McGraw-Hill Yearbook of Science & Technology

1985

GCA

TABLE OF CONTENTS

International Editorial Advisory Board

Editorial Staff

Sybil P. Parker, Editor in Chief

Jonathan Weil, Editor
Betty Richman, Editor
Alexander Hellemans, Editor

Edward J. Fox, Art director

Ann D. Bonardi, Art production supervisor
Sherelle Ramey, Art production assistant

Joe Faulk, Editing manager

Ann Jacobs, Editing supervisor
Frank Kotowski, Jr., Editing supervisor

Patricia W. Albers, Senior editing assistant
Nancy E. Grimes, Editing assistant
Barbara Begg, Editing assistant

Art suppliers: Eric G. Hieber, EH Technical Services, New York, New York; James R. Humphrey, Nanuet, New York; E. T. Steadman, New York, New York

Typesetting and composition by The Clarinda Company, Clarinda, Iowa

Printed and bound by R. R. Donnelley & Sons Company, the Lakeside Press at Willard, Ohio, and Crawfordsville, Indiana

Consulting Editors

Consulting Editors (continued)

Contributors

A list of contributors, their affiliations, and the articles they wrote will be found on pages 477–481.

Preface

The 1985 *McGraw-Hill Yearbook of Science and Technology*, continuing in the tradition of its 22 predecessors, presents the outstanding recent achievements in science and technology. Thus it serves as an annual review and also as a supplement to the *McGraw-Hill Encyclopedia of Science and Technology*, updating the basic information in the fifth edition (1982) of the Encyclopedia.

The Yearbook contains articles reporting on those topics that were judged by the consulting editors and the editorial staff as being among the most significant recent developments. Each article is written by one or more authorities who are actively pursuing research or are specialists on the subject being discussed.

The Yearbook is organized in two independent sections. The first section includes five feature articles, providing comprehensive, expanded coverage of subjects that have broad current interest and possible future significance. The second section comprises 159 alphabetically arranged articles.

The *McGraw-Hill Yearbook of Science and Technology* provides librarians, students, teachers, the scientific community, and the general public with information needed to keep pace with scientific and technological progress throughout the world. The Yearbook has successfully served this need for the past 23 years through the ideas and efforts of the consulting editors and the contributions of eminent international specialists.

SYBIL P. PARKER
EDITOR IN CHIEF

McGraw-Hill Yearbook of Science & Technology

1985

Archeoastronomy

MICHAEL ZEILIK

Michael Zeilik is an associate professor of astronomy at the University of New Mexico. He serves as a member of the Education Board of the American Astronomical Society and the Editorial Board of the "Journal of College Science Teaching." His research interests center on star birth, binary systems with solar-type stars, and archeoastronomy in the Southwest. Recipient of a doctorate in astronomy from Harvard University, he is the author of two books.

Archeoastronomy is a young scientific field that attempts to determine how much astronomy prehistoric people knew and how it influenced their lives. It involves the work of many disciplines—astronomy to chart the heavens, archeology to probe the cultural context, engineering to survey sites, and ethnology to provide clues to the cultural past. Born in the controversy over Stonehenge, modern archeoastronomy has yet to reach maturity. At its best, it amalgamates the talents of diverse experts; at its worst, it degenerates into controversies over intellectual turf. Despite these problems, archeoastronomy has prompted valuable insights into the astronomy of the past—even to revolutionizing some of the models of prehistoric cultures.

This article highlights some fruitful work in archeoastronomy. It touches on the Old World and New World, especially the southwestern part of North America. This restricted focus does not imply that archeoastronomers do not work throughout the world. Researchers in Africa, for instance, have just started to put together the astronomical past of that continent. Any prehistoric or preliterate people has their story to discover.

Of course, it can never be known for certain if archeoastronomers' ideas about the past are right. That is the ultimate weakness of all archeology. But researchers stand on firmer ground if ethnographic information is available to bolster the archeology. That is where the New World has a distinct advantage over the Old: here still live the descendants of people who were observing the sky before the arrival of Europeans. Their culture, although suffering from pressures to change, serves as a caution and guide to interpreting the past. Clearly, to gain a real understanding of ancient people, one needs to know their cultural heirs.

NAKED-EYE ASTRONOMY

Before plunging into the astronomy of the past, one must review briefly the kinds of observations of celestial cycles that can be made without a telescope—the same as were made by ancient people. For the purposes of this article, only the Sun and the Moon will be considered.

Sun. The Sun will be discussed first, since it is the easier of the two. Most people are aware that the height of the Sun in the sky at noon changes with the seasons: highest in the summer, lowest in the winter, and in between during spring and fall. The Sun reaches its highest noon point on the summer solstice (around June 22); drops to its lowest on the winter solstice (December 22); and is at the middle at the equinoxes (March 21 and September 23).

There is, however, less familiarity with the Sun's seasonal motion along the horizon. On the day of the summer solstice, for example, the Sun rises the farthest north of east that it will get for the year (Fig. 1). On the equinoxes, it rises due east. And on the winter solstice, it reaches its farthest point south of east. (The same occurs, mirror-reflected, at sunset.)

Thus, from summer to winter, the sunrise point moves to the south; from winter to summer, to the north. The rate at which the sunrise point moves on a day-to-day basis varies during the year. At the solstices, the sunrise points do not noticeably move for a few days. The Sun appears to "stand still" (the meaning of the word solstice). In contrast, at the equinoxes, the sunrise points move at their fastest rate—by almost the Sun's own diameter in a day at midlatitudes.

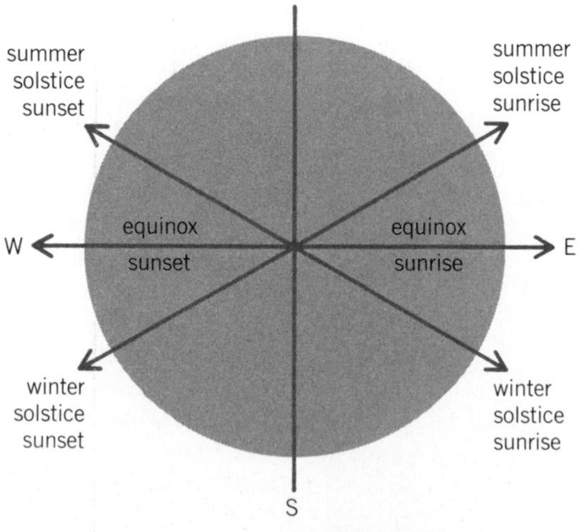

Fig. 1. Angular swing of the Sun along the horizon at rising and setting from solstice to solstice. The angle is for a latitude of 36°, that of the United States' Southwest.

This seasonal voyage of the Sun along the horizon differs with latitude with respect to the size of the solstice-to-solstice swing along the horizon. At 36°—the latitude of the Southwest—the swing amounts to 60°, one-sixth of the total horizon circle (Fig. 1). Farther north, the arc is greater; at the latitude of Stonehenge, it is about 80°. At more southerly latitudes, the arc is less; at 20° (middle of Mexico), it varies a total of 50°. So the Sun's seasonal dance appears most dramatically at far-northern latitudes.

Moon. The Moon's most obvious change is that of its phases, referring to how much the side turned to the Earth is illuminated. At new moon, that side is dark. At first and last quarter, half the side is illuminated. At full moon, the entire side is bathed in sunlight. The month of phases (synodic month) is simply the time from one phase of the Moon to the repetition of that phase, say from full moon to full again. It averages 29.5 days.

Suppose the point of moonrise is observed for a month. It would be seen that the moonrise point varies from a point farthest south to one farthest north during the month. In other words, the moonrise motion mimics the sunrise motion but occurs at a much faster rate, about 12 times as fast. Depending on when the observations are made, the moonrise arc may be larger than, the same as, or smaller than the sunrise arc. This is because the Moon's path in the sky with respect to the stars is not the same as the Sun's, but is inclined at about 5°, crossing the Sun's path at two points. Thus, the Moon can appear as much as 5° below the Sun's path, 5° above it, or right on it.

Complication. The matter is complicated in that the two points where the Sun's and Moon's paths cross (the nodes) move with respect to the stars, taking 18.6 years to circle the sky once. The result is that when the Moon's path reaches its highest point above the Sun's, the Moon's horizon swing is greater than the Sun's. When the two line up, the swings are the same. When the Moon's path falls below the Sun's, the total arc is less.

How much greater or less can be considerable. At a latitude of 36°, the Moon moves through a maximum arc of 70° and a minimum arc of 45° during the 18.6-year cycle (Fig. 2). In analogy to the Sun standing still at the solstices, the two extremes of the Moon's positions are also called standstills: major standstill for the maximum angle and minor standstill for the minimum, with 9.8 years between. Again in analogy to the Sun, these angular changes are more pronounced at more northern latitudes.

Much of the time while the Moon moves from minor to major standstill, the Sun's arc encompasses that of the Moon's. During this time, any alignment that works for the sun will work for the Moon as well. So the best way to ascertain that an alignment tracks the Moon is to see if it works when the Moon's swing lies outside the angle of the Sun. Then the alignment can relate to the Moon but not the Sun, while the Moon is near major standstill.

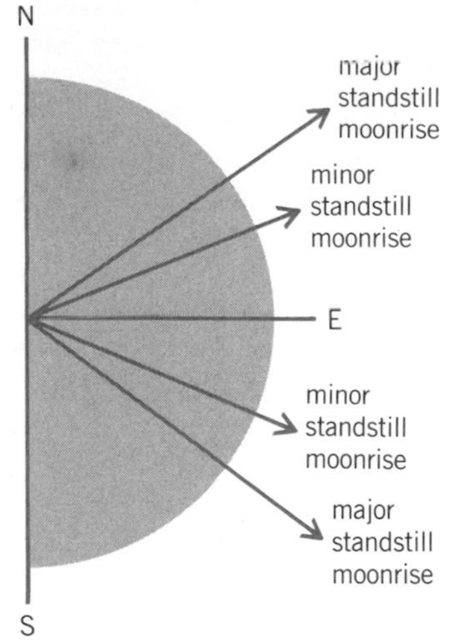

Fig. 2. Monthly angular swing of the Moon along the horizon at moonrise, for times of maximum angle (major standstill) and minimum angle (minor standstill). The latitude is 36°.

Horizon-marking system. It can now be seen how a simple horizon-marking system is set up. One must first find a spot to stand with a clear view of the horizon, which should have at least a few prominent features over the angular range of the sunrise (or sunset). Then it is necessary to return to this spot daily and note the rising positions of the Sun throughout the year at significant times: the solstices, when the Sun does not move, the equinoxes (halfway between the solsticial points), and, perhaps, important times to plant crops. One then has a basic solar calendar. Since the Sun's positions at various dates along the arc remain fixed for a long time, once established the calendar will be good for many years.

With this basic solar and lunar astronomy in mind, the relics of the past that may have had astronomical uses can be examined.

THE OLD WORLD: STONEHENGE AND OTHER MEGALITHIC SITES

In a direct sense, Stonehenge created the interest in archeoastronomy—revealing its inventive strengths, highlighting its weaknesses, and starting a confrontation between astronomers and archeologists that has only recently settled into a creative interaction. Stonehenge exemplifies the problems and potential of the archeoastronomical enterprise.

When one hears of Stonehenge, the massive upright stones that form a central horseshoe and circle (some 65 ft or 25 m in diameter) in the center of the site are envisioned (Fig. 3). Such large stones are commonly called megaliths in Great Britain; this term has come to be applied to all sites where stones, even fairly small ones, are arranged in some pattern. The horseshoe opens out on the main axis of Stonehenge, called the Avenue. Some 260 ft (80 m) from the center, within but not in the center of the Avenue, sits the tilted Heel Stone. This main axis of Stonehenge aligns roughly with the summer solstice sunrise, a fact noted more than 200 years ago. "Roughly" is emphasized because the Sun rises somewhat to the left of the Heel Stone as seen by an observer at the center of the structure.

Despite earlier interpretations of the astronomical

Fig. 3. The inner great trilithons of Stonehenge. (*Courtesy of O. Gingerich*)

use of Stonehenge, the modern controversy developed in the 1960s. Gerald Hawkins, an astronomer, applied the brute-force calculational power of a then-novel electronic computer to search for astronomical alignments to the Sun, Moon, stars, and planets for the main features of the site. He found them for the Sun and the Moon, including moonrise and moonset during major and minor standstills. Later, he proposed that the site could even have been used to warn of the times of possible eclipses. That assertion contradicated the rather confident claims of archeologists concerning the primitive nature of the society of the times.

Radiocarbon dates indicate that Stonehenge was built over a span from 3100 to 1000 B.C. in three separate stages. These cover the Neolithic era to the Bronze Age. The muddle over the astronomical use of Stonehenge comes, in large part, from the fact that it is a mosaic of structures, most likely built by different people, perhaps for different reasons. The great stones were erected between 2000 and 1500 B.C.; it is not clear what their cultural connection was to the earlier structure.

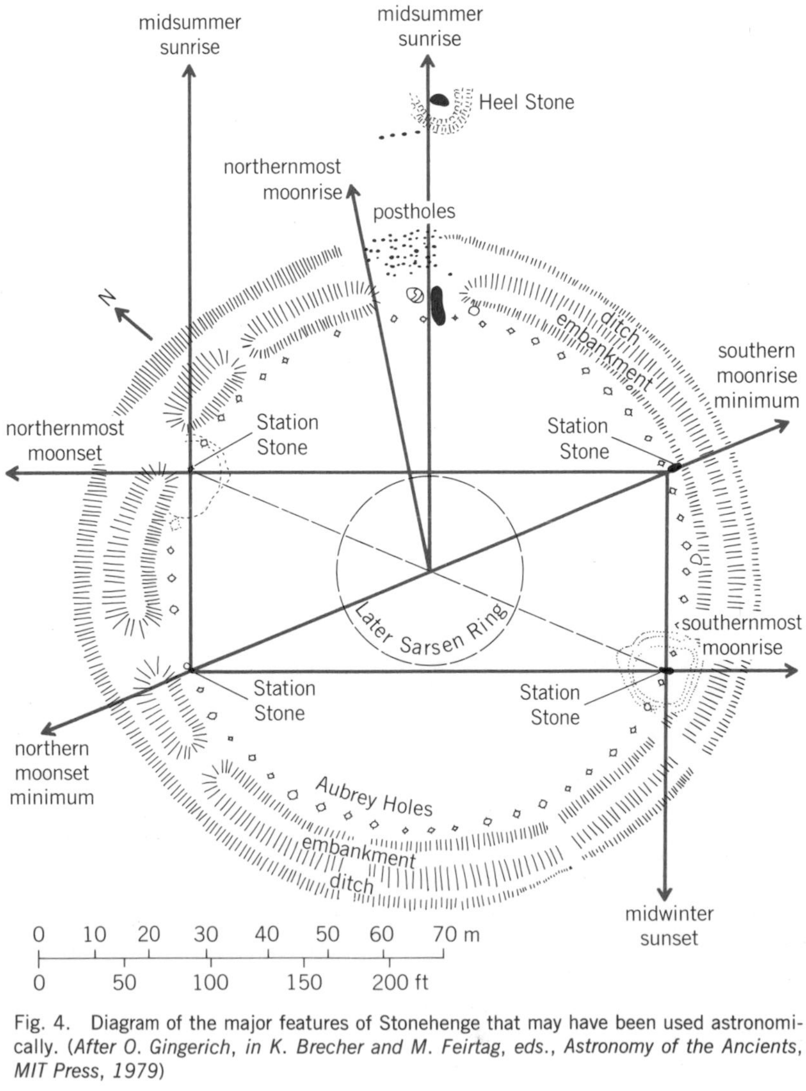

Fig. 4. Diagram of the major features of Stonehenge that may have been used astronomically. *(After O. Gingerich, in K. Brecher and M. Feirtag, eds., Astronomy of the Ancients, MIT Press, 1979)*

It is the earliest parts of Stonehenge, constructed between 3100 and 2100 B.C., that have the most astronomical promise (Fig. 4). These comprise the outer earthwork ring and ditch (about 330 ft or 100 m in diameter and 7 ft or 2 m high) broken only in the direction of the Heel Stone; a ring of 56 holes (the Aubrey Holes) that were dug and then quickly filled with chalk; an array of postholes near the opening to the Heel Stone; and the four Station Stones that lie along the circle of the Aubrey Holes.

Now with just these elements, lunar and solar observing can been done (Fig. 4). The four Station Stones form a fairly good rectangle. From its center, the summer solstice Sun rises along the opening to the Heel Stone. The short sides of the rectangle are parallel to this line, so they point to the summer solstice sunrise and winter solstice sunset. The long sides of the rectangle and its diagonals line up the moonrises and moonsets at the major and minor standstills.

Hawkins also contended that the inner megaliths of the horseshoe sighting outward through the ring around them also aligned to important settings and risings of the Sun and Moon. Hawkins then argued that the Aubrey Holes were used to indicate "danger seasons" when eclipses might occur. One lunar eclipse cycle (it is not the only one) takes 56 years (three times the 18.6-year standstill cycle). In this picture, the Aubrey Holes were used as a counter to keep track of the years within these cycles.

Does this all work out? Yes and no; both astronomical and archeological criticism can be applied. First, Stonehenge can be criticized as a lunar eclipse anticipator. A 56-year cycle does exist, but once worked out, it fails to apply after a few cycles. Also, to work out the cycle in the first place requires hundreds of years of careful observing and a preserved record (probably oral) from which to infer the cycle. Given problems with bad weather hiding eclipses and the difficulty of preserving a nonwritten record for such a long time, the establishment seems highly impractical. Second, the purported alignments with the inner megaliths are also questionable; the gaps are rather wide and, depending on where one stands, can cover a large angle on the horizon. Their crudeness suggests that alignments attributed to them are probably accidents of the layout.

The inner rectangle seems much better astronomically; because of its fairly large size (112 by 260 ft or 34 by 79 m), it results in fairly accurate sightlines. These also contain nice symmetries, which increase their appeal. However, the archeologist R. J. C. Atkinson notes that only two of the four stones actually survive; of these two (91 and 93), one has fallen and one seems to be a later replacement. So the original positioning of the stones is not known with accuracy.

All told, the older parts of Stonehenge make a reasonable solar and lunar observatory. The sloppiness (of about a degree of arc) should not be considered offensive, for the modern fetish with accuracy may be a cultural trait that was absent among Neolithic cultures. The extensive symmetries are very appealing and more important than the accuracies of the sightlines. Finally, even archeologists admit that the summer solstice alignment, more than any other, appears as an intention of the original construction.

The basic problem in all of this is that even if the astronomy works, one cannot be sure of the cultural context. Horizon watching can be used simply to tell the time of year or more forcefully to set a ritual calendar. Since megalithic societies have left few clues about their thinking—it can only be inferred from their material remains—the only hope of guessing about the importance of astronomy is to look at other sites along with Stonehenge.

That has been done for some hundreds of sites (in Great Britain and France) by Alexander Thom, an engineer. Long before the Stonehenge controversy, Thom started very carefully surveying prehistoric sites in the British Isles. He first found indications of alignments for the solstices and equinoxes, then for the lunar standstills. More recently, he has promoted the idea that megalithic astronomers made extremely accurate observations of the Moon (using very distant foresights, tens of kilometers long) so as to pick out very small, long-term variations of the Moon's motions.

The pervasiveness of the sightlines to astronomically important observations implies that megalithic cultures knew about the astronomy and deemed it important enough to construct numerous observing stations. More so than Hawkins's efforts, Thom's work forced archeologists to account for the astronomy in these cultures. The validity of precise lunar observations remains to be shown. They have been questioned, and different analyses lead to the other conclusion: the Moon was observed, but not with the precision inferred by Thom. From the view of cultural necessity, it is unclear how such precision would benefit megalithic people in terms of simple survival value. Even anticipating eclipses has dubious value—at most, a device to enhance the predictive dimension of priestly power.

Still, despite disputes over fine points, the archeoastronomers have forced a reexamination of the standard picture of megalithic life. That certainly marks one of the positive aspects of the field. Certainly astronomy was important, even if it is not known exactly how or why. Before the 1960s, such a concept was ignored for the most part by archeologists.

SKYWATCHING IN THE NEW WORLD

Compared to the Old World, the New World archeoastronomer has the advantage of the survival of remnants of the cultures from pre-Columbian times. Even the great destruction wielded by the Spanish in Mesoamerica—especially their burning of Mayan books that contained much astronomy—could not wipe out completely the astronomy inherent in that culture.

Turning northward, the Spanish marched on a fruitless search for the fantastic Seven Cities of Cíbola, said to be made of gold. They found none. But they did encounter the adobe villages, which they called *pueblos*, of the native peoples who had lived in them at least a thousand years prior to the arrival of the Spanish. Many of these pueblos disappeared in historic times (from 1540 onward); those that survived are the cultural connection to the people called the Anasazi, who occupied a vast area in the Southwest, centered on the Four Corners area (where New Mexico, Arizona, Utah, and Colorado now meet). Here stand ruins deserted from A.D. 1000 to 1400, stone and adobe constructions that provide some insight into the life of the Anasazi.

The Hopi pueblos (in Arizona) and Zuñi (in New Mexico) provide the best clues to the past because these villages were touched only lightly by the Spanish (in contrast to the submission demanded of the pueblos along the Rio Grande). Ethnographers worked here at the turn of the century and gathered cultural information before the severe pressures on the part of Anglos occurred. It is inferred that the Hopi and Zuñi are cultural descendants of the Anasazi (although it is not known from which specific Anasazi sites). So these pueblos preserve a remarkable cultural continuity with prehistory.

At Hopi and Zuñi, astronomy plays a central role in the agricultural and ceremonial life. The seasonal cycle of the Sun sets the ritual calendar and determines the times of specific crop plantings and harvestings. The dry Southwest demands an observant farming, for raising crops is a marginal activity; in the past, failed crops could mean death. So solar astronomy carries a practical weight as well as a religious one. The counting of months by lunar phases plays a secondary role in tracking the ritual calendar.

The observing is invested in a religious office, usually called the Sun Chief. He watches daily from a special spot within the pueblo or not far outside of it. The Sun Chief carefully observes sunrise (or set) relative to the horizon features. He knows from past experience what points mark the summer and winter solstice and the times to plant crops. These he announces within the pueblo, usually ahead of time so that ritual preparations can be carried out. The winter solstice—called Soyal at Hopi and Itiwanna at Zuñi—marks the heart of the ritual year. For the Hopi, each month was named, and the passing of a month sometimes was used to set the time for a ceremony.

Along with horizon features, the Zuñi Sun Watcher, called Pekwin, used a natural pillar to chart the seasons. When the shadow cast by the pillar lined up in a special fashion, Pekwin knew that the summer solstice would soon occur. Also, within the pueblo, special windows and portholes allowed sunlight to hit special plates or markings on the walls at significant times of the year. So light and shadows, along with horizon features, made up the basis of the puebloan solar astronomy. The ancestors are believed to have done much the same. (Pekwin no longer observes at Zuñi; the man in that office left the pueblo in the 1950s. However, another religious officer, the North Priest, now carries out the critical sun watch. This transfer of office attests to the importance of sun watching in the ritual life and to the continuity engendered by its importance.)

ANASAZI ASTRONOMY

Around A.D. 1000, the Anasazi prospered in the San Juan Basin and other regions of the Colorado Plateau. In this harshly beautiful landscape, they build community houses that were four or five stories high, contained hundreds of rooms, and many large and small kivas—round, underground rooms used for ritual purposes.

Chaco Canyon in northwestern New Mexico grew to be a center of Anasazi culture. By 1130. eight large villages were located within 9 mi (15 km) of the central canyon. Perhaps a few thousand people lived here in the large and small villages. These Chacoans faced climatic conditions similar to those of today; like the historic pueblos, they probably also had Sun Chiefs and seasonal solar calendars.

At least four places at Chaco may have been used for sun watching: Pueblo Bonito, Fajada Butte, Casa Rinconada, and a site not far from the pueblo of Wijiji. All have evidence in art or architecture as possible sun-watching stations.

Pueblo Bonito. This is a D-shaped apartment house of over 800 rooms built close to the north wall of the canyon. Within it are a number of corner doorways and windows, which are rather unusual in Anasazi architecture: Pueblo Bonito contains over half of the known examples. The archeologist Jonathan Reyman noted that two of the windows in rooms in the southeast part of the ruin have a clear view of the winter solstice sunrise (Fig. 5)—if, in fact, an outer wall did not obstruct the view. The use of windows is a necessary strategy in Chaco because the horizon is a flat expanse viewed from within the canyon—not at all good for a horizon calendar.

Casa Rincoñada. Directly across the canyon from Pueblo Bonito lies Casa Rincoñada (Fig. 6*a*). Archeologists regard it as a great kiva used for community ceremonies. It is almost 100 ft (30 m) in diameter—one of the largest kivas known. Standing in its center, one can see a window in the northeast corner. On the morning of the summer solstice, about a half hour after sunrise, a beam of sunlight through the window strikes one of six niches (Fig. 6*b*) that are spaced irregularly around the inside wall of the kiva. As the Sun climbs the sky, the beam moves downward until it hits the floor. The beam continues to strike the niche for about a week around the summer solstice.

Although visually arresting, it is not certain that this effect was intended by the Anasazi. First, the kiva has been reconstructed, and it is not known exactly where the opening was originally placed or its overall size. (It is clear in photos taken before the reconstruction that an opening did exist in the northeast side.) Second, a wall or room outside the window may have blocked out the sunlight. Third, kivas have roofs; Casa Rincoñada has stone circles that mark the positions of the four massive posts needed to support the heavy roof. The one in the northwest corner seems to just block the sunlight from hitting the niche on the summer solstice. Fourth, the historic pueblos do not use kivas for sun watching (although they did observe stars from them). Overall, the likelihood of actual use of the site by the Anasazi is shaky at best.

Fajada Butte. This butte (Fig. 7*a*) thrusts upward at the eastern end of the canyon, the lone dramatic break in the landscape. Atop the butte lies a sun marker that tracks the seasons. Within 30 ft (10 m) of the summit, three rock slabs lie against the butte's southeast face (Fig. 7*b*). The slabs are a few

Fig. 5. Pueblo Bonito. (*a*) View of the southeast area. The two corner windows may have been used to anticipate and confirm the winter solstice. (*b*) The light from the winter solstice sunrise streaming into the corner of room 228. (*Photos by M. Zeilik*)

Fig. 6. Casa Rincoñada. (*a*) View of the interior, showing a "window" at the right and a series of niches along the inside wall. The door in the middle is the north entrance. (*b*) The summer solstice sunlight entering through the window about 30 min after sunrise and illuminating a niche in the wall. (*Photos by M. Zeilik*)

meters long and spaced with about 25-in. (10-m) gaps between them. They shield the rock face on which they rest from the Sun except at times before local solar noon. Then the edges of the slabs allow sunlight to strike the rock face, on which are carved two spirals: a large one (almost 1.5 ft or 0.5 m wide) right behind the slabs, and a smaller one below and to the left of the larger.

The spirals mark the Sun's yearly cycle by light patterns visible late in the morning. On the summer solstice, a dagger of light materializes above the large spiral at about 11 a.m. In about 20 min, it

Fig. 7. Fajada Butte in Chaco Canyon. (*a*) View of entire butte (*photo by M. Zeilik*). (*b*) Rock slabs near the top of the butte. They rest against the rock surface on which two spirals are pecked. Their upper edges cause sunlight late in the morning to play upon the rock face on and around the spirals (*photo by R. Elston*). (*c*) The summer solstice sunlight cutting through the large spiral at about 11:13 a.m. (*photo by W. Wampler*).

descends and slices through the heart of the spiral design (Fig. 7*c*). On the winter solstice, two shafts of light appear on the outside of the large spiral and pass through its outer edges at about 10 a.m. At both equinoxes, two shafts appear, one shorter and to the left of the other. The little shaft cuts through the center of the small spiral, while the larger one drops through one side of the large spiral.

In addition to the solar markings, Anna Sofaer and her colleagues argue that it was also used to mark the 18.6-year lunar standstill cycle at moonrise for phases between full and waning crescent.

Is this a formation that the Anasazi intended for solar and lunar observations? The slabs are the result of a natural rockfall; they were not moved to their present location. People of historic pueblos tend to use natural features, with little modification, for sun watching. The spirals may have been placed behind the slabs after the natural play of light was noted for a year. The fact that the shafts do not appear at a particularly significant time of day (late morning) reinforces the view that the rocks were in a natural formation. The site certainly does confirm the solstices and equinoxes in a visually stunning way.

However, this site does have problems in a cultural context. First, almost all sun-watching stations of historic pueblos are within or close to the pueblo. Fajada Butte is a few kilometers from the nearest large pueblo and hard to climb. For daily sun watching, it is a rather impractical site. Second, spirals are not known to be sun symbols in historic puebloan culture—in fact, they are water signs. Why are they used here? Were spirals connected to the Sun in prehistoric times but not now? Third, as far as lunar observations are concerned, the historic pueblos certainly tracked the phases of the Moon. But little or no evidence has been found that they had any interest in the 18.6-year standstill cycle. So although tempting, the Fajada Butte site is not yet well proven as an intentional construct of the Anasazi for calendrical sun or moon watching.

Wijiji. This is a pueblo ruin built about A.D. 1100 late in Chaco's history. About a kilometer to the east of it, a large rincon (valley) opens up in the mesa. On the northwest side of the rincon runs a narrow ledge, which can be reached by climbing a prehistoric staircase. Here on the wall is painted a large four-pointed symbol (Fig. 8) that resembles the Zia Pueblo sun sign now used as the official seal of the State of New Mexico. North of the symbol, three boulders rest on the ledge; the largest has a double-spiral carved in its surface. The design and technique are clearly Anasazi. Eastward from the ledge, a large rock pillar, across the rincon, rises above the horizon. From a spot a few meters south of the boulder with the double spiral, the winter solstice Sun rises behind the pillar.

Was this the intention of the Anasazi? The area of the ledge contains much rock art, some of it Navajo. Ray Williamson has argued that the white "sun" symbol is Navajo in origin and that the site was used by Navajo (perhaps in the late seventeenth century) for sun watching. That may be the case, but there are cetainly Anasazi relics here, too. So the site may have been first used by the Anasazi and then adopted by the Navajo for ceremonial purposes. If so, it is the only site in Chaco so far that uses the horizon for sun watching, the most common practice in the historic pueblos, rather than the manipulation of light and shadow.

By 1250, Chaco was deserted, the great villages emptied of their human life. No one knows why. Some of these people settled the pueblos that are known today, taking with them the astronomy that was part of the tool kit of their culture.

Fig. 8. The painted, white sun symbol above the ledge near the Wijiji ruin in Chaco Canyon. The emblem resembles that of the Zia pueblo sun symbol and a sun symbol on a war shield from Jemez pueblo. (*Photo by M. Zeilik*)

OTHER PLACES AND CULTURES

Indigenous astronomy did not develop only in Great Britain and North America. In fact, every prehistoric or preliterate culture appears to have developed its own astronomy. The traditional navigators of the Pacific, both Micronesians and Polynesians, used naked-eye observations of the stars to sail thousands of miles successfully. The Carib people of northern South America developed a calendar that relied on the positions of the stars relative to each other and to the Sun at times of rising and setting. Incised pieces of bone hint that Ice Age people kept track of the phases of the Moon. The Ngas of Nigeria, Africa, predict the position of the new moon so that they can "shoot the moon" with a spear or arrow propelled by song. In the Eddas, the primary works of Nordic mythology, the motions of the Sun, Moon, and planets are presented in a mythic guise.

The worldwide study of archeoastronomy shows that every culture pays attention to different aspects of the sky, depending on its local environment and needs.

The sites discussed above have been placed in a cultural context. Without that, finding astronomical alignments may be no more than coincidence guided by the bias of wanting to find them and knowing where to look for "significant" results. It is usually possible to find an astronomical alignment at a site—the real question is whether or not the builders intended it as such. Looking for alignments must be done, but it must be guided by some idea of what it was like to be the Sun Priest of the day. The greatest danger here is the imposition of modern knowledge of astronomy upon an alien culture of the past.

In the past 10 years, archeoastronomy has blossomed with the germane interactions of experts who rarely wandered outside the narrow domains of their own specialties. Now that the professional ice has been broken, archeoastronomy may solidify its advances as its premises and practices are sharpened by public scrutiny and productive skepticism. As it reaches its deserved maturity, archeoastronomy will appear less like a special field and more like a broad template for understanding cultures of the past. [MICHAEL ZEILIK]

Bibliography: A. F. Aveni (ed.), *Archaeoastronomy in the New World*, 1982; A. F. Aveni and G. Urton (eds.), Ethnoastronomy and archaeoastronomy in the American tropics, *Ann. N.Y. Acad. Sci.*, vol. 385, 1982; D. C. Heggie (ed.), *Archaeoastronomy in the Old World*, 1982; R. A. Williamson (ed.), *Archaeoastronomy in the Americas*, Ballena Press, Los Altos, California, 1981.

Bioengineering

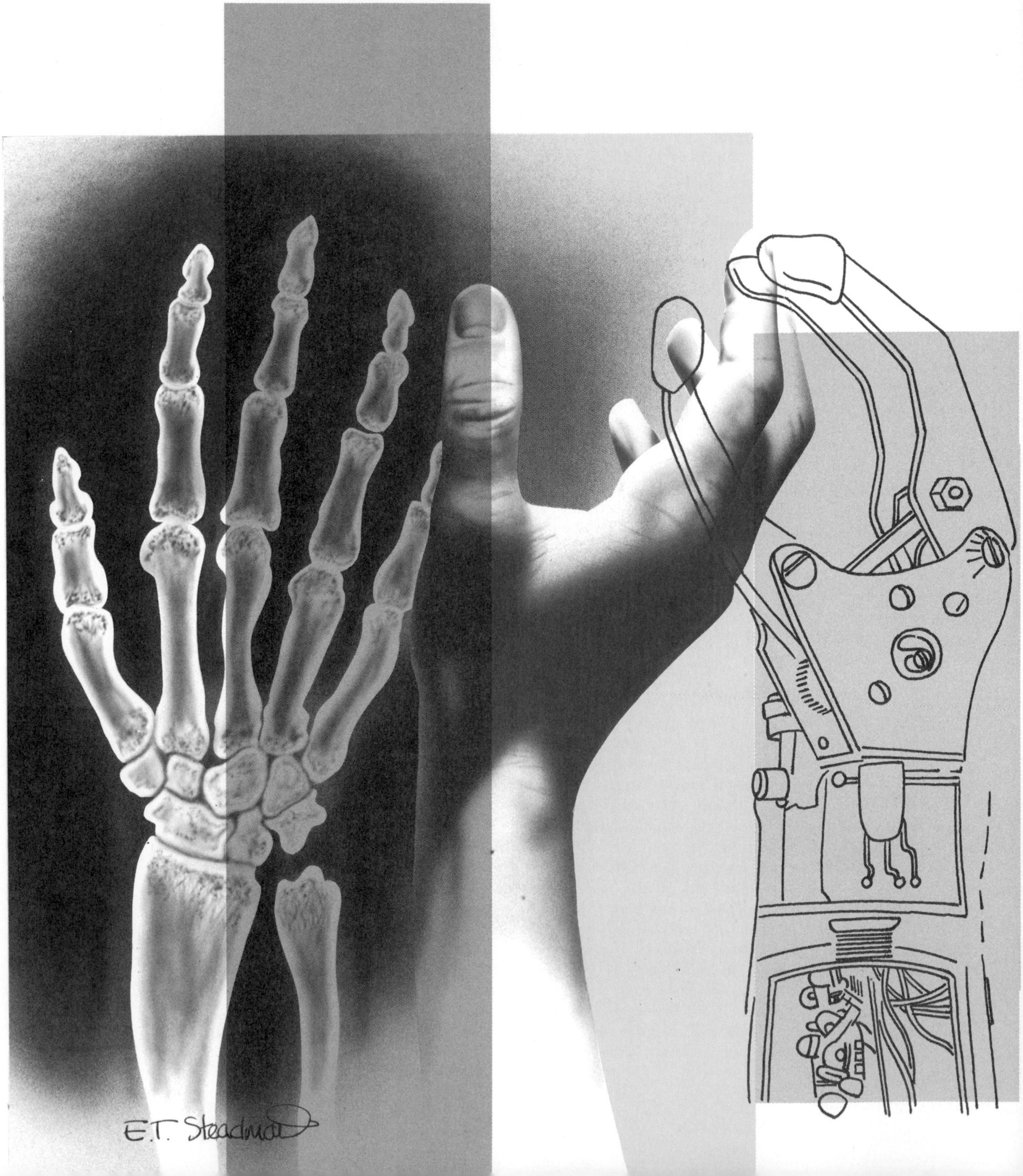

EUGENE F. MURPHY

Eugene F. Murphy, Ph.D., is a mechanical engineer, registered as a professional engineer in California and Illinois. A fellow of the Rehabilitation Engineering Society of North America, he has received many awards for his achievements in biomechanics, bioengineering, and rehabilitation engineering. During his 35-year career with the Veterans Administration, he pioneered in research on prosthetics, orthotics, and wheelchairs and automotive equipment. Upon his retirement, the Veterans Administration presented him with its Distinguished Career Award.

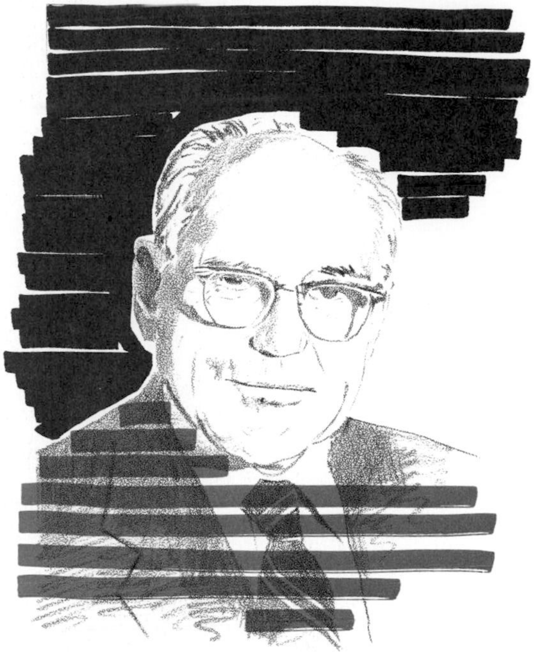

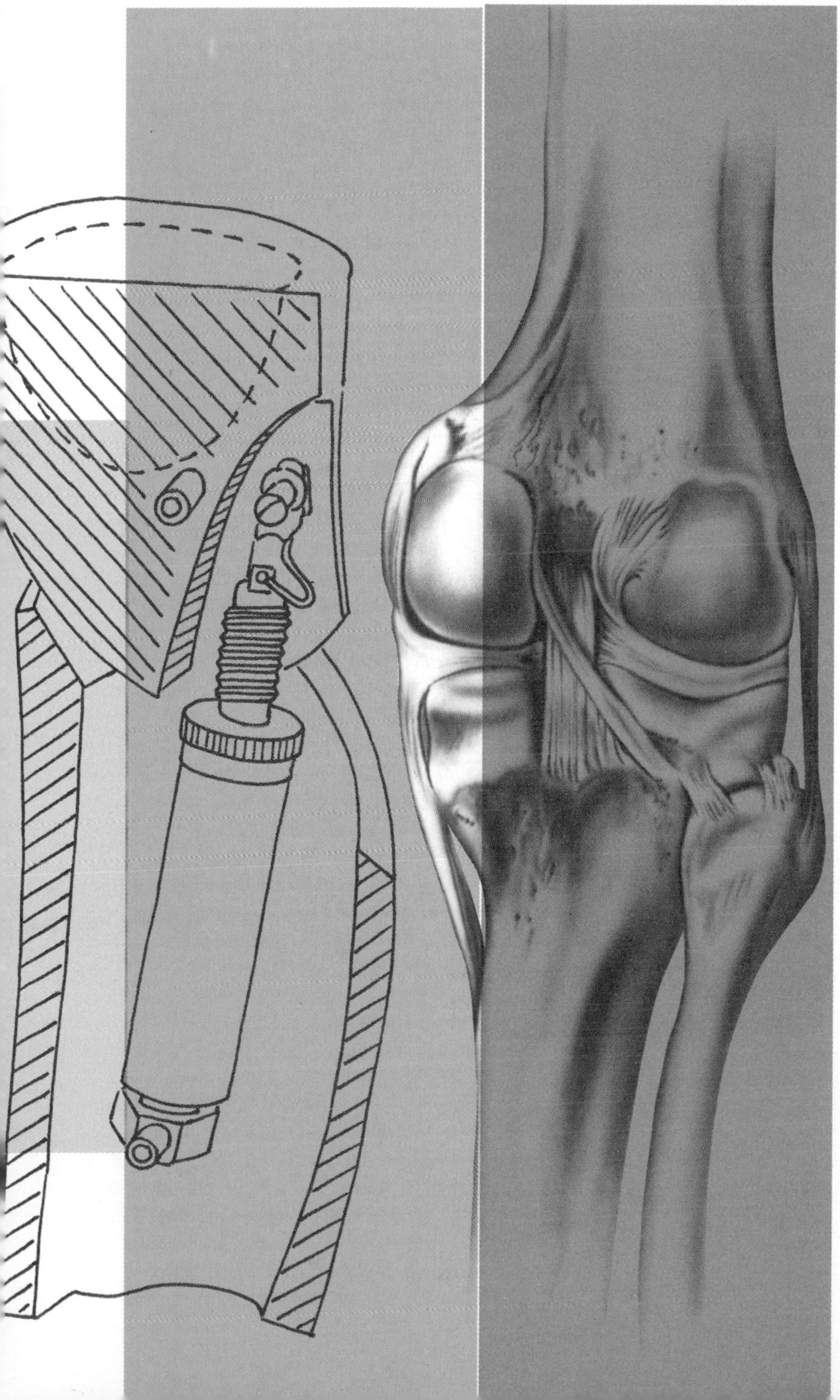

In the fall of 1982 the world was fascinated by the drama of the first artificial heart implanted in a human being for lifetime use. With major portions of his natural heart removed, leaving only the badly damaged left and right atria (intake chambers), Barney Clark was completely dependent on the artificial heart (Fig. 1). It was the seventh designed by Robert K. Jarvik, a member of the team assembled at the University of Utah by Willem Kolff, who himself had developed the first artificial kidney (dialysis machine). This artificial heart was a medical milestone, and although the patient's life was extended by only months it was recognized that the great triumph was the demonstration in a human being of the real success of innumerable medical and engineering decisions and compromises. These were based on a long period of laboratory, animal, and short-term human trials of components as well as systematic research on principles.

Complex and dramatic as they are, artificial organs are but one of many sectors of the interdisciplinary field of bioengineering. The following discussion summarizes progress in a number of representative areas and examines some underlying principles and problems.

For decades both the popular media and the professional journals have carried articles on bioengineering achievements—iron lungs or respirators, improved artificial limbs (prostheses) and braces (orthoses), artificial kidney machines, cardiac pacemakers, and external heart-lung machines (pump-respirators)—and hope exists for other organs. Life support systems, now commonplace in elaborately instrumented inten-

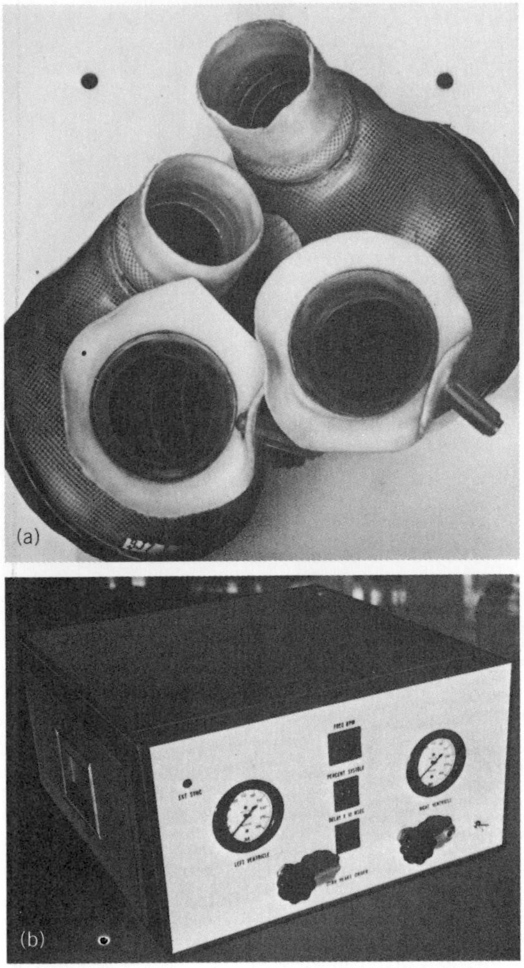

Fig. 1. Two components of the artificial heart system. (*a*) The Jarvik-7 artificial heart. The pumping diaphragm is composed of medical-grade polyurethane formed over a hemispherical mold. (*b*) The Utah heart driver. *(University of Utah)*

sive care facilities and increasingly complex operating rooms, improve the chances of survival. Some medical applications of technology permit better diagnosis or spectacularly save lives, but many are used rather prosaically for months or years to improve the level of rehabilitation and the quality of life for people left disabled long after an initial crisis. Indeed, even the advances in sanitary and environmental engineering have had profound impacts on the prevention of disease and injury, on avoidance of loss of life, or on the need for rehabilitation.

BIOLOGICAL AND PHYSICAL SCIENCES: DIVISIONS AND LINKAGES

Bioengineering overlaps portions of biology and medicine on one side and the physical sciences and engineering on the other, but also bridges these disciplines in numerous areas that formerly were separated. Bioengineering may also be considered to comprise sectors such as biomedical engineering, for the restricted interaction of human health problems with engineering; biomechanics, for the study of forces, accelerations, and stresses affecting animal and particularly human tissues; or biorheology, for the study of flow of blood and other body fluids. Bioengineering can be interpreted in a restricted sense to cover application of biological processes to engineering purposes. The applications of living organisms, or their biological systems or processes, to the manufacture of useful products (fermentation, genetic engineering, enzyme technology, conversion of wastes to fuels, and so forth) has been described as biotechnology. In contrast, biotechnology has also been used to describe the application of biological, physiological, and psychological principles in a wide variety of conventional engineering fields—involving design of handles and controls, tolerance to hot or cold environments, or illumination—as well as in design and control of artificial arms for amputees.

Biomedical engineering is a burgeoning field. In addition to mathematical modeling of physiological systems or of entire treatment systems, biomedical engineering involves numerous types of devices. Hospital engineering is particularly concerned with selection, maintenance, and safety of the very complex technology typical of modern hospitals. Clinical engineering is concerned with a variety of engineering problems associated with the direct care of patients, usually in the acute phase. In the last few years the interdisciplinary field of rehabilitation engineering has been particularly concerned with the application of technology to improve the quality of life for chronically disabled persons.

Obviously these subspecialties and interest areas share many common problems with basic sciences supporting either medicine or engineering (such as physiology or materials science), and clinical or engineering disciplines (such as surgery and rehabilitation medicine, or electrical or mechanical engineering), as well as with long-established hybrid fields like biophysics. The typical medical practitioner is initially bewildered by the quantitative, mathematical approach of the typical engineer. In turn, the engineer often encounters difficulty in adopting the "clinical" or "bedside manner" approach to patients (perhaps seriously ill) with emotions, hopes, and fears. Many of the physiological and anatomical facts and principles important in biomedical engineering are likewise crucial in sound solutions of seemingly more conventional engineering problems. For example, the relations between force, length, and holding time before fatigue for human muscle are as important in design of handles and tools as they are in the design of artificial limbs. Speed and accuracy of control of a jet plane by an able-bodied pilot is analogous to the control of an electric wheelchair. The human factors are as important as the stable and rapid response of the control system. Rational design of heating, ventilating, and air-conditioning systems must depend on physiology and on psychological perception of comfort as well as on fluid flow in pipes and ducts or on thermostats, hygrometers, and other controls. Thus, elementary training in the principles of biotechnology may be important for all engineers, just as mathematics and physics may be as useful as chemistry or biology for premedical students.

Studies have shown that the number of persons with major impairments in the United States is growing because of increasing success in preserving life. Severely injured or gravely ill patients nowadays are much more likely to survive, probably with the aid of biomedical engineering as well as surgical and medical skills, better drugs and antibiotics, and perhaps more rapid and skilled transfer to a medical center. Some impairments are particularly likely in elderly persons, notably problems with vision and hearing, but also amputation of lower limbs for peripheral vascular disease, paralysis of one side (hemiplegia) and perhaps additional problems with speech and vision after strokes, broken hips after falls, and painful joints from arthritis. Impairments (other than paralysis) of the back and spine are particularly likely in the working years. Spinal cord injury, particularly quadriplegia, is a special risk for young people because of automobile, motorcycle, and diving accidents.

Also, not only numbers of impairments but rates per 1000 persons may shift with time. (Census definitions and methods may vary, though successive studies in a given series are normally comparable.) The continuing increase in the total population of the country, as well as the increasing proportion of elderly, leads not only to shifting total numbers of certain impairments upward but to increasing numbers per 1000 total population.

In these studies, the unit is an impairment, so a single person with three categories of impairments would be counted three times; an estimate of the total number of selected impairments would not be an estimate of the total number of individuals affected. There are, of course, other acute conditions and other chronic impairments, for example mental ones, which are important but not directly concerned with bioengineering. Also, some conditions are not considered, including some cardiac problems, which in some cases may be treated by such bioengineering approaches as intensive care units, heart-lung machines, artificial heart valves, or an implanted artificial heart.

PROSTHETICS

The centuries-long history of efforts to replace missing extremities shows major interest in improvement during times of war but little sustained effort in peacetime. In World War I there were major programs on amputations and prostheses both in Germany and its allied countries and in Belgium, France, and Great Britain. Late in World War II the National Research Council in the United States was asked to conduct a research and development program to select and standardize the best artificial limbs for the various major levels of amputation. Still later, NRC advised the Veterans Administration and other agencies supporting such research and development. The program remained effective for about 30 years, terminating about 1976. The National Institute of Handicapped Research, created in 1978, now has expanded responsibilities for tasks that include research, demonstration projects, and related activities to aid rehabilitation of handicapped persons. Another role is leadership of an Interagency Committee on Handicapped Research to coordinate efforts of a number of federal agencies.

Lower-limb prosthetics. A major impact of the research program has been the increased interest of the medical profession in the total problem of amputee rehabilitation. There are now more precise methods of blood-flow diagnosis to avoid amputation in some cases, and surgery may be performed at lower levels to conserve human joints (particularly the knee in elderly patients with vascular problems). The use of rigid dressings instead of elastic bandages in the weeks after amputation seems to control edema or swelling more effectively, reduces pain in many cases, and sometimes (particularly in otherwise healthy individuals) permits the attachment of a temporary extension and artificial foot. The amputee is thus able first to stand and then to begin to walk much earlier, with both physiological and psychological benefits. In more debilitated cases, one or two changes of rigid dressings may be used to allow shrinkage of the residual limb and recuperation of the patient before a plaster trial socket and temporary prosthesis are applied. In either case a definitive prosthesis and appropriate therapy can be supplied about a month after amputation, much earlier than has been typical with prior methods.

Critics of the rigid dressing concept point out that a skilled person is needed to apply it immediately after the operation, at each cast change, and perhaps in the middle of the night in crises. Careful monitoring is needed, and the method is not practical for small hospitals. Advocates, however, argue for the therapeutic and economic advantages of immediate postoperative fitting of a rigid dressing on the residual limb, early prosthetic rehabilitation and ambulation, and geographically distributed amputation centers.

Since World War II there have been important gains in the understanding of the anatomical and biomechanical principles for fitting the socket to the residual limb and for aligning the socket, joint, and foot. The recent growth in use of transparent or translucent thermoplastics (for example, polypropylene) in both prosthetics and orthotics has extended to increased use of such materials for trial or check sockets, permitting direct observation of contact between skin and socket wall, reddening or blanching of skin in areas of excessive pressure, and response to changes of alignment. These thermoplastics also are used to construct lightweight prostheses. The total contact socket for above-knee amputations and the patellar tendon bearing socket or one of its variants for below-knee amputees are generally used.

A major result of the trend toward more distal amputations, as noted above, has been the shift from the predominance of above-knee amputees. For generations, a major concern of above-knee amputees, prostheses developers, and professionals was buckling of the single-axis knee joint. In standing, the axis was aligned slightly behind the vertical line through the center of gravity of the patient, so the corresponding floor reaction line tended to extend the knee against a mechanical backstop, preventing further hyperextension. Rocking forward on the forefoot created additional knee extension moment until the knee passed ahead of the toe. At, and immediately after, heel contact, however, the artificial knee joint tended to buckle, especially if the artificial heel bumper or cushion was too stiff. The situation was made worse if the axis of the socket was nearly vertical, not accommodating the apparent hip flexion of the residual limb caused by the anterior bowing of the upper portion of the femur (without the compensating backward curve of the amputated lower portion). True hip flexion contracture from prolonged sitting and inadequate therapy often further aggravated the risk of knee buckling at heel contact, especially descending hills.

Numerous locks, brakes, and controls have been invented to cope with knee buckling, but control is a major problem. However, excessive stability late in stance phase or undue resistance to knee motion during swing phase would be undesirable. Fortunately, better understanding and codification of the principles of fitting and alignment, and teaching of these concepts to prosthetists, have greatly reduced the risk of knee buckling. However, partly as byproducts of efforts to design improved knee joints for stance phase, a number of hydraulic and pneumatic controls for swing phase were developed over 20 years ago. Several have gained continuing substantial use to improve ability to walk gracefully at a wide range of walking speeds.

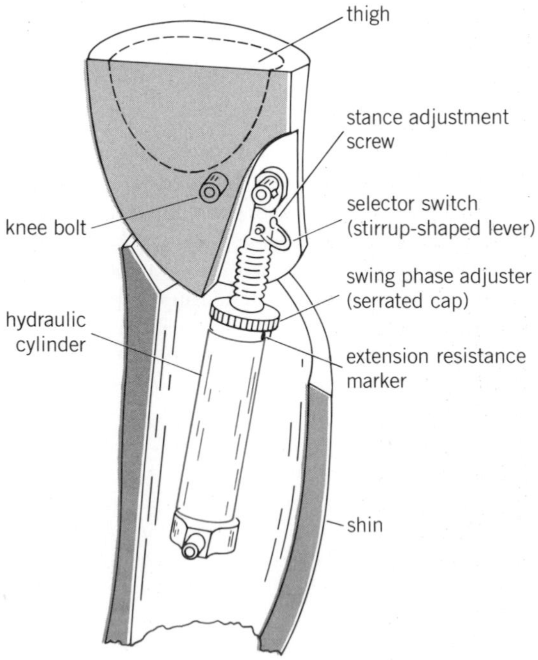

Fig. 2. Diagram of the Mauch SNS (swing and stance) knee.

The Mauch SNS (for swing and stance) knee control (Fig. 2) is the only surviving design from many projects on knee locks. Its unique control provides adjustably high resistance to knee flexion at all times except after prolonged hyperextension moment—the safe condition which occurs naturally when the amputee is on the forefoot of the prosthesis late in stance phase and is about to begin the rapid knee flexion of early swing phase. A separate system within the same hydraulic cylinder provides independently adjustable resistances to knee flexion and extension during swing phase, programmed to match fundamental data on normal walking and automatically increasing with walking speed. Additional features permit the amputee to select full lock against flexion for working at a bench or standing in a moving vehicle or, conversely, absence of locking for such activities as bicycling. Variants of the knee control and the long-awaited Mauch hydraulic ankle became available in 1984.

Upper-limb prosthetics. Absence of fingers or thumb, reported for 1,545,000 persons in the census for 1977, was far more common than minor amputations of toes only (300,000 persons) or absences of major extremities (264,000 persons missing lower extremities and 91,000, upper extremities). Distribution of levels was also quite different from the changing pattern noted in the lower extremity. Major absences included 28,000 amputations at or above the elbow, 24,000 below elbow and above wrist, and 38,000 hands (except digits only). Injury was by far the largest cause, particularly for minor absences, emphasizing the importance of safety engineering and education.

Unilateral upper-limb patients frequently learn to function reasonably well without a prosthesis, particularly in the cases with minor absence or, conversely, with such high-level absence as to be difficult to fit. Many tasks can be performed with the remaining hand alone or with hand plus body motions, assistance of any residual limb, or technical aids or adaptive equipment. The sensory feedback of the bare residual limb may be more valuable to the amputee than the better grip and more normal appearance afforded by conventional prostheses.

Conventional major amputations of the upper limb are wrist disarticulation (relatively uncommon), below-elbow, elbow disarticulation (relatively uncommon), above-elbow, shoulder disarticulation, and forequarter with loss of clavicle and scapula as well as the entire arm. Fortunately, the last two are uncommon; often at least a small portion of the humerus can be retained to round out the shoulder and perhaps provide voluntary control motions.

Prostheses are available for each level. Conventional body-powered prostheses are most frequently used. They involve plastic laminate exoskeletal construction and a Bowden cable with a fabric harness across the back and an anchoring at the opposite shoulder (or the opposite prosthesis in the case of a bilateral amputee). The cable and harness permit limited but substantial feedback of information on position and force. The amputee may use an artifi-

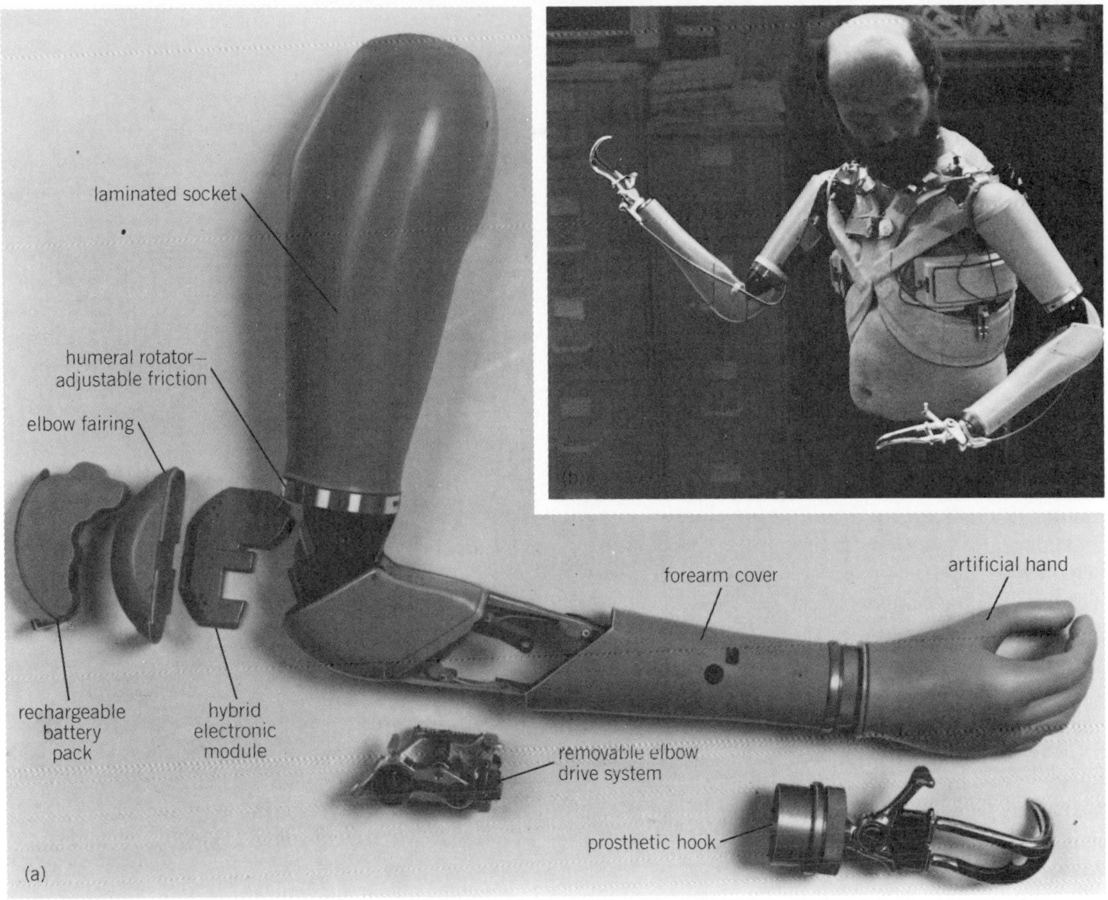

Fig. 3. The Utah artificial arm. (*a*) A clinically available, externally powered arm. Tiny myoelectric signals from remnants of elbow flexors and extensors are picked up through the skin by surface electrodes and used to drive the electric motor, causing elbow flexion and extension. The arm can be interchangeably fitted with a myoelectric hand or a split mechanical hook controlled by a cable from a shoulder harness. (*b*) A fitting for a high bilateral amputee; hooks are usually preferable. (*Motion Control, Inc., Salt Lake City, Utah*)

cial hand for social occasions and office work and a split mechanical hook for heavier work and, by choice, for some clerical jobs, tasks of every day living, and sports. Though an artificial hand with a cosmetic glove is not noticeable in casual observation, a split hook permits the amputee to function more skillfully and gracefully. Bilateral amputees, particularly, typically use hooks for dexterity.

External power for artificial arms has been studied since World War I, when both a compressed-air hand and a solenoid-actuated electric hand were designed. The Otto Bock electric hand, originating in West Germany, is probably the most widely used externally powered device, either with its own myoelectric control system controlled by forearm muscles or with adaptations to other control systems.

Another device is a self-contained, self-suspended prosthesis for the below-elbow amputee. With this device there is no need for either the shoulder harness or belt-supported battery pack of some earlier prostheses because the skin electrodes are built into the socket wall over the appropriate forearm muscles of the residual limb, the electronics and batteries are built into the prosthetic forearm, and the entire prosthesis is held in place by the fitting of the brim just above the bulging condyles at the elbow. The amputee can easily attach or remove the prosthesis, somewhat like a slipper. Cosmetic appearance and comfort are improved, and risk of clothing damage is reduced.

Electric elbows for above-elbow amputees have attracted considerable attention for many years. The early designs, using either switch or myoelectric control, were rejected by high proportions of amputees for various reasons, including weight, slow speed, noticeable noise while operating, poor appearance, and inconvenience of frequent battery recharging. The Boston elbow emphasizes myoelectric control of the elbow motor for flexion and extension, with a cable-controlled hook used for prehension and passive adjustment (by the remaining hand usually) of wrist rotation.

The Utah electric arm (Fig. 3) is ultimately intended to supply independent, myoelectric control of several joints (hand, wrist, and elbow). A simple version is now commercially available, which uses myoelectric signals from the biceps and triceps to control elbow flexion and extension; when the signals remain steady but above those of relaxed muscles for a brief period, the elbow automatically locks and the same myoelectric signals then become available to control closing and opening of the Otto Bock electric hand. An added circuit board allows independent adjustment of the sensitivity to signals for elbow and for hand. The wrist and turntable rotation are controlled manually against adjustable friction. A simpler harness pattern is said to allow better fitting of high-level above-elbow patients. Feedback of elbow position is provided indirectly by force patterns on the residual limb related to the bending moment caused by forearm, hand, and load.

The Veterans Administration Prosthetics Center (VAPC) electric elbow was produced in substantial quantities for field trials and fitted on volunteer veterans in a number of clinic teams throughout the country. The original design was noticeably noisy in quiet surroundings; an improved but more expensive flexible spline gear reduction system was quieter. The VAPC elbow is commercially available. There is also a VAPC modification of the Viennatone electric hand, equipped with a safety breakaway to release grip in emergencies.

The Northwestern University (NU) synergetic hook features two small motors operating simultaneously. One drives a rapidly moving finger until it meets the object, regardless of size, and then stalls; a special circuit cuts off power. The second motor operates through gearing with far more mechanical advantage to drive its finger against the opposite side of the object relatively slowly but with-gripping force steadily increasing. Neither motor-gearing system can be back-driven by external force. By controlling the brief time the second motor is allowed to operate, the amputee can vary the gripping force applied to the object. The mechanical system maintains grip with the power off. When the amputee gives the signal, the hook opens immediately at any load level.

Early externally powered arms were provided with on–off control. The automatic gear shift in the Bock hand and the NU synergetic devices were attempted improvements. In addition, there are attempts to provide proportional control, as in the Utah arm, and research on feedback.

Most artificial arms have been made with exoskeletal construction in plastic laminate, but in one design a tubular endoskeletal structure is made with matching elbow, wrist, and other components and a resilient foam for cosmetic covering. In both lower- and upper-limb prostheses, the choice of structural type requires the designer to compare strength, stiffness, ability to adjust alignment as desired and perhaps to readjust occasionally (yet securely fasten the parts as long as desired), and ability to provide satisfactory appearance at any joint position without risk of pinching clothing.

Sports and other activities. Recent developments for both lower- and upper-limb prostheses have tended to assist sports and recreation activities to supplement the traditional emphases on daily living and vocational tasks. Sturdy prostheses with special alignment for skiing, waterproof limbs and fins attached to sockets for swimming, and attachments or adapters to permit arm amputees to hold sporting equipment are examples.

SPINAL CORD INJURY AND ADAPTIVE EQUIPMENT

As a result of vastly improved medical care, antibiotics, and in some cases biomedical engineering advances, persons with spinal fractures and spinal cord injury have much better chances of survival. Biomedical engineering is contributing electronic methods for diagnosing the condition of the injured spinal cord, as well as increasingly sophisticated biomechanical principles, instrumentation for research, numerous types of rehabilitative devices, and a growing hope for functional electrical stimulation of paralyzed muscles in useful ways.

WHEELCHAIRS AND AUTOMOTIVE EQUIPMENT

Many rehabilitation engineering devices, such as wheelchairs, have long been available for patients with many types of disabilities. Recently, however, new designs have been developed, including models for sports, and there has been more interest in careful prescription of chair and equipment to meet individual needs. Electrically powered chairs, sometimes with powered reclining backs, and control systems requiring only slight forces and motions have become more widely used for quadriplegics and other severely disabled cases.

Various types of automotive adaptive equipment have been available for decades to increase comfort, convenience, and safety for normal drivers. Electric starters, windshield wipers, and similar aids are taken for granted. These and conveniences such as automatic transmission, power-assisted steering and brakes, and power windows may be crucial to permit disabled persons to drive. In addition, an increasing variety of other devices are available as custom or commercial equipment. Levers and cables, for example, permit hand operation of foot pedals, and there is increasing attention to electronic, hydraulic, and pneumatic servo systems.

The use of vans as passenger vehicles has considerably simplified travel for wheelchair users, either as passengers or as drivers. Entry and exit by ramp or power-operated lift while sitting in the chair may be much simpler than sliding from the chair to the seat of a sedan with the aid of a bridgelike transfer board. Once in the van, the disabled person may transfer to a conventional seat with seat belt, designed for resistance to accidents. Crash tests with dummies show clearly that the typical tubular-frame wheelchairs restrained by seat belts or bolts to the van deform dangerously. Experimental wheelchairs intended to permit deceleration loads required of automobile seats have been designed but are not yet readily available.

ENVIRONMENTAL CONTROLS AND MANIPULATORS

The term environmental controls has been used for some years in rehabilitation to designate devices by which severely disabled people such as high-level quadriplegics can voluntarily control not only temperature and air motion in their immediate environment but radio, television, attendant call, or a number of other features by exerting low-force, low excursion commands such as puffing or sipping. Originally relays were used, but solid-state controls were later applied.

Manipulators to perform tasks for severely disabled persons may be mounted on a worktable (Fig. 4) or, less commonly, on a wheelchair. To reduce the load on the user in commanding the various motions and combinations, a microprocessor works out details of the necessary path from simple commands given by the user via a retractable chin control. The chin control drives the wheelchair when it is moving, but communicates with the robotic arm or manipulator by infrared when the wheelchair is docked at the worktable. The manipulator can be used for eating, moving equipment on the worktable, changing diskettes in a computer, changing paper in a typewriter, and selecting reading material. A two-length mouth stick in a motorized holder can be used in the short form as a drinking straw and at full length to operate an electric typewriter, a personal computer keyboard, or a telephone keypad fitted with buttons.

FUNCTIONAL ELECTRICAL STIMULATION

The term functional electrical stimulation (FES) is used broadly for a variety of types of electrical stimulation of nerves or muscles, sometimes including the widely used transcutaneous electrical neural stimulation (TENS) to relieve pain, or assist efforts to control the bladder and urinary sphincter. Commonly, though, FES is used for the rapidly developing field of electrical stimulation intended to provide motion of paralyzed muscles. Over a decade ago there was considerable effort to assist hemiplegics to walk without the usual spring-loaded orthosis by electrically stimulating the anterior tibial muscles to lift the toes and forefoot during the swing phase of walking. While some users appreciated elimination of the metal or plastic orthosis, others found specific aspects objectionable. These included a tingling sensation, dangling wires leading to breakage with external systems using skin electrodes, or the surgery involved in implanting electrodes on the common peroneal nerve while still requiring an electronics box, battery, and induction loop on the body and a heel switch in the shoe.

After years of fundamental work, FES is now actively being studied, on muscles paralyzed by

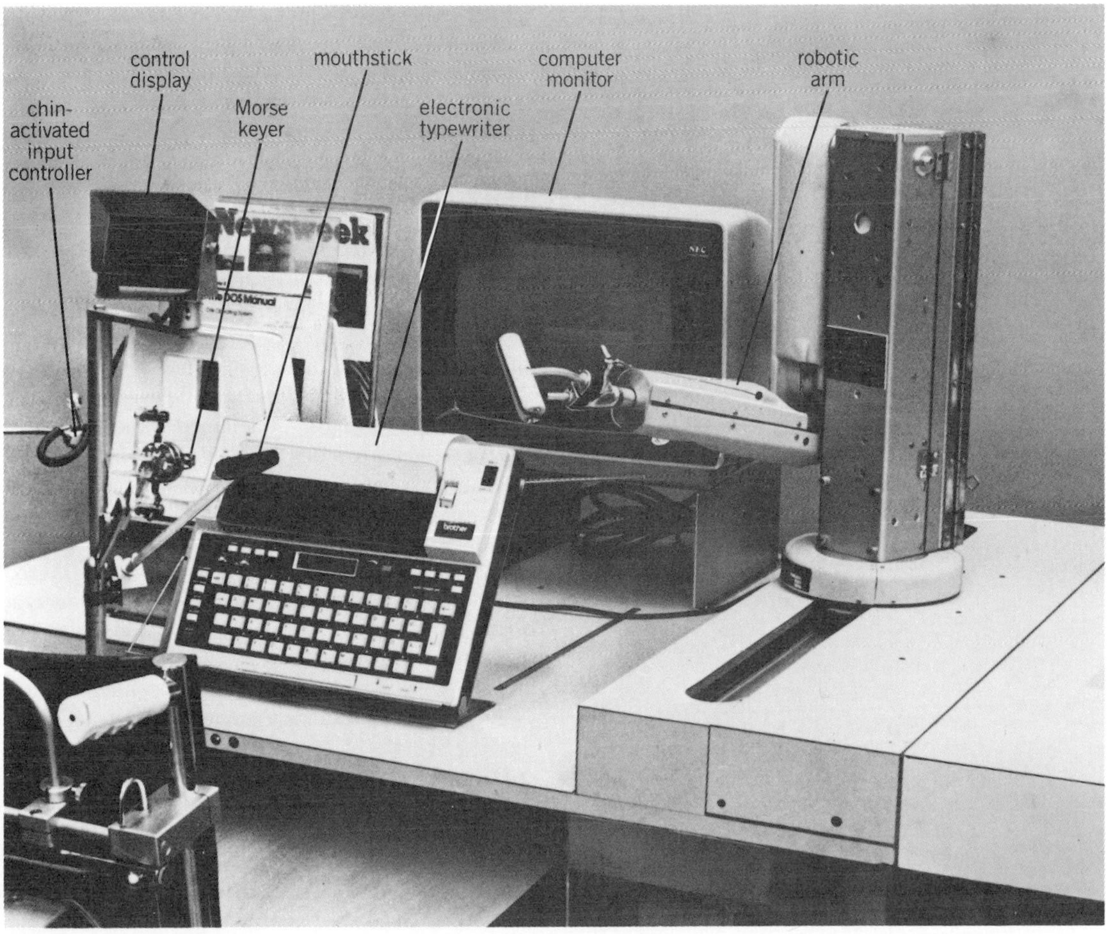

Fig. 4. Worktable configured to use a small portable electronic typewriter with a mouth stick. The robotic arm or manipulator may be controlled directly by slight motions and feeble force, such as chin motion of a quadriplegic, but computer assistance is available to reduce the load upon the user. (*Johns Hopkins Applied Physics Laboratory*)

spinal cord injury, at a number of centers in Yugoslavia, France, and the United States. Finger flexors have been stimulated in quadriplegics by shoulder control of a stimulator and percutaneous fine-wire electrodes left in place for months at a time. Leg muscles of paraplegics have been stimulated to allow standing and walking with support from a wheeled "walker," with voluntary control by finger-operated switches. Figure 5 shows a paraplegic walking outdoors by functional electrical stimulation involving a roller walker carrying battery-powered equipment and thumb switches in the handles. Though the patient exhibited excessive pelvis shifting, he demonstrated good stability on one leg while flexing and advancing the other. He walked very slowly, but could do so for 20 or more minutes without readjustments.

At the present time, a paralyzed person can move across a flat surface more easily and efficiently in a wheelchair than by FES. Thus there are important developments and projects on mobility and functional aids to improve conventional care, and very substantial hopes for at least limited applications of FES while further research and development continue.

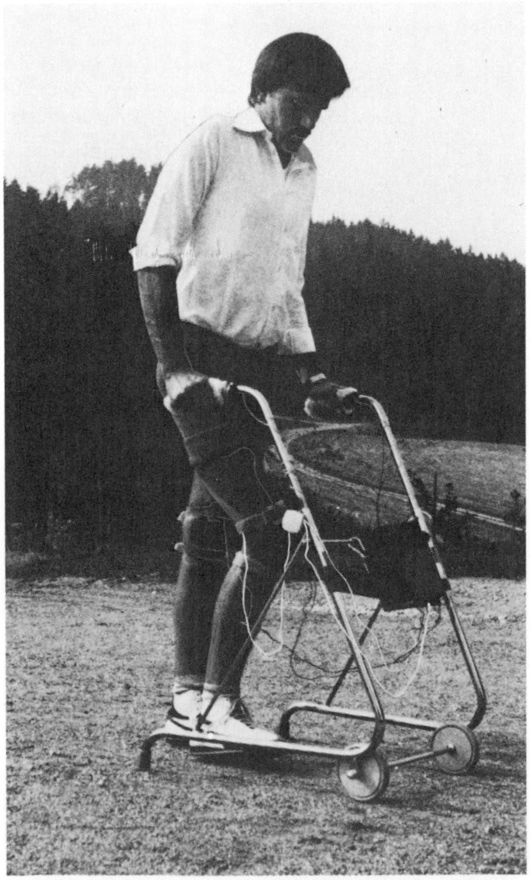

Fig. 5. Functional electrical stimulation walker for a paraplegic. (*From A. Kralj et al., Gait restoration in paraplegic patients: A feasibility demonstration using multichannel surface electrode FES, J. Rahab. Res. Dev., vol. 20, no. 1, July 1983*)

ORTHOTICS

The term orthotics was coined, by analogy to prosthetics, to describe the art and science of prescribing and fitting devices to support a part of the body. Formerly called a brace or orthopedic appliance, such a device is now termed an orthosis.

Orthotics is a much larger field than prosthetics, yet in many ways much more complex and difficult to improve. Orthoses may be useful for a wide range of conditions affecting only one or a few tissues, possibly remote from the visibly affected part. The condition may be temporary, long-term, or lifelong, and it may improve with time and perhaps treatment, remain stable, or deteriorate steadily or with remissions. Orthoses are worn outside the body, which may be atrophied or deformed, so appearance and possible clothing damage are always important considerations, and fitting is sometimes unusually difficult. Mechanical joints are typically opposite the bulbous bony condyles of the bones, so the orthosis may appear even more clumsy if the paralyzed muscles between joints are atrophied. Unlike the hollow interior of an exoskeletal prosthesis, there is no convenient hiding place for locks or power elements.

In many cases orthoses form only part of a total acute treatment and long-term rehabilitation plan, which may also include surgery, drugs, therapy, functional electrical stimulation, and such orthotics-related devices as wheelchairs, walkers, crutches, or canes. Orthotics management thus requires an interdisciplinary team.

Materials. A major trend in recent years has been the use of plastics instead of metal structures with leather cuffs and straps. Originally, thermosetting plastic laminates were adapted from prosthetics, but increasing amounts of thermoplastics have been used. Sheet material may be heated until soft, then draped, stretched, or vacuum-formed over a plaster model. The cooled rough structure is trimmed to shape and smoothed. The three-dimensional shape may provide adequate strength and stiffness. If necessary, a structure can be reinforced and stiffened by corrugations or by adding carbon fibers embedded in acrylic matrix or tapes of glass fibers to resist tensile loads. Because impact and repeated fatigue loads are particularly likely to cause breakage, some flexibility is desirable, and stress concentrations at sharp corners, notches, and scratches should be avoided.

Polycarbonate, ionomer, and polymethyl methacrylate are transparent thermoplastics at room and body temperatures, an advantage for checking pressure spots at fitting and for inconspicuous appearance of the finished appliance. Polycarbonate, though the strongest, must be thoroughly dried before molding, and polymethyl methacrylate is relatively weak. Ionomer, a relatively new material, is increasing in popularity.

Polypropylene and polyethylene, both semicrystalline materials, are not transparent, though some grades may be translucent in the thicknesses needed. Polypropylene is used extensively in many types of orthoses, particularly because its excellent fatigue resistance minimizes breakage of an entire structure and allows deliberate design of thin strips as hinges in place of moving joints. Sometimes the strip also serves as return spring. Polyethylene is not as resistant to fatigue but is useful in body jackets. Acrylonitrile-butadiene-styrene (ABS) has both high rigidity and high impact strength, and it is available in a wide range of colors. It is used for spinal supports and custom-molded wheelchair seats.

Plastic foams are increasingly used. Expanded polyethylene, with or without small holes for ventilation, is widely used for padding and liners. Sheet polyethylene, when heated and formed over expanded polyethylene pads placed on the model, bonds perfectly.

Plastic materials may stick to themselves or to a model when hot. However, with experienced handling, this property can be used to advantage for sealing when desired, but avoided when necessary by separating materials. Some thermoplastics can also be welded. Unlike the sharp melting point of metals, the plastic softens over a wide temperature range, so a different technique is used.

A synthetic balata which softens in boiling water and can be shaped comfortably on the body has been used widely by therapists for arm and hand splints and also for fracture splints allowing relatively early walking. The thermal diffusivity, related to the thermal conductivity and density, is so low that the rapidly evaporating water and radiation cool the surface before it touches the patient. The heat remaining in the slab escapes so slowly that neither the patient nor the operator is burned while the appliance is shaped as desired, then held (often with moist elastic bandage) until cool and firm. Unfortunately, the desirable ability to relieve pressure spots by local reheating carries the risk of inadvertent damage from high temperature, and the device is relatively weak compared with those from the materials discussed above.

Recognition is growing that many thermoplastic devices can be substantially prefabricated in a few sizes over typical wooden models, then quickly and easily adjusted with a heat gun to fit an individual patient.

Devices. Lower-limb orthoses are needed primarily for support of body weight by stabilizing joints despite paralyzed neuromuscular control, by relieving load or limiting motion in arthritic joints, or by supporting healing fractures or nonunions. Foot-ankle orthoses are also widely used to prevent toe drop and assist dorsiflexion in hemiplegics. Rigidly constructed bilateral orthoses designed to permit hip flexion and extension but block adduction have been used to prevent "scissoring" in certain cerebral palsy cases. Even if a patient cannot expect to walk, orthoses are sometimes prescribed to control spasticity while sitting in a wheelchair. Upper-limb orthoses range from temporary supporting splints to complex externally powered orthoses or perhaps a manipulator.

Orthopedic shoes, shoe modifications, custom inlays in stock shoes, and various supports should also be mentiond. Often these are used by persons with otherwise normal limbs, but sometimes they are used in conjunction with orthoses.

Spinal orthoses vary from simple belts and corsets to complex structures. Some act by restraining the abdominal contents, causing the torso to act as a beam when lifting loads, thereby reducing tension in the back muscles and compression on the discs between the vertebrae. Others serve partially or entirely as reminders to a wearer with normal sensation and muscles to avoid slumping into undesirable positions. Obviously, the reminder concept would not be applicable for quadriplegics. Some designs deliberately provide biofeedback to train the user's reflexes and muscles. An important aspect of spinal orthotics is the treatment of curvatures in the spines of teenagers, particularly girls.

SURGICAL IMPLANTS

Surgical implants, primarily for treatment of fractures, have been used for decades. Improvements in stainless steels and cobalt-chromium alloys and the introduction of pure titanium and titanium-aluminum-vanadium alloy allowed surgeons to obtain reasonably adequate compromises of mutual tolerance between host and implant; strength under static, impact, and fatigue loading even in the corrosive body fluids; stiffness or flexibility as needed; and ability to form complex shapes, and sometimes to modify in the operating room.

Bone plates, screws, and wire sutures are used widely for internal fixation of fractures. Cups to separate the surfaces of arthritic hip joints, partial joints, and in recent years total prosthetic replacements of joints have become widely used.

Ultrahigh-molecular-weight polyethylene, typically is now used for sockets of total hip prostheses. Methyl methacrylates have been widely used in contact with mucous membranes in dentistry. A major factor in the development of total joint prostheses was the introduction of polymethyl methacrylate as a grout (not really a cement, though often loosely termed that) between the stem of the prosthesis and the reamed medullary cavity of the bone. Porous surfaces of ceramics (or ceramics applied to metals) have been used to permit gradual bone ingrowth to anchor the implant stem to the bone. Medical grades of silicone rubber are widely used, for example, as finger joints.

In addition to orthopedic surgery, implants are used in other surgical specialties. Silicone rubber hydrocephalus shunts (to drain excess fluid from the skull) and metal plates and clips are widely used in neurosurgery. Plastics are used in cardiovascular, otological, and maxillofacial surgery, and are crucial in artificial organs.

In ophthalmology there is long experience with

plastic contact lenses on the cornea. Recently, implantable plastic lenses have been inserted immediately after surgical removal of the human lens.

For over 20 years there have been interdisciplinary efforts to develop standards for medical devices, especially implants. These range from basic materials, through specifications for numerous devices, to labeling and packaging. The American Society for Testing and Materials has published numerous voluntary consensus standard specifications, test methods, and recommended practices for medical and surgical devices, particularly surgical implants. The Association for the Advancement of Medical Instrumentation has worked particularly with cardiac pacemakers and other equipment involving electrical activity. The American National Standards Institute continues to sponsor a few standards-writing groups, including one on ophthalmic lenses (conventional eyewear, contact lenses, and recently implanted lenses), but mainly coordinates American efforts with international work. The Food and Drug Administration can impose legal standards, but frequently encourages development of and compliance with voluntary standards.

ARTIFICIAL ORGANS

The term artificial organs covers a diversity of approaches to aid persons with serious, usually life-threatening conditions affecting the kidneys, heart, or other vital organs. A spectrum of components and devices are involved. Some, like prosthetic heart valves, were highly experimental about 25 years ago, became reasonably satisfactory, and are still widely used. Intraaortic balloon pumping to assist a weakened heart temporarily is attempted in about 5000 cases a year in the United States with perhaps 70% success. The cardiac pacemaker to provide tiny electrical impulses to stimulate the heart muscle was experimental 25–30 years ago, but there are now said to be 60,000 or more recipients a year. At the other extreme, there are a handful of experimental subjects for research on prosthetic vision and hearing, animal research on various substitute organs, and basic research on such concepts as organized tissue regeneration.

Work on artificial organs or components may involve temporary, long-term, or lifetime application. The heart-lung machine or pump-oxygenator was originally used only to assist open-chest surgery, though now patients can be maintained for relatively brief periods.

Pacemakers can be used as tools in the hospital, then removed, but other designs are surgically implanted to function on demand, or continuously for several years at a time until the special battery begins to deteriorate. Originally replacement was routinely scheduled at perhaps 18 months when mercury batteries were considered safe for at least 2 years. Methods for testing pacemaker performance remotely via telephone and availability of longer-life lithium batteries have lengthened the intervals between surgery for many. Some work has been done on a nuclear isotope power source intended to last for years and perhaps for the remaining lifetime.

The artificial kidney, particularly as the dialysis machine, is a very widely used artificial organ. While it is sometimes used in hospitals for acutely ill in-patients, a widespread use is for periodic (usually three times a week) dialysis of persons with end-state renal disease, usually on an outpatient basis with the patient able to carry on normal activities for 2 days at a time between sessions. Selected patients with suitable training and perhaps home modifications to assure proper power and water supply may be able to dialyze themselves at home.

Another important device is the "wearable" artificial kidney with rechargeable batteries and a small plastic bag that serves as a tank for 15 minutes of true mobility, though normally it is connected to a 21-quart (20-liter) tank of dialysate for most of the customary 4-hour session. By careful planning and liaison with other artificial kidney centers, though, a patient using conventional stationary dialysis equipment can achieve considerable mobility.

Other approaches to kidney disease include peritoneal dialysis, hemofiltration, and hemoperfusion. Like hemodialysis, each involves compromises of medical, engineering, economic, and human factors. Access of blood to the machine or of dialysate to the body with minimal risk of infection and discomfort, minimal blood loss as tubes are disconnected, medical benefit including removal of appropriate molecules with minimal loss of others, first costs and operating costs (usually to a third-party payer), mobility, and convenience are among the considerations. Daily or even continuous home treatment might be desirable physiologically to avoid sharp changes in body chemistry and to simplify controls. On the other hand, thrice-weekly treatment at a major hospital-based artificial kidney center might assure more medical supervision and justify more sophisticated automatic controls of the equipment to respond to the individual's physiological condition.

Attempts are being made to aid diabetics by developing an artificial pancreas. The goal is to encapsulate insulin-producing cells in synthetic materials which will allow inward diffusion of nutrients and outward diffusion of waste products and insulin, but will not allow the host's body to perceive the cells as foreign or to attack them.

An ambivalence, and possibly a dilemma, in studying some artificial organs involves the goal in relation to transplantation. Is the organ a temporary life-support system and perhaps a tool for therapy or rehabilitation to improve the condition of the patient before receiving a transplant from a suitable donor (often hard to find)? Is even longer support needed until immunosuppressive drugs are further improved, donor finding and perhaps tissue and organ banks improved, and transplantation surgery further refined? Or are transplants the relatively temporary expedients until artificial organs are further improved or even invented? In principle the supply of donor organs is capable of great expansion but likely to remain limited, whereas any needed number of artificial organs could be constructed, with decreasing costs in mass production. Is a large transplant better or worse than a large foreign body? Can inert materials have sufficient fatigue life to last for many years without the metabolism and self-repairing properties of living tissues? What about growth if organs are to be considered for children?

[EUGENE F. MURPHY]

Bibliography: Alliance for Engineering in Medicine and Biology, Proceedings of various annual conferences on engineering in medicine and biology; American Society for Artificial Internal Organs, Transactions; A. M. Cook and J. G. Webster (eds.), *Therapeutic Medical Devices*, 1982; A. Kralj et al., Gait restoration in paraplegic patients: A feasibility demonstration using multichannel surface electrode FES, *J. Rehab. Res. Dev.*, vol. 20, no. 1, July 1983.

Integrated pest management

LARRY P. PEDIGO

Larry P. Pedigo is professor of entomology and chairman of the Pest Management Curriculum at Iowa State University. He is author or coauthor of over 70 articles related to integrated pest management and is active in teaching and research in this area.

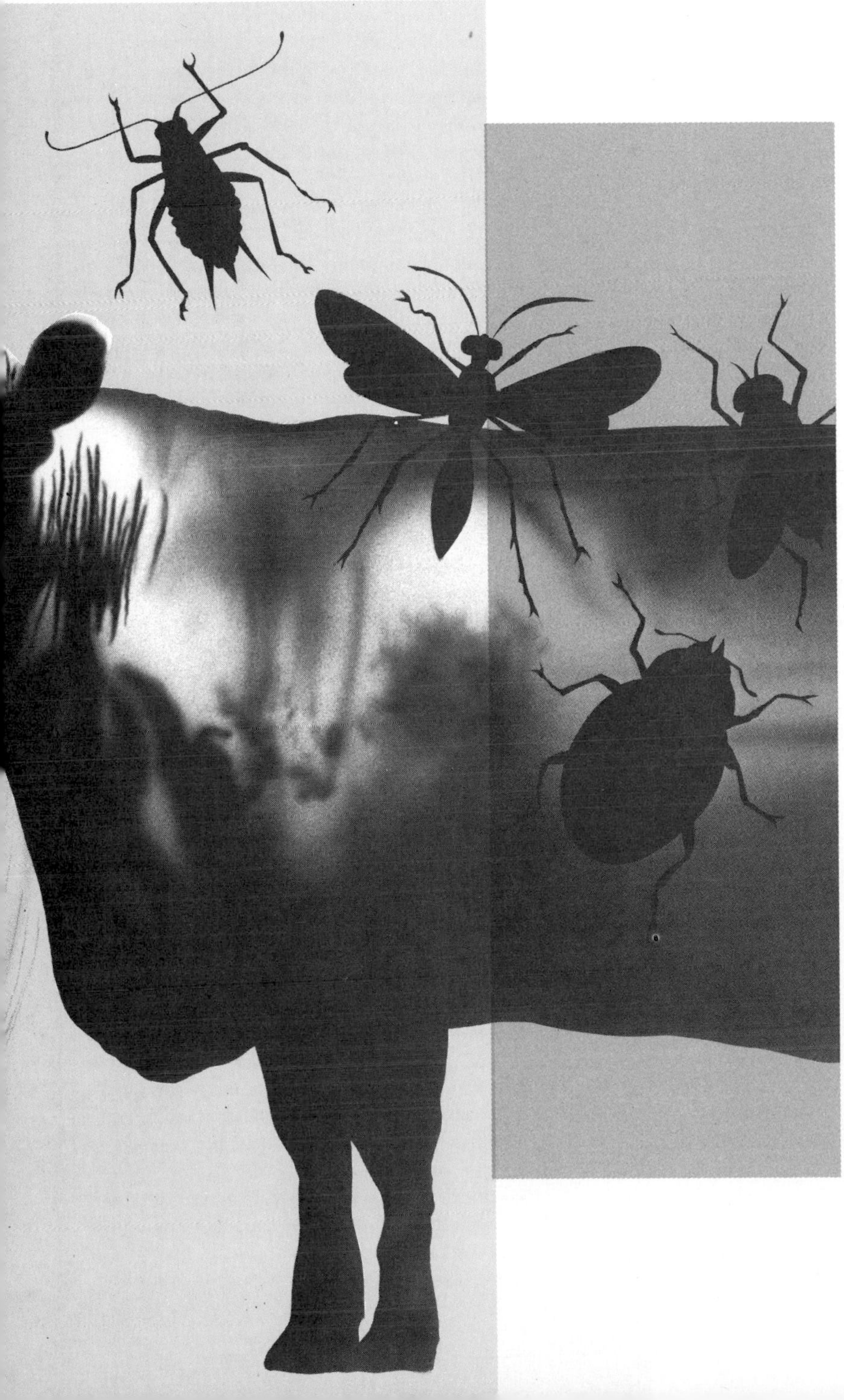

For the better part of its existence, agriculture has been seriously impeded by the ravages of pests. On a worldwide basis, an estimated 35% of potential food production is lost, and in many crops, dealing with pests is a major preoccupation.

Various approaches to alleviate pest problems have been attempted through the years. Probably no approach has been more pervasive or more popular among scientists than integrated pest management. Today, the practice of integrated pest management is supported at all levels of United States agriculture, including federal (Department of Agriculture; Environmental Protection Agency) and state (university research and extension) agencies, private crop consulting firms, and grower organizations. In fact, integrated pest management was ordered to be used wherever practicable by United States presidential decree in 1977, and is mandated by California law to be used when feasible.

Why has integrated pest management been so widely accepted? What has made the concept appealing not only in theory but also in practice? The answer is that integrated pest management is currently one of the few approaches to pest control that is effective and ecologically sound, yet is also based on economic realities.

DEFINITION

Many definitions have been proposed for integrated pest management, with each emphasizing various aspects. The following definition attempts to be broadly descriptive. Integrated pest management is a holistic approach to pest control that uses combined means to reduce the status of pests to tolerable levels while maintaining a quality environment.

In this definition there are several implications that set the approach apart from all others. First, integrated pest management seeks to deal with pests in the context of the whole production system, rather than as a separate collection of individual problems. It proposes that several tactics be integrated to alleviate problems, rather than relying on a single tactic.

The objectives of integrated pest management are also clear from this definition. The object of integrated pest management is to reduce pest status. Although reducing status may be done by killing pests, killing certainly is not the objective; protecting the commodity is. Consequently, pest status also may be reduced by avoiding or repelling pests or by reducing reproductive rates of pests.

Another implication of the definition is to produce tolerable levels of the pest. Here, tolerance means that humans live in the presence of pest species, albeit at levels that are not economically important. This aspect admits that complete elimination of pests may not be feasible or even desirable. Particularly, tolerance of pests sets integrated pest management apart from earlier approaches to pest control.

Finally, the keystone objective of integrated pest management is to maintain environmental quality. Conservation involves not only the quality of nonagricultural environments and their elements (water, soil, air, wildlife, and plant life) but also the quality of the cropping environment itself. With past approaches, environmental quality of the crop was largely ignored, excepting the attention paid to crop vigor and soil elements supporting the crop. With integrated pest management it is understood that cropping systems behave similarly to "natural" ecosystems, with a diversity of interacting elements such as natural insect and weed enemies that impinge directly on system stability. Therefore, maintaining a quality system, referred to as an agroecosystem, assists stability and therefore tends to reduce pest problems.

Achieving all of these objectives in every instance is difficult and may not always be feasible. Nevertheless, the integrated pest management system design attempts to satisfy these objectives and thus achieves more lasting and adroit solutions to pest problems. The final objective, therefore, is to provide for increased agricultural productivity while protecting the quality of the environment.

SCOPE

To achieve its objectives, the scope of integrated pest management is very broad. Integrated pest management addresses problems caused by many kinds of pests, including insects, mites and other arthropods, plant-parasitic nematodes, microbial and viral plant pathogens, weeds, and vertebrates. The focus for these organisms is on population density or infection frequency and the cause of variation in the numbers of pests in cropping and other environments. There is a preoccupation with numbers because, if number variation in time and place can be understood, predictions of problems can be made and populations can be manipulated more effectively.

Integrated pest management is sometimes equated to plant protection. However, in some ways its scope is broader than plant protection because it addresses pest problems with livestock, urban dwellings, and landscape, and includes certain aspects of human health (for example, mosquitoes and other vectors of human disease).

The wide scope of integrated pest management is also shown in the many suppressive tactics that may be utilized. An integrated pest management system may employ any number of tactics, including pesticides, host-plant or host-animal resistance, tillage, sanitation, and biological control. To use these tools effectively, the approach relies on an intimate understanding of the agroecosystem, quantification of pest activity, and appropriate integration of the information.

HISTORICAL ASPECTS

Historically, integrated pest management is an entomological concept. However, many elements of the approach were also used previously in other disciplines, particularly plant pathology. Integrated pest management developed during the late 1950s and 1960s from increasing problems with a single-component, mainly insecticidal, approach to insect control.

Insecticidal control. This single-factor approach developed out of the early successes with synthetic organochlorine insecticides, the best known being DDT (dichlorodiphenoltrichloroethane). Prior to DDT, insecticides were not efficient compounds for insect control. Most were inorganic compounds, and many, like paris green (copper acetoarsenite), were expensive, difficult to apply, toxic to plants, and not always effective. With the discovery of the insecticidal properties of DDT by Nobel laureate Paul Müller in 1939, a host of new and efficient synthetics were developed and widely used. The low cost and high performance of these compounds fostered the abandonment of earlier combined tactics for suppression, for example, using insecticides along with tillage and sanitation procedures. The insecticidal approach relied only on insecticide killing power.

As the practice of total insecticidal control grew, the goal of total eradication (complete elimination of insects) was developed, requiring increasingly frequent applications at greater dosages. With more and more dependence on insecticides, serious problems began to arise—problems related both to the nonagricultural environment and to the insect control program itself.

These problems were brought to public attention with the publication of Rachel Carson's book *Silent Spring* in 1962. This publication made effective indictments against pesticides, citing environmental degradation and chains of reactions caused by them. The result was a public outcry for environmentally safe tactics for pest control. Consequently, a United States Presidential Science Advisory Committee on Environmental Quality was appointed to study the problem, and a report recommending improved pest control practices was issued in 1965.

Pesticide misuse. Problems with the insecticidal approach in the cropping system also began to occur with shocking regularity at the dawn of integrated pest management. These problems have been referred to as the "three R's" of pesticide misuse: resistance, resurgence, and replacement.

Resistance is probably the best-known and most widespread problem. It occurs when a pesticide is applied repeatedly, initially causing heavy mortality in the pest population (Fig. 1). Because a few individuals in the population are predisposed to detoxifying the poison, they survive to produce a population made up of their own kind, resistant to the pesticide. Then, no matter how often and at what rate the pesticide is applied, it is ineffective. The first instance of resistance to modern insecticides was reported in 1946, with houseflies and DDT in Sweden and Denmark. Today, more than 400 cases of resistance have been documented.

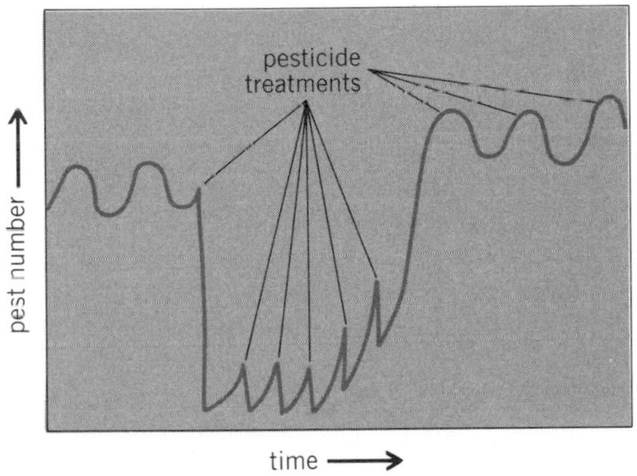

Fig. 1. Development of resistance in a pest population, showing that the pesticide is effective at first, followed by increasing ineffectiveness.

Resurgence (Fig. 2) occurs when an insecticide is applied to the pest population and the population is depressed momentarily; then, after a brief period, pest numbers return to a level higher than before the treatment. The primary cause of this phenomenon has been shown to be the destruction of arthropod natural enemies along with the pest. Once released from the limiting effects of predators and parasites, the pest can, at least temporarily, return in greater numbers. By the time integrated pest management concepts were developed, more than 50 species of plant-feeding arthropods had been documented as showing resurgence.

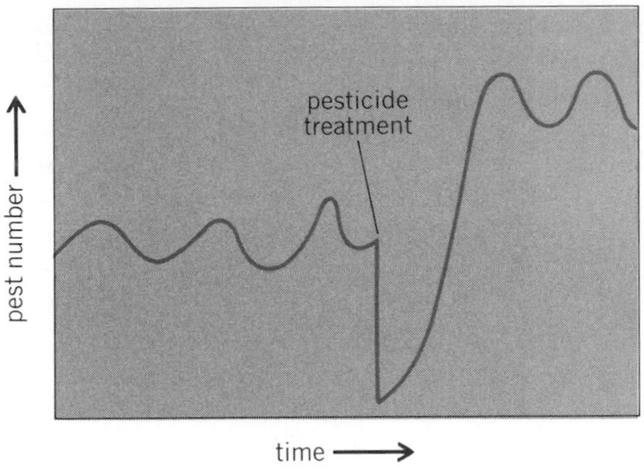

Fig. 2. Resurgence of a pest following depression of the population after pesticide treatment.

Replacement (Fig. 3), or secondary pest development, is similar to resurgence in that it is generally believed to be caused by the destruction of arthropod natural enemies. With replacement, however, the target pest (A) is kept in check with repeated insecticide applications, but it is replaced by another pest (B) formerly present at insignificant levels. In this instance, the replacing pest is not susceptible to the insecticide, but its natural enemies are strongly affected, allowing an increase in population levels. A good example of this phenomenon occurred in California in 1946–1947 when citrus was treated with DDT to control citricola scale. Along with suppression of citricola scale, populations of a specific predator, the vedalia beetle, were reduced and cottony cushion scale became a major pest within a few months.

Most of the problems with the "three R's" have been prevalent where pest problems are perennial and insecticide application is frequent. Usually, the most severe pest problems and, therefore, insecticide problems have occurred in crops with high market values, namely, cotton, citrus, deciduous fruits, vegetables, and ornamentals.

Before integrated pest management, the most common response to these problems was to switch to other insecticides when resistance occurred, to apply more frequent treatments for resurgence, or to use combinations of insecticides for replacement. These practices simply prompted ever-increasing in-

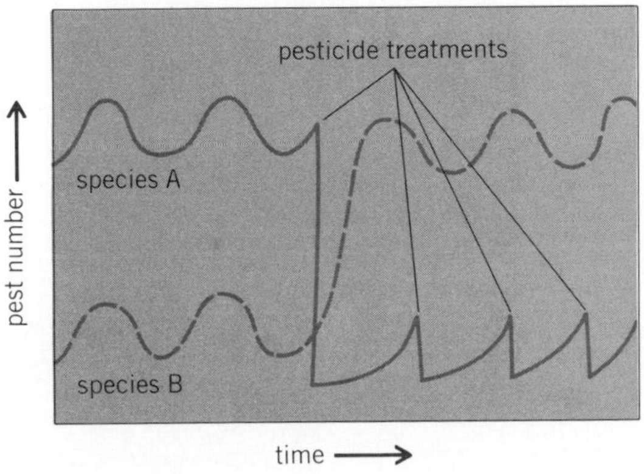

Fig. 3. Replacement of one pest, species A, by another, species B. The pesticide remains effective against A but has no effect on B.

secticide applications, with protection in the short term at best.

Pesticide treadmill. The practices eventually led to the so-called pesticide treadmill or pesticide syndrome (Fig. 4). The condition represents an addiction of the agroecosystem to pesticides most often caused by a reduction of natural enemy abundance and diversity. It represents human encumbrance of the density-regulating function of natural enemies and encourages a greater need to intervene in the system. The pesticide treadmill, in particular, was instrumental in fostering the integrated pest management philosophy and concept.

Biological and integrated controls. The integrated pest management concept has its roots in earlier theories of biological control—the use of natural enemies or introduced enemies to control insect and weed pests. Early successes and lasting solutions with devastating pests, for example, cottony cushion scale in California citrus (1888), *Opuntia* cacti in Australia (late 1920s), and Klamath weed, or Saint-John's-wort (*Hypericum perforatum*) in the western United States (late 1940s), encouraged the development of biological control and emphasized the significant impact of effective natural enemies in pest control.

With many pests, however, natural enemy controls were undependable for economic containment. Subsequently, the concept of integrated control was developed during the late 1940s and 1950s. Integrated control emphasized the selective use of insecticides so that natural enemies were conserved in the cropping system. This "integration" of control tactics was expanded in later years to include other tactics, for example, resistant hosts and sanitation, but natural enemy conservation was always the primary axiom of integrated control.

Pest management. With this background it was Australian entomologists L. R. Clark and P. W. Geier who clearly stated the principles of IPM in 1961. They suggested the terms protective population management or pest management for these ideas. Pest management differed from earlier approaches in its holistic viewpoint, its synthesis of ideas, and its inclusion of basic population theory in its design.

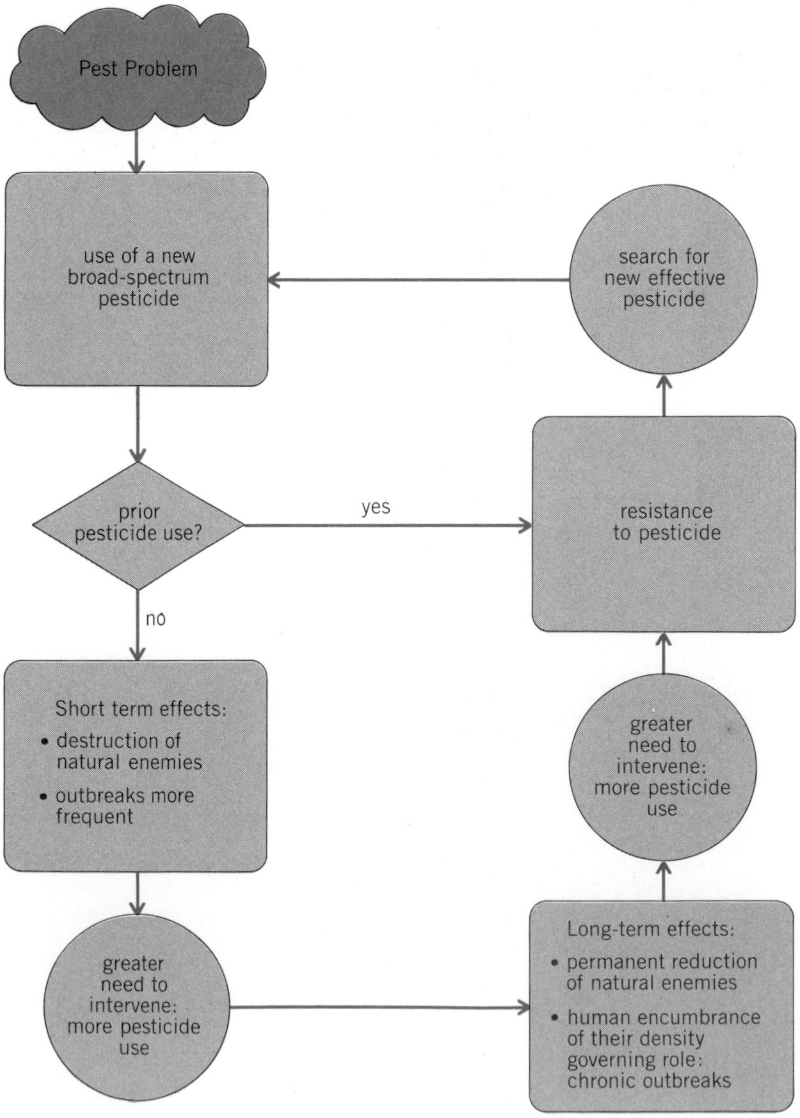

Fig. 4. Flow chart of the pesticide treadmill.

CONCEPTS OF PEST AND PEST STATUS

When is an organism a pest? Answering this question is central to the attainment of integrated pest management objectives.

Pests. Pest species are those that interfere with human activities. According to Geier, the quality of being a pest is anthropocentric and circumstantial. Termites feeding on dead wood in a forest serve an important ecological function, one of degradation in the process of returning nutrients to the soil. Clearly, these are not pests; indeed, they are beneficial to humankind. The same species, performing the same ecological function but in the environment of a human home, is a pest. An understanding of the often important ecological roles played by "pest" species in nonagricultural environments often gives insights in how to deal with them, and can aid in developing more tolerant attitudes to their presence.

Pest problems encountered in production systems are usually problems of numbers or intensity and are not determined by the mere presence of a species. Therefore, an agricultural pest may be described as any living species whose activities, enhanced by population numbers, cause economic losses.

The economics of loss plays a prominent role in the pest concept and helps define the circumstances under which a species is really a pest. In integrated pest management, a response is called for only when pest status demands it.

Pest status. This concerns the ranking of a pest relative to the economics of dealing with the species. Pest status is usually variable for a pest, depending mainly on the crop involved and the environmental matrix in which the pest–crop interaction occurs.

The major factors contributing to pest status are shown in Fig. 5. Among these, market values and management costs are among the most variable. Holding all other factors constant, a pest can assume higher or lower levels of importance because of changing economics. For example, a European corn borer population attacking corn at $2.10 per bushel carries a lower status than the same population attacking corn at $3.75 per bushel.

Susceptibility of the crop to pest damage or loss from weed competition is another important, but little understood, factor. Variable weather factors (particularly moisture) and cultural procedures (such as fertilization) can profoundly affect crop vigor. Under optimal conditions, pest status can be lowered because of compensatory growth or increased competiveness compared with adverse growing conditions.

But it is the environment, including the human social environment, that is the major cause of change in the factors governing pest status. These factors mediate pest status, which determines the measure and constraints of an integrated pest management program.

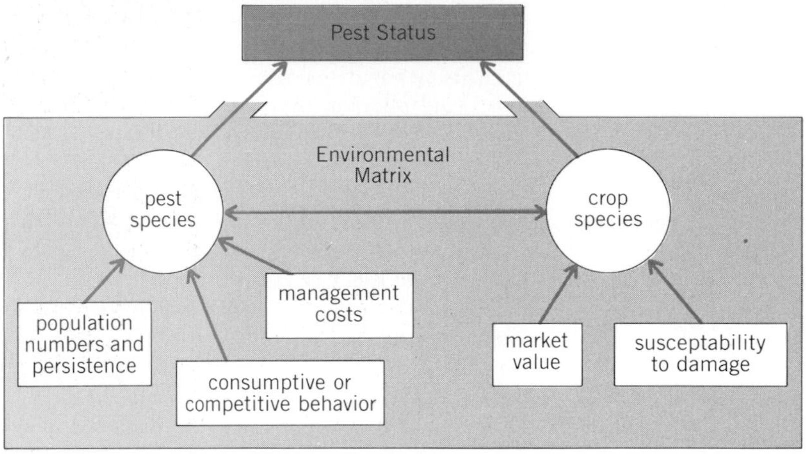

Fig. 5. Relationships between major factors influencing pest status.

PROGRAM COMPONENTS

The practice of integrated pest management is dependent on a series of well-designed programs that apply to key pests in the production system. Each program, or so-called subsystem, can be visualized as a bridge that can be crossed to avoid significant pest losses (Fig. 6). The bridge is composed of a foundation arch (information and techniques), several vertical pillars (tactics), and a road surface (avoidance of pest losses).

Information and techniques. The foundation arch, made up of several elements, represents the basic information that should be known and techniques that are needed in order to manage a pest. In Fig. 6, the elements to the left of center stand for biological information about a pest, and those to the right for procedures needed to obtain this information.

Biological information includes how individuals feed, grow, develop, reproduce, and disperse, as well as habitat requirements. Particularly, integrated pest management seeks weak points in the life cycle that can be exploited. The seasonal cycle is studied along with pest developmental rates to allow predictions of pest occurrence and relationship to biological events (phenology) of the crop.

Population dynamics is another critical element in the biological information needed for integrated pest

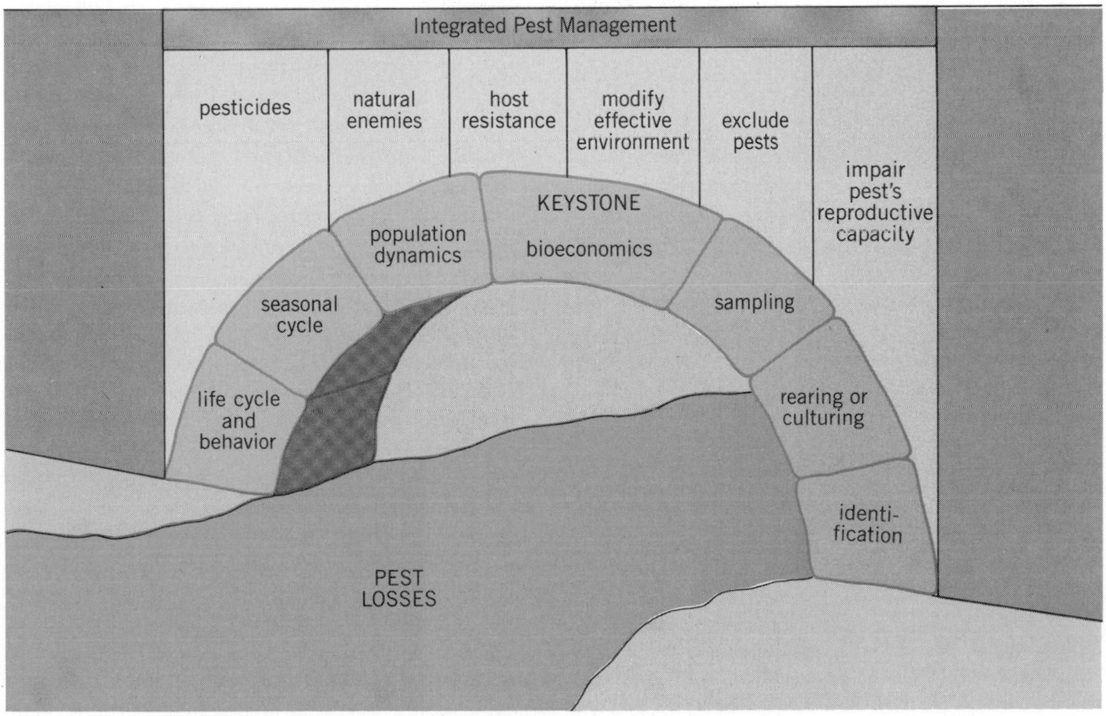

Fig. 6. The integrated pest management bridge, a schematic comprising the components in an integrated pest management subsystem.

management. It seeks to explain the forces (for example, natural enemies, weather, food, and habitat) causing population change, by quantifying these forces and allowing the description of numbers in terms of mathematical models. Such models, consisting of component equations, have often been organized to represent an ecological system and have been programmed for computer simulation. Such simulations are believed useful for understanding the nature of population dynamics in various environments. Population dynamics information builds upon the understanding of behavior, life cycle, and seasonal cycle, and allows predictions of infestations and outbreaks. It also gives indications as to environmental factors that may be exploited to reduce pest status.

The techniques in the foundation arch include species identification, rearing or culturing, and sampling. Identification involves assigning the correct name to a species and knowing the taxonomic classification. Since every species behaves differently, it is critical that accurate identifications be made. Such identifications are a key to the published literature about a pest. They aid in research and development of integrated pest management programs and serve as a reference for delivering such programs to practitioners. Not only is the ability to rear, culture, or colonize a pest useful for obtaining biological information about a species, but also laboratory and field experimentation with control tactics can proceed more rapidly than by having to depend completely on natural pest occurrence.

Sampling techniques and programs play a particularly important role in integrated pest management. Such techniques allow the estimation of numbers for population dynamics studies and later serve to assess the population for integrated pest management decisions. Lacking precise and economical sampling programs, it is doubtful that an effective integrated pest management program can be designed.

The keystone of the integrated pest management foundation arch is bioeconomics, that is, the relationship of pest numbers to losses. Bioeconomics ties together biological information of the pest and

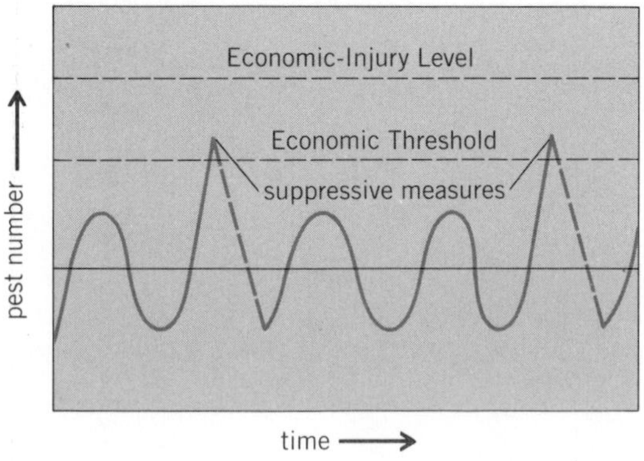

Fig. 7. The relationships between the economic-injury level, economic threshold, and a fluctuating pest population. (After *V. M. Stern et al., The integrated control concept, Hilgardia, 29:81–101, 1959*)

host crop, sampling programs, and economics. Determination of bioeconomics is greatly dependent on a knowledge of pest–plant (or pest–animal) relationships and how the crop responds to pest damage. Coupling known crop responses with pest densities and accounting for market values and pest management costs allow decision indices to be developed. The most definitive of these were proposed by V. M. Stern and other entomologists in the late 1950s. These indices, the economic-injury level and the economic threshold, denote the economic break-even point and the level at which pest management activities should be implemented, respectively (Fig. 7). According to F. L. Poston and others, these levels have formed the backbone of progressive insect control and integrated pest management development for the past 20 years. Far from being static, these levels change with changes in the variables shown in Fig. 5. Intensive efforts are presently under way to improve upon and better define bioeconomics, particularly from the standpoint of multiple pest problems and variable crop vigor.

Tactics. The foundation arch subsequently supports several pillars of the integrated pest management bridge. These pillars represent tactics that are employed, in combination, to reduce pest status or manage a pest. Those shown in Fig. 6 do not include all tactics but are some of the most useful.

Pesticides. These comprise mostly herbicides, acaricides, insecticides, fungicides, and rodenticides. They are used to kill pests outright. If applied properly, these materials are economical and efficient for short-term reduction of pest status. Today, judicious pesticide use forms the bulwark of integrated pest management programs against weed and many insect problems. If they are applied improperly, environmental damage, as discussed above, can result. Integrated pest management seeks to use pesticides thoughtfully and in moderation. The approach is to use selective pesticides (physiological selectivity) or apply broad-spectrum pesticides in a manner that is selective (ecological selectivity). Using reduced rates, restricting the area of treatment, and reducing the number of treatments during a season are all part of this selective strategy. In particular, integrated pest management seeks to conserve natural enemies with selective applications and avoid problems of pest resistance. This selective approach was most aptly stated by A. W. A. Brown, who said that people should "use insecticides (pesticides) as a stiletto, rather than as a scythe." Most important in this philosophy is tolerating residual levels of pests and using pesticides as indicated by the economic threshold, and not using them in attempts at eradication. An exception to this guideline might be with newly introduced pests in limited areas, for example, the Mediterranean fruit fly in southern Florida and California.

Natural enemies. The use of natural enemies is the prevalent tactic in almost all integrated pest management programs. Natural enemies include predators, parasites, and disease-causing organisms of pests. A complex of these often plays a role in the population dynamics and regulation of most species, including pests. Harboring or augmenting those natural enemies is often a cost-effective and environmentally desirable means of reducing pest status. Particularly important is the integration of this tactic with pesticide applications and environmental modification, a procedure that might give the natural enemy an advantage over the pest. The introduction of exotic natural enemies has great potential for long-term solutions to introduced-pest problems (classical biological control) and has shown dramatic success, particularly in perennial crops (for example, pastures, forests, and orchards).

Host resistance. This is another widely used tactic that has been combined with natural enemies and pesticides to reduce pest status. It is the primary tactic for dealing with problems with many plant diseases. Resistance to damage may be expressed in crops by: (1) outright destruction of an attacking pest or impairment of pest abilities (antibiosis); (2) being unattractive to a pest (antixenosis); or (3) tolerating injury without an appreciable yield loss (tolerance). Most host resistance has been developed in annual plants because of the relative ease of breeding techniques and shorter time involved. Resistance also has been accomplished in livestock, primarily cattle and sheep, by using conventional techniques of animal husbandry. It has been predicted, however, that with new technology, genetic makeup can be engineered at the molecular level to allow optimum efficiency in developing host resistance.

Environmental modification. Modifying the effective environment of a pest includes any procedure that removes or masks requisites (for example, habitat sites, overwintering refugia, or food supply) or in other ways creates an unfavorable microclimate for reproduction and survival. By removing or masking these requisites, the pest suffers greater than usual mortality from exposure to adverse weather and natural enemies, it starves to death, or it is lured away from the crop. To achieve this, changes in plant culture are usually made. Such techniques as sanitation, tillage, crop rotation, change in planting time, addition of trap crops, and change of row spacing are commonly used. Something can usually be done to incorporate this tactic in almost all integrated pest management programs, and, when feasible, the procedure may be one of the least expensive and most effective tactics, with few environmental risks.

Excluding pests. Exclusion, as an integrated pest management tactic, includes all procedures to prevent the introduction and spread of pests or, on a small scale, to block a pest from reaching its host. The main objective of exclusion is to prevent pest injury, as opposed to some of the previously discussed tactics where curing a problem is the major objective. Many of the exclusion activities include quarantines enforced by governmental agencies. The main idea is to prevent the spread of pests by legally restricting the movement of infested or infected commodities. This aspect of exclusion is probably the most important and is responsible for the reduction of countless potential losses. Although this approach seeks to exclude pests from the entire production system, attempts to exclude may also be used for individual plants or animals in the presence of pests. Examples of this production-level exclusion include cardboard collars to exclude cutworms from garden plants and screened cages to exclude codling moths from laying eggs on dwarf apple trees.

Impairment of reproductive capacities. This is a procedure that has been mostly developed with insects. The tactic is most often employed on an area-wide basis and aims at reducing numbers of new individuals produced. Commonly used methods include sterility induction (autocidal control) and pheromone applications. The idea behind sterility induction is to produce sterile insects, usually males, treated with gamma radiation or chemosterilants, and inundate the natural population with these. The sterile insects mate with fertile insects, resulting in unfertilized eggs. With continued inundations the pest population falls off and may be completely eliminated from a region. The most dramatic use of this tactic has been the eradication of the screwworm fly, an important parasite of cattle, from the Florida peninsula in 1959. However, biological problems (including multiple matings), reduced competitiveness of sterilized insects, and the costs of development have limited the potential of this tactic for many insects.

Pheromones are chemicals that are emitted by an individual and induce a response in another member of the same species. Sex pheromones usually attract members of the opposite sex, resulting in mating. The identification and synthesis of these materials, or production of imitations of them, has allowed their use in integrated pest management programs. Currently, the greatest success has been in sampling and monitoring populations with pheromone-baited traps. Such traps have been very important in the detection of new infestations of gypsy moth and in the timing of sprays in apple orchards for codling moth. The main idea for use of pheromones as a suppression tactic is to release the material into the atmosphere to "confuse" the receiving insect (usually males) as to the exact location of a mate. In theory, mating in a treated area would be greatly reduced and a population decline would result because of a much reduced birthrate. Tests in limited areas using microencapsulation and thermoplastic hollow fibers for constant pheromone release have met with some success. Some of the most promising uses probably lie with insect pests in forests, apples, and cotton. Although technical problems remain, it has been predicted that pheromones and other chemical attractants will become an increasingly important component in many integrated pest management subsystems.

STRATEGIES

Current practice in integrated pest management is to develop subsystems made up of the components

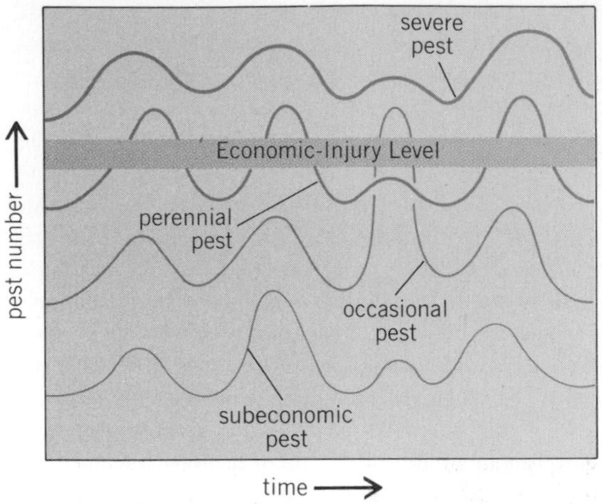

Fig. 8. Types of pests in relation to the economic-injury level.

discussed for pest species or complexes of related species. These subsystems then are combined, in a harmonious way, with other cultural activities of the production system. Therefore, the species subsystems are at the heart of the overall integrated pest management program.

The strategy and complexity of a subsystem are greatly dependent on the status of a pest. Four pest types based on status have been designated by Stern and others. These types include noneconomic populations (subeconomic pests) and occasional, perennial, and severe pests (Fig. 8). This classification is based on the relationship of their average density and highest fluctuations to the economic-injury level.

Subeconomic pest. The subeconomic pest may seem to be a contradiction of terms. However, it is a pest in the true sense, albeit one that causes insignificant losses; that is, trying to reduce these would cost more than the losses they inflict. Many insects and plant pathogens fall within this category. An example is the alfalfa caterpillar in Iowa. The insect is a defoliator and causes direct losses of forage. However, densities are low and market values modest, so that the most appropriate action is to take no action, at least from a control standpoint. The strategy for such pests is to monitor activity levels for changes that may alter the pest's status.

Changes either in the pest itself or in the crop can greatly alter pest status and turn a subeconomic pest into one of devastating importance. Such was the situation with *Helminthosporium maydis*, a fungus causing southern corn leaf blight. Before 1968, the organism was known to infect corn leaves but was of little consequence to the yield of corn in the Midwest. In 1968, a new race was discovered that infested Texas cytoplasm male sterile corn, whose character had been bred into most corn hybrids. In 1970, a combination of favorable weather and widespread planting of susceptible corn allowed the southern corn leaf blight epidemic to be one of the most widespread in the history of plant pathology. This case exemplifies the dangerous potential of subeconomic pests and the need to understand them and keep them under surveillance.

Occasional pest. The occasional pest is probably the most common type of pest among insects and plant pathogens. It has an average density substantially below the economic-injury level, but the highest fluctuations occasionally exceed this level. More often than not, these pests are not economic problems; therefore, integrated pest management seeks to deal with them in a curative mode, involving early detection, prediction of impending outbreaks, and employment of integrated pest management tactics when the economic threshold is exceeded. The objective is to dampen outbreak peaks, with no need to generally change the long-term average density. Therefore, integrated pest management subsystems in this category are less complex than more serious pests, often relying on only two or three tactics in the subsystem. Today, probably most integrated pest management subsystems are of this type.

Pesticides, including microbial insecticides, fit best in the strategy for occasional pests and, when used judiciously, may serve as long-term answers for them. This is because pesticide application is infrequent and usually does not lead to the undesirable side effects mentioned earlier (the "three R's" of pesticide misuse).

An example of this type of pest is the green cloverworm, an occasional defoliator of midwestern soybeans. This insect is present each year in all soybean fields, but unusually heavy moth flights from the South in early summer set the stage for midseason larval problems. The insect is rarely a problem in late season because of regulation by a disease-causing fungus, an important natural enemy. Efficient grower-based sampling plans, predictive indices, and definitive economic thresholds have been developed for this pest. Tactics of suppression are used only when necessary and can include insect-selective microbial insecticides (biological control agent *Bacillus thuringiensis*) or low-rate conventional insecticides. With the green cloverworm subsystem, natural enemy conservation is practiced, environmental problems have been averted, and grower profits have been sustained.

Perennial and severe pests. These types of pests cause the most serious and difficult problems in integrated pest management. Most weeds, many plant pathogens, and a few insects belong in these categories. Problems created by these pests are caused by relatively high market values of the crop and pest populations with high density. In this category are insects that attack the harvested produce directly and may cause blemishes on produce unacceptable to the consumer, for example, apple scab on some apple varieties.

These categories are characterized by having an average density that is very high in relation to the economic-injury level. With the perennial pest, the average density is below, but so close to, the economic-injury level that economic damage occurs most years. With the severe pest, the average density is above the economic-injury level, causing the pest to be a constant problem.

Unlike the occasional pest, the main integrated pest management objective for perennial and severe pests is to reduce the average density of the population over the long run. In the past, pesticide applications have been used heavily and frequently to accomplish this objective. Such usage, at least with insects and plant pathogens, has resulted in prime examples of adverse side effects and environmental problems. Because of these problems, pesticides alone are not a viable tactic for most perennial and severe pests, except as a stop-gap measure in the short run. Instead, more complex programs are required to produce lasting solutions. These programs are usually composed of several tactics to lower the environmental carrying capacity for, and consequently the average density of, the pest species.

Such a program was developed for the spotted alfalfa aphid in California's Central Valley. This aphid problem resulted from their accidental introduction from the Middle East in 1954, and the species subsequently spread to California and other states. Early attempts at suppression with insecticides showed success, but resistance to these chemicals developed rapidly. Particularly in fields where resistant aphids were found, repeated sprays reduced natural enemies, and aphids resurged.

An integrated program subsequently was developed, and it still offers effective protection from losses today. The program involved: (1) the introduction and establishment of exotic wasp parasites; (2) strip harvesting (leaving uncut strips on a given harvest date) to conserve and enhance natural enemies; (3) development of useful economic thresholds and implementation of sampling procedures for aphids and predators; and (4) use of a selective insecticide, Demeton, that was less destructive of natural enemies. The later development of aphid-resistant alfalfa largely eliminated the need for insecticide sprays in the program. The lasting value of such an integrated program lies in its reliance on several tactics rather than one.

FUTURE OUTLOOK

The development of integrated pest management as an approach to pest control has been an important step in the ability to deal with pests in a manner compatible both with the environment and with production economics. Further advances, however, will be required before optimal implementation is possible.

Foremost of these needed advances is a better understanding of agroecosystems and their characteristics with regard to the production system. The agroecosystem is biological, and the production system is socioeconomic. Changes in one influence changes in the other. Researchers must clearly delineate what changes can be made in production systems and the consequences of these changes, as well as delineate the factors beyond control. This delineation requires the perception of integrated pest management as an integral part of the production system and argues for an interdisciplinary approach to research and development. Without a strong commitment to interdisciplinary research, in both philosophy and resources, it is doubtful that integrated pest management can obtain its major objectives.

Probably most crucial in an interdisciplinary approach is an understanding of the crops (plants or animals) being produced and the reaction of these to pest attack or interference. Trying to understand host plant or animal reaction is the vehicle for bringing together many specialists in the same experiment. Such interdisciplinary experiments should attempt to explain interactions between insects, weeds, plant pathogens, and other pests, and reactions to stresses caused by combinations of these in practical situations. Such information would allow a more detailed modeling of the agroecosystem, giving a stronger theoretical base for system manipulations. The integration of pest management tactics then could move from a mode of empiricism to one of analytical optimization based on sound theory.

New advances in genetics also hold promise for most pest problems. The use of classical breeding techniques in developing host resistance has made great inroads on many pest problems, but with recent technological breakthroughs in genetic engineering it would seem that the heretofore slow and plodding breeding process may be greatly accelerated. Therefore, host resistance should attain greater prominence in future integrated pest management programs than in the past. This is particularly true with slow-growing woody plants.

Finally, before optimal implementation of integrated pest management programs can occur, many public attitudes must change. The idea that "the only good bug is a dead bug" must give way to a greater appreciation of the ecological role of pest species and acceptance of their presence. This is especially true of commodities that most consumers feel should be blemish-free. This attitude, termed the cosmetic effect, is responsible for application of thousands of tons of pesticides annually, with no or little gain in yield or nutritional quality. The same is true of esthetic pests (those that are unsightly) in the home, on ornamentals, and in yards. The economic-injury level for many of these is purely psychological. Problems of attitude must be rectified ultimately with education, particularly in public school and college classes.

[LARRY P. PEDIGO]

Bibliography: M. L. Flint and R. van den Bosch, *Introduction to Integrated Pest Management*, 1981; P. W. Geier, L. R. Clark, and D. T. Briese, Principles for the control of arthropod pests, 1. Elements and functions involved in pest control. *Protect. Ecol.*, 5:1–96, 1983; D. Pimentel (ed.), *CRC Handbook of Pest Management in Agriculture*, vols. 1–3, 1981; F. L. Poston, L. P. Pedigo, and S. M. Welch, Economic injury levels: Reality and practicality, *Bull. Entomol. Soc. Amer.*, 29:49–53, 1983.

Natural waste treatment

JAY BENFORADO

Jay Benforado is a Senior Scientist in the Office of Research and Development, U.S. Environmental Protection Agency. His responsibilities include planning and managing water-quality research at EPA Environmental Research Laboratories.

ROBERT K. BASTIAN

Robert K. Bastian works as an environmental scientist in EPA's Office of Water. He currently is serving as the principal technology expert on a special EPA Sludge Management Task Force.

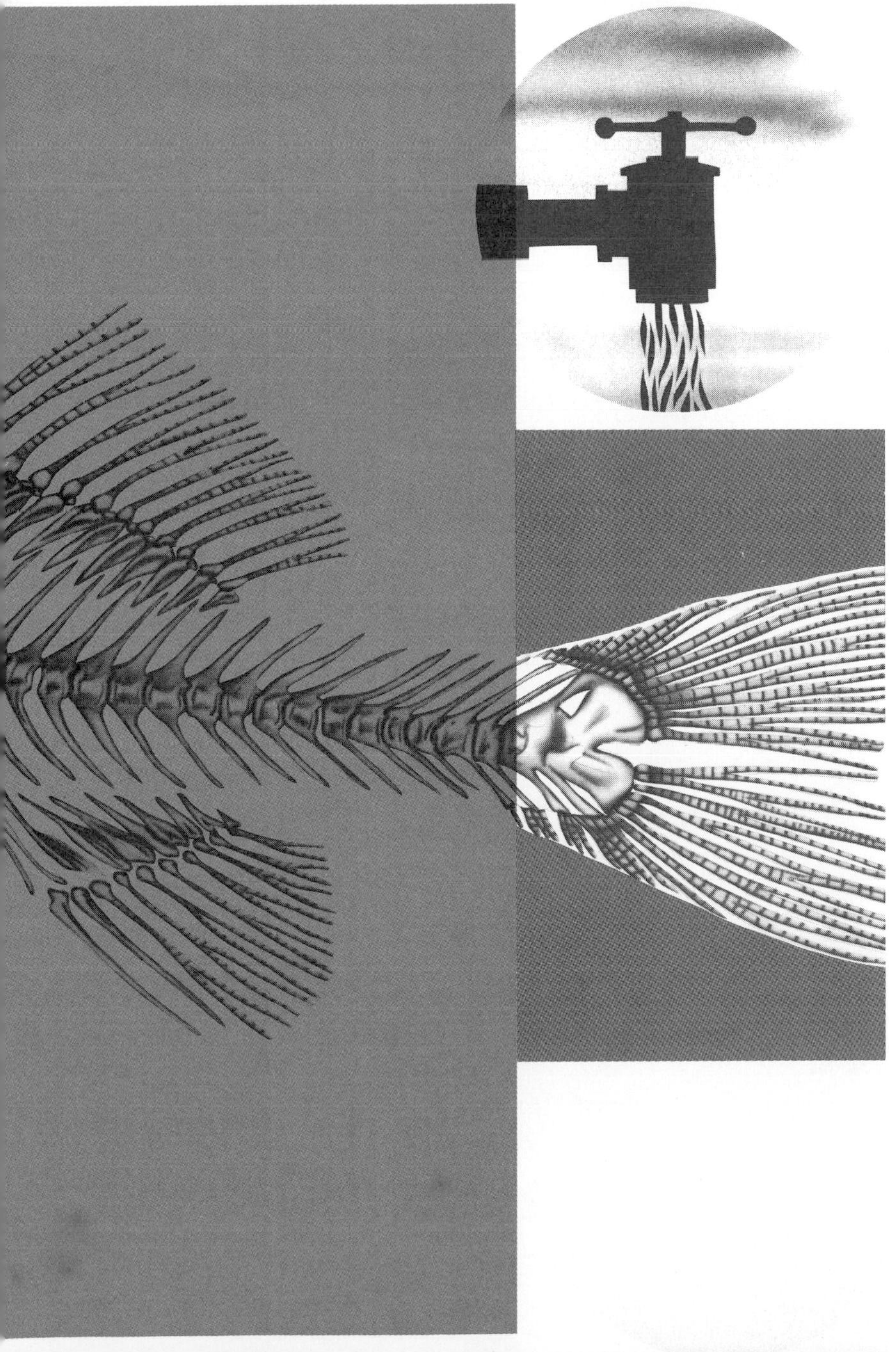

Innovative treatment techniques for wastewater and sludge that harness ecological processes found in natural ecosystems minimize the need for sophisticated engineering technology and can help reduce the cost of water pollution control. Although the landscape is littered with examples of inappropriate waste disposal—Love Canal, New York, and Times Beach, Missouri, are well known—the environment does have a finite capacity to assimilate pollutants and undergo "self-purification."

The successful use of natural environmental processes for treatment of biodegradable wastes has ranged from intensive management of an ecosystem where waste renovation is maximized at the expense of the natural qualities of the system, to the recycling of certain wastes to help enhance, restore, or create desirable ecosystems. The physical, chemical, and biological processes responsible for the breakdown and recycling of wastes in natural systems can also be reproduced in artificially constructed treatment facilities. Systems that are constructed expressly for treatment but capitalize on the use of the same environmental functions are less controversial, and can be advantageous in some circumstances because of better process control and greater system reliability.

A number of obstacles impede implementation of natural treatment techniques. In some systems, researchers are still uncertain about the eventual fate and long-term impacts of some contaminants in wastewater and sludge. Scientifically valid regulations and guidelines are needed that protect human health and the environment but are flexible enough to encourage innovation in treatment practices. A variety of practical engineering problems need resolution to allow for broader use of natural treatment systems by sanitary engineers and public officials that prefer to use the conventional technologies.

Waste treatment and disposal is a complex task, which when done improperly can adversely affect the air, land, water, and ultimately people. The use of properly managed natural ecosystems, although no panacea, offers an ecologically acceptable way to deal with many pressing water pollution problems at a reasonable cost.

CONCEPT AND CHARACTERIZATION

The disposal of waste materials has become one of the most critical problems facing the nation today; unsafe disposal practices can damage the environment and affect human health. Sources of waste include domestic wastewater and sewage sludge, industrial waste, dredged material, and solid waste, as well as urban runoff, mining wastes, and agricultural and silvicultural runoff. The size of the problem is significant—vast quantities of wastes are being generated that must be recycled into useful products or disposed of somewhere.

This article focuses on municipal sewage. Some 15,000 municipal sewage treatment plants in the United States handle more than 2.6×10^{10} gal (9.9×10^8 m^3) of wastewater each day, serving about 150 million people and at least 87,000 industries. Much of the treated wastewater—called effluent—flows after treatment into streams, lakes, and coastal waters, often to their detriment. There are five general types of pollutants that can degrade these aquatic ecosystems: oxygen-demanding wastes; suspended particulates; plant nutrients; synthetic organic chemicals; inorganic chemicals and mineral substances. The treatment plants also generate more than 17,000 tons (15,000 metric tons; dry weight) of sludge, the material removed from the wastewater during treatment, every day. Sludge disposal on land, in the ocean, or by incineration also poses environmental concerns that create economic problems for many communities.

What is more, because population is growing, municipal treatment plants are serving more industries; new processes in manufacturing are producing new, complex wastes that sometimes defy present pollution control technology; and federal and state regulations require more thorough treatment of sewage. Treatment plants in 1985 will handle about 15% more wastewater and sludge than they did in 1980, and treatment and disposal will also cost more as the prices of labor and energy continue to rise. At the same time, water pollution remains a serious national concern, and supplies of fresh wa-

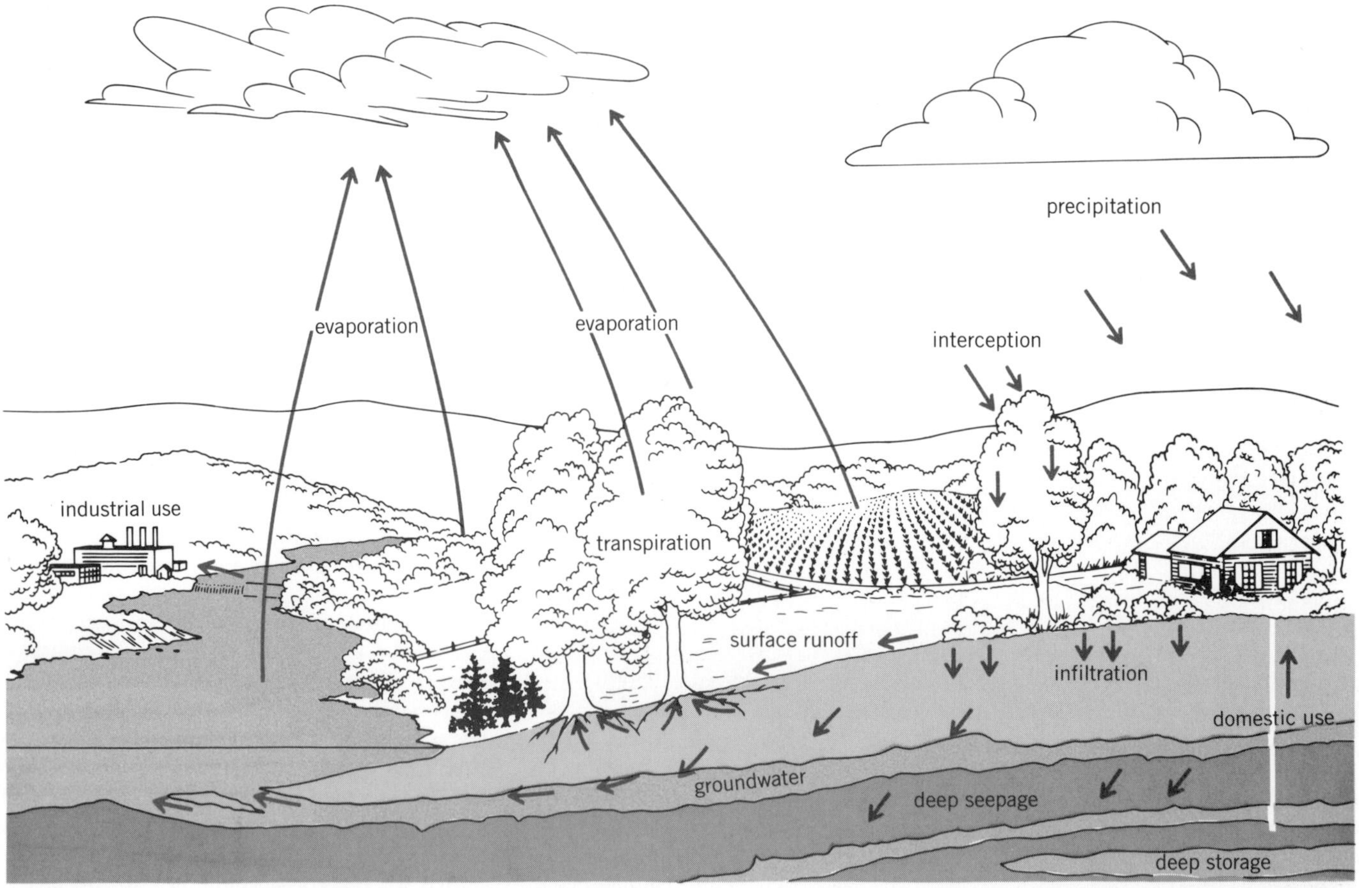

Fig. 1. The interchange of water between air, lanα, water, and living organisms is accomplished through the water cycle.

ter for irrigation, recharge of groundwater, and industrial and municipal uses are becoming scarce in some regions of the country.

The concept of natural waste treatment is not new. Nature has been recycling elements from one form to another since the world began. The idea of consciously recycling wastewater may seem distasteful to some, but this is a naturally occurring event linked to the Earth's hydrologic (water) cycle. As illustrated in Fig. 1, water evaporates, condenses to form rain, falls to the earth, is absorbed by or runs off the ground, and eventually reemerges in streams which flow into lakes or oceans, where again it evaporates, condenses to form rain, and so on. As a result, all water supplies, even rain or the purest wilderness streams, have been recycled. The same water is used over and over again. Also, effluents from many sewage treatment facilities, as well as urban and agricultural runoff, is presently discharged into streams and lakes that then serve as water sources for other municipalities downstream.

The nation's commitment to protecting the environment is an ambitious undertaking that relies on limiting the kinds and amounts of wastes that enter the environment. This is embodied in a network of federal environmental legislation, including the Clean Water Act; Clean Air Act; Marine Protection, Research and Sanctuaries Act; Safe Drinking Water Act; Toxic Substance Control Act; Resource, Conservation and Recovery Act; and the National Environmental Policy Act (Fig. 2). Although treatment of wastes prior to release into the environment is a central theme in pollution control, treatment does not totally purify waste material. The current approach is to allocate waste discharges based on acceptable waste characteristics and amounts, and an estimate of the ability of the receiving media (air, land, water) to assimilate the partially treated wastes. Natural ecosystems, in effect, provide a part of the actual waste treatment of the discharged wastes. Finding the proper balance between treating a waste before allowing a natural system to take over is a complex and long-standing issue.

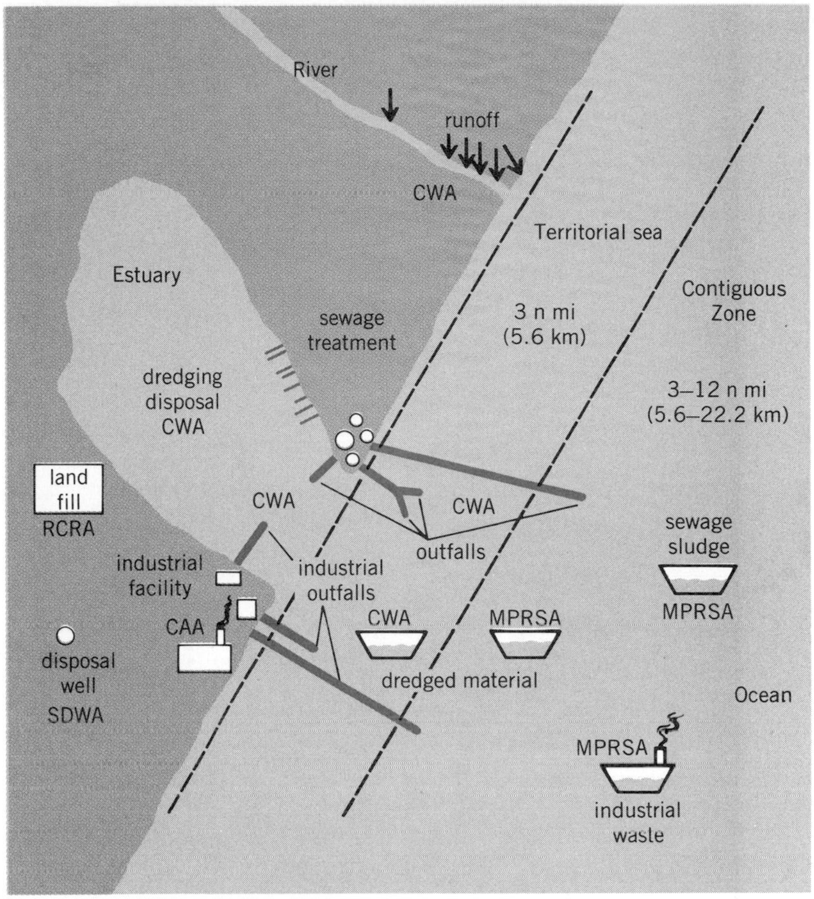

Fig. 2. Jurisdictional boundaries of key environmental laws. CAA — Clean Air Act; CWA = Clean Water Act; MPRSA = Marine Protection, Research and Sanctuaries Act; RCRA = Resource Conservation and Recovery Act; SDWA = Safe Drinking Water Act.

Historical perspective. Since the beginning of time, civilization has been faced with waste disposal problems—especially what to do with human wastes. A wide variety of technologies have been developed to deal with this problem. Some are simple to build and operate, while others are quite complex and utilize sophisticated equipment, large amounts of energy and chemicals, and skilled labor. They range from the once common backyard privy, septic tank and drainfield, and relatively simple land application systems, to advanced physical-chemical wastewater treatment systems and even the use of accelerated electrons or gamma rays from radioactive isotopes to condition and disinfect sewage sludge. Due to wide acceptance by the engineering community and public health officials, conventional treatment systems—typically, activated sludge and trickling filters—now are widely employed in wastewater treatment. These systems are generally followed by discharge of treated effluent into surface waters and disposal of sludge produced during wastewater treatment by incineration, land application, landfilling, or ocean disposal.

Only a century ago there was little regard for health hazards or environmental degradation caused by indiscriminant waste disposal. Rivers, lakes, estuaries, oceans, and the land were freely used as sewers for dumping wastes. Many of the centralized municipal wastewater collection and treatment systems of major urban areas were developed during the late 1800s and early 1900s as a means of protecting public health. At first these systems simply provided a means of flushing raw wastes out of the cities; later, facilities were added to provide primary treatment to remove large objects, floating materials, and settleable solids in raw sewage and eventually disinfection by chlorination (Fig. 3*a*).

As the public grew increasingly concerned about environmental problems, cities in heavily polluted areas gradually turned to higher levels of wastewater treatment. Secondary treatment breaks down organic matter and removes many of the suspended particles left after primary treatment (Fig. 3*b*). Most communities are now required by federal and state regulations to provide secondary treatment. And in recent years some communities have been required to do even more. Advanced or tertiary treatment goes after

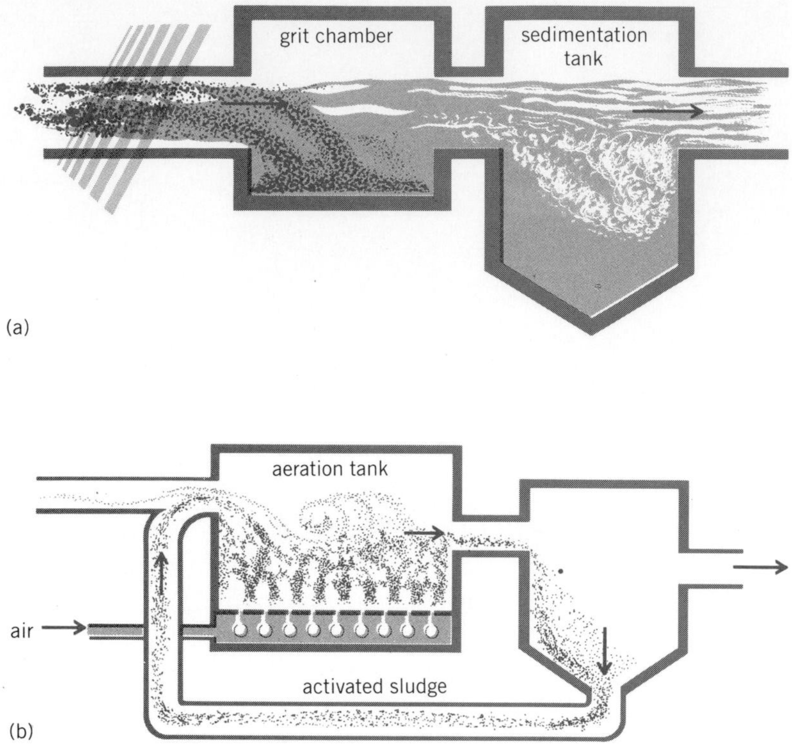

Fig. 3. Sewage treatment stages. (*a*) Primary treatment, which removes solids that will float or settle. Screens remove floating objects, and settling tanks remove heavy material. (*b*) Secondary treatment, the second step in most waste treatment systems, in which bacteria consume the organic parts of waste. It is accomplished by bringing together waste, bacteria, and oxygen in trickling filters or in the activated sludge process.

the oxygen-consuming materials and suspended solids remaining after secondary treatment, as well as compounds dissolved in wastewater. These range from nutrients, especially nitrogen and phosphorus which if discharged in large quantities can cause rampant growth of algae or other water-quality problems, to toxic heavy metals and an ever-increasing number of synthetic organic compounds, including PCBs and pesticides.

Shift toward natural ecosystems. Since the early 1970s, the U.S. Environmental Protection Agency (EPA) has pursued a technology-based approach to controlling water pollution. Municipalities and industries have been required to use certain kinds of treatment processes to meet specific levels of wastewater treatment. Most current approaches to wastewater treatment have evolved as a series of "add-on" technologies to the basic ideas developed at the turn of the century.

However, the emphasis on add-on technologies to achieve incremental improvements has locked many communities into costly, relatively inflexible, technologically complicated systems. Many of these systems use large amounts of energy—the treatment plant in a small town may require more energy than any other public service—while energy costs have more than tripled since 1970. As wastewater is treated to ever higher levels of quality by the more conventional technologies, sludge quantities increase. About half the volume of sludge that is produced in conventional advanced wastewater treatment represents the chemicals (such as alum) that are added during treatment.

In theory, natural and constructed ecosystems, if properly managed, can function as efficient and inexpensive waste treatment and recycling systems, but they must be approached in a different way than highly engineered facilities. Coupling engineered systems with natural systems is an iterative process; the donor and recipient systems must be considered as an integrated unit. The waste-producing side must be carefully managed to control the quality and amount of waste so that inputs can be assimilated without disrupting the natural system—for example, toxic pollutants that might damage the system must be tightly controlled if not entirely eliminated from the waste stream. The receiving side must be monitored to ensure adequate treatment and long-term treatment capabilities. There must be sufficient stability in the natural system to allow for adaptation. In many cases they should be allowed to change within certain limits so that the waste inputs can become an integral part of their metabolism.

The characteristics of the wastewater are of particular importance in the selection and design of a natural treatment system because of the sensitivities of both plants and animals. The nature of the wastes depend on the waste-producing source and the prior treatment. Many waste materials are rich in important nutrients: in the United States a year's worth of municipal wastewater and sludge contains an estimated 1.4×10^6 tons (1.3×10^6 metric tons) of nitrogen, 300,000 tons (270,000 metric tons) of phosphorus, and 600,000 tons (540,000 metric tons) of potassium. This represents roughly 10% of the amounts of these nutrients now supplied by commercial fertilizers nationwide—worth about $1 billion at current prices. Wastewater and sludge also contain many micronutrients that are essential for plant growth, organic matter that is valuable as a soil conditioner, and enough water to supply one-sixth of the nation's irrigated land. These potentially valuable resources, together with the major contaminants of concern (see Table 1), require careful attention. At the concentrations typically found in domestic wastewaters, of greatest immediate concern are suspended solids, pathogens, and biodegradable organic compounds. Other contaminants, such as heavy metals, can cause more subtle, long-term problems, and in the past decade pretreatment and control at the waste source have been implemented to reduce the concentrations of these compounds.

Evaluating natural ecosystems. There are several criteria by which to judge the efficacy of a natural ecosystem for waste recycling. The overriding consideration is effectiveness of treatment. Toward this end, specific waste characteristics such as contaminant levels are quite important, because there are no apparent solutions for removal of all constituents in one system. For example, the use of spray irrigation to volatilize trace organic chemicals may pose concerns about the potential for dispersal of pathogens by aerosols; similarly, the properties of wet-

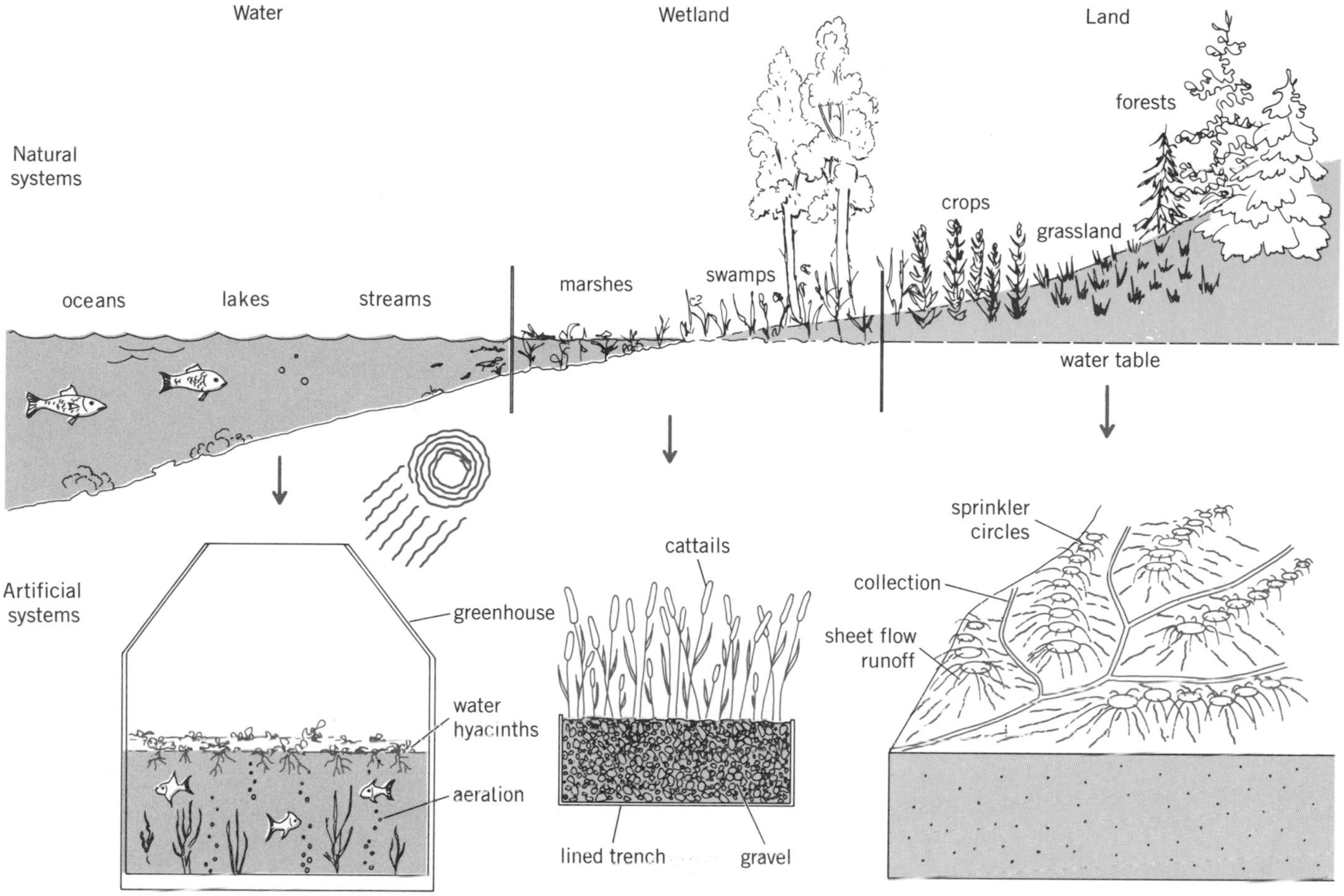

Fig. 4 A variety of different natural and constructed ecosystems can provide wastewater treatment.

land soils that promote precipitation and removal of heavy metals from wastewater also prolong the survival of viruses. A second criterion is the ease with which the system can be manipulated, managed, and controlled. Some environments are quite sensitive to change, while others can accommodate certain alterations with minimal effects. A third consideration involves the values and functions of the candidate ecosystem in its current state: can the ecosystem be changed or sacrificed in order to preserve other ecosystems? What needs to be evaluated are the treatment objectives, availability of the particular ecosystem type, and the ecologic, economic, and social roles that the system plays. The task be-

Table 1. Principal contaminants of concern in wastewater

Contaminants	Reason for concern
Suspended solids	Suspended solids can lead to the development of sludge deposits and anaerobic conditions in the receiving water.
Biodegradable organics	Composed principally of proteins, carbohydrates, and fats, biodegradable organics are measured most commonly in terms of biochemical oxygen demand (BOD) and chemical oxygen demand (COD). If discharged to the environment, the biological stabilization of these organics can lead to the depletion of natural oxygen resources and to the development of septic conditions.
Pathogens	Bacteria and viruses capable of causing communicable disease can be transmitted by water routes.
Nutrients	The nutrients essential for growth include carbon, nitrogen, phosphorus, and trace elements. When discharged to the aquatic environment, these nutrients can lead to excessive growths of undesirable aquatic life.
Refractory organic compounds	These synthetic organic compounds tend to be toxic in relatively low concentrations. Some may also accumulate in the environment, biologically and on adsorptive surfaces, concurrent with the slow decay of these compounds. Typical refractory organics are surfactants, phenols, and agricultural pesticides.
Heavy metals	Heavy metals are often toxic in relatively low concentrations. These contaminants are elemental, that is, environmentally conservative. They tend to accumulate biologically and on adsorptive surfaces. Typical examples are mercury, lead, and cadmium.
Dissolved inorganic salts	Inorganic constituents such as calcium, sodium, boron, and sulfate may have to be removed if the wastewater is to be reused.

comes one of matching a specific ecosystem with a specific waste and an environmentally sound management plan.

When using natural treatment processes as part of renovating and purifying wastewater and sludge, a choice must be made between a natural or a constructed system (Fig. 4). Each has advantages and disadvantages which depend upon individual circumstances, but selecting a constructed system often reflects a desire to protect existing natural systems.

Keeping this general ecosystem perspective in mind, an examination of the available information on technologies involving naturally occurring and artificial ecosystems that can effectively provide waste treatment follows. The technologies are grouped along a hydrologic continuum which grades from water to wetlands to terrestrial environments.

WATER

The extensive wastewater-induced changes in aquatic ecosystems caused by chronic overloading of wastes and the substantial costs of waste treatment needed to avoid these adverse environmental effects are issues of national importance. Congress addressed these concerns in passing both the 1972 and 1977 Amendments to the Federal Water Pollution Act (commonly called the Clean Water Act). This legislation, and EPA's policies adopted pursuant to it, stimulated construction of sewage treatment plants that restrict the discharge of pollutants into aquatic ecosystems, and encouraged use of other alternatives for treatment and disposal of sewage effluents.

Purification processes. The natural purification processes vary according to the physical, geochemical, and biological characteristics of a water body. The main physical mechanisms are dilution, adsorption, and sedimentation. Dilution is a fairly easy to understand process that involves diffusion and mixing. The old adage that "the solution to pollution is dilution" worked reasonably well when the waste quantities were small and readily degradable, and human population pressures were minimal. In terms of adsorption, natural waters contain an assortment of suspended particles that can attract a broad range of chemical pollutants from solution or suspension. The most important particles in this regard are clays which, because they have a negative electrical charge, sorb cations such as trace metals to their surface. Sedimentation removes pollutant particles that are suspended in water at rates that depend on the particle size, shape, and density, and the mixing and flow patterns of the waterway. The bottom sediments can often serve as a sink for many pollutants. Each year, rivers deliver billions of tons of dissolved and suspended matter to the oceans, which represent the ultimate destination for flowing waters.

The chemical reactions that affect aquatic pollutants are complex and strongly influenced by the biota, and thus not well understood. Certain types of reactions involved in the chemical breakdown of inorganic contaminants are best known. Acid-base reactions are important in the assimilation of acid and alkaline wastes. Nonbiologically mediated oxidation-reduction reactions are also of great significance for many hydrocarbons and heavy metals. Metal ions may be removed from solution by the formation of insoluble complexes. Some naturally occurring organic complexing agents such as humic and fulvic acids strongly bind heavy metals. Synthetic chelating agents such as nitrilotriacetic acid (NTA) and ethylenediaminetetraacetic acid (EDTA), which are used in detergents and industrial processes, are often discharged to surface waters, where they too may form metallocomplexes. Additional mechanisms for the removal of dissolved chemicals include coprecipitation, volatilization, hydrolysis, and, for some organics, phototransformation reactions on surface films.

Microbiological purification is generally associated with the assimilation of oxidizable organic wastes, which includes, for example, domestic sewage effluent and various industrial effluents such as those from paper manufacturing and food processing. The degradation of many hydrocarbons also proceeds to some extent by microbial action. A pollutant that is subject to decomposition by microorganisms is termed biodegradable. Microbial-mediated decomposition is a key component of the process of nutrient cycling. In nutrient cycles, organic wastes are broken down and inorganic nutrients released (for example, nitrogenous wastes are converted to nitrate, nitrite, and ammonia). Biological purification is a slow process, and exact rates are difficult to determine because of the compositional and mixing complexities of most waterways, as well as the intimate connections among biological, physical, and chemical processes.

The assimilative capacity of many water systems has been drastically exceeded, but when this stress is removed the systems do begin to recover. Flowing water systems generally are less susceptible to permanent damage by a given biodegradable waste than is a still-water body such as a lake. Groundwater appears to have a very limited capacity for biological purification in comparison with surface waters because environmental factors such as darkness limit the variety of microorganisms. However, in terms of treatment potential, this is, in part, offset by its high degree of physical purification.

Natural water bodies. The long-standing tradition of discharging sewage and other wastes into rivers, lakes, estuaries, and other waters stresses these natural environments and can affect the human uses of them. Even partially treated domestic sewage effluent can impact aquatic ecosystems by nutrient and organic matter loading, and can also impact public health by increasing human exposure to various contaminants. Nitrogen and phosphorus compounds stimulate the growth of blue-green algae and other undesirable plants in a process called eutrophication. Oxygen that is dissolved in the water is often depleted as organic matter is decomposed, and the resulting anoxic conditions are unsuitable for the survival of fish and other aquatic life. Bacterial and viral pathogens contaminate drinking water and affect contact water recreation like swimming. Other contaminants, like synthetic organic chemicals and heavy metals, although often present in minute quantities, can bioconcentrate or biomagnify in aquatic food chains and thus also create additional ecological and human health problems.

Many characteristics of natural aquatic ecosystems make them inappropriate sites for processing wastewater. The rapid flow of water in streams exports wastes downstream without processing. In lakes, the water residence time is often long, but there are usually insufficient outputs or internal sinks for nutrients. Lake eutrophication is common: algal growth, stimulated by excess nutrients, creates oxygen demand and results in unsuitable habitat for fish and other biota. Estuaries, located where rivers enter the sea, are characterized by insufficient flushing rates.

And finally, the oceans are the receiving ecosystem for millions of tons of waste each year, and a growing number of waste generators are seeking ways to dispose of greater quantities. The oceans do have a large capacity to absorb, recycle, or dilute waste material, but this capacity is unknown. Waste disposal and recycling at sea is controversial, and disposal practices are currently restricted.

Managed aquatic systems. Artificially constructed aquatic systems can be managed for water quality improvement. For example, shallow ponds, usually 3–5 ft (0.9–1.5 m) deep, can economically treat raw wastewater or remove additional contaminants from partially treated sewage (Fig. 5). Such stabilization ponds are already widely used by small communities and account for well over one-third of the municipal wastewater treatment facilities in this country. In the ponds, sunlight, algae, bacteria, and oxygen interact to help treat wastes, often to levels of secondary treatment. Bacteria play the major role by decomposing the organic matter in the wastes; and algae provide oxygen for bacterial respiration and use inorganic compounds (nitrogen and phosphorus) released by decomposition. Often disinfection of treated effluent is not critical because the long detention time significantly reduces pathogen levels. Some pond systems are aerated mechanically to allow greater loadings of wastewater than could otherwise be handled, and to reduce odors.

Other systems that also rely on the cultivation of aquatic organisms can be attractive alternatives to conventional methods of treating wastewater because they can clean the effluent to about the same level but often can be considerably cheaper. Managers of relatively small treatment systems, especially those handling less than 10^6 gal (4000 m^3) per day, commonly report savings of 25–75% in construction, operation and maintenance, and energy costs. Although aquatic systems can require less space than some land treatment methods, their use by large cities may still be limited by the cost or availability

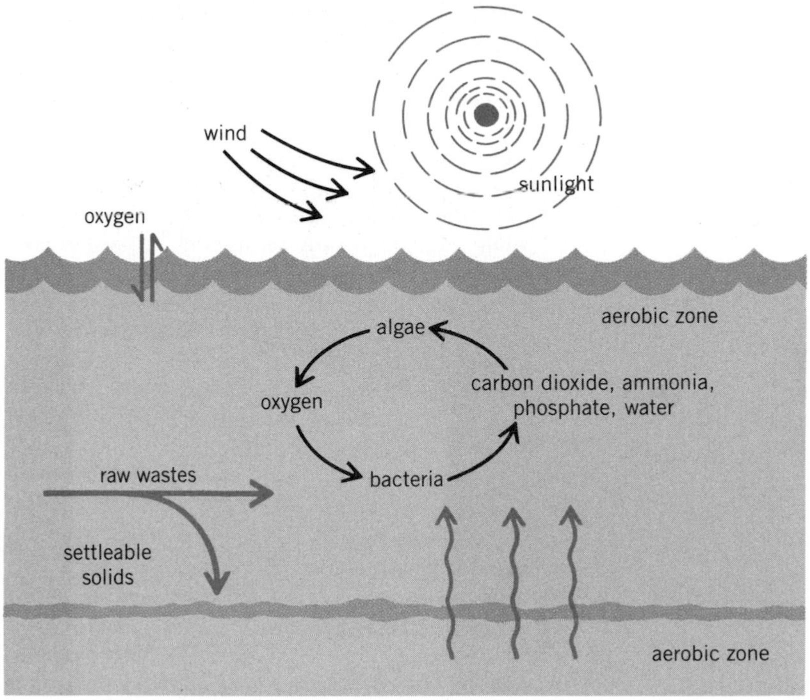

Fig. 5. Oxidation–stabilization ponds use sunlight, algae, and bacteria in pollutant removal.

of land. Full-scale aquatic treatment systems, as well as numerous research and development projects, are operating or under construction in at least 15 states as well as in several foreign countries.

Even though the use of aquaculture techniques for food production has been under way for years and wastewater treatment by aquatic plants and animals has been an accepted method in Europe and Asia for half a century, aquatic waste treatment is a fairly recent subject of research in this country. Aquaculture as applied to wastewater treatment pertains to the use of aquatic plants and animals in a controlled environment for the purposes of achieving treatment objectives. Aquaculture techniques have also been used to clarify turbid river water used as a drinking-water supply and in the demineralization of brackish groundwater.

Aquatic treatment is fundamentally different from conventional sewage treatment systems in that instead of rapid processing in a highly managed environment, treatment occurs at a comparably slow rate in essentially unmanaged systems. Aquaculture requires more land area but less construction and mechanization than conventional treatment facilities. Aquatic treatment is also subject to less operator control and greater environmental influence.

The use of aquaculture in wastewater treatment includes both aquatic plant systems in which the presence of free-floating plants such as water hyacinths, duckweed, and algae contribute to treatment, and animal systems that also use fish, mollusks, and crustaceans.

To date, the primary emphasis has been on plants, while the use of aquatic animals has generally been restricted to insect control and sometimes further polishing of treated effluent. Some attempts have been made with operations that combine more than one aquaculture component in a single unit or that are combined with other conventional units, but these have met with only limited success, especially in terms of economic feasibility.

Aquatic plant systems. Treatment operations that use aquatic plants typically consist of multiple-cell ponds. The initial cells are used for primary or secondary treatment, and the plants are often used only in the latter cells of the system. Although the main objectives are usually removal of nutrients (nitrogen and phosphorus) and additional removal of biochemical oxygen demand (BOD) and suspended solids, some success has been achieved in removal of heavy metals and trace organic chemicals.

The use of aquatic plants has also been effective in upgrading existing treatment ponds. Studies indicate that introduction of certain plants improves effluent quality while eliminating the need to increase hydraulic capacity or detention time in the ponds. Free-floating aquatic plants such as water hyacinths and duckweed can also help minimize nuisance growth of filamentous algae because they shade the water.

There are two general categories of plants used in aquaculture treatment: floating and submerged. Floating plants such as water hyacinths and duckweed have their photosynthetic parts at or just below the water surface with roots extending below the surface. With floating plants, the penetration of sunlight into the water is reduced and the transfer of gas between the water and atmosphere is limited. As a consequence, floating plants tend to keep the water free of algae and essentially anaerobic. Submerged plants such as algae may be suspended in the water column or rooted to the substrate. During the sunlight hours these plants add oxygen to the water.

As the plants grow, periodic harvesting is typically required to maintain high system performance; depending upon their rate of growth, plant harvesting intervals typically range from 2 to 4 months. Potential beneficial use of harvested plants, for such things as compost, animal feed, and methane production, is technically feasible but has often not been economically feasible. Development of improved recovery and reuse technologies may allow for plant uses to partially offset aquaculture treatment operational costs in the future.

The water hyacinth is a fast-growing fresh-water plant that is very efficient in removing nutrients and other materials. About 15 acres of water hyacinths, grown in ponds under controlled conditions, can treat 10^6 gal (4000 m^3) of wastewater to high levels per day. The plants take up nutrients and other chemicals as they grow, and their roots foster the growth of bacteria and higher organisms that assist in treatment. And by creating a blanket of vegetation that reduces wind and wave action on the water, they reduce unwanted growth of algae and help suspended solids settle to the bottom more quickly. Harvesting the hyacinths may be required to keep the system performing properly; 20 to 40 tons of plant material (dry weight) can be harvested per acre (44 to 88 metric tons per hectare) after about 3 months' growth. The hyacinths can be digested to produce methane fuel or processed to produce organic soil conditioner or animal feed, but such recycling systems are not yet economical.

The use of water hyacinths was pioneered by scientists at NASA's laboratory in Bay St. Louis, Mississippi, where the plants have been cleaning domestic and chemical wastewaters since 1975. Systems are also being used or tested in other southern and western areas, especially Florida, Texas, and California. For example, Walt Disney World near Orlando, Florida, has installed a water hyacinth treatment system where the production of energy and soil conditioner from harvested plants is being evaluated. And San Diego is building a prototype system that will treat 10^6 gal (4000 m^3) of wastewater per day and will be used to study the potential of water hyacinths for large-scale renovation systems.

Animal systems. Aquaculture systems that both treat wastes and produce valuable aquatic plants and animals have potential but are farthest from widespread use. In one particularly ambitious project, scientists at Woods Hole Oceanographic Institute tested a marine polyculture system that used municipal wastewater. They grew marine algae in a mixture of sea water and sewage effluent, and then fed the algae to shellfish, including oysters, clams, scallops, and mussels. The algae removed nutrients from the wastewater, and the shellfish removed the algae. Lobsters and fish such as flounder were then fed the wastes produced by the shellfish, and commercially valuable seaweed provided the final polishing of the effluent. The cost effectiveness of the total system is questionable, but it does appear that the seaweed unit by itself may prove attractive economically.

In other projects, a variety of fish species have been grown in wastewater stabilization ponds. In Arkansas, for example, buffalo fish, channel catfish, and several species of Chinese carp were raised in the last four stabilization ponds of a six-pond series in which municipal wastewater flowed from one to another. The fish got no other food. More than 3000 lb of fish per acre (3500 kg/ha) were harvested after 8 months, and the quality of the water discharged from the system was improved.

Another possibility is to use the cooling water discharged from power plants and industrial boilers, combined with various agricultural and industrial wastes, to support a commercial aquaculture system. Public utilities in at least 10 states are exploring this prospect. However, it appears that aquaculture systems cannot be optimized for both food production and waste treatment in the same unit; systems involving higher forms of animals seem less efficient in treating wastewater and are more difficult to manage. But in some cases it should be possible to combine waste treatment with aquaculture systems to help decrease net costs.

Design and operation. Because many aquatic plants that are well suited for treatment purposes are sensitive to cold temperatures, operation in cold climates may require the use of a protected enclosure like a greenhouse cover or, alternatively, limiting operation to a seasonal basis. Control of insects, particularly mosquitoes, is normally required, and mosquito-eating fish such as *Gambusia* sp. have proven to be an effective control measure. Toxic wastewater constituents need to be identified and adequately controlled in the influent wastewater. Because aquaculture systems utilize higher life forms that are less diverse than the microbial biota present in conventional biological treatment systems, they tend to be more susceptible to longer-term process disruption. It is therefore important to evaluate and control the effects of fluctuations in climate and wastewater characteristics.

Initial investigations of the use of aquatic plant aquaculture systems show this treatment to be very competitive when compared with the costs of constructing and operating other treatment alternatives. Actual construction costs are strongly influenced by local land availability and characteristics, climate, pretreatment methods, effluent quality standards, and plant harvesting and disposal methods. Of these factors, climatic conditions may have the greatest impact because the cost of providing a protective environment can be substantial. At present, information on operation and maintenance costs is limited but appears to be low compared with other treatment alternatives. A significant portion of labor and energy requirements are related to the harvesting and disposal of plants.

WETLANDS

The term wetland is relatively new and encompasses what for years have been referred to as marshes, swamps, and bogs, or by regional terms such as potholes, playa lakes, and pocosins. Wetlands occur in a wide range of physical settings typically at the transition between terrestrial and aquatic ecosystems. Because of this position in the landscape, some wetlands have been subjected to wastewater discharges from municipal and industrial sources, as well as agricultural runoff and irrigation return flow, and urban stormwater discharges. It is only in the past few decades that the planned use of wetlands for wastewater treatment has been studied and implemented.

Wetlands appear to be a major ecosystem category that can tolerate the conditions associated with wastewater input (that is, oxygen depletion associated with BOD removal and eutrophication) and effectively remove nutrients from wastewater. Stillwater wetlands (both palustrine and lacustrine) provide the best treatment if wastewater is allowed to percolate through peat before reaching groundwater or surface waters. This organic liner represents a significant nutrient sink. Wetlands that are adjacent to streams and rivers and intermittently flooded may also provide treatment, although some of the nutrients are subject to dilution in receiving waters. In both cases, because wetland plants tend to be preadapted to long periods of standing water and waterlogged soils, as well as reducing conditions, the impact of wastewater addition is generally less than on other terrestrial or aquatic plant communities.

Water quality improvement has been recorded in a variety of natural wetlands such as fresh-water marshes, northern bogs, southern swamps, bottomland hardwood forests, brackish and salt-water marshes, as well as artificially constructed wetland systems. Although the environmental conditions across the whole range of wetland types are quite diverse, most wetlands can effectively remove nitrogenous and phosphorus nutrients from wastewaters. Suspended and particulate matter is filtered out by plant roots or settles on the plant and soil surfaces. Hydrocarbons accumulate on the soil surface and may be actively broken down by microbial activity. Other contaminants that may be reduced are coliform bacteria and oxygen-demanding wastes.

Pollutant removal mechanisms. The pollutant removal mechanisms that operate in wetlands are quite similar to those found in aquatic ecosystems. Pollutants enter a wetland in dissolved, emulsified, and particulate forms and are removed by three main routes: loss to the atmosphere, incorporation into sediments or biota, and degradation. Some products of pollutant degradation may be inert or nontoxic (for example, oxidizable organic compounds are quickly converted to carbon dioxide and water), while others continue to pose environmental hazards (for example, some pesticides yield stable degradation products—they flow out of the wetland, volatilize, or remain in the wetland fixed in biota or sediment).

Wetland ecosystem properties that seem to provide this ability to degrade and eliminate contaminants in wastewaters are high plant productivity and nutrient needs, high decomposition rates, large adsorptive areas in sediments, and low oxygen content of sediment. The tendency for wetland sediments to become anaerobic is probably the main factor involved in the retention of various compounds. Reducing environments allow the conversion of heavy metals into relatively insoluble sulfides and also promote the removal of nitrate nitrogen through denitrification.

Sedimentation of pollutant particulates, or pollutants adhering to the surface of particles, can be the primary mechanism for the removal of these substances from the water column. This process is important for particulate nitrogen, oils, chlorinated hydrocarbons, and many metals. The nature of flow patterns through a wetland strongly influences deposition of suspended matter and overall pollutant removal effectiveness. High flow rates associated with storm events or seasonal runoff like snowmelt may resuspend previously deposited and decomposing organic matter and wash it out of the wetland system.

Emulsions are finely dispersed colloidal mixtures of two or more liquids that normally do not blend. In natural systems, emulsifying occurs most commonly in the presence of minute droplets of oil and, once formed, may persist in water bodies such as wetlands. Emulsion colloids are important to the fate of pollutants in wetlands because, aside from containing toxic oils, they tend to accumulate environmentally significant chemicals that are sparsely soluble in water. This is important for pollutants like mercury, polychlorinated biphenyls (PCBs), and some pesticides.

A principal means of removal of dissolved pollutants in wetlands is by adsorption to suspended solids or bottom sediments. Researchers have found that extensive contact between wastewater flows and wetland soils allows these interactions to occur. The most effective results are attained in seepage wetlands, where a large percentage of the flow passes through the soil complex prior to entering groundwater or discharging from the wetland. In wetlands having poor soil permeability, such as peatlands, shallow water depths and long resident times promote greater contact and increased pollutant removal efficiencies. Many substances adsorb to solids under conditions found in wetlands, including organic compounds, hydrocarbons, ammonium, phosphorus, heavy metals, bacteria, and viruses.

Wetland plants increase the overall capacity of the system to remove or retain pollutants, through interaction with the soil, water, and air. Although the primary mechanisms for pollutant removal in wetlands are usually the physical and chemical interactions discussed above, plant uptake of pollutants, particularly from the sediments, frees more soil exchange sites for further pollutant interaction and accumulation. Plants also provide surfaces for bacterial growth.

Wetland environments present ideal conditions for nutrient cycling and removal, particularly for nitrogen. The aerated water column and aerobic upper sediment layer promote nitrification and the formation of insoluble phosphorus–metal complexes. Reducing (anaerobic) sediment conditions and the interface between the aerobic and anaerobic layers promote ammonification and denitrification.

Natural wetlands. Wetlands exist primarily in flat to gently sloping terrain, are periodically flooded, and have groundwater at or near the surface of the ground for a major portion of the year. They occur in a wide range of climatic, topographic, geologic, and hydrologic settings, encompassing a diversity of ecosystem types. Examples include northern peatlands, bogs, and cattail marshes; bottomland hardwood forests and swamplands of the South; salt- and fresh-water marshes of coastal regions; and the marshes associated with high-energy rivers such as the Mississippi. Hydrology plays an important role in controlling wetland ecosystem characteristics and processes, and a clear understanding is needed to assess their potential utility in the assimilation of water-borne pollutants.

Natural wetlands vary widely in their ability to remove nitrogen and phosphorus. At low loading rates, good phosphorus removal efficiencies are common for nearly all wetland types, but this efficiency drops off at higher loading rates. A similar pattern occurs for nitrogen removal. Although removal rates are generally inversely correlated with loading rates, continued input of nutrients through time appears to shift wetland systems toward poorer removal efficiencies.

Various attempts have been made to correlate nutrient removal with wetland characteristics, but few generalities emerge that apply to all wetlands. This is due in part to the wide variability of wetland types and to the large number of factors affecting nutrient retention. A summary of natural wetland types and

Table 2. Summary of nutrient removal processes in different wetland ecosystems

Wetland type	Nitrogen	Phosphorus
Northern peatlands; *good removal*	Denitrification Vegetational uptake NH_4^+ adsorption to peat	Adsorption to peat Vegetational uptake
Nontidal fresh-water marshes; *variable removal*	Denitrification Macrophytic uptake Periphytic uptake	Chemical precipitation Macrophytic uptake Periphytic uptake Adsorption to substrate
Tidal fresh-water marshes; *poor or no removal*	Denitrification Vegetational uptake Tidal transfer	Litter uptake Vegetational uptake Tidal transfer
Brackish or salt marshes; *variable removal*	Denitrification Macrophytic uptake Periphytic uptake Tidal transfer	Macrophytic uptake Periphytic uptake Adsorption to peat Tidal transfer
Southern swamps; *good removal*	Denitrification Macrophytic uptake Periphytic uptake	Macrophytic uptake Periphytic uptake Adsorption to substrate
Sawgrass marshes; *poor removal*		Adsorption to substrate Litter uptake

nutrient uptake processes is shown in Table 2. In comparing nutrient removal of wetland ecosystems, a few generalizations can be made. Conditions favoring phosphorus removal are organic soils in poor nutrient regimes, plants that are limited by phosphorus supply, and the presence of aluminum and iron compounds. Conditions favoring nitrogen removal are a reduced soil–water interface (to promote denitrification) and plants that are nitrogen-limited. In addition, low-energy subsidy from tides, wave action, or stream flow favors removal of both nitrogen and phosphorus.

There is an adequate scientific basis for natural wetland treatment in only a few types of wetlands; for example, northern peatlands (Fig. 6) and cypress swamps have been well studied. Monitoring them under various conditions will provide "real-world" operating information and data needed for developing reliable design criteria.

Not all wetlands are good candidates for wastewater treatment—some are too valuable in their undisturbed condition, some are too sensitive to change, and some do not provide much treatment. The EPA is currently evaluating several kinds of wetlands in the Midwest and the Southeast to determine their suitability for use in wastewater treatment.

Artificial wetland treatment. The use of artificial wetlands for wastewater treatment seeks to take advantage of many of the same principles that apply in a natural system, but does so within a more controlled environment. Small-scale wetlands have been created expressly for the purpose of providing wastewater treatment, while some large-scale systems have been implemented with multiple-use objectives in mind (for example, using treated sewage effluent as a water source for the creation and restoration of marshes for wildlife use and environmental enhancement).

Artificial wetlands are less restricted by user conflicts and potential environmental concerns than natural wetlands. They can be constructed almost anywhere—including on lands with limited alternative uses—and they offer greater flexibility in design and operation that can lead to superior treatment and reliability. They can be built in natural settings, or they may entail extensive earthmoving, construction of impermeable barriers, or building of containers such as tanks or trenches. Wetland vegetation is typically established on a substrate such as gravel or peat. Some systems are set up to recycle a portion of the wastewater and to direct the final effluent into the soil for recharge of groundwater. Others act as flow-through systems, discharging final effluent to surface waters.

Applications of the artificial wetland concept are diverse and have been used across the country and around the world. Often, constructed wetlands are used in combination with other treatment systems. A meadow–marsh–pond system was constructed on Long Island, New York, where reed canary grass, cattails, and duckweed flourished in the different environments. Cattails and reeds appear to be ideal plants—they are hardy and widespread, and they propagate easily and grow quickly. In Ontario, for example, an experimental cattail marsh built on heavy clay soils provides efficient year-round treatment. There is only a slight decrease in performance during the winter, when ice forms on the marsh surface but wastewater continues to flow underneath.

Fig. 6. A large state-owned peatland located near Houghton Lake, Michigan, provides for advanced treatment for the discharge from a 10^6-gal-per-day (3800 m^3) municipal treatment plant.

Historical loss of marshes in the San Francisco Bay area has recently given rise to public interest in restoration of wetlands. The objective has been realized in one bayside city, Martinez, where secondarily treated effluent has been used to create a freshwater marsh. In addition to helping to meet stringent wastewater discharge limits to the bay, the newly established marsh provides habitat for numerous species of birds, aquatic invertebrates, and other animals.

The advantages of constructed wetlands are many, and include: flexibility in site location; optimized size for projected waste load; construction of topographic features such as channels, shallow bars, and levees to improve pollutant removal and facilitate management; less rigorous influent criteria, if the wetland is considered primarily to be a treatment system; and no alteration of natural wetlands. Perhaps the most beneficial feature is their potential to accept higher wastewater application rates. Concurrently, there are some disadvantages to using constructed wetlands relative to natural ecosystems. The cost and availability of suitable land and construction costs for grading the site are added expenses. In addition, the sites are unavailable during the construction period, and reduced performance can be expected during the period in which the vegetation becomes established. Other possible constraints are the costs of plant biomass harvesting and disposal and the fact that artificial wetlands, like their natural counterparts, may provide breeding habitat for nuisance insects or disease vectors, and may generate odors.

LAND TREATMENT

Upland ecosystems, for the purpose of this discussion, are considered to be all land areas that are not wetlands. Here, the water table is not near the surface of the ground during most of the year, and conditions are not favorable for the formation of hydric soils and the growth of water-loving plants. Uplands include rangelands, croplands, pastures, and forested lands.

Many efforts have been made during the past several years to advance the practice of utilizing upland soil and vegetation for the treatment and disposal of various waste effluents and sludges. A wide range of land reclamation and biomass production projects have been investigated and employed to date. Controlled application of wastes onto the land surface can achieve a designed degree of treatment in the soil–water–plant matrix. Such techniques, frequently called land treatment, have evolved from the time-honored practice of recycling animal manures and agricultural residues to the land and also crop irrigation processes.

The ecological processes involved in treating wastewater and sludge on land are largely physical and biological, although some chemical degradation does occur. The soil column plays an important role in the removal of contaminants. It acts as a site for physical interactions such as filtration and adsorption; chemical interactions such as precipitation, adsorption, and formation of organic complexes; and biological interactions such as bacterial metabolism, plant metabolism, and absorption. Chief factors governing the movement of pollutants are the soil pH, oxidation–reduction potential (redox potential), and the mineral composition of the soil. The cation exchange capacity (CEC) apparently influences the long-term accumulation of pollutants such as heavy metals and some organic compounds. Soil

Table 3. Key design and operation features in land treatment of wastewater

	Rapid infiltration	Slow rate	Overland flow
Application techniques	Sprinkler Surface flooding (most common)	Sprinkler Surface flooding	Sprinkler Surface flooding
Process objectives	Wastewater treatment Groundwater recharge Recharge surface streams	Wastewater treatment Crop production Groundwater recharge Recharge surface streams	Wastewater treatment Crop production (minor) Augment surface streams
Annual application rates, feet (meters) per year	20–400 (6–120)	2–20 (0.6–6)	10–70 (3–20)
Typical loading rate, inches (centimeters) per week	4–95 (10–240)	0.5–4 (1.3–10)	2–16 (5–41)
Land area required, acres per 10^6 gal per day (hectares per 1000 m^3)	5–65 (0.5–6.5)	50–500 (5–50)	15–110 (1.5–11)
Preapplication treatment (minimum)	Primary clarification	Primary sedimentation	Screening and grit removal
Slope	Suitable for basin construction	Less than 20% on cultivated land Less than 40% on uncultivated land	2–8%
Soil permeability	Moderate to rapid (sandy loam and sands)	Moderate	Slow (clay, silt, and soils with impermeable barriers)
Depth to groundwater	3 ft (0.9 m) during flooding cycle; 5–10 ft (1.5–3 m) during drying cycle	2–3 ft (0.6–0.9 m) (minimum)	Not critical
Climatic restrictions	Storage usually not required, even in cold climates	Storage usually required for cold weather and heavy precipitation	Storage usually required during extremely cold weather
Vegetation	Not critical	Required (agricultural crops, old fields, forests, and so on)	Grass crop required
Fate of applied wastewater	Primarily deep percolation to groundwater and eventually surface water	Evapotranspiration Percolation	Surface runoff Evaporation

reaeration is important for the maintenance of physical, chemical, and biological interactions.

Because metals tend to concentrate in sludges, land application of sludges can add appreciable amounts of trace metals to soils. High pH and low redox potential lead to the formation of insoluble metal complexes, which effectively prevent plant uptake of metals. As pH decreases and redox potential increases, a situation that can happen in flooded soils, many metal ions are released and become more available for plant uptake. In alkaline soils, most metal ions can be immobilized as precipitates.

Wastewater. Land treatment of wastewater is a versatile practice in which a variety of ecological processes are used to achieve a number of different objectives. The purposes for which this technique has been employed include water quality protection, wastewater reclamation and reuse, groundwater recharge, nutrient recycling, and crop production. The three principal methods of land treatment of wastewater in common use are slow-rate infiltration, rapid infiltration, and overland flow. Illustrations of these methods are provided in Fig. 7; design and operational considerations are given in Table 3.

Slow-rate land treatment. Similar to conventional crop irrigation, this relatively popular alternative recycles nutrients and produces a potentially marketable crop while reclaiming wastewater. Partially treated wastewater is applied to vegetated lands, and the soil, plants, and microorganisms clean the water as it soaks in and gradually percolates to the groundwater. Systems used to irrigate farmlands, old fields, and forests are currently operating successfully in many parts of the country. The operation of slow-rate land treatment systems is affected by the environmental conditions at a specific site. The local soil types and the plant communities that are present directly affect the methods that may be used to apply the effluent and the degree of renovation that is attainable. Land management practices such as field tillage and cultivation, crop planting and harvesting, as well as the installation of field tiles, recovery wells, or surface drainage systems also impact the amount of water that can be applied to a site and its infiltration into the soil after irrigation. Climatic conditions affect system operations since extended periods of precipitation or long periods of subfreezing temperatures can limit the period of irrigation.

The Clayton County Water Authority, located just south of Atlanta, Georgia, operates one of the largest totally forested spray irrigation systems in the United States. This year-round slow-rate system, in operation since 1982, can treat up to 2×10^7 gal (76,000 m^3) of wastewater per day by applying it to a 2400-acre (970-ha) forest of loblolly pines. The total irrigation network consists of about 250 mi (400 km) of buried pipeline and approximately 1800 sprinklers. The forest land is divided into seven sections, and when the design flow of 2×10^7 gal (76,000 m^3) per day is reached, only five sections will be required for operation, leaving the remaining area available for harvesting, system repair, and contingency irrigation. Once a week, each section receives 2.5 in. (6.3 cm) of wastewater over a period of 12 h. The trees, to be sold as pulpwood to help offset operating costs, will be harvested on a 20-year rotation cycle.

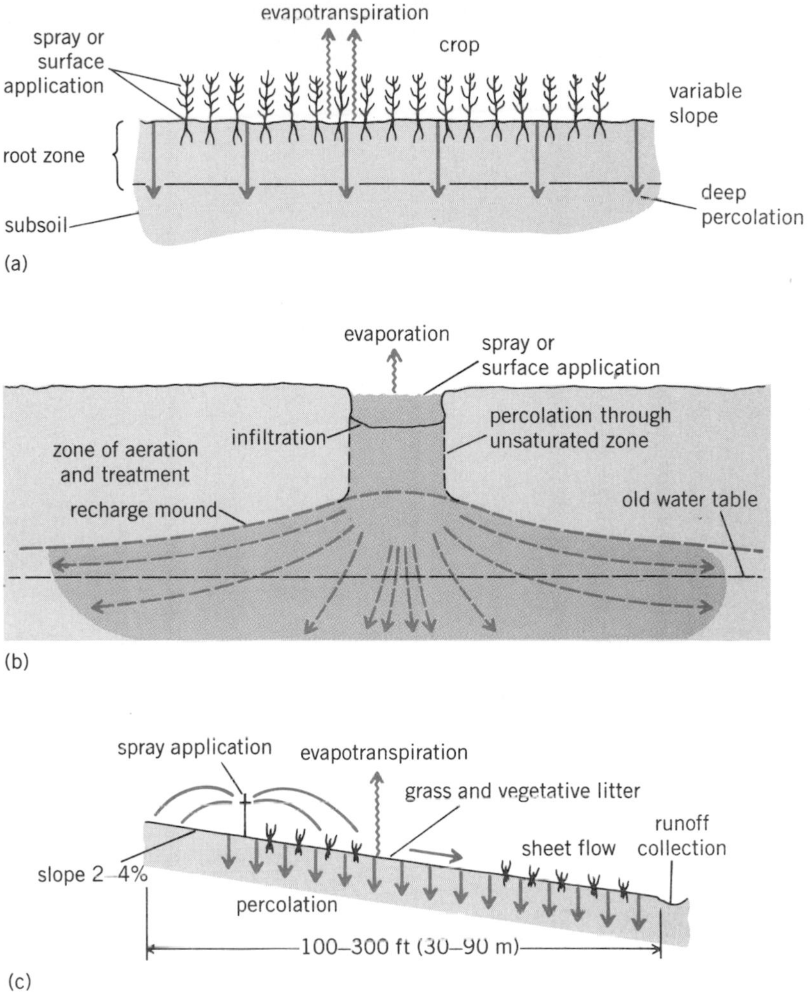

Fig. 7. Principal methods of land application of wastewater. (*a*) Slow-rate infiltration. (*b*) Rapid-rate infiltration. (*c*) Overland flow.

Because this site lies in the watershed of the county's water supply reservoir, the renovated wastewater eventually becomes part of the drinking-water supply. Recycling should help augment the available water supply, which has historically been limited during dry periods. To ensure that the groundwater and surface water leaving the site meet primary and secondary drinking-water standards, a network of groundwater observation wells and stream sampling points has been established.

Other relatively large slow-rate systems—each treating more than 2×10^7 gal (76,000 m^3) per day—have been installed in Michigan and Texas, while smaller systems operate in at least 40 states.

Overland-flow land treatment. Wastewater is applied at the top of grassy slopes which are underlain by a relatively impervious soil layer. The water flows in a thin film down the slope and is cleaned by the soil, vegetation, and microorganisms present at the soil surface. The vegetation is important as an integral part of the treatment process, in addition to protecting the slopes from erosion. Treated wastewater is collected in open ditches at the bottom for reuse or discharge to surface waters.

In terms of construction, the slopes must be steep enough to prevent ponding of the runoff, yet gentle enough to prevent erosion and allow the water enough time to be purified. The slopes must also be carefully graded and kept free from gullies to prevent channeling and allow uniform distribution of the water over the surface. Naturally rolling terrain is easily adapted to the network of slopes and terraces needed for an overland-flow system, minimizing land preparation costs. Cities can install such systems to preserve green belts and open spaces. In other locations, the systems may be used to produce forage grass for feeding cattle. Perennial grasses with long growing seasons, high moisture tolerance, and extensive roots are best for this purpose. Overland flow can be used to polish secondary effluents, such as the flows from oxidation ponds, or to provide secondary treatment. Under certain conditions, comminuted raw wastewater has been successfully treated to a secondary level with this process. In general, wastewater treatment is maximized in these systems, although minor crop production is possible.

The city of Davis, California, has installed one of the largest municipal overland-flow systems, which treats about 5×10^6 gal (19,000 m^3) per day. An industrial food-processing waste treatment system which has about the same capacity is located in Paris, Texas. Smaller or experimental systems are operating or are under construction in at least 15 states, including New Hampshire, South Carolina, Mississippi, Texas, and Illinois. The first systems were built in relatively warm climates, but recent advances in construction and design and greater operating experience have made them useful in cooler regions as well. Although overland flow has only recently been used to treat municipal wastewater, such systems have proved their worth in treating food-processing wastes for over 25 years.

Rapid infiltration. This system is usually used independent of agriculture and forestry operation. Vegetation plays a minor role and in many cases is not used at all. Wastewater is given an appropriate level of pretreatment and then applied by sprinklers or surface flooding to a constructed basin. As the basin is flooded, the water percolates downward through the soil profile, where physical, chemical, and biological processes cleanse the water, which eventually enters the groundwater. As in other land treatment, alternating wet and dry cycles are used to optimize microorganism activity, prevent waterlogging, and keep the soil from becoming clogged with suspended solids. For this reason, several basins are needed within a single system in order to have continuous treatment.

Rapid infiltration can be used only at certain sites, and design criteria tend to be quite site-specific. Moderate to highly permeable soils (such as sandy loams or sands) are needed, and the construction of infiltration basins must not disrupt this important site characteristic. Hydrologic testing at proposed sites is essential to determine realistic hydraulic loading rates.

Rapid infiltration is capable of providing high levels of wastewater treatment. BOD and suspended solids are removed almost completely by filtration and biological degradation at or near the soil surface. Removal of viruses, bacteria, and phosphorus varies according to the depth and composition of the soil. Nitrogen removal varies according to the operating scheme that is used. High nitrification rates are typical, and by careful system design and operation, rapid infiltration systems can be managed to achieve microorganism denitrification (that is, nitrate nitrogen is transformed to nitrogen gas, which escapes to the atmosphere), which prevents nitrate pollution of the groundwater.

Well-designed and -operated rapid infiltration systems offer several advantages: treatment performance can be equal or better than comparable conventional treatment facilities; land requirements are generally lower than other land treatment practices; construction costs, energy requirements, and operating costs are all low; systems can be operated in cold and wet weather so that storage lagoons are not needed; and, finally, system operation is quite simple. A secondary benefit of rapid infiltration systems is the recharge of groundwaters and augmentation of surface stream flows.

These systems are gaining in popularity for the treatment of both municipal and industrial wastewaters, but are still less widespread than the slow-rate systems. This is due in part to the somewhat unusual combination of land features that is needed. Certain sites contain soils that are well suited for the use of rapid infiltration but also have limitations that restrict its effective use (for example, high groundwater table, or underlying shallow fractured bedrock soils of low permeability). However, if the surficial soils are of sufficient depth, it is often possible to use subsurface wells or drainage systems to overcome many limitations.

Phoenix, Arizona, uses a large rapid infiltration system. Smaller systems are operating in about half of the states including New York, Massachusetts, and New Jersey, as well as a number of western states, including Montana, Wyoming, Colorado, Nevada, and California.

Costs. Land treatment systems can clean wastewater to a purity that equals or exceeds some of the most sophisticated conventional treatment processes. The systems also frequently cost less to build and operate than more highly engineered systems—if enough land is conveniently available at reasonable cost. For small treatment plants handling less than 10^6 gal (4000 m^3) per day, a 25% savings in construction costs and 50% savings in operation and maintenance costs (partly because they use much less energy) have been fairly common. In larger metropolitan areas with very large volumes of wastewater, these systems may be less practical because of land availability problems. However, large slow-rate systems may be attractive in regions where water is needed to irrigate crops. A slow-rate system usually requires 50–500 acres per 10^6 gal (5–50 ha per 1000 m^3) of wastewater handled daily. Rapid-infiltration and overland-flow systems require 5–65 acres and 15–110 acres per 10^6 gal per day (0.5–6.5 ha and 1.5–11 ha per 1000 m^3 per day), respectively. Both the land requirements and performance of similar projects can vary considerably due to local factors such as soil properties, wastewater characteristics, application rates, crops grown, climate, and the desired level of treatment.

Sludge. Processing and disposing of sludge frequently accounts for 50% or more of the costs of operating a typical sewage treatment plant. And with the usual disposal methods limited by lack of acceptable sites, rising costs, environmental problems, and legal restrictions, some communities have turned from a philosophy of disposal to one of reuse. Indeed, well over 20% of the nation's sewage treatment facilities currently rely to some degree upon land application as a sludge management practice. Sludge reuse projects are under way in many large metropolitan areas, including Milwaukee, Chicago, Denver, San Diego, Seattle, Philadelphia, and Washington, D.C., as well as in thousands of smaller cities and towns across the country, especially in the Midwest. The sludge is being recycled as a soil conditioner and organic fertilizer on cropland, pastures, sod farms, golf courses, parks, forests, and disturbed areas.

Raw sludge is 96–98% water; only 2–4% consists of solid material. It contains many microorganisms that can cause disease and material that can produce objectionable odors. Therefore, the sludge usually must be processed to stabilize it before it is applied to land. Sludge can be stabilized in a variety of ways, including long-term lagooning, composting, heat treatment, chemical treatment, and digestion. Anaerobic digestion, in which sludge is decomposed by bacteria in the absence of oxygen, is the most widely used.

Stabilized liquid sludge can be sprayed directly onto the soil surface, incorporated into the upper soil layer by plowing, or injected beneath the soil surface with specially designed injection systems. However, sludge is frequently dewatered in drying beds exposed to the sun or with the aid of vacuum filters, centrifuges, filter presses, belt presses, and even thermal dryers. This concentrates the sludge solids and lowers transportation costs. Dewatered sludge is spread on the soil surface and often disked in. Once the sludge is applied to the soil, natural biological systems take over, breaking down the sludge and incorporating its nutrients and organic matter into the soil matrix.

Three common natural treatment systems used for sludge management involve agriculture, forestry, and land reclamation. These practices are summarized in Table 4 and described below.

Agriculture. Sludge is most commonly applied to agricultural land. Many farmers long ago discovered the benefits of participating in sludge reuse programs: their fertilizer costs decreased, crop yields increased, and soil quality improved. Today, sludge from several Ohio cities is being supplied to farms in neighboring counties, under the direction of the Ohio Farm Bureau and other agricultural groups. Under a grant from the EPA, researchers have measured the resulting effects on crop yields, and have monitored human and animal health as well. Improvements in yields and the lack of any observed health problems stimulated greater interest among both farmers and county agricultural exten-

Table 4. Key design and operation characteristics of sludge treatment in upland systems

	Agricultural utilization	Forestry utilization	Land reclamation utilization
Application techniques	Surface (with or without incorporation) Subsurface injection	Surface (generally without incorporation)	Surface (with or without incorporation)
Process objective	Recycle nutrients and organic matter Enhance crop production Improve soil properties Disposal of sludge	Recycle nutrients and organic matter Enhance forest production Disposal of sludge	Establish vegetative cover Recycle nutrients and organic matter Improve soil properties Disposal of sludge
Application rates			
Range, dry tons per acre (metric tons per hectare)	1–30 (2–60)	4–100 (8–200)	3–200 (6–400)
Typical rates, dry tons per acre (metric tons per hectare)	5 (10)	20 (40)	50 (100)
Frequency	One time, infrequently or yearly	One time or 3–5-year intervals	Usually one time
Available technical literature and experience	Extensive literature Hundreds of large and small full-scale projects	Limited literature A few demonstration and small-scale projects	Limited literature A few full-scale and demonstration projects

sion agents. With such success stories becoming more common, farmers may someday routinely aproach municipal treatment plants to buy sludge for use as organic fertilizer.

In Madison, Wisconsin, a high percentage of the wastes comes from meat-processing plants, so the sludge is rich in nutrients, minerals, and organic matter. Sludge reuse dates back to the early 1930s, when the city sold or gave stabilized dry sludge to residents for fertilizing lawns and gardens. In the 1970s, city officials concluded that applying sludge on farmlands would be the most practical, economical, and environmentally sound method. Today, the Madison sewage treatment plant generates about 5500 tons (5000 metric tons) of stabilized sludge—which the city has named Metrogro—per year, which is distributed to farms located within 10 mi (16 km) of the treatment plant. Metrogro is transported in 3500-gal (13-m^3) truck spreaders equipped with flotation tires, which spray the sludge on the soil surface or inject it into the ground.

Forestry. There is also growing interest in the potential for using sludge to increase productivity in forests (Fig. 8). For example, researchers in the Pacific Northwest applied about 40 tons of sludge solids per acre (88 metric tons/hectare) to a 50-year-old stand of Douglas fir located on relatively poor soil. Two years later, they recorded a 60% increase in growth over untreated sites. The researchers also predict that the improvement in productive capacity of these forests will last at least 5 years and possibly longer. Sludge can benefit intensively managed tree plantations as well, often dramatically. At the Savannah River Laboratory near Aiken, South Carolina, researchers have found that applying sludge to loblolly pines may mean that three cuttings of pulpwood—rather than the normal two cuttings—can be harvested in a 20-year period.

However, not all species of trees appear to respond equally well to sludge. Red cedar and hemlock seedlings show high mortality rates when planted on recently treated sites, while ponderosa pine seedlings survive but their growth does not increase significantly. Grasses and weeds in newly clear-cut areas, as well as low-growing understory plants in established forests, also grow more rapidly when sludge is applied. Thus, unwanted plants may have to be controlled to reduce competition with desired tree growth.

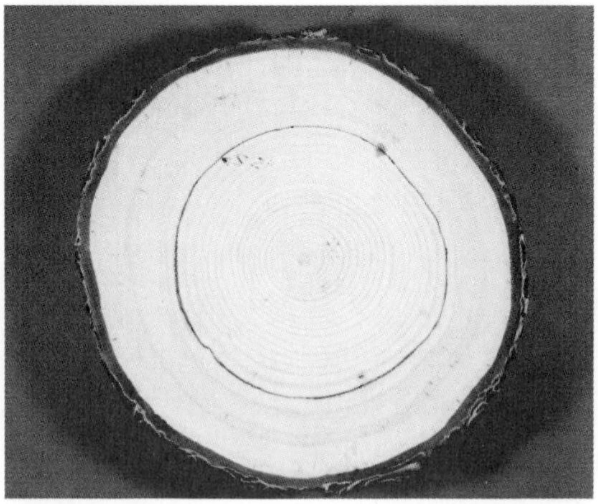

Fig. 8. A number of forest tree species respond to sludge additions with a dramatic increase in growth.

Land reclamation. Compelled by federal laws and court orders to curtail sludge dumping in the Atlantic, Philadelphia developed a master plan for managing sludge in 1975. The plan included using sludge to revegetate strip-mined lands, and the city is now reclaiming numerous spoil sites in western Pennsylvania. Because of initial opposition in mining areas, the city began with a pilot project, combined with a public information program, to show local communities that the method was environmentally sound.

In June 1978, the state approved the application of sludge to a 10-acre barren site in Somerset County. Following grading and liming, 50–60 tons per acre (110–190 metric tons/ha) of sludge solids were spread and mixed into the soil. The area was then seeded with a mixture of grass and legumes, and a lush green plant cover grew rapidly. Researchers monitored sludge, soil, and groundwater before, during, and after application, turning up no adverse effects on the environment. The city has since expanded the operation each year. About 1100 acres (445 ha) were reclaimed in both 1981 and 1982, using nearly 140,000 tons (130,000 metric tons) of sludge annually.

The need for reclaiming abandoned strip mines is great: there are more than 250,000 acres (101,000 ha) of despoiled lands in Pennsylvania alone and several million acres nationwide. The costs of transporting sludge for such projects—a potential drawback—can be minimized by backhauling the sludge in the same trucks or railcars that bring coal from the mining areas to the city. Chicago, Birmingham, and Seattle are already using sludge to reclaim despoiled lands, while Knoxville, Pittsburgh, Tulsa, Baltimore, and several other cities are seriously considering similar sludge management programs.

Other uses. One of the oldest sludge reuse operations is run by the Los Angeles County Sanitation Districts. More than 100 tons (110 metric tons) of sludge solids per day are composted after stabilization by digestion and then sold to a local company that screens, blends, and bags the material. The company, which has been in business for over 50 years, markets the sludge-derived product for home garden and horticultural use and to commercial nurseries. Demand is so great that the company must ration the material among selected customers. The selling or give-away of bagged or bulk-processed sludges to be used as organic fertilizers or soil conditioners has been practiced for many years in several other metropolitan areas, including Milwaukee, Houston, and Chicago, and more recently, Philadelphia, Missoula, Montana, and Washington, D.C.

Managing sludge in natural ecosystems. While the

benefits of recycling municipal sludge are well documented, there are also a number of concerns, especially for agricultural use. Most important is the possibility that pathogens or toxic chemicals which may be present in the sludge may contaminate the soil and the crops produced. If such contaminants accumulate beyond the soil's ability to assimilate them, the land could deteriorate rather than improve, groundwater quality could be degraded, surface water could be jeopardized, and crops could become unfit for consumption.

A number of factors must be considered in determining whether a particular site is suitable for sludge application and, if it is, for subsequent project design. Site characteristics of most importance are soil type, soil nutrient and organic matter content, slope, susceptibility to flooding, depth to groundwater, permeability of the most restrictive soil layer, vegetative cover, and cropping patterns.

Once a site has been selected and the treatment objectives defined, proper sludge-loading rates can be determined. This process involves characterizing the waste for a number of constituents; of most concern for municipal sludges are pathogens, nitrogen, phosphorus, cadmium, copper, nickel, lead, and zinc. The allowable loading rate of each constituent is calculated, and the actual sludge-loading rate is based on the most limiting constituent. Frequently sludge application is dictated by the nitrogen (or alternatively phosphorus) loading that meets expected crop needs. However, for persistant compounds that do not readily biodegrade such as heavy metals or synthetic organic chemicals (such as PCBs), the useful life of land application sites can usually be based on the cumulative amount applied. Of the five heavy metals mentioned above, the EPA currently regulates only cadmium applications to agricultural soils. The EPA has issued guidance on recommended cumulative loading limits for the other metals. The cadmium limit was established to protect human health by limiting cadmium loadings to fields used for the production of food-chain crops, and is considered by many scientists to be very conservative. Specific federal, state, and local requirements have been issued which restrict sludge application rates or prescribe specific management practices when some other compounds, such as PCBs, are present. [JAY BENFORADO; ROBERT K. BASTIAN]

Robotics

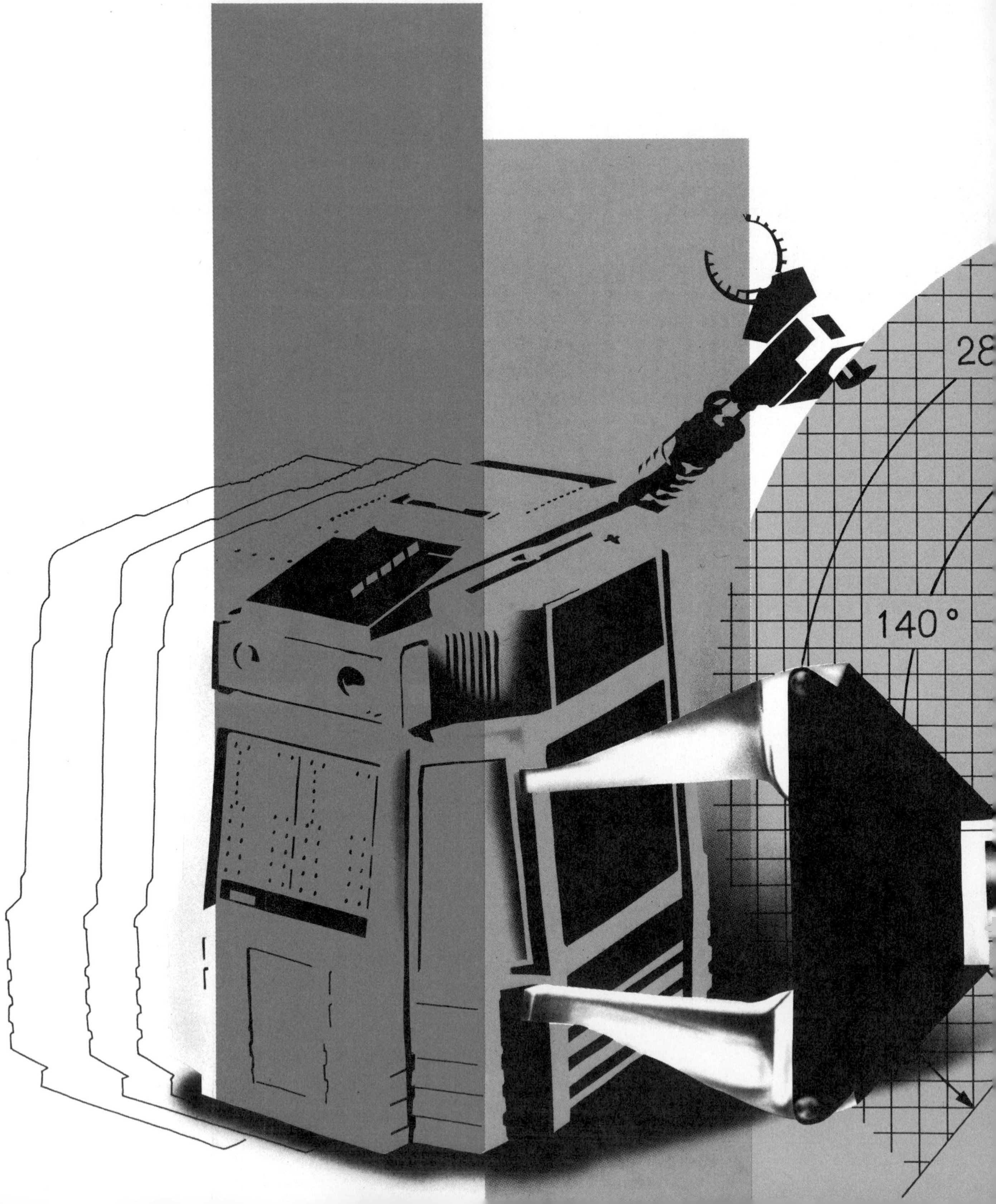

VINCENT M. ALTAMURO

Vincent M. Altamuro is the president of Management Research Consultants and the president of Robotics Research Consultants. Formerly on the faculties of the graduate schools of engineering of Columbia University and The City University of New York, he has conducted seminars and served as a consultant for manufacturing companies for over 25 years.

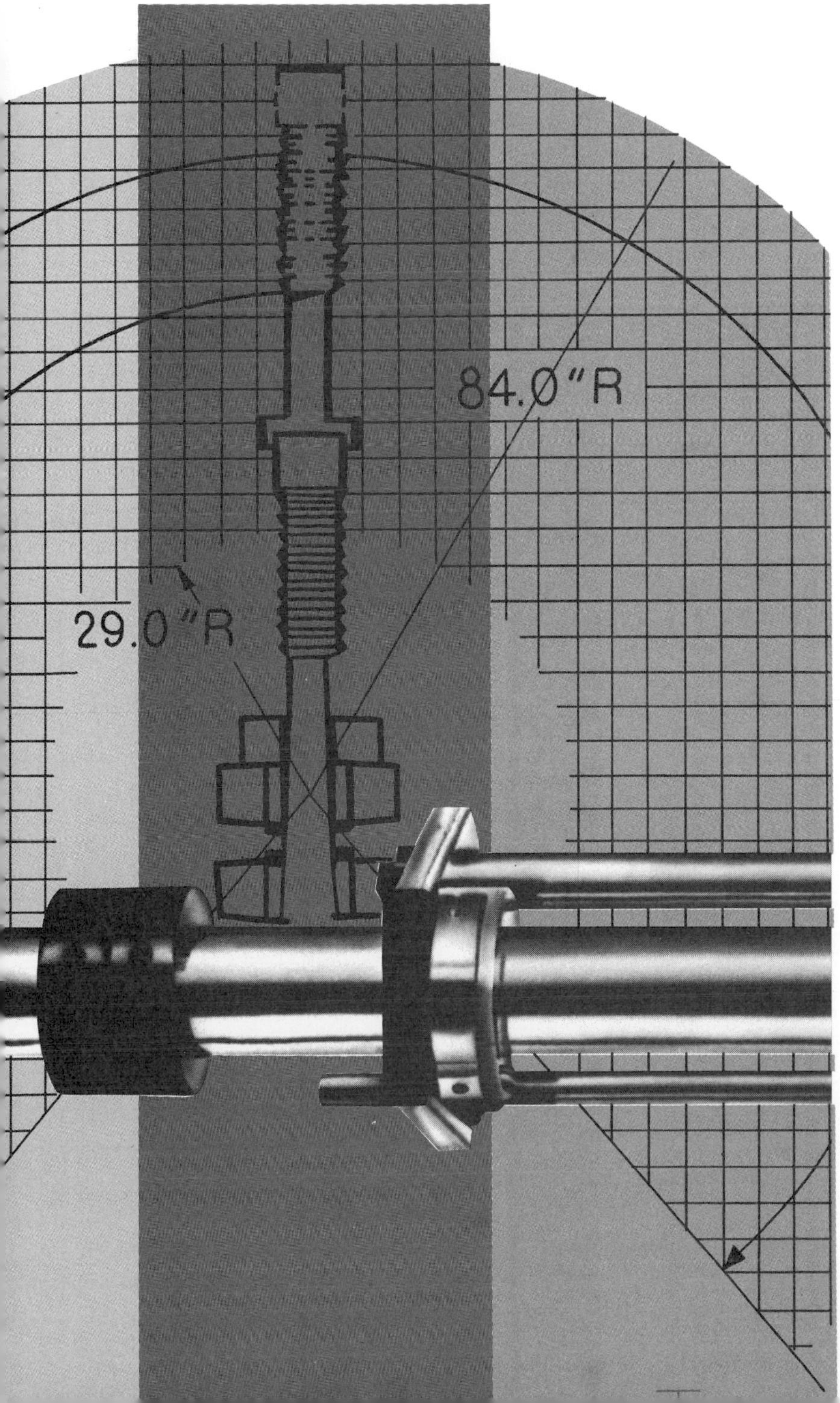

Robotics is the science of dealing with the characteristics, utilization, and interfaces of robots as individual machines and as components of larger systems.

History. Robotics is a new science, but robots as individual devices have ancient predecessors. Depending upon how loosely the word "robot" is defined, their ancestors can be traced back to devices found in early Egypt, where the priests used steam-activated mechanisms to open the temple doors and thereby impress the people with their mystical powers. The ancient Greeks constructed articulated statues to study human motions and to demonstrate physical principles. Some oracles were jointed figures and, in the *Iliad*, Homer referred to Hephaestus, the god of fire, metalworking, and mechanics, who was served by two golden working female statues. The ancient Chinese and Ethiopians also built moving statues powered by water or steam. The Byzantines and some early Arabians had clepsydras (water-powered clocks), combined with various forms of automata. In Europe, the French, Germans, and English used articulated toys in games and to amuse their royalty. In the 1700s, in Bruges, Belgium, a carillon was constructed to have 47 bells played by hammers operated by trip wires, which in turn were struck by an array of pins that could be placed ("programmed") in patterns into 30,000 holes on a huge wheel rotated by a clock. At about the same time, two Swiss brothers, Pierre and Henri Jacquet-Droz, advanced jaquemarts (mechanical mannikins that move to strike the hours on ornate clock bells) to the point where these spring-powered automata

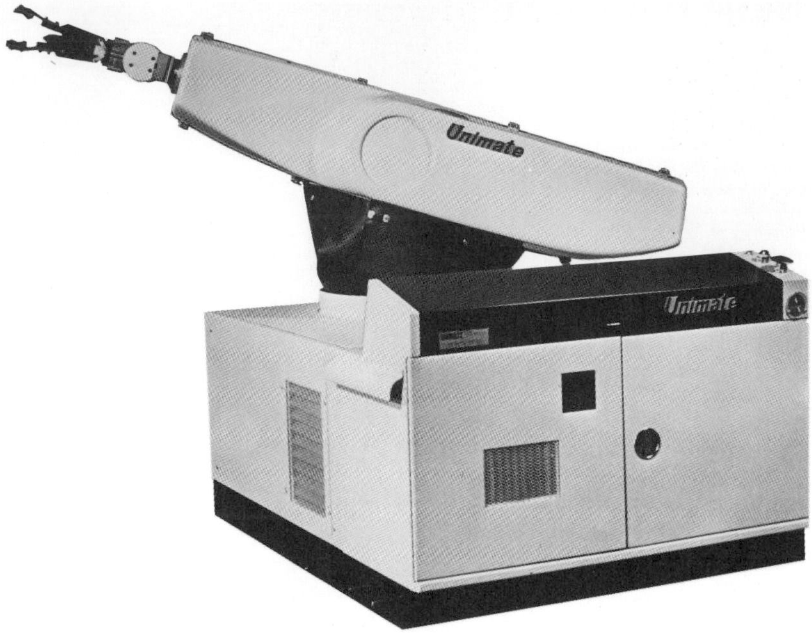

Fig. 1. Unimate 2000 industrial robot. (*Unimation Division, Westinghouse Electric Corporation*)

could play musical instruments and take pen to paper and write and draw pictures. In 1801, Joseph Jacquard constructed a loom controlled by punched cards and reprogrammable by changing the cards. In 1892, in the United States, Seward Babbitt built a motorized crane that had a hinged gripper so as to be able to reach into a furnace, grasp a hot ingot of steel, remove it, and deposit it where desired.

None of these were robots, as the term is used today, but each step advanced the technology in some respect—powering a mechanism, jointing and articulating members (links), programming and controlling motion, developing a mechanical hand (gripper), and so forth.

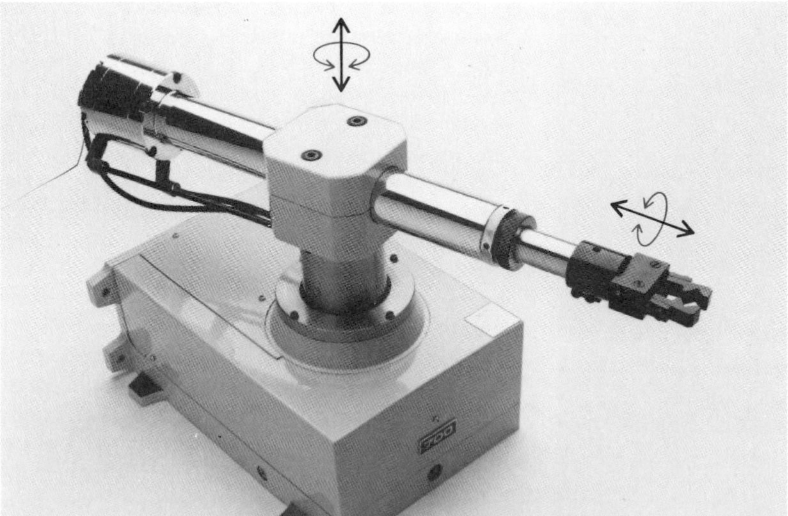

Fig. 2. Pneumatic robot. (*Seiko Instruments U.S.A., Inc.*)

Development. Further advances introduced additional capabilities which culminated in the robot of today. In addition to the foregoing technology, the modern robot also includes servomotors, optical encoders, feedback correction and control, microelectronics, computers, high-level language programming, and sophisticated sensors, all packaged into a fast-acting precision machine that operates reliably alone or integrated in a larger system.

Much of this development was done in the 1940s and 1950s, with some of it arising from the needs of World War II. In 1946, J. Presper Eckert and John Mauchly built the first large electronic computer (ENIAC), which in its minicomputer form of the 1970s and microcomputer size of the 1980s made the computer-based robot possible. During the same early period, the basis for the robot's manipulator (its arm, or more precisely, the mechanical linkage capable of varied controllable motions relative to its base) was developed in response to a need to handle radioactive nuclear materials safely. Master–slave telecherics ("hands at a distance") were developed at the Argonne National Laboratories. These had direct mechanical links through a wall so that a human, at a safe distance, could move the master side of it and have the slave side do dangerous work. Much of what was learned then about joints, axes, and degrees of freedom is used in the modern robot. Then, in the 1950s, at Lincoln Laboratories, Massachusetts Institute of Technology, touch sensors and feedback were incorporated in telecherics to give the master side a "feel" and the slave side a softer contact. These, too, are in today's robot. Another type of telecheric, the teleoperator ("operator at a distance") was also developed; this replaced the mechanical link between the master and slave with a connection via wires or radio signals sent from a control box equipped with switches, buttons, or a joy stick. These features exist in the modern robot's teach pendant. At that time, however, a robot was still not available, as all of the foregoing required continual control by a human. In the 1960s, researchers at Stanford University replaced the human at the master side of a teleoperator (their "Rancho Arm") with a computer to move and control an articulated manipulator.

Other technological advances that developed rapidly during the same time and that aided robotics engineering included orthotics, braces, and exoskeletal devices, prosthetics, bionics, numerical control (N/C) motion programming, high-level computer languages, and the beginnings of machine vision.

In 1946, George Devol developed a magnetic process controller by which the desired motion and actions of a machine could be programmed, stored, and reprogrammed, and which when played caused the process to be performed repetitively, as taught, until told to stop. It was the beginning of programmable industrial machinery, and in 1954 he applied it to a manipulator equipped with end-of-arm tooling (EOAT includes grippers and special tools, such as welding tips and paint spray guns) to create the first programmable industrial robot. At about the same time, prototypes of nonindustrial robots were being developed.

The first commercial robot to be used in industry was the Unimate 2000 (Fig. 1), in 1961, to tend a die-casting machine. It was a heavy-duty hydraulic-powered robot that gained wide use, first in Japan and later in the United States. Shortly thereafter, AMF introduced their Versatran line of robots, and then others followed suit. At approximately the same time, in Japan, Seiko widely used a simple, but accurate and reliable, small pneumatic robot (Fig. 2) to assemble wristwatches. All of the foregoing were "teach and repeat" robots—that is, they could only play back a previously programmed routine and had to be stopped and reprogrammed in order to do a new job. However, in 1973, Cincinnati Milacron introduced the T³ ("The Tomorrow Tool") robot, which was controlled by a minicomputer and could be reprogrammed via the computer. Now robots have microprocessors dedicated to each axis, plus extra microprocessors to handle vision, high-speed mass calculations, safety, and system balancing, all under the control of a master microcomputer. They have servo-controlled motors and encoders on each axis to inform the computer of their position 50 or 60 times each second. They have an array of sensors that are becoming more sophisticated and less expensive. Robots now have the senses of sight, feel, and touch. They can tell the distance, size, and temperature of an object, and do many other things faster, more accurately, and more consistently than a human. They can walk, climb stairs, float, hear (and follow spoken instructions), sing, and talk in any language.

Definitions. Robots have progressed rapidly in recent years and continue to advance, making it difficult to define them in a way that will be valid in the future.

A robot has been variously defined as "a heartless automaton," "a manufactured or machine-made man," and "a machine that looks like a human being and performs various complex acts (as walking or talking) of a human being." These are not good definitions. There is no requirement that the robot look like a human. It need not be anthropomorphic or anthropometric. In fact, in some cases, such characteristics would be a disadvantage. Some better definitions include: "a mechanism guided by automatic controls," and "an automatic apparatus or device that performs functions ordinarily ascribed to human beings or operates with what appears to be almost human intelligence."

The Robot Institute of America has defined a robot as "a reprogrammable, multifunctional manipulator designed to move material, parts, tools, or specialized devices, through variable programmed

ROBOT DESIGNATIONS

Capability in ascending order	If it does not look like a human, it is called:	If it looks like a human (is anthropomorphic) and is sized like a human (is anthropometric), it is called:
can only play back prerecorded programs	automaton	automan
and also is mobile	mechanoid	mandroid
and also has on-board computers and sensors	android	humanoid
and also is adaptive and heuristic	cyborg	syman
and also has extrahuman physical and mental capabilities	hyborg	supersyman

Fig. 3. Robot designations in ascending order of capability level. (*Courtesy of V. M. Altamuro*)

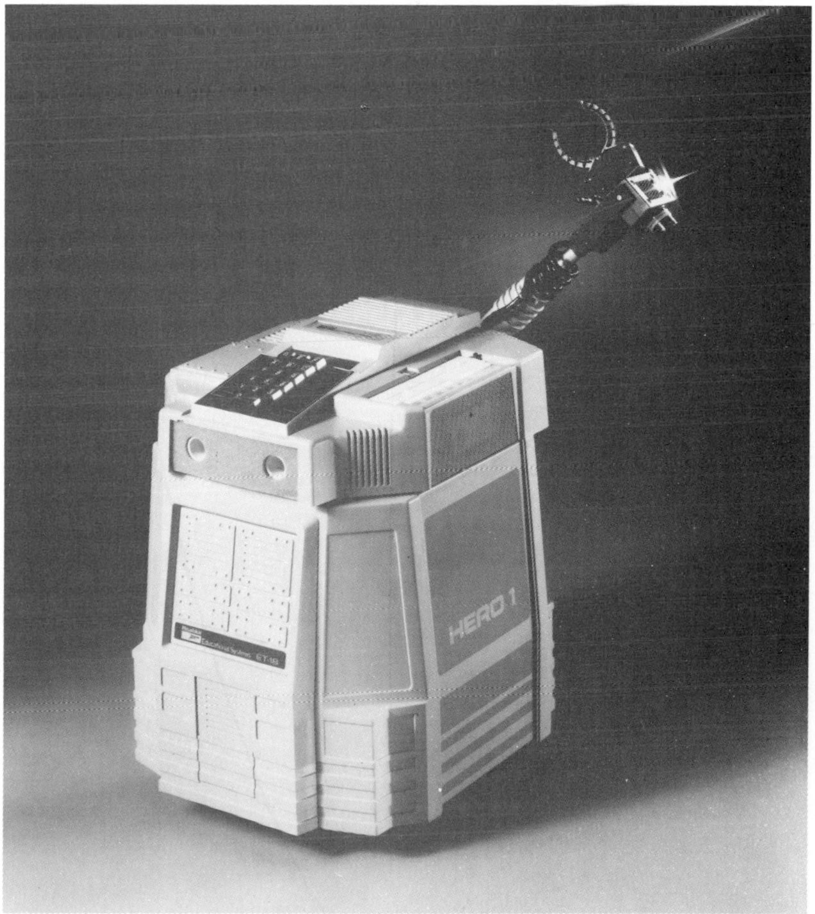

Fig. 4. HERO-1, an android robot. (*Heath Company, Division of Zenith Radio Corporation*)

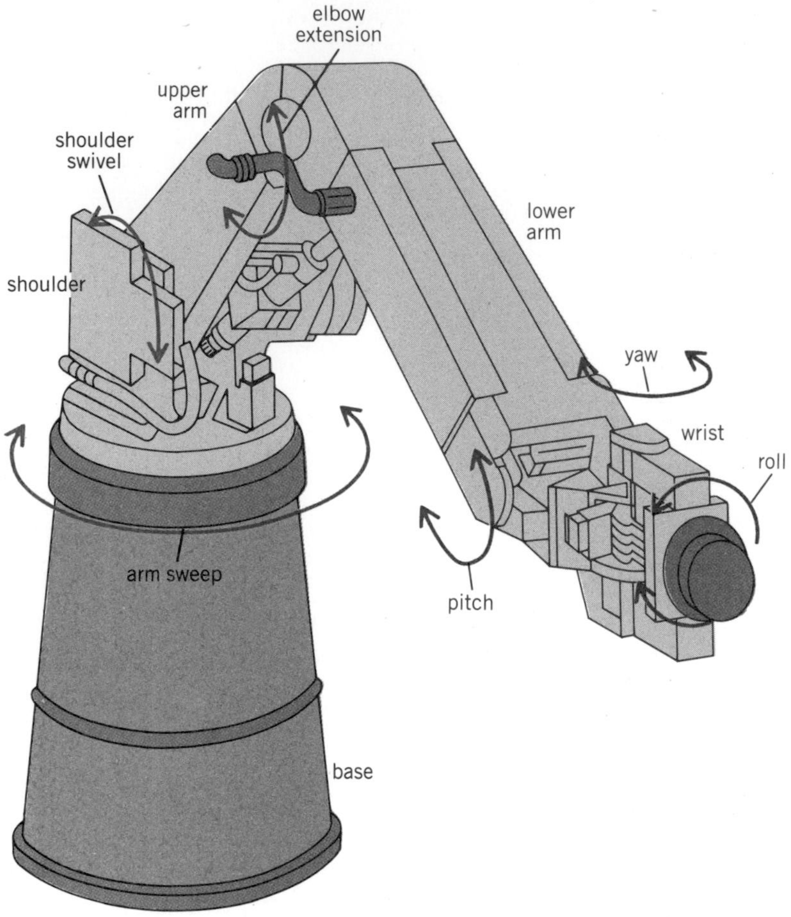

Fig. 5. Anatomy of a robot. (*Cincinnati Milacron, Inc.*)

motions for the performance of a variety of tasks." This definition was written several years ago, and it may have to be revised to recognize the recent advances in robotics. Any new definition should recognize that a robot must be highly articulated (jointed). Its arm should have at least three degrees of freedom, and it should have a wrist and gripper with additional articulations. A true robot should have "intelligence" in the form of a controller, computer, memory, sensors, and a closed-loop feedback system. It should be able to operate automatically and autonomously, without the need for human intervention. It should be more than a mechanism; it should be a machine capable of doing useful work.

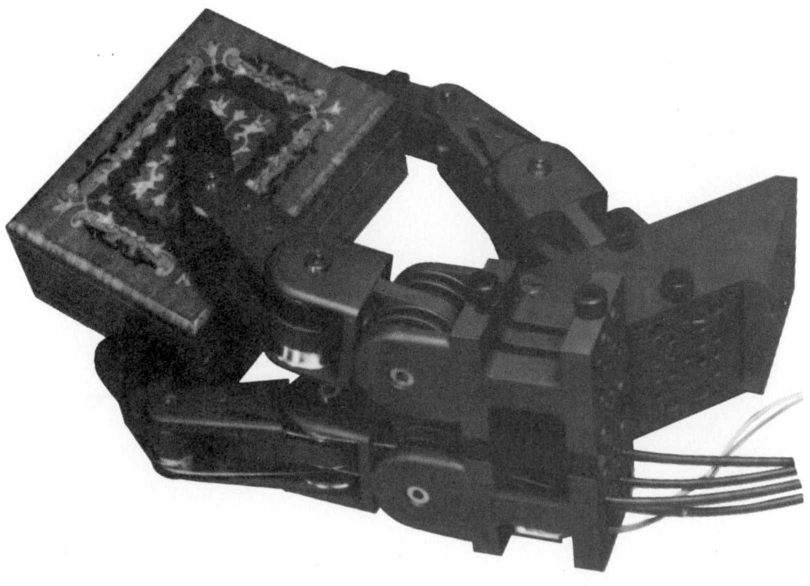

Fig. 6. Experimental gripper which has more degrees of freedom (9) and greater dexterity than the typical robot gripper. (*Salisbury Robotics, Inc.*)

So as not to eliminate a large portion of the existing robot population with new definitions to include the newer, more sophisticated robots, it will be necessary, in the future, to use modifiers rather than just the word "robot." Thus, a nonservo, noncomputer, pick-and-place pneumatic, four-degrees-of-freedom device may fit the definition of a "simple" robot, but not qualify as an "intelligent" or "sophisticated" robot. Also, the early robots were presented as general-purpose (as opposed to special-purpose) tools. But as new models are introduced, some are designed to do specific jobs. It is now proper to categorize a robot as a painting robot, welding robot, material-handling robot, electronic component assembly robot, and the like, to more clearly describe it. Further, each robot can be given ascending levels of capabilities. Names can be given to each such robot, with distinctions made between those which are also anthropomorphic or anthropometric and those that are not (Fig. 3; not all of the types of robots designated exist yet and some may never be created). The HERO (Fig. 4) is a good example of an android. Introduced in 1983 and improved in 1984, it is a completely self-contained, mobile robot with an on-board computer and sensors to enable it to interact with its environment in carrying out its program. It can sense 256 levels of sound and 256 levels of light, detect the motion of an object about the size of a human at a distance of about 15 ft (4.6 m), ultrasonically range distance to an accuracy of about 0.4 in. (1 cm) up to 8 ft (2.4 m) away, speak by synthesizing 64 phonemes, lift 0.5 lb (0.23 kg) with its arm extended, and "walk" (traverse on wheels) at about 1 mi (1.6 km) per hour. It does not look like a human. As yet, there are no humanoids, cyborgs (cybernetic organisms), symans (synthetic humans), hyborgs (hypercyborgs), or supersymans.

Robot anatomy. All robots have certain basic sections, and some have additional components to give them more capabilities. All robots must have a power source (mechanical, electrical, electromechanical, pneumatic, hydraulic, or a combination of these) and a means of applying the power (drives, gears, cams, bars, belts, chains, motors, pumps, actuators, and so forth) to move its manipulator (its assemblage of links, sometimes called its body, shoulder, arm, and wrist) (Fig. 5). Some robots have an immovable base which connects the manipulator to the floor, wall, column, or ceiling. Some have mobility components such as wheels, tracks, rollers, or "feet." All robots require a control sec-

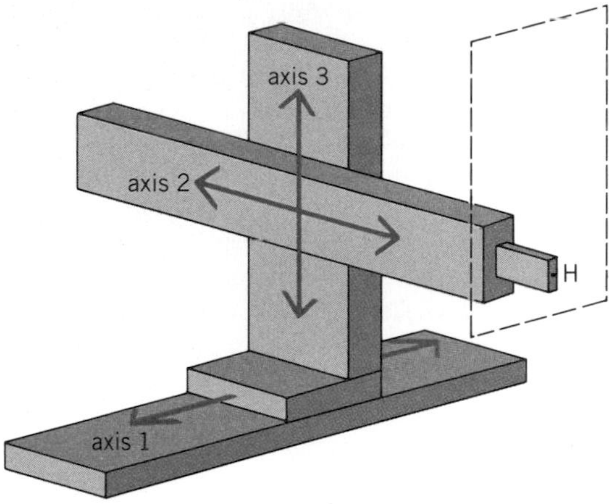

Fig. 7. A rectangular coordinate robot; H designates the hand, or end point. (*After Ken Susnjara, A Manager's Guide to Industrial Robots, Corinthian Press, 1982*)

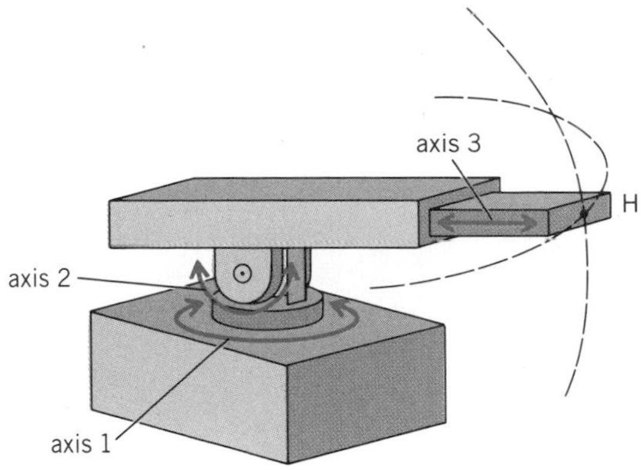

Fig. 10. A robot of spherical configuration; H designates the hand, or end point. (*After Ken Susnjara, A Manager's Guide to Industrial Robots, Corinthian Press, 1982*)

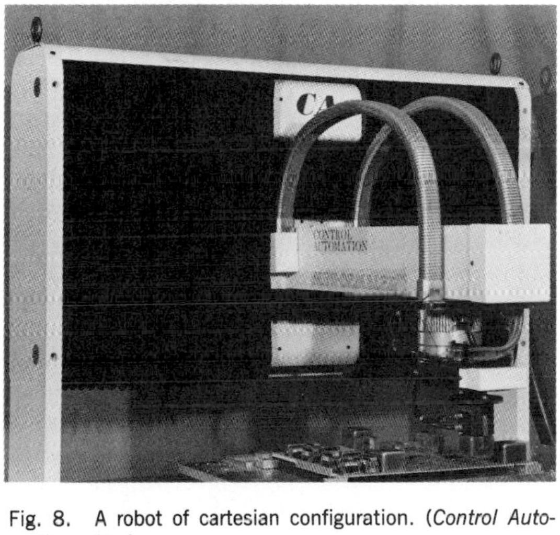

Fig. 8. A robot of cartesian configuration. (*Control Automation, Inc.*)

tion, which may be a relatively simple device in a "dumb" robot up to a very complex system in a "smart" robot that has the capability of continuously interacting with varying conditions and changing environment. Such control systems may include computers of various types, memory media, many input/output ports, absolute and incremental position encoders, signal feedback circuitry to provide real-time corrections, and all of the switches, interconnections, electronics, and wiring required. Teach pendants, keyboards, panels, cathode-ray tubes, and other devices (such as speech recognition and speech synthesis modules) are also needed to program the robot and monitor its performance. In addition, robots require end effectors or end-of-arm tooling to enable them to interface with the configurations of the objects to be handled. Grippers can

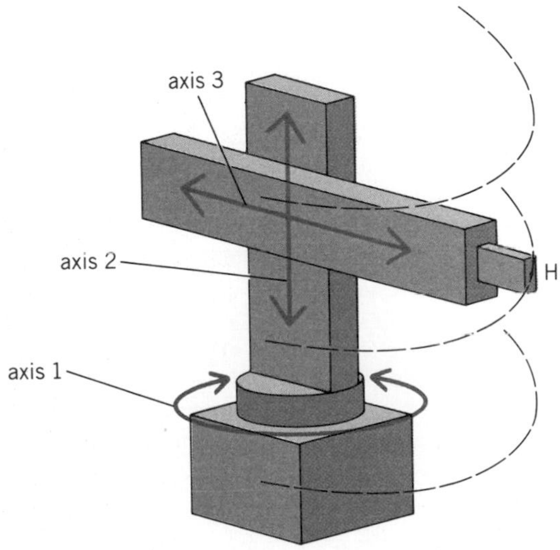

Fig. 9. A robot of cylindrical configuration; H designates the hand, or end point. (*After Ken Susnjara, A Manager's Guide to Industrial Robots, Corinthian Press, 1982*)

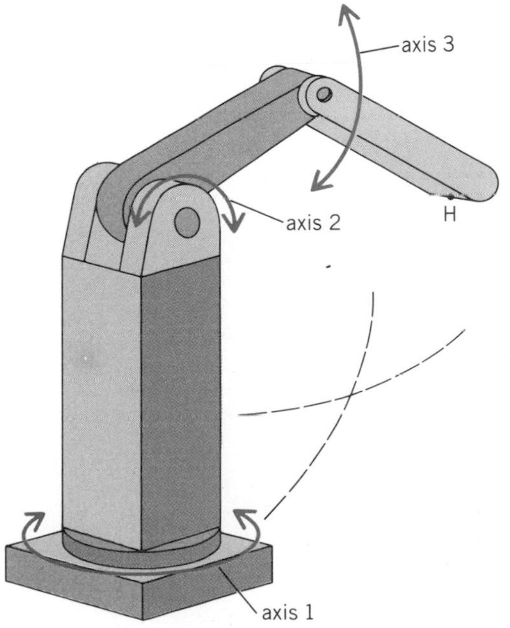

Fig. 11. A jointed arm robot; H designates the hand, or end point. (*After Ken Susnjara, A Manager's Guide to Industrial Robots, Corinthian Press, 1982*)

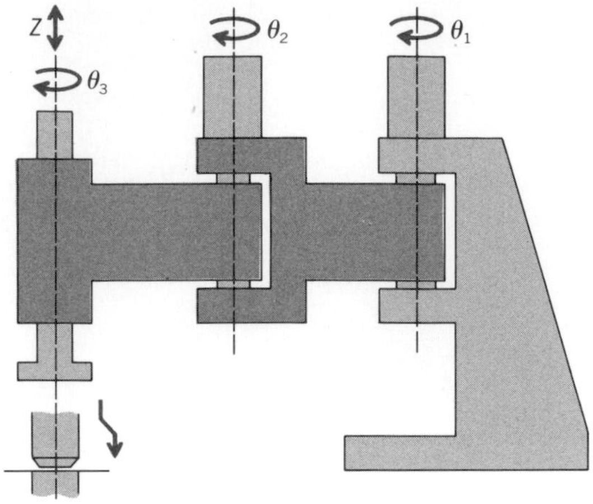

Fig. 12. Design of a robot of SCARA configuration.

become quite complex as their designs approach the capabilities of the human hand (Fig. 6). Further, the incorporation in them of sensors makes them more complex, heavier, and more expensive. Robot sensors help give the robot information regarding its position (where its links, axes, and tools are in space), presence (what is around the robot or coming toward it), and the environment (what is the ambient temperature, humidity, light level, and so on). Finally, the robot and the cell into which it is installed must contain adequate safety devices to prevent it from harming people, other equipment, the tooling, the product, and itself.

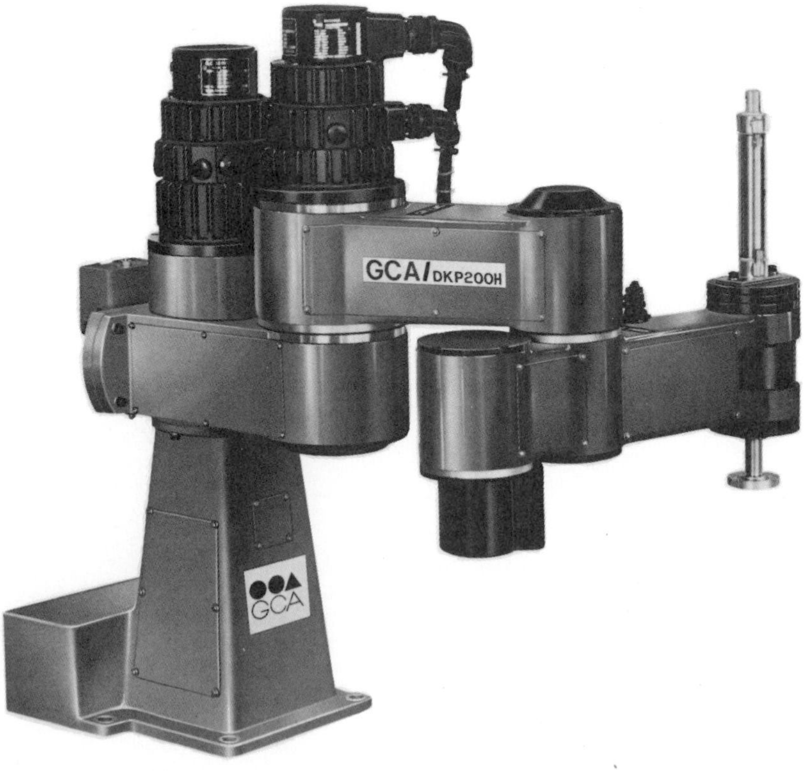

Fig. 13. A SCARA-type robot. (*GCA Corporation*)

Specifications of robots. Robots are designed in a wide range of specifications, so that a prospective user can match the features of a robot to the needs of a job. The important distinctions and definitions are as follows:

Configuration. The first three links of the manipulator can be designed to join and move in various ways. They can form a rectangular or cartesian configuration (Figs. 7 and 8); a cylindrical configuration (Figs. 2 and 9); a spherical or polar configuration (Figs. 1 and 10); a revolute or jointed arm configuration (Figs. 5 and 11); or the SCARA (selective compliant assembly robot arm) configuration (Figs. 12 and 13).

Workspace. The extent of each robot's reach in each direction is, of course, dependent upon its configuration, articulations, and size of its components (links and other members). The solid geometric space created by subtracting the inner (fully contracted) from the outer (fully extended) possible positions of a defined point (for example, wrist flange, center of gripper, and tip of tool) is called the robot's workspace or work envelope (Fig. 14). For a mobile robot, this space is greatly expanded—being limited only by physical barriers or programming restrictions—and is called the robot's probability shell.

Payload. The payload is the weight that the robot is designed to lift, move, and position repeatedly with accuracy, precision, and reliability. It varies, of course, with the degree to which it achieves each of these attributes, as well as whether it is handling the load with its manipulator fully extended or retracted.

Speed. Speed, too, is a specification that depends upon the point measured, whether extended or retracted, and the distance over which measured. From rest, the robot must accelerate, go to full speed, then decelerate before stopping. Clearly, if the distance to be moved is short, it will not have time to get up to its full rated speed. Further, many operations require positioning a load with precision, and the robot is usually slowed down to do that.

Accuracy. The accuracy of a robot is the difference between where a particular point (usually the hand or tool tip) goes and where it was directed (programmed) to go.

Repeatability. The repeatability or precision of a robot is the variance in successive positions each time the hand or tool returns to a taught position.

Resolution. The resolution of a robot is the smallest incremental change in position that the robot can make or its control system can measure.

Reliability. The reliability of a robot is the degree of probability that it will peform, as intended, for a given period of time, under normal conditions of use.

Robot motion control. While the robot's degrees of freedom and dimensions of its work envelope are established by its mechanical configuration and size, the paths it travels within its workspace are established by its control system.

Trajectory. There are several ways by which the robot's motion path can be categorized. It can move point to point (wherein its control system moves it from point A to point B without regard for the path, or trajectory, it takes to get to B) or continuous path (wherein the trajectory from A to B is controlled by the establishment of "way" points along the path through which the robot tip passes).

Another way to distinguish robot motions is by the fidelity of its actual trajectory to the desired or ideal trajectory. If the total desired path is thought of as a series of points, the robot's controller can calculate and implement its moves based upon individual points. (It could go to a point, stop, then go the next point—a jerky motion; or it could use its momentum to go through a point, recover, and go the next point—a jagged path.) It can use feed-forward control (wherein it gives consideration to a few points instead of its current position and passes slightly to the left or right of points so as to get in a better position to approach future points), or it can use total path planning (wherein it considers all of the points in the desired trajectory and plans its movements, before starting, to result in path with high fidelity).

Coordination. Still another way to distinguish the several levels of robot motion control is according to the amount of coordination among its parts as they move. Consider the example of a highly articulated revolute, or jointed-arm, robot with the task of moving a glass of water, filled to the brim, from one surface to a new location, a surface at a different height. If it moves the glass sequentially (one link starts, goes to its new position, stops, then the next link moves, and so on), the water will spill. If its links move independently (all start at the same time and each stops when it gets to its new position), the water will also spill, as some links will still be moving after others have stopped. If its links move interdependently (all start at the same time and the computer calculates the distance each has to move and causes the velocity of each to vary so that all stop at the same time), the water will still spill because there was no provision to keep the glass level. The robot must move in a coordinated manner (interdependent link movements, plus a master control instruction to maintain gripper orientation, at all times) for the water not to spill. Had the original or target surfaces been moving, that dynamic data would have to be incorporated in the robot controller's calculations to achieve a synchronized level of control. If something else were placed on the target surface just before the robot was to set the glass down, it would have adaptive abilities if it could sense that condition and switch to an alternate target; if, over time, the original target was occupied most of the time, it might go directly to the alternate target and would be at the heuristic, or ability-to-learn, level. When computers with artificial intelli-

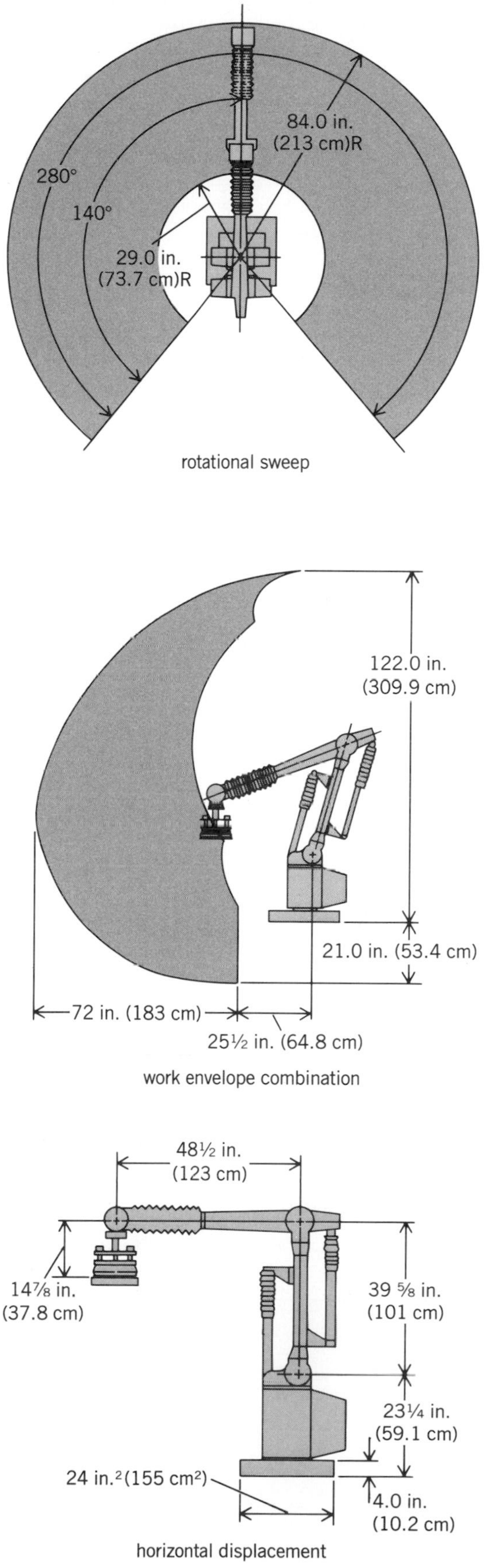

Fig. 14. Dimensions and work envelope of a robot in top and side views. (*Thermwood Corporation*)

gence are used to control robots, their step-by-step algorithms will be replaced by the more humanlike ability to use uncertain and incomplete information, symbolic reasoning, and judgment to determine the robots' actions.

Programming robots. The methods of programming robots, the number of languages used, and the levels of those languages have all increased as the robot hardware has advanced. The early robots were all of the "teach and repeat" type; that is they had static programs that, once loaded into the memory medium, could only be played back as recorded. Current robots have dynamic programs which can be altered while the robot is operating.

Robot programs can be created by setting end stops, switches, pegs, cams, wires, and so forth, on rotating drums, patch boards, or control panels; or by using a teach pendant (Fig. 15) with buttons, switches, or a joy stick tethered to the robot; or by physically grasping the end of the robot and moving it through the desired positions; or by using a keyboard to type in textual language instructions and data. All but the last are restricted to on-line programming, wherein the device must be attached to the robot; but the keyboard method may be done online or off-line, in which case the device is not attached to the robot. The advantage of off-line programming is that the robot can be doing job #1 while job #2 is being designed, programmed, analyzed, simulated, and debugged, with no loss of the robot's production time.

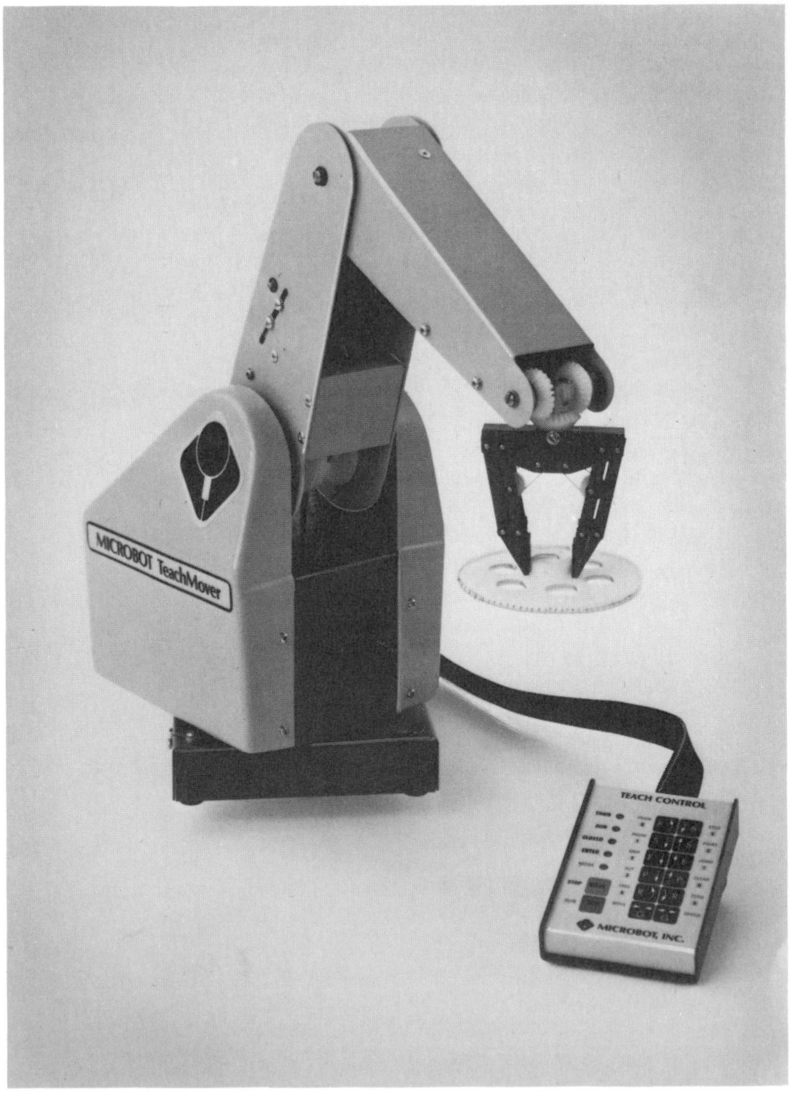

Fig. 15. The teach control pendant connected to a TeachMover robot. (*Microbot, Inc.*)

There are many robot programming languages in use—too many; around 20, counting variations of basic languages; and the number is still growing. The reason is that many robot manufacturers feel it is desirable to develop a special language for their robot rather than adopt one used by another manufacturer. Ultimately, standards must be developed and accepted so that all robot languages and programs are universal (will work on other robots), portable (can be transmitted from robot to robot), modular (constructed with interchangeable segments and sections that can be used or not, improved, replaced, and so forth), and resilient (able to sustain unexpected or emergency conditions and recover). Also, most robot programming languages in use today are explicit; that is, their instructions are at the level of such direct action commands to the robot as "open gripper," "close," and "rotate left." Future languages will be of a higher level and more implicit; that is their instructions will be at the task level: "assemble 25 of product model A-5," "walk the dog," "get me my pipe and slippers." Such languages are also called world-modeling languages and model-based systems, and are being developed by several universities, many robot manufacturers, and a few governments. These languages will be combined with the newer input methods of spoken commands and signals from a wide array of sensors to create robots that are able to operate very effectively and safely in the unstructured real world.

Robot applications. Robots may be operated alone or in systems. They may be used for industrial, commercial, medical, agricultural, military, home, hobby, education, and other purposes. Because of their many degrees of freedom, interchangeable end-effectors, and reprogrammability, they are close to universal machines and their range of potential applications is very wide. The major thrust of their recent development, however, has been for industrial applications, and nonindustrial uses are just getting started.

The first industrial applications of robots involved having them merely load and unload such machines as die casters, presses, and lathes. Next, they were used for materials handling, painting, welding, and other operations. Their most recent use is in assembly work, including high-precision electronics components assembly. In industrial applications, the robot can hold the work, the tool, or inspection probes. In the future it will transport the work as it processes it, which will bring industry closer to a continuous-flow factory.

The industrial robot is only one component of a total robot system, and that system is, in turn, only one part of a total autofacturing system which also includes the proper design of the part–product–process to accommodate handling by the robot or other automated equipment, computer-aided design, a common data base, an automated materials-handling system, and an information network linking the several work cells and people together to form an integrated, balanced system which operates in a real-time feedback, correction, and control mode.

Within the robot cell (Fig. 16) the robot or robots interface with conventional machines, to carry the work into the cell, position and hold it, and carry it away. In addition, the robot cell contains a computing or control capability, a multitude of sensors and interlocking switches, circuitry and power hook-ups, and "no harm" safety measures installed based upon a "worst case" simulation scenario.

The so-called nonindustrial applications of robots are growing rapidly. As androids with the interactive sensory capabilities of HERO (Fig. 4) are made more reliable, more rugged, and stronger, applications which were fantasy are now becoming possible. Working robots are already available that can act as bank vault tenders—they take a customer's key, verify it, fetch the correct safe deposit box, and return it when the customer leaves; act as nurses' aides—they fetch magazines and drinks for patients and bring the proper pills, on schedule, to each bed; aides to the handicapped—upon a voice command from a paralyzed person, they get a frozen dinner from a freezer, put it in a microwave oven, remove it at the proper time, and feed the person; act as guards—they rove the aisles of stores, warehouses, and such to sense moving objects, challenge them, or call the police; do the same in the home, and also sense for smoke, fire, and gas and water leaks at night and when the owner is away. They can also water plants and feed pets on schedule. They can answer the doorbell and the phone. Soon, they will be able to start the dinner at a preprogrammed time or when told to via a telephone call. They will be able to sense and maneuver around objects, vacuum the rugs, and mow the lawn.

Future uses of robots include assisting the police, fire fighting, and sanitation work. They can aid the military by doing reconnaissance and some types of warfare.

On the farm, they can pick and sort citrus fruit. They can pick up to 20 tons per day and sort it into six size categories, each with four color grades, and reject those with excessive defects. They soon will be able to plant crops; spread insecticides, herbicides, and fertilizer; pull weeds; gather, grade, sort, and pack the harvest. They will gather eggs, feed livestock, milk cows, and shear sheep. They will cut and carry timber, mine minerals (in the ground and at the bottom of the ocean), repair submerged ocean cables, work in nuclear power plants, and operate in space.

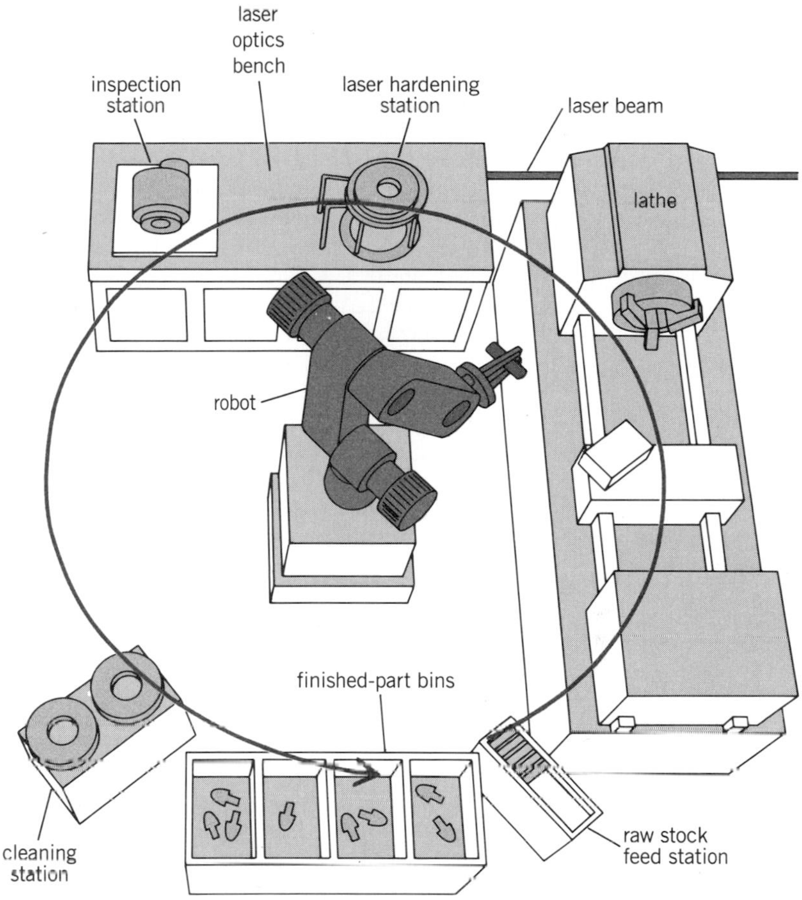

Fig. 16. A robot work cell that interfaces with production machines, inspection and cleaning stations, and a supervisory computer (not shown). (*After High Technology Magazine, July 1983*)

Advantages of robots. Robots can free humans from tasks which are dangerous. They can do jobs which are dirty, dull, and demeaning for people to do. They will do these things consistently, tirelessly, and without complaint. In some cases, they can exceed human performance. They can lift heavier loads, continually and without getting hurt. They can handle very small, delicate parts without damaging them. They can work in areas made hazardous by asbestos, paint fumes, noxious chemicals, and radiation. They can work in heat, cold, and dark. They can act as guards, sentries, fire and burglar alarms, inspectors, and monitors—relentlessly, reliably, and tirelessly. In some cases they work less expensively than humans in that they do not require overtime pay, vacations, cafeterias, rest rooms, lockers, and parking spaces. They are trained (programmed) quickly and easily and can be retrained (reprogrammed) to be flexible and adaptable to changing requirements. They do not become obsolete. When their first job is done, they can be used to do another. Their utility is limited only by their user's needs and imagination.

Disadvantages of robots. A robot's present capabilities lie between those of a human and a high-speed dedicated machine. Robots are not as good as humans in picking out stimuli against high noise backgrounds, picking out patterns which are varied or complex, applying prior experience to new problems, acting in emergencies with no new programming required, recovering from disruption, jiggling parts together, assembling a nut and bolt, or doing work which requires a "feel" and getting better with practice. With respect to high-speed dedicated machines, they are slower, not as economical for large outputs, and not as strong, durable, rigid, accurate, or precise.

The future. A science and technology of robotics now exists. It is in its earliest stages, and what is now regarded as advancement will be viewed as primitive by future generations. Robots have already relieved humans of some undesirable tasks; but they have also displaced workers who did not want to be displaced. Whether the robot will replace people, as the automobile replaced the horse, or will create new industries, more jobs, and a better quality of life, as did the computer, only time will tell. But they will advance. They will become more highly articulated, stronger, more reliable, more intelligent, and less expensive. Their capabilities will grow, perhaps to the point where certain abilities will be withheld from them—at the point where humans agree to control technology.

[VINCENT M. ALTAMURO]

Bibliography: J. S. Albus, *Brain, Behavior and Robotics*, BYTE Publications, Peterborough, New Hampshire, 1981; V. M. Altamuro, Working safely with the iron collar worker, *Nat. Safety News*, National Safety Council, July 1983; D'Ignazio, *Working Robots*, 1982; *Robotics Industry Directory*, Technical Data Base Corp., Conroe, Texas, 1985.

A-Z

Acid rain

Acid precipitation is the result of a complex and incompletely understood set of chemical reactions. Sulfur and nitrogen oxides emitted into the atomosphere react to form sulfuric and nitric acids, which are deposited on soil, water, and vegetation in either wet or dry form. Large and apparently expanding areas of Europe and eastern North America receive acidic precipitation. Within these areas are ecosystems that, because of the high acidity or low buffering capacity of soils, waters, or geologic materials, may be adversely affected by acid additions.

Just as the steps between emission and deposition are not clearly understood, the link between deposition of acid materials and environmental impact is uncertain. A great deal of data exists connecting decreased productivity of aquatic ecosystems to greater acidity and increased soluble aluminum concentrations. However, the link between acid precipitation and adverse effects on terrestrial ecosystems is tenuous despite recent research suggesting that root-damaging concentrations of soluble aluminum generated by acid precipitation are causing a decline in forest productivity.

Sensitive areas. In the absence of conclusive cause-and-effect evidence, much effort has been expended to delineate acid-sensitive areas. Classification parameters have included water alkalinity and acidity, neutralizing capacity of geologic materials, and acidity and neutralizing capacity of soils. In most United States ecosystems, soils are the key intermediate providing the root environment for vegetation and controlling runoff and drainage water quality. A number of workers have proposed criteria to divide soils into groups of varying acid sensitivity. These relate to the soil's ability to maintain its

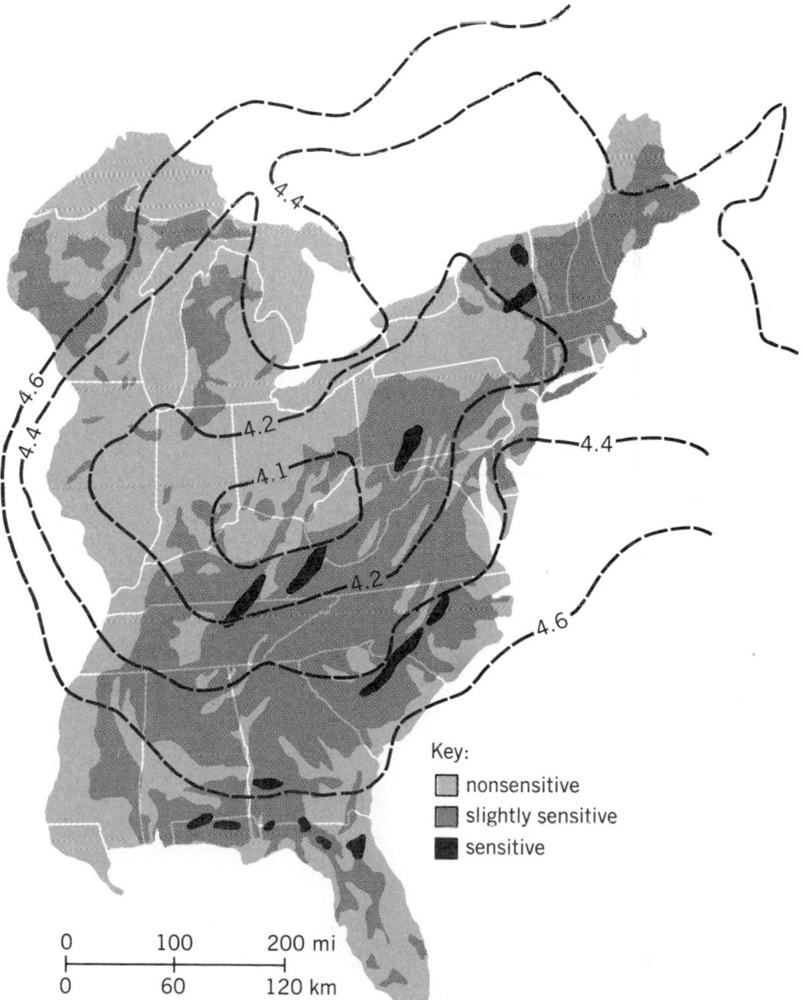

Fig. 1. Acid-sensitive soils of the eastern United States and annual average precipitation pH for 1980 (broken lines).

productivity against acid additions and to cause changes in the composition of water passing through it.

A number of reactions occur in soil which buffer changes in pH, lessening the impact of acid additions on both the soil and drainage waters. If sources of carbonate, such as limestone, exist in the soil, they react with acidic solutions, raising the pH of drainage water while maintaining a high soil pH. Clays and organic materials in soil have a net negative charge (cation exchange capacity) which attracts and absorbs cations. In soils near neutrality (pH in the range 6.0 to 8.0), the absorbed cations are primarily the basic cations Ca^{2+} and Mg^{2+}. However, when acids are added, the absorbed cations are replaced by hydrogen ions. This exchange process removes hydrogen ions from solution, thus reducing the acidity of drainage water while increasing soil acidity. Eventually the neutralizing capacity of carbonates and cation exchange capacity is exhausted. Amorphous aluminum and iron hydroxides and crystalline soil minerals then react with hydrogen ions to buffer the soil. At this point, usually pH 5.2 or less, considerable aluminum may be released into the soil solution as coatings on soil particles and clay minerals dissolve.

The carbonate content and the cation exchange capacity, both readily available soil data, are the characteristics most commonly used to classify acid-sensitive soils. Cation exchange capacity was the characteristic used to classify acid-sensitive soils in Fig. 1, which shows areas with acid-sensitive soils that also receive low-pH rainfall. The management imposed on a soil, however, can greatly modify its susceptibility to acid additions. For this reason, intensively managed or farmed soils are generally mapped as nonsensitive. The oxidation of ammonia- or urea-based fertilizer, commonly applied to cropped soils, produces many times more acid than that which falls to the ground as acid precipitation. The safety margin used to routinely amend agricultural soils with ground limestone, to maintain their productivity, is sufficient to neutralize the added effect of acid precipitation. Much of northeastern

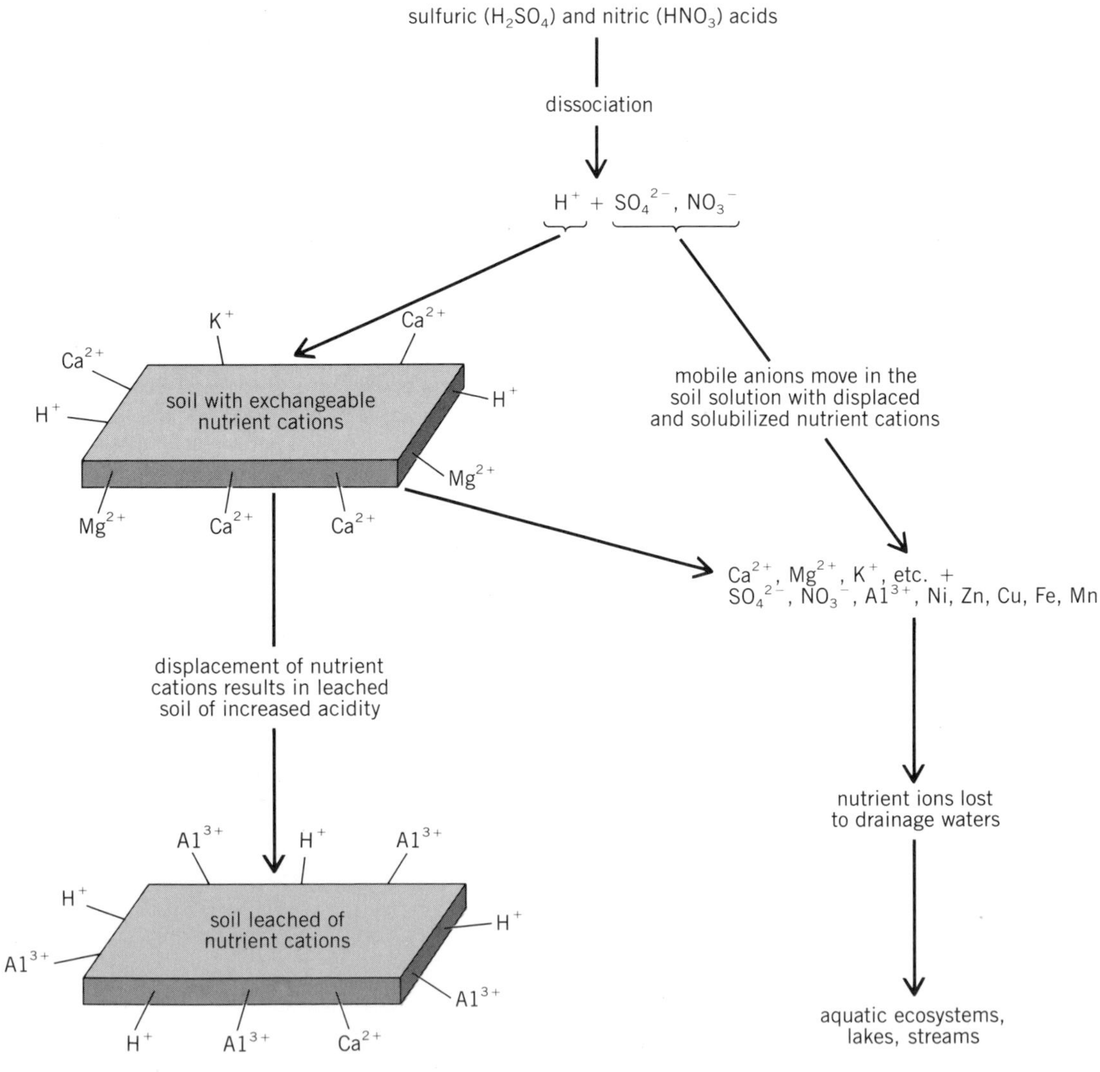

Fig. 2. Soil acidification and loss of soil nutrients following acid inputs.

North America, however, is nonagricultural, forested land which is not routinely limed; for example, Pennsylvania is 58% forested.

Acid precipitation is not a suitable medium for the growth of most aquatic organisms. However, when acidic solutions flow through soil, their composition may be modified. The reactions which alter the composition of drainage water are the same reactions which buffer the soil. Shallow soils with a low pH, low cation exchange capacity, and high permeability are least able to favorably modify solutions moving through them; therefore lakes and streams fed by such soils are most susceptible to degradation by acid precipitation. In general, those soils which remain productive despite acid additions also improve the quality of water moving them.

Direct entry of acid precipitation into lakes and streams causes chemical reactions analogous to those in the soil system. However, instead of soil, the carbonate-bicarbonate system buffers the solution. As with soils, waters of low pH and bicarbonate concentration are most sensitive. As the solution pH is decreased to about 4.5, bicarbonate is essentially depleted, and continued reductions in pH mobilize metals from suspended solids and the lake or stream bed.

Effects of acid additions. In humid regions, where precipitation exceeds evapotranspiration, soil acidification is a natural process, and the application of fertilizers can further acidify the soil. Thus, while the impact of acid precipitation on soil has only recently begun to receive attention, the consequences of soil acidification have been studied for many years.

Fertile soils near neutrality have negative charge sites on clay and organic matter balanced by basic cations necessary for plant growth. They support a diverse and vigorous population of soil microorganisms which decompose plant residues and convert organic nitrogen and sulfur compounds to plant-available inorganic forms. Soil minerals of neutral soils are relatively insoluble, and the solution concentrations of trace nutrients and other metals are maintained in nontoxic ranges. As a soil becomes more acid, the basic cations absorbed to the soil surface are replaced by hydrogen ions or solubilized metals. The cations now in solution can be leached through the soil (Fig. 2). The soil becomes less fertile and further reduced in pH. Decreases in soil pH result in reduced, less active populations of soil microbes, slowing decomposition of plant residues and cycling of the essential plant nutrients. Additionally, the solution concentration of phosphorus and trace elements changes as pH decreases. The concentration of plant-available phosphorus, controlled by aluminum and iron phosphate at low pH, decreases as pH decreases, while the concentrations of trace metals (Cu, Fe, Zn, B, Mn), with the exception of molybdenum, increase, sometimes to toxic levels. Aluminum concentrations sufficiently high to inhibit root growth have been found in soil solutions at low pH. When these solutions drain to stream and lake waters, they can also be toxic to aquatic organisms.

In the longer term as soils weather in a humid, acid environment, basic cations are leached from the soil, and minerals containing these elements dissolve, leaving behind a soil system rich in silica, iron, and aluminum oxides which has a lower pH, has lower cation exchange capacity, and is inherently less fertile. Since soil acidification is a natural phenomenon in humid environments, the major consequence of acid precipitation will likely be to accelerate this process.

While it is important to know what reactions occur in response to acid deposition on soil, the rates at which the reactions proceed are also important. It is the combination of neutralization capacity, reaction kinetics, and flow dynamics which governs the acidification rate and pattern of a soil system. The extent and location of weathering in the soil profile and the resultant change in solution composition depend on the specific flow rate and pathway taken by acidic solutions as they pass through the soil.

For background information *see* pH; SOIL CHEMISTRY in the McGraw-Hill Encyclopedia of Science and Technology.

[RONALD R. SCHNABEL; HARRY B. PIONKE]

Bibliography: W. W. McFee, *Sensitivity of Soil Regions to Acid Precipitation*, EPA-600/3-80-013, 1980; G. H. Tomlinson, Air pollutants and forest decline, *Environ. Sci. Technol.*, 17:246–256a, 1983.

Acoustic signal processing

Acoustic signal processing is the discipline that deals generally with the extraction of information from signals in the presence of acoustic noise. To obtain the best performance from acoustic signal processing, it is necessary to know, and take advantage of, the exact nature of this background acoustic noise. Thus, there is ongoing interest in determining the probability distribution of acoustic noise under various physical circumstances, and in deriving corresponding acoustic signal processing configurations or algorithms to obtain the maximum performance for each type of acoustic noise. Recent research is pointing the way to achieving this end.

It has been traditional to synthesize and analyze signal-processing systems on the basis of additive noise whose probability distribution is gaussian, or normal. The first-order probability density function for gaussian noise, whose mean, or average, value is zero, is given by Eq. (1), where n represents the

$$p_n(n) = \frac{1}{\sqrt{2\pi}\,\sigma}\, e^{-n^2/2\sigma^2} \qquad (1)$$

instantaneous value of the noise, and σ^2 is the variance of the noise (proportional to the noise power).

There are two principal reasons for assuming a gaussian distribution for acoustic noise:

1. Much of the naturally occurring noise encountered in acoustics, and other physical domains as well, truly is gaussian by virtue of the central limit

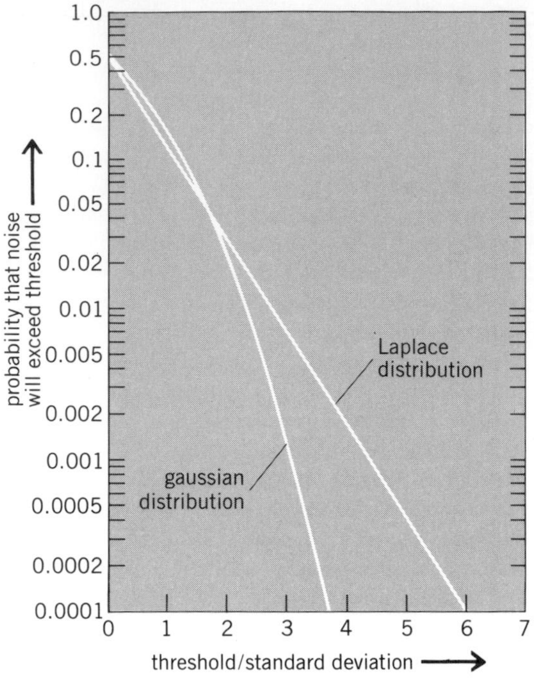

Fig. 1. Comparison of nongaussian noise governed by Laplace distribution with gaussian noise.

theorem, which states that, under a rather wide set of conditions, the distribution of the sum of a large number of independent random variables tends to the gaussian distribution. This corresponds to the physical mechanism for the generation of much—but not all—noise encountered in acoustics.

2. The mathematical analysis flowing from the gaussian assumption is tractable and well worked out. For this reason, even when the noise turns out not to be gaussian, it is tempting to continue to use signal-processing systems designed for gaussian noise. This practice, however, can result in failure to achieve optimal performance from the signal processing, as will be seen.

Nongaussian acoustic noise. This is a designation for any noise whose distribution is not governed by the probability density function given by Eq. (1). A common example of nongaussian noise is impulsive noise, where the probability of large excursions relative to the standard deviation (square root of the variance) is significantly higher than it would be for gaussian noise. Another way of expressing this idea is to say that the tails of the distribution are heavy. Figure 1 is a comparison of the probabilities of exceeding a given threshold (normalized by the standard deviation) for gaussian noise, on the one hand, and noise governed by the Laplace, or double-exponential, distribution on the other. This latter probability density function is given by Eq. (2).

$$p_n(n) = \frac{1}{\sqrt{2}\,\sigma}\, e^{-\frac{\sqrt{2}}{\sigma}|n|} \qquad (2)$$

This distribution has heavier tails than the gaussian, and can be used as one possible and tractable model for a nongaussian distribution.

Models for nongaussian noise experienced in the domain of radio reception have been developed in recent years. Current research is being directed to determining the applicability of these models to the acoustic domain. For example, it has recently been shown that under-ice acoustic noise is nongaussian with similar characteristics.

Two-sensor acoustic detection. The ideas presented in this article can be clarified by applying them to a specific example: Suppose sensors are positioned at two separate points in space such that a desired signal appears identically in the outputs of the two sensors (that is, perfectly correlated), but the accompanying acoustic noise is totally uncorrelated from one sensor to the other, and uncorrelated with the signal as well. What is the best way to process the received data from the two sensors in order to detect the wanted signal?

The most obvious approach is the "commonsense" approach: the signal differs from the noise in the degree of intersensor correlation. Guided by the mathematical definition of correlation, a configuration commonly known as an analog correlator (Fig. 2) is set up. The sensor outputs are separately bandpass-filtered to a bandwidth W, then multiplied together, and the product time-averaged over the observation time T. The resulting output is compared

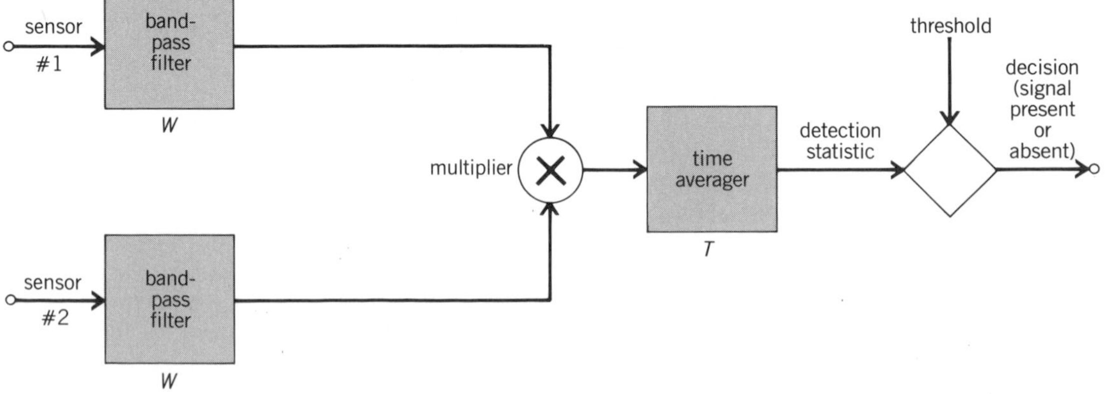

Fig. 2. Block diagram of an analog correlator.

with a threshold for detection of the presence or absence of the signal.

This can be a successful type of processor, but it suffers a drawback: it is difficult to know where to set the detection threshold, since the output is sensitive to the absolute level of the input. Some method of normalization is required. Automatic gain control can be used, but can be less than perfect in its implementation, especially if the input level changes rapidly over a wide dynamic range as can happen in acoustic detection systems.

One type of normalization which has achieved considerable success is that provided by the polarity-coincidence correlator, or clipper correlator (Fig. 3). This correlator differs from the analog correlator in the addition of a "hard clipper" in each sensor output channel. The hard clipper is a device which delivers a positive constant voltage (which can be represented by a $+1$) when the input voltage is positive, and a -1 when the input is negative: a one-bit quantizer.

Both types of correlator are further discussed below.

Optimization. Over the years, mathematical techniques have been developed to determine how to obtain the theoretical maximum of performance in signal processing. These techniques have been applied to the two-sensor acoustic detection problem, with interesting results. The highlights of the analysis will be sketched in here to give an appreciation for what is involved, followed by a statement of the results.

A receiver whose function is to detect the presence or absence of a signal can be thought of as an implementation of a statistical test of the null hypothesis, H_0, that the signal is absent, versus the alternative hypothesis, H_1, that the signal is present. The best such test under the Bayes minimum-risk criterion is one that forms the "likelihood ratio" and compares it with a threshold. The likelihood ratio, Λ, is defined by Eq. (3), where $\mathbf{v}$ is the input

$$\Lambda(\mathbf{v}) = \frac{p_1(\mathbf{v})}{p_0(\mathbf{v})} \tag{3}$$

data, written here in vector form; p_0 is the joint probability density function of the data under hypothesis H_0 (signal absent); and p_1 is the same, but under hypothesis H_1 (signal present).

For the problem on hand, it is convenient to work with $2WT$ statistically independent time samples in each sensor output channel. The samples can then be written as in Eqs. (4), where v_{ij} is the i-th sample

$$\begin{aligned} v_{i1} &= s_i + n_{i1} \\ v_{i2} &= s_i + n_{i2} \end{aligned} \tag{4}$$

of the voltage in the j-th channel; s_i is the i-th sample of the signal (same in both channels); and n_{ij} is the i-th sample of the additive background noise in the j-th channel.

The joint probability density function of the data under hypothesis H_0 (signal absent), p_0, is simple to establish, due to the assumed statistical independence of the noise, not only from time sample to time sample but also from channel to channel. It is simply the product of all the individual background noise probability density functions, $p_n(v)$. The joint probability density function of the data under hypothesis H_1 (signal present), p_1, is more involved. As would be expected, it depends on the signal probability density function. Also, because the signal appears identically in both channels, there is a channel-to-channel linkage, complicating the mathematics. Naturally, the behavior of the optimum signal processor is bound to depend in general on the nature of the signal, as well as that of the background noise.

A simplification does result, however, if it is recognized that an important application of an acoustic signal processor is to the situation where the signal-to-noise ratio is low—where the observer is "straining to hear" the signal. Under this circumstance, the individual conditional probability density function of the noise given a particular signal voltage can be expanded in a Taylor series, keeping the first two terms, as in Eq. (5), where the prime sig-

$$p_n(v - s) \cong p_n(v) - s \cdot p_n'(v) \tag{5}$$

nifies differentiation with respect to v. Following this approximation, the mathematical analysis can proceed, finding an approximate expression for the likelihood ratio Λ, Eq. (3). As is usual in optimization problems of this sort, the logarithm of the likelihood ratio turns out to be more convenient. Once this is written, a further small signal approximation can be used to express the logarithm by one term in its series expansion, leading to the sufficient detection statistic given by Eq. (6). The ratio

$$\sum_{i=1}^{2WT} \frac{p_n'(v_{i1})\, p_n'(v_{i2})}{p_n(v_{i1})\, p_n(v_{i2})} \tag{6}$$

p_n'/p_n can also be expressed as the derivative of the logarithm of the probability density function of the background acoustic noise. The dependence on signal characteristics is gone, due to the low signal-to-noise ratio assumption.

When Eq. (6) is specialized to the case of gaussian noise, where $p_n(v)$ is given by Eq. (1), the result, after disregarding multiplicative constants, is simply given by expression (7), which is a mathe-

$$\sum_{i=1}^{2WT} v_{i1} \cdot v_{i2} \tag{7}$$

matical description of the analog correlator (Fig. 2). Thus, the best way to process a weak signal in gaussian noise in this two-sensor context is with an analog correlator. The commonsense approach led to the right answer in this case. A word of warning: The arbitrary factoring-out of constants which include information about absolute input level tacitly assumes that the level is known. The normalization problem has been neglected. This is discussed below.

On the other hand, when Eq. (6) is specialized to the case of Laplace noise, where $p_n(v)$ is given by

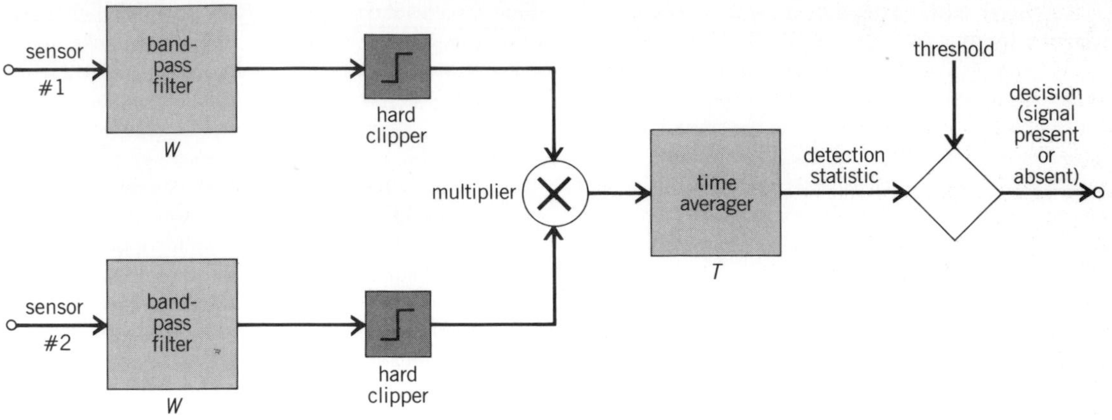

Fig. 3. Block diagram of a polarity-coincidence correlator.

Eq. (2), the result is given by expression (8), which

$$\sum_{i=1}^{2WT} \frac{v_{i1}}{|v_{i1}|} \cdot \frac{v_{i2}}{|v_{i2}|} \qquad (8)$$

is a mathematical description of the polarity coincidence correlator (Fig. 3). Thus, the best way to process a weak signal in Laplace noise in this two-sensor context is with a polarity-coincidence correlator.

Performance. In order to draw more complete conclusions, it is necessary to calculate the performance. The methodology here is basically to assume that the number of independent time samples, $2WT$, is large; and, again, that the input signal-to-noise ratio is low. Under the circumstances, the central limit theorem mentioned above will hold, causing the detection statistics at the output to approach the gaussian. Then a universal receiver operating characteristic plot (ROC curves) of detection probability versus false-alarm probability for various values of input signal-to-noise ratio can be drawn. A parameter called the detectability index is used for the purpose, and in itself becomes a measure of goodness. It is the square of the shift of the mean output caused by the presence of the signal, divided by the variance of the output (which does not change much when the input signal-to-noise ratio is low). The detectability index resembles an output signal-to-noise ratio, and in a sense it does serve this function. More importantly, it determines achievable values of the probabilities of detection and false alarm.

When the detailed calculations are completed, it is found that when the noise is gaussian, the performance of the analog correlator is 2½ times as good as that of the polarity-concidence correlator. When the noise has a heavy-tailed Laplace distribution, the performance of the latter is four times as good as that of the former. The absolute performance of the analog correlator is the same in both types of noise. From this point of view, the analog correlator appears to be more robust with respect to the characteristics of the background acoustic noise.

Threshold sensitivity. The above theoretical treatment assumes known input noise level, thus allowing a definitive threshold setting. The real world is not so obliging, as discussed above. To take this factor into account, a quantity called threshold sensitivity is now defined: the derivative of the logarithm of the standard deviation of the detection statistic (processor output) with respect to the logarithm of the input noise power. For such a complicated definition, the result is simple: the threshold sensitivity of the analog correlator is unity (very sensitive), while the threshold sensitivity of the polarity-coincidence correlator is zero (totally insensitive). Thus, the truly robust correlator is the polarity-coincidence correlator. In fact, its output variance, under these low input signal-to-noise ratio conditions, is equal to the reciprocal of the number of independent samples in each sensor output: $(2WT)^{-1}$, independent of the distribution or level of the background noise. Therefore, if the input bandwidth is maintained constant, a desirable constant false-alarm rate (CFAR) condition is achieved.

For background information *see* ACOUSTIC SIGNAL PROCESSING; DISTRIBUTION (PROBABILITY); PROBABILITY in the McGraw-Hill Encyclopedia of Science and Technology.

[JAMES F. BARTRAM]

Bibliography: R. F. Dwyer, A technique for improving detection and estimation of signals contaminated by under ice noise, *J. Acoust. Soc. Amer.*, 74:124–130, 1983; A. D. Spaulding and D. Middleton, Optimum reception in an impulsive interference environment—Part I: Coherent detection, *IEEE Trans. Commun.*, COM-25:910–923, 1977: A. A. Winder, Sonar system technology, *IEEE Trans. Sonics Ultrasonics*, SU-22:291–332, 1975; S. S. Wolff and J. L. Gastworth, Robust two-input correlators, *J. Acoust. Soc. Amer.*, 41:1212–1219, 1967.

Acquired immune deficiency syndrome (AIDS)

As the 1970s will be noted in medicine for the sudden emergence of two new syndromes, legionnaire's disease (LD) and toxic shock (TS), the 1980s will be associated with the new acquired immune deficiency syndrome (AIDS), a far more devastating and mysterious disease that has already claimed more lives than LD and TS combined.

AIDS describes an apparently lethal disease of undetermined etiology which has become identified with increasing frequency since 1980. It shows a predilection for homosexual and bisexual males, male and female intravenous drug abusers, and male and female Haitians. The aforementioned groups, as well as recipients of blood products, newborn infants of addicts, and heterosexual partners of AIDS patients, are apparently being exposed to an as-yet unidentified agent. AIDS seems likely to be caused by an agent transmitted by intimate sexual contact, contaminated needles, or less commonly, by percutaneous inoculation of infectious blood or blood products. The etiologic agent, whether a previously undescribed microorganism or a mutant of a known one, is presumed to have a proclivity for attacking and damaging human lymphocytes, culminating in a profound depletion of the helper-inducer T-lymphocyte population. Lacking these critical helper T lymphocytes, the immune system is crippled. In this immunocompromised state, the host becomes susceptible to developing multiple, severe, and recurrent opportunistic parasitic, viral, fungal, and mycobacterial infections, as well as malignancies. As of August 1983 the Centers for Disease Control (CDC) in Atlanta have reported over 2000 cases of AIDS in the United States, of which over 800 are known to have died. Since there is no effective therapy for the underlying acquired immune deficiency, most AIDS patients die from overwhelming infections within 2 to 3 years of the initial appearance of symptoms.

History and epidemiology. Cases of AIDS were first reported in 1981 by investigators from New York and California who were suddenly treating previously healthy homosexual men for a rare and unusual protozoal pneumonia caused by *Pneumocystis carinii*, and an equally rare and unusual tumor called Kaposi's sarcoma. These homosexual men had profound deficiencies in their immune response, yet careful clinical evaluations and autopsies failed to reveal a malignant neoplasm or any other immunosuppressive entity that might explain their immunodeficiency. From a clinical standpoint, the immunodeficiency had developed during adulthood since no patient had experienced any unusual infection or tumor during the first several decades of life. Thus, the immunodeficiency was acquired rather than congenital. The east and west coast investigators recognized that the *P. carinii* infections and the Kaposi's sarcoma tumors were secondary manifestations of an underlying acquired major defect in cellular immune function. Subsequent investigations discovered that these immunodeficient AIDS patients were susceptible to numerous opportunistic infectious agents including *Candida* species, *Herpes simplex* virus, cytomegalovirus, Epstein-Barr virus, *Cryptococcus neoformans*, *Toxoplasma gondii*, *Mycobacterium avium-intracellulare*, and *Cryptosporidium*.

As clinicians became aware of these occurrences in previously healthy individuals, unusual clinical events of all types in homosexuals, drug abusers, Haitians, and recipients of blood products became the focus of considerable interest. Cases of Epstein-Barr-virus-associated Burkitt's lymphoma, non-Hodgkin's lymphoma, colonic carcinomas, oropharyngeal carcinomas, immune complex nephropathies, generalized lymphadenopathy, and autoimmune phenomena were reported, often associated with opportunistic infections. Since no laboratory marker of AIDS has been discovered, defining AIDS in a manner that distinguished the truly new clinical occurrences from the heightened awareness of unusual disease processes in populations at risk for AIDS became a real problem. To determine the magnitude of AIDS and its true epidemiologic pattern, the CDC has adopted the following strict definition that is unlikely to be inflated by unrelated disease processes.

A. Patient with no known cause for immunosuppression
B. Disease predictive of a defect in cellular immune function, including:
   1. Meningitis, encephalitis, and pneumonitis due to:

Aspergillus	*Nocardia*
Cytomegalovirus	*Toxoplasma*
Strongyloides	*Cryptococcus*
Atypical mycobacteria	Pneumocystis
Candida	Zygomycosis

   2. Esophagitis due to *Candida*, *Herpes simplex*, cytomegalovirus
   3. Progressive multifocal leukoencephalopathy
   4. Chronic enterocolitis due to cryptosporidiosis
   5. Extensive herpes simplex persisting more than 5 weeks
   6. Kaposi's sarcoma
   7. Non-Hodgkin's central nervous system lymphoma

As more is learned about AIDS, the definition may need to be expanded or to become more sophisticated. This defintion is clinical rather than based on immunologic parameters, which do not yet clearly separate AIDS patients from those with other disease processes or from individuals who are clinically well.

Of the more than 2000 AIDS patients reported, 75% are homosexual or bisexual men; 17% are intravenous drug abusers; 5% are Haitians; and 1% are hemophiliacs. The metropolitan areas in the United States that have been identified as significant cluster areas include New York, San Francisco, Los Angeles, Miami, Newark, Houston, and Chicago. Cases of AIDS have been recognized in 40 states within the United States and 20 foreign countries. Two to four new AIDS cases are being reported to the CDC each day.

Opportunistic infections. The initial manifestations of AIDS can be either an opportunistic infection or Kaposi's sarcoma. Many patients who initially show evidence of Kaposi's sarcoma will ultimately develop opportunistic infections, but only a minority of patients who show evidence of opportunistic infections later develop Kaposi's sarcoma.

There is a typical grouping of clinical infectious syndromes and opportunistic infections in AIDS patients. Many AIDS patients experience malaise, fevers, anorexia, and weight loss for weeks, months, or years prior to the documentation of their initial opportunistic infection. These symptoms are nonspecific. AIDS patients may initially develop localized dermatomal herpes zoster (shingles) or oral candidiasis (thrush). Extension of oral candidiasis can lead to esophageal erosions, complaints of difficulty in swallowing, and a burning sensation behind the sternum. Both primary and recurrent *Herpes simplex* virus infections appear as painful vesicular lesions in oral, genital, and perineal areas.

Fever and weight loss. As patients' fevers and malaise persist, they become progressively debilitated and may lose up to 50% of their weight. The fevers may be low-grade and persistent, or episodic and spiking up to 104°F (40°C) with associated chills and perspiration. The etiology of their wasting syndrome is probably multifactorial. Nearly all the AIDS patients have had both Epstein-Barr virus and cytomegalovirus isolated from multiple sites. Epstein-Barr virus has been isolated from their throats and peripheral blood lymphocytes; cytomegalovirus has been isolated from their throat washings, urine, or blood. Granulocytopenia, lymphocytopenia, thrombocytopenic purpura, maculopapular rashes, interstitial pneumonia, chorioretinitis, encephalitis, and ulcerative gastrointestinal lesions can also occur with cytomegalovirus infections in AIDS patients. Many patients with AIDS have antibody against retrovirus in their blood. It is possible that the retrovirus is the cause of AIDS.

Mycobacterial pathogens. The bacterium *Mycobacterium avium-intracellulare* can often be isolated from bone marrow, lymph nodes, organ biopsies, and blood of AIDS patients. It is a ubiquitous environmental saprophyte that in the past had rarely been documented as a cause of disseminated disease in other groups of immunosuppressed patients. However, it has been strikingly common in AIDS patients, suggesting that they have an unusual immunologic lesion which selectively predisposes them to this heretofore rare mycobacterial pathogen. Other atypical mycobacteria have not been commonly found to cause disseminated disease in AIDS patients. However, *M. tuberculosis* has been a common infection in Haitians with AIDS. Like Epstein-Barr virus and cytomegalovirus, *M. avium-intracellulare* is relatively resistant to all forms of therapeutic agents, perhaps accounting for many AIDS patients' persistent febrile wasting conditions. Nevertheless, it is unclear whether the aforementioned pathogens are mainly responsible for the fever and constitutional symptoms, or whether as-yet unidentified pathogens or pathologic processes are responsible. A complete work-up is mandatory as other opportunistic pathogens, including *Pneumocystis carinii*, *Cryptococcus neoformans*, and *Toxoplasma gondii*, must be ruled out.

Diffuse pneumonitis. This is another common manifestaton of AIDS. AIDS patients frequently develop cough, shortness of breath, and fever, symptoms that can develop acutely over several days or insidiously over weeks or months. The syndrome of diffuse pneumonitis may be caused by cytomegalovirus, *Cryptococcus neoformans*, *Mycobacterium avium-intracellulare*, and even Kaposi's sarcoma when such patients' pulmonary tumors cause hemorrhage within alveoli. However, the most common cause of diffuse pneumonitis is *Pneumocystis carinii*. Pneumocystis pneumonia in AIDS patients is characterized by a subacute and insidious onset as the patient complains of a mild cough or chest discomfort of 2–10 weeks' duration. Unlike other groups of immmunosuppressed patients, AIDS patients suffer frequent relapses of pneumocystis pneumonia. Furthermore, pneumocystis pneumonia frequently appears in the setting of other concomitant opportunistic infections. Thus, prompt and aggressive diagnostic evaluations are necessary to determine the etiologic agents of pneumonia in AIDS patients.

Neurologic dysfunction. Central nervous system disease affects many AIDS patients, and includes progressive encephalopathies, relapsing cryptococcal meningitis, mass lessions due to *Toxoplasma gondii*, and central nervous system lymphoma.

The encephalopathies are characterized by a slowly progressive dementia that often becomes incapacitating. Cryptococcal meningitis is a common complication of AIDS patients. Headache, fever, and personality changes occur with this fungal infection. Cryptococcosis in AIDS patients is often a disseminated process with positive blood and bone marrow cultures. Relapsing disease is a major contributor to AIDS patients' morbidity and mortality. *Toxoplasma gondii* and lymphoma are the most common causes of mass lesions in the central nervous system. AIDS patients with *Toxoplasma* infection have fever and focal neurologic signs.

Chorioretinitis. Inflammation of the choroid and retina is also common in AIDS patients. Although occasional AIDS patients are seen with *Toxoplasma* retinitis, the most common cause of progressive chorioretinitis is cytomegalovirus. Initially patients are asymptomatic, but later, vision becomes compromised. At autopsy, numerous cytomegalovirus inclusion cells are demonstrated within the necrotic retinal, choroidal, and optic nerve tissues.

Diarrhea. Persistent or recurrent diarrhea that can reach 4 gallons (15 liters) per day is a frequent problem among AIDS patients. Homosexuals with AIDS may have a range of bowel complaints due to the enteric organisms that cause symptomatic disease in the general gay population, including *Entamoeba histolytica*, *Giardia lamblia*, and *Shigella*, *Salmonella*, and *Campylobacter* species. Appropriate antimicrobial therapy that eliminates these pathogens often fails to eliminate the copious watery diarrhea. Some patients with persistent watery stools have cryptosporidiosis, a major untreatable problem for AIDS patients. Many AIDS patients with persistent diarrhea have no demonstrable pathogen despite careful stool examination, endoscopy, small bowel biopsy, and autopsy.

Kaposi's sarcoma and other tumors. Kaposi's sarcoma is a vascular tumor of endothelial cell origin and appears as firm red or violet nodules involving the skin and mucous membranes; lesions are often multiple and may involve any portion of skin or any mucous membrane; body surfaces may eventually be covered by these multifocal tumors. Unfortunately, some patients develop life-threatening visceral involvement due to massive tumor infiltration of the lung or gastrointestinal tract; since the tumor is vascular, bleeding is not uncommon. It is unclear why Kaposi's sarcoma is so common in AIDS. There is evidence that it is associated with cytomegalovirus infection; perhaps the immunosuppressed state of AIDS patients allows cytomegalovirus to transform endothelial cells into a vascular sarcoma.

Over a dozen cases of Burkitt's lymphoma have been reported in young homosexual men. The number appears to be unusually high for this age group. Furthermore, an unusually high percentage of these cases have been associated with Epstein-Barr virus. The association of Kaposi's sarcoma with another herpes virus, cytomegalovirus, has stimulated speculation about the potential role that immunosuppression might play in permitting these viruses to cause tumors.

Colonic carcinoma, oropharyngeal carcinomas, and non-Hodgkin's lymphomas have been reported in homosexual men who appear to have the same immunosuppressed state and clinical profile as those with AIDS. Non-Hodgkin's lymphoma of the central nervous system has the strongest association with AIDS.

Immunologic profile. There is neither a serologic marker nor a diagnostic immunologic profile for AIDS. However, a characteristic pattern shows a modest anemia, leukopenia, and a normal or depressed platelet count. The differential count may reveal a severe lymphocytopenia. Most patients have antibody to one or more hepatitis viruses. Circulating immunoglobulin levels are usually normal or elevated.

Immunologically, patients with AIDS demonstrate lymphopenia due to a selective depletion in the helper-inducer (T4+ or Leu-3+) subset of T lymphocytes. This depletion of T4+ lymphocytes leads to a decrease in the ratio of helper to suppressor cells, the so-called T4/T8 ratio. This decrease in the number of T4+ lymphocytes is accompanied by profound abnormalities of T-cell function both in the body and in laboratory cultures. These immunologic abnormalities are not specific for AIDS and have been described not only in other disease states (predominantly viral infections) but in healthy homosexual men as well. There is currently no reliable blood test for AIDS, although an increased serum level of thymosin (a substance produced by the thymus) and an increased level of acid-labile alpha-interferon are being investigated as possible tests.

Questions. Successful treatment of AIDS would logically be based on reconstitution of immunologic function in addition to the treatment of specific infections and tumors, which will continue to develop as long as the patient is immunosuppressed. Current investigations involve the use of interferon, interleukin-2, cell transfers, bone marrow transplantation, and plasmapheresis, often in combination with antiviral drugs.

For background information *see* CELLULAR IMMUNOLOGY; OPPORTUNISTIC INFECTIONS; ONCOLOGY in the McGraw-Hill Encyclopedia of Science and Technology.

[ABE M. MACHER]

Bibliography: A. Friedman-Kien et al., Disseminated Kaposi's sarcoma in homosexual men, *Annu. Intern. Med.*, 96:693–700, 1982; M. Gottlieb et al., *Pneumocystis carinii* pneumonia and mucosal candidiasis in previously healthy homosexual men: Evidence of a new acquired cellular immunodeficiency, *N. Engl. J. Med.*, 305:1425–1431, 1981; H. Masur et al., An outbreak of community acquired *Pneumocystis carinii* pneumonia: Manifestations of cellular immune dysfunction. *N. Engl. J. Med.*, 305:1431–1438, 1981; B. Sonnabend, S. Witkin, and D. Purtilo, Acquired immunodeficiency syndrome, opportunistic infections and malignancies in male homosexuals: A hypothesis of etiologic factors in pathogenesis, *JAMA*, 249:2370–2374, 1983.

Aerodynamics

Numerical aerodynamic simulation is a computational tool for predicting the aerodynamics and fluid dynamics of aerospace vehicles by solving a set of mathematical equations with a high-speed digital computer. Researchers working in this area are required to be knowledgeable in four technical areas: aerodynamics, fluid dynamics, mathematics, and computer science. Numerical aerodynamic simulation, which is sometimes referred to as computational fluid dynamics, provides the aerospace vehicle designer with a relatively new but formidable tool to be used in the design and analysis process.

Analysis vs. simulation. To heighten understanding of the aerodynamics and fluid dynamics pertinent to the early stages of the flight-vehicle design process, the aerodynamicist currently has available three standard tools: analytic methods, experimentation, and numerical aerodynamic simulation. Analytic methods provide quick, closed-form solutions but require unduly restrictive assumptions, can treat only simple configurations, and predict only the idealized aerodynamics. Through experimentation, represenative or actual configurations can be tested and complete aerodynamic data can be produced. However, experimentation is costly in terms of both the model and actual tunnel test time. In addition, the limited range in flow conditions of a wind tunnel compared with anticipated real flight conditions severely restricts the scope of the experimental program.

Compared to analytic techniques, numerical aerodynamic simulations require very few restrictive assumptions and can be used to analyze relatively complicated flight-vehicle configurations. They also have few flow condition restrictions and result in complete surface and exterior flow field definition. Most important, they are far more cost-effective than wind tunnel testing because the cost of computer simulations is continually decreasing as result of improved numerical procedures and advances in computer technology.

The optimum design approach combines numerical aerodynamic simulation and wind tunnel testing. Numerical simulation can be used to complement experimentation by filling the gap in experimental test points quickly and cost-effectively. In addition, numerical simulation can be used to aid in the interpretation and understanding of the experimental data, and in its extrapolation to actual flight data. Perhaps the most potent use of numerical aerodynamic simulation lies in the fact that these computational tools can be combined with optimization procedures to arrive at a refined design, based on imposed design constraints governed by aerodynamics, performance, or structures.

Software. The design and application of software for numerical aerodynamic simulation involves basically four phases, as depicted in Fig. 1: software formulation, programming, validation, and utilization. Each phase is composed of facets requiring decisions by the software designer.

Software formulation. This phase is composed of nine facets. The first of these, problem definition and input, requires decisions regarding such factors as whether the flow problem is steady or unsteady, inviscid or viscous, and subsonic, transonic, or supersonic. In addition, required program control input parameters such as grid size, integration stepsize or relaxation factors, and smoothing parameters are established.

In the second facet, solution methodology and discretization topology are constructed; that is, such methods as multigrid, shock capturing, or shock fitting are examined, and grid generation concepts such as single-block, multiblock, component-adaptive, or interfering are explored.

The third facet deals with the equation set to be solved. Various approximation levels for the governing equations exist, ranging in complexity from the linearized inviscid equations to the exact full Navier-Stokes equations. Within that range are the small-disturbance, full-potential, Euler, and Reynolds-averaged Navier-Stokes equations. All are partial differential equations which do not possess closed-form solutions and hence require solution by numerical procedures using digitial computers.

To solve the governing partial differential equations, a decision must be made as to the numerical algorithm to be employed. This is required in the fourth facet. Numerical methods such as explicit or implicit procedures for steady and time-accurate problems are available. Under each of these classifications there are numerous algorithms from which a selection of the most appropriate can be made.

The fifth facet of the software formulation process is configuration definition, or the process by which

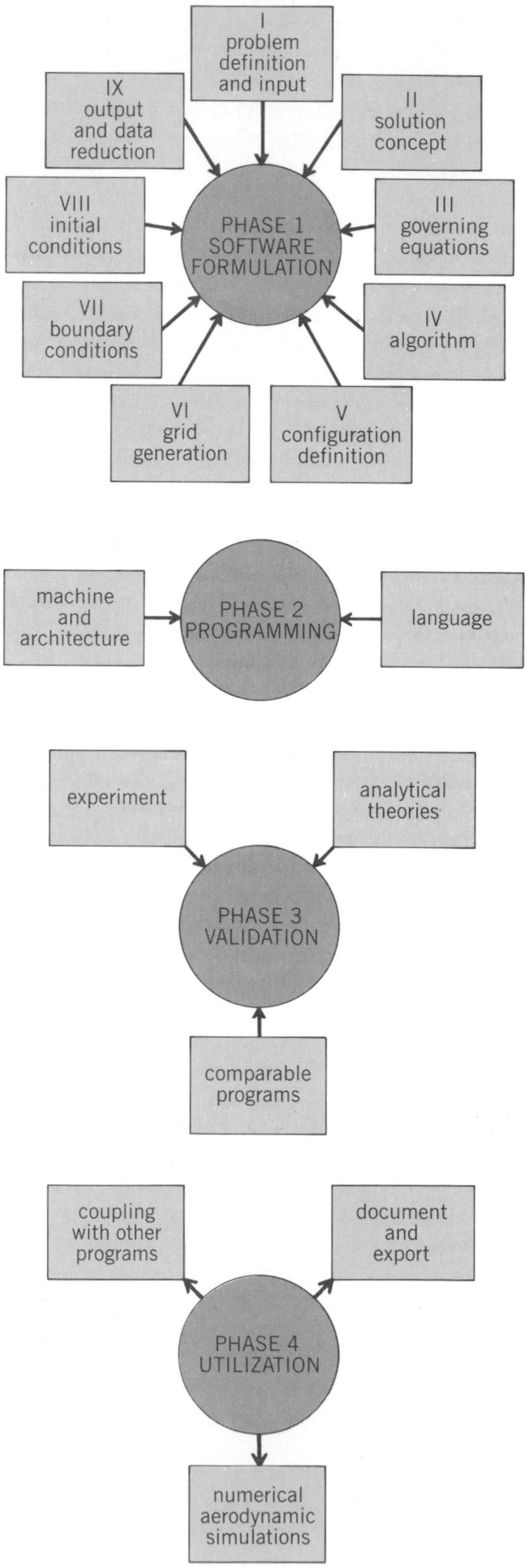

Fig. 1. The four phases of numerical aerodynamic simulation, including the facets of each phase.

the geometry of the flight vehicle is supplied to the computer. There are basically two approaches available for doing this, numerical and analytic, both of which result in the necessary information required by the computer program, that is, the body shape and its derivatives (slope and curvature).

The sixth facet, and one of the most important in the software formulation process, is grid generation. Because a numerical solution of the governing equations is required, the flow region of interest must be smoothly divided or discretized so that finite differences of the spatial derivatives of the governing equations can be obtained. To do this, there are two basic approaches: algebraic or analytic approaches, which include both conformal and nonconformal procedures; and differential or numerical approaches, which include elliptic, parabolic, and hyperbolic procedures.

The seventh facet of the software formulation process, and probably the most important, concerns boundary conditions. In order to properly signal the external flow field of the presence of the flight vehicle and the fact that the discretized flow region is finite, numerical boundary condition procedures are required at each face of the computational volume for three-dimensional problems. The boundaries of the computational domain can be classified as either permeable or impermeable; that is, there is either mass flow or no mass flow through the boundaries. The surface of a flight vehicle is an example of an impermeable boundary, while the bow shock wave encompassing that vehicle in supersonic flight is an example of a permeable boundary. To simulate the various permeable and impermeable boundaries that arise in numerical aerodynamic simulation, numerous procedures are available with diverse ranges of sophistication and complexity.

To begin a numerical aerodynamic simulation on the computer requires that the dependent variables of integration be initialized, that is, be assigned numerical values. This eighth facet is probably the easiest in the software formulation process, for in most instances these dependent variables are assigned values equal to the free stream. In all other cases, a guess as to their correct value is made and the integration process to determine their actual values is initiated.

The last facet involves the output and data reduction of the numerical solution that has been generated. Depending on the needs of the various users, the numerical data can be manipulated to satisfy all user output requirements. These data can be displayed graphically either on hard copy or on a cathode-ray tube. In addition, display media such as color still and motion pictures are available to aid in the understanding and interpretation of the numerical solution.

Programming. Phase two of the numerical simulation process is programming. For scientific programming, the most popular computer language is FORTRAN. Some computers have special architectures that permit certain mathematical operations to be performed more efficiently than on other computers. The programmer is responsible for taking advantage of these computer characteristics so that the programs utilize the computer as efficiently as possible.

Validation. To verify that numerical solutions being produced by the software are correct, the software designer's numerical solutions are compared with experimental data, analytic theories, or other computer programs. It is during this phase that the capabilities and limitations of the program are determined.

Utilization. In this phase of the simulation process the aerodynamic simulation program is first documented, and then, upon request, exported to be used by the computational fluid dynamic community. In addition, it is used to satisfy the designer's simulation requirements and, if applicable, it

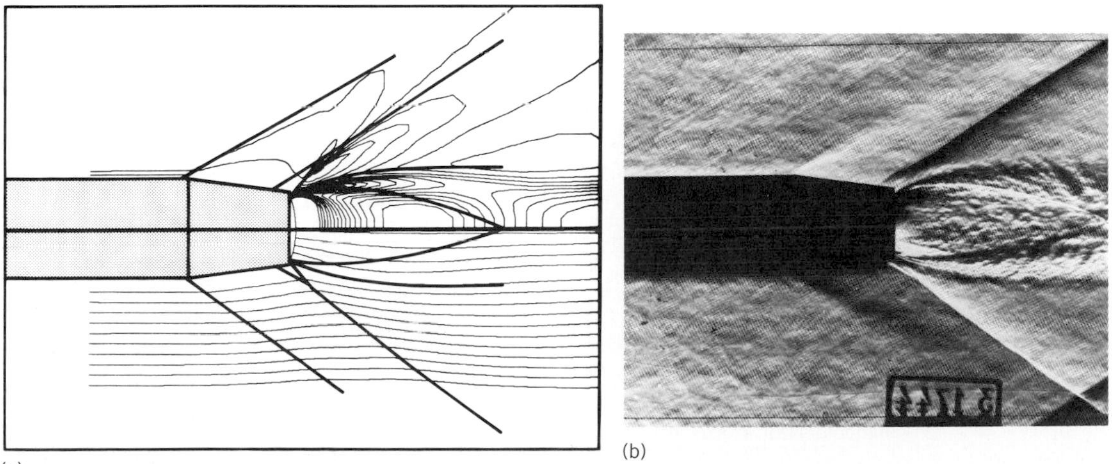

(a) (b)

Fig. 2. Computed and experimental flow features. (*a*) Computer simulation of density contours (top) and streamlines (bottom) for 8° conical missile afterbody with 20° nozzle; experimental shock-wave features (heavy lines) from schlieren photograph are superposed. (*b*) Experimental schlieren photograph; Mach number = 2. *(From G. S. Deiwert, A Computational Investigation of Supersonic Axisymmetric Flow over Boattails Containing a Centered Propulsive Jet, American Institute of Aeronautics and Astronautics, AIAA–83–0462, 1983)* is coupled with other programs such as optimization routines or structural response programs to enhance its overall versatility.

Applications. This process has great potential for solving complicated fluid dynamic problems. Two excellent examples generated from computer programs designed for numerical aerodynamic simulation are the problem of the conical afterbody with centered propulsive jet and the three-dimensional shock-on-shock problem. The solution of the first problem resulted in an understanding of a complicated fluid physics problem, while the solution of the second problem aided in the design of a missile system and eliminated the need for further experimental fluid dynamic testing.

The aft portion of jet engines and rocket motors produces undesirable aerodynamic drag. This drag is a result of not only the afterbody geometry but also the interaction of the exhaust plume with the external flow. In order to understand the fluid dynamics associated with this problem and to possibly minimize the aerodynamic drag, a computer code was designed to simulate this flow problem numerically. A typical numerical result from that program is shown in Fig. 2. The flow field features of the numerical solution and the experimental result compare quite well.

The flow field generated by the interaction of an incident shock wave with a missile traveling at supersonic speeds is of interest to the vehicle designer because of the structural and vibrational responses of the configuration under such a load. The occurrence of such a flow field is shown in Fig. 3*a*. To determine the flow field, approximate theories were developed. In addition, a rather elaborate experimental program was established (Fig. 3*b*). Because of the discrepancies between the two results, a computational effort was initiated. The results of all three prediction procedures in the form of peak surface pressures are shown in Fig. 3*c*. In essence, the numerical simulation verified the experimental data.

With the promise of improved numerical procedures and the planned advances in computer technology, numerical aerodynamic simulation should play a more dominant role in aerospace research and development and the flight vehicle design process.

For background information *see* AERODYNAMICS; DIGITAL COMPUTER PROGRAMMING; FLUID DYNAMICS; NAVIER-STOKES EQUATIONS; SUPERSONIC FLIGHT in the McGraw-Hill Encyclopedia of Science and Technology.

[PAUL KUTLER]

Bibliography: G. S. Deiwert, *A Computational Investigation of Supersonic Axisymmetric Flow over Boattails Containing a Centered Propulsive Jet*, American Institute of Aeronautics and Astronautics, AIAA–83–0462, 1983; P. Kutler, *A Perspective of Theoretical and Applied Computational Fluid Dynamics*, AIAA–83–0037, 1983; P. Kutler and L. Sakell, Three-dimensional, shock-on-shock interaction problem, *AIAA J.*, 13:1360–1367, October 1975.

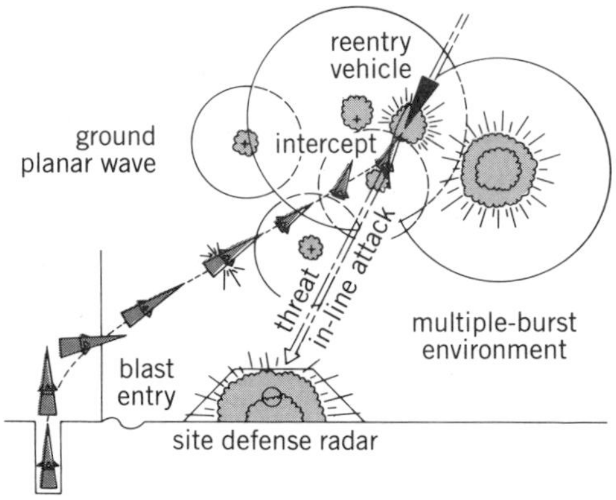

(a)

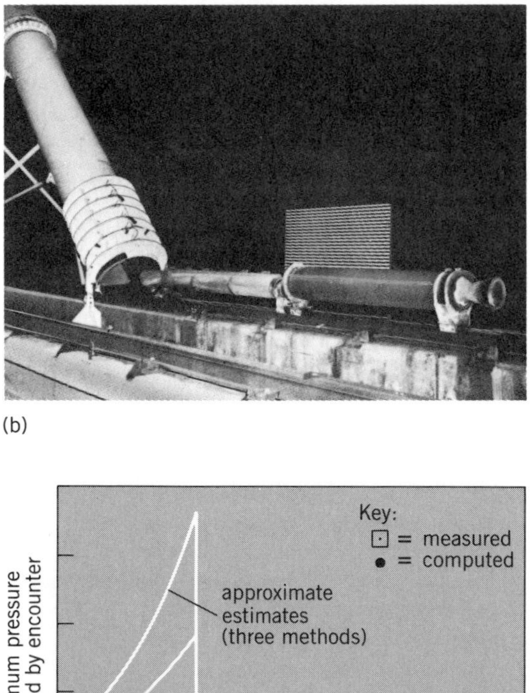

(b)

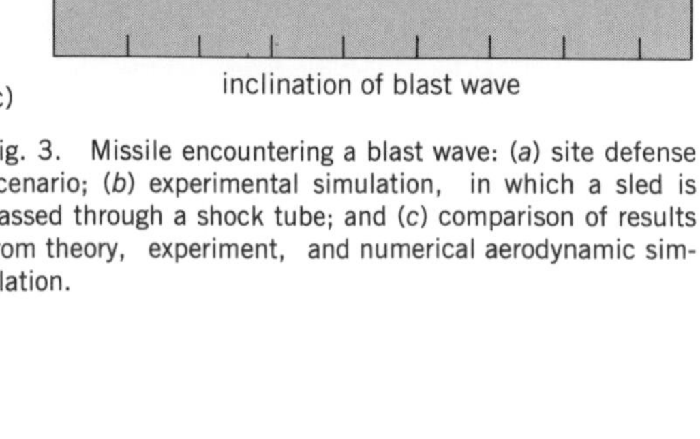

(c)

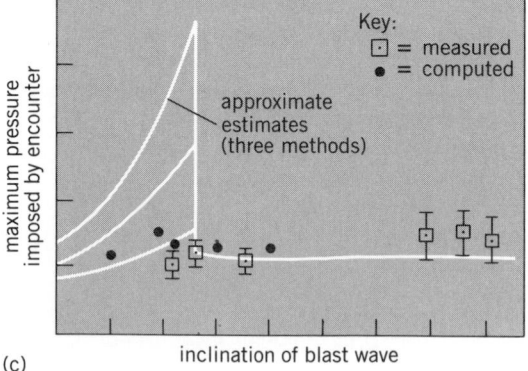

Fig. 3. Missile encountering a blast wave: (*a*) site defense scenario; (*b*) experimental simulation, in which a sled is passed through a shock tube; and (*c*) comparison of results from theory, experiment, and numerical aerodynamic simulation.

Agroecosystems

Agroecosystem research comprises analyses of the basic biology of agriculture in which economic production is explained as only one aspect of biological responses to the environment and to inputs of energy and materials. Unlike the empirical disciplines of agronomy and animal science that direct primary attention to increasing economic production, agroecosystem research considers all aspects of the biology of an agricultural system. This broader approach becomes more useful as production methods that have been developed through experimental methods begin to approach the limits of the biological processes. Then it becomes more important to know what biological processes determine yield and to learn how these processes act in determining the biological limits to yields. The chemistry of nutrient cycling must also be considered in order to determine whether these cycles can sustain the high yields of intensive agricultural production based on the use of chemicals.

Agroecosystem models. Agroecosystem research uses the methods of ecosystem analysis to measure the material and energy entering plant and animal populations and to explain how these inputs affect the physiologic processes determining growth and maintenance. A biologically complete agroecosystem (see illustration) has three basic components

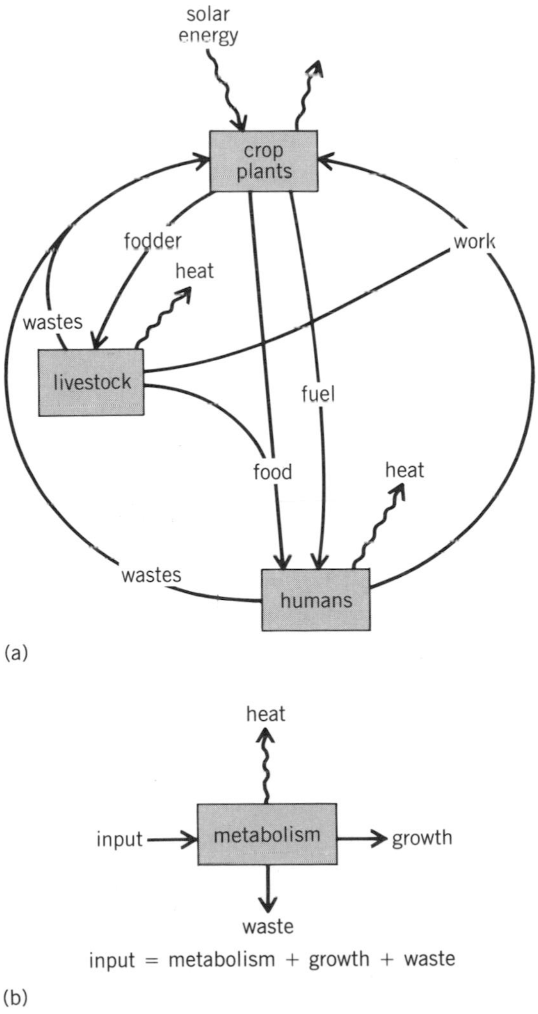

Energy and material flow in agroecosystems. (*a*) A biologically complete agroecosystem. (*b*) Flows of a single species population, either a part of an agroecosystem or the extreme of a specialized agroecosystem.

and a pattern of energy flow that is nearly the same in all agroecosystems. All ecosystems are driven by solar energy fixed as organic carbon by green plants. The accumulation of organic carbon fixed by plants in agroecosystems can be divided into three categories: coarser materials, such as fodder, eaten by livestock; grains, fruits, vegetables, and so on, eaten by humans and livestock; and dense stems and leaves which may be used for fuel and constructing shelters or utensils. The patterns of energy and material flow between the plants and the two consumers, livestock and humans, follow simple and consistent patterns that are similar in most agricultural systems. Unlike other ecosystems, which have an immense variety of species with patterns of interaction that form complex food webs, agroecosystems have relatively simple cycles.

The development of specific agroecosystem models from this general scheme has been very rapid because the three components are biologically very uniform. Nearly 60% of all crop production is from three species of grasses—wheat, rice, and maize—while the livestock component is largely made up of cattle, swine, and chickens, and *Homo sapiens* is the third component. The basic physiology of these seven species is understood in great detail, and the basic physiology, expressed as an energy budget (see illustration), sets limits to the consumption and use of energy by a species. Thus, the rates of energy flow through the three components of an agroecosystem can be determined from the general energy budget.

It is not possible to measure all these components in field situations, but the practical measures should be related to this equation because it is a standard point of reference.

Crop growth rates. There are no chemical byproducts from the input of solar energy to photosynthesis other than oxygen. Independent estimates of crop metabolism are extremely uncertain; hence, crop performance is given as net production, which is the organic carbon entering an agroecosystem. Three ways of measuring net production are given in Table 1. The maximum growth rate is an estimate of the physiological limit, the highest short-term rate obtainable under the most favorable conditions. Crops grow from seeds under conditions that are rarely optimal for very long, so the average growth rates over the season are well below the physiological potential. The average crop growth rate seems to be independent of the maximum photosynthetic potential, which leads some crop physiologists to argue that interactions within the crop community and between the plants of the community are as important as physiological processes of photosynthesis in determining the crop production.

The early progress in agronomy was based on an evaluation of yields and selection for the highest yields. Economic yield is a fraction of aboveground net production represented by the grain, and is designated as the harvest index (Table 1). Recent progress in developing high-yielding varieties is based

Table 1. Measures of ecosystem and agronomic production

	Ecosystem measures				
	Growth rate, oz/ft² day (g/m² · day)		Aboveground net production, ton/acre (metric ton/ha)	Harvest index	Economic yield, ton/acre (metric ton/ha)
	Maximum	Average			
Wheat	>0.07 (>22)	0.03–0.06 (8–18)	4.0–5.7 (9–13)	0.45	1.8–5.3 (4–12)
Rice	0.18 (55)	0.02–0.06 (7–17)	3.5–5.3 (8–12)	0.50	1.8–2.6 (4–6)
Maize	0.17 (52)	0.06–0.07 (18–23)	8.8–1.3 (20–30)	0.43	4.0–5.7 (9–13)

on theoretical models for the ideal distribution of aboveground biomass, and the yields of newer varieties were increased by altering the harvest index from levels of 0.25 or 0.30 to nearly 0.50. The gain in yield resulting from an increased harvest index was developed through the application of basic theories for the interception of light by plant communities to determine the structure of individual plants that would develop the most efficient canopies for the production of grain.

The production of plant biomass in agroecosystems can be converted to energy by using energy constants for grain and straw. The range of variation in these constants is much less than the yearly variation in yields; therefore general models can reasonably be based on extrapolations from the constants for energy per unit biomass given in standard agricultural references.

Animal energy budgets. The metabolism, food consumption, and growth of all domestic animals is so well known that reliable energy budgets can be specified for animals of any age or function in the agroecosystem (Table 2). If there is a simple census of the livestock in an agroecosystem, the numbers of animals can be multiplied by the values in the energy budget to determine the energy flow in the system. The energy requirements of humans is easily determined from standard nutritional tables as well.

Agroecosystem structure. With data on crop production and standard energy equivalents, a reasonably accurate estimate of the energy and material flow can be determined for any agroecosystem. Some agricultural villages have all the attributes of an ecosystem, while specialized farms may be comparable to a single species population in an ecosystem. Nearly all the farms in developed countries specialize in the production of a particular crop or animal and are more like a single ecosystem component isolated from, and independent of, the rest of the ecosystem components.

Village agroecosystems. The farming villages in many developing countries are largely self-sustaining communities based on renewable (biological) resources with nonessential or minor links to other ecosystems. The agricultural systems of self-sufficient farming villages are closed interdependent communities in which the number and kind of crops planted are determined by the local demands of livestock for fodder and of humans for food, fodder, and fuel. The work needed to manage the land and water resources and regulate the crops is developed by humans and animals from the energy stored in the harvest from earlier crops. The community is both an ecosystem unit and a discrete economic unit. The primary economic goal is self-sufficiency, with profit as a secondary goal. The farmers are acutely aware of the biological relations that are shown in illustration *a*. The primary goal of long-term self-sufficiency can be achieved only by regulating the components on the agroecosystem.

Specialized farms. The farms in a centralized economy tend to be dedicated to the production of only one or two species for an economic profit. Large-scale production units for a few crops can operate when land is abundant relative to the demands of the local population, fossil carbon is a cheap source of energy and feedstock for chemicals, and

Table 2. Energy budgets for the livestock in agroecosystems

	Energy budget in MJ/day (Mcal/day)				Feed efficiency (for dry weight)
	Intake =	Metabolism +	Growth +	Waste	
Meat					
Cattle	123.0	78.2	7.0	39.2	0.04
(at 660 lb or 330 kg)	(29.4)	(18.7)	(1.67)	(9.36)	
Swine	61.5	25.4	22.9	13.2	0.09
(at 176 lb or 80 kg)	(14.7)	(6.07)	(5.48)	(3.15)	
Chickens	2.15	0.65	1.04	0.45	0.05
(3.3 lb or 1.5 kg)	(0.51)	(0.16)	(0.25)	(0.11)	
Animal products per day					
Cow (1100 lb or 500 kg)	189.0	89.5	46.9	48.5	0.16
[33 lb or 15 kg milk]	(45.2)	(21.4)	(11.2)	(11.6)	
Hen (4.4 lb or 2 kg)	1.40	0.74	0.20	0.46	0.09
[0.6 egg]	(0.33)	(0.18)	(0.05)	(0.11)	

an inexpensive transportation network allows for an easy exchange of products. The consumption of products in a centralized agricultural economy follows the same paths as in a village agroecosystem, but production is concentrated in areas determined by production costs rather than the location of the sites of consumption (urban areas) so that the flow of materials and energy between compartments varies with commodity prices. For example, the economics of a national commodities market determine how much of the production of green plants will be diverted into cattle, hogs, chickens, or human food at any given time. Under such conditions, the decisions about what to grow and where to grow it are largely independent of the biological linkages required in a village agroecosystem. The functional biological systems then become isolated populations, such as a flock of chickens or a crop of wheat, and the physiology of growth for the population of one species is the principal, if not the only, biological factor affecting the viability of a farm.

Conclusion. The biological functions of a farm are dependent on inputs of fossil carbon or fuel or chemicals, and it is uncommon for the food or fodder produced at the farm to be cycled within it. Even reproduction of the populations has become isolated from growth, as the seed for future crops and the sperm for animal reproduction is purchased from specialists. Dairy farming and grazing systems are the only major two-trophic-level agroecosystems commonly left in developing countries, and much dairy production now comes from farms that produce little of the necessary feed. The agroecosystems in a centralized economy consist of a set of isolated populations regulated by economic factors and linked by a transportation network.

For background information *see* AGRICULTURAL SCIENCE (ANIMAL); AGRICULTURAL SCIENCE (PLANT); AGRICULTURE; AGRONOMY; BIOLOGICAL PRODUCTIVITY; BIOMASS; ECOSYSTEM; ECOSYSTEMS ENERGY; FOOD CHAIN; PHOTOSYNTHESIS in the McGraw-Hill Encyclopedia of Science and Technology.

[ROGER MITCHELL]

Bibliography: L. T. Evans, *Crop Physiology*, 1975; National Academy of Sciences, *Nutritional Energetics of Domestic Animals*, 1981; D. Pimentel and M. Pimentel, *Food Energy and Society*, 1979; C. R. W. Spedding, *The Biology of Agricultural Systems*, 1975.

Airplane

Textile materials have made a significant contribution in bringing the Boeing 757 and 767 models to their advanced status. The large-volume, nondecorative fabrics for both exterior and interior parts include the high-technology fibers and fabrics of glass, Nomex, Kevlar, and carbon (graphite) in knitted and woven laminate or composite configurations. These materials impart the necessary properties of high strength, high modulus, light weight, and enhanced fire safety. The decorative materials—carpets, drapery, upholstery—incorporate the newest low-smoke flame-retardant finishing technology and have been structured to impart a weight savings of 20–25% over previous materials.

Interior uses. Interior uses of textile materials may be classified as nondecorative or decorative.

Nondecorative uses. Predominant interior applications are in panel laminates. Figure 1 illustrates the basic materials and the layering configuration of the 757 and 767 panels. The principal interior paneling component, known as honeycomb core, was introduced with the 747 in the late 1960s. Multiple layers of prebonded Nomex aramid paper, coated with phenolic resin, expand to a predetermined cell size: ⅛, ¼ or ⅜ in. (3.2, 6.4, or 9.5 mm). One or more layers of resin-impregnated glass or Kevlar in knit and woven fabric form are bonded to the core. The surface seen by the passenger is then covered with the decoratively printed Tedlar surface. The specific fabric selected and its orientation onto the panel are dependent upon the calculated tensile, peel-strength, and compression requirement of a given part plus esthetic criteria.

Interior panels are categorized as either crushed-core or sandwich panels. Sidewall panels, ceiling panels and frames, and stowage bins are the primary interior framing part incorporating contours. The contoured shape necessitates the crushing of the honeycomb core in matched metal tools.

Acoustic panels, bulkhead partitions, lavatory and galley walls, and floor panels are referred to as sandwich panels. Additional applications of textile-covered panels include the ducting, cargo liners, and cargo flooring.

A new application of textiles is the encapsulation of the polyurethane seat cushioning to retard flammability. This use requires a lightweight, flexible, durable, compact fabric composed of a thermally stable polymer. Woven, knitted, felted structures in a variety of fiber types have been evaluated and candidates identified.

Small quantities of industrial textile materials used on commercial airplanes include: rubber-coated nylon for escape slides and rafts; nylon webbing as baggage restraints in the cargo hold; and knit or woven, porous or compact polyester structures in approximately 50 sizable rubber seals. The fabric imparts greater strength and increased flexibility. In areas with large openings such as a cargo door, two or more layers of fabric are added to increase the stiffness of the elastomeric seal.

The 757 and 767 models introduced a new concept for carpet underlay, needled Nomex felt. The felt provides superior acoustic damping and wear properties while eliminating the problems of moisture absorption and tearing when the carpet is cleaned or replaced.

Decorative uses. While the primary fiber type for most decorative textile materials is 100% wool with low-smoke Zirpro flame retardant finish, 90% wool–10% nylon is commonly used as upholstery fabric. Inherently flame-retarded polymers of rayon and of polyester have had limited service use. The flame

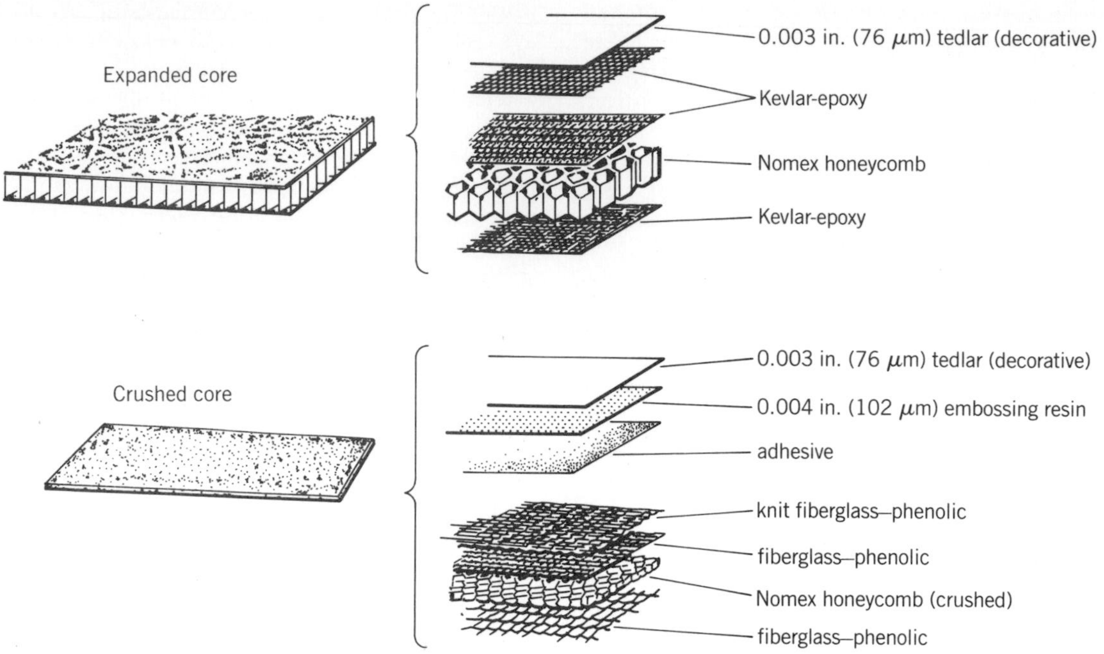

Fig. 1 Construction of interior sidewall and sandwich panels on the Boeing 757 and 767 aircraft.

retardant polyester may be formulated to be launderable or dry cleanable, or both. Whereas flame retardant rayon retains a better appearance with dry cleaning, intimate blends or combination structures with wool provide the esthetic qualities of wool. Fiber types used for upholstery are also suitable for drapery applications. Fabrics containing polymers of polyvinyl chloride and acrylonitrile are not considered suitable for commercial airplane usage due to the high levels of smoke generated as tested in the National Bureau of Standards smoke chamber.

The class of thermally stable synthetic polymers with backbones containing nitrogen, while possessing superior fire properties, possess color-related problems. Either the natural color is difficult to bleach out, or the color fastness range is limited to three or four dark, select colors. Until a feasible method of coloration or a printing technique is developed to alleviate this limitation of color palette, these fibers will not be satisfactory in decorative applications. New uses for these high-temperature polymers in interior applications are under consideration.

The design and development of a woven lightweight [52 oz/yd^2 (1763 g/m^2) total weight; 25 oz/yd^2 (847 g/m^2) pile weight] carpet with improved flammability parameters, pleasing esthetic qualities, and adequate service wear regarding foot traffic and food carts was particularly difficult. The weight constraint, plus smoke and toxicity guidelines, necessitated changes in backing yarn fibers and sizes as well as the quantity and chemical formulation of the backcoating. Current construction parameters include a low-level loop wool pile with polyester chain and stuffer plus preshrunk linen or Dref spun glass backing yarns coated with 12–15 oz/yd^2 (407–508 g/m^2) of ethylene–vinyl acetate copolymer or styrene butadiene latex.

To simulate flight service conditions to which the decorative materials are subjected, specialized testing equipment was designed to measure the multifaceted aspects of wear.

Exterior uses. On the 757 and 767 models, exterior applictions of advanced composites are numerous, including elevators, rudders, spoilers, ail-

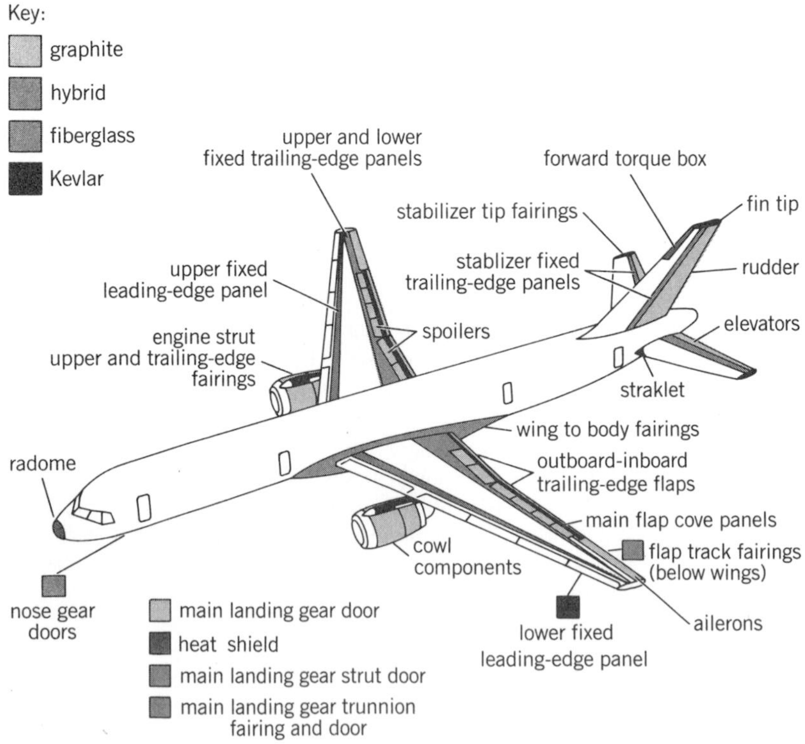

Fig. 2. Applications of advanced composite materials on the Boeing 757 aircraft.

erons, flaps, engine cowling components, fairings, landing gear doors, and the radome. Composite materials can be generally defined as a mixture of two dissimilar materials to enhance the properties of the individual components.

These lightweight, nonmetallic systems made of high-strength fibers are embedded in a polymeric matrix. Besides their outstanding properties of high tensile strength, high modulus, and low density, the graphite composites offer superior thermal stability, acoustic damping, and resistance to fatigue, wear, and corrosion as compared with the metals they are replacing. On the 757 model, approximately 3% of the structural weight has been replaced with advanced composites for a weight savings of nearly 1500 lb (680 kg).

Two fundamental resin-impregnated fiber arrangements are used in a composite consisting of tape (a strip of unidirectional fiber bundles) and fabric in three basic weave structures: 8 harness satin, 3 × 1 satin, and 1 × 1 plain weave. These three basic weaves are represented in 21 glass, 2 Kevlar, and 4 graphite structures; and each one has a specific application. Graphite fiber bundles or yarn size are referred to as 3K or 6K tow; the number denotes the quantity per thousand individual filaments in a fiber bundle or yarn.

During manufacturing, the fabric or tape fibers are stacked in multiple layers in precisely calculated orientations. These exacting orientations and the number of layers required are determined by the known stresses or loads to which a specific part is exposed in service. On an engineering drawing, each ply or layer is referenced as 0°, 90°, ±45°, or other angle as necessary. This notation specifies the direction of each ply in the lay-up (stacking) process. Figure 2 schematically locates the advanced composite applications on the 757 model. Approximately 3340 lb (1515 kg) of graphite, Kevlar, and hybrid materials are used, resulting in a weight savings of approximately 1490 lb (675 kg).

Nontraditional yarn-spinning systems and alternative fabrication techniques such as stitch bonding, noncrimp weaving, and air layering are being evaluated. The exploration into and transfer of this innovative technology to transportation applications are in the infancy stage. Though substantive progress is still needed, especially in the fabricating or structural aspects, textiles have become an important support system for the aerospace industry.

For background information *see* AIRFRAME; COMPOSITE MATERIAL; TEXTILE in the McGraw-Hill Encyclopedia of Science and Technology.

[SALLY A. HASSELBRACK]

Aluminum alloys

Although the aluminum industry is concerned with the development of many different types of alloys, including those used in applications such as packaging, automotive parts, and architectural products, the most active area—or what might be considered the new frontier in aluminum alloy research—is found in the aerospace industry. There suppliers, users, and researchers in both the private and public sectors are cooperating in an effort to create more fuel-efficient aircraft. Since aluminum constitutes 70–80% of the weight of an airframe, metallurgists have been pursuing aluminum alloy development programs directed toward producing materials characterized by stronger, stiffer, lighter-weight properties. In addition, the titanium alloys of aircraft gas turbines are also prime targets for replacement by lighter-weight alloys. Researchers have developed improved aluminum alloys using alloying techniques involving the addition of lithium, the lightest metal, to aluminum; and three processing technologies—powder metallurgy, mechanical alloying, and a process which involves blending aluminum alloy powders and silicon carbide fibers to form a composite. Since aerospace alloys must be thoroughly tested to meet weight criteria and a stringent balance of strength, ductility (the ability to deform before fracturing), toughness (resistance to the propagation of a crack), and corrosion resistance, these candidate aircraft alloys are still in the experimental stage, and will continue to be until the late 1980s. Although aircraft designers consider these new alloys most promising, the alloys are more costly initially than those now used. However, these high-strength alloys may provide full cost savings over the lifetime of the aircraft.

Aluminum-lithium alloys. Lithium is the lightest metal found in nature. For each weight percent of lithium added to aluminum, there is a corresponding decrease of 3% in the alloy's weight. Thus, aluminum-lithium alloys offer the attractive property of low density. A second beneficial effect of lithium additions is the increase in elastic modulus (stiffness). Also, as the amount of lithium is increased, there is a corresponding increase in strength due to the presence of very small precipitates which act as strengthening agents for the aluminum. Increasing the amount of lithium increases the number of precipitates, so that in a given volume of material the spacing between the precipitates decreases, making it stronger. The combination of the precipitate size and the spacing controls the alloy strength. As the precipitates grow during heat treatment, the strength increases to a limit, then begins to decrease.

Aluminum-lithium alloys thus come under the classification of precipitation-strengthening alloys. Since the size and distribution of the precipitates can be controlled by heat treating, these alloys are also referred to as heat-treatable. Since the addition of lithium to aluminum has also been shown to result in an alloy with unacceptable levels of ductility, which is a necessary engineering property for many aerospace applications, other elements such as copper, magnesium, and zirconium have been added to offset the loss in ductility. Though these alloy additions improve the ductility, they also have the deleterious effect of increasing the density, particularly in the case of copper. Consequently, the schemes of alloy development have been focused on balancing the various positive and negative attributes of the different elements and to arrive at a composition which will give a suitable combination of strength, ductility, and density. Figure 1 is a schematic relating yield strength and elongation for a variety of aluminum-lithium alloys. The alloy 7075-T651, a commonly used alloy for such applications, is included as a reference point. In order to be competitive, the new alloys should have strength-ductility relations equal to or better than 7075.

Powder metallurgy. As early as 1947, investigators showed interest in the development of high-modulus, high-temperature aluminum alloys. Elements such as iron, cobalt, and nickel have been added to aluminum in large quantities to produce alloys which have relatively stable structures at elevated temperatures and higher elastic moduli than conventional aluminum alloys. However, the casting techniques available to metallurgists of that day resulted in extremely brittle alloys. Nevertheless, investigators have again been spurred by the aircraft industry's desire to replace the titanium in the fan sections of gas turbines, where operating temperatures are up to 450°F (235°C), with lighter-weight materials. For example, the substitution of a suitable aluminum alloy for titanium could result in a 31% reduction in weight for that part. However, existing aluminum alloys do not have sufficient strength to withstand the operating temperatures of even the cooler portions of a gas turbine and to remain thermally stable. Consequently, researchers are considering alloy additions to aluminum which move slowly (have low diffusivities) in it, so that the characteristics of that structure do not change with time at the service temperature. Unfortunately, the elements which have low diffusivities in aluminum also have very limited solubilities; therefore, processes other than conventional ingot casting must be utilized.

An alternative for producing a candidate alloy is the powder metallurgy process, which is a rapid solidification process. Rapid solidification processing involves the transformation of finely divided liquid in the form of either droplets or thin sheets which become solid at high solidification rates (on the order of 10^4 K/s). Solidification occurs as heat is removed from the molten metal. The primary mechanisms of heat transfer can be divided into two broad categories: conductive and convective. Certain rapid solidification process forms, such as ribbons and fibers, are cooled primarily by conduction. The heat is transferred away by a substrate, the surface on which the liquid is cast. The substrate is generally made of copper and can be either a rotating drum or a rotating wheel. On the other hand, another rapid solidification process form, powder, is cooled primarily by convection. The heat is transferred away from the fine liquid droplets in a gas stream. Regardless of the mechanisms of heat transfer, the resulting structures are more homogeneous and are on a finer scale than those formed by conventional ingot-casting techniques. The rapidly solidified particulates are then consolidated by forging or extrusion into a final shape. Even after consolidation, the various rapid solidification process forms are more homogeneous and have finer microstructures than a corresponding ingot-cast and worked product.

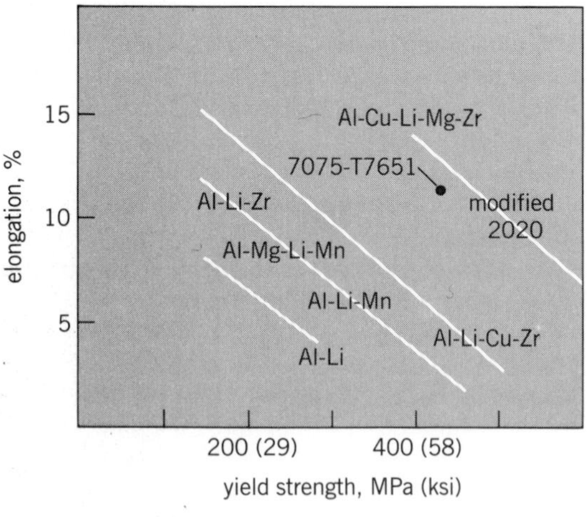

Fig. 1. Schematic of strength-ductility relationships for various Al-Li-X alloys.

Recent work has centered on utilizing the natural benefit of rapid solidification processing, producing metastable microstructures which decompose and precipitate during consolidation and working. Thus, elements which have limited solubility and low diffusivity in aluminum such as iron, cobalt, and nickel (the transition metals) in conjunction with elements such as molybdenum or cerium can be used to maximum advantage. However, the use of these alloying elements necessitates control of thermal history of the particulate both in its formation and upon subsequent processing. For example, in aluminum-transition metal alloys, the precipitates can form directly from the melt rather than from the solid. Consequently, the precipitates in these systems tend to be coarser than those in aluminum-lithium system.

As in the case of aluminum-lithium alloys, the yield strength is affected primarily by the interparticle spacing and the fineness of the precipitates which form. The ductility and fracture toughness are affected by macroparticles that concentrate stress, thereby reducing the toughness. Since ductility and fracture toughness are important properties, rapid solidification processing alloys will be limited until a control of the volume fraction of coarse particles can be achieved.

Recently, rapid solidification process aluminum alloys containing iron and cerium have shown promise. Alloys containing both of these elements appear to result in an alloy which has a refined particle-size distribution and improved high-temperature stability. The new powder metallurgy alloys appear competitive with titanium up to 375°F (191°C). It is clear, however, that more work must be done to achieve higher strengths and better thermal stability in the temperature range 450–500°F (230–260°C).

Mechanical alloying. The mechanical alloying process circumvents the limitations of conventional ingot casting. Blends of powders are mixed in a ball mill. A drum is mounted horizontally and half-filled with steel balls and blends of elemental metal powders. As the drum rotates, the balls drop on the metal powder. The degree of homogeneity of the powder mixture is determined by the size of the particles in the mixture. If the powder's particles are too coarse, the different constituent elements in the blend will not interdiffuse during consolidation. Consequently, high rates of grinding must be used. The rate of grinding is proportional to the speed of rotation. There is a limit, however, to the speed. At high speeds the steel balls are forced to the wall of the drum, at which point grinding stops. A mill with the capability of higher energies has the drum mounted vertically with a series of impellers. Still higher grinding rates can be achieved on a small scale with a high-speed shaker. However, such a mill produces only a few grams of material.

In a high-energy mill the particles of metal are flattened, fractured, and rewelded. Every time two balls collide, they trap powder particles between them. The impact creates atomically clean fresh surfaces. When these clean surfaces meet, they reweld. Since such surfaces readily oxidize, the milling operation is generally conducted in an inert atmosphere. During the initial stages of the mechanical alloying process, the particles are layered composites of the constituent powders, and the starting constituents are easily identifiable within the composite particle. The composition and shape of the individual particles are different. As alloying progresses, the particles are further fractured and rewelded. The intimate mixture of the starting constituents decreases the diffusion distance and the particle tends to become more homogeneous. Metastable and stable phases are beginning to form in the individual particles. Diffusion of the elements is accelerated by the strain introduced during the milling. During the final stage of processing, the individual particles became similar in composition, and similar to one another. Completion of the process occurs when the tendency to fracture is balanced by the tendency to weld and the particle size distribution is constant. The relationship between the thickness and layers within the composite is schematically illustrated in Fig. 2.

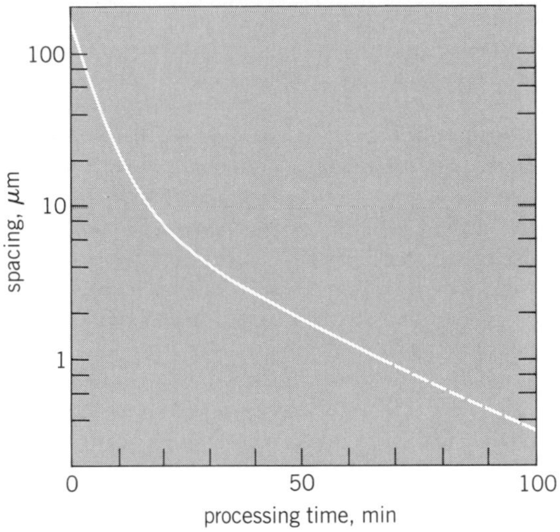

Fig. 2. Graph illustrating the effect of processing time on the spacing between the layers of mechanically alloyed powders. The smaller the spacing, the more homogeneous the powder particle.

A variety of aluminum alloys have been developed and evaluated, and a few are now commercially available. Currently, extensive research is being conducted to mechanically alloy other aluminum alloy compositions, including those containing lithium.

Aluminum–silicon carbide composites. A composite is a material in which two or more constituents are combined to result in a material which has properties different from those of either constituent. Typical composites are from materials in which one of the components has very high strength and modulus and the other has high ductility. Their properties generally follow a rule of mixtures. For example, if elastic modulus is the property of interest,

Data for several loading fractions of silicon carbide*

	Tensile properties of SiC/2024 Al-T4 composites†; vol. % SiC			
	0	15	20	25
Ultimate tensile strength, MPa (ksi)	400–434 (58–64)	NA‡	455–524 (66–76)	552–641 (80–93)
Elongation, GPa (Msi)	73 (10.6)	89–97 (13–14)	97–117 (14–17)	117–151 (17–22)
Elongation to failure, %	19–20	1–2	1–2	1–2

	Tensile properties of SiC/2024 Al composites, as extruded§; vol. % SiC		
	0	20	25
Ultimate tensile strength, MPa (ksi)	186 (27)	331–372 (48–54)	400–448 (58–65)
Elongation, GPa (Msi)	73 (10.6)	97–117 (14–17)	117–151 (17–22)
Elongation to failure, %	20–22	1–2	1–2

*MPa = megapascals. GPa = gigapascals. ksi = kips (10^3 pounds) per square inch. Msi = 10^3 pounds per square inch.
†18 specimens.
‡Not available.
§20 specimens.

the elastic modulus of the composite is approximately the weighted sum of the elastic moduli of the constituents.

Historically, the cost for reinforcing fibers has been high; however, in the last 10 years new filaments with excellent mechanical properties have been developed. For example, silicon carbide (SiC) was made available by the inexpensive process of pyrolyzing rice hulls. The SiC whiskers can be mixed with aluminum alloy powder, compacted under pressure at elevated temperatures, and extruded or forged into a final product. The very-high-modulus SiC is incorporated in the ductile aluminum alloy matrix. The resulting properties depend on the volume fraction of SiC, and to a large degree on the fabricating methods. If the composite is incorrectly fabricated, the SiC whiskers will be broken, reducing their effectiveness, and internal defects will be produced which reduce the ductility.

The primary advantage of an aluminum–silicon carbide composite is the high elastic modulus and strength. The tensile properties in a composite of SiC/2024 with three different volume fractions of silicon carbide are summarized in the table. The corresponding tensile properties for the aluminum alloy powder are included. These data demonstrate the advantages of the SiC fibers. However, work still must be done to improve the ductility.

For background information *see* ALUMINUM; COMPOSITE MATERIALS; LITHIUM; METALLIC GLASSES; POWDER METALLURGY in the McGraw-Hill Encyclopedia of Science and Technology.

[T. H. SANDERS, JR.]

Bibliography: A. P. Divecha, S. G. Fishman, and S. D. Karmarkar, Silicon carbide reinforced aluminum: A formable composite, *J. Met.*, 33:12–17, September 1981; P. S. Gilman and J. S. Benjamin, Mechanical alloying, *Annu. Rev. Mater. Sci.*, 13:279–300, 1983; S. L. Langenbeck, *Elevated Temperature Aluminum Alloy Development*, Interim Technical Report for April–September 1981, U.S. Air Force Rep. LR 29977, October 1981; E. A. Starke, Jr., T. H. Sanders, Jr., and I. G. Palmer, New approaches to alloy development in the Al-Li system, *J. Met.*, 33:24–33, August 1981.

Alzheimer's disease

First described in 1907 by Alois Alzheimer, a German physician, Alzheimer's disease is an adult-onset neurological disorder of unknown etiology manifest by loss of memory, impaired thought processes, and abnormal behavior. When the illness begins before the age of 65, the disease is termed Alzheimer's disease; when onset is after 65, the disorder is referred to as senile dementia of the Alzheimer's type. Approximately 5% of the United States' population over 65 years of age have severe dementia; an additional 10% have a mild-to-moderate impairment in memory and cognition. Of these demented individuls, approximately 40–50% have Alzheimer's disease, making this disorder the most common cause of dementia in middle and late life.

Affected individuals are, at first, forgetful. As the memory disorder gradually worsens, these individuals, although able to recall occurrences in the distant past, are unable to remember recent events. Subsequently, speech, the ability to calculate, visuospatial orientation, judgment, and social behavior become progressively abnormal. Eventually, the individuals become profoundly demented, and frequently die of intercurrent infection. The disease is more common in women, and the average duration of disease varies from 3 to 8 years. The diagnosis of Alzheimer's disease is usually made on the basis of clinical history and neurological examination since, other than brain biopsy (now rarely performed in the United States), there are no specific tests which definitively establish a diagnosis.

Brain pathology. At autopsy, the brains of patients with Alzheimer's disease are usually slightly smaller than normal for their age. Microscopic examination discloses three characteristic pathological features: neurofibrillary tangles, neuritic plaques, and loss of specific populations of nerve cells. Neurofibrillary tangles are fibrillar inclusions within the cell bodies of neurons, particularly in the cerebral cortex. These tangles are composed of unusual paired helical filaments thought to be cross-linked protein polymers perhaps derived from neurofilaments normally existing in nerve cells. Tangles may be associated with abnormalities of intracellular transport within nerve cells. Neuritic plaques, the principal histological hallmark of Alzheimer's disease, are located within the cortex and are composed of several elements: abnormal neurites (axons and synaptic terminals); amyloid, an accumulation of extracellular fibrillar material in a β-pleated sheet configuration; and nonneuronal reactive cells. The presence of plaques appears to be important in the clinical expression of disease in that their relative abundance correlates with the presence of dementia and with severity of loss of certain neurotransmitter markers, particularly cholinergic enzymes. A third feature of Alzheimer's disease is dysfunction and death of certain populations of nerve cells in the cortex, hippocampus, amygdala, basal forebrain, and specific brainstem nuclei.

Neurochemical studies have shown that the brains of individuals with Alzheimer's disease exhibit a selective reduction in enzyme markers for specific neuotransmitter systems. Specific populations of nerve cells use specific neurotransmitters; for example, certain nerve cells in the basal forebrain use acetylcholine as a neurotransmitter and contain choline acetyltransferase, the enzyme which synthesizes acetylcholine. When this population of neurons is stimulated, these cells release acetylcholine at their nerve terminals (synapses). The transmitter can excite, inhibit, or modulate the activity of target neurons, depending on the nature of the receptors on these nerve cells. In Alzheimer's disease, alterations in neurotransmitters have been established by measuring the transmitter or its synthesizing enzyme, and there is evidence of abnormalities in several neurotransmitter systems, including those using acetylcholine, norepinephrine, and somatostatin. The most consistant abnormality in Alzheimer's disease is reduction in the activity of choline acetyltransferase in the cortex, and this abnormality has been attributed to dysfunction and death of basal forebrain cholinergic neurons which are known to innervate the cerebral cortex. Similarly, changes in the norepinephrine and somatostatin systems may result from degeneration of neurons using these transmitters.

Etiology. The etiology of Alzheimer's disease is unknown. The disease is associated with increasing age but is not an invariable concomitant of aging. Although many cases appear to be sporadic in nature, available evidence suggests that a significant number of affected individuals have inherited a predisposition to develop the disease: in certain families, the disease is inherited as an autosomal dominant; in addition, individuals with trisomy-21 (Down's syndrome), a chromosomal disorder, have, if they live into middle life, a very high incidence of Alzheimer's disease. Infectious agents and toxins have been suggested to play a role in the pathogenesis of Alzheimer's disease, but there is no conclusive evidence that the disease is due to a transmissible agent or is caused by a toxic process.

Treatment. There is no specific treatment for Alzheimer's disease. In recent years, with the demonstration that there are cholinergic abnormalities in the brain, cholinomimetic drug therapy has been tried, but these drugs have not produced any sustained remission of signs and symptoms. Because Alzheimer's disease is so common and because of the increasing number of individuals reaching the at-risk period of life, Alzheimer's disease has been called an approaching epidemic. Although no specific treatment is yet available, the application of new approaches to this disorder should clarify the biology of the disease and eventually allow the design of rational therapeutic approaches.

[DONALD L. PRICE]

Bibliography: J. T. Coyle, D. L. Price, and M. R. DeLong, Alzheimer's disease: A disorder of cortical cholinergic innervation, *Science*, 219:1184–1190, 1983; D. L. Price et al., Basal forebrain cholinergic neurons and neuritic plaques in primate brain, *Biological Aspects of Alzheimer's Disease*, Banbury Rep. 15, pp. 65–77, 1983; R. D. Terry and P. Davies, Dementia of the Alzheimer type, *Annu. Rev. Neurosci.* 3:77–95, 1980.

Antibiotic

Antibiotics have been defined as compounds which are produced by microorganisms and in high dilution inhibit the growth of other microorganisms. The practical use of antibiotics is to control infection and disease caused by microorganisms. Aminocyclitol antibiotics, also known as aminoglycoside antibiotics, are an important group of antibacterial compounds, and include the clinically useful streptomycins, gentamicins, kanamycins, neomycins, and their semisynthetic relatives such as amikacin.

Ideally, aminocyclitol antibiotic research will lead to compounds which are nontoxic and will kill all pathogenic bacteria. Whether this will ultimately happen is open to conjecture. Progress has been made toward this goal, including establishment of the mechanism of action, resistance mechanisms, and toxicity. This has resulted in a better understanding of the biochemistry associated with these compounds. While still incomplete, this knowledge, combined with empirical approaches, has permitted attempts at preparing better antimicrobial compounds.

Characteristics of aminocyclitols. All aminocyclitols have the common feature of possessing an aminocyclitol unit, which is a cyclic alcohol containing amino functionalities. Examples include 2-deoxystreptamine (I), the most commonly occurring aminocyclitol; streptidine (II), found in streptomycin; fortamine (III), found in the fortimicins; and validamine (IV), isolated from validamycins. These

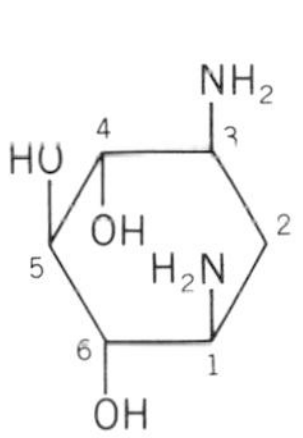

(I)

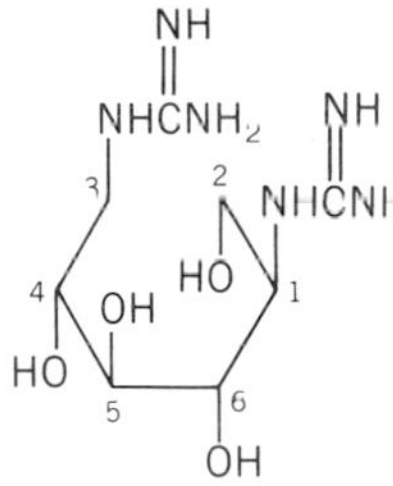

(II)

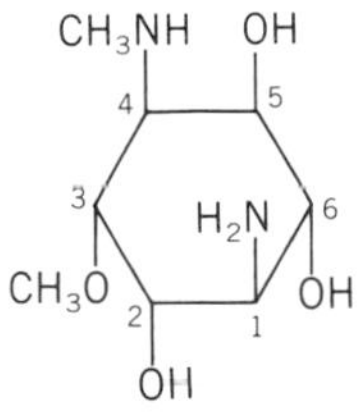

(III)

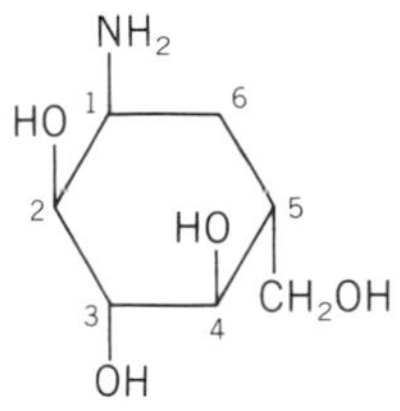

(IV)

units are glycosylated to form the antibiotic from which they are isolated. Thus kanamycin (see illustration) contains 2-deoxystreptamine which is glycosylated with aminoglucose derivatives at positions 4 and 6, whereas neomycins contain 2-deoxystreptamine glycosylated at positions 4 and 5. Streptomycin contains streptidine glycosylated only at positive 4, and fortimicins contain fortamine which is also monoglycosylated but at position 6. Validamycin contains two aminocyclitol units and a glucose moiety. All of these antibiotics appear to be biosynthesized by the producing organisms using glucose as the starting material. Some of the reactions involved in the biosynthesis are known, but generally little experimental data are available.

The aminocyclitol mechanism for inhibiting the growth of bacteria seems to be by interfering with ribosomal protein biosynthesis. The lethal nature of

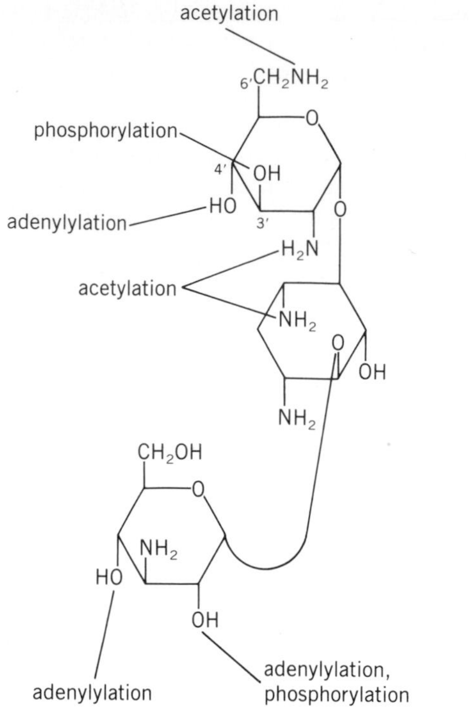

The sites of modification and types of reactions involved in the inactivation of kanamycin B by various bacteria.

this interference is probably via inhibition of protein synthesis, but there are experimental results which demonstrate that "misreading" of the nucleic acid template also occurs, which would result in "wrong" proteins being produced. Since bacteria and mammalian protein biosynthetic processes involve substantially different ribosomes, these antibiotics can kill bacteria without harming the host.

Clinical problems. Clinically two of the biggest problems associated with aminocyclitol antibiotics are antibiotic resistance, and nephro- and ototoxicity.

Resistance. Aminocyclitol antibiotic resistance in a selection of organisms has been studied, and this has led to a useful understanding of the various resistance mechanisms involved. A frequently occurring resistance mechanism is that in which the target organism enzymatically modifies the antibiotic to render it inactive. The modifications described until now include *N*-acetylation, *O*-adenylylation, and *O*-phosphorylation. The enzymes which catalyze these inactivation reactions differ depending on the microbial source, but generally any one enzyme can inactivate a variety of antibiotics. The enzyme which acetylates the 6′-amino group of kanamycin modifies similar positions in neomycin, tobramycin, and amikacin. The illustration shows the different sites for antibiotic inactivation of kanamycin.

A further problem associated with these resistance-conferring enzymes is that often the genetic information for the enzyme synthesis is coded within an extrachromosomal element called a plasmid. Furthermore, these are often transferrable from one bacteria to another; during this plasmid transfer a bacteria is generated which can synthesize the inactivating enzymes, and is therefore as antibiotic-resistant as the first, and can possibly transfer the plasmid to yet another bacteria.

One of the aims of the antibiotic medicinal chemist has been to synthesize modified compounds which cannot be inactivated. A classic example of this is the synthesis in Japan of kanamycin, in which the 3′ and 4′ hydroxyl groups (see illustration) have been replaced with hydrogen. The new antibiotic 3′,4′-dideoxykanamycin, also known as dibekacin, lacks the sites which were previously susceptible to phosphorylation or adenylylation. 3′,4′-Dideoxykanamycin is active against kanamycin-resistant organisms which possess this mechanism. More recent examples have been described, but none perhaps as striking as the first.

Toxicity. The mechanism for oto- and nephrotoxicity, which results in damaged hearing and kidney function respectively, is not well understood, and an empirical approach has been employed to find novel aminocyclitol antibiotics which have lower toxicity. Fortimicins are a relatively recently discovered group of antibiotics produced by *Micromonospora olivoastereospora*, and have been shown to have much lower toxicity in experimental animals. However, as yet they have not become available for general clinical use in the United States.

Novel aminocyclitols. The trail leading to the final compound of this discussion was first described in 1969 by K. L. Rinehart and coworkers. Starting with a neomycin-producing organism which was exposed to a chemical mutagen, mutants were generated which were unable to synthesize neomycin unless 2-deoxystreptamine (I) was present in the growth medium. When chemically modified 2-deoxystreptamines were added to these cultures instead of the regular 2-deoxystreptamine, neomycins containing the novel aminocyclitol were biosynthesized in some instances.

One of the most clinically promising antibiotics to be developed using this technique is hydroxygentamicin. This was biosynthesized by using a mutant of the gentamicin-producing organism, *Micromonospora purpurea*, which required 2-deoxystreptamine to synthesize gentamicin. When the organism was grown in medium containing streptamine (V), hydroxygentamicin was produced. Hydroxygentamicin is approximately as active as gentamicin in terms of antibacterial activity, but is one-half to one-fifth as toxic in experimental animals. Hydroxygentamicin has also been prepared by using *myo*-inosose (VI)

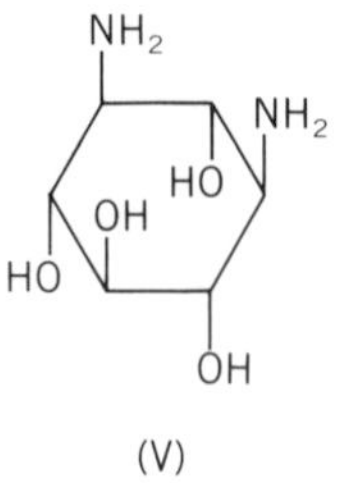

(V)

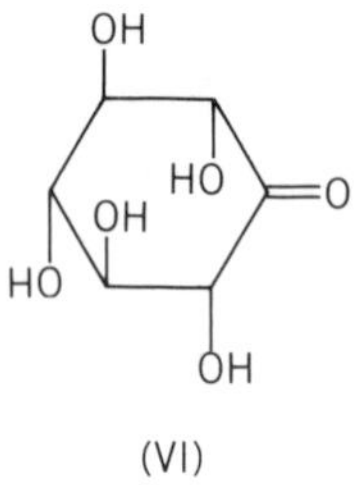

(VI)

instead of streptamine. In this case *M. purpurea* converts the *myo*-inosose to streptamine, presumably using that part of the deoxystreptamine-biosynthetic route which is still present, and then converts streptamine to hydroxygentamicin.

For background information *see* ANTIBIOTIC; BACTERIAL GENETICS; GENTAMICIN; MEDICAL BACTERIOLOGY; NEOMYCIN; STREPTOMYCIN in the McGraw-Hill Encyclopedia of Science and Technology.

[CEDRIC PEARCE]

Bibliography: C. J. Pearce and K. L. Rinehart, Jr., Biosynthesis of aminocyclitol antibiotics, in J. W. Corcoran (ed.), *Antibiotics*, pp. 74–100, 1981; J. Reden and W. Durckheimer, Aminoglycoside antibiotics, chemistry, biochemistry, structure-activity relationships, *Topics Curr. Chem.*, pp. 105–169, 1979; K. L. Rinehart, Jr., and T. Suami, *Aminocyclitol Antibiotics*, 1980; H. Umezawa and I. R. Hooper, *Aminoglycoside Antibiotics*, 1982.

Antimicrobial agents

A major development in medicine during this century was the discovery of antimicrobial agents. Recent research has shed light on the possible effects of large doses of antibiotics on the immune system and on the mechanisms of antibiotic resistance in bacteria resulting from the widespread use of antibiotics.

The first section of this article discusses the effects of antibiotics on immunodefense capabilities of the host. The second section discusses penicillin resistance in pneumococci.

EFFECTS OF ANTIBIOTICS ON IMMUNODEFENSE CAPABILITIES

In the absence of the immunodefenses, the human body deteriorates rapidly; without polymorphonuclear cells antibiotics can hardly add more than 2 weeks of survival time. It is, therefore, important to know the effects of antibiotics on the highly complex immunodefense system. It is also important to know the effects of antibiotics on bacteria, and how these antibiotic-modified bacteria challenge the immunodefense system.

Effects on antibodies. The only antibiotic which has been shown consistently to have some effect on antibodies is rifampicin. In rabbits and guinea pigs, rifampicin causes a dose-dependent suppression of antibody response to bovine serum albumin, but suppression in rabbits was reversible and was obtained only with high doses. Studies in humans have shown that seroconversion following vaccinia vaccination was suppressed if the vaccine was administered simultaneously with local rifampicin. However, orally administered rifampicin had no effect. It has been shown that a year or more of rifampicin therapy in tuberculosis patients has no effect on the antibody response to influenza vaccine. Most of the evidence on the effect of rifampicin on the humoral response is obtained from cell cultures and animal experiments. There is no conclusive evidence that rifampicin shows any immunodepressing activity with clinical consequences.

Effects on lymphocytes. Lymphocyte transformation is affected by almost all antibiotics. The most consistent effect has been produced by amphotericin B, an antifungal antibiotic, followed by rifampicin and tetracycline. Again, there is no evidence that these drugs affect lymphocytes in humans with any significant clinical effects.

Effects on polymorphonuclear cells. The effects of antibiotics on polymorphonuclear cells, which are the front line of defense at the various points of entry of microorganisms into the body, have been the subject of many detailed investigations. These effects can be divided into those on chemotaxis, and on phagocytosis and intracellular killing.

Migration of polymorphonuclear cells in laboratory cultures is inhibited by all therapeutically used aminoglycosides. The inhibition is reversible as washed polymorphonuclear cells become active, which is an indication that inhibition is dependent on the continued presence of the antibiotic. The chemotactic response of polymorphonuclear cells to *Escherichia coli* endotoxin can be significantly inhibited in cell cultures by gentamicin at concentrations that are easily attainable in serum. Amikacin appears to cause a similar but moderate inhibition of chemotaxis. Tetracycline and its derivatives are the second most common agents which have been shown to inhibit chemotaxis in cell cultures. The serum of volunteers treated with tetracycline inhibits chemotaxis of normal polymorphonuclear cells in cell cultures. Doxycycline has a more profound effect than that of other tetracyclines, probably because of its higher lipid solubility. Rifampicin and sulfamethoxazole have also been shown to inhibit chemotaxis in cell cultures.

For each antibiotic there is at least one study which contradicts the other findings, and since there is little animal experimentation and no clinical evidence that these antibiotics would significantly inhibit chemotaxis during infections in humans, the inhibition of chemotaxis of polymorphonuclear cells by antimicrobial agents remains at most an academic subject rather than a phenomenon of medical importance.

Phagocytosis with intracellular killing is slightly inhibited in cell cultures by the presence of aminoglycosides and by tetracycline. A new cephalosporin, however, has been shown to enhance both chemotaxis and phagocytosis by polymorphonuclear cells. Again, the clinical significance of these findings remains to be demonstrated.

Effects on bacterial virulence. Antibiotics affect bacterial virulence and, therefore, alter the manner in which bacteria challenge the immunodefense system. Antibiotics at concentrations equal to the minimum bactericidal concentration eliminate the virulence of any given bacteria since these concentrations kill the offending organism. At concentrations equal to the minimum inhibitory concentration, antibiotics inhibit the growth of bacteria and, therefore, also eliminate the virulence. If antibiotics are successful in influencing factors of virulence and the synthesis and release of substances which enable the bacteria to invade the host, the concentration of the antibiotic or the time of exposure to the antibiotic must be less than that which kills the bacteria or even that which prevents growth. Therefore, only concentrations below the minimum inhibitory concentration can be considered to play a role in influencing the virulence of bacteria, and therefore, their challenge to the immunodefense system.

Exposure of various species of Enterobacteriaceae or of gram-positive cocci to various beta-lactam antibiotics at concentrations below the minimum inhibitory concentration results in replication without separation. Long filaments of the respective bacilli or large round cells are produced. The problem arises as to whether polymorphonuclear cells can phagocytize and kill these filaments or large cocci which are often longer and larger than the diameter of a polymorphonuclear cell.

In order to solve this problem, strains of *Escherichia coli* were exposed to ampicillin or mezlocillin at concentrations one-half to one-third of the minimum inhibitory concentration, and the resulting filaments were incubated with polymorphonuclear cells. The process of phagocytosis was observed under phase-contrast microscopy and, in some cases, recorded on film. The filaments as well as the bacilli were harvested, planted for colony-forming unit counts, dried, and weighed. The total weight divided by the number of colony-forming units yields the mean weight of an individual bacillus or filament. A single polymorphonuclear cell is able to phagocytize about 2–40 normal bacilli or 1–2 filaments (Fig. 1). In rare cases up to 5 filaments are phagocytized.

The mean weight of an *E. coli* bacillus is approximately 3×10^{-9} mg and, the mean weight of an *E. coli* filament is approximately 65×10^{-9} mg. Since a filament is about 20 times heavier than a bacillus, and since both, when planted on agar, produced only one colony, the bactericidal effect of polymorphonuclear cells on bacilli and filaments was evaluated both by the number of organisms and by the weight of organisms killed. The bactericidal effect of polymorphonuclear cells on bacilli and filaments was comparable after 3 h of incubation, when the weight of the phagocytized bacteria was approximately 30×10^{-9} mg or less, or about 10 bacilli or 1 filament per polymorphonuclear cell.

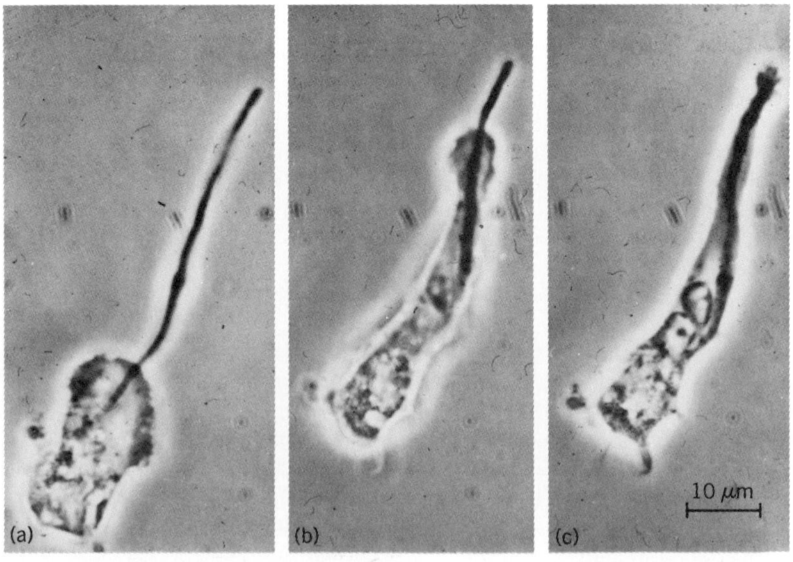

Fig. 1. Steps *a–c* in phagocytosis of a long filament by a single polymorphonuclear cell.

Increased amounts of either bacilli or filaments were handled with more difficulty but, considering their weight rather than their number, filaments were killed more readily. In the living organism, the chance of a polymorphonuclear cell finding and killing 20 bacilli is less when they are randomly dispersed than when they are together in the form of a single filament. Since the number of bacteria in infections rarely exceeds the number of polymorphonuclear cells at the site of infection, the presence of filaments should be of little consequence to the efficacy of the immune system of an uncompromised host.

Antibiotics alter the synthesis and excretion of numerous substances produced by bacteria. *Staphylococcus aureus* produces at least four hemolysins. In the presence of various beta-lactam antibiotics at concentrations below the minimum inhibitory concentration, staphylococci increase significantly their output of alpha-hemolysin. Pneumococci are known for their production of alpha-hemolysin. In the presence of beta-lactam antibiotics and vancomycin, over 100 strains of pneumococci tested produced beta-hemolysin. In contrast to the effects produced by beta-lactam antibiotics, other drugs, mostly those interfering with ribosome activity, inhibit the production of hemolysins as well as of other enzymes generated by staphylococci. *Clostridium difficile* has been implicated as an etiological agent in pseudomembranous colitis in humans. The organism produces an exotoxin which, in the presence of clindamycin at concentrations below the minimum inhibitory concentration, is synthesized at a higher rate, sometimes up to 128 times the normal amount in a period of 4 days (Fig. 2).

Conclusion. Antibiotics have been shown to affect the immunodefense system directly as well as indirectly through their effects on bacteria and have been shown to produce a large variety of alterations in laboratory cultures. Some of these alterations were also demonstrated in animal experiments. However, from all the years of antibiotic therapy of patients with infections as well as from prospective studies regarding the effects of antibiotics on the immunodefense system, there is no evidence that the administration of antibiotics at their recommended dosage in patients with normal immunodefenses has any clinical significance. When, however, large doses of antibiotics are given to immunodeficient patients, some adverse reactions could be observed,

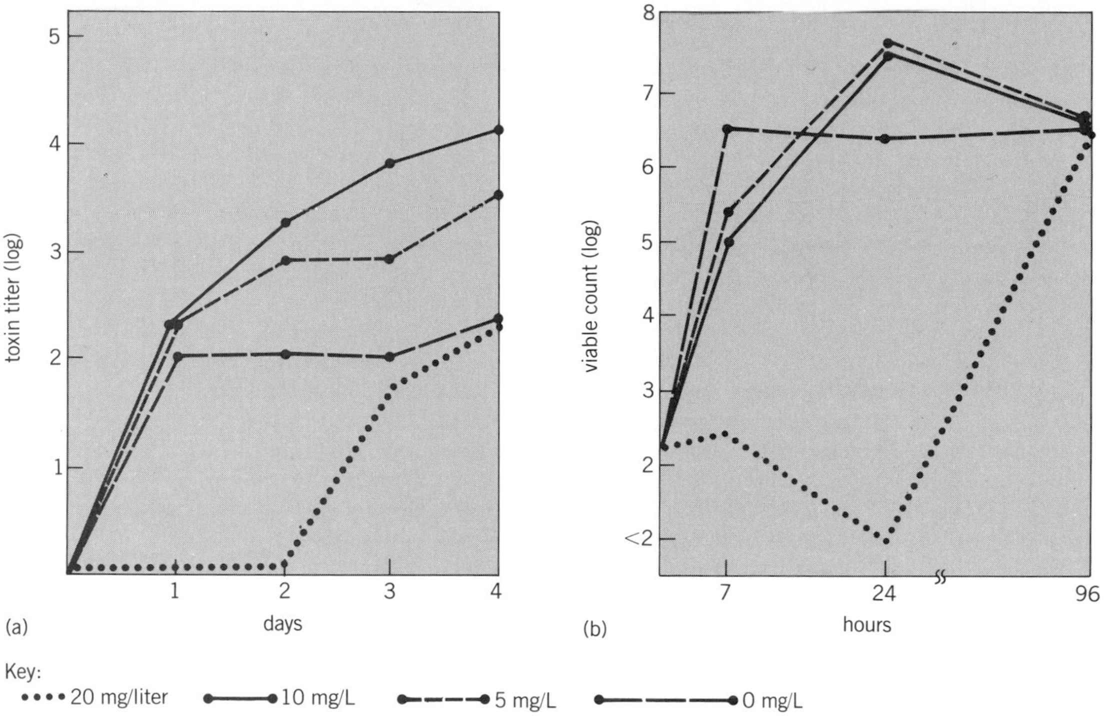

Fig. 2. Effect of clindamycin at different concentrations on *Clostridium difficile*, strain 23289. (*a*) Toxin production. (*b*) Growth. (*After R. H. George et al., Induction of Clostridium difficile toxin by antibiotics, Curr. Chemother.,* 955, 1980)

but little can be done since the administration of these antibiotics is indispensable for survival.

[VICTOR LORIAN]

PENICILLIN TOLERANCE IN PNEUMOCOCCI

Antibiotic tolerance is a unique, rather newly recognized phenomenon in which bacteria gain resistance against the irreversible antibacterial effects of a drug, but remain sensitive to the growth-inhibitory (reversible) effects.

The phenomenon of tolerance was discovered in 1970 during an analysis of the phenotype of some pneumococcal mutants and physiological variants isolated in the laboratory. The pneumococci in question were defective in the activity of a cell wall–degrading enzyme (autolysin), an *N*-acetylmuramic acid L-alanine amidase. This enzyme can hydrolyze the amide bonds connecting the *N*-acetylmuramic acid and L-alanine residues in the peptidoglycan and degrade pneumococcal cell walls to soluble fragments. When growing cultures of such lysis-defective (LYT^-) bacteria were exposed to various concentrations of penicillin, a peculiar phenomenon was observed. Upon receiving the antibiotic, the cultures underwent the normal course of residual growth, followed by a halt in growth, with a time course indistinguishable from that of lysis-prone (LYT^+) wild-type cultures. Furthermore, the minimum concentrations of penicillin needed for the inhibition were the same in the lysis-defective and lysis-prone cultures. However, the two kinds of bacteria exhibited one striking difference in their response to penicillin treatment. In the wild-type cells, the onset of growth inhibition was followed by a rapid loss of viability and lysis. In the mutant bacteria, loss of viability was slow and lysis did not occur at all. Cell wall–synthesis inhibitors other than beta-lactam antibiotics (for example, vancomycin, D-cycloserine, or fosfonomycin) could not cause lysis of the mutant bacteria either, while these inhibitors have still caused inhibition of culture growth at drug concentrations that were identical or very close to the corresponding minimal inhibitory concentration values for the wild-type cells.

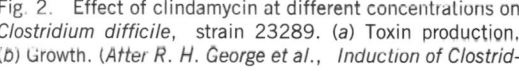

Subsequent studies showed that the initial course of penicillin action was indeed identical in the LYT^- and LYT^+ cells. Both types of pneumococci had the same set of identical penicillin-binding proteins (cell wall–synthetic enzymes that form covalent complexes with penicillin). During treatment with penicillin these penicillin-binding proteins have reached the same degrees of saturation in the two types of bacteria, and treatment with the antibiotic has induced the same set of inhibitory effects in cell wall synthesis, in protein synthesis and, eventually, in cellular growth. It seemed that the difference between the penicillin responses of LYT^+ and LYT^- cells was connected to some events secondary to the inhibition of the biochemical targets of the antibiotic (that is, the penicillin-binding proteins). The central agent in these secondary (terminal) events was the autolytic enzyme. In the LYT^+ cells, which contained high specific activity of autolysin, the interference with cell wall synthesis was shown to lead to the mobilization or triggering of uncontrolled autolytic activity, causing rapid loss of viability and lysis. In the LYT$^-$ cells, containing only a fraction (about 0.1%) of the autolytic activity of the LYT$^+$ cells, these terminal events did not take place and the final consequence of penicillin action for the bacteria was, primarily, an inhibition of growth. That this penicillin-tolerant phenotype was the consequence of lowered autolytic activity was more rigorously established by the demonstration that genetic transformation resulting in the LYT$^-$ property has invariably brought along antibiotic tolerance as well.

Other penicillin-tolerant mutants. The basic findings described for pneumococci have by now also been confirmed in a number of other bacteria, including mutants of *Bacillus subtilis*, *B. licheniformis*, *Escherichia coli*, and *Streptococcus faecium*. While each one of these penicillin-tolerant mutants was defective in one or more autolytic enzymes, individual strains have differed in the types of catalytic activity that appeared to be associated with antibiotic tolerance. The lowered autolytic activity was that of an amidase in *B. subtilis*, amidase and glucosamidase in *B. licheniformis*, transglycosylase (and endopeptidase) in *E. coli*, and a muramidase in *S. faecium*.

A common feature of all these mutants, including pneumococci, was that the low autolytic activity did not seem to cause any gross defect in the ability of the cells to grow. Thus, one is forced to conclude that the enzymatic activity responsible for the penicillin-induced lysis (and, at least in part, for the loss of viability) is not essential for the multiplication and growth of bacteria. This conclusion has an important bearing on the mode of action of penicillin and on the way that hydrolytic and synthetic reactions may be integrated during replication of the bacterial cell wall. Why and how interference with cell wall synthesis leads to the apparent triggering (that is, upset regulation) of cell wall–degrading activity is not understood.

The complexity of the mechanism of antibiotic tolerance is indicated by the isolation of bacterial mutants that show antibiotic tolerance and yet appear to have normal or only slightly decreased autolytic activities. Such mutants have been described in pneumococci and in *E. coli* by selecting for bacteria surviving various arbitrary doses of exposure to beta-lactam antibiotics. An unusual feature of these LYT$^+$-tolerant mutants was that their tolerance was drug-specific, and also antibiotic dose–dependent, in contrast to the "blanket" resistance of LYT$^-$ mutants to lysis by many wall inhibitors applied even at very high doses. For instance, pneumococcal mutants isolated by selection for survival in the presence of penicillin were found to show tolerance to penicillin and cycloserine, but could still be lysed by some other cell wall inhibitors (vancomycin or bacitracin). Still another novel feature of these mutants is that their tolerance is a quantitative phenomenon affecting the rates and degrees of lysis of the cells. These findings point to the complexity of autolysin regulation in bacteria.

Nonlytic death of bacteria. As it has been pointed out, none of the tolerant mutants isolated so far is completely resistant to the bactericidal effect of cell wall inhibitors, and the mechanism of bacteriological death (that is, inability of a cell to give rise to a clone of progeny) need not always involve damage to the bacterial cell wall. For instance, the rapid loss of viability during penicillin treatment of group A and B streptococci is not accompanied by cell wall degradation and cellular lysis. The mechanism of this process may involve irreparable autolytic damage limited to the hydrolysis of a relatively few covalent bonds. However, alternative mechanisms, for example, damage to the plasma membrane, cannot be excluded at the present time. The analysis of tolerant group A and B streptococci may eventually provide answers.

Multiple mechanisms of penicillin tolerance. Once the notion is accepted that the irreversible effects of penicillins are caused by an indirect mechanism, a number of different mechanisms that increase bacterial survival without loss of penicillin sensitivity in bacteria can be visualized. Besides low autolytic activity, such mechanisms may involve overproduction of an autolysin inhibitor, production of a "hardier" plasma membrane or altered autolysin sensitivity of the cell wall and, possibly, the existence of more complex mutants affected in the regulation of autolytic activity within the cell cycle. Clearly, as long as the selection is for bacterial survival, any one of these (and additional) mutants may be picked up and may show a common, tolerant phenotype.

Tolerant bacteria among clinical isolates. Since the late 1970s, an increasing number of reports have described penicillin tolerance among clinical isolates of human pathogens. Many of these originated from patients with relapsing infections in spite of antibiotic therapy. Most of these were recognized on the basis of a dissociation between the minimum inhibitory concentration and minimum bacterial concentration values. The bacterial species include clinical isolates of *Staphylococcus aureus*, *S. sanguis*, *S. mutans*, group A and B streptococci, *Listeria*, *S. faecalis* and *S. faecium*, and the *viridans* group of streptococci and pneumococci. A particularly striking fact is the frequency of tolerant bacteria among pathogens often involved with infections at sites of impaired host defense (valvular endocarditis, osteomyelitis). Whether or not tolerance presents problems for the chemotherapy of such pathogens remains an unsettled question.

Phenotypic tolerance. The striking resistance of nongrowing bacteria against the irreversible effects of cell wall inhibitors has been known for a long time, and this forms the basis of a frequently used laboratory method for the isolation of auxotrophic mutants. The phenotypic tolerance of nongrowing bacteria is also likely to be the basis of antagonism observed between bacteriostatic agents and penicillins. Selective suppression of the irreversible effects of cell wall inhibitors, without action on their growth-inhibitory effects, may be achieved by a variety of environmental influences (for example, the shifting of the pH of the medium, or the addition of lipids or lipoteichoic acids or certain enzymes or antibody against the autolytic enzyme to the media). The mechanism of these phenomena is not clear. Such tolerance-providing conditions may exist in microenvironments at the sites of bacterial infection and may contribute to difficulties in chemotherapy.

For background information *see* ANTIBIOTIC; ANTIBODY; CHEMOTHERAPY; DRUG RESISTANCE; IMMUNITY; LYSIN; PENICILLIN; VIRULENCE in the McGraw-Hill Encyclopedia of Science and Technology.

[ALEXANDER THOMASZ]

Bibliography: H. Fischer and A. Tomasz, Coordinate control of murein and teichoic acid biosynthesis in *S. pneumoniae*, FEMS Symposium: The Murein Sacculus of Bacterial Cell Walls, Architecture and Growth, Berlin, March 1983; D. Horne and A. Tomasz, Tolerant response of *Streptococcus sanguis* to beta-lactams and other cell wall inhibitors, *Antimicrob. Agents Chemother.*, 11:888–896, 1977; V. Lorian (ed.), *Antibiotics in Laboratory Medicine*, 1980; V. Lorian and B. Atkinson, Effect of serum and blood on Enterobacteriaceae grown in the presence of subminimal inhibitory concentrations of ampicillin and mecillinam, *Rev. Infect. Dis.*, 1:797–806, 1979; L. D. Sabath (ed.), *Action of Antibiotics in Patients*, 1982; L. D. Sabath et al., A new type of penicillin resistance of *Staphylococcus aureus*, *Lancet* 1:443–447, 1977; A. Tomasz, The mechanism of the irreversible antimicrobial effects of penicillins: How the beta-lactam antibiotics kill and lyse bacteria, *Annu. Rev. Microbiol.*, 33:113–137, 1979; A. Tomasz, Penicillin tolerance and the control of murein hydrolases, in M. Salton and G. D. Shockman (eds.), *Beta-Lactam Antibiotics: Mode of Action, New Developments, and Future Projects*, 1981; A. Tomasz, E. Zanati, and R. Ziegler, DNA uptake during genetic transformation and the growing zone of the cell envelope, *Proc. Nat. Acad. Sci. U.S.A.*, 68:1848–1852, 1971.

Archeological chemistry

With the exception of glycine, the amino acids found in proteins can exist in two isomeric forms called the D- and the L-enantiomers. The bulk of the amino acids in living organisms consist of only the L-enantiomers, although there are a few microbial peptides which contain D-amino acids. Under conditions of chemical equilibrium, equal amounts of the D- and L-enantiomers should be present. Living organisms thus maintain a disequilibrium state through a system of enzymes that selectively utilize only the L-enantiomer. When an organism dies, however, the enzymatic processes that have maintained chemical disequilibrium are no longer active, and the amino acids are subject to a racemization reaction in which the L-amino acids are converted into an equilibrium (racemic) mixture having equal amounts of the D- and L-enantiomers.

The racemization reaction of amino acids can be written as reaction (1), where k_i is the first-order

$$\mathrm{L} \underset{k_i}{\overset{k_i}{\rightleftharpoons}} \mathrm{D} \qquad (1)$$

rate constant for the interconversion of enantiomers. The rate equation for the racemization reaction is given by Eq. (2), where D/L is the ratio of the two

$$\ln\left[\frac{1 + \mathrm{D/L}}{1 - \mathrm{D/L}}\right] - \ln\left[\frac{1 + \mathrm{D/L}}{1 - \mathrm{D/L}}\right]_{t=0} = 2k_i t \qquad (2)$$

enantiomers for a particular amino acid at time t. The term $t = 0$ is necessary to account for some racemization that occurs during sample preparation.

Amino acid dating. During the last decade, the use of the amino acid racemization reaction as a geochronological tool for directly dating fossil bones has been extensively investigated. Fossil bones have been found to contain both D- and L-amino acids, and the D/L ratio of a particular amino acid in fossil bones is observed to increase with the increasing geological age in a manner that is described by Eq. (2). The value of k_i for a fossil bone from a particular site is determined principally by the environmental temperature of the locality. Since racemization is a chemical reaction, the value of k_i is temperature-dependent and increases with increasing temperature. The racemization rate constant for a locality can be estimated by using a "calibration" procedure, wherein the D/L ratio in a fossil of known age from the area of study is substituted into Eq. (2) and the local k_i value is calculated. This calibrated rate constant is an average value that integrates variations in temperature and other environmental parameters of the locality over the time period represented by the age of the calibration sample. Following calibration, the calculated k_i value can be used, with certain limitations, to date other samples from the general area.

Amino acid racemization dating has been found to be particularly useful in dating bones which are too old (that is, more than 35,000 years) for conventional radiocarbon techniques. Even within the period datable by radiocarbon, racemization can be used both to verify radiocarbon ages and to date samples that are too small for conventional radiocarbon methods. Other radiometric techniques are generally not applicable to bones, since either they are based upon the complex incorporation of a radionuclide during the burial history of the bone (as in uranium series dating) or they require closed-system behavior with regard to loss of the daughter isotope (as in potassium-argon dating).

Application to anthropology. A good example of the application of the racemization dating method to an important anthropological site is provided by studies at Olduvai Gorge in the north-central Tanzanian Rift Valley of East Africa. A large number of important hominid fossils have been found here. The geology of the region has been extensively studied. The ages of the various stratigraphic sections range from the recent back to about 3.5 million years.

Upper Pleistocene samples from both Olduvai Gorge and the Nasera rock shelter, located approximately 25 mi (40 km) northwest of Olduvai, have been dated by radiocarbon. By using these ages and the measured D/L aspartic acid ratios in the samples, it is possible to calculate the first-order rate constant for interconversion of the aspartic acid enantiomers (k_{asp}) with Eq. (2). By using these k_{asp} values, other Upper Pleistocene samples from the Olduvai region can be dated; some representative results are given in Table 1. The aspartic acid ages obtained for the other levels in the Nasera rock shelter are in good agreement with the radiocarbon ages of the samples. The aspartic acid ages for the Upper Ndutu Beds at Olduvai are also consistent with the finding that the radiocarbon ages of this stratigraphic unit are more than 29,000 years old.

The k_{asp} values obtained for the Upper Pleistocene samples in the Olduvai Gorge region suggest that after a period of 80,000 to 100,000 years aspartic acid should be totally racemized; that is, the D/L aspartic acid ratio should equal 1.0. However, analyses of bone samples from older stratigraphic units in the Olduvai Gorge region indicate that this is not the case. For example, a fossil bone from the Masek Beds whose age ranges from 450,000 to 600,000 years, was found to have a D/L aspartic acid ratio of 0.75. Samples from older stratigraphic units showed lower D/L aspartic acid ratios. These various results indicate that after a period of 50,000 to 100,000 years, fossil bones in the Olduvai region contain such low quantities of indigenous aspartic acid that secondary aspartic acid introduced from surrounding soil and percolating groundwaters introduces significant aspartic acid contamination, thus yielding D/L ratios that are lower than expected.

The aspartic acid results from the older samples at Olduvai were originally interpreted as indicating that beyond 80,000 to 100,000 years the bones were so badly contaminated that they were effectively undatable by racemization. However, some recent results suggest that in tooth enamel, isoleucine (iso) still may be relatively free of contamination and the racemization (actually epimerization) of isoleucine to alloisoleucine (allo) can be used to date the older stratigraphic sections at Olduvai. Some of the isoleucine racemization results are given in Table 2. As can be seen, the allo/iso ratio steadily increased with age of the older stratigraphic sections at Olduvai. Moreover, in the enamel fraction from a fossil rhinoceros tooth from the oldest stratigraphic section, the 3.5-million-year-old Laetolil Beds (approximately 25 mi or 40 km southwest of Olduvai), the allo/iso ratio is close to the allo/iso equilibrium ratio of 1.3. (Since isoleucine has two optically active centers, this amino acid is diastereomeric, and as a result the equilibrium ratio is not 1.0 but about 1.3–1.4.) This result suggests that in the Olduvai region isoleucine racemization in teeth is effectively a closed system with respect to the introduction of secondary isoleucine contamination for periods in excess of 3–4 million years. Using the sample from the base of Bed I for calibration of the isoleucine racemization rate (k_{iso}) yields ages (Table 2) for the other stratigraphic units which are in reasonable agreement with the ages estimated from stratigraphic and radiometric age determinations.

The isoleucine reaction was used to calculate the ages of some deposits around Olduvai which have been difficult to date by other means. Bones from an archeological site on the shore of Lake Ndutu (about 19 mi or 30 km southwest of Olduvai) yielded an allo/iso ratio of 0.6, which is the same value as for the bone from the lower unit of the Masek Beds (age 450,000 to 600,000 years) at Olduvai Gorge. This result suggests that the Lake Ndutu site is comparable in age to that of the lower unit of the Masek Beds at Olduvai, a conclusion that has also been reached from geological evidence. The Lake Ndutu site has yielded an important hominid fossil. Isoleucine racemization provides an important tool for at least establishing an approximate age for this locality.

Although a large number of investigations have demonstrated that racemization-based ages of fossil

Table 1. Aspartic acid racemization ages for Upper Pleistocene bones from the Olduval Gorge region, Tanzania

Sample location	D/L Aspartic acid	Radiocarbon age, years*	Aspartic acid age, years†
Nasera Rock Shelter			
Level 4	0.406	—	22,000
Level 5A	0.428	21,600 ± 400‡	$k_{asp} = 1.8 \times 10^{-5}\ yr^{-1}$
Level 6	0.486	22,900 ± 400	26,000
Olduvai Gorge			
Naisuisui Beds	0.32	17,500 ± 1000‡	$k_{asp} = 1.5 \times 10^{-5}\ yr^{-1}$
Naisuisui Beds, stratigraphically below UCLA 1695	0.57		39,000
Upper Ndutu Bed	0.72	>29,000§	56,000

*Organic fraction (collagen?).

†Calculated from Eq. (2), using the indicated samples for the determination of k_{asp} and D/L asp = 0.07 to calculate $t = 0$.

‡Calibration sample.

§Determined on calcareous fossils and calcrete deposits.

SOURCE: From J. L. Bada, Racemization of amino acids in fossil bones and teeth from the Olduvai Gorge region, Tanzania, East Africa, *Earth Planet. Sci. Lett.*, 55:292–298, 1981.

Table 2. Alloisoleucine/isoleucine (allo/iso) ratios and estimated ages of fossil teeth (enamel fraction) from the Olduvai Gorge region, Tanzania

Locality	Age, years	Allo/iso	Allo/iso estimated age, years*
Olduvai Gorge			
Upper Bed IV	600,000–700,000	0.82	700,000
Bed III/IV	~800,000	0.92	800,000
Upper Bed II	$\sim 1.3 \times 10^6$	1.00	1×10^6
Lower Bed II	1.65×10^6	1.17	1.5×10^6
Top of Bed I	1.72×10^6	1.09	1.2×10^6
Base of Bed I†	1.8×10^6	1.21	$k_{iso} = 1.3 \times 10^{-6}\ yr^{-1}$
Laeotili			
Laetolil Beds	$\sim 3.5 \times 10^6$	~1.30	—

*Calculated from $\ln\left(\frac{1 + \text{allo/iso}}{1 - 0.8\ \text{allo/iso}}\right) = 1.8\ k_{iso}t$.

†Calibration sample.

SOURCE: From J. L. Bada, Racemization of amino acids in fossil bones and teeth from the Olduvai Gorge region, Tanzania, East Africa, *Earth Planet. Sci. Lett.*, 55:292–298, 1981, and J. L. Bada, unpublished results.

bones agree with the ages deduced from other geochronological methods, the validity of racemization dating has been questioned by some who claim that environmental factors, such as groundwater leaching, pH, humidity, and variations and exposure to extreme temperatures, could potentially affect the rate of racemization in fossils. In particular, a major controversy has developed about the validity of some racemization ages based on aspartic acid results for some Paleo-Indian skeletons from California. The racemization ages suggest that human beings had migrated into the Americas by at least 40,000 years ago. These ages contrast with the more traditional view that people first populated the New World only about 12,000 to 15,000 years ago. The accuracy of the racemization ages of the California Indian skeletons is currently being investigated by using a new particle-accelerator–based radiocarbon dating technique.

For background information *see* AMINO ACIDS: ARCHEOLOGICAL CHEMISTRY; CHEMICAL DATING; FOSSIL MAN; RACEMIZATION; RADIOCARBON DATING in the McGraw-Hill Encyclopedia of Science and Technology. [JEFFREY L. BADA]

Bibliography: J. L. Bada, Racemization of amino acids in fossil bones and teeth from the Olduvai Gorge region, Tanzania, East Africa, *Earth Planet. Sci. Lett.*, 55:292–298, 1981; J. L. Bada, Racemization of amino acids in nature, *Interdiscipl. Sci. Revs.*, 7:30–46, 1982; G. Belluomini, Direct aspartic acid racemization dating of human bones from archaeological sites of central southern Italy, *Archaeometry*, 23:125–137, 1981; P. E. Hare, in A. K. Behrensmeyer and A. P. Hill (eds.), Organic geochemistry of bone and its relation to the survival of bone in the natural environment; *Fossils in the Making*, pp. 208–219, 1980.

Artificial skin

Artificial skin is a synthetic double-layered material consisting of a temporary silastic outer layer and, in the deeper layer, a permanent collagen–glycosaminoglycan fiber open mesh, resembling the pattern of normal dermal connective tissue fibers. This material is used as a skin replacement to cover surgically excised areas of deep burn when the individual's own skin is not available for autografting.

The clinical need for a synthetic skin replacement has steadily increased during the past 10 years because an increasing number of burn patients are surviving and require more extensive areas of burn wound to be grafted than can be covered by donations from the patient's own, sometimes very limited, unburned areas. In order to cover the large area of burned skin, the standard medical approach has been to remove the burned skin surgically and apply partial thickness skin autograft, removed from the healthy, unharmed areas of the body, over the previously excised sites. The importance of wound closure has led researchers to the remarkable development and use of skin allografts, the perfection of methods to freeze and preserve skin in the viable state, the organization and operation of viable frozen-skin banks, the use of human amnion as a temporary skin substitute, and extensive use of skin xenografts. All have proven temporary, and although successful over a short period, do not fill the need for a permanent skin replacement. Grafts of cadaver skin from unrelated donors do substitute for the patient's own skin and have been used as coverings in some cases, but skin allografts are useful only on a temporary basis since they are soon immunologically rejected. Many attempts have been made to provide a temporary skin closure using synthetic polymers as sheets or polymer composites. Again, these have acted as temporary dressings and behaved as adjuncts to the treatment of the patient, but have not solved the problem of wound closure in a permanent way.

Artificial skin has been developed in response to the shortage and critical need for permanent burn-wound closure with a healthy skin replacement which is not immunologically rejected. The process of healing in all types of injuries involves the laying down of new collagen to close a defect. The final result of this process is scar formation, and the scar is always associated with contraction and a loss of normal function. Since skin which has been destroyed below the level of the dermis and skin appendages has lost its ability to regenerate, an alternative form of cover must be created for the denuded area. Further, there is a need to explore the possibilities of obtaining tissue repair with minimal scar formation in the treatment of burned patients. Artificial skin has to be capable of healing when it has been injured, as well as to resist bacterial invasion through normal antibacterial activity. For these reasons, artificial skin has been designed to permanently replace the dermal, and secondarily, the epidermal portion of skin. It should provide a mechanism of healing of a burn wound which resembles regeneration and avoids scar formation.

Characteristics. With the idea of achieving these goals, two major objectives for artificial skin were formulated: (1) The capability to induce the migration of connective tissue cells into a fibrillar collagen dermal portion of the artificial material without inflammation or immunologic rejection, but with synthesis of a normal connective tissue matrix. (2) The capability to direct this incoming connective tissue matrix into a pattern resembling normal dermis and not scar, and at about the same rate to provide for controlled biodegradation of the manufactured artificial dermis, creating a "self-neodermis" which would function more as dermis than as scar.

To achieve these the following assumptions were made: (1) A collagen–glycosaminoglycan composite, made into an open fibrillar lattice would induce the migration of mesenchymal cells and the synthesis of new connective tissue without inducing inflammatory reactions or immunologic rejection. (2) An artificially constructed fibrillar lattice constructed in a manner similar to the three-dimensional distribution of connective tissue fibers in normal dermis would allow incoming mesenchymal cells and microvasculature to use these artificial fibers as a template or scaffolding, and would lay down newly synthesized connective tissue matrix in a pattern resembling normal dermal structure rather than the structure of scar. (3) A neodermis whose fibrillar structure resembles normal dermal architecture would function both physically and cosmetically more as dermis in the clinical setting than scar tissue. (4) The newly formed neodermis, if seeded with the patient's own based epidermal cells, should produce a functioning cornified epidermis. The neodermis and epidermis should function permanently as normal skin with conventional antibacterial and reparative properties. It should remain intact through the same type of remodeling processes found in normal dermis and epidermis for the rest of the individual's life.

The development of artificial skin incorporated a two-phase concept: First, a mechanical phase provides the necessary immediate protection of the wound as the patient's own cells are being induced to rebuild new dermis and epidermis. This function is supplied by the silastic epidermis and the artificial dermis which, when applied to the excised burn area, provide a barrier to the invasion of bacteria and prevent loss of heat, fluids, electrolytes, and protein. Second, a biologic phase permits the patient's mesenchymal cells to migrate into the artificial collagen–glycosaminoglycan fibrillar scaffolding, biodegrade it, and replace it with a neodermis. This is followed by the outgrowth of epidermis from the wound edge and the seeding of the neodermis in areas far removed from the wound edge with the patient's own epidermal cells to produce a confluent epidermis. This type of two-phase wound repair is designed to provide permanent skin replacement without, or with minimal, scar formation. It involves the initial use of a biodegradable synthetic template which can be manufactured commercially and stored without special precautions, and which organizes the incoming cellular and connective tissue component into a dermallike structure that supports a mature autoepithelium and, in coordination with it, functions as normal skin.

Biomedical engineering considerations. The present methodology for the production of artificial skin consists of the precipitation of purified collagen from a bovine hide in an acid environment by addition of chondroitin 6-sulfate (a glycosaminoglycan) to form a composite fiber. The relative amount of the glycosaminoglycan in the coprecipitate is important and is controlled by the amount added and by the pH. Since biological systems can dissolve these substances in their pure form, and they have a low mechanical stability, they are unsuitable for surgical material. It is important to stabilize them both chemically and physically. This is done through control of the chemical composition of the fibers themselves and by cross-linking the carefully controlled collagen–glycosaminoglycan fibers with two methods—treatment with gluteraldehyde and the use of high-temperature vacuum dehydration. These treatments do not alter the configuration of the collagen triple helix in the membranes.

This fibrous material allows connective tissue cell migration into it with synthesis of new connective tissue and without inflammation or a demonstratable local immunologic reaction. Further, these fibers, when cast into an open porous membrane whose structure resembles normal dermis, provide a template for the creation of a neodermis with the newly produced connective tissue.

The configuration of this artificial template is of the utmost importance to the final configuration of the fibers in the new tissue. Pore size, structure, and orientation are critical and determine both the cell migration path and rate of progress into the artificial membrane. In addition, the artificial membrane ensures the shaping of the newly synthesized fibers into a dermallike pattern. Therefore, it must be biodegraded into nontoxic components at a rate that is approximately equal to the rate of synthesis of new connective tissue. Biodegradation, as noted above, is partly controlled by glycosaminoglycan content and cross-linking. Small pores were found to be associated with little cell migration into the artificial mesh and with early formation of a dense fibrous tissue capsule at the membrane–tissue interface. Methods for controlling pore size are confined to purely physical processes which do not require additives or chemical reagents. A carefully designed freeze-drying process yields membranes with the biologically correct mean pore size. Finally, the thin silastic covering on the outer surface of the artificial dermis, with a water transmission rate close to normal skin, provides a complete artificial skin.

Clinical applications. Artificial skin has been employed in a number of cases of extensive burn injuries since 1980. It has been successful in closing up to 60% of the body surface in both adults and children who had total burn injuries of 80% or more of their body. At this time the artificial material has only been used following prompt excision of the burn and immediate grafting with the skin substitute. Histologic study has demonstrated that following grafting, the deeper layers of the artificial skin are invaded with mesenchymal cells with the prompt synthesis of a connective tissue matrix. This is well advanced in the first 4 days following grafting. Blood vessels migrate from the wound bed into the newly synthesized connective tissue close behind the advancing cellular edge. The fibers of the artificial template are slowly biodegraded over a period of approximately 60 days, and they are replaced at approximately the same rate by newly formed connective tissue.

Following vascularization at between 14 and 21 days, the temporary silastic epidermis is removed at a time that is clinically convenient and the newly forming neodermis is immediately covered with a thin (0.002–0.003 in. or 0.05–0.07 mm) epidermal graft harvested from the patient. This graft is meshed and spread evenly over the developing neodermis so that although the epidermal cells do not form a continuous cover, there are no large areas without epidermal cells. An occlusive dressing is applied, and when removed in 14 to 16 days, a newly formed epidermis is present and beginning to cornify over the neodermis. This skin replacement retains many of the anatomic characteristics and much of the mechanical behavior of normal skin, thus offering the promise of immediate wound closure for patients with severe burn injuries.

For background information *see* CONNECTIVE TISSUE; INTEGUMENT; TRANSPLANTATION BIOLOGY in the McGraw-Hill Encyclopedia of Science and Technology.

[FREDERICK ACKROYD; JOHN F. BURKE]

Atomic structure and spectra

Hydrogen, being the simplest of all the atoms, has long been considered an excellent proving ground for the quantum-mechanical theories that underlie present understanding of matter and of the interaction of radiation with matter. One of the outstanding examples of the improvement in understanding of the interaction of radiation with matter is the finite energy difference between the two $n = 2$ electron orbit levels, the $2s_{1/2}$ and $2p_{1/2}$ levels, predicted to be exactly zero by standard quantum mechanics. The small difference results from the quantization of the electromagnetic field described by quantum electrodynamics (QED). The electron in the $n = 2$ orbit can be pictured as interacting with the zero point fluctuations of the quantized radiation field. The interaction is due to the virtual absorption and emission of photons, and affects the spherically symmetric $s_{1/2}$ state more than the antisymmetric $p_{1/2}$ state, leading to an upward shift in the energy of the $2s_{1/2}$ state with respect to that of the $2p_{1/2}$ state. Willis Lamb, after whom the energy shift is named, did the first experimental work on hydrogen and set the stage for an explosion of experimental and theoretical studies which have continued to the present.

A nucleus of atomic number Z with $Z - 1$ of its electrons removed should have the same quantum-mechanical description as the neutral hydrogen atom (described above), substituting Z for 1 as the electric charge of the nucleus. Thus, the phenomenon of Lamb shift which has been used as a yardstick of how well quantum electrodynamics describes the hydrogen atom should still be a good tool for these exotic hydrogenlike atoms (in reality, ions). Recent work and the resurgence of interest in the Lamb shift in heavy hydrogenic systems are due to both experimental and theoretical advances. The

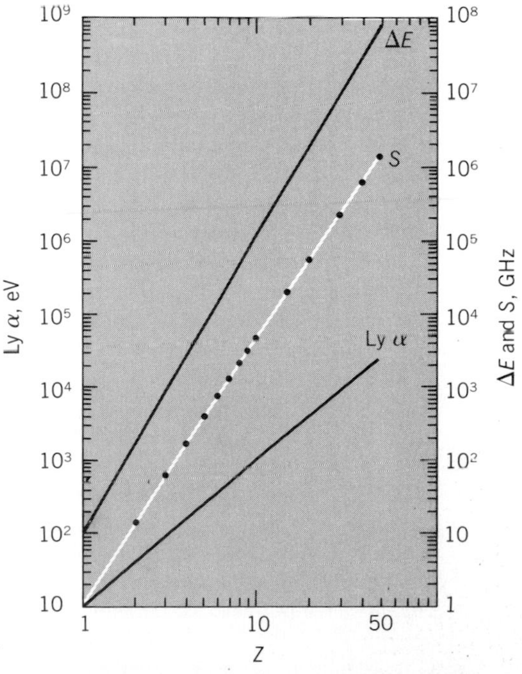

Fig. 1. Predicted scaling of energy splitting in hydrogenic ions. S is the Lamb shift, ΔE is the $2p_{1/2}$–$2p_{3/2}$ separation, and Ly α is the $n = 2$ to $n = 1$ energy difference. Ly α is measured in electronvolts on the left-hand scale, and ΔE and S are measured on the right-hand scale in terms of the frequency corresponding to the energy splitting in gigahertz.

Lamb shift calculation within the framework of quantum electrodynamics yields an exact prediction for the quantity which rapidly increases with the charge Z. Experimental verification of the predictions can serve to confirm the theory and give confidence in the newer field theories such as quantum chromodynamics (QCD) which attempt to describe fundamental particles and their interactions. The Lamb shift varies approximately as Z^4, and thus it is thousands of times larger in heavy systems than in hydrogen (Fig. 1). Tests of quantum electrodynamics predictions for Lamb shift in different energy regimes could lead to the discovery of new interactions as well.

The technological advances which allow experiments in heavy systems are the development of new heavy-ion accelerators to produce the hydrogenic species and concurrent availability of high-power single-frequency lasers to probe the magnitude of the energy splitting. The experimental techniques developed for Lamb shift studies led to innovative designs for high-power laser–particle beam interactions for fundamental studies, and are of interest as well for laser-induced fusion, directed energy devices, and short-wavelength laser developments.

Nonlaser experiments. In addition to the Lamb shift between the $s_{1/2}$ and $p_{1/2}$ levels, other fundamental properties of the two states differ. An electron in the $p_{1/2}$ state rapidly drops to the $1s_{1/2}$ ground state, emitting one x-ray photon while an electron in the $2s_{1/2}$ state can remain in that level for a relatively long time, finally decaying to the ground state by emitting one or two x-rays. The exact values for the lifetimes and decay energies are dependent on Z and are illustrated for the case of $Z = 17$ in Fig. 2.

A basic Lamb shift experiment is carried out by first producing the hydrogenic system, for example, by stripping $Z - 1$ electrons from a neutral atom. As inner atomic electrons are bound with higher and higher energies, accelerators capable of energies of order 10 to 15 megaelectronvolts per atomic mass unit (MeV/amu) are necessary for moderate atomic numbers (less than 20) and even higher for higher Z systems. Then, the longer-lived $2s_{1/2}$ level must be populated and its energy relative to the $2p_{1/2}$ state precisely determined.

One way to indirectly determine the energy splitting is by applying an intense constant electric field to ions in the $2s_{1/2}$ level which serves to "mix" the $2s_{1/2}$ wave function with the $2p_{1/2}$ wave function. The observable effect of this mixing is a change in the average lifetime of the ions, with the lifetime change being directly related to the energy separation. Experiments have been carried out on C^{5+} and O^{7+}.

In these experiments, an accelerated ion beam was fully stripped of electrons and the $2s_{1/2}$ level was populated in a single-electron pickup interaction. The electric field was produced by a transformation of a static magnetic field to the frame of reference of the moving ion, and the lifetime was determined by detecting the rate of x-rays emitted as a function of position in the magnetic field. Since the ion velocity, about 10% of the speed of light, is fixed by the accelerator energy, position information is equivalent to timing information. These experiments confirmed the prediction of quantum electrodynamics to the level of about 1%.

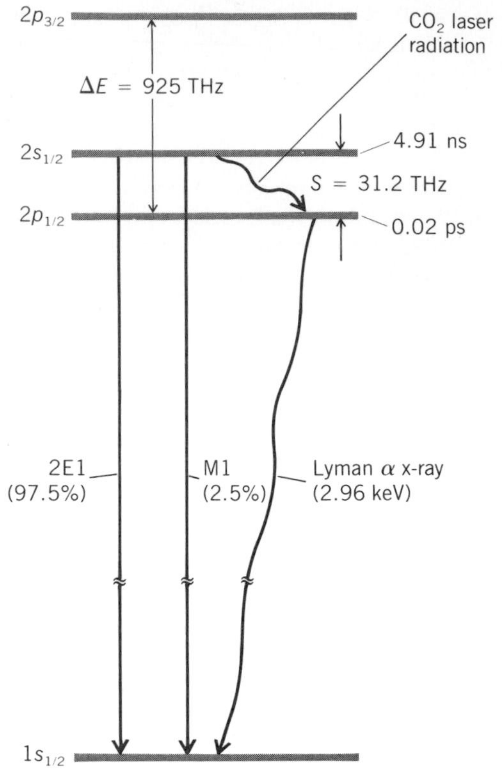

Fig. 2. Energy-level diagram, with energy splitting and lifetime values of the hydrogenic system with $Z = 17$, Cl^{16+}. For other hydrogenic systems, the structure remains unchanged, but the various parameters scale with Z, as shown for the Lamb shift values in Fig. 1.

Laser resonance experiments. A direct high-precision measurement of the Lamb shift splitting can be accomplished by using a high-power laser whose frequency exactly matches the $2s_{1/2}$–$2p_{1/2}$ energy difference. Ions in the $2s_{1/2}$ state can resonantly absorb laser photons and make a transition to the $2p_{1/2}$ state. From the $2p_{1/2}$ state an x-ray will immediately be emitted as the system decays to the $1s_{1/2}$ ground state.

The most precise of such measurements reported in the past few years was carried out on the system Cl^{16+}. For this experiment a three-stage electrostatic accelerator was used to accelerate chlorine ions to 190 MeV in order to have sufficient energy for complete ionization of the chlorine atom. The beam of chlorine nuclei then passed through an ultrathin carbon foil where a small fraction of the chlorine nuclei capture an electron to form the $2s_{1/2}$ state of hydrogenic Cl^{16+}, which normally lives for 5 nanoseconds before emitting one or two x-rays. Since the ions are traveling at approximately 10^8

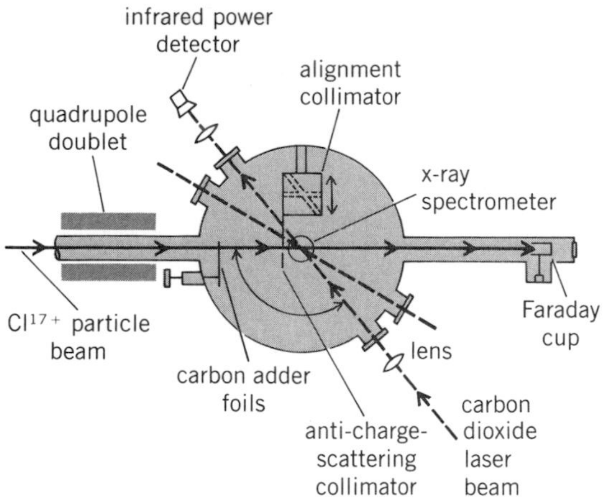

Fig. 3. Schematic diagram of the apparatus for measuring Cl^{16+} Lamb shift. Details of the accelerator and the laser system are omitted. (*After O. R. Wood, II, et al., Phys. Rev. Lett., 48:398–402, 1982*)

ft/s (3×10^7 m/s), the x-rays are not emitted, on average, until they are 0.5 ft (15 cm) past the carbon foil. Prior to this point a specially designed carbon dioxide laser is focused to a spot less than 0.02 in. (0.5 mm) in size on the ion beam whose diameter is also less than 0.02 in. (0.5 mm; Fig. 3). When the laser frequency is tuned to the precise energy difference between the two energy states, a laser photon is absorbed, exciting the Cl^{16+} ion from the $2s_{1/2}$ state to the $2p_{1/2}$ state. A chlorine ion in the $2p_{1/2}$ state instantaneously emits an x-ray photon and returns to the ion ground state. The laser-induced x-rays are detected with precision x-ray spectrometers, and the change in x-ray intensity as the laser frequency is changed yields the Lamb shift.

In order to be able to precisely change the laser frequency as well as maintain the necessary high power (more than 1.5 kW) and small laser spot size, a unique laser system was necessary. A high-power carbon dioxide laser originally designed for cutting and drilling was modified to act as an amplifier for a smaller laser with a diffraction grating used for frequency changes near the normal 10.6-micrometer output wavelength of the carbon dioxide laser. The laser oscillator and amplifier were synchronously turned on and off to achieve maximum average power output, and the grating angle was controlled by a minicomputer which also processed signals from the x-ray spectrometers, laser power monitors, and particle-beam intensity monitors.

The result of the measurement yielded the Cl^{16+} Lamb shift with a precision of 0.7%, confirming one of two present quantum electrodynamics calculations. This result is important for interpreting related experiments on other hydrogenic systems, for assessing relativistic calculations of energy levels in heavy and possible superheavy atoms, and for extending the understanding of fundamental quantum electrodynamics.

Experimental and theoretical work continues to extend this work to heavier hydrogenic atoms in order to test the high-energy limits of this fundamental theory. In addition to the laser resonance experiments, very high-precision x-ray spectroscopy can be used for ions with Z larger than about 20.

For background information *see* ATOMIC STRUCTURE AND SPECTRA; LASER; LASER SPECTROSCOPY; QUANTUM ELECTRODYNAMICS in the McGraw-Hill Encyclopedia of Science and Technology.

[D. E. MURNICK; C. K. N. PATEL]

Bibliography: H. Gould and R. Marrus, Lamb shift in the lifetime of the $^2s_{1/2}$ state of hydrogenlike argon ($Z = 18$), *Phys. Rev. A*, 28:2001–2025, October 1983; H. W. Kugel and D. E. Murnick, The Lamb shift in hydrogenic atoms, *Rep. Prog. Phys.*, 40:297–343, March 1977; W. E. Lamb, Jr., and R. C. Retherford, Fine structure of the hydrogen atom IV, *Phys. Rev.*, 86:1014–1022, June 1952; C. K. N. Patel, High-power carbon dioxide lasers, *Sci. Amer.*, 219(2):22–33, August 1968; O. R. Wood, II, et al., Measurement of the Lamb shift in hydrogenic Cl^{16+}, *Phys. Rev. Lett.*, 48:398–402, February 1982.

Biomimetic chemistry

A term coined by R. Breslow in 1972 to describe the many diverse investigations that mimic either the result or style of biochemical reactions. This definition is necessarily broad because the field itself is still developing. Every aspect of a biological system is potentially a subject for mimetic study, causing this branch of chemistry to be extremely interdisciplinary, traversing both bioorganic and bioinorganic chemistry. Various topics, each a specialized field in itself, are collected under the title biomimetic chemistry because of their unifying approach of imitating biological process. The term is used in many different applications, even those which bear little structural similarity to any biological system other than to produce similar chemical transformations.

Over the course of billions of years, an astounding variety of biochemical reactions and processes have evolved; even the most thoroughly studied is still incompletely understood. Just as in organic chemistry, where the traditional criterion for "knowing" the structure of a molecule is to synthesize it, the necessary criterion for complete understanding of a biochemical process must be to reproduce it artificially. This is the rationale of biomimetic chemistry, and model systems are its core. Most insights into the functioning of coenzymes, particularly vitamins, stem from the study of models. The model system can be either a derivative of the natural one or a completely artificial one incorporating only those features thought key to the natural function. The goal is a complete understanding of the structural and mechanistic aspects which control the natural system.

Once a biological process is at least partially understood, it can serve as a model for the creation of useful chemical processes. The widely different molecular structures produced by biological processes provide, in essence, a catalog of possibilities from which a chemist can choose. This biomimetic approach has been particularly efficacious in the medical sciences. Drugs can be designed by mimicking the structures of naturally occurring molecules, keeping those features which are effective against a disease and either adding features which will enhance that effectiveness or removing features which have adverse side effects. This approach was used in the development of serum antibodies and antibiotics even before the elucidation of the method.

Since enzymes are natural catalysts, remarkable by chemical standards for their rate and selectivity, they have two aspects of great interest to the chemical industry. While even the simplest of enzymatic reactions is still incompletely understood, it is one of the ultimate goals of biomimetic chemistry to extend the kinds of rates and selectivities found in enzymatic reactions to reactions of industrial importance for which an enzyme counterpart may be created. Because of their potential economic importance, there is a great deal of research directed at the development of industrial mimics of processes such as nitrogen fixation and water splitting which occur efficiently only inside living cells. The design of such so-called artificial-enzyme systems is only now becoming a possibility.

The goals of biomimetic chemistry are thus twofold: first, to elucidate the mechanism of biochemical reactions through the investigation of model systems; second, to develop hitherto-unknown compounds modeled on and mimicking biological systems. The following examples serve to illustrate this diversity.

Cyclodextrin chemistry. Cyclodextrins are a class of cyclic polyglucosides that have six or seven glucose molecules linked in a toroidal ring. This arrangement provides a hydrophobic cavity large enough to contain and surround molecules of a certain size and geometry, which can then undergo a reaction catalyzed by the close proximity of the cyclodextrin. This structure mimics, in essence, the environment provided by an enzyme for the binding of a substrate and subsequent reaction. For instance, the substrate *m-t*-butyl phenyl acetate binds within the cavity and then transfers an acetyl group to the cyclodextrin at a rate 250 times faster than a similar reaction without the cyclodextrin. This very modest rate enhancement has led to the conclusion that not only must the substrate bind the enzyme but there must also exist a proper geometric relationship between the enzyme and substrate during the transition state when the substrate is converted to product. The situation has been likened to finding a "key" that fits into the cyclodextrin "lock" during this transition state.

The converse of this, finding a lock that fits the key, is also possible. A cyclodextrin derivative was prepared that contained a "floor" at the end of the cavity. This caused the substrate to sit higher in the cavity, and subsequently it was bound more weakly. However, this also brought the substrate closer to the geometry of the needed transition state and thus caused the rate of acetyl transfer to increase by a factor of 9.

Substrates have also been found that bind closely to the necessary geometry. The nitrophenylester of ferrocinnamic acid was predicted to be a better substrate and, indeed, its reaction with cyclodextrin was found to be 750,000 times greater than the control.

While none of these reactions are of any particular utility in themselves, they demonstrate that the geometric relationship of the substrate to the enzyme during the reaction state is of critical importance for catalytic activity. Understanding this lock-and-key relationship points the way to the design of artificial enzymes for which no natural equivalents exist.

Vitamin B_{12}. Vitamin B_{12} acts as a coenzyme in a number of enzymatic rearrangements whose general characteristic is the interchange of a hydrogen atom and a group (X) on adjacent carbon atoms, as shown in the reaction below. In addition to performing this

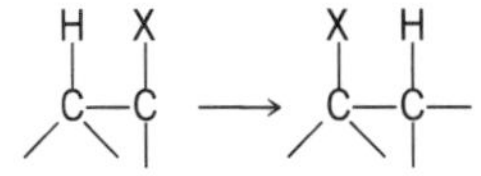

rearrangement, vitamin B_{12} is also of interest as the first known naturally occurring organometallic compound, containing a stable Co-C sigma bond. How the structural features of vitamin B_{12}, particularly those involved in the Co-C bond, participate to perform this rearrangement has been an area of active study since its characterization in 1958. The complex structure of vitamin B_{12} (see illustration) does not lend itself to any easy rationalization of its mechanism of action, so model studies of structural analogs have played a role in understanding the process.

Modifications of region 1 of the molecule show that this region is important for binding of the coenzyme to the enzyme, and not for catalytic activity. Modification of the outer substituents of region 2 show that they also assist in binding. Alterations to the inner ring structure have important effects on the electronic interaction with the cobalt nucleus and therefore with axial ligation. Alterations in region 3 have profound effects on the activity, leading frequently to complete inactivation. This result has led to the conclusion that the hydrogen bonding of this region to the enzyme activates the Co-C bond to cleavage, which, along with other evidence, implicates this as an important first event in vitamin B_{12}'s action.

Flavins. Flavins are coenzymes involved in a number of biological oxidation-reduction processes. They serve as key intermediates in one-electron transport systems such as the cytochromes; two-electron transfers in organic redox reactions; and activation of molecular oxygen without the aid of a metal in order to insert O and O_2 into a substrate. One interesting enzyme utilizing a flavin in the latter process is bacterial luciferase, which chemilumi-

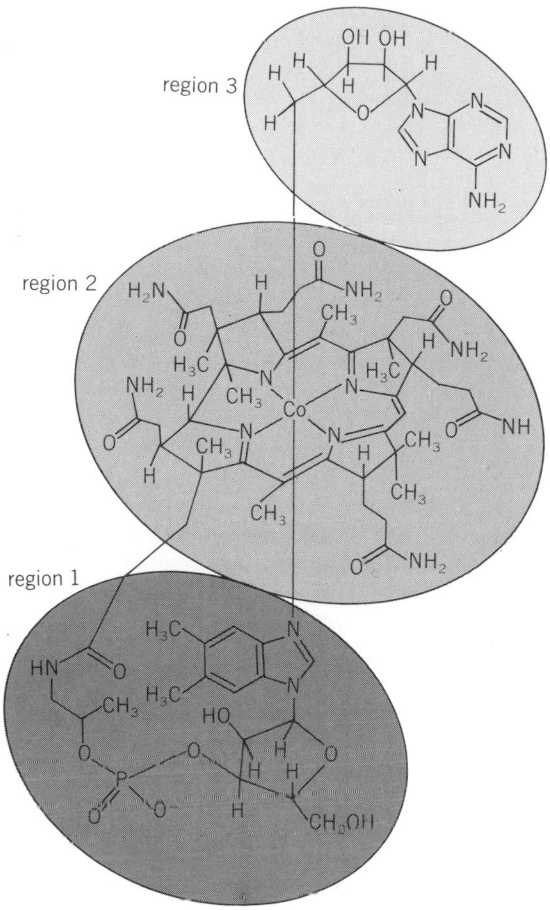

Structure of vitamin B_{12}.

nesces during the oxidation of an aldehyde to a carboxylic acid.

The flavin's ability to act as a stable one- and two-electron reservoir is due to the structure (I) which

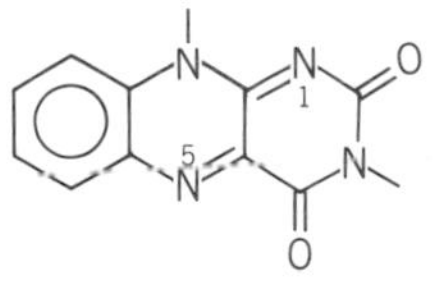

(I)

allows for extensive delocalization of electrons over the whole molecule. From the structure, the nitrogens at positions 1 and 5 were predicted to have key roles in the coenzymatic function. The significance of these positions was studied by the synthesis and evaluation of flavin analogs where these nitrogen atoms were replaced by carbon atoms.

Substitution at position 5 yielded a flavin that is restricted to only two-electron processes. Direct hydrogen transfer from substrate to this position was also shown in this analog, thus proving it is the site of electron entry. In contrast, carbon substitution at position 1 yields a catalytically active analog capable of both one- and two-electron processes. The importance of these model studies has been underscored by the discovery of a naturally occurring, five-position-substituted flavin that is used by a bacterium which produces methane from CO_2 and H_2.

Heme proteins. The importance of the heme proteins stems from the particular ability of the heme unit to interact with oxygen, and for this reason hemes are found in a number of very important oxygen-utilizing systems in nature. For example, hemes are involved in the following: reversible binding of O_2 for transport and storage (hemoglobin and myoglobin); activation of O_2 in order to incorporate one or two atoms of oxygen into a substrate, even those as unreactive as hydrocarbons (cytochrome P450, tryptophane dioxygenase); functioning as terminal enzyme in the respiratory redox chain that reduces O_2 to water (cytochrome and oxidase); and utilization of H_2O_2 (catalase and peroxidase).

How the heme groups' reactivity is modulated by the particular environment of each of these proteins has been an area of active research, with many potentially useful applications. Model systems which mimic these processes in some specific manner have been instrumental in probing the causes of the heme's reactivity.

The heme group is composed of an iron atom surrounded by the four nitrogens of a porphyrin ring (II). This leaves two sites on the iron atom open for

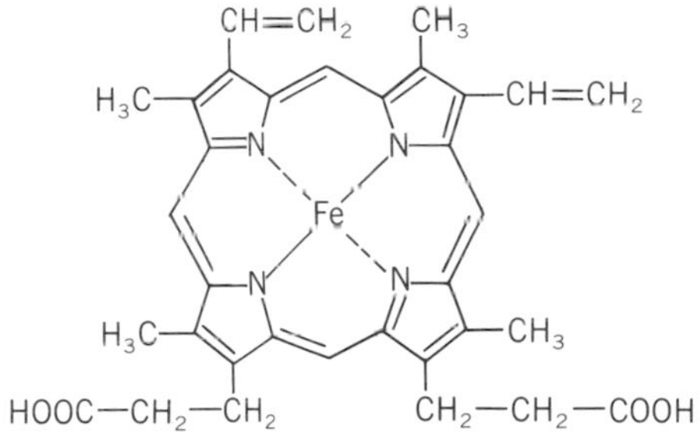

(II)

coordination by O_2 and another ligand group on the protein, such as histidine.

A number of model systems have been used to elucidate the role of hemes in hemeproteins. One approach that has been extensively employed is to substitute a metal other than iron into the porphyrin cavity. The different steric and electronic properties of these metals result in altered reactivity of the heme when reconstituted with protein, yielding information on the role of the central metal ion. Other metals are also frequently more suitable for study with a particular physical method (such as nuclear magnetic resonance, electronic paramagnetic resonance, or magnetic susceptibility), yielding information that is difficult to obtain with an iron heme. A method yielding information of the heme-O_2 interaction is the substitution of O_2 with other small molecules (CO, CN, NO) capable of binding to the heme's metal.

Another approach is to structurally modify the heme itself so that the modified heme contains elements which formerly were supplied by the protein. For example, hemes which have a covalently at-

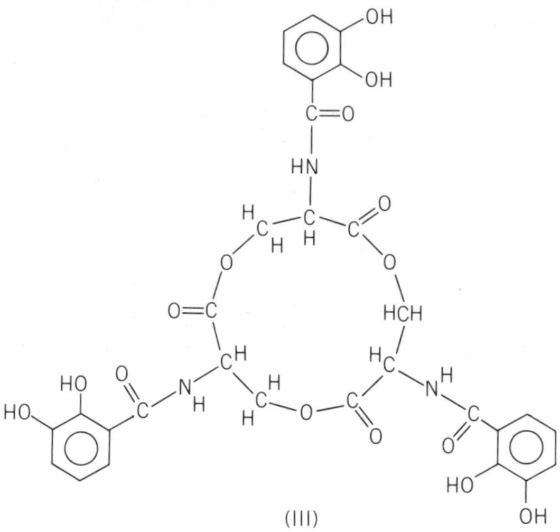

(III)

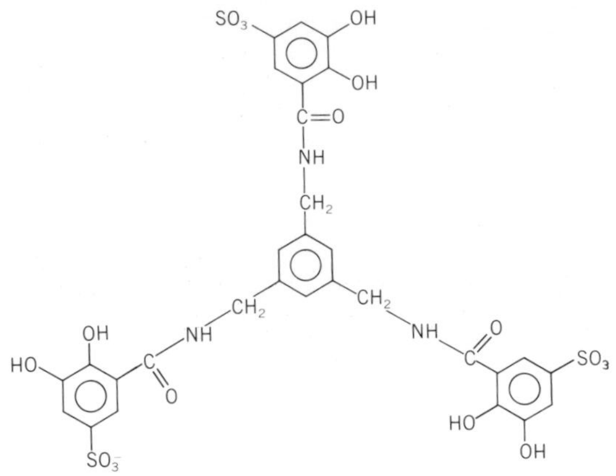

(IV)

tached imidazole or steric enclosure on one side have been synthesized, and in some respects mimic the environment provided by the globin in hemoglobin. Such synthetic models bind O_2 and CO_2 in a manner which mimics that of hemoglobin.

Iron chelation drugs. β-Thalassemia is a genetic disease affecting approximately 5 million people worldwide; treatment requires constant transfusions on a regular basis. Iron is an essential element for which there is no normal bodily mechanism for removal; thus, the body retains the hemoglobin iron of each transfusion, eventually accumulating toxic levels of the metal. For a drug to be capable of removing this excess iron, it must be specific for Fe^{3+}, since removal of any other metals, particularly calcium and magnesium, leads to undesired side effects. To promote rapid iron removal, it should have a higher affinity for Fe^{3+} than the human iron transport protein, transferrin, probably the kinetically important source of excretable iron.

The biomimetic approach is one of the most promising approaches to the design of such a drug, since compounds with similar properties evolved under natural conditions over a billion years ago. In order to obtain iron from the environment, microorganisms secrete low-molecular-weight iron chelating agents, called siderophores, which complex the iron and then facilitate transport across the cell membrane for use. Several of the siderophores have extremely high affinities for Fe^{3+}, high enough to remove it from transferrin. The problem of specificity for Fe^{3+} was achieved by employing principally either catechol groups

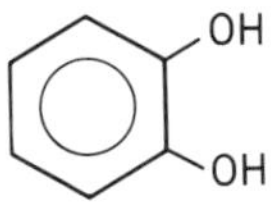

or hydroxamate groups

which bind iron selectively via two oxygen atoms.

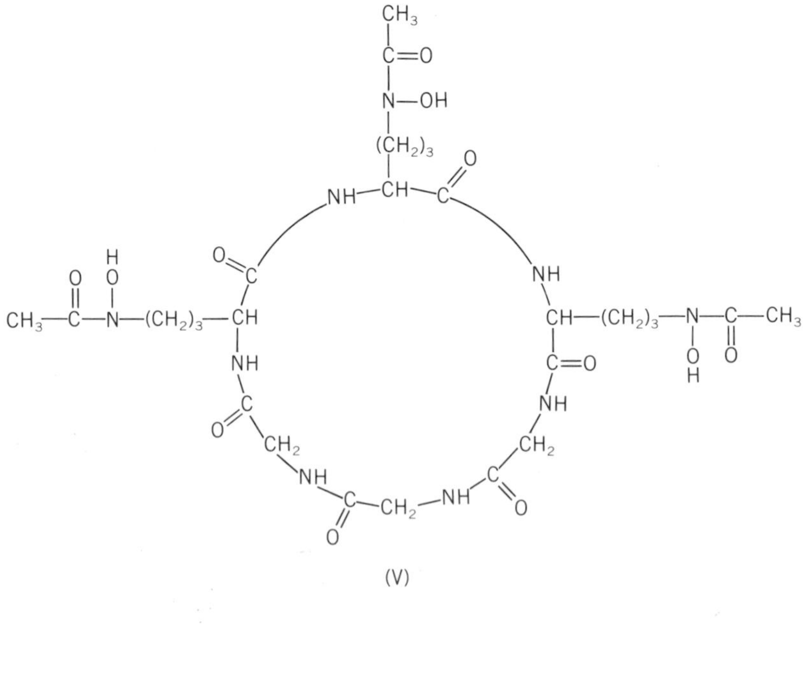

(V)

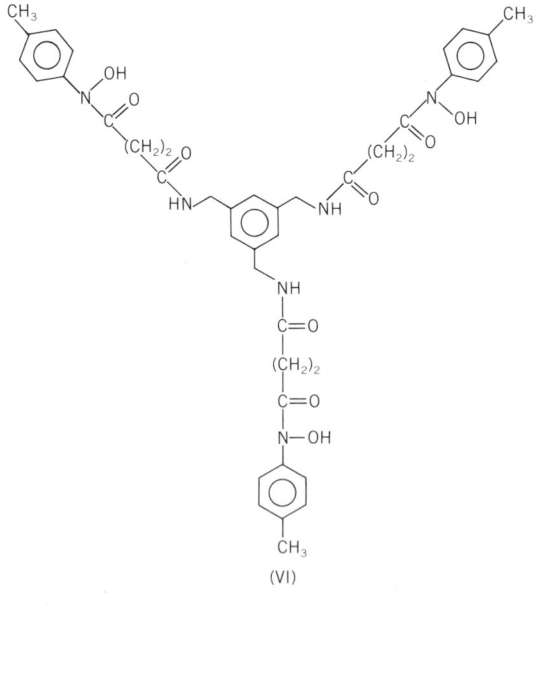

(VI)

While the siderophores themselves are not well suited for drug use, their structures serve as ideal models on which to base the synthesis of analogs. The analogs can be designed to retain the best of the siderophore properties—high affinity and specificity—coupled with other properties more suitable for drug use.

The siderophore enterobactin (III) consists of three catechols suspended from a platform constructed of a cyclic triester of *l*-serine. Although enterobactin has the highest iron-binding affinity of any iron chelator, natural or synthetic, it suffers as a drug from low availability and instability and insolubility in water. The synthetic analog called MECAMS (IV) retains the best feature of enterobactin: three catechols suspended from a central platform. The platform here, however, is a benzene ring which is stable to hydrolysis, unlike the serine ester linkages of enterobactin. Investigations with MECAMS have shown that it possesses a high iron-binding affinity and a rate of iron removal from transferrin comparable to enterobactin.

Ferrichrome (V) is a siderophore containing three hydroxamates suspended from a cyclic hexapeptide platform. An analog called MEDROX (VI) has been synthesized and is constructed of three hydroxamates suspended from the same benzene platform as MECAM.

Both catechols and hydroxamates have separate properties which render them attractive for a biomimetic design of an iron-removing drug. Hydroxamates (in the form of desferrioxamine B) have proven themselves at least partially successful in iron removal. Catechols have shown themselves to be more effective at iron removal from transferrin than hydroxamates are. Thus the attachment of a catechol to the amine terminus of desferrioxamine B might be expected to improve its iron removal properties. Such a molecule has been synthesized, and indeed does remove iron from transferrin at a rate 100 times greater than desferrioxamine B itself.

Summary. The explosion in knowledge of the biological sciences in the last few decades has provided the chemist wth a seemingly endless variety of interesting subjects which challenge imitation. Model studies of biological processes are only now yielding chemical information on the complex and subtle nature of these processes. The knowledge gained from such systems will ultimately allow the creation of completely artificial systems, designed for a specific purpose, which perform functions similar to the enzymatic reactions found in nature.

For background information *see* BIOINORGANIC CHEMISTRY; CHELATION; ENZYME; VITAMIN B_{12} in the McGraw-Hill Encyclopedia of Science and Technology.

[KENNETH N. RAYMOND; STEVEN J. RODGERS]

Bibliography: R. Breslow, Biomimetic chemistry, *Chem. Soc. Rev.*, 1:553, 1972; D. Dolphin et al. (eds.), *Biomimetic Chemistry*, 1980; A. Martell, W. Anderson, and D. Badman (eds.), *Development of Iron Chelators for Clinical Use*, 1981.

Birds

Earlier in this century scientists generally assumed that birds genetically inherited their song and call patterns. In support of the genetic hypothesis, investigators found that the "songs" of a few nonsongbird species were indeed genetically coded. For example, techniques such as hybridization, foster rearing of young birds by adults of another species, or isolation of young birds from all sounds revealed that the "coos" of several dove species and the calls of domesticated chickens are genetically coded.

However, beginning as early as the late 1930s a series of experiments demonstrated that some species of songbirds do not inherit their songs intact; rather they acquire their song patterns by learning from adults. It is now generally held that in many of the evolutionarily less advanced birds, songs develop normally without learning, but that song patterns are often learned in species belonging to evolutionarily more advanced groups, such as parrots, hummingbirds, and in particular the more advanced passerines (the songbirds). Futhermore, it is now known that many passerine and nonpasserine species can modify their call patterns by learning. The evolution of this capacity for vocal learning does not necessarily preclude a genetic basis for the songs of these more advanced species. For example, when tutored with the appropriate tape recordings, the white-crowned sparrow (*Zonotrichia leucophrys*) learns its own species song pattern and rejects the alien song patterns of closely related sparrow species. This form of selective learning, as well as other constraints that affect song development, is believed to be genetically based. Recently, researchers have begun to examine these constraints on learning and development, and some investigators are beginning to identify those features of passerine songs that are genetically heritable.

Hybridization experiments. Behavioral geneticists often use hybridization experiments to document the genetic heritability of behavioral traits. The analysis focuses on similarities and differences between the behaviors of the hybrid and the parental species. If the hybrid is fertile, it may be backcrossed with one or both parental species, and the behavior of these second-generation hybrids can be included in the analysis. In one of the earliest hybridization experiments, the songs of five different dove species and six different hybrid crosses were analyzed. Preliminary study had shown that these doves did not learn their coos. The hybridization experiments revealed that hybrid doves inherited the tonal quality of the coos of both parental species, but the rhythmical pattern of the hybrid songs was almost always disorganized and did not resemble that of either parental species. However, backcrossing some of the hybrids with one or the other parental species eliminated most of this rhthmic disorganization; that is, the backcross hybrids uttered coos that closely resembled those of the dominant parental species.

A more recent hybridization experiment focused on the song of the canary (*Serinus canaria*), a species that learns its song. The genetic basis of canary song has been analyzed by hybridizing domestic canaries with greenfinches (*Carduelis chloris*) and European goldfinches (*C. carduelis*). Analysis of the songs of the parental species and their hybrids shows that the precise form of the syllables, the basic building blocks of song patterns, are learned, but the features responsible for the characteristic canary temporal organization (that is, the introduction, the intratrill interval, the intertrill interval, and the interphrase interval) are genetically inherited. It is notable that doves, which do not learn their songs, and canaries, which do learn their songs, both have the rhythmical organization of their songs genetically coded in some way. Temporal features, such as song duration, have been identified as species-typical constants for other songbirds known to modify their songs by learning, such as the wren (*Troglodytes troglodytes*).

Although the canary learns the specific form of its syllables, a recent study has shown that even this syllable learning is directed by gene-based constraints on learning. Three different genetic strains (Roller canaries, Border canaries, and Roller-Border hybrids) were tested in a choice learning experiment. The birds were individually reared in acoustic isolation and then were tutored with recordings of both Roller and Border songs. Roller canaries learned only Roller syllables and trill sequences; Border canaries learned only Border syllables and trill sequences; but the hybrids learned Roller and Border syllables and sequences. This shows that different genetic strains of canary have different learning predispositions, and these learning differences are genetically heritable. Thus the genetic constitution of a canary influences what syllables and syllable sequences it will learn, as well as determining the rhythmical organization of its song.

Learning predispositions. Another recent experiment examined the song-learning predisposition of another songbird, the swamp sparrow (*Zonotrichia geogiana*). Its song learning is guided by features of the syllable, and an alien rhythmical pattern imposed on the tutor song does not prohibit learning. Fourteen male swamp sparrows were reared in acoustic isolation, then tutored with a number of different synthetic songs. These artificial songs consisted of either swamp sparrow or song sparrow (*Z. melodia*) syllable types. Swamp and song sparrows are closely related, but their songs differ in temporal pattern. Both the simpler swamp sparrow tempo and the more complex song sparrow tempo were represented in the set of artificial tutor songs. The young male swamp sparrows rejected all song sparrow syllable types and learned only swamp sparrow syllable types regardless of whether they were presented with song sparrow timing or swamp sparrow timing. Thus the swamp sparrow was highly selective in its learning. It exhibited a marked learning predisposition for swamp sparrow syllables, and the temporal pattern of the tutor songs did not appear to influence the learning. It is notable, however, that if a swamp sparrow syllable was presented in a tutor song with a song sparrow–like tempo, the form of that syllable but not the song's tempo was acquired. The young male swamp sparrows transposed the tempo and sang these syllables with typically simple swamp sparrow timing.

This shows that the syllable types of these two closely related sparrow species are not equivalent stimuli to the swamp sparrow. It is suggested that this may be due to species-typical perceptual predispositions—gene-based perceptual predispositions that are attuned to features in swamp sparrow syllables. Such a mechanism would guide a young bird to learn from adults of its own species.

Developmental contraints. Bird songs generally do not appear fully formed at first utterance. Song development involves several stages: subsong, plastic song, and finally crystallization into the full song pattern of the adult. Depending on the species, the period for song learning may precede, or overlap, these stages. Many species that learn their song may acquire parts of other species songs during the sensitive period for song learning. These alien syllables are then sung in the plastic song stage. But when the young male crystallizes his songs, these alien syllables are often lost and are therefore eliminated from the full song pattern. Alien timing is also lost at crystallization. For example, a young male house finch (*Carpodacus mexicanus*) will learn a long canary song from his canary foster parent. House finch songs are usually 3 seconds long, but when in plastic song this male house finch can sing precise imitations of the canary model that are 10 seconds or more in duration. When that bird crystallizes these songs, many canary trills are abruptly lost, and that male house finch thereafter sings only a 3-s portion of the original canary song as an adult; these short songs no longer sound canarylike even though they are composed of canary syllables. Similarly, male swamp sparrows sing very long songs and have repertoires of over a dozen different syllables when in plastic song, but after crystallization many of the syllable types are lost and the song is abruptly shortened. These sparrows finish development with a species-typical repertoire: two or three 2-s-long songs, each composed of one repeated syllable type. These examples of development reveal that biological (gene-based) constraints operate during song development. These developmental processes often screen out alien syllables or alien temporal features from the adult song.

Not all passerine birds follow the developmental sequence just described. The alder and willow flycatchers (*Empidonax alorum* and *E. traillii*) are two very closely related species. They look so much alike that they are usually identified only by their respective songs, which are similar on a voiceprint but sound distinctively different. Very young alder flycatchers tutored with a willow flycatcher song show no evidence of learning it. Instead they develop a surprisingly normal alder flycatcher song, and they produce this adultlike alder flycatcher song at a very early age, just after fledging at about 12 days of age. There is apparently no long developmental sequence of subsong, plastic song, and finally full song. Furthermore, wild populations of alder and willow flycatchers exhibit essentially no geographic variation in their respective song patterns. All of this evidence indicates that these flycatchers do not learn their songs, that their songs are genetically determined, and that the song development program is highly constrained. Although flycatchers are passerines, they represent a relatively primitive family and are not true songbirds.

Summary. Over 9100 bird species are now living on Earth, and over half of these are songbirds. Clearly there are many differences among these species in how genes code for their call and song patterns. A general model of gene action is beginning to emerge from studies such as those mentioned above: (1) genes in some way code for the overall temporal pattern of complex vocalizations such as songs, and for species with highly constrained song development (for example, doves and flycatchers) the form and sequence of the constituent syllables are genetically determined as well; (2) other genes appear to code for perceptual constraints that guide song learning in species with a capacity for vocal learning; (3) developmental processes like song crystallization modify what has been learned and impose more limited species-typical constraints on the final full song product, and these processes are presumably gene-based too. [PAUL C. MUNDINGER]

Bibliography: H. R. Güttinger, The integration of learnt and genetically programmed behavior: A study of hierarchical organization in songs of canaries, greenfinches, and their hybrids, *Z. Tierpsychol.*, 49:285–303, 1979; D. E. Kroodsma and E. Miller (eds.), *Acoustic Communication in Birds*, Vol. 2, 1983; P. Marler and S. Peters, Selective vocal learning in a sparrow, *Science*, 198:519–521, 1977.

Brain

The fact that the human cerebral hemispheres control the opposite sides of the body was probably one of the earliest discoveries of anatomists and physicians. The idea that specific areas of the brain are responsible for particular functions is considered to have originated with the French physician Paul Broca, who in 1861 established the relationship between the absence of expressive speech and a lesion in an area of the left cerebral hemisphere. Broca stated that expressive speech was to be found in this area for all humans, but almost immediately after his pronouncement, cases appeared where speech was affected following lesions in the right hemisphere. Since individuals showing this pattern were almost invariably left-handed, the notion developed that left-handedness represented a mirror image of the cortical function to be found in the right-handed. This belief persisted for quite some time despite contrary evidence. It is now known that speech functions are located in the left hemisphere in approximately 95% of all individuals. Those persons having speech located either in the right hemisphere or in both hemispheres are more often left-handed, although the relationship is far from perfect. Probably the majority of left-handed have speech located in the left hemisphere, identical to the right-handed.

The contemporary study of specialization of hemisphere function has received the greatest impetus from the studies of the psychologist Roger Sperry. Sperry and his students carried out systematic studies of a small group of patients who had undergone a cerebral commissurotomy—a radical surgical procedure undertaken as a last-resort life-saving procedure. In a commissurotomy, the corpus callosum—the large tract of nerve fibers connecting the two cerebral hemispheres—is almost completely severed. This procedure, which has the effect of containing life-endangering epileptic seizures, leaves the two hemispheres of the brain with almost no intercommunication, each hemisphere functioning as a separate entity. Sperry, in studying these patients, established specific functions for each hemisphere. The results of his work stimulated many investigations into the nature and function of cerebral specialization, including cortical differences associated with handedness.

Classification of handedness. It is now well established that handedness is of genetic origin. A convenient classification of handedness is: (1) right-handed with no family history of left-handedness; (2) right-handed with a family history of left-handedness; (3) left-handed with no family history of left-handedness; (4) left-handed with a family history of left-handedness. A classification of this type is useful in the study of handedness and cerebral asymmetry, since differences are found between the left-handed who have a family history of left-handedness and the left-handed who occur as isolated individuals in entirely right-handed families.

Cerebral asymmetry in normal individuals. Two methods are often used by psychologists to study differences in cerebral specialization in normal individuals: dichotic listening in the study of audition, and visual half-field presentation for vision.

In monaural listening by the left ear, as illustrated in Fig. 1*a*, a sound (the syllable "ba") presented to the left ear is sent to the right hemisphere by way of contralateral nerve fibers (connecting the left ear to the right cerebral hemisphere) and to the left hemisphere by way of ipsilateral nerve fibers (connecting the left ear to the left cerebral hemisphere). The syllable "ba" is reported accurately.

In monaural listening by the right ear (Fig. 1*b*), a sound (the syllable "ga") is presented to the right ear and is sent to the left hemisphere by way of contralateral nerve fibers and to the right hemisphere by way of ipsilateral nerve fibers. The syllable "ga" is reported accurately.

In dichotic listening (Fig. 1*c*), an individual hears two sounds, the syllables "ga" and "ba," si-

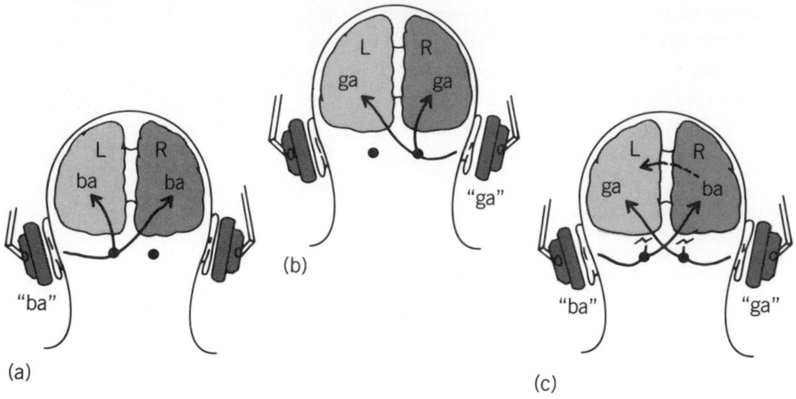

Fig. 1. Model (developed by D. Kimura) of dichotic listening in normal subjects. (*a*) Monaural listening by the left ear. (*b*) Monaural listening by the right ear. (c) Dichotic listening. (*After S. Springer and G. Deutsch, Left Brain, Right Brain, W. H. Freeman, 1981*)

multaneously through stereo earphones. Since there is functional evidence that the contralateral nerve fibers take precedence over the ipsilateral fibers, verbal material (the syllable "ga") presented in the right ear should be more easily detected than verbal material (the syllable "ba") presented in the left ear. In these studies, the subjects repeat "ga" more accurately than "ba." This procedure indicates that verbal processing is carried out primarily in the left hemisphere.

Visual half-field studies are possible because the anatomical arrangement of nerve fibers in the human visual system allows presentation of visual material initially to only one cerebral hemisphere. In humans, nerve fibers on the nasal half of the retina cross over to the opposite hemisphere. Those fibers on the temporal half of the retina pass directly back to the same-side hemisphere.

When subjects focus on a defined fixation point, information can then be shown to the right or left of that fixation point and be transmitted along the nerve paths leading to a given hemisphere. Although the technique has its limits since material can be shown for only 150 milliseconds before eye movement begins, it does allow the study of differential hemisphere functions in vision (Fig. 2).

Application of these techniques to the study of cerebral specialization has produced some interesting differences as related to handedness. Right-handed individuals characteristically recognize verbal material more easily when it is presented to the right ear in a dichotic listening task, and recognize verbal material more accurately and more quickly when it is presented in the right visual field leading directly to the left hemisphere. These results are entirely consistent with verbal processing being located in the left hemisphere. By contrast, some left-handed persons respond identically to the right-handed, showing left-hemisphere superiority, while other left-handed are equally accurate, regardless of ear or visual half-field presentation, showing no difference in hemispheric specialization. Those persons who were left-handed within families of entirely right-handed showed response patterns identical to the right-handed, while those persons with a family history of left-handedness had symmetrical patterns of response, being equally good with either ear or in either visual half-field.

Handedness and brain damage. The study of behavioral effects of cortical damage has produced evidence concerning differential effects related to handedness. A summary of the clinical experience follows.

About 24% of right-handed persons with left-hemisphere cortical lesions develop language disorders, as compared with 6.7% for right-handed with right-hemisphere lesions. The left-handed with left-hemisphere cortical lesions show language disorders in 22.4%, and with right-hemisphere lesions, 13.7%.

If such differences are examined by comparing the frequencies of performance deficits in relation to hemisphere lesions in the right- and left-handed, a much higher relationship is found in the right-handed between location of lesion and the behavioral deficit or problem. In the left-handed, disorders specific to one hemisphere occur much less frequently. H. Hecaen and J. Sauget assessed 50 types of behavioral performance deficits in relation to lesion site—right or left hemisphere—on right- and left-handed patients. For their right-handed patients, they reported 47 out of 50 between-hemisphere differences as significant at a probability level of .05 or less. The corresponding number of between-hemisphere differences for the left-handed patients was 4 out of 50. There is a strong relationship between site of location of cerebral injury and a corresponding performance deficit for the right-

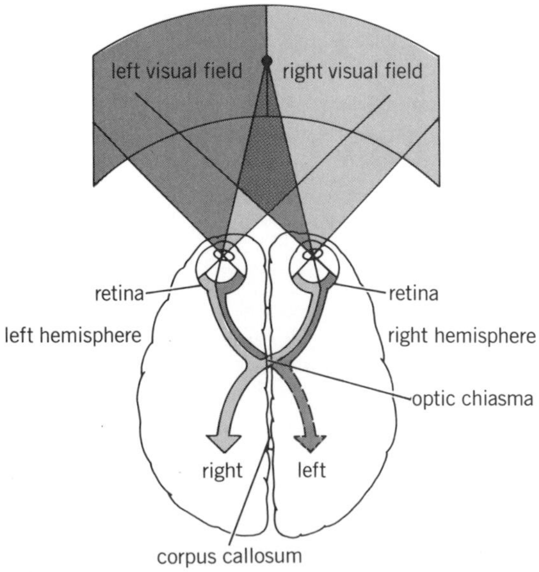

Fig. 2. Visual pathways to the hemispheres. When fixating a point, each eye sees both visual fields, but information about the right visual field is sent only to the left hemisphere and information about the left visual field only to the right hemisphere. (*After S. Springer and G. Deutsch, Left Brain, Right Brain, W. H. Freeman, 1981*)

handed. For many of the left-handed, this relationship appears to be at a chance level.

When these data are examined in relation to the handedness classification given earlier, the differences are even more noticeable. There are no differences in the verbal performance abilities of the familial left-handed who have suffered either right or left cortical lesions—the frequency of language disorders is virtually identical for right- and left-hemisphere lesions. Those patients who are the left-handed from families with no history of left-handedness show the same high relationship between lesion site and disorder found in the right-handed.

Recent developments. A systematic study of handedness in patients undergoing computerized axial tomography scanning for investigation of brain damage has revealed an interesting structural difference in head shape between the familial left-handed and the right-handed. In the right-handed, the skull shape tends to have a right frontal protrusion and a left occipital (rear) protrusion, resulting in an asymmetrically shaped skull. By contrast, the familial left-handed show a much more symmetrical head shape (Fig. 3).

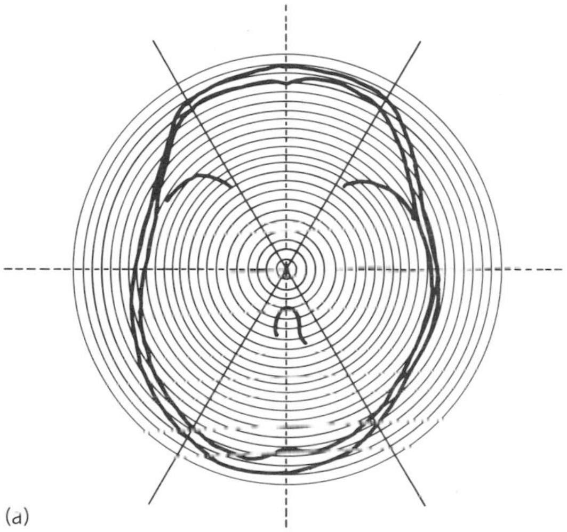

(a)

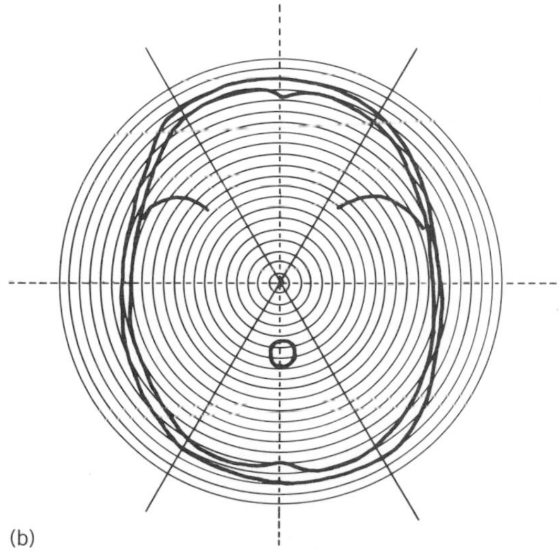

(b)

Fig. 3. Asymmetrical skull shape as shown in base views of skulls; obtained by computerized axial tomography scanning; circular graphs are superimposed. (*a*) Adult right-handed male. (*b*) Adult male who writes with his right but is otherwise left-handed. His sister is left-handed. (*After M. LeMay, Asymmetries of the skull and handedness: Phrenology revisited, J. Neurol. Sci., 32:243–253, 1977*)

Although the behavioral and anatomical differences between left- and right-handed persons are now well established, the relationship to other characteristics is not known, with one exception. For many years, much psychological research on handednesses was devoted to attempts to link left-handedness with other kinds of deficits such as stuttering, awkwardness, mental deficiency, emotional instability, and alcoholism. There is no evidence that handedness is related to any kind of deficiency, just as there is no evidence that handedness is related to any unusual talents or abilities.

For background information *see* BRAIN; HEARING (HUMAN); INTERHEMISPHERIC INTEGRATION; VISION in the McGraw-Hill Encyclopedia of Science and Technology.

[CURTIS HARDYCK]

Bibliography: C. D. Hardyck and L. F. Petrinovich, Left-handedness, *Psychol. Bull.*, 84:385–404, 1977; C. D. Hardyck, L. F. Petrinovich, and R. D. Goldman, Left-handedness and cognitive deficit, *Cortex*, 12:266–279, 1976; H. Hecaen and J. Sauget, Cerebral dominance in left-handed subjects, *Cortex*, 7:317–328, 1971; D. Kimura, Functional asymmetry of the brain in dichotic listenings, *Cortex*, 3:163–178, 1967; M. LeMay, Asymmetries of the skull and handedness: Phrenology revisited, *J. Neurol. Sci.*, 32:243–253, 1977.

Cactus

The cactus family (Cactaceae) consists of about 1600 species of perennial stem succulents native to the Western Hemisphere. Cacti occur naturally from southern Canada to Patagonia, the West Indies, and the Galápagos Archipelago; but one epiphytic genus, *Rhipsalis*, occurs in tropical America, Africa, and islands in the Indian Ocean. Because most cacti are killed by freezing, they are most common in hot, dry deserts and tropical and subtropical habitats, especially in Mexico (over 700 species). However, numerous species occur in temperate and montane localities, and a few, for example, caespitose species of *Tephrocactus*, occur in the Andes of Peru and Bolivia at elevations above 13,200 ft (4000 m).

Morphology. Cacti have sessile, showy solitary flowers (see illustration) possessing many free perianth parts that are borne on a short or long floral tube, numerous stamens, and an inferior, multicarpellate, uniloculate ovary with a single style and several to many stigma lobes. Placentation is parietal, and the ovary develops into a many-seeded fleshy berry, which is sometimes shed as a dry fruit

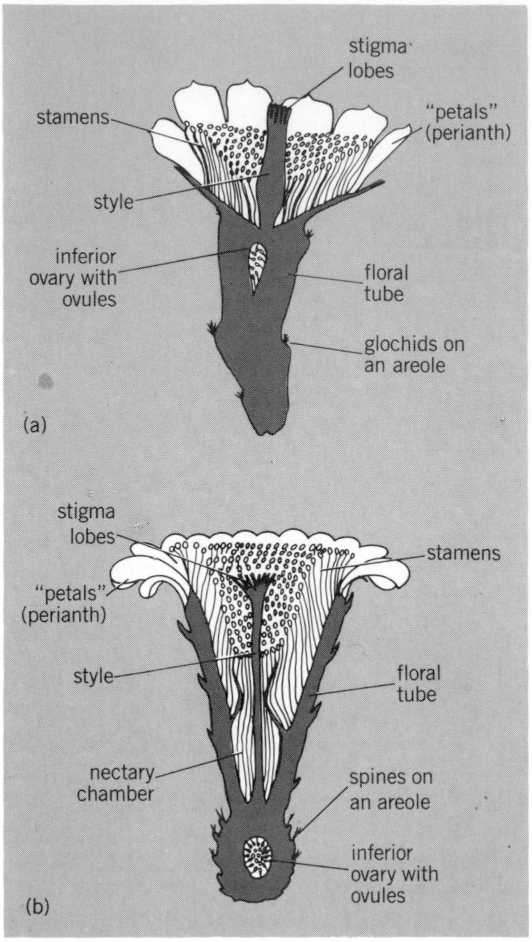

Cactus flowers. (*a*) Prickly pear (*Opuntia*, subfamily Opuntioideae). (*b*) Organ-pipe cactus (*Stenocereus thurberi*, subfamily Cactoideae).

or may be dehiscent by valves. In the Cactaceae, leaves are small, vestigial, or absent, and the succulent, water-storing stem has become the chief photosynthetic organ. Concomitantly, the axillary bud has been modified as a short shoot, called an areole, on which are borne stiff spines instead of foliar primordia. Areoles are produced strictly in helically alternate phyllotaxy, but these succulent stems are frequently modified so that the spiny areoles project above the stem on cylindrical tubercles, called podaria, or on elongate ribs, formed by the vertical fusion of podaria. In platyopuntias (*Opuntia* spp.), the flattened stems are called cladodes.

Certain species of *Pereskia* are the principal exceptions to the general description of Cactaceae. These small trees and shrubs have fairly large simple, entire leaves, poorly developed areoles with few spines, and relatively nonsucculent stems. Moreover, *P. sacharosa* has a superior, multicarpellate ovary with axile placentation and a low number of stamens. *Pereskia* is considered to be the group with the most primitive vegetative and reproductive features and therefore most similar to nonsucculent dicotyledons.

Classification. For many years Cactaceae was classified with other families having parietal placentation, such as begonias (Begoniaceae), loasas (Loasaceae), and passion flowers (Passifloraceae). Gradually phylogeneticists realized that parietal placentation had evolved independently several times, and cacti were ultimately reassigned to a new order, the centrosperms (Chenopodiales or Caryophyllales).

Centrosperms have a unique set of derived embryological and seed features, including a strongly curved embryo surrounding a diploid nutritive tissue, termed perisperm, which replaces triploid endosperm. In the last 25 years, placement of cacti in this order became certain when investigators discovered three additional shared derived characteristics: presence of betalains (15-carbon, nitrogenous pigments), which are absent in all other angiosperm orders; presence of an unusual form of plastids in phloem sieve tubes; and occurrence of an advanced type of pollen morphology and exine sculpturing. Cacti, and in particular the pereskias, seem to be most closely related to portulacas (Portulacaceae), the xeromorphic Didiereaceae, and certain pokeweeds (Phytolaccaceae) that have alternate phyllotaxy and normal secondary growth.

Subfamilies. Cacti are grouped into three subfamilies: Pereskioideae, Opuntioideae, and Cactoideae.

Pereskioideae. This subfamily includes species which have leaves that persist through a full growing season, the genus *Pereskia* (12–15 species) which possesses the primitive features of the family, and *Maihuenia* (2 species), comprising highly modified cushion plants of the Andean foothills of Argentina and Chile.

Opuntioideae. This subfamily includes about 250 species, all of which have two specialized features, a white bony aril covering the seed, and glochids, which are thin deciduous spines on the areoles. *Pereskiopsis* of North America and *Quiabentia* of South America have large leaves, a condition which shows the origin of opuntioids from pereskia-like ancestors. Other members of this subfamily have reduced or ephemeral leaves. Species of *Opuntia* are often very difficult to identify because populations are polymorphic and at the same time related species can interbreed successfully and thus create a mixed population of hybrids (hybrid swarms).

Cactoideae. This subfamily, formerly known as Cereoideae, is the largest, consisting of 1300 or more valid species, all of which lack prominent leaves and have no glochids or arils. Here one finds an amazing diversity of growth forms and habits, ranging from massive arborescent candelabra forms (*Pachycereus weberi*) and solitary columnars as much as 50 ft (15 m) tall (*Neobuxbaumia mezcalaensis*) to small trees and erect shrubs, barrel cacti, creeping forms, and small cylindrical and globular forms. The tiniest species is *Blossfeldia liliputana* of Argentina, which lives in rock crevices and forms a cluster of small shoots, each less than 0.8 in. (2 cm) across. The subfamily also contains about 15 genera of epiphytes, including such commonly cultivated forms as the Christmas cacti (*Schlumbergera truncata*), *Epiphyllum*, and *Rhipsalis*.

Over the years plant taxonomists have in general avoided the study of Cactoideae because the plants do not lend themselves to standard herbarium techniques and their succulent tissues require special preparation. Gross morphological features of flowers, fruits, and shoots have proved by themselves to be unreliable indicators of relationship because morphological convergence and parallelism have repeatedly occurred. To complicate matters, many species have marked phenotypic variation, and many of the extreme variants have at some time been given scientific names by amateur collectors without much critical study and based on insufficient materials.

Each tribe of Cactoideae has a clear geographic range. Three tribes (Cacteae, Echinocereeae, and Pachycereeae) are primarily Mexican but have a few species on the margins of the range; and four tribes (Browningeae, Cereeae, Notocacteae, and Trichocereeae) are primarily South American with some outliers in Central America and the West Indies. Tribe Leptocereeae, which includes the primitive features of the subfamily, occurs in northwestern South America, Central America, and the West Indies, and tribes Hylocereeae, which has many epiphytes, and Nyctocereeae occur in North America and the West Indies. From this information emerges a general phylogenetic and biogeograhic model for the subfamily in which Mexican tribes evolved independently and isolated from the southern tribes. Another tribe is Pachycereeae from Mexico.

One fascinating taxon of small Mexican cacti, which includes *Ariocarpus*, *Aztekium*, *Lophophora* (peyote), and *Obregonia*, includes forms from very dry areas with cryptic coloration and low form. These cacti have abundant alkaloids, the most famous of which is mescaline, a derivative of tyramine, which has been used as a potent hallucinogen in native religious ceremonies and which has become a controversial drug in western cultures.

Economic importance. The major economic importance of cacti is their use as garden and house plants. Cacti are cultivated widely throughout the world because of their beautiful and bizarre vegetative forms and showy flowers. Because collectors have removed large numbers of cacti from native habitats, many species are now uncommon or rare in the wild. Consequently, many countries have signed an international agreement to limit sharply the export and import of cacti, and now many cactus nurseries propagate species mainly from seeds or by cuttings. Species with large, sweet fruits, especially platyopuntias and columnar cacti, either are cultivated around homes and cultivated fields or are collected from the wild. Moreover, the young pads of platyopuntias and young shoots of chollas are used in native recipes. Straight columnar forms are frequently used as living fences in Latin America, especially in Mexico; and in desert communities, where wood is scarce, trunk wood of large columnars is used as support for primitive dwellings.

For background information *see* CACTUS; CARYOPHYLLALES; PLANT PHYLOGENY; PLANT TAXONOMY in the McGraw-Hill Encyclopedia of Science and Technology.

[ARTHUR C. GIBSON]

Bibliography: L. Benson, *The Cacti of the United States and Canada*, 1982; N. H. Boke, The cactus gynoecium: A new interpretation, *Amer. J. Bot.*, 51:598–610, 1964; A. C. Gibson and K. E. Horak, Systematic anatomy and phylogeny of Mexican columnar cacti; *Ann. Missouri Bot. Gard.*, 65:999–1057, 1978.

Cell (biology)

The Earth's first cells were probably micrometer-sized microbes that fed on organic compounds produced by electrochemical reactions on the primitive planet. It is unknown when these first heterotrophic, bacterialike prokaryotes appeared, for it is unlikely they would have been preserved. However, there exists a fossil record of microbes and the biosedimentary structures that they might have produced, called stromatolites, in rocks radiometrically dated at 3.5×10^9 years. In slightly metamorphosed sedimentary rocks of the early Archean Warrawoona Group in Western Australia, several types of filamentous microbial fossils are found preserved in finely laminated carbonaceous cherts (Fig. 1). These microfossils constitute the oldest direct evidence of the existence of life on Earth, and also indicate that by 3500 Ma (Ma = 10^6 years) ago, microbial life was diverse, complex, and possibly capable of a primitive photosynthetic metabolism.

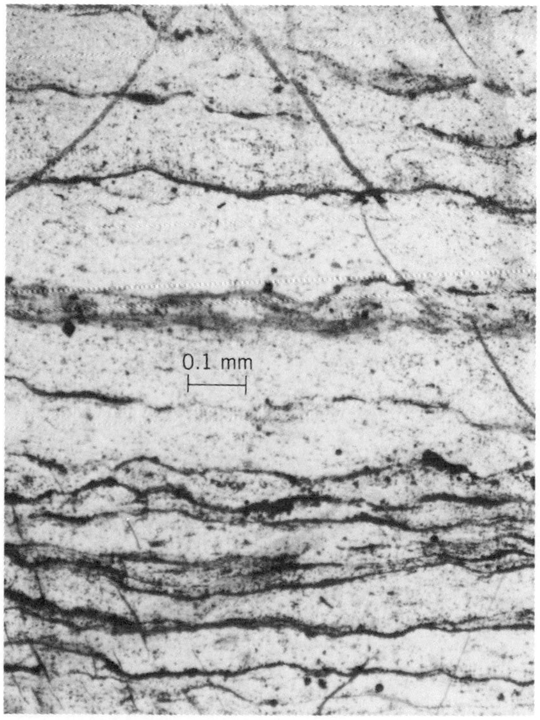

Fig. 1. Carbonaceous stromatolitelike lamination of microfossil-bearing Warrawoona chert.

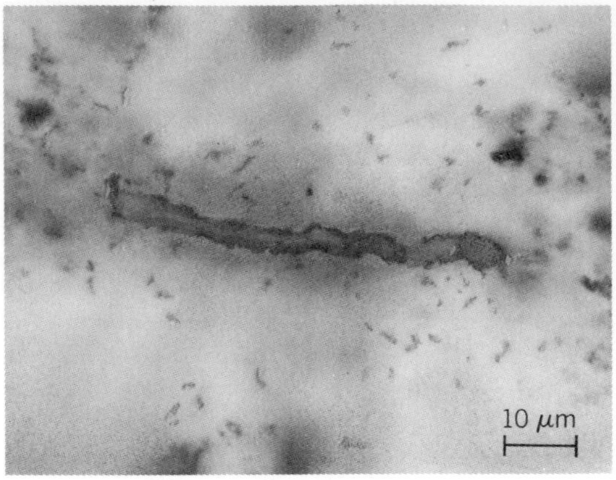

Fig. 3. Cylindrical, tubular microfossil found within a carbonaceous lamina.

Fossil-containing succession. The Warrawoona Group is a thick (upwards of 7 mi or 12 km), predominantly volcanic succession composed mainly of extrusive ultramafic and mafic volcanics with interstratified sedimentary units. Chert (cryptocrystalline quartz) and chert-barite units <160 ft (50 m) thick are common sedimentary rocks. Most of the sequence was deposited in shallow water. Although the group has undergone a complex geologic history since deposition, significant parts of the 20,000-mi^2 (60,000-km^2) outcrop area have undergone little thermal alteration. Tectonically, the region underwent uplift and faulting about 3400–2750 Ma ago, producing a variety of fractures and weathering surfaces.

The age of the Warrawoona Group has been established by a variety of isotopic-age dating techniques: Sm-Nd dates of 3556 ± 32 Ma have been obtained from volcanics found in the lower third of the group; U-Pb dates on zircon from lavas near the middle of the group have given a 3452 ± 16 Ma age; and model lead ages from galenas yield about 3400–3500 Ma. The age of the group is generally regarded to be about 3500 Ma.

Microfossils. Four types of filamentous microfossils have been detected in petrographic thin sections of finely laminated carbonaceous chert. The first type are <1-micrometer, threadlike filaments up to 160 μm long; they are the most abundant microfossil type. These filaments appear to be composed of organic material, are sinuous, they generally lack orientation with respect to lamination and bedding, and are more common in the thicker (10–500-μm), clear, organic-poor layers than in the thinner (5–10-μm) carbonaceous laminae. Another type of small filament is tubular, 0.8–1.1 μm in diameter, possibly with partitions (septae). These filaments are also found in the clear layers and lack orientation with respect to lamination and bedding. A third type consists of large, 4.0–6.0-μm-wide, septated filaments also found in the clear layers; the specimen illustrated in Fig. 2 has one of the <1-μm, type-1 filaments wrapped around it. Cylindrical, tubular microbial fossils 3.0–9.5 μm in diameter and up to several hundred micrometers long constitute the fourth type. These tubular microfossils are found within, parallel to, and partly defining thin, 5–10-μm carbonaceous layers (Fig. 3).

Because of their small size, relatively simple morphology, and lack of preserved reproductive structures, the affinities of these filamentous microbes remain unclear, although they are in all probability prokaryotes. Morphologically they resemble a wide variety of modern prokaryotes such as sheathed bacteria, oscillatoriacean cyanobacteria, pigmented and colorless gliding bacteria, and green sulfur bacteria. Somewhat more revealing as to a possible affinity are the 3.0–9.5-μm-diameter empty tubes (Fig. 3) oriented parallel to and within carbonaceous laminae. They might be some kind of proto-cyanobacterium. This kind of microbe-within-lamina organization is suggestive of stromatolitic activity, with the tube representing the empty, discarded sheath of a photoresponsive (photosynthetic) microbe. There is a term, $^{13}C_{PDB}$, which expresses the ratio of two stable isotopes of carbon, ^{12}C and ^{13}C. Photosynthetic and other autotrophic organisms preferentially utilize the lighter isotope of carbon. Isotope ratios are expressed in parts per thousand (‰) in terms of deviation from a standard, a fossil belemnite from the Cretaceous Pedee Formation (C_{PDB}), according to the following equation:

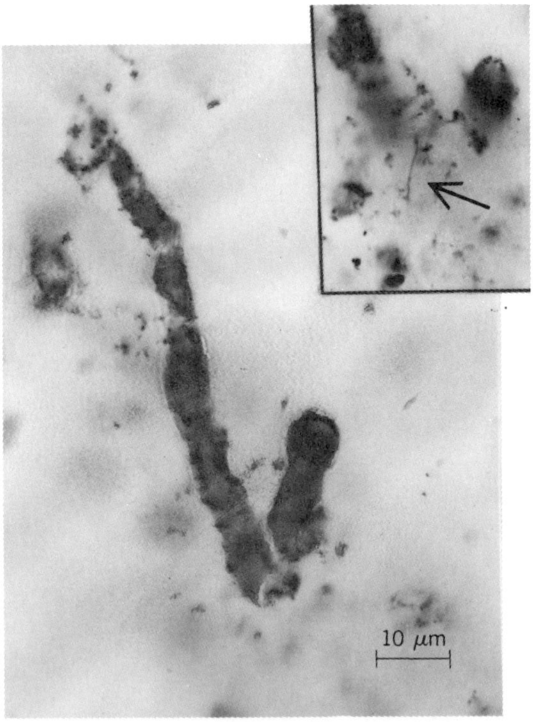

Fig. 2. Partitioned filament in Warrawoona chert; reconstructed from photomicrographs taken at different focal depths. Inset at upper right shows the small, <1-μm-diameter filament (arrow) wrapping around a larger filament, as mentioned in the text.

$$\delta^{13}C(‰) = \frac{(^{13}C/^{12}C \text{ sample} - {}^{13}C/^{12}C \text{ standard})}{(^{13}C/^{12}C \text{ standard})} \times 1000$$

Stable carbon isotope analyses of Warrawoona chert have yielded $\delta^{13}C_{PDB}$ values of −35.4‰, suggesting that some microbes in the Warrawoona biota may have been capable of autotrophic CO_2-fixation. Thus, a possible scenario has some of these microbes living at the sediment-fluid interface, responding to sunlight, autotrophically fixing CO_2 (not necessarily with the release of O_2), and contributing to the production of a laminar fabric (Fig. 1). Centimeter-sized, domal, laminated structures interpreted as stromatolites have been discovered at several other localities of the Warrawoona Group, supporting this scenario.

Micropseudofossils. Associated with the microbial fossils that are considered syngenetic with the Warrawoona chert are a variety of micropseudofossils—structures that superficially resemble microfossils but were produced by nonbiological processes. Spheroids 4.0–14 μm in diameter with carbonaceous outer boundaries superficially resemble coccoid microbes. But careful study of rock thin-sections reveals that they were produced within a fracture in the chert where mobilized, particulate organic matter collected around some sort of bubble, forming a spheroid. This clearly postdates deposition of the Warrawoona chert. Another postdepositional phenomenon are trails, several micrometers across, left behind as small pyrite grains moved through the chert producing filamentous micropseudofossils. Radiating clusters (up to 100 μm across) of 0.5–0.75-μm solid filaments are also probably pseudomicrofossils. The radiating habit of the filaments is somewhat similar to fabrics in radial fibrous chalcedony, a polymorph of quartz.

Significance. The Warrawoona contains a variety of microscopic structures covering the spectrum of those which are assuredly microfossils to those demonstrably micropseudofossils. Because the outcrops from which the Warrawoona microfossiliferous cherts were collected are faulted, are intruded by chert dikes, are weathered, contain fractured chert, and have been subjected to a few episodes of postlithification dissolution and silicification, questions have been raised as to whether the four types of microfossils detected in the thin sections are syngenetic with deposition of the Warrawoona rocks. Possibly they are younger microfossils, 2750-Ma-old microbial contaminants introduced along fractures developed parallel to bedding during uplift of the region; they might be microbial contaminants, such as iron bacteria, introduced during phases of fracturing and weathering in the late Archean or the Tertiary; or perhaps they are even more recent.

Research has demonstrated that the four types of microbial fossils are found within the first generation of chert, of Warrawoona age. None of the microfossils described here has been found in fractures or in cherts that formed in voids or during weathering. Hence, the interpretation most consistent with and constrained by the data is that these microbes are the oldest remains of cells yet discovered.

Implications. The ancient, diverse, morphologically sophisticated, metabolically "advanced" microfossils found in Warrawoona rocks, as well as the presence of stromatolites, mean that by 3500 Ma ago, prokaryotes were well established in shallow water environments. Either the origin of life appreciably predated the Warrawoona fossils or, once life appeared a few hundred million years before the Warrawoona, it evolved rapidly along prokaryotic lines.

For background information *see* CELL (BIOLOGY); CHERT; LIFE, ORIGIN OF; MICROPALEONTOLOGY; STROMATOLITE in the McGraw-Hill Encyclopedia of Science and Technology.

[S. M. AWRAMIK]

Bibliography: S. M. Awramik, J. W. Schopf, and M. R. Walter, Filamentous fossil bacteria from the Archean of Western Australia, *Precambrian Res.*, 20:357–374, 1983; D. I. Groves, J. S. R. Dunlop, and R. Buick, An early habitat of life, *Sci. Amer.*, 245:64–73, 1981; D. R. Lowe, Stromatolites 3,400-Myr old from the Archean of Western Australia, *Nature*, 284:441–443, 1980; J. W. Schopf (ed.), *Origin and Evolution of the Earth's Earliest Biosphere*, 1983.

Cell differentiation

The processes leading to cytodifferentiation, or cell differentiation, begin with a poorly identified process frequently termed determination. When determined, cells are restricted to a pathway of development which leads to a specific differentiated state. In spite of the universality of the process of determination, little of the molecular mechanisms for it is understood. It is assumed that the unidirectional nature of development requires the stable establishment of cell type and the orderly progression of cellular states leading to differentiation. In most instances the time of cellular determination, that is, the point at which cells become fixed in their developmental potential, occurs fairly early in development. In many cases this process requires a series of restrictive steps, each further limiting the developmental potential of the cell type.

A thoroughly documented description of cell differentiation has been obtained from studies of the development of the mammalian pancreas. Following a poorly understood early determinative step, the tissue primordium passes through an early differentiated stage in which the specific gene products, characteristic of the differentiated state, first appear at low levels during a period of rapid cell proliferation. Then, following the cessation of mitosis, the features of the differentiated pancreas cell appear. Similar sequences of cytodifferentiation have been demonstrated for a wide range of tissue types, such as lens fiber cells, muscle, and erythrocytes. In spite of these thorough molecular documentations of the sequence of events associated with cytodifferentiation, there is still little knowledge of the nature of the first step in the process during which a cell lineage becomes committed or determined for a specific developmental fate.

Germ cell lineage. One of the prime experimental systems for studying the events leading to cellular determination has been the establishment of the germ cell lineage. In some insects and vertebrates, primordial germ cells appear before or simultaneously with the determination of the basic germ layers of the embryo. In those instances where germ cells appear before other cell types, experimental studies have demonstrated that the appearance of germ cells is dependent upon cytoplasmic regions localized in the egg during oogenesis. In a number of holometabolous insect groups a cytologically detectable region of the egg has been associated with the appearance of primordial germ cells. In the fruit fly *Drosophila*, a critical proof that this region contains true germ cell determinants has been achieved by transplanting a small portion of this cytoplasm to ectopic locations in the embryo (see illustration). Subsequently, during cell formation, primordial germ cell-like spherical cells formed at this ectopic site. To demonstrate that these induced cells are bona fide germ cells, they were transplanted to a region of the embryo where they could reach the embryonic gonad. By the use of genetic labels to follow the fate of cells, it has been possible to show unequivocally that the induced cells were fully able to function as germ cells by producing functional gametes. Thus, these experiments demonstrated that at the posterior tip of such insect embryos a region of ooplasm is preformed which is sufficient in itself to induce the stable determination of germ cells.

Germ plasm. Further experiments have documented a number of additional facts concerning germ plasm. This cytoplasm forms during oogenesis and becomes fully functional prior to the completion of oogenesis. This is the sole demonstration of an ooplasmic localization responsible for a specific cellular determination. It is not surprising that true maternal effect mutations have been found which eliminate completely the presence of the distinctive posterior region. These are termed "grandchildless" because the first generation lacks germ cells and consequently no second generation is possible. In addition, function germ cells have been produced employing cytoplasm from different *Drosophila* species, indicating that the controlling molecules may have been conserved during evolution.

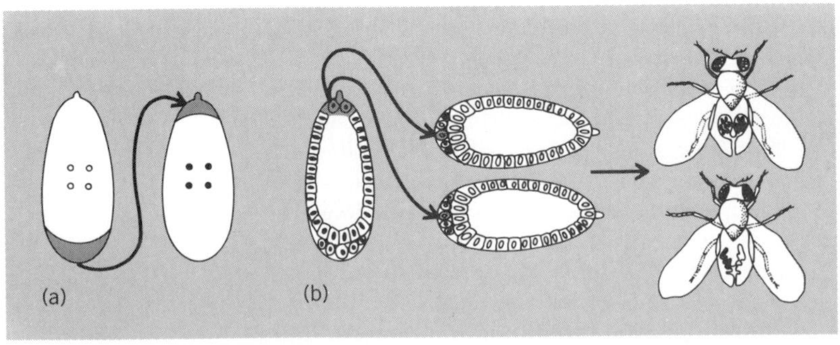

Schematic illustration of germ cell determination. *(a)* Posterior cytoplasm from the donor egg of *Drosophila* is transplanted to the future anterior tip or head region of a recipient embryo. *(b)* This embryo produces cells at the anterior tip identical in morphology to those formed at the posterior tip, and these induced cells are shown to be germ cell precursors by transplanting them to their normal location in a genetically marked host embryo where they produce germ cells after metamorphosis. The gonads of the flies on the right side of the figure have germ cells of two different genotypes, showing that the cells induced at the anterior tip are functional germ cells.

Molecular basis for determination. The molecular basis for germ cell determination is not known. In *Drosophila* the cytological organelles restricted to the germ plasm, termed polar granules, have been isolated in sufficient purity to identify a 95,000-dalton basic protein as a major constituent of the organelle. Cytochemical observations have indicated that ribonucleic acid (RNA) is associated with these organelles prior to germ cell formation and that after the appearance of polysomes with the organelles the RNA disappears. Furthermore, the production of germ cells has been known for 50 years to be sensitive to ultraviolet irradiation. Recently, it has been shown in *Smittia* that the ultraviolet effect can be reversed by irradiation with visible light. Such photoreversions have been demonstrated to be due to excision of pyrimidine dimers formed by the ultraviolet light. These experiments support the hypothesis that RNA is important in the functioning of germ plasm. It has been proposed that the polar granule is a repository for a maternally provided messenger RNA which is utilized for producing one or more proteins necessary for germ cell determination. A recently discovered grandchildless mutation eliminates these organelles. Molecular analysis of this mutation may enhance understanding of cellular determination in germ cell differentiation.

While there are many other examples of stable determinations during early development, little is known about the causes of determination. The basic test to discover whether cells have been determined is to challenge them with the possibility to differentiate as another cell type. For example, while most tissue types in a regenerating amphibian limb reform from the corresponding tissue types, muscle cells, for example, can contribute to both cartilage and muscle lineages. A critical test for the determined state requires not only that a cell be unable to transform into a new cell lineage when challenged, but that in spite of the challenge the cells remain fixed in their developmental fate. While there are many examples of stable determinations, a few examples will suffice.

Primordial germ cells are not regenerated when they fail to form. Similarly, they do not differentiate into other cell lineages when placed ectopically. Many types of experiments have demonstrated that epidermal structures of insect segments are determined for specific portions or compartments of the organism. However, the following critical experiment provides the clearest demonstration of a specific determination. After precursor cells for the first thoracic leg were transplanted to the region of the third thoracic leg at the inception of gastrulation, these cells became integrated into the third leg, but nevertheless formed tissue that was first leg in character. Thus, these cells retained their determined cellular fate in spite of becoming integrated into a different structure.

For background information *see* CELL DIFFERENTIATION, SENESCENCE, AND DEATH; DEOXYRIBONUCLEIC ACID (DNA); EMBRYONIC DIFFERENTIATION; MOLECULAR BIOLOGY; RIBONUCLEIC ACID (RNA).

[ANTHONY P. MAHOWALD]

Bibliography: K. Illmensee, A. P. Mahowald, and M. Loomis, The ontogeny of germ plasm during oogenesis in *Drosophila*, *Developmental Biology*, 49:40–65, 1976; A. P. Mahowald and R. E. Boswell, in A. McLaren and C. Wylie (eds.), *Current Problems in Germ Cells*, 1983; N. K. Wessells, *Tissue Interactions and Development*, 1977.

Cell walls (plant)

The increase in size of plants is due to the enlargement of undifferentiated cells arising from cell division in meristematic regions. The primary cells walls of these plants are a major factor in controlling the rate of cell enlargement and thus also the rate of growth of the plant. Such enlargement can be striking, resulting in some cases in as much as a 2000-times increase in length of a cell. Cessation of enlargement is characterized by the onset of deposition of a thicker, additional layer of cell wall called the secondary cell wall. An understanding of the structure of the primary cell wall, and particularly the interconnections between the polymers in this wall, is now recognized as essential for understanding the mechanism of cell enlargement in plants.

Recent research in this field has focused on several areas: (1) Studies of the structure of the various polymers which compose the primary cell wall have revealed that the structures are highly complex. (2) Studies on the nature of the interconnections between the polymers have resulted in the discovery of several potential sites of cross-linking, containing special bonds which could control the extensibility of the wall and which need to be broken and reformed in order for irreversible wall enlargement to occur. (3) Studies on the role of hormones in promoting cell enlargement have continued, but a satisfactory explanation remains elusive.

Wall enlargement. In most plant cells undergoing enlargement, growth occurs uniformly throughout the entire lateral surface of the wall (surface growth). Exceptions in higher plants are root hair and pollen tube cells which elongate only by an increase in area at the tip (tip growth). During cell enlargement the thickness of the wall remains relatively constant, and so growth is coupled to the concomitant synthesis of new cell wall. In the general case of surface growth, the new wall polymers must be deposited throughout the surface of the existing wall.

The driving force for wall extension is the turgor pressure of the cell—a force which keeps the plasma membrane firmly pressed against the cell wall. It is generally believed that the initial events in wall growth involve modifications in wall structure which, in response to the force created by turgor, cause the polymers of the wall to shift in position so that the wall increases in surface area. Subsequently, new cell wall polymers are laid down to maintain constant wall thickness, and new solutes are accumulated within the cell to maintain constant turgor. Although, for simplicity, these events are presented as sequential in time, the overall process is a continuous one, and all of these events are occurring simultaneously within the growing cell.

Structure. Since primary cell walls are the type which exist in growing cells, this discussion will concentrate on what is known about the structure of these walls. In very general terms, the primary cell wall consists of a fibrillar network of cellulose fibrils enmeshed in an amorphous matrix of polysaccharides and proteins. At least some cellulose is present in almost all primary walls, and primary walls from different types of plants differ chiefly in the composition of the matrix components. Not too many years ago, it was proposed that only a few distinct polymer types existed in primary walls, and that primary walls from all dicotyledonous plants (dicots) are quite similar to each other while those from monocotyledonous plants (monocots) differ from dicot walls but are similar to each other in structure. It is now known that this model, while correct in a general way, is an oversimplified view, and that the polymers are quite complex and heterogeneous. In fact, the extent to which minor differences in structure exist among even closely related cell types is not known, nor to what extent such differences might imply differences in mechanism of cell wall expansion. Nevertheless, it is possible now to describe the general features of the major wall polymers of primary cell walls. However, it should be recognized that the descriptions provided below do not reflect all the subtle differences in structure reported by various investigators.

Cellulose. Cellulose is a polymer of β-1,4-glucan, that is, a homopolymer, the units of which consist entirely of glucose residues linked in such a fashion as to create extended chains. These linear glucan molecules are held together by hydrogen bonding to form fibrils which are highly insoluble and partially crystalline. In the primary cell wall, the fibrils have an average diameter of about 3.5 nanometers and are called elementary fibrils. At the beginning of cell elongation, the fibrils are randomly oriented, but the process of extension apparently causes a passive reorientation of these fibrils so that they become transverse to the direction of wall extensions.

Matrix polysaccharides. These polymers are usually divided into two classes which are defined operationally by their presence in fractions obtained by sucessive chemical extraction of isolated cell walls. One fraction, the pectic polysaccharides, is obtained by extracting walls with boiling water, chelating agents, or dilute acid. Subsequent extraction of the insoluble residue with alkali solubilizes the second fraction, called the hemicellulosic polysaccharides. These are crude definitions based upon long-standing extraction techniques which certainly result in some degradation of the wall polymers during extraction.

In primary cell walls, the major types of hemicellulosic polymers are the xyloglucans, the glucuronoarabinoxylans, and the mixed-linked β-glucans. In primary walls derived from cells of dicotyledonous plants, the major polymer is the xyloglucan. This polymer resembles cellulose in that it possesses a backbone structure of extended chains of β-1,4-glucan; it differs in that many of the glucose residues have a pentose sugar, xylose, attached as branches by β-glycosidic linkages to carbon-6 of glucose. Some the xylose residues are further substituted with galactose and fucose. These branches on the glucan backbone prevent the association of xyloglucan chains into fibrils, but there is evidence that hydrogen bonding of xyloglucan chains to cellulose fibrils can occur, as will be discussed later.

The primary walls derived from cells of monocotyledonous plants, in particular the cereals which have been most extensively studied, also contain some xyloglucan. However, in these walls the quantitatively more dominant related hemicellulosic polymer is a glucuronoarabinoxylan. Here, the backbone consists of a β-1,4-xylan with side chains attached which can vary somewhat from species to species. The most common side chains are a single arabinose residue attached to carbon-3 of backbone xylose residues. Other common side chains are glucuronic acid or 4-*O*-methyl glucuronic acid residues attached as single units to carbon-2 of backbone xylose residues.

Primary walls derived from cells of monocotyledonous plants contain an additional prominent hemicellulosic polymer apparently not found at all in dicotyledonous plants. This polymer is a mixed-linked β-glucan, a linear polymer of glucose containing a mixture of 3-linked and 4-linked glucosyl residues in an average ratio of 1:2.

The primary cell walls of dicot, but not monocot, cells are characterized by a relatively high content of pectic polysaccharides. Recent work indicates that substantial complexity exists in these polymers. The negatively charged rhamnogalacturonans are considered to be the backbone chains of pectic polymers. The basic structure consists of a chain of α-1,4-linked galacturonic acid residues in which 2-linked rhamnose residues are interspersed. Some of the rhamnose residues are substituted with arabinose and galactose residues. A very complex and highly branched variation of this polymer also exists, and it contains the rarely observed sugars apiose, 2-*O*-methylxylose, and 2-*O*-methylfucose. Some pure α-1,4-linked galacturonan free of rhamnose can also be extracted from these walls. In addition, neutral polymers are found in this fraction, such as the highly branched arabans containing arabinose residues in α-1,3 and α-1,5-linkages as well as β-1,4-galactans.

Cell wall protein. In addition to certain enzymes which exist in cell walls and apparently play no structural role, the primary cell walls derived from cells of dicots (and to lesser extent from monocots) contain a unique hydroxyproline-rich protein possessing, in addition to other amino acids, a common repeating sequence of serine followed by four hydroxyproline residues. Most of the hydroxyproline residues are substituted with three or four arabinose units, and many of the serine residues are substituted with a single galactose residue. This protein has long been thought to play a role in cell wall growth, and for this reason it is called extensin.

Interconnections between wall polymers. The existence of several interconnections of polymers is reasonably well accepted by now, as described below. The polymers xyloglucan (found predominantly in dicots) and glucuronoarabinoxylan (found predominantly in monocots), while incapable of self-association into fibrils, can associate via strong hydrogen bondings to the surface of cellulosic fibrils of the wall, forming a coating around these fibrils, and perhaps also create an interconnecting network of fibrils.

Primary cell walls are known to contain Ca^{2+}. Model studies with isolated pectic polysaccharides which contain negatively charged galacuturonic acid residues indicate that Ca^{2+} interacts with these residues to promote stable structures. Calcium has long been known to confer rigidity on cells walls, and it also plays a role in stabilizing interchain interactions between negatively charged wall polymers. Some variable amount of methyl esterificaiton of the galacturonic acid residues exists in these walls, and it is expected that this would alter the degree of calcium-induced stabilization.

The primary cell walls of monocots (and, to a lesser extent, of dicots) contain phenolic compounds, primarily ferulic and *p*-coumaric acid. Recent work indicates that these residues are esterified to arabinose or galactose residues of acidic and neutral pectic polymers (in dicots), or to glucuronoarabinoxylan (in monocots). It is suggested that these polymers can be cross-linked by a peroxidase-mediated oxidative coupling, forming diferulic acid cross-bridges. In support of this theory are the observations that walls do contain peroxidases, and that diferulic acid has been isolated from cell walls. The ester links to the polymers are alkali-labile, and the ability of alkali to solubilize many wall polymers could be explained by the breakage of these potentially crucial alkali-labile cross-links.

Recent work indicates that when a procedure which hydrolyzes all of the cell wall polysaccharides was employed with isolated primary cell walls derived from cultured plant cells, a network of hydroxyproline-rich protein (extensin) was left behind as an insoluble network which retained the original shape of the cell wall. Other recent work provided evidence that extensin molecules are cross-linked to each other via isodityrosine residues, which could be formed by a peroxidase-mediated oxidative reaction between tryosine residues on adjacent chains of extensin.

There is evidence that the neutral pectic polysaccharides, araban and galactan, are covalently attached to the acid pectic polysaccharide rhamnogalacturonan. It is not yet clear whether there are covalent associations between pectic and hemicellulosic polymers.

Mechanisms of wall loosening. Understanding of the mechanism of wall loosening is crucial for an understanding of the mechanism of cell enlargement. A further important fact is that the plant hormones indoleacetic acid, ethylene, and gibberellic acid are all known to stimulate cell expansion. The effectiveness of these different hormones varies with cell and tissue type. In elongating stem tissues, indoleacetic acid and gibberellic acid stimulate the longitudinal extension of cells, whereas ethylene induces a radial expansion of such stem tissues. In most tissues where these effects are observed, one can also observe a hormone-induced increase in the extensibility of the wall as measured by mechanical tests on isolated cell walls. The most extensive studies have involved indoleacetic acid–induced elongation; in the case of this hormone, it is clear that the effect of indoleacetic acid is rapid (effects are observed within 10 min) and is accompanied by an increase in proton secretion from the cell which results in a lowering of a cell wall pH. Because the extensibility of walls can also be enhanced in some cases by a similar lowering of external pH, it has been proposed that wall loosening is somehow regulated by a change in cell wall pH. One proposal involved a possible effect of low pH on disrupting the hydrogen bonding between xyloglucan or glucuronoarabinoxylan and cellulose; this is now not considered a plausible mechanism since it has been shown experimentally that changes in pH similar to those observed in living tissue have no effect on the association between these polymers.

Since maximal wall loosening occurs around pH 4.7, mechanisms involving a nonenzymatic acid-induced cleavage of covalent bonds seems unlikely as well. Current thinking focuses on the idea that this pH would be optimal for enzyme-catalyzed modification in wall structure which could lead to wall loosening. One suggestion involves a hypothetical transglycosylase type of reaction which would break some glycosidic link between polysaccharides and reform the link between similar polysaccharides in a different location. This could explain why tension (turgor pressure) is required for wall extension. Under tension, when bonds are broken, the polymers can slide to new positions before reformation of new bonds with new partners, whereas in the absence of tension the bonds would reform in the same position.

An alternate proposal considers the transesterification reaction involving diferulic acid ester cross-links as a likely candidate. Esterases are known to exist in cell walls, but it is not yet known if these have the capacity for transesterification involving phenolic ester linkages. Nonenzymatic acyl migration of fatty aids on lipids is known to occur, so it is not entirely implausible that such a transesterification could alternatively occur by a nonenzymatic reaction. Another possible transfer reaction which should now be considered is the possibility of transpeptidation between different portions of cross-linked extensin chains, particularly in view of the recent results cited earlier which indicate that cross-linked extensin could be the backbone network for the entire wall. Another indication that isodityrosine cross-linking of extensin or diferulic acid cross-linking of polymers may be of some importance is that the hormones indoleacetic acid and gibberellic acid have long been known to influence the levels of secreted peroxidase—a key candidate for catalyzing the synthesis of these cross-links.

It is worth noting that, as a general rule, dicot primary cell walls are rich in extensin but low in ferulic and *p*-coumaric acid, and monocot primary cell walls show the opposite distribution. If these cross-links are related to the degree of extensibility of the walls, it may be that different mechanisms of wall loosening will occur in these distinct types of cell walls. In this regard, involvement in cell extension of the noncellulosic β-glucan, found only in monocot walls, has also been proposed. Degradation of this glucan is closely correlated with cell extension, being optimal at the time of the maximal rate of elongation. Furthermore, indoleacetic acid stimulates the activity of an enzyme which can degrade this glucan, and an inhibitor of this enzyme, nojirimycin, also inhibits indoleacetic acid–induced elongation. However, if growth is inhibited by reduction in turgor, glucan degradation still occurs, but the walls do not increase in extensibility. At this time, the possible role of this glucan in mediating extensibility is not clear. If it does play a role, it must be restricted to monocot cell walls.

Dicot walls show a different type of glucan degradation in response to indoleacetic acid treatment—a solubilization of portions of xyloglucan which is stimulated by indoleacetic acid. Indoleacetic acid is shown to stimulate production of certain cellulases which are now known to effectively hydrolyze xyloglucan; however, as for the monocot β-glucan, the role of xyloglucan degradation in effecting wall loosening still remains unclear.

Thus, no single mechanism of wall loosening has been confirmed to be involved in the process of hormone-stimulated cell enlargement. Future research must attempt to prove or disprove the possible mechanisms described above or to provide an alternate explanation for the process.

For background information *see* CELL WALLS (PLANT); PLANT CELL; PLANT GROWTH; PLANT HORMONES in the McGraw-Hill Encyclopedia of Science and Technology.

[DEBORAH P. DELMER]

Bibliography: R. E. Cleland, Wall extensibility: Hormones and wall extension, in W. Tanner and F. A. Loewus (eds.), *Encyclopedia of Plant Physiology*, new series, vol. 13B, pp. 255–273, 1981; A. Darvill et al., The primary cell walls of flowering plants, in N. E. Talbert (ed.), *The Biochemistry of Plants*, vol. 1, pp. 92-162, 1980; S. C. Fry, Isodityrosine, a new cross-linking amino acid from plant cell wall glycoprotein, *Biochem. J.*, 204:449–455, 1982; K. Kato, The ultrastructure of the plant cell wall: Biochemical viewpoint, in W. Tanner and F. A. Loewus (eds.), *Encyclopedia of Plant Physiology*, new series, vol. 13B, pp. 29–46, 1981.

Chaotic behavior

Many systems governed by completely determinate differential equations can exhibit apparently chaotic behavior. An obvious example is turbulence in fluid flow, and this behavior is associated with the existence of nonlinear terms in the equations of motion which not only makes these equations analytically intractable but also makes their solution depend critically on minute changes in the initial, or boundary, conditions. Although the behavior of nonlinear systems has been studied since the early nineteenth century, there has been a recent revival of interest stimulated by the discovery that very similar behavior is exhibited by simple systems of recurrence relations such as Eq. (1), which deter-

$$x_{n+1} = 4\lambda x_n(1 - x_n) \qquad (1)$$

mines the $n+1$-th value of a variable x in terms of its nth value. For $\lambda < 3/4$ this series converges to a unique value, but as λ is increased from $3/4$ toward 1 the solution first exhibits period doubling with the values of x for odd n being unequal to the values for even n. This is followed by a sequence of further period doublings until at $\lambda_c = 0.892\ldots$ the behavior becomes chaotic. For $\lambda_c<\lambda<1$ there are narrow windows in the chaos in which periods other than 2^n appear. The discovery by M. J. Feigenbaum in 1981 that there are definite regularities in this behavior which are identical for a broad class of recurrence relations has raised hopes that the investigation of the recurrence relations associated with differential equations can illuminate the nature of their solutions. The connection is, in principle, simple, though in detail very complex. Consider Eq. (2). If $\omega t_n = 2n\pi + \phi$, the values of $\dot{q}$ and q

$$\ddot{q} + f(q) = V \cos \omega t \qquad (2)$$

at t_n determine the values at t_{n+1} so that a dual recurrence relation given by Eq. (3) is satisfied. The

$$q_{n+1} = g(q_n, \dot{q}_n), \ \dot{q}_{n+1} = h(q_n, \dot{q}_n) \qquad (3)$$

behavior of this sequence will exhibit all the features associated with period multiplication and chaos even if it does not give a complete description of the motion.

As a result there has been extensive mathematical investigation, mainly using digital and analog computers, of the properties of both nonlinear differential equations and recurrence relations. This has been accompanied by an experimental search for chaotic effects in a variety of physical and electronic systems. Since many of the equations of physics are only approximately linear and the active devices used in electronics are highly nonlinear, these effects are expected to be widespread. An interesting example is due to the nonlinearity of the forces binding ions in a crystal. When the piezoelectric, mechanical resonance of a Rochelle salt crystal is driven hard, it exhibits both chaotic behavior and period multiplication.

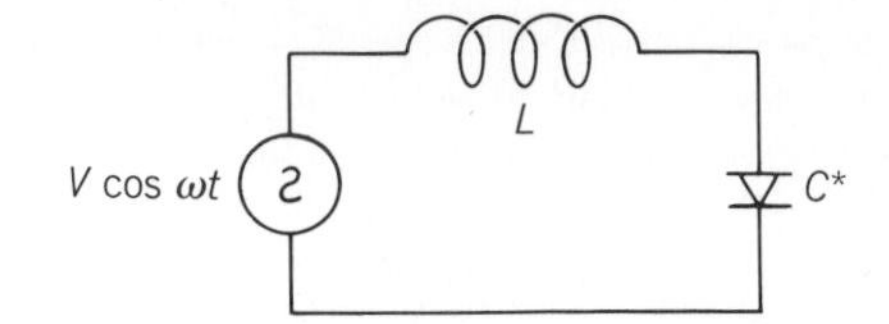

Fig. 1. Series-tuned circuit with nonlinear diode capacitor.

Varactor diode circuit. Perhaps the clearest examples, however, have been found in the behavior of the simple series-resonant tuned circuit shown in Fig. 1, in which the nonlinear capacitance C^* is a varactor diode. The charge q on the capacitor (with an appropriate change of units) obeys an equation of the same form as Eq. (2) and, if the diode is reverse-biased, this can be written as Eq. (4). A tra-

$$\ddot{q} + \omega_0^2 q + \gamma q^2 = V \cos \omega t \qquad (4)$$

ditional approach to this equation would be to begin with a small signal approximation and ignore γq^2 to obtain a first-order linear solution given by Eq. (5).

$$q_1 = q_{10} \cos \omega t = (\omega_0^2 - \omega^2)^{-1} V \cos \omega t \qquad (5)$$

This solution would then be inserted as q^2 and the second-order correction q_2 found as the solution of linear equation (6), which gives a term correspond-

$$\ddot{q}_2 + \omega_0^2 q_2 = -\tfrac{1}{2}\gamma q_{10}^2(1 + \cos 2\omega t) \qquad (6)$$

ing to a direct current and a term corresponding to an alternating current of frequency 2ω. Suppose, however, that an extra term is retained in q^2, giving Eq. (7). Then the equation satisfied by q_2 is Eq. (8).

$$q^2 \sim q_1^2 + 2q_1q_2 \qquad (7)$$

$$\ddot{q}_2 + \omega_0^2 q_2 + 2\gamma q_{10} q_2 \cos \omega t = -\tfrac{1}{2}\gamma q_{10}^2(1 + \cos 2\omega t) \qquad (8)$$

Without the term on the right-hand side this is Mathieu's equation which, for finite γq_{10} and ω near $2\omega_0$, has a solution of frequency $\omega/2$ which grows exponentially and, in the absence of any resistive loss, finally dominates the behavior. This is the behavior of a degenerate parametric oscillator. The question of what happens for large V remains unanswered. A mechanism for period doubling has been found, but apparently this does not lead to more complex effects as long as the diode remains reverse-biased. But in the experiments investigating the behavior of the circuit, the diode was not biased and was driven hard into forward conduction. The circuit is now described by a more complicated equation but still, essentially, one of second order. As V was increased, the period went through a sequence of doublings which could be followed up to 2^5, and thereafter the response exhibited chaotic behavior in which narrow windows showing other periods, for example period 5, appeared. The most remarkable feature of experimental results, however, was that if λ in Eq. (1) was identified with V and x with either the diode voltage or the diode current, the behavior reproduced in great detail that of recurrence relation (1). This is somewhat surprising since the governing equation including forward conduction is very complex and there is some evidence that minority carriers stored during conduction play a role in the process, as Schottky diodes with no charge storage do not yet yield these effects although they have a very similar reverse-biased voltage-charge relation.

Synthesized nonlinear circuits. It is also possible to synthesize a nonlinear voltage-charge (V-q) relation using fast field-effect transistor switches and a discriminator. Two possible examples are shown in Fig. 2. If the element described by Fig. 2*a* is

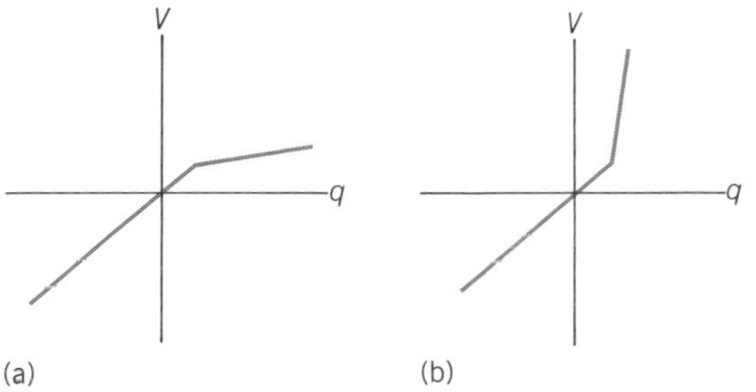

Fig. 2. Synthesized nonlinear voltage-charge (V-q) relations. The behavior of series-tuned circuits with elements described by *a* and *b* is discussed in the text.

inserted in the series-tuned circuit, then as the drive is increased, a sequence of period doublings leads to chaos. However, as the drive is further increased very broad windows appear (separated by chaotic regions) in which coherent responses of periods 2, 3, 4, 5, and up to at least 15 arise in sequence. The element described by Fig. 2*b* leads to quite different behavior. As the drive V is increased, a very complex, high-period, possibly chaotic, motion begins as soon as q reaches the knee. A further increase in V leads to simpler and lower-period motions. A similar effect may be observed in the motion of a ship in a rough sea when it is loosely moored to a jetty. It appears, therefore, that although it may be possible to relate the behavior of nonlinear systems to recurrence relations, the connection will not always be as direct as in the remarkable experiments on the varactor diode circuit.

More complex systems. So far, only systems governed by ordinary differential equations have been considered, and even there dissipation, which has a significant effect, has been ignored. Many of the most interesting effects, for example, turbulence, convective instabilities, and atomic and electronic motions in solids, involve nonlinear, partial-differential, wave or diffusion equations, and here much more complicated relations must be expected. For example, the relations between spatial and temporal chaos have been investigated for a system obeying the sine-Gordon wave equation, which has been widely studied in connection with solitons.

Although since the late 1970s there has been considerable progress in understanding of nonlinear systems, and chaotic behavior and speculation has been rife that such diverse phenomena as turbulence, low-frequency, or l/f, noise, melting, atmospheric instability, and even population dynamics may soon yield their secrets, the subject is still in its infancy. In view of the very difficult problems associated with the numerical solution of realistic nonlinear differential equations, theoretical advances will, to a great extent, depend on clear-cut experimental results.

For background information *see* DIFFERENTIAL EQUATION; FLUID FLOW; PARAMETRIC AMPLIFIER; PIEZOELECTRICITY; SOLITON; VARACTOR in the McGraw-Hill Encyclopedia of Science and Technology. [F. NEVILLE H. ROBINSON]

Bibliography: A. R. Bishop et al., Coherent spatial structure versus time chaos in a perturbed sine-Gordon system, *Phys. Rev. Lett.*, 50:1095–1098, 1983; J. P. Eckmann, Roads to turbulence in dissipative dynamical systems, *Rev. Mod. Phys.*, 53:643–654, 1981; D. R. Hofstader, Strange attractors: Mathematical patterns delicately poised between order and chaos, *Sci. Amer.*, 245(5):22–43, November 1981; E. Ott, Strange attractors and chaotic notions of dynamical systems, *Rev. Mod. Phys.*, 53:655–671, 1981; J. Testa, J. Pérez, and C. Jeffries, Evidence for universal chaotic behavior of a driven nonlinear oscillator, *Phys. Rev. Lett.*, 48:714–717, 1982.

Chemostat

The chemostat is a convenient device for the continuous cultivation of bacterial populations in a constant, competitive environment and so can be used to study natural selection. A systematic experimental study of natural selection should lead to a better understanding of the principles of biological evolution.

Basic principles. The chemostat provides a constant competitive environment by limiting the concentration of one nutrient in the medium and then slowly adding this medium to a growing culture of bacteria and siphoning off used medium plus bacteria at the same rate. Initially the bacteria added to a chemostat grow more rapidly than they are lost from the siphon, and so they use up the nutrients in the medium. The lowering concentration of the limited nutrient causes the bacteria to grow more slowly until a steady state is reached in which the rate of growth equals the rate of removal. The bacteria compete for this limiting nutrient. The concentration of the limiting nutrient in the fresh medium determines the density of bacteria growing in the chemostat at steady state. The growth rate of the bacteria is determined by the rate at which the me-

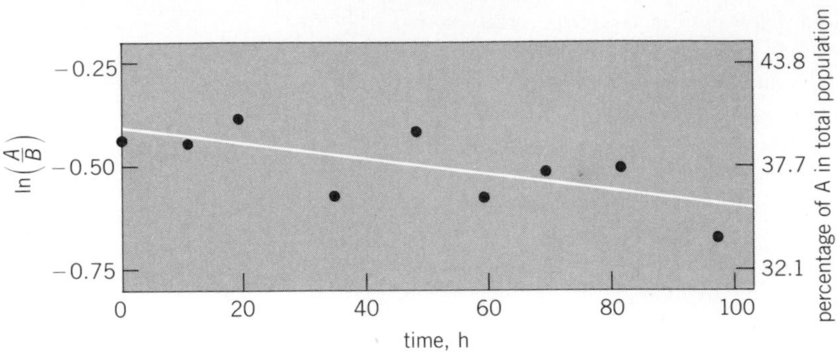

Selection against strain A in an energy-limited chemostat. The generation time is 2 h.

dium is pumped into the chemostat. At steady state the concentration of the various molecules in the medium remains constant, and the bacteria stay in the same physiological state for extended periods of time.

Experimental results. Because of this extended constancy, small selective differences between strains can be measured. The selection rate s is estimated by using linear regression from the equation

$$\ln \frac{X_1(t)}{X_2(t)} = \ln \frac{X_1(0)}{X_2(0)} + st$$

where $X_1(0)$ and $X_2(0)$ represent the number of the two strains at time $t = 0$, and $X_1(t)$ and $X_2(t)$ represent these numbers at time t (conveniently measured in hours). The illustration shows an example where the selection rate is 0.0021 ± 0.0007/h against strain A. This selection is significantly different from zero, and when the experiment is repeated there is again significant selection against A.

In chemostats the growth rate and the density of a population are independent parameters. This independence gives some interesting experimental results. Two pairs of chemostats were run, all at the same density, but one pair had a growth rate twice as fast as the other. In the chemostats with the faster growth rate, twice as many cells were produced. After 500 h the populations were removed from the chemostats and allowed to compete against each other. All the populations were about equally fit whether they competed at the faster growth rate or at the slower growth rate, even though so much evolution had taken place that the strain of bacteria used to initiate these experiments was strongly selected against when it competed against the evolved strains. This experiment implies that evolution is constant per unit time rather than per generation.

Artificial vs natural selection. Natural selection should be distinguished from artificial selection. In artificial selection a population with a specific character is obtained by selecting for that character. For example, in selecting for an increase in the number of bristles on the fruit fly, *Drosophila melanogaster*, a scientist counts the number of bristles on a certain number of flies and selects those with the greatest number of bristles to breed and produce the next generation. Then the selection is repeated, and after some generations the flies have acquired more bristles. This is artificial selection. Conversely, the scientist may be able to construct an environment in which an increased number of bristles helps the flies live longer and produce more offspring so that after a few generations the flies have an increased number of bristles. Then the scientist is studying natural selection, which happens in, and because of, the environment, even if the environment is not "natural." The important component of natural selection as distinguished from artificial selection is the interaction of the organism with the environment.

Environmental variables. With the chemostat, experiments can be repeated by changing only a single environmental variable to ascertain how these changes affect natural selection. Even small changes in the environment can affect the selection rate. Strains of *Escherichia coli* which are constitutive for the lactose operon have a large selective advantage over those strains which are normally regulated when they compete in chemostats where lactose is the limiting nutrient and sole source of energy. Constitutive strains synthesize the lactose enzymes irrespective of presence or absence of lactose. The normally regulated strains synthesize the enzymes only when lactose is present. Isopropylthiogalactoside (IPTG) causes the lactose enzymes to be synthesized in the normally regulated cells just as lactose does, but IPTG cannot be metabolized. As the table shows, increasing the amounts of IPTG

Selection rates for lactose constitutive strains in lactose-limited chemostats as a function of IPTG concentration; average generation time 2.19 ± 0.10 h

IPTG concentration	Selection/h
0	0.172 ± 0.013
$6 \times 10^{-7}\,M$	0.086 ± 0.004
$2 \times 10^{-6}\,M$	0.027 ± 0.002
$6 \times 10^{-6}\,M$	0.008 ± 0.003
$6 \times 10^{-5}\,M$	−0.002 ± 0.003

added to the chemostats (in concentrations of less than one to a few parts per million) causes a decreasing selective advantage for the constitutive strain until at $6 \times 10^{-5}\,M$ IPTG the selection becomes zero. Thus addition of a nonmetabolizable chemical at low concentrations can profoundly affect natural selection. In chemostats the concentration of lactose is too low for the normally regulated bacteria to maintain proper induction of the lactose enzymes. The IPTG maintains this induction. In other environments, as when the sugar maltose is the limiting nutrient, the lactose constitutive cells are selected against.

Genetic differences. In the same environment different mutations of the same gene can confer different degrees of fitness if they have different physiological effects upon the cell. The *trpE* gene in *E. coli* codes for an enzyme which is the first one in the pathway for making tryptophan. In a chemostat where glucose is the limiting nutrient and excess tryptophan is provided, a missense mutation and a polar nonsense mutation of the *trpE* gene are equally fit, and both are selected for over the normal gene. However, when indole, an intermediate in tryptophan biosynthesis, is provided instead of tryptophan, the two mutations are no longer equally fit. The missense mutation is still selected over the normal gene; but the normal gene is selected over the polar nonsense mutation. The missense mutation only affects the *trpE* gene, while the polar nonsense mutation also affects the other tryptophan genes, reducing the amount of these enzymes in the bacterium. Presumably, because of the reduced amounts of these enzymes, not enough indole can be converted into tryptophan, causing the mutant cells to be partially starved for tryptophan and to grow more slowly.

Thus natural selection can be studied experimentally. Its properties can be discovered, and will depend upon the interaction between certain factors of the environment and the physiological effects of the genetic differences between the strains. Genetic differences that do not produce functional differences will be selectively equivalent. Phophoglucose isomerase is a very important enzyme when *E. coli* is growing on glucose. In a glucose-limited chemostat, there is very strong selection against a strain carrying a missense mutation of *pgi*, the gene for this enzyme. Yet naturally occurring genetic variants (electromorphs) of this gene are selectively equivalent and seem to be biochemically equivalent.

Transposable elements. Various predictions can be tested with chemostats. Transposable elements are specialized segments of DNA that can increase their numbers within a cell by autonomously duplicating and inserting a new element into a new site in the cellular DNA. It has been proposed that transposable elements are selfish or parasitic DNA. Selfish DNA has two properties: it uses cells as the environment in which it behaves like an individualized entity by evolving separately from the cell, and it has no beneficial phenotypic effects upon the host cells.

Transposable elements have the individualization required to be selfish DNA. Whether or not they also give a phenotypic advantage to the bacterium has been tested twice, with Tn*10* and with Tn*5*, both in *E. coli*. In both cases the strain carrying the transposable element was selected over the isogenic strain without the element in glucose-limited chemostats. The mechanism of selection is different in the two cases, and in neither case is it simple.

It has been shown that Tn*10* confers the selective advantage by transposition of a part of Tn*10*, the IS*10*, to a certain site in the bacterial chromosome which gives the cell an advantage. This change is not achievable by simple mutation. The selection is characterized by a long lag of up to 140 generations (300 h), during which time the strains are selectively equivalent, followed by rapid selection for the strain carrying Tn*10*. The selection appears to be frequency-dependent in that at least 1 out of every 50,000 cells must carry Tn*10* before the selection for the Tn*10* strain is seen. However, it also depends upon the numbers of Tn*10*-bearing cells. When the Tn*10*-bearing cells are too rare, transposition to the advantageous site does not happen before the other strain evolves another advantageous adaptation and eliminates the Tn*10*-bearing strain.

It has also been shown that Tn*5* confers an immediate selective advantage which seems to last for only about 25 generations (50 h). Transposition does not cause the selective advantage; rather the proteins produced by the transposable element help the cell to adapt physiologically more quickly to the new environment of the chemostat.

Thus in two experiments transposable elements had a beneficial phenotypic effect. Therefore it is unlikely that transposable elements are selfish or parasitic, but have evolved to help the cell adapt to changing environments.

For background information *see* CHEMOSTAT; GENE; MUTATION; OPERON in the McGraw-Hill Encyclopedia of Science and Technology.

[DANIEL DYKHUIZEN]

Bibliography: S. W. Biel and D. L. Hartl, Evolution of transposons: Natural selection for Tn*5* in *Escherichia coli* K12, *Genetics*, 103:581–592, 1983; L. Chao et al., Transposable elements as mutator genes in evolution, *Nature*, 303:633–635, 1983; D. E. Dykhuizen and D. L. Hartl, Selection in chemostats, *Microbiol. Rev.*, 47:150–168, 1983.

Chromosome

Chromosomes are structures in the cell nucleus which contain genetic information coded in the molecules of deoxyribonucleic acid (DNA). In addition to their role as carriers of heredity information, some chromosomes have other cellular functions. Recent research has focused on the structure and function of lampbrush and *B* chromosomes.

LAMPBRUSH CHROMOSOMES

The largest known chromosomes, larger even than the giant polytene chromosomes of the fruit fly *Drosophila*, are found in developing oocytes of various frogs and salamanders. They have a rather diffuse or fuzzy appearance when viewed at low magnification with a microscope, and they were long ago dubbed lampbrush chromosomes because they resembled the brushes used to clean lamp chimneys. The fuzziness is due to hundreds of paired loops extending laterally from the main axis of each chromosome (Fig. 1). The lateral loops are the sites of intense ribonucleic acid (RNA) synthesis, or transcription. Most recent work on lampbrush chromosomes has focused on the nature and fate of the RNA transcribed on the loops.

Although lampbrush chromosomes are best known from studies on frogs and salamanders, they actually occur in the oocytes of many other animals, both vertebrate and invertebrate, including birds, reptiles, sharks, insects, crustaceans, and mollusks. Spermatocytes of many animals pass through a diffuse stage that has been termed lampbrush, but it is not clear that the chromosomes are in the same

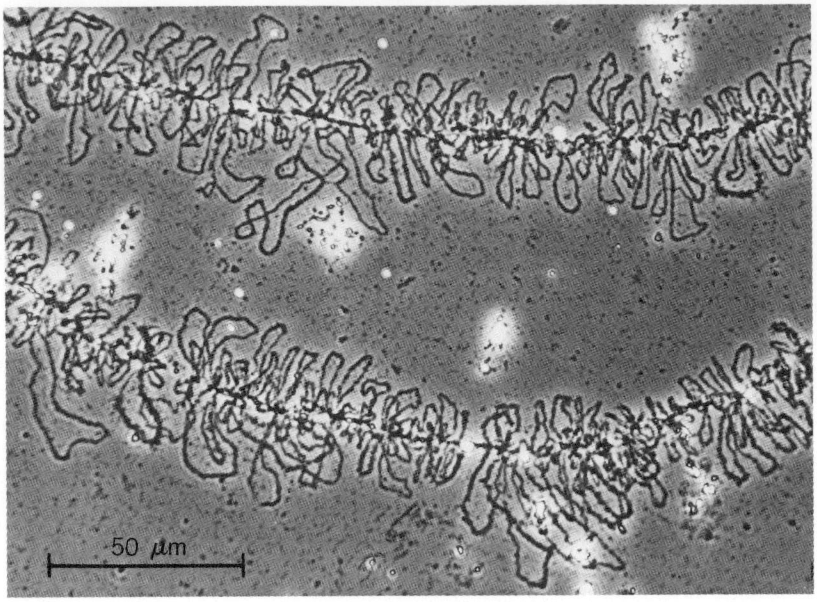

Fig. 1. Two homologous lampbrush chromosome segments from an oocyte of the newt *Notophthalmus* viewed by phase contrast microscopy. Lateral projections in the form of loops extend from the main axis of each chromosome.

active state seen in oocytes. However, a few pairs of typical lampbrush loops do occur on the Y chromosome in spermatocytes of *Drosophila* and related flies. Finally, well-developed, although small, loops are found on the chromosomes of the alga *Acetabularia*, proving that the lampbrush condition is not limited to animal cells.

Structure. Observations made on salamander oocytes have clarified the major structural features of lampbrush chromosomes. Before these features are outlined, a few words on oocytes will help place the chromosomes in context. An oocyte is a single cell that gives rise to the female gamete, or egg. During its major growth phase it increases enormously in volume. For instance, a mature frog or salamander oocyte may be 0.04 in. or more (one to several millimeters) in diameter, depending on the species. Also, the yolk of a hen's egg is one gigantic cell filled with nutrients for the embryo. During the entire growth period of the oocyte, its nucleus, or germinal vesicle as it is often called, remains in the prophase of the first meiotic division. Accordingly the lampbrush chromosomes within the germinal vesicle are dual structures, consisting of paired homologous chromosomes. These homologs are, of course, corresponding members of the two chromosome sets derived from the parents of the individual under consideration. They are held together at one or more points called chiasmata. Viewed through the light microscope, each homolog consists of a series of axial granules, the chromomeres, from which extend one or more pairs of lateral loops (Fig. 2*a*). A typical loop displays an asymmetrical structure, being very thin at one insertion into the chromomere and progressively thicker around the loop to the other insertion. Alternatively, there may be two or more such thin-to-thick regions in the same loop.

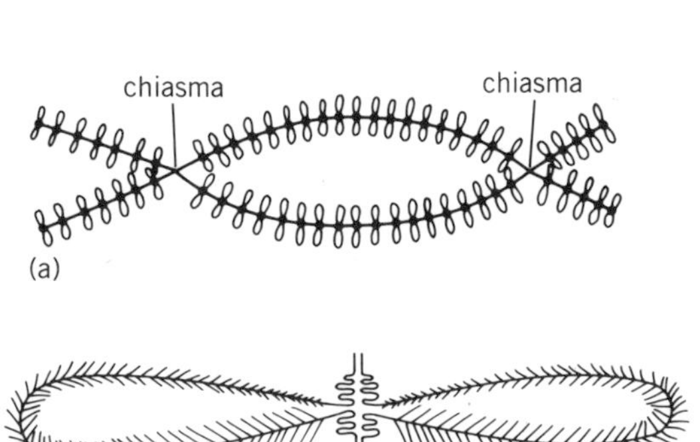

Fig. 2. Structure of lampbrush chromosomes. (*a*) Two homologous lampbrush chromosomes held together at their chiasmata. It is believed that genetic exchange occurs at or near the chiasmata. The major distinguishing features of lampbrush chromosomes are the paired lateral loops arising from axial granules called chromomeres. (*b*) Chromomere with an associated pair of lateral loops. Each loop has in its axis a very thin DNA fiber which is coated with numerous ribonucleoprotein filaments. These filaments consist of partially completed RNA transcripts associated with protein.

For many years the loops were a source of confusion from both a structural and functional standpoint. Staining experiments demonstrated that they consisted largely of RNA and protein, with no detectable DNA. Furthermore, the RNA was known to be synthesized in the loops themselves because of the very rapid incorporation of RNA precursors. Despite the evidence for the importance of RNA (and protein), studies have demonstrated that the structural integrity of the loops depends on DNA. Researchers treated lampbrush chromosomes with the enzyme DNase and found that the loops were fragmented extensively. In comparable experiments with RNase and proteases, the loops lost most of their mass, but remained as continuous strands. From these and other experiments it was concluded that the loops contain a DNA fiber that runs from one end of the chromosome to the other. The loops are paired because there are two identical sister chromatids in each homolog (Fig. 2*b*).

RNA synthesis. Current understanding of lampbrush chromosome loops stems from electron microscope observations, showing that the unusual morphology of loops, particularly their thin-to-thick gradients of RNA and protein, is directly related to RNA synthesis. It was found that each loop consists of an extraordinarily thin DNA axis from which arise numerous lateral filaments (Fig. 3). The filaments, which are graded in length along the thin-to-thick regions, contain RNA and protein. It was suggested that the thin-to-thick regions are, in fact, RNA transcription units; that is, the numerous RNA polymerase molecules attach to the DNA axis at the thin regions and move along to the thick regions, each synthesizing an RNA copy of the DNA as it proceeds. The short filaments contain recently initiated RNA transcripts, whereas the longer filaments contain nearly completed transcripts ready to leave the chromosome.

Aspects of this model have been tested by the

method of nucleic acid hybridization. In this procedure, radioactive DNA or RNA molecules representing specific genes are synthesized biochemically for use as gene probes. Lampbrush chromosomes attached to a glass microscope slide are then bathed in a solution containing the radioactive molecules. If the RNA on one or more loops contains sequences complementary to the probe, the probe will form a molecular hybrid with the RNA; the hybrid can then be detected by coating the slide with a thin film of photographic emulsion. Quite a few specific RNA transcripts have been identified in this way, and the detailed patterns of hybridization confirm the hypothesis that each thin-to-thick region corresponds to a functional transcription unit.

The most unusual feature of the lampbrush chromosome transcription unit is its extraordinary length. It appears that transcription does not terminate at the end of each gene, but instead proceeds long distances along the DNA after the gene sequence has been transcribed. The biological significance of this feature of transcription is presently obscure.

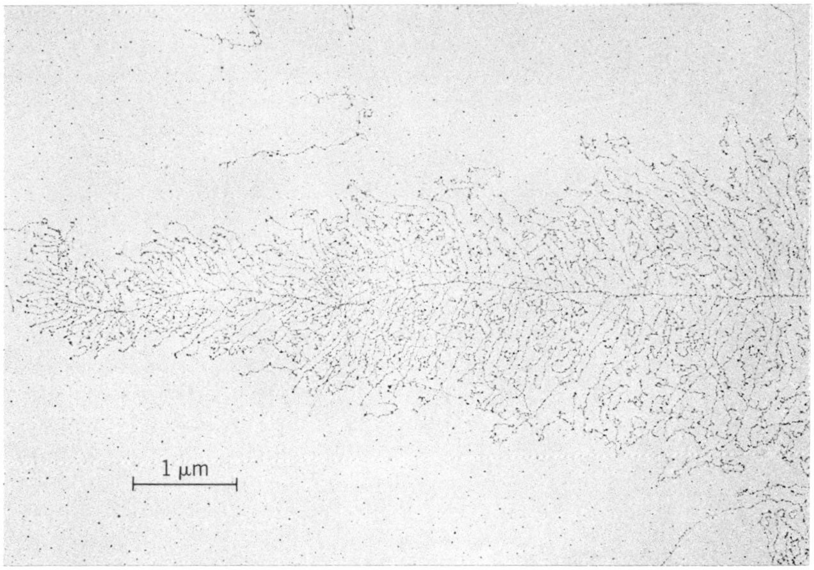

Fig. 3. Electron micrograph showing a small portion of a lampbrush chromosome loop from the newt *Notophthalmus*. Note the central DNA axis with its projecting ribonucleoprotein filaments. The filaments increase in length from one end of the transcription unit to the other. (*From O. L. Miller, Jr., B. R. Beatty, and B. A. Hamkalo, Nuclear structure and function during amphibian oogenesis, in J. D. Biggers and A. W. Schuetz, eds., Oogenesis, copyright © 1972 by University Park Press*)

Function. The intense RNA transcription of lampbrush chromosomes has led to much speculation about their functional significance. Not only are the transcription units unusually long, but the number of transcription units active at any one time is very large. It has been estimated that at least 5000 pairs of loops occur on the chromosomes of the newt *Triturus*, and since many of these have multiple transcription units, the total number of active transcription units must be still higher. Thus lampbrush chromosomes pose two possibly unrelated questions about RNA synthesis: Why are the transcription units so large, and why are there so many transcription units?

Almost all discussions of the functional significance of lampbrush chromosomes have been influenced by the fact that they have been studied primarily in oocytes of tailed amphibians. There has been a strong presumption that the lampbrush chromosome state has something to do with development of the egg into an embryo, specifically that lampbrush chromosomes supply the egg with a large number of preformed messenger RNA molecules to handle the complex events of embryogenesis. That lampbrush chromosomes synthesize essential RNA molecules for the eggs in which they occur is undeniable, but it is clear that they are not a universal feature in all species. For instance, many animals have small non-yolky oocytes, and in these a lampbrush chromosome stage is difficult to demonstrate or may be absent altogether. In other species, large yolky oocytes may not develop lampbrush chromosomes if they are associated with nutritive or nurse cells. This is true, for instance, of *Drosophila* oocytes. Here it is quite possible that the nurse cells take over the essential functions of the lampbrush chromosomes, since it is known that they supply RNA to the egg cytoplasm through large intracellular bridges. These examples demonstrate that the lampbrush condition is not an absolute requirement for growth of an oocyte and subsequent embryonic development. The fact that lampbrush chromosomes occur primarily in large yolky oocytes that lack accessory nutritive cells may be a clue to their functional significance: their intense synthetic activity is perhaps more an adaptation to the general RNA needs of a gigantic cell than to specific involvement in developmental events.

[JOSEPH G. GALL]

B CHROMOSOMES

The presence of supernumerary *B* chromosomes is one of the forms of numerical chromosome variation commonly found among species of higher organisms. The other kinds of change in number are those due to polyploidy and aneuploidy: in aneuploids the extra chromosomes are duplicates of individual members of the set, rather than whole sets, as in polyploidy. The name *B* chromosomes was first used by the American geneticist L. F. Randolph in 1928 in order to distinguish them from extra members of the basic complement (as in aneuploidy), which he called the *A* chromosomes. A classic example of aneuploidy is Down's syndrome (mongolism) in humans, due to an extra copy of chromosome 21. In this case, as in most instances of aneuploidy, the consequences of the unbalanced number are disastrous to the individual.

B chromosomes are extra chromosomes, in excess of the normal diploid or polyploid complement, but they are not duplicates of any of the members of the basic set and their presence has little, if any, recognizable effect upon the phenotype. They are found in variable numbers among some of the individuals of a population, while others are without them. In other words, they constitute a chromosome polymorphism. The *B* chromosomes are additional unique members of the genome which are nonessential for the normal processes of growth and development. It is not known how they originate or what their role is within the genetic system. One school of thought argues for some adaptive significance, and another says that they are a "selfish" or "parasitic" form of DNA.

Although *B* chromosomes (or simply *B*s) are something of an enigma, their importance is undeniable because of their widespread occurrence. They have turned up in almost every major group of higher plants and animals that have been investigated. In total, they are now known in more than 1000 species of flowering plants, including 10 gymnosperms, and 260 species of animals, most widely among the Insecta, and including 10 species of amphibians and 19 species of mammals, including humans.

Structure and genetic organization. In somatic cells, for example from rye (*Secale cereale*; Fig. 4*a*), *B* chromosomes can often be recognized by their smaller size and by the difference in morphology. The distinction between *A*'s and *B*'s is much clearer, however, at meiosis, when homologous partners pair together. *B* chromosomes reveal their structural uniqueness at this stage because they never show any homology (that is, never pair) with any members of the basic set. When a single *B* is present, it will remain unpaired, as a univalent (Fig. 4*b*). When two or more are present in the same individual, they are all seen to be structurally identical and to synapse among themselves. In many animals, and to a lesser extent in plants, they are out of phase with the *A* chromosomes in their cycle of coiling and replication, and their chromatin takes on a distinctive heterochromatic form in the prophase I stage of meiosis. They can often be distinguished from the *A* chromosomes, as well, by their specific patterns of Giemsa banding at mitosis. (Giemsa banding is a specialized form of chromosome staining which shows patterns of DNA organization within the chromosomes. Heterochromatic regions appear as darkly stained bands interspersed with lighter euchromatin.) *B* chromosomes appear to be devoid of major genes, and in species such as maize (*Zea mays*), which has been intensively studied by genetic analysis, they are not represented in the chromosome linkage map. Despite this notable lack of genes, they do not appear to differ from the *A* chromosomes in their DNA composition. The only certain knowledge that exists about their genetic organization is that they contain some elements (possibly genes) which control aspects of their own heredity, and some elements which affect pairing and recombination among the *A* chromosomes in some species; these latter effects are principally known in maize.

Inheritance. *B* chromosomes are less stable than the *A*'s and show an irregular and nonmendelian pattern of inheritance. At mitosis they are characteristically constant in some species, while in others they are subject to defects in movement during nuclear division, to nondisjunction, and consequently, to somatic mosaicism. When nondisjunction takes place at mitosis, it often leads to a situ-

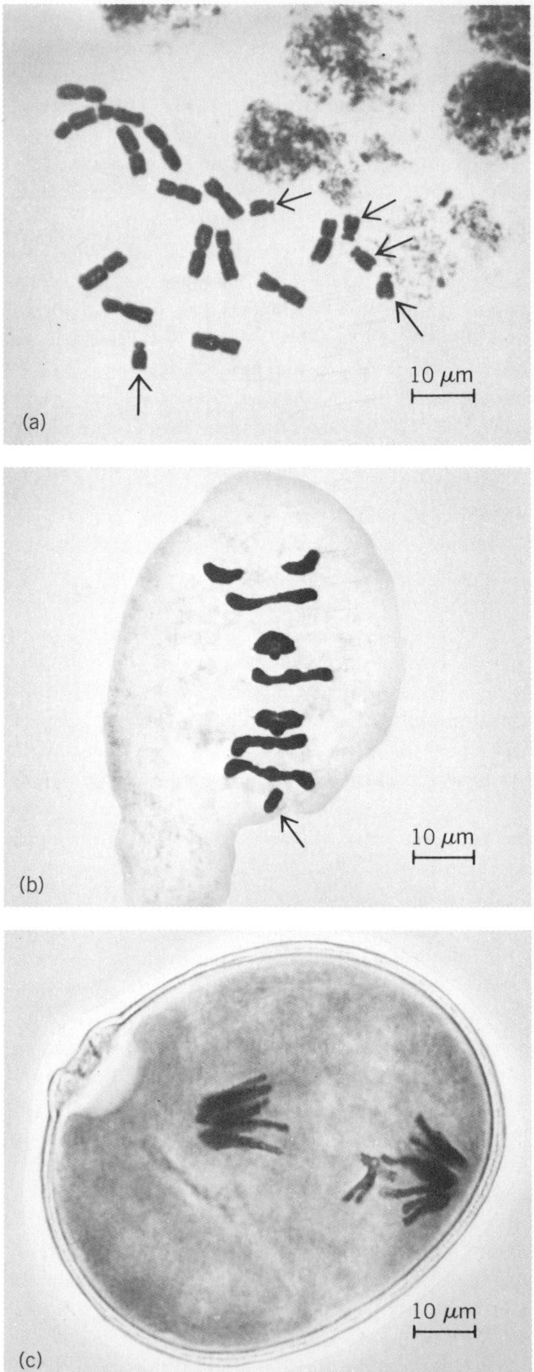

Fig. 4. *B* chromosome of rye (*Secale cereale*). (*a*) Metaphase of c-mitosis in a root meristem cell from a plant with five *B* chromosomes (*arrows*). (*b*) Metaphase I of meiosis in a 1*B* plant with an unpaired univalent (*arrows*). (*c*) First pollen grain mitosis showing a *B* chromosome undergoing directed nondisjunction. The centromere has divided, but the two chromatids have remained attached together in two sensitive regions adjacent to the centromere and are passing undivided into the generative nucleus, and hence, the germ line.

ation where one of the daughter cells has a deficiency of *B*'s, and the other a surplus of *B*'s compared with the original number. A cell with one *B*, for example, can have two daughter chromosomes passing into one cell and none to the other. This frequently happens in the Orthoptera (grasshoppers and locusts) during multiplication of the primary spermatocytes, and the cells with the higher number of *B*'s then appear to be at an advantage in contributing to the germ line in the male. The instability thus serves as an accumulation mechanism.

At meiosis there is often a lack of pairing among the *B*'s and a certain degree of elimination of univalents through failure to be included in the telophase nuclei. Rarely, as in the grasshopper *Myrmeleotettix maculatus* and the flowering plant *Lilium callosum*, there is meiotic drive in female meiosis, and unpaired *B*'s are preferentially distributed to the egg cell. (Meiotic drive is a nonrandom assortment of chromosomes leading to their preferential movement to one pole of the cell at anaphase I of meiosis).

In plants accumulation mainly takes place after meiosis by a drive mechanism based on directed nondisjunction in the male gametophyte, that is, at the first pollen grain mitosis (Fig. 4*c*). The situation in rye is exceptional in that the boosting of numbers occurs on both the male and female side. In maize there is nondisjunction at the second pollen grain division, and the gain in *B*'s among the progeny of crosses occurs due to preferential fertilization of the egg cell by the sperm containing the doubled-up number of *B*'s.

Experimental crosses between pairs of individuals with low numbers of *B*'s invariably show an increase in their number in the progeny as a result of the various accumulation processes. At the population level the gain in numbers eventually levels off when an equilibrium is reached between the forces of gain and loss.

Phenotypic effects. *B* chromosomes carry no known major genes and cause no obvious qualitative variation in the phenotype—with rare exceptions. Their effects are quantitative and undetectable in most species. High numbers of *B*'s seem disadvantageous to vigor and particularly to fertility. In plants the most serious effect is reduction in seed set.

At the level of the nuclear phenotype, the addition of the extra chromosomes, and consequent increase in DNA quantity, has some measurable effects. Cell size and division time are increased, and there are disturbances to RNA and protein metabolism.

One of the most interesting and significant activities of the *B*'s is the influence which they have upon the *A* chromosomes at meiosis. They bring about changes in the frequency and distribution of chiasmata and thereby impose some regulation over the genetic system of variation. Some workers consider this to be one reason for their maintenance within natural populations.

Natural populations. Extensive studies have been made on the *B* chromosome polymorphism in natural populations. In some species distinct clines and obvious patterns of distribution have been found, and some of these correlate with variations in the climate or soil-type factors. Generally speaking, the *B*'s are most abundant under conditions which are most favorable for the species concerned. It has not been possible as yet, however, to demonstrate any clear-cut selective advantage to their presence, except under certain extreme conditions of experimentation. This leaves the question about the reason for their existence unanswered.

One possibility is that they are selfish chromosomes which perpetuate and spread themselves in populations by the capability which they have to replicate faster than the *A* chromosomes.

A computer simulation model has been made of the population dynamics of the *B* chromosome system in rye. The model was based on certain parameters which determine the transmission rate of the *B*'s from one generation to the next, such as meiotic pairing and elimination rates, nondisjunction rate, and the reproductive capacity of different *B* chromosome classes of plants. Essentially it was found that the drive mechanism, based on directed nondisjunction in the male and female gametophytes, leads to an accumulating situation within the range of the parameter values which are normally found in wild populations. An equilibrium is ultimately reached when the average *B* frequency is around the level of two *B*'s per plant. In rye, at least, the *B*'s are self-perpetuating and are equipped with their own genetic information which specifies their transmission characteristics and ensures their survival in the population without any involvement of natural selection based on host plant phenotypes.

In some respects these findings are not unexpected since it is now known that much of the DNA in the basic *A* chromosome complement of higher organisms has no apparent useful function either, and that much of it may also be redundant and selfish. For background information *see* CHROMOSOME; DEOXYRIBONUCLEIC ACID (DNA); GENE; MEIOSIS; MITOSIS; RIBONUCLEIC ACID (RNA) in the McGraw-Hill Encyclopedia of Science and Technology.

[R. N. JONES]

Bibliography: H. G. Callan, Lampbrush chromosomes, *Proc. Roy. Soc. London*, B214:417–448, 1982; J. G. Gall et al., The transcription unit of lampbrush chromosomes, in S. Subtelny and F. C. Kafatos (eds.), *Gene Structure and Regulation in Development*, pp. 137–146, 1983; R. N. Jones, Are *B* chromosmes selfish?, in T. Cavalier-Smith (ed.), *DNA and Evolution: Natural Selection and Genome Size*, 1983; R. N. Jones and R. B. Matthews, Selfish *B* chromosomes in rye, *Kew Conference II*, pp. 183–190, 1983; R. N. Jones and H. Rees, *B Chromosmes*, 1982; H. C. Macgregor, Recent developments in the study of lampbrush chromosomes, *Heredity* 44:3–35, 1980.

Clover

Among the true clovers, subterranean clover (usually referred to as subclover) is unique because of

Fig. 1. Single-spaced plants of cultivar Tallarook.

the development of seeds in burs below or on the soil surface. Subclover is a persistent annual and regenerates well in pastures. Australia grows 40 million acres (16 million hectares), and over 1 million acres (400 hectares) have been sown on hill lands in Oregon and California. Use is increasing in the southeastern United States.

Recent studies have placed commercial subclover into three species. In addition to the most important *Trifolium subterraneum*, there are *T. yanninicum*, more suitable on wet, waterlogged soils, and *T. brachycalycinum*, adapted to alkaline soils. Speciation is on the basis of cytological, genetic, and morphological discontinuity.

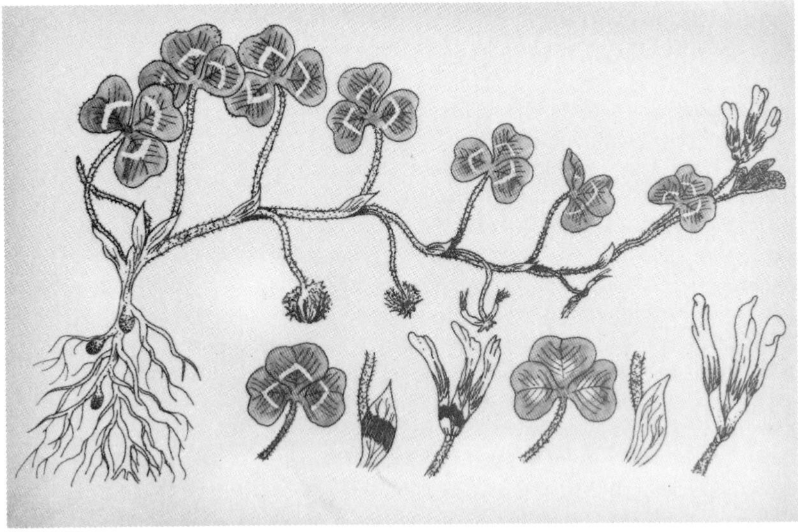

Fig. 2. Plant parts of subterranean clover showing variation of marking on leaflets and calyxes.

Adaptation. All species are relatively tolerant to acid soils and are productive wherever other winter annual clovers are grown. Persistence is better in summer-dry areas. Subclover behaves as a winter annual, but the many cultivars from Australia provide a great range in length of growing season (rainfall) and temperature and photoperiod requirements for flowering. Short-season cultivars (4-month season) are used in areas with rainfall as low as 12 in. (300 mm). The late-season cultivars for higher-rainfall areas will grow during an 8-month season.

Description. Stems develop as prostrate runners. Spaced plants may have runners 18 in. (45 cm) long (Fig. 1). Flowers, usually three or four to a cluster, are cream color to white. Seeds are purplish black, except for white seeds of *T. yanninicum*, and are the largest of the clover species. Pubescence varies with species and cultivars. All the species are self-pollinated and cleistogamous.

Seed burial and regeneration. Subclovers are geotropic: after pollination, the florets deflex and the peduncle turns downward and elongates, similar to peanuts, pressing the fruiting body onto the soil (Fig. 2). The cultivated species are also geocarpic. Two mechanisms involved are the exertion of a downward push on the developing fruiting body by the elongating peduncle, and the pulling action of the sequential whorls of sterile calyxes which grow downward from the apex of the inflorescence and then deflex backward through the soil to form the protective covering of the bur. Early-maturing cultivars (Dwalganup) are more strongly geocarpic, burying up to 80% of the burs. Those cultivars are grown in areas with hot, dry conditions during seed maturation. It is likely that bur burial modifies the microenvironment in which seed development takes place. The later-season cultivars (Tallarook and Mt. Barker) bury few burs, even when subjected to harsher conditions (Fig. 3). Burial, in all cases, is reduced in soils that set hard from rain or compaction by animals.

Having some seeds inaccessible to grazing animals is important for regeneration of the clover. Seed for the following year is obtained while grazing the plants during the reproductive period. In addition, there are mechanisms for delayed germination: physiological dormancy reduces germination for a period after maturity; the amount of hard seed (seedcoats impermeable to water, common among legume species) is relatively high in most cultivars; and seeds may remain in the soil several years before germinating. These mechanisms are most useful in areas with summer rainfall, and provide for regeneration when seed set is prevented for a season by excessive spring drought or by winter-killing of plants.

Nitrogen fixation. Similar to other pasture legumes, subclover fixes atmospheric nitrogen symbiotically. However, specific strains of rhizobia are required, and nodulation failure results either be-

Fig. 3. Seed burs developing in soil surface.

cause inoculum is misused or because inoculated seeds are often sown in acid soils or exposed to desiccation by surface sowing, with detrimental effects on the bacteria. Such problems have been largely overcome by the use of lime-pelleted seeds, by the use of large quantities of inoculum, and by the covering of seeds or sowing just before a rain.

Nutritive value. Subclover is comparable with other clovers in quality. Spring growth that is not used for grazing, hay, or silage is dead and dry in the summer, but it is still useful for livestock maintenance. Green plants in the spring may be as high as 30% in crude protein content, but on drying in summer the protein is reduced to about 7%. The seeds in burs contain up to 40% crude protein and 15% fat. Thus, the dry material provides the above-minimum requirements for maintenance of livestock.

Like other clovers, there is potential for causing bloat in ruminants. The animal stress factor of most importance in some areas is phytoestrogenic activity in some cultivars, resulting in loss of fertility in breeding animals. The activity is the result of isoflavone content, principally formononetin and is mainly an inherited characteristic. New cultivars are selected for low content, although there is some environmental influence on content. Isoflavones are reduced by wilting and drying.

Management. Yield increases with a more dense stand. One objective for the first year is to provide seed for succeeding years, and moderate grazing pressure the first year, especially by sheep, may be required. The prostrate clover develops best with grazing, which reduces competition for light provided by taller-growing companion grasses. To ensure good regeneration, it is necessary to fully utillize all old growth before the autumn rains. The clover fails to reestablish well if uneaten forage and debris are left on the surface.

Seed production. Australia has been the main source of seeds. Harvest was first done with rakes and sweeps. Then sheepskin rollers were used; burs attached to the wool on the roller were brushed onto adjacent containers. Most seed harvest at present is done with the suction-type combine, developed in Australia. The vacuum action collects burs, plant material, and soil, and blowers and screens allow separation of the burs for threshing. Some seed is grown in western Oregon by using imported combines. There is potential for up to 1 ton of seed per acre, but 500–600 pounds per acre (560–675 kg per hectare) is usual. For background information *see* CLOVER; LEGUME FORAGES; METABOLISM IN RUMINANTS; NITROGEN FIXATION in the McGraw-Hill Encyclopedia of Science and Technology.

[W. S. MCGUIRE; W. E. KNIGHT]

Bibliography: K. P. Barley and P. J. England, Mechanisms of bur burial in subterranean clover, *Proceedings of the 11th International Grassland Congress*, 1970; N. C. Brady (ed.), *Advances in Agronomy*, 1982; J. Katznelson, Biological flora of Israel, *Israel J. Bot.*, 23:69–108, 1974; N. L. Taylor (ed.), *Clover Science and Technology*, 1984.

Comet

New instrumentation, better computer technology, and the placing of several astronomical observing artificial satellites into orbit above the Earth's atmosphere have contributed to many advances in cometary astronomy during the past decade. Three recent developments in this field concern the 1985–1986 apparition of Halley's Comet, three recently discovered Sun-grazing comets in a 2-year span, and two comets having very close approaches to the Earth in less than a month. The higher-quality instrumentation now available, as well as that which is planned, should give much information in the near future concerning the true nature of the enigmatic cometary nucleus.

Halley's Comet. Astronomers throughout the world are currently planning programs for observing Halley's Comet, the most famous of such objects, as it once again approaches the inner solar system on its regular 76-year orbit around the Sun. Recovered near its predicted position by astronomers using the 200-in.-aperture (5.08-m) reflecting telescope at Palomar Observatory in southern California in October 1982, periodic comet Halley was then far fainter than any previously detected comet. A newly designed charge-coupled device (CCD) was employed to detect the comet as it was about 11 astronomical units (AU), or 1.02×10^9 mi (1.65×10^9 km), from the Sun—somewhat further than the distance to the planet Saturn.

Four space probes are now scheduled to fly by Halley's Comet in early March 1986, a few weeks after perihelion (its closest approach to the Sun) on February 9. The Soviet Union has two such artificial satellites, *Vega 1* and *Vega 2*, planned for encounters on March 5 and 9, respectively. The European Space Agency's *Giotto* will fly by the comet on March 13, while Japan's *Planet-A* encounter will occur on March 7.

Much criticism of the United States' failure to allocate funding for a probe to Halley's Comet has led to two employment plans for existing satellites to study comets. One probe, *ISEE 3*, orbiting the Sun and studying the ejected solar wind particles, has been temporarily sent in a path which will take it near periodic comet Giacobini-Zinner, much fainter than Halley's, in September 1985. *ISEE 3* and other satellites studying the solar wind, such as *Pioneer 7*, can be used indirectly to study Halley's Comet, since the charged particles of the solar wind are thought to be the major influence on a comet's behavior as it nears the Sun.

A second American probe which will be scheduled to observe directly Halley's Comet is the *Pioneer Venus* spacecraft which is orbiting Venus. This satellite will observe the comet in early February 1986, just prior to perihelion, as the comet passes 2.5×10^7 mi (4.0×10^7 km) from the orbiter. A spectrometer aboard the spacecraft will observe Halley's Comet in ultraviolet light to determine various atomic and molecular species that are released from the comet's icy nucleus into the coma and tail. *Pioneer Venus* should detect hydrogen, carbon, nitrogen, oxygen, sulfur, and other species typical of the icy components making up the cometary nucleus.

Unfortunately for Earth-based observers, Halley's Comet is making its most unfavorable apparition in millennia, for it will be on the far side of the Sun at perihelion. Northern Hemisphere observers, in particular, will not have the best of views, as the comet will be low on the horizon when at its peak brightness. Southern Hemisphere observers, however, will have comet Halley placed almost overhead in the morning sky at the beginning of twilight during late March and the first week of April.

The best view of the comet from northern latitudes will come in mid-to-late March 1976, when the comet will be only some 10° above the southeastern horizon at the break of dawn. Halley's Comet will be two or three times higher in the evening sky in early January and late April, but will also be somewhat fainter and will not exhibit as long a tail. The appearance of this comet for observers in North America is expected to be quite similar to that of Comet Kohoutek in 1973–1974.

The comet is best viewed as far from city and suburban lights as possible with small- to moderate-sized binoculars. Its brightness at this apparition likely will be comparable to a third- or fourth-magnitude star from late January until mid-April. It should thus be faintly visible to the unaided eye in a dark, clear sky. The comet will gradually pull away from the inner solar system and fade from view, although the largest telescopes, including the forthcoming Space Telescope, may follow Halley's Comet until perhaps 1989. The next chance to view this comet will not be for another 76 years or so, around its next perihelion in July 2061.

Sun-grazing comets. Images of three probable comets have been found on digital exposures taken with an instrument aboard an artificial satellite. The three comets were observed either to collide with the Sun or to completely disintegrate in the outer solar atmosphere, for none of them was found to reappear following their closest approaches to the Sun. Two of the three Sun-grazing comets found to date with the *P78–1* Earth-orbiting satellite were observed in 1981, and the third apparently collided with the Sun in August 1979. In each case, the U.S. Naval Research Laboratory's SOLWIND coronagraph on board the satellite, an instrument designed to study the Sun's outer atmosphere, or corona, by using a disk to hide the Sun itself, observed the comets for no longer than about 4 h each. The little information derived from the observations was supplemented by the knowledge of a family of comets, known as the Kreutz Sun-grazing comet group, to determine orbital characteristics of each object.

Evidently none of the three comets was observed from the ground, owing to a combination of poor placement with respect to the Sun and small size with correspondingly faint brightness. It is thought that, if these objects are indeed members of the Kreutz group, they are part of what was originally a much larger comet, which broke up at a much earlier approach to the Sun. The past 200 years has seen several of the Kreutz comets swing around the Sun, but the SOLWIND coronagraph findings surprised astronomers for two reasons. First, the discovery of three such comets in less than 2 years hinted at an unexpectedly large number of possible such objects orbiting the Sun, and, second, these included the first known cases of objects that actually collided with the Sun.

The first SOLWIND comet discovery, known as Comet 1979 XI, had a perihelion distance substantially less than the solar radius. Uncertain figures place this distance at only 0.35 solar radius, while Kreutz Sun-grazing comets typically have perihelia near 1.5 solar radii. While Comet 1979 XI did not reappear after its closest approach to the Sun, a diffusing of material appeared on the opposite side of the Sun and persisted for several hours. This material was probably matter strewn out from the cometary nucleus and dissipated over a large area. The perihelia distances of the 1981 Sun-grazing comets, known as Comets 1981 I and 1981 XIII, seem to have been 1.05 and 0.92 solar radii, respectively. The discovery of additional Kreutz members will add more pieces to the puzzle.

Close-approaching comets. George E. D. Alcock, an amateur astronomer living near Peterborough, England, discovered a new comet on May 3, 1983. Another amateur in Japan, Genichi Araki, had independently found the same comet a few hours earlier.

The Earth-orbiting *Infrared Astronomical Satellite (IRAS)* had actually located the comet as early as April 25, and the comet was named IRAS-Araki-Alcock for the order in which the discoveries were made. Astronomers soon had enough observations of this new object to learn that it would be making the closest approach to the Earth of any comet in over 200 years—0.03 AU or 2.8×10^6 mi (4.5×10^6 km) away—on May 11. *See* INFRARED ASTRONOMY.

The news of Comet IRAS-Araki-Alcock reached the scientific community less than a week before closest approach, when the comet was already of naked-eye brightness. Large telescopes and orbiting satellites such as the *International Ultraviolet Explorer (IUE)* were trained on the large diffuse celestial object as it traced a path through the northern constellations Draco, Ursa Minor (the Little Dipper), and Ursa Major (the Big Dipper). At its brightest, one day after closest approach, many observers reported the total apparent magnitude to be equal to that of a first- or second-magnitude star, brighter than the individual stars of the Big Dipper.

The comet's nucleus, perhaps 0.6 mi (1 km) in diameter, typical of an ordinary comet, released icy and dusty gases due to effects of solar radiation, producing an outer atmosphere, or coma, which exceeded 60,000 mi (100,000 km) in diameter. At closest approach to the Earth, the apparent visual size of Comet IRAS-Araki-Alcock was 4° in extent, or eight times the apparent size of the Moon.

What made this comet extraordinarily important to astronomers was indeed its proximity to the Earth. This permitted detection for the first time of diatomic sulfur in a comet, with the *IUE* satellite. Radar echoes were successfully received with large radio telescopes in Puerto Rico and California, revealing texture and the most concrete data on a comet's nucleus to date. Reduction of the data will probably yield reliable information concerning the nuclear size and rotation rate. While some 10 or 20 comets can be followed in an average year with large telescopes, very little is known about the true nature of the cometary nucleus. F. Whipple's theory of the dirty snowball consisting of a mixture of various ices and dust material, announced in 1950, still appears to be the best available.

A second close-approaching comet was discovered on May 8, 1983, by three Japanese amateurs and named Comet Sugano-Saigusa-Fujikawa. Its encounter with the Earth brought it to a geocentric distance of 0.06 AU, or about 5.9×10^6 mi (9.5×10^6 km) in early June. Two close cometary encounters in less than a month is an exceedingly rare occurrence. Both comets were among the 16 closest-approaching comets in history. Successful radar contact was also made with Comet Sugano-Saigusa-Fujikawa, and initial results placed its nuclear size as substantially smaller than that of the May comet.

For background information *see* COMET; HALLEY'S COMET in the McGraw-Hill Encylopedia of Science and Technology.

[DANIEL W. E. GREEN]

Bibliography: B. G. Marsden and D. W. E. Green, 1983d: May's surprise comet, *Sky Telesc.*, 66:26–29, 1983; C. S. Morris and D. W. E. Green, The light curve of periodic comet Halley 1910 II, *Astron. J.*, 87:918–923, 1982; Z. Sekanina, The path and surviving tail of a comet that fell into the sun, *Astron. J.*, 87:1059–1072, 1982; D. K. Yeomans, *The Comet Halley Handbook*, NASA JPL 400–91, 1981.

Communications satellite

During the early 1980s the number of communications satellites has increased substantially. To a great extent this is due to the rapid growth of television services which are primarily of an entertainment nature. The services have expanded to include multiple KTV services offering 24-h-a-day movies without commercials or interruption. Because of the existence of different time zones within a country, it is often necessary to use multiple satellite repeators to accommodate the different viewers. Specialized (narrow-casting) television applications have led to the evolution of specialized transponders offering full-time weather, health, women's programs, foreign movies, education, congressional debates, and so forth. Over 100 regularly scheduled television transponders are in use. All of these are receivable on fairly simple and relatively inexpensive earth stations.

Radio and other audio services have continued to expand, and telephone and data requirements have continued to increase. Nearly one-third of the satellite capacity is now consumed by telephone services. It is expected that this number will continue to increase at a rate even faster than video, which is reaching a temporary saturation point.

Looking into the future, even greater capacities and demands for satellite communications are seen. New direct-to-home broadcasting services are being initiated. This may completely change the character of how television is distributed on a nationwide basis. *See* DIRECT BROADCAST SATELLITE SYSTEMS.

Just as in the United States, there is an emerging boom in the rest of the world. Vast improvements have been brought about in telecommunications in areas such as Indonesia, Alaska, northern Canada, and within certain areas of South America and Africa. Further examples of this explosive growth are given below.

Communications satellites are usually placed in a geostationary orbit, so that they appear stationary from the Earth. The individual repeaters aboard the satellites are called transponders.

AUSSAT. The Australian National Communications satellite system is being managed by AUSSAT Proprietary, Ltd., and will consist of at least two domestic satellites providing coverage to Australia and Papua New Guinea. Five major coverage areas are included, with sufficient satellite power being provided for direct-to-home television broadcasting. The equivalent radiated power is approximately 50,000 W per beam. The home antenna will be under 5 ft (1.5 m) in diameter. The satellites may also be used for voice, teletype, and network video to connect the cities in back areas of Australia. A total of 15 transponders will be available on each stellite.

Two launches are planned on the space shuttle in July and October 1985. These satellites will be spin-stabilized, with an initial mass of approximately 1430 lb (650 kg). The anticipated lifetime is at least 7 years.

ARABSAT. ARABSAT is operated by the Arab Satellite Communications Organization, headquartered in Riyadh, Saudi Arabia. It provides a regional communications satellite service, over the region extending from the Atlantic across North Africa and eastward to Saudi Arabia.

Two launches are planned in 1984 and 1985 (one each on the European Ariane rocket and the space shuttle). Each satellite will be body-stabilized and will contain 25 transponders in the 4-GHz band, plus one direct broadcasting transmitter operating at 2.5 GHz. The transmitter for the direct broadcasting service provides an equivalent isotropically radiated power (eirp) of about 25,000 W. The 10-ft-diameter (3-m) earth station antennas may be fabricated from screen and wire mesh because of the low frequency. This frequency was first used in the Rocky Mountain States Experiments on NASA's *Advanced Technology Satellite 6* and is being used in India with INSAT. The portion of the Arab world that will be covered by this satellite is approximately 5000 mi (8000 km) east-west and 2800 mi (4500 km) north-south.

CS 2. The Japanese *Communications Satellite 2* (also called *Sakura,* which means cherry blossoms) is the second generation of a satellite operating in the 4-GHz and the 20–30-GHz satellite frequency bands.

The *CS 2* series consists of two satellites: *CS 2a* and *CS 2b*, which were launched in 1983. These are upgraded versions of the original experimental communications satellite (*ECS*, *CS*, or *Sakura 1*). Both satellites were launched from the Tanegashima Space Center in Japan using the N-2 launch vehicle. The second generation is being used for operational telephone traffic among the Japanese islands. The satellites weigh approximately 770 lb (350 kg) each and are placed in the geostationary orbit. Each satellite carries two transponders receiving at 6 GHz and retransmitting at 4 GHz. It also has six transponders capable of receiving at 30 GHz and retransmitting to the Earth at 20 GHz.

ECS. The *European Communications Satellite* (also known as EUTELSAT) was first launched in early 1983 on the Ariane rocket. A series of these satellites will be launched to provide both telephony and cable services series to Europe. The satellites are body-stabilized and are produced by a European manufacturing consortium. These satellites operate in the new 14- and 11-GHz bands providing coverage throughout Europe.

INMARSAT. INMARSAT (of the International Maritime Organization) provides the world's ships and offshore platforms with telephone, telex, facsimile, and data communications via a network of satellites. More than 85% of the world's merchant ships' gross tonnage belongs to nations that are members of INMARSAT. The two largest owners are the United States' COMSAT General Corporation (23.4% ownership) and the Soviet Union (14.1%).

The INMARSAT system is a successor to the COMSAT General MARISAT network, which consists of three satellites in the Atlantic, Indian, and Pacific oceans. INMARSAT has added a *Marecs* satellite in the Atlantic. The *Marecs* is the maritime derivative of the *European Communications Satellite*. INTELSAT, the International Telecommunications Satellite Organization, provides additional satellites for use in the Indian Ocean. These are the INTELSAT V satellites, which also contain a maritime communications subsystem package.

While the present generation of satellites is not especially powerful or capable of handling many voice circuits simultaneously, it only requires relatively modest ship terminals. There are now over 1500 such terminals with typical antenna diameters of 4 ft (1.2 m). In the near future there will be still smaller ship stations with antennas of approximately 8 in. (0.2 m). These will be for telex and 2400-baud data services only. By 1985 there will be 24 shore stations (used for transmitting to and receiving from satellites), including the 3 from the old MARISAT system.

INSAT. An ambitious satellite program constructed in the United States for the government of India started with disaster when the first of the three INSAT satellites failed to operate after reaching geostationary orbit. Modifications have been made to the second and third satellites in this series to resolve the problems encountered. The second one has been successfully launched and placed in operation.

The INSAT satellites are an example of the beginnings of the Integrated Satellite System (formerly called the Orbital Antenna Farm or Geostationary Platform) of the future. In the case of Insat, three basically different missions were combined on the same spacecraft. In addition to the fixed communications satellite service, there also was a direct-to-village broadcasting service at 2.5 GHz, which could be received on fairly simple antennas made of screening. The third mission involved meteorological sensors for detection of weather patterns around the Indian subcontinent.

INTELSAT VI. The International Telecommunications Satellite Organization has entered into a contract with Hughes Aircraft Company and its subcontractors in Canada, France, Italy, Japan, the United Kingdom, and West Germany for 16 spacecraft. These satellites will be of the double-drum spinning variety capable of generating over 2 kW of electrical power at the end of life.

Each INTELSAT VI will be 39 ft (11.7 m) tall. Multiple antennas will permit precisely shaped antenna beams at each of two frequency band pairs (at 4–6 and 11–14 GHz). These new satellites will be capable of being operated in a satellite-switched time-division multiple-access mode. They contain various techniques for cross-strapping of the frequency bands. For example, an uplink at 6 GHz may become an 11-GHz downlink. The first launch is anticipated in early 1986.

Morelos. The Mexican Communications System (called Morelos or Sistema Mexicano de Satellite) is being designed for launch in 1985. This hybrid system uses two pairs of frequencies, 4–6 and 12–14

GHz. These satellites will be capable of providing both telephone and television distribution services, in addition to some educational direct-to-village capability.

Palapa-B. The Indonesian Domestic Satellite System, called Palapa-A, consists of two satellites launched in 1976 and 1977. This system has grown into a regional satellite system, with eventually three satellites to be used throughout the entire Association of Southeast Asian Nations region.

The new satellites will have twice as many transponders as the earlier version (that is, 24 vs 12), each with double the power, and yet the launch vehicle has grown only modestly. The original Palapa system had 40 earth stations, and an additional 200 or more will be constructed for the new system.

SBS 3. Satellite Business Systems' satellites have now grown to three in orbit plus plans for at least three more. *SBS 3* carries 10 transponders operating in the 12-GHz band. These have been primarily used for voice and data services, although it is planned to make use of and a portion of *SBS 4* for direct-to-home video services.

Telstar 3. The first of this series of satellites was launched in 1983. These satellites, which will succeed the Comstar series, have been constructed for the AT & T Communications (Long Lines) to carry voice, data, and leased video services (primarily for television networks). The power amplifiers (each satellite has the capability of operating 24 transponders) may be either traveling-wave-tube amplifiers or solid-state amplifiers.

Whereas Comstar carries 750 full circuits per transponder, the new Telstars can carry at least 900 using conventional modulation. This number can be increased to 3900 full circuits using companded single-sideband amplitude modulation. Fully and optimally loaded, a *Telstar 3* could carry as many as 93,600 full circuits.

For background information *see* COMMUNICATIONS SATELLITE in the McGraw-Hill Encyclopedia of Science and Technology.

[WALTER L. MORGAN]

Computer-integrated manufacturing

In a computer-integrated manufacturing (CIM) system, individual engineering, production, and marketing–support functions of a manufacturing enterprise are organized into a computer integrated system. Functional areas such as design, analysis, planning, purchasing, cost accounting, inventory control, and distribution are linked through the computer with factory floor functions such as materials handling and management, providing direct control and monitoring of all process operations.

Computer-integrated manufacturing may be viewed as the successor technology which links computer-aided design (CAD), computer-aided manufacturing (CAM), industrial robotics, numerically controlled machine tools (NCMT), automatic storage–retrieval systems (AS/RS), flexible manufacturing systems (FMS), and other computer-based manufacturing technology which has been developed over the last 20 years. Computer-integrated manufacturing is also known as autofacturing and integrated computer-aided manufacturing (ICAM). Autofacturing includes computer-integrated manufacturing, but also includes conventional machinery, human operators, and their relationships within a total system.

Development. Three main economic and technical factors have contributed to the development of CIM.

1. It has been found that even the most successful investment in the technology necessary to improve productivity in a single function of manufacturing can often lead to extreme inefficiencies in both predecessor and successor functions. For example, when using a CAD terminal the design engineer can create, at a very efficient rate, many designs that may not be able to be fabricated; numerically controlled machine tools can process parts that may not be able to be made as well by conventional machining, but numerically controlled machines can be tedious to program and cannot easily be made to interface with or interact with their environment; inventory storage systems can efficiently store and retrieve parts, but most of this inventory has been found to be unnecessary.

2. Industrial robots have evolved to the point where they can serve as reliable, fast, and flexible materials transfer devices between processes. An example is a robot loading and unloading machining center. Robots can also perform some of the processes. Examples are drilling or routing by robots, or an assembly operation by robots.

3. Computers and computer technology, including both large-scale computer-based information system (CBIS) and microprocessors, have continued to evolve. The information systems allow data bases from typical manufacturing functions such as design, planning, and scheduling to be shared by all persons responsible for performing these functions. The microprocessors permit design of shop floor controllers that not only control the operation of their own machines but also allow machines to communicate with one another and to the original source of their programs virtually in real time. The "instant" effect of decisions can thus be known at all times by the decision makers, enabling corrective action to be taken before problems occur. This principle has been referred to as "on-time" decision making.

Factory of the future. The term computer-integrated manufacturing was first popularized by Joseph Harrington in 1975. He predicted that future manufacturing success would be based more on effective management of data and information flow rather than on how efficiently either piece parts or finished products were manufactured. He postulated that in the CIM factory of the future there would be areas of "departmentalized" decision making and "nondepartmentalized" decision making. "Departments" would be defined as being process ("hard technology") based, that is, involving processes such as milling, drilling, routing, and grinding.

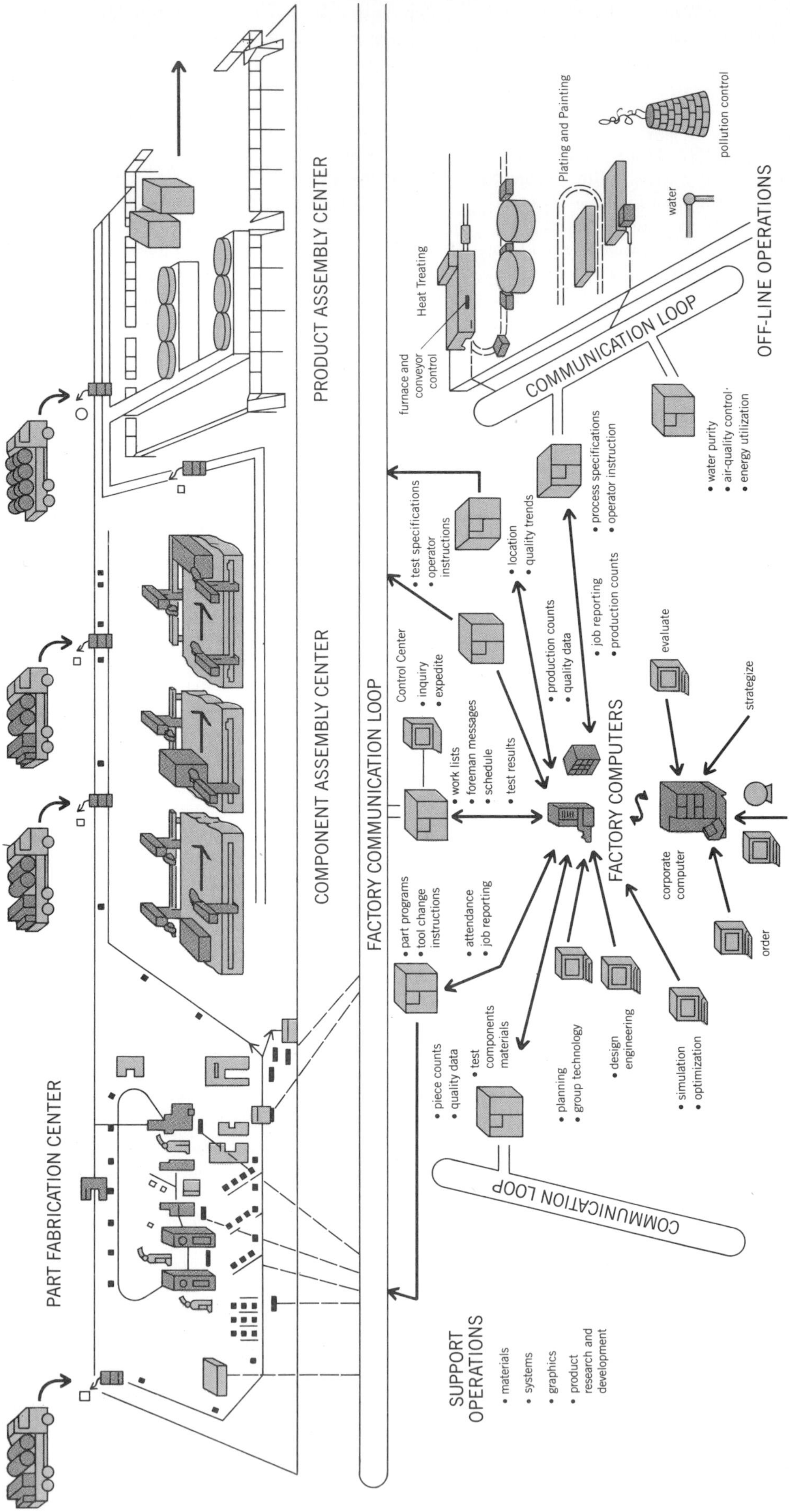

Fig. 1. Factory-of-the-future concept.

"Nondepartments" are defined as being information ("soft technology") based, such as design engineering, process planning, and inventory management, the basic premise being that they are generic throughout an organization. For example, process planning considered as the function "planning" applies equally to all processes in the factory, whether they be machining of metal or lay-up of composite material. Investment made to computerize the function of process planning based on the assumption that it is generic will result in reuse of this investment many times throughout the organization. The computer-based process planning function will not change at all; but the data on which it operates will be specific to the process of the moment. Computer-aided design systems are the best example of the concept, but the technology is still primitive.

The CIM factory of the future will look quite different from the factory of today. It will integrate the traditional hard or process-based technology factory of today (departments) with emerging software of systems-based technology (nondepartments). Figure 1 shows such a factory concept. The goal of this factory is to be efficiently integrated and continuously flexible and economical in the face of change. It will be effectively organized to achieve maximum productivity along with consideration for personnel.

Effectiveness and efficiency. In this scenario, efficiency is most often associated with transfer-line or assembly-line technology. The best-known developer of this technology was Henry Ford. His Model T factory could produce cars at an unprecedented rate, but the consumer had no choice at all about the specifications of the product. Flexibility is most often associated with robotics and other reprogrammable automation. The Japanese excel in flexible automation. One company (Toyota) has more flexible manufacturing systems than are known to exist in all other companies combined. Effectiveness is most often associated with the soft technology and organizational support structure of the enterprise. Effectiveness is the concept of "doing the right thing." Efficiency is the concept of "doing the thing right." Effectiveness seeks at all times to balance the functions of the factory so that they are working with, not opposing, one another.

Economies. Optimization of the CIM factory is based on different criteria from that used to evaluate today's factory. The primary criterion for assessing today's factory is efficiency. It aims to produce as many parts in a given time as the process technology will allow. Today's factory strives for economy of scale; that is, more and more volume of a standard product should lead to reduced unit cost (cost for each part or product) and therefore more profits.

The CIM factory, by considering efficiency and flexibility and effectiveness, enables economy of scope. In economy of scope, flexibility is at least as important as efficiency because the factory is seen to be an ever-changing dynamic environment which must always quickly respond to the needs of the marketplace, if not actually lead the need. Alvin Toffler coined the term global "prosumer" economy to describe an economy where the producer and the consumer are inextricably linked together; the consumer believes that he or she has the right to demand a custom-built product at an affordable price. The producer is therefore obligated to manufacture

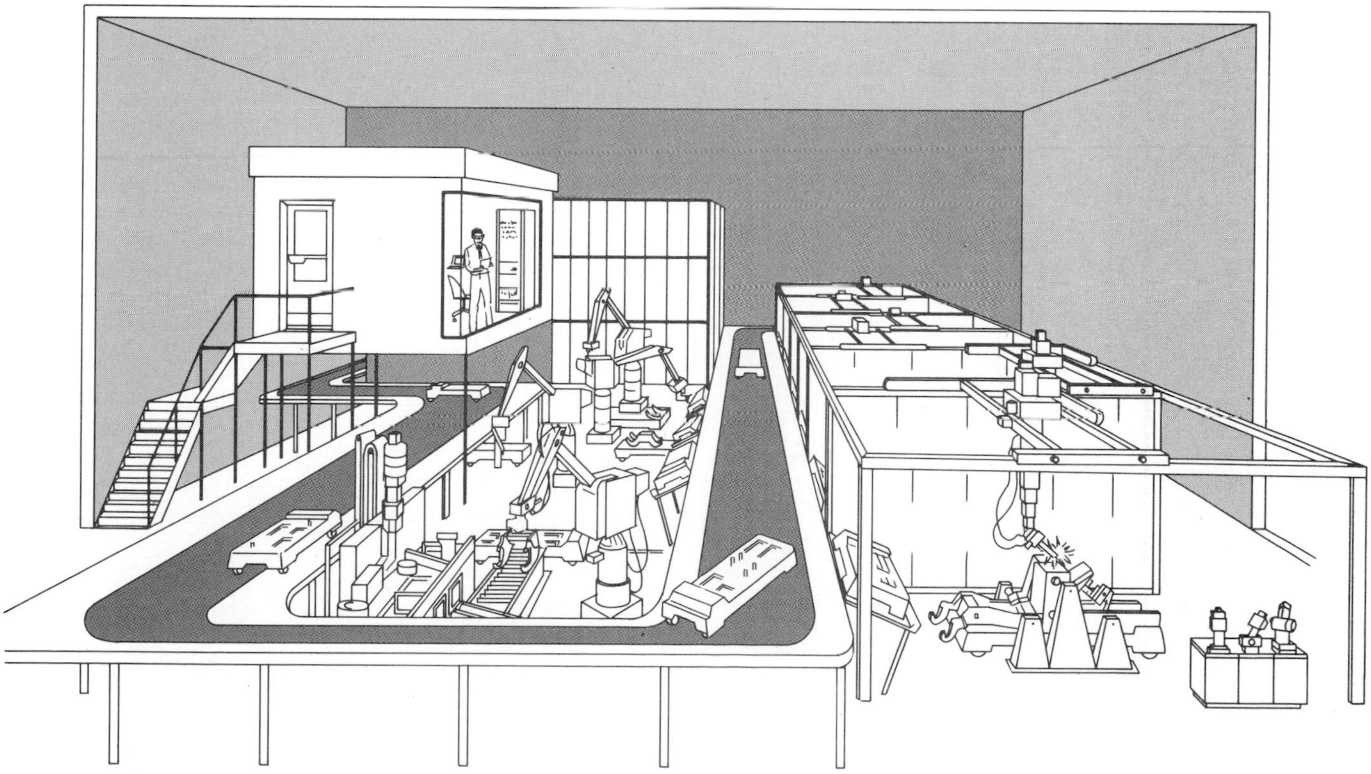

Fig. 2. Typical CIM work cell.

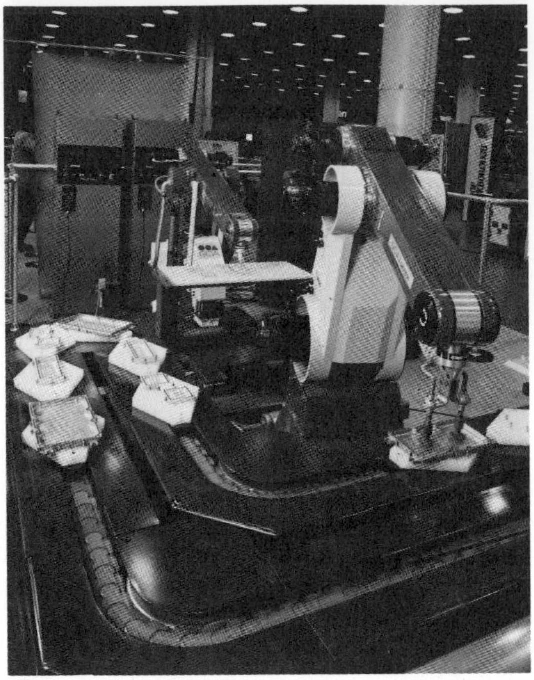

Fig. 3. Typical CIM work station.

this product virtually on demand. Clearly, the traditional factory with both rigid transfer-line technology and a rigid organizational structure cannot satisfy this objective; it strives for maximum volume of each product. In the ideal case, the CIM factory provides for profit with an order quantity of one, that is, when no two products are identical.

Organization. The CIM factory concept is illustrated in Fig. 1: hard technology is shown in the upper half and soft technology in the lower half. Soft technology can be thought of as the intellect or brains of the factory, and hard technology as the muscles of the factory. The type of hard technology employed depends upon the products or family of products made by the factory. For metalworking, typical processes would include milling, turning, forming, casting, grinding, forging, drilling, routing, inspecting, coating, moving, positioning, assembling, and packaging. More important than the list of processes is their organizaton.

The CIM factory is made up of a part fabrication center, a component assembly center, and a product assembly center (Fig. 1). Centers are subdivided into work cells, cells into stations, and stations into processes. Processes comprise the basic transformations of raw materials into parts which will be assembled into products. In order for the factory to achieve maximum efficiency, raw material must come into the factory at the left end and move smoothly and continuously through the factory to emerge as a product at the right end. No part must ever be standing; each part is either being worked on or is on its way to the next work station. In today's factories, a typical part is worked on about 5% of the time.

In the part fabrication center, raw material is transformed into piece parts. Some piece parts move by robot carrier or automatic guided vehicle (AGV) to the component fabrication center. Other piece parts (excess capacity) move out of the factory to sister factories for assembly. There is no storage of work in process and no warehousing in the CIM fac-

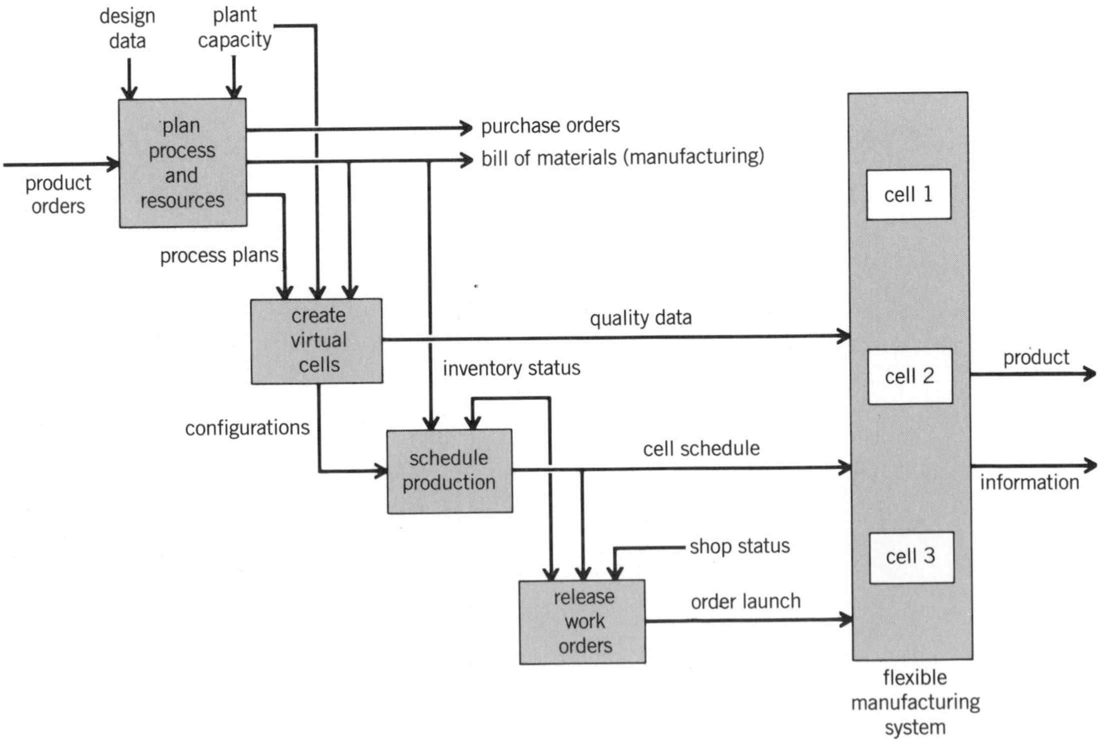

Fig. 4. Diagram of the architecture of manufacturing. (*Designed by D. Shunk, GCA Corporation*)

tory of the future. To accomplish this objective, part movement is handled by robots of various types. These robots serve as the focus or controlling element of work cells (Fig. 2) and work stations (Fig. 3). Each work cell contains a number of work stations. The station is where the piece part transformation occurs—from a raw material to a part, after being worked on by a particular process.

Components, also known as subassemblies, are created in the component assembly center. Here robots of various types, and other reprogrammable automation, put piece parts together. Components may then be transferred to the product assembly center, or out of the factory (excess capacity) to sister factories for final assembly operations there. Parts from other factories may come into the component assembly center of this factory, and components from other factories may come into the product assembly center of this factory. The final product moves out of the product assembly center to the end user.

Integration. The premise of CIM is that a network is created in which every part of the enterprise works for the maximum benefit of the whole enterprise. Independent of the degree of automation employed, whether it is robotic or not, the organization of computer hardware and software is essential (lower half of Fig. 1). The particular processes (upper half of Fig. 1) employed by the factory are specific to the product being made, but the functions performed (lower half of Fig. 1) can be virtually unchanged in the CIM factory of the future no matter what the product. These typical functions include forecasting, designing, predicting, controlling, inventorying, grouping, monitoring, releasing, planning, scheduling, ordering, changing, communicating, and analyzing.

It must be recognized that independent, optimum performance of each function is not as important as their integration with one another, or their integration with the factory floor itself. This integration is brought about in two ways. The first is through an architecture of manufacturing which specifies precisely how and when each function is integrated with the other (Fig. 4). The second is through a cell controller and cell network.

The cell controller ensures integration of data between each machine, robot, automatic guided vehicle, and so forth, of the cell. The cell network (Fig. 5) actually performs this communication within the cell, and connects cells together into centers. In the cell network, communication is possible between each machine of the network and between different networks. The cell controller itself serves to perform such tasks as: downloading part programs from the CAD system to each machine in the cell, monitoring actual performance of each machine and comparing this performance to the plan, selecting alternate routing for a part if a machine is not operable, notifying operators of pending out-of-tolerance conditions, archiving historical performance of the cell, and transmitting to the center level on an exception basis the cell performance compared to

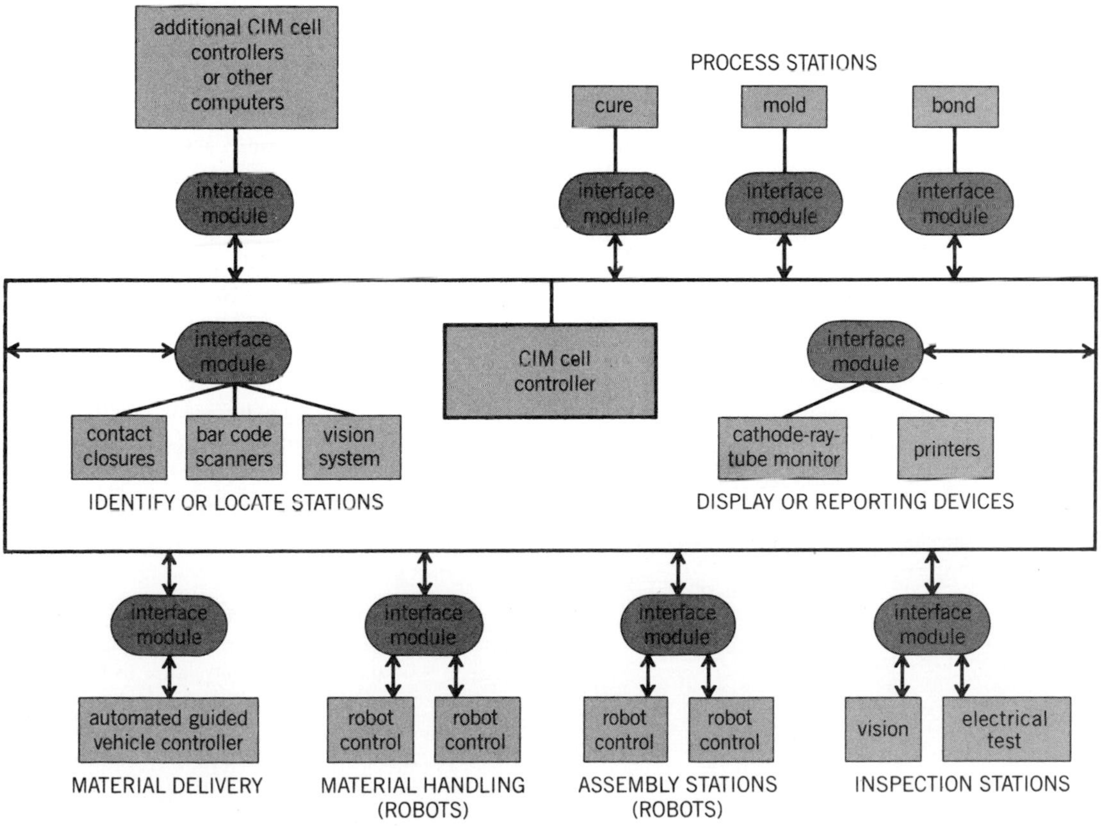

Fig. 5. CIM production cell network.

plan. The center level of the CIM factory of the future is all of the policies and procedures that run the factory. These policies and procedures are embodied in computer software, which is in turn based upon an overintegration plan or architecture of manufacturing.

The purpose for the architecture of manufacturing (Fig. 4) is to provide a blueprint for employing CIM. While there are as many variations of the details of the architecture of the CIM factory of the future as there are factories, the structure of the architecture is generic. The focus is the flexible manufacturing system (FMS), which is made up of cells, stations, and processes. The information to be fed back to functions where decisions are made is more important than the product itself. In this example, the functions are defined very broadly and are used only to illustrate the absolute necessity for interaction between functions if the CIM factory is to be made to work efficiently, flexibly, and effectively.

Analysis of the first block of Fig. 4 illustrates the idea. Here input to the model is product forecast. The input is changed into output depending upon the status of the controls coming into the top of the function box. In this case, one control is design data. It should not be possible for an order to be accepted when design does not exist. A second control in plant capacity; an order cannot be accepted if it cannot be filled by the plant within cost and schedule.

In the ideal CIM factory of the future, before any real decision is made, it is tested on a computerized model of the architecture to determine its feasibility. Thereby, any mistakes are made in theory, not in practice. Several simulation software packages are routinely used for this purpose. Typical output of an integrated decision support software (IDSS) model is shown in Fig. 6 for a small manufacturing cell. It is clear that for the true CIM factory of the future to be realized it will be necessary to build a complete simulation model.

Future development. For today, all of the principles involved are well accepted and are implemented in some way. The factory-of-the-future's un-

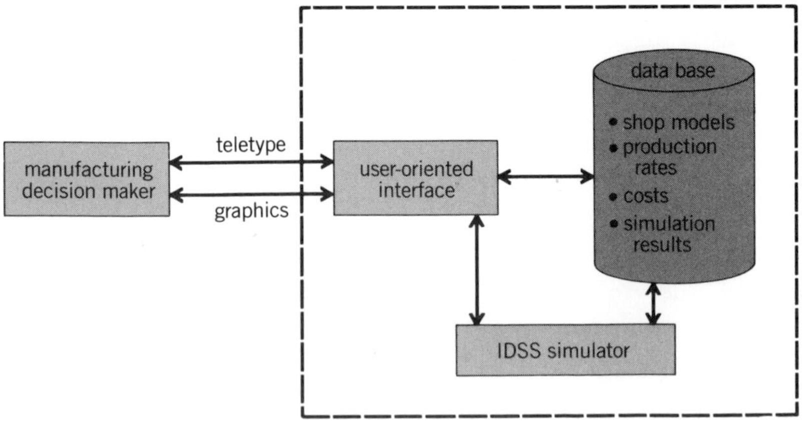

Fig. 6. Integrated decision support software (IDSS) model.

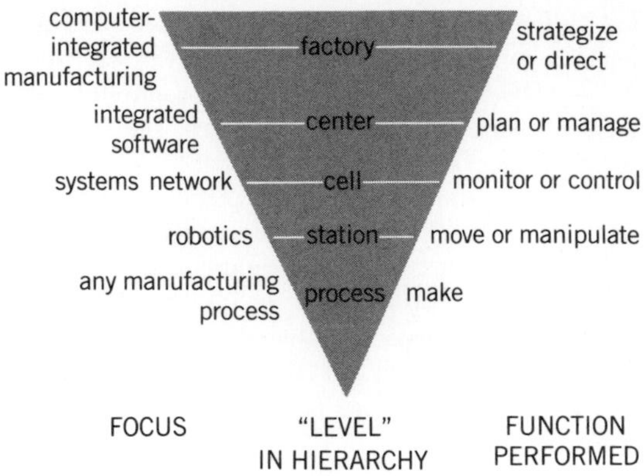

Fig. 7. Factory of the future. (*Designed by M. Eastwood, GCA Corporation*)

derlying structure of center, cell, station, and process (Fig. 7) is the basis for the control scheme being used by principal CIM companies in the United States. This structure serves both as the physical integration mechanism and as the highest-level architecture for integration of functions, the purpose of CIM. The benefits from both a hard technology and soft technology point of view are significant. For example, from a systems viewpoint advantages include: phased development and implementation planning which permits pay-as-you-go financing; capability of managing change in lot size, workload, and technology; just-in-time materials management; operation from factory control; and integration of business systems. From a process viewpoint, advantages include computer-aided design (CAD) integration, independent and multiple work cells, flexible routing, flexible automation of parts handling and processes, and integration of all hardware.

For background information *see* COMPUTER-AIDED DESIGN AND MANUFACTURING; COMPUTER GRAPHICS; GERT; INDUSTRIAL ROBOTS in the McGraw-Hill Encyclopedia of Science and Technology.

[DENNIS E. WISNOSKY]

Bibliography: J. Harrington, Jr., *Computer Integrated Manufacturing*, 1975, reprint 1979; R. M. Metcalf and D. R. Boggs (Xerox PARC), Ethernet: Distributed switching for local computer networks, *Commun. ACM*, 19:395–404, July 1976; A. B. Pritsker and C. D. Pegden, *Introduction to Simulation and SLAM*, 1979; C. M. Savage, Preparing for the factory of the future, *Mod. Mach. Shop*, October 1983.

Computer vision

Computer (or machine) vision deals with the use of digital computer techniques for extracting, characterizing, and interpreting information in visual images of a three-dimensional world. Visual sensing technology is receiving increased attention as a means to endow machines with the capability of exhibiting a greater degree of "intelligence" in dealing with their environment. Thus, a robot or other machine that can "see" and "feel" should be easier to train in the performance of complex tasks, while requiring less stringent control mechanisms than preprogrammed machines. A sensory, trainable system is also adaptable to a much larger variety of tasks, thus achieving a degree of universality that ultimately translates into lower production and maintenance costs.

The computer vision process may be divided into five principal areas: sensing, segmentation, description, recognition, and interpretation. These categories are suggested to a large extent by the way in which computer vision systems are generally implemented. It is not implied that human vision and reasoning can be so neatly subdivided, nor that these processes are carried out independently of each other. For instance, it is logical to assume that recognition and interpretation are highly interrelated functions in a human. These relationships, however, are not yet understood to the point where they can be modeled analytically. Thus, the subdivision of functions discussed below may be viewed as a practical (albeit limited) approach for implementing computer vision systems, given the level of understanding and the analytical tools presently available.

Visual sensing. Visual information is converted to electrical signals by the use of visual sensors. The most commonly used visual sensors are vidicon cameras and solid-state arrays. Vidicons are the usual vacuum-tube cameras used as television imaging devices. An input video signal is digitized and transferred to a computer as an image of a size ranging typically from 64 × 64 to 512 × 512 discrete image elements, depending on the resolution requirements of a given application.

Solid-state devices are available as linear and area arrays. If the scene to be imaged is in continuous, uniform motion (as in belt conveyors), a linear array can be used to scan a line across the conveyor, and the motion of an object in the direction perendicular to scan produces the desired two-dimensional image. In a one-dimensional array each element is read every N time intervals, where N is the number of image elements in the line, whereas in a two-dimensional array each element is read every N^2 time intervals; therefore, the two-dimensional array maintains a higher output data rate while allowing a long integration time for noise reduction. Linear arrays with resolution exceeding 2048 elements are available.

In order for a vision system to be able to interact with its environment, a geometric relationship between the real world and the images in the picture seen by the system must be established. This relationship is a transformation of measurements from a three-dimensional coordinate system to the image coordinate system, or vice versa. Essential to the derivation of the object-image relationship is a precise mathematical description for the camera (that is, a camera model). The use of two or more cameras and their mathematical models allows extraction of three-demensional information, such as depth.

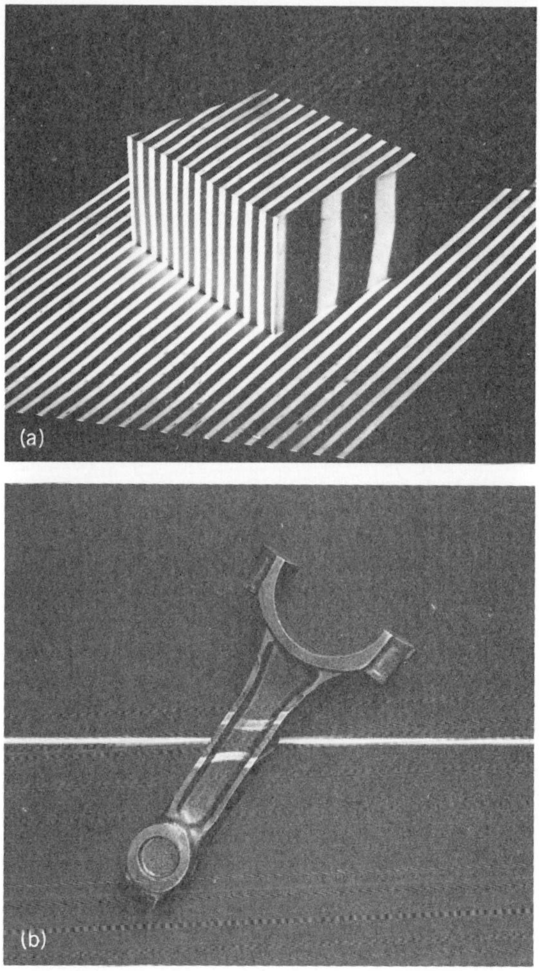

Fig. 1. Two examples of scene illumination by structured lighting. (*a*) Illumination of a block by series of parallel strips of light (*from F. Rocher and E. Keissling, Methods for analyzing three-dimensional scenes, Proceedings of the 1975 International Conference on Artificial Intelligence, pp. 669–673, used by permission of International Joint Conference on Artificial Intelligence, Inc.*) (*b*) Illumination of a workpiece by two strips of light which coincide at a background surface (*from W. Myers, Industry begins to use visual pattern recognition, Computer, 13(5):21–31, 1980*).

Illumination of a scene is an important factor affecting the complexity of vision algorithms. Arbitrary lighting of the environment often results in low-contrast images, specular reflections, shadows, and extraneous details. When control of the illumination is possible, the lighting system should illuminate the scene so that the complexity of the resulting image is minimized, while the information required for analysis is enhanced. For instance, the so-called structured lighting approach used in some industrial vision systems projects points, stripes, or grids onto the scene. The way in which these features are distorted by the presence of an object simplifies the computer interpretation of the scene. Two examples of this approach are shown in Fig. 1.

Segmentation. Segmentation is the process that breaks up a sensed scene into its constituent parts or objects. Hundreds of segmentation algorithms have been proposed over the past 15 years. This is still an active area of research because of its importance as the first processing step in any practical computer vision application. Although image sementation has proved to be a difficult task in unconstrained situations such as automatic target detection, the problems encountered in industrial applications can be considerably simplified by special lighting techniques such as those discussed above.

Segmentation algorithms are generally based on one of two basic principles: discontinuity and similarity. The principal approach in the first category is edge detection. The principal approaches in the second category are thresholding and region growing. Most edge detection techniques for industrial applications are based on the use of spatial convolution masks in order to reduce processing time. The idea is to move a mask over the entire image area, one image element location at a time, and, at each location, to compute a measure proportional to discontinuity (for example, the gradient) in the image area directly under the mask. Thresholding is by far the most widely used approach for segmentation in industrial applications of computer vision. There are two reasons for this. First, thresholding techniques (in their simpler forms) are fast, and in addition, they are quite straightforward to implement in hardware. Second, the lighting environment is usually a controllable factor in industrial application; this results in images that often readily lend themselves to a thresholding approach for object extraction. Region growing techniques are applicable in situations where objects cannot be differentiated from each other or the background by thresholding or edge detection. Although region growing has been used extensively in scene analysis, it has not found wide applicability in industrial applications because this method is usually impractical from a computational or hardware implementation point of view, and many of the problems which would require region growing for segmentation can usually be handled by special lighting or other enhancement techniques. A considerable amount of work has dealt with techniques that attempt to incorporate contextual information in the segmentation process. This includes the use of relaxation, plan-guided analysis, and the use of semantic information.

Description. The description problem in computer vision is one of extracting features from an object for the purpose of recognition. Ideally, these features should be independent of object location and orientation and should contain enough discriminatory information to uniquely differentiate objects. Descriptors for computer vision are based primarily on shape and amplitude (for example, intensity) information. Shape descriptors attempt to capture invariant geometrical properties of an object. Approaches for shape analysis and description are generally either global-oriented (region-oriented) or boundary-oriented. Global techniques include principal-axes analysis, texture, two- and three-dimensional moment invariants, geometrical descriptors such as P^2/A (where P is the length of the perimeter and A is the area), and the extrema of a region, topological properties, and decomposition into primary convex subsets. Boundary-oriented techniques include Fourier descriptors, chain codes, graph representations, of which strings and trees are special cases, and shape numbers. Boundary feature extraction is often preceded by linking procedures which fit straight-line segments or polynomials to the edge points resulting from segmentation. An example of object description by extraction of its boundary and principal directional axes is shown in Fig. 2.

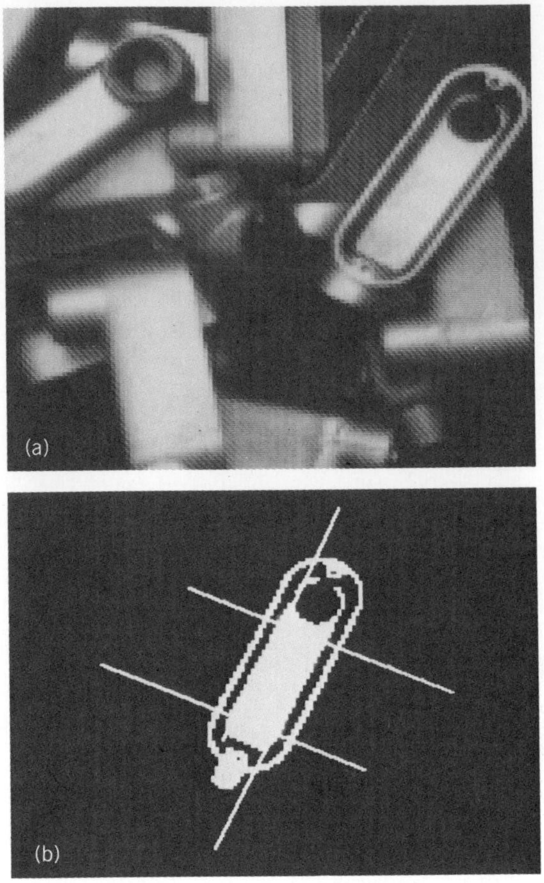

Fig. 2. Description of an object by extraction of its boundary and principal directional axes. (*a*) Bin of workpieces. (*b*) Description of one of the segmented objects. (*From J. Birk, R. Kelley, and H. Martins, An orienting robot for feeding workpieces stored in bins, IEEE Trans. SMC, SMC-11:151–160, 1981*)

Recognition. Recognition is basically a labeling process; that is, the function of recognition algorithms is to identify each segmented object in a scene and to assign a label (for example, wrench, seal, or bolt) to that object. Recognition approaches presently in use may be subdivided into two principal categories: decision-theoretic and structural. Decision-theoretic techniques are based on the use of decision (discriminant) functions. Given M object classes, $\omega_1, \omega_2, \ldots, \omega_M$, the basic problem in decision-theoretic pattern recognition is to identify M decision functions, $d_1(x)$, $d_2(x)$, . . . , $d_M(x)$, with the property that, for any pattern x^* from class ω_i, $d_i(x^*) > d_j(x^*)$, for $j = 1, 2, \ldots, M$, $j \neq i$. The objective is to find M decision functions such that this condition holds for all classes with minimum error in classification.

Decision-theoretic methods deal with patterns on a quantitative basis and largely ignore structural interrelationships among pattern primitives. Structural methods of pattern recognition attempt to describe fundamental relationships among pattern primitives via discrete mathematical modes. Here, the most widely used method is syntactic pattern recognition in which concepts and results from formal language theory provide the basic mechanisms for handling structural descriptions. The existence of a recognizable and finitely describable structure is essential in the success of the syntactic approach. Basically, a formal grammar is developed to generate elements

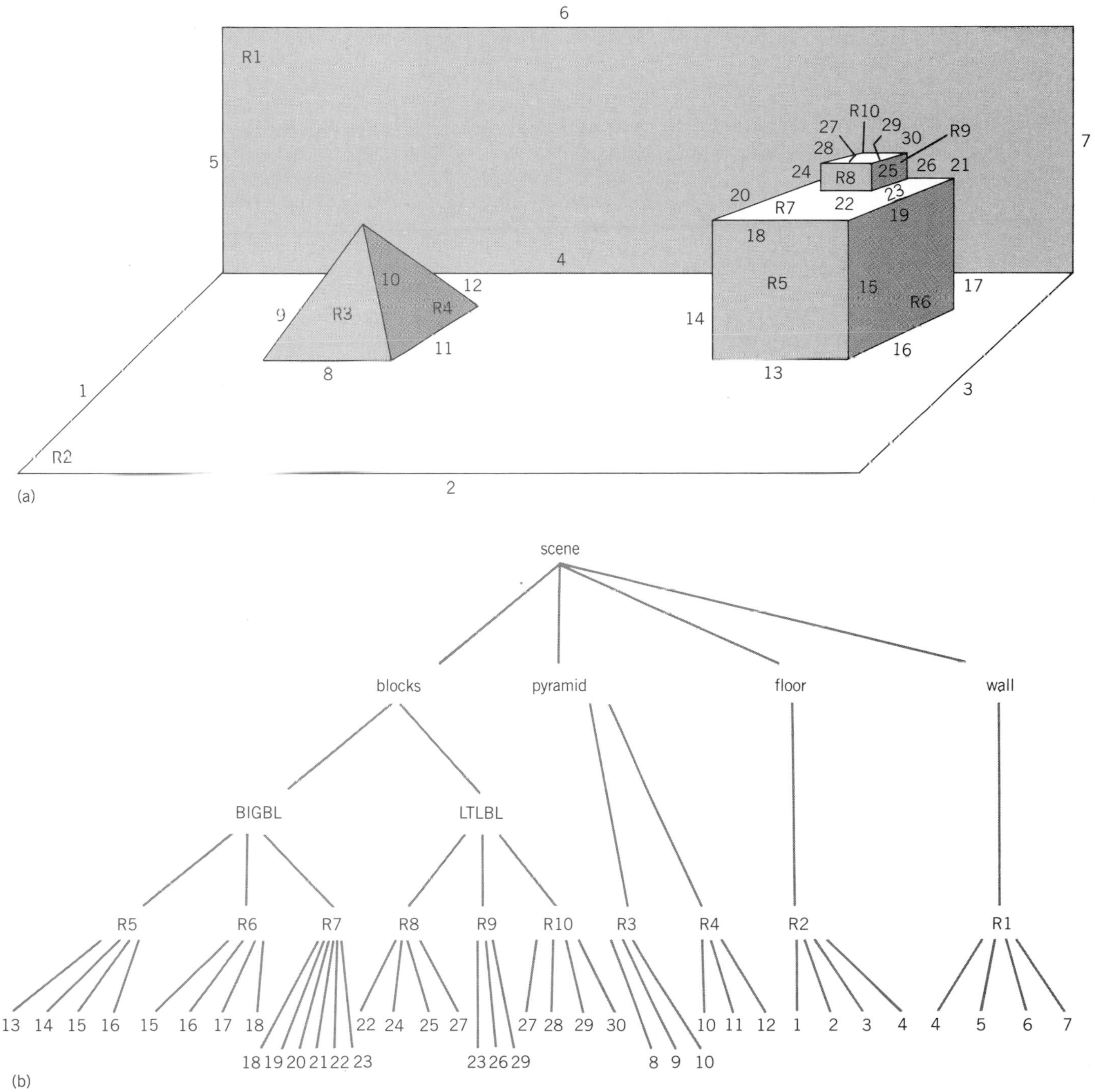

Fig. 3. Hierarchical representation of a scene. (*a*) Scene with blocks, pyramid, floor, and wall. (*b*) Hierarchical representation. (*After M. G. Thomason and R. C. Gonzalez, Database representation in hierarchical scene analysis, in L. Kanal and A. Rosenfeld, eds., Progress in Pattern Recognition, Elsevier-North Holland, 1982*)

of a language that defines a pattern class, and an automaton (or equivalently, a parsing algorithm) is developed to recognize precisely that same language. Such a language may consist of strings of primitives and relational operators (for example, directed line segments along the boundary of a polygonal representation of a workpiece) or of higher-order data structures such as trees, graphs, and webs.

One of the most significant recent extensions of syntactic techniques has been the inclusion of semantic evaluations simultaneously with syntactic analysis by means of attributed grammars. In this approach, a pattern primitive is defined by two components: a token or symbol from a finite alphabet, and an associated list of attributes consisting of logical, numerical, or vector values. The syntactic rules provide the basic structural description, while the semantic rules assign meaning to that description.

Interpretation and the use of models. In this discussion, interpretation is viewed as the process which endows a vision system with a higher level of conception about its environment. Sensed information, tasks to be performed, and types of parts to be handled are all essential items in establishing the level of competence and adaptability of a vision system. Given the limited state of development in "truly intelligent" vision systems, careful definition of a constrained set of operating conditions is essential. This is usually accomplished via the use of models.

The structure and complexity of a model depends on the stage of visual processing in which it is used. Computer vision techniques may be divided into three basic levels of processing: low-, medium-, and high-level vision. Although this division is somewhat arbitrary, it does provide a convenient method for categorizing the various processes that are inherent components of a computer vision system.

Low-level vision techniques attempt to extract "primitive" information from a scene. Examples of the use of models in low-level vision procedures range from modeling the characteristics of incident and reflected light properties of a body, to the detection of edge segments in a scene by modeling an edge as an abrupt change in intensity amenable to detection by gradient operators. Medium-level vision refers to procedures which use the results from low-level vision to produce structures that somehow carry more meaning than the elements extracted by the low-level vision process. Medium-level vision processes include edge linking, segmentation, description, and recognition of individual objects. High-level computer vision may be viewed as the process that attempts to emulate cognition. At this level of processing, the present knowledge and understanding of a suitable model is considerably more vague and speculative. While models for low- and medium-level vision tend to be rather specific in nature, a model for high-level vision encompasses a considerably broader spectrum of processing functions, ranging from the actual formation of a digital scene through interpretation of interrelationships between the objects in a scene. Figure 3 illustrates modeling of a scene by decomposing it into successively simpler elements. The simplest element considered in this case is an edge. Thus, regions are composed of edges, objects are composed of regions, and the scene is composed of objects.

For background information *see* ARTIFICIAL INTELLIGENCE; CHARACTER RECOGNITION: COMPUTER GRAHICS; DECISION THEORY; INDUSTRIAL ROBOTS; TELEVISION CAMERA in the McGraw-Hill Encyclopedia of Science and Technology.

[R. C. GONZALEZ]

Bibliography: H. G. Barrow and J. M. Tenebaum, Computational vision, *Proc. IEEE*, 69:572–595, 1981; G. G. Dodd and L. Rossol, eds., *Computer vision and Sensor-Based Robots*, 1979; K. S. Fu, R. C. Gonzalez, and C. S. G. Lee, *Introduction to Robotics*, 1984; R. C. Gonzalez and R. Safabakhsh, Computer vision techniques for industrial applications and robot control, *Computer*, 15(12):17–32, 1982.

Conodont

Conodonts are represented in the fossil record by toothlike phosphatic microfossils which range in age from Cambrian to Triassic, a time span of 300 million years. Being hard and durable, they are readily preserved. They are often very abundant, and within the last 25 years they have proved to be of major importance in correlating sequences of sedimentary rocks. While conodont skeletal elements have been known for 125 years, the nature of the evidently soft-bodied animal that bore them has been unknown until very recently. The nature and affinities of the conodont animal have been the subject of considerable speculation and debate, and indeed Klaus Müller wrote in 1981 that "the origin of conodonts is considered to be one of the most fundamental unanswered questions in systematic paleontology." A specimen from the Lower Carboniferous rocks near Edinburgh described by Derek Briggs, Euan Clarkson, and Richard Aldridge in 1983 provides the first reliable evidence for the nature of the animal which bore conodont elements.

The elements are normally found as isolated, discrete specimens and are often extracted by dissolving rock samples in weak acids. For many years, conodont taxonomists treated individual element types as separate species, but the discovery of "natural assemblages" on bedding surfaces provided direct evidence that each animal possessed a skeletal apparatus consisting of several different kinds of elements. These assemblages were first described in 1934, independently by H. Schmidt and H. W. Scott, and several hundred similar associations, each representing the skeletal remains of an individual animal, have now been found. Most are from shales, as their preservation requires that the soft parts of the animal decay undisturbed by currents, so that the skeleton remains intact. Some elements may also become fused into "clusters" during the

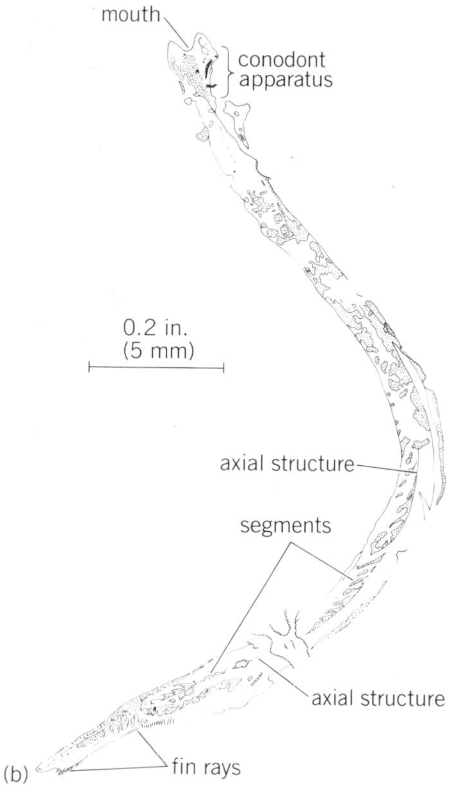

Fig. 1. The unique specimen of the conodont animal discovered in Lower Carboniferous rocks in Edinburgh. (*a*) Photograph of specimen. (*b*) Outline of the specimen for comparison. (*From D. E. G. Briggs, E. N. K. Clarkson, and R. J. Aldridge, The conodont animal, Lethaia, 16:1–14, 1983*)

diagenesis of the sediments in which they occur, and these, too, give an indication of the composition of the conodont apparatus. Although most conodont species are still known only from the elements scattered in the sediment, it is now common procedure in conodont taxonomy to reconstruct their skeletal apparatuses by using morphological and statistical criteria. However, although the skeletal apparatuses of some conodont animals had been known for almost 50 years, the nature of the organism itself had remained an enigma.

Morphology of conodont animal. In the single specimen known, the body is elongate, about 1.6 in. (40 mm) long and 0.08 (2 mm) wide (Fig 1*a* and *b*). The only skeletal structures are the conodont elements, which occur in the head region (Fig. 2). No distinct features are evident in the anterior part of the trunk, probably due to the nature of the preservation. There is a distinct axial line in the posterior two-thirds of the trunk, however, on one side of which are posteriorly inclined structures which may represent segments. The posterior extremity of the trunk preserves evidence of fin rays indicating the presence of a caudal and a posterolateral fin.

The head region, apart from the conodont apparatus, is relatively poorly preserved (Fig. 2). Two lobes, one on each side of the head, project anteriorly beyond the conodont apparatus. They flank a median opening which may lead to the mouth. The conodont apparatus is composed of three separate groups of elements. The first, anterior, group consists of about eight comblike (ramiform) elements, in most cases with the denticles angled posteriorly. Like the other elements, they show characteristic white matter in the denticles. The second group consists of a pair of arched (ozarkodinoform) ele-

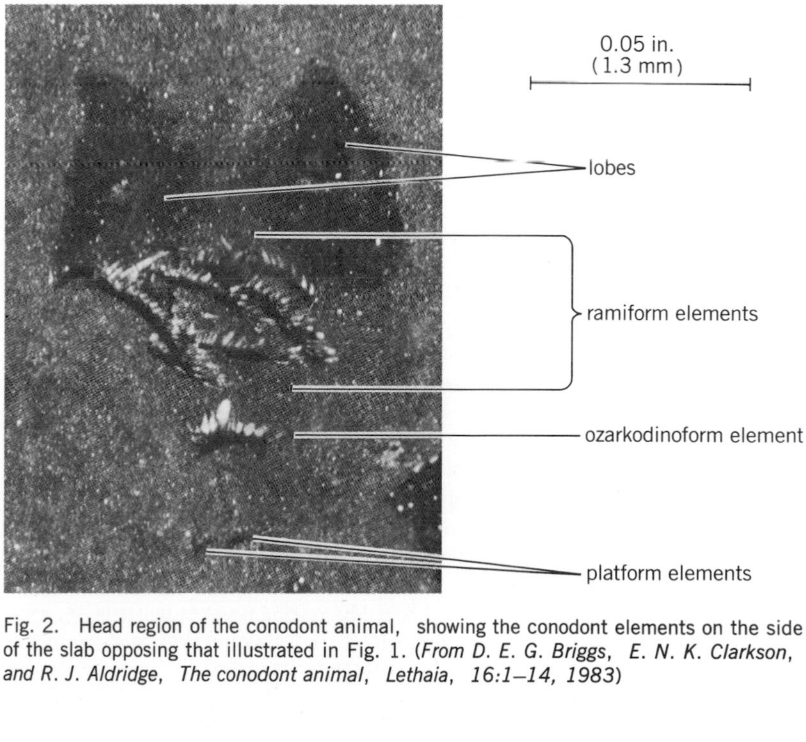

Fig. 2. Head region of the conodont animal, showing the conodont elements on the side of the slab opposing that illustrated in Fig. 1. (*From D. E. G. Briggs, E. N. K. Clarkson, and R. J. Aldridge, The conodont animal, Lethaia, 16:1–14, 1983*)

ments, the slab having split between them so that one is preserved on the part, the other on the counterpart. The third group consists of a pair of platform elements, which are not as well displayed as the rest because they are deeply embedded in matrix. It has been possible to expose them partially, however, by careful preparation. Unfortunately, it is the platform element that is most characteristic and provides the name for the apparatus (and therefore for this conodont animal); because it is partially concealed, the identification can only be tentative, that is, *Clydagnathus*? cf. *cavusformis*.

Affinities of conodont animal. Two fossil organisms have previously been described as possible conodont animals. The first is the soft-bodied "conodontochordate" described from the Upper Carboniferous of Montana in 1973 by W. G. Melton and Scott. In this the conodonts occur in the gut and are not present as apparatuses—the conodontochordate unfortunately turns out to have eaten the conodont animal. Such a possibility seems remote in the case of the newly discovered specimen from Scotland in which the arrangement of the conodonts is characteristic of those in previously reported bedding plane assemblages, and the apparatus is in the expected position in the head region.

The second possible conodont animal was described in 1976 from the Middle Cambrian Burgess Shale of British Columbia by Simon Conway Morris. This flattened annulated animal, *Odontogriphus*, is quite different from that described here. It possesses about 20 toothlike structures which may represent cone-shaped conodont elements, arranged in a double loop around the mouth. They are interpreted as the supports for the tentacles of a lophophore. If these structures are indeed conodonts, this reinforces the suggestion that at least some Cambrian "conodonts" are unrelated to later forms. In this case they appear to belong to the lophophorates.

The morphology of the Scottish conodont animal suggests the possibility of affinity with either of two living phyla, the chordates and chaetognaths (arrow worms). The chordates have received more attention then any other phylum in discussions of conodont affinity, which have included the possibility of relationship to agnathans (including cyclostomes), selachians, ostracoderms, placoderms, and different kinds of primitive vertebrate. In 1982, however, Hubert Szaniawski showed that the grasping spines of the living chaetognath *Sagitta* are morphologically identical to some Cambrian conodontlike elements, thus renewing interest in the possibility of chaetognath affinity.

Of the chordates, the conodont animal shows greatest similarity to the textbook examples of primitive morphology: amphioxus (*Branchiostoma*) and the ammocoete larva of lampreys, both of which are eellike, are laterally flattened, and bear fins. Chaetognaths, which also have fins, are dorsoventrally flattened. Unfortunately it is not possible to reconstruct the three-dimensional appearance of the conodont animal based on a single flattened specimen. Hence the important difference between chordates and chaetognaths in the plane of flattening of the body cannot be used in assessing the likely affinity of the conodont animal.

It is tempting to interpret the divisions of the trunk in the conodont animal as V-shaped and to draw the obvious parallel with the muscle blocks (myotomes) of amphioxus and fish, but the evidence for this is inconclusive. There is no structure in the chaetognaths, however, comparable to this apparent segmentation. The series of eggs which may occur in the ovaries is only superficially similar, and the cuticle of chaetognaths is so thin that the "segmentation" is unlikely to represent cuticular banding. The axial trace might compare with a number of different structures in a chordate. It appears too insubstantial to represent the notochord, however, and the gut would normally terminate anterior of the tail. It resembles the horizontal septum dividing the segments in chordates above the cyclostome level, but the other evidence is not consistent with the animal being a gnathostome. The axial structure might equally represent the dorsoventral mesentery which subdivides the trunk and tail coelom in chaetognaths.

It is the conodont elements themselves that provide the greatest obstacle to arguing a chordate or chaetognath affinity for the animal. Had there been an obvious living homolog for the conodont apparatus, there would not have been such wide-ranging speculation and discussion about conodont affinities. The conodonts might have functioned as teeth or as supports for the tentacles of a branchial basket. No chordate, however, has teeth arranged in a configuration similar to the conodont apparatus. It has been suggested that the conodonts might be equivalent to the lingual teeth in the living hagfishes, but these consist only of horny cusps and lack skeletal supports. The comblike ramiform conodonts provide possible candidates for the internal supports of filtering tentacles in a primitive gill basket. In this case, however, it is difficult to suggest a complementary function for the paired arched and platform conodonts behind them.

The Cambrian conodontiform elements to which the grasping spines of chaetognaths show such a striking resemblance belong to a group known as protoconodonts. They were secreted internally in a manner distinct from the euconodonts or "true" conodonts, which include all post-Ordovician examples. The relationship between protoconodonts and later conodonts is not clear. Thus even if the anterior ramiform conodonts in the conodont animal were used for grasping, a relationship to the chaetognaths remains conjectural.

Phylum Conodonta. Determination of the affinities of the Carboniferous conodont animal from Edinburgh is hampered to a degree by the discovery of only one specimen to date. The evidence available, however, demonstrates neither chordate nor chaetognath affinity and emphasizes the uniqueness of conodonts. The animal is preserved in an intertidal stromatolite, an unusual environment for conodonts, where it occurs in association with a fauna of crustaceans, probable hydroids, worms, and palaeoniscid fish. Isolated conodonts and some fused clusters have been extracted from the same bed, but until additional specimens showing the soft tissues of the animal are discovered there is no apparent reason to disagree with those who have assigned the conodonts to a separate phylum, the Conodonta. Although the function of the conodonts remains equivocal, Briggs, Clarkson, and Aldridge consider it most likely that they were teeth, the anterior group of comblike ramiform conodonts grasping prey which was impaled and pulled into the mouth to be processed by the arched and platform pairs.

For background information *see* CHAETOGNATHA; CHORDATA; CONODONT in the McGraw-Hill Encyclopedia of Science and Technology.

[DEREK E. G. BRIGGS; EUAN N. K. CLARKSON; RICHARD J. ALDRIDGE]

Bibliography: D. E. G. Briggs, E. N. K. Clarkson, and R. J. Aldridge, The conodont animal, *Lethaia*, 16:1–14, 1983; D. E. G. Briggs and E. N. K. Clarkson, The Lower Carboniferous Granton "shrimp-bed," *Spec. Pap. Palaeontol.*, 30:161–177, 1983; K. J. Müller, *Treatise on Invertebrate Paleontology*, W, Suppl. 2, *Conodonta*, pp. 478–482, 1981; H. Szaniawski, Chaetognath grasping spines recognized among Cambrian protoconodonts, *J. Paleontol.*, 56:806–810, 1982.

Coulomb excitation

Coulomb excitation provides the traditional means of measuring the electromagnetic moments of nuclei, and information collected with this probe has been instrumental in establishing and refining detailed models of nuclei since the early 1950s. The recent development of accelerators providing energetic and precisely defined beams of very heavy nuclei has made possible the study of Coulomb excitation in a new regime—one in which the Coulomb forces between the accelerated and struck nuclei are so strong that qualitatively new phenomena emerge.

Process. Coulomb excitation of atomic nuclei occurs as a result of the Coulomb forces acting between an accelerated projectile and a target nucleus as the two undergo a collision. These forces are exerted between the positively charged protons bound within the respective nuclei. Prior to the collision, the nuclei are in their energetically lowest, or ground, states, but the Coulomb interaction produces transitions that promote one or both of the interacting nuclei into excited states. The process can be visualized most easily when the target nucleus has a relatively rigid but nonspherical shape, as is the case for many rare-earth and actinide nuclei. As shown in Fig. 1, the torque exerted on the ^{238}U target by the Coulomb field of the incident projectile tends to induce rotational motions corresponding to the possible quantized excitations of the struck nucleus. When the kinetic energy of the projectile is not high enough for the Coulomb repulsion between the colliding nuclei to be completely overcome, the short-range nuclear forces do not come into play, and consequently the excitation occurs solely as a result of the Coulomb forces. Under these circumstances the measured probabilities for transitions to excited states can be related to the electromagnetic properties and shapes of the nucleus with a high degree of accuracy.

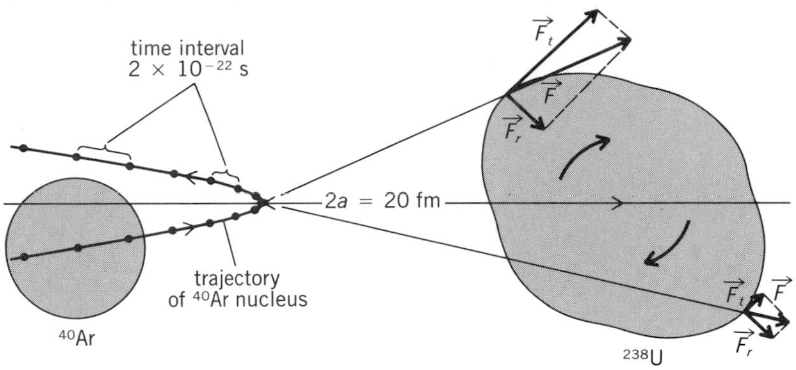

Fig. 1. Simple picture of the Coulomb excitation of a ^{238}U nucleus by the scattering of a 140-MeV ^{40}Ar projectile through an angle of 160°. The resolution of the electrical force $\vec{F}$ exerted by the ^{40}Ar nucleus into a radial component $\vec{F}_r$ and tangential component $\vec{F}_t$ is shown for two points on the ^{238}U nucleus. The tangential components of the electrical force exert a net torque on the ^{238}U nucleus which gives it a rotational excitation. (*After F. K. McGowan and P. H. Stelson, Coulomb excitation, in J. Cerny, ed., Nuclear Spectroscopy and Reactions, pt.C, pp. 3–54, Academic Press, 1974*)

In an approximation based on first-order perturbation theory, the semiclassical cross section $\sigma_{E\lambda}$ for Coulomb excitation when a projectile with atomic number Z collides with relative velocity v with a target nucleus can be written as in the equation below,

$$\sigma_{E\lambda} = (Ze^2/hv)^2 \, a^{-2\lambda+2} \, B(E\lambda) \, f_{E\lambda}(\xi)$$

where h is Planck's constant. The quantity λ represents the multipolarity of the transition and takes on integer values, for example $\lambda = 1$ for dipole excitation, $\lambda = 2$ for quadrupole, and so forth. The quantity a denotes one-half the distance of closest approach of the two nuclei in a head-on collision. The equation illustrates that for the typical case of quadrupole excitation ($\lambda = 2$) the cross section increases as the closest approach decreases. As a consequence, particle accelerators capable of producing beams with sufficient energy to reduce this distance by partially overcoming the Coulomb repulsion between the projectile and target nuclei are required for efficient Coulomb excitation measurements. The function $f_{E\lambda}(\zeta)$ in the equation depends on the excitation energy of the level populated in the collision, and describes the fact that the excitation probability decreases with increasing excitation energy. The remaining quantity, $B(E\lambda)$, is the reduced transition probability. It contains the information about the electromagnetic properties of the nucleus. The determination of $B(E\lambda)$, together with additional electromagnetic properties that are neglected in the approximations leading to the above

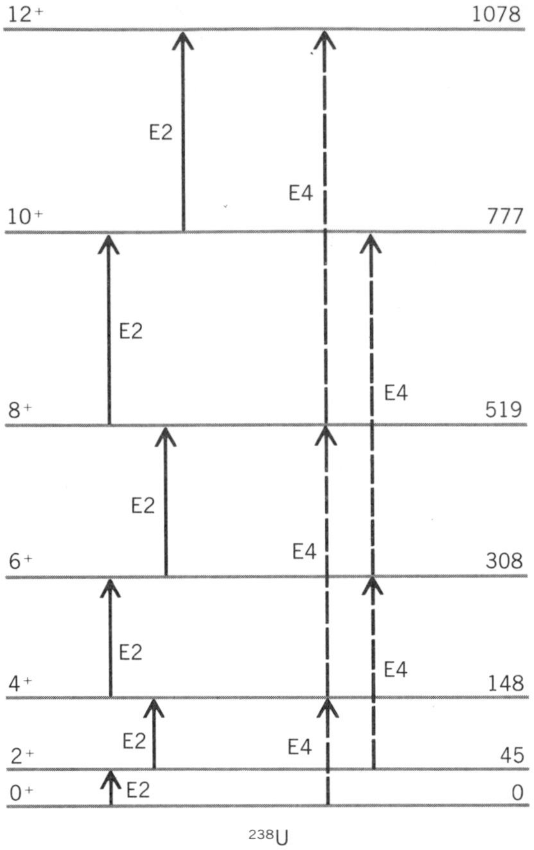

Fig. 2. Multiple Coulomb excitation of several excited states of ^{238}U. The angular momenta and intrinsic parities of the states are labeled at the left, and excitation energies in units of kiloelectronvolts appear at the right. Quadrupole (E2) and hexadecapole (E4) excitations are indicated.

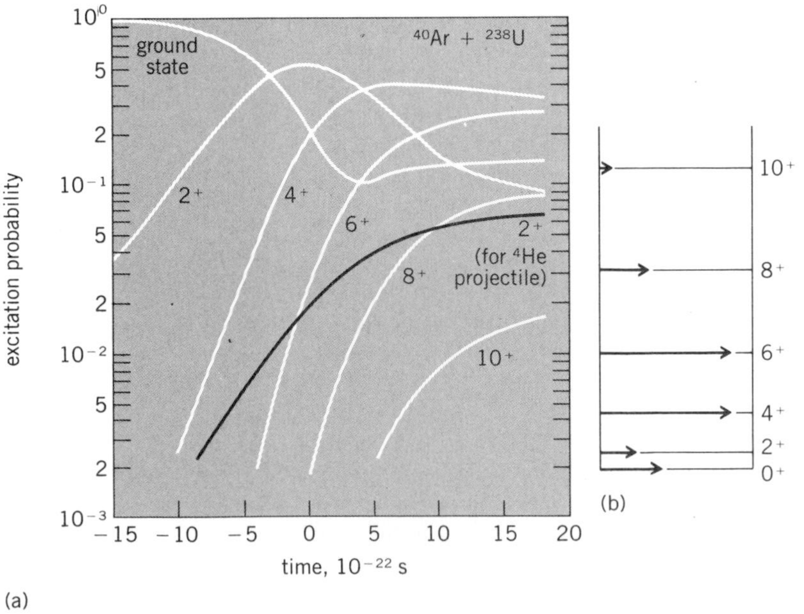

Fig. 3. Excitation of the ground-state rotational band of ^{238}U for the situation illustrated in Fig. 1. (*a*) Excitation probabilites as a function of time. (*b*) Level diagram in which the final relative probabilities for excitation of the different rotational states are shown by the lengths of the arrows. (*After F. K. McGowan and P. H. Stelson, Coulomb excitation, in J. Cerny, ed., Nuclear Spectroscopy and Reactions, pt. C, pp. 3–54, Academic Press, 1974*)

equation, constitutes the traditional goal of Coulomb excitation measurements.

Heavy projectiles. Recent developments in accelerator technology have resulted in the production of beams of essentially all nuclear species—from protons through uranium nuclei—having sufficient energy to carry out Coulomb excitation measurements. When heavy nuclear projectiles are employed in place of the traditional protons or alpha particles, the cross sections increase dramatically owing to the greater electrical charge carried by the heavier projectiles. One important consequence is that multiple Coulomb excitation processes can compete effectively with the normal single excitation mechanism (Fig. 2). Combinations of multiple quadrupole excitation ($\lambda = 2$) occur readily and lead to the population of states of higher angular momentum than can be achieved via single excitation. Bombardments of ^{238}U with ^{208}Pb projectiles have extended the pattern illustrated in Fig. 2 to states having angular momenta in excess of $I = 30\ \hbar$ (where $\hbar$ is Planck's constant divided by 2π). Such experiments take maximum advantage of the projectile Z^2 dependence of Coulomb excitation contained in the above equation. They have opened a new region of the nuclear spectrum to precise exploration and have revealed how the nucleus responds to the enormous centrifugal stresses associated with high angular momenta. The nucleus can carry angular momentum principally in two ways: by rotation of the nucleus as a whole, if the nucleus is deformed; and by the alignment of the orbital motion of individual nucleons along the rotation axis. Coulomb excitation experiments with heavy projectiles have revealed in detail how the Coriolis force in a rotating nucleus tends to counteract the tendency of nucleons to pair off. With increasing angular momentum the force occasionally succeeds in breaking apart a pair of high-spin nucleons which then align themselves as noted above. The abrupt alignment causes an irregularity in the sequence of excited states which is known as backbending.

Population inversion. Another important consequence of the enhanced Coulomb excitation obtained with heavy projectiles is the population inversion of energy levels that occurs in the colliding nuclei. As the colliding pair approach and then recede, Coulomb excitation takes place continuously. The results of a calculation describing Coulomb excitation of ^{238}U by a 140-MeV beam of ^{40}Ar nuclei are shown in Fig. 3. The ^{238}U nucleus enters the collision in its ground state, but as time proceeds the excited energy levels characterized by angular momentum $I = 0, 2, 4$, and so forth, become progressively more populated. After the nuclei have receded from one another (Fig. 3*b*), the nucleus is left predominantly in excited configurations for the particular trajectory considered. These calculations are in excellent agreement with experimental studies. The excited states decay back to the ground state through emission of gamma radiation, but this occurs on a time scale which is long compared to the reaction time. Thus, Coulomb excitation with heavy projectiles achieves a population inversion.

Rearrangement reactions. The existence of the population inversion is particularly significant at the distance of closest approach, which corresponds to time = 0 on the abscissa of Fig. 3. It is at this point that rearrangement and other nuclear reactions achieve their greatest likelihood, which can be appreciable if the bombarding energies are raised to correspond to the top of the Coulomb barrier. A typical rearrangement reaction involves the transfer of one or more nucleons—protons or neutrons, or both—from one nucleus to the other under the influence of the nuclear, or strong, force. When traditional projectiles such as protons and alpha particles are used, these reactions are governed by structure overlap integrals involving the target ground state. With heavier projectiles the population inversion associated with the strong Coulomb excitation dictates that overlap integrals involving excited states also contribute to the reaction process. The measured cross sections depend on the overlap integrals, and the latter reflect important features of nuclear structure. Thus, population inversion from Coulomb excitation can provide experimental access to aspects of nuclear structure that are inaccessible with light-mass projectiles.

In the presence of strong Coulomb excitation, rearrangement reactions can proceed simultaneously through the ground and several excited states of the target nucleus. Quantum mechanics dictates under such circumstances that the reaction amplitudes associated with the various routes combine coherently and thereby interfere with one another. The measurement of the interference pattern provides a very sensitive test of the models for both the nuclear structure and the reaction process. Reactions in which two neutrons are transferred between the interacting nuclei in the presence of moderate Coulomb excitation have been measured by using carbon and oxygen projectiles. The detailed theoretical reproduction of the observed interference patterns shows that the reaction mechanism and the nuclear structure of low-lying levels of permanently deformed nuclei are well described by current theories. The new generation of accelerators just now becoming available for experimentation will permit future studies with comparable precision but with much heavier projectiles. These experiments will provide precision tests of nuclear models at much higher excitation energies and angular momenta.

For background information see COULOMB EXCITATION; COULOMB'S LAW; NUCLEAR REACTION; NUCLEAR SPECTRA; NUCLEAR STRUCTURE; SCATTERING EXPERIMENTS (NUCLEI) in the McGraw-Hill Encyclopedia of Science and Technology. [KARL A. ERB]

Bibliography: K. Alder and A. Winther, *Coulomb Excitation*, 1966; R. Bass, *Nuclear Reactions with Heavy Ions*, 1980; F. K. McGowan and P. H. Stelson, Coulomb excitation, in J. Cerny (ed.), *Nuclear Spectroscopy and Reactions*, pt. C, pp. 3–54, 1974.

Crown gall

Crown gall is a cancerous disease that afflicts plants. It is characterized by proliferating tumorous outgrowths that commonly appear on the crown of the infected plant—that is, at the juncture of the root and the shoot. The disease occurs in hundreds of different species of higher plants, including the flowering plants and conifers and their relatives. Remarkably, one major group of plants, including grasses, palms, and orchids, is not susceptible to the disease.

In 1907 Erwin F. Smith first showed that a specific bacterium, *Agrobacterium tumefaciens*, is responsible for the crown gall disease. It is still the only bacterium that has been identified as a direct causal agent of cancer in any higher organism.

Cell transformation. Crown gall poses a question that is fundamental to all growing organisms and to cancer growth in particular. How does *Agrobacterium* succeed in transforming normal cells—cells that have already become differentiated and that would ordinarily undergo little or no further division—into cells that divide uncontrollably and have the capacity for essentially perpetual division? The bacterium does not itself enter the plant cells it infects. Consequently, it must produce a tumor-causing agent, called the tumor-inducing principle, which enters the plant cells and alters their growth.

When it became possible to grow plant tissues in a sterile culture medium, it was shown that cultured tumor cells differ from cultured normal cells in that tumor cells will grow on simple media consisting only of salts and sugar. Normal cells require hormones and often additional organic compounds for growth. Based on these data, it was proposed that the transformation from normal cell to cancerous cell results from the appearance of gene-specified reactions for the chemical conversion of nutrients into certain organic compounds in cells that would ordinarily not produce these compounds. These changes enable the cancerous cell to produce hormones and other organic substances, the absence of which normally limits cell growth and division.

This theory can account for the cancerous state in all organisms, and considerable evidence to support the concept has since been found in animal tumor systems. This remarkable and fundamental insight has pointed for the past 30 years to gene control as the principal altered feature at the crux of all abnormal cell growth. But how these alterations are brought about and maintained remains unknown. Also, scientists have yet to determine the compounds and chemical pathways in the cell that are ultimately responsible for uncontrolled growth.

Plasmids. When the new techniques of molecular biology became available, a correlation was established between the presence of a certain plasmid in *Agrobacterium* and the ability of the bacterium to induce tumors. Bacterial plasmids are small circular molecules of deoxyribonucleic acid (DNA). Plasmids functon in bacteria exactly as small chromosomes do; though they are not essential, they may allow the bacteria to grow in adverse environments, to use unusual organic compounds, or (as in the case of *Agrobacterium*) to cause disease. In 1976 it was shown that crown gall tissue cultures contained *Agrobacterium* plasmid DNA. Since these tissue cultures were free of the bacterium itself, the source of the plasmid DNA had to be the tumor cell.

It is now known that through the action of *Agrobacterium* a specific piece of the plasmid molecule (about 10% of the intact plasmid) is inserted and becomes stably integrated into the chomosomal DNA of the host plant. Thus, all or part of the plasmid must move from the bacterium to the plant cell during the cell's transformation from a healthy to a cancerous state. Once this piece of the plasmid molecule has become part of the plant's DNA, it is then replicated at each tumor cell division right along with the DNA that constitutes the genes of the normal plant.

The identity of the tumor-inducing principle is now established as a piece of *Agrobacterium* plasmid DNA, although rigorous proof that this DNA alone is entirely responsible for the transformation is still lacking. Evidence does show that recovery of organized growth in plant tissue occurs with loss of this foreign DNA, which therefore must be present for the tumorous growth to continue. Thus, it is apparent that *Agrobacterium* is a natural genetic engineer, able to introduce genetic material into plant chromosomes.

Role of amino acids. Since it is known that plasmids typically convey useful properties to their bacterial host, the question naturally arises as to what possible benefit this tumor plasmid could have for *Agrobacterium*—the more so because plasmid-free (and therefore harmless) strains of the bacterium are a common and successful component of the bacteria in ordinary soil. At least a partial explanation for this unique association between bacterium, plasmid, and host plant emerged from studies which began in the 1950s. A new amino acid was discovered in bacteria-free cultures of crown gall. This compound was shown to be a derivative of lysine, a common protein amino acid. A family of similar amino acid derivatives, all isolated from crown gall tumors and mostly based on another amino acid, arginine, has since been found. These amino acid derivatives were not detected in the normal plant and were of unknown function.

The tumor-causing ability in *Agrobacterium* is highly correlated with the ability of the bacterium to use these amino acid derivatives as a source of energy. This was the first physiological activity of the virulent bacteria shown to be correlated with the capacity to induce tumors. These compounds also promote tumor growth when they are added to small tumors soon after their inception, indicating that they function to promote tumor development. However, the bacteria themselves have never been shown to make these compounds; it is only the tumors they induce which exhibit this capacity. The ability of agrobacteria to utilize these amino acid derivatives is a plasmid determined process. Moreover, a second plasmid gene determines if the tumor will produce the amino acid derivatives. Although both genes occur on plasmids of tumor-forming bacteria, either or both may be lost without loss of the capacity to induce tumors; thus, they are not themselves essential for transformation.

The nutritional value of these tumor products to the bacterium seems ample reason for *Agrobacterium* to have evolved and maintained the tumor-causing plasmid. However, these amino acid derivatives also facilitate the transfer of plasmid DNA from bacteria that carry the plasmid to related bacteria that do not carry it. In effect, the altered amino acids act on *Agrobacterium* by stimulating this molecular genetic exchange. The crown gall tumor thus has an even more important role in the life of the bacterium. It constitutes an environment where related but nontumorigenic bacteria multiplying in the same wound as the tumor-producing bacteria are impregnated with the plasmid, possibly accompanied by some of the donor bacterium's chromosomal DNA. The tumor then may be viewed both as the plasmid's way of colonizing new bacterial hosts and as the bacterium's way of accomplishing genetic exchange and the potential adaptive benefits that accrue.

Genetic engineering. *Agrobacterium* as a natural genetic engineer for plants is well advanced beyond what humans using DNA and plant cells can presently accomplish in the laboratory, and practical use can be made of this natural engineer. After isolating the *Agrobacterium* plasmid and splicing into it a new gene, one can then transfer the plasmid into a plasmid-free *Agrobacterium*. When this bacterium infects a plant, it will then transfer the plasmid DNA to the plant cell. This plasmid DNA carries the newly inserted DNA sequence of the spliced gene, along with the tumor-causing DNA, into the plant chromosome. If the normal protein-synthesizing apparatus of the cell correctly understands this new gene and correctly transcribes and translates it, the result is the appearance of a new protein that can change some essential characteristic of the cell. The cell will have the new properties desired by the investigator in the original selection of the DNA to be inserted into the plasmid. This sort of experiment has actually been accomplished, and resulted in the production of a bean-seed protein in a sunflower tumor cell, the so-called sunbean. The remaining problem is how to suppress the tumorous character of these cells so that organized growth can occur; no one wants to cultivate a field of tumors.

Bacterial adherence. In addition to the tumor-inducing principle, there is another requirement for successful infection of plant cells by agrobacteria. Early models of the initiation of crown gall tumors presented a straightforward picture. Virulent agrobacteria, containing the tumor-inducing principle, could be found in the watery wound of a plant's ruptured cells; the bacteria released their tumor-inducing principle into the aqueous medium, whereupon the plant cells took it in and became tumorous. But results obtained in the 1970s showed that to cause a tumor the bacteria must locate specific infection sites and adhere to them.

This kind of surface adherence or attachment is necessary before many infectious agents can cause disease. Thus *Agrobacterium* provides a model system to determine the nature of these adherence reactions which, unlike the adherence of insects to flypaper, exhibit selective specificity. When virulent agrobacteria are mixed with harmless related bacteria that compete with it for adherence sites, tumor initiation is reduced. Nonrelated bacteria do not have this effect. When the harmless competing bacteria are added to a wound medium even a few minutes after virulent bacteria have been introduced, tumor initiation is not reduced. This shows that virulent bacteria must locate and attach themselves to the available sites soon after they enter a wound.

This demonstration that specific *Agrobacterium* adherence is essential for tumor initiation in plants preceded most of the related studies in animals and is currently one of the best-characterized cases of host adherence by a cellular pathogen. Adherence involves a specific component of the surface of the bacterial cell envelope and an acid polymeric sugar in the outer portion of the cell wall of the plant. It is clear that a carbohydrate-carbohydrate interaction is responsible, although the exact structures of these surface carbohydrates in plant and bacterium have still to be determined.

The specificity of adherence of *Agrobacterium* to its host suggests complementary, intermeshing shapes and charges, a sort of lock-and-key relationship. This is exactly the kind of system that might help to determine host range and host specificity, that is, the variety of hosts susceptible to a particular pathogen and the ability of a pathogen to select a particular host to infect. By plasmid modification of adherence characteristics, this possibility has been demonstrated experimentally, and it was also shown that *Agrobacterium* does not adhere to a large group of plants which are nonhosts.

Since the cereal grains all belong to this latter group, the use of *Agrobacterium* to genetically modify corn, wheat, rice, or barley is now impossible. It may become possible if the bacterium could be modified so that it would adhere to these plants or, alternatively, if the cereal strains could be altered, either genetically or by direct chemical treatment, so that the bacterium would adhere to them. The value of such a procedure, if successful, makes efforts to extend the host range of *Agrobacterium* worthwhile.

Agrobacterium can adhere to the walls of cells taken from roots, stems, and leaves of host plants, but it does not adhere to cell walls taken either from embryonic parts of the plant or from crown gall tumors on the plant. This suggests that there is a change in cell wall structure and metabolism early in plant development, and that the cells that form these walls may revert to the embryonic state in the process of transformation from healthy to diseased state. The specific nature of these changes is still to be resolved, but may have a parallel in animals since the production of embryonic forms of several proteins has been observed in many animal tumors.

Unanswered questions. Although much is now known about the tumor-inducing factors of crown gall, there are still many unanswered questions. The mechanism of the bacterium-to-plant DNA transfer has still to be determined, and since this knowledge may help in the design for placing new genes in plants, it is very desirable. How the crucial piece of plasmid DNA both causes uncontrolled growth and activates several biosynthetic reactions can only be guessed at present. This information may be of use in developing effective chemical means to modify plant growth. Also, the questions of whether the altered amino acids, which seem thus far to be found only in these tumors, exist in the normal plant, and the nature of their functions there, have still to be resolved. Clearly much research and exciting potential applications of the results lie ahead.

For background information *see* CROWN GALL; GENETIC ENGINEERING; ONCOLOGY; PLANT PATHOLOGY in the McGraw-Hill Encyclopedia of Science and Technology.

[JAMES A. LIPPINCOTT]

Bibliography: K. A. Barton and W. J. Brill, Prospects in plant genetic engineering, *Science*, 219:671-676, 1983; M.-D. Chilton, A vector for introducing new genes into plants, *Sci. Amer.*, 248:50-59, 1983; J. A. Lippincott, B. B. Lippincott, and J. J. Scott, Adherence and host recognition in *Agrobacterium* infection, in *Current Perspectives on Microbial Ecology*, C. A. Reddy and M. J. Klug (eds.), American Society for Microbiology, in press; E. W. Nester and T. Kosuge, Plasmids specifying plant hyperplasias, *Annu. Rev. Microbiol.*, 35:531-565, 1981.

Cyclotron

Superconducting cyclotrons are accelerators that combine the technologies of precision isochronous cyclotrons and superconducting coils of the type used in large bubble chamber magnets. The result is small, cost-effective machines for accelerating heavy ions to tens of MeV per nucleon, because the large $B\rho$ product required to bend the ions is obtained from the high magnetic induction B generated by the high current density in the superconductor rather than by a large radius ρ. Not only the capital cost of the accelerator itself is reduced, but also building, shielding, and operating costs. The small size essential for these savings has imposed stringent requirements on some of the cyclotron subsystems, but this challenge has been overcome by the ingenuity of the designers. The first superconducting cyclotron is now in operation, four more are in various stages of construction, and several others are in the proposal stage.

Funded superconducting cyclotron projects

Laboratory	Midplane induction, teslas	Extraction radius, in. (cm)	First beam	Source of ions
Michigan State University phase 1	4.8	26 (67)	1982	Internal ion source
Chalk River	5.0	26 (65)	1984	Radial injection from 13-MV tandem
Milan University	4.9	34 (87)	1985	Radial injection from 16-MV tandem
Texas A&M	4.8	26 (67)	1986	Internal ion source
Michigan State University phase 2	5.3	39 (100)	1986	Radial injection from phase 1 cyclotron

The table gives some selected parameters for the funded superconducting cyclotron programs listed in order of the actual or expected dates for first beam. All have a midplane induction of approximately 5 teslas, roughly three times greater than that common to nonsuperconducting cyclotrons. Several advantages other than reduced cost accrue from this high midplane induction. The pole steel is saturated over the entire operating range, and this traditionally nonlinear component of the magnetic induction is reduced to a constant, greatly simplifying the calculation of the midplane field. As a result, there is excellent agreement between the measured and calculated fields, and only minor shimming has been required to achieve the desired profiles. Furthermore, the saturated steel significantly reduces variations in the magnetization; the magnets do exhibit a small hysteresis effect, but providing the operating point is approached from a lower excitation, the resettability of the midplane field is remarkably good. This simplifies reestablishing an ion beam that has been previously accelerated.

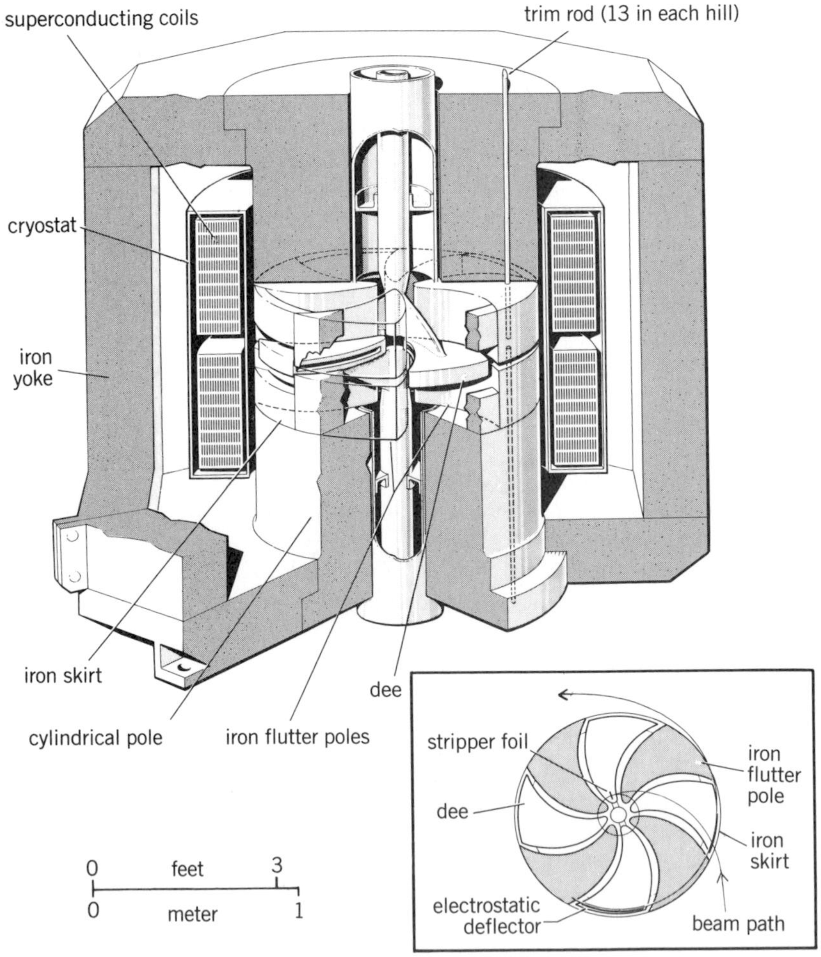

Fig. 1. Cutaway view and plan view (inset) of the Chalk River (Ontario) superconducting cyclotron.

Structure. Common to all these cyclotrons is a cryostatically stabilized niobium-titanium coil (typically about 5 megampere-turns) cooled by immersion in a bath of boiling helium. In a cryostatically stabilized coil, the conductor consists of a number of superconducting filaments within a copper matrix. If a short portion of the superconductor should go normal, the copper shares the current and has sufficient area exposed to liquid helium to remain below the critical temperature of the superconductor, permitting the filaments to regain their superconducting state and resume carrying the current. The cryostat for this superconducting coil restricts radial access to the midplane. As a consequence, in all these cyclotrons the geometry is such that the accelerating dees are supported by resonators that extend axially through holes in the poles, rather than the more conventional geometry of resonators supporting the accelerating structure at the outer circumference. This is illustrated in Fig. 1. The combination of multiple accelerating gaps (eight in Fig. 1), high dee-to-ground voltage (typically 100 kV), and high charge state of the accelerated ions gives a large energy gain per turn that minimizes deterioration in beam quality by rapidly accelerating the beam through troublesome resonances and also eases the extraction problem.

Beam extraction. Extracting the beam from all of these superconducting cyclotrons follows the same sequence used in many lower-field cyclotrons, namely, an orbit perturbation to allow the penultimate turn of the beam to clear the inside electrode of an electrostatic deflector that directs the beam into a magnetic focusing channel to compensate for the radial defocusing at the pole edge. This sequence was one of the earliest concerns in the design studies on superconducting cyclotrons because: the small size reduces clearances on the final orbits; practical considerations do not allow the electric field in the deflector to be scaled up with the increased magnetic field; and radial defocusing is more severe in the steep gradient as the beam escapes the magnetic field. The combination of high energy gain per turn, careful tailoring of the magnetic profile at the edge of the pole, and novel combinations of electric and magnetic fields was necessary to solve this problem.

Production of highly stripped ions. The maximum energy of an ion beam from a cyclotron is described by the equation $E = KQ^2/u$, where E is the energy, K is a constant proportional to the square of the $B\rho$ product, Q is the ionic charge, and u the atomic mass of the accelerated ion. In superconducting cyclotrons, K is made large by the high magnetic field, but to take best advantage of this large K it is still necessary to accelerate as high a charge state as possible. Presently used internal ion sources are not capable of producing adequate currents of highly stripped ions of the heavier elements, but fast ions that pass through a thin foil do emerge highly charged because the faster an ion moves the easier it is to strip its orbital electrons. Three of the cyclotrons listed in the table take advantage of this feature by preaccelerating the beam in either a tandem Van de Graaff or another cyclotron before stripping the beam in a thin carbon foil near the center of the cyclotron; the other two are accelerators preceding such a foil stripper. One promising idea being actively pursued at Milan is to inject axially into the cyclotron from an electron cyclotron resonance source that can produce very highly stripped ions by sequential ionizations in a hot electron plasma. This combination holds promise for making an even more cost-effective heavy-ion accelerator.

Axial focusing and maximum energy. In an azimuthally varying field cyclotron, the axial focusing varies as the flutter, $F \sim \Delta B^2/\langle B\rangle^2$, where $\langle B\rangle$ is the average midplane induction and ΔB is the maximum azimuthal induction difference. In a superconducting cyclotron, the pole steel is saturated over the entire operating range, and ΔB remains unchanged at approximately 1.6 teslas. Hence, F is proportional to $1/\langle B\rangle^2$, unlike lower-field cyclotrons that have a constant flutter. As a result, for all but the heaviest ions the maximum energy from a superconducting cyclotron is limited not by its bending strength but by axial focusing. This leads to the characteristic specific-energy (energy per nucleon) curves shown in Fig 2 that increase as the mass (and the mass-to-charge ratio) of the accelerated ion decreases.

Applications. Although superconducting cyclotrons are well suited for research in atomic physics with highly charged heavy-ion beams, their principal application is in heavy-ion nuclear physics research where there is an increasing interest in the energy range that they span. The Chalk River (Ontario), Michigan phase 2, and Milan facilities will all be capable of producing ion beams that are comfortably above the Coulomb barrier for any projectile-target combination. This will permit the study of a wide variety of exotic new nuclei, new states in nuclei, and new nuclear phenomena, and will also lead to an improved understanding of heavy-ion reactions. Since projectile velocities will be comparable to that of Fermi motion within the target nuclei, experiments will be possible that span a transition region from collective nuclear behavior to a regime where individual nucleons act more independently. The wide range of experimental studies made possible by the high specific energies available for essentially all masses of projectiles will enhance knowledge of nuclear structure with an even more exciting prospect of uncovering unforeseen phenomena.

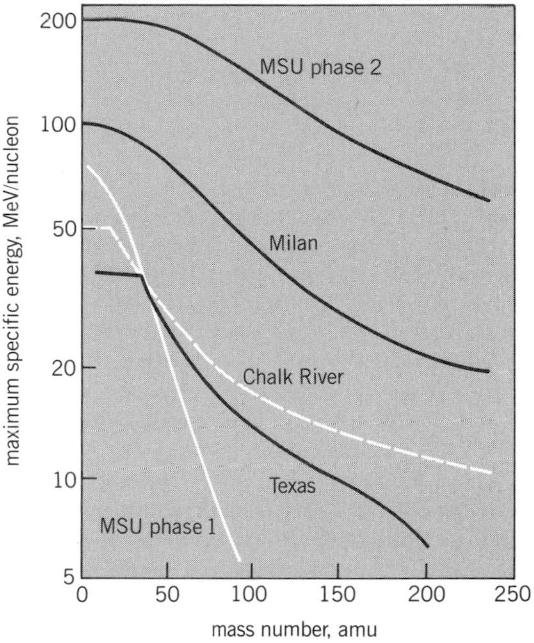

Fig. 2. Maximum specific energy as a function of mass of the accelerated ions for funded superconducting cyclotrons. The Texas A&M curve is for the beam from the facility's existing normal-coil cyclotron injected from the superconducting cyclotron.

For background information *see* ION SOURCES; MAGNETIZATION; NUCLEAR REACTION; NUCLEAR STRUCTURE; PARTICLE ACCELERATOR in the McGraw-Hill Encyclopedia of Science and Technology.

[JOHN H. ORMROD]

Bibliography: G. Bellomo et al., *Nucl. Instrum. Methods*, 206(1):19–46, 1983; H. G. Blosser et al., *Proceedings of the 7th International Conference on Cyclotrons*, pp. 584–593, 1975; J. H. Ormrod, *IEEE Trans. n.s.*, 28(3):2062–2066, 1981.

Data communications

The purpose of Digital Electronic Message Service (DEMS) is to provide efficient means for two-way high-speed data communications, transfer of graphic material images (facsimile), and teleconferencing between cities and within a city environment.

DEMS was authorized by the Federal Communications Commission (FCC) on April 17, 1981. The authorization included the use of Digital Termination Systems (DTS) to provide intracity digital radio links between a central serving office and subscribers. DTS is a digital microwave system, designed to operate in the previously little-used 10.6-GHz frequency band. It can send and receive information in digital form at speeds up to 2,100,000 bits per second (2.1 Mbps). While the terms DEMS and DTS have separate meanings, these acronyms are frequently used interchangeably. From a practical point of view, DTS is the equipment that makes the DEMS service operational.

Prior to 1968, the only two nationwide networks available for moving data electronically were the American Telephone and Telegraph Company and Western Union. It was illegal for subscribers to connect their own equipment to the telephone lines, and the lines themselves were designed for voice or low-speed data transmission. This situation was drastically changed with the 1968 FCC Carterphone ruling, which permitted the Carter Electronics Corp. to connect its mobile radio equipment to the telephone network. This ruling and others that followed opened the door to a new interconnect and private telecommunications market within the United States. *See* MOBILE RADIO.

Nationwide networks. DTS can be used in connection with multicity domestic satellite or terrestrial microwave networks. Information can travel cross-country over these facilities at the rate of millions of bits per second, but this flow historically gets constricted within cities by inadequate, unavailable, or prohibitively expensive data channels between the communications carrier's terminal and the user's premises. This is frequently referred to as the "last mile" problem.

DTS was designed to distribute effectively high-speed data to the users, transmit business data back to the regional or national communications carriers, or exchange information in digital form among subscribers within a city or metropolitan area. This new communications alternative also has been called telephone bypass, implying that it bypasses the local phone lines. While DTS can effectively bypass the phone lines (also called local loops) that were designed primarily for voice service, it also offers opportunities to the telephone companies for new services.

Service advantages. There are three general reasons why DTS is considered superior to the traditional telephone wires: improved quality of transmission, improved availability of service, and reduced service costs.

Transmission quality. It is generally recognized that transmissions, in analog form, over wire networks are subject to noise and distortions. Analog signals vary in intensity over a wide range, as opposed to a digital signal in which information is converted into one of two coded information bits or distinct electrical pulses. While usually acceptable to voice conversations, wire networks are generally not suitable for handling digital data in the quantity, quality, and speeds required by business organizations.

Digital transmission quality is usually defined in terms of bit error rates (BER). Wire networks that have been designed for data transmission usually have a bit error rate of 10^{-6}, or 1 error in 10^6 bits transmitted. DTS networks operate at a conservative 10^{-8} BER, which corresponds to the probability of 1 error in 10^8 bits transmitted. This is analogous to transmitting error-free a 50-page report or the contents of a 5000-page encyclopedia.

Service availability. In many areas, adequate telephone company facilities are not readily available, especially where data rates above 9600 bits per second (9.6 kbps) are required. New 56-kbps line installations typically take between 2 and 6 months. In contrast, a DTS subscriber station can be added to an established DTS network in as short a time as 3 days, and operate at data rates as high as 1.544 or 2.1 Mbps, depending on the DTS system installed.

Economic benefits. DTS can be economically attractive for three reasons: potentially lower communications system costs, new business opportunities to communications carriers, and increased productivity to users.

As a consequence of the reorganization of American Telephone and Telegraph, it is generally expected that the charges for telephone company local distribution facilities will increase appreciably. Charges for DTS services are expected to be lower.

The conventional data channel transmission rates are predetermined. The user has a choice of either ordering a data channel that can transmit high data rates at proportionally higher cost or using a less expensive channel that has the capability of moving data at a slower rate. In the first case, the user may be paying for unused transmission capacity, while in the second case productivity may be adversely affected. DTS facilities, by contrast, have remote network management capability, so that a subscriber's data transmission rate can be changed quickly in response to a telephone call at a computer terminal keyboard. Large information-processing jobs can be completed rapidly, and the system then returned to a lower and economical data rate for routine operations. The subscriber only pays for the capability that is being used.

DTS can accommodate higher data rate channels and greatly increase the productivity of subscriber organizations. Most business machines now work at only half of their top design speeds, or slower. The magnitude of this data bottleneck is indicated in the table. The efficiency levels of some computer host-to-host data transfer operations have dropped as low as 8%.

The production speed of business machines, such as line printers, depends directly on the communi-

Comparison of traditional operating speeds and capabilities of various data terminals

Application	Terminal	Traditional operation, kbps	Terminal capability, kbps
Inquiry and response	IBM 327X	2.4–9.6	19.2
Host-to-host	IBM 370X FEP	9.6–56	128–230
Remote job entry (RJE) station	Harris 1600	4.8–9.6	19.2–56
Computer-aided design and manufacturing (CAD/CAM)	—	9.6	56
Bank, point-of-sale (POS) terminals	IBM 3600/4700	2.4–9.6	19.2
Remote printing	Xerox 2700/5700	9.6–19.2	96

cations channel bit rate. For example, the printing time of a 100,000-line document is slightly more than 3 h if data are transmitted at 9.6 kbps, requires 1½ h at 19.2 kbps, and decreases to approximately ½ h at 56 kbps.

Digital Termination Systems. Most of the proposed DTS will use radio transmission as the communications medium, as opposed to cable or lightwave technology. By the end of 1983, there were only three established DTS equipment manufacturers: Local Digital Distribution Company (LDD), NEC America, Inc., and TXR, Inc.

The LDD point-to-multipoint system, called RAPAC, provides system subscribers with synchronous digital channels that allow facsimile machines, video conferencing equipment, computer terminals, and other business machines to be connected to a nationwide or local telecommunications network. Figure 1 shows three rack-mounted RAPAC units which connect a subscriber's business machines to the DEMS network. In this example, the top unit contains a modem and a control circuit card assembly, and two printed circuit cards, called port cards. Each port can accommodate a business machine, such as a printer or a computer terminal. The two lower units can each connect four additional business machines to the network. Thus, the total capacity of the arrangement shown is 10 business machine connections.

Radios at the central station typically transmit to subscriber stations within a 120° sector at a distance of 6–10 mi (10 to 16 km). Three rooftop antennas, each the size of a traffic light, can cover a 360° "cell." The subscriber stations are equipped with 2.5-ft-diameter (0.76-m) dish antennas that can be mounted on a rooftop or placed inside a window, as long as there is line of sight to the central station antenna.

The radio at the central station continuously transmits information to all subscriber stations within the assigned sector by using time division multiplexing (TDM) coding technique. Each subscriber station monitors this common transmission, but processes only that portion of the broadcast that has been addressed to it.

Each subscriber within the sector transmits back to the central station on a common frequency, and avoids interference by transmitting information in bursts in preassigned time intervals. This technique is called time division multiple access (TDMA). Both the TDM and the TDMA transmission bursts are measured in milliseconds.

Transmission capacity can also be shared by using the demand assigned multiple-access (DAMA) technique. It gives the central station the capability to monitor user needs and assign capacity based on demand for service.

Within each 120° sector, the RAPAC radio transmits and receives digital signals at a rate of 0.8, 1.8, or 2.1 Mbps, depending on the bandwidth that has been allocated. Up to 254 subscriber stations can be accommodated within each sector. The RAPAC system is composed of many small cells or multipoint radio equipment, operating at a specific set of frequencies, overlaid on a metropolitan area to provide a wide area of coverage. The allocated frequencies can be reused to obtain extended coverage (Fig. 2).

The subscribers can have assigned to them individual channels with data rates ranging between 1.2 kbps and 1.544 Mbps. (A number of these individual channels can be used by various subscribers as long as the total data rate does not exceed the overall channel capacity.) Changes in data rates can be easily accomplished by an operator at the keyboard. In situations where a subscriber has a sustained

Fig. 1. Subscriber equipment for the RAPAC Digital Termination System.

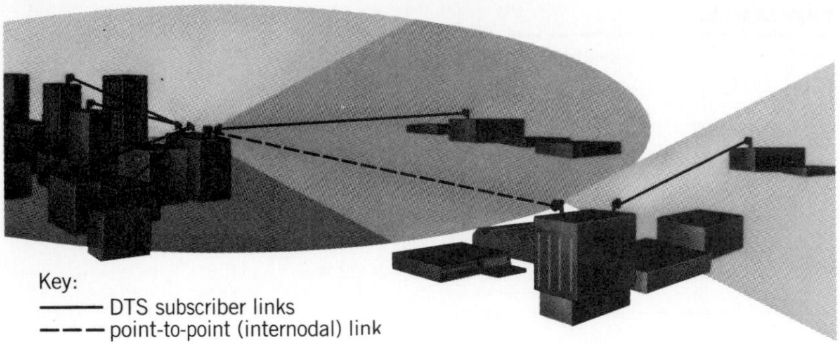

Fig. 2. Portion of a RAPAC network. One 360° cell, centered on a central station, is connected by a point-to-point microwave link to a one-sector Digital Termination System central station.

need for information transmission at high rates, DTS can be supplemented by a high-density, narrow-beam microwave link that can be integrated within the system, and can function as a supplement to DTS.

Using RAPAC equipment, the entire DTS network, which may consist of several interconnected station cells, can be controlled by one operator who can switch to each central station without leaving the computer keyboard of a Network Management Facility. From a remote location, the operator can change the network operation, add and delete subscribers, change channel data rates, and troubleshoot the system in case of a malfunction.

DEMS operational status. The feasibility of the DEMS and DTS concepts was first successfully demonstrated in a 3-month cross-country hookup completed in late 1981. Using a 1.5-Mbps two-way satellite link between New York City and San Francisco, seven communications users tested various service features. Both radio and coaxial cable was used for local distribution. During this test, the DTS radio repeatedly operated at bit error rates of 10^{-11}, which is 10^5 times better than conventional wire network performance.

In 1983 the first operational DTS systems went into service in New York City, San Francisco, Chicago, and London.

For background information *see* DATA COMMUNICATIONS; ELECTRICAL COMMUNICATIONS in the McGraw-Hill Encyclopedia of Science and Technology.

[UMBERTO RAFAELS]

Bibliography: M. Edwards, Local loop bypass paves way for wideband services, *Commun. News*, 20(9):50–54, September 1983; Gartner Group, Inc., *Digital Termination Services: A New Communications Option*, 1983; Telecommunications—Everybody's favorite growth business: A battle for a piece of the action, *Bus. Week*, no. 2760, pp. 60–70, October 11, 1982.

Dental materials

In recent years dentistry has benefited from advances in the science of materials. The materials must be biocompatible for the intended application, and corrosion- and tarnish-resistant because they are expected to function in the mouth for many years. Often, however, other properties, such as strength and elastic modulus, are important. In addition, the materials must satisfy esthetic requirements, especially when anterior (front) teeth are restored or replaced. To meet physical, chemical, and esthetic requirements simultaneously, two or more materials may be used together. For example, a restoration may be made from a gold alloy with porcelain bonded to it. The gold alloy provides strength and ductility, while the porcelain meets the esthetic requirements by providing a surface whose color blends with that of neighboring natural teeth.

Amalgams. In restorative dentistry the decayed part of a tooth is removed and replaced with an appropriate material. Posterior teeth toward the back of the mouth are often filled with dental amalgam. The amalgam is prepared by mixing a powder of an alloy containing silver, tin, and other elements with mercury, a process known as trituration. When it is freshly prepared, the amalgam is plastic and can easily be pressed into a cavity and carved to the desired shape. It then sets to provide a more or less permanent restoration.

Over the past several decades, there have been some advances in the formulation of amalgam. First, attention was focused on the shape of the silver-tin-based alloy particles, and it was found that the use of spherical particles gave higher strengths. More recently it was found that including 12–30% copper in the alloy provides even greater strength and an improvement in corrosion resistance. These new alloys have resulted in a considerable increase in the life of amalgam restorations.

Plastics. Since the 1940s plastics have been used for filling anterior teeth because they are more pleasing esthetically. For about 20 years acrylic resins were used, but these were not entirely satisfactory because they tended to stain, wear, and dissolve away. Decay often developed in adjacent regions of a tooth with this type of restoration. Following acrylic resins came resins containing particles of a filler to improve their strength and durability. A recent development is the use of resins which can be mixed, inserted into a cavity, carved to shape, and then hardened by the application of an intense beam of light or ultraviolet radiation.

Crowns. When decay is extensive, a tooth is usually restored with a full or partial coverage crown. Traditionally, crowns for posterior teeth have been made from 14–18-karat gold alloys. The increase in the price of gold in recent years has tended to bring about the use of lower gold contents or alloys containing two or more of the metals silver, platinum, and palladium but no gold. Base-metal alloys containing various combinations of the metals cobalt, chromium, nickel, and such have also been developed. All of these alloys are satisfactory clinically but differ in their corrosion rates, mechanical strength, and ease of fabricability. Base-metal alloys, for example, tend to have higher melting points than gold alloys, which makes them more difficult to cast.

Esthetic considerations are important in constructing crowns for anterior teeth, and therefore simple metal castings are seldom used. Frequently an anterior crown is constructed by bonding a porcelain or plastic veneer to a metal coping which fits over the prepared tooth. Less commonly a jacket crown made entirely from porcelain is used. The techniques used for bonding porcelain to the metal coping are complex. First the coping has to be carefully heat-teated in order to bring certain elements, such as tin and indium, to the surface and convert them to oxides. Then an opaque layer of porcelain, which typically contains an oxide of hafnium or zirconium, is fired onto the surface. This is followed by the addition of successive layers of porcelains of different compositions until the required dimensions and color are achieved. The result should be a restoration which matches the anatomy and color of adjacent natural teeth precisely.

Prosthodontics. The area of dentistry which deals with the replacement of one or more complete teeth is known as prosthodontics. When a patient has lost all teeth, they are replaced with a full denture consisting of a tinted acrylic resin base into which are bonded acrylic or porcelain artificial teeth. While the materials used for full dentures have steadily improved over the years, progress has not been dramatic. Nevertheless, the performance of full dentures has been very satisfactory.

When a patient still has one or more natural teeth, reconstruction is achieved with a partial denture. A fixed partial denture, also known as bridgework, consists of crowns, fitted over remaining teeth, with replacement teeth attached. The materials used are much the same as those for single crowns, but the fabrication procedures are more complex.

Removable partial dentures are made from a cast metal frame, with attached tinted plastic to simulate gums, and teeth made from plastic or porcelain. They are attached to remaining teeth by metal clasps. As in other areas of dentistry, the increasing price of gold has led more and more to the use of base-metal alloys. Cobalt-chromium- and cobalt-nickel-based alloys have been used extensively, but the latter are less favored because there is some evidence that nickel causes an allergic reaction in some patients. In the past there have been problems with the relative brittleness of cobalt-chromium alloys, but in recent years changes in their compositions have improved their ductility. It is no longer common for a clasp to fracture when a small adjustment in its shape is made as the denture is being fitted.

Implants. One of the frontiers of dentistry, where much work is still experimental, is the use of implants for the retention of dentures. A post of an alloy, or some other material, is screwed into, or otherwise retained in, the jawbone. The denture is then attached to the post, or more than one post, thus providing better retention than is usually achieved. This procedure is expensive because it requires surgery to achieve implantation of the posts, and it is therefore not used extensively. It is employed with patients in whom pressure tends to produce pain. When implants are used, the stresses of mastication are borne by the jawbone rather than the gums or remaining teeth. The materials used for implants include cobalt-chromium alloys, titanium alloys, resins, carbon-impregnated resins, and carbon. As dental science advances, implants may well be more widely used.

Orthodontics. For esthetic reasons or to improve a patient's occlusion (bite), the dentist may decide to straighten or move one or more teeth. With the techniques of orthodontics, a brace is devised which applies a permanent set of stresses to bring about the required straightening and migration of the teeth. The process, which may take months or years, is carried out in stages at each of which the orthodontist adjusts or replaces the brace. The brace must be strong and so it is made from wrought wire, frequently made of stainless steel or a cobalt-chromium alloy. A recent development is the use of a titanium-nickel intermetallic compound, TiNi, which exhibits a shape memory effect when it is in the annealed condition. That is, it can be bent into shape at room temperature and then, when warmed, caused to regain its original shape. This effect will doubtless find wider applications in dental science, and TiNi seems certain to be one of the dental materials of the future. *See* ALLOY.

For background information *see* ALLOY; AMALGAM; DENTISTRY; POLYACRYLONITRILE RESINS in the McGraw-Hill Encyclopedia of Science and Technology.

[J. A. VON FRAUNHOFER; A. A. JOHNSON]

Developmental biology

All multicellular animals and plants use certain of their cells for specific purposes. Even the simplest types of sponges have different types of cells: some sponge cells line the interior canals, while others set up water currents or sift out food. As multicellular organisms become increasingly complex, the number of different types of cells with different types of functions increases dramatically. A conservative estimate of the number of different cell types in the human body is of the order of 200. However, having a large number of cell types is only one factor. In order for a plant or animal to be able to function, the cells must be arranged together to form tissues and organs. In turn, tissues and organs must be organized in a certain way to make a whole body that can carry out a series of actions. Living systems are in this respect analogous to machines; a car engine will not run if it is in bits or is built incorrectly. The method and design of tissue and organs, and the way in which different cell types are arranged, are crucial aspects of the functioning of multicellular organisms.

Form and pattern. The construction of tissues and organs and the spatial arrangement of cell types within them during development are the subjects of studies of morphogenesis (the creation of form) and of pattern formation. Clear-cut examples of patterns of structures can be seen in higher plants. For example, leaves in certain higher plants are arranged in specific positions on the stem (Fig. 1*a*). Each leaf has a certain shape and a highly ordered pattern of different cell types within it. The tissues within the stem and root are also highly ordered. The organization of vascular cells, which transport water and soluble salts and sugars, is particularly striking: the phloem (solute-conducting tissue) is always found outside the xylem (solute-conducting tissue; Fig. 1*b*). In combination, these tissues form specific patterns within the stem or root.

Pattern formation has also been examined in animals. The vertebrate limb is a fine example. The human hand has five fingers which are arranged in a specific pattern. The thumb is on one side and the little finger is on the other. Each finger is different from the others and has a characteristic form. The essential features of this simple description, which seems obvious and even inconsequential, encapsulates the central importance of the study of pattern formation. But, the simple question of how such patterns of cell types are generated during development still has to be answered. For example, it is not understood what causes the thumb to be different from the little finger. Until now it has not been possible to fully elucidate any particular pattern formation of cells in an organ of a plant or animal. However, some of the general principles of pattern formation have been established. These principles, and the cellular and molecular processes which underlie them, have been gleaned from studies on many different systems. These general principles can, therefore, be illustrated with examples from different types of animals and plants.

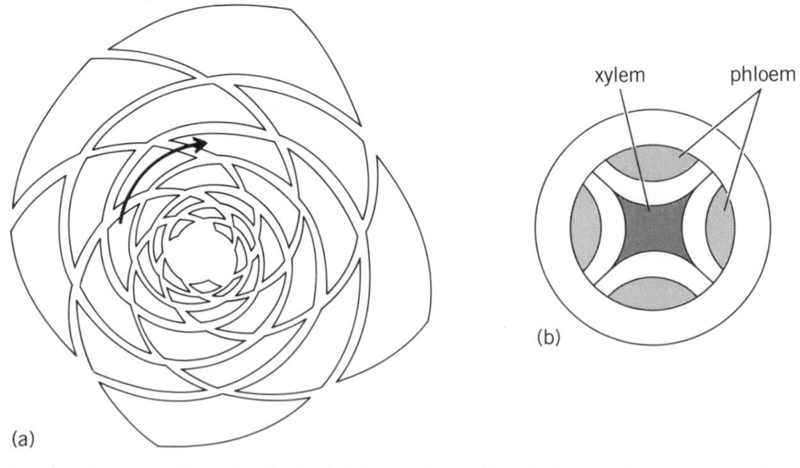

Fig. 1. Simple patterns in plants. *(a)* The pattern of newly formed leaves as seen in a transverse section of an idealized shoot tip. The young leaves are arranged in a series of spirals, one of which is shown by the arrow. *(b)* The vascular pattern in a typical dicotyledonous root. The phloem lies in four groups outside and between the four central xylem strands.

Positional specification. A central concept to the study of pattern formation is that a cell will have its future fate decided by the position it happens to occupy within a group, or field, of developing cells. In order to act upon this information, some mechanism must exist for "informing" each cell, or group of cells, what its position, or address, is within the field. Once this positional information is available, the cell will interpret it and follow an appropriate pattern of cytodifferentiation. The exact manner in which a cell interprets its positional information will depend upon its previous developmental history and the genes it contains.

At the tissue level of development, considerable effort has been focused on attempting to demonstrate the existence of mechanisms which specify cellular position. At the molecular level these have been singularly unsuccessful. However, their existence remains undoubted. An experiment will illustrate the point. The developing chick limb is a small tongue-shaped mass of homogeneous cells when it first forms. However, after several days, the pattern of bones typical of the adult wing is evident within the egg. If a small piece of tissue is taken from the posterior side of one early-stage wing before any obvious cell differentiation has occurred, and is grafted into the opposite side of the wing of another embryo of comparable age, a remarkable result is obtained. The resulting wing has an additional set of digits formed on the side of the wing in which the grafted tissue was placed (Fig. 2). Thus the pattern of the skeleton in the operated wing has been drastically altered by the movement of a tiny piece of tissue from one part of the wing field to another.

It can be assumed that this result is produced by an interaction between the grafted posterior limb cells and the surrounding anterior limb cells. The point is, even though the limb cells all look alike, they are in fact different in some way, and this difference reflects the position in the limb from which the grafted cells are taken. If the exact position of origin from which the graft cells are taken is altered, the character of the abnormal limb which results is changed in a defined and characteristic manner. This means that cells in different limb positions have different characters; they are somehow aware of their position within the limb and are able to interact with other limb cells in a manner which is defined by their position.

This type of evidence for positional specification has been gained from comparable experiments in many different vertebrates, but the importance of cellular position is also evident in invertebrates. For example, a single *Hydra* polyp can regenerate two whole *Hydra* individuals when it is cut in half. Observation of this process shows that pattern formation in the polyp, which has a simple head and tentacles at one end and a sticky foot at the other, follows similar rules to pattern formation in the chick limb. Following bisection, the head end of the *Hydra* always reforms a new foot and the foot end always reforms a new head. Thus each part of the *Hydra* is aware of the position it normally holds within the whole, complete animal.

Cellular interpretation. The specification of position can only be part of the process of pattern formation. Once this aspect of the process is complete, the cells of the field, a chick wing or a *Hydra*, must turn this cryptic information into an overt pattern. Little is known of the cellular machinery responsible for this aspect of pattern formation. Its existence is again a logical deduction rather than an experimental proof. Another experiment with the chick limb provides the evidence. The chick leg and wing have different patterns of skeletal and muscular components. Its four toes are different from its three fingers. If a small piece of tissue is grafted from the future thigh region of the leg to the future finger-forming region of the wing, the grafted cells form toes. Thus the leg cells have changed their fate. They respond to the positional cues of the wing and form digits, but they form leg digits, that is, toes, rather than wing digits. It can be deduced, therefore, that leg cells are subtly different from wing cells. This information, which can be obtained only during the early stages of chick development, determines how the positional cues available in the limb field are to be interpreted as leg or wing. This kind of result demonstrates the existence of positional specification and its cellular interpretation. Little progress has been made, however, in the search for the real mechanisms underlying these phenomena by using tissue grafting techniques. Nonetheless, in recent years considerable advances in the understanding of cellular interpretation have come from the genetic analysis of the development of the fruit fly *Drosophila*.

Genetic mechanisms. The genetics of *Drosophila* is better understood and documented than any other multicellular organism. Among the enormous collection of known mutations are a small series which have remarkable morphological effects. These mutations can change one structure, such as a leg or an antenna, into another body part normally located at another position on the fly: for example, there is a mutation that changes an antenna into a leg. It is assumed that these types of genes, which are called homeotic genes, somehow control the interpretation of positional cues which map out the anatomy of the various cuticular organs of the fly. Numerous other experiments have demonstrated that the same, or closely similar, mechanisms exist in all of the developing fields for all the cuticular parts of the fly. The homeotic genes somehow control the interpretation of this positional information and ensure that an antenna and a leg appear as different structures during normal development.

The best-understood set of genes which control fly morphology is the bithorax complex, a complicated group of at least eight genes. Particular genes in this complex have different morphological consequences if they are mutated. In any genetic analysis, the most effective method for analyzing the function of a particular gene or gene complex is to remove it and to examine the consequences. With regard to the abdominal segments of *Drosophila*, the removal of the bithorax complex produces a remarkable result. In the larva of a normal fly, there are 12 segments which are all anatomically different. At the front end is the head, followed by 3 thoracic segments and 8 abdominal segments. If the whole bithorax gene complex is removed, all of the 8 abdominal segments develop in identical fashion and form the character of the second of the thoracic segments (the mesothorax), even though 12 segments are still formed in the abnormal larva. Parts of the bithorax complex can then be added back into the genome, and as different genes are added, the different segments reappear. Thus the positional information present in each abdominal segment is interpreted differently as the result of the action of a series of products produced by the bithorax gene complex.

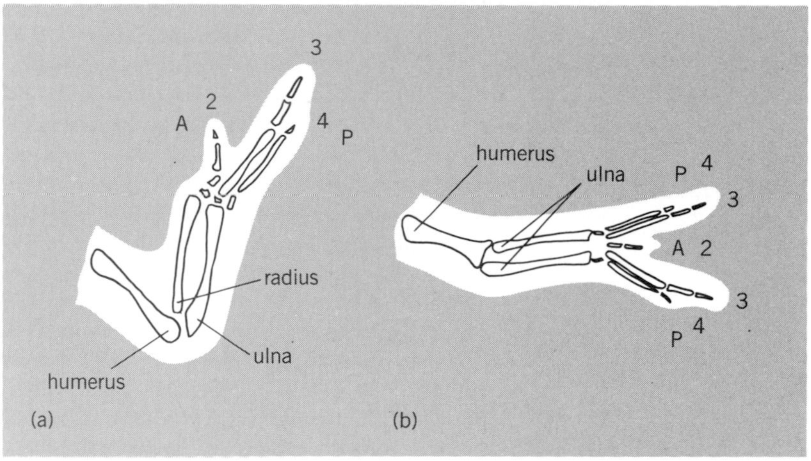

Fig. 2. The pattern of digits in the vertebrate limb is controlled by the establishment of cellular position and the subsequent interpretation of this information. The specification of cellular position is demonstrated with reference to an experiment performed on the chick wing. *(a)* The skeleton of a normal wing, which has three digits (2, 3, and 4 from anterior (A) to posterior (P). *(b)* If a piece of limb bud tissue is grafted from the posterior side of one wing bud to the anterior side of another, a wing with mirror image symmetry of the skeletal parts is formed.

Plants. Pattern formation in plants follows different processes. It is also known that not all animal patterns rely on the same two-step positional information process. The main difference in pattern formation lies in the types of cells that are produced within a developing field. In the vertebrate limb, numerous different cell types form within a single developing field. The same cell types, such as cartilage or muscle cells, form in different limb positions. For example, cartilage cells form in both the developing radius and the developing ulna in the wing or arm, yet these bones are clearly different structures. Thus, in the wing the same cell types can appear at different wing positions. It is assumed, therefore, that cartilage cells can result from the cellular interpretation of different positional values. These types of patterns may be termed nonequivalent.

In plants, and in some animal patterns, identical structures appear to be arranged in space, such as hairs on the skin, leaves on a stem, or stomata on a leaf. In these cases, it is assumed that each element in the pattern, each hair or leaf, forms as a result of a peak concentration of a single stimulus. The arrangement of these elements in space therefore corresponds to the arrangement of a series of equivalent peaks of the stimulus. Such patterns can be thought of as equivalent or prepatterns. As is the case with nonequivalent patterns, no example of a prepattern has been examined in sufficient detail to allow the identification of molecular mechanisms forming it.

Conclusion. Pattern formation is a critical process in the development of plants and animals. To date, the mechanisms underlying the spatial arrangement of cell types in developing organs and tissues have remained elusive. If the secrets of development are to be revealed, the fascinating aspects of tissue level controls of patterning must be analyzed with continuing energy. These studies will not be completed until studies at the molecular, cellular, and organismic levels of organization become merged.

For background information *see* ANIMAL MORPHOGENESIS; DEVELOPMENTAL BIOLOGY; EMBRYONIC DIFFERENTIATION; GENETICS; PLANT MORPHOGENESIS; REGENERATION (BIOLOGY) in the McGraw-Hill Encyclopedia of Science and Technology.

[NIGEL HOLDER]

Direct broadcasting satellite systems

Direct broadcasting satellites (DBS) provide the latest means of distributing television signals, which are transmitted directly from the satellite to the home over superhigh-frequency radio waves. Both performance and economic advantages may be realized by this direct method. This article describes the basic DBS system and compares it to the existing means of television signal distribution. An assessment is made of some of the problems and the means for their solution. The status of DBS programs in the United States, Europe, and Asia is presented. The variety of programming made possible by the economics of DBS system is surveyed. Finally, the future of DBS systems is forecast.

System configuration. A DBS system consists of a television repeater station in geostationary orbit, a ground station that transmits program signals to the spacecraft, and the consumer terminals that receive the signals from the satellite and convert them to a format compatible with existing television sets (see illustration). These system components are derived from existing equipment used for satellite distribution of television signals to homes, hotels, cable systems, and network stations. The DBS systems will provide, in addition to television, high-fidelity audio, teletext and other services. Some DBS systems will also scramble the signals at the program generation facility, thus requiring descramblers at the consumer terminals (see illustration).

To reduce the cost and installation problems of the user equipment, commonly known as homesat terminals, the satellite transmitter provides as much as 100 times the signal strength used by existing satellite transmission systems. As a result, the homesat terminal antenna (a dish type) can be made as small as 2 ft (0.6 m) in diameter as compared with the 10–15-ft (3–4.5-m) diameters of existing earth terminal equipment. The consumer cost for equipment and its installation is reduced to affordable levels through this reduction in antenna size.

The increase in transmitter power does, however, limit the number of channels that can be transmitted from a particular spacecraft. A single spacecraft of the same size and weight as today's communications satellites can provide 3 such higher-power transmission channels compared with 24 lower-power channels. This limitation requires that many such satellites be collocated to provide the 32 channels available in the designated frequency band for each orbital location. Alternatively, larger satellites can be employed. There are eight such locations allocated for the United States. The grouping or clustering of spacecraft at such an orbital location allows the consumer's antenna to be fixed in direction and receive the maximum number of channels available.

Each of the eight orbital locations in the United States will typically provide coverage of a single time zone, although some can provide two-zone coverage. Signals from each satellite are transmitted to all locations within that coverage area, with high-quality reception capability and with no dependence on proximity to a major city, where terrestrial transmitters are normally located. Because a single transmitter is employed per channel for each coverage area, the common interference between terrestrial stations located in adjacent cities is eliminated. The problems of interfering signals from adjacent satellites is reduced by the rejection capability of the homesat antenna pointed at its proper spacecraft.

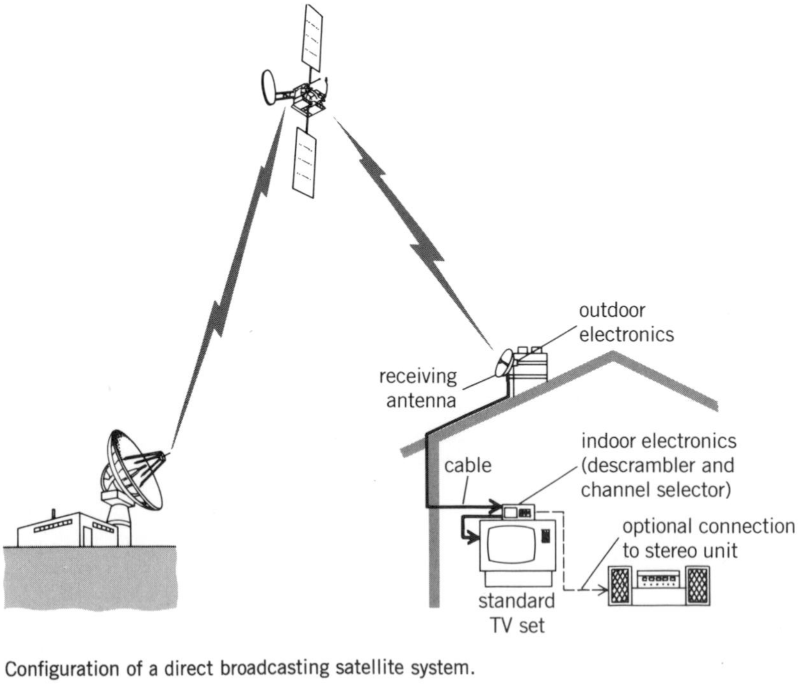

Configuration of a direct broadcasting satellite system.

Interference-free reception may be, therefore, provided for all locations within the country.

Comparison with existing TV distribution. Television signals are presently distributed to the individual homes via five means: the terrestrial broadcasting station (TBS), cable TV (CATV), satellite master antenna TV (SMATV), low-power TV (LPTV), and multipoint distribution system (MDS). The most common are the TBS and CATV systems.

The terrestrial broadcasting station has the most widespread use and provides signals to the majority of the viewing community. Its principal drawbacks are the limited number of channels available to any area and the quality of fringe-area reception. The channel authorization plan for all areas was configured by the Federal Communications Commission to eliminate interference between adjacent cochannel stations. This plan, which is necessary due to interference, limits the number of stations available to any location. There is also a line-of-sight requirement between the transmitter and receiver antennas for strong signal reception, causing limitations of the range of coverage. This limits the number of viewers per transmission station and increases the cost to the viewer due to sometimes elaborate home antenna installations.

These restrictions gave birth to the cable TV industry. Through the use of elaborate antenna installations in the fringe areas, good-quality reception from existing terrestrial stations could be received and rebroadcast over cable transmission systems to the homes. Later CATV systems supplemented the reception of terrestrial broadcast signals with signals from satellites, again using rather costly antennas. Because the cost of these antennas can be distributed over all viewers served by a CATV network, the system can be profitable. Current CATV systems offer up to 100 channels in some areas. CATV systems are not profitable, however, where there are large distances between viewers, because of the cost of installing the transmission cables. Picture quality is generally superior to that of direct terrestrial broadcast.

The satellite master antenna TV system is an outgrowth of the CATV system and is in fact a miniature CATV system. It is designed to provide high-quality transmission to hotel and apartment complexes, with signals generally received directly from satellites. One advantage of the SMATV system is that it is privately owned and operated and is not subject to the community's administrative and political restrictions normally imposed on operators of CATV systems. These installations require on the order of 300–500 viewer terminals to become profitable.

Low-power TV stations are similar to the terrestrial broadcast station, with the major exception that they are restricted in transmitter power to alleviate the interference problem from adjacent stations. Reception quality is generally good only if the viewer is close to the transmitter.

The multipoint distribution system is comparable to the LPTV system in that it is short-range. MDS uses a frequency band different from the UHF and VHF bands for terrestrial TV transmission. It provides a good signal quality within its range, and its implementation cost is lower than that for CATV. However, MDS can provide a maximum of only 4 television transmissions.

All of these systems require a separate physical installation to provide services to each local area. Each installation requires a staff for operation and maintenance, and, therefore, the total operating cost is not insignificant. By comparison, the DBS system requires a single staff for operation and maintenance and only one transmitter station per time zone. However, the DBS system has a disadvantage in that local-event coverage is not possible due to the wide coverage area of a time zone.

Technical problems. Problems encountered in the implementation of a DBS system include rain attenuation, impact of high-power radio-frequency satellites, initial implementation cost, and establishment of a broad user base.

Rain attenuation is the degradation of received signal strength from the satellite when rain is present in the line of sight from the satellite to the homesat terminal. The signal strength may be reduced by a factor of 3 to 4 or more over clear air transmissions and may be severe enough in dense rain conditions to cause total loss of reception. The two solutions for this problem are more transmitter power in the spacecraft and a larger homesat terminal antenna. For very high-rain areas, such as the southeastern United States, homesat antenna diameters will most likely be 3¼–4 ft (1–1.2 m) compared with 2 ft (0.6 m) in average areas.

The increased transmitter power in the spacecraft causes a reduction of the number of channels per spacecraft. This in turn causes a requirement for multiple spacecraft at a given orbit location. This problem is resolved by management of precise orbit location of each spacecraft to avoid possibility of collision and radio interference in the control signals to the spacecraft. Other technical problems are achieving the desired lifetime of the high-power satellite transmitter tubes (7–10 years) and dissipating the thermal energy they create.

The cost to implement the DBS system is much less than that of a terrestrial distribution system per square mile of coverage area. However, the DBS system is implemented over an entire time zone instantaneously, and the total cost to implement becomes very large. When the first DBS system is implemented, the system operator has the problem of incurring the cost of transmitting hardware (the satellite and control stations) with a very small user population due to the initially small number of homesat terminals.

This leads to the last problem of establishing a user base of adequate size for the systems operator to recoup the investment. Most potential systems operators have indicated that the break-even level of homesat terminals installed is, on a nationwide basis, 5,000,000 sets.

Current status. The most advanced DBS system is in Japan. In 1978 the first experimental direct broadcast satellite was launched and operated on an experimental basis for approximately 24 months until satellite transmitter failure terminated the experiment. During this time, many experiments were conducted, and plans were made to build and launch the replacement spacecraft, still on an experimental basis. This spacecraft, due to be launched in 1984, has a two-channel capacity. Japan's third-generation spacecraft is now in the conceptual design phase and will provide four channels to that country, with a forecast launch date of 1989.

Germany, the United Kingdom, and France all have programs in progress to develop and manufacture direct broadcast satellites. These are to be launched during 1985–1986. The transmitter power in these spacecraft is double that of Japan's due to the increase in required coverage area and the desire to have a high margin against rain attenuation. Other European countries, such as Luxembourg and the Scandinavian countries, are currently defining their DBS requirements.

In the United States, several companies have DBS construction permits authorized by the Federal Communications Commission, and some of these have contracted for spacecraft.

Programming considerations. Of great importance to the success of a DBS system is the programming. DBS systems, because of the possibility of vast numbers of viewers served, have the opportunity to provide special-interest broadcasting. Anticipated movie fare will include children's matinees, classics, foreign films, horror films, and adult films. Real-time entertainment will come from legitimate theater, sports, charitable events, and political events such as conventions. The economics of the broadcasting of such specialty programming becomes profitable when, as in a DBS system with a large user base, a small percentage of the total viewing community can provide the necessary revenue. For the terrestial-based systems, each station must provide broad-interest programming to maintain a large percentage of the viewer population to achieve profitability.

In addition to specialty broadcasting, the DBS system can provide both normal wide-area advertiser-supported and pay-for-view programming. Some of the prospective DBS system programmers have envisioned the offering of an open satellite university, electronic games, teletext service, interactive communications and services, and multilingual audio service to accompany the video programming.

Prospects. The present trend in satellite technology is toward larger, more efficient satellites. Continuation of this trend will result in as few as one spacecraft supplying the 32 channels authorized at each orbital location. Advances in technology are also foreseen in the design of higher-power transmitters on the spacecraft, thereby providing greater margin against weather degradation of received signals and permitting less expensive home equipment.

Perhaps the most exciting advance is the possibility of improving the video and audio quality of the transmission. High-definition television with photographic clarity and stereophonic sound would be only one possible application of this development.

For background information *see* CLOSED-CIRCUIT TELEVISION; COMMUNICATIONS SATELLITE; TELEVISION in the McGraw-Hill Encyclopedia of Science and Technology. [ROBERT F. BUNTSCHUH]

Earth

Diamond may melt to a more dense, metallic liquid along a melting curve whose temperature decreases with increasing pressure. Pressures and temperatures along the melting curve are not accurately known, but some data suggest that the curve may intersect Earth's geothermal gradient so that the stable form of native carbon in the deep mantle and core may be a metallic liquid.

Structure and stability of diamond. In the structure of diamond, each carbon atom is bonded, by overlapping sp^3 hybrid orbitals, to four nearest neighbors situated at the corners of a tetrahedron. The strength of the structure is indicated by diamond's unexcelled hardness and resistance to melting. In fact, the melting of pure diamond has never been confirmed, and it has been suggested that, instead of melting on increase of pressure, diamond transforms to a metallic solid. There are theoretical and experimental reasons to suppose, however, that diamond does melt and that, like silicon, germanium, gray tin, and ice, its melting temperature decreases with increasing pressure.

Diamond, silicon, germanium, and gray tin have similar structures and bonding. The phase diagrams for silicon, germanium, and tin (Fig. 1) are very similar. In each the low-pressure solid is a semiconductor with a diamond-type structure; this diamond-type solid melts to a metallic liquid along a curve whose slope is negative with respect to temperature and pressure; and below the melting curve the diamond-type solid transforms, with increasing pressure, to a metallic solid with the white tin structure.

Diamond, an electrical insulator, is expected to behave like silicon, germanium, and tin in that, upon melting, the strongly directional covalent bonds will be destroyed and, at any point along the melting curve, the carbon atoms are expected to be more closely packed in the melt than they are in the solid diamond. Near the melting curve, therefore, the melt will have a smaller molar volume and higher density than solid diamond. Structurally, molten diamond may resemble the solid carbon phase that is stable at still higher pressures. By analogy with silicon, germanium, and tin, this phase may be metallic with a white tin structure.

Neither molten diamond nor the metallic solid phase has yet been confirmed, although in experiments with attained shock pressures of 1.9 to 2.2 megabars (1.9 to 2.2 $\times$ 10^2 gigapascals) a transformation from diamond (density = 2.03 oz/in.^{3} or 3.52 g/cm^3) to a denser form (2.28 oz/in.^{3} or 3.95 g/cm^3) has been observed which may have been the metallic phase.

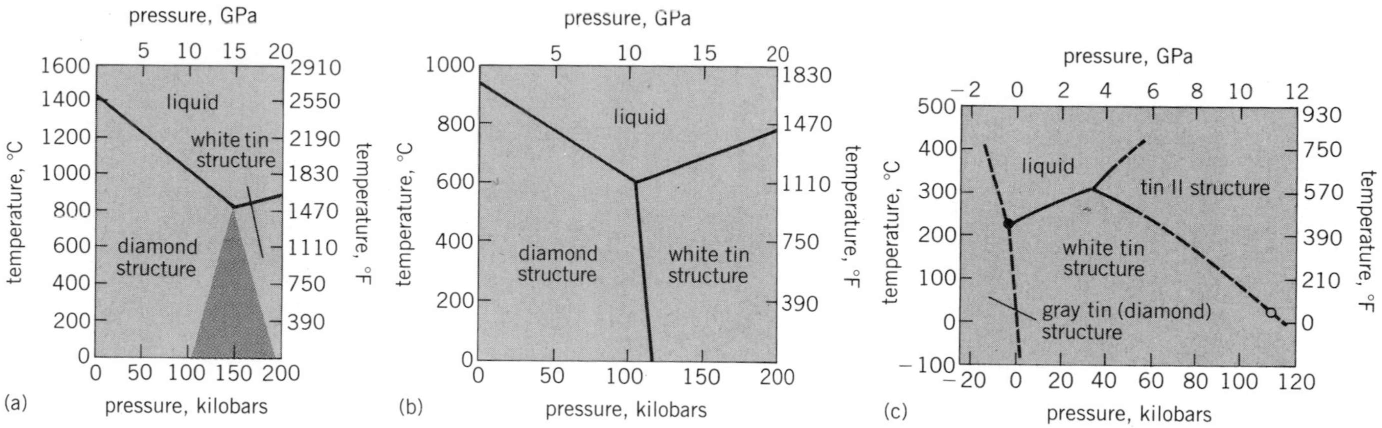

Fig. 1. Phase diagrams of (*a*) silicon, (*b*) germanium, and (*c*) tin. Less-certain boundaries are shown by shaded areas (silicon) and broken lines (tin). The crystal modifications (polymorphs) on each diagram are labeled according to the structure type.

A suggested pressure-temperature diagram for carbon, showing regions of phase stability, has been drawn by F. P. Bundy, who first synthesized diamond. This diagram (Fig. 2) is based upon experimental data from various sources, including flash-heating graphite-to-diamond conversion experiments and shock-wave experiments. Bundy was unable to melt pure diamond (because the charge was electrically insulating), but he did melt boron-doped diamond at 140 kilobars (14 GPa). In that experiment he observed a marked decrease in electrical resistivity, suggesting the presence of metallic liquid. The charge quenched to a rounded, glistening mass of graphite and black diamond crystals, but there was no evidence of an amorphous phase (diamond glass).

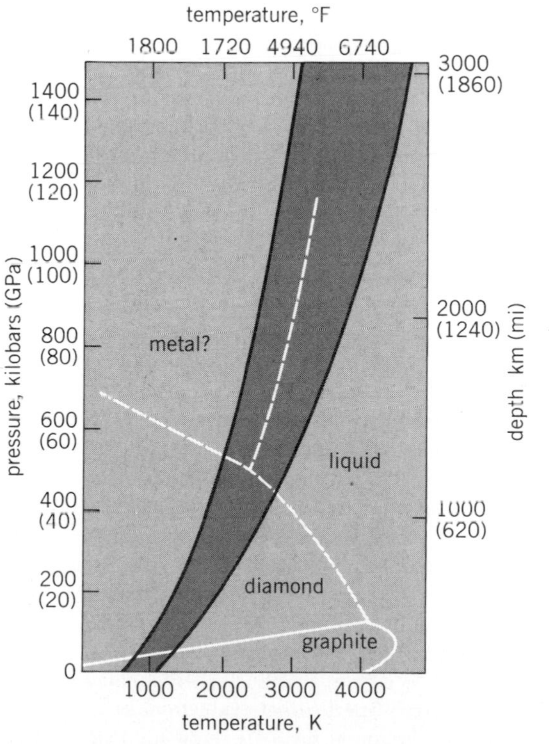

Fig. 2. Phase diagram for carbon suggested by F. P. Bundy. The shaded area shows the range of estimates of temperature inside the Earth according to B. A. Bolt. (*After J. S. Dickey, Jr., et al., Liquid carbon in the lower mantle?, Geology, 11:219–220, 1983*)

In new experiments to melt diamond and define its melting curve, the charge is compressed between two diamond anvils and heated by a Q-switched Nd:Yag laser. Run conditions exceed 100 kilobars (10 GPa) and 5400°F (3000°C). Cooling is so rapid (nannoseconds) that it may be possible to quench the melt to glass.

Implications for the Earth. Superimposed upon Bundy's carbon-phase diagram in Fig. 2 is a shaded area showing the range of estimates of temperature variation within the Earth. The estimated diamond melting curve intersects the estimated temperature band between approximately 450 kilobars (45 GPa) [4490°F (2750 K)] and 400 kilobars (40 GPa) [4040°F (2500 K)]. These conditions correspond to depths of 777 mi (1250 km) and 696 mi (1120 km). If the Bundy melting curve is correct, molten diamond may be the stable form of native carbon in more than three-quarters of the Earth.

Whether molten diamond in the Earth's interior would be of geologic importance depends upon the quantity of carbon present. Carbon is the fourth most abundant element in the solar system and the sixth most abundant element in C1 meteorites, which may represent the less-volatile ingredients of the nebula where the terrestrial planets accreted. Judging by the compositions of the crust and upper mantle, however, the Earth appears to contain only 0.00001 times as much carbon as the nebula. Such an enormous depletion is puzzling, especially if the early condensates which accreted to form the Earth included refractory compounds such as carbides and graphite. An alternative possibility is that most of the Earth's original carbon was dissolved in the molten iron core.

The solution of carbon in iron at low pressures is

interstitial and does not significantly affect the density of the liquid. At core pressures, which exceed 1 megabar (10^2 GPa), however, metallic carbon may substitute for iron atoms in the melt. It is conceivable that carbon helps to reduce the density of the outer core from that of molten nickel-iron (at core pressures) to the observed outer core density, which is about 10% less.

Given the large difference in atomic radii between iron and carbon, even under core conditions, iron- and carbon-rich immiscible liquids might be created. (Comparable liquid immiscibility is observed between iron and tin at 1 bar, or 10^5 pascals.) It has been suggested that carbon-rich fluids, emanating locally on the core surface, may create diapiric plumes of hotter, lighter, and less-viscous matter in the lower mantle, and that these may be responsible for local areas of anomalous melting and other thermal phenomena in the upper mantle.

For background information *see* CARBON; DIAMOND; EARTH INTERIOR; PHASE TRANSITIONS in the McGraw-Hill Encyclopedia of Science and Technology. [JOHN S. DICKEY, JR.]

Bibliography: F. P. Bundy, The P, T phase and reaction diagram for elemental carbon, *J. Geophys. Res.*, 85:6930–6936, 1980; J. S. Dickey, Jr., et al., Liquid carbon in the lower mantle?, *Geology*, 11:219–220, 1983; K. I. Kondo, and T. J. Ahrens, Shock compression of diamond crystal, *Geophys. Res. Lett.*, 10:281–284, 1983.

Ecological interactions

Rocky shores and reefs, with their large variety of inhabiting species, are ideal areas for ecological studies. Their accessibility allows both direct observation and experimentation by the artificial exclusion of certain species. This article discusses the effects of asteroid predation on benthic (bottom-dwelling) communities and the nature and consequences of interactions between limpets and barnacles on rocky shores.

STARFISH

Although early naturalists knew that most starfish (also called seastars or asteroids) were predators of marine benthic invertebrate animals, the importance of their effects on community structure, that is, distribution, abundance, and diversity of their prey, was not demonstrated until the mid-1960s. Studies on starfish effects now suggest that starfish predation is a major factor in structuring cold-water marine communities. Starfish are present in warm-water habitats but have relatively minor roles as predators compared to fishes.

Predation in North America. Most studies on the significance of starfish predation have been concentrated in rocky intertidal habitats, that is, the area between high and low tidemarks, of North America. Salient features of community structure in this system are: sharply delineated horizontal bands or zones of organisms, with barnacles in the high, mussels in the middle, and seaweed on the low shores; low numbers of species in the two upper zones and high numbers in the low; and the presence of starfish as "top" predators (species which prey on other species but are not usually preyed upon themselves).

The effects of feeding by these starfish (members of the genera *Pisaster* and *Asterias*) on prey populations, such as mussels and barnacles, were determined by removing the starfish from specified plots and preventing their return. Results were dramatic. Within 6 months to 3 years, mussels had invaded the lower shore from the middle intertidal and had outcompeted and thereby excluded all other sessile species, including the marine plants. These experiments displayed several important features of processes in this community which were not previously known. First, although mussels are not always the most frequently eaten prey of the starfish, the lower edge of the mussel band owes its apparent constancy to continuous foraging by starfish. Second, a by-product of this foraging is the maintenance of many competitively inferior species (since the superior competitor, the mussel, is prevented by the starfish from excluding them). Third, mussels are simultaneously the dominant competitor for space and the favorite food of starfish in these communities.

Effects on other rocky shores. These investigations spurred others to evaluate the importance of starfish as predators and mussels as space competitors in other parts of the world. Results to date from such places as England, New Zealand, and Chile indicate that the dynamics of rocky intertidal communities in cold- but not warm-water habitats in both Northern and Southern hemispheres are similar to those described above. Although species in different locations have evolved independently, the interaction between starfish and mussels is of major importance in determining rocky intertidal community structure in the zones they inhabit.

Effects in subtidal habitats. A different picture emerges from studies in cold-water subtidal (habitats which are below the low tide mark and continuously submerged) regions. Compared to intertidal shores, community structure is more complex: there are more species, more complicated patterns of distribution and abundance of sessile organisms, and more intricate relationships between predators and prey. Notably, there are usually more starfish species, and although all are predators, some eat non-starfish prey, and some starfish species actually prey on other starfish. Further, mussels are rarely abundant in subtidal regions; conspicuous sessile species include kelps and other marine plants, sponges, sea pens, sea cucumbers, and in muddy habitats, clams and other bivalve mollusks. Corresponding to this increased complexity is an increased difficulty in studying subtidal communities. Although few experimental evaluations are presently available, indirect methods indicate that starfish play roles comparable to those documented for intertidal regions in many, but not all, cold-water subtidal habitats.

Crown-of-thorns and coral reefs. Starfish are ubiquitous and often abundant in the tropics but, with the partial exception of coral reefs, community dynamics in tropical subtidal habitats are poorly known. Available studies indicate that despite the popular view that plagues of crown-of-thorns (*Acanthaster*) starfish threaten the existence of coral reefs, this and other starfish are of minor importance compared to fishes. Although *Acanthaster* outbreaks can devastate reefs, this effect depends on the composition of the coral community. In some areas (for example, Hawaii) the favorite coral prey species of *Acanthaster* are scarce, and the starfish has little effect. Only where favored species of branching coral are abundant do outbreaks of this starfish have much effect. Further, plagues are infrequent, and recent evidence indicates that their occurrence is restricted to mountainous islands with bays or lagoons.

Additional limitations to outbreaks are that a particular combination of biological and atypical meteorological conditions must coincide. Specifically, when the crown-of-thorns starfish reproduces simultaneously with the occurrence of heavy precipitation after a dry period, runoff to the sea leads to high nutrient input, which produces high abundances of phytoplankton in bays and lagoons. The larvae of the starfish feed on these microscopic plants, and consequently many more survive than usual. After 3 years of feeding and growth in cracks and crevices of the reef, the starfish suddenly emerge in astonishing numbers and, if the favored corals are abundant, may devastate the reef. Because these conditions are rarely met, *Acanthaster* plagues are not frequent. The apparent increase in outbreaks in recent decades seems attributable to the coincidental emergence and popularity of scuba diving for recreation and scientific research, and increases in tourism in the South Pacific.

Conclusion. Studies of the importance of starfish in the dynamics of benthic marine communities have been in the forefront of recent advances in community ecology. Characteristics influencing the preeminent effect that starfish have on cold-water benthic marine communities include the potential to grow orders of magnitude larger than the size at which they reach sexual maturity, a generalized morphology (including their shape, possession of tube feet, and extraoral digestion of prey) permitting them to attack prey in almost all taxonomic groups and of all sizes, and the ability to survive long periods without food. However, as mentioned earlier, starfish are not dominant predators in all benthic marine habitats. Decapod crustaceans (lobsters, crabs), mammals (sea otters), and especially fishes are now known to control community structure in many marine habitats. Nonetheless, the most widespread groups of predators which regulate their communities are starfish in cold-water habitats and fishes in warm-water habitats. Future research will hopefully provide a much greater fund of basic knowledge on the dynamics of a greater variety of marine benthic communities, especially in subtidal and tropical habitats.

[BRUCE A. MENGE]

LIMPETS AND BARNACLES

On the majority of rocky shores throughout the world, limpets and barnacles are prominent members of the community. Barnacles are often the dominant space occupiers on the rock surface over a broad zone covering much of the mid and upper shore, and limpets are the major grazers of algae. Until quite recently it was assumed that for both of these groups, and indeed for intertidal species in general, the limits of distribution were basically set by tolerance of environmental conditions. Understanding is now increasing that interspecific interactions, such as competition and grazing, are also major determinants of distribution, and hence of community structure. Limpets and barnacles exhibit a complex spectrum of both adverse and beneficial interactions, some acting directly and some through the effects which both groups exert upon the intertidal macroalgae.

Adverse effects of barnacles on limpets. The most obvious effect is that a cover of barnacles interferes with the grazing abilities of the limpets. There are no direct observations of reductions in the rate of food intake, but there is considerable circumstantial evidence of lower feeding efficiency. Thus, with heavy grazing pressure, sporelings of the brown algae *Fucus* survive only in the crevices between barnacles, and there is also improved survival of juvenile barnacles in such refuges. This reduced grazing efficiency results in either a lower food intake or a prolonged foraging period for the limpets; the former will restrict growth, the latter increases mortality. Growth rate is inversely related to barnacle cover in a number of species, such as *Collisella digitalis*, *Patella granularis*, and *P. vulgata* (Fig. 1). Changes in growth rate have been observed both in natural communities with different barnacle cover and in experiments where the barnacle density has been manipulated. Increased mortality in dense barnacle cover has been observed in *C. digitalis* and *P. vulgata*. The cause may be increased susceptibility to dislodgement or predation while foraging over the barnacle matrix, reduced

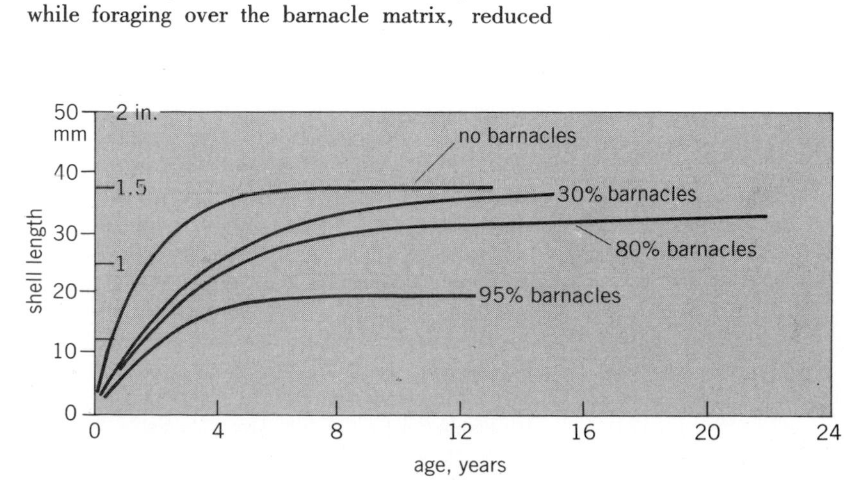

Fig. 1. Growth rates of four natural populations of *Patella granularis* from Kalk Bay, South Africa, growing on different degrees of barnacle cover. (*After G. M. Branch, Interspecific competition experienced by South African Patella species, J. Anim. Ecol., 45:507–529, 1976*)

tolerance to stress because of limited food resources, or an inability to form a proper seal to the rock surface as a defense against desiccation.

Beneficial effects of barnacles on limpets. The interference with grazing discussed above may operate differentially between size classes and species of limpet. Although acting as an adverse influence in general, barnacle cover may benefit some limpets by reducing competition. Large limpets are effectively excluded from the spaces between barnacles, while small individuals or species may exploit that resource which would otherwise not be available. Thus the small limpet *Patelloida latistrigata* shows reduced mortality in the presence of barnacles, because it is protected from competition with the larger limpet *Cellana tramoserica.*

Barnacles can also be beneficial by facilitating limpet recruitment. Very small limpets have a high surface area–to–volume ratio, which makes them far more vulnerable to desiccation than larger specimens. Barnacles retain water at low tide and so raise the humidity near the rock surface, and also reduce air movement in the boundary layer. Thus, recruitment and survival of young limpets are promoted in various species (for example, *Collisella digitalis*, *Patella granularis*, and *P. vulgata*), resulting in a situation in which mortality is raised and growth slowed, yet the population density is higher. The interaction is obviously complex, since there are other studies where highest limpet settlement was observed in barnacle-free areas. An experimental study of *P. vulgata* resulted in higher settlement on a small-scale mosaic of bare rock and barnacles than on either a 100% cover of barnacles or totally bare rock. The mosaic probably provided the optimum conditions on bare rock for attachment as well as humidity for survival. An indirect effect of barnacles arises from their interference with limpet grazing. This permits the occasional "escapes" of fucoid algae which develop into scattered clumps, and these clumps provide humid conditions and are favored sites for limpet settlement.

Adverse effects of limpets on barnacles. As a result of moving and feeding, limpets kill or ingest large numbers of cyprids and young metamorphosed barnacles. Thus, in natural communities there can be a negative correlation between settlement density of barnacles and the numbers of adjacent limpets, a relationship demonstrated for *Semibalanus balanoides* and *Patella vulgata.* When limpets are experimentally excluded, the survival of young barnacles is improved (Fig. 2*a*). In these experiments, timing is critical; exclusion permits algal growth, and if carried out too long before settlement, the algal growth itself inhibits barnacle settlement (Fig. 2*b*). Less damage is inflicted by the limpets on uneven rock than on a smooth surface, and similarly damage is reduced by the presence of a dense population of old barnacles. That limpets may be more harmful to some barnacles than others confers a competitive advantage upon the more resistant species: *Acmaea* prevents total dominance being established by the larger *Balanus*, thus permitting the coexistence of the smaller *Chthamalus.*

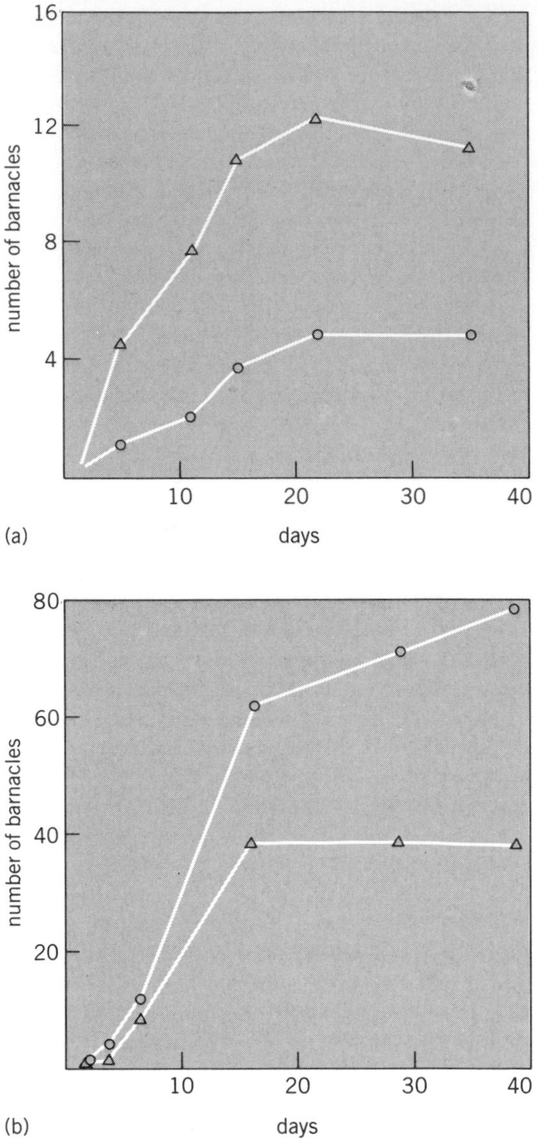

Fig. 2. Number of metamorphosed *Semibalanus balanoides* per scraped settlement square in two experiments, with circles indicating the controls to which limpets have access, and triangles the experiments from which limpets are excluded. (*a*) Experiment commenced during settlement. (*b*) Experiment commenced one month before the start of settlement, allowing algal growth. (*After S. J. Hawkins, An experimental study of the interactions of Patella and macroalgae with settling Semibalanus balanoides, J. Exp. Mar. Biol. Ecol., 1983*)

Beneficial effects of limpets on barnacles. Benefits conferred by limpets on barnacles are indirect, and arise from the fact that in most barnacle-dominated areas the algal cover is absent because of the grazing activities of the limpets. The presence of significant quantities of algae has several adverse effects on barnacles. Small algae such as filamentous greens form a layer on the rock surface which inhibits the settlement and establishment of barnacles, an effect clearly demonstrated when limpets are ex-

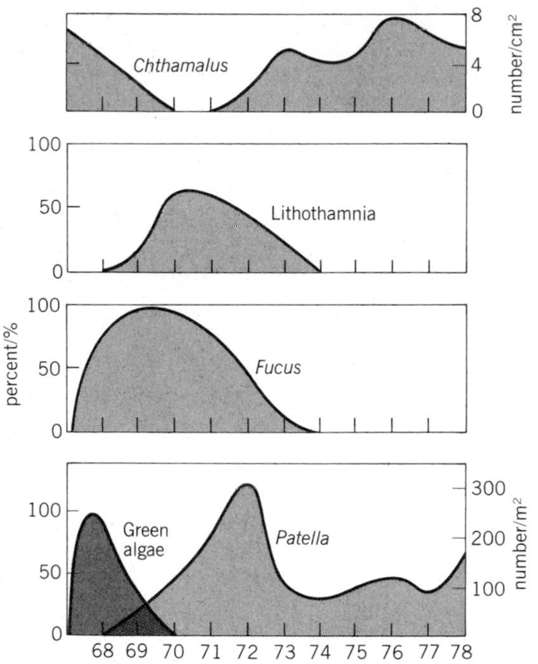

Fig. 3. Changes in the community at midtide level in an area of Southwestern Britain following the death of all limpets in 1967 from the effects of dispersants used after the Torrey Canyon oil spill. (*After A. J. Southward, in E. Naylor and R. G. Hartnoll, eds., Cyclic Phenomena in Marine Plants and Animals, 1979*)

cluded from the area a month before settlement (Fig. 2*b*). Larger algae exert a sweeping effect which reduces barnacle density. This reduction is apparent in the vicinity of natural algal clumps, and also in experiments where algal plants were transplanted to previously algal-free areas. In dense stands the algal holdfasts themselves will overgrow and smother many of the barnacles. The most striking demonstrations of this limpet-algae-barnacle interaction have occurred in the large-scale ecological disaster resulting from major oil spills. The sequence of events following the Torrey Canyon incident in 1967 is shown in Fig. 3. The dispersants used in cleaning up the oil effectively killed all the limpets, and this was followed by the development of 100% algal cover, which eliminated the barnacles. The barnacles reappeared only when the limpets recovered and grazed down the algae.

Conclusion. It is apparent that the success of both barnacles and limpets on rocky shores is based upon a dynamic rather than a static equilibrium. Although each group affects the other adversely, there is a counterbalance formed by other advantageous interactions.

Barnacles inhibit movement and feeding of limpets, but to offset this they promote limpet recruitment, and permit limited algal growth under conditions of high grazing pressure. This algal growth acts as a food reserve, and perhaps more importantly, as a reservoir for algal recruitment to the grazed rock. In the absence of barnacles, limpet grazing could be so effective as to lead to severe food shortages, and limpet recruitment so poor as not to sustain the population. Extreme population fluctuations would result, fluctuations which are damped in the presence of barnacles.

The benefits which barnacles receive from the activities of limpets immeasurably outweigh any adverse effects. While recruitment is reduced by the activities of the limpets, in the absence of limpets the algae would proliferate and the barnacle population would decline and could be eliminated. It seems fair to conclude that the large areas of barnacle-dominated intertidal rock which occur over much of the world are maintained only by the continual grazing activity of herbivores. The dominant herbivores, in the midtidal regions at least, are characteristically limpets, and barnacles depend upon them for the provision of space. The structure and habit of intertidal barnacles is not only a reaction to the physical environment, but renders all but the smallest specimens resistant to damage from the grazing of the limpets upon which they ultimately depend.

For background information *see* ECOLOGICAL INTERACTIONS; ECOLOGY; MARINE ECOSYSTEM; POPULATION DYNAMICS in the McGraw-Hill Encyclopedia of Science and Technology.

[RICHARD HARTNOLL]

Bibliography: C. Birkeland, Terrestrial runoff as a cause of outbreaks of *Acanthaster planci* (Echinodermata: Asteroidea), *Mar. Biol.*, 69:175–185, 1982; G. M. Branch, The biology of limpets: Physical factors, energy flow, and ecological interactions, *Oceanogr. Mar. Biol. Annu. Rev.*, 19:235–380, 1981; S. J. Hawkins, Interactions of *Patella* and macroalgae with settling *Semibalanus balanoides* (L.), *J. Exp. Mar. Biol. Ecol.*, 71:55–72, 1983; S. J. Hawkins, and R. G. Hartnoll, Grazing of intertidal algae by marine invertebrates, *Oceanogr. Mar. Biol. Annu. Rev.*, 21:195–282, 1983; P. Jernakoff, Factors affecting the recruitment of algae in a midshore region dominated by barnacles, *J. Exp. Mar. Biol. Ecol.*, 67:17–31, 1983; B. A. Menge, Effects of feeding on the environment: Asteroidea, in M. Jangoux and J. M. Lawrence (eds.), *Echinoderm Nutrition*, pp. 521–551, 1982; N. A. Sloan, Aspects of the feeding biology of asteroids, *Oceanogr. Mar. Annu. Rev.*, 18:57–124, 1980; G. M. Wellington, Depth zonation of corals in the Gulf of Panama: Control and facilitation by resident reef fishes, *Ecol. Monogr.*, 52:223–241.

Electric power systems

The essential requirements placed upon electric utility systems are to provide customers with safe, adequate, and continuous service at reasonable cost. This places an obligation upon electric utility systems to produce, transport, and distribute electrical energy in a manner which maintains a high quality of service to consumers and preserves the integrity of the interconnected system.

Indices of performance. While these broad requirements have been in effect nearly as long as ac systems have existed, the absolute quantitative expression of these characteristics has evolved as public needs and expectations have changed. Quality of service established itself as a factor in commercial competition between alternative energy supplies for illumination and power early in the history of electric utility systems. Present requirements place high priority on service continuity and upon fast restoration following loss of service. As a result, the availability of service realized by electric utility systems is high. For many years, North American systems have achieved Average Service Availability Indices of 0.9997 to 0.9998. Indices for describing service performance are illustrated in the table, which uses data for a hypothetical system.

Causes of service interruption. The principal causes of sustained interruption of service to customers served by the distribution system are weather-related, such as extreme winds, lightning, snow, and ice storms. Equipment-related faults, outages, and misoperations, public interference, and operations-related problems and troubles within the customer's own system are also important causes of interruption. Seventy to ninety-five percent of events resulting in sustained interruption of service to customers arise in the distribution system.

An important fraction of bulk power disturbances resulting in interruption of service has involved misoperation of protective relaying and supervisory control systems.

Variation in reliability. In general, interruption performance varies with load density and the size of load served. Low-load-density areas (suburban and rural) are usually supplied from radial distribution feeders. Higher-load-density areas may be served by feeders, with switching to permit sectionalizing and transfer of loads. Service to higher-load-density areas, commercial and urban, and to essential public services may be provided with redundant facilities such as spot networks and secondary networks;

System reliability indices*

Base system (example)	
(1) Customers served	8.3×10^6
(2) System peak demand	35 GW
(3) Annual production	156 TWh
I. Sustained interruption data (annual)	
(4) Number of interruption reports	77,500
(5) Customer sustained interruptions	12.4×10^6
(6) Customers affected	7.5×10^6
(7) Customer-minutes interruption	843×10^6
II. Reportable bulk power interruptions 100 MW or more for 15 or more minutes	
(8) Number of bulk power interruptions	5
(9) Sum of interrupted loads	1290 MW
(10) Sum of interruption durations	200 min

Customer indices

(11) System interruption frequency index: $\frac{(5)}{(1)} = 1.49/\text{yr}$

(12) Customer average interruption duration: $\frac{(7)}{(5)} = 68$ min

(13) Customer interruption frequency index: $\frac{(5)}{(6)} = 1.65/\text{yr}$

(14) System annual interruption duration: $\frac{(7)}{(1)} = 102$ min/yr

(15) Average service unavailability: $\frac{(14)}{525600} = 0.00019$

(16) Average service availability: $1.0 - (15) = 0.99981$

(17) Customers interruption per event: $\frac{(5)}{(4)} = 160$

Bulk power indices

(18) Loads interrupted per event: $\frac{(9)}{(8)} = 258$ MW

(19) Bulk power interruption index: $\frac{(9)}{(2)} = 0.037$

(20) Bulk power interruption duration: $\frac{(10)}{(8)} = 40$ min

*Data are given for a hypothetical system. Each quantity listed is labeled with a bracketed number, and the various indices are calculated from the data as indicated. For example, the system interruption frequency index, which is labeled (11), is obtained by dividing the customer sustained interruptions, labeled (5), by the customers served, labeled (1).

service to major industrial loads and to large secondary networks may be from the bulk power system. As a result, the reliability of service to high-density-load areas, essential public services, and large commercial and industrial loads is correspondingly higher than for the lighter-load-density areas. Historically, the level of service reliability provided has corresponded closely to the value placed upon service continuity by the type of customer and public function. Surveys show consistently that industrial operations and commercial functions put higher value on continuity of service.

Nature of interruptions. The nature of interruptions caused by disruptions and failure in the bulk power system differs in extent, in frequency, and in effect from interruptions arising in distribution systems. The number of customers interrupted per case of interruption in distribution systems is usually small, ranging from a few customers to a few hundred and averaging around 150 customers. Disturbances arising in the bulk power system can affect large numbers of customers and can have widespread social and economic impact. Major failures of the bulk power system have resulted in interruption of hundreds of megawatts of load and interruption of service to hundreds of thousands of customers for periods of more than a day. Public and political reaction is influenced strongly by the severity of the power failure, by the consequential damages, and by the type of facilities affected.

The public has shown tolerance to localized outages arising in the distribution system if the interval between outages is reasonably long and service restoration proceeds within a short time (less than 4 h). However, public reaction to widespread interruptions has been strong, particularly if the source of disturbance cannot be identified readily with widespread and severe natural causes, such as hurricanes, thunderstorms, tornadoes, or ice storms. Civil regulatory and legal actions have resulted from prolonged interruption of service, particularly when widespread failure interferes with public facilities, including water, sewage, health, safety and protection, transportation services, and lighting. On the other hand, the public has responded to appeals for voluntary curtailment and has accepted emergency load-relief measures to protect the power system, provided the restoration of service is prompt.

Categories of failure. Many degrees of failure are possible within the power supply system. At least three categories of failure should be identified: service deficiencies, emergencies in the supply system, and power system interruptions.

Service deficiencies include events of service voltage excursions, such as unacceptably low voltage and voltage dips, unacceptably high voltage and voltage surges, excessive harmonics, and unbalance in polyphase power supplies. Such effects may arise within the customer's own system. Consumer processes or equipment may be susceptible to voltage excursions, which may result in process upsets (printing processes and paper production, for example), resulting in poor-quality product or production interruption. Consumer equipment may be vulnerable to abnormal voltage, harmonics, or unbalanced conditions.

Power system emergencies arise when demand exceeds supply capability. Emergencies have occurred from widespread shortfall of resources. Deficiencies have also occurred from generating-capacity shortfall or transmission-system transfer capability limitations. System responses to emergencies include appeals for voluntary load reduction, the curtailment of interruptible loads, and reduction of service voltage to customers or through load management control. Load shedding and isolation of areas from interconnected systems (controlled separation) are permitted in major emergencies.

Power system interruptions result in sudden and uncontrolled interruption of service to consumers. Typically, widespread interruptions have resulted from a combination of causes including two or more of the following: (1) a heavily stressed system, (2) critical facilities on maintenance, (3) high hazard conditions causing multiple failures, (4) protection system failure, and (5) operator error.

Design and operating philosophy. The design and operating philosophy for interconnected bulk power systems has been to provide adequate reserves to minimize the risk of power supply emergencies and to provide system strength to withstand specific classes of disturbance. Bulk power system reliability has been assessed in terms of two basic attributes termed adequacy and security.

Adequacy relates to the provision of sufficient facilities to satisfy customer load demand in the presence of scheduled and unscheduled outages of generation, transmission, and distribution facilities necessary to transport energy to the actual customer load points. Adequacy is assessed on both deterministic (contingency) and risk bases in terms of static generating capacity reserves and transmission and distribution load-carrying capability.

Security relates to system capability to withstand disturbances arising within that system. Design and operating philosophy has adopted generic performance tests (umbrella tests) which will assure that the system has the capability and flexibility to maintain secure operation and bulk power system integrity.

Good system design requires the application of testing procedures to establish margins to failure. Margins can be measured in adequacy terms, such as load-carrying or transfer capability, and in security terms such as critical clearing times and insensitivity to protective relaying and control misoperations. Performance testing must involve explicit consideration of overload cascading, instability, and voltage sensitivity, including voltage collapse. Selection of contingencies for performance tests must include common-mode and other multiple transmission outages that are considered credible during severe hazard conditions.

Strategic reliability considerations can have a major impact on the actual performance of the bulk power supply system. Some of the strategic factors that should be considered at the time of system design are: avoiding excessive dependence on one type of fuel or resource, avoiding dependence on a single generic type of generating capacity, providing sufficient tie capacity to adjacent power systems, developing emergency energy allocation procedures, and providing flexibility to handle changes in environmental, transportation, or regulatory constraints.

For background information *see* ELECTRIC POWER SYSTEMS; ELECTRIC POWER SYSTEMS ENGINEERING in the McGraw-Hill Encyclopedia of Science and Technology.

[ROBERT J. RINGLEE]

Bibliography: R. Billinton et al., *Power System Reliability Indices, Philosophy and Techniques*, 10th Annual Engineering Conference on Reliability, Availability, Maintainability for the Electric Power Industry, Tutorial Lectures, Montreal, May 24, 1983; *Definitions of Customer and Load Reliability Indices for Evaluating Electric Power System Performance*, IEEE Pap. A75–588–4, 1975; *The National Electric Reliability Study*, Department of Energy, Tech. Study Rep. E/P-005, 1981; *Operating, Design, and Planning Criteria for Bulk Power Systems of New York State*, Final Report, vol. 1, *Criteria Reviews, Recommendations, and Findings*, Power Technologies, Inc., Schenectady, March 4, 1981.

Electrical utility industry

Three unusual developments marked the year 1983 for the electrical utility industry. All were financial in nature, rather than technological.

First, 1983 saw the first default on indebtedness in the modern history of the industry. The Washington Public Power Systems (WPPS), a consortium primarily of public power entities in Washington and Oregon, defaulted on payment for bonds issued to secure construction costs of two of five nuclear power stations. When construction costs soared just as regional demand for electric energy dropped precipitously, making unnecessary the plant's capacity, the members of the consortium sought and received relief from the courts from their "take or pray" contracts with the Bonneville Power Authority. These payments, according to those contracts, had to be made even if the units were canceled, and were to be used to pay off the bonds in such case. The units were canceled, saddling the consortium with $2.25 billion in bonds to service. The court decision, which said that the utilities did not have authority to enter into such a contract and therefore need not pay, led to default.

Second, a regulatory agency refused for the first time to permit a utility to place a new generating unit into its rate base. The Long Island Lighting Company's Shoreham nuclear plant, with a capitalized cost of about $3.2 billion, would normally have entered the rate base upon completion, permitting the company to earn its allowed rate of return on the asset. That, however, would have meant that the utility's customers would have had to absorb an immediate 50% rate increase. Regulators are now seeking a solution in some type of phase-in of the plant. This problem threatens to become widespread as large numbers of nuclear and fossil units are scheduled to enter commercial service over the next few years.

The third unique event occurred when the operating license of the Utah Power and Light Company's 2.5-MW Bountiful hydroelectric plant came up for renewal by the Federal Electric Regulatory Commission (FERC) after 50 years of operation. The City of Bountiful, Utah contested the renewal on the basis that public power utilities had, by the Federal Power Act, the right to preference in renewal. In effect, the City of Bountiful claimed the right to take over ownership of the station, paying Utah Power and Light Company its original cost less depreciation. The FERC held for the city, initiating thereby a series of court cases leading to a petition by a group of private utilities to the U.S. Supreme Court to overturn the FERC decision. A number (490) of other hydroelectric licenses are ultimately involved, and this will be a fiercely contested issue.

Spurred by extreme and long-lasting summer heat throughout the Midwest and Southeast, peak demand for the industry rebounded from its historic drop in 1982 to climb 6.6% in 1983. This rise, in the face of a still-sluggish economy, lends support to industry consensus forecasts of future load growth in the order of 2.5–3.0% per annum over the next decade.

Ownership. Ownership of electrical utilities in the United States is pluralistic, being shared by investor-owned corporations; customer-owned cooperatives; and public bodies on the municipal, district, state, and federal levels of government. The industry, however, is dominated in all essential measures by the investor-owned sector. Investor-owned utilities serve 76.5% of the 95.94 million electric customers in the United States. Municipal, district, and state utilities serve 13.6% of total customers, slightly more than do the cooperatives, which serve 9.9%. Although federal utilities do serve some few customers on the retail level, they are primarily wholesalers, selling to other utilities.

Investor-owned utilities also own and operate 77.3% of the nation's installed generating capacity. Publicly owned utilities own 10.2% of all installed capacity, and cooperatives own only 2.9%. Federal entities own and operate 9.6% of the installed capacity.

The large discrepancy between the percentage of customers served by the cooperatives and the much smaller percentage of capacity owned by them comes about because such organizations are mostly distribution companies which buy their power from others at wholesale rates and distribute it to their member customers. Some cooperatives do generate, and some, such as the Tri State G&T Association of Denver, Colorado, do not serve distribution customers at all, but produce and transmit wholesale power to distribution cooperatives.

The extraordinary financial pressures on investor-owned companies in recent years and antitrust rulings by courts have led investor-owned companies to enter into joint ownership of new large generating units with cooperatives and publicly owned utilities. These two types of utilities are eager to buy into these large efficient units so as to have access to the lower-cost energy produced by the economies of scale of these units. Although their financial resources generally could not permit their building of such units, especially nuclear ones, their access to lower-cost funding makes them an attractive partner. For example, Duke Power Company sold a 75% share of a 1250-MW nuclear unit to a consortium of 10 cooperatives from North Carolina and 5 from South Carolina. Duke will construct the plant and operate it under contract from the other owners.

Some cooperatives and municipals, unregulated and with assess to capital at lower interest rates, see the financial plight of the privately owned companies as an opportunity to build large units themselves and sell the power to the investor-owned companies. Seminole Electric Cooperative in Florida, for example, with no retail customers, is currently building two 600-MW coal-fired units, the first of which is due to be in service in 1984.

Capacity additions. Utilities had a total generating capacity at the end of 1983 of 661,000 MW, having added 11,465 MW during the year. This produced a capability—that is, the actual ability to generate power at the time of peak demand—of 605,700 MW. The difference comprises capacity unavailable because of maintenance, failures, or derating from nameplate capacity (see table).

The composition of the capacity additions made during 1983 was 390 MW of conventional hydroelectric, 6530 MW of fossil-fired steam, 254 MW of pumped-storage hydroelectric, 4040 MW of nuclear power, and 260 MW of diesel and combustion turbines.

The composition of total plant by type of generation at the end of 1983 was 457,973 MW, or 69.3% fossil-fired steam; 64,250 MW (9.7%) conventional hydroelectric; 14,349 MW (2.2%) pumped-storage hydroelectric; 67,182 MW (10.2%) nuclear; 51,992 MW (7.9%) combustion turbines; and 5199 MW (0.8%) internal combustion engines, essentially all diesels (see illustration).

Fossil-fueled capacity. All units entering service in 1983 were coal-fired in accord with the provisions of the Fuels Use Act of 1974. There are a considerable number of older oil-fired units in operation that have the potential to be converted to coal firing. These are heavily concentrated (62%) in the Northeast, with some in the East Central (22%) and Southeast (13%). Utilities plan to convert 8700 MW from oil to coal in the 1982–1990 period.

Because of the environmental problems associated with coal burning, utilities have launched a number of developmental projects to perfect fluidized-bed combustion. This is a technique in which the coal is burned together with limestone in a bed that is kept turbulent, or fluidized, by passing air through it. This provides efficient combustion and pollution control simultaneously. The largest such installation in the United States is a 40-MW unit on the Tennessee Valley Authority system.

A number of developmental processes to gasify coal and use the clean gas as a boiler fuel are also under way. A 40-MW prototype gasifier based on gasifying high-sulfur Illinois coal in a rotating kiln similar to those used in cement production began pilot operation at the Wood River Plant of the Illinois Power Company in 1983.

Utilities spent $10.5 billion for construction of fossil-fired units in 1983. Investor-owned utilities spent $7.6 billion; state and municipal units, $1.7 billion; cooperatives, $1.3 billion; and federal agencies, only $2.3 million.

Nuclear power. The year 1983 marked the fifth straight year that no new orders for nuclear units have been placed, and industry analysts now feel that new orders cannot be expected until after 1990

United States electric power industry statistics for 1983*

Parameter	Amount	Increase compared with 1982, %
Generating capacity, $\times 10^3$ kW		
Conventional hydro	64,250	0.6
Pumped-storage hydro	14,349	1.8
Fossil-fueled steam	457,973	1.4
Nuclear steam	67,182	6.4
Combustion turbine and internal combustion	57,191	0.5
TOTAL	660,945	1.8
Energy production, $\times 10^6$ kWh	2324.4	1.8
Energy sales, $\times 10^6$ kWh		
Residential	735,100	0.8
Commercial	539,500	5.0
Industrial	777,000	0.8
Miscellaneous	78,800	−1.0
TOTAL	2,130,500	1.8
Revenues, total, $\times 10^6$ dollars	129,325	6.8
Capital expenditures, total, $\times 10^6$ dollars	39,855	−0.9
Customers, $\times 10^3$		
Residential only	85,300	1.3
TOTAL of all classes	95,940	0.9
Residential usage, kWh/customer	8,623	−0.8
Residential bill, cents kWh (average)	7.0	3.7

*From 34th annual electric utility forecast, *Elec. World*, 197(9):55–62, September 1983. Extrapolations from monthly data of the Edison Electric Institute; and 1983 annual statistical report, *Elec. World*, 197(3):61–84, March 1983.

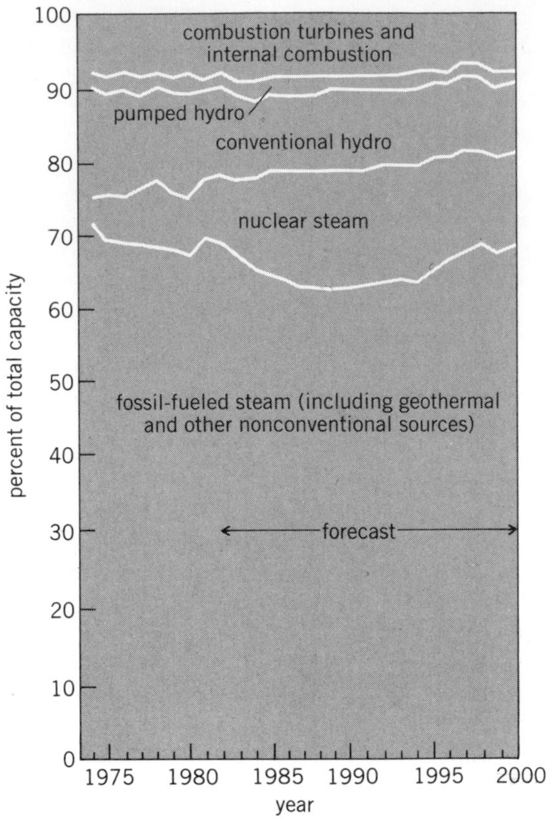

Probable mix of net generating capacity. (*After 34th annual electric utility forecast, Elec. World, 197(9):55–62, September 1983*)

at the earliest. The cancellation of units already announced or under construction continued, with four units totaling 4810 megavolt-amperes (MVA) joining the list of those canceled. Because the average completion time from date of Nuclear Regulatory Commission (NRC) approval to entrance into commercial service has now stretched out to more than 10 years, units ordered and begun in previous years continue to come on line. During 1983 four units with an aggregate capacity of 4040 MVA began commercial service. The total capacity of all nuclear units now in service in the United States is 67,182 MW, provided by 81 units in 56 individual generating plants.

One of the units entering service in 1983 was a boiling water reactor (BWR), and three were pressurized water reactors (PWR). This brings the totals in the United States to 28 BWRs and 53 PWRs.

Utilities currently have under construction, but less than 50% complete, a total of six nuclear units aggregating 6915 MW. There are an additional 38 under construction but more than 50% complete, aggregating 39,100 MW.

With the refusal by the United States Congress to fund further construction of the liquid metal fast breeder reactor, the project was terminated for practical purposes. This ended the 30-year effort by the consortium of 750 utilities and the Department of Energy to build the breeder, for which site preparation had started, engineering had been 95% completed, and more than $100 million spent on component fabrication. A minimal effort in breeder research and development will continue through operation of the fast flux test facility at Hanford, Washington, and through the Department of Energy's small experimental breeder reactor (ERB) program.

Utilities spent $16.8 billion on nuclear construction in 1983. Investor-owned spending accounted for $13.4 billion, state and municipal utilities for $2.8 billion, and cooperatives for $536 million.

Combustion turbines. Combustion turbines have historically been installed by utilities for use at times of peak demand to supply up to 10–15% of that peak. The low capital cost of such machines, currently about $290/kW compared to $660/kW for coal-fired units, more than offsets their high fuel consumption when used only for about 200 h per year in this way. Combustion turbines, which use either natural gas or oil as fuel, have heat rates of about 13,000 Btu/kWh (3.8 joules of heat per joule of electric energy) compared with 9000 Btu/kWh (2.6 joules of heat per joule of electric energy) for a modern fossil unit. Further, they have quick-start capability which permits them to be brought up to full load in 3 to 8 min and provides flexible capability for emergency conditions. They are also used to provide startup power for generating stations that have experienced a complete shutdown, as during a blackout of an entire area or system.

Because of government restrictions on fuel, and because there are already excess reserves at peak periods, the percentage of peak represented by gas turbines is decreasing, and currently stands at 8%.

Although individual units as large as 125 MW have been installed, units most favored are in the 25–50 MW range. In 1983, utilities added 200 MW of combustion turbine capacity to raise the total industry capacity to 51,992 MW, which is 7.9% of total capacity installed.

Combustion turbines are also used in combination with steam turbines in highly efficient combined-cycle operation. In this mode, the 900–1000°F (482–538°C) gas turbine exhaust produces steam in a heat recovery boiler which supplies a steam turbine. Efficiencies of this cycle may run as high as 60%, compared with 34–35% for the most efficient current steam cycles. In this mode the combustion turbine will contribute about 70% of the combustion turbine output.

Utilities spent $98.5 million on combustion turbines in 1982.

Hydroelectric installations. Utilities brought 390 MW of conventional hydroelectric capacity into service in 1983, raising the total now installed to 64,250 MW. Hydro units of this type, that is, hydroelectric turbines driven by impounded water or by the natural flow of a river, constitute 9.7% of total installed capacity of all types of utility systems in the United States.

Future plans call for an additional 3590 MW of capacity to be built over the next 10 years, though locations for the required dams are becoming increasingly difficult to find and to license. The few major sites suitable for high dams are, in general, in areas where the environmental effects of the resulting lake are unacceptable. Utilities spent $552 million on this type of installation during 1983. Investor-owned companies spent $356 million; state and municipal utilities, $137 million; cooperatives, $12.2 million; and federal agencies, $46.4 million.

The future of hydroelectricity may rest with small-scale projects that either can be incorporated into existing dams or can be powered by the normal flow of the river. These units carry ratings of 1.5–50 MW. Such units have become especially popular in the West, where the extensive irrigation canal systems often have locks that create heads of 15–25 ft (4.5–7.5 m), adequate to power small tube- or bulb-type hydroelectric units.

Pumped storage. Pumped storage represents one of the few possible methods for storing large amounts of energy from electrical generators. Water is pumped from a body of water on which the generating unit is located into a reservoir some distance above. This is normally done during off-peak periods when large, efficient base-load units that would ordinarily be shut down for lack of demand are available. During the subsequent peak demand period, this water is released through the pumps, which can be reversed to act as turbines, recovering only about 65% of the energy originally expended but reducing the need for the equivalent capacity to be provided at peak.

Utilities installed an additional 254 MW of pumped storage capacity during 1983, raising total installed capacity in the United States to 14,349 MW, or about 2.3% of the nation's total capacity.

Utilities spent $145 million for pumped storage facilities in 1983. The investor-owned companies spent $124 million, the major portion of that total. Federal agencies spent only $2.7 million; cooperatives, $2.5 million; and publicly owned utilities, $15.9 million.

Renewable energy sources. Utilities have actively sought energy sources that provide alternatives to the combustion of coal, oil, gas, and uranium. At the end of 1983, utilities had a total of 1690 MW of capacity other than the conventional types listed above. These include geothermal, wind, solar, waste, and refuse-fired units. Utilities have planned an additional 3383 MW of such units for the 1983–1992 period. These consist of 2040 MW of geothermal, 357 MW of solar, 273 MW of wind, and the rest waste- or refuse-fired capacity. More than 87% of all such new capacity planned is located in the state of California.

Rate of growth in demand. Some credible forecasters have begun to project extremely low or even negative growth prospects over the next 10–20 years. These projections are predicated to a large degree on the rise in electricity prices as rates adjust to absorb the large additions to the rate base that the new expensive units represent as they come into service. For example, the just-completed Shoreham unit of the Long Island Lighting Company would raise rates roughly 50% upon entering the rate base. Coupled with this negative price elasticity, stable prices for competitive energy sources such as coal and oil will also exert downward pressure on use. Others, however, point to the fact that the normalized rate of growth, even throughout the past 10 years, has tracked the growth in the gross national product. These are predicting growth of peak demand in the range of 2.5–3.0%.

A long-term decreasing trend naturally arises from the mix of demographic factors that characterize a maturing society such as the United States. Population is growing at a decreasing rate, and passing beyond the years of peak consumption. This effect will be compounded by price-induced conservation. Because of the high technological content of the utility industry's plant, the preponderance of skilled labor employed, the high cost of capital engendered by unresponsive regulation, and escalating fuel costs, electricity prices should rise slightly more than inflation over the coming decade, somewhat moderating growth.

Industry and commerce have also embraced the concept of demand control. In this technique, electrical equipment is computer-controlled to minimize peak demand on the entire manufacturing plant or building within the constraints of required production. This reduces the demand charge, which is a substantial part of total energy cost. Load shedding agreements have become popular between utilities and groups of industrial customers acting as a load bloc which, in return for a reduced rate for electricity, agree to reduce demand at peak upon request.

Distribution utilities that purchase power wholesale from generating companies can derive immediate benefits from restraining peak demand, since a major element of their charge is for demand. These utilities have begun to employ controls on air conditioners and electric water heaters to cut them out of service at peak periods. Cooperatives such as Georgia's Oglethorpe Electric have already installed tens of thousands of such devices.

The overall national declining pattern of growth (from 7.2% before 1974 to an average of 3.4% in the period 1974–1983) has a major effect on reserve margins—that is, the excess of installed capacity over demand—on a national basis. A rule of thumb is that average national reserve margin should be about 25%, and it is now about 35%. Though utilities are delaying the construction of or canceling many major generating units, some plants started years ago will be finished and will come into service. Margin will not decline to the 25% level until 1990.

Usage. Sales of electricity, after a drop of 2.6% in 1982, rose 1.8% in 1983. Total national usage was 2.130×10^{12} kWh. The rise was due primarily to extraordinary heat throughout the Midwest.

Commercial sales held up best again in 1983, rising 5.0% to a total of 539.5 × 10^9 kWh.

Industrial sales rose slightly, about 1%, reflecting the sluggish economic recovery, and reached about 777 × 10^9 kWh consumed.

Residential sales also rose only slightly from 1982, recording a rise of 0.8%, with 735.1 × 10^9 kWh consumed. This usage is dependent on housing starts, which continued at a very low annual level of 1.2 million units.

Electric heating for residences continues to make gains. More than half of all new homes constructed in the United States in each of the last 10 years have been electrically heated, and in 1983 heating energy sales topped 164.4 × 10^9 kWh. The emergence of the heat pump, now taking almost a third of new electric installations, has made electric heating competitive in most areas with competitive fuels. Electric heating is now used in 18.6% of all housing stock in the United States, second only to gas.

Residental use per customer dropped slightly to 8623 kWh from 1982's use of 8696 kWh. Continued rate increases, however, boosted the average residential rate to 7.0 cents/kWh and the annual average bill per residential customer to $603.

Total revenue for all classes of service for the entire industry was $129.3 billion.

Fuels. Consumption of coal by utilities in 1982, the last year for which good figures are available, actually dropped 0.5% to 594.1 × 10^6 tons (538.8 × 10^6 metric tons). Oil dropped sharply 28.9% to 249.7 × 10^6 bbl (39,600 m^3). Essentially all the shift from oil was absorbed by nuclear and hydroelectric units since consumption of natural gas also dropped 1.1% to 3.2266 × 10^{12} ft^3 (91.3 × 10^9 m^3).

The same type of shift was seen in actual energy generated. Total generation output dropped from 2.1507 × 10^{12} kWh, to 2.0936 × 10^{12} kWh in 1982, while that portion generated by other than hydroelectric dropped from 2.0341 × 10^{12} kWh to 1.9325 × 10^{12} kWh. Coal generated 1.1925 × 10^{12} kWh of the total; oil accounted for 146.6 × 10^9 kWh; gas was used to generate 305.3 × 10^9 kWh; and nuclear power put out 282.8 × 10^9 kWh. The rest was primarily hydroelectric, with a minor contribution from other sources such as geothermal.

Distribution. Distribution capital expenditures for 1983 amounted to $7.05 billion, and an additional $1.981 billion was spent maintaining existing plants. During the year, 14,800 mi (23,840 km) of three-phase equivalent overhead lines and 6060 three-phase equivalent mi (9760 km) of underground lines came into service at voltages ranging from 4.16 to 35 kV. The majority of this mileage was at 15 kV, which accounted for 9290 three-phase equivalent miles (15,721 km) of overhead and 4042 three-phase equivalent miles (6510 km) of underground circuitry. The percentages for overhead construction held by other voltage classes were 14.8%, 11.8%, and 2.5% for 35-, 25-, and 4-kV, respectively. For underground construction, the equivalent percentages are 13.1%, 17.4%, and 2.8%. During 1983, utilities energized 12,667 MVA of distribution substation capacity and expended $708 million for substation construction.

Transmission. Utilities spent $4.2 billion in capital accounts for transmission lines in 1983. During the year, they spent $2.02 billion for overhead lines at 345 kV and above, and $884 million for overhead circuits of 220 kV and below. For underground transmission construction, which can cost on an average eight times more than equivalent overhead construction, capital expenditures amounted to $58.5 million at voltages of 220 kV and higher, and $20.9 million for circuits at 161 kV and below. Utilities installed 1260 mi (2030 km) or overhead lines at 345 kV and above, but 4060 mi (6540 km) at 220 kV and below.

The picture is completely different in the underground sector, because current cable technology costs favor the lower voltages. In 1983, only 6 mi (9.7 km) of cables operating at or above 230 kV came into commercial service, and 18 mi (29 km) at or below 161 kV.

Utilities brought 37.4 gigavolt-amperes of transmission substation capacity into service in 1983 and spent $1.15 billion for substation construction. Maintaining existing transmission plant cost $600 million.

Capital expenditures. Total capital expenditures in 1983 dropped slightly from 1982's $40.2 billion to $39.9 billion as a result of utility efforts to curtail construction. Of this total, $28.2 billion went for generating facilities, $4.2 billion for transmission, $5.5 billion for distribution, and $2.0 billion for miscellaneous facilities such as headquarters buildings and vehicles. Total assets held by the investor-owned segment of the industry were $322.4 billion at the end of 1982. Municipals held about $30 billion in assets and cooperatives about $33 billion.

For background information *see* ELECTRIC POWER GENERATION; ELECTRIC POWER SYSTEMS; ENERGY SOURCES; TRANSMISSION LINES in the McGraw-Hill Encyclopedia of Science and Technology.

[WILLIAM C. HAYES]

Bibliography: Edison Electric Institute, *Statistical Yearbook of the Electric Utility Industry*, 1982; 1983 annual statistical report, *Elec. World*, 197(3):61–84, 1983; 34th annual electric industry forecast, *Elec. World*, 197(9):55–62, 1983; 23d steam station cost survey, *Elec. World*, 197(11):49–60, 1983.

Electromagnetic compatibility

Electromagnetic compatibility denotes the condition in which electrical and electronic equipment functions as intended by the designer; no part or component of a system interferes with another. In addition, the system functions compatibly with its electromagnetic environment; that is, it neither causes interference nor is the victim of interference. Interference may take the form of surges or other disturbances on a power or control line, or of radio noise or other radiated disturbances.

Recent developments in the field of electromagnetic compatibility stem from the expanded usage of electronic equipment in the areas of cable television, citizens' band radio, computers, and military communications. In addition, concerns about the compatibility of electromagnetic radiation with the human body, or the lack thereof, have resulted in new standards for limiting the exposure of the general public to such radiation. In the ongoing effort to achieve electromagnetic compatibility, especially for newer and more sensitive components and systems, recently developed shielding techniques have become important, especially because of the growing usage of plastic enclosures. For secure transmission, the standards have been revised to assure that equipment handling such transmissions will not produce compromising emanations.

CB radio interference with television. The tremendous popularity of citizens' band (CB) radio has resulted in demands upon manufacturers that the third-harmonic output of these radios be kept to extremely low levels. The third harmonic of CB channel 1 falls just 64 kHz above the television channel 5 color subcarrier; CB channels 29 and 30 produce third harmonics that fall, respectively, only 15 kHz below and above the TV channel 5 sound carrier. Many CB radios operate in very close proximity to television receivers in the same or a nearby household. Accordingly, manufacturers endeavor to keep their spurious emissions to a level on the order of 1 microwatt or less.

CATV interference with aeronautical radio. Cable TV (CATV) systems now are built to provide as many as 55 different channels to their subscribers on a single coaxial cable, with some systems having a capability of twice that number through the use of dual cables. Until the early 1970s, many CATV systems carried only the 12 very-high-frequency (VHF) channels. Thus the channels on the cable were only at the TV frequencies used over the air. Stations on ultra-high-frequency (UHF) channels were carried by CATV on a VHF cable channel that was not being used otherwise. Expansion of CATV systems to carry more than 12 VHF channels means using very high frequencies, on the cable that are not used over the air for TV. This is necessary since many of the UHF-TV channels would suffer high losses if attempts were made to transmit them via cable at their normal frequencies.

Among the additional very high frequencies used on cable for extra channels are the 108–136 MHz frequencies used for aeronautical radio and the 225–400 MHz frequencies used for aircraft landing systems as well as for various military purposes. Some CATV systems have been found to leak slightly, especially at connectors joining the cables to repeater amplifiers. The leakage has caused interference with aeronautical radio navigational and voice circuits as well as with the Forestry Service and amateur radio operators.

The Federal Communications Commission (FCC) now requires that cable systems within 66 mi (110 km) of aeronautical radio stations must offset their frequencies by 100 kHz (plus tolerance) in the air communications bands and by 50 kHz (plus tolerance) in the air navigational bands.

As the number of cable operators increases, the likelihood of interference increases. Cable operators are required to eliminate any harmful interference. Most of the problems have resulted from cable levels above 100 microwatts and without frequency offset. Thus care on the part of the cable operators with respect to their levels, frequency offset, and connector tightness are seen as the solutions to this problem.

Interference from computing equipment. The FCC in its Rules and Regulations established Part 15, Subpart J, to deal with interference from digital computing equipment. Actually any equipment not otherwise covered by the Rules and Regulations comes under the computing equipment heading if it uses pulses at a rate of 10,000 per second or more in its operation. Tests are performed by the manufacturer of such equipment to provide assurance that it meets the FCC's requirements. The FCC specifies conducted emission measurements in the 450 kHz–30 MHz band and radiated emission measurements in the 30–1000 MHz band. To satisfy many of the measurement requirements, radiated emission measurements often are performed at an open-air test site to avoid having to compensate for shielded enclosure effects and to avoid the use of costly anechoic chambers.

New radiation hazard standards. The American National Standards Institute (ANSI) has revised its radio-frequency protection guide (RFPG) to lower values than those of its previous (1974) standard. The RFPG is defined as the radio-frequency field strength or equivalent plane-wave power density that should not be exceeded without careful consideration of the reasons for doing so, careful estimation of increased energy deposition in the human body, and careful consideraton of the increased risk of unwanted biological effects. Values averaged over a 0.1-h period are given in the table, where E is the electric field strength in volts/m, H is the magnetic field strength in amperes/m, and P is the power flux density in milliwatts/cm^2.

The American Conference of Government Industrial Hygienists (ACGIH) extends the low-frequency limit from 0.3 MHz down to 0.01 MHz at the same levels as those for 0.3–3 MHz.

One of the ANSI provisions is that the values may be exceeded if the average whole-body specific absorption rate is less than 0.4 W/kg, with the peak

Radiation hazard standards specified by radio-frequency protection quide (RFPG)*

Frequency, MHz	E^2, V^2/m^2	H^2, A^2/m^2	P, mW/cm^2
0.3–3	40,000	2.5	100
3–30	4000/(900/f^2)	0.25(900/f^2)	900/f^2
30–300	4000	0.025	1.0
300–1500	4000/(f/300)	0.025(f/300)	f/3000
1500–100,000	20,000	0.125	5.0

*f = frequency.

spatial specific absorption rate, averaged over 1 g of tissue, less than 8 W/kg. The specific absorption rate is defined as the time rate at which radio-frequency electromagnetic energy is imparted to an element of mass of a biological body.

The foregoing recommendations apply to both nonoccupational and occupational exposures, but do not apply to the intentional exposure of patients under the direction of physicians for healing purposes.

For fields consisting of a number of frequencies for which there are different values of the RFPG, the fraction of the RFPG incurred within each frequency interval must be determined; the sum of all such fractions should not exceed 1.0.

At frequencies between 300 kHz and 1000 MHz the RFPG may be exceeded if the radio-frequency input power of the device producing the radiation is 7 W or less.

Conductive coatings. Nonmetallic materials are being used increasingly throughout industry. They appear in the form of structural foam plastics as housings for relatively small appliances and systems; in addition, they appear in the form of fiberglass-reinforced plastics for aircraft structural and fuel system components. Two types of problems result: such materials provide no radio-frequency shielding effectiveness, and static charges can accumulate on the plastic surfaces. Sudden discharges then produce noise bursts and possible component failure or system malfunction.

The use of conductive coatings on plastic surfaces becomes important not only to provide shielding and to prevent static discharge problems but also to provide for grounding. An important characteristic of any conductive coating is its surface resistance R_s measured in ohms per square. The surface resistance for a rectangular area is given by $R_s \leq \rho/d$, where ρ is the resistivity of the material in ohm centimeters and d is the thickness of the material in centimeters.

For shielding, R_s generally should be 1.0 ohm/square or less, with shielding effectiveness improving as R_s decreases. For grounding, an R_s value of 10 ohms/square often is adequate, whereas for electrostatic dissipation an R_s value of 50–150 ohms/square is desirable so that the static charges are dissipated slowly, thereby minimizing the production of noise.

Techniques for applying conductive coatings include vacuum metallization, which can be used on any plastic with a controlled film thickness and thus a controlled attenuation, but which is costly, requires a base coat, and requires a vacuum chamber large enough to accommodate the part being coated. The resulting aluminum film is also susceptible to corrosion in moist atmospheres.

Another technique for applying a conductive coating is flame spraying. This results in a hard, dense coating, often of zinc, but the plastic to be coated first must be sandblasted, and the heat from the flame may warp the plastic. The equipment is expensive, the process is hazardous, and the zinc coating may separate from the plastic with temperature changes.

Sputtering involves the use of a high-vacuum atmosphere in which the metal target is bombarded by electrically excited argon ions that sputter (dislodge) metal atoms. These metal atoms then are deposited on the desired parts. Generally chrome is applied to the plastic, followed by a layer of copper (for its conductivity) and then another layer of chrome (for corrosion protection). The film thickness is on the order of 0.5 micrometer (0.00002 in.).

Nickel acrylic and urethane coatings can be sprayed on a surface to a thickness of 50 to 60 μm (0.0020 to 0.0024 in.). This thickness provides good shielding without the cracking and adhesion problems associated with thicker coatings.

For background information *see* ELECTROMAGNETIC COMPATIBILITY; RADIO SPECTRUM ALLOCATIONS in the McGraw-Hill Encyclopedia of Science and Technology.

[BERNHARD E. KEISER]

Bibliography: *Electromagnetic Radiation with Respect to Personnel, Safety Level of*, ANSI Stand. C95.1–1982; *Interference Technology Engineers' Master, 1983*, R & B Enterprises, Pennsylvania; J. P. Wong, Cable signal leakage: Where is it all going?, *Technical Papers, Cable '83*, pp. 23–29, National Cable Television Association.

Electromagnetic pulse (EMP)

The electromagnetic pulse is a transient electromagnetic signal produced by a nuclear explosion in or above the Earth's atmosphere. Though not considered dangerous to people, the EMP represents a potential threat to many electronic systems.

This article describes the historical evolution of the concept of EMP, including how it is generated and what electrical characteristics are usually ascribed to it. Its coupling to typical systems is discussed, as is the manner in which it causes upset or damage to such electronic components as semiconductors.

Discovery. Though predicted by some scientists involved with the early development of nuclear weapons, EMP was not then considered to be a serious threat to people or equipment. However, in the early 1960s some of the high-altitude nuclear tests conducted in the Pacific led to some strange occurrences many miles from ground zero. In Hawaii, some 800 mi (1300 km) from the Johnston Island test, EMP was credited with setting off burglar alarms and turning off street lights. In other tests conducted in Nevada, significant EMP-induced signals were coupled to cables.

Initial nuclear radiation. In a typical nuclear detonation, parts of the shell casing and other materials are rapidly reduced to a very hot, compressed gas, which upon expansion gives rise to enormous amounts of mechanical and thermal energy. At the same time the nuclear reactions release tremendous amounts of energy as initial nuclear radiation (INR).

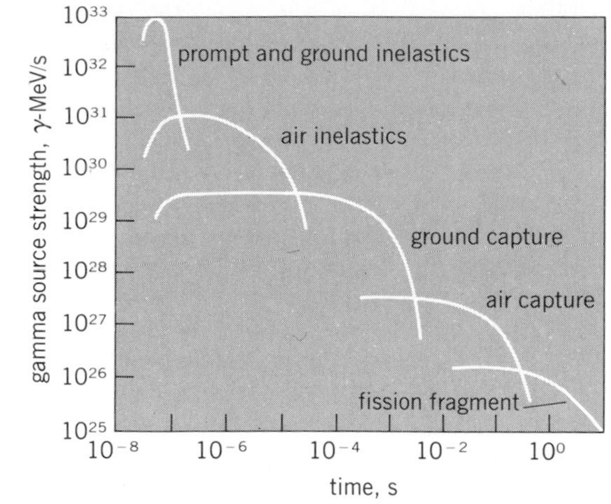

Fig. 1. Total gamma source strength versus time for nominal 1-megaton surface burst. Both horizontal and vertical scales are logarithmic. (*After C. L. Longmire, On the electromagnetic pulse produced by nuclear explosions, IEEE Trans. Antennas Propag., AP-26(1):3–13, 1978*)

This INR is in the form of neutrons and high-energy electromagnetic radiation, called gamma rays. About a minute after the detonation, the radioactive decay of the fission products gives rise to additional gamma rays and electrons (or beta particles), known as residual nuclear radiation (RNR). The distribution of the total explosive energy of a hypothetical fission detonation in the atmosphere below an altitude of 6 mi (10 km) is 50% blast, 35% thermal, 10% RNR, and 5% INR. At higher altitudes where the air is less dense, the thermal energy increases and the blast energy decreases proportionately.

EMP is associated with the INR output, which is a small percentage of the total explosive energy. Nevertheless, EMP is still capable of transferring something of the order of 0.1–1 joule per square meter (0.1–0.01 ft-lbf/ft^2) onto a collector, more then enough to cause upset or damage to normal semiconductor devices.

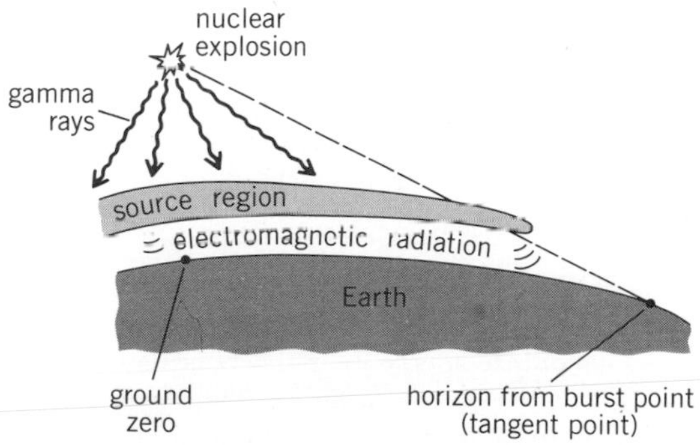

Fig. 2. Schematic representation of the EMP in a high-altitude burst. The extent of the source region varies with the altitude and the yield of the explosion. (*After S. Glasstone and P. J. Dolan, eds., The Effects of Nuclear Weapons, U.S. Department of Defense and the Energy Research and Development Administration, 3d ed., 1977*)

Early research developed the physics of high-altitude EMP. In 1978 a consistent explanation was set forth of how EMP is generated, regardless of the detonation height above ground. Figure 1 identifies six contributions to the gamma source for a hypothetical 1-megaton surface burst. As the detonation height is elevated, contributions from the ground and air sources decrease; for the high-altitude detonation (higher than 37 mi or 60 km) the EMP gamma source becomes essentially the prompt gammas from the nuclear burst. The gamma-source wave shape then approaches a smooth curve, approximating a double exponential, and since EMP is generated from the gamma source, it too approaches a double exponential wave shape.

EMP generation from a high-altitude burst. What occurs to the prompt gammas moving away from a high-altitude nuclear detonation? In Fig. 2, those gamma rays moving toward the Earth penetrate a more dense region of the atmosphere called the source or deposition region. In this region the highly energetic gamma rays interact with the air molecules to form 1-MeV Compton electrons and less energetic gamma rays, which then proceed in the same general direction as the original gamma rays. The fast Compton electrons slow down by stripping electrons from air molecules to form secondary electron-ion pairs. (Though these secondary electrons and ions do not contribute to the generation of the EMP, they do cause the region to become highly conductive, and therefore play an important role in determining the EMP wave shape and amplitude.) The very intense, short-duration pulse of Compton electrons is also deflected by the Earth's geomagnetic field, according to Eq. (1), where the deflection force $\vec{F}$ is

$$\vec{F} = q(\vec{v} \times \vec{B}) \qquad (1)$$

perpendicular to the geomagnetic field $\vec{B}$ and the velocity $\vec{v}$ of a Compton electron q. The Compton electrons then spiral about the geomagnetic lines as they slow down.

Characteristics for a high-altitude burst. For a high-altitude nuclear detonation the radiated EMP observed at large distances from the source region can be represented in time t as an electric field $E(t)$, and a magnetic field $H(t)$, given by Eqs. (2) and (3), with $E_o = 5.2 \times 10^4$ V/m (1.6×10^4

$$E(t) = E_o(e^{-\alpha t} - e^{-\beta t}) \qquad t \geq 0 \qquad (2)$$
$$H(t) = H_o(e^{-\alpha t} - e^{-\beta t}) \qquad t \geq 0 \qquad (3)$$

V/ft), $H_o = 1.4 \times 10^2$ A/m (4.2×10 A/ft), $\alpha = 4.0 \times 10^6\ s^{-1}$, and $\beta = 5.0 \times 10^8\ s^{-1}$. (In air, E and H are related by the impedance of free space, 377 ohms.) If the observer is directly below the detonation (ground zero), the polarization of both fields is predominantly horizontal; if the observer is at the horizon, the fields can have both horizontal and vertical components. The magnitude of these field components varies according to the latitude and longitude of the observer and the direction of the burst.

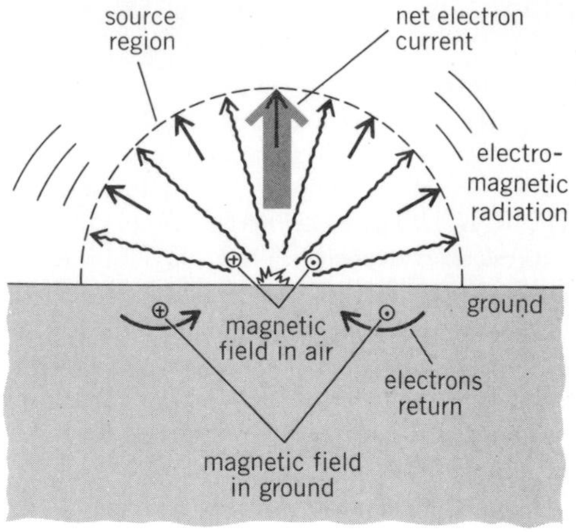

Fig. 3. Schematic representation of the EMP in a surface burst. (*After S. Glasstone and P. J. Dolan, eds., The Effect of Nuclear Weapons, U.S. Department of Defense and the Energy Research and Development Administration, 3d ed., 1977*)

EMP in a surface burst. Should the nuclear detonation occur closer to the Earth, the EMP generation process becomes far more complex and the electric and magnetic fields become very complicated. The most dramatic change occurs with a surface burst (Fig. 3).

When the observer is somewhere within the source region of a near-surface burst, where the air conductivity varies between 10^{-4} and 10^{-2} siemens/m (3×10^{-5} and 3×10^{-3} siemens/ft), the resultant electric field is predominantly vertical, and the resultant magnetic field is polarized perpendicular to the plane of the figure. The electric field polarization is due to the Compton-electron and ion-charge-separation fields which tend to become perpendicular to the conducting earth, leaving a resultant vertical electric field near the ground. As these Compton electrons move radially away from the detonation, they curve earthward and return to the detonation point through the conducting earth. This Compton current loop then gives rise to a resultant magnetic field. When the observer is far from the detonation (outside the source region), the electric and magnetic fields begin to approximate fields from a vertical dipole, decaying with distance (r) as $1/r$.

Internal EMP. It is possible for INR to directly interact with systems, causing EMP signals internal to structures. This phenomenon has been called internal or system-generated EMP (IEMP or SGEMP, respectively) and is potentially a serious problem for satellites and electronics in metallic enclosures. These forms of EMP are generated by gamma rays impinging on the enclosures, producing currents of Compton electrons internally that then produce internal electromagnetic waves. They are very dependent upon the nuclear detonation, the system topology, and the relative position of one to the other.

Coupling. The coupling of these rather wide-band (10^4 to 10^8 Hz) signals to systems of different topologies can be significant. For example, it is not unusual to predict voltage and current levels of hundreds of thousands of volts and a few thousand amperes coupled by high-altitude EMP onto extended systems. The exact level of coupling depends upon the size of the system and its orientation with respect to the incident field, and upon whether or not it is near an earth ground. For ground-based systems the incident field components can actually be enhanced or degraded, depending upon the polarization. Solving Maxwell's equations at the interface between an ideal conductor and a dielectric shows that the net horizontal electric field is zero at the interface and the net vertical electric field doubles. (For the magnetic field components the opposite is true.) Since the earth is not a perfect conductor, the field components do not completely cancel or quite double in amplitude.

An estimate of about 1 joule (0.7 ft-lbf) of EMP coupled energy is considered reasonable for many systems. Even if the coupling onto circuits is inefficient, as little as 10^{-13} J can upset some semiconductor devices and 10^{-6} J can cause damage. The potential for such upset and damage in critical electronic systems has led to the development of semiconductor devices which are "hardened" or protected against EMP. This development is particularly advanced in communications systems whose disruption by EMP is considered an important civil and military vulnerability.

For background information *see* COMPTON EFFECT; ELECTROMAGNETIC RADIATION; NUCLEAR EXPLOSION in the McGraw-Hill Encyclopedia of Science and Technology. [ROBERT A. PFEFFER]

Bibliography: V. Gilinsky and G. Peebles, *J. Geophys. Res. Space Phys.*, 73(1):405–414, 1968; S. Glasstone and P. J. Dolan (eds.), *The Effects of Nuclear Weapons*, U.S. Department of Defense and the Energy Research and Development Administration, 3d ed., 1977; W. J. Karzas and R. Latter, *Phys. Rev.*, 157(5B):1369–1378, 1965; A. S. Kompaneets, *Sov. Phys.—JETP*, 35(8),6:1076–1080, 1959; C. Longmire, *IEEE Trans. Antennas Propag.*, AP-26(1):3–13, 1978.

Electronic warfare

In recent years, advances in electronic warfare technology have taken place in the development of devices and systems for the self-protection of aircraft, and in the design of electronic support measures which give an overall picture of the battlefield and aid in tactical decision making.

AIRCRAFT SELF-PROTECTION

Since the early days of aeronautics, the governments of the world have commissioned their engineers not only to oppose their opponent's aircraft but also to contrive means to protect their own. Recognition of the enemy, countermoves to evade and avoid, attack, and camouflage were the means of protection adopted during World War I. However, aircraft were not perfected to the extent that they could form a credible threat.

World War II developments. During World War II, planes became more expensive and plentiful, and, consequently, means to defend against their intrusion became more formidable. Listening devices and radar were developed. By 1943, Germany had perfected the gun-laying radar to sight height, range, speed, and direction. It was developed because American forces had developed the "box formation," which provided an adequate defense against German fighters. In addition, fighters accompanied the bombers and provided protection.

The American-British team came up with two methods of confusing the radar image. One involved a device that transmitted gross amounts of noise. When tuned to the same frequency as the ground receivers, it blocked out the radar. This "Carpet" (APT-2), in volume production by the end of World War II, had a 5-watt output at around 520–560 MHz. A second method to reduce the effectiveness of electronic trackers was "window," invented by the British, also known as chaff by the Americans, who improved the manufacture. The device consisted of aluminum strips 11.5 × 0.045 in. (292 × 1.14 mm) and 0.0009 in. (23 micrometers) thick. They were manufactured in a V configuration, packed 6000 strips in a bundle. These bundles were placed aboard heavy bombers. In addition, "path finder" squadrons flew in front of the formations, seeding the targets. The chaff was very effective against the tremendous concentration of German 88-mm flak guns employing proximity fuses. Each strip appeared as a plane.

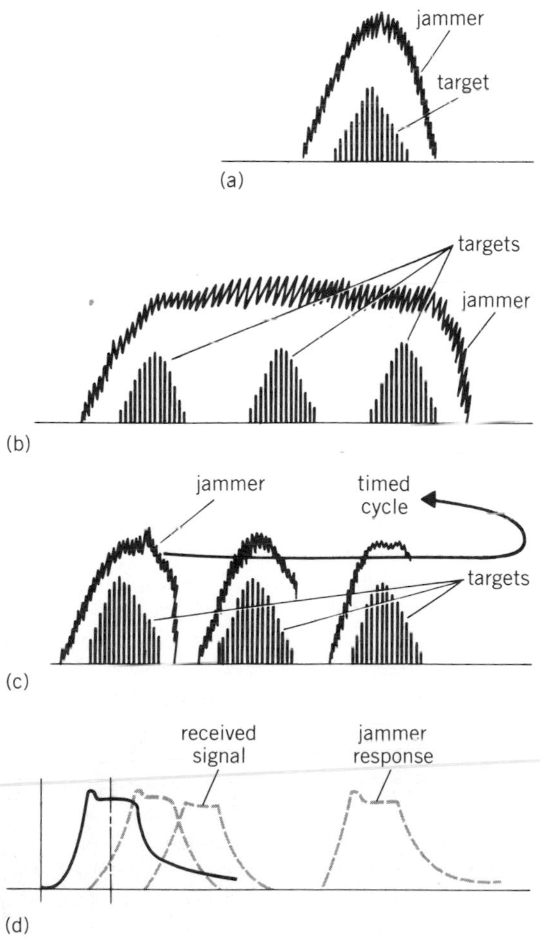

Fig. 1. Active jamming techniques. (*a*) Spot jammer. (*b*) Barrage jammer. (*c*) Swept-spot jammer. (*d*) Repeater.

As the end of the war approached, techniques improved. The German radars could be turned to any of several frequencies. Consequently bombers were fitted with jammers tuned to various frequencies to afford a carpet to protect themselves. Ravens, or radio countermeasure officers, often flew on the aircraft to cover ranges not covered by the fixed installations. These officers used special receivers tuned from 20 to 1000 MHz.

Protection of modern aircraft. Modern aircraft, in addition to opposing other aircraft, must contend with antiaircraft guns, ground-to-air missiles, and air-to-air-missiles. All of these weapons can have various types of explosives, heat-seeking systems, and on-board radar or television guidance systems, which increase the complexity of defensive avionics.

The modern aircraft requires means of warning its occupant and setting up evasive action to avoid missiles. The aircraft must carry some receiving device which can identify a threat and, more important, sort the threats to ensure that the most perilous is counteracted.

Modern aircraft contain various types of radar which provide the range, angle (azimuth and elevation), and velocity of an impending threat. The signal received by the attacker must have the same parameters to allow the triggering device or final closing mechanism contained on board the missile to be effective. The means of protecting the aircraft can be either a deception or a noise device. Deceptive devices include chaff, and noise devices include jammers, described above. They are still the mainstays of aircraft self-protection.

Noise and deception techniques. The noise generated by electronic weapons systems can be one of three types: spot, barrage, or swept-spot. These electronic countermeasure (ECM) devices can be assigned to one of three classes: active radiators, medium modifiers, and reflectivity modifiers.

Active radiators. An active jammer or radiator can, as explained above, execute spot jamming. It picks up the signal from the threat and actively deceives the attacker's receiver by overshadowing the signal on the attacker's set with a stronger reply (Fig. 1*a*).

The barrage-type jammer responds to a situation in which the same spot occurs in several time sequences; that is, the attacker's receiver plots multiple spots. The target's on-board computer receives these spots and commands the jammer to cover the spots in much the same manner as the spot jammer, but to cover all the jammer spots simultaneously (Fig. 1*b*). The disadvantage of this type of jammer is that the spread of the energy reduces the density or power of the noise available to a shorter bandwidth.

The swept spot (Fig. 1*c*) combines the wide-band blankets of the barrage jammer and high density of the spot jammer by executing a timed cycle (Fig. 1*c*).

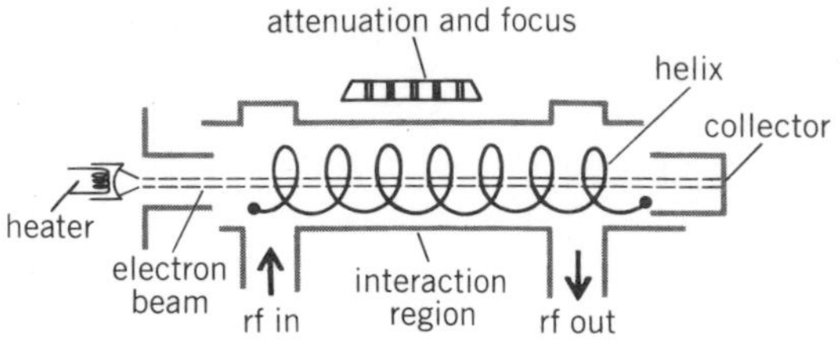

Fig. 2. Traveling-wave tube used in jamming.

Another type of active jammer is the repeater. It receives a signal, reproduces it, and returns it out of phase or time-delayed (Fig. 1*d*) to deceive the attacking vehicle, causing it to aim at the wrong target, or at unoccupied air space.

These simplified "active" jamming devices can be multiplied in their effectiveness and efficiency, the objective being to confuse the attackers. Other types of active deception are the creation of false targets and inverse-gain inverse-square-wave main-lobe squint jammers.

Medium modifiers. Another type of jamming consists of medium modifiers, such as chaff, which reproduces the signal many times and creates a group of targets that the incoming vehicle will confuse unless programmed to recognize a definite profile. Aerosols are also used as a deceptive device by clouding air space near or offset from the position of the target aircraft. The missile seeks the most prominent target.

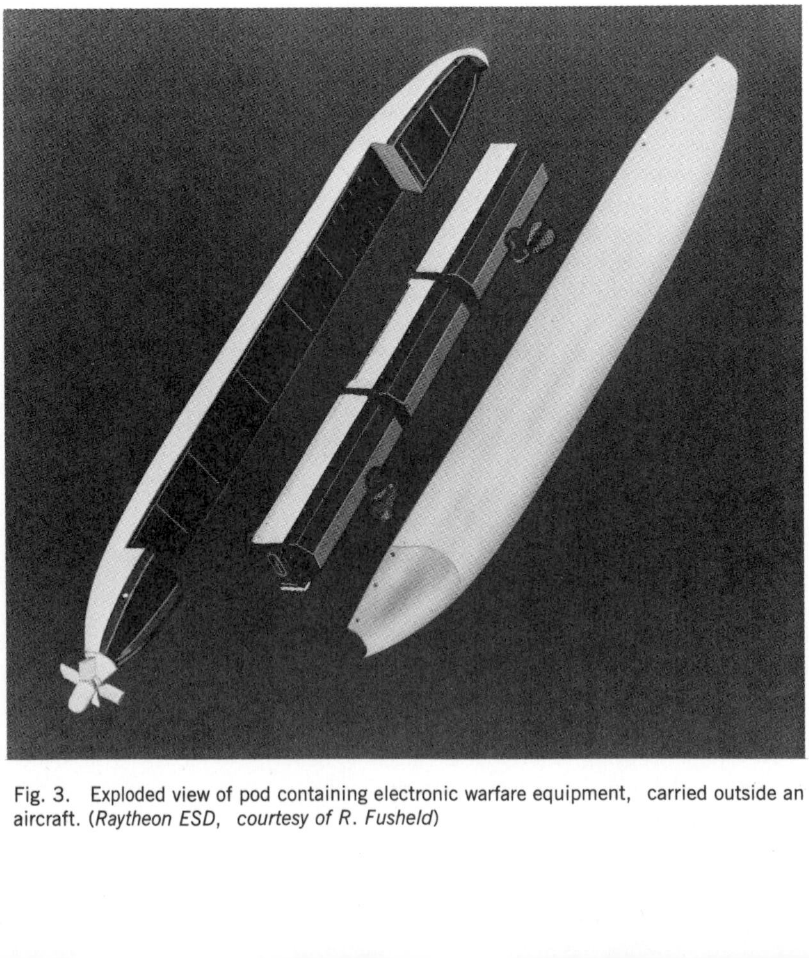

Fig. 3. Exploded view of pod containing electronic warfare equipment, carried outside an aircraft. (*Raytheon ESD, courtesy of R. Fusheld*)

Reflectivity modifers. Reflectivity modifiers change the shape or reduce the profile of the return signal. Most "targets" have particular reflective signature; that is, they will reflect a certain pattern of the return signal. The interdicting aircraft can use a microwave absorption material to reduce the power of the return signal. Another technique is to cause the reflected signal to deviate from the desired return path so that it appears to pass through decoys which can be dropped from the host aircraft and present a better target.

Protection against closing vehicles. The various systems described above deal with protection of the target aircraft during the phases of initial acquisition, that is, locating the target, tracking, obtaining the range, and directing the attacking aircraft toward the target. During the final few seconds of the chase, the attacker sometimes uses radar or some other radiation-receiving device, such as low-light-level television, which picks up the target and directs the attacking vehicle to attack the target. Infrared or heat seekers are also a closing-vehicle means of last-stage target acquisition. A certain signature of the target's radar signal, carried in the memory of the missile, can enable it to literally latch onto this signal.

The target aircraft may be protected against these devices by an image modifer coupled to a disposable jammer fitted with a device which forces the oncoming target seeker to vote for a stronger signal, that is, a device which presents, as it were, a target that is more like the target than the target itself. The decoy or disposable device can be made to separate from the host and have the missile attack it by presenting a stronger image than the host.

Generally, attacking missiles are propelled by rockets and can be approaching at speeds of mach 3 or higher. The on-board device must be capable of tracking the missile to enable another on-board device to fire a counterprojectile or deploy a decoy.

The host vehicle can also have a passive type of defense. This is often a warning device which emits a signal that identifies the oncoming threat and warns another device or a person on board. The pilot can then exercise a maneuver to evade the oncoming device.

Computing equipment. In general, a modern aircraft must possess the technology to identify a threat, recognize the type of threat, and deploy countermeasures which increase the probability of completing its mission. This requires a sophisticated receiver-signal processor and an extremely fast computer. Such a computer requires an extremely complex memory to enable all of the logic functions to operate and respond effectively.

Standoff jammer. Sometimes one aircraft is designated as a standoff jammer. It patrols the target air space and engages in high-power jamming of both the acquisition or tracking devices and the closing vehicles. These aircraft carry powerful transmitters excited by traveling-wave tubes. A high-power high-frequency amplifier will send out a selected frequency to jam the radar devices.

An example of one such microwave device is shown in Fig. 2. The traveling-wave tube uses velocity modulation to amplify an input radio-frequency (rf) signal. Electrons emitted at the cathode are forced through the interaction region. The collector removes the electrons. The rf signal that is to be amplified is coupled at the input port to the periodic structure or helix. A wave traveling along this helix travels at the speed of light, but the velocity of propagation in the direction of the electron beam is much smaller because the length of the wire is about 15 times the axial length of the helix. The wave travels along the beam with one-fifteenth the speed of light, and the electrons will be in synchronism with the wavefront if they are accelerated by about 1500 volts. The synchronism between the electromagnetic wave and the electrons traveling axially results in amplification.

System requirements and configuration. The modern aircraft must carry equipment that can receive signals warning of an oncoming threat, and analyze, process, and react to them with sufficient rapidity to permit the pilot or the radio operation officer to evade the threat. The avionics equipment must also permit the aircraft and its occupants to distinguish a real threat while maintaining communication with the parent organization.

The modern systems contained in aircraft can be stowed on pods to be added externally (Fig. 3) or internally (Fig. 4). Figure 5 depicts the architecture of a typical airborne system.

[HORACE J. CORIGLIANO]

Fig. 4. Bomb bay of a bomber holding electronic warfare equipment. (*Raytheon ESD, courtesy of R. Fusheld*)

ELECTRONIC SUPPORT MEASURES

Electronic support measures (ESM) is that portion of electronic warfare (EW) pertaining to the problems of detection, identification, localization, and processing of radiated electromagnetic energy from radars or communication systems. These systems communicate with countermeasure systems, controls, displays, central processors, and recording devices. The principal purpose of ESM is to provide a near real-time (within tens of seconds) picture of the electronic order of battle (EOB) for tactical decision making. The mission requirements of ESM differ from those of Elint/Surveillance systems and radar warning receiver (RWR) systems. Elint systems gather intelligence data for non-real-time-processing, while RWR systems are designed to fulfill the self-protection requirements of the platform. Different implementation approaches result from these distinct mission requirements.

Advances in several technology areas have had dramatic impacts on the nature, uses, performance, and cost of present and future ESM systems. These areas in order of impact are computer systems, microwave integrated circuits, microprocessors, digital signal processing, antenna technology, and optical technologies. Specific changes created or caused by ESM requirements in each of

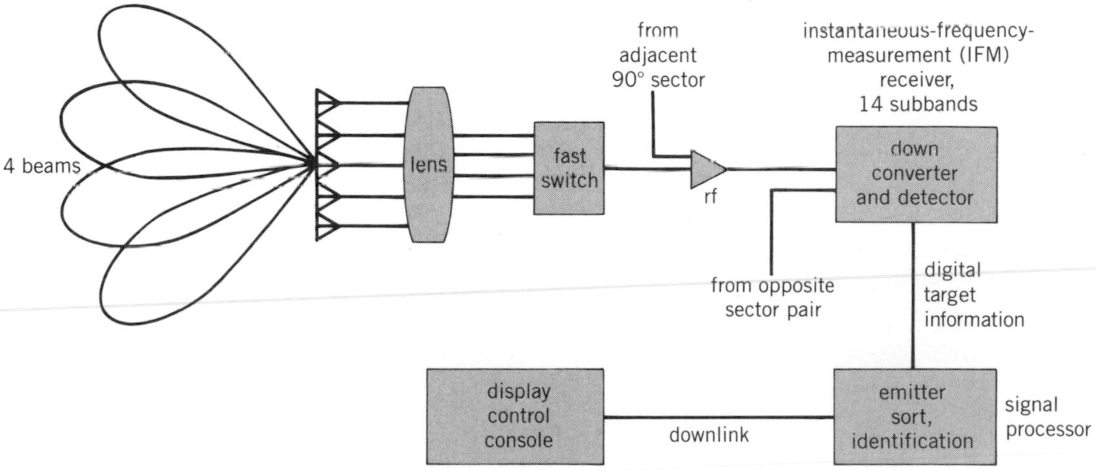

Fig. 5. Architecture of a typical airborne electronic warfare system, the light airborne multipurpose system (LAMPS).

these technology areas will be discussed. The increasing complexity of radar systems and electromagnetic environments has also motivated ESM system developments.

Missions and requirements. ESM has found application on land, marine, underwater, space, and airborne military platforms. The missions of these systems include support for three activities: determination of the electromagnetic order of battle; defensive and offensive tactical operations; and multisensor data correlation for enemy intent determination.

To fulfill these missions, the ESM system must carry out the timely interception of emissions with a high degree of confidence, and report the intercepts with parametric data that support other platform system needs. These platform needs include the transfer of this data to displays, central processors, weapon systems, and so forth. As the above missions imply, ESM equipment is designed to be an integrated part of a combat system used as a distinct and crucial adjunct to the tactical decision-making functions of the complete system. The generic requirements for ESM commensurate with this use fall in the following areas: needed spectral and spatial coverage; needed system response time; necessary parameter measurement; required signal and target detection ranges; and specific interface to subsystems.

These requirements are translated by system engineers, trained in electronic warfare, into the technical characteristics appropriate for each of the embedded electronic subsystems. These characteristics are those needed to satisfy the stated missions. A generic modern ESM system with its major subsystems is depicted in Fig. 6. These systems contain wide-bandwidth (several octaves) sensitive microwave receivers, precision antenna arrays and systems, analog-to-digital (A/D) converters, digital signal processors, general-purpose digital processors, and interface devices. The requirement areas defined above are applied to each subsystem where applicable to generate detailed specification. These subsystem specifications are tied back to the overall ESM system specification needed to fulfill the mission objectives. The correspondence of requirement areas and subsystems is indicated in Table 1.

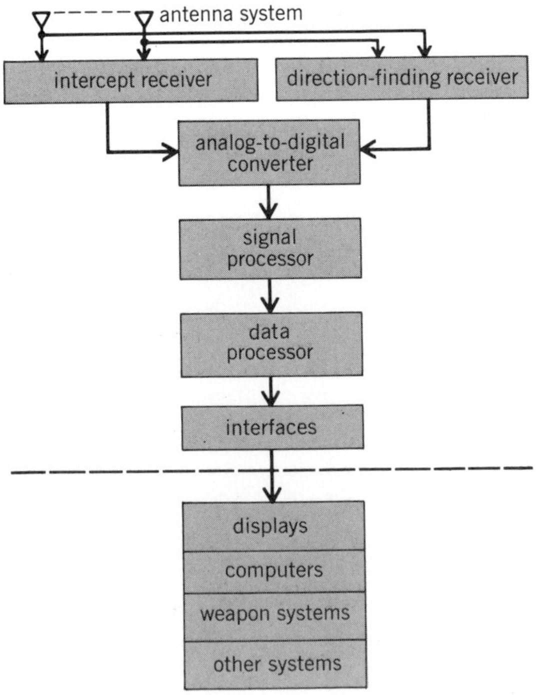

Fig. 6. Electronic support measures (ESM) system.

Table 1. ESM requirements and corresponding subsystems

Requirement area	Subsystem
Spatial coverage	Antenna array Platform aperture available
Spectral coverage	Antenna array Intercept receiver Direction-finding receiver
Response time	Intercept receiver Signal processor Data processor
Parameter measurement	Intercept receiver Direction-finding receiver Signal processor Analog-to-digital converter
Detection range	Intercept and direction-finding receiver Antenna array Signal processor
Interfaces	Data processor Interfaces

Although the ideal ESM system would provide detection ranges to and beyond the horizon, and instantaneous viewing both at all frequencies and in all directions, technology combined with cost and size constraints has not permitted such an ideal system to be developed. Trade-offs in system architectures are made to optimize system performance for the specific platform and mission.

The ESM problem can be viewed as a classical stochastic communication problem. The ESM system must intercept, detect, identify, and locate radiated energy without definitive prior information as to the exact characteristics of the signals to be detected. The problem is one of search of the potentially viable spatial and spectral regions for radiated signals that may be of potential interest to the platform containing the ESM system. The specific values for these parameters are discussed below.

Radar signals. The wide application of radar systems and the need for counter-countermeasure techniques have led to the generation of an ever-increasing variety and number of radar-type signals. These signals include: the conventional types which are characterized by high-energy pulse generation at a predictable constant rate pulse repetition frequency, or PRF; frequency-modulated continuous-wave (CW) radars; pulse Doppler types; multiple-beam types; frequency and PRF agile types; and spread-spectrum types.

Each class of radar can operate at any frequency

Table 2. Radar parameter ranges

Parameter	Range
Pulse width	100 nanoseconds to 100 microseconds (continuous-wave and pulse Doppler types may differ)
Pulse repetition frequency	100 to 3000 pulses per second (pulse Doppler types are higher and may be 50,000 to 300,000 pulses per second)
Scan rate	10 scans per minute to 100 scans per second
Scan type	Circular, sector, raster, conical, fixed, electronic, and so on

within a range of several hundred megahertz centered around a microwave band frequency. ESM systems must therefore maintain threat file data defining possible parameter ranges of operation for each class or type of radar. Radars of potential interest to platforms carrying ESM systems operate at radio frequencies normally falling in the 500-MHz to 18-GHz spectrum, and millimeter waves have been increasingly used for some applications. Radar parameter ranges are generically large although reasonably constrained for a given radar system. Some parameter ranges are given in Table 2.

Radar scan characteristics (scan rate and scan type) are an important design consideration for ESM systems. These scan characteristics can aid in determination of radar type, but they can also make intercept difficult.

System characteristics. ESM system capabilities have evolved since the 1950s as the utility of this type of information has been realized. The first type of ESM systems was characterized by rotating antenna systems (similar to a radar except covering much broader spectral regions) feeding narrow-band superheterodyne receivers. (A multiple number of receivers was used to cover needed frequency regions.) The receiver outputs were generally input to analog displays for operator interpretation. These systems were plagued by the "scan-on-scan" problem. Scan-on-scan refers to the situation that at long ranges intercept can occur only if the radar of interest is pointing at the ESM system platform, the ESM antenna is pointing at the radar, and the radio frequency of the ESM superheterodyne receivers is at that of the radars. For electromagnetic environments containing small numbers of signals, reasonable response times may be achieved, although tens of minutes may be needed for intercept. The receiver tuning strategy used was typically one of scanning from the low frequency to high frequency of operation at a selectable rate.

The use of superheterodyne receivers provided selectivity and good sensitivity (detection range). This type of receiver, although limited due to its relatively narrow bandwidths, continues to be utilized, although in more advanced forms, in modern ESM systems. As the complexity and numbers of radars increased, technology was needed to address the need for rapid response times with low MTTI (mean time to intercept). Several new types of receivers, as well as new system architectures, have been introduced. These receiver types provide selectivity, wide instantaneous bandwidth, and reasonable sensitivities. Examples of this new breed of ESM receiver types are given in Table 3.

These receiver types have increasingly become the intercept receiver function for ESM systems. System architectures have evolved that separate the direction-finding (DF) function from the signal-intercept function. In these architectures the intercept receiver provides rapid intercept of the radars operating over a broad spectral region (as much as an octave of frequency), while the parameter measurement including direction finding is provided by narrow-band superheterodyne receivers tuned to the correct frequency by the intercept receiver. This system concept has been made possible by the advent of digital and computer technology growth. The emergence of large-scale integrated (LSI) circuits in the digital domain resulted in a dramatic movement to fully automatic ESM systems with minimal operator interface for the multitude of processing functions required.

The need for automatic and high-speed digital processing was necessitated by the use of wide rf bandwidth intercept receivers. These receivers made possible the simultaneous interception of many operating radars, needed for rapid system response, as well as the detection of parameter-agile-

Table 3. New types of ESM receivers

Type	Positive features	Disadvantages
IFM (instantaneous frequency measurement)	Approximately 4000 GHz bandwidth Better than 10 MHz resolution Large dynamic range Small size Low cost	Time-synchronous signal measurement errors
Microscan	Approximately octave bandwidth High resolution Instantaneous narrow bandwidth	Very high processing rate required 30–40 dB dynamic range High cost
Channelized (surface acoustic waves)	Octave bandwidths Better than 20 MHz resolution	Relatively high cost Relatively large
Channelized (acoustooptics)	Multioctave bandwidths Better than 20 MHz resolution Minimum amount of microwave hardware	Present-technology limitations on lasers, Bragg cells, and detectors

Table 4. ESM pulse processing functions and requirements

Function	Requirement
Parameter measurement	Encoding of pulse width, pulse amplitude, pulse frequency, pulse unique characteristics, and pulse direction of arrival characteristics
Pulse train de-interleaving (separaton)	Separaton of multiple signals exhibiting similar parameters on the basis of multiple pulse time of arrival characteristics
Pulse repetition frequency (PRF) and PRF type	Encoding of time sequence of pulses for each signal in terms of rates and characteristics, that is, pulse groups
Scan type and rate (if needed)	
Receiver hardware control	Real-time control of receiver system assets to optimize spectral and spatial search.

type radars. ESM systems with such receivers have been designed to simultaneously process hundreds of operating radars, resulting in millions of pulses of data per second that must be automatically processed by the processing subsystem. The typical processing functions required for these systems fall into two classes: the pulse processing requirements and signal processing requirements. The functions required and definitions of the requirements are given in Tables 4 and 5.

In the early automatic systems, the pulse processing functions were implemented in special-purpose logic, and signal processing was implemented by machine language software operating in general-purpose computers adapted for military use. As technology advances have occurred, distributed microprocessors using firmware programs have taken over the pulse data-processing functions, while signal-processing functions have been implemented in high-order languages (HOL) in military general-purpose digital computers executing millions of operations per second.

Antenna systems. As the applications of ESM grew, the need for precision line-of-bearing direction-finding measurement became increasingly important. With direction-finding performance of better than 1°, the rapid set-on of weapon systems (missiles or guns) became feasible and even essential to own-force defense. Sophisticated multioctave interferometer array systems have emerged to meet this need. Systems using archimedian spiral antenna elements have evolved, providing small arrays (less than 3–4 ft or 0.9–1.2 m) with greater than 4:1 microwave bandwidths. Whereas interferometry utilizes phase difference measurements for direction-finding determination, other techniques using amplitude are also used for ESM system applications that do not require precision direction finding. Lens and phased array techniques have also been used to support the ESM direction-finding function.

Microwave receiver technology. A consistent goal in the development of ESM systems has been to provide increased performance with reduction in equipment size and weight. This has helped push technology toward new techniques for miniaturization of microwave receiver systems. ESM systems use microwave integrated circuit (MIC) technology to reduce receiver sizes by orders of magnitude. In addition, the use of the field-effect transistor (FET) in microwave applications such as amplification and oscillators has provided enhanced-sensitivity microwave receivers. Microstrip technology coupled with precision etching and cutting has provided MIC receiver systems that are producible in volume at reasonable cost and have improved performance over those of their predecessors.

Significant research and development efforts are occurring in microwave receiving systems using both digital and optical technologies. The applicable digital technologies include gallium arsenide (GaAs) digital devices and Josephson junction devices. The gallium arsenide devices provide picosecond switching speeds, enabling multigigahertz operations needed for ESM-type applications. When very high-speed digital devices become available (perhaps in the late 1980s), real-time powerful digital processing algorithms such as convolution and correlation will become possible even for very dense electromagnetic environments.

In the optical processing area, investigation into the practicality of the use of acoustooptic techniques for channelized receiver operation is in process. This technology, made possible by coherent light sources (lasers), makes use of the phenomena of light interacting with high-frequency sound waves in certain crystalline materials. This Bragg cell phe-

Table 5. ESM signal processing functions and requirements

Function	Requirement
Signal identification	Identification of the class and type of signal from measured characteristics
Signal tracking	Provision for retaining and tracking of detected and identified (even if unknown) signals from intercept to intercept of same signal
Signal location	Some systems use sophisticated triangulation algorithms to compute signal location; multiple intercept measurements of direction are needed over a few degrees of arc
System interface	Format and transmission of needed data to on-board systems
System self-test	Automatic or semiautomatic system self-test and diagnosis

nomenon causes the light source to be deflected through the crystalline material as a function of the high-frequency sound wave produced in the Bragg cell material. Developments in lasers, photosensors, and Bragg cell device fabrication promise receiver systems with 2–4-GHz rf bandwidths for 100-nanosecond-pulse-width signals, 20-MHz frequency resolution, and 60-dB dynamic range. This technology also has applicability to direction-finding receivers and high-speed processing subsystems for optical correlation and convolution.

For background information *see* ACOUSTOOPTICS; ELECTRONIC COUNTERMEASURES; INTEGRATED CIRCUITS; MICROWAVE SOLID-STATE DEVICES; RADAR; SURFACE ACOUSTIC-WAVE DEVICES; TRAVELING-WAVE TUBE in the McGraw-Hill Encyclopedia of Science and Technology.

[ALAN S. SOCK]

Enzyme

Enzymes are powerful and selective catalysts which are important in virtually all physiological functions. They consist of globular macromolecules with molecular weights ranging from 10,000 to 500,000, and are composed of chains of α-amino acids in the L configuration. The exploitation of enzymes for practical purposes has been of interest to researchers in many fields, including food and agriculture, pharmaceuticals, chemical synthesis, and chemical analysis.

In many instances, the use of soluble enzymes is not feasible because of high cost. Immobilization of the enzymes offers the benefit of reuse of the catalyst, thus reducing expenses, and has a second possible advantage of enhanced reaction stability.

Enzyme immobilization. The principal techniques for enzyme immobilization are adsorption, covalent attachment, and entrapment within polymer gels (Fig. 1). Many enzymes have a preponderance of either positive or negative amino acid side chains on the surface, and electrostatic attraction at multiple points between a solid support and an enzyme leads to a stable complex. Hydrophobic forces may also contribute to stability of noncovalent complexes of enzyme with water-insoluble materials. Covalent attachment requires reagents that create raw chemical linkages (between enzyme and support). For entrapment, the enzyme is mixed with polymerizable elements, such as vinyl monomers or "prepolymers." The polymcrizaton yields a gel containing the enzyme entrapped within it.

Properties of immobilized enzymes. In most cases, enzymatic activity is reduced as a result of attachment of enzymes to solid supports. Loss of activity may occur either because of chemical reaction of a functional group at or near the enzyme active center or because of physical denaturation. When charged supports are used, microenviromental effects result in the perturbation of the pH-rate profile and the apparent reactant-binding capability of the enzyme. Both effects are reduced by increased ionic strength of the bulk solution. The transport of reactant to the surface of an enzyme particle can be limited by the unstirred (Nernst) layer surrounding the particle. The supply of reactant to enzyme molecules in the interior of the supporting particle may also be rate-limiting.

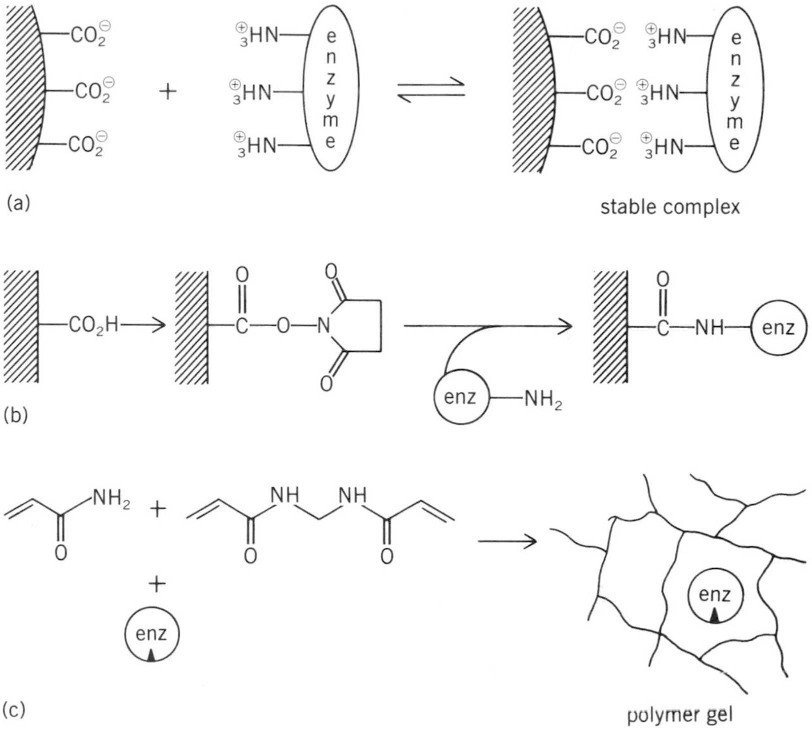

Fig. 1. Enzyme immobilizaton by (*a*) adsorption; (*b*) covalent attachment; (*c*) entrapment.

The effect of immobilization in multienzyme systems has been examined in the two-step conversion of glucose to gluconolactone-6-phosphate. Hexokinase catalyzes the phosphorylation of glucose to glucose-6-phosphate. This product is the reactant for the second enzyme, glucose-6-phosphate dehydrogenase. When both enzymes are attached to the same matrix particle, the lag time prior to steady-state production of final product is significantly reduced. A reasonable explanation is that the reactant for glucose-6-*P* dehydrogenase is produced nearby, that is, the apparent concentration of it is large compared with the case where the enzymes occur free in solution. Enhanced reaction rates are observed in biological cells when a large number of enzymes associated with the sequential reactions of a metabolic pathway are fixed on membranes within these cells.

Immobilized enzyme reactors. The attachment of enzymes to insoluble supports permits their use in reactors which resemble those employed in the chemical process industry. Tank reactors are often preferred for intermittent, small-scale production. Viscous reactant solutions and enzyme particles with low activity might also dictate the use of tank reactors. For large-scale processes, continuous packed-bed reactors are frequently selected. Upward-flow reactors are used when the enzyme parti-

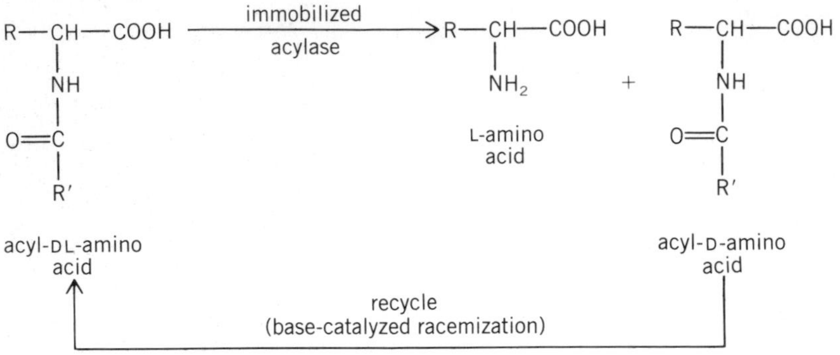

Fig. 2. Production of optically pure amino acids.

cles are compressible and when gaseous products are produced. The fluidized-bed reactor, where the enzyme particles are kept in suspension by the upward flow of reactant, can be used with viscous reactant solutions or when gaseous products are produced.

The development of ultrafiltration membranes and hollow-fiber devices has led to the construction of reactors which depend on the fact that the enzyme (of high molecular weight) can be retained by the membrane, while the product (of low molecular weight) is allowed to pass. Such reactors are not appropriate for processes in which proteases are present by choice or as contaminants. Proteases are often subject to autolysis, which means that they digest themselves as well as the reactant. Contaminating proteases can severely limit the lifetime of an enzyme in this type of reactor. In contrast, when the enzyme preparation is covalently immobilized on particles, the contaminants are also immobilized (or washed through), and proteolytic degradation is eliminated.

Application of immobilized enzymes. A large-scale process based on immobilized enzyme technology is the production of high-fructose corn syrup. In the United States more than 5 billion pounds (2.3 billion kilograms) are produced annually, much of which is used as a sweetener for soft drinks. In this process, glucose syrup (made enzymatically from corn starch) is passed over a bed of immobilized glucose isomerase; about 50% of the glucose is isomerized to fructose. Because fructose is sweeter than glucose, high-fructose corn syrup is sweeter than the original glucose solution; its sweetness is equivalent to a solution of sucrose (table sugar) of comparable concentraton.

In the production of 6-amino penicillanic acid from penicillin G, as shown in reaction (1), the phenylacetyl moiety is released from the amide linkage, but the lactam amide is left intact. The transformation occurs quantitatively under very mild conditions. This application is an excellent example of the exploitation of enzymatic specificity.

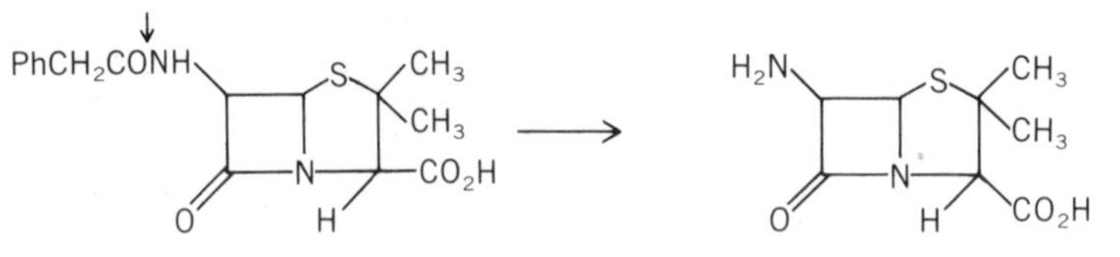

Penicillin G 6-Amino penicillanic acid (1)

In the production of optically pure amino acids from a racemic mixture with immobilized acylase, the enzyme exhibits preference for one optical isomer; only the L enantiomer is deacylated (Fig. 2). The resulting L-amino acid crystallizes, and the acyl-D-amino acid remains. The process can be run continuously by reracemization of the D isomer to the D,L mixture.

Chemical analysis is a very important area of application for immobilized enzymes. Here again, enzyme specificity is exploited. Enzymes can selectively transform a single compound in a complex mixture. The immobilized enzyme may be coupled with a thermal detector or an ion-selective electrode. Coating the latter with an enzyme-polymer layer results in an "enzyme electrode." An even simpler approach is to cover the tip of a conventional electrode with an ultrafiltration membrane which contains the desired enzyme solution. A cation-selective electrode immersed in a solution of glutamate dehydrogenase has been successfully used for the detection of glutamic acid.

Medical applications also appear promising. For example, recently a deheparinization method for blood has been proposed. The use of extracorporeal devices for blood oxygenation during bypass surgery or kidney dialysis requires heparinization of the blood to prevent clotting. However, heparin treatment can lead to hemorrhagic complicatons. To prevent these problems, an immobilized heparinase reactor is placed in the external circuit, prior to the point of blood reentry (Fig. 3). Heparinase in the column completely eliminates the heparin anticoagulant activity, and heparinizaton is restricted to the external circuit. Immobilized heparinase may find applications in other extracorporeal devices used in enzyme therapy.

Certain types of leukemia are asparagine-dependent; when the tumor cells are deprived of asparagine, proliferaton slows dramatically. The enzyme L-asparaginase degrades asparagine as in reaction (2). Depletion of asparagine by direct injection of

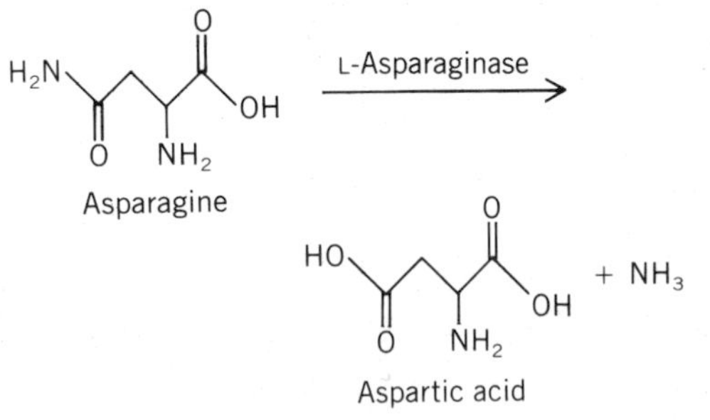

(2)

asparaginase into a patient is accompanied by two problems which relate to the fact that the asparaginase most commonly used is of bacterial origin. The reticuloendothelial system recognizes the enzyme as

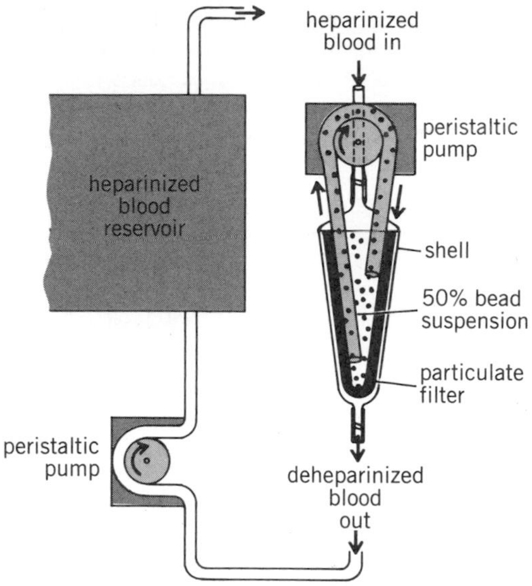

Fig. 3. Extracorporeal heparinase reactor. (*After R. Langer et al., Science, 217:261–263, 1982*)

alien and clears it from the bloodstream. Also, life-threatening allergic reactions may result following administration of the foreign protein asparaginase. To immobilize the enzyme within biocompatible capsules would appear to be a potential solution to the problems. Unfortunately, immobilized enzymes are not generally good catalysts at low concentration of reactants. Since the tumor cells can survive at low levels of asparagine, effective enzyme therapy requires reduction of the asparagine concentration to very low levels. The consequence is that leukemia therapy based on the use of immobilized asparaginase remains in the experimental stage.

For background information *see* ENZYME; ION-SELECTIVE MEMBRANES AND ELECTRODES in the McGraw-Hill Encyclopedia of Science and Technology.

[G. P. ROYER]

Bibliography: P. W. Carr and L. D. Bowers, *Immobilized Enzymes in Analytical and Clinical Chemistry*, 1980; R. Langer et al., *Science*, 217:261–263, 1982; K. Mosbach and B. Mattiasson, *Acta Chem. Scand.*, 24:2093–2096, 1970; G. P. Royer, *Fundamentals of Enzymology*, 1982.

Ergonomics

Ergonomics is synonymous with biomechanics, the study of muscles, bones, and organs and how they are affected by work, and human-factors engineering, a field dealing with all factors of the physical environment and its effects on workers.

A recent issue of *Occupational Safety and Health Reporter* contained statistics released by the National Institute for Occupational Safety and Health (NIOSH) stating that in the United States an estimated 3.2 million cases of occupational injuries were treated in hospital emergency rooms in 1982. Of these, 580,000 cases were identified as musculoskeletal disorders. The National Safety Council also reported the cost of all industrial accidents, conservatively estimated, to total $55 billion. The rising number of injuries and worker compensation costs has led managements to be concerned with reducing injuries sustained by their employees. Ergonomic principles and their application offer an engineering approach to injury reduction with a corresponding reduction in compensation costs. Additional benefits include improvement in worker morale and an increase in productivity levels.

Within the past two decades ergonomics has moved from military and aerospace applications to private industry. The first applications were directed toward the identification of, and improvements in, stressful jobs in warehouses and manufacturing shop areas. With the increased interest of concerned management for the safety and health of employees, ergonomic applications multiplied and now are seen in all areas of industry. The principles of ergonomics are equally useful in the office environment. Furniture manufacturers are actively engaged in the design of office equipment planned for improved compatibility with the worker and the new office technology.

Design rationale. In considering the total system of any task, ergonomists collect dimensional data on the universes of people who will populate the work assignment; similar data are also considered for the consumer. The information is used in the design of work space. This design will be affected by: the size of the product to be produced; the work classification (precision, light, heavy); the motion inventory of the task (important for safe operation); and the strength forces and pressures necessary to accomplish the job requirements. Human physical limits must be considered in both design and placement of controls.

The operational flow should be smooth. Feedback responses, which must be considered in the information transfer between machine and operator, utilize tactile, visual, and sensory systems. Tactile response is dependent on the size of a control and its designed resistance. Visual response is feedback through a given instrumentation or lighting of a given unit. Sensory feedback may include auditory alarms or signals that may be felt. In any system, ergonomists must avoid overloading the human operator. Operator frustrations and accidents occur when the system overload exceeds operator capability.

Ergonomic principles. In general, optimum design in industrial environments involves a number of ergonomic principles. The aim should be for movement; static work should be avoided. The best mechanical advantage should be provided. Extremes in the position of body joints, overloads of the muscles, and unnatural postures should be avoided. Work conditions that permit posture changes should be provided with alternate sitting and standing activity, including provision for proper seated posture. Variability in operator size must be accommodated. Balanced motion activity for both hands and prepositioning of tools in their order of use should be provided. Workers should be trained to use the facility correctly in work spaces planned for employee safety.

Guidelines for hand tools. The following ergonomic principles have been developed for the design and use of hand tools.

Axis of hand work. In general, manipulative work of the hand is best performed when the hand is in line with the axis of the forearm, rather than with the wrist flexed, extended, or in ulnar deviation. The use of the body members in their midrange and functional position usually results in the best performance and with the least fatigue.

Precision vs power grasps. Tools held in the hand are generally directed toward the radial side when speed and precision movements are desired, and toward the ulnar side for slower and stronger types of movements. This is due to the sensory capabilities of the regions of the hand and the manipulative abilities of the fingers. Examples are a precision grasp in the use of a sewing needle and a power grasp in the use of a screwdriver.

Action axis of forearm-hand. There is a different line of action between the application of force and the application of rotation in the forearm-hand. The force axis runs approximately through the elbow and digit 2 (on the radial side), whereas the rotation axis runs from the elbow to the base of digit 4 (ulnar side). Should a task require supination or pronation, the best axis for this movement is along the line of rotation. Force movement is best applied along the axis of force.

Movements to avoid. A work task or tool should not be designed so as to require the wrist to flex and the arm to be pronated at the same time. This action stretches the tendons and may result in the condition known as tennis elbow. A task or tool should not be designed that will require the distal phalanx to be flexed prior to or without flexing the middle phalanx, as this results in a constriction on the tendons leading to the fingertips. A movement which flexes the wrist restricts the ability to grasp, due to insufficient length of the extrinsic muscles, and should be avoided.

Horizontal use of hand tools. Ulnar deviation is to be avoided when using plier-type tools in a horizontal position. This posture provides a very fatiguing position that puts an undue amount of stress at the point of intersection of the carpals and the ulnar. This problem can be remedied by means of appropriate jigs and fixtures to orient the work or by application of biomechanically designed pliers.

Vertical use of hand tools. When using pliers in a vertical position, it is generally preferable to grip them in a manner such that the jaws are pointing downward rather than to hold them in a normal manner, to maintain the action axis of the forearm.

Vertical operation of hand tools such as ratchet screwdrivers create undesirable biomechanical conditions, as either a single finger or the entire hand is cupped over the end of the screwdriver. The addition of a flange on the top and bottom of the ratchet handle helps eliminate the undesirable pressure from the palm arch, which passes nerve endings to fingers, or the end of the index finger.

The common screwdriver operated in a vertical position involves static loading of the muscles of the shoulder. A better method is to grasp the handle with the arm hanging down to the side and then use the second hand to guide the direction of the screw. Excessive movement or turning of the screw causes overstretching at the wrist joint and overloading of the extensor muscles.

Handle design. The handle on a set of plier-type tools must be long enough so that the end of the handle does not terminate in the palm of the hand. A short handle allows undesirable concentration of high pressure on the sensitive nerves and blood vessels of the hand.

A handle with a wide span can present special problems to the female worker with a small hand. The larger span of the tool may demand that the distal phalanx of the finger be flexed. Repetitive action of this type is conducive to strain and a trigger finger reflex.

It is desirable to cover the handle on hand tools with a rubber or plastic material to relieve high pressure points at the tool surface, to provide good heat and electrical insulation, and to provide a surface less likely to slip from the hand. A material should be selected that does not absorb chemicals and is resistant to the embedding of grit or chips from the work.

A spring should be incorporated on the plier handle to open the jaws automatically. The opening of the plier by insertion of a finger between the handles induces cramped posture of the hand and may cause pinches or trigger finger.

Flutes or ridges should be provided on the handles of tools such as a screwdriver if high torque is required. A small diameter and smooth front end of such a tool allows better control and manipulation, provided consideration be given to the diameter of the handle. Studies have shown that a T-shaped handle provides the highest torque capability.

Heat protection. When handling hot materials with hand tools or when using a hot tool such as a soldering iron, both comfort and safety require a protective heat shield or flange on the tool. The source of heat should be moved farther from the surface of the hand, but without leading to additional muscular effort or unbalancing the tool.

Form-fitting tools. Form-fitting tools incorporating handles such as pistol grips on power tools or on heavy-duty pliers should be matched carefully with the population who will use them. A tool which matches the fiftieth percentile of the population will not provide characteristics desirable for a person with a large hand. If the fingers are stretched to fit a grip, the hand will be penalized by a loss in power. Therefore it is better to avoid form-fitting tools.

Control buttons or triggers. Buttons or triggers

should be located in a convenient position relative to the normal functional position of the hand. Some latitude in the location and posture of the hand should be allowed so that a single muscle group will not be called on repetitively to operate the button. The index finger is commonly used, but is somewhat ineffective, for tripping a control button. A trigger converted into a long lever arm which fits the curvature of the hand will provide a control that is easy to actuate.

Arm rests. The use of arm rests where hand work is being performed generally reduces the static loading of the back muscles and enhances ability to perform precision work at the bench.

Tool prepositioning. It is desirable to design for the prepositioning of tools in the work place. For example, a tool can be mounted overhead, put into a holder, or into any specific place so that it is available when needed.

Combination tools. Design considerations should include the possibility of more than one function for a single tool. Examples include the nut-holder driver and tweezer, bladed screwdriver, and needle-nose and cutter pliers. Combining the tools will often ease manipulative ability without the necessity of placing aside the first tool to grasp the second one. The design of such tools should consider the anatomy of the hand, sensory limitations, strength requirements, and safety.

Design principles for video terminal work spaces. The video display terminal environment includes many factors that can affect the health, comfort, and productivity of a worker: noise, distracting glare, visual fatigue, improper lighting, and excessive noise.

In the design of work space for video display terminals, it is essential to consider all elements early in the planning. The key elements of the video display terminal system are the terminal support platform, the terminal keyboard, the terminal monitor, and the dimensional data on the population who will be assigned to use the terminal.

Text-entry or "regular" typing requires the equal use of both hands for keying, so the keyboard is placed directly in front of the operator. Data entry is usually performed by one hand, with the other hand used for reference position for the source document. Usually, for data entry, the keyboard is placed to favor the keying hand.

There are ergonomic principles helpful in establishing video display terminal work stations that provide operator compatibility with minimal discomfort.

Keyboard position. The most desirable position for the keyboard is on the support table 25–28 in. (63–70 cm) over the floor. Comfortable keyboard height is dependent on arm position. The operator's elbow should be level with the home-base key row.

A palm rest should be provided to increase the manipulative ability of the fingers. Arm and hand motions are dependent on one group of muscles to originate the movement, while a second group opposes the activity. Both groups of muscles act to maintain a position. The palm rest permits the use of minimal force by the arm muscles for the keying activity.

In addition, a document holder should be provided for the operator. The holder should be equipped with a small task light, placed to prevent its beam from striking the monitor screen.

Furnishings. Comfortable seating that can be adjusted easily by the operator is necessary. Ergonomically designed seating is available that incorporates nitrogen gas–filled cylinders for fast and simple height adjusting. The backrest angle should be adjustable. The chair should adjust vertically to provide support for the operator's back, essential for good working posture. Proper seating height should permit the operator's feet to rest comfortably on the floor, with adequate space to permit leg movement between 90 and 105°.

The furniture should have a matte finish in neutral colors. Strong contrasts can contribute to operator visual fatigue.

Terminals should not be placed in front of windows. Drapery or blinds should permit control of external light sources. Bright outside light sources are the most prevalent cause of eyestrain.

Monitor. The top of the monitor should be placed at eye level, so that the operator's head is in the natural center range of movement.

Phosphor selection is important. While a green phosphor is commonly used, new research indicates an orange phosphor increases screen readability.

High illumination levels can cause screen glare. Low light levels cause the operator to strain to read the screen characters or the source document. The latest research indicates light levels between 50 and 70 footcandles (500 and 700 lux) is best for video display terminal areas. Indirect lighting offers a sensible approach for the video terminal workplace.

Terminal monitors should have matte screen surfaces or be equipped with filter screens. These features reduce the veiling reflections on the screen surface. Shielding the screen surface from overhead light sources also minimizes the veiling reflections. Extensive tests conducted by the U.S. Bureau of Radiologic Health of the Food and Drug Administration and NIOSH clearly indicate that video display terminals pose no radiation hazard to the operators.

For background information *see* ANTHROPOMETRY; BIOMECHANICS; HUMAN-FACTORS ENGINEERING in the McGraw-Hill Encyclopedia of Science and Technology.

[HARVEY C. FOUSHEE]

Evaporative casting

The evaporative casting process has been developed only in the past 30 years. Early developments served to establish the suitability of this process for the production of one-of-a-kind castings, such as machinery bases, dies, and works of art. In the 1970s, the ground vehicle industry became interested in the process as a potential source of mass-produced iron and aluminum castings. As a result of these efforts, the evaporative castings process is today regarded as an economical production process for both high-volume casting and unique castings. This flexibility, coupled with the ability to make aluminum, iron, brass, or steel castings, suggests that this process will displace older casting technologies which require significantly higher capital investment and are relatively inflexible.

Process technology. The evaporative casting process is a blend of two distinctly different technologies: creating polystyrene patterns of replicas of the part to be cast, and casting of the molten metal so as to assure complete displacement of the polystyrene replica. For the production of one-of-a-kind castings, the polystyrene pattern can be carved from board stock or from large polystyrene blocks which are the source of the board stock. For production of tens to hundreds of identical castings, temporary tooling may be used. For larger production runs, or if dimensional tolerances are tight, a high-quality aluminum die is recommended. Because this die is used only for molding polystyrene patterns at relatively low temperatures and pressures, die life is virtually unlimited.

The raw materials used in the process are polystyrene beads which contain 5–8% pentane by weight. These beads are available in different sizes. The smaller beads allow for better dimensional control and lower overall pattern density, a significant casting advantage; however, they tend to be more expensive than the larger beads. In the first process step, the raw beads are transferred by air into an expander and heated to 212°F (100°C). If greater bead expansion (low density) is desired, a vacuum expander may be used. In the expansion step, the polystyrene bead softens and the pentane volatilizes and inflates the beads from an initial bulk density of 0.37 oz/in.^{3} (0.64 g/cm^3) to 0.0092 oz/in.^{3} (0.016 g/cm^3). This expansion is accompanied by a roughly 3½-fold increase in bead diameter. After expansion, the beads (sometimes referred to as prepuff) are transferred to holding containers or are blown directly into the die mold.

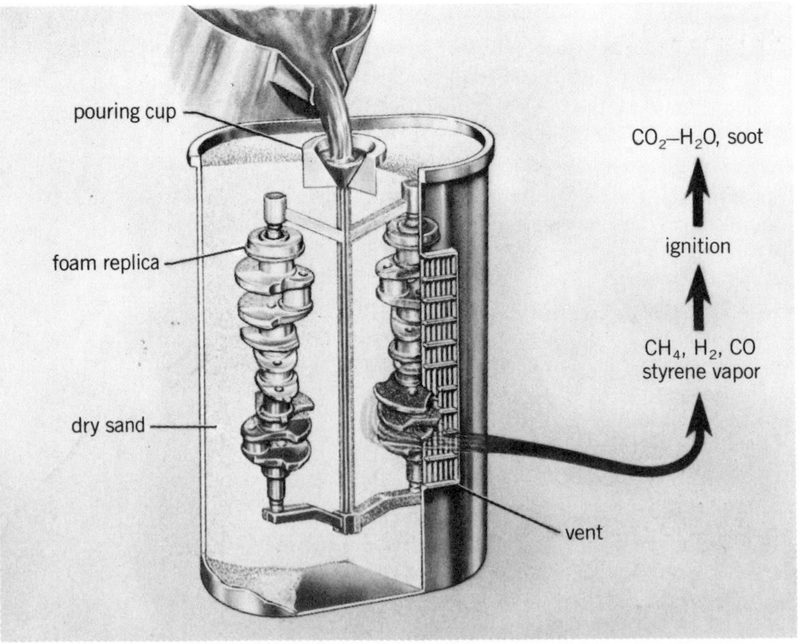

Schematic of the evaporative casting process. The part shown is an experimental hollow crankshaft for a V-8 engine. (*After S. A. Weiner et al., Evaporative casting process: Some metallurgical considerations, Met. Prog., 122(8):21–26, December 1982*)

Prior to transfer of the prepuff into the mold, the die is closed and preheated by low-pressure steam. After the prepuff is blown into the mold, additional steam is applied to fuse the beads and further expand them to assure that the pattern fills the mold. This latter step is critical for pattern dimensional reliability and surface finish, even though the overall expansion is less than 5%. After fusion, the die is water-cooled, the press is opened, and the pattern is carefully removed because it is still soft and pliable. In fully automated production, molding press cycle times are on the order of 15 s.

Molding presses are available with platen sizes up to 6½ by 6½ ft (2 by 2 m). However, because of the pliability of the polystyrene pattern immediately after molding, it is best to keep the pattern size as small as possible. Despite this concern, it is advantageous to mold the pattern with integral gating both to simplify assembly and to improve cluster handling characteristics. In some cases, it may even be desirable to design additional structure into the pattern for mechanical stability prior to assembly.

Once the outer skin of the pattern has dried, pieces may be joined to create the final casting assembly. This casting assembly should be an accurate replica of the part to be cast along with the gating and risering. After joining, excess polystyrene material such as replica stiffeners may be removed simply prior to casting. Depending upon pattern shapes, especially along the joint face, replicas may be bonded by plastic hot-melt resins, adhesives, vibration bonding, or hot wire joining.

After the replica is assembled, a refractory wash coat is applied. As in conventional molding, the primary function of the wash coat is to prevent sand burn-in in subsequent casting operations.

After the refractory coating has dried, the assembly is placed into a molding flask, and a backup medium, typically unbonded sand, is compacted around the replica. Compaction may be via fluidized bed or vibration. The molding flask should be vented or equipped for vacuum pouring in order to facilitate the removal of polystyrene off-gases.

Basic pouring practices required to produce good castings with conventional sand molds also apply to evaporative casting. The heat from the molten metal decomposes the polystyrene, allowing the metal to flow into the void created by the departing polystyrene decomposition products. The amount and nature of the decomposition products depend on the metal temperature. In the progression from pouring aluminum through iron to steel, the breakdown of the polystyrene is more complete and more rapid. Thus, venting and gating are more critical for steel castings than for aluminum castings made by evaporative casting. In casting iron or higher-temperature metals, the polystyrene off-gases are ignited. Because of ignition and the initial low amount of organic material, evaporative casting produces significantly fewer organic emissions than conventional sand casting.

After cool-down, the casting is removed, usually by simple inversion of the flask. This is possible because the typical backing medium is unbonded sand. Occasionally, residual polystyrene decomposition products will ignite on shakeout. After shakeout, the casting is cleaned by either a light sand or shot blast. The sand is allowed to cool and treated to remove fines and spills prior to reuse. A schematic of the process is shown in the illustration. The part shown is an experimental hollow crankshaft to be cast in nodular iron for use in a 5.0-liter V-8 engine.

Application. The evaporative casting process is still used for production of one-of-a-kind castings such as works of art and machine cases because the process is compatible with bronze, steel, aluminum, and iron castings. However, major applications involve the high-volume production of cast iron and aluminum components for the ground vehicle industry.

In general, microstructural characteristics of castings made by evaporative casting tend to be the same as those of castings made by conventional sand casting. For example, nodular iron castings made by evaporative casting have the same carbide content and nodularity (graphite nodule count) as those made by green sand molding, but cool at a somewhat slower rate and exhibit a slightly coarser grain structure with a lesser amount of pearlite. Microstructures of aluminum components made by evaporative casting are comparable to microstructures of sand-cast aluminum parts. Shrinkage rules for various aluminum alloys and iron types are the same for evaporative casting and conventional sand molding. Thus, the presence of the polystyrene foam (and its decomposition products) does not alter casting dimension and has little effect on microstructure or metal chemistry.

Extensive work with both iron and aluminum components and test bars revealed that casting dimensions and tolerances were determined by the dimensions and tolerances of the polystyrene foam replicas. By controlling the pentane level in the prepuff (or in the final replica), tolerances of ± 0.005 in. (± 0.01 cm) over a 7.25-in. (18.4-cm) length in a complex configuration were obtained in a 1000-piece run of 5.8-liter V-8 crankshaft replicas conducted by a commercial polystyrene foam molder. Other cast iron components made by evaporative casting include front spindle supports, connecting rods and caps, cylinder block sections, flywheels, ring gears, water pumps, rear axle housings, and exhaust manifolds.

More recently, interest has centered on production of highly cored aluminum castings made by evaporative casting, in particular intake manifolds and aluminum cylinder heads. These components are in production using the semipermanent mold casting process with either gravity or low-pressure feeding of aluminum into the mold. However, the trend is toward using evaporative casting for such components. Turn-key production systems are available from traditional foundry equipment suppliers. Arrangements can be made to produce the polystyrene replicas in-house or to obtain them from commercial polystyrene foam molders. Aluminum water pumps and cylinder block sections have also been made by evaporative casting.

Advantages. The major advantages of evaporative casting are reduced casting costs and greater overall flexibility. Cost savings occur both in reduced capitalization and in lower variable costs. Compared to conventional core-making operations, polystyrene foam molding presses, dies, and raw materials are all significantly less expensive. Furthermore, polystyrene foam tooling tends to outlast the design life of the part. In addition, the following aspects of the process significantly reduce foundry operating costs: sand handling and reuse are greatly simplified (and cost is reduced) because of the absence of binders; shakeout is made less complex because of the absence of cores; gating and risering can be incorporated into the polystyrene replica; and the use of replicas tends to minimize casting fins which result from conventional parting lines and cores. For complex cored aluminum castings made by the semipermanent mold process, evaporative casting eliminates the need for the expensive carousel pouring tables and the multiple production dies required for high-volume production.

In terms of process flexibility, evaporative casting offers advantages both from design flexibility and from the ability to respond rapidly to changes in the marketplace. The greater design flexibility results from there being virtually no requirement for draft angles. Also, process requirements such as parting lines, transitions from thin to thick sections, and locators are less severe in evaporative casting than in conventional molding.

In terms of response to the marketplace, evaporative casting is compatible with many different metals and thus greatly reduces the cost in going from iron to aluminum for a given component. Furthermore, the low tooling cost and the reduced numbers of dies required for production relative to semipermanent mold aluminum casting allow for quicker and much less expensive product changes. The absence of multiple dies in particular greatly decreases the cost and lead times required for new product launch. Thus, the evaporative casting process is suitable for producing quality castings of complicated designs in a variety of metals over a wide range of production volumes.

For background information *see* METAL CASTING in the McGraw-Hill Encyclopedia of Science and Technology.

[S. A. WEINER]

Bibliography: P. G. Kohler, Evaporative pattern process: A new dimension for the foundry, *Mod.*

Cast., 71(8):36–37, August 1981; S. A. Weiner et al., Evaporative casting process: Some metallurgical considerations, *Met. Prog.*, 122(8):21–26, December 1982; S. A. Weiner and C. D. Piercecchi, Dimensional behavior of polystyrene foam shapes, Annual Meeting of the EPS Division of the Society of the Plastics Industry, March 7, 1980, *AFS Transac.*, publication pending.

Fertilization

Fertilization is the interaction between sperm and egg which ultimately results in the fusion of the two cells, the comingling of their genetic material, and the triggering of embryonic development. The process involves dramatic changes in the morphology and physiology of both gametes. This article discusses recent research on the mechanism of these changes in animal gametes.

Major alterations occur in the sperm as they are released from the testis and approach the egg; these include changes in the rate of sperm motility and changes in the surface of the sperm so that they can bind to egg envelopes, pass through these envelopes, and finally fuse with the plasma membrane of the egg. When sperm and egg meet, there are two major changes in the egg. The first involves a rapid change in the receptivity of the egg to the sperm, referred to as the block to polyspermy. This change ensures that only one sperm will fuse with the egg. The second change is the triggering of embryonic development, which includes the preliminary events of the migration and fusion of the sperm and egg nucleus in the egg cytoplasm and the important events leading to metabolic changes which somehow initiate cell division and embryogenesis.

Much of the research on fertilization has been carried out on simple organisms, such as gametes of sea urchins, since they provide convenient models for studying early development. Recently, development of techniques for the study of fertilization outside the organism has allowed significant advances to be made in the understanding of fertilization in higher forms, such as mammals.

Sperm motility. Recent research has focused on the mechanisms by which sperm are kept nonmotile in the testis and how motility is initiated upon release of sperm from the testis, as during ejaculation in mammals, or upon dilution in sea water, as in marine organisms. This work suggests that the dormancy of the sperm is a consequence of a low intracellular pH (pH_i), which apparently prevents the flagellar apparatus from operating. Upon release of semen from the testis, or dilution of the semen, the pH_i rises (to about 0.5 pH unit in the sea urchin), and this increase is sufficient to activate the motility apparatus. The mechanism involved in the pH_i rise requires extracellular sodium, and it may operate by a Na^+–H^+ exchange process.

Sperm motility increases in the vicinity of the egg, and these changes probably are related to soluble substances released by the egg (as in sea urchins) or occur upon binding to the outer envelopes of the egg (as in mammals). The mechanism of this motility change has been best studied in sea urchins, where a low-molecular-weight peptide, termed speract, increases the motility of the sperm. Here too, sodium is required for the action of speract, and it is thought that speract somehow lowers the pH optimum for the motility increase.

The motility change may relate to the passage of the sperm through the outer egg coat. In sea urchins the outer egg coat is an acidic mucopolysaccharide, and the change in the external pH optimum might permit the sperm to pass through this acidic coat. In the mammal the change in motility is probably related to the capability of the sperm to pass through the thick outer egg covering, known as the zona pellucida.

Acrosome reaction. In order to bind to the outer egg coverings and pass through them, the sperm must first undergo an alteration referred to as the acrosome reaction (see illustration). This alteration involves changes in a vesicle (acrosomal vesicle) near the anterior end of the sperm; the vesicle contains substances involved in species-specific binding between sperm and egg (referred to as bindins) and also enzymes or proteins which somehow digest the outer egg envelopes so that the sperm can pass through them and ultimately fuse with the egg plasma membrane.

The acrosomal reaction or opening of this acrosomal vesicle is a typical exocytosis, requiring extracellular calcium. In sea urchins the trigger for the acrosome reaction is the jelly coat of the egg or possibly substances on the vitelline layer. The trigger in the mammal has not been identified, although fertilization experiments outside the organism show that the acrosome reaction can be triggered by amines such as taurine and possibly neurotransmitters such as epinephrine. These substances somehow cause an influx of calcium into the acrosomal vesicle, thus inducing the exocytosis. Recent studies also indicate an increase in the pH_i of the acrosomal vesicle prior to exocytosis. In many invertebrates this pH_i increase may be related to a simultaneous polymerization of actin which results in the formation of an acrosomal filament at the apical end of the sperm; however, actin polymerization is not seen in mammals.

The mechanism by which the sperm passes through the outer egg coats probably involves proteases contained in abundance in the acrosomal vesicle. In some organisms, notably in the mollusks, a noncatalytic mechanism is involved. Here the acrosomal vesicle is very large and contains special proteins which somehow allow the sperm to pass through the outer egg chorion via a nonenzymatic mechanism. One speculation is that these proteins somehow destabilize the chorion so that the sperm becomes able to pass through it.

Sperm-egg binding. The binding of sperm and egg is species-specific and is related to interactions between surface receptors on both. In sea urchins

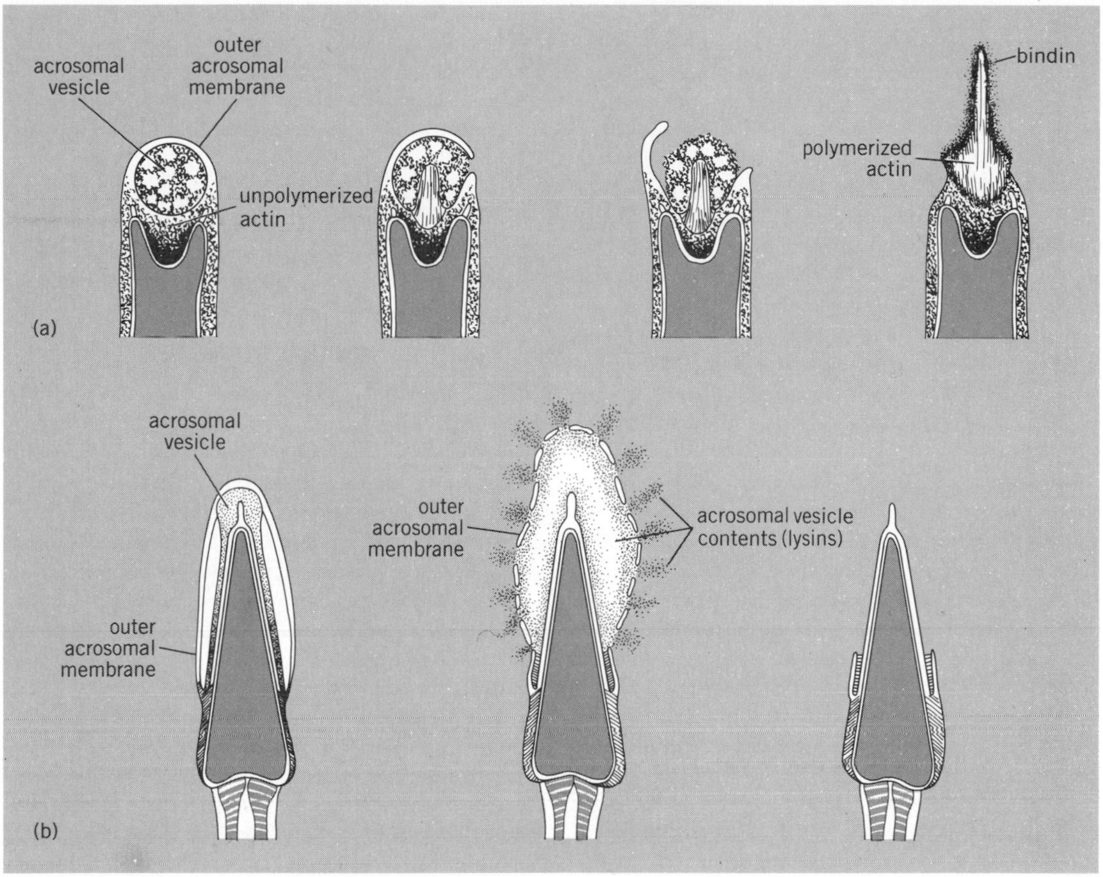

The acrosomal reaction in (*a*) echinoderm sperm and (*b*) mammalian sperm. In both cases lysins are contained within an acrosomal vesicle and become exposed by the fusion or breakdown of the outer acrosomal membrane. The lysins are now able to digest a passage through the outer egg envelope. The polymerization of actin, resulting in the formation of a filament or process in front of the sperm, is a major difference seen in echinoderms (and many other invertebrates). The filament becomes coated with the contents of the acrosomal vesicle. (*After David Epel, Fertilisation, Endeavour, 4(1):26–31, 1980*)

and oysters, there are large amounts of proteinaceous substances in the sperm (the bindins) which attach in a species-specific manner to the egg surface. Although these bindins have not yet been found in sperm of other organisms, they may be present. Little is known about sperm-binding substances located on the egg surface. However, in ascidians and mammals it appears that the sperm receptor on the egg is a glycoprotein that is rich in fucose. The detailed structure is not known, but presumably specificity must reside in more than just the fucose moiety, most probably in the nature of the fucose-containing macromolecule.

Block to polyspermy. An egg could, theoretically, be inundated by hundreds or even thousands of sperm unless there were mechanisms to prevent polyspermy. Work on several aquatic organisms, notably sea urchins and frogs, indicates that a rapid change in the membrane potential (depolarization) is involved in the block to polyspermy. This sort of electrical change is not seen in eggs of fish and mammals. In the eggs of fish, a physical mechanism allows only one sperm to enter. The sperm must pass one by one through a tiny channel, referred to as the micropyle, and when a sperm fuses with the egg, the micropyle becomes impermeable to other sperm. This change results from a secretory event in the egg (the cortical reaction) which releases products that clog the micropyle channel. The mechanism of polyspermy prevention in the mammals is not yet completely understood. Perhaps there are very few sperm at the site of fertilization, so that a rapid electrical depolarization mechanism is not essential. In these cases, the cortical reaction (a secretory event in which numerous granules embedded in the cortex secrete their contents outward and alter the egg coats) alone may be responsible for this block to polyspermy. In organisms which employ depolarization as the primary block to polyspermy, the cortical reaction is often a secondary block.

Mechanisms of egg activation. Classical research on artificial parthenogenesis revealed that the egg contains the program of early embryonic development and that this program is somehow triggered upon fertilization. The triggering mechanism apparently involves an increase in intracellular calcium; in fact, changes in free calcium (Ca^{2+}) levels have now been directly demonstrated in eggs of sea urchins, starfish, fish, and mammals. This Ca^{2+} increase is the controlling factor since one can regulate the postfertilization changes by manipulating the intracellular calcium levels with calcium buffers or by triggering development directly with ionophores which increase intracellular calcium. Most eggs can be activated by ionophores in Ca^{2+}-free media or even by sperm in Ca^{2+}-free media (if the Ca^{2+}-requiring acrosome reaction is induced before adding sperm to the egg). This suggests that the sperm somehow causes release of Ca^{2+} from an intracellular store. The exception to this rule may be found in mollusks and perhaps other protostome organisms where the sperm induces a depolarization which then allows calcium to enter from outside the cell.

Regardless of the calcium source, it now seems clear that a Ca^{2+} increase triggers development, and work on some of the changes that occur after fertilization indicates that the calcium receptor is calmodulin, a Ca^{2+}-binding protein which appears to be involved in the regulation of numerous cellular processes. For example, in sea urchins a role for the calcium-calmodulin reaction has been seen in the activation of nicotinamide adenine dinucleotide (NAD) kinase (an enzyme activated at fertilization) and for the cortical reaction.

In some forms, the calcium increase is followed by a fairly large increase in pH_i. This has been especially well described in the sea urchin egg, where determination of pH_i by a number of methods reveals a pH_i increase of about 0.4 pH unit. This pH_i change requires extracellular Na^+, and proceeds most likely by a Na^+–H^+ exchange mechanism. In the sea urchin, this pH_i change is necessary for initiating protein and DNA synthesis and also the early movements of sperm and egg nucleus within the egg cytoplasm.

Such pH_i changes may not be a universal phenomenon in fertilization. For example, although changes in pH_i have also been seen after fertilization of frog eggs, they are not observed following fertilization of starfish eggs. Perhaps the pH_i increase relates to the relative metabolic dormancy of the egg, but more comparative studies are needed. However, it does seem certain that a calcium increase is the primary trigger, and future research could reveal how the sperm-egg contact causes the Ca^{2+} increase, and how the Ca^{2+} increase then triggers cell division and ultimately development.

For background information *see* FERTILIZATION; REPRODUCTION (ANIMAL); in the McGraw-Hill Encyclopedia of Science and Technology.

[DAVID EPEL]

Bibliography: C. G. Glabe et al., Carbohydrate specificity of sea urchin sperm bindin: A cell surface lectin mediating sperm-egg adhesion, *J. Cell Biol.*, 94:123–128, 1982; S. Hagiwara and L. A. Jaffe, Electrical properties of egg cell membranes, *Annu. Rev. Biophys. Bioeng.*, 8:385–416, 1979; B. M. Shapiro, R. W. Schackmann, and C. A. Gabel, Molecular approaches to the study of fertilization, *Annu. Rev. Biochem.*, 50:815–843, 1981; M. J. Whitaker and R. A. Steinhardt, The relation between the increase in reduced nicotinamide nucleotides and the initiation and maintenance of DNA synthesis in the egg of the sea urchin *Lytechinus pictus*, *Cell*, 25:95–103, 1981.

Food engineering

This article surveys some recent advances in food engineering that involve the use of novel materials for processing and packaging.

MEMBRANE PROCESSING

Membrane separation processes are based on the ability of semipermeable membranes of the appropriate physical and chemical nature to discriminate between molecules primarily on the basis of size and, to a lesser extent, on shape and chemical composition. The main role of a membrane is to act as a selective barrier. It should permit passage of certain components but retain certain other components of a fluid mixture. By implication, either the permeating stream or the retained phase should be enriched in one or more components.

The illustration shows various membrane processes and examples of the type of solutes that can be separated. Membrane separation processes cover a wide range of particle sizes, and are matched in versatility only by centrifugal processes. However, an absolute requirement for centrifugation is the existence of a suitable density difference between the two phases to be separated, in addition to their being immiscible. Membrane separation processes have no such requirements, and indeed, the real value of ultrafiltration and reverse osmosis processing is that they permit separation of dissolved molecules down to the ionic range, provided a suitable membrane is used.

Process characteristics. The nature of the membrane itself controls which component permeates and which component is retained. The table shows the characteristics of various membrane processes. The driving force for osmosis is chemical potential differences between the water on either side of the membrane. With an ideal semipermeable membrane, only water should permeate the membrane. The common laboratory technique of dialysis, on the other hand, is primarily a technique for purifying macromolecules, such as desalting of proteins, and the primary driving force is the difference in concentration of the permeable species between the solution in the dialysis bag and outside the bag. Electrodialysis relies mainly on electromotive force and ion-selective membranes to effect a separation between charged ionic species. What distinguishes the three common membrane processes—microfiltration, ultrafiltration, and hyperfiltration–reverse osmosis—is the application of hydraulic pressure to increase the rate of the transport process.

Ideally, reverse osmosis membranes should retain all components other than the solvent (water) itself, while ultrafiltration membranes retain only macromolecules or particles larger than about 1–20 nanometers. (It is customary to refer to molecular weight cutoff instead of pore size when attempting to classify membranes within ultrafiltration itself; thus ultrafiltration covers "particles" or molecules that range in molecular weight from about 100 to 1,000,000 daltons.) Microfiltration processes, on the other hand, are designed to separate particles in the so-called micrometer range, that is, suspended particles of 0.10–10 micrometers. Particles larger than 10 μm are best handled by conventional macrofiltration or centrifugal processes. The distinction between the various membrane processes is somewhat arbitrary and has evolved with usage and time. In its broadest sense, reverse osmosis is essentially a dewatering technique; ultrafiltration is a method for simultaneously purifying, concentrating, and fractionating macromolecules or fine colloidal suspensions; and microfiltration is a clarification process that separates molecules and particles on the basis of size and solubility. Generally, particles on the order of >0.05 μm are retained by microfiltration membranes.

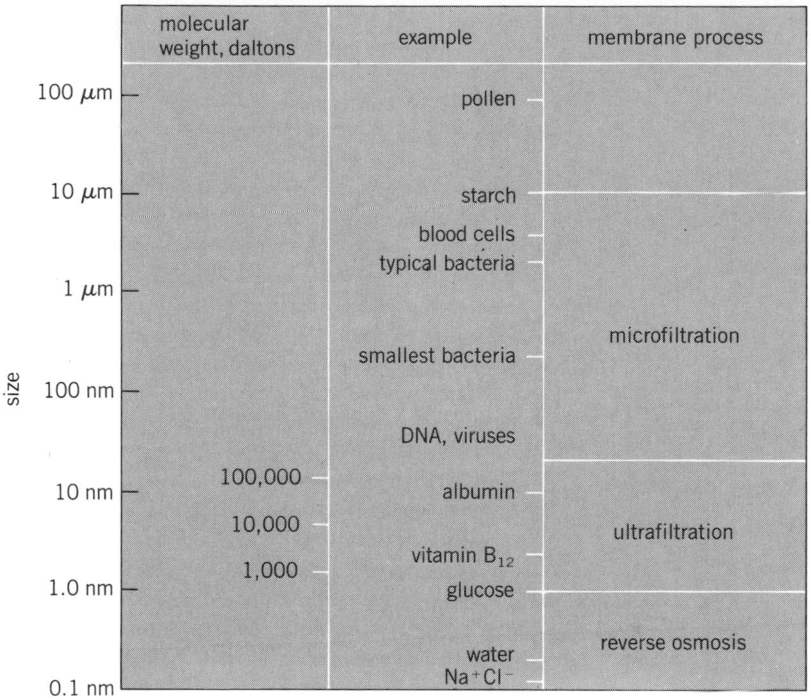

Examples of compounds separated by membranes.

Advantages of membrane processes. Ultrafiltration and reverse osmosis constitute perhaps the first continuous molecular separation processes that do not involve either a phase change or interphase mass transfer. In their simplest form, reverse osmosis, ultrafiltration and microfiltration consist merely of pumping the feed solution under pressure over the surface of a suitably supported membrane, of the appropriate chemical nature and in the optimum physical configuration. The removal of water is accomplished without a change in phase or state of the solvent. Evaporation, freeze concentration, or freeze drying are common dewatering techniques used in the food, pharmaceutical, and biological processing industries. Evaporation requires the input of about 1000 Btu/lb of water evaporated (2.3×10^6 J/kg), while freezing requires about 144 Btu/lb of water frozen (3.3×10^5 J/kg), merely to effect the change in state of water from liquid to vapor and liquid to solid, respectively. Since membrane processes do not require a change in state of the solvent to effect a dewatering, this should result in considerable savings in energy, even when comparing membrane processes to multiple-effect and mechanical vapor-recompression systems, up to certain feed concentration limits. A less obvious advantage is the fact that no complicated heat-transfer or heat-generating equipment is needed, and the membrane operation, which requires only electrical energy to drive the pump motor, can be situated far from the prime power-generating plant. No additional steam capacity need be installed to handle the ultrafiltration–reverse osmosis unit. In addition, no condensers (or huge condenser-cooling water supply) are needed, thus avoiding related problems like thermal pollution and overloading of sewage treatment systems.

Another feature of membrane processes is that they can be operated at ambient temperatures. Thus thermal or oxidative degradation problems, common to evaporation processes, can be avoided. However, there may be occasions when operation at considerably lower temperatures is necessary (for example, to prevent microbial growth problems or denaturation of heat-sensitive components) or at higher temperatures (for example, to minimize microbial growth problems or to lower the viscosity of the retentate, thus lowering pumping costs and improving mass transfer). Finally, since small molecules should normally pass through ultrafiltration membranes freely, their concentration on either side of the membrane should be the same during processing and about equal to the original feed so-

Characteristics of membrane processes

Process	Driving force	Permeate	Retentate
Osmosis	Chemical potential	Water	Solutes
Dialysis	Concentration	Water plus small molecules	Large molecules
Ultrafiltration	Pressure	Water plus small molecules	Large molecules
Reverse osmosis	Pressure	Water	Solutes
Electrodialysis	Electromotive force	Water plus ionic solutes	Water plus nonionic solutes
Microfiltration	Pressure	Water plus dissolved solutes	Large suspended particles

lution. Thus there should be minimal changes in the microenvironment during ultrafiltration, that is, few changes in pH or ionic strength.

Limitations. There are some limitations to ultrafiltration and reverse osmosis processes since neither can take the solutes to complete dryness. In fact, membrane processes are quite limited in their upper solids limits. In reverse osmosis it is frequently the osmotic pressure of the concentrated solutes that limits the process. In ultrafiltration it is rarely the osmotic pressure of the retained macromolecules but rather the low mass-transfer rates of concentrated macromolecules and the high viscosity that make pumping of the retentate difficult. As an example, current technology permits skim milk to be concentrated by multiple-effect evaporation to about 50% total solids, while the upper limit using reverse osmosis is about 25% total solids and using ultrafiltration about 42% total solids. Other problems that plagued early membrane applications were the fouling of membranes, poor cleanability of some early modules, and restricted operating conditions, although some of these problems have been overcome through the development of superior membrane materials and improved module design.

Applications. The most important applications of the various membrane processes in the 1960s and 1970s were for desalination, water treatment, and recovery of paint and chemicals. Since that time the food engineering industries have developed a variety of additional applications.

Those involving reverse osmosis include concentration of milk and cheese whey, recovery of lactose from ultrafiltered cheese whey permeate, protein recovery from soy whey, recovery of carbohydrates and protein from potato-processing water, recovery of dye and sugar from confectionery waste, and recovery of anthocyanins from cranberry pulp wastes.

Applications involving ultrafiltration include separation of enzymes and proteins from fermentation broths, preconcentration of milk for manufacture of dairy products, protein recovery from cheese whey, removal of glucose and concentration of egg white, effluent treatment of potato and corn wastewaters, separation of soy bean concentrates and isolates, and clarification and chill proofing of clear fruit juices.

Applications of microfiltration include pretreatment of water, harvesting microbial cells for fermentation, removal of yeast from wine or beer, sterilization of liquids, and concentration of emulsions or suspensions. It is anticipated that major advances will take place in the 1980s in development of membrane technology applications in food processing as well as biotechnology.

[MUNIR CHERYAN]

FOOD PACKAGING

Since food represents the market for more than half of all commercial packaging, far more effort is generally focused on food packaging than on any other category. Recent federal regulations have dealt with tamper resistance for over-the-counter drug products, and food packagers have been concerned that the regulations could be applied to them. However, because proprietary drugs represent only about 2 billion packages annually and a few hundred manufacturers, compared to the 300–400 billion packages and 26,000 manufacturers for food, tamper-resistant packaging for food is not being forced by government at this time.

Carbonated-beverage and beer packaging represent the largest fraction of food packaging, accounting for more than a third of the units used. The size and economic value of this market have made it the target of major supplier developmental programs. Thus many food-packaging developments are initiated for the beverage industries and subsequently transferred to other sectors of the food industry.

With the 1981 Federal Food and Drug Administration (FDA) issuance of a regulation on the maximum residual hydrogen peroxide permitted after chemical sterilization of polyethylene surfaces, aseptic packaging employing this technique became commercially acceptable in the United States. The result has been an explosive growth of aseptic packaging for consumer, institutional, and industrial sizes of food products, the major progress being achieved in institutional sizes.

Paperboard packaging. Recent progress in this area has involved single-wind paperboard cans and coextruded coatings.

Single-wind paperboard cans. The manufacture of paperboard cans by erection of laminated or coated laminated paperboard into cylindrical or rectangular, solid shaped containers from flat blanks has been attempted for several years. A number of in-line paperboard can-body-making machines have been installed to package dry foods in place of spiral-wound composite cans, and dairy products in place of off-line fabricated paper cups. There is one complete packaging system introduced (Cekacan) that makes, fills, and closes single-wind paperboard cans with high water-vapor and oxygen-barrier performance. By using extrusion-coated and extrusion-laminated paperboard, single-wall semirigid packages are made in-line with packaging operations at speeds, efficiencies, and protection functions paralleling those employed with conventional canning lines.

Coextruded coatings. A mainstay of paperboard for frozen and other food packaging has been low-density polyethylene extrusion coating on one or two sides of the paperboard to improve water-vapor and grease resistance of the base material. At the opposite extreme has been lamination to high-barrier materials such as aluminum foil. With the technical success of plastic coextrusions in film formation, the concept has been translated to paperboard coating. Rather than attempt to achieve function with single layers or multiple applications of single layers, different functional materials are applied simultaneously through a single die to obtain optimum function. For example, ionomer and polyethylene are coextruded to effect bond to the substrate, water-vapor resistance, and effective package heat sealing.

Plastic. This packaging technology has advanced with the development of linear low-density polyethylene, crystallized polyester, high-barrier plastics, comaterials, and metallized film.

Linear low-density polyethylene. Although this material with its toughness and extended heat-seal range has been used for milk pouches in Canada for more than 10 years, broad-scale use in the United States has taken place only since 1979 with the development of commercial low-pressure polyethylene production. Increased tensile strength at lower cost has made this film and injection-molding material attractive for bag-making and bundling.

Crystallized polyester. Polyester materials expanded greatly with their application to carbonated beverage bottles after a Federal Food and Drug Administration ban on polyacrylonitrile bottles in 1977. Use of polyester materials in extrusion coating of paperboard to impart heat, liquid, and fat resistance was a natural outgrowth of polyethylene extrusion coating of paperboard. These boards are being used in both conventional and microwave oven heating.

All-polyester constructions may be heat-set at temperatures below their glass transition temperature, but still well above the operating temperatures of food heating. Solid white frozen-food trays to replace aluminum foil are being made by compression-molding mineral-filled polyester. Less expensive trays and dishes for the same purpose are made by extruding thermoplastic polyester sheet, thermoforming at elevated temperature, and cooling under restraint. These partially crystallized polyester trays are being introduced to compete against plastic-coated ovenable paperboard being used for frozen-food packaging.

High-barrier plastics. High-oxygen-barrier constructions in semirigid materials for use as possible metal or glass replacements are being made by coextrusion processes. The relatively expensive high-oxygen-barrier plastics such as polyvinylidene chloride or ethylene vinyl alcohol copolymer are sandwiched between polyolefin layers to provide single formable sheets. Thus, the oxygen-barrier materials are protected from abrasion or moisture, and high-water-vapor-barrier polyolefins are present to function synergistically.

Multilayer plastics incorporating high oxygen barriers are being used in Japan for bottles and jars to contain catsup, mayonnaise, edible oil, and wine. Such packages have been introduced in the United States for catsup and barbecue sauce. In the United States, thermoform–fill–seal aseptic packaging systems require multilayer materials for long-term ambient-temperature distribution.

Comaterials. In addition to coextrusion, multiple materials are being joined in injection-molding parisons and in one process of thermoforming and blow-molding independent sheets to produce bottles and jars with high barrier to contain hygroscopic foods.

Metallized film. Vacuum deposition of thin (micrometer) layers of aluminum on polyester, polypropylene, and even polyethylene has resulted in films with orders-of-magnitude increases in oxygen- and water-vapor-barrier performance. These films have been used to make pouches for roasted and ground coffee, snack foods, wine, and processed meats in place of aluminum foil.

Bottles. In this type of food packaging, advances have been made using polyester, coated polyester, and aluminum.

Polyester. Injection blow-molded polyester bottles for carbonated beverages were the greatest commercial packaging success of the 1970s. These have been followed by copolyesters with lower performance characteristics but significantly better fabrication properties. Thus, the materials may be thermoformed or extrusion-blow-molded to provide clear and colorless packaging for food products with modest barrier requirements such as sugar syrups.

A plant is being constructed in the United States to produce polyester bottles capable of being used for hot filling products such as juices. This factory is employing processes developed in Japan, where the products are already commercial.

Coated polyester. Although polyester bottles have been commercially acceptable, their oxygen barrier is marginal in smaller sizes with their lower surface-to-volume ratios and for highly oxygen-sensitive products such as beer. By externally coating the parison or finished bottle by dip, spray, or roller coating, the oxygen barrier may be increased several times. Beer bottles made from polyester coated with polyvinylidene chloride (PVDC) have been introduced commercially in the United Kingdom. Wine and single-service-size PVDC-coated polyester bottles have been introduced commercially in the United States.

Aluminum. A three-piece aluminum package for carbonated beverages, designated the aluminum bottle, has been introduced into a test market. A drawn cylindrical body is affixed to a conical aluminum top. The bottle-shaped container is closed with a plastic-plug fitment capable of reclosure.

Cans. Recent developments in cans involve both plastic and metal.

Plastic. The notion of displacing tin-plated steel cans with their weight and cost variables has fostered development of numerous types of plastic cans. An extruded tube of polypropylene to which is bonded a lamination of aluminum foil and polypropylene film was developed in Sweden. Ends, fabricated by injection-molding polypropylene and laminating aluminum foil, are affixed to the square cross-section bodies by mechanical clinching plus induction sealing. This package is designed for terminal sterilization of particulate low-acid food products.

A thermoformed polyester can for carbonated beverages which has now been extended to beer and wine has been developed in the United Kingdom.

The can body is a tapered cup-shaped unit, and closure is by double-seaming an easy-open aluminum end. A similar can has been introduced in Italy.

Also in the United Kingdom, one company is employing insert injection-molding methods to produce cans claimed to be suitable for both terminal sterilization and for dry foods. Aluminum foil–polypropylene laminations are inserted in the injection mold, and the can body is formed by injection and simultaneous fusion bonding to the film. The polypropylene imparts water-vapor barrier and the aluminum foil, oxygen barrier. The package is closed by double seaming an easy-open aluminum end.

A Japanese company has developed an insert injection-molded can including ethylene–vinyl alcohol copolymer as a high-oxygen-barrier material in a coextruded coating on paperboard. The tapered paperboard/plastic can is closed by heat-sealing a compatible closure. The can is then sterilized by microwave–thermal techniques.

Metal. Welded side-seam steel cans and two-piece aluminum cans are rapidly displacing traditional soldered side-seam tin-plated cans for food packaging.

Although tens of billions of two-piece aluminum cans are used annually for beer and carbonated beverages, part of their applicability is due to the presence of internal pressure from the carbon dioxide content. Cans for food require an absence of oxygen. This is usually achieved by vacuumization which, for thin-wall aluminum cans, would lead to paneling. Recently, aluminum can makers in the United States have introduced two-piece aluminum cans in which liquid nitrogen is introduced after filling to provide an inert environment and a positive internal pressure to maintain can shape. This system has been used commercially to can wine.

Processes. Among the newer processes being used by modern food packagers are aseptic packaging, high temperature–short time (HTST) sterilization, terminal sterilization, insert injection molding, shrink film casing, computer-assisted design, and controlled-atmosphere packaging.

Aseptic packaging. In this process the product and packaging are sterilized independently of each other and then brought together in a sterile environment. Sterilization of fluid products outside of the package permits HTST techniques, which lead to better-quality food. Independent sterilization of the packaging allows for less stringent demands on the materials and permits the use of lower-cost materials. Canning using an aseptic system has been used in the United States since the 1940s. Aseptic packaging in bulk cans has been commercial since the 1960s, and in paperboard and plastic laminations has been commercial in western Europe, Latin America, and Asia for more than 15 years. The 1981 FDA decision on hydrogen peroxide sterilant opened the way for commercialization of consumer-sized aseptic packaging in the United States.

A system known as the Brik Pak vertical form–fill–seal system can run roll-stock paperboard–polyethylene–aluminum-foil laminations through hot aqueous hydrogen peroxide solutions. After evaporation of the sterilant, sterile liquid food such as juices, juice drinks, or milk are introduced, and the pouches are heat-sealed through the contents to eliminate head space.

The Bloc Pak system, developed first in West Germany, uses preadhered tubes made of much the same lamination as that used by Brik Pak. The tubes are erected and conveyed into the sterile section, where hydrogen peroxide is sprayed into the open-top package. After thermal evaporation, the packages are filled and heat-seal-closed, sometimes with inert-gas injection prior to sealing.

A number of deposit–fill–seal and thermoform–fill–seal aseptic packaging systems for all-plastic cup-shaped configurations have been introduced by several European firms. Packaging sterilization methods include hydrogen peroxide, steam, and the heat required to extrude the base plastic sheet.

In larger sizes, 1–5 gal (3.8–19 liters) or greater, sealed pouches made from metallized polyester are presterilized with ionizing radiation. Sterile pouches are aseptically attached to aseptic filling units and filled with wine, sterile milk, or sterile high-acid purees such as tomato or fruit. In corrugated fiberboard cases, these filled pouches may be distributed at ambient temperature for institutional or industrial applications. In the United States, this process is being used for pouches, drums, and tanks.

HTST terminal sterilization. Often referred to as retort pouch packaging, which is only a single segment of the total market, HTST terminal sterilization is now being performed in thin cross-section containers fabricated from steel, aluminum, plastic, or flexible laminations. The aim is to increase the surface-to-volume ratio to accelerate the heat penetration and to effect a reliable seal. Shallow-drawn steel trays may be double-seam-closed. Shallow-drawn extrusion-coated or laminated aluminum-foil trays are heat-seal-closed in several western European processes. Coextruded PVDC–polypropylene sheet is thermoformed, filled with low-acid particulate food, and vacuum-sealed, with the resulting package steam-sterilized under counterpressure. Smooth-surface flat-sealing areas free from food particles permit more reliable, high-speed sealing than is practical with pouches. Large multigallon aluminum foil pouches are being filled, heat-sealed, and thermally sterilized. These pouches are designed to displace the institutional No. 10 tin-plated steel can.

Insert injection molding. A system known as the VersaForm insert injection-molding process bonds paperboard and plastic structural members to produce package components intended to be optimum in function and economics. The decorative advantages and inextensibility of paperboard are blended with the cost and strength of plastic. In the United States, commercial applications of the VersaForm process include primary closure, can overcaps, tubs and cups, and beverage bottle multipackers.

Shrink film casing. Packaging for distribution performed by carbonated beverage packagers need not comply with government regulations, which largely dictate the use of all corrugated fiberboard case constructions to protect the contents. Further, a limited number of regulatory permits have been issued for reducing the amount of corrugated fiberboard and replacing it with shrink film for common-carrier transport. Cans, bottles, jars, cartons, and even pouches may be unitized in a shallow corrugated fiberboard tray which is then overwrapped in polyolefin film. Applying brief heat to the film shrinks it and creates a tight bundle capable of withstanding vertical compression, vibration, and drop as well as corrugated fiberboard, but at lower cost. The film wrap may be complete or may be partial by adhering it to the side wall of the tray.

Shrink film casing is used by most carbonated beverage canners and by a few brewers and food canners for distribution packaging.

Computer-assisted design. The ability of computer memories to store and of computers to retrieve vast quantities of data has led to their use for multiple repetitive computations and tasks. Geometric shapes may be displayed graphically by storing either the mathematical formula or the visual depiction. Thus, the task of drafting two-dimensional representations of three-dimensional objects such as packages may be facilitated by computer. A library of component shapes and dimensions may be retrieved in part and fitted together on a computer output display to generate package design trials. When the designer is satisfied, hard-copy drawings may be made, and with them, the numerical control tapes to produce laser-cut dies or molds. Thus, sample packages may be made rapidly, and production dies and molds are precise reproductions of approved samples and of each other. Even three-dimensional representations of the package may be generated on the flat computer-output screen. In the United States commercial applications of computer-aided design and subsequent laser die-cutting are in production of paperboard folding cartons.

Controlled atmosphere packaging. Vacuum and inert-gas packaging for processed meats and cheeses has been commercial since the 1960s. Application of this concept to respiring or dynamic biological systems has been limited to wholesale red meats, where the issues of product color control are minimal. The Cryovac system for vacuum packaging of prime-cut red meats in high-oxygen-barrier pouches is now used for more than half of all red meat in the United States. A small number of meat packers and retailers have packaged retail cuts in sealed packages under vacuum, which leads to dark coloration from myoglobin pigment, or under high carbon dioxide–oxygen atmosphere, which permits the typical color of oxymyoglobin pigment.

A number of western European bakers have been employing high–carbon dioxide internal environments in pouches or thermoforms to extend the shelf life of their soft baked goods such as breads and cakes.

In both North America and western Europe, low-oxygen packaging of both fresh and frozen seafoods is being employed to permit long-distance distribution of these products.

For background information *see* DIALYSIS; FOOD MANUFACTURING; ION EXCHANGE; OSMOSIS; PLASTICS PROCESSES; ULTRAFILTRATION in the McGraw-Hill Encyclopedia of Science and Technology.

[AARON L. BRODY]

Bibliography: C. J. Benning, *Plastic Films for Packaging*, 1983; S. D. Brock, *Membrane Filtration; A User's Guide and Reference Manual*, 1983; M. Cheryan, *Ultrafiltration Handbook*, 1984; R. H. Cramm and W. R. Sibback, *Coextrusion Coating and Film Fabrication*, Technical Association of the Pulp and Paper Industry, 1983; J. F. Hanlon, *Handbook of Package Engineering*, 1984; P. Meares, *Membrane Separation Processes*, 1976; F. A. Paine, and H. Y. Paine, *A Handbook of Food Packaging*, 1983; S. Sacharow and R. Griffin, *Food Packaging*, 2d ed., 1982.

Fuel

Coal–water mixtures have potential use as burnable industrial fuels, and therefore are the subject of research and development projects in several industrialized nations, including the United States. This approach to the need for low-cost fuels as substitutes for fuel oils and natural gas in industrial fuel-intensive applications is considered an excellent "bridge-over" fuel system. This concept groups industrial fuels into two categories: those fuels that are used extensively today, in particular, fuel oils and natural gas; and fuels known as futuristic, such as solar energy, wind power, biomass, geothermal energy, oil shale, tar sands, and coal gasification. Although energy researchers think futuristic fuels have great promise, such fuel alternatives will not be widely available to industry at competitive rates for another 15–20 years.

In the intervening time, coal–water mixtures appear to be an alternate fuel type that is available immediately to fuel users for both industrial applications and electric power generation. Uncertainties regarding the preparation, storage, transportation, and combustion of coal–water mixtures have been almost entirely resolved.

Fluid coal background. In the early 1960s, a project team successfully transported (via barge and railroad tank car), stored, pumped, and burned coal–water fuel in a cyclone-type furnace boiler. Similar experimental projects were completed in the Soviet Union and Germany later in the 1960s. An interesting aspect of these projects was the fact that, in all three cases, coal–water mixtures were prepared at coal mine sites, pumped through coal slurry pipelines, and burned on an as-received basis in industrial boilers. However, despite these successful experiments, coal–water mixtures were not widely adopted by industrial fuel users, primarily due to economic factors. Fuel oils and natural gas are preferred, clean fuels, and their prices, adjusted for inflation, were actually declining during the 1960s and early 1970s. That circumstance was sharply reversed following the OPEC oil embargo in late 1973 and early 1974; many industrialized, oil-dependent nations hastily initiated alternate energy programs aimed at the rapid commercialization of new technology for energy.

Coal slurries were the focus of immediate and intense study by energy researchers. Coal–water mixtures were not the first object of attention, however, as researchers investigated coal–oil mixtures as the fastest route to relief from skyrocketing oil prices. It was not until the late 1970s that a thorough economic analysis of coal–oil mixtures revealed that this fuel alternative would not produce a saving for industrial fuel users. In the early 1980s, coal–water mixtures were finally recognized as the most cost-effective approach to fluid coal technology.

Fuel specifications. Coal–water mixtures encompass a variety of formulations of finely pulverized coal, water, and, in some cases, highly volatile fuels or various types of chemical additives. Generally, coal constitutes 55–60% or more by weight of these mixtures, which include coal–water mixtures; coal, oil and water mixtures; and coal, methanol, and water mixtures.

Coal–water formulations are usually prepared with one or more chemical additives for specific applications. As the amount of coal in the mixture is increased above 50%, viscosity increases sharply, and commercially available pH adjustment, stabilization, and viscosity-reducing chemical agents may have to be added in order to achieve a mixture that can be stored for long periods of time, pumped, atomized by an industrial burner nozzle, and burned successfully.

Since coal–water mixtures are suspensions and not chemical solutions, there is a tendency for the coal particles to settle toward the bottom of storage vessels, transportation equipment, or fuel pipes. Extensive work has been performed on the use of stabilizing chemical agents to retard this settling effect, but agitation equipment (that is, mechanical mixers, compressed air lances, or pumps) is still required.

Agents for controlling pH, for retarding slagging and fouling, and for flame enhancement are commercially available. Their use is dictated by the site-specific circumstances of each user. Very little, if any, work has been done on the addition of SO_x abatement chemicals, such as lime, nahcolite, or trona, to coal–water mixtures. Researchers anticipate that this approach may solve an SO_x emission problem resulting from the use of high-sulfur coal.

The value of chemical additives cannot be generalized for all fuel users or even specific classes of fuel users, for the following reasons. First, coal is a complex mixture of several chemical elements that vary in amount from one coal field to another and even within the same mining area. Thus, the use of additional chemical additives depends on an analysis of the coal utilized by each user. Second, the type and specifiction of the equipment in which coal–water mixtures are combusted are variable. Further differences must be noted within combustion equipment classes, such as coal-fired boilers compared with oil- and gas-fired boilers. Third, the addition of one or more chemicals to a coal–water mixture raises the cost of the fuel to the user according to the type and amount of each additive. Fuel preparers could also charge the user an additional premium for special or proprietary formulations that may or may not enhance product performance.

Coal–water mixture preparation. Coal–water mixtures are prepared by wet milling, dry grinding and mixing, or a combination of both approaches.

Wet milling, a technique derived from the cement industry, involves the use of ball mills that are suitable for the processing of mineral aggregates in combination with water as a fluidizing medium. Grinding data reveal that this wet milling approach results in a 25% reduction in horsepower compared with dry grinding for the same coal type, throughput requirement, and particle sizing.

In the drying grinding and mixing method, coal is pulverized in a dry state in a variety of equipment types (attrition mill, bowl mill, roller mill, cage mill, ball mill, and so on) and combined with water in a mixer tank.

The combination approach involves a coal–water mixture production process in which there is provision for both wet and dry grinding. This provides a mixture with a measure of micronized (that is, 100% passing 325 mesh) coal particles produced through a dry grinding circuit and blended with the coal produced through a wet milling circuit.

The proper approach to coal–water mixture preparation is dictated by the desired particle-size distribution of the coal and the cost of the plant equipment required to provide a usable fuel.

Handling, storage, and transportation. There are four major considerations in handling, storing, and transporting coal–water mixtures. Coal–water mixtures are suspensions of finely ground, abrasive particles in water; handling and storage system designs must consider equipment erosion, particularly in pumps and piping layouts. For instance, diaphragm-type and progressive-cavity pumps with abrasion-resistant parts are typically used in this application. In piping arrangements, pipe T joints are usually replaced with long-radius elbows; abrasion-resistant flexible hose connections are used where applicable; and larger-diameter, thick-walled piping may be specified in order to slow down material flow velocity and reduce system maintenance.

The tendency of the suspended coal particles to settle to the bottom of storage vessels and transportation equipment requires special equipment design considerations. One or more of several approaches may be specified: Mechanical agitation, an established technology with regard to refluidizing particulate suspensions (for example, paints and chemicals), is available. It involves the use of electric motor–driven, rotating-propeller agitation devices. Compressed-air sparging is a second approach in which, as an alternative to or in connection with mechanical agitation, perforated piping is injected with compressed air for refluidization. Finally, in the pump-around approach to refluidization, heavy-duty pumps are used with a piping system designed to remove coal–water mixtures from the bottom of storage vessels and inject them into the top of the vessels.

Given sufficient time and exposure to air, water will evaporate from a coal–water mixture, leading to an increase in viscosity and, ultimately, caking of the coal particles. Fortunately, this problem can easily be rectified by adding more water in combination with agitating in order to rejuvenate the mixture.

System design must address the tendency toward freezing of coal–water mixtures in cold weather. First, it must be recognized that these mixtures have to be exposed to subfreezing temperatures before there is a real threat of ice formation. Second, mixtures subjected to an agitation process or flow conditions are less prone to freeze than those in a static condition. Third, system design provisions for heat and insulation may be considered adequate if they maintain mixture temperatures in the 50–70°F (10–20°C) range. The type of high-level heat inputs required by residual oils are not required for coal–water mixtures.

Coal–water mixture combustion. The combustion requirements for coal–water mixtures are similar in most respects to the requirements for heavy-oil firing, but there are some important differences.

Storage and pumping circuitry. Each combustion location must have a surge tank adjacent to the burner in order to ensure a constant supply of fuel and to receive the fuel recirculated from the burner. Also, there should be a screening device in the input supply line to the burner to remove oversize material. The fuel pump must be capable of achieving high pressures at the burner nozzle in the range of 100–300 psi (700–2100 kilopascals); and long-radius elbows must be substituted for pipe T joints and abrasion-resistant flexible hose must be used.

Atomization. Two-fluid atomization with either compressed air or steam air as the second fluid is required for coal–water mixtures. As with heavy oils, these mixtures must be converted into a finely divided mist without significant agglomeration of the particles for high-efficiency combustion (that is, greater than 90% carbon conversion). There are several burner nozzle designs that closely resemble those used for heavy oils, except that coal–water mixture nozzles must contain highly abrasion-resistant steel, steel alloy, or ceramic exit ports.

Combustion. Once the mixture has been atomized, the surface moisture on each particle must be quickly evaporated before the coal will burn and release heat. This is accomplished primarily by means of the radiant heat from a fire brick or castable refractory-lined combustion chamber located immediately adjacent to the burner nozzle. Secondarily, evaporation is achieved by heat released from the coal already burning in the combustion flame zone. The heat required for water evaporation is known as evaporative heat loss, which amounts to an approximate 4–8% loss in the heating value of the coal.

Secondary fuels. Users who choose to switch from fuel oil or natural gas to coal–water mixtures will require dual-fuel capability. First, it is prudent for a user to have a backup fuel in case of emergencies. Second, lower-quality coals or those coals with a very low volatile rating may require supplementary or cofiring assistance from fuel oil or natural gas. Third, industrial equipment that normally goes through startup and shutdown procedures requires preheating of the combustion equipment with fuel oil or natural gas prior to the combustion of coal–water mixtures.

Burner controls. A properly designed system, and one that will comply with NFPA safety standards, will include modern computerized burner controls based on a programmable controller or other suitable industrial-grade computer technology. The safety and optimization of the combustion process, including the dual fuel capability, dictate the use of several analog feedback loops (temperature, pressure, flow, and so on) monitored by a computer as well as the control over motors and valves.

Coal–water mixture economics. Industrial users will switch to coal–water mixtures only if it can be proved that they are a viable, low-cost alternate to fuel oil or natural gas. Average energy prices on a per-million-Btu (10^6 Btu $= 1.1 \times 10^9$ joules) equivalent basis in the United States as of January 1983 are shown in the table.

By comparison, cost estimates for coal–water mixtures, which must include the costs for mixture ingredients and fuel preparation range from $3.00 to $4.50 per 10^6 Btu. The user's savings would be in the range of a low of 3% to a high of 63%, depending on the oil or gas product displaced.

Another important consideration for users is the need for conversion equipment, such as baghouses for particulate emission control and coal fly ash–handling equipment. Those industrial users who require this equipment must analyze each combustion site on a case-by-case basis, and generally will reduce their coal–water mixture fuel savings estimates

Average energy prices in January 1983

Fuel	Price per 10^6 Btu*
Industrial steam coal (delivered)	$ 2.18
Industrial natural gas	$ 4.62
Industrial propane	$ 5.51
Industrial residual fuel	$ 5.60
Industrial distillate fuel	$ 8.04
Industrial electricity	$14.01

*10^6 Btu $= 1.1 \times 10^9$ joules.

by approximately 50%. The fuel users who do not need conversion equipment, such as those operating kilns and mineral aggregate dryers, can expect sizable energy savings from coal–water mixtures.

For background information *see* COAL; FOSSIL FUEL in the McGraw-Hill Encyclopedia of Science and Technology.

[WILLIAM K. FOX]

Gardnerella

Gardnerella vaginalis is currently reognized as the etiological agent most frequently responsible for cases of bacterial or "nonspecific" vaginosis. Signs and symptoms of this infection include pruritus (itching) and a malodorous vaginal discharge. *Gardnerella vaginalis* is now considered to be a rare cause of neonatal septicemia, urinary tract infection, and cystitis. Recent research on this bacterium has been directed at either finding new treatment regimens for *G. vaginalis* infections or uncovering the relationship between clinical manifestations of nonspecific vaginosis and *G. vagninalis* colonization (for example, the pathogenic mechanisms).

Treatment. Numerous drugs and treatment regimens have been recommended for the cure of vaginosis caused by *G. vaginalis*. Early recommendations included sulfonamides, tetracyclines, penicillin, or polymyxins by either oral or intravaginal routes. It is now recognized that these antimicrobials usually failed to provide a cure because either *G. vaginalis* in culture is totally resistant to some of these drugs (such as polymyxins) or the bacteria show strain-related resistance (for example, to tetracyclines). Interestingly, although *G. vaginalis* cultures have been shown to be resistant to sulfonamides, many investigators demonstrated clinical cures with this type of antimicrobial agent. This apparent paradox has recently been explained with the reported results from a study that demonstrated inhibition of cultures of some strains of *G. vaginalis* by high levels of sulfonamides (25,000 micrograms per milliliter of sulfacetamide). These high levels are probably reached when triple sulfa vaginal suppositories are used for treatment.

The most effective treatment for *G. vaginosis* at the present time appears to be metronidizole. Recent studies have reported 100% cure rates with this antimicrobial. Irrespective of drug regimen chosen, simultaneous treatment of consorts has been reported to improve cure rates. This finding probably reflects the fact that greater than 50% of sexual consorts of infected women are asymptomatic urethral carriers of *G. vaginalis* and that untreated sexual partners may be a source of reinfection.

Pathogenic mechanisms. Recent reports in the literature might now explain the relationship between *G. vaginalis* infection and the clinical manifestations of nonspecific vaginosis. It has been found that seven amines (methylamine, isobutylamine, putrescine, cadaverine, histamine, tyramine, and phenylamine) were present in vaginal washings of women with clinical symptoms of vaginitis but absent in washings from asymptomatic women. It is postulated that *G. vaginalis* produces high concentrations of pyruvic acid and amino acids which are then decarboxylated by the vaginal anaerobic flora into the corresponding amine. Amines are known to cause skin irritation, erythema, blistering, and increased vascular permeability. Thus, it is conceivable that the presence of these amines might be responsible for epithelial cell shedding and the discharge found in patients with nonspecific vaginosis.

Because vaginal anaerobes are postulated to play a role in the pathogenesis of nonspecific vaginosis, an in-depth study of vaginal fluid has also been undertaken. With the combined use of quantitative anaerobic cultures and gas-liquid chromatography, it has been demonstrated that lactate is the predominant acid in normal vaginal fluid. However, in symptomatic women, lactate was decreased, with a concomitant increase in succinate, acetate, butyrate, and propionate. In addition, streptococci and lactobacilli are the predominant species in vaginal fluid from normal women, while the predominant flora of symptomatic women were *G. vaginalis*, *Bacteroides* species, and *Peptococcus* species. These findings are consistent with the hypothesis that both *G. vaginalis* and anaerobic bacteria are necessary to produce nonspecific vaginosis.

Future developments. Future work on *G. vaginalis* and nonspecific vaginosis no doubt will involve studies on both the epidemiology of infection and new diagnostic methods for use by laboratories and clinicians. Both of these areas require additional work before the pathogenicity of *G. vaginalis* is fully explained.

For background information *see* ANTIBIOTIC; VAGINAL DISORDERS in the McGraw-Hill Encyclopedia of Science and Technology.

[J. R. GREENWOOD]

Bibliography: K. C. S. Chen et al., Amine content of vaginal fluid from untreated and treated patients with nonspecific vaginitis, *J. Clin. Invest.*, 63:828–835, 1979; J. R. Greenwood and M. J. Pickett, Transfer of *Haemophilis vaginalis* (Gardner and Dukes) to a new genus, *Gardnerella*: *G. vaginalis* (Gardner and Dukes) comb. Nov., *Int. J. Systemat. Bacteriol* 30:170–178, 1980; C. A. Spiegel et al., Anaerobic bacteria in nonspecific vaginitis, *New Engl. J. Med.* 303:601–607, 1980.

Garnet

Garnets occur as characteristic minerals in certain rocks. They consist predominantly of oxygen (O), silicon (Si), aluminum (Al), iron (Fe), calcium (Ca), magnesium (Mg), and manganese (Mn), but may also contain titanium (Ti), chromium (Cr), and zirconium (Zr). The garnet structure displays a very dense packing of oxygens with tetrahedral cavities filled by silicon. These SiO_4 tetrahedra share oxygens (corners) with oxygen octahedra containing aluminum, ferric iron (Fe^{3+}), or chromium. The oxygen tetrahedra and octahedra are linked to poly-

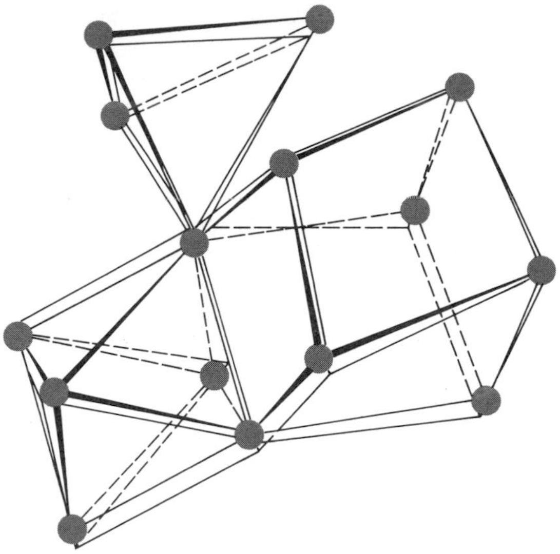

Fig. 1. The portion of the garnet structure showing the three principal oxygen coordination polyhedra for grossular (light lines and broken lines) and hydrogrossular (heavy lines and broken lines). The upper tetrahedron shares a corner with an octahedron which in turn shares an edge with a distorted cube. The circles represent oxygen anions (O^{2-}) in the cube, whose center site (not shown on the drawing) is occupied by calcium, and in the octahedron, whose center is occupied by aluminum. In the tetrahedron, the circles represent $(OH)^-$ anions, and refer to hydrogrossular containing no silicon. Note that the edges for grossular do not interesect at the position of the anions of the hydrogrossular. Their configurations display the principal coordination polyhedra but have different dimensions.

hedra having eight oxygens as corners which can at best be described as distorted cubes. Their center site is occupied by calcium, ferrous iron (Fe^{2+}), or manganese. The three principal coordination polyhedra for specific garnets are shown in Fig. 1. The generalized garnet formula can be written as expression (1), where { } represents eightfold, [ ] sixfold,

$$\{Ca,Mg,Fe^{2+},Mn\}_3[Al,Fe^{3+},Cr]_2(Si)_3O_{12} \qquad (1)$$

and ( ) fourfold oxygen coordinations. Eight cations (for example, 3 Ca, 2 Al, and 3 Si) and 12 anions (oxygen) represent one eighth of the cubic unit-cell having a space symmetry of Ia3d. Certain combinations of cations occur in nature, such as, Ca, Fe^{2+}, Mg, and Mn in eightfold oxygen coordination, and Al, Fe^{3+}, and Cr for the sixfold coordination. This reflects certain physical and chemical conditions of their formation within the Earth's crust and upper mantle.

Role of water. Recent research has elucidated the role played by water in garnet formation. Except for the titanium-bearing andradites and the zirconium garnet kimzeyite with a Si $\rightleftharpoons$ Fe^{3+} and Si $\rightleftharpoons$ Al substitution, respectively, garnets are considered to have silicon in their tetrahedral site according to the generalized garnet formula (1). However, it has been found that silicon content in some garnet systems is less than 3.000 and that considerable amounts of water are present. Experimental work involving the synthetic garnet system has revealed that the presence of a water vapor phase has an effect on the garnet composition. The H_2O vapor interacts with the garnet and alters its composition, so that a substitution of $Si^{4+} \rightleftharpoons 4H^{1+}$ [in effect a structural substitution of $(SiO_4)^{4-} \rightleftharpoons (OH)_4^{4-}$] has taken place. The spatial dimensions of the structure will change and with it the physical properties of the garnet. The silicon–hydrogen exchange can formally be written as reaction (2), in which Me^{2+} stands for

$$\underset{\text{Garnet}}{\{Me^{2+}\}_3[Me^{3+}]_2(Si)_3O_{12}} + \underset{\text{Water}}{6H_2O} \rightleftharpoons \underset{\text{Hydrogarnet}}{\{Me^{2+}\}_3[Me^{3+}]_2(OH)_{12}} + \underset{\text{Quartz}}{3SiO_2} \qquad (2)$$

Ca, Fe^{2+}, Mg, or Mn, and Me^{3+} for Al, Fe^{3+}, or Cr, respectively. A garnet–hydrogarnet solid solution will be formed as a function of bulk chemical composition, temperature, and pressure, yielding reaction (3). Generally speaking, all rock-forming

$$\underset{\text{Garnet}}{\{Me^{2+}\}_3[Me^{3+}]_2(SiO_4)_3} + \underset{\text{Water}}{nH_2O} \rightleftharpoons \underset{\text{Hydrogarnet solid solution}}{\{Me^{2+}\}_3[Me^{3+}]_2(SiO_4)_{3-n/2}(OH)_{2n}} + \underset{\text{Quartz}}{n/2\ SiO_2} \qquad (3)$$

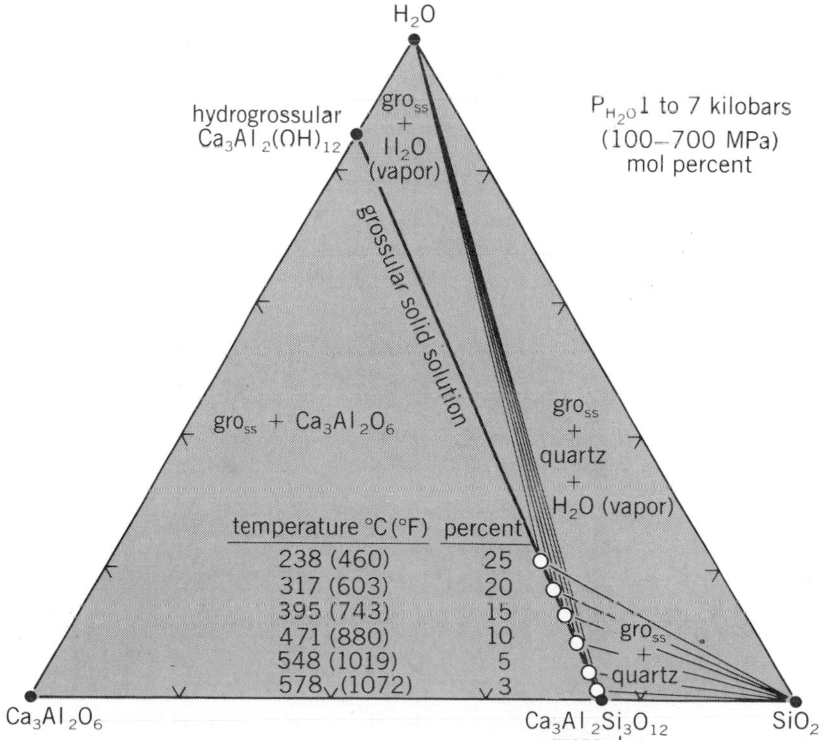

Fig. 2. The H_2O–SiO_2–$Ca_3Al_2O_6$ system with the intermediary grossular solid solution series between grossular ($Ca_3Al_2Si_3O_{12}$) and hydrogrossular ($Ca_3Al_2(OH)_{12}$). Open circles along the grossular solid solution series refer to temperatures at which the garnet composition was determined chemically and physically. Note that only garnet compositions of assemblages of gro_{ss} + quartz + $H_2O_{(vapor)}$ are indicative of temperature and can be used for thermometry, regression equation (4) in text. Assemblages of gro_{ss} + quartz and of gro_{ss} + $H_2O_{(vapor)}$ are not temperature-dependent. gro_{ss} = grossular–hydrogrossular solid solution. Not shown on the diagram is the solution of SiO_2 in H_2O. Numbers refer to mol % hydrogrossular in grossular at a given temperature.

garnets (grossular, $Ca_3Al_2Si_3O_{12}$; andradite, $Ca_3Fe_2^{3+}Si_3O_{12}$; spessartine, $Mn_3Al_2Si_3O_{12}$; almandine, $Fe_3^{3+}Al_2Si_3O_{12}$; uvarovite, $Ca_3Cr_2Si_3O_{12}$; and also the Ti-bearing andradite, $Ca_3Ti^{4+}Fe_2^{3+}Si_2O_{12}$, as well as pyrope, $Mg_3Al_2Si_3O_{12}$) appear to react that way. The result of the hydration is an increase of the unit-cell volume (or unit-cell parameter) accompanied by a decrease of the refractive index and density as a function of the hydrogarnet content.

The hydration effect can best be demonstrated by the Ca-Al garnet grossular and the Ca-Fe^{3+} garnet andradite. These are the end members of the grandite solid solution series and are characteristic minerals of impure limestones which have undergone metamorphism, particularly in aureoles around mafic intrusions. Depending upon the temperature (and pressure) of their formation, they may incorporate a hydrogarnet component in small-to-moderate amounts. Plazolite (= hibbschite) is a grossular-rich hydrogarnet solid solution which exhibits the maximum substitution, 1 $(SiO_4)^{4-}$ by 1 $(OH)_4^{4+}$, ever found in garnets of natural rocks. Members of the grossular ($Ca_3Al_2Si_3O_{12}$)–hydrogrossular ($Ca_3Al_2(OH)_{12}$) have been synthesized in the laboratory, and the phase relationships between H_2O, grossular, and hydrogrossular were studied experimentally. Although there is complete solid solution between grossular and hydrogrossular, the solid solution is incomplete at any given temperature and pressure.

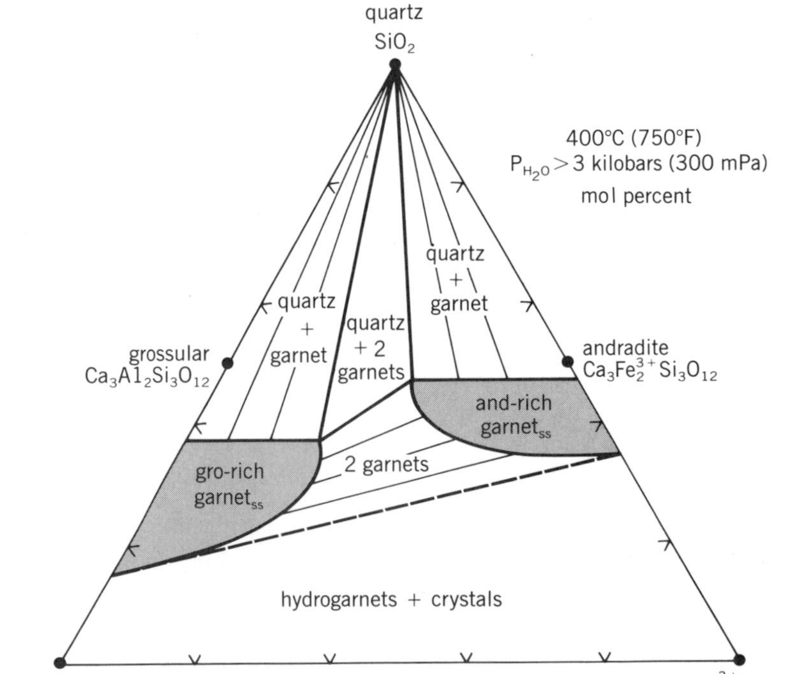

Fig. 3. The grossular ($Ca_3Al_2Si_3O_{12}$)–andradite ($Ca_3Fe_2^{3+}Si_3O_{12}$)–hydrogrossular ($Ca_3Al_2(OH)_{12}$)–hydroandradite ($Ca_3Fe_2^{3+}(OH)_{12}$) quadrilateral including quartz (SiO_2) as a stable phase. The quadrilateral displays phase relations at 750°F (400°C) and H_2O pressures of less than 3 kilobars (300 MPa). Solid circles for garnets refer to end-member compositions and are not stable as phases. At the given temperature of 750°F (400°C) and at pressures of 3 kilobars (300 MPa) and above, the grossular-rich (gro-rich) garnet solid solution series and the androdite-rich (and-rich) garnet solid solution series cover two distinct portions of the quadrilateral (shaded areas). The composition of coexisting garnets in the 2-garnet assemblage (solvus), and in the two different garnet (gro-rich) + quartz assemblages are given by their connecting tie lines (light lines). Compositions of the garnets in the quartz + 2-garnet assemblage are indicative of the temperature and pressure (750°F or 400°C at 3 kilobars or 300 MPa and above). These compositions will change as a function of temperature (and pressure for the and-rich portion) and can be applied to garnet thermometry. Note that all assemblages are coexistent with a H_2O vapor phase (not shown on the diagram).

At a given pressure and temperature, only a certain portion of the solid solution series is formed. High temperature favors grossular whereas low temperature favors that of hydrogrossular component in the garnet. The unit-cell volume changes from 1.6619 nm^3 for grossular to 1.9885 nm^3 for hydrogrossular, caused by an expansion of the oxygen polyhedra as shown in Fig. 1. The refractive index and the specific gravity decrease from 1.733 and 3.60 respectively for grossular to 1.605 and 2.52, for hydrogrossular. The silicon–hydrogen exchange reaction (3) equilibrates the garnet in the presence of excess H_2O vapor to a specific grossular–hydrogrossular solid solution whose composition is solely a function of temperature in the pressure range of 1 to 7 kilobars (100 to 700 megapascals). Thus, the amount of the hydrogrossular component incorporated in the grossular is indicative of the temperature range at which the garnet was formed.

Thermometry. If the hydrogrossular content of the grossular can be determined chemically or physically, the temperature can be obtained from the mean of regression equation (4), where X is given in mol %.

$$T(°C) = 623.2 - 14.988\, X_{\text{hydrogrossular}} - 0.017\, X^2_{\text{hydrogrossular}} \pm 10°C \quad (4)$$

Grossular–hydrogrossular solid solution thermometry is illustrated in Fig. 2 for the chemically pure system H_2O–SiO_2–$Ca_3Al_2O_6$ with the grossular–hydrogrossular solid solution series as intermediary phase.

Solid solutions. Grossular–hydrogrossular solid solutions consisting only of silicon, aluminum, calcium, and water very rarely occur in nature. The most important associated element is (ferric) iron, and the solid solution of grossular and andradite with its waterbearing end members is more characteristic for rocks. The incorporation of a hydroandradite component, $Ca_3Fe_2^{3+}(OH)_{12}$, in adradite, $Ca_3Fe_2^{3+}Si_3O_{12}$, is dependent on temperature, pressure, and oxygen fugacity of the chemical system. The latter must maintain the iron in the trivalent state. In contrast with the grossular–hydrogrossular solid solution, which is nearly pressure-independent, the incorporation of hydroandradite at a given temperature is more extensive at pressures of 1–2 kilobars (100–200 MPa) than at pressures of 3 kilobars (300 MPa) and above. Andradite and hydroandradite do not form a complete solid solution series. At a temperature of 400°C (750°F) and a pressure of less than 3 kilobars (300 MPa), the solid solution series is restricted to the portion between andradite$_{95}$hydroandradite$_{05}$ and andradite$_{70}$hydroandradite$_{30}$. Incomplete solid solutions between grossular, andradite, hydrogrossular, and hydroandradite occur in nature and are very common in mafic igneous rocks metamorphosed in an anhydrous environment at low temperature (less than 400°C or 750°F) and at high pressure (less than 5 kilobars or 500 MPa). The solid solution, however, is not complete. Because of the differences in the nature of hydration of grossular and andradite, a solvus is formed within the grossular–andradite–hydrogrossular–hydroandradite quadrilateral, separating a grossular-rich garnet solid solution from an andradite-rich garnet solid solution. The range of the garnet solid solution is shown for 400°C (750°F) and pressures of less than 3 kilobars (300 MPa) in Fig. 3. If the amount of the hydrogarnet component present in these two garnets in coexistence with quartz and excess H_2O as vapor can be determined chemically or physically, garnet–hydrogarnet thermometry can be applied.

For background information *see* GARNET; GEOLOGIC THERMOMETRY; MINERALOGY; SOLID SOLUTION in the McGraw-Hill Encyclopedia of Science and Technology.

[HANS G. HUCKENHOLZ]

Bibliography. E. P. Flint, H. F. McMurdie, and L. S. Wells, *J. Res. Nat. Bur. Stand.*, 26:13–33, 1941; H. Bartl, *N. Jb. Miner. Mh.*, pp. 404–413, 1969; H. G. Huckenholz and K. T. Fehr, *N. Jb. Miner. Abh.*, 145:1–33, 1982; P. Ribbe (ed.), *Review in Mineralogy*, vol. 5: *Orthosilicates*, 2d ed., Mineralogical Society of America, 1982.

Genetics

One of the major tasks of the evolutionist is to explain the detailed manner in which new species are formed. No biological individual exists alone; each is part of a larger community of individuals that share in a common pool of genes. These populations of individuals that are recognized as species must have arisen as a genetically altered, descendant subdivision of a preexisting, ancestral species. Extraordinary new techniques are now available for measuring genetic variability in populations, and application of these techniques to the population genetics of the speciation problem have led to significant theoretical reassessments. One of these relates the origin of species to a shift in the genetic basis of sexual behavior.

By almost any criterion, the production of progeny is the crucial event in the life of an organism. Failure to reproduce signals the end of many hundreds of millions of generations of prior success by ancestors. With the death of each unproductive individual, a line of descent reaching back to the very origin of life is finally terminated. Accordingly, then, in each generation any genetic variant that enhances successful reproduction will be naturally selected.

Organisms that reproduce sexually are abundantly represented in the existing biosphere, a fact that attests to the evolutionary success of the lineages that utilize this system. In most cases, the ultimate union of the gametes comes only after the interaction of a series of complex behavioral or comparable physiological forces that are an integral part of the genetic endowment of the species. These factors are equally as important in evolution as those that determine the developmental fate of the zygote after it is formed.

Natural selection for sexual behavior. As a number of modern evolutionists have stressed, natural selection operates not so much through simple survival as through the capacity of certain members of a species to outbreed others and thus to differentially perpetuate the genetic variations that they carry. This means that natural selection will tend to continually maximize all those processes and behaviors that contribute to the bringing of the sexes together and ultimately the uniting of the gametes. Over time, these forces come to be built into an all-pervading reproductive drive that characterizes each sexually reproducing species.

Viewed in this light, sexual interaction and behavior can be seen as the result of an intimate and highly specific genetic coadaptation of the sexes. Thus, each sex not only has special characteristic structures but also employs specific sensory cues that facilitate the meeting of the sexes.

Sexual selection. In a significant number of cases, sexual attraction and behavior serves a larger function in reproduction than the mere conjunction of one male and one female drawn at random from the breeding population. Darwin was the first to emphasize the process called sexual selection, in which the high fitness of both partners is assured. There are two types of sexual selection, generally referred to as intrasexual and epigamic. Intrasexual selection involves particularly males. Some individuals have behavior and armaments that confer on their carriers the "power to conquer other males in battle," as Darwin put it. Still other hereditary characters confer on their male carriers the "power to charm the female." Thus, in epigamic selection, the female becomes an especially active selective agent whose mating preferences impinge on male characters. Essentially she assumes the role of making a choice from among a field of genetically variable males of her own species. The results of these kinds of selection may be seen in the spectacular displays of male territoriality in such colonial animals as seals or in the extraordinary male sexual displays of the birds of paradise. In these birds selection has in some cases become "runaway," producing elaborate results, driving the male characters to such extremes as to make the male increasingly vulnerable to predation. The outcome appears to be a balance of these opposing forces.

Recently, both the biological function and the genetics of sexual selection have been studied in several insects. Certain sexually dimorphic species of *Drosophila* flies, such as those endemic to the Hawaiian Islands, have been used. (*Drosophila* provides an opportunity for detailed genetic and behavioral analyses under controlled laboratory conditions.)

Sexual behavior and species formation. Considerable discussion has recently centered on genetic changes that may occur when a new population is founded in a previously unoccupied area or niche. In certain cases, the number of "founders" of such a population may be very small. They are likely to carry to the new population an incomplete or unbalanced portion of the gene pool of the ancestral species. It is these newly founded populations that sometimes show characteristics suggesting that they may be on the genetic path leading to the formation of a new species.

Most large (ancestral) populations have a complex and balanced genetic system that tends to support extensive polygenic polymorphism. Among these polymorphisms are those underlying the sexual selection systems mentioned above. When a population is founded from one or a few individuals, the new gene pool formed from these individuals may undergo disorganization, shifting the genetic basis of sexual selection. As the new population is built up, the theory suggests that a novel round of sexual selection is instituted. The sexual system of signals and responses may thus shift and be reorganized around a novel mode.

Especially when the system depends on female choice, the adjustment forged in the new population may be built into a unique system. Under such circumstances, the new system may show an incipient premating isolation from other such groups.

The appearance of sexual isolation may be merely one of the incidental outcomes of sexual selection. Reproductive isolation does not have to be the direct result of an ad hoc selective process. The scheme seems to be applicable to the exuberantly speciated Hawaiian *Drosophila*. In these flies, shifts appear to particularly involve characters related to sexual selection. This disorganization-reorganization hypothesis of species formation, however, may be much more widely applicable among sexually reproducing plants and animals, especially those that have other complex balanced adaptive systems that are susceptible to disorganization when a population is temporarily reduced to a very small number of individuals.

For background information *see* POLYMORPHISM (GENETICS); POPULATION GENETICS; REPRODUCTIVE BEHAVIOR; SPECIATION; SPECIES CONCEPT.

[HAMPTON L. CARSON]

Bibliography: J. Alcock, *Animal Behavior*, 2d ed., 1979; H. L. Carson, Speciation as a maor reorganization of polygenic balances, in C. Barigozzi (ed.), *Mechanisms of Speciation*, 1982; R. Lande, Models of speciation by sexual selection of polygenic traits, *Proc. Nat. Acad. Sci. USA*, 78:3721–3725, 1981; A. R. Templeton, The theory of speciation via the founder principle, *Genetics*, 94: 1011–1038, 1980.

Geotectonic imagery

The *Seasat* satellite was launched in June 1978 and suffered a total power failure in early October of the same year. Among the instruments on board *Seasat*, which orbited at an altitude of 480 mi (800 km), was a radar altimeter capable of measuring the shape of the ocean surface to a precision of ±0.2 ft (±.05 m). Although its lifetime was brief, *Seasat* obtained continuous, precise measurements of the ocean surface along nearly 1000 suborbital paths uniformly distributed between latitudes 72°S and 72°N (Fig. 1). The accuracy of the measurements ensured the recovery of information on gravitational field perturbations at spatial scales of several tens of miles. Analysis and reduction of this data set has led to the development of geotectonic imagery (GTI) or maps depicting gravity field variations over the oceans. Geotectonic imagery provides an ideal means for assessing visually the information obtained by *Seasat* and promises to advance significantly understanding of the nature of the sea floor.

Examination of geotectonic imagery has already led to the identification of a large number of previously unmapped features of the sea floor. A catalog of discoveries includes seamounts, fracture zones, ridges, and peculiar zones of regional oceanic crustal deformation. The newly discovered features come in a variety of shapes and sizes, from seamounts 30 mi (50 km) wide and 300 mi (500 km) high to mountains which rise 13,000–16,000 ft (4000–5000 m) above the surrounding sea floor to fracture zones which can be traced for 600 mi (1000 km) or more. Analysis of geotectonic imagery will lead eventually to improved maps of the sea floor, more accurate knowledge of the history of the relative movement of lithospheric plates (for example, the relative positions of continents through geologic time), and a better general understanding of the physical processes which govern the formation and evolution of the oceanic crust and lithosphere.

Satellite altimetry. The physics of gravitational fields requires the existence of a nonuniform gravity field where mass is nonuniformly distributed. Mountains and valleys, therefore, cause small perturbations from the spherically symmetric gravity field of the Earth, as do variations in the distribution of mass within the Earth. Over the past two centuries, these perturbations have been measured by a variety of instruments. The latest development in the effort to measure terrestrial gravity was the introduction of the radar altimeter to the satellite program in the 1970s.

A radar altimeter measures the altitude of an orbiting satellite by bouncing a microwave pulse off the Earth's surface. Precise determination of the position of the satellite permits these measurements to be transformed into elevations relative to a fixed reference surface. The most advanced altimeter manufactured to date was among the instruments carried by *Seasat*. Each of the altitude measurements over the oceans obtained by *Seasat* represented an area several miles in diameter and was accurate to about ±0.2 ft (±.05 m) (Fig. 2). Measurements of this quality are of considerable interest to marine geodesists and geophysicists because they provide such detailed information about the marine gravity field.

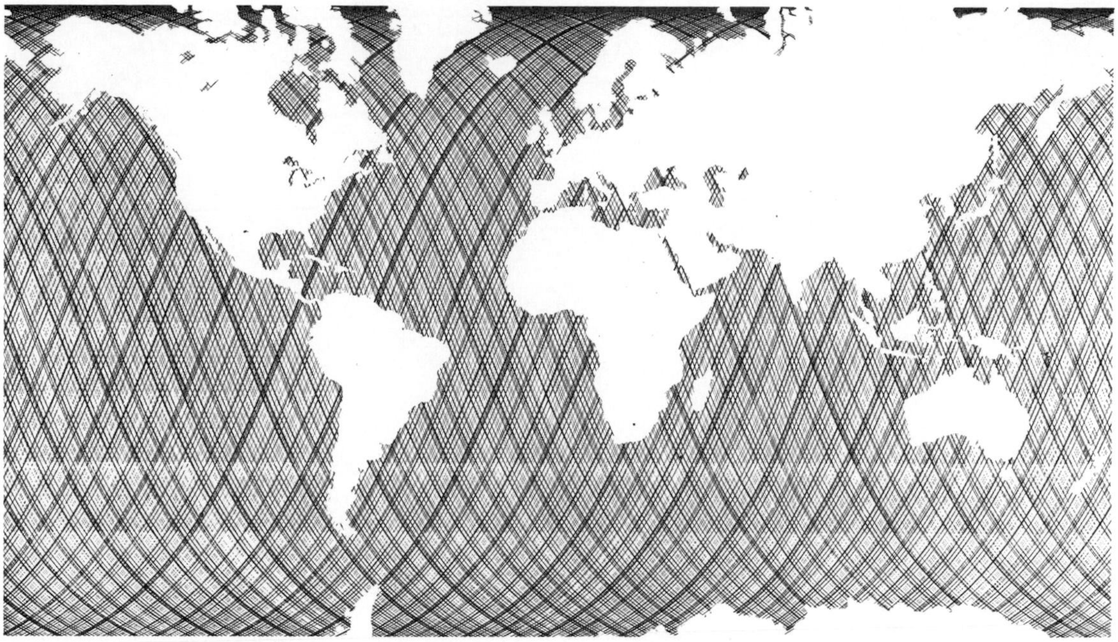

Fig. 1. Suborbital satellite tracks along which *Seasat* obtained radar altimeter measurements of the sea surface.

To a good approximation, the ocean surface is a level surface; that is, the surface is everywhere perpendicular to the gravitational force vector. In the vicinity of a massive object, the force vector will show an anomalous component pointing toward the object, causing a rise in sea surface toward the object. The magnitude of an object's influence on the sea surface is proportional to its mass; a large seamount, for example, can cause an anomalous swell of 16–33 ft (5–10 m) in the ocean surface. The accuracy of the *Seasat* radar altimeter was sufficient to recover the 0.7-ft (0.2-m) swell that would be associated with a seamount 30 mi (50 km) wide and 1600 ft (500 m) high. In remote areas, where the distribution of surveys by research vessels can be very sparse, altimeter data may be the only available source of information on the sea floor.

Seasat operated for less than 3 months during the late summer and early fall of 1978. Had it completed its planned mission, it would have sampled the sea surface along a dense grid of orbital paths with adjacent tracks separated by less than 12 mi (20 km). Its premature demise prevented the achievement of this goal, leaving tracts of unsampled sea floor up to 120 mi (200 km) wide. Therefore, a discrepancy exists between the resolution permitted by the instruments' accuracy and the sampling distribution. The analytic procedures which eventually led to geotectonic imagery were developed with the intention of preserving as much as possible of the detail present in the altimeter data while preparing a two-dimensional representation of the information.

Data processing. In order to fully exploit the information provided by the *Seasat* altimeter, it was necessary to correlate and combine the individual measurements obtained along separate orbits. A two-dimensional representation based on the entire data set was produced. The most profitable way to accomplish this was to estimate sea-surface elevations at the intersections of a regular grid. By doing this, the spatial characteristics of the data could be visually perceived and digitally analyzed. Had *Seasat* fulfilled its mission, the gridding procedure

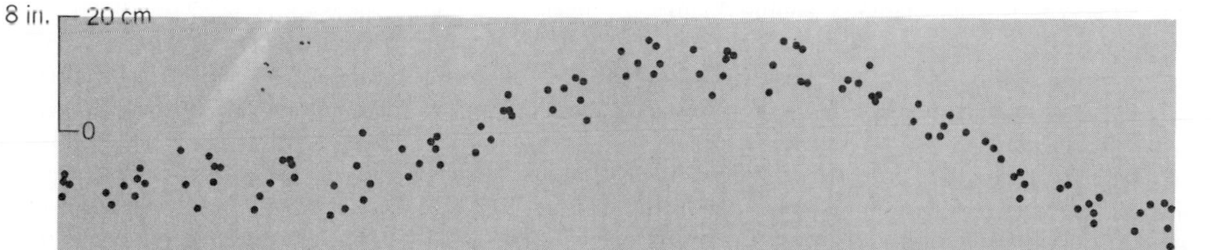

Fig. 2. Radar altimeter measurements at points along seven coincident *Seasat* tracks crossing the Reykjanes Ridge (south of Iceland).

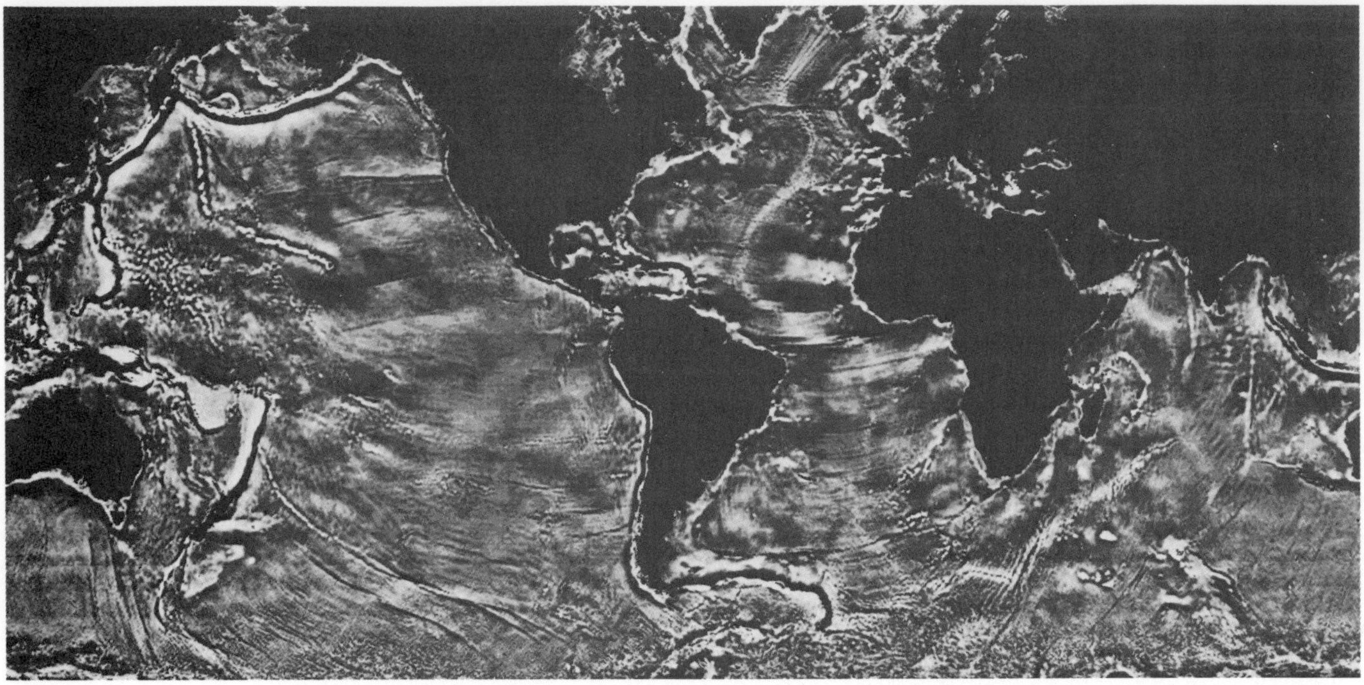

Fig. 3. Geotectonic image of the ocean basins between latitudes 67°S and 67°N. Dark shades represent areas of low gravitational pull and light shades represent gravity highs. (*Courtesy of W. F. Haxby*)

could have been easily accomplished, since each grid point would never be further from measurement points than the intrinsic spatial resolution of the data. However, gridding the existing data set required estimating sea-surface elevation at points separated from the nearest *Seasat* pass by as much as 60 mi (100 km). The technique used to accomplish the gridding took advantage of the fact that sea-floor physiography is dominated by topographic features which are strongly lineated. Accordingly, the data obtained along adjacent *Seasat* orbits were examined to identify prominent trends through a particular area. If a trend could be established, the interpolation of data between tracks would be along that trend. The grid points at which sea-surface elevations were assigned were separated by about 12 mi (20 km) at the Equator and were closer at high latitudes, where *Seasat* tracks were more closely spaced.

The final step in processing the data was the transformation of sea-surface elevations, which are directly related to the gravitational potential, into estimates of gravitational force, which is the radial derivative of the gravitational potential. This computation has the effect of enhancing gravity field perturbations at small spatial scales and subduing the broad regional variations in the field. The computation of force perturbations (or gravity anomalies) was also desirable because the result is a quantity routinely measured by surface research ships and therefore verifiable.

Once the altimeter data had plotted on a grid and transformed, all that remained was to display the result. Modern computer graphic techniques permit a range of display possibilities which are likely to revolutionize conventional cartography. These techniques enable the display of topographic maps that look like photographs of landscapes with illumination possible from any direction. Besides providing an esthetically pleasing display which even a casual observer can appreciate immediately (not possible with conventional contour maps), an advanced graphics capability can be exploited to display far more information than conventional maps. The use of shading, for example, can enhance subtle textural information which would otherwise be lost. The image displayed in Fig. 3 depicts gravity anomalies by assigning a shade of gray to a range of anomaly values. Dark shades represent gravity lows, and light shades gravity highs. Seamounts and ridges therefore appear white, whereas deep sea trenches appear black.

Most of the features which appear in the geotectonic image in Fig. 3 are associated with bathymetric features of the sea floor. However, mass inhomogeneities beneath the surface also influence the gravity field and therefore appear in the image. Thus, because midocean ridges are underlain by a zone of low-density upper-mantle material, their influence on the gravity field is lessened. The East Pacific Rise, the dominant bathymetric feature in the southern and eastern equatorial Pacific Ocean, is 9800 ft (3000 m) in elevation and several thousand miles wide, but is nearly invisible. On the other hand, some features in the image, such as the texture depicted over continental shelves, are due to buried geologic structures which show no surface expression whatever. Some of the features in the map are the result of components of the sea surface which are not due to gravity perturbations; some of the structure in the western North Atlantic Ocean adjacent to the North American continental margin is due to the presence of the Gulf Stream.

Discoveries. Of all the ocean basins surveyed by *Seasat*, the southern Indian Ocean is perhaps the least surveyed by ship. Hundreds of miles may separate ship surveys in this remote region. Although geotectonic imagery exposes new features in all the oceans, the map of this area reveals more in terms of size and quantity than any other. Due south of Madagascar, at a latitude very near that which intersects the southern tip of South America, is a group of very large seamounts, thousands of feet in elevation and over 60 mi (100 km) in diameter. Four of these massive seamounts, forming most of the eastern part of the group, had escaped detection until finally mapped by geotectonic imagery. To the east of this seamount group, a linear gash trending northeast–southwest and extending more than 600 mi (1000 km) across the sea floor reveals a previously unmapped fracture zone which formed as the Indian subcontinent separated from Antarctica during the breakup of Gondwanaland. Still farther to the east, a ridge extending southeast from the Kerguelen Plateau is newly exposed. On bathymetric maps, this ridge is manifested as a single isolated seamount. These features, and many others which have been identified all over the globe, help in assessing the level of volcanic activity in the ocean basins, provide information on the history of continental drift, give clues as to the dynamical nature of the Earth's interior, and contribute to the understanding of the process of crustal deformation.

In the future, the radar altimeter will figure prominently in several planned satellite missions. The United States Navy will launch *Geosat* in the fall of 1984. The first 18 months of *Geosat* will yield altimeter measurements along the dense grid of orbital tracks which *Seasat* was meant to provide. Later in this decade NASA's Topex (topography experiment) mission will measure the ocean surface with more than double the accuracy of *Seasat* and will provide unprecedented information on global ocean dynamics (currents and tides). These and future missions as yet unplanned promise to unlock further the secrets of the oceans and the Earth beneath.

For background information *see* CARTOGRAPHY; GEODESY; GEOPHYSICAL EXPLORATION; RADAR; SCIENTIFIC SATELLITES, in the McGraw-Hill Encyclopedia of Science and Technology.

[W. F. HAXBY]

Geothermal power

Hydrothermal, or "natural," geothermal reservoirs are found where groundwater has contacted rock formations heated by sources lodged at shallow depths in the Earth's crust. In a few of these locations, large circulating groundwater convection systems have formed within porous, fractured rocks at depth. The hydrothermal reservoirs, usually capped by an impervious layer, are characterized by relatively high heat flux at the surface. They are sometimes associated with leakage of water via hot springs, geysers, and fumaroles, but also may be "blind" resources having none of these surface indicators. These rare hydrothermal systems, once discovered, drilled, and exploited, are excellent sources for power generation and direct heat uses. The current growth rate of geothermally fueled electric power capacity is approaching 15% per year worldwide. The obvious value of such developments has motivated research into advanced geothermal energy extraction techniques. There are several active projects at present.

Hot dry rock energy. In contrast, many areas of the Earth's surface have above-average heat flow and are underlain by rocks that are heated by near-surface heat sources; but they have insignificant quantities of water and exceedingly low permeability. These areas of subsurface hot dry rock have excellent potential for development as hydrothermal reservoirs by artificially increasing permeability and the heat transfer area in the subsurface.

Extraction systems. The extraction of usable energy from a hot dry rock reservoir can be accomplished by fracturing hot, essentially dry rock between two deep drill holes with water pressure, and then circulating water through the wells and fractured reservoir in a closed circulation loop. This technology has been successfully demonstrated at Fenton Hill, New Mexico, by the Los Alamos National Laboratory.

The energy extraction concept of the Los Alamos Hot Dry Rock Geothermal Program, as first constructed, consisted of drilling a test system of two holes at Fenton Hill (see illustration) to a depth of approximately 10,000 ft (3 km) where the temperature is 392°F (200°C), then connecting the boreholes with a large-area vertical fracture zone. Water circulates down one borehole, is heated by the hot rock, and rises up the second borehole to the surface, where the heat is extracted and the cooled water is reinjected into the underground circulation loop. This system operated for a cumulative 416 days during engineering and reservoir testing experiments. An energy equivalent of 3–5 MW of thermal power was produced without adverse environmental problems. During one test, an electric generator installed in the circulation loop produced 60 kW (electrical).

A second-generation engineering evaluation system, recently drilled to 15,000 ft (4.5 km) at Fenton Hill, entails creating multiple fracture zones between a pair of inclined boreholes (see illustration). The higher, 608°F (320°C) temperature and larger reservoir size of this advanced system should produce 5–10 MW (electrical) for 20 years and be commercially feasible.

Applications. Depending on the location and depth of future hot dry rock reservoirs, the extracted energy may be either high-temperature (for generation of electricity) or low-temperature (for

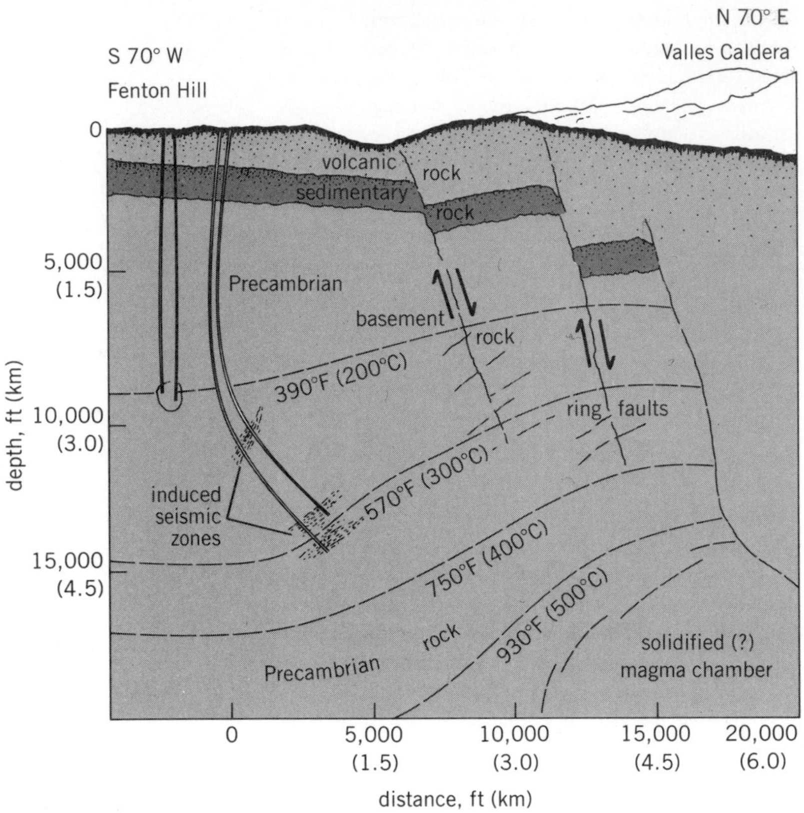

Schematic cross section of Hot Dry Rock (HDR) Geothermal Test Site at Fenton Hill, New Mexico, showing relative positions of the two energy extraction systems.

such direct uses as space heating and food processing). The circulating hot water can also be used to augment energy production from other energy systems, for such diverse uses as boiler feedwater preheat, process heat for petroleum refining, or stimulating bacterial growth in cold climates as might be required for more rapid digestion in sewage treatment plants or landfill dumps.

Conventional geohydrothermal resources have been successfully integrated into existing energy supply networks in many places throughout the world. There are other aspects of geothermal integration with other alternative energy sources to complement and increase the usefulness of each. When the hot dry rock technology of drilling and fracturing in crystalline rock is coupled with solar energy production, excess summertime heat from solar collection facilities can be transferred and stored in artificial underground reservoirs for wintertime withdrawal utilization. The same technology can provide huge, but easily accessible, heat sinks for reject industrial heat, thus creating many options for industry, municipalities, and district heating organizations to integrate energy demands with heat disposal requirements.

Forced geoheat recovery. Heat extraction systems that use subsurface geologic structures in near-surface regions of the Earth, where the temperature increases downward at an average rate (normal geothermal gradient), have been termed forced geoheat recovery resources. Potential reservoirs are: permeable sediments in deep basins; fracture zones with sufficient natural permeability; formation contact areas, such as lava beds within extensive basalt flows; and some regions below volcanoes. The normal to slightly active regions of the Earth's crust create in-place temperatures of 176–302°F (80–150°C), at depths of 3000–7000 ft (1–2 km). These resources are adequate for many direct energy uses, and at the upper end can be expected to produce electric power through the use of special low-temperature power plants. The space heating applications in the Paris Basin are examples of forced geoheat extraction. Heat extraction, fluid flow, and economic analyses have been performed for typical subsurface reservoir models. The results show that such systems require pumping and can be economically attractive in near-normal thermal gradient reservoirs that can produce a few megawatts of thermal power, that is, can supply a space heating load equivalent to that required for a town of a few thousand population.

Magma energy. Nearly every volcanic eruption involves large amounts of thermal energy. After eruptions, slowly cooling molten or near-molten rock remains below the surface as pod-shaped magma bodies, in volcanic conduits, and as networks of sheetlike dikes and sills. A 7-year study of the feasibility of extracting energy from magma bodies was funded by the U.S. Department of Energy.

It has been estimated that the magma energy resource of the United States is greater than 50,000 Quads (1 Quad = 10^{15} British thermal units). Locations of these resources are readily identified with existing geological and geophysical prospecting techniques. Multiple prospecting techniques were used during a 6-year study of the near-surface Kilauea Iki lava lake in Hawaii. A combination of seismic, gravity, magnetic, and geologic techniques is most useful for identification of drilling sites.

Small-scale drilling experiments in the Kilauea Iki lake indicate that drilling into rock with temperatures as high as 1870°F (1020°C) is possible with water-cooled drill bits and insulated drilling strings. Deeper, production-size holes for energy production have not yet been tested.

Laboratory experiments involving melted rock and single-tube boilers and computer models indicate that useful energy extraction rates can be achieved with closed heat exchangers in lower-viscosity magmas (basalts, andesites) and open heat exchangers in magmas of high viscosity; the open heat exchanger is 10–100 times more effective than closed heat exchanger systems.

Extraction of energy from magma is scientifically possible. However, problems concerning the hazards of power plants located on active or dormant volcanoes have not been addressed.

For background information *see* COMFORT HEAT-

ING; DISTRICT HEATING; ELECTRIC POWER GENERATION; GEOTHERMAL POWER; MAGMA in the McGraw-Hill Encyclopedia of Science and Technology.

[JOHN ROWLEY; ROLAND PETTITT; GRANT HEIKEN]

Bibliography: G. Bodvarsson and G. M. Reistad, Forced geoheat recovery for moderate temperature uses, *J. Volcanol. Geotherm. Res.*, 15:247–267, 1983; J. L. Colp, *Final Report: Magma Energy Research Project*, Sandia Nat. Lab. Rep. SAND 82–2377, 1982; Z. V. Dash et al., Hot dry rock geothermal reservoir testing: 1978 to 1980, *J. Volcanol. Geotherm. Res.*, 15:59–99, 1983; J. F. Hermance and J. L. Colp, Kilauea Iki lava lake: geophysical constraints on its present (1980) physical state, *J. Volcanol. Geotherm. Res.*, 13:31–61, 1982.

Goat

A small ruminant mammal, widely distributed around the globe under greatly differing climatic conditions. It is especially adapted to dry and mountainous regions, and productive even where cattle and sheep have difficulty surviving. The domesticated goat is closely related and similar in size to sheep, but has many anatomical and physiological differences. It is an important provider of meat, milk, and fiber (called mohair and cashmere), income, fertilizer, and energy (draft and heating). Domesticated goats of the major breeds have been developed in, and are mainly derived from, five geographical areas: Swiss mountains (for milk); Indian, Arabian, and northeast African drylands (for meat and milk); west African lowland (for meat and milk); south African prairie (for meat); Turkish highland (for mohair production).

Accompanying colonizing immigrants and released from the live-food supplies of early merchant ships, goats became established in the Americas, and today make up a small (compared to cattle, swine, poultry, and sheep) but increasingly visible and important part of agriculture in the United States. Mohair production from Angora goats is of economic significance in Texas and ranks first in the world, followed by that of Turkey. Dairy goats number about 1 million in the United States, with more than half in California. Six dairy goat breeds are recognized in the United States: Alpine, LaMancha, Nubian, Oberhasli, Saanen, and Toggenburg.

Breeds. Alpine goats are of Swiss origin. They have many colors, especially faded shades of white into black, a straight face, erect ears, and sickle-shaped horns, when present. Alpine does (mature females) may weigh at least 135 lb (60 kg), measure 30 in. (75 cm) at the withers (top of shoulder), and are (with Saanen and Nubian breeds) the biggest and heaviest dairy goats. Alpine goats rank second in milk-producing ability, behind the Saanen breed.

LaMancha goats are of Spanish origin but were developed in California. They have many colors, a straight face, but characteristically no external ears. They are smaller than Alpine goats and produce less milk, but their offspring are fleshier. They are very calm goats, and persistent milk producers in the winter when the other breeds often dry up.

Nubian goats are mostly of Indian origin, but were developed in England. They too have many colors, but characteristically arched faces and long pendulous ears. Nubian goats produce less milk than the Swiss breeds, but their milkfat level is the highest. They are the most popular in the United States and are the most heat-tolerant. They are also more fleshy. Horns in Nubian, as in LaMancha, goats are rarely seen. Although anatomically present, they are routinely and permanently removed in early age to facilitate animal handling. Horns in all Swiss breeds are of the upright sickle shape, while Indian, Spanish, and Turkish breeds have more sideways-growing horns with varying degrees of spiraling. Goats can be selected for the genetic trait of hornlessness, but this is only partially advisable since the trait is genetically linked to sterility, and a certain percentage of intersex or hermaphrodite offspring can be expected when both parents carry the hornless trait.

Oberhasli goats are of Swiss origin. They are related to Alpine goats and similar to them, except that they are often colored solid red or black. They are smaller and lower in milk production than Alpine goats, but very popular in Europe where they are well adapted to high mountain grazing and long-distance traveling. They have also been called Chamoisee, Brienz, or Swiss Alpine goats in contrast to the French Alpine and British Alpine breeds which are today just called Alpine goats.

Saanen goats (Fig. 1) are of Swiss origin. They are only white and have straight faces and erect ears. They are rated as the top milk producers around the world and have been used for the upgrading of many native breeds in underdeveloped countries. All dairy goat breeds in the United States are generally short-haired, although some Swiss strains may have partially long hair which is of some advantage to them in alpine climates.

Toggenburg goats are also of Swiss origin. They are very popular in the United States and rank in milk-producing ability next to the Alpine and Saanen goats. They have only brown color with characteristic white facial markings and white ear and leg

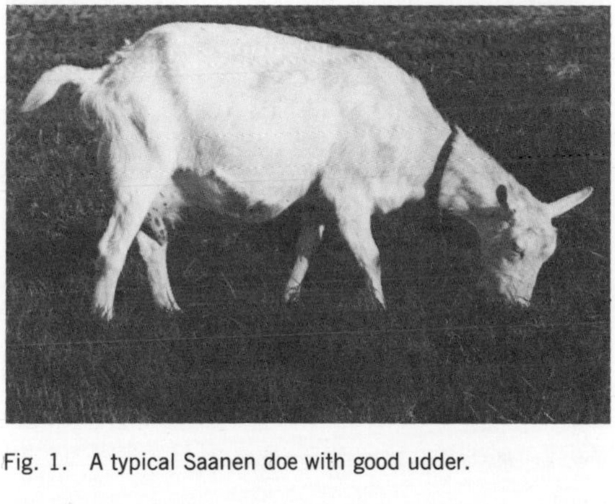

Fig. 1. A typical Saanen doe with good udder.

stripes. They are the smallest of the Swiss breeds, approximately 120 lb (54 kg) for mature does and 26 in. (65 cm) height at withers. They have straight faces and erect ears.

In addition to the six American dairy breeds above (and the Angora), there are also two major goat populations in the United States for mixed purposes, mostly meat, brush clearance, pets, or laboratory use: the Spanish goat, mainly kept on open range land, and the West African Pygmy.

Milk. Milk is the main product of dairy goats in the United States (Fig. 2). There are two manufacturing plants for dry goat milk, one in California and one in Arkansas. Interest, mostly of limited local scale, is increasing in goat cheese production: soft curd (similar to cottage-ricotta), cheddar, and feta. Impetus comes from European and Near East traditions and imports, especially from France, which has a thriving, sophisticated chèvres (soft goat cheese) industry. Liquid raw goat milk can be officially marketed in only 14 states in the United States, those with public health laws permitting such sales. In all other states, goat milk must be pasteurized. There is a strong interest nationwide in raw goat milk by people with malnourished children, by persons with cow milk allergies and gastric problems, and by those interested in natural foods. This situation has led to a wide dispersion of mostly small goat herds throughout the United States to satisfy needs on a family level.

Goat milk is valued for raising orphaned foals and puppies and for the production of prime veal. Biochemically goat milk differs in several important aspects from cow milk. The major casein in cow milk (alpha-s-1) is found in neither goat nor human milk. The casein type in goat milk makes a softer curd and is more easily digested. The fat in goat milk consists of smaller globules than in cow milk and is more easily digested. The goat milk fat consists characteristically of much more short-chain fatty acids than cow milk.

There are also some consistent differences in vitamin, mineral, and enzyme contents between goat and cow milk, although gross compositions are more similar when compared with sheep milk, which has a much higher solids content and smaller daily yield. Daily production of good dairy goats in the United States is at least 1 gallon (3.8 liters) with average content of 3.8% fat, 3.4% protein, 4.6% lactose, and 0.8% minerals, totaling 12.6% solids.

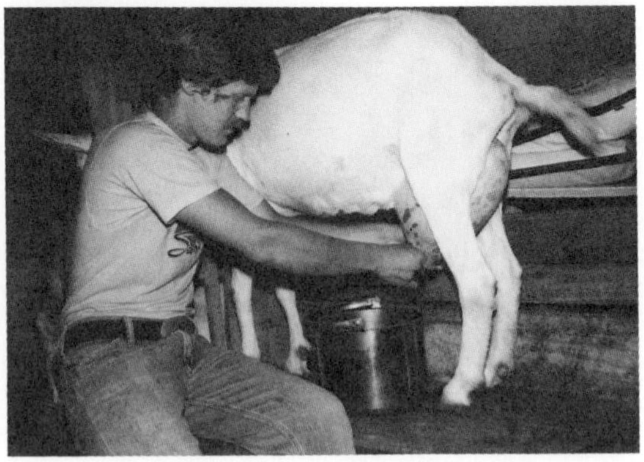

Fig. 2. Milking dairy goats on elevated stands.

Supervised production records are obtained and processed by a subdivision of the Dairy Cattle Herd Improvement Program of the United States Department of Agriculture (USDA) for the purpose of statistics, improved herd management, and sire proving for genetic improvement. Participation is voluntary, with a monthly fee. Most goat herd participation is found in California, Oregon, Wisconsin, New York, Ohio, Washington, Arkansas, and Arizona (in descending order). Average milk production per doe per year in 1981 was 1907 lb (866 kg) for Saanen goats and 1522 lb (691 kg) for Nubian goats. The other breeds have production rates between those two values; the national average is 1715 lb (779 kg). Average milk fat contents range from 3.44% for Toggenburg goats to 4.51% for Nubian goats, with a national average of 3.79%. Three-quarters of all dairy goat herds in the United States have 10 or less milking does. A world goat milk record was made by a Saanen goat in Australia with 7714 lb (3502 kg) in 365 days.

Breeding. Breeding of dairy goats is similar to that of sheep. Goat kids, born in the spring, are ready for breeding in the fall of the same year and will have their next generation 150 days after conception, usually the next spring. This cycle coincides with the seasonal changes of vegetation; the spring with new pasture growth favors milk production and provides greater survival opportunity. Thus, the dairy goats (for example, the Swiss breeds) of the Northern Hemisphere with its distinct seasons evolved as seasonal breeders. The disadvantage is the consequent shortage of goat milk during winter months. Goats of Mediterranean, Near Eastern, and Indian origin show less seasonality. Seasonal breeding is governed physiologically by day length; and by artificially reducing daylight in controlled housing, dairy goats can be brought into earlier estrus during the summer. This is a practical and successful method for obtaining an even goat milk supply throughtout the year.

Breeding of dairy goats is still mostly by natural service. Bucks produce, in the scent glands on top of their head, odors attractive to does but obnoxious to people. Estrus lasts 1–2 days and may recur in 21 days if impregnation fails. Artificial insemination with frozen semen stored in liquid nitrogen tanks at −320°F (−196°C) is gaining popularity since it is the best and most economical avenue of genetic improvement. Pregnancy testing is possible with ultrasonic scanning and radiography, but more practical by routine testing of milk samples for progesterone levels at 3 weeks after breeding.

Kidding in dairy goats is usually easier than in cows or sheep because of their generally lean condition and the narrow bone structure of kids. Two or three kids per pregnancy, not necessarily identical, is the rule rather than the exception, including backward presentations. Rebreeding is not inhibited by nursing, as in some other mammalian species, but rather by the increasing day length of spring.

New developments are estrus synchronization of does with hormone implants or injections to overcome estrus detection problems and to facilitate artificial insemination, and embryo transfer in conjunction with estrus synchronization to enable wider use and greater number of offspring of genetically superior does. Embryo freezing is possible even now, and the sorting between different sex embryos as well as cloning from single superior embryos will be practical in the near future, thus opening exciting new methods of dairy goat improvements. Considering that in many parts of the world, where the majority of the 450 million goats are found, the productivity per animal usually averages only 1/10 to 2/10 of the average goat milk production in the United States, the new breeding techniques in developed countries are contributing to the solution of malnutrition problems in the third world.

Feeding. Feeding of dairy goats is similar to that of cows and sheep, except that goats need and thrive on browse like deer, and some breeds may go without water for days. High-producing dairy goats need scientifically balanced rations as do high-producing dairy cows.

Diseases. Facilities and equipment are not strongly developed for goats specifically, but—as with housing, milking machines, and feed storage—medications against parasites, diseases, and metabolic disorders are adapted from, or patterned after, those for cattle. Unlike in dairy cows, the largest problem in dairy goats is internal parasites, which appear to adapt often to certain standard medications. Also, unlike in dairy cows, tuberculosis and brucellosis are not a problem in dairy goats in the United States, although annual testing continues to be advocated for safety reasons. On the other hand, there are viral diseases causing arthritis, bacterial infections producing abscesses, and intestinal bacterial disorders which are more serious in dairy goats than in dairy cows—particularly because veterinary research, education, and care in the United States are more focused on cattle than on goats.

Overall, American dairy goats are among the world leaders, even with their limited organization and government support. Once that organization materializes in the United States, dairy goats may become an industry as successful as the dairy cattle industry.

For background information *see* CASHMERE; GOAT; MOHAIR in the McGraw-Hill Encyclopedia of Science and Technology. [GEORGE F. W. HAENLEIN]

Bibliography: J. L. Ayers and W. C. Foote, Proceedings of the 3d International Conference on Goat Production and Disease, *Dairy Goat J.*, 1982; G. F. W. Haenlein et al., Proceedings of the International Symposium on Dairy Goats, *J. Dairy Sci.*, 63(10):1591–1781, 1980; G. F. W. Haenlein and D. L. Ace, *Goat Extension Handbook*, University of Delaware, 1983; P. Morand-Fehr et al., Nutrition and systems of goat feeding, *Proceedings of the International Symposium at Tours*, Institut National de Recherche Agronomique, Paris, 2 vols., 1981; National Research Council, *Nutrient Requirements of Goats: Angora, Dairy, and Meat Goats in Temperate and Tropical Countries*, 1981; S. N. Singh and O. P. S. Sengar, *Final Technical Report: Studies on the Combining Ability of Desirable Characters of Important Goat Breeds for Meat and Milk Separately and in Combination*, Department of Animal Husbandry and Dairying, Raja Balwant Singh College, Bichpuri (Agra), India, 1979.

Gypsum

The mineral gypsum ($CaSo_4 \cdot 2H_2O$) can occur as the result of several processes; however, in association with sedimentary deposits, it is generally considered to have formed as an evaporite. The recent finding of gypsum in deposits that did not form under evaporating conditions has led to the conclusion that gypsum can also form diagenetically. In this diagenetic process, iron sulfide (pyrite) is oxidized by oxygen-rich and calcium-rich water to form iron oxide (limonite) and calcium sulfate (gypsum). Thus, paleoenvironmental interpretation of ancient sedimentary deposits must recognize that not all gypsum-bearing strata indicate semiarid to arid evaporating conditions; gypsum can form in cool, humid environments through the diagenetic alteration of sulfide-rich sediments.

Evaporative formation. Evaporites form where salts dissolved in water are concentrated through evaporation to the point that the solution becomes oversaturated and salts are precipitated. For gypsum to precipitate from sea water, the salinity must exceed 124 parts per thousand (3.35 times that of normal sea water), and the sea water must evaporate to 0.2 its original volume. (If sea water is evaporated to 0.1 its original volume, the mineral halite, NaCl, will also be precipitated.) There are two conditions, both of which require semiarid to arid conditions, for the formation of gypsum as an evaporite. A constant but restricted influx of sea water into a restricted basin in an arid region, such as a lagoon or land-locked sea, could lead to the precipitation of very thick layers of gypsum. Similarly, sea water flooding tidal flats to shallow depths in arid regions could lead to the formation of concentrated brines in the tidal flat sediments, and gypsum crystals would precipitate. Under such conditions other minerals, such as calcite, aragonite, dolomite, halite, polyhalite, and sylvite, can also be precipitated to varying degrees depending on the severity of evaporation.

Diagenetic setting and process. In Ohio, sand and clay of deltaic deposits which formed in glacial lakes during the late Pleistocene contain both single crystals and thin layers of gypsum. The sequence of deposits consists of basal gray prodelta overlain by red sand and gravel. The contact between basal mud and overlying sand is gradational, consisting of alternating thin beds of sand and mud. The basal mud deposits contain clay minerals as well as finely ground rock particles and finely crystalline pyrite (FeS_2). At the contact between the basal mud and overlying sand, the sediment displays a reddish-brown stain, in striking contrast to the gray appearance of the sequence, which is due to the presence of iron oxides, probably limonite.

As the delta was built forward into the glacial lake, streams carried sand and gravel over the earlier deposited gray prodelta mud. The muds being deposited in deeper reducing water contain iron sulfide (pyrite). The overlying sand and gravel deposits are permeable and contain abundant limestone fragments. As a result, water that moves through these deposits is rich in calcium bicarbonate. Where calcium bicarbonate–rich water comes in contact with pyritic muds, pyrite is oxidized to iron oxide (limonite) and gypsum (see illustration). If this alteration occurs in the mud layers, clear tabular, occasionally twinned, gypsum crystals grow and displace enclosing muds, with little, if any, mud included in the crystals. However, if crystal growth occurs in interbedded sand layers, thin, dovetail, generally twinned, crystals form. Commonly, thin sheets of intergrown crystals occur along the upper contact of sand layers. In contrast to the displacive crystals formed in the mud layer, gypsum crystals in sand layers include sand grains (gypsum sand crystals), and there is little evidence of any displacement of sand. This difference can be explained by comparing the compact, grain-supported sand deposits with the loose, water-saturated muds. As gypsum crystals grow, mud can be pushed aside easily because of its gel-like nature, while the more rigid sand deposits do not allow for movement of sand grains. Therefore, sand grains remain in place and gypsum crystals continue their growth into the interstitial spaces between grains, cementing sand into crystals.

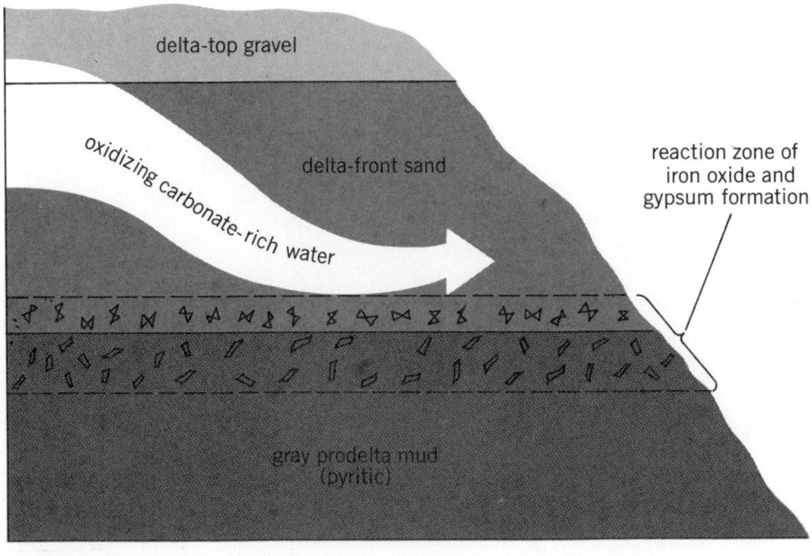

Model for diagenetic nonevaporative gypsum formation. The dovetail gypsum sand crystals form in sand, and the tabular gypsum crystals form in mud.

Significance. It is important to recognize that at least some of the gypsum in the sedimentary record may be the result of diagenetic alteration of sulfides. The setting and process involved are simple, and conditions suitable for alteration are widespread in nature. All that is required is a setting where initial reducing conditions produce sulfur-rich deposits, and that these deposits are overlain by carbonate-rich sediments. During and after deposition, oxidizing water would migrate through this sequence and gypsum could form. Potential settings that could give rise to conditions other than the glaciolacustrine delta described above include: transgressive carbonate beach deposits which are deposited over lagoonal muds; carbonate reef debris which is deposited over forereef, basinal sediments; and river deltaic sands deposited over prodelta muds or interdistributary muds. In the absence of other indications of high evaporation rates such as redbeds, halite, restricted fauna, and dolomite, the diagenetic origin of gypsum might be the most reasonable explanation for the occurrence of gypsum.

For background information *see* AUTHIGENIC MINERALS; DIAGENESIS; SALINE EVAPORITE in the McGraw-Hill Encyclopedia of Science and Technology.

[ROGER J. BAIN]

Bibliography: W. E. Dean and B. Charlotte Schreiber (eds.), *Marine Evaporites*, Soc. Econ. Paleontol. Geol. Short Course 4, 1978.

Heavy-ion accelerator

In 1982 the Bevalac accelerator at the University of California's Lawrence Berkeley Laboratory became the first machine ever to accelerate uranium ions to relativistic velocities—velocities so close to the speed of light that the kinetic energy of the projectile becomes comparable to or greater than its rest energy (the energy equivalent of its rest mass). A beam of uranium ions with a kinetic energy of approximately 1 GeV/nucleon (238 GeV total kinetic energy) was extracted and used for atomic and nuclear physics experiments. In one such experiment, it was shown that these high-energy ions could easily be stripped of all their remaining electrons, leaving bare uranium nuclei in the 92+ charge state ($^{238}U^{92+}$); this was the first time that bare nuclei of the heaviest elements have been produced in the laboratory. With these milestone achievements, the Bevalac opened important new vistas in the still-young field of relativistic nuclear collisions.

Accelerator complex. The Bevalac (Fig. 1) consists of a complex of ion sources and accelerators. The main accelerator, the 328-ft-circumference (100-m) Bevatron, is a synchrotron, and was the highest-energy proton accelerator in the world when it was built in 1954. It was used to create and study antiprotons and antineutrons as well as for the discovery of many of the new particles and resonances now familiar in high-energy physics. Through frequent upgrading, the Bevatron has continued to be a forefront research tool, but in fields other than elementary particle physics.

Fig. 1. Lawrence Berkeley Laboratory's Bevalac accelerator, the result of linking the SuperHILAC—an Alvarez-type linear accelerator (top)—with the Bevatron—a nuclear synchrotron (foreground)—by means of a beam-transfer line. The path of the beam is shown by a broken arrow.

In 1971 it was demonstrated that the Bevatron could accelerate nitrogen ions produced in small quantities in the proton source. In 1973 a transfer line was constructed from the SuperHILAC (an Alvarez-type linear accelerator) to provide a more powerful source of heavier ions. The SuperHILAC accelerated all ions up to lead to a kinetic energy of 8.5 MeV per nucleon. Some of these (all ions up to iron) could be fully stripped, injected as bare nuclei into the Bevatron, and accelerated further, to 2.1 GeV per nucleon. With the linking of the Bevatron and the SuperHILAC—called the Bevalac—an international program of research on relativistic nuclear collisions was launched in 1974. Nuclear physics, astrophysics, atomic physics, and radiobiology experiments were conducted with beams ranging from protons to iron ions. Of these programs, the latter has progressed from early work on cells and animals to clinical research in large-field radiotherapy, with heavy-ion treatments delivered to as many as 20 cancer patients per day.

As progress was made in relativistic heavy-ion physics, it became apparent that acceleration of the heaviest ions would be highly desirable, and a major upgrade of Bevalac to a uranium-beam capability was completed in 1982. This required upgrading the SuperHILAC to increase its beam intensity and improving the Bevatron vacuum system to permit the acceleration of these heaviest ions, which can be produced by the SuperHILAC only in a partially stripped state.

Accelerating heavy ions. The main difficulty encountered in accelerating heavy ions results from the acceleration being proportional to the charge-to-mass ratio, q/A. For a proton, this ratio is unity, but for a heavy atom such as uranium ($A = 238$), from which it is difficult by standard methods to remove more than a few electrons, q/A typically remains very small. This means that a linear accelerator would have to be immensely long to overcome this limitation. No such problem exists for the Bevatron, however, where the ions, constrained by the strong magnetic field to a circular orbit, pass the accelerating electrodes about 10^6 times. But here another limitation arises: the momentum p that can be held in a circular orbit of radius r in a magnetic field B is given by the formula $p = qBr$. Once again, the low charge state of a typical uranium ion poses a limitation on the maximum momentum that can be achieved.

The method adopted at the Bevalac to overcome these problems is to accelerate the ions in several stages. After each stage, the beam is passed through a thin foil or a vapor, and a few more electrons are stripped off. As the energy is increased, the acceleration and stripping processes both become more efficient, until finally the desired energy and charge state are produced.

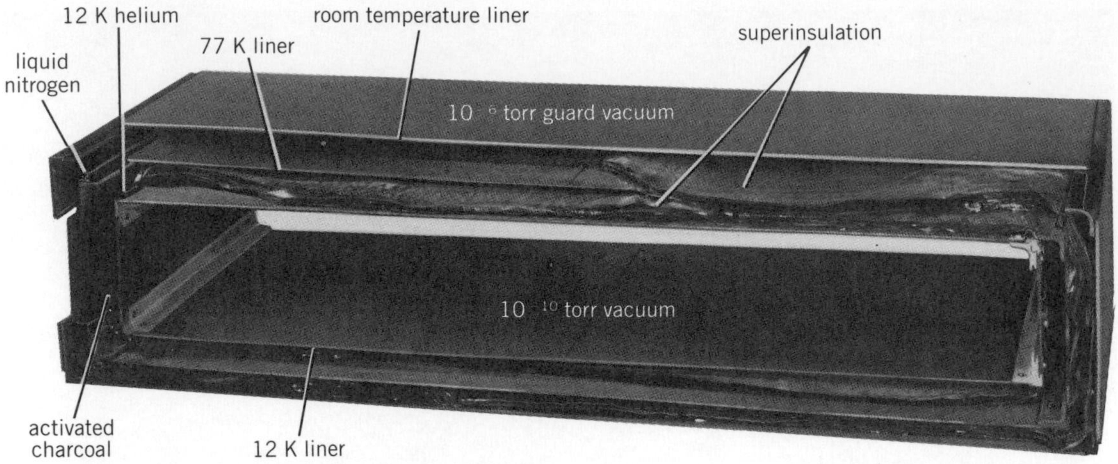

Fig. 2. Cross section of the cryogenically pumped high-vacuum liner installed in the toroidal vacuum tank of the Bevatron to permit the acceleration of relativistic heavy ions. 12 K = −438°F = 22°F above absolute zero; 77 K = −321°F = 139°F above absolute zero. 1 torr = 0.02 lbf/in.^{2} = 133 Pa.

The process begins at the SuperHILAC in a newly constructed heavy-ion preinjector called Abel (to go with Adam and Eve, the two existing preinjectors), which consists of a 0.75-megavolt Cockcroft-Walton accelerator followed by a Wideröe linear accelerator. Weakly ionized uranium produced by a Penning ion gauge is accelerated to 112 keV per nucleon and then passed through a fluorocarbon vapor stripper to produce as many ions as possible in the 13+ charge state. These enter the main Alvarez linear accelerator and are accelerated in two stages (separated by a carbon-foil stripping to the 40+ charge state) to 8.5 MeV per nucleon. After yet another carbon-foil stripping, the 68+ charge state is selected for injection into the Bevatron via the transfer line. Only 0.24% of the original ion beam survives to this stage, the rest having been lost as unwanted charge states in the three stripping stages. This optimized procedure represents a complex balance between accelerator costs and the desirability of high energies as opposed to high intensities. Higher intensity at lower energy can be achieved by omitting one of the stripping stages.

Improved Bevatron vacuum. The need for an improved vacuum in the Bevatron can be understood from atomic collision arguments. The total flight path of ions during acceleration in the Bevatron is of the order of 3 × 10^8 ft (10^8 m)—incomparably greater than the 656 ft (200 m) traversed by the ions in the SuperHILAC and transfer line. In the Bevatron's earlier vacuum of 10^{-9} lbf/in.^{2} (10^{-7} torr or 10^{-5} pascal), each ion underwent more than 1000 collisions; if it gained or lost an electron in any one of these collisions, it was lost from the synchronous orbit. At Bevatron energies, electron capture by the ions is rare (except near injection), and electron loss dominates. Thus, ions injected in an already fully stripped state, which was possible with ions up to iron, could be accelerated in the existing vacuum, although often with poor intensities. A thousandfold improvement of the vacuum to 10^{-12} lbf/in.^{2} (10^{-10} torr or 10^{-8} Pa) was needed to achieve higher intensities and to permit acceleration of the partially stripped ions, particularly $^{238}U^{68+}$.

To achieve this goal, a cryogenically pumped liner was installed inside the vacuum tank, enclosing about one-quarter of its original volume (Fig. 2). This innovative device consists of nested boxes of printed circuit boards, coated with copper strips to conduct heat and separated by multilayered superinsulation (striped aluminized Mylar blankets). The innermost box is cooled to 22°F above absolute zero or −438°F (12 K) by helium gas, and the middle box to 139°F above absolute zero or −321°F (77 K) by liquid nitrogen; the outermost box, made of fiberglass, is at room temperature and provides support and protection for the liner. The size of the central chamber is 6 in. × 43.3 in. × 328 ft (15.2 cm × 110 cm × 100 m).

The extensive use of organic materials in a high-vacuum system may appear surprising; however, at 22°F above absolute zero (12 K) the outgassing rate of these materials is not even measurable, and the total air-pumping speed of the helium-cooled surfaces is many millions of liters per second. Pumping for hydrogen and helium is provided by activated charcoal panels on the helium box and by small auxiliary diffusion pumps. The average pressure in the machine is now 2 × 10^{-12} lbf/in.^{2} (1 × 10^{-10} torr or 1.3 × 10^{-8} Pa), and the total heat load at 22°F above absolute zero (12 K) is 150 watts.

Measurement of such low pressures is very difficult by conventional means. In the Bevatron, a beam of $^{12}C^{4+}$ ions at 7.2 MeV per nucleon was allowed to circulate for extended periods while the beam loss was monitored. From previously measured cross sections, the average gas density experienced by the ions could be calculated.

In the first year of operation, many new heavy-ion beams have been used for experiments, including $^{56}Fe^{24+}$, $^{84}Kr^{33+}$, $^{93}Nb^{37+}$, $^{139}La^{47+}$, $^{197}Au^{61+}$, and $^{238}U^{68+}$, at intensities of 10^6–10^7 particles per pulse (one pulse every 5 s), and kinetic energies of 0.96–1.70 GeV per nucleon. No significant difficulties have been encountered; in fact, the improved vacuum has greatly eased many procedures, and the overall reliability of the Bevalac has improved.

Relativistic uranium beams. The availability of relativistic uranium ions made it possible, for the first time, to produce bare uranium nuclei in the laboratory. In an experiment to test the effect of a copper foil on the charge-state composition of a uranium beam of 962-MeV per nucleon (originally 68+), it was found that a foil of only 4×10^{-5} in. (1 micrometer) thickness is sufficient to strip all the ions to the 89+ charge state or higher. With a 8×10^{-4} in. (20-μm) foil, more than 80% of the ions are fully stripped (92+), while the remainder are hydrogenlike (91+). A variety of tests of quantum electrodynamics are in progress for hydrogenlike and heliumlike uranium ions, including measurement of the Lamb shift.

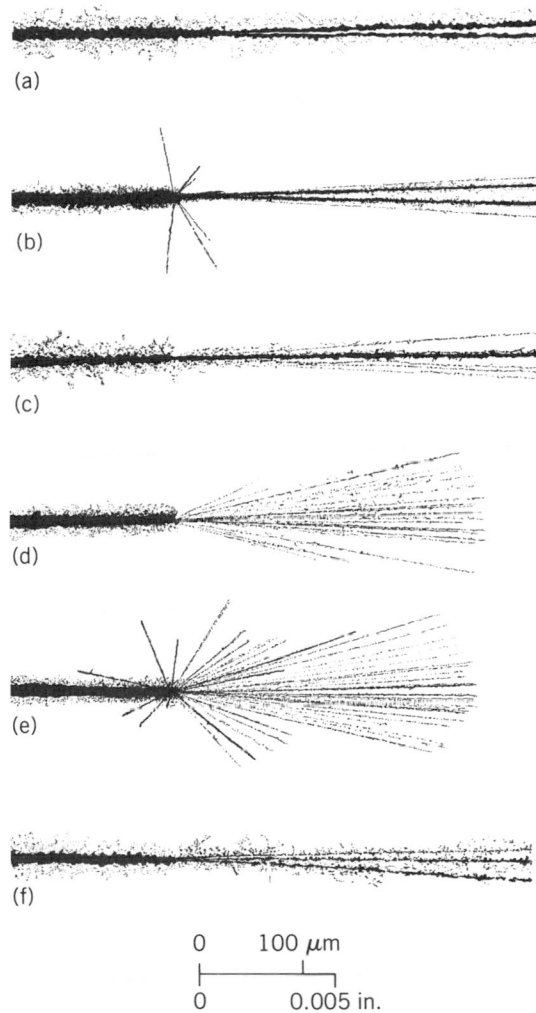

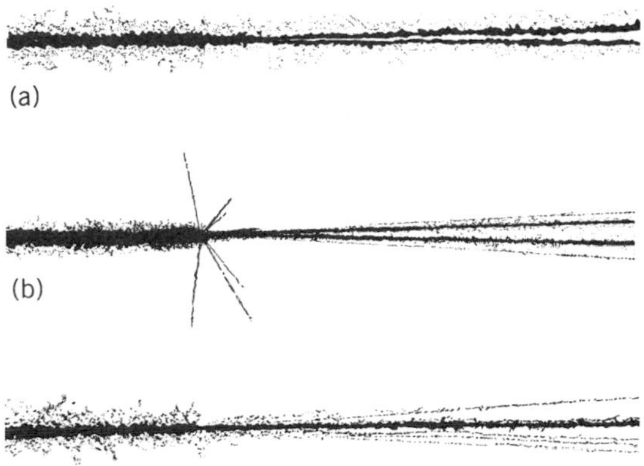

Fig. 3. Microprojection drawings of collisions of relativistic uranium nuclei (about 1 GeV per nucleon) with nuclei in a photographic emulsion. (*a*) "Clean" binary fission of the projectile induced by a grazing collision with a nucleus in the emulsion (no tracks from the target nucleus). (*b*) "Dirty" binary fission accompanied by light particles from the projectile and from the target nucleus. (*c*) Probable fission of the projectile, with destruction of one of the fission fragments into lighter particles. (*d*) Head-on collision with a light target nucleus, in which the projectile is completely fragmented. (*e*) Extremely violent head-on collision with a heavy target nucleus (silver or bromine), with fragmentation of both nuclei. (*f*) Ternary fission of the projectile.

The first nuclear physics experiments with relativistic uranium beams were done by bombarding photographic emulsions (Fig. 3). The dynamics of such collisions are being studied with many new detection devices that have recently come into operation—especially the Heavy-Ion Spectrometer System (HISS); the Plastic Ball-Wall, an electronic detector covering a solid angle of 4π steradians (coverage in all directions) with 991 individual modules; and the Streamer Chamber, which provides visual information comparable to the emulsion technique but which can be triggered to select events of special interest.

Nuclear thermodynamics. Preliminary physics results show that even at the highest energies of the Bevalac, the projectile nucleus can be stopped completely in the target nucleus, briefly creating a thermodynamic equilibrium state at enormous temperatures (100 MeV, or about 2×10^{12}°F or 10^{12} K), from which hundreds of nuclear fragments and created particles emerge. In this equilibrium state, densities of more than four times normal nuclear density are reached—a condition believed to exist in nature only in the center of neutron stars.

Now that fully stripped uranium ions have been produced, there appears to be no further obstacle (except cost) to nuclear accelerators of even higher energy. The Lawrence Berkeley Laboratory has already made plans for a higher-energy accelerator, to be called the Tevalac because the projectile energies will be in the TeV range. With such an accelerator, it is believed that nuclear matter can be raised to temperatures and densities so high that its constituent quarks and gluons will be released. This would create, for an instant, a "quark-gluon plasma" similar to that which is believed to have existed about 1 microsecond after the big bang.

For background information *see* BIG BANG THEORY; PARTICLE ACCELERATOR; QUARKS; RADIOLOGY. [HOWEL G. PUGH]

Bibliography: J. R. Alonso et al., Acceleration of uranium at the Bevalac, *Science*, 217:1135–1137, 1982; E. M. Friedlander, H. H. Heckman, and Y. J. Karant, Nuclear collisions of uranium nuclei up to ~1 GeV/nucleon, *Phys. Rev. C*, 27:2436, 1983; H. Gould, in *Proceedings of the 6th High-Energy Heavy-Ion Study, Lawrence Berkeley Laboratory, June 28–July 1, 1983*, LBL Tech. Rep. 16281, December 1983; *The Tevalac: A National Facility for Relativistic Heavy-Ion Research to 10 GeV per Nucleon with Uranium*, LBL Publ. 5081, Lawrence Berkeley Laboratory, December 1982.

Higgs boson

The recent discovery of the *W* and *Z* weak bosons, the carriers of the weak force, is a milestone in confirming gauge theory unification of the electromagnetic force with the weak force. A final vital ingredient of the highly successful electroweak theory remains to be found—a particle known as the Higgs boson. In a gauge theory of fundamentally massless interacting particles, Higgs boson effects give masses to the weak bosons, and to the quarks and leptons, through a spontaneous symmetry-breaking mechanism. The theory does not predict the mass of the Higgs boson, but the interactions of this boson are completely specified. It likes to couple to heavy particles, such as the W, the Z, and the t quark. Unfortunately the Higgs boson remains elusive due to its very small couplings to light quarks and leptons, and formidable experimental efforts will be required to detect it.

Gauge theories. A vision of particle physics is the ultimate unification of the four known basic forces in nature—strong, electromagnetic weak, and gravitation—into one all-encompassing theory. In the last two decades major progress has been made toward that goal. Gauge theories have been proposed to describe the first three of these forces by a unified theory, and many experiments are being carried out to test the validity of the predictions.

The first unification of forces (the electric and the magnetic), and the first gauge theory, was the classical theory of electric and magnetic fields introduced in 1868 by James Clerk Maxwell. In the quantum extension of the Maxwell theory, quantum electrodynamics (QED), developed in the twentieth century, the carrier of the electromagnetic force is a massless vector particle, the spin-1 (in units of Planck's constant) photon which interacts with a spin-½ electron field. The observables of the theory are independent of the unmeasurable phase of the electron field at any space-time point, a property known as a local gauge symmetry. QED has survived exacting experimental tests, and served as a prototype for unified theories embodying local gauge symmetry.

At first sight, electromagnetism and the weak force appear to have little similarity. Electromagnetism holds atoms together and describes light, while the weak force accounts for beta-decay radioactivity. The electromagnetic force is infinite in range, while the weak force has a very short range. At low energies the interaction strengths are vastly different. Nevertheless, there was a remarkable parallel in that both forces appeared to be interactions mediated by the exchange of spin-1 particles. This led Sheldon Glashow in 1961 to propose a gauge theory to unify the two forces. However, this theory could not account for the fact that the carrier of the weak interactions had to be heavy to explain the short range of the weak force, since the local gauge symmetry required the weak bosons to be massless, like the photon.

Spontaneous symmetry breaking. The development of theories with spontaneous symmetry breaking eventually resolved the weak boson mass difficulty. An example of such a symmetry breaking is the bending of a thin column (for example, a darning needle) under a compression force (Fig. 1).

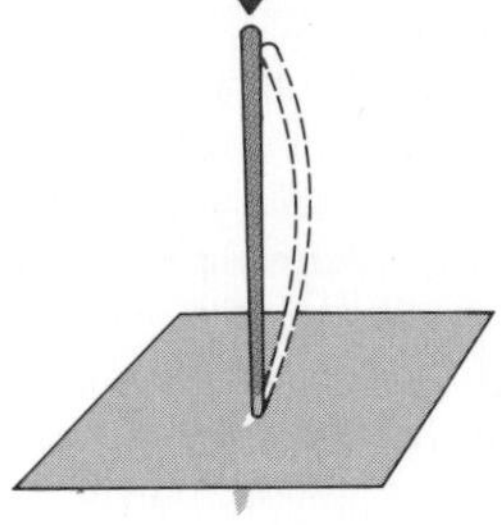

Fig. 1. Bending of a darning needle by a compressive force, an example of spontaneous symmetry breaking.

Even if the column is symmetric about its axis and the force is along the axis, a particular direction is selected when it bends under application of a sufficiently large force. The direction of bending is arbitrary, however, and, on the basis of physical principles, the bending could have occurred along any direction. The original symmetry is actually not lost in spontaneous symmetry breaking, but becomes hidden.

Spin-0 (scalar) particles were introduced by Peter Higgs in 1964 as a means of causing spontaneous symmetry breaking in gauge theories. The so-called Higgs mechanism using these particles was the vital development needed to generate masses for the weak bosons without destroying the local gauge symmetry. These spin-0 particles are normally referred to as Higgs bosons.

Weinberg-Salam theory. In 1967 Steven Weinberg and Abdus Salam, working independently, formulated a unified theory of electromagnetic and weak interactions based on gauge symmetry. This theory has so far accounted for all experimental measurements of electromagnetic and weak interactions. In the unbroken theory the massless gauge bosons form a triplet W^+, W^-, W^0, and a singlet B^0 in a so-called weak-isospin space. Here the superscripts denote the electric charges in units of the electron charge e. The W bosons interact with the members of lepton and quark doublets which are left-handed (that is, their spins are antiparallel to their momenta). For instance, the lightest lepton and quark doublets are given by notation (1). This

		Electric charge	
Neutrino	$\begin{pmatrix} \nu_e \\ e \end{pmatrix}$	0	
Electron		-1	
Up-quark	$\begin{pmatrix} u \\ d \end{pmatrix}$	$^2/_3$	(1)
Down-quark		$-^1/_3$	

type of interaction is needed to explain left-right asymmetries (parity violation) observed in beta decays. The B^0 interacts with both left- and right-handed quarks and leptons, with a coupling strength proportional to a quantum number known as weak hypercharge. In group theory language the gauge symmetries are SU(2) for the W interactions and U(1) for the B.

A left-handed doublet of Higgs fields, with electric charges shown in notation (2), is introduced to

$$\begin{pmatrix} \phi^+ \\ \phi^- \end{pmatrix} \qquad (2)$$

cause spontaneous symmetry breaking. The potential energy of the Higgs fields is such that it has a minimum when the neutral field ϕ^0 has a finite value v, instead of zero as would happen in usual field theories (Fig. 2). This gives a preference for a particular orientation in the weak isospin space, breaking the original SU(2) × U(1) symmetry. The resulting masses of the weak bosons are proportional to the vacuum expectation value v, which is determined by its relation to the weak coupling to have a value of about 250 GeV.

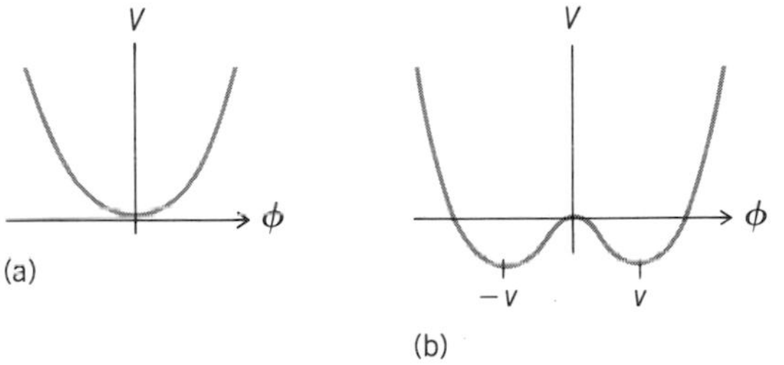

Fig. 2. Potential energy V of a real Higgs boson ϕ. (a) Usual field theory; minimum energy occurs for $\phi = 0$. (b) Higgs mechanism; minimum energy occurs for $\phi = \pm v$.

The physical gauge bosons of the spontaneously broken SU(2) × U(1) theory are W^+, W^-, Z, and γ, where Z and γ are linear combinations of the primordial W^0 and B^0 gauge bosons given by Eqs. (3). The single new parameter of the theory is

$$\begin{aligned} Z &= \cos\theta_W\, W^0 - \sin\theta_W\, B^0 \\ \gamma &= \sin\theta_W\, W_0 + \cos\theta_W\, B_0 \end{aligned} \qquad (3)$$

$\sin^2\theta_W$, which is simply related to the SU(2) and U(1) couplings of the gauge bosons to quarks and leptons. The photon γ couples to particles according to their electric charges and is massless, which gives the electromagnetic force infinite range, as required.

The theory predicted a new class of weak process called neutral currents (Fig. 3) involving no charge transfer. Neutral current processes were subsequently discovered, and extensive studies show that they conform precisely to the predictions of the theory. Analysis of the data on weak neutral currents determines the value of $\sin^2\theta_W$ to be about 0.215.

Massless vector bosons have only two observable spin orientations, while massive vector bosons have three. In the Higgs mechanism the scalar fields are the source of the additional spin orientations of the W^+, W^-, and Z^0. Of the four degrees of freedom

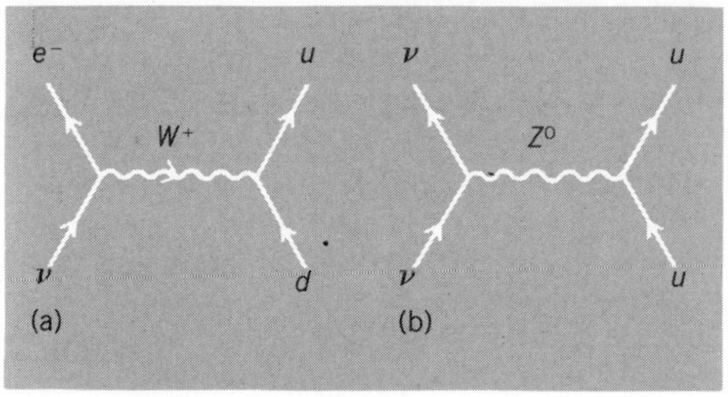

Fig. 3. Representative weak interactions of leptons and quarks. (a) Charged current process $\nu_e d \rightarrow e^- u$ mediated by W-boson exchange. (b) Neutral current process $\nu_e u \rightarrow \nu_e u$ mediated by Z-boson exchange.

of the scalar fields (ϕ^+ and ϕ^0 are both complex), three disappear to become vector boson components; the massless vector particles are said to "eat" the Higgs bosons in acquiring their masses. The remaining real scalar particle, denoted by H^0, is the observable Higgs boson.

The predicted $W^\pm$ and Z^0 masses can be expressed in terms of the electric and weak coupling strengths at low energy and the parameter $\sin^2\theta_W$. The predicted values are given by Eqs. (4), with

$$\begin{aligned} M_{W^\pm} &= \frac{38.5\ \text{GeV}}{\sin\theta_W} = 83\ \text{GeV} \\ M_{Z^0} &= \frac{M_{W^\pm}}{\cos\theta_W} = 94\ \text{GeV} \end{aligned} \qquad (4)$$

uncertainties of order 2 GeV mainly associated with uncertainties on the measured value of $\sin^2\theta_W$. An antiproton-proton colliding beam accelerator was constructed at CERN in Geneva, Switzerland, with the express purpose of finding the weak bosons. The fusion of quarks and antiquarks in the incident beams produces weak bosons which subsequently decay, for example, through reactions (5), where

$$\begin{aligned} u\bar{d} &\rightarrow W^+ \rightarrow e^+\nu_e \\ d\bar{u} &\rightarrow W^- \rightarrow e^-\bar{\nu}_e \\ u\bar{u}, d\bar{d} &\rightarrow Z^0 \rightarrow e^+e^- \end{aligned} \qquad (5)$$

the bars denote antiquarks. The expected W and Z production rates were very small, about one part in a million of the total proton-antiproton cross section.

The experimental discovery of the $W^\pm$ and Z^0 weak bosons by the CERN experiments was reported in early 1983. The measured masses in notation (6)

$$\begin{aligned} M_{W^\pm} &= 80.9 \pm 2.0\ \text{GeV} \\ M_{Z^0} &= 93.0 \pm 2.0\ \text{GeV} \end{aligned} \qquad (6)$$

are in accord with the theoretical prediction. This agreement has been hailed as the crowning achievement of 50 years of weak interaction research by several generations of physicists. *See* INTERMEDIATE VECTOR BOSON.

Higgs boson properties. The Higgs mechanism ensures the renormalizability of the theory; all infinities appearing in perturbation theory can be adsorbed into a redefinition of physical parameters such as masses and coupling constants. The discovery of the Higgs boson is an important test of the renormalizability.

The single physical Higgs particle H^0 has an unknown mass. However, there is a strong theoretical prejudice that it must have a mass greater than 10 GeV and less than 1000 GeV. This lower bound is the mass due to radiative corrections to the theory when the mass in lowest order is taken to be zero. The upper bound comes from the stipulation that the Higgs boson self-interactions do not become strong, for in that case perturbation theory breaks down. One might therefore expect that the Higgs boson mass is within an order of magnitude of the $W^{\pm}$ and Z^0 masses.

In contrast to its mass, the Higgs boson couplings to other particles are uniquely determined. Its couplings to fermions (either lepton or quark pairs) are of a trilinear Yukawa form $H^0 f\bar{f}$ with coupling strengths m_f/v that are in direct proportion to the fermion mass m_f. As a consequence the Higgs boson decays predominantly into the highest-mass fermion pair that is energetically allowed. For instance, a Higgs boson of mass 20 GeV would decay mainly into a $b\bar{b}$ pair, where b is the b quark of mass m_b approximately equal to 4.6 GeV. Since the b quark is long-lived, with lifetime of order 10^{-12} s, it may be possible to detect the b and $\bar{b}$ through their decay paths and then reconstruct the mass of the Higgs boson from the energy and momenta of these decay products.

The Higgs boson couplings are small, even on the scale of the weak interactions, making its production exceedingly rare. The only feasible observations result from its bigger couplings to heavier particles, like the t quark and weak bosons. One proposed source is the decay of a bound state of t and $\bar{t}$ quarks, called toponium, to a Higgs boson and a photon (Fig. 4). If the Higgs mass were 10 GeV and the t-quark mass 30 GeV, toponium would decay via this mode about 1% of the time. At a future electron-positron colliding ring accelerator, with the energy tuned to sit on a toponium resonance, the

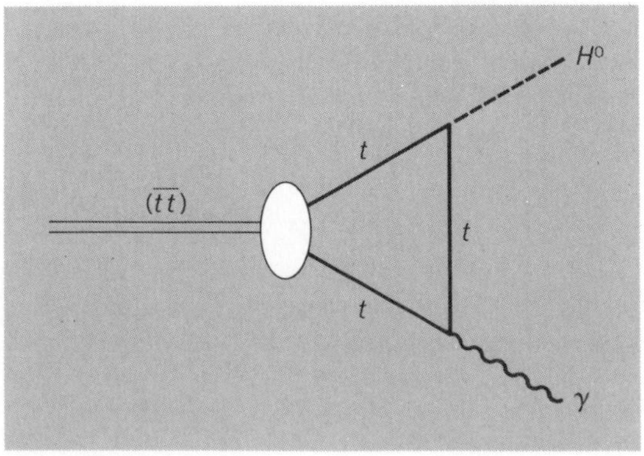

Fig. 4. Decay of a toponium state (a $t\bar{t}$ bound state) into a Higgs boson and a photon.

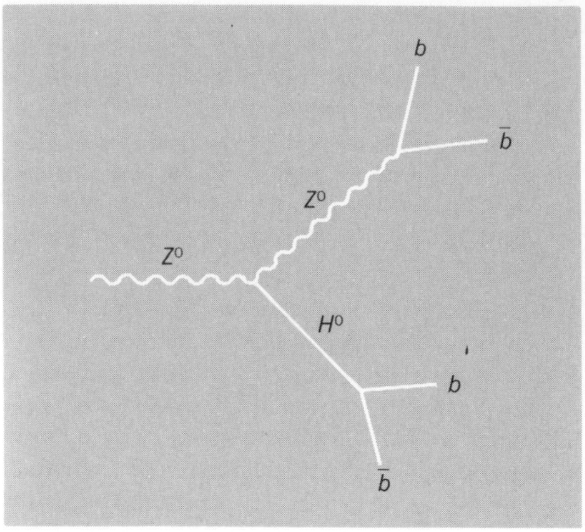

Fig. 5. Radiation of a Higgs boson from a Z^0 boson, with both H^0 and Z^0 decaying to $b\bar{b}$.

monochromatic energy from the photon emitted in this decay could provide a unique signature for Higgs boson hunting. *See* QUARKONIUM.

Another promising source of Higgs boson is radiation from a Z^0 boson with both decaying to a pair of long-lived particles (Fig. 5). This final-state configuration of four heavy long-lived particles, of which two determine the Higgs boson mass, provides a spectacular signal. The current developments of experimental techniques for heavy-particle identification (τ, c, b, t) and availability of higher-energy accelerators may make it possible to detect this elusive particle which plays a critical role in electroweak unification. *See* PARTICLE ACCELERATOR.

For background information *see* ELEMENTARY PARTICLE; FUNDAMENTAL INTERACTIONS; QUARKS; SYMMETRY LAWS (PHYSICS); WEAK NUCLEAR INTERACTIONS in the McGraw-Hill Encyclopedia of Science and Technology. [VERNON D. BARGER]

Bibliography: R. Donaldson, R. Gustafson, and F. Paige (eds.), *Proceedings of the 1982 DPF Summer Study on Elementary Particle Physics and Future Fascilities, Snowmass, Colorado, 1982*, American Institute of Physics, 1983; J. Ellis and M. K. Gaillard, Physics of intermediate vector bosons, *Annu. Rev. Nucl. Part. Sci.*, 32:443–497, 1982; W.-Y. Keung, Higgs hunting, in V. Barger, D. Cline, and F. Halzen (eds.), *Proton-Antiproton Collider Physics—1981: AIP Conf. Proc.*, 85:186–195 (1982); UA1 Collaboration, Experimental observation of isolated large transverse energy electrons with associated missing energy at $\sqrt{s}$ = 540 GeV, *Phys. Lett.*, 122B:103–116, 1983, Experimental observation of lepton pairs of invariant mass around 95 GeV/c^2 at the CERN SPS Collider, *Phys. Lett.*, 126B:398, 1983; UA2 Collaboration, Observation of single isolated electrons of high transverse momentum in events with missing transverse energy at the CERN $\bar{p}p$ collider, *Phys. Lett.*, 122B:476, 1983, Evidence for $Z^\circ \rightarrow e^+ e^-$ at the CERN $\bar{p}p$ collider, *Phys. Lett.*, 129B:130, 1983.

High-altitude illness

Nonacclimatized lowland dwellers develop a variety of physiologic responses during acute exposure to altitudes above 10,000 ft (3000 m). The initial circulatory responses serve to facilitate oxygen delivery to vital organs. Occasionally, however, maladaptation occurs, and intravascular fluid may shift into the brain, lung, or other tissues. Resulting organ dysfunction is recognized as acute mountain sickness (AMS), high-altitude cerebral edema (HACE), high-altitude pulmonary edema (HAPE), or high-altitude retinal hemorrhage (HARH). These disorders frequently coexist, and progression from mild AMS to severe HACE or HAPE is not uncommon. There is great individual variation regarding susceptibility. Drugs which depress ventilation, such as sedatives or alcohol, or diseases which impair oxygen transport, including emphysema, sickle-cell disease, or organic heart disease, place individuals at increased risk. Further research involving the control of ventilation during sleep, endogenous vasoactive substances (arachidonic acid metabolites, prostaglandins, and leukotrienes) and the biochemical events controlling capillary membrane integrity should lead to advances in altitude physiology as well as in diseases characterized by lack of oxygen or an alteration in cellular membrane permeability.

Physiologic responses to hypobaric hypoxia. At increasing altitudes barometric pressure falls, reducing the partial pressure of oxygen. The illustration shows the relationship of barometric pressure to the partial pressure of oxygen (pO_2) in various body compartments. From sea level to 10,000 ft (3000 m) the arterial oxygen tension falls from 100 to 60 mmHg (13.3–8.0 kilopascals). This degree of hypoxia stimulates a variety of complex physiologic responses.

The rate and depth of ventilation increases when arterial pO_2 falls below 60 mmHg (8 kPa). Respiratory alkalosis develops as arterial pCO_2 tension decreases, reducing the hypoxic drive to increase ventilation. Alkalosis diminishes after 2–3 days as the kidneys excrete sufficient bicarbonate, which results in increased ventilation characteristic of "acclimatized" individuals. Ventilation decreases during sleep at high altitude, resulting in lower pO_2 levels. This is analogous to an altitude gain during sleep (see illustration). Those individuals with more vigorous respiratory responses to hypoxia (awake, asleep, or during exercise) appear to be less susceptible to altitude illness and perform better at extreme altitudes in excess of 24,000 ft (7200 m). The increased renal production of erythropoietin requires 4–12 weeks before red blood cell mass increases.

Direct measurement of human cerebral blood flow is difficult, but observations of the retinal vessels with fundoscopy has increased understanding of cerebral vascular responses to hypoxia and alkalosis, and of the transudation of fluid across capillary membranes. The cerebral and retinal blood vessels dilate when arterial pO_2 is less than 60 mmHg (8 kPa). This vasodilatation, however, is countered by hypocapnia (lowered pCO_2) resulting from increased ventilation. Brain volume may increase because of increased cerebral blood flow or fluid transudation from cerebral vessels into brain tissue (cerebral edema). Failure of the cerebral autoregulatory vasoconstrictor response could facilitate the development of cerebral edema, analogous to hypertensive encephalopathy. A rise in intracranial pressure secondary to brain swelling is facilitated by the rigid cranial vault. Symptoms related to cerebral edema include headache, nausea, and deterioration of cerebral function, such as coma or death, if the process is not reversed.

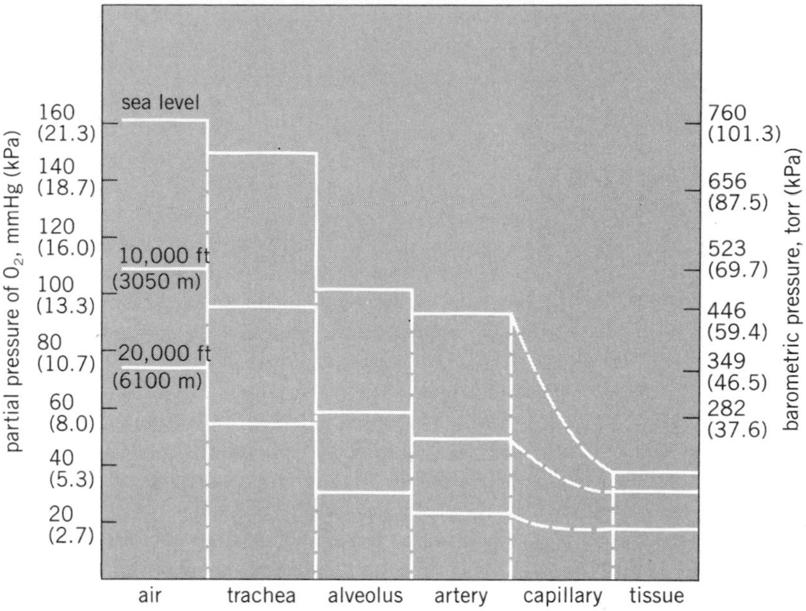

Oxygen tension gradients at sea level and 10,000-ft (3050-m), and 20,000-ft (6100-m) elevation. (*After R. T. Meehan and D. C. Zavala, The pathophysiology of acute high-altitude illness, Amer. J. Med., 73:395–403, 1982*)

Hypoxia raises pulmonary vascular resistance. Pulmonary blood flow increases further as a result of increase in cardiac output secondary to catecholamine-induced increased heart rate. Exercise further augments this increase in pulmonary blood flow. The opening of intrapulmonary shunts may lead to overperfused capillary beds. Injury or increased endothelial cell permeability allows the leakage of protein-rich intravascular fluid into the alveoli, which interferes with gas exchange. Thus, pulmonary edema aggravates hypoxemia, and a vicious cycle develops. The precise mechanism of the hypoxic pulmonary vasoconstrictor response is unknown, but a variety of chemical mediators are currently under study, including histamine, serotonin, angiotensin, catecholamines, prostaglandins, thromboxanes, leukotrienes, and calcium fluxes which modulate pulmonary vasomotor tone. Some of these vasoactive substances may alter capillary endothelial permeability and promote the transudation of fluid into the alveoli.

Body fluid homeostasis at high altitude remains

difficult to study and as a result is poorly understood. The factors responsible for an initial decrease in intravascular fluid volume are unknown. The pathogenesis of peripheral edema which is common at high altitude and is associated with AMS susceptibility also remains unexplained. Cold-induced reflexes, orthostatic hypotension, alkalosis, and exercise, in addition to hypoxia, may all contribute to these fluid shifts. Dehydration at high altitude is common, and symptoms are similar and often mistaken for acute mountain sickness. Decreased humidity and exercise, plus increased ventilation, all contribute to a substantial increase in insensible fluid loss.

Classification of altitude illness. The table outlines a classification of altitude-related disorders, ranging from mild AMS and asymptomatic HARH to life-threatening HAPE and HACE.

Twenty-five percent of lowland dwellers will develop AMS symptoms if transported rapidly above 8000 ft (2400 m). The incidence and severity of illness increase with greater altitude. Modern rapid transportation places increasing numbers of skiers, trekkers, climbers, and military personnel at risk for developing altitude illness. Symptoms of AMS develop 6–8 h after arrival and usually subside over the next 24–36 h. Sleep may aggravate AMS as ventilation decreases.

Maintaining adequate hydration, slowing the rate of ascent, and sleeping at lower altitudes reduce the frequency and severity of AMS. Individual variation regarding susceptibility, however, is great. Acetazolamide remains the only drug proven to reduce the severity of AMS, but it does not prevent severe high-altitude pulmonary edema, high-altitude cerebral edema, or high-altitude retinal hemorrhage. It may be used by persons who are extremely susceptible to AMS or by rescue or military personnel who must ascend rapidly.

High-altitude cerebral edema may be an exaggerated manifestation of AMS with a similar pathogenesis. It is unproven, however, whether AMS is secondary to mild cerebral edema. A recent study supports this hypothesis since 4 mg of dexamethasone every 6 h prevented AMS symptoms in all subjects during an acute exposure to 15,000 ft (4500 m) of altitude for 36 h. Increased cerebral blood volume secondary to vasodilatation from hypoxemia, or sleep-depressed ventilation with a rise in pCO_2 facilitate development of cerebral edema. Autopsy studies of HACE victims have revealed severe brain edema and hemorrhage. Sedatives or "subclinical" pulmonary edema, which lower arterial pO_2, may place individuals at increased risk, especially during sleep. The coexistence of HAPE and HACE is common, and they may be related. The success of dexamethasone treatment for HACE is anecdotal, but its effectiveness in other types of vasogenic cerebral edema is well recognized. Oxygen should also be utilized if available.

High-altitude pulmonary edema is a noncardiogenic form of pulmonary edema which may be more common in individuals with an exaggerated pulmonary vasoconstrictor response to hypoxia. Subclinical pulmonary edema detected by the presence of audible rales with a stethoscope can be discovered

Classification of acute altitude-related illnesses

Type	Occurrence	Altitude	Symptoms	Management
Acute mountain sickness (AMS)	Common	Above 7000–8000 ft (2100–2400 m)	Headache Nausea Vomiting Insomnia Shortness of breath Lethargy Weakness	Rest Complete recovery usual Analgesics for headache Descent if symptoms are severe Prevention by slower rate of ascent, avoidance of sedatives, and acetazolamide prophylaxis
High-altitude cerebral edema (HACE)	Rare	Usually above 12,000 ft (3600 m)	Severe headache Vomiting Unsteady gait Incoordination Impairment of judgment Hallucinations Severe neurological defects Coma Death	Immediate descent O_2 if available Dexamethasone
High-altitude pulmonary edema (HAPE)	Rare; common subclinically	Usually above 9000 ft (2700 m)	Progressive shortness of breath; may be present at rest Cough; may produce bloody sputum Coma Death	Immediate descent O_2 if available
High-altitude retinal hemorrhage (HARH)	Rare below 14,000 ft (4200 m)	Usually above 14,000 ft (4200 m)	Usually no symptoms Visual blurring or blind spots, which rarely are persistent	Descent if visual symptoms are present

in 20% of climbers who reach 14,000 ft (4200 m). The much lower incidence (1%) of overt HAPE suggests that a continuum exists between mild and severe forms of HAPE. Postmortem studies reveal widespread pulmonary edema similar to changes of the acute respiratory distress syndrome. Individuals at greatest risk are young high-altitude natives who reascend after a stay at lower altitudes. Death from progressive hypoxemia may develop within 8 h of initial symptoms unless descent is undertaken. A rapid descent of only 2000–3000 ft (600–900 m) may be life-saving. Most victims recover completely with descent alone; however, in a hospital setting bedrest and oxygen are sufficient treatment.

A variety of other disorders are precipitated or aggravated by hypobaric hypoxia. Sickle-cell crisis or spontaneous thrombus formation in either the systemic or the pulmonary circulation are well documented at high altitude. Peripheral edema is common among trekkers at high altitude and appears to be associated with AMS. The use of a mild diuretic may be warranted if symptomatic peripheral edema has already developed. High-altitude retinal hemorrhage has been observed in 56% of climbers at 17,500 ft (5250 m). This contrasts with 4% in trekkers, suggesting the importance of heavy physical exertion in the pathogenesis of this vasculopathy. Fluorescein leakage from retinal vessels has been demonstrated during vigorous exercise at 17,500 ft (5250 m). Spontaneous retinal hemorrhages are usually asymptomatic unless the macular area is involved. However, visual defects including scotomas may be permanent. HARH correlates poorly with AMS or sleep ventilatory responses to hypoxia, and is not prevented by the use of acetazolamide or prostaglandin synthetase inhibitors, such as naproxen.

Future research. The factors governing individual susceptibility to altitude illness remain unknown. The precise process by which lowland dwellers acclimatize to acute exposure to altitude is under investigation. Ongoing investigation of the central nervous system control of ventilation, mechanisms of blood vessel dilatation and constriction, and control of capillary endothelial membrane permeability should provide insight into the pathogenesis of altitude illness and a variety of disorders including sleep apnea syndrome, pulmonary hypertension, vasospastic disorders, and adult respiratory distress syndrome. Improved treatment should follow these advances in better understanding of the control of vascular tone and membrane permeability at the cellular level.

For background information *see* CIRCULATORY SYSTEM; RESPIRATION in the McGraw-Hill Encyclopedia of Science and Technology.

[RICHARD T. MEEHAN]

Bibliography: G. English, Physiologic adaptations to high altitude, *Univ. Wash. Med. J.*, vol. 5, no. 2, summer 1978; C. S. Houston, Going higher! The story of man and altitude, Charles Houston, Burlington, Vermont; C. Lenfant and K. Sullivan, Adaptation to high altitude, *New Engl. J. Med.*, 284:1298–1309, 1971; R. T. M. Meehan and D. C. Zavala, The pathophysiology of acute high-altitude illness, *Amer. J. Med.*, 73:395–403, September 1982.

Hydropower

The redevelopment of many dams in the United States has stimulated advances in the smaller hydropower field. It has been estimated that 100,000 sites in the United States could be developed with this new technology.

For many years the systems which delivered power to the public utility network were primarily designed as synchronous generating power plants. In such systems, the speed of the generator must be matched to the frequency of the network, and the peak of the generator sine wave must be in exact timing with the peak of the power system before the breakers can be closed to deliver power to the networks. This required expensive synchronizing equipment. On smaller units, below 150 kW, the synchronizing equipment cost often exceeds the cost of the generator, and in sizes below 25 kW the synchronizing equipment may exceed the cost of the complete turbine-generator package.

Induction motor-generator. A recent advance is based on the fact that any induction motor becomes a generator when connected to its normal power supply and forced to run above its synchronous speed. For all 60-cycle applications the design synchronous speed can be determined by dividing the number of poles in the motor into 7200. Thus a four-pole motor would have a design synchronous speed of 1800 rpm (revolutions per minute). When load is applied to an induction motor, its speed will fall below the synchronous speed until maximum allowable load is reached; the reduction in speed at that point is known as the slip. Conversely, when the motor is operated as a generator and the turbine connected to it forces it to operate above the design synchronous speed with the same amount of "positive" slip that it experienced at maximum load in motor mode, the output to the network will be the same amount of delivered energy as the motor formerly consumed in motor mode. For the small hydroplant operator this induction generating system is very cost-effective, being about 50% of an equivalent synchronous generator cost.

Single-phase induction motors are normally built in sizes up to 7.5 kW (10 hp). This would produce approximately a 7-kW generator output. However, three-phase induction motors are available in any sizes required by small hydropower developers. Such motors may then be operated as single-phase generators and still be less costly than synchronous generators or custom-built single-phase induction generators. As long as the "live" legs of a given unit receive no greater amperage as a single-phase motor, operation will be completely safe, and output is about 60% of nameplate rating for the same temperature rise on the active windings. Single-phase induction generators in larger sizes are available, but their high cost may prevent redevelopment of the small existing dams as well as new run-of-river sites.

Low-head hydraulic turbines. Advances in the design of low-head hydraulic turbines for small hydroplants have paralleled advances in generator design. Full-Kaplan designs in the so-called micro-size turbines are now available commercially. These very small units are capable of maintaining essentially constant efficiencies over a wide range of flows, from about 100% rates to about 20% of design flow rates. This permits greater utilization of year-around water flows using a single unit with single piping, controls, and such. Previously, very few attempts had been made to utilize full-Kaplan capability below about 5 MW. Micro-size units in full stainless steel construction have been installed as mini-elbow-tube turbines down to 1 kW at 20 ft (6 m) of head. Full-Kaplan capability has been extended downward to 10 kW. The smallest turbine has a 4-in. (10-cm) inlet and a 2-in. (5-cm) runner. These low-head units find ready application on irrigation canal "drops" or on small dams with heads ranging down to 4–6 ft (1–2 m). Many of them have been installed on run-of-river installations in remote locations in third-world countries. All sizes work between 6 ft (2 m) and 45 ft (14 m) of head.

Generator–load controller system. When the more economical induction generator cannot be used for a small hydroplant in a specific location because of the lack of a utility network within a reasonable distance, low-cost equipment is available that permits the generation of synchronous power without the use of expensive and high-maintenance speed-control systems such as hydraulic-oil-relay governors. In applications where a widely varying load is mated with widely varying water flows, the Lima MACR (a model classification indicating heavy duty) generator with an electronic load controller has been used successfully (see illustration).

The Lima MACR generator is self-excited and self-regulated. The excitation system is a three-phase ac system revolving past two flux-field sources. The first is a permanent-magnet flux field to provide initial current flow to the main rotating field. The second is a stationary winding in series, with the load being drawn from the stationary field in which the power is generated. Both sources produce ac power in the excitation circuit. The first source is at fixed level, but since the second varies with load demand, the excitation power delivered to the rotating field also varies directly, and constant voltage is produced at the generator output terminals even with varying load. Since the excitation power is ac, diodes rated at 10 times normal root-mean-square voltage are placed in the circuit converting the power to dc at the rotating pole windings. The

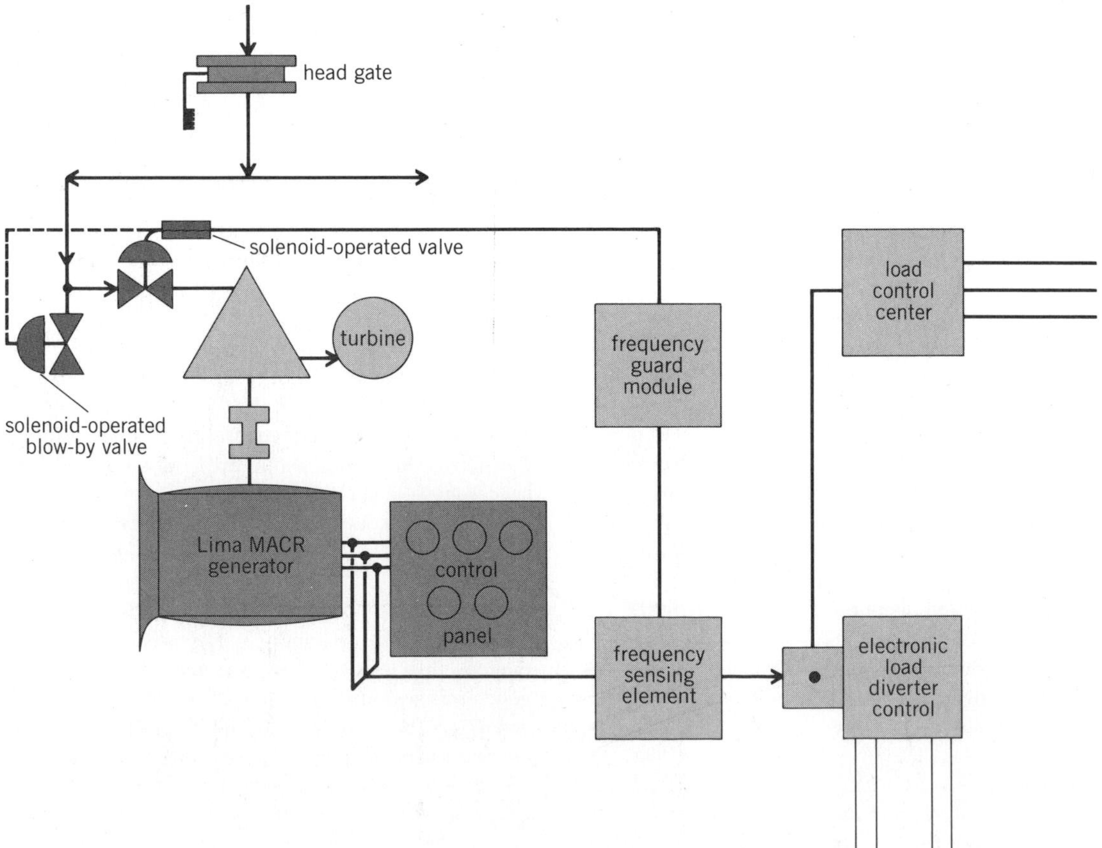

Schematic diagram of a system using a Lima MACR generator with an electronic load controller. *(GSA International Corp.)*

diode assembly rotates with the excitation winding and is wired directly to the rotating dc field; thus the system is brushless, self-excited, and self-regulated.

As shown by the turbine power equation below

$$P = \frac{Q\,he}{11.8}$$

(Q = flow, h = head, and e = efficiency), power remains constant as long as flow and head are held constant at the turbine. If the load at the generator terminal is used as a base, the point at which the system is in equilibrium can be determined. The electronic load diverter controls speed (frequency) by controlling load, once the inlet valve to the turbine has been set to satisfy the full-load demand of the generator.

Small changes in load at equilibrium conditions will produce small changes in speed. This system can instantly detect the smallest change in frequency of the generator output; thus the sensing element is connected to the output of the generator at the terminals. The control is stepless throughout the first 10 kW of load change (either increase from zero percent to 10 kW load or decrease from 10 kW to zero percent load). What actually happens is the stepless substitution of an outside load (such as a hot water heater) in any amount required to control the frequency at 60 cycles whenever the normal demand for power is reduced. The reverse procedure is used to avoid lowering the frequency whenever the demand is increased. When the stepless diversion is inadequate to absorb the generator output, a relay applies a fixed step of load of any magnitude less than 10 kW, and the stepless load instantly and automatically readjusts its diversion to hold speed constant. By having enough sink loads to absorb the full output of the generator, it is possible to allow normal load demand to go to zero, and have frequency remain constant.

It is now possible for the small hydroplant operator to sell power to the network without the expense of autosynchronizing equipment by using an induction system having greatly simplified controls at a fraction of the cost of synchronous systems. Also, the use of the electronic load diverter makes it possible for the small operator to operate an isolated synchronous system with low-cost controls.

For background information *see* HYDROELECTRIC GENERATOR in the McGraw-Hill Encyclopedia of Science and Technology.

[KENNETH M. GROVER]

Immunological phylogeny

All vertebrates can recognize and respond to nonself molecular configurations on microorganisms, cells, or organic molecules by utilizing a complex recognition system termed the immune response. Essentially two general types of cellular systems are involved in immune recognition and responsiveness (see illustration). The first comprises phagocytic cells (macrophages) which are not intrinsically spe-

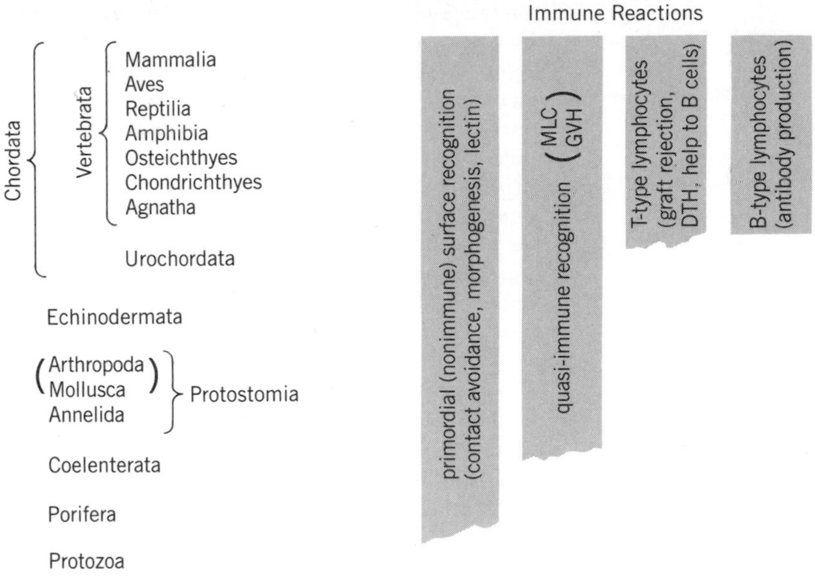

Distribution of immune reactivities within the animal kingdom. MLC = mixed leukocyte culture response; GVH = graft-versus-host reaction; DTH = delayed-type (contact) hypersensitivity. Data are not available for Arthropoda and Mollusca.

cific for foreign antigen but are capable of interacting with cells of the lymphoid series, degrading antigen, and presenting it to the lymphocytes. The lymphocytes have cell surface receptors for foreign antigen, and not only carry out the initial specific recognition role but differentiate into plasma cells which secrete antibodies (B-lymphocyte series), proteins which are specifically induced in large quantity following contact with foreign antigen. Other lymphocytes (T lymphocytes) can differentiate into cytotoxic killer cells which can destroy foreign tissue or cells. T lymphocytes also differentiate into cells termed helpers or suppressors which can affect in either a positive or negative sense the production of antibody. The two general types of specific lymphocytes differ in surface markers and differentiation patterns. One of the major means of distinguishing the two cell populations is that the B cells have readily detectable immunoglobulins (the family of related serum proteins which possesses antibody activity), namely of the IgM and IgD classes, on their surfaces, whereas the T cells lack these markers but have characteristic determinants such as the Thy-1 alloantigen.

Recent studies have focused upon the presence and characterization of cells of the T and B lymphoid series in lower vertebrate species; the nature of the antibodies produced by lower vertebrate species and their relationship to well-characterized immunoglobulin classes of mammals; the extent of diversity in the antibodies of lower vertebrates with particular respect to the degree of variable region heterogeneity within lower species compared with that exhibited by mammals; and the deoxyribonucleic acid (DNA) sequence of genes encoding variable regions of immunoglobin of lower species. The presence of lymphocytes and circulating antibodies has now been documented in all extant vertebrate species, ranging from the ostracoderm-derived cyclostomes, such as hagfish and lampreys, to mammals. However, the existence of induced, specific reactions directly homologous to the immune repertoire has not been clearly established in invertebrates. Studies designed to pinpoint the origin of immune diversification are now being actively pursued in the protochordates, such as tunicates, and there is also considerable interest in the defense mechanisms of invertebrate phyla not directly related to the evolutionary stream which gave rise to the chordates.

Cellular immune responses. The presence and role of phagocytic cells in engulfing foreign pathogens has been documented in virtually all metazoan organisms. Phagocytic cells possess limited capacity to discriminate self from nonself, and this is due in part to the presence of lectins (molecules capable of binding specifically to various sugars) on their surface. This conclusion has been directly shown for phagocytic cells of the oyster, and apparently correlates with the fact that most invertebrates possess lectins of various specificities circulating in their hemolymph.

Although there is no evidence at present to suggest that invertebrate lectins and vertebrate immunoglobulins are homologous structures, sufficient diversity exists within lectins of certain species to indicate that these types of molecules and their cellular expression on phagocytes might serve as a primitive and universal recognition mechanism. Colonial tunicates (protochordates) have shown the capacity to reject allografts, a phenomenon comparable to T-cell-mediated graft rejection of true vertebrates, although the existence of true lymphocytes in protochordates has yet to be firmly established. All true vertebrates possess cells clearly recognizable as lymphocytes and can carry out T-cell functions, such as graft reject, and show the capacity of B cells to synthesize and secrete immunoglobulins. The teleost fish and amphibians possess lymphocytes of distinct functional and cell surface marker phenotypes which resemble T and B lymphocytes of birds and mammals. In addition, functional assays indicate that teleost fishes and amphibians possess helper T lymphocytes which are capable of interacting specifically with B cells in the production of circulating antibody. Lampreys and hagfish lack discrete lymphoid organs characteristic of higher vertebrates, but contain diffuse collections of lymphoid cells in the area of the gills, which may correspond to a primitive thymus, and in the gut, which may correspond to the spleen of higher vertebrates. By contrast, all placoderm-derived vertebrates (sharks, bony fishes, amphibians, reptiles, birds, and mammals) have clearly demarcated thymuses and spleens. True lymph nodes are not present in vertebrate species more primitive than mammals, but birds possess aggregates of lymphoid tissue probably serving a function similar to that of lymph nodes. The thymus is a primary lymphoid organ which generates cells of the T-lymphocyte series. In birds, a discrete organ, the bursa of Fabricius, is the primary source of B cells. The precise anatomical location of the bursa is not known in mammals or vertebrate species distinct from birds, but it is clear that some sort of functional equivalent must exist.

B cells of teleosts and amphibians have recently been shown to express surface immunoglobulin, and this surface immunoglobulin has been isolated and characterized. It resembles the 7SIgM (IgMm) which is associated with the membrane of B lymphocytes of mammals in overall properties. Mammalian B lymphocytes express, in addition, IgD membrane immunoglobulin. This has not yet been conclusively shown on the surfaces of B lymphocytes of teleost and amphibian species. The nature of the antigen receptor on T lymphocytes is not exactly known at this time, but studies with T lymphocytes of teleost fish suggest that a molecule related to immunoglobulin might serve as the specific antigen receptor on these cells.

Immunoglobulins of lower species. Humans possess five major classes or isotypes of immunoglobulin: IgG, IgM, IgA, IgE, and IgD. These classes are defined by the type of heavy chain (γ, μ, α, ϵ, or δ) which occurs in disulfide-bonded association with light chains (either of κ or λ type). The usual structure of an immunoglobulin consists of two heavy chains joined to two light chains; in order to form the combining site to antigen, a light chain must be associated with the heavy chain. The combining site for antigen is made up by interaction of the variable (V) regions of a light chain in association with that of a heavy chain. A molecule consisting of two light (L) and two heavy chains would thus have two combining sites. The variable regions comprise approximately the first 100 N-terminal residues (amino acids at the amino terminus of the peptide chain) of light and heavy chains, and the remainder of the molecules are common (or constant) within a given isotype.

Some immunoglobulins, notably IgM (immune macroglobulin), exist as polymers of the basic four-chain unit. The IgM molecule normally consists of a cyclic pentamer of $L_2\mu_2$ subunits. This is the first immunoglobulin to appear in ontogeny, and the first to appear in phylogeny. Immunoglobulins of cyclostomes, elasmobranchs (sharks and rays), and many teleost fish consist only of IgM polymers (5-mers in elasmobranchs; 4-mers in teleosts) and in some cases IgM subunits (μ_2L_2). Immunoglobulins possessing heavy chains distinct from the μ-like heavy chains of that species are present in some lungfish (Dipnoi) and in anuran amphibians (frogs and toads). Dipnoi have a low-molecular-weight non-IgM immunoglobulin (termed IgN) that possesses a heavy chain with a molecular weight of approximately 36,000 daltons. The μ chain has a mass of approximately 70,000 daltons, and that of the γ chain is approximately 50,000 daltons. Anuran amphibians show a classical progression of antibody activity from high-molecular-weight molecules (IgM) to molecules of approximate mass 180,000 daltons which is apparent with time following immunization. The low-molecular-weight immunoglobulin is not IgM, and it does not correspond directly to IgG or other of the non-IgM immunoglobulins of mammals. Amino acid sequence data are required to allow definitive conclusion regarding its direct homology to mammalian classes, but the present data suggest that the heavy chain of the low-molecular-weight immunoglobulins of anuran amphibians represents a parallel duplication of a heavy-chain gene distinct from that which generated the γ chain of mammals. Birds possess IgM and IgA immunoglobulins, but also possess a non-IgG immunoglobulin similar to that of amphibians as their major immunoglobulin class. This immunoglobulin has been termed IgY. IgG immunoglobulins containing γ chains clearly homologous to those of humans and of true mammals are found only within the three subclasses of living mammals, namely, eutherians, metatherians (marsupials), and monotremes (for example, the echidna).

Duplication of constant-region genes appears to be an essential feature of immunoglobulin evolution both on the large scale of emergence of vertebrate classes and on the smaller scale of speciation where distinct subclasses still arise from the major classes which might have emerged early in vertebrate evolution. The constant regions consist of domains or internal homology units of approximately 100 amino acids, and contain an internal disulfide bond. These domains are specified by individual gene segments or exons occurring within the heavy-chain chromosome. For example, the μ chain consists of five domains, one variable and four constant, each of which is encoded by a separate exon. The emergence and evolutionary variation of the constant regions in immunoglobulin has resulted from duplication or deletion of gene segments corresponding to the domains of the actual proteins. Selective forces have acted upon the proteins produced because each domain carries out a particular effector function, such as fixation of complement or binding to cell surfaces.

Variable-region diversity in lower vertebrates. Although all vertebrates can produce recognizable immunoglobulin, and sharks and higher vertebrates possess immunoglobulin clearly similar to that of mammals, the degree of diversity within the V-region compartment of lower vertebrates is most probably less than that of mammals. Sharks, for example, can make specific antibodies to a wide variety of antigens, including small organic molecules (haptens), viruses, and foreign cells, but the affinity of these antibodies is low and does not increase with length of immunization. Although the size of the antibody can shift from IgM pentamer to IgM monomer with prolonged immunization, only the IgM class is represented and the affinity does not increase. By contrast, immunization of a common mammal, such as the rabbit, with a hapten usually generates first a primary response which is predominantly IgM with low affinity in the combining site. However, with prolonged immunization, the predominance of antibody activity is shifted to the IgG class, and the affinity of the antibody for the hapten may increase by three or four orders of magnitude. Failure of affinity maturation is also a property of IgM antibodies of higher vertebrates and may reflect a lack of somatic mutation in the V_H genes that become associated with the μ constant regions. In addition, analyses, such as isoelectric focusing (which provides a means of counting the number of distinct antibodies on the basis of charge) and amino acid sequence analysis, indicate that in the shark and in the anuran amphibian *Xenopus* the number of antibodies capable of responding to a hapten, such as dinitrophenol, is much less than that of mammals. The evidence suggests that the degree of heterogeneity of amphibian and fish antibodies is less than that of mammals. Although the basic event underlying immune recognition must be ancient and possibly preceded the origin of vertebrates, subsequent developments for increasing diversity, such as somatic mutation, apparently do not occur in lower vertebrates.

The primal event in the emergence of the classic immune system was the generation of V-region genes, probably by tandem duplication of a precursor and the joining of one of these genes to a constant-region gene in order to allow formation of an immunoglobulin heavy or light chain. It is possible that V-region genes and their products preceded constant regions which are necessary for effector functions because specific alloreactive lymphocytes, comparable to T cells, occur in species, such as tunicates, which lack the capacity to form circulating antibodies. Substantial evolutionary conservation has been found by comparing the sequences of heavy-chain variable-region genes of a reptilian species, the caiman, with those of human and mouse. It is interesting that the caiman gene sequence clearly fits into the third heavy-chain variable-region subgroup of humans. This indicates that the generation of these subgroups preceded the events of speciation which lead to the demarcation between reptiles and mammals.

Although the precise nature of the precursors of the specific elements of the immune system in evolution remains to be determined, the genetic and cellular events which lead to the capacity for specific immune recognition, diversification, and reactivity occurred early in vertebrate evolution. V- and C-region genes and a mechanism for joining them must have emerged prior to the emergence of true vertebrates, although somatic diversification mechanisms resulting in amplified repertoires and affinity maturation may not have arisen until relatively late in vertebrate evolution corresponding to the phylogenetic levels of birds and mammals. The specific immune elements were grafted onto a primitive and universal mechanism involving phagocytic cells and lectins.

For background information *see* CELLULAR IMMUNOLOGY; IMMUNITY; IMMUNOGLOBULIN; IMMUNOLOGY; LYMPHATIC SYSTEM; PHAGOCYTOSIS; TRANS-

PLANTATION BIOLOGY in the McGraw-Hill Encyclopedia of Science and Technology.

[JOHN J. MARCHALONIS]

Bibliography: N. Cohen and M. M. Sigel (eds.), *The Reticuloendothelial System*, vol. 3: *Phylogeny and Ontogeny*, 1982; L. Du Pasquier, Antibody diversity in lower vertebrates—why is it so restricted? *Nature*, 296:311-313, 1982; G. W. Litman and J. J. Marchalonis, Evolution of antibodies, L. N. Ruben and M. E. Gershwin (eds.), *Immune Regulation*, pp. 29-60, 1982.

Immunology

Adjuvants are substances that increase the effect of administered antigens on the immune system; that is, they potentiate the immunogenic activity of antigens. Typically, the effect of adjuvants is observed as an increased amount of antibody that is formed by an animal injected with a mixture of an antigen and adjuvant, in contrast to the lesser amount of antibody that is formed in response to the antigen alone. However, an adjuvant can also affect other immune properties, such as cell-mediated immunity, and it can be used to enhance the likelihood of specific antiorgan reactions, such as arthritis, encephalitis, uveitis, and orchitis, or to increase resistance to infectious agents or tumors.

Adjuvant effects. The adjuvant effect can be manifested by various substances. The mechanism of action of different adjuvants varies. Manipulation of the antigen itself is not considered to fall under the category of adjuvant effects. For instance, most substances of low molecular weight, termed haptens, are typically not immunogenic. However, they can be linked to a protein carrier, and the compound is then able to induce antibody formation against the low-molecular-weight substance. Large-molecular-weight substances, such as proteins, may fail to induce an antibody response under certain circumstances—for instance, if all aggregated material is removed by ultracentrifugation. The remaining soluble material is generally unable to induce an antibody response, but is capable instead of producing suppression or tolerance. Aggregation of the soluble material by heating or chemical linkage, or introduction of the soluble material with an adjuvant usually restores the ability to induce specific antibody formation.

The action of adjuvants is not completely understood. They can act by modifying the physical characteristics of the immunogen, slowing its dispersal from the site of adminstration, or altering in some way the ability of the immune system to react to the presence of the immunogen.

The last category includes various processes, acting either singly or in combination. These processes range from an alteration in the properties of the cell membrane that initially encounters the immunogen, or the direct enhancement of intracellular processes, to the release of factors (monokines, lymphokines) from one set of cells (for example, T lymphocytes, macrophages) that promote the activity of other cells (for example, B lymphocytes).

Types. The earliest described adjuvants include lipids, especially waxes, and protein precipitants, such as alum. These materials are used to delay the dispersal of the immunogen. Waxes and lipids do so by retaining the immunogen within the lipid, thereby allowing only a low rate of diffusion through the oil-water interface into the animal's body fluids. A common adjuvant of this type is called Freund's incomplete adjuvant, which consists of light paraffin oil and a small amount of an emulsifer such as mannide monooleate (Arlacel A). Alum is used as an adsorbent that slowly releases the adsorbed immunogen. A complete Freund's adjuvant is formed by the addition of mycobacteria to the incomplete adjuvant, for example, heat-killed *Mycobacterium tuberculosis*, *My. butyricum*, or live BCG (bacillus Calmette-Guérin).

The action of the mycobacteria appears to be largely reproducible by a polymer of one or more disaccharides containing *N*-acetyl glucosamine and *N*-acetyl muramic acid attached to a basic peptide. The adjuvant activity is reproducible with synthetic analogs such as *N*-acetylmuramyl-L-alanyl-D-isoglutamine (MDP) or *N*-acetylmuramyl-L-alanyl-D-glycine. A similar effect may also be produced by other agents, for example, *N*,*N*-dioctadecyl-*N'*,*N'*-bis(2-hydroxyethyl) propanediamine (CP20961). It is likely that this compound, as well as the mycobacterial and synthetic glycopeptides, chiefly acts on macrophages which then release substances that act upon T and B lymphocytes, but a direct action on lymphocytes can also occur. Probably the initial action of this type of adjuvant is to induce a marked acute inflammatory reaction followed by an enhanced immunological response mediated by lymphokines. It should be noted that an excessive or repeated administration of complete Freund's adjuvant, though not of MDP, results in a profound disorganization of the local lymph nodes and a subsequent reduction of the immune response which may be attributed to the lymph node disorganization, as well as to the intense stimulation of suppressor cells evoked by adjuvants of this type. Freund's adjuvant, as well as some other procedures used to stimulate increased humoral antibody production (for example, multiple intradermal injections), is also highly effective in eliciting delayed hypersensitivity mediated by T cells and the formation of immunoglobulin E, which is responsible for immediate hypersensitivity and anaphylaxis. The ability to affect a variety of immunological responses is termed immunomodulation; the ability to increase the responses is called immunopotentiation.

Immunopotentiation. Marked immunopotentiation can occur with bacteria other than the mycobacteria, as well as with bacterial products. *Corynebacterium parvum* appears to act on macrophages but not T cells, and it does not stimulate delayed hypersensitivity. *Bordetella pertussis*, the agent of whooping cough, releases a protein lymphocytosis-promoting factor, which acts on T cells; the organism may also release other active materials. The heat-inactivated organism is especially effective in promoting IgE antibody. Endotoxins, or lipopolysaccharide extracts of gram-negative bacteria (for example, *Escherichia coli*), in which the active constituent is termed lipid A, act on B lymphocytes and macrophages. Lipopolysaccharides are very efficient B-lymphocyte mitogens. The toxic and immunostimulatory properties of lipopolysaccharides appear to coexist. There are many other materials with immunological activity, including a phospholipid fraction of gram-negative bacteria, and a methanol-acetone extract of mycobacteria. Intrinsic immunopotentiators such as thymic hormones, lymphokines, or the transfer factor (a soluble product of leukocytes which has the ability to transfer specific cell-mediated immunity from individuals who possess it to ones who do not) can have an effect similar to adjuvants, but are not so termed. Pharmacological and miscellaneous agents with adjuvantlike or immunopotentiating effects include polyribonucleotides (A;C and I;C polyribonucleotide made up of adenosine and citidine, and of inosine and cytodine, respectively) and levamisole, as well as bestatin (an enzyme inhibitor from *Streptomyces olivoreticuli*), carrageenan, colchicine, derivatized (chemically modified) nucleosides, isoprinosine, polyacrylic acid, pyran, saponin, and tilorone. These agents are generally noted to have immunostimulatory or immunomodulatory properties in the course of their usage—for example, for therapeutic purposes. Levamisole has been used as an antihelminthic agent and isoprinosine as an antiviral one. The agents vary in their effects on the components of the immune system, and these effects have been studied for each agent. Generally, the effect is to enhance macrophage activity, T- or B-cell proliferation, or the release of intrinsic immunopotentiators.

For background information *see* ANTIBODY; ANTIGEN; CELLULAR IMMUNOLOGY; HYPERSENSITIVITY; IMMUNITY; IMMUNOGLOBULIN in the McGraw-Hill Encyclopedia of Science and Technology.

[ALEXANDER BAUMGARTEN]

Bibliography: A. C. Allison, Mode of action of immunological adjuvants, *J. Reticuloendothelial Soc.*, 26:619, 1979; J. F. Bach, in P. J. Lachmann and D. K. Peters (eds.), *Clinical Aspects of Immunology*, 4th ed., 1982; P. Jolles and A. Paraf, *Chemical and Biological Basis of Adjuvants*, 1973; J. W. Osebold, Mechanisms of action by immunologic adjuvants, *J. Amer. Vet. Med. Ass.*, 181:983, 1982.

Immunotherapy

Immunotherapy is defined as the treatment of cancer by improving the ability of a tumor-bearing individual (the host) to reject the tumor immunologically. There are molecules on the surface of tumor cells, and perhaps in their interior, that are recognized as different from normal structures by the immune system and thus generate an immune response. These are tumor-associated antigens, clearly demonstrated with animal tumors and strongly suggested with human tumors. The two components of the immune response are cell-mediated and antibody-mediated immunity, which must work in concert to overcome tumor cells. One type of thymus-derived lymphocyte (also called a cytotoxic T cell) can destroy tumor cells directly, while another recruits other white blood cells, macrophages, that do the killing. Natural killer cells and perhaps other white blood cells may also participate. Antibodies produced in reaction to a tumor probably work mostly by binding to white blood cells at one end and to tumor cells at the other, bringing the two cells into close proximity and facilitating destruction of the tumor cells. However, elements that normally regulate immunity such as suppressor T cells are stimulated excessively by the tumor, and this leads to an immune response that is deficient and unable to reject the growing tumor. Thus, the strategy of immunotherapy is to stimulate within or transfer to the tumor-bearing individual the appropriate antitumor elements, while avoiding further stimulation of suppressor elements.

There are four broad categories of immunotherapy: active, adoptive, passive, and restorative.

Active immunotherapy. Active immunotherapy attempts to stimulate the host's intrinsic immune response to the tumor, either nonspecifically or specifically. Nonspecific active immunotherapy utilizes materials that have no apparent antigenic relationship to the tumor, but have modulatory effects on the immune system, stimulating macrophages, lymphocytes, and natural killer cells. A large and rather heterogeneous group of agents have been studied, the best known of which are microbial agents, such as Bacillus Calmette-Guerin (BCG), and also pyran, interferon, and the vitamin A derivatives called retinoids.

Clinical trials with nonspecific active immunotherapy alone have largely failed to improve the survival of cancer patients, with only a few exceptions, such as in nodular lymphoma and in ovarian cancer. Interferon-alpha is a low-molecular-weight protein produced by leukocytes in response to viral infections. Initial studies with partially purified extracted interferon-alpha reported several remissions in patients with multiple myeloma and nodular, poorly differentiated lymphocytic lymphoma, and a prolonged survival of patients with osteosarcoma treated prophylactically after amputation. Partial responses were also observed in late-stage breast cancer patients. Recent trials of recombinant interferon-alpha have noted significant responses of patients with renal cell (kidney) carcinoma, multiple myeloma (a bone marrow tumor), and Kaposi's sarcoma, a tumor found in patients with AIDS. However, it is uncertain whether the nonspecific immunomodulatory effects of interferon were responsible, since interferon also has a direct antitumor effect. Studies on retinoids were prompted by epidemiological data showing a decreased incidence of lung and bladder cancer associated with a high intake of vitamin A. While retinoids do augment the immune response, like interferon they too have direct effects on neoplastic cells. Retinoids also promote the maturation of preneoplastic cells toward normal. This promising class of agent is now undergoing early clinical trials. *See* ACQUIRED IMMUNE DEFICIENCY SYNDROME.

Specific active immunotherapy attempts to stimulate specific antitumor responses with tumor-associated antigens as the immunizing materials. Tumor cells from the same patient or from another with the same type of tumor, irradiated or altered in an artificial environment to make them more antigenic, have been used in the past to try to augment specific immunity. Presently, the potential usefulness of antigenic extracts, both soluble and particulate, is being investigated in such dissimilar diseases as malignant melanoma (malignant skin mole) and lung cancer. Rapid progress in identifying important tumor-associated antigens is being made with these and several other tumors, especially through the use of monoclonal antitumor antibodies. The use of extracts of tumor antigens for immunotherapy must be approached with great caution, since soluble tumor antigens can generate suppressor T cells when combined with a small amount of host antibodies. This could lead to enhanced tumor growth. The content of antigens in specific tumor vaccines, the form in which antigens should be presented to the patient (such as particulate, or together with adjuvant materials), and the schedule and route of administration are all critical questions.

Adoptive immunotherapy. This denotes the transfer of immunologically competent white blood cells or their precursors into the host. Bone marrow transplantation, while performed principally for the replacement of hematopoietic stem cells, can also be viewed as adoptive immunotherapy. Some success in the treatment of acute leukemia has been noted by allogeneic (second-party) bone marrow transplantation after sublethal irradiation of the patient. Interestingly, the results from syngeneic (twin) bone marrow transplants have been less encouraging, perhaps because the identical twin's lymphocytes do not recognize the leukemic cells as sufficiently different from "self" to reject them. In the near future, transfusion of specifically "educated" T lymphocytes in the test-tube will be feasible. It is now possible to grow T cells in artificial environments with the aid of T-cell growth factor (interleukin 2), stimulate (or restimulate as the case may be) them with tumor antigens, and transfuse the cells into the patient. This has been effective therapy in mice and could soon be applied to patients. Transfusion of leukocytes "armed" with antitumor antibodies is another approach that appears feasible in the near future for the treatment of human cancers.

Restorative immunotherapy. This comprises the direct and indirect restoration of deficient immunological function through any means other than the direct transfer of cells. Repletion of competent effector cell populations is one form of restorative immunotherapy. Thymic hormones, such as thymopoietin and thymosin, can convert T-cell precursors into helper T cells, a property shared by the antihelminthic agent Levamisole and several of its congeners. In a study of small-cell carcinoma of the lung, thymosin increased the median survival of patients who had sustained a complete response from chemotherapy. The subgroup that benefited most was the one with a deficient complement of T cells. Deficient functioning of T cells may be a problem in most patients with a tumor. Deficient percentages of T cells relative to other leukocytes have been noted in several types of tumor, the most noteworthy of which is Kaposi's sarcoma, a malignant blood vessel tumor occurring mostly in patients with a kidney transplant or AIDS. Repletion of number or function is an important strategy which might by itself be curative in certain tumors. Antagonism of suppressor influences is another form of restorative therapy. Low doses of the antitumor drug cyclophosphamide selectively inhibit suppressor T cells in mice. The exact dose effective in humans remains to be determined. Inhibitors of prostaglandin synthesis, such as indomethacin, may be useful in overcoming other suppressor cells of the macrophage series, which act principally through prostaglandins.

Passive immunotherapy. This means the transfer of antibodies to tumor-bearing recipients. This approach has been made feasible by the development of hybridoma technology, which now permits the production of large quantities of monoclonal antibodies specific for an antigenic determinant on tumor cells. Monoclonal antibodies made against tumor-specific antigens, if they in fact exist, or against tumor-associated antigens, which may be shared by certain normal cells, are of potential importance therapeutically. As noted above, these antibodies probably work through the arming of white blood cells, especially macrophages, of the host. They may be most effective in treating dispersed tumors such as leukemia, and ovarian cancer in ascites (fluid in the abdominal cavity). The potential for enhancement of tumor growth as a result of the formation of antigen-antibody complexes and the excessive stimulation of suppressor cells exists and must be carefully monitored. Currently, mouse monoclonal antibodies to human melanoma and colon cancer are undergoing trials in humans, as are anti-T-cell antibodies in T-cell lymphomas and leukemia, with encouraging early results. The development of human monoclonals, which is now progressing, should obviate some of the allergic reactions that have occurred with mouse antibodies, and promises still greater specificity of attack on tumor cells.

For background information *see* CELLULAR IMMUNOLOGY; IMMUNITY; INTERFERON; ONCOLOGY; TRANSPLANTATION (BIOLOGY) in the McGraw-Hill Encyclopedia of Science and Technology.

[J. KAN-MITCHELL; M. S. MITCHELL]

Bibliography: M. S. Mitchell, The advent of monoclonal antibodies, *Consultant*, 23:127–134, 1983; M. S. Mitchell and J. H. Bertram, Tumor immunology: Current concepts and therapeutic applications in human cancer, in P. Calabresi, P. S. Schein, and S. A. Rosenberg (eds.), *Medical Oncology*, 1983; W. D. Terry and S. A. Rosenberg (eds.), *Immunotherapy of Human Cancer*, 1982.

Inertial guidance system

A low-accuracy Doppler-aided strapdown inertial navigator is an integrated navigation system which consists of a low-cost low-accuracy inertial navigation system, augmented by a Doppler radar. The integration of these systems yields a hybrid navigator whose performance is superior to that of either system when acting alone. The main contenders for the installation of Doppler-aided strapdown inertial navigators are military helicopters, in which Doppler radar systems are standard equipment.

Inertial navigation systems (INS) are autonomous systems which supply position, velocity, and attitude information (in the ensuing discussion this information will be referred to as the navigational information). By their nature, INS are unstable in that the error in their indicated output navigational information diverges with time. To check this divergence, it is possible to use a very accurate INS or some external partial navigational information to aid the INS. Although the second approach requires an additional navigational aid, it is quite often less expensive than the first approach because of the high cost of a high-grade INS. A typical example of the second approach is a low-accuracy Doppler-aided strapdown inertial navigator. It consists of a low-cost low-accuracy INS which belongs to the strapdown category, a Doppler radar system, and a processor (Fig. 1). The processor has three tasks: it performs the strapdown INS computations; it resolves the Doppler velocity vector measurement; and it processes, in some optimal way, the data obtained from both the INS and the Doppler subsystems to yield an estimate of the navigational information as well as the subsystem error sources.

Low-accuracy Doppler-aided strapdown INS is particularly suitable for helicopters in which a Doppler navigator is usually standard equipment. Therefore, when a low-cost INS and an inexpensive processor are added to the original Doppler navigator, a high-performance navigational system results.

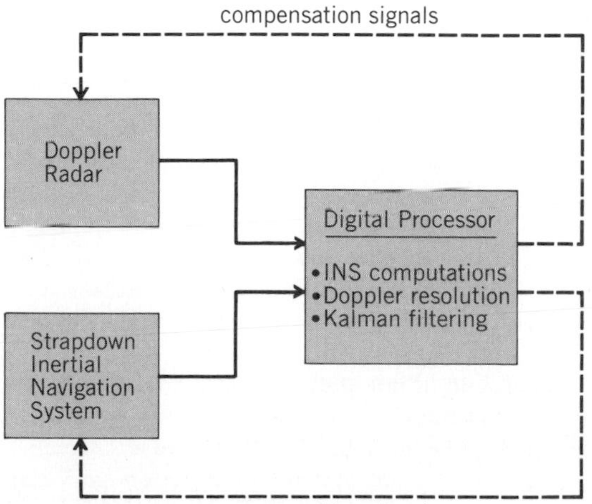

Fig. 1. Schematic diagram of the principle of operation of a Doppler-aided strapdown inertial navigator.

The INS which is most suitable for such application is a strapdown system, principally because its cost is low while its accuracy is adequate. Many fixed-wing aircraft also carry a Doppler navigator as a standard system, but they are usually also equipped with a high-cost high-accuracy INS. The two systems can be integrated in much the same way as the low-accuracy strapdown INS and the Doppler system are integrated on a helicopter, but this is not a low-cost solution and the improvement is not as remarkable as that of the low-cost Doppler-aided strapdown INS.

Several key factors arose at the beginning of the 1970s to enable the fabrication of an efficient low-accuracy Doppler-aided INS. First, the miniaturization of electronic circuits made it possible to manufacture light, relatively small Doppler navigators with low power consumption. Second, a new efficient estimation algorithm known as Kalman filter was introduced at the beginning of the 1960s. This algorithm enables the integration of the Doppler navigator and the INS into one efficient system. Third, the emergence of fast inexpensive processors allowed real-time navigation and Kalman filter computations. Finally, the need which emerged in the 1970s for midcourse guidance of tactical standoff weapon systems and new breakthroughs in strapdown-gyro technology promoted the development of the low-cost low-accuracy strapdown INS. The concurrence of these developments made possible the realization of a low-cost Doppler-aided INS.

A general description of the three key elements in a low-accuracy Doppler-aided strapdown navigator (that is, the strapdown INS, the Doppler radar, and the Kalmon filter) is presented below.

Strapdown INS. Strapdown INS is an autonomous navigation system which provides the navigational information of the vehicle on which it is installed. The system consists of an inertial measuring unit which contains accelerometers and gyroscopes, and a processor. Unlike a gimbaled INS, where the inertial measuring unit is isolated from the rotational motion of the vehicle by a set of gimbals, in a strapdown INS the inertial measuring unit is strapped to the vehicle. The accelerometers measure the specific force vector which operates on the INS; then this vector is transformed into a reference coordinate system in which the gravity vector is added to change the specific force vector into the acceleration vector of the vehicle. Compensation for Earth rotation and curvature finally yields expressions for the rate of change of the velocity vector of the vehicle. One integration of the latter, together with the initial velocity, produces the instantaneous velocity of the vehicle. Division by radii of curvature, together with another integration added to the initial position, yields the geographic position as well as the height of the vehicle. The gyro outputs are processed to keep track of the angular motion of the vehicle. The result of this processing is a transformation matrix which transforms vectors from the vehicle, and hence the inertial measuring unit coordinate system, to the reference system. This matrix is used to transform the specific force from the vehicle to the reference coordinate system as described earlier. The transformation matrix supplies the attitude information of the INS. Research into strapdown INS started as early as the late 1950s, but it was only in April 1970 that it came into prominence. At that time the Strapdown Abort Guidance System of *Apollo 13* performed a crucial role in returning its crew safely to Earth following a rupture of a fuel-cell oxygen tank in the service module which aborted the lunar mission.

Two classes of strapdown INS are of interest: the accurate and expensive kind which is installed on the latest vintage of commercial jets, and the low-cost low-accuracy systems which are designated for use in tactical weapons. The latter is of interest in this article.

INS initial alignment. Strapdown INS, like gimbaled INS, has to undergo an initial alignment stage during which the initial attitude of the inertial measuring unit coordinate system (body system) with respect to the reference system is determined. This can be done external of the INS by using optical means, or internally by using the IMU sensors. The latter is known as self-alignment. The self-alignment about the level axes is based on the inertial measuring unit accelerometers' measurement of the gravity vector and is accomplished very quickly. The self-alignment in azimuth, although it relies on the gyros' readings of the Earth's rate of turn, which is a very weak signal. Consequently this process is time-consuming. It is known as gyrocompassing. In gimbaled INS, gyrocompassing accuracy is bounded by the constant error of the east-pointing gyro. In strapdown INS, the accuracy is bounded by the constant error of the equivalent east gyro. (The equivalent east gyro error is the sum of the projections of the three strapdown gyro errors on the east direction.) This bound is expressed by the relation below,

$$\phi_{\min} = \frac{\epsilon_E}{\Omega \cos L}$$

where $\phi_{\min}$ is the lower bound on the azimuth error in gyrocompassing, ϵ_E is the absolute value of the equivalent east gyro constant-error component, Ω is the Earth turning rate, and $\cos L$ is the cosine of the latitude angle at which the INS is aligned. A by-product of the self-alignment is the calibration of the north gyro error in gimbaled INS, or the equivalent north gyro error in strapdown INS. (The definition of the equivalent north gyro error is similar to that of the equivalent east gyro error.)

Doppler radar. The Doppler radar continuously measures the vehicle velocity vector with respect to ground by utilizing the Doppler effect. The three components of the velocity are obtained in the antenna coordinate system. The attitude of the antenna with respect to the navigation reference coordinate system has to be known in order to determine the velocity components in that reference system. Attitude is usually determined by using a vertical, and a directional gyro (or a compass). However, in a Doppler-aided strapdown system the strapdown INS provides the necessary attitude information. To measure the ground velocity of the vehicle, a radar transmitter-receiver is mounted on the vehicle and radiates electromagnetic energy toward the Earth's surface by means of several beams. Some of the energy is backscattered by the Earth and is received by the radar receiver on the vehicle. When there is a relative velocity between the vehicle and the Earth's surface, the received frequency differs from the transmitted frequency. This difference is proportional to the relative velocity. This frequency difference, which is known as the Doppler effect, is detected, and the velocity is extracted from it. In typical microwave Doppler radars the value of this Doppler shift is on the order of 20 to 30 Hz per knot of speed (40 to 60 Hz per m/s of speed).

Kalman filter. Kalman filter is an algorithm which provides an estimate of the random state of a linear system. This estimator is unbiased and is optimal in the sense that the estimation error is a minimum variance error. The role it plays in a low-accuracy Doppler-aided strapdown navigator is to blend together the strapdown INS and the Doppler radar into an integrated system whose performance surpasses that of either system.

The way that the filter operates can be roughly described as follows. The propagation model of the INS error due to its error sources is stored in the filter. The velocity vector computed by the INS is compared with that measured by the Doppler radar. The difference is considered by the filter as an indication of the velocity error generated by the INS. Through using the INS error model, the filter attributes this error to the various INS error sources and assigns them numerical values. Having estimated the values of the errors, the processor takes them into account and compensates for them in future INS computations so that when these error estimates are exact, the error sources are fully compensated. In fact, the filter also estimates some of the Doppler radar errors in the same way. It should be emphasized, however, that error sources which have no distinct influence on the INS velocity error are said to be unobservable and cannot be estimated by the filter. In the process of estimating the errors, the filter assigns different weights to the INS and to the Doppler radar data in accordance with the accuracy of each system. This way the filter optimally weighs the information obtained from each system. Kalman filters were introduced in 1960 and 1961, and in 1968 a Kalman filter was used for the C-5 aircraft INS. Nowadays practically every INS utilizes some form of a Kalman filter for initial alignment and midcourse updates.

Operation of the integrated system. By using the vehicle compass, the strapdown INS is coarsely aligned prior to take-off. This coarse alignment yields a poorly aligned INS whose errors grow rapidly; however, since the Kalman filter blends the INS with the Doppler radar, the INS errors are

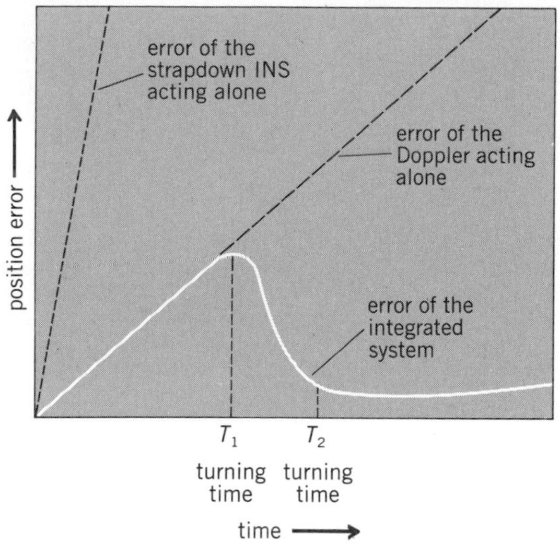

Fig. 2. Typical position error growth as a function of time for the strapdown INS when it operates alone, for the Doppler radar when it operates alone as a navigator, and for the integrated low-cost Doppler-aided strapdown INS.

bounded by the Doppler. Figure 2 presents the growth of the INS position error, as well as that of the Doppler radar when the latter is used as a navigator. It is seen that up to a time T_1 (when the inertial measuring unit is rotated about its vertical axis), the position error of the augmented system is bounded by the Doppler. However, after time T_1, the augmented system operates better than either system.

This phenomenon is explained as follows. At the beginning of the flight, the coarsely aligned INS begins a self-alignment stage using the velocity information supplied by the Doppler radar. (This alignment is being accomplished by the Kalman filter.) Consequently a fast level alignment takes place but, as explained earlier, the azimuth alignment is bounded by the inaccuracy of the gyros and remains poor. More precisely, the gyrocompassing operation is bounded by the equivalent east gyro error (see equation above) which cannot be calibrated out. However, as was mentioned before, during this gyrocompassing stage the equivalent north gyro error is being calibrated. Then, at time T_1 when the inertial measuring unit has rotated 90° about its vertical axis, the calibrated equivalent north gyro turns into an equivalent east gyro. Consequently the bound on azimuth alignment decreases, and a new gyrocompassing stage commences which increases azimuth alignment accuracy. Furthermore, the Kalman filter yields better estimates of the vehicle position. After T_1, the present equivalent north gyro error (which prior to T_1 was the east equivalent gyro error) is calibrated. Then, at T_2 when the inertial measuring unit is rotated again, the newly calibrated equivalent north gyro becomes the equivalent east gyro. At this point the equivalent horizontal gyros are well calibrated, the strapdown INS is well aligned, and many other system error sources are calibrated as well. It is seen, then, that the rotations are necessary for transforming the low-accuracy strapdown INS into an accurate inertial navigation system by virtue of the good alignment and calibration. Moreover, if the Doppler-carrying vehicle also turns, the slowly varying Doppler radar errors are calibrated too. Since the calibrated error sources do not generate entirely constant errors, occasionally turning the inertial measuring unit, or the whole vehicle, helps to maintain the calibration and the high accuracy of the augmented system. In this manner the joint operation of the low-cost strapdown INS and the Doppler radar yields a relatively high-accuracy hybrid navigation system.

If there is enough time to perform the alignment phase prior to take-off, the same sequence of operations can be carried out on the ground without the use of the Doppler system. Basically, there is no difference between this case and the one just described. In fact, this case is a special case of the prior one in which the velocity is precisely known to be equal to zero. In the ground-alignment situation, the vehicle takes off with an already well-aligned and -calibrated INS. Here, too, occasional in-flight turns of the inertial measuring unit, or the whole vehicle, help to maintain the high accuracy of the system.

For background information see DOPPLER RADAR; ESTIMATION THEORY; INERTIAL GUIDANCE SYSTEMS in the McGraw Hill Encyclopedia of Science and Technology.

[I. Y. BAR-ITZHACK]

Bibliography: I. Y. Bar-Itzhack, D. Serfaty, and Y. Vitek, Doppler-aided low-accuracy strapdown inertial navigation system, *AIAA J. Guid. Control and Dynamics*, 5:236–242, May-June 1982; A. Edwards, Jr., The state of strapdown inertial guidance and navigation, *Navigation: J. Inst. Navig.*, 18:386–401, Winter 1971–1972; M. Kayton and W. R. Fried (eds.), *Avionics Navigation Systems*, pp. 207–280, 1969; P. S. Maybeck, *Stochastic Models, Estimation and Control*, vol. 1, 1979.

Infrared astronomy

In January 1983 NASA launched into space a new type of telescope. The *IRAS (Infrared Astronomy Satellite)*, a joint development by British, Dutch, and American astronomers, is a cryogenically cooled instrument designed to provide the first sensitive all-sky survey in the infrared region of the spectrum. Initial results from *IRAS* suggest that this satellite should provide astronomers with important new information bearing on the process of stellar birth and the energy sources powering emission from galactic nuclei.

Infrared astronomy before IRAS. In the early 1960s, advances in technology provided astronomers with detectors of sensitivity sufficient to begin studies of the universe at wavelengths between 1 and 1000 micrometers. To make successful observations in this wavelength range from the surface of the Earth is challenging for two reasons: (1) Light can penetrate the Earth's atmosphere in only a few,

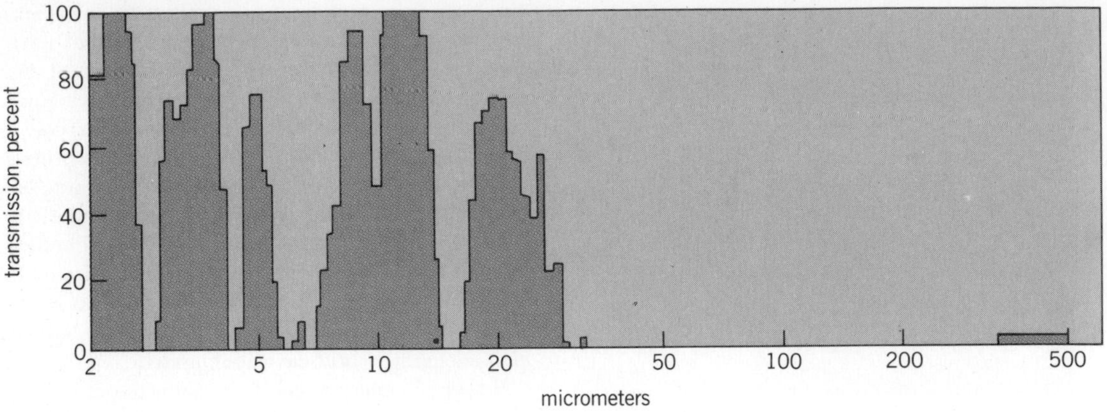

Infrared transmission of the Earth's atmosphere. It is almost completely opaque between about 30 and 1000 μm. (*After S. E. Strom, Infrared and submillimeter astonomy from space, Sky Telesc., 65(4):312–315, 1983*)

relatively narrow-wavelength "windows" (see illustration). (2) Both the Earth's atmosphere and the telescope are "warm" and consequently radiate at a characteristic temperature near 300 K (80°F); the peak emissivity of objects at this temperature occurs at a wavelength near 10 μm. At this wavelength, the background signal from the telescope and from the Earth's atmosphere far exceeds the typical signals from celestial sources by orders of magnitude.

Despite these formidable difficulties, infrared measurements have provided astronomers with a rich harvest of discoveries over the past 20 years. First, infrared measurements have permitted astronomers a glimpse into the birthplaces of new stars—dense, dust-laden molecular cloud complexes. The dust grains contained within these cloud complexes are typically 0.1 to 0.5 μm in diameter and consequently are efficient scatterers of light in the optical region of the spectrum (centered at a wavelength of 0.5 μm). At wavelengths longward of 1 μm, the effective opacity of the grains is much lower, making it possible to search these stellar wombs for newly formed stars. Just after birth, stars are relatively large and cool (typical temperatures of 100 to 1000 K or −280 to 1300°F). Hence, young stellar objects emit most of their light in the infrared, which is another reason why astonomers seek to study the stellar birth process in this region of the spectrum. Later, when they are hotter, but not yet fully mature hydrogen-burning main sequence objects, young stars are often surrounded by dust envelopes which absorb optical radiation and reemit the absorbed energy in the infrared. Hence, even these "adolescent" stars are most easily studied in the infrared. From the ground, astronomers have scanned nearby molecular cloud complexes, succeeding in imaging newly formed stellar clusters and beginning the spectroscopic studies which promise detailed physical understanding of young stellar objects. However, only the brightest such objects within these clouds can be studied from the ground.

Second, infrared observations have revealed that in some galaxies the luminosity measured at long wavelengths (greater than 10 μm) equals or exceeds the luminosity observed in the optical spectral range. The sources of this infrared luminosity have not been identified with certainty. In many cases, it is believed that huge clusters of newly formed stars having a combined luminosity exceeding 100 million suns account for the observed flux. These clusters are, however, obscured from view by thick clouds of submicrometer dust particles. At present, there is no generally accepted explanation for the relatively high frequency with which such "IR galaxies" are observed. It is not possible for an IR galaxy to sustain its current power output for times comparable to the age of the universe. It is conjectured, therefore, that galaxies are infrared-bright only for relatively brief periods during their evolutionary history. However, without a complete census of infrared galaxies, it is impossible to assess whether this explanation is correct or whether a new type of powering "engine" is perhaps responsible for the behavior of these galaxies.

Third, the infrared has already proved itself to be an essential tool in studies of galaxy evolution. In nearby spiral galaxies similar to the Milky Way, infrared measurements provide a dust-free view of the underlying distribution of old stars (the bulk of the luminous mass in a galaxy), whereas optical pictures are most often dominated by the light from the youngest stars. Study of the relationship between the underlying matter distribution, as revealed by infrared "pictures," and the star-forming history of a galaxy is currently one of the more important problems in galactic evolution studies.

The potential of infrared observations for providing keys to the study of galactic evolution is even greater. Edwin Hubble first described the relation between the recession velocity of a galaxy and its distance: the greater the distance of a galaxy, the greater its velocity away from the Earth. Recession velocities are estimated from the Doppler shift of selected spectral features; the greater the shift to the red of a given feature, the greater the velocity of a galaxy away from the observer on Earth. The light

Characteristics of the IRAS survey detector array

Central wavelength, μm	Field of view, arc-minutes	Wavelength interval, μm	Detector material	Detection limits, janskys*
12	0.75 × 4.5	8.5–15	Si:As	0.7
25	0.75 × 4.6	19–30	Si:Sb	0.65
60	1.5 × 4.7	40–80	Ge:Ga	0.85
100	3.0 × 5.0	83–120	Ge:Ga	3.0

*A source is considered "detected" if its flux exceeds 10 times the standard duration of a single measurement. 1 jansky = 10^{-26} W/(m^2·Hz).

emanating from galaxies receding from the Earth at velocities approaching the speed of light is so red-shifted that the peak of the spectral energy distribution for a galaxy, normally at a wavelength near 0.5 μm, is shifted far into the infrared. Hence, to observe light from the most distant galaxies, astronomers must carry out their searches in the infrared.

IRAS. Despite the important contributions of infrared astonomy over the past two decades, knowledge of the infrared sky is extremely incomplete. The first attempt to provide an unbiased survey in the infrared was made at the California Institute of Technology in the mid-1960s by Robert Leighton and Gary Neugebauer. Their 2-μm survey resulted in the discovery of several thousand infrared stars—comparable to the numbers of stars visible to the naked eye on a clear, dark night. Later, the Air Force Geophysics Laboratory conducted two rocket surveys of the infrared sky at wavelengths of 5, 10, and 20 μm. Neither the Cal Tech nor Air Force survey achieved very deep limits. Although both contributed enormously to initial infrared studies of young stellar objects and to the detection of very evolved stars in the process of shedding their outer envelopes, only a few extragalactic sources were detected. Consequently, most infrared observations to date have been made of a limited number of preselected objects. It was recognized that an unbiased, deep survey of the infrared sky would be an essential next step in achieving a full understanding of the cool or dust-shrounded constituents of the universe.

About a decade ago, a group of American, Dutch, and British astronomers undertook the challenge of developing an Explorer satellite to carry out such a survey. They decided to develop a cryogenically cooled space telescope. By going to space, they would avoid the obscuration and thermal emission from the Earth's atmosphere. By cooling the telescope with a reservoir of liquid helium, they would greatly reduce the warm background against which celestial sources are usually viewed from the ground. The technical challenge of developing a long-lived, cool telescope was formidable. After several years of disappointment and delay, the product of their efforts, *IRAS*, was launched by a NASA Thor-Delta rocket into a 370-mi-high (600-km) polar orbit in late January 1983.

The telescope has a beryllium primary mirror 22 in. (57 cm) in diameter, of focal ratio *f*/9.6, cooled to a temperature of 10 K (18°F) above absolute zero. *IRAS* has three kinds of instruments arrayed in its focal plane. Most important is the grid of detectors used to carry out the survey. The table summarizes their characteristics. Basically, these detectors, cooled to 2.5 K (4.5°F) above absolute zero, were designed to carry out a survey in four wavelength bands from 8 to 120 μm. The initial performance of *IRAS* exceeded all the expectations of the experimenters. During each day of operation nearly 1000 sources—both within the Milky Way Galaxy and external to it—were detected. The initial survey at all four bands was completed in late 1983, resulting in the discovery of several hundred thousand heretofore unknown infrared sources. The experimenters published a catalog of all the extended infrared sources and unresolved objects discovered in the survey along with a "picture" of the infrared sky in each of the four survey wave bands.

The second focal-plane instrument is a photometer having somewhat higher spatial resolution than does the "grid" of survey detectors. The photometer was designed to make careful wide-band measurements of selected sources at wavelengths of 52 and 100 μm.

The third instrument is a low-resolution spectrograph which was designed to obtain spectral energy distributions and atomic and molecular feature strengths for the brighter sources discovered by *IRAS*.

In addition to the all-sky survey, the experimenters pointed *IRAS* at a variety of astronomical objects which, at least from current beliefs and prejudices, appeared attractive candidates for detailed study. Foremost on their list were the molecular clouds thought to be the current birthplaces of stars. Their survey located the obscured stellar population of these clouds; future analysis of these heretofore-unseen sources should lead to a better understanding of the process of star formation. They may also have caught a few clouds in the very act of forming a star. The search for large, cool protostars was one of the most important quests of the *IRAS* experiment team. While astronomers have thus far pieced together a fairly convincing picture of how prestellar material might have been gathered together into molecular cloud complexes (stellar wombs) and how stars look a few hundred thousand years after birth, very little is known regarding the intervening, embryonic phases of stellar development. Yet it is just during the evolutionary phase between fragmentation of a molecular cloud into protostellar units and the appearance of a stable, near-equilibrium young stellar object that solar systems are supposed to have formed. When analysis of the data is complete, *IRAS* observations may provide some of the missing clues necessary for understanding these critical phases of stellar development.

IRAS also carried out a survey of the plane of the Milky Way Galaxy. Freed of the obscuring effects of dust, the survey should eventually provide a much clearer picture of the shape of the galactic disk and a better understanding of the nature of stellar populations in different locations in the Milky Way.

"First-look" data suggest that *IRAS* has detected several thousands galaxies, many more than the experiment team originally estimated. One of these systems (a galaxy known as MCG + 1–11–013) has been shown to emit more than 10 times as much energy in the infrared beyond 60 μm as compared with its optical luminosity. The source of this optically invisible luminosity is still a mystery. The *IRAS* survey should provide a much better idea regarding the frequency of infrared-bright galaxies and hence the probability that all systems pass through one or more such phases.

When complete and analyzed, the *IRAS* survey may contribute as much to knowledge of the sky as previous landmark mappings such as the Henry Draper Catalog and the National Geographic Society–Palomar Observatory Sky Survey. Certainly, *IRAS* promises to make fundamental contributions to the study of how stars form and how galaxies evolve.

During its lifetime, *IRAS* discovered a number of heretofore-unknown comets. Among the first was Comet IRAS-Araki-Alcock, which approached within a few million miles of Earth during May 1983. Because of its close approach, astronomers were able to study the comet's nuclear region with unprecedented spatial resolution. From their observations of Comet IRAS during its brief encounter with Earth, astronomers hope to learn how these mile-size (kilometer-size) balls of dirty ice are constructed and what factors control the ejection of nuclear gas and dust as a comet approaches the Sun. *See* COMET.

Perhaps the most dramatic event during *IRAS's* lifetime was the discovery of a glowing infrared cloud consisting of relatively large (greater than 1/3-inch or 8-mm diameter) dust particles surrounding the bright nearby star, Vega. The cloud was discovered by accident—Vega was thought to be an ideal "standard" for calibrating the *IRAS* detectors rather than an object of intrinsic interest. The *IRAS* observations suggest that this relatively young (age less than 1 billion years) star many be the host for a solar system still in the process of forming.

The infrared cloud is thought to be a flat disk of large particles, glowing feebly (at a temperature of 90 K or −280°F) as a result of energy absorbed from Vega. The dimension of the disk—comparable to that of the solar system—as well as the presence of large dust grains—unusual in well-studied astronomical objects—were the keys which led the *IRAS* team to propose that Vega was the expectant sire of a family of planets.

Instruments after IRAS. The apparent success of *IRAS* has led astronomers to urge the rapid construction of a permanent infrared observatory. One approach is SIRTF, the Shuttle Infrared Telescope Facility. Astronomers conceive SIRTF to be a 3.3-ft-diameter (1-m), cryogenically cooled telescope, launched into a polar orbit similar to *IRAS* but with the ability to drop to a lower orbit in order to be serviced by the space shuttle. SIRTF would permit astronomers to look at individual regions of the sky with spatial resolution 10 or more times that of *IRAS*. Because SIRTF would point and track for long periods of time, it would permit detailed imaging and moderate-resolution spectroscopic study of some exciting sources discovered by *IRAS*.

Later, astronomers hope to construct a giant (33-ft or 10-m diameter or greater) telescope capable of providing arc-second angular resolution pictures of the sky at a wavelength of 100 μm. Such a telescope is so large that it must be erected or deployed in space. Initially named the LDR (Large Deployable Reflector), such an instrument would permit the earliest phases of stellar evolution to be seen and resolved. It also should permit observation of the feeble, redshifted light from galaxies and galaxy clusters in their earliest formation phases. Hence, LDR promises to provide the ultimate tool for probing galaxies and stars at their very birth, perhaps providing an answer to the most fundamental question: Where do we come from?

For background information *see* GALAXY; GALAXY, EXTERNAL; INFRARED ASTRONOMY; STELLAR EVOLUTION in the McGraw-Hill Encyclopedia of Science and Technology. [STEPHEN E. STROM]

Bibliography: G. Rieke and M. Lebovsky, Infrared emission of extragalactic sources, *Annu. Rev. Astron. Astrophys.*, 17:477, 1979; S. E. Strom, Infrared and submillimeter astronomy from space, *Sky Telesc.*, 65(4):312–315, 1983; S. E. Strom, K. M. Strom, and G. L. Grasdalen, Young stellar objects and dark interstellar clouds, *Annu. Rev. Astron. Astrophys.*, 13:187, 1975; C. G. Wynn-Williams, The search for infrared protostars, *Annu. Rev. Astron. Astrophys.*, 20:587, 1982.

Insect

The physiology and ultrastructure of insect ovaries have been studied intensively during the last 15 years. From the result of comparative investigations of ovariole development among diverse insect species, it is now possible to outline the main evolutionary paths followed in hexapod oogenesis. This article describes insect ovaries in general, and shows how different types of ovarioles arise by different programs controlling the formation of germ cell clusters. Based on these differences, a phylogenetic tree of insect ovarioles can be constructed.

General anatomy. In all insects the ovaries are subdivided into a group of polarized tube-shaped organs, the ovarioles (Fig. 1). Each ovariole is cov-

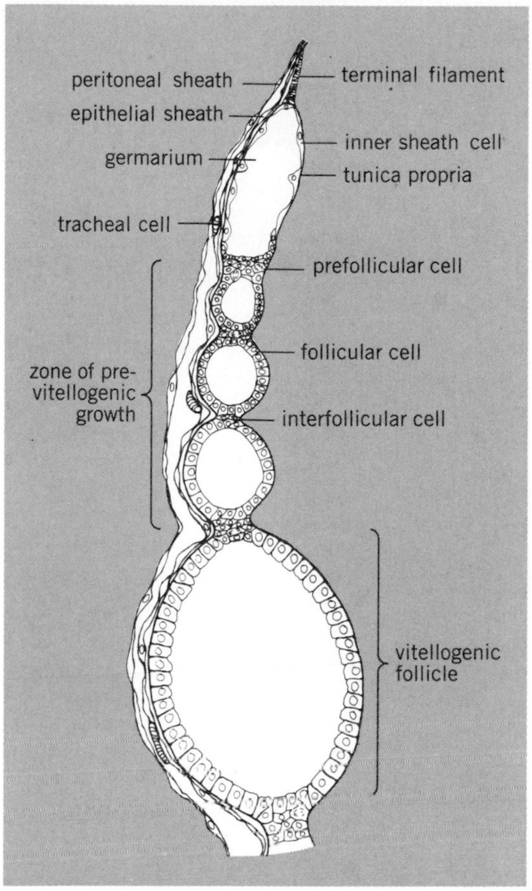

Fig. 1. General morphology of an insect ovariole. Germ cells and their descendants, as well as the outer epithelial sheath of the right side of the ovariole, are not shown.

ered by a peritoneal sheath bearing tracheoles and muscles. A terminal filament attaches each ovariole dorsolaterally in the thoracic or first abdominal segments. Next to the terminal filament, a thin sheath of somatic cells surrounds a group of primary germ cells or their descendants. This region is called the germarium. The more posterior inner sheath cells differentiate into the prefollicular cells which generate the follicular cells that surround each growing oocyte. This takes place in the region of the ovariole called the vitellarium. In its anterior portion the growth of oocytes is directed only by genes of the germ line cells, and this period of oocyte growth is called previtellogenesis. Later, during vitellogenesis, yolk is deposited in the egg from protein precursors mainly synthesized in the fat body. At the end of vitellogenesis, internalization of yolk precursors is prevented by the synthesis of the egg shell. The ripe eggs leave the ovariole and enter the oviduct, where they are often stored.

Panoistic ovary. In panoistic ovaries (lineage *a* in Fig. 2) all germ cells undergo subsequent differential mitoses, producing both stem line oogonia and oocytes. Prefollicular cells are scattered in small nests among young oocytes, and these somatic cells form a monolayered follicular tissue about each oocyte at a later stage during previtellogenesis.

Meroistic ovary. In germaria of meroistic ovaries, oogonial cells (stem cells) divide into stem cells and cystoblasts. These normally undergo a limited series of mitoses, each followed by incomplete cytokinesis. Thus, germ cell clusters in which the sibling cells are connected to each other by persisting intercellular bridges will be formed. The arrangement of cell divisions, as well as the process of oocyte–nurse cell determination, leds to a further subdivision in the classification. In polytrophic meroistic ovaries (lineage *b* in Fig. 2) each germ cell cluster develops as an isolated physiological unit, whereas in telotrophic merositic ovaries (lineages *e*, *h*, and *i* in Fig. 2) all nurse cells of all clusters are combined in one tropharium which nourishes all oocytes simultaneously.

Germ cell cluster formation. All types of insect ovarioles have the same somatic tissue constituents, and therefore the structural diversity is dependent on the manner in which germ cell clusters are formed. In the most common type (lineage *b* in Fig. 2) of polytrophic meroistic ovrioles, 2^n cells arise by a series of synchronous divisions. In most insects the number of divisions (n) is species-specific and ranges from 1 (yielding 2 interconnected cells) to 6 (yielding 64 cells). The germ cells in a given clone form a branching chain of cells. This is because the division spindles are oriented in such a way that, during a given cycle, one sister cell retains all previously formed cytoplasmic bridges while the other receives none. Once the divisions terminate, one of the cells with the largest number of intercellular bridges starts to receive a stream of cytoplasmic components from the other cells of the clone. These function as nurse cells, and the intercellular bridges serve as canals through which the oocyte is nourished. The oocyte enters meiosis, while the nurse cells begin a series of endomitotic DNA replications. As a result, each nurse cell may come to contain thousands of times as much DNA as the oocyte it serves.

A second type of polytrophic meroistic ovary has evolved in the earwigs (Dermaptera; lineage *c* in Fig. 2). In the upper part of the germarium lies a clone of interconnected oogonial cells which undergoes asynchronous divisions. Some cells pinch off the clone and behave as cystoblasts. Following the same method of germ cell cluster formation described above, clones of eight interconnected cells develop, all of which enter the prophase stages of meiosis. The four cells bearing three or two intercellular bridges become oocytes, and the others develop into nurse cells. The cluster will then divide into four subclones, each containing an oocyte and a nurse cell.

A third type of polytrophic meroistic ovary has evolved in one flea family (the mole fleas, Hystrichopsyllidae; lineage *n* in Fig. 2). In each germarium a large clone of germ cells exists which splits into subclones. It is not clear what determines which cells become oocytes, and nurse cells do not seem to replicate their DNA endomitotically.

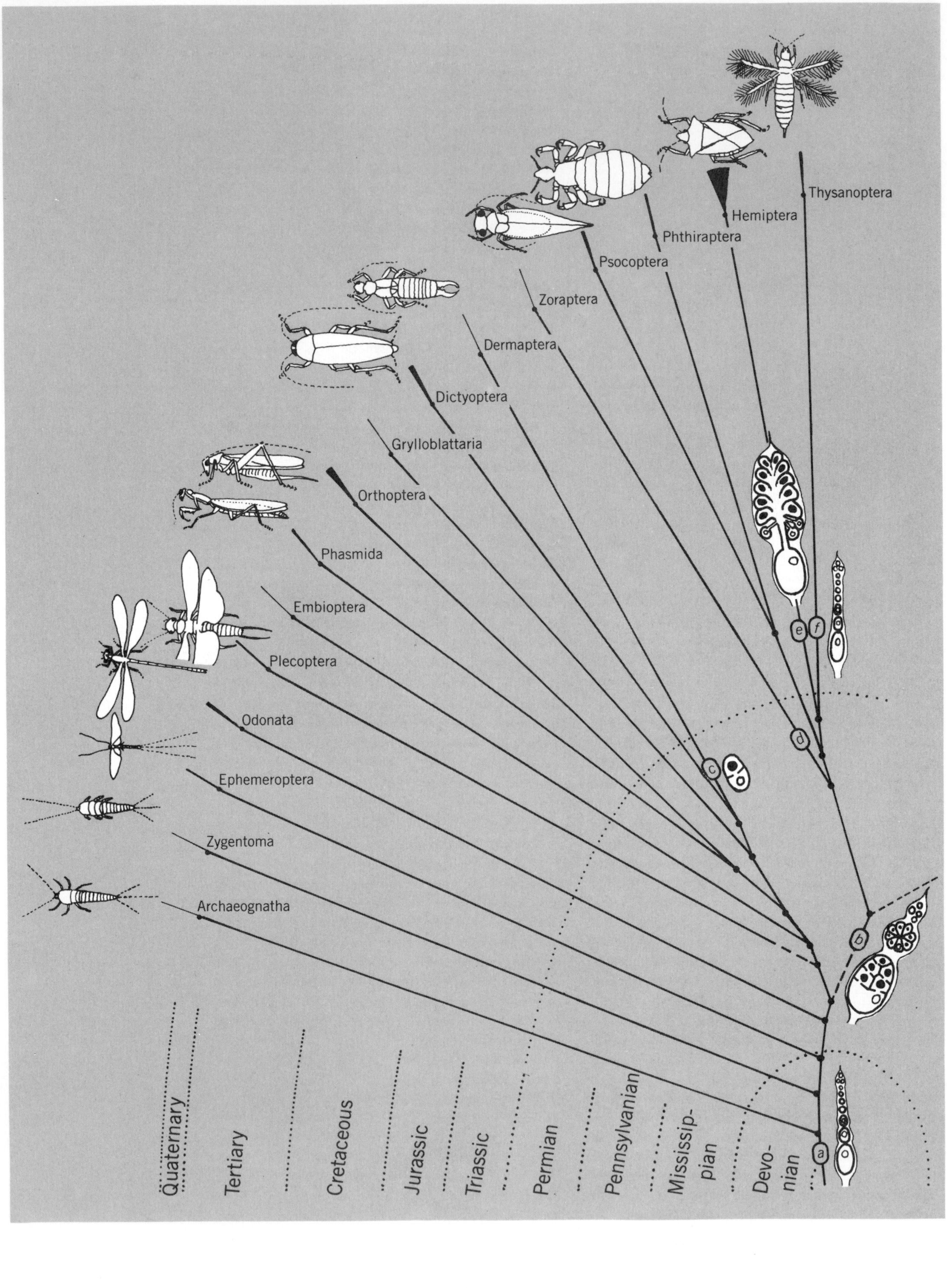

Thysanoptera
Hemiptera
Phthiraptera
Psocoptera
Zoraptera
Dermaptera
Dictyoptera
Grylloblattaria
Orthoptera
Phasmida
Embioptera
Plecoptera
Odonata
Ephemeroptera
Zygentoma
Archaeognatha
a
b
c
d
e
f
Quaternary
Tertiary
Cretaceous
Jurassic
Triassic
Permian
Pennsylvanian
Mississip-pian
Devo-nian

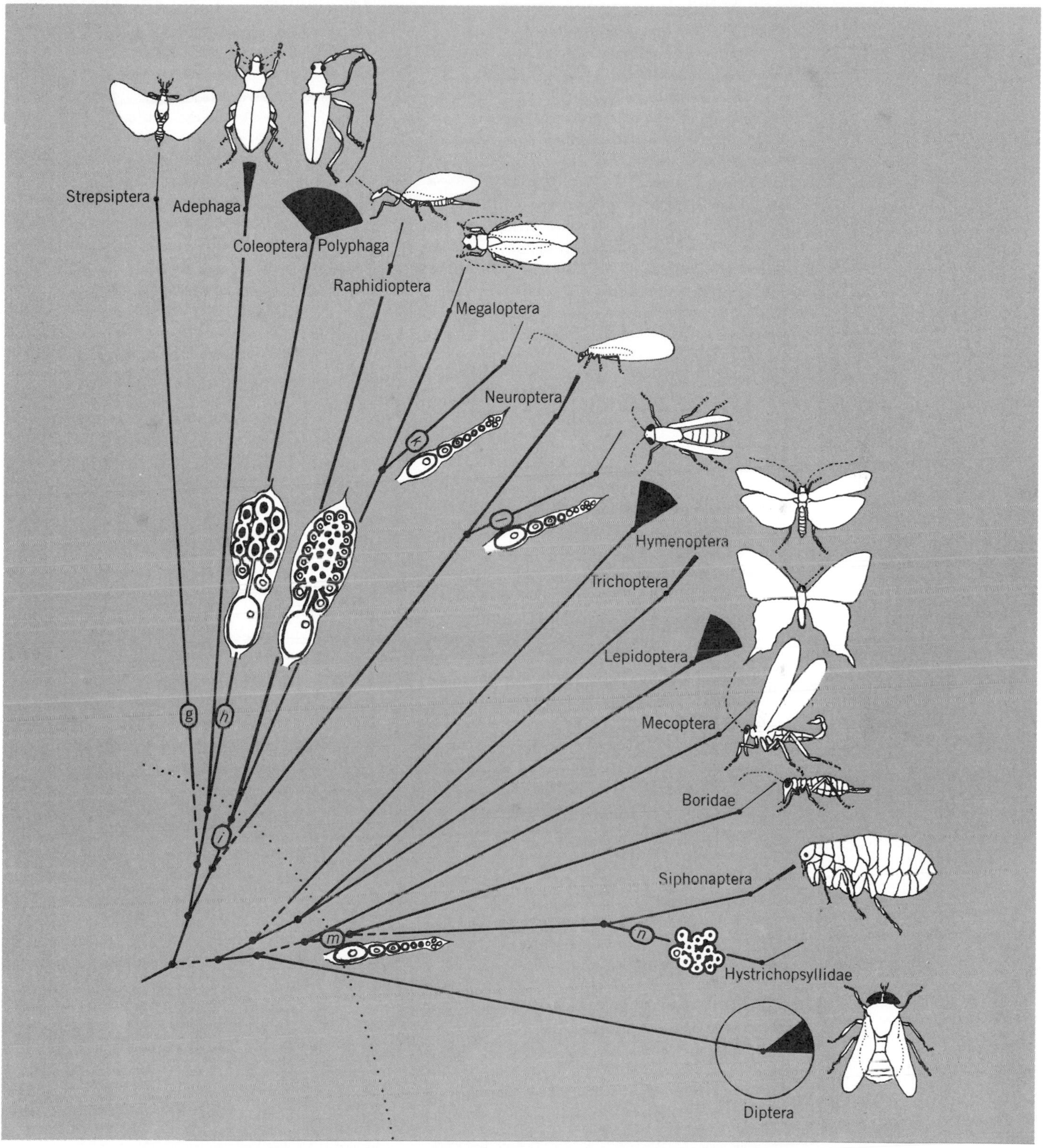

Fig. 2. Phylogenetic tree of insect ovarioles showing insect orders and their relationships. The bifurcation points (black dots) are correlated roughly with geological periods. The encircled letters indicate changes in the formation of germ cell clusters in females. At the base of the insect line, the panoistic ovary had already developed (*a*). During phylogenesis this panoistic ovary was replaced by the polytrophic meroistic ovary in the order of earwigs (*c*) and for a second time among the common ancestors of holometabolous (all orders from the Strepsiptera to the Diptera) and paraneopteran (all orders from the Zoraptera to the Thysanoptera) insects (*b*). A switch back to the panoistic ovary occurred in the order of Thysanoptera (*f*), in at least one species each of the orders of Megaloptera (*k*) and Neuroptera (*l*), and in the common ancestors of the Boreidae and Siphonaptera (*m*). In one flea family (Hystrichopsyllidae) a third type of the polytrophic meroistic ovary has evolved (*n*). Three telotrophic meroistic ovaries developed independently in the order of Hemiptera (*e*), in the polyphage Coleoptera (*h*), and in the Megaloptera/Raphidioptera (*i*). The ovariole structure of the Zoraptera (*d*) and Strepsiptera (*g*) is unknown. Amount of species of each order is symbolized by the black sectors of a circle standing for all insect species as indicated in the order Diptera. Stippled lines indicate relationships still in question.

Germ cell cluster formation is still enigmatic in the three types of telotrophic meroistic ovaries found in the Insecta. In hemipterans (true bugs, cicadas, aphids, and so on) only one germ cell cluster exists per ovariole (lineage *e* in Fig. 2). The first, incomplete division produces one preoocyte and one prenurse cell. In all species the preoocyte has a limited program of further divisions, yielding a constant number of definitive oocytes (2^n in aphids). Except in cicadas and bugs, the same pattern of incomplete divisions is found in prenurse cells, yielding a constant number of nurse cells (2^n in aphids). In cicadas and bugs some apical prenurse cells retain their dividing capacity and establish a gradient of differentiating nurse cells from top to the bottom of the tropharium. In all hemipteran species the intercellular bridges are concentrated in a central area, the central core, to which all nurse cells are connected, as well as all oocytes, via their nutritive cords.

The polyphage Coleoptera (lineage *h* in Fig. 2) have evolved a second type of telotrophic ovariole. Unbranched or slightly branched germ cell clusters of 2^n cells arise by synchronous divisions. The terminal cell next to the prefollicular tissue develops into the oocyte, and all other cells of each cluster become nurse cells. All clusters are packed side by side, and in many species the cell membranes separating adjacent nurse cells disappear.

A third type of telotrophic meroistic ovariole developed in the common ancestors of alder flies (Megaloptera) and snake flies (Raphidioptera; lineage *i* in Fig. 2). Oogonial stem cells invade the young ovarioles. From these stem cells small clusters arise. In central areas the sibling cells fuse, whereas peripheral cells of each cluster remain unaffected. A central syncytium results, surrounded by a monolayer of persisting cells. Next to the somatic prefollicular cells some of the persisting cells develop into oocytes; all others function as nurse cells. Endomitotic polyploidization of nurse cell nuclei does not occur.

Phylogenesis of insect ovarioles. About 95% of all insect species living today have ovarioles containing clusters of interconnected germ cells. Furthermore, the formation of germ cell clusters is not restricted to insect females. Throughout the animal kingdom clusters of interconnected, developmentally synchronized spermatocytes are found. Therefore, germ cell cluster formation is an ancient evolutionary invention. In females, germ cell clusters are present in many phylogenetically diverse groups (for example, crustaceans, mammals, reptilians, tardigrades, and annelids). Within the Hexapoda, the Entognatha (containing the Collembola, Protura, and Diplura) is the sister subclass to the Insecta. Germ cell clusters are found in females of collembolans and some diplurans, but the proturans and the other diplurans have panoistic ovaries as do all "primitive" insect orders (lineage *a* in Fig. 2).

The question remains as to why cluster formation is reduced or eliminated in females belonging to many taxonomic groups, but not in males. Perhaps the answer lies in the basic differences between gametogenesis in the two sexes. In males a large number of small, functionally equivalent gametes are produced. Cluster formation multiplies the gamete population and ensures the synchronous development of the sibling cells during the time their cytoplasms are being loaded with the RNAs to be utilized during spermiogenesis. On the other hand, the formation of clusters of gametes in females is counterproductive. Here, small numbers of giant cells are needed, and this presumably is the reason that the meiotic divisions are not accompanied by cytokinesis and that only one of the heploid nuclei is retained per cell.

A switch back and forth between repression (lineages *a*, *f*, *k*, *l*, and *m* in Fig. 2) and derepression (lineages *b*, *c*, and *n* in Fig. 2) of germ cell cluster formation in females has occurred several times. The great breakthrough (lineage *b* in Fig. 2), that is, the origin of the meroistic ovary, requires at least three further inventions: a method of selecting certain sibling cells as oocytes, and others as nurse cells; a mechanism for gene multiplication (endomitotic polyploidization) of the nurse cells, and the development of an effective system for transporting macromolecules from nurse cells to oocytes. All these events must have happened among the common ancestors of paraneopteran and holometabolous insects (lineage *b* in Fig. 2). The aberrant polytrophic meroistic ovaries of earwigs and mole flies (lineage *c* and *n* in Fig. 2) are believed to have arisen independently.

Telotrophic meroistic ovarioles must have developed independently three times (lineages *e*, *h*, and *i* in Fig. 2). In polytrophic ovaries RNA synthesis by the oocyte nucleus is reduced in many species, whereas in all three types of telotrophic ovaries the oocyte genomes are fully active. Possibly, RNA synthesis of telotrophic oocytes and nurse cells must be qualitatively different because all nurse cells nourish all growing oocytes at the same time and differential gene activation during oogenesis may be needed to establish maternal pattern formation in eggs.

For background information *see* GAMETOGENESIS; INSECT PHYSIOLOGY; INSECTA in the McGraw-Hill Encyclopedia of Science and Technology.

[JÜRGEN BÜNING]

Bibliography: W. Hennig, *Insect Phylogeny*, trans. and ed. by A. C. Pont with revisionary notes by D. Schlee, 1981; E. Huebner, *The Ultrastructure and Development of the Telotrophic Ovary*, in R. C. King and H. Akai (eds.), *Insect Ultrastructure*, vol. 2, 1984; R. C. King and J. Büning, *The Origin and Functioning of Insect Oocytes and Nurse Cells*, in G. A. Kerkut and L. I. Gilbert (eds.), *Comprehensive Insect Physiology, Biochemistry and Pharmacology*, vol. 1, 1984; N. P. Kristensen, Phylogeny of insect orders, *Annu. Rev. Entomol.*, 26:135–137, 1981.

Intermediate vector boson

The W and Z particles were discovered in 1983 in very high-energy proton-antiproton collisions. These elementary particles, which are also called intermediate vector bosons, are the fundamental particles tht transmit the weak force. (An example of a weak interaction process is nuclear beta decay.) It is through the exchange of W and Z bosons that two particles interact weakly, just as it is through the exchange of photons that two charged particles interact electromagnetically. The intermediate vector bosons were postulated to exist many years ago; however, their large masses have prevented their production and study at accelerators until recently. Their discovery is a key step toward unification of the weak and electromagnetic interactions.

Production and detection. The W and Z particles are predicted by electroweak theory to be very massive, roughly 100 times the mass of a proton. Therefore, the experiment to search for the W and the Z demanded collisions of elementary particles at the highest available center-of-mass energy. Such very high center-of-mass energies capable of producing the massive W and Z particles were achieved with collisions of protons and antiprotons at the laboratory of the European Organization for Nuclear Research (CERN) near Geneva, Switzerland. This was accomplished through conversion of the existing 4-mi-circumference (6-km) superproton syncrotron (SPS) into a proton-antiproton ($p\bar{p}$) collider. The first step in the conversion of the SPS into a $p\bar{p}$ collider was the creation of a dense source of antiprotons. The antiprotons were created by collision of an intense beam of protons with a stationary nuclear target. The antiprotons, which were produced with varying momenta, were then injected into a specially constructed ring called an antiproton accumulator. In the antiproton accumulator, the momentum spread of the antiprotons was narrowed by using an innovative technique called stochastic cooling. Dense bunches of antiprotons could then be injected into the SPS along with protons circulating in the opposite direction. After injection into the SPS, the protons and antiprotons were accelerated together to an energy of 270 GeV each. The beams were stored for several hours at this energy, with the desired $p\bar{p}$ collisions occuring when the beams crossed one another. This made a center-of-mass energy of 540 GeV per $p\bar{p}$ collision, which was about a factor of 10 higher than previously achieved with any other accelerator.

The $p\bar{p}$ collisions were monitored in two underground experimental areas (UA1 and UA2). Experiment UA1 consisted of a large, 2000-ton magnetic spectrometer completely surrounding the $p\bar{p}$ collision region. The inner portion of the detector contained a precision device for charged-particle tracking. This detector (central detector) was made up of about 6000 sense wires in a gas-filled cylindrical volume. Each sense wire was equipped with its own sophisticated electronics. When charged particles passed through this chamber, the ionization that they created in the gas was carefully collected in order to produce a three-dimentional image of the particle trajectory. The entire central detector was in a homogeneous magnetic field so that accurate measurement of the curvature of each track gave the particle momentum. Immediately surrounding the central detector were devices (calorimeters) for measuring the energies of particles, both charged and neutral. These calorimeters were built of modules of

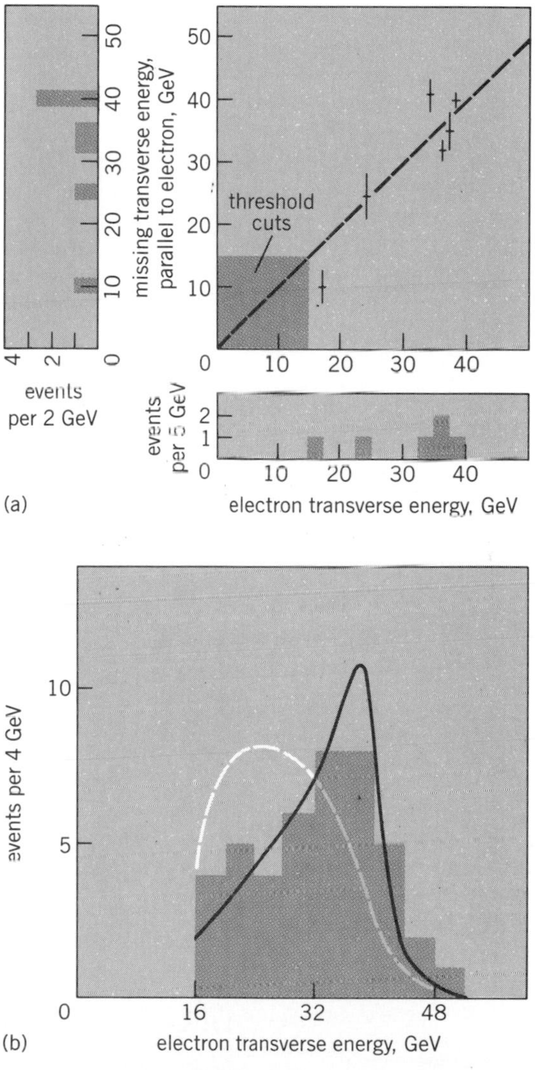

Fig. 1. Observation of the W particle in the UA1 data. (*a*) Scatter plot of electron (or positron) transverse energy versus the transverse energy imbalance (due to a neutrino) in the whole event, from a sample of six events. Threshold cuts select events with large electron transverse energy and missing transverse energy (*after G. Arnison et al., Experimental observation of isolated large transverse energy electrons with associated missing energy at $\sqrt{s} = 540$ GeV, Phys. Lett., 122B:103–116, 1983*). (*b*) Electron transverse energy distribution from a larger sample of 43 events. The solid curve is the expected distribution for the decay, $W \rightarrow e\nu$, while the broken curve is the distribution expected for the decay of a different particle "X" into an electron and two neutrinos (*after G. Arnison et al., Further evidence for charged intermediate vector bosons at the SPS collider, Phys. Lett., 129B:273–282, 1983*).

lead and plastic scintillator and iron and plastic scintillator. Particles entering the calorimeter were completely absorbed, with a nearly constant fraction of the energy being converted into light. This light was carefully collected and monitored by about 2400 photomultiplier tubes. The outermost portion of the detector consisted of several layers of ionization chambers for charged-particle detection. These chambers detect charged particles that pass completely through the calorimeter without interacting. There is only one known charged particle that has such properties, the muon (μ). Neutrinos (ν) leave no direct trace in the detector.

The UA2 detector was a smaller, 200-ton nonmagnetic detector. Its principal components were inner charged-particle tracking chambers followed by calorimeters. However, the main difference between this detector and the UA1 detector was that there was no magnet (for charged-particle momentum measurement), and the calorimetry (for particle energy measurement) did not cover the total solid angle.

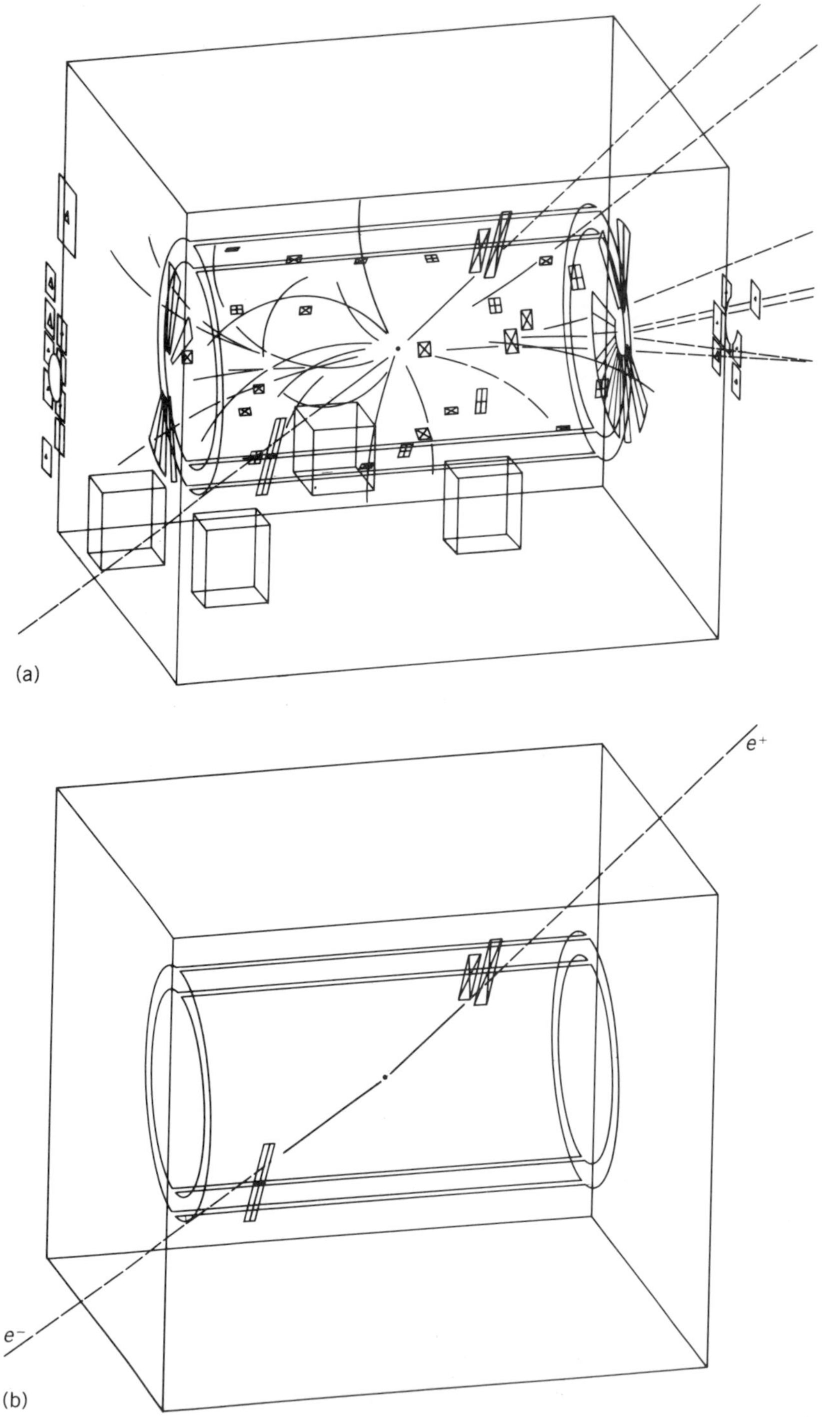

Fig. 2. Graphics display of a decay $Z^0 \rightarrow e^+e^-$ in the UA1 detector. (*a*) All charged-particle trajectories (curves) and energy deposits in the calorimeters (rectangles and boxes) are displayed. (*b*) Only charged tracks and calorimeter hits with transverse energy greater than 2 GeV are displayed, leaving just the Z_0 decay products, an electron and positron of energy about 50 GeV each. (*After G. Arnison et al., Experimental observation of lepton pairs of invariant mass around 95 GeV/c^2 at the CERN SPS collider, Phys. Lett., 126B:398–410, 1983*)

Discovery of the W particle. The intermediate vector bosons have very distinctive signatures in the UA1 detector. The *W* particle (which comes in two electric charges, W^+ and W^-) is identified through its decay into positron (e^+) and neutrino (ν), $W^+ \rightarrow e^+\nu$, or electron (e^-) and antineutrino ($\overline{\nu}$), $W^- \rightarrow e^-\overline{\nu}$. The trajectory of the electron is measured by the trail of ionization that it leaves in the central detector. The energy of the electron is measured by the calorimeters. The details of the energy deposition of the electron in the calorimeters distinguish it from heavier, strongly interacting particles, such as pions and protons. Checks are made to see that the position of the energy deposition in the calorimeter agrees with the position of impact measured from the central detector. A further check is made to see that the momentum of the electron, as measured from the curvature of its trajectory in the magnetic field of the spectrometer, is compatible with the amount of energy deposited in the calorimeter. The neutrino has such a small interaction probability that it leaves no direct traces in the detector. However, strong evidence for the presence of a neutrino exists in the *W* events. This evidence comes from an apparent lack of momentum conservation analogous to the decay $n \rightarrow p\, e^-\overline{\nu}$. In fact, it was the study of such neutron decays that originally led to the hypothesis of the existence of the neutrino.

Data were recorded in November and December 1982 from about 10^9 $p\bar{p}$ interactions. After careful analysis of these data, the UA1 group found six events with an isolated electron or positron of very large transverse momentum (momentum perpendicular to the colliding beam axis) together with large missing transverse momentum (Fig. 1*a*). The UA1 group could find no explanation for these events other than the production of new massive charged particles (W^+ and W^-), which decayed into an electron (or positron) and neutrino. This discovery was announced in January 1983 by C. Rubbia (who was also instrumental in converting the SPS into a $p\bar{p}$ collider). Confirmation of the UA1 result was soon announced by the UA2 group.

Discovery of the Z particle. Another run of the CERN $p\bar{p}$ collider was made in the spring of 1983. For this run, the performance of the machine was improved to obtain nearly 10^{10} $p\bar{p}$ collisions. The existence of the charged boson W was readily confirmed with about 10 times the origianl statistics (Fig. 1*b*).

Also, for the first time, events were seen that contained both a high-energy electron and a high-energy positron. Four such events were found in the UA1 data with such an e^+e^- pair (Fig. 2). All four events have a common value of invariant mass (about 95 GeV/c^2 where c is the speed of light) within experimental resolution. No other high-mass e^+e^- events were found in this sample, and all known possible background contributions were calculated to be negligible. Events were also detected that contained a positive muon (μ^+) and a negative muon (μ^-) with invariant mass of about 95 GeV/c^2.

These data are clear evidence for the existence of a new massive particle that decays (with yet unmeasured branching fraction) into e^+e^- and $\mu^+\mu^-$. The only known interpretation of this particle which fits the measured properties is that it is the neutral intermediate vector boson Z^0. The discovery of this particle was announced by the UA1 group in May 1983. Confirmation of this result came later from the UA2 group.

Properties of the W and Z particles. Striking features of both the charged W and the Z^0 particles are their large masses. The charged boson (W^+ and W^-) mass is measured to be 80.9 ± 1.5 GeV/c^2, and the neutral boson (Z^0) mass is measured to be 95.6 ± 1.5 GeV/c^2 from the UA1 experiment. (For comparison, the proton has a mass of about 1 GeV/c^2.) There is an additional systematic error on these mass values of 3%, mainly due to calibration of the calorimeter. The W and Z^0 mass values reported by the UA2 experiment are in agreement with the UA1 results. Prior to the discovery of the W and the Z, particle theorists had met with some success in the unification of the weak and electromagnetic interactions. The electroweak theory as it is understood today is due largely to the work of S. Glashow, S. Weinberg, and A. Salam. Based on low-energy neutrino scattering data, which in this theory involves the exchange of virtual W and Z particles, theorists made predictions for the W and Z masses. The actual measured values are in agreement (within errors) with predictions. The discovery of the W and the Z particles at the predicted masses is an essential confirmation of the electroweak theory.

Only a few intermediate vector bosons are produced from 10^9 proton-antiproton collisions at a center-of-mass energy of 540 GeV. This small production probability per $p\bar{p}$ collision is understood to be due to the fact that the bosons are produced by a single quark-antiquark annihilation. Thus, to make a boson from a $p\bar{p}$ collision, one of the quarks in the proton must annihilate with one of the antiquarks in the antiproton. Furthermore, the energy of the quark (q) plus antiquark ($\bar{q}$) in the $q\bar{q}$ center-of-mass system must just equal the boson mass. The other production characteristics of the intermediate vector bosons, such as longitudinal and transverse momentum distributions (with respect to the $p\bar{p}$ colliding beam axis), all support the theoretical picture of production through simple quark-antiquark annihilation.

The decay modes of the W and Z are well predicted. The simple decays $W^+ \rightarrow e^+\nu$, $W^- \rightarrow e^-\bar{\nu}$, $Z^0 \rightarrow e^+e^-$, and $Z^0 \rightarrow \mu^+\mu^-$ are all spectacular signatures of the intermediate vector bosons. However, these leptonic decays are expected to be only a few percent of the total number of W and Z decays. There will be much experimental activity in the coming years to find other decays of the W and Z. The W particle also has an identifying feature in its decay, which is due to its production through quark-antiquark annihilation. The quark and antiquark carry intrinsic angular momentum (spin) that is conserved in the interaction. This means that the W is polarized at production; its spin has a definite orientation. This intrinsic angular momentum is conserved when the W decays, yielding a preferred direction for the decay electron or positron (Fig. 3). This distribution is characteristic of a weak interaction process, and it strongly favors the assignment of the spin of the W to be 1, as expected from theory. (Spin-1, or more generally integer-spin, particles are called bosons; another example is the photon.)

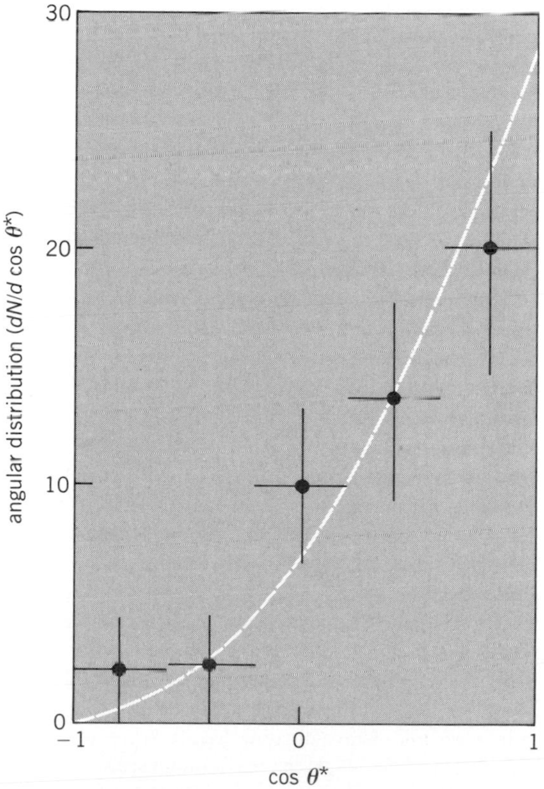

Fig. 3. Angular distribution of the emission angle θ^* of the e^- with respect to the proton direction (or the e^+ with respect to the antiproton direction) from W^+ and W^- events (in the rest frame of the W). The experimental data are given by the crosses. The curve is the expectation for a spin-1 particle produced by simple quark-antiquark annihilation followed by the weak interaction decay, $W^+ \rightarrow e^+\nu$ or $W^- \rightarrow e^-\bar{\nu}$. (*After G. Arnison et al., Further evidence for charged intermediate vector bosons at the SPS collider, Phys. Lett., 129B:273–282, 1983*)

For background information *see* ELEMENTARY PARTICLE; FUNDAMENTAL INTERACTIONS; PARTICLE ACCELERATOR; QUARKS; WEAK NUCLEAR INTERACTIONS in the McGraw-Hill Encyclopedia of Science and Technology.

[JAMES W. ROHLF]

Bibliography: G. Arnison et al., Experimental observation of isolated large transverse energy electrons with associated missing energy at $\sqrt{s}$ = 540 GeV, *Phys. Lett.*, 122B:103–116, 1983; G. Arnison et al., Experimental observation of lepton pairs of invariant mass around 95 GeV/c^2 at the CERN SPS collider, *Phys. Lett.*, 126B:398–410, 1983; M. Banner et al., Observation of single isolated electrons of high transverse momentum in events with missing transverse energy at the CERN $p\bar{p}$ collider, *Phys. Lett.*, 122B:476–485, 1983; C. Rubbia et al., Producing massive neutral intermediate vector bosons with existing accelerators, *Proceedings of the International Neutrino Conference*, Aachen, West Germany, pp. 683–687, 1976.

Interplanetary matter

Each year as the Earth orbits the Sun it collides with over 10,000 tons of interplanetary matter. The smallest particles are abundant and settle to the Earth's surface at a rate of one 10-micrometer particle per square meter per day. Larger objects become increasingly rare, and those of around a half-mile or a kilometer in size impact at a rate of one per 300,000 years for the entire Earth. As evidenced by impact craters seen on ancient lunar rocks, this impact rate has been maintained for most of the Earth's history. Nearly all of the impacting material is debris from comets and asteroids, objects believed to be relic planetesimals not greatly altered since the formation of the solar system 4.6 billion years ago. The analysis of this debris provides detailed information on early processes, that occurred in the solar system, and in some cases the studies yield useful information on terrestrial processes as well.

Meteoritics, the study of collected samples of interplanetary matter, has in the past dealt almost exclusively with meteorites large enough to be noticed and found at random on the surface of the Earth. Recent advances have now extended this field to include smaller, often microscopic, particles popularly referred to as cosmic dust. Although the extraterrestrial dust samples are small and difficult to work with, they are important for a variety of reasons, the most important of which is that they provide a means of investigating fragile types of material that are too weak to survive atmospheric entry in sizes large enough to be recognized on land as meteorites. Dust particles decelerate from cosmic velocity near 60 mi (100 km) altitude and are not subjected to high dynamic stress. In contrast, hand-sized meteorites retain their high initial velocities down to altitudes of well below 30 mi (50 km), where the stress due to friction with dense air disintegrates all but the strongest materials.

Collected dust samples provide new insight into a long-standing dilemma concerning the origin of meteorites. Trajectory analyses of meteoroid entry paths in the atmosphere indicate that the bulk of interplanetary matter that collides with the Earth is debris from comets. Most meteoroids disintegrate during entry into the atmosphere, but objects that do survive and are found on the ground appear to be material from asteroids. Analyses of conventional meteorites indicate that most and perhaps all have had thermal histories that appear to be inconsistent with origin from comets, bodies that presumably have been cold since their origin. The study of the atmospheric deceleration of meteoroids from known comets has shown that cometary materials are fragile. It is possible that cometary materials survive atmospheric entry only as dust.

Collection. Nondestructive collection of cosmic dust is not practical in space because the high impact velocities usually vaporize the sample during collection. The atmosphere, however, is an ideal collector because it decelerates the particles slowly enough that, in some cases, they are not severely heated. Dust particles decelerate at altitudes between 60 mi (100 km) and 30 mi (50 km), depending on the particle size and density, and then gently settle toward the surface of the Earth, a process that can take days to months depending on particle size.

Particles smaller than 50 μm (0.002 in.) are abundant enough that they can be collected directly from the atmosphere. To reduce terrestrial contamination problems, the collections are made in the stratosphere above 12 mi (20 km) by using balloon- or aircraft-borne collectors. The concentration of 10-μm (0.0004-in.) particles is only one particle per 110,000 ft^3 (3000 m^3) of air, so that clean, high-volume air-sampling collectors must be used. Since the first successful collections with the "Vacuum Monster" balloon collector were made in 1970, over 500 extraterrestrial dust particles have been collected from the atmosphere. Most were collected on oil-coated impaction plates lowered beneath the wings of NASA U-2 aircraft. For the 2–50-μm (0.00008–0.002-in.) particles collectible in the stratosphere, most are not heated to their melting points during entry, and are called micrometeorites as defined by Dr. Fred Whipple in the 1950s.

Particles larger than 100 μm (0.004 in.) are strongly heated during atmospheric entry and usually melt, forming so-called ablation spheres. These larger particles produce sufficient ionization along their entry paths that they are detectable by radar reflection, and those larger than a millimeter (0.04 in.) become visual "meteors," the luminous phenomena resulting from entry. Rare particles as large as 1 mm (0.04 in.) can survive entry without melting if they enter at very low angles and hence dissipate most of their energy at high altitude. Particles larger than 100 μm (0.004 in.) are much too rare to be directly collected from the atmosphere, but they can be found in terrestrial sediments that have long collection times and only minor "contamination." The best location on Earth for such collection is the floor of the Pacific Ocean at depths greater than 13,000 ft (4000 m) and at locations far from continents or islands. Sediments at these locations accumulate at a rate of only 3.3 ft (1 m) per million years, and extraterrestrial particles are more highly concentrated than in any other location on Earth. Ablation spheres constitute less than 1 part per million of the bulk sediment weight, but for the subset of magnetic particles larger than 300 μm (0.01 in.) they can constitute as much as half. Nearly all ablation spheres are magnetically recoverable because they contain magnetite (Fe_3O_4) crystals produced during the entry process. Known as cosmic spheres, they were first collected over a century ago on the *Challenger* voyages when a hand magnet was passed through sediment samples dredged from the sea floor. The initial clue that the spheres were extraterrestrial was that some contained metallic nickel-iron, a natural mineral that is exceedingly rare in small terrestrial particles. In modern collections, particles are collected from the ocean floor with magnetic sleds.

Stratospheric micrometeorites. The majority of micrometeorites collected in the stratosphere are easily distinguished from the various types of contaminants that are inevitably included in collection experiments. The elemental compositions of most of the particles closely match the cosmic abundance pattern seen in the Sun and primitive meteorites. For the elements that can be contained in stony materials, this pattern is equivalent to what is believed to be the bulk composition of the interstellar dust and gas from which the solar system formed. The composition in descending order of abundance is O, Mg, Fe, Si, C, S, Ca, Al, Ni, Na, P, Cr, Mn, and Ti. Ninety percent of the particle mass is contained in the first six elements. Fortunately no known natural or synthetic terrestrial materials match this cosmic composition, at least in the form of small particulates. Micrometeorites are also collected that are composed primarily of one mineral and hence cannot have a cosmic abundance pattern, because no mineral can accommodate the elements in the appropriate proportions. The most common of these particles are composed of the minerals olivine, pyroxene, or iron sulfide. Most of these particles were previously embedded in a matrix of cosmic proportions, and they still retain small bits of this material on their exteriors. Small patches of this material only a micrometer in size can readily be seen in a scanning electron microscope and analyzed for elemental composition by measuring the characteristic x-rays emitted due to bombardment by the focused electron beam.

Meteorites whose major elemental composition is close to cosmic abundances are called chondrites. A rare subgroup of these, the carbonaceous chondrites, contains relatively high abundances of volatile elements such as C, S, and Bi. They are believed to be the most primitive meteorites and those that formed at the lowest temperatures. Remarkably, over 90% of the collected dust particles have elemental compositions similar to this meteorite group. Carbonaceous chondrites are evidently very common in space. Their rarity as conventional meteorites is probably a result of their friable nature and tendency to crumble during atmospheric entry. In mineralogy and structure, some of the dust particles appear to be identical to established classes of carbonaceous chondrites. The majority of particles, however, differ in significant ways. Many of the particles are porous aggregates of submicrometer grains and morphologically resemble a cluster of grapes (Fig. 1). Some of the particles have porosities greater than 50% and bulk densities much lower than the most porous meteorites. This structure is similar to models inferred for cometary meteoroids from the basis of their entry behavior into the atmosphere. If the particles were indeed cometary, the pore spaces would have originally been filled with ices. Although the elemental composition is the same, these particles are mineralogically quite different from established carbon-rich meteorite types. The most striking difference is that the particles are anhydrous, while most of the mass of carbon-rich chondrites is in silicates that contain over 10% bound water.

Several new types of chondritic composition materials have been recognized from studies of micrometeorites. Some of these are likely to be materials from comets. It is possible that a reliable identification of some samples as cometary debris can be made on the basis of in-place particle analyses performed on the European and Soviet missions to Halley's Comet in 1986. Comets are believed to have formed at the outer fringes of the planetary system by accretion of primordial ice and dust. They may contain both materials formed in the early solar system and grains formed in the interstellar medium

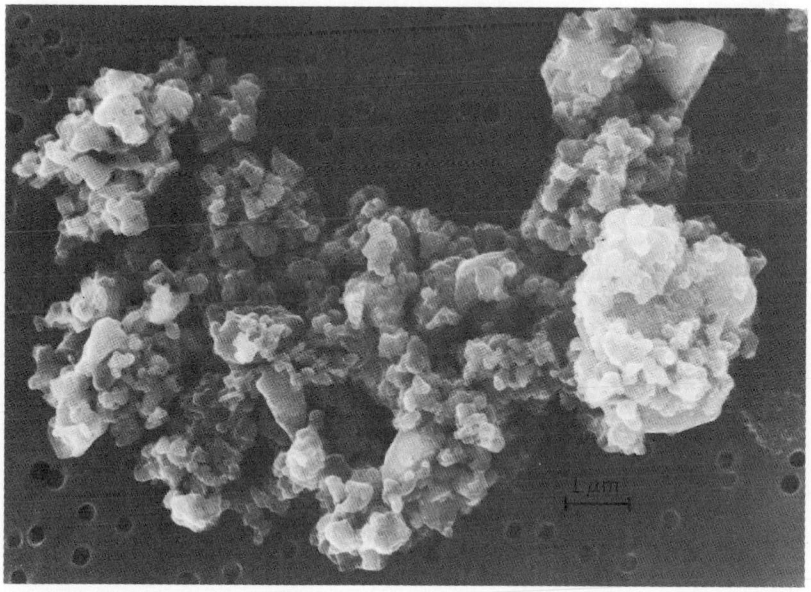

Fig. 1. Scanning electron micrograph of a chondritic composition micrometeorite collected in the stratosphere.

that actually predate the solar system. The collected dust samples are being extensively studied to determine what information they may contain on early and perhaps presolar environments and processes. The particles are being studied by practically all analytical techniques applicable to small samples. These include mass spectroscopy, infrared spectroscopy, neutron activation, and analytical and high-resolution electron microscopy.

Cosmic deep-sea spheres. The deep-sea spheres that have been collected range in size from less than 50 μm (0.002 in.) to over 2 mm (0.08 in.) in diameter. The specimen shown in Fig. 2 is 300 μm (0.012 in.) in diameter. Deep-sea spheres are believed to be solidified melt droplets produced by the atmospheric melting of millimeter and larger meteoroids. The spheres can be classified into two groups first recognized over a century ago on the *Challenger* voyages. The abundances vary with size, but roughly half the spheres are stony and half are iron. The iron spheres commonly consist of a eucentric FeNi metal core surrounded by a iron oxide shell of the minerals magnetite and wustite. Nickel in the spheres is usually concentrated in the metal core. The stony spheres are composed of olivine, glass, and magnetite, and they usually have quench textures indicating formation by rapid cooling from a molten silicate droplet. The elemental composition of the stony spheres is chondritic except for volatile elements like S and Na and highly siderophile (iron-loving) elements like Ni. The observed depletions of nickel are probably due to the loss of a metal droplet while the sphere was molten and either spinning or decelerating. Some stony spheres are found with a Ni-rich metal bead at one end, suggesting that at least some of the iron-type spheres probably were derived from stony spheres and not iron meteoroids. Some of the metal is actually produced during atmospheric heating due to reduction from iron-bearing silicates by finely dispersed carbon.

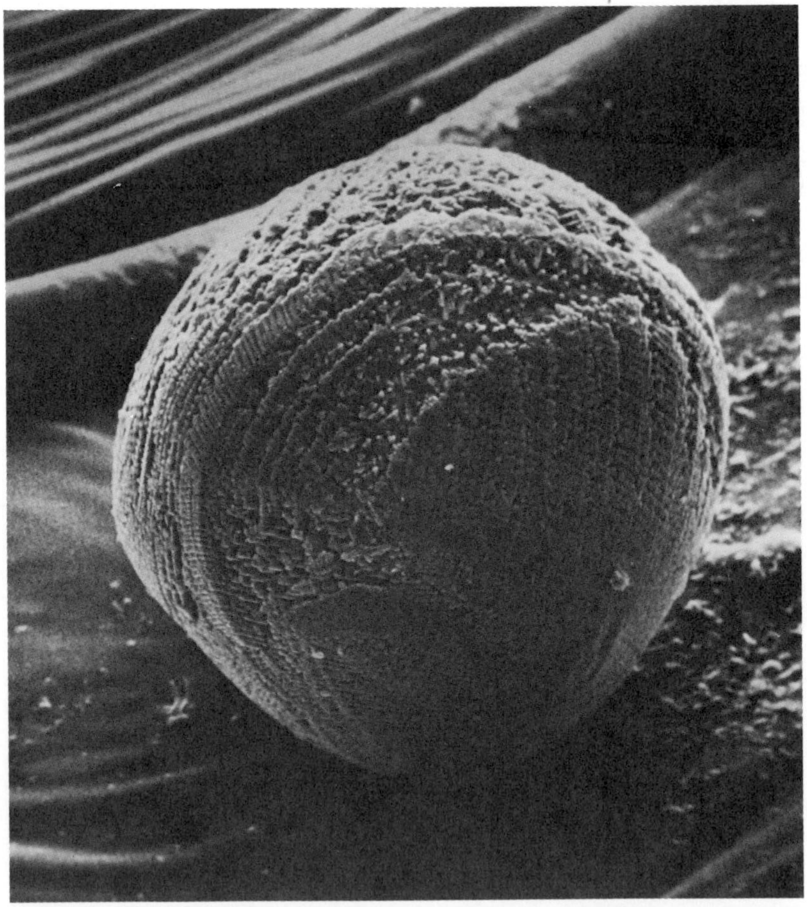

Fig. 2. Scanning electron micrograph of a 300-μm-diameter (0.012-in.) stony deep-sea sphere collected on the floor of the Pacific Ocean. Etching by sea water has removed glass from the surface, leaving olivine and magnetite crystals standing out in positive relief.

The deep-sea spheres are by no means pristine samples because of severe heating during atmospheric entry followed by long-term burial in the ocean floor. They are important, however, because they do preserve certain elemental and isotopic abundance ratios of their parent materials. The abundances of the moderately refractory elements such as Al, Ca, and Ti clearly indicate that over 80% of the stony spheres are materials from parents with compositions like those of carbonaceous chondrites. Presumably, most of the spheres were formed from millimeter-sized pieces of the same materials that also formed the 10-μm-sized (0.0004-in.) particles collected in the stratosphere as micrometeorites. In terms of analysis, the spheres offer some significant advantages over stratospheric particles in that they can be collected in large sizes. The largest spheres contain more than a million times as much mass as the typical collected atmospheric particle. Large mass allows abundance measurements to be made on trace elements and precise isotopic measurements to be made on a significant number of elements. For example, measurements of the amounts of the isotopes ^{26}Al, ^{53}Mn, and ^{10}Be have recently shown that the typical particles were exposed to cosmic rays for roughly a million years. Prior to this time, the samples must have been shielded inside bodies more than 3 ft (1 m) in diameter. Measurements of the isotopes of Sr permitted setting a rough formation age of the parent materials consistent with the age of the solar system.

Potentially the most valuable aspect of the deep-sea spheres is that they are deposited in a stratigraphic sequence. Reasonably well preserved stony spheres have been found in 40-million-year-old sediments, and it is anticipated that both stony and iron types can be recoverd in much older deposits. In contrast, most stony meteorites on land weather beyond recognition in much less than a million years. Samples from sediment cores can be studied to determine changes in the amount or nature or extraterrestrial materials that have impacted the Earth in the past. They could, for example, record past visits of giant comets or passage of the solar system through a dusty region of the Galaxy. If spheres can be found in deposits several hundred million years old, their oxidation state might provide clues on the time of emergence of high levels of oxygen in the atmosphere. The spheres may also provide a means of detecting past variations in the cosmic-ray flux.

For background information *see* COMET; INTERPLANETARY MATTER; METEOR; METEORITE; MICROMETEORITE in the McGraw-Hill Encyclopedia of Science and Technology.

[DONALD E. BROWNLEE]

Bibliography: D. E. Brownlee, in *The Sea*, vol. 7: *The Oceanic Lithosphere*, ed. by C. Emiliani, 1981; P. Fraundorf, D. E. Brownlee, and R. M. Walker, Laboratory studies of interplanetary dust, in L. L. Wilkening (ed.), *Comets*, 1982; I. Halliday and B. A. McIntosh (eds.), *Solid Particles in the Solar System*: *IAU Symposium 90*, 1980; J. A. M. MacDonnell, *Cosmic Dust*, 1978.

Irrigation (agriculture)

Nearly 15% of the cropland in the United States is irrigated. Irrigated lands, the majority of which are found in the western United States, account for approximately 25% of the total agricultural production in the United States. These lands produce a major share of such important crops as sugarbeets, processing tomatoes, strawberries, lettuce, peaches, grapes, and navel oranges, and substantial proportions of the nation's output of cotton, rice, almonds, and vegetables. In the past, irrigated agriculture has been supported by groundwater sources and surface supplies that were developed relatively inexpensively. Today, declining groundwater tables and competition for water from municipalities, from industries, and for esthetic uses foreshadow changes in the availability of water for agriculture. These changes will almost certainly limit future expansion of irrigated agriculture in the semiarid West and will probably lead to changes in the way that irrigation water is managed.

Constraints on water supplies. The growth of irrigated agriculture in the United States has been enhanced by substantial private and public investment in surface-water impoundment and conveyance facilities. Developed surface-water supplements ground water supplies and provides water for areas with no groundwater resources. Future development of additional surface-water supplies will be more costly than in the past. There are several reasons for this. First, virtually all relatively inexpensive supplies have been developed. Remaining supplies will be considerably more costly to develop because they may be far removed from sites of use and because the unit costs of developing these less desirable sites will be sharply higher. Second, the costs of constructing water development facilities have risen at rates higher than the rate of inflation, making any construction project relatively more costly now than it would have been in the past. Third, the cost of energy required to pump and convey water has risen more than threefold over the last decade. Thus, water that must be conveyed over long distances or elevation changes will be more costly to deliver than it once was. Agricultural water users state that they are unable to defray these higher costs, and public willingness to subsidize new projects through taxes appears low.

The costliness of new supplies has implications beyond the obvious dampening effects on the growth of irrigated agriculture. In some areas, irrigation is supported by the practice of overdrafting groundwater (that is, the rate at which the groundwater is pumped exceeds the rate at which it is recharged or replenished). While overdrafting is justifiable in some situations, it can be continued only until groundwater tables decline to uneconomical pumping depths. Overdrafting is a temporary and self-limiting practice, and ultimately irrigation that is supported by it will cease without supplementary sources of water. The U.S. Water Resources Council estimated in 1978 that as much as 10% of the groundwater pumping in the western United States entails overdraft. Clearly, some irrigated lands will be converted to less profitable dryland farms or taken out of production altogether without additional sources of supply.

Calls for additional water supplies are not limited to the agricultural sector, however. Increases in population and in industrial activity throughout the West, together with emerging development of western coal and other energy resources, translate into burgeoning demands for nonagricultural water. Since all sectors face similar constraints with respect to development of new supplies, increasing attention has been devoted to the possibilities of reallocating water from the agricultural sector, which uses between 85 and 90% of the developed supply in the West, to nonagricultural uses. The extent of such reallocations may not be large because of legal restrictions, and because the demands of other sectors are relatively small compared with total levels of agricultural water use. Yet even small reallocations combined with the declining availability of groundwater and the difficulties of developing new supplies imply that irrigated agriculture will have to adapt to increasing water scarcity.

Potential for water saving. There is considerable controversy over estimates of the quantity of water that could be saved by agriculture and made available for other purposes. Estimates range from 2 to 20% of current levels of use. The difference in estimates is largely attributable to differences in the definition of saving. Normally, most water applied to irrigated fields is used consumptively by the crop in processes that contribute to growth and production. If the crop has insufficient water, production declines. The water not used by the crop either percolates through the soil profile or runs off the field to adjacent areas. Often, percolating water and runoff are recaptured through groundwater pumping and drainage facilities and subsequently used for further irrigation or other purposes. Less frequently, percolating or runoff waters are irretrievably lost, either to saline sinks, such as the sea, or to depths from which retrieval is not economically possible.

Virtually all estimates of potential physical savings include the waters that are irretrievably lost. To the extent that these waters can be saved, they represent new increments of supply whose use would not affect the extent or productivity of irrigated agriculture. On the other hand, the physical saving or conserving water that is subsequently reused represents a reallocation of water away from some existing use or user to another use or user. The differences in estimates of potential physical savings is primarily explained by the fact that some estimates count all runoff and percolation water as a potential savings and others do not.

While estimates of physical savings may measure the ultimate potential of water conservation measures, they are not useful indicators of the quantities of water that may actually be saved. Prevailing water laws and pricing policies have served to keep the price of most irrigation water relatively low. As a result, users of this water have little incentive to bear the costs of new technologies or sophisticated management techniques in order to economize on the use of relatively cheap water. Only when water supplies are absolutely constrained or priced at levels higher than current levels will users be induced to economize further on its use. Physical savings will be realized only to the extent that it remains economical.

The evidence suggests that prices must rise substantially from current levels to induce significant declines in water use. For example, throughout the Western United States the average price of irrigation water is currently about $25 per acre-foot. (An acre-foot is the amount of water required to cover 1 acre to a depth of 1 ft. It is equivalent to 325,851 gallons or 1,238,230 liters, the average quantity used by a family of four in a year.) A 10% increase in this price would induce less than a 7% reduction in current levels of use. Price increases for sources of water which must be pumped are inevitable since the price of fuels and electricity required for pumping is rising. Energy-related price increases will induce some cutback in agricultural water use, but it seems unlikely that these will be sufficient to offset both the ultimate shortfall in groundwater supplies and increased demands from the nonagricultural sectors.

Markets for water. One proposal that is receiving attention involves the idea of a water market. The principal advantage of markets is that they permit the allocation of limited resources among competing demands so that the resources are put to their highest-valued uses. Markets also rely on voluntary exchange, and buyers and sellers are never coerced to participate. In theory, then, water markets may be an attractive solution to the current water scarcity problem in the West, since they would permit water to be moved from some of the low-valued uses (generally, certain agricultural uses) to higher-valued uses (generally, urban and industrial uses). Moreover, sellers would be induced to sell only if the offered price was higher than the return they would obtain be using it themselves. Water markets would also result in changes in the effective price of water. The higher emergent prices would induce users to engage in further water-economizing practices. There is evidence that even limited water transfers between agriculture and other sectors would be profitable to both buyers and sellers and would, in many instances, obviate the need to build costly new water supply projects.

Although current western water laws inhibit water trading, a number of transfers have been successfully consummated in Colorado, Utah, and New Mexico. These sales have involved transfers from the agricultural sector to the energy industry and municipalities. The advent of water markets will undoubtedly reduce the amount of water available to agriculture and reduce irrigated acreage. However, markets would ensure that those selling their water would be compensated for reducing water use. Additionally, markets would induce agricultural water users to shift from low-valued to high-valued crops where appropriate, and adopt water-saving irrigation technologies and water management practices. The full potential of water markets cannot be realized, however, without changes in western water law which would facilitate the transfer of water while protecting the rights of those who use return flows productively.

For background information *see* IRRIGATION (AGRICULTURE); LAND DRAINAGE (AGRICULTURE); WATER CONSERVATION in the McGraw-Hill Encyclopedia of Science and Technology.

[HENRY J. VAUX, JR.]

Biblography: R. E. Howitt, D. E. Mann, and H. J. Vaux, Jr., The economics of water allocation, in E. Engelbert (ed.), *Competition for California Water*, 1982; R. E. Howitt, W. D. Watson, and R. M. Adams, A reevaluation of price elasticities for irrigation water, *Water Resources Res.* 16(4):623–628, 1980; U.S. Water Resources Council, *Second National Assessment*, 1978.

Land reclamation

In the past decade much stricter state and federal environmental regulations have been passed which require industry to reclaim mined lands to a level at least as productive as they were prior to mining. One of the problems in the reestablishment of vegetation can be the highly acidic conditions found in some spoil material.

Acidic environment. Extremely acidic conditions of the spoil material are generally the result of oxidation and hydrolysis of iron sulfide minerals such as pyrite and marcasite. Prior to disturbance of the area, reducing conditions in the overburden prevail. Under these reducing conditions the pyritic minerals are not oxidized, and therefore sulfuric acid is not produced. However, removal of the overburden exposes iron sulfides to the air. In the presence of water, a series of chemical reactions are initiated with the pyritic minerals, resulting in the production of sulfuric acid. This is a very strong acid and has the potential for drastically lowering the pH of the spoil and topsoil material. A high acid concentration (low pH) is toxic to vegetation.

Elemental toxicities to plants can result from an

increase in the plant availability of many elements in an acidic environment. Another potential problem is the restriction of root penetration into the acidic spoil, resulting in limited nutrient uptake and increased susceptibility of the plants to drought. Low pH also leads to a reduction in the populations of free-living and symbiotic nitrogen-fixing soil microorganisms. There is a corresponding increase in the population of microorganisms that oxidize iron and sulfur minerals. These microorganisms thrive in an acidic environment and therefore maintain the acidic conditions. Such problems are directly or indirectly due to the high acid concentration of disturbed pyritic-bearing overburden and can be relieved by increasing the pH of the material. Flyash and calcium carbonate (agricultural lime) can be used as soil amendments to reduce the acidity associated with the presence of the iron sulfide minerals.

Amendments. There are certain advantages in using flyash as a treatment for increasing the pH of acid spoils. Flyash is a waste product of coal-burning power-generating plants and is already creating waste disposal problems. It generally has a high pH (approximately 11–12) and therefore can help to neutralize acidic spoil material. Also, flyash contains many essential plant nutrients, and often the particles are in the size range of very fine sand to silt which may improve the moisture-holding capacity of coarse-textured spoil material. A disadvantage of flyash is that, depending upon the quality of the coal being utilized by the power-generating plants, boron toxicity to the vegetation may occur.

The advantages of using calcium carbonate for increasing the pH of acidic spoils include: its effectiveness in ameliorating the acidic conditions of the spoils; its lack of toxic levels of other elements; and its availability. However, the costs of transporting the lime to the mine sites could be quite high.

In a greenhouse study, five levels of flyash and five levels of agricultural lime were tested as spoil amendments in conjunction with three topsoil depth treatments. The results of this study suggest that topsoil addition should be enough to correct this problem of revegetating acidic mine spoils. However, this is only a short-term remedy since the reactions of hydrolysis and oxidation on the untreated pyritic spoils would continue just beneath the soil surface until the topsoil itself would become too acidic to support vegetation. A layer of treated spoil material with topsoil applied may give the system time enough to prevent the free exchange of oxygen and moisture to the untreated spoil material. This would result in limited acid production and a more optimum pH for plant growth.

The aboveground biomass and root distribution of barley (*Hordeum vulgare*) were used to evaluate the effectiveness of the amendments in the greenhouse study. Water-soluble boron and pH were measured in the amended spoils to evaluate the effectiveness of the flyash and the agricultural lime for ameliorating the acid problem.

Aboveground biomass. Spoil amendments and topsoil treatments resulted in large increases in aboveground biomass over the untreated spoil (Table 1). Aboveground biomass was consistently higher where 4 or 8 in. (10 or 20 cm) of topsoil was applied to either the agricultural lime or the flyash-amended spoil material. In most cases the agricultural lime amendments resulted in slightly greater biomass production than the flyash amendments.

Root distribution. This is an important factor to consider when evaluating the success of revegetation of disturbed lands, especially for those areas subject to severe conditions of wind or water erosion. The rooting system of a plant serves many functions, including the uptake of nutrients and water, anchoring the plant to the soil, and stabilizing the soil against erosion. The root distribution was measured by determining root biomass per 4-in. (10-cm) depth increment of the soil. The results indicate that the roots have a definite tendency to penetrate into the spoil material treated with either flyash or agricultural lime. Furthermore, the data show that the

Table 1. Comparison between lime and flyash amendments on aboveground biomass of barley

Lime			Flyash		
Tons/acre (metric tons · ha^{-1})	Topsoil depth, in. (cm)	Aboveground biomass, tons/acre (metric tons · ha^{-1})	Tons/acre (metric tons · ha^{-1})	Topsoil depth, in. (cm)	Aboveground biomass, tons/acre (metric tons · ha^{-1})
0	0	0.0	0	0	0.0
9.8 (22)	0	0.98 (2.2)	9.8 (22)	0	0.71 (1.6)
20 (45)	0	1.03 (2.3)	20 (45)	0	1.07 (2.4)
40 (90)	0	1.07 (2.4)	30 (67)	0	0.98 (2.2)
80 (180)	0	1.07 (2.4)	40 (90)	0	0.94 (2.1)
0	4 (10)	1.12 (2.5)	0	4 (10)	1.12 (2.5)
9.8 (22)	4 (10)	1.34 (3.0)	9.8 (22)	4 (10)	1.25 (2.8)
20 (45)	4 (10)	1.34 (3.0)	20 (45)	4 (10)	1.29 (2.9)
40 (90)	4 (10)	1.38 (3.1)	30 (67)	4 (10)	1.34 (3.0)
80 (180)	4 (10)	1.34 (3.0)	40 (90)	4 (10)	1.38 (3.1)
0	8 (20)	1.43 (3.2)	0	8 (20)	1.43 (3.2)
9.8 (22)	8 (20)	1.29 (2.9)	9.8 (22)	8 (20)	1.47 (3.3)
20 (45)	8 (20)	1.47 (3.3)	20 (45)	8 (20)	1.34 (3.0)
40 (90)	8 (20)	1.51 (3.4)	30 (67)	8 (20)	1.15 (2.6)
80 (180)	8 (20)	1.25 (2.8)	40 (90)	8 (20)	1.38 (3.1)

Table 2. Boron and pH of spoil as affected by amendments

Amendment, tons/acre (metric tons · ha^{-1})	Boron,* $\mu g \cdot g^{-1}$	pH*
	Topsoil	
	3.6	7.3
	Untreated spoil	
	3.6	2.8
$CaCO_3$		
9.8 (22)	3.6	6.6
20 (45)	4.2	6.8
40 (90)	8.2	7.1
80 (180)	14.0	7.1
Flyash		
9.8 (22)	56.8	5.0
20 (45)	52.3	6.6
40 (67)	65.5	7.6
80 (90)	77.0	7.7

*Boron and pH amendment level means for all 2-in. (5-cm) increments, excluding the topsoil increments.

roots will not penetrate into the untreated acidic spoil material, but simply spread out in the limited topsoil available.

pH data. Examination of the pH data (Table 2) for spoil material showed that the untreated spoil material was very acidic (pH 2.8), which prevented almost all plant growth (Table 1). Increasing the amount of agricultural lime resulted in a corresponding increase in pH; however, the difference in pH between the highest and the lowest agricultural lime rate is only 0.5 pH unit. This may be due to the fact that not all the agricultural lime has reacted, especially at the higher amendment rates. The change of the pH with increasing flyash amendment levels was much greater. This was not unexpected since the flyash is finer in texture than the agricultural lime, thus allowing more immediate solubilization and reaction with the acidic spoil.

Boron. In evaluating the boron concentrations of the two spoil amendments at their respective levels, it was found that the topsoil and the untreated spoil material contained 3. 6 micrograms per gram of boron. Although the boron concentration increased to 14 $\mu g \cdot g^{-1}$ at the highest concentration of agricultural lime, it certainly is not toxic to the native plants used to revegetate mined lands in the western United States. However, the boron concentration in the flyash-treated spoil material ranged from 52 to 77 $\mu g \cdot g^{-1}$. Some of the plants grown on spoils treated at the higher levels of flyash expressed limited symptoms of boron toxicity; however, there was little or no reduction in aboveground biomass. As compared to native grasses of the western coal regions, barley is relatively sensitive to boron.

Conclusion. Agricultural lime and flyash can be used to ameliorate acidic conditions caused by the oxidation of pyritic materials in mine spoils. Although lime is an effective method of increasing the pH of acidic soils, the expense of transporting it to the mine sites could be prohibitive. Flyash is also an effective treatment for increasing the pH of acidic spoils and, because power-generating plants frequently are in proximity to the mines supplying their coal, the transportation costs of the flyash should not be excessive. Flyash is a waste product that currently requires special handling techniques in its disposal, and therefore any use of flyash would reduce these disposal problems.

For background information *see* pH; PLANT MINERAL NUTRITION; SOIL CONSERVATION; SOIL MICROBIOLOGY; STRIP MINING in the McGraw-Hill Encyclopedia of Science and Technology.

[E. M. TAYLOR, JR.; GERALD E. SCHUMAN]

Bibliography: A. C. Chang et al., Physical properties of flyash-amended soils, *J. Environ. Qual.*, 6:267–270, 1977; J. W. Doran and D. C. Martens, Molybdenum availability as influenced by application of fly ash to soil, *J. Environ. Qual.*, 1(2):186–189, 1972.

Laser alloying

Laser alloying is a material processing method which utilizes the high power density available from focused laser sources to melt metal coatings and a portion of the underlying substrate. Since the melting occurs in a very short time and only at the surface, the bulk of the material remains cool, thus serving as an intimate heat sink. Large temperature gradients exist across the boundary between the melted surface region and the underlying solid substrate. The result is rapid self-quenching and resolidification.

Transitions. The sequence of schematic cross sections in Fig. 1 illustrates the transitions occurring during and following an individual laser exposure. In Fig. 1*a*, the metal substrate (B) coated with a thin metal film (A) is irradiated with a laser pulse. A fraction of the incident laser light is absorbed by free carriers within the electromagnetic skin depth of 10^{-7} to 10^{-6} in. (10^{-6} to 10^{-5} cm). For metal surfaces and most laser wavelengths, a significant fraction of the incident light will be specularly or diffusely scattered away (the reflectance process is shown in Fig. 1). The absorbed energy is "instantaneously" (10^{-12} s) transferred to the lattice. The near-surface region very rapidly reaches the melting point, and a liquid-solid interface starts to move through the film (Fig. 1*b*). In Fig. 1*c*, the liquid-solid interface has swept through the original thin film–substrate interface. Interdiffusion of the film and substrate elements starts. The laser pulse is nearly terminated, and the surface has remained below the vaporization temperature. In Fig. 1*d*, the maximum melt depth has been reached, and interdiffusion continues. The resolidification interface velocity is momentarily zero and then rapidly increases. In Fig. 1*e*, the resolidification interface has moved approximately half way back to the surface from the melt depth. Interdiffusion in the liquid continues, but the resolidified metal behind the liquid-solid interface cools so rapidly that solid-state diffusion may be neglected. In Fig. 1*f*, the material is completely resolidifed, and a "surface alloy" of A in B has been produced. What makes laser surface alloying both attractive and interesting is the wide variety of chemical and microstructural states that can be retained because of the rapid quench from the liquid phase. These include chemical profiles where the "alloyed" element A is highly concentrated near the atomic surface and decreases in concentration over shallow depths (hundreds of nanometers), and uniform profiles where the concentration of A in B is the same throughout the entire melted region. The types of microstructures observed include extended solid solutions (the concentration of A in B greatly exceeds equilibrium values), metastable crystalline phases (high-temperature phases retained because of the rapid return to room temperature), and metallic glasses.

Surface alloying. In general the output of a laser source must be focused in order to achieve sufficient energy density to induce melting of a metal surface. This focused spot is characterized by its optical spot size. The size of the melted spot (effective spot size) resulting from the laser exposure will generally be smaller than this optical spot size. For a continuous-wave laser source, a melt strip will be produced as the focused beam and metal surface are moved relative to one another. The dwell time, that is, the time that the continuous-wave laser beam irradiates a particular surface point, will strongly influence the depth of the melting. As is schematically shown in Fig. 2*a*, individual melt strips must be partially overlapped to produce area coverage. The effective spot size, degree of overlap, and relative scanning speed will thus determine the area per unit time which can be produced.

For a pulsed or Q-switched laser (high-power, short-pulse laser) source, there also is an optical and effective spot size, and the "pulse length" is the time of exposure. The laser pulses are emitted in a train of pulses characterized by a repetition rate. As is schematically shown in Fig. 2*b*, this train of pulses is raster-scanned across the metal surface to produce area coverage. For these lasers, the effective spot size, degree of overlap, and repetition rate determine the area per unit time which can be produced.

For all the laser sources (continuous-wave, pulsed, and Q-switched), the exposure time (dwell time or pulse length) strongly influences the depth that will be melted. Longer exposure times result in deeper melting. Since deeper melting means a longer total time in the molten state, that means more time available for diffusion of the one or more alloying elements into the molten portion of the substrate. Deeper melting and longer melt times therefore result in more dilute surface alloys, while shallow melting and shorter melt times result in more concentrated surface alloys. It is also evident that in some instances convection, surface tension, and plasma effects can enhance the mixing within the liquid state and drive the melt toward homogenization.

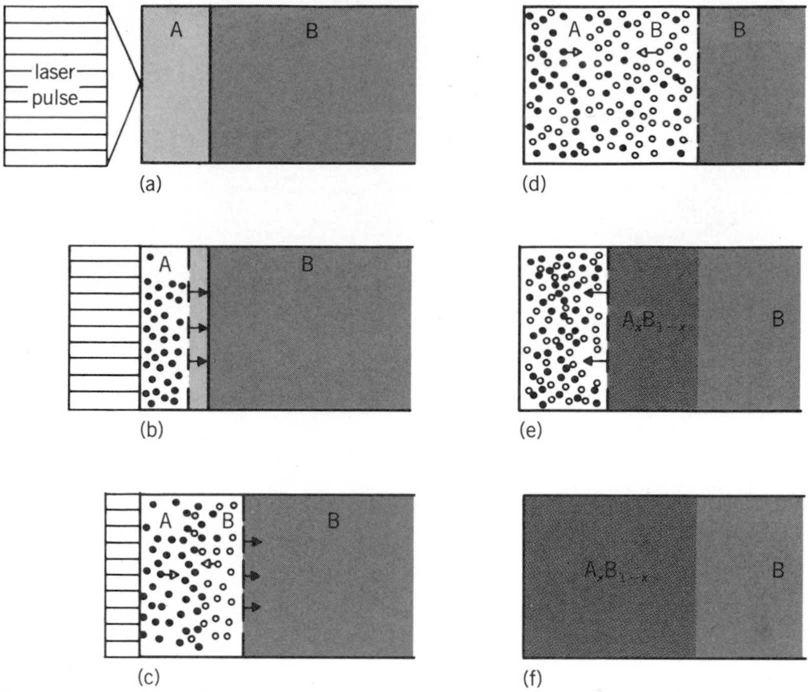

Key:
• = film elements
∘ = substrate elements
→ = movement of interface
→ = movement of film/substrate elements

Fig. 1. Sequence of schematic cross sections for laser alloying, with time increasing from *a* to *f*. A_xB_{1-x} represents surface alloy (with composition fixed by *x*) of film elements A in substrate elements B.

In making laser alloys, many other processing variables need to be considered. In addition to the exposure time just discussed, these include the laser power, the thickness of the film put down prior to laser melting, and in some instances the nature of the gaseous ambient during the laser processing. The processing variables are interrelated, and one variable cannot be freely changed without affecting another. Another consideration is that laser alloying is a liquid state–rapid quenching phenomenon. The near-surface region must be melted and yet vaporization avoided. Different minimum and maximum energy densities are thus defined for each laser exposure time. In addition to these processing constraints, there are certain properties of matter which strongly influence whether or not certain element combinations may be laser-alloyed. For example, it is not possible to laser-alloy a low melting point–high vapor pressure element like zinc (Zn) into a high melting point–low vapor pressure metal substrate such as tungsten (W). The Zn would vaporize before the underlying W could be melted. Since liquid-state intermixing is required, suitable systems must exhibit miscibility in the molten state. Binary systems like silver-nickel (Ag-Ni), iron-lead (Fe-Pb), copper-molybdenum (Cu-Mo), and aluminum-bismuth (Al-Bi) have miscibility gaps in the liquid state spanning nearly all compositions. Such systems cannot be laser-alloyed.

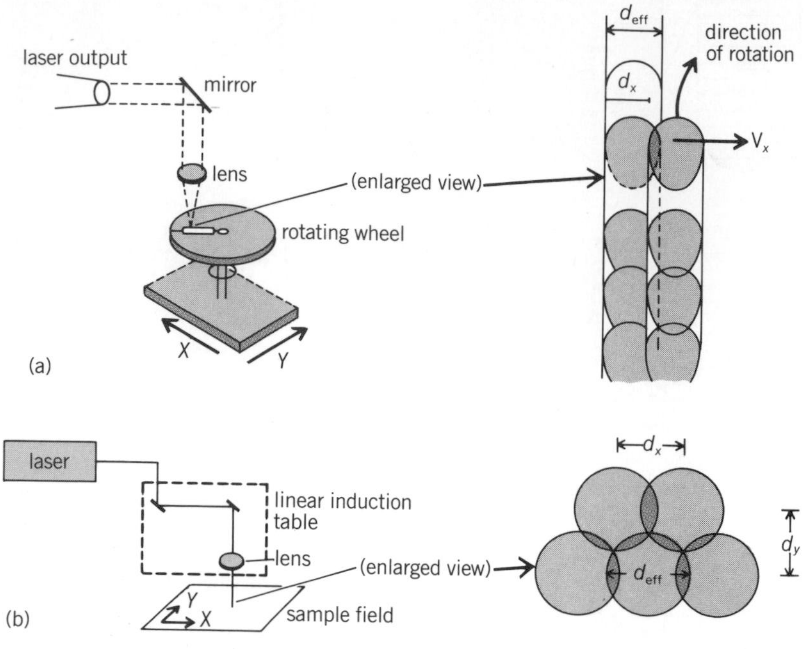

Fig. 2. Schematic drawings and enlarged views of material processing for (*a*) continuous and (*b*) pulsed or Q-switched laser system. d_{eff} = diameter of melt spot or melt strip; V_x = velocity of sample movement in *x* direction; d_x, d_y = displacement between subsequent laser melt strips or spots resulting from relative motion of laser and sample.

Advantages. Surface alloys, particularly laser surface alloys, have advantages that include tailoring surface properties, conservation of materials, and creation of new metal surfaces.

Surface alloying allows alteration of the metal surface in order to achieve characteristics best suited to the service environment. For example, in an ordinary saw blade the material requirements for the cutting teeth are very different from those of the length of the blade. In many applications requiring corrosion or oxidation resistance, that resistance is needed only on the external surface of the material. It is possible to design the bulk of the material with characteristics most suitable for structural needs or ease of fabrication and to tailor the metal surface for interface requirements. Material conservation is another reason often cited for considering surface alloying. In the case of stainless steels and superalloys, the elements added to impart the special properties are often strategic elements. That is, they are elements that the United States consumes in great quantities, with sparse domestic supplies. For example, chromium (Cr) and nickel (Ni) may not be necessary on the inside of knives, forks, and spoons. Further, precious metals are expensive and in relatively short supply. Thus surface alloying can be very cost-effective.

These first two advantages—tailoring for surface properties and material conservation—are generally important considerations in almost any coating technology. However, laser alloying differs from coating technologies in that the near-surface region is a continuous extension of the interior of the metal. There is no interface between bulk and "coating"; the laser alloy is a mixture of bulk and surface elements. Also, problems in regard to porosity and adherence do not exist. Synergistic benefits may be derived which are not realized by a coating alone.

The prospect of creating new metal surfaces intrigues the largest number of investigators involved in laser alloying research. Laser alloying involves very large temperature gradients and quenching from the liquid state. In this way it is much like other rapid-solidification technologies. The thermodynamic constraints which limit the conventional metallurgist do not necessarily apply. It is anticipated that laser alloying will produce many new metal alloys which could not have been produced by conventional methods.

Current research. Scientists at Rockwell International and the Naval Research Laboratory have been producing surface alloys of Cr and Ni in carbon steels by the use of continuous CO_2 lasers to melt steel coated with thin films or powder mixtures. The alloys produced have a very uniform composition throughout the melted near-surface region, which has ranged from 10^{-6} to 10^{-1} in. (hundreds of micrometers to several millimeters) in depth. In order to demonstrate the corrosion resistance of these surface alloys, electrochemical polarization studies were performed. Figure 3 contains electrochemical results on both 29% Cr–13% Ni surface alloy (carbon steel substrate) and a commercial bulk 304 stainless steel. The electrochemical medium was deaerated 1*N* H_2SO_4. As can be seen by inspection of the polarization curves, the surface alloy, in relation to bulk 304 stainless, has a comparable current density within the passive region and an order-of-magnitude lower critical current density for onset of passivation. Thus, surface-alloyed stainless steels can readily be prepared by laser processing.

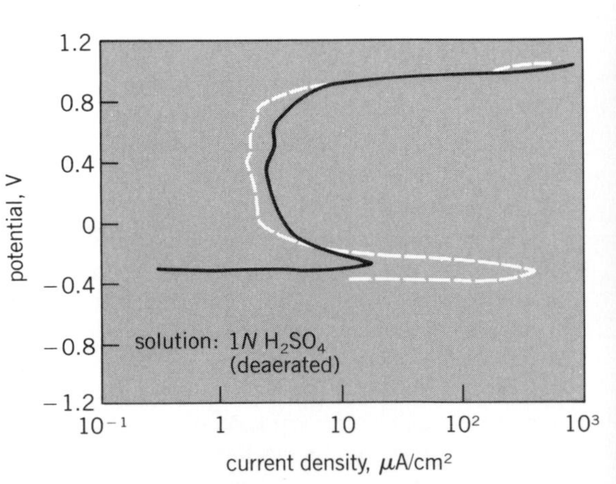

Fig. 3. Electrochemical polarization curves for surface alloy (solid curve) and for commercial bulk stainless steel (broken curve). Potential measured with reference to a standard calomel electrode. 1 μA/cm² = 6.45 μA/in.² (*After J. B. Lumsden, D. S. Gnanamuthu, and R. J. Moores, Corrosion resistance of AISI 4140 steel laser surface alloyed with Cr and Ni, in C. R. Clayton and C. M. Preece, eds., Corrosion of Metals Processed by Directed Energy Beams, Metallurgical Society of AIME, 1982*)

Numerous scientists in the United States, Germany, Italy, and the Soviet Union are using continuous, pulsed, and Q-switched lasers to produce metallic glasses via laser alloying of films and substrate. Laser alloying has two inherent advantages over the usual methods of producing amorphous metals (splat cooling, melt spinning, and so on): higher cooling rates are accessible because of the self-quenching nature of the mechanism; and laser alloying can be performed on thicker, bulkier material while the other methods generally produce thin foils of limited usefulness.

Other innovative material developments through laser alloying are expected. Commercial applications have begun. In the not-too-distant future, laser alloying may be as common a practice as decorative coating, ion implantation, or electroplating.

For background information *see* ALLOY; LASER; METAL COATINGS in the McGraw-Hill Encyclopedia of Science and Technology.

[CLIFTON W. DRAPER]

Bibliography: C. W. Draper, Laser surface alloying: A bibliography, *Appl. Opt.*, 20(18):3093–3096, September 1981; C. W. Draper, Laser surface alloying: The state of the art, *J. Met.*, 34(6):24–32, June 1982; D. S. Gnanamuthu, Laser surface treatment, *Opt. Eng.*, 19(5):783–792, 1980; D. K. Sood, Metastable surface alloys produced by ion implantation, laser of electron beam treatment, *Radiat. Effects*, 63(1):141–167, January 1982.

Launch vehicle

The shuttle is the primary United States space launch system for providing space transportation. Its reusability and versatility, its ability to carry multiple payloads from space, and its ability to deliver large payloads to low Earth orbit make it the national space transportation system for the future.

There are two families of expendable vehicles used to support and extend the capability of the shuttle. The first of these families is known as the Shuttle Upper Stages, and consists of stages which are designed to be launched from the shuttle payload bay. The second is the expendable launch vehicles which have been used to provide space transportation for the last 20 years and which will continue to be used in the transition period while payloads are being designed and developed for the shuttle. The expendable vehicles launch their payloads directly into orbit independently of the shuttle.

SHUTTLE UPPER STAGES

Shuttle Upper Stages are used to place payloads into orbits or trajectories of higher energy than the shuttle alone is capable of attaining. The capabilities of these stages range from the PAM-D, which will place 2760 lb (1250 kg) into a geosynchronous transfer orbit, to the STS/Centaur, which can insert up to 13,030 lb (5910 kg) into a geosynchronous orbit. Some of the Shuttle Upper Stages are entirely new developments, while others are adaptations of existing stages from expendable launch vehicles with a long history of successful usage.

The Shuttle Upper Stages may be further subdivided into perigee stages, whose primary purpose is to provide only perigee thrust to move the payload from the shuttle orbit and place it into an elliptical geosynchronous transfer orbit as shown in Fig. 1 (for these applications, the apogee thrust needed to place the payload into geosynchronous orbit is usually furnished by an apogee motor which is combined with the payload), and those stages that are capable of providing both perigee and apogee thrust to place the payload into geosynchronous orbit.

The United States has four perigee stages either operational or under development. These are the Payload Assist Modules (PAM): PAM-D, PAM-A, and PAM-DII; and the Transfer Orbit Stage (TOS). There are three stages which have geosynchronous orbit capability: Inertial Upper Stage (IUS), Centaur-G, Centaur-G Prime. In addition, there are two "integrated" systems which combine the perigee motor, the apogee motor, and the payload in a single package. These are the SYNCOM IV and INTELSAT VI integrated systems. These total systems place the payloads (SYNCOM IV and INTELSAT VI) into geosynchronous orbit.

The Shuttle Upper Stages can obviously be used for other than geosynchronous transfer orbital or geosynchronous orbital applications; for example,

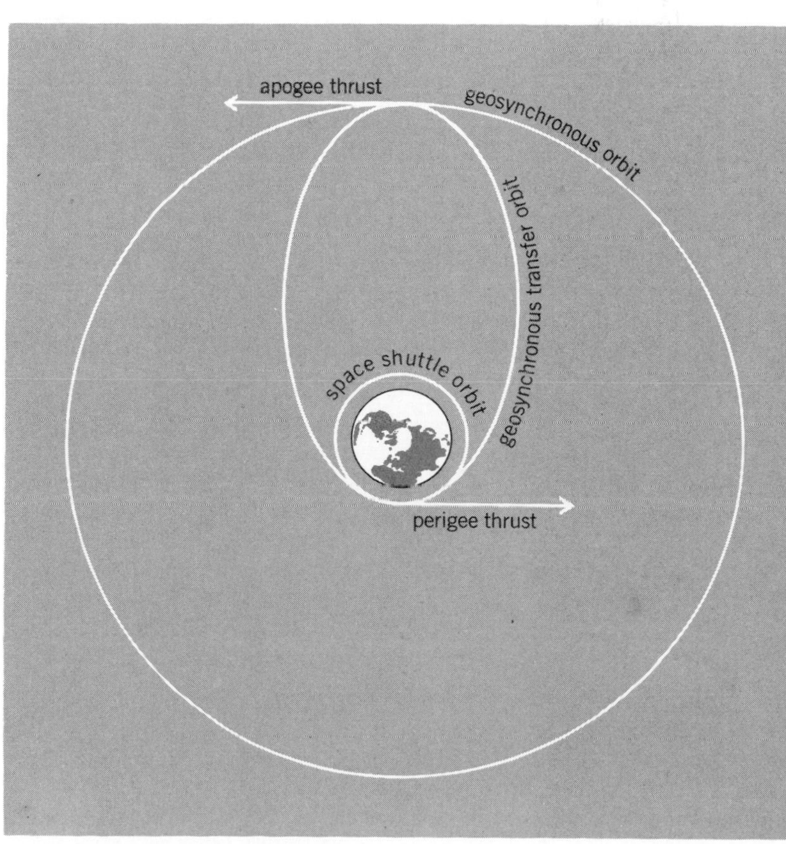

Fig. 1. Geosynchronous orbit transfer.

Table 1. Characteristics of Shuttle Upper Stages

Upper stage	Length, ft (m)	Weight, lb (kg)	Diameter ft (m)	ASE weight, lb (kg)	Payload to geosynchronous transfer orbit, lb (kg)	Payload to geosynchronous orbit, lb (kg)	Operational capability, year
PAM-D	6.2 (1.9)	4800 (2180)	3.9 (1.2)	2500 (1140)	2760 (1250)	1380 (625)	Current
PAM-A	7.5 (2.3)	8500 (3860)	4.3 (1.3)	4200 (1910)	4400 (2000)	2200 (1000)	1984
PAM-DII	6.2 (1.9)	8100 (3680)	5.2 (1.6)	3420 (1550)	4100 (1860)	2050 (930)	1985
TOS*	11.2 (3.4)	16,420 (7450)	7.5 (2.3)	3300 (1500)	6020-13,030 (2730-5910)	3000-6500 (1360-2950)	1986
IUS	16.4 (5.0)	32,560 (14,770)	9.5 (2.9)	7360 (3340)	—	5000 (2270)	Current
Centaur-G	19.7 (6.0)	36,270 (16,450)	14.1 (4.3)	7340 (3330)	—	10,000 (4540)	1987
Centaur-G Prime	29.2 (8.9)	42,770 (19,400)†	14.1 (4.3)	8050 (3650)	—	13,030 (5910)	1986

*Estimated values.
†Off-loaded for shuttle limitation of 65,120 lb (29,540 kg).

the Centaur-G Prime is specifically designed for planetary applications. However, the payload capability to geosynchronous or geosynchronous transfer orbits is a convenient way to compare the performance capability of the various stages. The geosynchronous transfer orbit considered in the following discussion is that resulting from a due-east launch of the shuttle from Cape Canaveral, Florida, at an inclination to the Equator of 28°. The perigee stage is then launched into a transfer orbit at an inclination to the Equator of 26°; at this inclination, the weight placed in transfer orbit is about twice that of the useful payload placed in geosynchronous orbit, that is, the weight of the apogee motor is approximately equal to the weight of the useful payload.

Fig. 2. First launch of Shuttle Upper Stage PAM-D from the bay of the space shuttle orbiter in November 1982. The axis of PAM-D is perpendicular to the longitudinal axis of the orbiter.

The following is a brief description of the characteristics of each of the Shuttle Upper Stages mentioned above. Table 1 is a summary of the stage characteristics and payload capabilities.

PAM-D. The PAM-D is a spin-stabilized solid-propellant perigee stage specifically designed to place payloads, which are equivalent to those launched on the expendable Delta launch vehicle, in geosynchronous transfer orbit. It will place 2760 lb (1250 kg) in geosynchronous transfer orbit when launched from the shuttle payload bay in a 183-mi (295-km) orbit. The PAM-D is carried to orbit in specially designed airborne support equipment (ASE) which restrains the stage and payload during the shuttle launch, spins the stage and payload at selected rates between 30 and 100 revolutions per minute to provide stabilization, and launches it from the shuttle payload bay by means of ejection springs.

After sufficient separation between the stage and orbiter is attained (by maneuvering the orbiter and allowing an inactive 45-min coast period), the stage is fired by a timer, placing the satellite and its apogee motor on geosynchronous transfer orbit.

The PAM-D system occupies approximately 6 ft (2 m) of the orbiter payload bay length—up to four PAM-D's can be accommodated on a single shuttle flight. Figure 2 is a picture of a PAM-D being launched from the orbiter bay. Note that the PAM-D is transported in the orbiter, with its axis perpendicular to the longitudinal axis of the orbiter.

The PAM-D is a commercial development by the McDonnell Douglas Astronautics Company. The first operational flight of this system from the shuttle orbiter was in November 1982. The PAM-D also flies as the third stage of the expendable Delta vehicle. The first flight on the Delta was in November 1980.

PAM-A. The PAM-A is a solid-propellant perigee stage designed to place payloads, which are equivalent to those launched on the expendable Atlas Centaur launch vehicle, in geosynchronous transfer orbit. It will place 4400 lb (2000 kg) on geosynchronous transfer orbit when launched from a 183-mi (295-km) altitude, 28°-inclination shuttle orbit. As with PAM-D, the ASE serves to support and restrain the stage and payload during shuttle launch and ascent. The PAM-A with its spacecraft is launched with its longitudinal axis parallel to that of the shuttle. The stage and payload are then spun up to provide stability and launched from the orbiter payload bay by ejection springs. After 45 min, the PAM-A motor is ignited and places the payload and apogee motor in geosynchronous transfer orbit. The PAM-A motor is a slightly modified Minuteman III third stage.

The PAM-A system, with payload, occupies approximately 30 ft (9 m) in the orbiter payload bay—two PAM-A's can be accommodated on a single shuttle flight.

The PAM-A is a commercial development by the McDonnell Douglas Astronautics Company and may be available for flight as early as 1984.

PAM-DII. The PAM-DII is a spin-stabilized solid-propellant perigee stage whose capability lies between those of the PAM-D and PAM-A discussed above. The system uses ASE similar to that of the PAM-D. The PAM-DII will be capable of placing approximately 4100 lb (1850 kg) into transfer orbit and will be designed to be offloaded to a capability of about 2760 lg (1250 kg). The system is operationally similar to the PAM-D and PAM-A. The system is also a commercial development of the McDonnell Douglas Astronautics Company. First flight is planned for mid-1985.

TOS. The TOS is a solid-propellant perigee stage which will be capable of placing payloads ranging from 6020 to 13,030 lb (2730 to 5910 kg) in geosynchronous transfer orbit when launched from the orbiter payload bay.

The motor to be used for this application is currently planned to be the same basic motor used on the Inertial Upper Stage which is discussed below. This motor has demonstrated capability at 50% offload. It will be used in the three-axis-stabilized mode of flight.

The TOS will be mounted longitudinally in the orbiter payload bay on ASE, which may be a modification of that used for the IUS—two TOS systems can be accommodated on a single shuttle flight. The first flight is planned for 1986.

IUS. The IUS is a two-stage solid-propellant vehicle capable of placing 5000 lb (2270 kg) in geosynchronous orbit when launched from the shuttle orbiter, or 4200 lb (1910 kg) in geosynchronous orbit when launched from the Titan-34D expendable launch vehicle. It is a three-axis-stabilized vehicle which employs a highly sophisticated avionics system having a predicted probability of mission success of greater than 98%.

The IUS is launched with its longitudinal axis parallel to that of the shuttle. It is rotated upward at an angle of 60° in preparation for deployment. As with the systems discussed above, deployment is accomplished by springs. The stage maintains attitude control for 45 min before ignition of the first stage to place the payload and the IUS second stage in geosynchronous transfer orbit. The second stage is fired to place the 5000-lb (2270-kg) payload in geosynchronous orbit.

The IUS was developed by the Boeing Company under an Air Force contract. The first flight on the Titan was in 1982; the first shuttle launch was in 1983. The IUS second-stage failed due to a control anomaly during the latter flight. After review of the anomaly and modifications to prevent its recurrence, it was expected that operational flights of the IUS will resume in late 1984 or 1985.

Centaur-G. The Centaur-G vehicle is a liquid hydrogen–liquid oxygen fueled stage capable of placing 10,000 lb (4540 kg) into geosynchronous orbit. Its length is limited to 20 ft (6 m) in order to accommodate long (up to 40 ft or 12 m) payloads in the orbiter cargo bay.

The Centaur-G is an adaptation of the expendable Centaur stage used on the expendable Atlas Centaur vehicle. The stage will fly in the orbiter, with its longitudinal axis parallel to that of the orbiter. The erection and release mechanisms are similar in principle to those of the IUS.

The Centaur-G is being developed by the General Dynamics Corporation. The development is being jointly funded by NASA and the Air Force. The Centaur-G will be ready for flight in 1987.

Centaur-G Prime. The Centaur-G Prime vehicle is a longer (almost 30 ft or 9 m) version of the Centaur-G. This vehicle is being developed by NASA for use in planetary missions. In a shuttle orbiter which is capable of carrying 65,120 lb (29,540 kg) to low Earth orbit, the Centaur-G Prime is capable of delivering 13,030 lb (5910 kg) to geosynchronous orbit. When shuttle cargo weight capability increases, the Centaur-G Prime will be able to deliver approximately 17,950 lb (8140 kg) to geosynchronous orbit. The Centaur-G Prime will be used on the Galileo and International Solar Polar Mission in 1986.

SYNCOM IV and INTELSAT VI. The SYNCOM IV and INTELSAT VI integrated systems are commercial developments by the Hughes Aircraft Corporation. These propulsion systems are specifically designed to place their respective payloads (SYNCOM IV and INTELSAT VI) in geosynchronous orbit. The SYNCOM IV system has a capability of about 2910 lb (1320 kg) into geosynchronous orbit, and the INTELSAT VI system will place about 5000 lb (2270 kg) into geosynchronous orbit. The SYNCOM IV is scheduled to be available for launch in 1984 and the INTELSAT VI in 1986.

EXPENDABLE LAUNCH VEHICLES

The family of expendable launch vehicles maintained by the United States is graduated in size and capability to complete a variety of missions demanded by space activity. The family is now operating in a transition phase, having evolved to where five systems are being procured and maintained in active status. These are the Scout, Atlas Centaur, Atlas, Delta, and Titan systems. Their characteristics are shown in Table 2.

Scout. The Scout launch vehicle was developed in the late 1950s by Vought, Inc., to launch small payloads as probes, to place them in low Earth orbit, and to serve as a launch vehicle for reentry studies. The current Scout is a four-stage solid-propellant vehicle to which a fifth stage can be added to provide highly elliptical orbits. Scout vehicles are launched from the Wallops Flight Facility at Wallops Island, Virginia; the Western Space and Missile Center at Vandenberg Air Force Base, California; and the San Marco Test Range off the coast of Kenya, Africa. The San Marco Range is operated by the Italian government in cooperation with NASA. Scout has an open-loop inertial guidance system located in the third stage. The fourth stage is spin-stabilized, after having received proper orientation from the control exerted by the first three stages. The optional fifth stage has an attitude correction system.

Atlas Centaur. Atlas Centaur, developed by General Dynamics/Convair, is a 2½-stage vehicle intended as a booster for high-energy missions, particularly lunar and planetary exploration missions and geosynchronous orbit missions. The Atlas first stage is powered by three Rocketdyne engines. At an appropriate time of flight, two outboard engines and support structures are discarded, allowing the remainder of flight to be completed with one engine. This separation practice gives rise to the reference

Table 2. Characteristics of United States expendable launch vehicles

Vehicle	Stages	Propellants*	Thrust, 1000 lbf (kN)	Maximum diameter × height, ft (m)	Maximum payload, lb (kg)† 115-mi (185-km) orbit	Maximum payload, lb (kg)† Geosynchronous transfer orbit
Scout	1. Algol IIA	Solid	108.9 (484.5)	3.6 × 75 (1.11 × 22.9)	560 (255)	—
	2. Castor IIA	Solid	64.1 (285.2)			
	3. Antares IIIA	Solid	18.7 (83.1)			
	4. Altair IIIA	Solid	5.8 (25.6)			
Atlas Centaur	1. Atlas booster and sustainer	LOX/RP-1	430.1 (1913.0)	10.0 × 131 (3.05 × 39.8)	13,330 (6045)	5210 (2365)
	2. Centaur	LOX/LH_2	32.8 (146.0)			
Atlas	1. Atlas booster and sustainer	LOX/RP-1	387.1 (1722.0)	10.0 × 92 (3.05 × 28.1)	4600 (2090)‡	—
	2. TE-364-4	Solid	14.8 (65.8)			
Delta (3900)	1. Thor + 9 TX-562-2	LOX/RP-1 Solid	205.0 (912.0) each	8.0 × 116 (2.44 × 35.4)	6710 (3045)	—
	2. Delta	N_2O_4/Aerozine	9.9 (44.2)			
	3. TE-364-4 or PAM	Solid	14.8 (65.8)			2810 (1275)
		Solid				3090 (1400)
Titan III	0. Two half-segments, 3.05 diameter	Solid	2599.9 (11,564.8)	10.0 × 162 (3.05 × 49.4)	32,900 (14,920)	
	1. LR-87	N_2O_4/Aerozine	5319.0 (23,660.3)			
	2. LR-91	N_2O_4/Aerozine	112.2 (499.3)			
	3. Transtage or IUS	N_2O_4/Aerozine	15.7 (69.8)			3550 (1610)
		Solid	62.0 (275.8) 26.0 (115.7)			4080 (1850)

*Solid = solid propellant combining a single mixture both fuel and oxidizer. LOX/RP = liquid oxygen and a modified kerosine. LOX/LH_2 = liquid oxygen and liquid hydrogen.

†Due east launch unless otherwise specified.

‡Polar (north-south) launch.

of Atlas being a 1½-stage system. The Centaur is a high-energy second stage using liquid hydrogen and oxygen as its propellants. This stage utilizes two Pratt and Whitney RL-10 engines.

The Atlas Centaur is capable of either direct ascent (using one burn of the Centaur stage) or a parking-orbit flight path (using two burns). Its guidance system, located in the Centaur, uses a Minneapolis Honeywell inertial reference system and a Teledyne computer to provide steering signals to both stages. A launch complex including two pads and logistic support is maintained at the Eastern Space and Missile Center at Cape Canaveral to support Atlas Centaur launches.

Atlas boosters. The Atlas used as a space booster is much the same as the Atlas used on Atlas Centaur. These vehicles were originally built as early ballistic missiles, but are now updated and refurbished under a modification program to be used as space launch vehicles. They are also powered by three Rocketdyne engines, two of which along with their supporting structure are dropped in flight with the continuation of flight maintained by a single engine. The Atlas booster does not have restart capability and delivers payloads to low-altitude orbital injection points. On-board spacecraft kick motors are used to achieve the final injection into the desired orbit. A General Electric radio guidance system is used in the Atlas boosters which are launched from the Western Space and Missile Center.

Delta. The Delta vehicle was developed for intermediate-size spacecraft and is presently operational with two- and three-stage configurations. The current configuration consists of an 8-ft-diameter (2.4-m) first stage provided by McDonnell Douglas and utilizing a Rocketdyne RS-27 engine augmented by nine Thiokol Castor IV strap-on motors and a 6-ft (1.8-m) second stage powered by an Aerojet AJ10-118 engine. The Delta Inertial Guidance System (DIGS) is mounted atop the second stage and consists of a redundant inertial measurement system and guidance computer. Third stages include Thiokol solid-propellant motors with spin stabilization. Missions have been launched by the two-stage vehicle with PAM-D described above as part of the mission payload. Deltas are launched from the Eastern Space and Missile Center into inclined orbits and from the Western Space and Missile Center to obtain polar or nearly polar orbits, that is, north-south orbits which cross the Earth's poles.

The Delta has launched scientific, meteorological, and communications satellites for United States government agencies, for foreign governments, and for domestic and foreign private corporations on a reimbursable basis. It is the workhorse of United States launch vehicles, having completed over 170 launches with a 95% success record.

Titan IIIC. The Titan IIIC is a four-stage solid- and liquid-propellant launch vehicle developed by Martin Marietta for the Air Force for military space applications. The central core of the Titan IIIC is composed of two liquid stages which use Aerojet engines. Two 120-in.-diameter (3.05-m) five-segment solid-propellant United Technology Corporation motors are added as a "0 stage." The final or fourth stage, also developed by Martin, is called the Transtage and contains an inertial guidance system and an attitude control system. The Transtage has a multistart capability and provides the propulsive maneuvers for achieving a variety of circular and elliptical orbits. It can orbit multiple payloads to the same or different orbits on a single launch.

Versions of the Titan III, less the Transtage, have been used by itself or with alternate upper stages to launch payloads into space. When used with Centaur as an upper stage, it provides a high-energy capability for planetary missions. This vehicle launched the Viking spacecraft which landed on Mars, and the Voyager mission to the outer planets. Another version was used to launch the first flight of the IUS, discussed above.

For background information *see* LAUNCH COMPLEX; ROCKET ENGINE; SPACE SHUTTLE; SPACE TECHNOLOGY; SPACECRAFT STRUCTURE in the McGraw-Hill Encyclopedia of Science and Technology.

[JACK W. WILD; JOSEPH B. MAHON]

Machining operations

Water jet cutting, sometimes referred to as hydrodynamic machining, is a relatively new development for industrial use. In this process, water at extremely high pressure, up to 60,000 psi (400 megapascals), is directed through very small nozzles to cut a variety of nonmetallic materials.

This concept was developed some years ago as a university experiment. Because of pump limitations, pressure had to be built up slowly and then released in a surge. Even though the pressure could be maintained only for very short periods, the potential was recognized, and industry soon developed special continuous-duty high-pressure pumps. This permitted employing the water jet cutting method on a commercial basis.

Basic operation. Most water jet units use a 50–100-hp (37–75-kW) hydraulic pump to operate a reciprocating pump called an intensifier, which pumps the water to the 60,000-psi (400-MPa) level. The high-pressure water is piped to the point of use by small-diameter heavy-wall stainless steel pipe. After passing through a valve (for on and off operation without turning off the pump), the water is directed at the workpiece through jewel nozzles, usually sapphire, because such nozzles resist the abrasive force of the water and are relatively inexpensive. The typical orifice of the nozzles ranges from 0.005 in. (0.125 mm) to 0.012 in. (0.3 mm); some of the larger pumps can operate a 0.016-in.-diameter (0.4-mm) nozzle. The advantages of these small nozzles are that they produce a very narrow kerf or width of cut, and that water usage is low despite the high velocity (up to Mach 3). It is important to maintain nozzles in good condition, as worn nozzles reduce the quality of the cut edge. Figure 1 illustrates a typical water jet in operation.

Although some water jet units use water with

Fig. 1. Water jet cutting with stationary nozzle. *(Hydroforce Systems Inc.)*

added copolymer to obtain a long cohesive stream, most units do not, and fairly well-collimated streams can be maintained by using good nozzles. Generally, for the first inch (2.5 cm) from the nozzle the jet is clear and very cohesive. The jet starts to diverge after that but is still effective for 4 to 5 in. (10 to 13 cm). Insulation material 6 in. (15 cm) thick can be cut, but of course the cut edge is not critical. Thus, to obtain the best-quality cut and for safety reasons the nozzle should be as close to the workpiece as practicable.

Fig. 2. Kevlar-epoxy part being trimmed with water jet by guiding mold line of part against catcher tube. *(Lockheed-California Company)*

In installations where the nozzle is held in a fixed position and the work passed under it, a catcher drain tube is usually used. The water passes through the material being cut and into the catcher tube. This tube is approximately 6 in. (15 cm) long and is placed directly under the nozzle. It is in the shape of a tee so that the water is actually drained from the side of the tube, and the lower part of the tube (closed at the end) remains filled with water. This column of standing water serves to absorb the excess energy from the jet. In installations where the nozzle is moved back and forth in a straight line, a slot is used to capture the water after cutting.

Types of installations. There are a number of installations where the material is hand-fed under a fixed nozzle much like a pin router or a band saw (Fig. 2). Then the catcher drain tube can be fitted with a removable bushing that can be used to guide a standard router block tool which holds the part to be cut.

There are also numerous water jet cutting systems which are operated by numerical control or computer control. There are other units which use an optical tracer; that is, an electric eye follows a line on a drawing and transmits the motion to the water jet. In these cases where the nozzle is moved, special high-pressure swivel fittings are used, and frequently the thick-wall tubing is formed in several large coils to provide some flexibility in movement.

Materials cut. Water jet cutting is primarily used for nonmetallics material (see table); it will not cut metals except for the thinnest foils. One of the first uses of water jet cutting was in the forest products area, such as for cutting corrugated board. Today, it is successfully used in the aircraft industry to cut fiberglass, graphite, and Kevlar materials (cured and uncured). It has a particular advantage in cutting Kevlar, which tends to fray and produce a ragged edge when cut with conventional saws and routers. Water jet cutting is also effective in cutting both hard and soft rubbers, including foam. The water jet exerts so little force on the part that the material is not compressed and therefore retains its shape. This is particularly noticeable when cutting fiberglass insulation batts and soft foam rubber. The water jet is sometimes used for cutting honeycomb structure and fiberglass face sheets. It is marginal in cutting honeycomb, particularly the heavier types, because the water stream column is disturbed after cutting through the top face sheet and consequently the lower face sheet is cut with a rather rough and ragged edge.

Care must be taken in cutting laminates such as fiberglass-epoxy or graphite-epoxy because if thickness is in excess of 0.25 in. (6.3 mm) delamination tends to occur. In such cases it is helpful to reduce the cutting rate and to use a quality nozzle.

Characteristics. Water jet cutting has some unique characteristics, which require consideration when this type of machining is to be used.

Dust. Most conventional cutting methods create some dust and are particularly troublesome with materials such as graphite or asbestos. Water jet cut-

Parameters for water jet cutting of various nonmetallic materials*

Material	Thickness, in. (cm)	Nozzle opening, in. (cm)	Feed rate,† in./min (m/min)
Plywood	0.25 (0.6)	0.012 (0.3)	200 (5)
	0.37 (0.95)	0.012 (0.3)	100 (2.5)
	1.0 (2.5)	0.012 (0.3)	60 (1.5)
Foam rubber	0.5 (1.27)	0.008 (0.2)	1000 est. (25.4)
	4.0 (10.16)	0.012 (0.3)	1000 est. (25.4)
Neoprene	0.5 (1.27)	0.012 (0.3)	600 est. (15.2)
50 Shore rubber	2.0 (6.08)	0.012 (0.3)	100 (2.5)
Gray silicone rubber, 40 Shore	0.6 (1.52)	0.012 (0.3)	500 est. (12.7)
Fiberglass-epoxy	0.25 (0.6)	0.012 (0.3)	120 (3)
Fiberglass-phenolic	0.25 (0.6)	0.012 (0.3)	150 (3.8)
Circuit board	0.090 (0.23)	0.008 (0.2)	60 (1.5)
Nomex core	0.75 (1.9)	0.012 (0.3)	340 (9)
Nomex core only	2.0 (5.08)	0.012 (0.3)	300 (7.6)
Kevlar-epoxy	0.25 (0.6)	0.012 (0.3)	100 (2.5)
Graphite-epoxy	0.25 (0.6)	0.012 (0.3)	40 (1)
ABS	0.090 (0.23)	0.012 (0.3)	300 (7.6)
Polycarbonate	0.092 (0.24)	0.012 (0.3)	340 (8.6)

*At a water pressure of 55,000 psi (3800 mega pascals).
†These feed rates can be extrapolated inversely on a nearly linear basis as the material thickness is reduced. That is, if a 0.25-in. (0.6-cm) thickness can be cut at 200 in./min (5 m/min), a 0.125-in. (0.3-cm) thickness can be cut at 400 in./min (10 m/min).

ting offers a great advantage, as almost all the cutting residue is carried away in the water. However, the venturi vacuum created by the water at high velocity is so strong that the material is difficult to move across a flat-table working surface; this can be overcome by cutting grooves in the table to break the vacuum.

Noise. Generally the noise from water jet cutting is well within acceptable limits. In most cases the hydraulic pump makes more noise than the water jet itself. However, noise from the water jet increases sharply as the catcher tube is moved farther from the nozzle. Depending on the pressure used, the noise becomes uncomfortable when the catcher tube is more than about 6 in. (15 cm) from the nozzle. In cases where a slot type of catcher is used (with numerical control operations, for example), the noise is significantly greater. In the few applications where no catcher is used, the noise is unbearable and the operation must be done in a closed area with a robot or some automated device.

Safety. Inasmuch as the water jet can quickly cut off a finger, one of the operating problems with using this technique is psychological because water is generally thought of as harmless; people must be educated as to the danger of the high-pressure jet. A basic practice that should be observed, particularly on hand-fed units, is to lower the water jet nozzle (before turning on the jet) so that it just clears the material to be cut. This also has the benefit of positioning the material close to the nozzle where the best cut is obtained. The occasional splashing of water that occurs is not harmful.

Future trends. Development work is being conducted to incorporate abrasive cutting with water jet. This concept, which uses additional nozzles, usually carbide, to introduce a fine abrasive material into the jet stream after the water has left the conventional sapphire nozzle, would provide the capability to cut hard metals and thick laminates.

For background information *see* MACHINING OPERATIONS in the McGraw-Hill Encyclopedia of Science and Technology.

[DALE L. ZUELOW]

Magnetic monopoles

On February 14, 1982, a prototype superconductive detector used to search for magnetic monopoles observed a single-candidate event. Even though a spurious cause has not been ruled out, the possible existence of superheavy magnetically charged particles passing through the Earth's surface would have profound significance for elementary particle physics as well as cosmology. Thus, the experiement has produced enormous interest within the scientific community. Larger superconductive detectors have been built and are now operating, and many conventional ionization detector arrays are being used in the search.

Serious interest in magnetic monopoles began in 1931, when P. A. M. Dirac proposed the existence of magnetically charged particles to explain the observed quantization of electric charge. He showed that only integer multiples of a fundamental magnetic charge g (Dirac charge) are consistent with quantum mechanics. Many years of experimental searches produced no convincing candidates.

The last decade has seen enormous advances in theoretical attempts to unify the known forces of nature into a single consistent theory. Work on these unification theories has unexpectedly yielded strong renewed interest in monopoles. In 1974 G. 't Hooft and independently A. M. Polyakov showed that in theories which accomplish true unification (those based on simple or semisimple compact groups) magnetically charged particles are necessarily present. These include the standard SU(5) grand unification model (which attempts to unify the electromagnetic force, the weak force which accounts for beta decay, and the nuclear force). This modern theory for magnetic monopoles predicts the same long-range field and thus the same charge g as Dirac found; now, however, the near field is also specified, leading to a calculable mass. The standard SU(5) model predicts a monopole mass of approximately 10^{16} GeV/c^2, or 10^{-8} g (approximately the weight of an amoeba), far heavier than had been considered in previous searches.

Such supermassive magnetically charged particles would possess qualitatively different properties from those assumed in earlier searches. These are due to their necessarily nonrelativistic velocities, from which follow weak ionization and extreme penetration through matter. Thus such particles may very well have escaped detection in earlier searches.

Astrophysical limits. Cosmological theories based on grand unification lead to predictions for monopole particle flux limits which are impossibly high or unobservably low. However, the latter results are exponentially model-dependent and thus are not inconsistent with observable levels. Astrophysical arguments yield more concrete observational limits. In the discussion below, a particle mass of 10^{16} GeV/c^2 is assumed to obtain representative numbers. Then an absolute upper bound for the galactic monopole particle flux of 4×10^{-10} cm^{-2} s^{-1} sr^{-1} is obtained from the limits on the local galactic dark mass.

A much smaller upper bound of 10^{-15} cm^{-2} s^{-1} sr^{-1} has been obtained by assuming an isotropic flux from arguments based on the existence of the 3-microgauss (3×10^{-10} tesla) galactic magnetic field. However, several others have demonstrated that models incorporating monopole plasma oscillations allow a much larger particle flux, approaching in some cases the local galactic dark mass limit. All of these galactic bounds suggest particle velocities in gravitational virial equilibrium, thus very near 10^{-3} of the speed of light c.

An enhanced monopole density, gravitationally bound to the solar system, has been suggested. This would allow much smaller average galactic flux levels, and would lead to lower particle velocities, near 10^{-4} c. Although a mechanism for the formation of such a cloud remains obscure, its possible existence has not been ruled out.

Finally, it has been shown theoretically that the supermassive monopoles arising from grand unification theories will catalyze nucleon decay processes. If the cross section for such events is of the order of hadron-hadron cross sections, as has been suggested, then all attempts at direct detection of these monopoles may be doomed to failure. Arguments which are based on x-ray flux limits from galactic neutron stars and which assume a strong cross section lead to an upper bound for a magnetic particle flux of about 10^{-22} cm^{-2} s^{-1} sr^{-1}. However, there is much uncertainty regarding both the catalysis cross section and the astrophysical arguments based on incomplete understanding of neutron stars.

Conventional detectors. Perhaps the most important question regarding direct experimental detection is whether conventional ionization or scintillation devices, much larger in sensing areas than superconductive devices, can detect the passage of single Dirac charges with velocities of the order of 10^{-4} to 10^{-3} c. Several phenomenological theories suggest that they can; however, experimental tests of these theories are not possible. Recently, new calculations based on fundamental quantum-mechanical arguments have suggested that helium gas devices would provide such a sensitivity.

Searches with conventional detectors have seen no candidates for slow penetrating particles, setting upper flux limits several orders of magnitude lower than those of the prototype superconductive detector. These efforts are complementary, with the conventional detectors providing up to 1000 times larger sensing areas and convincing detection for any slow-moving electric or magnetic particles of greater velocity than about 10^{-3} c, and the superconductive detectors providing definitive identification of magnetic charge for any particle velocity.

Superconductive detectors. The remarkable theoretical similarities between flux quantization in superconductors and Dirac magnetic monopoles make superconductive systems natural detectors for these elusive particles. The magnetic flux emanating from a Dirac charge, $4\pi g = hc/e = 4 \times 10^{-7}$ gauss-cm^2 in gaussian units (or $h/e = 4 \times 10^{-15}$ weber

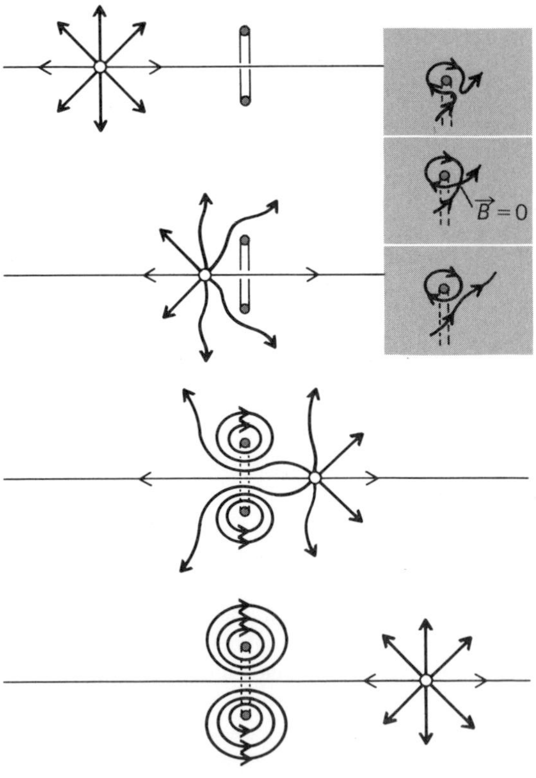

Fig. 1. Successive configurations of magnetic field lines as the monopole passes through the superconducting ring. Inset shows detail of a single magnetic field line leaving a closed loop around the ring; $\vec{B}$ is the magnetic flux density.

in SI units), is exactly twice the flux quantum of superconductivity, $\Phi_0 = hc/2e$ in gaussian units (or $h/2e$ is SI units). A magnetic charge g passing through a superconducting ring is shown schematically in Fig 1. No magnetic field line can pass through the superconductor; thus every field line emanating from the pole must leave a closed loop around the ring wire as the particle moves through. The magnetic flux change through the ring must exactly equal the total flux from the pole, $2\Phi_0$ for a monopole of unit Dirac charge. No net flux change results from trajectories which miss the ring.

Superconductive technologies, developed over the last two decades, have led naturally to very sensitive detectors for magnetically charged particles. These superconductive detectors directly measure the magnetic charge independent of particle velocity, mass, electric charge, and magnetic dipole moment. In addition, the detector response is based on simple and fundamental theoretical arguments which are extremely convincing. Because of their velocity-independent response, these detectors are a natural choice in searches for a particle flux of supermassive (thus slow) magnetically charged particles.

The prototype superconductive detector which observed the candidate event has operated for a total of 382 days. It consists of a 2-in. diameter (5-cm) loop made of superconducting niobium wire mounted inside a superconducting cylinder. The cylinder provides an ambient magnetic field of 5×10^{-8} gauss (5×10^{-12} tesla) and excellent isolation from external field changes. The current in the wire is continuously monitored with a SQUID (superconducting quantum interference device) magnetometer. The single-candidate event is consistent in magnitude with the passage of a Dirac charge within an uncertainty of $\pm 5\%$, but no other candidates have been seen. The sensitivity of the instrument to mechanical disturbances and the lack of redundant information have not allowed a spurious cause to be ruled out. These data have set an upper limit of 2.4×10^{-10} cm^{-2} s^{-1} sr^{-1} on any possible flux of monopoles.

A new detector based on the same principles as the prototype system has been constructed and is operating. It has improved mechanical stability, a greater sensing area, and much better spurious signal discrimination. It consists of three mutually orthogonal superconducting loops (Fig. 2), each 4 in. (10 cm) in diameter (twice that of the original device). In addition to a sevenfold increase in direct sensing area for trajectories passing through at least one loop, this new instrument is sensitive to near-miss trajectories which traverse the superconducting shield but do not intersect the three loops. The near-miss category provides an additional effective sensing area 42 times greater than that of the prototype detector for Dirac charges, and will further reduce the flux bound if no candidate events are seen. However, a near-miss candidate with its smaller signal levels would not be as definitive as one corresponding to a trajectory passing through at least one loop. No candidate events have been seen with the new detector, and the upper limit on the flux of monopoles obtained with the prototype detector has been reduced by more than an order of magnitude.

Fig. 2. Three-loop superconductive monopole detector.

Work is beginning on a yet larger detector with a sensing area approaching 10 ft^2 (1 m^2). During the next several years either definitive candidate events will be found or the upper flux limit will be reduced by an additional factor of 100 or more, relegating the uncorroborated candidate to the class of improbable nonreproducible results.

For background information *see* FUNDAMENTAL INTERACTIONS; MAGNETIC MONOPOLES; SQUID; SUPERCONDUCTIVITY in the McGraw-Hill Encyclopedia of Science and Technology.

[BLAS CABRERA]

Bibliography. B. Cabrera, First results from a superconductive magnetic monopole detector, *Phys. Rev. Lett.*, 48:1378, 1982; R. A. Carrigan and W. P. Trower (eds.), *Magnetic Monopoles*, 1983; W. P. Trower and R. A. Carrigan, Superheavy magnetic monopoles, *Sci. Amer.*, 246(4):106–118, 1982.

Marine navigation

Marine navigation systems have been developed in recent years which integrate various navigation functions in order to reduce the workload of ships' officers and increase efficiency and safety.

The MANAV (Marine Integrated Navigation) system originated with concern about the maneuvering of large tankers. HYCATS (Hydrofoil Collision Avoidance and Tracking System) was designed for high-speed ships.

MANAV

MANAV originated in 1967, when there was much public, governmental, and company concern about the stopping and maneuvering of large tankers, particularly when close to shore, in estuaries, or when berthing. From studies carried out by a number of national and international organizations, it became clear that the majority of accidents at sea were due to human error, and not mechanical failure as many had thought. For the engineer, the concept of human error is difficult to deal with, and it was some years before ergonomics began to make an impact on maritime affairs. *See* ERGONOMICS.

Terminology proved the first stumbling block in conceiving an integrated system which would provide the mariner with the required information. The initial concept of integrating equipment has now been replaced by the concept of routing information from remote sensors and presenting this information through visual display units sited on a central compact console. By centering design on silicon chip technology, the price can be reduced to an acceptable level. The original concept of retaining the paper chart, with its wealth of knowledge, has been maintained, although this is expected to eventually give way to digitized information.

Early studies. The first study in 1967 arose from work carried out by the Esso Tanker Company in a bid to develop a system to assist in ensuring the safe operation of its very large tankers (known in the Merchant Marine as very large crude carriers, or VLCCs), which were then coming into service. In 1969 Esso and the United Kingdom Ministry of Technology (MINTECH) decided that an industrial partner should be sought; the Decca (now Racal-Decca) Navigator Company was contracted to undertake a design study. This study began in January 1971, and although a wealth of supporting data has been collected in the intervening years, the objectives for MANAV as set by Decca and MINTECH have not radically changed.

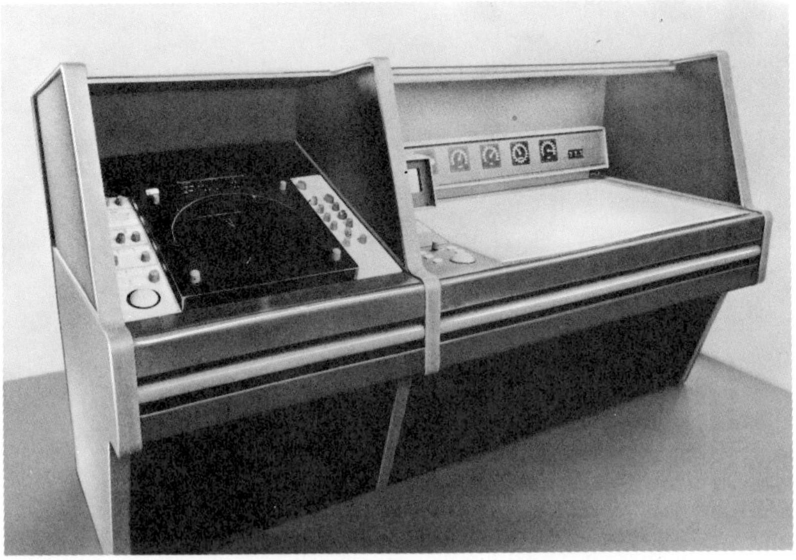

Fig. 1. Mock-up of MANAV system.

Aims. The aims of MANAV are threefold. The first aim is to supply, through a set of sensors, the required data for the safe conduct of the ship throughout each phase of its voyage. The second aim is to integrate the sensors and present data to significantly reduce the work load of the officer on the bridge and provide the officer with anticollision and ship-handling information in a clear and readily assimilable form; safeguard against gross errors; and provide better information to aid voyage management and passage economy. Finally, the MANAV system is designed to provide a ready and simple means for the mariner to communicate with the system, and to provide raw and processed data and differentiate clearly between them.

Equipment. MANAV consists of an automatic chart table, under which is sited a Racal-Decca System 500 computer. The chart table is butted to a 16-in. (40-cm) radar display. A 12-line interactive tabular display is mounted at the chart table. The whole system can be mounted on the bridge so as to give maximum possible visual field of view from the console (Fig. 1).

Automatic chart table. The automatic chart table accepts standard marine navigational charts laid on top of it in a manner similar to the use of any standard chart-table surface. The chart is clamped in position and then referenced by a quick and simple procedure. Thereafter, a computer-driven illuminated cursor, projected from beneath the chart, will accurately indicate to the deck officers various information. The position estimated by dead reckoning or the computed position may be displayed. Dead reckoning is based on log and compass inputs from the last known fix position. Computed position is based on the input from electronic navigation aids.

The automatic chart table can also be used to trace any position or contour, for example, objects for a fix, planned tracks, and navigation marks, from the chart to be digitized for various uses within the system. The facility enables synthetic charts to be compiled, showing, for example, depth contours, traffic zones, coastlines, hazardous areas, and offshore structures. When required, such synthetic charts can be superimposed on the radar screen. Also, the radar display rolling-ball marker can be effectively linked to the chart-table cursor, and vice versa, to provide direct chart–radar cross-referencing. Thus a chosen mark or feature on the chart can be pointed up on the display, or an echo on the display can be indicated and identified on the chart.

As chart information can be displayed on the radar screen at any stage in the voyage, synthetic radar coastlines and other fixed targets can be generated for on-board training exercises, such as practicing approaches to channels, when the ship is in "quiet" oceanic waters.

Data display. The main interface between mariner and machine is provided by the tabular display and a relatively simple control panel. The display presentation is in plain-language page form with a capacity of 12 lines, each of 40 characters. The control panel is used in conjunction with the data display for general system control, including the input of information such as waypoints, estimations of set and drift, chart referencing data, and position information obtained from visual fixes.

Navigation display. The system enables up to 20 targets to be tracked. These are manually acquired and tracked automatically by the computer with track numbers automatically allocated. Tracks can be canceled manually when they cease to be a threat, but are automatically canceled should the officer of the watch forget, at a distance of 50 nautical miles (93 km). If a track is lost by the computer, a visual warning is provided.

Computer. A Racal-Decca system 500 computer that is used in the system was specially chosen as being suitable for the rugged marine environment. It has been well proven by its reliability on various vessels, including offshore patrol vessels and dynamically positioned vessels operating in the North Sea. In addition to carrying out the system functions described, a considerable amount of computer time is dedicated to simplifying control procedures, including error checking, and carrying out real-time tests of data credibility and fault detection.

Pelorus unit. This facility enables the operator to acquire the bearings of up to four objects which, having been stored in the system, can be recalled at a later stage. The facility acts, in effect, as a four-channel electronic note pad to hold bearings and time information for transfer onto the chart. To use the facility, the azimuth ring is aligned on the required object; one of four channel-select buttons is then pressed to record bearing and time of bearing.

Sea-trial evaluations. During 1980, while the equipment was being developed and tested ashore, several preinstallation trips were made in order to obtain background information to carry out a "before" and "after" data-collection exercise on the study vessel. Bridge activities were tape-recorded, the recorded information noting time and nature of the activity undertaken by bridge team personnel. By working watches, observers were able to keep activity records for the complete duration of a voyage because the vessel's trading voyages were relatively short. One of the limitations of this method of data collection is in the difficulty experienced in observing the small detailed activities undertaken by the ship's officers; this was of particular concern when recording activity details when the officers were operating the MANAV system.

In addition to the data recorded, a questionnaire was used to obtain subjective opinions from the ship's officers. The questionnaire contained sections designed to solicit information on the experience of individual officers with the equipment through a detailed critique of 18 listed items. Comments on an officer's overall impression of the system and on each of the 29 "pages" of information available on the system display panel were also included.

Workload. Reduction of bridge workload is one of the claims for the MANAV system. Bridge activity workload, with or without MANAV, is high (Figs. 2 and 3). The bridge, being the nerve center of the ship, is a communications and business center as

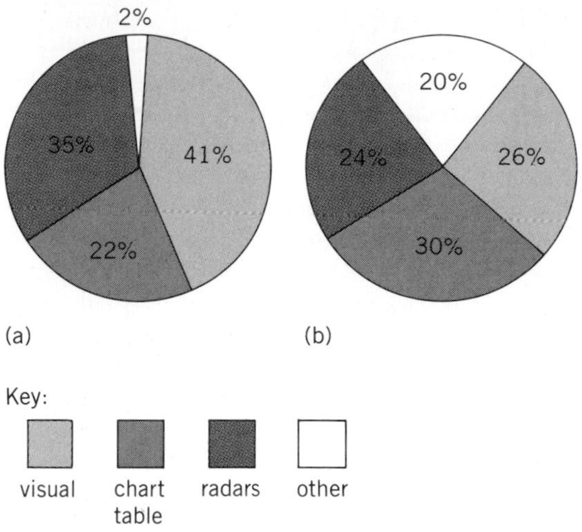

Fig. 2. Proportion of time spent in various activities by ship's officers on the bridge before the installation of the MANAV system. (*a*) Department of Industry bridge study data, 1975. (*b*) Pre-MANAV evaluation, 1980.

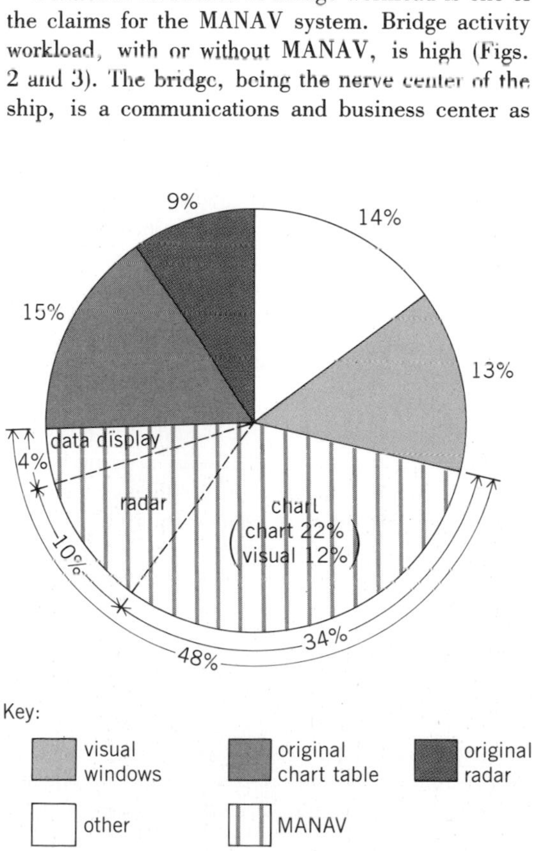

Fig. 3. Proportion of time spent in various activities by ship's officers on the bridge after the installation of the MANAV system.

well as a navigational platform. This complicates the task of assessing how much a system such as MANAV reduces workload. A comparison of the nighttime results may prove more realistic, as many of the "social" or nonnavigational elements of the workload are not present.

If a comparison is made of the time spent at or moving between the equipment making up MANAV for pre- and post-MANAV installation situations, a considerable reduction in workload is noted, with the total time being reduced from 54% (time on radar, Decca navigator, and chart) to 20–25%. This is almost certainly accounted for by the fact that the officer of the watch can collect all three items of information from a single piece of equipment.

Error reduction and voyage economy. A number of possible ways of examining track deviation were considered in order to provide the comparative information necessary to enable an assessment of error reduction and possible voyage economy. Inspection revealed that deviation from a planned course was generally small, whether MANAV was fitted or not. Furthermore, the course followed by the ship, whether laid down on the chart or not, is not sacrosanct, and officers are free to follow a course within a tolerable limit of the line.

Deviations along track are taken to be those deviations from the planned distance to be run over a given period of time. It was one of the aims of the MANAV study to determine whether MANAV provided enhanced voyage management through easier monitoring. Analysis of along-track errors, from the data collected, does not support this hypothesis.

When operating to a preplanned course, the point at which the heading of the ship has to be changed is an important factor in economy and voyage management considerations and can be critical in emergencies; it was one of the error reduction features examined. No significant difference between the sets of data could be deduced.

A measure was computed to provide an indication of the uncertainty in an officer's mind as to the ship's precise position. Before installation, the time interval between fixes varied between 4 and 27 min with a mean of 11 min. After installation, measurements gave a mean of 17.5 min with a range of 5–35 min. Not enough data were collected to make these results statistically significant. However, the confidence level of the officers using MANAV was significantly increased, in that the spotlight denoting position gave a ready and continuous reference to position.

Subjective opinions and conclusions. The general impression of MANAV as expressed by the ship's officers was that it was a useful, effective, and "user friendly" piece of equipment. The study revealed that ship's officers considered the automatic plotting table to be an extremely reliable and well-thought-out facility which improved their effectiveness.

The plasma alphanumeric display panel generally proved reliable, although one fault has occured during the time the equipment has been on board. It is considered that the procedure in paging information could be simplified, and the associated work load reduced, by employing a cathode-ray-tube visual display unit and touch-sensitive keyboard.

Automatic tracking, display of collision avoidance information, and the interaction facilities between plotting table and ARPA, and vice versa, all proved to be highly effective, especially in estuaries and channel transits. The clearway or closest-point-of-approach (CPA) display was particularly well liked. The facility was frequently used because the navigator could determine at a glance whether a course or speed change had been made safely or not, without having to resort to a trial maneuver, which is time-consuming.

Interfacing of MANAV to other bridge equipment, in particular to a Doppler speed log, is likely to improve the accuracy of track keeping and the accuracy to which dead reckoning and estimated position can be determined.

Navigational sensors. At a later stage, multisensors are envisaged, which will increase the accuracy considerably. Multisensor inputs require sophisticated mathematical filtering techniques, which demand a great deal of computing capacity. Whether the cost and acceptability to the mariner of these techniques are justified is not yet known.

Integration of shipboard information is a trend that appears to be on the increase, with the MANAV hardware being an advanced system in that trend. The trend toward further digital display of information is expected to continue.

[I. C. MILLAR]

HYCATS

HYCATS, an acronym for Hydrofoil Collision Avoidance and Tracking System, is a computer-based system designed to automate target tracking and navigation functions in order to increase the safety of high-speed ships. The developmental model, installed in USS *Pegasus* (PHM-1), was known as the High Speed Ship Collision Avoidance and Navigation System (HICANS); other systems are now being installed in the missile patrol hydrofoils. A brief description of the functions and the physical configuration follows.

Functions. Navigation performed by HYCATS primarily addresses coastal navigation and piloting where danger of grounding, particularly at high speeds, is great. The basic concept is the superposition of chart data over radar data, allowing precise positioning to be accomplished. Two different methods of displaying chart data are used in HYCATS: a television image of a standard nautical chart, and digitized chart data that is stored in and displayed by the computer.

The digital chart consists of selected chart data that are digitized, stored on magnetic tape, and loaded into the computer when needed. The types of data on the chart consist of navigation aids, coastlines, and other radar-prominent features,

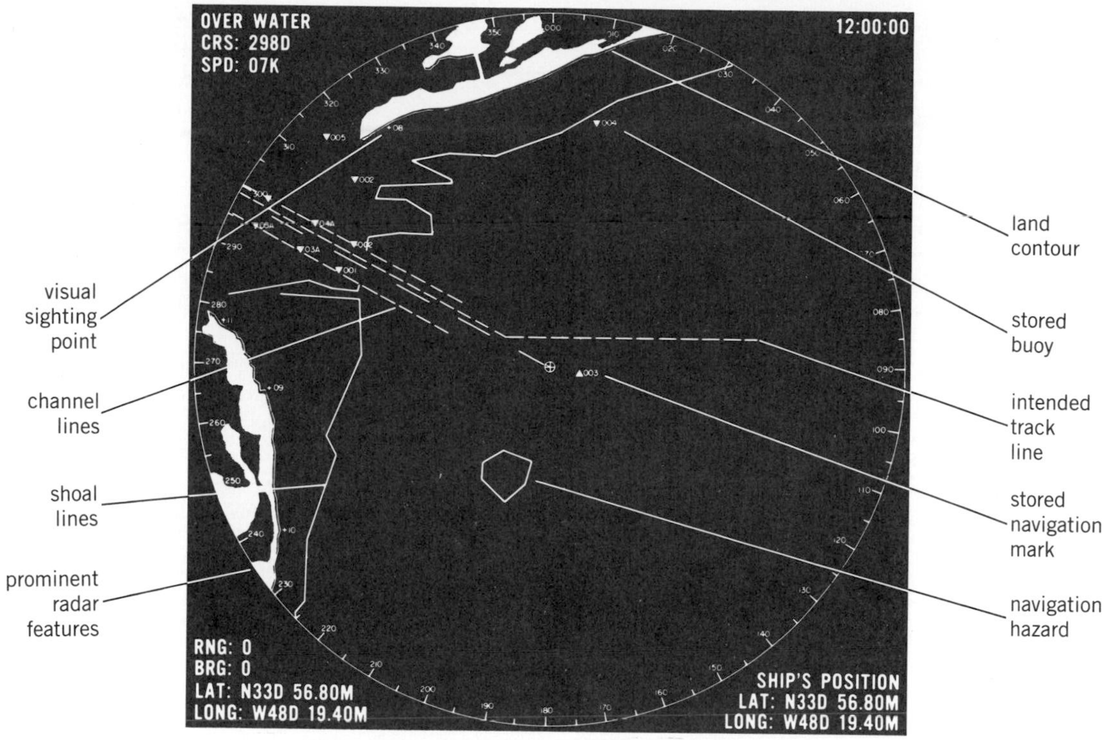

Fig. 4. Typical HYCATS digital chart display. The digital chart is superimposed on the radar video. Additional information is provided in the corners of the display.

shoal lines (which are selected to show shallow water relative to the draft of the specific ship), and intended track lines (Fig. 4). Ranges from 1 to 64 mi (1.6 to 103 km) can be selected for the chart and radar data to meet the particular circumstances. The initial position fix is accomplished by aligning the radar returns of coastlines, prominent radar features such as piers and bridges, and navigation aids with the corresponding chart data. The chart is then automatically dead-reckoned by using ship heading and speed to maintain alignment. Periodic minor adjustments may be required if effects of current are significant. Fully automatic position fixing is accomplished by designating radar-trackable navigation aids as fixed aids matched to their respective chart data. The computer then maintains alignment between the radar return from the aid and the chart data, thereby automatically fixing the ship's position on the chart. As a backup to radar position fixing, adaptors installed on the bridge peloruses allow visual bearings to be displayed as lines of position on the chart display. Using the intended track line, the system can assist the navigator by providing distance and time to the next course change as well as the course itself. This information is presented both on the graphic display and on the tabular display. *See* PILOTING.

Pictures of nautical charts are generated by a high-resolution television camera, mounted above a chart table on which the desired chart is mounted. The camera is moved along both the x and y axes on a carriage mechanism that is under computer control. Initial alignment and subsequent dead reckoning are similar to the digital chart. A zoom lens on the camera allows limited changes in range scales, depending on the scale of the chart being used. Radar positioning fixing and display of visual lines of position can be accomplished by using operator-entered digital navigation aids that are superimposed by the computer on the television chart.

Safe operation at high speeds is enhanced by the addition of collision avoidance data to the radar and chart display. HYCATS interfaces with the ship's standard radars. Using the radar data, it automatically tracks up to 45 surface targets, both radar aids (as discussed earlier) and ships and boats. Once acquired by the system, target position is updated on every sweep of the radar, and from this the course and speed of the target are calculated. The target symbol with identifying number is displayed over the radar video to provide a confirmation of proper operation. The system also provides a graphic display of those targets which pose a collision threat and the maneuvers which are safe. Points of possible collision (PPC) are computed for each target and are displayed if they occur within the selected range scale. These PPCs are the points along the target ship's course where a collision would occur if the tracking ship maintained its speed and selected a course to aim at the PPC. This is equivalent to the solution for firing a torpedo to hit the target if the torpedo has a speed equal to present ship's speed. Solutions for safe courses when a minimum closest point of approach is desired are displayed in the form of ellipses around the PPCs. These ellipses are called predicted areas of danger (PAD) and are computed on the basis of the operator-selected miss distance. Audible and visual alarms warn the operator of potentially dangerous situations. A trial-speed capability allows the operator to assess effects of changing speeds. This technique for displaying collision avoidance data is similar to that found in Sperry Corporation's commercial collision avoidance systems, which have been successfully used at sea for more than 10 years.

In addition to the graphic display, a tabular display of target data lists each target's number, range, bearing, course, speed, and closest-point-of-approach (CPA) range, bearing, and time.

Targets can be classified by the operator as friendly, unknown, or hostile, or can be classified as a formation guide. With this designation and station coordinates, the system will assist in maneuvering the ship to its assigned station in the formation.

Physical description. HYCATS has three operator positions. The primary position where all functions are controlled and performed is located in the Combat Information Center (CIC). The second position, which functions as a remote display of HYCATS data, is located on the bridge. The third position is the navigator position, which is used to initalize and control the paper chart. Figure 5 shows the bridge of a missile patrol hydrofoil with HYCATS installed.

The CIC console provides two 17-in. (43-cm) displays along with lighted pushbuttons to control the system and a trackball which is used to align the chart and to move the balltab for acquiring targets. The console also has numeric thumbwheels to set system parameters such as CPA distance, a numeric keyboard to enter data such as trackline number, and other system and display controls. The bridge console has two 14-in. (36-cm) displays along with a trackball and limited pushbuttons for display selection. The navigator's position has one 17-in. (43-cm) display with the specialized controls for the paper chart system.

Fig. 5. HICANS installed in USS *Pegasus* pilot house. HICANS with its two television displays occupies the left half of the console, while ship control functions occupy the right half.

The displays are high-resolution raster scan (television-type) monitors which provide the brightness necessary for the displays to be viewable under the high ambient light conditions experienced on the bridge. Red filters are used to protect the bridge watchstanders' night vision. The bridge and CIC displays are interchangeable to allow for different operational requirements and to provide redundancy in the event of failure. The navigator's display can be switched to display either the graphic or tabular data.

HYCATS software operates in the standard Navy shipboard minicomputer, which is a 16-bit computer with 65,000 words of core memory. Other digital electronic equipment interfaces (1) with the ship's radars for target tracking and television radar display, (2) for generation of the graphics and alphanumerics, (3) with the ship's heading and speed sensors, and (4) for control of the paper chart system. The paper chart system uses a high-resolution television camera with a zoom lens moved by an x-y carriage mounted over a chart table holding the nautical chart.

For background information *see* HUMAN-FACTORS ENGINEERING; MARINE NAVIGATION; PILOTING; RADAR in the McGraw-Hill Encyclopedia of Science and Technology. [LANNY PUCKETT]

Bibliography: *Code of Practice for Ships' Bridge Design*, Department of Industry, London, 1977; R. F. Hansford and J. Vickers, Navigation and the computer, *Seminar of Marine Computer Applications*, Institution of Electrical Engineers, London, November 1980; I. C. Millar and A. A. Clark, Recent developments in the design of ships' bridges, *Proceedings of the Symposium on the Design of Ships' Bridges*, RINA/NI, London, 1978; I. C. Millar and R. F. Hansford, The MANAV integrated navigation system, *International Congress of the Institutes of Navigation*, Paris, September 1982; L. Puckett, HICANS: Navigation for high speed ships, *Proceedings of the Nation Marine Meeting*, Institute of Navigation, October 27–29, 1982.

Marine sediments

Research on marine sediments continues to provide new insights into depositional processes in the ocean. Recent studies have involved the study of deep-sea fans using long-range side-scan sonar and the determination of sediment accumulation rates using radionuclides.

DEEP-SEA FANS

Deep-sea fans are thick, fan-shaped wedges of sediment that form along the base of the continental slope at locations where rivers discharge large volumes of sediment into the ocean. These sediments are transported down the continental slope to the continental rise through submarine canyons by gravity-induced bottom flows such as turbidity currents. The sediments are then deposited radially outward from the mouths of the canyons as fans that are somewhat similar in geometry to alluvial fans found at the bases of mountain ranges in arid regions.

Modern deep-sea fans vary widely in size; lengths and widths range from several miles to more than 600 mi (1000 km) and thicknesses range from a few hundred feet to more than 6 mi (10 km). The largest modern fans are developed on the continental margin of India off the Ganges and Indus rivers, off the Amazon River in Brazil, and off the Mississippi River in the Gulf of Mexico. Deep-sea fans of all sizes are important elements in building and shaping the continental rise.

Morphology. In recent years, modern deep-sea fans, as well as ancient fans now locked in continental rocks through the collisions of tectonic plates, have become important targets for oil and gas exploration. The fine-grained organic-rich sediments produce oil and gas when the fan is buried, and the coarse sands in the channels and at other sites in the fan act as reservoirs to trap the oil and gas. However, to predict where potential reservoir sands may occur within fans and thus exploit the hydrocarbon potential of both ancient and modern fans successfully, it is essential to understand in detail the morphology, the vertical and lateral sediment facies distributions, the sedimentation processes, and the growth patterns of present-day fans. During the past 25 years seismic reflection studies, bathymetric mapping, and sediment sampling of modern deep-sea fans have revealed a great deal of information about these parameters. However, the large size of most fans has limited the spacing of data lines and thus the amount of information that these conventional oceanographic techniques can provide. In general, these conventional data have proved to be too widely spaced to provide the detailed geologic information (for example, morphology, distributary channel trends and bifurcation patterns, or sediment facies distribution) necessary to fully understand the sedimentation processes and evolution of fans.

During the past few years, however, newer, more sophisticated instruments such as side-scan sonars and multi-narrow-beam echo sounders have become available for studying the deep-sea floor. Unlike conventional echo sounders which return only a one-dimensional profile of morphology and bathymetry along the ship track, these new instruments can map continuously a two-dimensional swath of sea floor from 0.3 to 36 mi (0.5 to 60 km) wide, depending on the instrument type and water depth. Thus, such instruments potentially give scientists the capability of mapping the entire surface of even the largest fan within a reasonable time frame (weeks).

GLORIA. At present the only instrument which can continuously map very large areas of the sea floor (for example, a swath up to 6 mi or 10 km wide) is the GLORIA long-range side-scan sonar operated by the Institute of Oceanographic Sciencies of Wormely, United Kingdom. GLORIA, or geological long-range inclined asdic (asdic is the British term for sonar), is towed just below the sea surface at speeds of 8–12 knots (4–6 m/s) and is capable of mapping continuously (with reflected sound waves) a swath of sea floor 9–37 mi (14–60 km) in width (4–19 mi or 7–30 km on each side of the ship), depending on the depth and acoustic characteristics of the water column. In 1982 oceanographers began to use GLORIA to conduct detailed studies of large deep-sea fans; both the Amazon and the Mississippi were surveyed. The results of a study of the Amazon Deep-Sea Fan illustrate how the GLORIA long-range side-scan sonar can be utilized to improve available knowledge of the development and evolution of deep-sea fans.

Amazon Cone. The Amazon Deep-Sea Fan (also called the Amazon Cone) is the world's third-largest modern deep-sea fan and extends from the continental shelf off the Amazon River mouth in northeast Brazil seaward for a distance of 435 mi (700 km) to depths of 16,000 ft (4800 m; Fig. 1). Various studies of the Amazon Fan have been conducted since 1970 with conventional seismic reflection techniques and piston cores. These studies revealed that despite its large size, the Amazon Fan is a typical deep-sea fan. The fan is fed sediments by a system of distributary channels, and these channels radiate outward across the fan surface from a single large channel that extends downslope from the Amazon Submarine Canyon at the head of the fan (Fig. 1). On the uppermost part of the fan these channels are quite large, up to 2 mi (3 km) wide and up to 820 ft (250 m) deep. The channels are perched atop levee systems that are 6 to 30 mi (10 to 50 km) wide. Channels become more numerous as they branch, and become progressively smaller downslope. They are only 30–90 ft (10–30 m) deep on the lower part of the fan (less than 13,000 ft or 4000 m in depth).

The GLORIA study of the Amazon Fan was an international cooperative program. The field work was conducted in 12 days, during which most of the upper and middle Amazon Fan (to depts of 13,000 ft or 4000 m) was covered with overlapping GLORIA sonar images. The GLORIA records resolved the positions and trends of the distributary channels in much greater detail than had ever been seen before. Individual channels were mapped continuously from the Amazon Canyon downslope to depths of 12,000 or 13,000 ft (3500 or 4000 m) on the middle fan. Figure 1 shows channels observed on GLORIA sonographs. The GLORIA sonographs, combined with conventional seismic records from precision depth recording (PDR) along shiptracks outside the limits of GLORIA coverage, show that the levees associated with these channels form two broad but separate levee complexes across the middle fan. The western edge of the Western Levee Complex and its

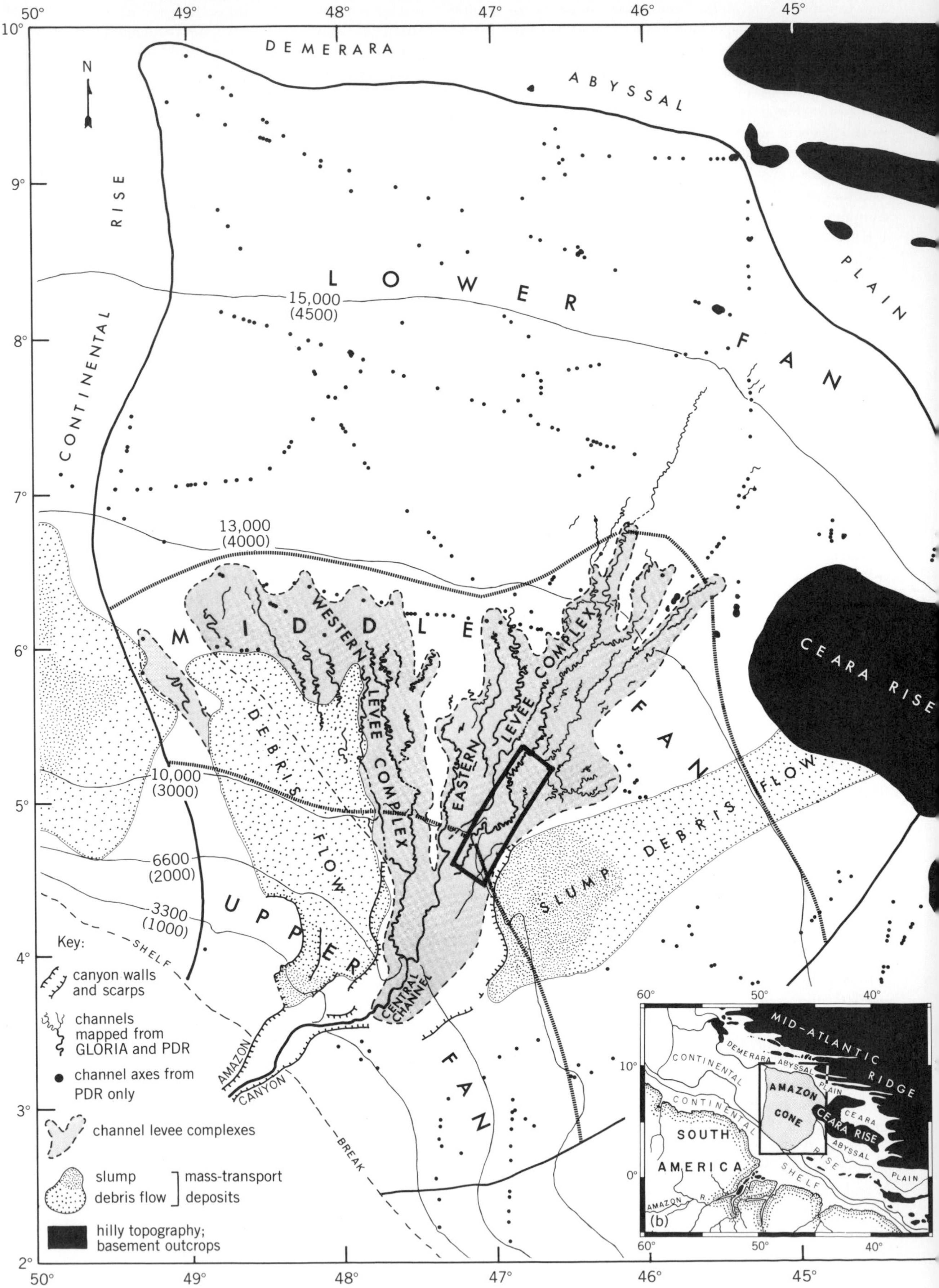

DEMERARA ABYSSAL PLAIN
CONTINENTAL RISE
LOWER FAN
MIDDLE FAN
UPPER FAN
CEARA RISE
WESTERN LEVEE COMPLEX
EASTERN LEVEE COMPLEX
DEBRIS FLOW
SLUMP DEBRIS FLOW
CENTRAL CHANNEL
AMAZON CANYON
SHELF BREAK
15,000 (4500)
13,000 (4000)
10,000 (3000)
6600 (2000)
3300 (1000)
Key:
canyon walls and scarps
channels mapped from GLORIA and PDR
channel axes from PDR only
channel levee complexes
slump
debris flow
mass-transport deposits
hilly topography; basement outcrops
(b)
MID-ATLANTIC RIDGE
DEMERARA ABYSSAL PLAIN
CONTINENTAL RISE
CONTINENTAL SHELF
AMAZON CONE
CEARA RISE
CEARA ABYSSAL PLAIN
SOUTH AMERICA
AMAZON R.

(a)

channels have been buried by a large debris flow which extends downslope to at least 12,000 ft (3700 m). Such debris flows and related mass-transport processes are important but poorly understood processes that contribute to fan morphology.

The most striking characteristic of the leveed distributary channels revealed by GLORIA is their highly developed, fluviallike meanders. Most channels show continuous, intricate, often recurving meanders (Fig. 2). Abandoned meanders, cutoffs, and related floodplainlike features are observed. Branching of channels to form additional distributaries generally occurs by breaching of levee walls, especially on the outside of meander loops. The meanders and associated floodplainlike features are comparable in size and morphology to those of late-maturity rivers on land such as the lower Mississippi.

Interpretation of sediment deposition. The surprising occurrence of such large, fluviallike meander systems on a deep-sea fan suggests that large-scale, relatively continuous downslope flow of sediment-laden turbidity currents through the channels is required in order to supply the enormous volumes of water and sediments necessary for maintenance and modification of such channel systems. This evidence for relatively continuous flow challenges the widely held traditional view that channel development and sediment deposition on deep-sea fans are the result of only intermittent or sporadic turbidity-current events. Additional studies are in progress to determine which concept is correct. In any case, downslope sediment transport was apparently accomplished only during Pleistocene and previous glacial periods when lowered sea level allowed the Amazon River to flow across the exposed continental shelf and continuously discharge sediments directly into the Amazon Submarine Canyon on the continental slope. In contrast, during high-sea-level stands such as at present (Holocene), and during previous interglacials, sediments have been unable to cross the wide continental shelf. The sediments thus cannot reach the fan channels, and the Amazon Cone becomes temporarily inactive.

Growth pattern. The GLORIA sonograms (such as Fig. 2) have enabled scientists to map continuously the trends of the distributary channels across the Amazon Fan. The stratigraphic position of each channel–levee system relative to the others was determined from a series of seismic reflection profiles down the fan; this information permitted determination of the relative age relationships between the various distributary channels. These age relationships suggest a growth pattern for the fan. The fan apparently grows through formation of a succession of broad levee complexes, such as the Western and Eastern complexes in Fig. 1. Each levee complex is composed of several individual distributary channels whose levees overlap or coalesce with one another. Probably only one levee complex is active at any given time; and on that levee complex, probably only one channel is active. Eventually this active channel is abandoned, probably through avulsion, and a new channel–levee system forms at an adjacent location, thereby expanding the levee complex. Eventually the entire levee complex becomes inactive through channel abandonment or relocation. In this manner, a succession of overlapping levee complexes develops through time and, as a result, the fan grows upward and radially outward downslope.

Conclusion. The GLORIA side-scan sonar study of the Amazon Cone demonstrates the usefulness

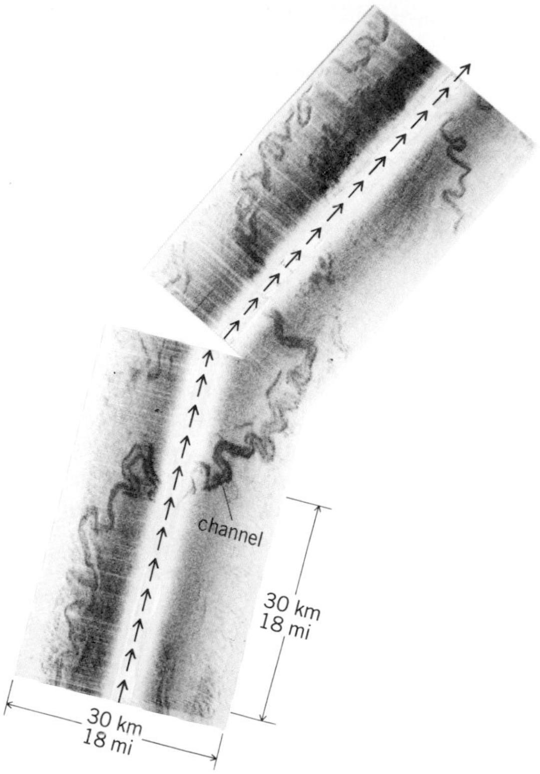

Fig. 2. Slant-range corrected GLORIA side-scan sonar record (sonograph) showing a portion of a large meandering distributary channel between depths of 9500 ft (2900 m) and 12,000 ft (3600 m) on the middle Amazon Fan (location of sonograph shown by black box in Fig. 1). Ship track is along the middle of the sonograph in direction of arrows (downslope). This channel is 160–330 ft (50–100 m) deep and up to 0.6–1.2 mi (1–2 km) wide. In this sonography, the dark patterns are regions of the sea floor (such as the channel) which reflect much of the acoustic energy sent out by GLORIA, whereas light regions are much less reflective (for example, smooth interchannel areas). The channels are highly reflective because of their rough walls and floors, whereas the channel levees and interchannel areas are less reflective because they are relatively smooth. *(Courtesy of John E. Damuth and Roger D. Flood)*

Fig. 1. Amazon Deep-Sea Fan. (*a*) Map showing the distributary channel system and other morphologic features. Bathymetric contours are in feet (meters). The Amazon Cone is outlined on the map. The three portions of the fan are divided by broken lines. The locations and trends of the distributary channels shown in the shaded areas on the upper and middle fan were mapped from GLORIA sonographs (example in Fig. 2). The black box shows the location of the GLORIA sonograph in Fig. 2. (*b*) Map showing location of fan in relation to northeast South America.

and scientific potential of long-range side-scan sonar for studying deep-sea sedimentation processes and growth patterns of submarine fans. Initial surveying of a submarine fan with GLORIA can thus provide a map of the total fan surface which can then be used to focus more detailed fan studies such as deep-towed side-scan sonar studies, quantitative bathymetric swath-mapping with narrow-beam echo sounders (Sea Beam), sediment sampling, and submersible studies on the most scientifically important locations of the fan. [JOHN E. DAMUTH]

SEDIMENT ACCUMULATION RATES

The determination of sediment accumulation rates in the marine environment has played an important role in understanding global and oceanic histories. Estuarine, continental margin, and deep-sea sediment columns provide a record of oceanic processes over various time scales which can be examined in detail by establishing a time-versus-depth relationship. Naturally occurring and bomb-produced radionuclides can be used to establish time-versus-depth relationships in marine sediments on time scales ranging from days to millions of years. For example, radiochemical chronologies have been used to examine seasonal changes in the rate of sediment accumulation near the mouth of the Yangtze River, as well as to examine changes in oceanic productivity during glacial and interglacial periods (100,000-year time scale). In addition to such studies, sediment accumulation rates from modern environments are useful in evaluating the depositional and sedimentological processes which form the preserved strata in the geologic record.

Radiochemical models. The determination of sediment accumulation rates is based on the fundamental radioactive decay relationship in Eq. (1),

$$A(t) = A_0 \exp \{(-\ln 2/T_{1/2})(t)\} \qquad (1)$$

where $A(t)$ and A_0 are the activities or concentrations of the radionuclide (disintegrations per minute per gram of sediment, abbreviated dpm/g) at time (t) and $t = 0$, respectively (years); and $T_{1/2}$ is the half-life of the radionuclide (years), which is the time required to reduce the activity (or decay rate) of the radionuclide in half. The symbol λ is commonly used in place of $\ln (2)/T_{1/2}$ and is called the decay constant. When one radionuclide decays to form another, the original isotope is called the parent and the newly formed isotope is called the daughter. Equation (1) describes the activity (A) of a radionuclide which is isolated from its parent. When the half-life of the parent is at least 100 times longer than that of the daughter, the A in Eq. (1) also describes the so-called excess activity, which is the activity of the daughter in excess of that of the parent.

Incorporating the relationship that the sediment accumulation rate (S) is equal to depth (z) in the sediment divided by the elapsed time (t) since deposition, Eq. (1) can be transformed into Eq. (2),

$$A(z) = A_0 \exp(-\lambda z/S) \qquad (2)$$

where z is the depth (cm) below the sediment-water interface or a designated horizon, A_0 is the activity (dpm/g) at depth $z = 0$, $A(z)$ is the activity z centimeters below the zero horizon, and S is the sediment accumulation rate (cm/year). The assumptions necessary to use Eq. (2) are: (1) the activity [or excess activity if there is coexisting parent activity] has been constant at depth $z = 0$ for a period of time equal to at least five half-lives of the radionuclide [corresponding to normal analytical detection limit]; (2) the sediment accumulation rate has been constant; (3) below the horizon where $z = 0$, no biological or physical reworking has altered the radiochemical profile; and (4) there has been no chemical migration of the parent or daugher radioisotopes through the sediment interstitial waters. Inherent in this technique is the assumption of a steady-state distribution for the radionuclides with respect to the sediment-water interface. Despite the fact that sediment is accumulating continually, the profile with respect to the sediment-water interface (or horizon where $z = 0$) remains constant. This technique for determining accumulation rates is called the constant surface activity model, and it is used with naturally occurring radionuclides which are continuously supplied to or produced in the marine environment.

In most marine sediments, surface particles are reworked by biological activities (such as feeding, burrowing, and escape from predators) as well as physical processes (such as tides, storms, and downslope debris flow) to a depth of 4–8 in. (10–20 cm). If sediment accumulation rates are determined in a region undergoing reworking and accumulation, Eq. (1) must be modified. Particle reworking is modeled as a diffuse process with the intensity of reworking quantified by incorporating a mixing coefficient, D (cm^2 year). The corresponding steady-state advection-diffusion equation is given in (3).

$$D \frac{\delta^2 A}{\delta z^2} - S \frac{\delta A}{\delta z} - \lambda A = 0 \qquad (3)$$

Based on the assumption of a constant surface activity ($A = A_0$ at $z = 0$) and decreasing activity at depth ($A \rightarrow 0$ as $z \rightarrow \infty$), the solution to Eq. (2) is given by Eq. (4).

$$S = \frac{\lambda z}{\ln\left(\frac{A_0}{A_z}\right)} - \frac{D}{z} \ln\left(\frac{A_0}{A_z}\right) \qquad (4)$$

Physical and biological reworking are not always diffusive in nature (infinite number of small random steps). However, when sufficient numbers of individual events are integrated over time, the reworking process simulates diffusion. If mixing and accumulation affect the distribution of radionuclides in a particular depth interval, two radionuclide profiles are needed to uniquely solve Eq. (4) for the two unknowns, the mixing coefficient (D) and the sediment accumulation rate (S).

In some marine environments there may not be

two naturally occurring radionuclides with appropriate half-lives for resolving the importance of mixing and accumulation. In such cases, impulse tracers can be used. The steady-state advection-diffusion equation cannot be used because impulse tracers are delivered suddenly to an environment rather than supplied continuously. The dispersion equation, used to model the depth of penetration of impulse tracers, is given in Eq. (5), where h is the

$$h = (2Dt')^{1/2} + St' \tag{5}$$

depth of penetration of the impulse tracer into the seabed, and t' is the elapsed time between tracer injection into the seabed and core collection. If the half-life of the impulse tracer is small relative to t', additional corrections for radioactive decay must be made. Using a naturally occurring tracer profile and an impulse tracer profile along with Eqs. (4) and (5) enables calculation of S and D, because there are two equations and two unknowns. If the sediment core penetrates sufficiently below the depth of physical and biological reworking, measurement of a single radionuclide at depth along with Eq. (2) can be used to establish the sediment accumulation rate.

Naturally occurring and bomb-produced geochronometers. An important consideration in establishing accumulation rates in the marine environment is time scale. The rate of sediment accumulation in estuarine and continental shelf deposits ranges from inches per year to hundredths of inches per year, whereas accumulation rates in continental slope, rise, and deep-sea sediments vary from hundredths of inches per year to hundredths of inches per 1000 years. The table shows a list of naturally occurring and bomb-produced radionuclides used to establish accumulation rates in various marine environments. Radionuclides with relatively short half-lives are used in continental shelf environments, whereas longer-lived tracers are used in continental slope and deep-sea environments.

The radionuclide ^{234}Th, which is produced from the decay of ^{238}U, is insoluble in sea water and is scavenged from the water column by settling particles. Particles arrive at the sediment–water interface with excess ^{234}Th activity because uranium is soluble in sea water. Thorium-234 is used to establish accumulation rates in continental shelf environments near large sediment sources because of its relatively short half-life (24 days). Figure 3*a* shows

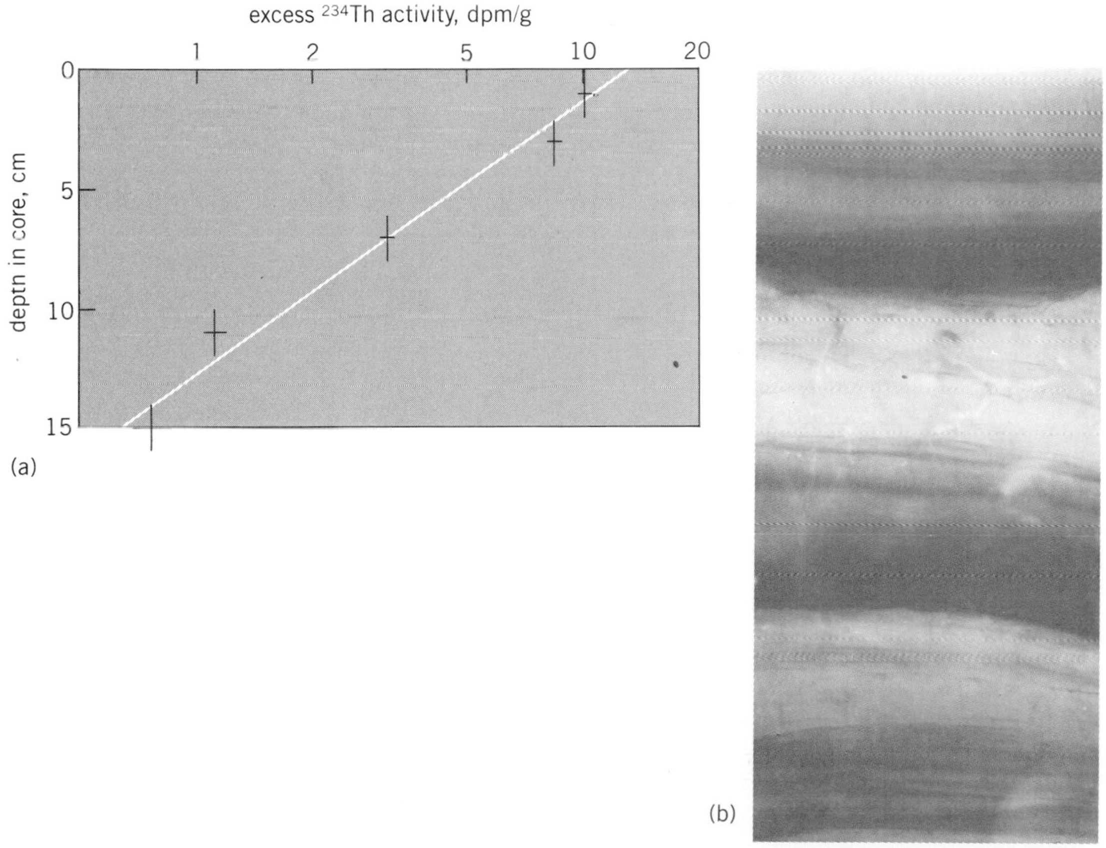

Fig. 3. Naturally occurring geochronometer. (*a*) ^{234}Th profile from the East China Sea showing a very rapid accumulation rate (100-day time scale) near the mouth of the Yangtze River. The vertical bars on the data points represent the intervals that were sampled, and the horizontal bars the error in analysis. (*b*) X-radiograph from the core. The integrity of the horizontal strata in the radiograph indicates that the ^{234}Th profile is controlled by accumulation and not particle reworking. Here dpm/g = disintegrations per minute per gram of sediment. (*From B. M. McKee, C. A. Nittrouer, and D. J. DeMaster, The concepts of sediment deposition and accumulation applied to the continental shelf near the mouth of the Yangtze River, Geology, copyright 1983 by Geological Society of America*)

Radionuclides used to determine marine sediment accumulation rates

Naturally occurring radionuclides	Source	Half-life	Environments where radionuclide is used to determine accumulation rate
^{234}Th	^{238}U	24.1 days	Continental shelf near major sediment source
^{7}Be	Cosmic-ray-produced	53 days	Continental shelf
^{228}Th	^{228}Ra	1.9 years	Continental shelf and rapidly accumulating basins in continental margin
^{210}Pb	^{226}Ra	22.3 years	Continental shelf and continental slope
^{32}Si	Cosmic-ray-produced	100 years	Siliceous continental shelf and continental margin
^{14}C	Cosmic-ray-produced	5,730 years	Continental shelf, continental slope, and deep sea
^{231}Pa	^{235}U	33,000 years	Deep sea
^{230}Th	^{234}U	75,000 years	Deep sea
^{10}Be	Cosmic-ray-produced	1,600,000 years	Deep sea
Impulse tracers	**Source**	**Duration of tracer injection**	**Environments where radionuclide is used to determine accumulation rate**
^{137}Cs	Nuclear bomb testing	30–40 years	Continental shelf and continental margin
^{239,240}Pu	Nuclear bomb testing	30–40 years	Continental shelf and continental margin

a ^{234}Th profile from a core collected near the mouth of the Yangtze River. Biological and physical reworking can be eliminated as important processes because the x-radiograph for this core (Fig. 3*b*) shows numerous horizontal strata. If diffusive processes were active in the sea bed, the sediment forming the horizontal strata would have become homogenized. Based on Eq. (2) and the slope of the excess ^{234}Th profile, the accumulation rate for this core is 21 in./year (53 cm/year), which is characteristic of the 100-day period (approximately five half-lives) prior to core collection.

The most commonly used geochronometer for continental shelf sediments is ^{210}Pb. The three sources of ^{210}Pb that contribute to the marine environment are: decay of ^{222}Rn in the atmosphere to ^{210}Pb, which is subsequently supplied to the oceans by rainfall; riverine sediment yield; and in-place production in the water column supported by ^{226}Ra. The radionuclide ^{210}Pb, which is insoluble in sea water, is scavenged from the water column by settling particles and arrives at the sea bed in excess of its effective parent, ^{226}Ra. As an example of ^{210}Pb geochronology, Fig. 4 shows a profile from a varved Gulf of California sediment core. The varves, which are alternating semiannual layers of biogenic silica-rich and biogenic silica-poor sediment, provide a check on the ^{210}Pb accumulation rate. Because of the preserved varve structures, sediment reworking can be eliminated. Equation (2) and the excess ^{210}Pb profile yield an accumulation rate of 0.075 in./year (0.19 cm/year), which is in good agreement with the average varve thickness of 0.079 in. (0.20 cm). The independent check on the accumulation rate confirms the ^{210}Pb technique and the necessary assumptions used in the steady-state constant surface activity model. In most continental shelf environments, sediment reworking and sediment accumulation are both important processes. Therefore, two tracers must be incorporated to resolve the accumulation rate from the effects of reworking. Because there is no naturally occurring radionuclide with a half-life and chemistry similar to ^{210}Pb, impulse tracers such as nuclear-bomb-produced ^{137}Cs and ^{239,240}Pu are used. The ^{210}Pb profile plus Eq. (4) coupled with an impulse tracer profile and Eq. (5) enable calculation of accumulation rates, as well as rates of reworking.

Carbon-14, which is produced by cosmic-ray bombardment in the atmosphere, reaches the marine environment via carbon dioxide exchange between the atmosphere and surface sea water. The ^{14}C activity is incorporated into calcareous as well as organic-rich biota. After normalizing the ^{14}C activity to the amount of ^{12}C and correcting for the effects of nuclear-bomb-produced ^{14}C and the burning of fossil fuels, surface plankton should have a

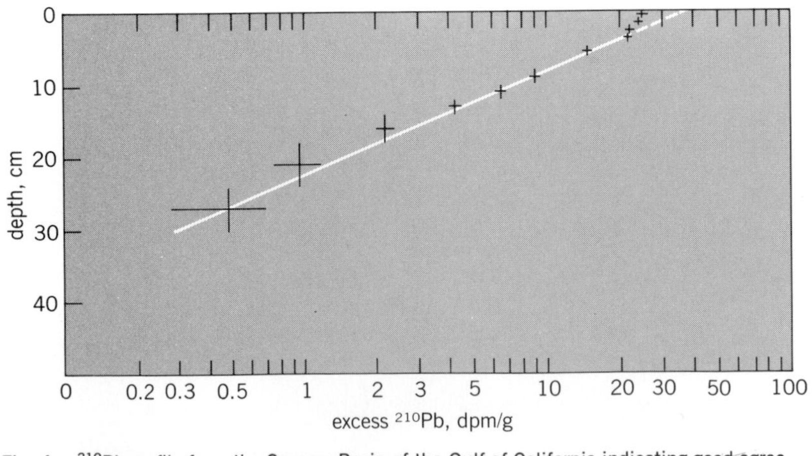

Fig. 4. ^{210}Pb profile from the Carmen Basin of the Gulf of California indicating good agreement between the radiometric accumulation rate (0.75 in. or 0.19 cm per year) and the varve chronology (0.79 in. or 0.20 cm per year). The data points are to be interpreted as in Fig. 3. Here dpm/g = disintegrations per minute per gram of sediment. (*After D. J. DeMaster, The half-life of Si-32 determined from a varved Gulf of California sediment core, Earth Planet. Sci. Lett., 48:209–217, Elsevier, 1980*)

zero age (relative to atmospheric ^{14}C). Under these conditions the activity-versus-time relationship can be established by using Eq. (1). Surface sediments in the marine environment, however, rarely have zero ^{14}C ages in the carbonate or organic fractions because: (1) detrital carbon [which is relatively depleted in ^{14}C] is supplied to ocean sediments from coastal or riverine environments in addition to the planktonic carbon; (2) particle reworking mixes newly deposited material with carbon from depth in the sediment column [depleted in ^{14}C]; and (3) due to upwelling and inefficient carbon dioxide exchange, surface waters are frequently out of equilibrium with respect to the atmospheric $^{14}C/^{12}C$ ratio. Accumulation rates can be determined by using ^{14}C; however, the ^{14}C age of the surface sediment must be assumed constant over the time period of interest (up to 30,000 years).

The radionuclides ^{231}Pa and ^{230}Th, which are both produced from uranium isotopes in sea water, are used to establish accumulation rates in the deep sea. Both radionuclides are insoluble in sea water and reach the sea bed via settling particles. Their relatively long half-lives (^{231}Pa, 33,000 years, and ^{230}Th, 75,000 years) enable determination of accumulation rates as low as hundredths of inches per 1000 years. An additional advantage of their long half-lives is that most of the radiochemical profile occurs below the zone of particle reworking (4 to 8 in. or 10 to 20 cm), and Eq. (2) can be used to determine the accumulation rate. The production of ^{231}Pa and ^{230}Th in the water column and the corresponding supply to the sea bed are controlled by the water depth because the uranium concentration in sea water is uniform. If the rate of sediment supply to a deep-sea environment changes, the activity of the sediment reaching the sea bed can change. Under these conditions a modification of the constant surface activity model can be used which utilizes the excess $^{230}Th/^{231}Pa$ to establish the time-versus-depth relationship.

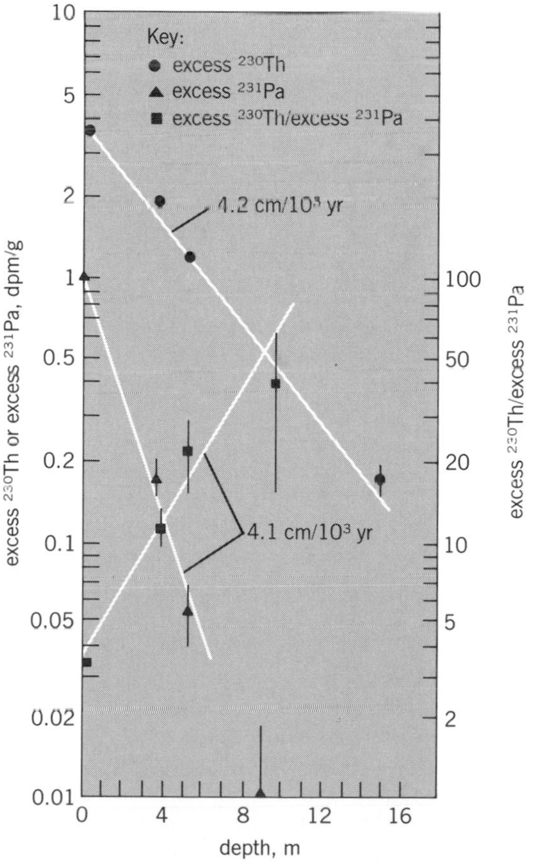

Fig. 5. ^{231}Pa, ^{230}Th, and $^{230}Th/^{231}Pa$ profiles from an Antarctic deep-sea sediment core showing concordancy between the three different radiometric techniques. Here dpm/g = disintegrations per minute per gram of sediment. (*After D. J. DeMaster, The supply and accumulation of silica in the marine environment, Geochim. Cosmochim. Acta, 45:1715–1732, Pergamon, 1981*)

The $^{230}Th/^{231}Pa$ ratio increases with depth because the half-life of ^{231}Pa is less than that of ^{230}Th. An advantage of this model is that if, for example, the deep-sea accumulation rate doubles, the excess ^{230}Th and ^{231}Pa activities will decrease to half (when the radionuclide flux remains constant). Despite the change in individual activities, the $^{230}Th/^{231}Pa$ ratio remains constant and can be used for geochronometry. Figure 5 shows excess ^{230}Th, ^{231}Pa, and $^{230}Th/^{231}Pa$ data from a rapidly accumulating deep-sea Antarctic core. The agreement among the ^{230}Th, ^{231}Pa, and $^{230}Th/^{231}Pa$ chronologies indicates that the surface radonuclide activities as well as the sediment accumulation rate have been constant.

For background information *see* GEOCHRONOMETRY; LEAD ISOTOPES (GEOCHEMISTRY); MARINE SEDIMENTS; RADIOCARBON DATING; SONAR; TURBIDITE; TURBIDITY CURRENT in the McGraw-Hill Encyclopedia of Science and Technology.

[DAVID J. DEMASTER]

Bibliography: J. E. Damuth et al., Age relationships of distributary channels on Amazon Deep-Sea Fan: Implications for fan growth pattern, *Geology*, 11:470–473, 1983; J. E. Damuth et al., Distributary channel meandering and bifurcation patterns on the Amazon Deep-Sea Fan as revealed by long-range side-scan sonar (GLORIA), *Geology*, 11:94–98, 1983; L. E. Garrison, N. W. Kenyon, and A. H. Bouma, Channel systems and lobe construction in the Mississippi Fan, *Geo-Mar. Lett.*, 2:31–39, 1983; E. D. Goldberg and K. W. Bruland, Marine geochronology, in *The Sea*, vol. 5 (ed. by E. D. Goldberg), pp. 450–487, 1974; A. S. Laughton, The first decade of GLORIA, *J. Geophys. Res.*, 86:11511–11534, 1981; K. K. Turekian and J. K. Cochran, Marine geochronologies with natural radionuclides, in J. P. Riley (ed.), *Chemical Oceanography* pp. 313–360, 1978.

Medical parasitology

Infections with parasites are typically of long duration, 40 years or more in the case of some malarias. These long-lasting infections have been interpreted by some as meaning that the protective immune response to parasites is either inadequate or nonexistent. However, it is now clear that in all parasitic infections there is an adequate immune response but it is foiled by the ability of parasites to evade its consequences. Long-lasting infections coupled with misdirected immune responses lead to gross immunopathology, of which the grotesque swelling of limbs in elephantiasis caused by filarial worms is one example. The difficulties inherent in unraveling the complex immune responses and the dangers of adverse consequences have rightly made parasitologists cautious about the prospects of developing vaccines. The tools of modern molecular biology have in many cases been used to dissect those parts of the immune response that are protective from those that are not, and the possibility of developing vaccines against major parasitic diseases is now a reality.

Evasion of the immune response. The best-known example of the ways in which parasites can evade the immune response is the antigenic variation shown by African trypanosomes. In the bloodsteam of their human or animal host, the trypanosomes are coated with a thick layer of glycoprotein which rapidly stimulates an antibody-mediated immune reaction. Before the parasites can be eliminated, however, trypanosomes with different kinds of glycoproteins appear, and as these are recognized by the immune system, they too are replaced by others with yet another kind of surface coat. The glycoprotein coat is called the variant antigen and is coded for by over 100 genes generated by mutation and duplication. This gives rise to an almost unlimited repertoire of distinct antigens that allow the parasite to remain one step ahead of the immune response. So efficient is this system that if an experimental animal is infected by the bite of a tsetse fly, the vector of trypanosomiasis, tens or hundreds of different variant types of trypanosomes can be recognized in its blood within a few days. In natural situations, there is a tendency to revert to a basic repertoire of antigens, and in the case of acute, or Rhodesian, sleeping sickness, the repertoire may be very limited. In heavily infested areas, cattle can eventually become immune to trypanosome infections, which suggests a gradual exhaustion of the repertoire of variant antigens and susceptibility to immune attack. Thus, despite antigenic variation, there is always the hope that a vaccine can be developed.

The presence of antigenic variation has been searched for in other parasites, and there is evidence for such variation in experimental malaria infections, but this has not been clearly demonstrated in humans.

Other parasites illustrate a variety of ways in which the immune response can be evaded. The blood flukes, *Schistosoma* spp., that cause bilharzia, survive in a potentially immunologically hostile environment by disguising themselves with host antigens, thus becoming immunologically invisible. They acquire these antigens when, as larvae, they bore through the skin and tissues on their way to blood vessels, where they mature. The antigens involved are glycolipids similar or identical to those of the red blood cell ABO series.

One of the most interesting ways in which parasites evade the immune response is by inhabiting macrophages or other phagocytic cells. These cells usually incorporate invading organisms in cell membrane–surrounded vesicles known as phagosomes and kill them by means of a respiratory burst and the release of toxic oxygen radicals or lytic enzymes contained in lysosomes. The lysosomes fuse with a phagosome to form a phagolysosome. The causative agent of Chagas' disease, *Trypanosoma cruzi*, lives freely in the cytoplasm of the cell and thus does not come in contact with the phagolysosome. *Toxoplasma gondii*, a ubiquitous parasite of cats and many other hosts including humans, is able to prevent the lysosomes from fusing with the phagosome in which the parasite lives. *Leishmania* spp., which cause appalling skin lesions and even inhabit the macrophages of the spleen and liver, actually survive in the phagolysosome.

All these special methods of evading the immune response are supplemented by a general immunodepression which is characteristic of parasitic infections. This immunodepression not only damps down a partially ineffective immune response and gives the parasites an increased chance of survival, but also leaves the host susceptible to opportunistic viruses and bacteria. The prevalence of Burkitt's lymphoma in Africa, a consequence of Epstein-Barr viral infection, is thought to result from immunodepression associated with malaria.

Protective immune mechanisms. Despite the fact that parasites are capable of evading the immune reaction, there is considerable evidence that in most cases there are protective responses which often consist of several distinct mechanisms operating together or in sequence. In malaria, the first response is directed against the sporozoites, the infective stages injected by a mosquito. An antigen consisting of a single polypeptide on the surface of the sporozoite seems to be the target recognized and neutralized by antibody. However, sporozoites quickly enter liver cells, where they are safe from immune attack. The sporozoites transform in the liver into another parasitic form, the merozoites, which emerge from the liver cells and invade red blood cells. The merozoites multiply in the red cells, and the new generation of parasites infects fresh red cells. Nevertheless, the parasites which leave a cell are susceptible to the action of antibody while they remain in plasma prior to their entry in another cell. Once within the red cells, the parasites are not susceptible to antibody but can be killed by nonspecific factors released from macrophages. Finally, the sexual stages (gametocytes) that are formed from merozoites in some red cells and are released into plasma can also be affected by antibody after they and the plasma have been ingested by mosquitoes. Antibody is therefore involved in at least three ways in the immune response to malaria, but because the antigens of the various stages are quite distinct, at least three different antibodies must be involved.

A similar situation occurs with the blood flukes. The adult worms are not affected but the larval forms, or schistosomula, are attacked during passage through the skin or tissue. The immune attack is complex. Various kinds of antibodies bind to the surface of the schistosomulum, and macrophages, eosinophils, mast cells, or platelets are bound to the surface of the worm by these antibodies. It is not clear exactly what happens, but signals from bound mast cells cause eosinophils to flatten out on the surface of the young worm and to release their contents. These disrupt the surface layer of the worm, allowing eosinophils to enter the parasite and thus bring about its destruction. Eosinophils are also involved in immunity to several other worms and usually attack migrating larval stages.

The macrophage-dwelling parasites are safe unless their host cells become activated as part of an immune response. The parasites then lose their protection and are killed. The skin-inhabiting leishmanias are attacked by a delayed-hypersensitivity type of reaction. Immune lymphocytes attack the infected macrophages, destroying them and the parasites they harbor. Interestingly, in this case, recovery, albeit slow, is the rule. Occasional absence of healing is attributable to the induction of suppressors cells which switch off or damp down the immune response. The organisms, *Theileria* spp., that cause the deadly East Coast fever in cattle in Africa and similar diseases elsewhere, live in lymphocytes. In immune animals, these infected cells are attacked by cytotoxic lymphocytes in a manner similar to that in viral infections.

Immunopathology. The long duration of parasitic infections provides every opportunity for the development of immunopathological reactions. In some malarias, antigen-antibody complexes block and damage the vessels of the kidney and brain. In trypanosomiasis, the whole pathology of the disease is consistent with an immune response that has got out of hand. This is because the variant antigens are shed and lodge in tissues, and the immune responses become directed against these antigens instead of the trypanosomes themselves. A comparable situation exists in infections with blood flukes. The adult worms produce eggs which lodge in the liver, and these are subject to immune attack of the delayed-hypersensitivity type. This results in the accumulation of cells which form large granulomas. These eventually compress blood vessels of the liver and may cause death. Elephantiasis, resulting from filarial infections, is also thought to have an immunological basis, although the mechanisms are not understood.

Autoimmunity can be considered a form of immunopathology, and autoimmune destruction of red blood cells occurs in malaria presumably as a result of the binding of parasite antigens to red cell membranes. In Chagas' disease, parasite antigens occur on nerve and heart cells, and these may be destroyed even at a distance from the site of infection. An additional problem in this disease is that the parasites have antigens in common with some nerve cells, which are destroyed instead of the parasites.

Vaccination. A comparative lack of safe and effective drugs and resistance to those that are in use, coupled with insecticide resistance on the part of the vectors, make the development of vaccines a prime aim of much parasitological research. For cattle and sheep, there is an effective commercially available vaccine against lungworm which consists of irradiated larvae. A similar vaccine against canine hookworm was extremely effective but has been discontinued because of its limited shelf life and low volume of sales. Irradiated-larva vaccines have also been used successfully against cattle blood flukes but are not available commercially. A vaccine consisting of attenuated parasites is widely used against the blood parasites that cause red water fever (babesiasis) in cattle in Australia. Vaccines of this kind are unlikely to be acceptable for human use, and a sign of the things to come is the identification of the sporozoite antigen of the malaria parasite and its successful cloning in *Escherichia coli.* With the eventual synthesis of this antigen, a vaccine against the most intransigent infectious disease of humans will become a real possiblity.

For background information *see* ANTIBODY; ANTIGEN; HYPERSENSITIVITY; IMMUNITY; IMMUNOPATHOLOGY; PARASITOLOGY; VACCINATION in the McGraw-Hill Encyclopedia of Science and Technology.

[F. E. G. COX]

Bibliography: O. O. Barriga, *The Immunology of Parasitic Infections*, 1981; A. R. C. Capron (ed.), Immunoparasitology, *Clinics Immunol. Allergy*, vol. 2, no. 3, 1982; S. Cohen and K. S. Warren (eds.), *Immunology of Parasitic Infections*, 1982; W. Frank (ed.), *Immune Reactions to Parasites*, 1982.

Metallurgy

Rapid solidification (RS) refers to the transformation from the liquid state to the solid state in a very short period of time (usually milliseconds or less). There is some question as to whether a minimum solidification or cooling rate is required in order to classify a transformation as rapid solidification. Most criteria for classifying a transformation as rapid solidification are based on the nature of the solidified material. The term rapid solidification applies if, when compared to conventionally processed materials, the solid exhibits decreased micro- or macrosegregation, increased solid solubility, decreased second-phase grain size, elimination of segregated phases, formation of metastable phases, and formation of amorphous phases. The minimum cooling rate needed to achieve these solid structures varies from material to material. Perhaps the most useful definition of rapid solidification is that it includes transformations with sufficiently high cooling rates that result in microstructural modifications and property changes in the solid that would not occur with conventional solidification. Based on this criteria, phenomena with cooling rates as low as 10^2 °C/s (1.8×10^2 °F/s) can be considered rapid solidification; however, most practical materials must be solidified at rates $>10^5$ °C/s ($>1.8 \times 10^5$ °F/s). Currently, the terminology has also been extended to transformations from the vapor and plasma states to the solid state.

Metals and alloys have been the primary focus of rapid solidification technology (RST) since the discovery, in the late 1950s and early 1960s by Pol Duwez and his students at the California Institute of Technology, that when solidifying Au-Si and Pd-Si alloys by splat quenching, it is possible to suppress normal crystallization and produce an amorphous microstructure. In an amorphous structure, the atoms are more or less randomly arranged and there is no long-range order or crystal structure. The consequence of such random structures is that the alloys have properties which are significantly different from crystalline alloys of the same composition. This new way of controlling materials' properties stimulated international research in rapid solidification technology.

Developments. Throughout most of the 1960s, rapid solidification technology was a laboratory curiosity, under study at a few universities. It was not until the 1970s that it began to emerge as a significant technology. Since then, research activity has grown at an exponential rate. Recent advances in rapid solidification technology have been as revolutionary to the technology as Duwez's discoveries were to classical metallurgy. Most of the developments in rapid solidification technology have resulted in process advances and new alloy compositions.

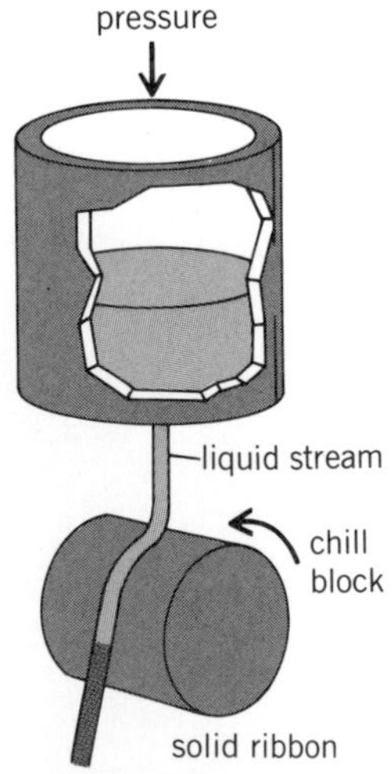

Fig. 1. Schematic of chill block melt spinning process. (*After R. E. Maringer, The New Metallurgy: Challenge and Opportunity, Battelle Technical Inputs to Planning, Rep. 1, Columbus, Ohio, 1977*)

The early rapid solidification processes produced very small amounts of material, usually only hundredths of ounces per hour. These are the so-called splat quenching techniques, the most popular of which are the ski slope and the piston anvil. In both of these techniques, small drops of molten metal are spread over a stationary substrate which extracts the heat and results in rapid solidification. The next generation of rapid solidification technology processes were capable of producing much larger quantities of rapidly solidified materials. Among these processes are melt spinning and melt extraction. Basically, these two processes involve solidification on a rotating substrate. The major difference in these processes is the manner in which the molten metal is brought in contact with the rotating substrate. In the melt spinning process, a molten stream is extruded from a crucible onto the rotating substrate which is usually a cylindrical drum (Fig. 1). The product of this process is a thin, rapidly solidified narrow ribbon. However, by changing the configuration of the surface of the drum, flakes can be made. In melt extraction, the rotating substrate is normally a disk with a rather sharp edge around the circumference which is brought into contact with a molten metal bath (Fig. 2). The melt extraction process usually produces a wire or filament, but if notches are put into the edge of the disk, the process makes fibers of a specified length. There are several variations of each of these processes, of which at least two have been commercialized for the manufacture of aluminum flakes and steel fibers.

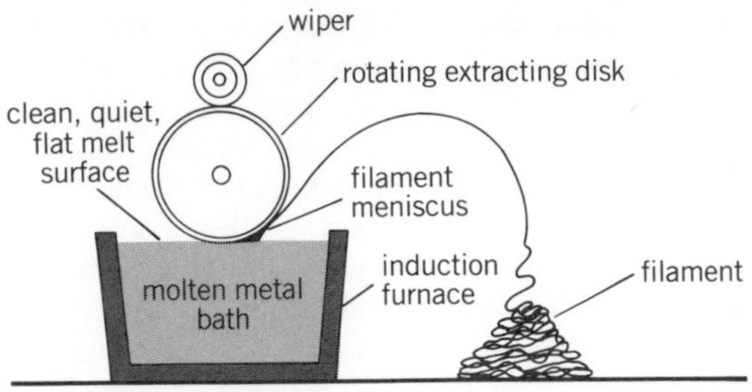

Fig. 2. Schematic of crucible melt extraction process. (*After W. K. Kinner, Rapid quenching: A materials bonanza?, Mater. Eng., pp. 32–36, January 1979*)

Currently, the most significant developments in rapid solidification technology are in direct strip casting, rapidly solidified powders, and new alloy compositions. The thrust of these developments is primarily for commercialization and, if totally successful, will revolutionalize several areas of materials processing and alloy development.

Direct strip casting. Extensive developments have been under way in direct strip casting for several years. Most of the effort has been directed to the production of amorphous magnetic alloy strip for use in distribution transformers, and of amorphous Ni and Cu alloys for brazing applications. Transformers made from amorphous magnetic alloys can reduce electrical transmission losses by as much as 80%, and the brazing alloys are reducing fabrication costs in jet engine assemblies. There are several direct strip casting processes, but they all involve feeding molten metal onto a single rotating drum or into the bite of two counterrotating drums. The molten metal is fed onto the drums in such a manner as to produce wide (presently up to 10 in. or 25 cm wide) thin strip. These processes can produce wide crystalline and microcrystalline, as well as amorphous, strip. The incentive to develop these processes for crystalline alloy strip is primarily for the economic savings in processing, whereas the amorphous and microcrystalline alloys are being developed for superior properties and performance. For example, the energy savings alone for making directly cast crystalline steel strip can amount to over 60%, when compared with conventional ingot or continuous casting methods. Additional savings would be realized from the reduced need for rolling mills and other handling equipment, maintenance of that equipment, and labor. The success of this development will have a revolutionary impact on the steel industry. Research on direct strip casting processes for crystalline materials is in the preliminary stage, but it is definitely one of the major trends in rapid solidification technology.

Rapidly solidified powders. Although not a new technology, powder metallurgy (P/M) is another area in which rapid solidification technology is under way on a large scale. The application of rapid solidification powders is very broad, ranging from tool steels, turbine engines, and aircraft structures, to batteries, catalysts, and bearings. The classification of a powder as "rapidly solidified" is somewhat arbitrary; however, it is generally accepted that most alloys that have practical rapid solidification technology applications have undergone solidification rates $>10^5$ °C/s ($>1.8 \times 10^5$ °F/s). When this is used as a basis, some of the smaller-size fractions of water- and gas-atomized powders can be considered rapidly solidified. However, there are several other techniques specifically designed for making rapidly solidified powders. Among these are the rapid solidification rate (RSR) centrifugal atomization process, the rotating electrode and plasma rotating electrode processes, modified melt extraction and melt spinning, ultrasonic gas atomization, and the recently developed rapid spinning cup process. Rapidly solidified powders are also being made by milling rapidly solidified strip and ribbon.

Only a few of these techniques produce amorphous powders, even though they all are considered rapid solidification processes. Most of the powders are microcrystalline, that is, with grain diameters under 5 micrometers. The emphasis on microcrystalline, rather than amorphous, powders is due mainly to the difficulty in retaining desired properties in consolidated amorphous alloys which have been heated to temperatures above the glass transition temperature T_g (crystallization point). There is also a great deal of difficulty in consolidating amorphous alloys without going to temperatures above T_g, since these alloys do not exhibit much plastic deformation below T_g. In fact, the consolidation of rapid solidification powders is a major research area currently under development.

Most of the recent developments in the rapidly solidified powder-making processes have been to control the process parameters to economically produce clean, uniform powders. Cleanliness and uniformity are the most important considerations in powder production. Recent improvements in the rotating electrode process have eliminated nearly all sources of contamination. The new plasma rotating electrode process has been set up to produce rapid solidification powders. The rapid spinning cup process, which has produced superalloy powders with relatively low O_2 concentrations (<80 ppm), is being modified to produce very fine, clean powder with particle diameters under 10 μm. In addition to developments designed to make very clean and uniform powders, these processes are being modified to produce larger quantities of rapid solidification powder. The centrifugal atomization process is being scaled up to handle 2000 lb (900 kg) of material per melt. This facility will produce 300 lb (140 kg) of rapid solidification powder per minute.

Alloy development. The potential for developing new alloys is one of the most significant aspects of rapid solidification technology. By rapidly solidifying most alloy systems, it is possible to increase the solid solubility composition range. In addition to increased solid solubility, most multiphase rapid solidification materials have microstructures that feature very fine, uniformly distributed precipitated phases. These factors make it possible to enhance the properties of established alloys and to develop alloys having new compositions not possible through conventional pressing. By combining rapid solidification technology with powder metallurgy or direct strip casting, alloys with improved properties could be produced at lower costs.

As an example, new rapidly solidified iron-base alloys are showing promise for improved bearings through increases in bearing lifetime. This allows increased rotational speeds in aircraft gas turbine engines, which means increased performance. The expected improvements in bearing properties, and in steel properties in general, are the result of refined microstructures and small carbide sizes developed in the rapidly solidified alloys. In conventional ingot metallurgy bearings, the sizes of the largest carbides present are often in the range of the critical flaw size calculated by linear elastic fracture mechanics. In these rapid solidification technology materials, there may be property gains from two sources: a decrease in the carbide sizes to below the critical flaw size, and an increase in facture toughness.

Rapidly solidified, compositionally modified tool steels have significantly better rolling contact fatigue life compared with conventional tool steels. In addition, a rapid solidification high-chromium (19% Cr) alloy is being developed for corrosion-resistant bearings. This high-chromium alloy, made from a narrow size fraction of atomized powder, also shows improved reliability in rolling contact fatigue life compared with current engine bearing materials. These developments, which utilize compositions and thermomechanical treatments tailored to take advantage of the microstructural control offered by rapid solidification, will produce alloys that have longer life and greater reliability than current bearing alloys.

In the area of light metals, rapidly solidified aluminum alloys are being developed with the aim of increasing the upper use temperature from 200°F (93°C) to 650°F (343°C). Accomplishing this would enable these aluminum alloys to replace titanium alloys in some gas turbine engine applications, such as fan blades and other low-temperature compressor components. The incentive here is savings in material and fabrication costs. To accomplish this aim, improvements in specific strength and stiffness approaching those of the titanium alloys are necessary. Extensive studies of rapid solidification aluminum alloys indicate that several transition element–containing ternary systems are promising candidates for development. One alloy, Al-8Fe-2Mo, has specific strength comparable to that of Ti-6Al-4V at 400°F (204°C). Other Al alloys under development are based on the Al-Li system. These alloys not only will have excellent high temperature properties, but will be lighter in weight because of the Li content.

It appears that rapid solidification aluminum alloys will significantly extend the useful upper temperature limit for aluminum alloys.

In contrast to the extensive investigations of rapid solidification aluminum alloys, rapid solidification titanium alloys are only beginning to be studied. One objective is to develop titanium alloys for use up to 1300°F (704°C). Toward this end, precipitation-hardened, titanium–rare earth binary alloys are being investigated. By oxidizing the rare-earth precipitates, it is possible to form dispersoids which are stable at high temperatures. There is presently not enough information to assess the potential for rapid solidification titanium alloys or the most fruitful directions for development; however, these issues should become clearer in the next few years.

Recently, a concept for a new type of rapid solidification material has been developed. This concept consists of making brittle RS amorphous strip or ribbon from hypoeutectoid metallic glass alloys, attriting it into particulate, then consolidating the particulate into bar by extrusion. The amorphous material crystallizes during consolidation into a fine homogeneous multiphased microstructure. The high boron and carbon content of these alloys results in a high content of hard borides and carbides. These impart high hardness and strength suitable for die and tool applications. Iron-, nickel-, and cobalt-base alloys have been made by this technique. Certain alloys have shown high resistance to sulfuric acid attack compared with stainless steel. Low-carbon modifications also show a response to heat treatment with hardnesses greater than R_c 68 (Rockwell C hardness scale). Further development of these alloys should expand their areas of possible application.

Rapidly solidified materials are usually made as particulate and strip. In these forms, their use is extremely limited. To be useful, the material must be consolidated into billets for secondary fabrication into shapes and parts or directly into near net shapes. Much of the consolidation of RS particulate has been accomplished by extrusion. Other possible methods are forging, rolling, vacuum hot pressing, hot isostatic pressing, and dynamic consolidation. The aim in consolidating rapid solidification materials is to maintain the consolidation temperature as low as possible to retain maximum latitude for subsequent manipulation of the microstructure and metallurgical phases. For this purpose, dynamic consolidation techniques using very high pressure (kilobars; 1 kilobar = 0.1 gigapascal) and low temperatures are being studied. These methods require further development before their practical possibilities can be assessed. The more practical approach is to use conventional methods, for example, extrusion, as much as possible. To do this, the alloys will have to be designed to respond as favorably as possible to the consolidation process selected.

Laser surface melting and alloying. The widespread use of lasers has spawned another area of rapid solidification technology, that is, laser surface melting and alloying. By focusing a laser beam onto the surface of a metal or alloy, it is possible to controllably melt the surface. If the melted layer is thin <0.02 in. or 0.5 mm), and the underlying material is thick enough to quench the melt layer in a few milliseconds, then the surface layer is a rapidly solidified material. With this process, it is possible to put fused, thin, highly alloyed coatings on less expensive bulk substrates or parts. These coatings can provide wear, erosion, and corrosion resistance while saving material costs. A problem which must be addressed for wider use of this technique is the residual surface tensile stresses often present in the resolidified surface layer. No commercial applications of laser surface melting or of alloying are known, but a number are being seriously investigated. These include surface melting of gray cast iron to form a hard, wear-resistant white cast iron surface, refining the surface carbides of tool steels and bearings by surface melting, and surface alloying for oxidation and wear resistance on steel substrates. In addition, the buildup of successive thin layers of rapidly solidified material into bulk form offers an alternative approach for obtaining parts or structures of rapid solidification material besides the powder or strip route.

Future outlook. It is now evident that rapid solidification technology is much more than a laboratory curiousity, and that it has established itself as a viable technology with the potential for extensive commercialization. However, in spite of the dynamic growth in research and the significant developments in processes and alloys, RST has not yet realized its full potential. In some respects, the technology may be viewed as a solution looking for a problem. There are many cases where the problem and the solution have been matched, and the results have been exciting and very beneficial.

The complexion of RST has changed dramatically in the last few years. Amorphous materials, which dominated interest earlier, are finding only limited use and interest today. Now the primary focus is on rapid solidification crystalline and microcrystalline materials, the processes for making rapid solidification materials economically, and new alloy compositions that are made possible by rapid solidification technology.

For background information *see* ALLOY; AMORPHOUS SOLID; METALLIC GLASSES; METALLURGY; POWDER METALLURGY in the McGraw-Hill Encyclopedia of Science and Technology.

[ROBERT S. CARBONARA]

Bibliography: R. S. Carbonara et al., *State-of-the-Art Review of Rapid Solidification Technology (RST)*, Metals and Ceramics Information Center, Columbus, Ohio, 1981; H. Jones, *Rapid Solidification of Metals and Alloys*, Institute of Metallurgists, London, 1982; T. Masumoto and K. Suzuki (eds)., *Rapidly Quenched Metals*, vols. 1 and 2, Japan Institute of Metals, 1982.

Meteorite

Meteorites provide the only samples of material outside the Earth-Moon system that are available for laboratory studies of composition and physical properties. Since 1969, when Japanese explorers first found meteorites in the Yamato Mountains of Antarctica, approximately 6000 Antarctic meteorites have been collected. Many of these are multiple fragments from larger meteorites, but before the discovery of Antarctic meteorites only about 2100 meteorites were known, and generally less than 10 new ones were recovered each year worldwide. The United States meteorite recovery program is a collaborative effort by several government agencies. The National Science Foundation provides support for Antarctic recovery expeditions, the National Aeronautics and Space Administration and the Smithsonian Institution are involved with meteorite processing and curation, and all three agencies sponsor research on these meteorites. In the last few years United States teams have discovered a number of Antarctic meteorites that have provided some startling views of meteorite parent bodies and the origin and early evolution of the solar system.

Occurrence. Most meteorites are collected on the Polar Plateau, a harsh environment of high elevation in the Antarctic interior characterized by ice thickness greater than 1 mi (1.6 km). The specimens in the United States collection were found in areas along the Transantarctic Mountains in Victoria Land. The meteorites are localized in certain areas, though the mechanism of concentration is not completely understood. One plausible hypothesis is that over many thousands of years meteorites fall into, and are trapped within, snow and ice in the continental interior. The gradually thickening glacial ice sheets flow to the edge of the continent, except where ice flow is impeded by obstacles, for example, mountains (nunataks). In such areas meteorites may be exposed by wind ablation and concentrated as successive layers of ice are removed (Fig. 1). Most antarctic meteorites have been found in such fields of exposed "blue" ice located in the inland sides of nunataks (Fig. 2).

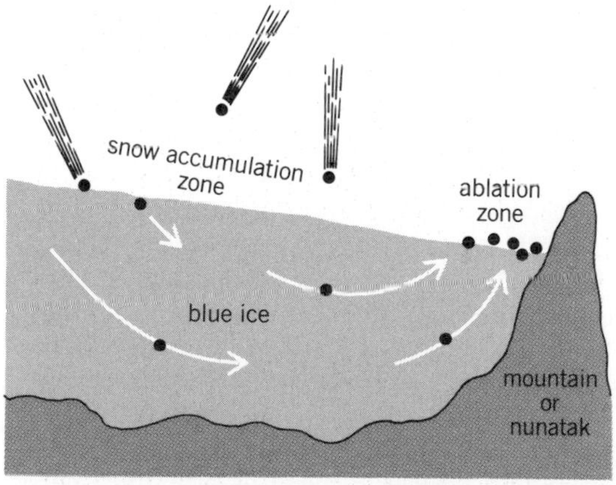

Fig. 1. Possible mechanism for concentrating meteorites into "blue" ice areas. Meteorites fall into snow accumulation zones, where they are entrapped in ice that moves toward the edge of the continent. Where ice movement is impeded, for example behind nunataks, ablation exposes meteorites as layer after layer of ice is eroded by ablation.

Fig. 2. Exposed meteorite (in foreground) spotted during a helicopter traverse. (*NASA*)

Carbonaceous chondrites. The meteorites returned from Antarctica have provided a wealth of scientific data and some surprises. One interesting example is the carbonaceous chondrites, rare types of meteorites whose numbers have been significantly increased by Antarctic specimens. They are commonly considered to be the most primitive types of extraterrestrial materials because their chemical compositions so closely match that of the Sun. Because the Sun contains almost all of the mass of the solar system, carbonaceous chondrites have average solar system composition, except for the most volatile elements such as hydrogen and helium. In fact, in some ways carbonaecous chondrites may record the ancient solar system composition even better than the present-day Sun. For example, significant quantities of lithium have been consumed in the Sun by nucleosynthesis during its evolution, but carbonaceous chondrites retain the original lithium values. These meteorites are thought to have been formed at the beginning stages of the solar system. They, as well as other types of chondrites, define the age of the solar system as 4.5 billion years, as determined by measurements of various radioactive isotopes with known decay rates.

Unlike most other chondrites, carbonaceous chondrites have escaped later heating and recrystallization and thus preserve an unparalleled record of early solar system processes. Many contain white inclusions that formed at high temperatures during the birth of the solar system. These consist mostly of highly refractory minerals that formed by condensation of the solar nebula gases or as residues from partial vaporization as the nebula was heated during its collapse to form the Sun. These inclusions have exotic isotopic compositions that suggest input from a nearby exploding star in which these isotopes were created. These inclusions were incorporated along with low-temperature dust and gases to form these meteorites.

Carbonaceous chondrites take their name from the

carbon-rich organic molecules which they contain. These consist mostly of straight to slightly branching chains or rings of carbon atoms connected to hydrogen and oxygen, as well as complex amino acids. In fact, most of the organic bases in deoxyribonucleic acid (DNA) and ribonucleic acid (RNA) have been found in carbonaceous chondrites. Detailed studies of these molecules have established that they are not biologic in origin. However, their occurrence in ancient meteorites suggests that most of the raw materials for life were present in the early solar system and were not restricted to the Earth.

Antarctic samples may be particularly important in the characterization of the chemical compositions and organic compounds in chondrites, because they have been preserved in a cold, sterile environment. Modern techniques are capable of analyzing extremely low concentrations of elements or isotopes. At these low levels, many meteorites collected by standard methods and stored in museums are already contaminated. However, great care is taken to prevent contamination of Antarctic meteorites, especially carbonceous chondrites. These are collected in sterile bags and shipped in a frozen state to the Meteorite Curation Facility at the NASA Johnson Space Center, Houston. There the meteorites are thawed and processed in controlled-atmosphere cabinets formerly utilized for lunar rocks returned by the Apollo missions.

The estimated chemical compositions of planets are very similar to carbonaceous chondrites and other types of chondrites. This suggests that chondrites are probably left-over building blocks for planets which somehow escaped planetary processing. It is thought that they were never incorporated into large bodies, but instead represent fragments of small asteroids that may have orbited between Mars and Jupiter. The huge gravitational field of Jupiter prevented these asteroids from being assembled into a larger planet, enabling them to escape melting and geological processing.

Achondrites. However, it is now recognized that not all meteorites come from small asteroids. This recent discovery is an exciting result of the study of Antarctic achondrites, meteorites that have been melted. One of the most intriguing meteorites of this type is Allan Hills A81005, the first known lunar meteorite (Fig. 3). It is a breccia consisting of angular fragments of rock in a fine-grained matrix. The rock types include species such as anorthosite and basalt that constitute much of the lunar highlands and maria, respectively. Chemical and isotopic compositions of these fragments provide an unmistakable lunar fingerprint. The meteorite is similar to other lunar soil breccias from highlands areas, but minor differences apparently preclude its derivation from a location near one of the sites already sampled by Apollo or Luna missions.

Fig. 3. Allan Hills A81005, the first meteorite from the Moon. (*NASA*)

Shergottites. Several examples of a rare group of meteorites called shergottites have also been recovered in Antarctica. These igneous meteorites crystallized about 1.3 billion years ago and have been suggested to be samples from the planet Mars. Although this age is old, it is significantly younger than other types of meteorites. The source of heat on small asteroids was probably the rapid decay of short-lived radioactive isotopes, but no known internal asteroidal heat source could extend melting until 1.3 billion years ago. However, Mars had volcanic activity at that time. The chemical composition of the shergottites is similar to that measured for Martian soil by *Viking*, and one of the Antarctic samples contains unusual trapped gases similar to those that *Viking* analyzed in the Martian atmosphere. Magnetic intensity measurements indicate that these meteorites formed on a parent body with a weak to nonexistent magnetic field, similar to present-day Mars. Furthermore, geochemical characteristics suggest that the mineral garnet may occur in the source regions that melted to produce shergottites, and this high-pressure mineral is stable only in the deep interiors of planets. None of these lines of evidence proves that shergottites are Martian rocks; however, all are consistent with this idea but not with what is known about small asteroids. Nevertheless, there are serious difficulties in extracting rocks from a planetary surface. It is presumed that large impacts into the surface would be required to accelerate material to escape velocity (about 3 mi/s or 5 km/s for Mars), and numerical models indicate that this is improbable. The question of the source for shergottites will probably not be answered until returned Martian samples are available for direct comparison.

Significance. Recovery of rare or unique meteorite types is only one of the scientific values of Antarctic meteorites. Meteorites found in other parts of the Earth are generally recent falls, because they are not preserved indefinitely due to weathering. However, most Antarctic meteorites fell between 30,000 and 750,000 years ago. The residence time on Earth can be determined by measuring the remaining radioactive isotopes produced while the meteorite was bombarded by cosmic rays in space. Because asteroidal fragments whose orbits cross that of the Earth are swept up by the Earth within fairly short times, it is probable that the proportions of meteorite types in the past may have been different from the present flux. Statistics on the ancient flux recorded by Antarctic meteorites should provide information to unravel this problem.

Besides the lunar meteorite A81005, some other Antarctic meteorites are also soil breccias composed of angular fragments of rock embedded in a fine-grained matrix. Such breccias were formed on or near the surface of their asteroidal parent bodies by impacts of other meteoroids which crushed and mixed the various target rocks. These breccias preserve a record of ancient solar wind and flares in the form of gases implanted during periodic overturn of the soil. Their magnetic properties also provide evidence for an early magnetic field in the solar system. Although similar regolith breccias are found on the Moon, the meteoritic breccias are older and extend the historical record of the Sun's evolution.

Finally, Antarctic meteorites have scientific value that is unrelated to their extraterrestrial origin. They can also provide information on geological processes in Antarctica, although this application is just beginning. Recent field parties have begun to collect ice samples associated with newly exposed meteorites, and analyses of both kinds of samples may place important constraints on the three-dimensional paths of moving ice sheets. The distribution of meteorite concentration fields in Antarctica, when they are more completely known, will also be useful for reconstructing continental ice flow and ablation patterns. Mineralogical and chemical studies of the alteration experienced by a few Antarctic meteorites will be an aid to understanding weathering in polar environments.

Antarctic meteorites are clearly an important new scientific resource for a number of reasons. Samples of unique or rare meteorites provide new kinds of extraterrestrial materials for laboratory studies. The long terrestrial residence times of Antarctic meteorites allow reconstruction of the relative abundances of objects in Earth-crossing orbits in the past. The samples are useful for precise chemical and isotopic analyses because they were preserved in an uncontaminated state. The meteorites also act as probes of Antarctic processes. Their value will increase in future years as additional scientific questions are formulated. In the meantime, this well-documented collection will continue to expand as more meteorite concentrations are discovered on the Antarctic ice.

For background information *see* ANTARCTICA; MARS; METEORITE; MOON; SOLAR SYSTEM in the McGraw-Hill Encyclopedia of Science and Technology. [HARRY Y. MCSWEEN, JR.]

Bibliography: Antarctic Meteorite Working Group, *Antarctic Meteorites: An International Resource for Scientific Research*, Lunar and Planetary Institute, 1981; W. A. Cassidy and L. Rancitelli, Antarctic meteorites, *Amer. Sci.* 70:156–164, 1982; H. Y. McSween, Jr., Are carbonaceous chondrites primitive or processed: A review, *Rev. Geophys. Space Phys.*, 17:1059–1078, 1979.

Micrographics

Micrographics is a technology that is widely used to store and retrieve document images and information. Developments that link micrographics equipment with computer technology have made micrographics an important information management tool.

Organizations that deal with large quantities of information are confronted with the need to retain massive volumes of paper documents. Despite widespread application of electronic office equipment and the rapid growth of electronically generated data, paper documents continue to grow in volume and value. Typically they contain far more information than users convert to digital form, largely because of the cost of data entry and storage. The source document often remains the repository of this additional information. Micrographics technology provides a cost-effective alternative to paper for storing and retrieving these needed documents.

Micrographics encompasses the technology associated with the production, handling, and use of microforms. A microform contains one or more microimages that are too small to read without magnification. These microimages are usually the images of documents, pages of text, or other information. The microforms themselves are typically film, either 16-mm roll microfilm or 105-mm microfiche "cards."

Roll microfilm lends itself to large-volume applications. These applications are best served by central look-up stations equipped with automated microimage retrieval terminals. Microfiche is commonly used for reports and information that receive widespread distribution in an organization. Here the information usually is read via manually operated fiche readers.

Either format offers users the ability to capture and store substantial amounts of information at low cost. A 4 × 4 × 1 in. (10 × 10 × 2.5 cm) magazine of 16-mm roll microfilm can hold the images of as many as 25,000 8½ × 11 in. (22 × 28 cm) documents.

Microfilming. There are two basic methods of placing microimages on microfilm. The first is a photographic method, commonly used to capture source document images, in which devices called microfilmers are used to photograph documents. Microfilmers range from microprocessor-controlled high-speed rotary devices to manually operated planetary cameras. Some high-speed devices can microfilm up to 225 letter-size documents per minute.

The second method employs a device called a computer output microfilmer, which directly generates computer output into either roll microfilm or microfiche. Typically, these devices are used to generate microimages of documents captured by a microfilm camera.

Document retrieval. Microfilm was once the tool of the archivist: old records, seldom needed again, were converted to microfilm for long-term storage.

On those rare occasions when they were required, an individual had to laboriously search through a roll of microfilm to find the desired images.

Today, however, microfilm is most often used with active files in a wide range of organizations. This change has been made possible by the linking of micrographics equipment with computer technology. With this combination, users today can rapidly and automatically find a desired document image among millions in a matter of seconds.

One of the fastest-growing uses of micrographics technology is for the computer-assisted retrieval of source documents. Incoming documents are photographed by a high-speed microfilmer. At the same time, the microfilmer automatically encodes each document and its image with a unique identification number and an image mark. This identification number becomes the address for each image in the microimage file.

The microimage is transferred into the microimage file, and the address number is entered into the organization's computer system, along with other key data. Thus the computer file becomes the index for source documents, which are retained on microfilm. The cathode-ray tube displays the microimage file address number for the document image, along with other key data; the cathode-ray-tube display tells the operator which magazine to select. The operator then loads the appropriate microfilm magazine into an intelligent retrieval terminal. Depending on how the system is configured, either the computer automatically directs the retrieval terminal to the right image, or the operator manually keys in the address number, enabling the retrieval terminal to locate the correct document image.

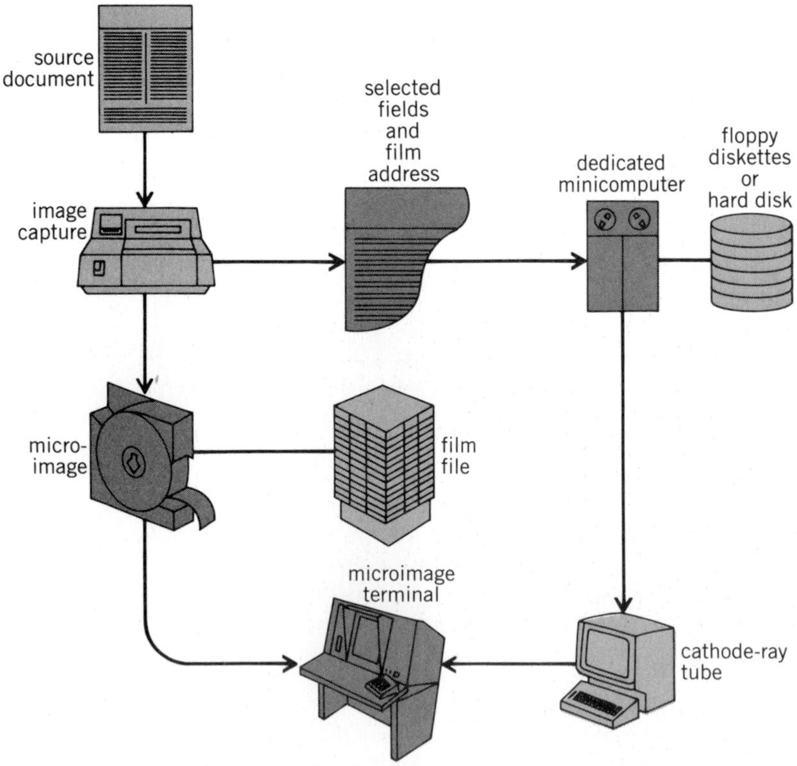

Fig. 1. Components of a stand-alone minicomputer-based microfilm system.

Computer-assisted retrieval began to proliferate some years ago as many organizations linked micrographics equipment to their large mainframe computers in the data-processing operation. However, in recent years the micrographics industry has paralleled an overall trend in the data-processing industry and has turned to minicomputers.

Many departments within larger organizations today employ minicomputers and microcomputers to meet various data-processing and information-handling needs. They have linked micrographics equipment to minicomputer systems to facilitate the storage and retrieval of paper documents. While the use of micrographics equipment with mainframe computers is still common, the application of this technology with minicomputers points to expanded use of micrographics in the coming years.

In fact, demand for minicomputer-based microfilm systems has been so sharp that a number of firms have offered total information systems as complete packages (Fig. 1). These systems are designed specifically for computer-assisted retrieval, complete with all the necessary hardware and software. The image is stored on microfilm in a microimage file, while selected key data and the microimage file address are entered into a minicomputer for magnetic storage. When the image is needed, an operator calls up the key data on a cathode-ray-tube screen at a computer terminal. The computer then directs an intelligent microimage terminal to the appropriate document image, which it displays on its viewing screen.

Users of computer-assisted retrieval generally find several major benefits. First, they eliminate the burden of retaining paper documents and can more efficiently manage the storage of source documents through microfilm. Second, they have a computerized index that retains the precise location of each image within the file. And third, they have rapid, automatic retrieval of any document in the system.

Application of computer-assisted retrieval include accounts payable and receivable, personnel records, customer credit charge slips, insurance claims processing, check processing, loan files, and correspondence, among others.

Computer output microfilm. Another popular use of micrographics technology involves computer output microfilm. Here devices called computer output microfilmers directly generate computer data onto either roll microfilm or microfiche.

Computer output microfilm devices typically print alphanumeric data onto microfilm, much as paper printers generate computer reports on paper. They are large high-speed systems designed for high-volume job streams; some can print up to 10,000 pages of output per hour. Also, some specialized computer output microfilm units can record line art and computer graphics onto microfilm.

The more advanced computer output microfilm units use a laser to literally write the computer data directly onto microfilm. The special film requires dry heat rather than chemical processing. The computer output microfilm actually processes its own film after exposure by the laser, delivering ready-to-use microfilm or microfiche. In addition, these units contain built-in minicomputers and operate on-line to most mainframe computers. In this manner, they function as true computer peripherals.

But even with the growing acceptance of microfilm, trends indicate that organizations must cope with an increasing paper flow. Every working day, for example, United States business alone generates an estimated 600 million pages of computer output, another 235 million photocopies, and 76 million letters. This is added to the estimated 21 trillion pages of paper already stored in file drawers across the country. To help manage this volume, a number of new technologies are being considered.

Microimage transmission. The next step for micrographics technology may be a concept called microimage transmission. Basically, this is the ability to electronically scan a photographic image stored on microfilm, convert that image to a digital signal, transmit that image to a remote location, and then display it on a video screen.

The present generation of computer-assisted retrieval microfilm systems cannot transmit images. To use the system, an individual must be at the microimage retrieval terminal. But with microimage transmission, users at remote locations would be able to access the central microfilm files and view images on their own video screens. The central file itself would be automated, with a robot selecting the appropriate microfilm magazine on demand and inserting it in a scanner. A computer would then select the appropriate image to be scanned, transmitted, and displayed.

The heart of any microimage transmission system would be the computer address index for individual images, much like the index already in place with current computer-assisted retrieval microfilm systems.

Optical disk. Beyond microimage transmission, optical disk technology may be a mass-storage media for images and information. These systems may be interactive and compatible with microimage transmission systems, giving users the ability to call up individual images that are stored on either microfilm or optical disks.

This total information system would operate much like present magnetic-disk-based computer systems. Here digitized information is encoded on a magnetic disk. With an optical-disk-base system, a light-sensitive disk would be optically encoded with digitized information.

In use, a document would be electronically scanned. The image would be digitized, and a laser would be employed to encode a special disk with the digital information. The laser would burn microscopic pits into the disk's optical surface, which is sensitive to certain wavelengths of light. To retrieve the image, a different-wavelength laser would read

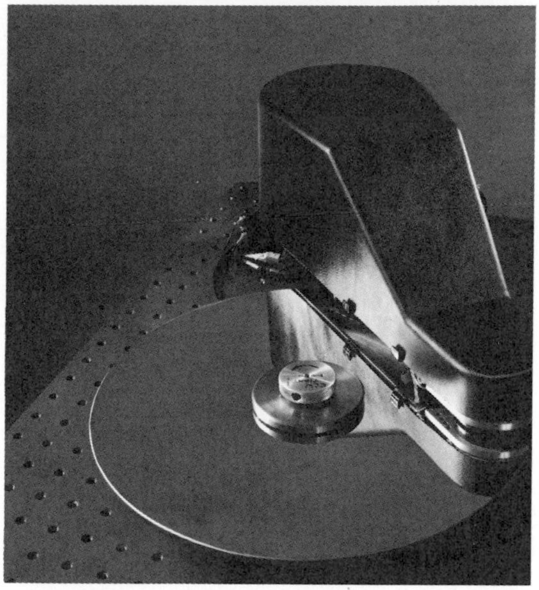

Fig. 2. Prototype optical disk system incorporating a laser record-read device. (*Eastman Kodak Co.*)

the encoded information from the disk and transmit the information to a high-resolution video display terminal for viewing. Research indicates that a single optical disk (Fig. 2), about the size of a long-playing phonograph record, could store more than 100,000 pages of information. Thus, a single disk might offer the mass storage potential of 50 magnetic tapes or 2000 floppy disks.

Because both optical disk and microimage transmission call for electronic scanning and transmission of images, it is assumed that they will be compatible and interactive. Indeed, they will probably even interact with magnetic storage systems, giving users the ability to manage document images and information on a total-systems basis. Both image and alphanumeric information will be stored, accessed, and retrieved by a system that blends the use of microfilm, magnetic storage, and optical disk.

For background information *see* COMPUTER; COMPUTER STORAGE TECHNOLOGY; DATA-PROCESSING SYSTEMS; MICROCOMPUTER; MICROFILMING; VIDEO DISK RECORDING in the McGraw-Hill Encyclopedia of Science and Technology. [JOHN A. LACY]

Bibliography: J. J. Hurley and D. Massengill, Micrographics and minicomputers, *Mini-Micro Systems*, 15(12):238 244, December 1983; M. Obenzinger, *Approaches to Information Storage: Micrographics Equipment and Magnetic Equipment and Media*, Lehman Brothers, Kuhn Loeb Research, 1982; L. J. Thomas, Future storage technologies for the information industry, *IMC J.*, 18(4):21–23, Fourth Quarter 1982.

Microholography

Physicists and life scientists around the world are now engaged in research that will ultimately allow three-dimensional imaging of living organisms with resolution and contrast far beyond the reach of optical microscopes. The impetus for this activity is the imminent availability of high-intensity coherent sources of electromagnetic radiation with wavelengths between 0.1 and 10 nanometers. Much of the study is concentrated on holographic imaging because it can eliminate the need for focusing elements which are difficult to fabricate with enough precision to achieve diffraction-limited resolution in the soft x-ray regime. Furthermore, several of these new sources promise extremely high intensity and subnanosecond pulses, and can circumvent the problem of killing and altering the specimen with the x-ray exposure by extracting an image from the specimen before it is obliterated.

X-ray sources. To be suitable for holography, the x-radiation must be monochromatic and have a relatively high degree of coherence—the property that enables two waves to interfere, constructively or destructively. Electrons bombarding heavy-element targets in conventional x-ray tubes generate bremsstrahlung, which is not monochromatic, and characteristic line radiation, which, although quasi-monochromatic, has insufficient coherence for holography. Synchrotrons using magnetic undulators can generate a narrow band of radiation at an intensity much higher than that obtained by x-ray tubes. Use of monochrometers and pinhole apertures can improve the coherence of this radiation at the sacrifice of intensity but with retention of sufficient intensity to image biological specimens on time scales from a few seconds to a few hours. X-ray sources driven by high-energy, short-pulse laser systems or gas-puff Z pinches can produce pulses that are short enough to preserve the biological integrity of a specimen during exposure. Although these sources do not produce radiation with enough coherence for holography, they will be very useful for shadowgraphy and conventional forms of x-ray microscopy.

The most promising sources are under development. Nonlinear optical frequency multiplication techniques produce intense picosecond pulses of tunable coherent radiation, and have reached wavelengths as short as 40 nm. Similarly, multiphoton excitation can "pump" atoms to higher energy levels that have lasing transitions at wavelengths much shorter than the excitor laser. Large cross sections for multiphoton excitation have been demonstrated for quantum energies as high as the ninety-ninth harmonic of a 6-eV excitor laser. X-ray lasers

<table>
<tr><th>Type</th><th>On-axis Fresnel transform (Gabor holography)</th><th>Off-axis Fresnel transform (Leith-Upatnieks holography)</th><th>Planar Fourier transform (Stroke holograpy)</th></tr>
<tr><td>Geometric configuration</td><td>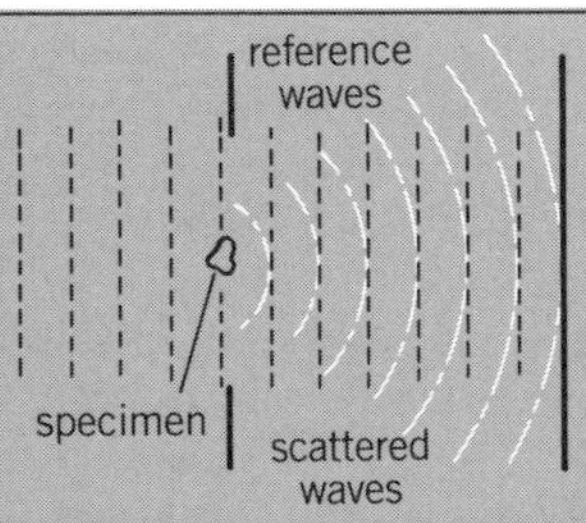
</td><td>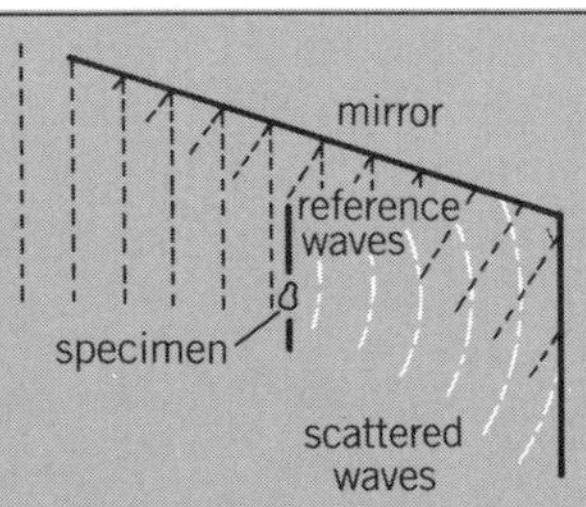
</td><td>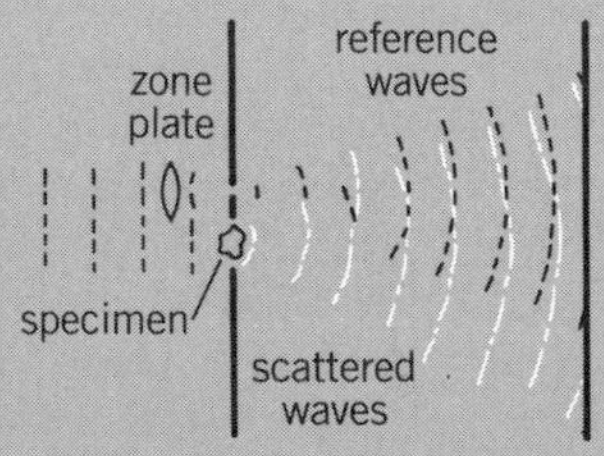
</td></tr>
<tr><td>Coherence requirements</td><td></td><td></td><td></td></tr>
<tr><td>spatial</td><td>$2\frac{b\lambda}{\delta\mu}$</td><td>$3\frac{b\lambda}{\delta\mu}$</td><td>$2D$ (not less than $\frac{3}{2}D$)</td></tr>
<tr><td>temporal</td><td>$\frac{b}{1-\xi}(\mu^{-1}-1)$</td><td>$\frac{2}{\pi}\frac{b}{1-\xi}(\mu^{-1}-1)$</td><td>$\frac{2D\lambda}{\pi\delta(1-\xi)}$</td></tr>
<tr><td>Maximum specimen volume</td><td></td><td></td><td></td></tr>
<tr><td>coherence limit</td><td rowspan="2">$\pi\frac{\delta^2\lambda(1-\xi)^2x^2}{\Delta^2(\mu-1)^2}$</td><td rowspan="2">$\frac{\pi^3}{4}\frac{\delta^2\lambda(1-\xi)^2x^2}{\Delta^2(\mu-1)^2}$</td><td>$\frac{\pi^4}{48}\frac{x^3\delta^2}{\lambda^2}(1-\xi)^3$</td></tr>
<tr><td>mapping limit</td><td>$(\mu^{-2}-1)\frac{\pi b^2\delta^4}{\Delta^2\lambda}$</td></tr>
</table>

Fig. 1. Geometry, coherence requirements, and maximum volume for microholographic techniques discussed in the text. $\mu = [1 - (\lambda/\delta)^2]^{1/2}$.

driven by nuclear explosives and by more conventional laboratory sources are under development. Stimulated emission has been reported for several x-ray transitions. The formidable obstacles to the realization of gamma-ray lasers are slowly being overcome. X-ray and gamma-ray lasers will be inherently short-pulse, high-intensity devices because: they will probably not have resonant cavities, so the radiation being amplified can make only a single passage through the active medium; and the creation and maintenance of a high density of excited atomic states of short lifetime and high quantum energy require enormous power, which terrestrial sources can supply only in the form of pulses.

Geometries. Holography uses wave interference to register information on both the amplitude and phase of a monochromatic wave at a recording surface. Once recorded, this information can be used to reconstruct the configurations of specimen features that contributed to the structure of the registered wave. Such wavefront reconstruction is accomplished either by coherent illumination of the hologram or by an equivalent computer analysis. Reconstruction gives rise to real and virtual images with magnifications determined by the ratio of the wavelengths of the original and the reconstructing illuminations.

Essential features of the principal geometries for holography are shown in Fig. 1. The Fresnel transform techniques use planar reference waves and have resolution limited by the grain size of the recording medium. The on-axis (Gabor) form is inherently simple but suffers from overlap of the real and virtual images. The off-axis (Leith-Upatnieks) modification reduces the image overlap problem but requires a mirror and a broadened beam for system illumination; both may be difficult at x-ray wavelengths. The Fourier transform (Stroke) geometries, using curved wavefronts, achieve large fringe spacings and are therefore less sensitive to grain size.

Coherence is characterized by the effective finite length of a photon wave train in the direction of its travel (temporal) and in the transverse direction (spatial). Two overlapping wave trains produce interference fringes or photon noise, depending on the extent of their overlap and on the constancy of their respective phases. Figure 1 shows the spatial and temporal coherence requirements as a function of wavelength (λ), the desired resolution (δ), the shortest distance from the specimen to the recording surface (b), the transverse dimension of the specimen (D), and the required signal-to-noise ratio (ξ).

Both coherence length and geometry limit the holographable volume of a specimen. Figure 1 shows the maximum volume, which depends on temporal coherence length (x) and recording medium resolution (Δ). The term "mapping limit" refers to the fact that a hologram is a homomorphic mapping of specimen volume onto the surface of the recording medium. For most specimens of biological interest, spatial and temporal coherence lengths of 10^{-3} to 10^{-1} cm are adequate.

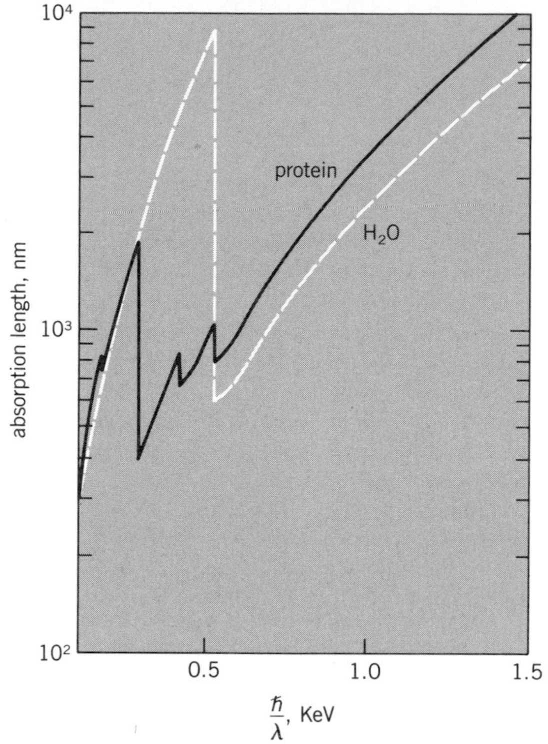

Fig. 2. Absorption lengths (distance over which x-ray intensity is attenuated by a factor of e ≅ 2.72) for protein and water.

Interactions of x-radiation. The interaction of x-radiation with matter is quite different from the interaction of visible light with matter. Whereas the extinction of a visible beam traversing matter is mainly due to scattering, the extinction of an x-ray beam is mainly due to absorption. X-rays can also be scattered, but usually the cross section for coherent scattering is very much smaller than for absorption. In the visible regime, holographic images are primarily formed by refraction or reflection, whereas in the x-ray regime they are dominated by diffraction. The greatest contrast in x-ray absorption between water, which composes most of the cytoplasm, and protein (or the nucleic acids) occurs between the *K* edges of oxygen and nitrogen. Figure 2 shows the absorption lengths for protein and water as a function of x-ray quantum energy. In Gabor holography, image overlap obscuration can be mitigated by requiring $b\lambda \gg D^2$, but in typical cases this implies extraordinary temporal coherence length. If the specimen has high contrast, this requirement is reduced as D can be replaced by the transverse dimension of the specimen's largest opaque feature.

Snapshot microholography. Existing x-ray sources, in particular, synchrotron radiation sources, have been used to make holograms. However, they require long exposures, limiting their usefulness for research on living specimens. More coherent sources may also be developed, but those of low intensity will be similarly limited, since ionization will have decomposed molecules, modified compositions, and altered biological functions before enough radiation can be received to form a useful hologram. Snapshots are essential for microholography of living specimens. Fortunately, it is likely that x-ray sources producing brief intense bursts will be developed in the near future.

With an intense pulsed coherent source (such as an x-ray laser), hydrodynamic expansion, initiated by sudden heating, rather than normal biological activity, chemical change, or thermal agitation, will limit the time during which recording of the hologram must be accomplished. Analytical expressions for the explosion of a semiopaque feature (such as a protein globule) are useful for estimating the radiation requirements for typical cases. They are based on the criterion that, to achieve a linear resolution δ, a specified minimum number of photons must have been coherently scattered in a volume δ^3 and that, during the exposure time Δt, no dimension of the specimen should have increased by more than δ.

To examine a feature within a specimen with resolution $\delta \ll d$, where d is the feature's transverse dimension, the maximum exposure time is shown in Eq. (1), where m is the mass of the feature, Σ_a is

$$\Delta t \cong \left[\frac{\delta^2 m}{FI\Sigma_a}\right]^{1/3} \qquad (1)$$

its absorption cross section, I is the x-ray intensity, and the dimensionless coefficient F is a function of the equation-of-state parameters and the rate at which transport processes remove energy from the exploding feature. Longer exposures result in blurring; that is, the resolution achieved will correspond to a larger value of δ. If, on the other hand, $\delta \gg d$ is acceptable, the exposure time given by Eq. (1) should be multiplied by the factor δ/d, after substituting a different dimensionless coefficient G. If the x-ray burst duration exceeds the value found above, some shutter mechanism must be provided for the incident radiation, or a time gate must be provided for recording.

The number of photons elastically scattered from a single resolution element and registered in the hologram during the maximum exposure time is given by Eq. (2), where ρ is density, Σ_e is the fea-

$$N \cong \frac{\epsilon\lambda}{\hbar m}\rho I\Sigma_e\delta^3\Delta t \qquad (2)$$

ture's elastic cross section, and ϵ is the quantum efficiency of the recording medium. This number must be statistically significant compared to the background photon noise. Combining Eqs. (1) and (2) yields Eqs. (3), giving the intensity required to

$$I = m\sqrt{\left(\frac{N\hbar}{\Sigma_e\rho\lambda\epsilon}\right)^3\frac{\Sigma_a}{\delta^{11}}} \times \begin{cases}\sqrt{F}, & \delta < 3d/8 \quad (3a)\\ \sqrt{G(d/\delta)^3}, & \delta \geq 3d/8 \quad (3b)\end{cases}$$

obtain a specified resolution in an exposure for the maximum time before unacceptable hydrodynamic blurring can occur. Note the strong (5.5- and 7-power) dependences of I on the linear resolution δ. To improve resolution by a factor of 2, one must increase intensity by two orders of magnitude, because the statistically significant number of useful photons would then have to be scattered in a shorter time by a smaller-volume element. For most biological specimens, intensities on the order of 10^{12} W · cm^{-2} with pulse lengths on the order of 10^{-11} s will be required to obtain a microhologram with resolution of 10 nm.

Recording. An x-ray hologram can be registered by radiation-induced prompt or latent chemical change, or by photoelectron emission. Photographic emulsion is unsatisfactory for Fresnel transform microholography as the resolution is limited by grain size—one would prefer a medium with grain size limited only by its atomic or molecular structure. If an electron microscope could be used to image the points of electron emission from a photocathode reference surface, time-gated holography might be possible. However, the continuous distributions in energy and in angles of emission of electrons from a photocathode preclude the formation of sharp electron-optical images, imposing a trade-off between quantum efficiency and resolution, unless image-deblurring analysis can be applied. Photoresists (materials that lose resistance to chemical etching at points exposed to radiation) have grain sizes that approach 5 nm, which is entirely adequate for microholograms with resolutions of 10 nm. To reconstruct a photoresist hologram, one could make a transmission electron micrograph, which is viewed with visible laser illumination, or one could use a transmission electron microscope to scan and digitize the photoresist for analysis by computation, which can also mitigate nonlinearities that may be troublesome in optical reconstruction.

By using Fourier transform microholography, it is possible to arbitrarily adjust the fringe spacing at the sacrifice of intensity and thereby record with common photographic emulsions. However, when compared with photoresists on the basis of number of quanta required to produce a developable speck, it is not clear that the greater sensitivity of photographic emulsion offers any advantage, and consequently there is no clear advantage of Fourier over Fresnel methods.

Practical considerations. The realization of microholography as a practical research tool still awaits the solution of some challenging technical problems:

1. Development of sources that can generate intense coherent radiation at the precise wavelengths to optimize contrast among specimen constituents. Perhaps nonlinear mixing with tunable visible radiation will be necessary.

2. Termination of exposure at the maximum time demanded by Eq. (1) with the corresponding minimum intensity given by Eqs. (3). Frequency multiplication techniques and multiphoton excitation la-

sers can achieve these short pulses because the optical laser driving them can be mode-locked. Corresponding schemes are difficult to envisage for x-ray or gamma-ray lasers, and their pulse lengths are likely to be much longer. A shutter or gate, somewhere in the system, that operates when full intensity is reached will be essential.

3. In principle, photoelectric recording could be time-gated. However, complexity and precision required of the electronics, and blurring associated with initial electron velocity distribution make this approach unattractive. On the other hand, exposure control in photoresist recording is not likely to be managed by a gate; therefore exposure control must be provided elsewhere in the system.

4. Leith-Upatnieks holography may be necessary to avoid image overlap obscuration, and this requires an x-ray mirror. A synthetic Bragg crystal may suffice, and thermal expansion, if sufficiently uniform, can provide automatically time-gated reflection.

For background information *see* HOLOGRAPHY; LASER in the McGraw-Hill Encyclopedia of Science and Technology.

[JOHNDALE C. SOLEM]

Bibliography: G. Baldwin, J. Solem, and V. Gol'danskii, *Rev. Mod. Phys.*, 53:644, 1981; H. Egger, H. Pummer, and C. Rhodes, Excimer lasers for the generation of extreme ultraviolet radiation, *Laser Focus*, 18–6:59, 1982; J. Kirz and D. Sayre, Soft x-ray microscopy of biological specimens, in H. Winick and S. Doniach (eds.), *Synchrotron Radiation Research*, 1981; J. Solem and C. Baldwin, Microholography of living organisms, *Science*, 218:229–235, 1982.

Mineral

Detailed knowledge of long-range atomic ordering in minerals is essential to an understanding of the relationships between mineral composition, structural state, physical properties, and ambient conditions of crystallization and equilibration. The experimental techniques whereby such ordering is characterized have proliferated in recent years. Traditional methods have undergone considerable development and are now capable of producing results of high precision and accuracy, while new methods have been developed to deal with those problems that were not solvable by conventional means.

All methods used for characterization of ordering can be divided into two categories, fully quantitative and qualitative. In some cases the qualitative methods (infrared and Raman spectroscopy, electronic absorption spectroscopy, conventional nuclear magnetic resonance spectroscopy, and electron paramagnetic resonance spectroscopy) merely need some technical development to become fully quantitative, but in other cases there are fundamental reasons that restrict the method to qualitative or at best semiquantitative results.

Single-crystal diffraction. In general, the chemical composition of a crystal used in a diffraction experiment is known, and thus the scattering factors are not generally considered as variable once the atoms have been correctly identified. The exception to this occurs when atoms are disordered over more than one crystallographically unique site in the mineral, when the scattering factors at those sites are composite averages of all the atoms occupying these sites in the crystal. This average scattering factor is determined by the diffraction experiment, and the problem is then to quantitatively determine the amounts of those atoms giving rise to this scattering. For binary site-occupancies, this has long been done by incorporating expressions for the site-occupancies directly into the least-squares structure refinement and using the known (externally determined) composition of the mineral as a constraint in the refinement procedure. Improvements in crystallographic technique have obviated the need for such a procedure, although it should be noted that "soft" constraints of a similar sort might lead to improved precision.

For binary order-disorder, diffraction techniques can determine both individual site-occupancies and total crystal composition, provided that significant differences exist between the scattering powers of the atomic species involved in the order-disorder. If the disorder involves more than two scattering species, individual site-occupancies cannot be determined from a single diffraction experiment, unless external information can be included in the analysis. This can be either partial site-occupancy data from a different experiment or the inclusion of an additional constraint equation based on the crystal-chemical characteristics of that specific structure type (for example, charge-balance constraint or mean-bond-length constraints).

Mössbauer spectroscopy. The Mössbauer effect is the recoil-free emission and absorption of gamma rays by a specific atomic nucleus. There are 30 or so isotopes that are sensitive to this effect, but nearly all quantitative site-occupancy studies in minerals have involved ^{57}Fe. Subject to the assumption that the recoil-free fraction of the absorber at energetically similar sites is the same, site-occupancies may be derived from a comparison of the relative intensities of the lines in the absorption spectrum, together with an externally derived chemical analysis. For most problems of interest in mineralogy, the Mössbauer spectra consist of a complex overlap of peaks that reflect differential occupancy of more than one crystallographic site. The precision with which site-occupancies can be determined is strongly a function of the degree of overlap of the constituent lines in the spectrum. Use of half-width and area constraints can alleviate this problem to some extent, but often such constraints are difficult to justify, and spectral complexity limits the application of this technique to simpler minerals. However, in such cases it is capable of quite high precision, and comparison with analogous diffraction results suggests that both methods are accurate to within the limits of the precision.

Nuclear magnetic resonance. In minerals, dipolar interaction between nuclear spins and anisotropy of the ^{29}Si chemical shifts leads to very wide nuclear magnetic resonance line widths and to weak, completely overlapping lines if conventional nuclear magnetic resonance methods are used. Rapid sample spinning at the magic angle ($3 \cos^2 \theta = 1$) to the external magnetic field gives greatly reduced line widths (~200 ppm → 0.5–3 ppm); combined with a high magnetic field, this leads to high resolution and good sensitivity, thus providing a direct probe for ^{29}Si in different environments. Initial studies show isotropic ^{29}Si chemical shifts to be strongly dependent upon structure type, varying from ~70 ppm in orthosilicates to ~110 ppm in tectosilicates (quoted relative to Me_4Si), a result of progressive increased diamagnetic shielding. Most subsequent work has concentrated on zeolites, framework aluminosilicates in which substitution of aluminium for silicon results in paramagnetic deshielding and a progressive decrease in ^{29}Si chemical shift from ~110 ppm for Si with four Si neighbors to ~85 ppm for Si with four Al neighbors. This has proved a powerful method for directly characterizing Al-Si order-disorder relationships in framework aluminosilicates, an area that has resisted investigation by other techniques. This has been further strengthened by the introduction of ^{27}Al magic angle spinning nuclear magnetic resonance, which provides complementary information on the coordination and distribution of aluminum.

Rietveld method. Many minerals of interest do not occur in crystals large enough to examine by conventional single-crystal diffraction techniques. Until recently, the structural characterization of such minerals (or synthetic materials) was ignored. However, the importance of microcrystalline minerals to such diverse topics as nuclear waste disposal, tooth enamel, and industrial catalysts has promoted the development of a diffraction method to handle these materials. The Rietveld method uses all of the information in a powder diffraction pattern (x-ray or neutron) to characterize the structure of the material examined. The structural parameters of the mineral, atomic coordinates, site-occupancies, and thermal parameters, together with various experimental parameters affecting the pattern, are refined by least-squares procedures to minimize the difference between the entire calculated and observed powder patterns. The method is superior for neutron diffraction as compared with x-ray diffraction because neutron scattering does not fall off rapidly with scattering angle as does x-ray scattering, and because the diffraction intensity profile is more easily described mathematically in the neutron case. Nevertheless, extremely useful results on microcrystalline material have recently been obtained by x-ray diffraction.

There is obviously a lot less information contained in a powder diffraction pattern than in a complete single-crystal diffraction data set. In recognition of this, initial work with the Rietveld technique tended to concentrate on high-symmetry structures with small unit cells. However, this has gradually been extended to more complex lower-symmetry structures in which order-disorder effects can be of importance. In more complex cases, information from external sources can be incorporated into the refinement procedure, allowing the derivation of information that could not otherwise be determined. A straightforward example of this concerns the incorporation of typical fixed temperature factors into a refinement when site-occupancy refinement is being carried out; more complex examples could include the incorporation of "soft" bond-length constraints derived from distance–least-squares refinement modeling.

Atom location by channeling-enhanced microanalysis. An electron beam, incident on a crystal oriented to produce Bragg diffraction, undergoes elastic scattering that causes a modulation of the electric current over the unit cell of the crystal. The variation in electric current across the unit cell results from interference between the incident and Bragg-diffracted beams within the crystal, and hence the modulation is particularly strong when the crystal is oriented so as to produce a strong low-order reflection. Atoms lying at points of high current density will be excited more than atoms lying at points of low current density. At particular crystallographic orientations of the sample, the resulting pattern of modulated intensity may have a maximum at a particular site on the unit cell, thus giving rise to an anomalously high production of characteristic x-rays. This effect may thus be used to derive site-occupancy information in single crystals of extremely small dimensions. For the crystal structure of interest, the sites to be characterized must lie on alternate crystallographic planes, and the projected charge density on these planes must be significantly different; these restrictions ensure that the majority of the current passes between the atoms of the crystal (channeling) and that the electron current can be made to have minima and maxima at the sites of interest. Also, there must be one atomic species that lies solely on one of the planes to act as a reference element for quantitative analysis; in principle, this is not necessary as the intensity distribution of the electron wave field within the unit cell can be calculated, but the experimental parameters for this approach are not easily determined, and the empirical approach is far more convenient. Used in conjunction with energy dispersive spectroscopy, it provides a rapid and simple method for site-occupancy characterization.

Additional useful information may be obtained by electron energy loss spectroscopy on the transmitted electron beam to derive valence states of ions at specific sites. By using localization enhancement such that only ionization events due to electrons passing within ~0.1 nanometer of the nucleus are registered, valence differences can be characterized by the chemical shift observed in the energy loss spectra.

Conclusion. Of all the techniques outlined above, single-crystal diffraction is the most general and provides the background information which allows the results of the other methods to be interpreted. However, the specific sensitivity of the other methods makes them a natural counterpart to both diffraction methods and other spectroscopic techniques, and the different methods should be regarded as complementary rather than competitive.

For background information *see* MINERAL; MÖSSBAUER EFFECT; NEUTRON DIFFRACTION; NUCLEAR MAGNETIC RESONANCE (NMR); X-RAY DIFFRACTION in the McGraw-Hill Encyclopedia of Science and Technology. [F. C. HAWTHORNE]

Bibliography: F. C. Hawthorne, Quantitative characterizaton of site occupancies in minerals, *Amer. Mineral.*, 68:287–306, 1983; E. Lippmaa et al., Structural studies of silicates by solid-state high-resolution ^{29}Si NMR, *J. Amer. Chem. Soc.*, 102:4889–4893, 1980; J. Taftø and Z. Liliental, Studies of the cation atom distribution in $ZnCr_x Fe_{2-x}O_4$ spinels using the channeling effect in electron-induced x-ray emission, *J. Appl. Cryst.*, 15:260–265, 1982; R. A. Young and D. B. Wiles, Application of the Rietveld method for structure refinement with powder diffraction data, *Adv. X-ray Anal.*, 24:1–23, 1981.

Mobile radio

In cellular mobile radio service the service area is subdivided into multiple cells or zones, each spanning no more than approximately 10 mi (16 km) from center to edge. The radio channel frequencies assigned to each cell differ from those assigned to any adjacent cell, but cells separated by a sufficient number of intervening cells may use the same frequencies. Cellular systems serve both portable units and mobile units installed in vehicles.

Origins. A proposal for a cellular system existed as early as 1947. During the 1960s, the overcrowding of mobile radio channels at 150 and 450 MHz in many of the more densely populated areas of the United States revealed a demand for mobile communications that exceeded the capacity of existing spectrum allocations and existing methods of providing service. The cellular approach offered an attractive method for meeting this demand.

The 1970s saw a dramatic increase in efforts by industry to develop practical cellular systems for public telecommunications use. In 1978 the Bell System began serving subscribers in the Chicago area on the nation's first operational cellular system. Bell Laboratories designed this developmental system, known as Advanced Mobile Phone Service (AMPS), and Western Electric manufactured most of the land-based equipment. In 1981 the Federal Communications Commission issued regulations governing the Domestic Public Cellular Radio Telecommunications Service. In 1983 Ameritech Mobile Communications Inc., established the first commercial cellular system in the United States. Located in Chicago, this system was distinct from the earlier developmental system and incorporated advances.

By 1983 cellular systems were operational in Australia, Bahrain, Denmark, Finland, Hong Kong, Japan, Norway, Qatar, Saudi Arabia, Singapore, Spain, Sweden, and the United Arab Emirates.

Fundamental properties. Certain basic properties characterize a cellular system. The system must contain more than one base station (also called land station or cell site). Base stations constitute the physical reality of a cellular system, whereas cells themselves are a conceptual convenience, which can be defined in several ways. A pragmatic definition of a cell is the geographical area in which a specific base station is the most likely one to serve calls. A cell in the interior of a service area can be defined in a different manner as the territory closer to one particular base station than to any others.

The ability to grow in terms of serving an expanded area or serving more subscribers within a constant area is another essential property of cellular systems. Conceptually, expansion of the service area implies adding more cells around the periphery. In practice, expansion is achieved by installing additional base stations in areas outside the previously existing service area. Growth in terms of serving more subscribers within a constant area may require cell splitting. Conceptually, cell splitting means that an area previously treated as a single cell is subdivided into several smaller cells. In practice, cell splitting is accomplished by installing additional base stations within the existing service area. Cell splitting enables a system to serve a higher traffic density (expressed in units such as calls per unit area), because a base station's entire complement of channels, which previously served a larger cell area, can now be devoted to a smaller cell area. The number of successive stages of cell splitting is limited only by practical difficulties of obtaining base station sites sufficiently closed to intended cell centers. A cellular system can adapt to variations in traffic density from one part of the service area to another by splitting cells in each part only to the extent that the traffic density warrants.

Frequency reuse gives cellular systems the ability to use spectrum far more efficiently than earlier types of mobile radio systems. Frequency reuse refers to the simultaneous use of the same frequency to serve a call at each of several base stations in different parts of a service area.

Associated capabilities. Cellular systems require certain supporting capabilities in order to realize the benefits offered by the fundamental properties discussed above. A vital capability is the transferring of the communication path used for a particular call from one base station to another as the mobile unit moves within the service area. In many systems, at least some of the base stations are equipped with multiple transmitter-receiver antennas which serve multiple regions that differ from one another in range or azimuthal direction or both. When such a base station serves a call, it sometimes transfers the call from one of its antennas to another. The transfer of a call to a different antenna or base station is commonly called a handoff.

In most cellular systems, the assignment of channel frequencies to base stations is static in the sense that a particular base station serves calls only on a fixed subset of the allocated frequencies. A completely dynamic channel assignment scheme would permit a given base station to use every frequency some of the time. In a system that employs static channel assignment, a handoff always requires that the mobile unit change from one frequency to another while the system reroutes the communication path through its facilities. A system employing some degree of dynamic channel assignment would sometimes reroute the communication path without requiring the mobile unit to change channels.

In order to complete calls between mobile units and land telephones, a cellular system needs a capability for interconnection with the land telephone network. A system must also make the connections necessary to complete calls between two mobile units. The capability to make various kinds of connections requires the presence of one or more switching machines to connect trunks together.

A cellular system executes call-control actions such as handoffs and trunk connections only after exercising its capability for decision making. This capability is implemented through computer processors. The processing capability may reside in various parts of the system. At one extreme, a system's total processing capability could be concentrated in a single central processor. At the other extreme, this capability could be distributed among all the base stations. Most systems distribute their processing capabilities in a way that lies somewhere between these extremes. Mobile units contain microprocessors to control many mobile-unit functions.

The proper coordination of system and mobile-unit actions requires an elaborate vocabulary of digital messages for the exchange of information and commands between base stations and mobile units. In the United States, the formats of basic messages have been standardized, as have the details of the encoding and modulation methods.

Hexagonal cellular geometry. In a service area the cells, no matter how they are defined, turn out to be irregular in shape and unequal in area. Nevertheless, system designers use the idealization of regular polygonal cells to simplify matters of channel assignment and cell splitting. Designers seek to place base stations in positions that jointly approximate a regular lattice.

Different regular lattice configurations would correspond to different cell shapes. Although other cell shapes are possible, regular hexagons are most often used to represent idealized cells. The lattice of base station positions shown in Fig. 1*a* corresponds to the array of regular hexagons shown in Fig. 1*b*. In Fig. 1*b*, interior cells are taken to be the territory closest to their corresponding base stations, and cells around the perimeter are assigned the same size and shape as the interior cells.

The number of cells per cluster in a cellular system refers to the number of contiguous cells which, taken together, are allowed to use all possible frequencies, although each frequency is permitted in only one cell of the cluster. The relationship between the number of cells per cluster, designated N, and the minimum distance between cochannel cells (cells that use the same frequencies) is given by Eq. (1) for a grid of regular hexagons. D is the

$$D/R = \sqrt{3N} \tag{1}$$

distance between corresponding points in cochannel cells, and R is the cell "radius," which is the radius of a circle that circumscribes the hexagon. Figure 2 shows an example for seven cells per cluster. The line representing the distance D points to the centers of two cochannel cells. (The center of a hexagon is the center of the circle that circumscribes the hexagon.) These and the other cells that are cochannel with them are shaded. The letters in the cells symbolize the mutually exclusive channel sets of a static channel-assignment scheme.

Only certain values of the number of cells per cluster are realizable. These values have the form given in Eq. (2), in which i and j are nonnegative

$$N = i^2 + ij + j^2 \tag{2}$$

integers. The parameters i and j, sometimes called

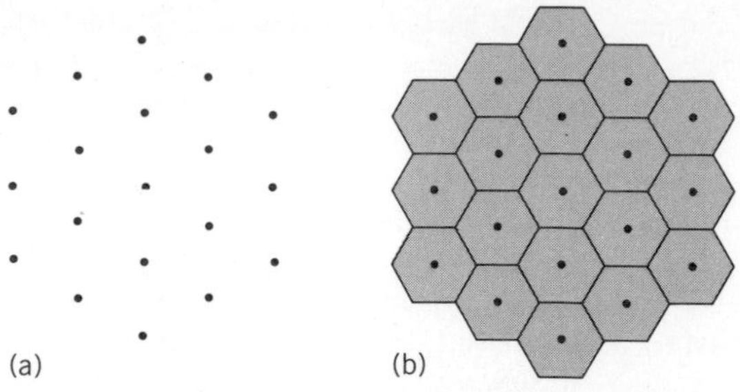

Fig. 1. Relationship between base stations and cells. (*a*) Lattice of base stations. (*b*) Base stations with associated cells.

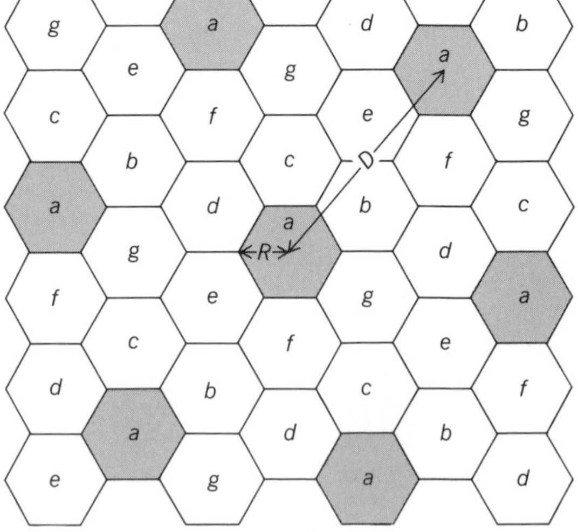

Fig. 2. Frequency reuse in a cellular system.

shift parameters, provide instructions for locating any cell's nearest cochannel cells. The instructions could be stated as follows: Start at the center of the cell of interest and move toward the midpoint of one of its sides. Continue in this direction, counting off i cells. After arriving at the i-th cell, rotate 60° counterclockwise and count off j cells in the new direction. The j-th cell and the cell of interest are cochannel cells. Instructions in which i and j are interchanged or in which the rotation is clockwise are also valid, as long as the same instructions are used for locating all cochannel sites.

United States rules and standards. In the United States, cellular service is authorized on frequencies between 825 and 845 MHz (mobile transmit) and between 870 and 890 MHz (land transmit), with a channel spacing of 30 kHz. The allocated spectrum is split into two subbands of equal width, so that two competing systems may serve the same service area. Compatibility standards specify the use of frequency modulation, with a frequency-deviation limit of 12 kHz for voice signals and a frequency deviation of 8 kHz for digital call-control messages exchanged between mobile units and base stations.

For background information *see* MICROPROCESSOR; MOBILE RADIO in the McGraw-Hill Encyclopedia of Science and Technology. [VERNE H. MacDONALD]

Bibliography: *Bell Syst. Tech. J.*, vol. 58, no. 1 (special issue on Advanced Mobile Phone Service), January 1979; *Cellular System Mobile Station–Land Station Compatibility Specification*, Electronic Industries Association, EIA Interim Stan. CIS-3-A, June 1983; *Cellular System Mobile Station–Land Station Compatibility Specification*, Office of Science and Technology, Bull. 53, April 1981; *Code of Federal Regulations*, Title 47, Part 22: Public Mobile Radio Service, Subpart K: Domestic Public Cellular Radio Telecommunications Service.

Molecular engineering

The properties of organic molecules floating at the interface between water and air have found applications for many centuries. The seagoing civilizations of the Persians and Greeks certainly employed oil spread on troubled waters to reduce wave motion. One of the earliest researchers to question why the film spread so far and what held it together was Benjamin Franklin. His simple experiments in measuring the area covered by small quantities of oil placed on a lake surface led other scientists to pursue the study of such phenomena. Lord Rayleigh repeated Franklin's experiment in England and, from an estimate of the number of molecules placed on the water surface and the area covered, deduced that the film was really only one molecule thick. Agnes Pockels followed up Rayleigh's work in a series of "kitchen-sink" experiments in Germany in the late 1800s. Using basic cooking utensils, she derived a surprisingly accurate value for the area occupied by each molecule of a simple saturated fatty acid dispersed on a water surface.

Beginning in the mid-1920s, Irving Langmuir and Katherine Blodgett devised techniques for manipulating the monolayers on the water surface and for removing the layers onto a solid substrate. Further, they attempted experiments to illustrate that such films might be useful in reducing sliding friction between moving surfaces, as antireflection coatings on optical components, and as model membranes in biological systems simulation. These films, now commonly called Langmuir-Blodgett films, form the basis for the emerging field of molecular engineering—the use of ordered arrangements of organic molecules, almost like two-dimensional crystals, in electronics, optics, lubrication, filtration, bioengineering, and many other areas.

Film production technique. Conceptually, the technique is simple. The organic molecule must be such that one end of the molecule has a hydrophobic termination while the other end is hydrophilic. The molecules are spread upon the water surface by dis-

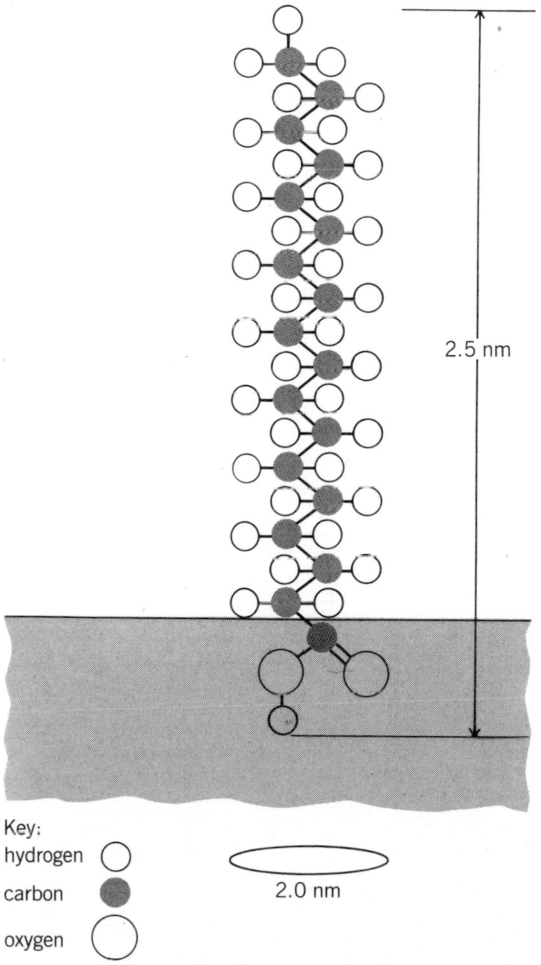

Fig. 1. Schematic of the molecular structure of stearic acid. The refractive index increases with decreasing CH_2 chain length, but increases if the termination in the water is replaced by a divalent metal ion. (*After C. W. Pitt and F. Grunfeld, Organic thin films in integrated optics, in J. R. Jacobson, ed., Thin Film Technologies: Proceedings of the Society of Photo-Optical Instrumentation Engineers, no. 401, pp. 156-164, (1983)*)

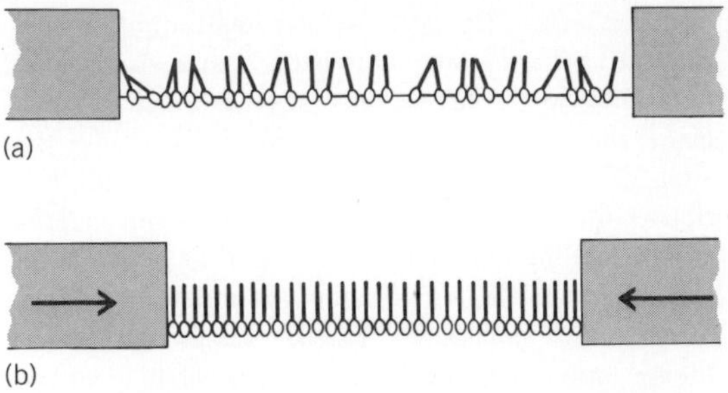

Fig. 2. Diagram of the arrangement of the compressed monolayer on the subphase. (*a*) Before compression. (*b*) After compression. When the solvent evaporates, the molecules are randomly oriented on the water surface. As the layer is compressed, an oriented monolayer is formed. (*After C. W. Pitt and F. Grunfeld, Organic thin films in integrated optics, in J. R. Jacobson, ed., Thin Film Technologies: Proceedings of the Society of Photo-Optical Instrumentation Engineers, no. 401, pp. 156-164, (1983)*)

solving them in a rapidly evaporating solvent; the diluted molecules arrange themselves on the water surface so that the hydrophilic termination is immersed while the hydrophobic termination remains in the air—provided that the main body of the molecule is sufficiently long to maintain low solubility in the water subphase. When the solvent has evaporated, a continuous monolayer may be created on the surface by applying a gentle lateral pressure to the layer in the plane of the subphase surface. Figure 1 shows a typical long-chain saturated fatty acid molecule of the type suited for Langmuir-Blodgett deposition; Fig. 2 shows the arrangement of the compressed monolayer on the subphase. The compression of the layer must be adequate to overcome the repulsive forces between the molecules, but not so great as to cause the layer to collapse and fold over on itself; a pressure of $2\text{–}4 \times 10^{-2}$ newtons per linear meter (20–40 dynes per linear centimeter) of monolayer has been found suitable for materials such as stearic acid, dyonic acid, and vinyl stearate. The film may be removed from the subphase by drawing a substrate suitably prepared with a polar surface through the monolayer as shown in Fig. 3*a*. The polar groups in the hydrophilic termination of the molecule adhere to the polar surface of the substrate, and a section of monolayer is coated onto the substrate. Upon reversing the direction of motion of the substrate, the hydrophobic exposed surface of the film on the substrate encounters the nonpolar terminations of the molecules on the subphase and adhesion occurs (Fig. 3*b*). The adhesion in this case is relatively weak since only nonpolar bonds are involved; however, it is sufficiently strong to enable a third layer to be deposited in a format of molecule tail to molecule tail (Fig. 3*c*). Thus, by sequential dipping of the substrate through the monolayer, a multilayer film can be built up on a substrate.

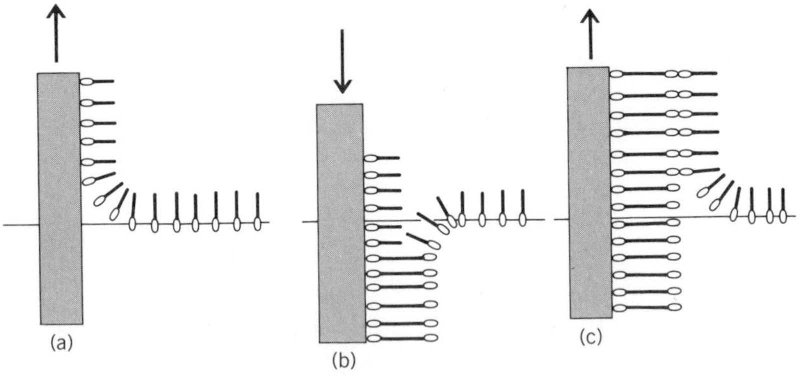

Fig. 3. Removing the film from the subphase. (*a*) The substrate is drawn out of the water and collects a layer. (*b*) Reimmersion collects a second layer. (*c*) A multilayer film can be built up by sequential dipping. (*After C. W. Pitt and F. Grunfeld, Organic thin films in integrated optics, in J. R. Jacobson, ed., Thin Film Technologies: Proceedings of the Society of Photo-Optical Instrumentation Engineers, no. 401, pp. 156-164, (1983)*)

Although the process is simple, a number of precautions must be observed if the purity and homogeneity of the films are to be maintained. The film material and the water of the subphase must be of very high purity; a few contaminant molecules per million film molecules are sufficient to weaken it so that it tears during removal, exhibits pinholes, or incorporates optical scattering centers. The temperature of the water is important: if it is too low, the film becomes rigid like a layer of ice and so is difficult to remove; if too high, the molecular layer is disrupted by thermal agitation. The removal rate of the film from the water surface affects the quality of the film: if the removal rate is rapid, the film may tear, or the compressive forces on the monolayer may not be able to maintain the layer fully compressed so that holes will appear in the multilayer. However, if these factors are carefully controlled, the deposition process is capable of producing homogeneous films of laboratory scale areas (square centimeters), albeit at modest deposition rates (thicknesses of 100 nanometers per hour).

Advantages. In addition to the normal chemical and physical properties of the organic molecules used in the Langmuir-Blodgett process, several unique advantages can be gained in certain applications by utilizing the peculiar characteristics of the films. Foremost is their thinness; for example, a one-molecule-thick dielectric film has real advantages in producing high values of capacitance in a parallel plate capacitor, or much-improved quantum efficiencies in an electroluminescent flat-panel display. The physical properties may also be tailored by selection of the hydrocarbon chain length and by chemical reaction in the water subphase. Possibly the simplest subphase reaction is the conversion of the fatty acids (arachidic, palmitic, stearic, and so forth) to their salts by the incorporation of divalent metal ions in place of the hydrogen atom in the hydrophilic group; such metal atoms act as cross-links to adjacent molecules, thus strengthening the films as well as modifying the electrical–optical characteristics. The mechanism is extremely simple: the degree of ionization of the fatty acid in the subphase

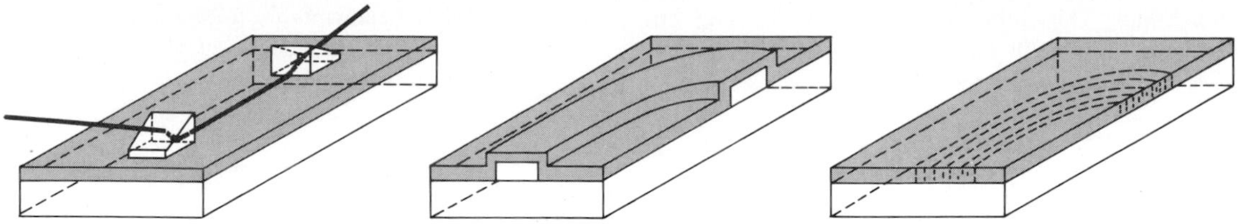

(a) (b) (c)

Fig. 4. Three examples of integrated-optics application of Langmuir-Blodgett layers. (*a*) Planar guide layer in which the laser beam is coupled into the film by prisms. (*b*) Overlay on a prefabricated guide to alter propagation characteristics. (*c*) Langmuir-Blodgett layer modified by radiation to form a stripe guide. (*After C. W. Pitt and F. Grunfeld, Organic thin films in integrated optics, in J. R. Jacobson, ed., Thin Film Technologies: Proceedings of the Society of Photo-Optical Instrumentation Engineers, no. 401, pp. 156-164, (1983)*)

depends on the pH of the solution. A strongly basic subphase will cause ionization and hence increase the conversion of fatty acid to salt: an increased film dielectric constant and refractive index result from the enhanced molecular polarizability produced by the metal ions included in the film. The refractive indexes of stearic acid can be changed by doping with metal ions. Further selection of the index may be obtained by choosing materials of similar long-chain structure but containing —C=C— or —C≡C— groups in the body. Such materials as the diacetylenes, with —C=C— in the body, and vinyl stearate, with —C≡C in the body, may be polymerized by exposing a monomer Langmuir-Blodgett film to ultraviolet light, x-rays, or an electron beam.

Applications. There is currently a wide range of application of Langmuir-Blodgett films under investigation in laboratories throughout the world. Devices reported recently include field-effect transistors (FET) in which the gate insulation layer has been replaced by a monolayer of organic materials. In this case, the monolayer is permeable to certain gases, so that the transistor reacts electrically to the presence of such gases; the FET has been adapted for direct sensing of the ambient.

Surface acoustic waves—microsound waves traveling along the surface of a solid—have found application in signal storage and processing because the propagation velocity of sound in a solid is low, enabling large amounts of data to be stored. The usual electrical-to-sound transducers employed in these devices are piezoelectric in operation, but recently a novel very-thin-film capacitor, using a Langmuir-Blodgett layer as the dielectric film, has been found to detect surface acoustic waves directly. The sound waves modulate the thickness of the molecular layer, causing the electrical capacitance to vary at the same frequency as the surface acoustic waves. Very fine-line pattern definition in thin metal films is a key requirement for fast integrated circuits in both silicon and gallium arsenide. High-resolution photoresists are needed to achieve this objective. Although pattern writing with focused electron beams is, in principle, able to meet the need, in practice the electrons are scattered in conventional resist-layer thicknesses, thus degrading the pattern resolution. By using very thin Langmuir-Blodgett layers of polymerizable materials, it has been demonstrated that line widths as fine as 60 nm can be written with electron beams.

Langmuir-Blodgett films have also made an impact in the recent field of integrated optics, in which light is manipulated in a waveguiding dielectric medium in order to perform signal transmission and processing. Figure 4 illustrates how laser light may be guided by an organic molecular multilayer of suitable refractive index and thickness, and how waveguides made of other materials may be improved by cladding with a Langmuir-Blodgett layer.

The future for Langmuir-Blodgett layer applications appears to lie firmly in the burgeoning area of molecular engineering, encompassing not only conventional physics and chemistry but also the biosciences. Thus, current research in magnetic Langmuir-Blodgett films, in photovoltaic structures, in phonon memory effects, in hyperfiltration membranes, in Josephson tunnel junctions, in catalysis, and in fluorescence and luminescence in Langmuir-Blodgett films will be extended by using the layers to simulate biological membranes. For example, the penetration of lipid membranes by proteins is an important area of investigation in topics as diverse as photoconduction in cells and studies of enzyme reactions. A particularly exciting area in which new work is expected to start is that of growing two-dimensional protein crystals. Data on the structure of protein crystals, provided by x-ray and electron diffraction, will, it is hoped, enable computer models of these complex structures to be developed. The sophistication of these models will be such as to enable the effects of modifying the protein structure—for example, by incorporating salt bridges to enhance the temperature tolerance of the protein—to be precisely predicted before building the altered structure. If such research is successful, in the next few decades molecular engineering may well match the achievements of solid-state physics over the past 40 years.

For background information *see* INTEGRATED OPTICS; MONOMOLECULAR FILM; SURFACE ACOUTIC-WAVE DEVICES in the McGraw-Hill Encyclopedia of Science and Technology. [C. W. PITT]

Bibliography: W. Barlow (ed.), *Langmuir-Blod-*

gett Films, Thin Film Science and Technology Series, vol. 1, 1980; G. G. Roberts and C. W. Pitt (eds.), *Langmuir-Blodgett Films, 1982*, Thin Film Science and Technology Series, vol. 3, 1983.

Molybdenum

Molybdenum (Mo) is a recently recognized essential trace element involved in many plant-animal relationships. The requirement for Mo, which was first discovered in Australia, is so small that seed treatment with Mo can correct Mo deficiency in most crops. The physiological role of Mo appears to be in protein synthesis in plants and symbiotic nitrogen (N) fixation in legumes. Typical visual symptoms of Mo deficiency are pale green or yellow leaves and stunted growth. In cauliflower, Mo deficiency causes whiptail. Toxicity of Mo is rare in plants.

Deficiency of Mo in livestock is, so far, unknown. The main concern for animals is toxicity related to high consumption of Mo which interacts with copper (Cu) metabolism resulting in Cu deficiency.

Functions. Enzymes, namely nitrate reductase and nitrogenase, require Mo, and are involved in protein metabolism in plants, and they govern the processes of nitrate reduction and fixation of N_2 from the air by legumes. Reduction of nitrate (NO_3^-) to ammonium (NH_4^+) is a significant step in the synthesis of proteins in the plants. The increased Mo requirement of most plants grown on NO_3^--N compared with NH_4^+-N can be almost completely accounted for by the requirement of Mo in the enzyme nitrate reductase. Molybdenum has been found to have a protective function in alleviating damage caused by the accumulation of NO_3^- in small plants. On soils containing low amounts of Mo, plants develop many nodules on their roots, but N_2 is not fixed. Besides increasing the protein content of plants, Mo enhances the synthesis of nucleic acids. A low concentration of ascorbic acid in a number of plant species is a characteristic of Mo deficiency.

Molybdenum is recognized as an essential element for animals, but a deficiency has never been developed or observed in cattle. This indicates that the minimum requirement for Mo is very low. It is apparent from studies with low-Mo diets that Mo is involved with enzyme activity and is required for particular metabolic processes in animals.

Molybdenum fertilization. Molybdenum can be applied to the soil, to plants as a foliar spray, or to seed in a coating. For soil application, 0.7–1.4 oz Mo/acre (50–100 g/hectare) for most agronomic crops and up to 5.6 oz/acre (400 g/hectare) for vegetable crops, such as cauliflower, are used. In New Zealand and Australia, Mo has been applied to soils in combination with superphosphate called molysuper. Residual effect of Mo applied to soils can last 2–5 years depending upon the rate, soil type, and geographical region. Due to the small rate of application, it is difficult to achieve uniform distribution if Mo is mixed in a fertilizer. Also, Mo may become fixed or rendered unavailable to plants when mixed with the soil. Foliar application of Mo for overcoming Mo deficiency has been a common practice in many crops. Rates of less than 1.4 oz/acre (100 g/ha) are needed when applied as a foliar spray. This method avoids the problem of Mo fixation in acid soils and is more desirable than soil applications under dry conditions.

Because of the extremely low requirement of crops for Mo, the most common method of overcoming a Mo deficiency is to treat the seeds with a Mo preparation (0.1–0.5 oz/acre or 7–35 g/ha). Seed treatment is effective in controlling a Mo deficiency in soybeans, cauliflower, Brussels sprouts, and other crops. Application of Mo with the lime on lime-pelleted legume seeds is a practical way of applying Mo in close proximity to the seeds and without detriment to the rhizobia in the inoculum.

Tissue contents. In general, plant species with a greater requirement for Mo, such as legumes, have higher Mo concentration than cereals and grasses. The sufficiency levels of Mo range from about 0.1 ppm for grasses and cereals to 0.3–0.5 ppm for alfalfa, soybeans, cauliflower, and clovers. In general, legumes contain more Mo than grasses, but exceptions occur. Climate may affect the Mo content of the same plant species; for example, the Mo content in pasture legumes in the tropics is about 0.02 ppm, but in the temperate zones it typically exceeds 0.1 ppm. In areas of the world where Mo-toxic soils are found, plant Mo levels may exceed 100 ppm.

Generally, leaf tissue is considerably higher in Mo than the stem. In wheat and barley, the Mo content of grain is lower than that of the vegetative tissue, while in soybeans the seed contains much higher amounts than do leaves, stems, or pod shells. Likewise, peanut kernels have a considerably higher Mo content than does the leaf petiole.

Among the factors affecting the plant Mo content, soil pH is of utmost importance, and the tissue Mo increases sharply with an increase in soil pH.

In animals, liver and kidney tissue contain higher concentrations of Mo than do muscle, lung, or brain tissue. For sheep and cattle grazing herbage normal in Cu and low in Mo, 6 micrograms Mo per 100 ml blood is considered normal.

Seed content. A relationship between Mo deficiency symptoms and seed reserves has been established. For example, Mo deficiency in corn is unlikely even on low-Mo soils if the Mo in planted seed exceeds 0.08 ppm, but is likely if it is below 0.02 ppm. Soybeans show no yield response to applied Mo when the seeds planted contain more than 2.6 ppm Mo. It has been suggested that Mo accumulation in the nodules of soybeans may be highly dependent during the early growing period on the Mo content in the seeds planted. Thus seed Mo may play an important role in the early growth and in the N_2 fixation by nodules. During the pod-filling stage, Mo in the roots and pod shells (and to a lesser extent nodules, leaves, and stems) is translocated to the seeds.

Molybdenum responses. Forage legumes (alfalfa and clover), soybeans, and crops of the genus *Brassicae* (for example, cauliflower, broccoli, and Brussels sprouts) frequently respond to Mo on soils low in Mo or low in pH. Deficiency of Mo has now been found in other crops such as grasses, cereals, and corn on acidic soils particularly. In Australia, Mo has been applied on clover primarily to secure symbiotic fixation of N_2 for permanent pastures and leys. Responses to Mo were found on wheat and oats on light sandy soils; on rice in Brazil; and on cow peas in Sierra Leone (western Africa). Since the response to Mo is more significant during early growth, the activity of nitrate reductase in producing nitrite and, thence, protein is the process primarily involved.

Since responses to Mo by plants are closely related to soil properties (that is, acidic, coarse, and leached soils frequently respond to Mo), patterns of Mo deficiency are related to certain geographical areas of North America, Australia, New Zealand, and eastern Europe.

Plant deficiency and toxicity symptoms. The symptoms associated with Mo deficiency are closely related to the metabolism of N and are expressed in young plants as white and necrotic areas extending back along the leaves from the tips. Other characteristic plant symptoms include whiptail of cauliflower and golden yellow coloration of older leaves plus short internodes and reduced foliage in wheat. Molybdenum deficiency in melons causes severe yellowing, stunting, and failure to fruit, while in citrus trees leaf burn and yellow spots related to nitrate accumulation occur. In sugarbeets inward curling of leaf margins, arrow-shaped leaves, and curling of lamina occurs.

Concentrations of up to 100–200 ppm Mo in a variety of plant species have not been found to be toxic to them. Molybdenum toxicity in plants marked by chlorosis and yellowing is induced only under extreme conditions.

Animal deficiency and toxicity symptoms. Under naturally occurring conditions, true Mo deficiency has never been reported in humans or farm animals. By using tungsten as a Mo antagonist, Mo deficiency was created in rats. This resulted in the loss of the enzymes sulfite oxidase and xanthine oxidase in the liver as well as a loss of total hepatic Mo.

Nutritional interest in the element is overwhelmingly concerned with its toxic effects on animals. Species differences in tolerance to Mo are substantial. Cattle are by far the least tolerant to Mo toxicity, followed by sheep, while horses and pigs are the most tolerant of farm livestock. High levels of Mo in forages can induce Cu deficiencies in animals, and this disorder is referred to as molybdenosis. This Mo toxicity occurs under natural grazing conditions in many parts of the world. All cattle are susceptible to molybdenosis, but milking cows and young stock are affected most. Toxicity symptoms include the development of harsh scouring and discolored coats. Typical teart (Mo-toxic) pastures contain 20–100 ppm Mo in plant tissue. Under such conditions very large Cu supplements are needed to control teart. On the other hand, Cu poisoning in grazing sheep in parts of Norway is caused by a combination of normal or subnormal levels of Cu and low or very low levels of Mo.

For background information *see* FERTILIZER; MOLYBDENUM; MOLYBDENUM INTOXICATION; NITROGEN FIXATION; PLANT MINERAL NUTRITION in the McGraw-Hill Encyclopedia of Science and Technology.

[UMESH C. GUPTA]

Bibliography: W. R. Chappell and K. K. Petersen (eds.), *Molybdenum in the Environment*, vol. 1, 1976, vol. 2, 1977; U. C. Gupta and J. Lipsett, Molybdenum in soils, plants and animals, *Adv. Agron.*, 34:73–115, 1981; J. Ishizuka, Characteristics of molybdenum absorption and translocation in soybean plants, *Soil Sci. Plant Nutr.*, 28:63–77, 1982; E. J. Underwood (ed.), *Trace Elements in Human and Animal Nutrition*, 1977.

Muscovite

Muscovite occurs in granite either through direct crystallization from hydrous silicate melts (magmatic muscovite) or as a later product of alteration in the solid state (subsolidus reactions). The two kinds of muscovite can occur in the same rock. Understanding the conditions of formation of magmatic muscovite is important because muscovite granites occur worldwide, and attain batholithic dimensions as both synorogenic and postorogenic intrusions.

Structure and chemistry of muscovite. Muscovite has an ideal composition of $KAl^{vi}_2Al^{iv}Si_3O_{12}H_2$ (superscript vi indicates aluminum in octahedral coordination; iv, aluminum in tetrahedral coordination). The silicon and Al^{iv} atoms are tetrahedrally coordinated to four oxygens; the tetrahedra share corners and form continuous sheets. The Al^{vi} atoms, octahedrally coordinated to six oxygens, form matching sheets of linked octahedra, with one out of every three octahedral sites empty. Six of the twelve oxygen atoms in the formula coordinate only the tetrahedra, and four coordinate both tetrahedra and octahedra. The other two oxygens are linked only to octahedra, to which the two protons are attached. Each sheet of octahedra is sandwiched between two sheets of tetrahedra. These so-called sandwiches have a residual negative charge which is balanced by interlayer potassium in twelvefold oxygen coordination.

Many ionic substitution schemes are possible; principal substitutions found in nature are summarized as follows:

1. A site (twelvefold oxygen coordination; interlayer cation site) for K: Na, Rb, Ba, H_3O, Cs, Ca.
2. Nonbridging oxygen (attached only to the octahedra) for OH: F, Cl.
3. Tetrahedral site (the "bread" of the sandwich) for Si: Al, Fe^{3+}, Ti(?).
4. Octahedral site (the "meat" of the sandwich) for Al^{vi}: Fe^{3+}, Fe^{2+}, Mg, Mn, Ti, Cr, V, Li, vacancy(?).

Some of the substitutions are simple, for example, K ⇆ Na (paragonite substitution), Fe^{3+} ⇆ Al^{vi} (ferrimuscovite substitution), or OH ⇆ F (fluormuscovite substitution); others are charge- or vacancy-coupled, for example, (Fe^{2+},Mg) + Si ⇆ Al^{vi} + Al^{iv} (leucophyllite substitution), (Fe^{2+},Mg) + Ti ⇆ 2 Al^{vi}, (Fe^{2+},Mg) + H ⇆ $Fe^{3+,vi}$ + □, or, probably, Ti + □ ⇆ 2(Mg,Fe^{2+}) [the symbol □ represents vacancies in the crystal]. Celadonite represents a linear combination of leucophyllite and ferrimuscovite substitutions; phengite represents a 50% reach toward the leucophyllite component.

Criteria for recognition of magmatic muscovite. Discussion of the nature, composition, and origin of magmatic muscovite depends on the identification of a given muscovite as magmatic. Even if consideration is confined to truly intrusive magmatic granitic rocks, muscovite that formed through subsolidus reaction must still be distinguished from muscovite that crystallized directly from a magma.

To differentiate muscovite produced by these two processes, both chemical composition and textural relation of the muscovite to the other minerals have been used. Since compositions of magmatic muscovite must be established first, only textural criteria can be considered. A review of the textural data led to the following suggested criteria for magmatic muscovite: (*a*) coarse crystals, grain size comparable to other, obviously magmatic minerals in the rocks; (*b*) clean crystal terminations, ideally subhedral or euhedral; (*c*) freedom from ragged inclusion relation to minerals that may be related to the muscovite by subsolidus alteration; and (*d*) absence of other indications of subsolidus rock alteration. These are all plausibility criteria. Insistence on euhedral intergrowth relation with biotite, especially in butt-end rather than interlayer configuration, could help in the second criterion, but even that is not conclusive. Figure 1 shows photomicrographs of six samples of muscovite in granite.

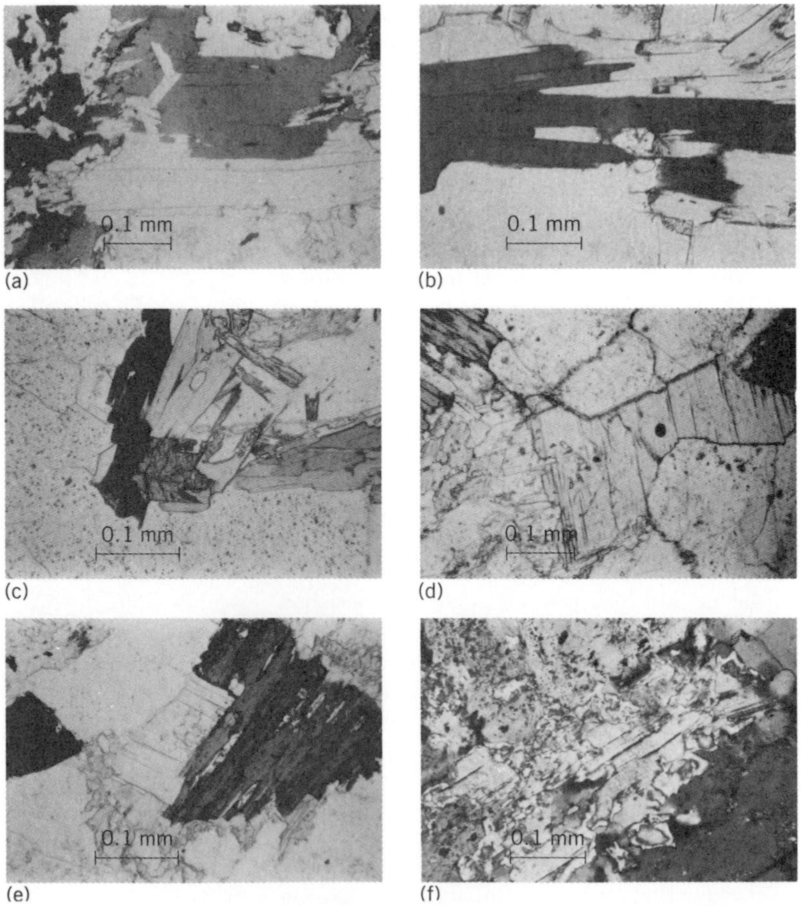

Fig. 1. Photomicrographs of muscovite in granite. Muscovite and biotite in euhedral butt-end intergrowth; muscovite is interpreted to be magmatic: (*a*) from the Pioneer Mountains, Montana, and (*b*) from Stone Mountain, Georgia. Muscovite in isolated euhedra or subhedra, genesis uncertain: (*c*) from Mount Airy pluton, North Carolina, and (*d*) from the Pioneer Mountains, Montana. (*e*) Muscovite associated with altered biotite, from the Pintlar Range, Montana. (*f*) Muscovite as tiny ragged crystals enclosed in altered plagioclase, from the Granite Mountains batholith, Wyoming. Both *e* and *f* are probably subsolidus alteration products.

Chemistry of magmatic muscovite. It has been determined that muscovite identified by criteria *a*, *b*, and *c* does not have the stoichoimetric composition of $KAl_2AlSi_3O_{12}H_2$. Principal characteristics of magmatic muscovite, identified on the basis of textural relations, are the presence of about 3.2 atoms of Si per formula, about 2.7 atoms of Al, about 0.4 atom of total Fe + Mg, and consistently about 0.04 atom of Ti, in addition to the expected K–Na and OH–F substitutions. The ratio of ferric iron to ferrous iron is generally not known, because chemical data on samples that satisfy the textural criteria are crystal-specific in a given rock, and methods used to determine ferric and ferrous iron directly, including Mössbauer spectroscopy, require bulk samples. Nevertheless, stoichoimetry of muscovite indicates that substitutions in magmatic muscovite do not require filling the vacant octahedral site.

With the above caveat, the bulk of individual chemical analyses of magmatic muscovite can be accounted for in terms of four reciprocal components: muscovite s.s. (s.s. represents solid solution) or Mu, composition $KAl_2AlSi_3O_{12}H_2$; celadonite or Ce, composition $K(Fe^{2+}Mg)Fe^{3+}Si_4O_{12}H_2$; leucophyllite or Lp, composition $K(MgFe^{2+})AlSi_4O_{12}H_2$; and ferrimuscovite or Fu, composition $KFe^{3+}AlAlSi_3O_{12}H_2$. Available data (Fig. 2) indicate that the principal substitution can be described as between Mu and Ce, with Mu ranging 70–80 mol %, Ce 25–15 mol %, and Lp or Fu seldom exceeding 5 mol %. These components do not include Ti, Na, F, or other elements listed above. The composition of texturally defined magmatic muscovite is believed to be distinct from subsolidus muscovite mainly in the higher Ti content; by the same token, it is also distinct from most phengitic muscovite of metamorphic rocks.

Magmatic muscovite and petrogenesis. Rocks that contain magmatic muscovite are peraluminous; that is, the atomic ratio R = Al/(Na + K + 2Ca) is greater than unity, such that there is more Al in the system than would be needed to take up all of the Na, K, and Ca into feldspar. Such peraluminous rocks may be expected to contain high-alumi-

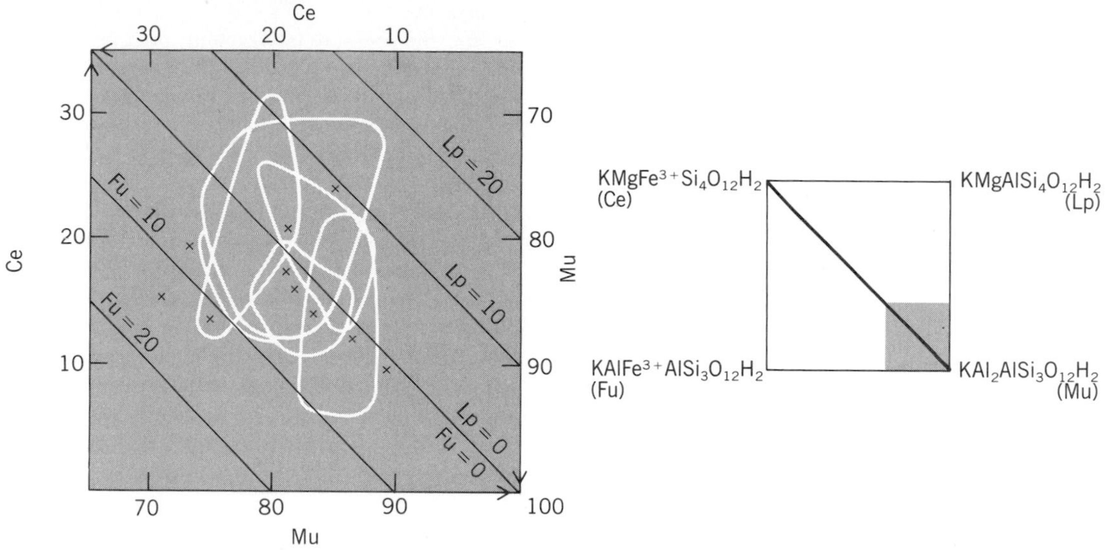

Fig. 2. Plots of magmatic muscovite using the reciprocal coordinates muscovite s.s. (Mu), celadonite (Ce), leucophyllite (Lp), and ferrimuscovite (Fu), as discussed in the text. Encircled areas define microprobe-analyzed magmatic muscovite from individual rock samples, Pioneer batholith, Montana. All iron assumed to be ferric. Crosses are selected data from the literature.

num minerals such as almandine garnet, cordierite, and the aluminum silicate polymorphs sillimanite and andalusite, in addition to muscovite. Indeed, all these minerals are found in peraluminous granites.

Peraluminous melt chemistry, in conjunction with other discriminants, often suggests derivation by partial or complete melting (anatexis) of sedimentary source material (so-called S-type granite), because many sedimentary rock sequences consist of clay-rich material, peraluminous according to the above definition. Which high-aluminum minerals appear in the resulting granite depends on the details of the rock chemistry and the condition of crystallization, but muscovite is the most common.

Not all peraluminous granite is S-type granite, however. A second possible origin of peraluminous magma is by crystallization of a magma that was not initially peraluminous but became so through continuous or episodic fractional separation of solids from the melt. If the solid phases thus separated have an aggregate atomic ratio (R) value less than that of the original melt, then the melt becomes enriched in alumina; sufficient operation of the process could lead to a peraluminous residual granitic melt. It has been proposed that fractional crystallization of hornblende, a major rock-forming mineral, could effect this passage; crystallization of other minerals having $R < 1$ would have a similar effect. However, controlled laboratory study of melt-related phase equilibria in this system is needed to demonstrate the efficacy of the process.

Experimental data. Laboratory hydrothermal stability studies indicate that stoichiometric muscovite (Mu), $KAl_3Si_3O_{12}H_2$, is stable in the presence of quartz (SiO_2), at temperatures that overlap the H_2O-saturated granite solidus. On a pressure–temperature diagram (Fig. 3), the intersection of the two curves that limit the thermal stabilitites of granitic silicate melt and of muscovite + quartz indicates approximately the minimum pressure for naturally occurring magmatic Mu muscovite to be stable. This pressure is about 4 kilobars (400 megapascals), which corresponds to a depth in the Earth's crust of about 9 mi (15 km). This postulated depth leads to a dilemma, because many granites that carry textur-

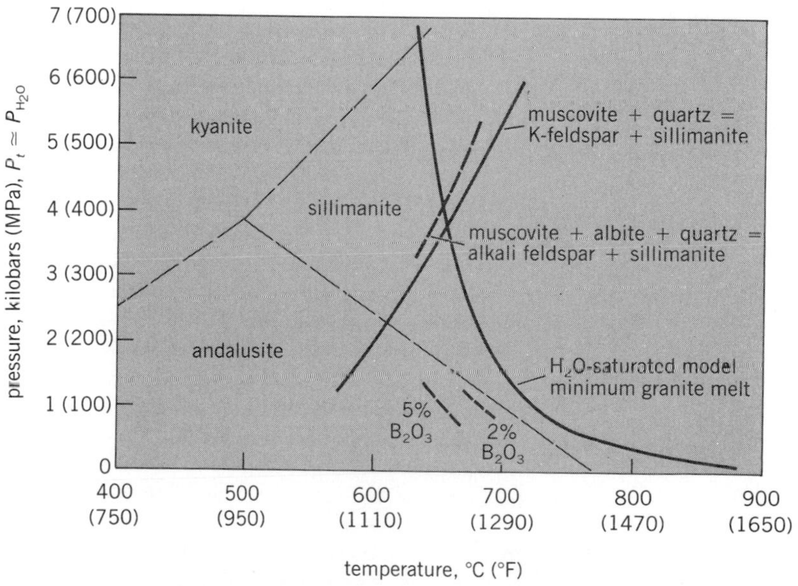

Fig. 3. Pressure–temperature diagram showing: upper thermal stability of stoichiometric muscovite + quartz and of muscovite + quartz + albite, the minimum melting curve for ideal granite, the effect of B_2O_3 saturation on the latter curve, and the effect of 1.5 wt % F on the same curve. The stability fields of the aluminum silicate polymorphs are added for reference. P_t = total pressure.

ally determined magmatic muscovite show geologic evidence of emplacement at much shallower depths. The phase equilibrium data obtained in the laboratory thus flatly contradict some geologic information.

Challenges. Reconciliation of these contradictions remains a major challenge to the understanding of granitic rocks that carry magmatic muscovite. Attempts to explain the difficulty have raised some questions.

Shallow-level intrusive muscovite. Despite the textural evidence, muscovite from shallow-level intrusions may not be magmatic but instead may be a subsolidus reaction product. While difficult to refute conclusively, the compositional distinction found in recent experiments casts doubt on this hypothesis for many rocks. Large, euhedral crystals of muscovite, especially those in butt-end intergrowth with obviously magmatic biotite, are difficult to interpret as being anything other than magmatic.

Nonstoichiometric magmatic muscovite. Because magmatic muscovite is not stoichiometric Mu, $KAl_2AlSi_3O_{12}H_2$, phase equilibrium relations determined on the stoichiometric material do not apply. Some workers believe that the celadonitic and other components in the phase may stabilize muscovite to the extent that the intersection of the muscovite + quartz thermal stability limit with the minimum melting curve of granite may be at a much lower pressure. However, no empirical evidence exists that the other components in magmatic muscovite do lead to greater thermal stability; scanty experimental data on the stability of celadonite and phengitic muscovite suggest that phases having these compositions should be thermally less stable, not more so.

Lowering of minimum melt temperature. The granitic melt may contain components that, at any given pressure, would drastically lower the temperature for the minimum melt composition. Halogens such as fluorine and chlorine were present in many melts, as shown by their occurrence in the muscovite and associated minerals; if the muscovite is magmatic, then these components could have lowered the temperature of the granite solidus to result in appreciable lowering of the minimum pressure for magmatic muscovite. The efficacy of fluorine in lowering the temperature of the granite minimum melt is known; the practical problem is to establish that a given melt contained enough fluorine to explain the crystallization of magmatic muscovite. For example, the fugacity of F in natural granitic melts may be controlled by the fluorine-bearing minerals fluorite or topaz, and the quantitative effects found in the laboratory using higher concentrations of F may be unrealistic. Another possible component is boron, which is significant because muscovite-bearing granite is commonly associated with the boron-rich mineral tourmaline. Boron partitions heavily into the melt and does lower the temperature of the solidus of the minimum-melting granite; at 1 kilobar (100 MPa) an H_2O-saturated melt having about 5 wt % B_2O_3 shows a reduction of the solidus temperature of about 75°C (170°F); if the muscovite is boron-free, the pressure intersection could be lowered to about 2 kilobars (200 MPa; see Fig. 3). However, saturation of boron in natural granitic systems is at most with respect to tourmaline, not B_2O_3, so again the quantitative effect found in the laboratory is probably unrealistic. Definitive data on tourmaline stability are scanty, and the efficacy of boron in stabilizing magmatic muscovite remains unclear.

Crystallization conditions. Magmatic muscovite may have crystallized in a geologic setting different from where it is actually found in a pluton. For many plutons, varied petrographic and geologic evidence indicates that the material was emplaced not as a simple liquid but as a mixture of crystals and liquid. These mushes may be expected to remain physically mobile until a reasonably high volume ratio of solid to liquid is reached. Thus, magmatic muscovite conceivably could have crystallized at deeper levels of the crust and then not reequilibrated at the final pressure of solidification. Igneous rocks commonly include minerals that did not equilibrate at the final site of consolidation; however, many such minerals show evidence of either chemical zoning (such as zoning in plagioclase feldspar), or breakdown into other phases (such as relict pyroxene with amphibole overgrowth). Magmatic muscovite does not generally show such textural evidence of disequilibrium.

Proposed research. Many questions about magmatic muscovite remain unanswered. There is a need for laboratory studies of muscovite stability as a function of its chemical composition and as a function of geologically plausible concentrations of components in the melt that would not be recorded directly in the solid phases (such as chlorine). Independent determination of conditions of formation of the muscovite should be attempted, for example, by use of fluid inclusions in this mineral or in other, coexisting minerals. Current research involving use of trace elements, for example, the rare-earth-element patterns of minerals in a granite, to test mutual thermodynamic equilibrium and to trace their crystallization sequence, might provide another helpful tool. If a muscovite is produced by subsolidus crystallization at a much later time, the difference in age might be detectable, on a crystal-by-crystal basis, by the use of the laser-heated $^{39}Ar/^{40}Ar$ technique that allows age determination on single crystals. If the subsolidus muscovite formed in contact with circulating groundwater, that fact might be shown by crystal-specific oxygen-isotope studies, perhaps by using the ion probe. Thus state-of-the-art technology may help to resolve some of the difficult problems in the near future.

Oxygen fugacity during crystallization. Many peraluminous granites contain ilmenite; these rocks formed under fairly reducing conditions. Other muscovite-bearing granites, however, contain magnetite. The muscovite–quartz–magnetite assemblage is more oxidizing than the chemically equivalent iron biotite–almandine garnet assemblage. In the presence of an alkali feldspar that is ubiquitous in granite, these two alternative assemblages meet along a univariant reaction curve that is a function of pressure, temperature, and the fugacity of oxygen. In the temperature range of 650–750°C (1200–1380°F), which is applicable to the late-stage solidification of granite, the oxygen fugacity of this buffer system is in range of 10^{-14} to 10^{-16} bars (10^{-9} to 10^{-11} Pa) at reasonable total pressures, approximately midway between the oxygen fugacities of the quartz–magnetite–fayalite buffer and the hematite–magnetite buffer. Thus relatively oxidizing conditions could enhance the stability of magmatic muscovite.

Conclusions. Laboratory calibration of conditions of formation of magmatic muscovite is needed. This work should be done in a system that is chemically a good model of natural materials; detailed study of the composition of the silicate melt and its physicochemical properties, as well as those of the solid phases, needs to be made as a function of the total pressure, the temperature, and the fugacity values of oxygen, H_2O, the halogens, and boron. The study is important not only because these rocks have worldwide distribution through the geologic ages and in different tectonic settings, but also because these rocks are genetically associated with many economically important mineral deposits, such as those of tin, molybdenum, and tungsten, in the form of veins, skarns, and greisen deposits that are synmagmatic or late magmatic. Understanding the genesis of peraluminous granite could illuminate the origin of these mineral deposits that are the objects of considerable contemporary exploratory and research activity.

For background information *see* CHEMICAL DATING; COORDINATION CHEMISTRY; FUGACITY; GRANITE; MUSCOVITE; ROCK AGE DETERMINATION; SILICATE MINERALS in the McGraw-Hill Encyclopedia of Science and Technology. [E-AN ZEN]

Bibliography: R. G. Cawthorn and P. A. Brown, A model for the formation and crystallization of corundum-normative calc-alkaline magmas through amphibole fractionation, *J. Geol.*, 84:467–476, 1976; D. B. Clarke (ed.), Peraluminous granites, *Can. Mineral.*, 19:1–216, 1981; D. A. C. Manning, The effect of fluorine on liquidus relationships in the system Qz-Ab-Or with excess water at 1 kb, *Contrib. Mineral. Petrol.*, 76:205–215, 1981; M. Pichavant, An experimental study of the effect of boron on a water saturated haplogranite at 1 kbar vapour pressure, *Contrib. Mineral. Petrol.*, 76:430–439, 1981.

Mutagens and carcinogens

Damage to deoxyribonucleic acid (DNA), the genetic material that stores hereditary information, has been implicated as a first step in a complex set of processes leading to cancer. Physical and chemical agents that cause DNA damage, which may permanently alter the genetic material, are called mutagens. They exist in the environment naturally or are produced by human activity. Since many mutagens are known to be carcinogens, short-term mutagenicity bioassays have been developed that allow for the rapid identification of a large number of chemicals that damage DNA. Because the structure of DNA is universal in all living organisms, microbial mutagenicity tests can be used to identify chemical mutagens. One such bioassay, the bacterial Ames *Salmonella* microsome assay, is probably the most widely used short-term bioassay because of its high predictive value for mammalian carcinogenesis.

Ames test. This test, developed by Bruce Ames, employs several specially constructed strains of *Salmonella typhimurium* that are unable to synthesize the essential amino acid histidine because of a defective (mutated) histidine gene. Unless the gene is made functional through a specific reverse mutation, no growth will occur in the absence of histidine. When about 1 billion bacteria of each of the tester strains are exposed to a test chemical on petri plates containing a selective minimal medium, only those bacteria that have reverted to histidine independence are able to grow and divide to such an extent that they form visible colonies after 48 h of incubation at 98.6°F (37°C). When several dose levels of a mutagenic chemical are tested (as is usually the case), a dose-related increase in the number of histidine-independent revertant colonies is generally obtained. Because bacteria mutate spontaneously, a control plate, containing the test organism only, is included in every assay to provide a base level for comparison. Figure 1 depicts the experimental procedure for testing chemicals in the Ames mutagenicity assay. A chemical is considered nonmutagenic when no increase in the number of revertant colonies above the spontaneous background colonies is observed.

One important aspect of the Ames test is the addition of an extract of liver, usually derived from

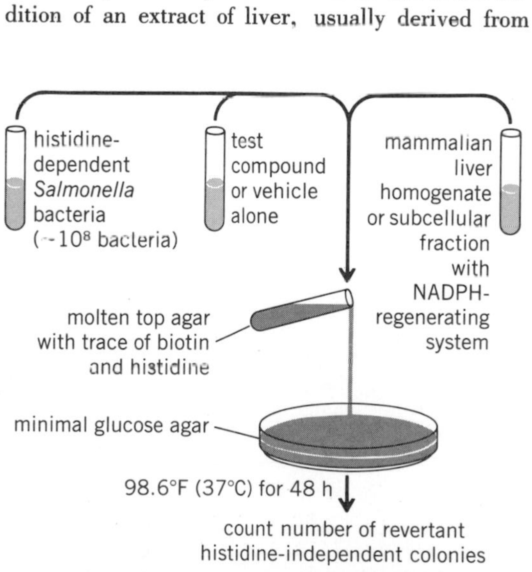

Fig. 1. Experimental procedure for testing chemicals in the standard Ames plate incorporation assay. (*After K. E. Mortelmans, Salmonella/mammalian-microsome test to detect chemical mutagens, J. Food Protec., 41(12):989–995, December 1978*)

rodents, to the petri plate before it is incubated. The liver homogenate mimics mammalian metabolism, which is known to transform many substances to excretable products, many of which are mutagenic. However, not all chemical substances need to be transformed enzymatically in order to exert a mutagenic effect. In some instances, the inclusion of the metabolic activation system may even decrease the mutagenic activity of a chemical.

The assay has been used to evaluate the mutagenicity of pure chemicals, complex mixtures (such as cigarette smoke and automobile exhaust emissions), and body fluids. Figure 2 depicts the mutagenic dose response obtained with aflatoxin B_1, a potent mutagen and animal and human liver carcinogen, and illustrates the differential sensitivity of some of the bacterial *Salmonella* strains.

Reliability of test. The Ames test has proved to be about 85% accurate in detecting certain classes of chemical carcinogens as mutagens and in predicting that a nonmutagenic compound is noncarcinogenic. Development and improvement of existing strains and assay conditions should increase the test's accuracy in predicting the potential carcinogenicity of chemicals. Caution is necessary when assessing the reliability of the Ames test, because there are several classes of chemical carcinogens that are not readily detected as mutagens in the Ames assay. The composition of a group of chemicals can therefore theoretically change the accuracy determination in a range between 0 and 100%. Examples of chemical classes that are not readily detected as mutagens in the Ames assay are polyhalogenated hydrocarbons (many of which are pesticides), hydrazines, nitrosamines, and metals. Examples of chemical classes that are readily detected as mutagens are alkylating agents and polycyclic aromatic hydrocarbons.

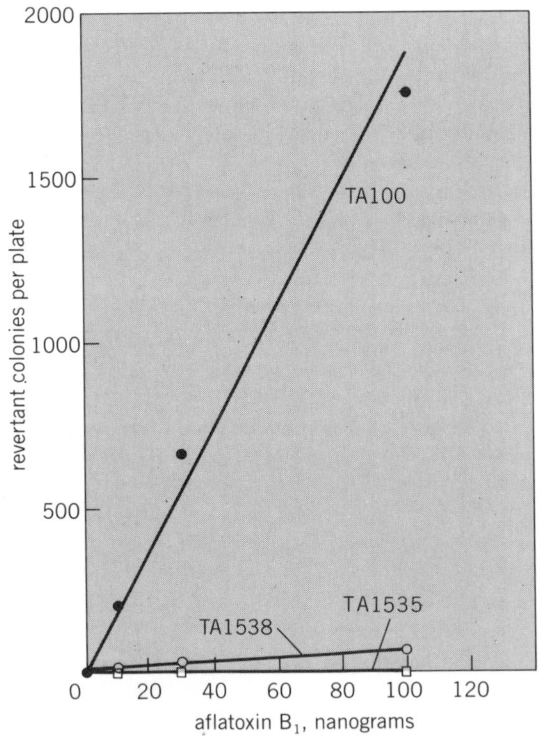

Fig. 2. Mutagenic activity in three *Salmonella* strains (TA100, TA1538, TA1535) of aflatoxin B_1 in the standard plate incorporation assay, in the presence of rat liver homogenate. The spontaneous revertant colonies have been subtracted. (*After J. McCann and B. N. Ames, The Salmonella/microsome mutagenicity test: Predictive value for animal carcinogenicity, in H. H. Hiatt, J. D. Watson, and J. A. Winsten, eds., Origins of Human Cancer, Cold Spring Harbor Laboratory, pp. 1431–1450, 1977*)

Use of test results. Because test results are obtained in 2 days (compared with 2–3 years in animal carcinogenicity studies, and with a latency period of 20–30 years for chemical carcinogenesis to be manifested in humans), a large number of chemicals have been tested in the Ames assay over the past 10 years. Mutagens (naturally occurring or produced by human activity) have been found in food, in drinking water, in a large number of industrial chemicals, in cosmetics, in urine of cigarette smokers, and in the air, to name a few sources.

Since it is impossible and impractical to ban all chemical mutagens from the environment, the results of the Ames assay are often used to establish priorities for further testing in more complex systems such as mammalian mutagenicity assays and animal studies. Over the past few years, attempts have also been made to establish reliable human risk-assessment models using short-term mutagenicity test results. The risk-assessment evaluations are based to a great extent on mutagenic potency. As a sole source, results obtained with the Ames test or with other short-term tests have not yet been used in the United States by government agencies to regulate the use of chemical substances. However, as human risk-assessment models based on short-term tests become more refined and reliable, they most likely will become important tools in decision making not only by government agencies but also by private industry.

For background information *see* BIOASSAY; MUTAGENS AND CARCINOGENS; MUTATION in the McGraw-Hill Encyclopedia of Science and Technology.

[KRISTIEN E. MORTELMANS]

Bibliography: B. N. Ames, Dietary carcinogens and anticarcinogens, *Science*, 221:1256–1264, 1983; B. N. Ames, J. McCann, and E. Yamasaki, Methods for detecting carcinogens and mutagens with the *Salmonella* mammalian-microsome mutagenicity test, *Mutat. Res.*, 31:970–973, 1975; D. R. Calkins et al., Identification, characterization, and control of potential human carcinogens: A framework for federal decision-making, *J. Nat. Cancer Inst.*, 64:169–1976, 1980; J. McCann and B. N. Ames, Detection of carcinogens as mutagens in the *Salmonella* microsome test: Assay of 300 chemicals—Discussion, *Proc. Nat. Acad. Sci. USA*, 73:950–954, 1976; J. McCann and B. N. Ames, The *Salmonella*/microsome mutagenicity test: Predictive value for animal carcinogenicity, in H. H. Hiatt, J. D. Watson, and J. A. Winsten (eds.), *Origins of Human Cancer*, 1977.

Neutron-antineutron oscillations

In order to understand the observed stability of proton and bound neutron against decay into known lighter elementary particles such as positrons (e^+), pions ($\pi^\pm$, π^0), and so forth, it is necessary to introduce the concept of baryon numbers (B). The proton and neutron are assigned a baryon number $B = +1$, and all particles that are lighter are assigned zero baryon number. Stability of matter is explained by the further assumption that, like electric charge, baryon number is conserved in all physical processes. Similarly, observed properties of weak interactions necessitate the introduction of another conserved attribute, called the lepton number (L) for particles such as electron, muon, and neutrino (all of which have $L = +1$). The antiparticles ($\bar{p}$, $\bar{n}$, e^+, and so forth) have opposite baryon and lepton number, that is $B(\bar{p}) = -1$, $L(e^+) = -1$.

Recent attempts to construct unified pictures of all matter (baryons and leptons) and all forces (weak, nuclear, and electromagnetic) have raised the possibility that there may be ultraweak interactions that do not respect the conservation of baryon and lepton number and can therefore lead to the instability of the proton and thus all nuclear matter.

Extensive ongoing searches for the decay of the proton and bound neutron, to date, however, have failed, and the results can be interpreted to give lower limits on the proton lifetime and the lifetime for nuclear instability of about 10^{31} years. This implies that the strengths of baryon-number-violating interactions are extremely weak and are weaker than nuclear forces by at least 30 orders of magnitude. See PROTON.

Once the conservation of baryon number is sacrified as a law of physics, various new baryon-number-violating physical processes other than proton (and bound-neutron) decays (to leptons) can be envisaged and experimentally searched for. All of them must, however, respect electric charge conservation, since it is a fundamental physical law and sensible theoretical exceptions are not known to exist. One such new process found to exist in a class of unified gauge theories by R. N. Mohapatra and R. E. Marshak in 1980 is the neutron-antineutron (N-$\bar{N}$) oscillation, which changes baryon number by two units ($\Delta B = 2$). From purely phenomenological considerations, this process had been noted independently by V. Kuzmin and S. L. Glashow. Neutron-antineutron oscillation has certain unique, interesting features from both theoretical and experimental viewpoints, and its discovery would have significant bearing on the future of theoretical physics.

The process and its detection. According to the laws of quantum field theory, the neutron (N) and its antiparticle, the antineutron ($\bar{N}$), have the same mass. Therefore, if there are interactions that violate baryon number by two units, they can take a free neutron into an antineutron. In the language of quantum mechanics, a free-neutron wave function at time $t = 0$ will after a time acquire an antineutron component proportional to t/τ (for t much less than τ), where τ is the characteristic oscillation time in which all neutrons would become antineutrons if they were otherwise stable. A sample of n free neutrons from a source would, after time t, be expected to generate $n(t/\tau)^2$ antineutrons. Since the neutron lives only about 1000 s before it undergoes beta decay into other particles, for the N-$\bar{N}$ oscillation to be observable at current reactor neutron sources (for which $n = 10^9$–10^{13}/s), the characteristic oscillation time τ should not be more than 10,000 years. In actual experiments, due to other practical limitations, however, only oscillation times shorter than 1–100 years can be observed.

At first sight it appears that an N-$\bar{N}$ oscillation time of about a year should be unacceptable, since the age of the universe is about 10 billion years, and it appears that all matter (or at least the neutrons in all matter) would have converted to antimatter, which would subsequently annihilate with matter into radiation, leaving an "empty" universe. Luckily, this would not happen, and, indeed, the stability of matter with a lifetime for matter, τ_{matter}, greater than 10^{31} years is quite consistent with $\tau_{N\text{-}\bar{N}}$ less than 1 year. The reason is that the neutron is present in nature not as a free particle but always in a nuclear force field. The nucler force field exerts a different force on neutrons than on antineutrons. For example, at extremely short distances, neutrons are repelled by other protons and neutrons, whereas antineutrons are always attracted. The difference between the two kinds of potential experienced by the neutron and the antineutron is roughly of the order of 100–1000 MeV. Therefore, unlike the case of a free neutron, a bound neutron has to overcome this potential barrier to become an antineutron. This provides a suppression of the neutron-antineutron conversion inside the nuclei. Careful calculations reveal that present limits on nuclear stability indeed imply a free-neutron oscillation time $\tau_{N\text{-}\bar{N}}$ of about a year. It is, therefore, a lucky coincidence that precisely a $\tau_{N\text{-}\bar{N}}$ of about 1–100 years is what is observable in current reactors.

In performing the actual experiment, reactors are used as the source of thermal neutrons, which travel with a velocity of about 1000 ft/s (300 m/s). These neutrons are allowed to travel for a time, after which they reach the detector. If any of the neutrons have converted into an antineutron, the antineutron would annihilate with protons or neutrons in the detectors and would give rise to pions, which can be detected. To use the entire lifetime of the neutron would require an experiment extending over 180 mi (300 km). Aside from just the size considerations that limit the flight time t, the neutron beam must not be allowed to spread too much at the end of the flight. There is also another restriction. The magnetic field of the Earth exerts opposite forces on the neutron and antineutron since their magnetic moments are opposite. This, like the nuclear force discussed above, has the effect of suppressing the N-$\bar{N}$ oscillation. But it was pointed out that, if this magnetic field is reduced (degaussed) a thousand times, the neutron can behave effectively as a free particle. This degaussing also limits the size of the flight path, thus reducing t. The flight times used or contemplated for experiments are about 10^{-2} s. Two experiments are being conducted in Europe at Laue-Langevin Institute (ILL) at Grenoble, France, and at the Pavia reactor in Italy. There are also proposed experiments at Oak Ridge, Tennessee. The ILL experiments have already provided a lower limit on $\tau_{N\text{-}\bar{N}}$ of 10^6 s. All these experiments when concluded would either observe an N-$\bar{N}$ oscillation or increase the lower bound to about 10^9 s.

Theoretical implications. The observation of neutron-antineutron oscillations would have implications for the basic laws of nature. The present state of understanding of the subatomic phenomena is based on identifying the fundamental constituents of matter—the building blocks that make up the protons, neutrons, and so forth—and the nature of the forces between them. It is believed that the basic constituents at the present level of experimental accuracy are six quarks known as u (up), d (down), c (charm), s (strange), t (top), and b (bottom), the t quark remaining to be discovered. Each quark comes in three colors. For each quark there exists a lepton, which does not participate in nuclear forces; thus there are six leptons (all discovered): ν_e (neutrino), e^- (electron), ν_μ (muon-neutrino), μ^- (muon), ν_τ, and τ^-. The everyday universe needs only the u and d quarks and the ν and e, and only these particles will be discussed in this article. The weak, electromagnetic, and nuclear forces between these quarks are believed to arise from exchange of spin-1 bosons associated with certain symmetries operating on them. This idea is an extrapolation of the extremely successful theory of electromagnetic phenomena, where the associated symmetry is a local phase transformation symmetry, called U(1) symmetry, which to be theoretically consistent needs a massless spin-1 boson, the photon.

The associated symmetry, as far as weak interactions go, is believed to be one that operates in the same way on the pairs of quarks (u,d) and leptons (ν,e^-). Since it operates on two states at a time [and thus is called SU(2) symmetry], there are three spin-1 bosons that carry the weak force, denoted W_L^+, W_L^-, and Z. These particles have recently been discovered in the high-energy machines at CERN at Geneva, Switzerland. There is, however, a difference between the electromagnetic force and the weak force, arising from the fact that the weak interactions do not respect mirror reflection symmetry whereas electromagnetism does. For the quarks and leptons, this means that only their left-handed helicity states carry the SU(2) symmetry. Belief in this hypothesis is strengthened by the observation that the neutrinos that emerge in all weak processes such as beta decay and muon decay carry mostly left-handed helicity. Since neutrinos participate only in the weak interactions, this has been interpreted as evidence that the neutrino, unlike the electron and all the quarks, comes only with one helicity. It appears somewhat unesthetic that the neutrino, similar in many ways to its brothers, the electron and the u and d quarks, would be so dissimilar to them by having only one helicity. *See* INTERMEDIATE VECTOR BOSON.

A key question in theoretical physics is to determine whether there are hidden clues that will give evidence for the existence of a neutrino with right-handed helicity (ν_R), thereby "completing" the symmetry between the quarks and leptons. If the ν_R exists, then the natural symmetry of electroweak forces becomes $SU(2)_L \times SU(2)_R \times U(1)$ rather than $SU(2)_L \times U(1)$, where the $SU(2)_R$ symmetry operates on the right-handed helicity paired states, (u_R,d_R) and (ν_R,e_R^-). This makes parity a fundamental symmetry of all forms of interactions, including weak interactions, a point of view that was first advocated in 1974. In a manner similar to the $SU(2)_L$ case, there will be three new spin-1 states, W_R^+, W_R^-, and Z_R. In 1980 it was realized that in the $SU(2)_L \times SU(2)_R \times U(1)$ theory there are forces associated with U(1) symmetry which at very high energies respect the $B - L$ (baryon minus lepton) symmetry and, when parity symmetry breaks down, as it must to conform with observations, the $B - L$ symmetry must break by two units. This, therefore, leads to $\Delta B = 2$, $\Delta L = 0$ or $\Delta B = 0$, $\Delta L = 2$ interactions. The $\Delta B = 2$ transition is the phenomenon of neutron-antineutron oscillation. Thus the observation of N-$\bar{N}$ oscillations would not only reveal a new kind of ultraweak force in nature but also provide a clue to new symmetries of particles.

For background information *see* FUNDAMENTAL INTERACTIONS; SYMMETRY LAWS (PHYSICS); WEAK NUCLEAR INTERACTIONS in the McGraw-Hill Encyclopedia of Science and Technology.

[RABINDRA N. MOHAPATRA]

Bibliography: S. L. Glashow, in *Proceedings of Neutrino '79*, Bergen, Norway, vol. 1, p. 518, 1979; M. Goodman et al. (eds.), *Proceedings of the Harvard Informal Workshop on Neutron-Anti-neutron Oscillations*, 1982; V. A. Kuzmin, *JETP Pisma Rf.*, 12:335–336, 1970; R. N. Mohapatra and R. E. Marshak, *Phys. Lett.*, 94B:183–186, 1980; R. N. Mohapatra and R. E. Marshak, *Phys. Rev. Lett.*, 44:1316, 1980.

Nobel prizes

For 1983 there were seven recipients of the Nobel prizes, awarded by the Swedish Royal Academy.

Medicine or physiology. Barbara McClintock, of Cold Spring Harbor Laboratory, Long Island, New York, was recognized for her discovery of mobile genetic elements—"jumping genes"—which move from one location to another on the chromosome. Her experiments with chromosome exchanges in *Zea mays* provided cytological proof for T. H. Morgan's theory of crossover (breakage and crosswise reunion of chromosomes). A jumping gene can cause changes that range from varicolored kernels on an ear of corn to inherited resistance of bacteria to an antibiotic. McClintock's work also holds promise for research into the causes of cancer.

Physics. This prize was shared by two astrophysicists. Subrahmanyan Chandrasekhar, of the University of Chicago, was recognized for his calculation of rules governing the collapse of stars greater than 1.4 solar masses. The concepts of the neutron star and the black hole were later developed from his work. The Chandrasekhar limit, a critical value of stellar mass first calculated by Chandrasekhar with reference to white dwarf stars, has long been a fundamental tool of astrophysics.

William A. Fowler, of the California Institute of Technology, was honored for his leading role in developing the theory of synthesis of the heavier elements in the cores of stars. He hypothesized a complex series of steps leading to the collapse of a star. In the ensuing supernova, neutron ferment synthesizes the very heavy elements. New stars that condense from the remnants of these explosions would consequently be rich in the heavy elements, which are lacking in the older stars.

Chemistry. Henry Taube, of Stanford University, was recognized for his work in the mechanisms of electron transfer reactions, especially in metal complexes. Taube's discoveries arose from experiments concerning inorganic reactions, a subject which has flourished only since the 1940s. His work with reactions in metallic solutions has applications in industry and also in biology—specifically, how metal ions such as zinc, iron, and copper affect enzyme catalytic function.

Economics. Gerard Debreu, who holds professorships in both mathematics and economics from the University of California (Berkeley), was chosen for his contributions to economics at the theoretical level. His work has laid the mathematical foundation for modern microeconomic theory, with emphasis on supply and demand, and he is regarded by his colleagues as a pioneer in the science of mathematical economics.

Literature. British author William Golding was chosen for "his novels which, with the perspicuity of realistic narrative art and the diversity and universality of myth, illuminate the human condition," according to the Nobel committee. They went on to praise Golding for creating "not only somber moralities and dark myths . . . [but] also colorful tales of adventure . . . full of narrative joy, inventiveness and excitement." Golding's best-known novel is *Lord of the Flies*, a story of shipwrecked children in which he deliberately set out to contradict what he calls the "freshfaced" image of innocent youth.

Peace. Lech Walesa, of Gdansk, Poland, was honored for his founding of the Solidarity union movement in Poland and subsequent efforts to conduct peaceful dialog with the authorities there. The Nobel committee cited Walesa's "personal sacrifice to insure the workers' right to establish their own organizations," and declared that "a campaign for human rights is a campaign for peace."

Nondestructive testing

There are four nondestructive acoustic testing techniques which are capable of detection of fatigue damage in a solid, namely, body-wave reflection, surface-wave reflection, ultrasonic attenuation, and acoustic emission. The ideal nondestructive testing technique would permit very early detection of fatigue damage, so that proper assessment of the severity and the rate of severity increase of the structural damage leading to failure could be made. Thus, the most sensitive systems would be capable of detecting motion, pile-up, and break-away of dislocations; the next most sensitive would be capable of detecting microcracks; the least sensitive would only be capable of detecting macrocracks.

It is practically expedient to have nondestructive techniques which can successfully detect fatigue damages in each of these regimes, since some structures can tolerate larger amounts of fatigue damage or larger crack sizes than others without serious concern for loss of structural integrity. The best nondestructive technique for detecting and sizing of macrocracks is ultrasonic reflection; the best nondestructive techniques for detection of microcrack formation and possible premicrocrack fatigue damage are ultrasonic attenuation and acoustic emission; a large number of nondestructive techniques for residual stress (or strain) measurements are candidates for extremely early fatigue damage detection, with current interest primarily focused on ultrasonic wave velocity measurements.

Crack detection survey. Historically, nondestructive testing techniques have primarily been exploited to detect the existence of cracks in structural materials. Of prime concern in this regard is the size of the smallest flaw which can be detected by each of the nondestructive testing methods. In a comprehensive statistical analysis of the detectability of artificially induced fatigue cracks in aluminum alloy test specimens, 118 test specimens containing a total of 328 fatigue cracks were evaluated. The cracks ranged in length from 0.007 to 0.5 in. (0.18 to 12.7 mm) and in depth from 0.001 to 0.178 in. (0.03 to 4.51 mm). The test specimens were evaluated in the "as-milled" surface condition, in the "etched" surface condition, and after proof loading, in a randomized inspection sequence. The nondestructive test methods used were x-radiographic, dye penetrant, eddy current, and ultrasonic. The 984 nondestructive observations taken with each method served as a sample base for establishment of high confidence levels. Based on the results of these measurements, it was concluded that x-radiography is the least reliable of the four test methods for detection of tight cracks and should not be considered as a sensitive method for this use. On the other hand, the ultrasonic method was shown to be the most reliable for crack detection as well as to be the most accurate in measuring crack dimensions.

Body-wave reflection. In this technique, a transducer, longitudinal or shear-wave, generates an ultrasonic pulse which propagates through the body of the test specimen. Upon striking a crack, due to the acoustic impedance mismatch, part of the acoustic energy contained in the incident pulse is reflected to the transducer. The transducer now acts as a receiver and detects that portion of the pulse reflected from the crack.

Body-wave reflection techniques as applied directly to the fatigue damage detection process itself were first reported in 1964, and have continued up to the present time. These techniques have experienced considerable success, particularly with regard to crack detection and sizing. They offer a big advantage over penetrant, eddy current, and surface-wave reflection techniques in that they can locate and measure flaws at large depths in most materials. However, body-wave reflection techniques are not sensitive to material changes which give warning of fatigue damage prior to macrocrack formation. The main reason for this is that in order for an easily detectable fraction of the incident ultrasonic energy to be reflected from a crack to the transducer, the crack must be relatively large, and often the structure will already be well on the way to fracture.

Surface-wave reflection. In this technique, a transducer generates an ultrasonic pulse which propagates along the surface of the test specimen. Upon striking a crack, acoustical impedance mismatching causes a portion of the incident pulse to be reflected to the transducer, where it is detected.

Although surface-wave reflection techniques have been used quite successfully since 1962 to detect fatigue cracks which are undetectable by other nondestructive testing methods, surface-wave reflection techniques are not sensitive to material changes which give warning of fatigue damage prior to macrocrack formation, that is, prior to a surface crack becoming large enough to reflect to the transducer an easily detectable amount of ultrasonic energy.

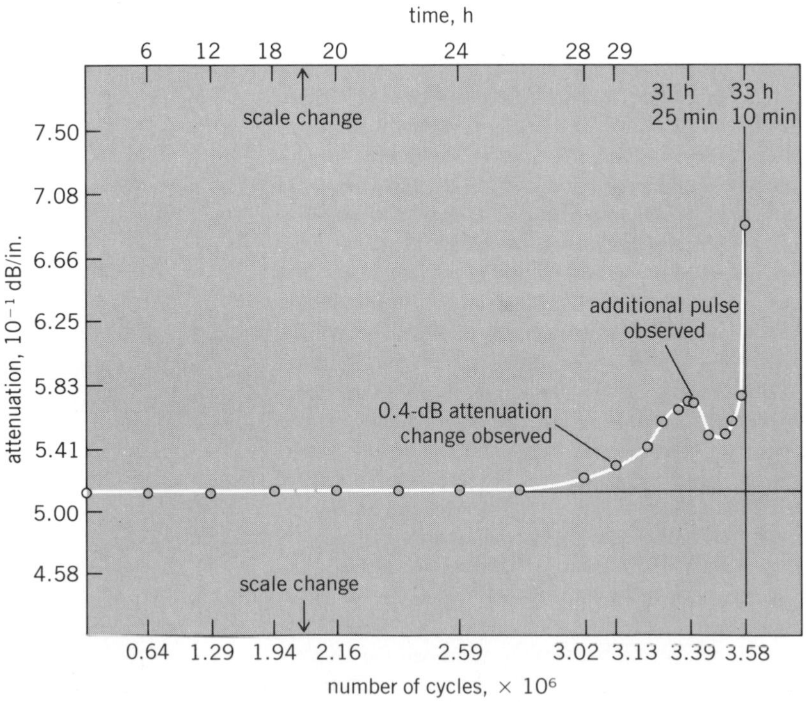

Fig. 1. Typical plot of average value of ultrasonic attenuation versus number of fatigue cycles for 6061-T6 polycrystalline aluminum. 10^{-1} dB/in. = 3.94 dB/m. (*After N. R. Joshi and R. E. Green, Jr., Ultrasonic detection of fatigue damage, J. Eng. Fract. Mech., 4:577–583, 1972*)

The condition of the workpiece and proper transducer attachment are special problems associated with the use of surface waves. Moreover, in many real materials, internal stress concentrations cause cracks to form in the interior of the structure and not at the surface where they can be detected by surface waves. Surface-wave reflection techniques are better suited to detection of surface-breaking macrocracks and crack surface-length measurement than to early detection of precrack fatigue damage.

Ultrasonic attenuation. In this technique, a transducer generates an ultrasonic pulse which propagates either through the body of the workpiece or along its surface. As a result of various mechanisms (in the case of fatigue, severely plastically deformed regions or microcracks), energy is lost from the incident pulse. Hence, the pulse which is reflected from the opposite side of the workpiece from which it was transmitted is detected to have a decreased amplitude in comparison with the incident pulse. Alternatively, a second transducer, affixed to the side of the workpiece opposite the sending transducer, can serve as a receiver, or for the surface-wave case, a receiving transducer can be affixed to the same surface of the workpiece as the generating transducer.

The first ultrasonic technique used to study the development of fatigue damage during stress cycling was the ultrasonic attenuation technique. As early as 1956, changes were observed in ultrasonic attenuation in the early stages of fatigue cycling on polycrystalline aluminum specimens. Similar measurements have continued up to the present time, since ultrasonic attenuation measurements and acoustic emission measurements have proved to be superior to all other nondestructive testing techniques for the early detection of fatigue damage.

It has been well documented that nucleation of fatigue cracks occurs as a result of inhomogeneous plastic deformation in microscopic regions of cyclically loaded metals. This inhomogeneous plastic deformation can be in the form of mechanical twinning, slip bands, deformation bands, or localized concentrations at grain boundaries and inclusions. Moreover, the mechanisms responsible for these regions of inhomogeneous plastic deformation are all based on dislocation interactions. In particular, dislocation interactions with point defects, with other dislocations, with stacking faults, with grain boundaries, and with inclusions are known to create regions of severe localized plastic deformation, which develop into microcracks, and these, in turn, grow to sufficient size to cause fracture. Therefore, since ultrasonic attenuation is extremely sensitive to dislocation motion and since dislocation motion is a prerequisite to fatigue damage, measurements of change in ultrasonic attenuation permit monitoring of dislocation motion and subsequent fatigue crack formation.

The extreme sensitivity of ultrasonic attenuation measurements for early detection of fatigue damage is illustrated by the results of experiments in which the change in ultrasonic attenuation measurements was used as a continuous monitor of fatigue damage during cyclic testing of aluminum and steel specimens. Ultrasonic pulses were generated by an x-cut quartz transducer firmly bonded to the clamped end of a polycrystalline aluminum or steel bar using Eastman 910 contact cement. The specimen shape, transducer size, and frequency used ensured that the entire specimen was filled with ultrasound in a guided wave mode. This mode of operation permitted the ultrasonic wave to detect both bulk and surface alterations in the test specimens. Each specimen was held in a horizontal plane and was displaced vertically at the opposite end by a fixture attached to an eccentric cam of a fatigue machine. In this manner, the specimen was fatigued in reverse bending as a cantilever beam to fracture at 30 Hz with a vertical displacement peak-to-peak which was set at a fixed value between 0.3 and 0.6 in. (7.5 and 15 mm).

Figure 1 shows the results of a typical experiment for a 6061-T6 polycrystalline aluminum specimen tested at a constant vibration amplitude peak-to-peak of 0.3 in. (7.5 mm). The upper abscissa indicates the test time in hours, while the lower abscissa indicates the number of fatigue cycles which elapsed before various events were detected. It can be seen that the attenuation began to increase at about 2.8 million cycles and proceeded catastrophically to increase up to fracture at about 3.5 million cycles. The 2.8-million-cycle mark occurred at about 26 h, and failure occurred after about 33 h. Thus the ultrasonic attenuation measurements gave a strong indication of the onset of fatigue failure approximately 7 h prior to the occurrence of the fracture itself. Conventional ultrasonic body-wave reflection monitoring was unable to detect any additional echoes due to energy reflected from the crack until about 2 h prior to failure. Thus, for this particular experiment, ultrasonic attenuation indicated that failure was imminent 5 h before conventional ultrasonic testing could indicate any crack formation.

Figure 2 summarizes the experimental results obtained from ultrasonic attenuation monitoring during fatigue tests on cold-rolled polycrystalline steel specimens. The entries represented by circles indicate the average time elapsed in three similar tests before an 0.4-dB change in attenuation was observed. The entries represented by triangles indicate the average time elapsed in three similar tests before detection of an additional ultrasonic pulse due to reflection of energy from a crack. The entries represented by squares indicate the average time elapsed in three similar tests before fracture occurred. These results are similar to those obtained with aluminum in that ultrasonic attenuation gave early warning of fatigue failure.

Acoustic emission. In this technique, elastic waves, some of which possess ultrasonic frequencies, are internally generated when a test specimen is subjected to a sufficiently high state of stress. A transducer affixed to the surface of the test specimen detects these internally generated waves.

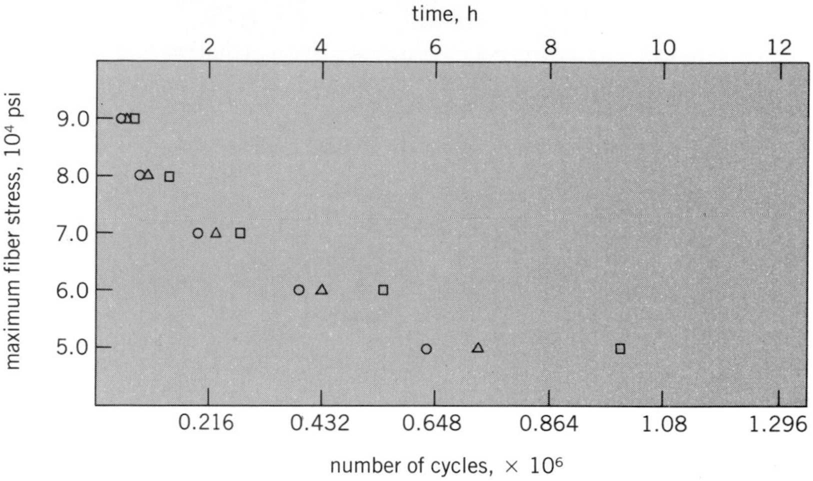

Fig. 2. A summary of experimental results obtained from ultrasonic attenuation measurements made during fatigue cycling of cold-rolled polycrystalline steel. Symbols explained in text. 10^4 psi = 69 megapascals. (*After N. R. Joshi and R. E. Green, Jr., Ultrasonic detection of fatigue damage, J. Eng. Fract. Mech., 4:577–583, 1972*)

Even though J. Kaiser is generally credited with "discovering" acoustic emission in 1950, it was not until 1964 when work was initiated in the ultrasonic range of frequencies that acoustic emission actually became an engineering tool. The amount of work performed in this field since that time is enormous, and many investigators have monitored acoustic emission successfully during fatigue testing, despite the difficulty encountered in separating the acoustic emission signals from background noise.

The extreme sensitivity of both ultrasonic attenuation and acoustic emission monitoring suggests that the greatest potential for fatigue damage detection lies in utilizing these two techniques simultaneously. Both monitoring techniques have been applied simultaneously to prismatic specimens of 7075 polycrystalline aluminum. The capability of making reliable acoustic emission measurements during cyclic loading was provided by use of an acoustic emission transducer-cable-amplifier system. This system, which combined a multishielded Mu-metal- and copper-screened cable with other special features, has a signal-to-noise ratio at least three orders of magnitude better than conventional acoustic emission monitoring systems.

Figure 3 shows the results of a typical experiment for an as-received 7075-T651 aluminum specimen tested at a constant vibrational amplitude of 0.35 in. (9 mm) peak to peak. Although there was very little change in ultrasonic attenuation during the early portion of the fatigue life, at 80% of the fatigue life the attenuation increased rapidly, quickly exceeding the sensitivity range monitored. On the other hand, the true root-mean-square voltage of the acoustic emission signal was observed to fluctuate during the first 20% of the fatigue life, followed by a portion in which it remained relatively unchanged, and then at 80% of the fatigue life it in-

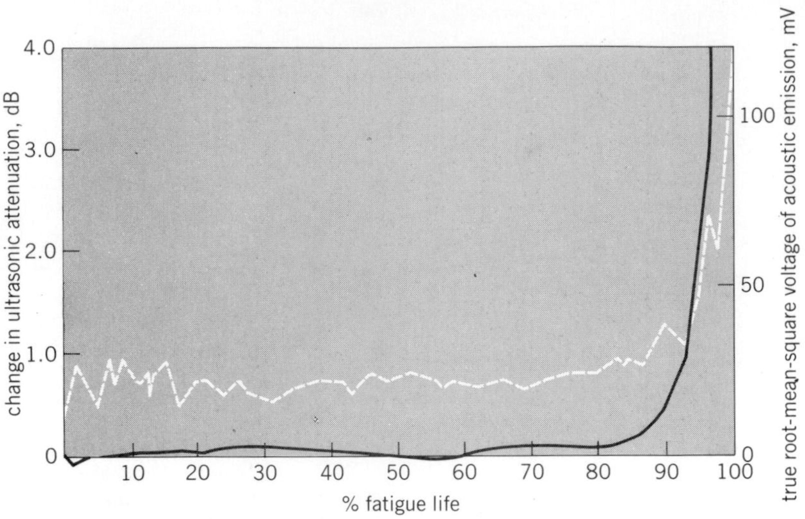

Fig. 3. Typical plot of change in ultrasonic attenuation (broken curve) and true root-mean-square voltage of acoustic emission (solid curve) versus fatigue life for as-received 7075-T651 aluminum. (*After J. C. Duke, Jr., and R. E. Green, Jr., Simultaneous acoustic emission and ultrasonic attenuation monitoring of the mechanical deformation of aluminum, Proceedings of 2d International Conference on Mechanical Behavior of Materials, pp. 1646–1650, 1976*)

creased rapidly with only minor fluctuations until failure.

For background information *see* ACOUSTIC EMISSION; METAL, MECHANICAL PROPERTIES OF; NONDESTRUCTIVE TESTING; PLASTIC DEFORMATION OF METAL; ULTRASONICS in the McGraw-Hill Encyclopedia of Science and Technology.

[ROBERT E. GREEN, JR.]

Bibliography: R. E. Green, Jr., Non-destructive methods for the early detection of fatigue damage in aircraft components, *Proceedings of AGARD/NATO Lecture Series No. 103*, AGARD-LS-103, Pap. 6, 1979; R. E. Green, Jr., Ultrasonic attenuation detection of fatigue damage, *Ultrasonics International 1973 Conference Proceedings*, pp. 187–193, 1973; W. D. Rummel et al., The detection of fatigue cracks by nondestructive test methods, *Mater. Eval.*, 32:205–212, 1974; D. O. Thompson et al., *Proceedings of DARPA/AFML Review of Progress in Quantitative NDE*, Ames Laboratory, Iowa State University, 1973–1983.

Nowcasting

Nowcasting is a form of very short-range weather forecasting that is emerging as a result of recent technological developments, especially in the fields of meteorological observation, data processing, and communications. The term nowcasting is sometimes used loosely to refer to any area-specific forecast ranging up to 12 h ahead that is based on very detailed observational data. However, nowcasting should probably be defined more restrictively as the detailed description of the current weather along with forecasts obtained by extrapolation up to about 2 h ahead. Useful extrapolation forecasts can be obtained for longer periods in many situations, but in some weather situations the accuracy of extrapolation forecasts diminishes quickly with time as a result of the development or decay of the weather systems.

Comparison with traditional forecasting. Much of the weather forecasting practiced today is based on the widely spaced observations of temperature, humidity, and wind, obtained from the worldwide network of balloon-borne radiosondes, supplemented by satellite soundings. These data are used as input to numerical-dynamical weather prediction models in which the equations of motion, mass continuity, and thermodynamics are solved for large portions of the atmosphere. The resulting forecasts are general in nature, and although these general forecasts for 1 or more days ahead have undoubtedly improved in line with continuing developments of the numerical-dynamical models, there have not been corresponding improvements in local forecasts for the period up to 12 h ahead. The trouble is that most of the mathematical models cope adequately with only the large weather systems such as cyclones and anticyclones. These are what the meteorologist refers to as synoptic-scale weather systems.

During the past decade nowcasting has been growing up alongside the more traditional weather forecasting approach. It involves the use of very detailed and frequent meteorological observations, especially remote sensing observations, to provide a precise description of the "now" situation from which very short-range forecasts can be obtained by extrapolation. Of particular value are the patterns of cloud, temperature, and humidity which can be obtained from geostationary satellites and the fields of rainfall and wind measured by networks of ground-based radars. These kinds of observation enable the weather forecaster to keep track of smaller-scale events such as squall lines, fronts, and thunder-

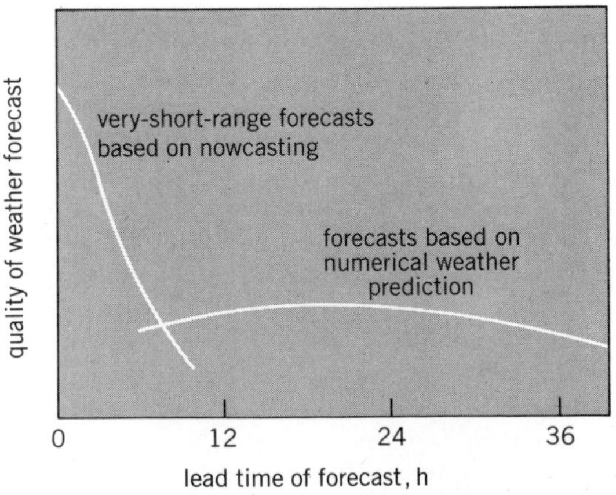

The quality of weather forecasts, defined as the product of the accuracy and detail achievable, depicted schematically as a function of lead time for two different forecasting methods. Traditional forecasts based on numerical weather prediction deal with synoptic-scale events such as cyclones and anticyclones; very short-range forecasts based on the nowcasting approach resolve so-called mesoscale events, such as thunderstorms, fronts, and squall lines.

storm clusters, and various terrain-induced phenomena such as land-sea breezes and mountain-valley winds. Meteorologists refer to these systems as mesoscale weather systems because their scale is intermediate between the large, synoptic scale cyclones and the very small, or microscale, features such as boundary layer turbulence.

The graph in the illustration summarizes the distinction between traditional forecasting and nowcasting in terms of the lead time and quality of the forecast. The two approaches are complementary; each has its place according to the lead time and detail required of the forecast. Conceptually, nowcasting is a simple procedure. However, vast amounts of data are involved, and it is only recently that the necessary digital data-processing, transmission, and display facilities have become available economically.

Role of information technology. A key problem in nowcasting is that of combining diverse and complex data streams, especially conventional meteorological data with remote sensing data. It is widely held that this combination should be achieved by using digital data sets displayed on interactive video displays—the so-called work station concept. By means of the advanced human-computer interaction techniques now being developed, the weather forecaster of the future will be able to analyze the merged data sets by using a light pen or a finger on touch-sensitive television screens. Various automatic procedures can be implemented to help the forecaster carry out analyses and extrapolation forecasts, but the incomplete nature of the data sets is such that the forecaster will usually be in the position of needing subjectively to fine-tune the products. The idea behind the forecasting work station is to simplify the routine chores of basic data manipulation so that the forecaster is given the maximum opportunity to exercise judgment within the context of a highly automated system.

Very short-range forecast products are by their nature highly perishable: they must be disseminated promptly if they are not to lose their value. Advances in technology now offer the means for the rapid tailoring and dissemination of digital forecast information and for presenting the material in convenient customer-oriented formats. In some cases dissemination may be by direct computer-to-computer link with the user's control system. In other cases the user will benefit from new methods of visual presentation, using such media as cable television or viewdata.

Mesoscale numerical models. There is some evidence of a gap in forecasting capability between nowcasting and traditional synoptic-scale numerical weather prediction. As shown in the illustration, this occurs for forecast lead times between about 6 and 12 h, that is, for periods when development and decay are beginning to invalidate forecasts by simple extrapolation. To some extent the forecaster can identify some of the likely developments by interpreting the nowcast information in the light of local climatologies and conceptual life-cycle models of weather systems. But the best way to forecast changes is to use numerical-dynamical methods. It would be natural to assume that an immediate way forward would be to incorporate the detailed nowcast data as input to numerical-dynamical models with a finer resolution than those models presently in operational use; these are the so-called mesoscale numerical models. Research into the use of detailed observational data in mesoscale models is actively under way, but there are significant technical difficulties which will make progress slow. The most rapid advances in the application of mesoscale models to very short-range forecasting are thus likely to be made in those circumstances where the initial state of the atmosphere does not need to be specified in detail. Those circumstances arise where the predominant forcing of the weather is by well-defined terrain features such as land-water boundaries and hills.

For background information *see* METEOROLOGICAL SATELLITES; METEOROLOGY; RADAR METEOROLOGY; WEATHER FORECASTING AND PREDICTION in the McGraw-Hill Encyclopedia of Science and Technology.

[KEITH A. BROWNING]

Bibliography: K. A. Browning (ed.), *Nowcasting*, 1982.

Number theory

Number theory has long been known for its beauty, purity, and uselessness. With the advent of computer technology, this situation has changed. Due to an increased emphasis on information theory, coding, and cryptography, number theory is losing its purity and is being applied.

The applications fall roughly into two groups: information processing codes and secret codes. The first type of "code" is simply a way of storing and transmitting information, the well-known Morse code providing an excellent example. Skillful coding can speed up the running of computer programs. The example given below provides some indication of the manner in which number theory has found a place among computer code techniques.

Congruence arithmetic. Number theory is the theory of whole numbers, of arithmetic operations performed without recourse to fractions or decimals. In computer terminology, number theory is the study of arithmetic in the integer mode. Its basic operations are addition, subtraction, multiplication, and division with remainder. The last operation is the most interesting; furthermore, for many purposes, the remainders are more important than the quotients.

The procedure of dividing and keeping only the remainder is called congruence arithmetic. Thus two integers a and b are said to be congruent modulo m, if a and b give the same remainder upon division by m. The congruence relation is written $a \equiv b \pmod{m}$. For example, $22 \equiv 1 \pmod{7}$; also $365 \equiv 1 \pmod{7}$; and hence $22 \equiv 365 \pmod{7}$. [The fact that $365 \equiv 1 \pmod 7$ explains why Christmas advances one day of the week (for example, from Tuesday to Wednesday) between one year and the next (if leap years are ignored). If only "the day of the week" is of concern, and not how many weeks intervene, then the only relevant fact about the 365 days in a year is that 365 divided by 7 gives a remainder of 1.]

The congruence relation, $\equiv$, behaves like an equals sign in many respects. For example, consider a calculation in which the modulus m remains fixed. If $a \equiv b$ and $b \equiv c$, then $a \equiv c$. Likewise, if $a \equiv a'$ and $b \equiv b'$, then $a + b \equiv a' + b'$, $a - b \equiv a' - b'$, and $ab \equiv a'b'$. Putting this into words: congruence is preserved under addition, subtraction, and multiplication.

Application to multiple precision. The fact that congruence is so preserved is one reason for its utility in machine computation. For example, consider the problem of multiple precision. Computers are normally designed to handle numbers consisting of a block of digits, this block being of fixed length. For purposes of illustration, imagine a machine which handles blocks of 13 digits (base 10). Suppose more precision, say 24 digits, is needed. One approach would be to break each 24-digit number into two halves (like breaking 6789 into 67 and 89). Unfortunately, this operation does not behave nicely with respect to addition or multiplication. To correct for this, a carry algorithm must be introduced, essentially like the method of "carrying" learned in school. Then this rather clumsy procedure will be executed every time an arithmetic operation is performed. Despite its awkwardness, this procedure has a certain logical simplicity and is frequently employed by programmers.

However, congruence arithmetic sometimes provides a better approach to the problem of multiple precision. Suppose 24 digits are needed. The computer holds 13-digit numbers, but its arithmetic is only accurate for numbers of about 6 digits (because multiplication doubles the number of digits). So four moduli, m_1, m_2, m_3, and m_4, of around 6 digits are chosen. These must be relatively prime, that is, no two of them should have a common factor. A good choice for m_1, m_2, m_3, and m_4 might be $2^{20} - 3$, $2^{20} - 1$, $2^{20} + 1$, and $2^{20} + 3$. These four numbers are relatively prime, and their product, $m_1m_2m_3m_4$, has 25 digits, giving a little more precision than needed. Now an ancient theorem (the Chinese remainder theorem) states that each integer x from 1 to $m_1m_2m_3m_4$ can be uniquely represented by its four congruent numbers, r_1, r_2, r_3, and r_4, that is, by its remainders upon division by m_1, m_2, m_3, or m_4, respectively.

As discussed above, the congruent numbers r_i behave naturally with respect to addition, subtraction, and multiplication. Thus the very fast internal algorithms of the machine, applied to the short 6-digit numbers and then combined via the Chinese remainder theorem, can be used for this arithmetic. Unfortunately, this procedures does not work well for division, or for telling which of two numbers is the larger. Hence the procedure is mainly useful for multiple-precision problems involving long chains of additions, subtractions, and multiplications. Such problems are not too common, but they are not altogether rare either.

Composite numbers with unknown factors. Congruence arithmetic also forms the basis for the following remarkable fact. A number n is called composite if it has nontrivial factors d, that is, if there are numbers d other than n or 1 such that d divides exactly into n. At present, there are many large numbers which are known to be composite but for which no factors are known; that is, factors can be proven to exist, but it is not known how to find them.

The following is typical of the way in which such results are proved. Recall that a number n, greater than 1, is called prime if it is not composite, that is, if n has no divisors other than itself and 1. Now a theorem of P. Fermat states: If p is prime, and a is not divisible by p, then $a^{p-1} \equiv 1 \pmod p$ [where $\equiv$ means congruence, as defined above]. Consider a number p whose primality is to be tested. Applying Fermat's theorem, it is possible to reason as follows. Take any number a not divisible by p and compute a^{p-1}. Fermat's theorem indicates that if p is prime, then $a^{p-1} \equiv 1 \pmod p$. Hence if it is found a^{p-1} is not congruent to 1 module p, it follows that p is not prime. The theoretical knowledge that p is not prime, that is, p has factors, does not provide any practical method for locating these factors. In fact, as noted above, there are many numbers which are known to be composite but for which no factors have been found.

When p is large, the computation of a^{p-1} requires special devices, and these too are based on congruence. The method can be illustrated by an example, $p = 2^{128} + 1$, which was proved to be composite 50 years before any factors were found for it. For the other number a in Fermat's theorem, it suffices to take $a = 3$. Now there are two steps which shorten the computation immensely:

(1) Since Fermat's theorem involves congruence (mod p), congruence arithmetic can be used. That is, whenever, at some stage in the calculation, a number larger than p is obtained, that number is divided by p, and only the remainder is retained. This prevents the numbers occurring in the calculation from becoming too large.

(2) To compute $a^{p-1} = 3^{2^{128}}$, one does not multiply 3 by itself 2^{128} times, since the run-time for such a procedure would exceed the age of the universe. Instead, one merely starts with 3 and then squares 128 times. After each squaring, the resulting number is reduced (mod p), as described above.

The converse of Fermat's theorem is false. Thus, if p is not prime, the equation $a^{p-1} \equiv 1 \pmod p$ might still hold, by accident, so to speak. Hence Fermat's equation can give a conclusive proof that certain numbers are not prime [when the equation $a^{p-1} \equiv 1 \pmod p$ fails, contradicting Fermat's theorem if p were prime]. But this method can only give probabilistic support to the statement that p is prime [when the equation $a^{p-1} \equiv 1 \pmod{p}$ holds]. In this latter case, the equation might simply hold by accident.

The problem of finding a fast and universal test for primality is still open. Considerable progress has been made in recent years, by L. M. Adleman, C. Pomerance, R. S. Rumely, and others. The related problem, of actually finding the factors of composite numbers, appears to be even more difficult.

Application to secret codes. This circumstance—that the factors of large composite numbers are very hard to find—has been suggested as a basis for secret codes. A typical one of these codes works as follows. The code maker takes two large prime numbers p and q (of, say, around a hundred digits in base 10) and multiplies them to produce the composite number $n = pq$. The number n is made public, but the factors p and q are kept secret. In the operation of the code, the encode procedure requires only a knowledge of n, which was made public. Hence these are sometimes called public key codes. However, the decode procedure requires a knowledge of p and q as well. The code maker knows the factors p and q, because he or she put them there, but a code breaker does not know these factors and presumably cannot find them. In brief: here is a code in which anyone can encode, but only the code maker can decode.

The details of this code are as follows. The message (say, in English) is first replaced by a string of digits, for example, by writing $A = 01$, $B = 02$, ... There is nothing secret about this part. Then this string of digits is broken into chunks of 200 digits, so that each chunk is a number smaller than n. The chunks are encoded separately, all by the same method.

The encode process involves, besides the number n mentioned above, a prime number r which does not divide exactly into either $p - 1$ or $q - 1$. The value of r, like n, is made public. Then the encode procedure is:

Start with x = the number corresponding to a "chunk."
Raise x to the r-th power, and reduce modulo n.

The resulting number y is the encoded form of the message. Thus x corresponds to the original chunk, and y is its encoded version.

To decode, it is necessary to go backward from y to x. Actually, this is easy if the factors p and q of n are known, precisely the information which the code maker has kept secret. The recipe for decoding is:

Set $\varphi = (p - 1)(q - 1)$.
Find a number s such that $rs \equiv 1 \pmod{\varphi}$.
Raise y to the s-th power, and reduce modulo n.

Here is a brief explanation. It is observed that the argument involves two different kinds of congruence: (mod n) and (mod φ). The number φ is known as Euler's phi function for $n = pq$. By a theorem of Euler, expression (1) is valid. (Actually, this only

$$x^{\varphi} \equiv 1 \pmod{n} \qquad (1)$$

holds if neither p nor q divides exactly into x. In practice, since p and q have about a hundred digits apiece, it is almost infinitely unlikely that either of them will divide x.)

Then, for any multiple $k\varphi$ of φ, expression (2) is valid.

$$x^{k\varphi} \equiv 1 \pmod{n} \qquad (2)$$

Since $rs \equiv 1 \pmod{\varphi}$, $rs - 1$ is some multiple $k\varphi$ of φ, so that $rs = k\varphi + 1$.

Now since $y \equiv x^r \pmod{n}$, and the purported "decode" $\equiv y^s \pmod{n}$, expression (3) is valid.

$$\text{decode} \equiv y^s \equiv x^{rs} \equiv x^{k\varphi+1} \equiv x \pmod{n} \qquad (3)$$

Thus the "decode" equals the original message x, as desired.

The code is not useful unless the operations listed in the "decode" procedure can be executed in practice. The computation of $\varphi = (p - 1)(q - 1)$ is just a multiplication, provided that p and q are known. The finding of a number s such that $rs = 1 \pmod{\varphi}$ is done by using an algorithm developed by Euclid. Finally, the raising of y to the s-th power (mod n) can be carried out by a slight extension of the method given above for determining the congruence of $3^{2^{128}}$.

However, if p and q are not known, the problem is very difficult. For this reason, the code has even been labeled unbreakable. This designation seems unwarranted. Difficult problems do get solved, and nobody can predict when, or by whom.

For background information *see* CRYPTOGRAPHY; NUMBER THEORY in the McGraw-Hill Encyclopedia of Science and Technology.

[J. IAN RICHARDS]

Bibliography: L. M. Adleman, C. Pomerance, and R. S. Rumely, On distinguishing prime numbers from composite numbers, *Ann. Math.*, 117:173–206, 1983; M. E. Hellman, The mathematics of public-key cryptography, *Sci. Amer.*, 241(2):146–157, August 1979; C. Pomerance, The search for prime numbers, *Sci. Amer.*, 247(6):136–147, December 1982; I. Richards, The invisible prime factor, *Amer. Sci.*, 70:176–179, March-April 1982.

Nutrition

Food is a complex mixture of many chemical compounds. Plants and plant products normally eaten by people contain nutrients, toxicants, and antinutrients. The nutritional quality of plant foods is affected by the amount and bioavailability of specific nutrients in the food and by the interaction of nutrients with antinutritive compounds.

Plants as sources of nutrients. Plants provide about 60% of the food available for consumption by people in the United States, and on a dry-weight basis, about one-third of the food consumed is derived from cereal grains (wheat, corn, rice, oats, barley) and legume seeds (beans, soybeans, peas, lentils). In less-developed countries, plants contribute more of the human diet.

Plants and plant products are valuable sources of nutrients and energy for people. Nutrients are those compounds or elements required for normal development, growth, or maintenance of bodily functions. Essential amino acids (usually incorporated into proteins), essential fatty acids, various vitamins, and mineral elements are some of the nutrients in plant foods. Plants also supply a substantial amount of the energy (calories), mostly in the form of carbohydrates (starches and sugars) and lipids (oils and fats), consumed by people.

Major foods of plant origin contribute more than 30% of the protein, approximately 40% of the fat, more than 90% of the carbohydrate, and about 60% of the energy available for consumption by people in the United States. These values are based on the amount of food available at the retail level, and they do not reflect potential waste of food or loss of nutrients during food preparation. Although amounts of nutrients in plant foods are important, additional factors must be considered to properly assess the nutritional value of specific foods. Protein quality and mineral bioavailability are examples of factors that affect the value of plants as sources of specific nutrients.

Protein quality. The digestibility and quality of plant proteins influence the nutritional value of plant foods. All whole-plant foods provide some protein, but cereal grains and legume seeds are the major sources of plant protein consumed by people. In general, plant proteins are less digestible than animal proteins. Additionally, when protein is consumed at the recommended level (about 0.01 oz/lb or 0.8 g/kg body weight for mature adults), plant proteins have less nutritional value (quality) than do animal proteins because of the amino acid composition of the proteins. Compared to animal proteins, plant proteins may lack sufficient quantities of one or more of the essential amino acids necessary to meet optimal nutritional requirements. Most cereal grains have low amounts of lysine, and the levels of tryptophan or threonine in some grains may also be limiting. Legume seeds are mainly deficient in methionine, an essential sulfur-containing amino acid. Requirements of people for amino acids can be met by eating mixtures of proteins from cereal grains and legume seeds. Moreover, plant protein quality, particularly that of certain cereal grains, has been improved by genetic selection of varieties with increased lysine content. Factors controlling the methionine content of legume seeds, particularly soybeans (which may contain as much as 40% protein), is currently being studied in several laboratories.

Mineral bioavailability. The total amount of a mineral nutrient in a food does not necessarily reflect the nutritional quality of that food. Rather, the amount of an element that is absorbed (available) from a food is important when determining the adequacy of specific plants as sources of essential minerals. Bioavailability of a mineral element in food refers to that portion of the total amount of the element present that is absorbable in a metabolically active form. Because many of the processes and factors that affect mineral bioavailability are not understood, predicting the available mineral content of plant foods currently is extremely difficult. Estimates of mineral bioavailability from a limited number of plant foods have been obtained for only a few elements. Some dietary constituents enhance mineral bioavailability, and other factors depress mineral bioavailability. Several of these constituents are shown in Table 1. Many of the inhibitors of mineral bioavailability listed in Table 1 are controversial, and additional research is needed before conclusive proof of their negative role in mineral bioavailability is obtained.

Antinutrients. Some of the many compounds occurring naturally in plant foods are considered to be antinutritives. These secondary plant metabolites

Table 1. Constituents of plant foods that possibly inhibit or promote mineral bioavailability

Constituent	An element affected	Source of constituent or dietary levels
		Inhibitors
Cadmium	Zinc	High cadmium levels in contaminated leafy vegetables or rice grain
Fatty acids (long-chain)	Magnesium	Oil seeds (possibly)
Fibrous carbohydrates	Zinc	Whole cereal grains
Goitrogens	Iodine	Some species of *Brassica* (cabbages), soybeans
Molybdenum	Copper	High molybdenum levels in some sorghum varieties
Oxalic acid	Calcium	Oxalate accumulators (spinach, rhubarb)
Phytic acid	Zinc	Mature seeds and cereal grains
Protein	Zinc	Zinc absorption is depressed when dietary protein levels are low
Tannins (polyphenols)	Zinc	Coffee, tea, some sorghum varieties
		Promotors
Ascorbic acid	Iron	Citrus fruits
Amino acids (selected)	Zinc	Histidine and cysteine promote zinc uptake
Vitamin D	Calcium	Common food plants lack high vitamin D levels

may impart, to the plant, resistance against plant pathogens, insects, or vertebrate herbivores. With respect to nutrition, antinutrients are constituents of plants that adversely affect the digestion, absorption, or utilization of a nutrient or are compounds that in some way increase nutrient requirements. In addition to factors that adversely affect mineral bioavailability, some examples of proposed antinutrients in selected plant foods are proteinase inhibitors (protease inhibitors), lectins (hemagglutinins, phytohemagglutinins, phytoagglutinins, and mitogens), and amylase inhibitors (so-called starch blockers). Some common plant foods containing these biologically active substances that may act as antinutrients are shown in Table 2.

Proteinase inhibitors. Those occurring in selected legume seeds and in some varieties of potato tubers have been studied more extensively than those in other plant foods. The major proteinase inhibitors in legume seeds are proteins that interfere with the action of serine proteinases, mainly trypsin and chymotrypsin, protein-digesting enzymes secreted by the pancreas. Multiple forms (isoinhibitors) of serine proteinase inhibitors usually occur within seeds of a given legume species. Inhibitors of other pancreatic proteinases are known. The serine proteinase inhibitor serves as a pseudosubstrate for the digestive enzyme; the inhibitor functions by binding to the enzyme and forming an enzyme-inhibitor complex that is relatively inactive at normal physiological conditions. When experimental animals are fed raw legume seeds or purified proteinase inhibitors, inhibition of the protein-digesting enzymes results in decreased protein digestion, increased secretion of digestive enzymes, excessive fecal loss of nitrogen and protein, and depressed growth. Since various experimental animals respond differently when fed raw legume seeds, extrapolation of results from animal species to humans must be done with care. Moreover, with the exception of potato tubers, relatively little is known about the nutritional significance of proteinase inhibitors in nonleguminous plant foods. However, much of the proteinase inhibitor activity in most plant foods is destroyed or markedly reduced by processing, moist-heat treatment, or ordinary cooking methods. The influence of prolonged consumption of low levels of proteinase inhibitors on health remains to be determined.

Lectins. While these are found mainly in seeds, particularly in legumes, some lectins occur also in roots, tubers, and leaves. Most lectins are glycoproteins that will agglutinate erythrocytes and other types of cells, and some lectins have antinutritional properties. Specific lectins, particularly those in some varieties of beans (*Phaseolus vulgaris*), decrease protein digestion, depress absorption of amino acids and sugars, and cause growth retardation and death when fed to experimental animals. Lectins with antinutritional significance apparently bind to receptor sites on intestinal epithelial cells, and this binding results in nonspecific interference with the absorption of nutrients. People normally do not eat plant species and varieties containing lectins that are highly toxic. Moreover, normal cooking procedures usually eliminate the hemagglutinating ability and antinutritional effects of those lectins present in common plant foods.

Amylase inhibitors. Starch is hydrolyzed (digested) initially to maltose by the action of salivary and pancreatic alpha-amylases. Maltose is further degraded to simple sugars by enzymes located in the brush border of intestinal epithelial cells. Cooked starch is hydrolyzed more rapidly than raw starch. Inhibitors of the alpha-amylases have been found in various cereal grains and legume seeds. The digestion of raw starch is impaired more by amylase inhibitors than is the digestion of cooked starch. By depressing the utilization of starch, amylase inhibitors effectively reduce the supply of energy (in the form of sugars) available for absorption from the alimentary tract. Purified amylase inhibitors have been used to control weight gain. However, any carbohydrate that for any reason escapes digestion and absorption in the small intestine has the capability of causing flatus. Excessive flatus production may cause nausea, diarrhea, and abdominal cramps and pain. Because some of these signs have occurred in some people who have purposely eaten high levels of amylase inhibitors, commercial preparations of amylase inhibitors are not presently available to the general public. Moist-heat treatment generally inactivates amylase inhibitors that occur naturally in plant foods, but some amylase inhibitory activity in wheat products may persist through the baking process.

Table 2. Common plant foods containing biologically active substances that may have antinutritional properties*

	Biologically active substance†		
Plant	Proteinase inhibitor	Lectin	Amylase inhibitor
Cereal grains			
Barley (*Hordeum vulgare*)	X	X	X
Corn (*Zea mays*)	X	X	X
Oats (*Avena sativa*)	X	—	X
Rice (*Oryza sativum*)	X	X	—
Rye (*Secale cereale*)	X	—	X
Wheat (*Triticum aestivum*)	X	X	X
Legume seeds			
Beans (*Phaseolus vulgaris*)‡	X	X	X
Lentil (*Lens esculenta*)	X	X	X
Peanut (*Arachis hypogaea*)	X	X	—
Pea (*Pisum sativum*)	X	X	X
Soybean (*Glycine max*)	X	X	—
Tubers			
Potato (*Solanum tuberosum*)	X	X	X

*Most antinutritional properties are destroyed by processing, moist-heat, or cooking methods.

†The proteinase inhibitor adversely affects digestion of protein. All lectins are not toxic; specific lectins impair protein digestion and depress absorption of sugars and amino acids. The amylase inhibitor impairs digestion of starch.

‡Several varieties: kidney, navy, pinto, wax, French, and garden beans.

Additionally, because they are proteins or glycoproteins, amylase inhibitors are generally inactivated by pepsin and trypsin digestion. However, when large quantities of amylase inhibitors are consumed, some amylase inhibitory activity may not be destroyed by proteolytic digestion.

Summary. The nutrient composition and ability of various plant foods to meet human nutritional requirements have been studied extensively. However, the significance to human nutrition of many of the antinutrients that occur naturally in plant foods is largely unknown. Most of the adverse effects of specific antinutrients have been determined in studies with experimental animals fed raw foods. The activities of various antinutrients in selected plant foods are markedly reduced or eliminated by proper processing, preparation, and cooking procedures. As normally consumed by people, common plant foods contain amounts of antinutrients lower than the levels reported to induce deleterious effects in experimental animals. Moreover, people generally eat a moderate amount of a variety of plant foods so that continuous exposure to any particular antinutrient is diminished.

For background information *see* AMINO ACIDS; ENZYME; NUTRITION; PROTEIN in the McGraw-Hill Encyclopedia of Science and Technology.

[WILLIAM A. HOUSE]

Bibliography: D. Duffus and C. Slaughter, *Seeds and Their Uses*, 1980; W. A. House and R. M. Welch, Effects of naturally occurring antinutrients on the nutritive value of cereal grains, potato tubers and legume seeds, in W. H. Gabelman and R. M. Welch (eds.), *Crops As Sources of Nutrients for People*, Amer. Soc. Agron. Spec. Publ. 46, 1984; I. E. Liener, The nutritional significance of plant lectins, in R. L. Ory (ed.), *Antinutrients and Natural Toxicants in Foods*, 1981; I. E. Liener and M. L. Kakade, Protease inhibitors, in I. E. Liener (ed.), *Toxic Constituents of Plant Foodstuffs*, 2d ed., 1980.

Oncology

The retinoblastoma gene is considered to be a model for a class of recessive human cancer genes which function as a suppressor or regulator of tumorigenicity. A primary mechanism for tumor development appears to be the loss or inactivation of both alleles of this gene. This mechanism contrasts with the molecular mechanism proposed as critical for the expression of putative human oncogenes. These latter genes may produce tumors after they are activated or otherwise altered. In addition, the extremely high incidence of second primary tumors among individuals who inherit one inactive retinoblastoma allele suggests that the retinoblastoma gene is directly responsible for the development of several other primary cancers.

Genetics of retinoblastoma. Retinoblastoma (Rb) occurs in hereditary, nonhereditary, and chromosomal deletion forms. Hereditary Rb refers either to cases with a positive family history or to individuals who have tumors in both eyes. Nonhereditary Rb applies to those individuals in whom only one eye is affected and who have no family history of developing this tumor. These patients, as well as the hereditary cases, have no chromosomal abnormalities in their peripheral lymphocytes. In the chromosomal deletion form of Rb, there is an inherited deletion of band 14 on chromosome 13 (region 13q14).

Possible genotypes at the Rb locus are outlined in the table. When only one eye is involved and there is no family history of tumor formation, Rb is usually nonhereditary, with a constitutional genotype of Rb+/Rb+. If both eyes have Rb or there is a positive family history of the tumor, it is assumed that there is a germinal mutation or inactivation of one allele, with the genotype being Rb+/rb−. As many as 5% of patients with retinoblastoma have the 13-deletion form of Rb (an inherited deletion of one chromosome 13 including region q14) and a genotype of Rb+/−.

Retinoblastoma genotypes at the Rb locus in chromosome 13 (13q14)

Genotype	Sporadic unilateral	Hereditary bilateral	13-deletion
Constitutional	Rb+/Rb+	Rb+/rb−*	Rb+/−
Tumor	rb−/rb− or rb−/−	rb−/rb− or −/rb−	rb−/− or −/−

*The genotype referred to when a patient is said to carry the retinoblastoma "gene."

Families with the hereditary forms of Rb show a classical dominant inheritance of tumor development. However, the inheritance of one inactive or deleted allele (Rb+/rb− or Rb+/−) is not sufficient for Rb to be produced. A second event ("hit") is required for Rb to develop. The near certainty that this second event will occur in at least one retinoblast accounts for this dominant pattern of inheritance.

Evidence for a common Rb locus. Recent chromosomal and gene mapping studies have documented that there is a common locus for both the hereditary and 13-deletion forms of Rb. The loci for each of these forms are closely linked to the locus for esterase D, which is a gene-dose-dependent polymorphic enzyme assigned to chromosomal region 13q14. Since both Rb loci are linked to esterase D and since both loci are responsible for the development of retinoblastoma, it is highly likely that there is only a single Rb locus in 13q14 common to hereditary and 13-deletion forms.

Although there is only circumstantial evidence to support the conclusion that the locus for nonhereditary Rb is also 13q14, this also is most likely the case. Deletions including 13q14 have been found in tumor cells from patients with both the hereditary and nonhereditary forms of Rb, suggesting that the same Rb locus is involved in the tumorigenic events responsible for nonhereditary as well as hereditary Rb.

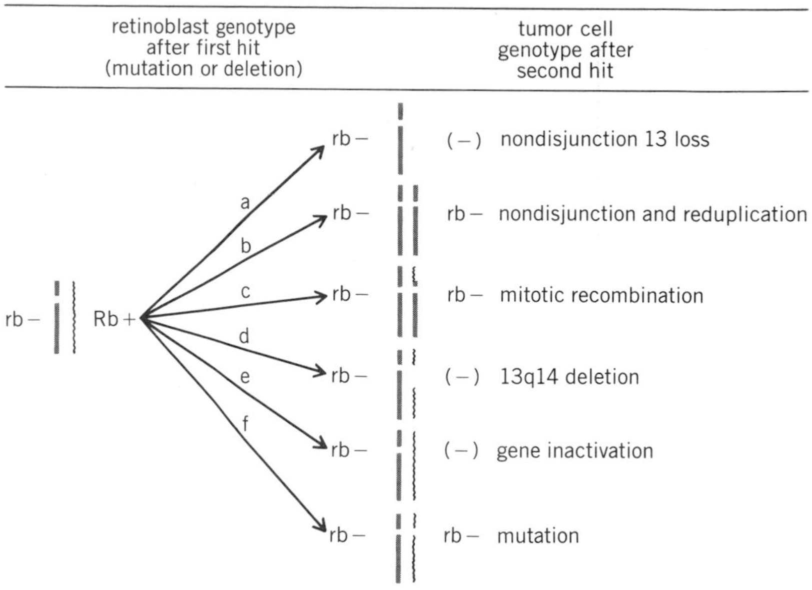

Several mechanisms that would produce hemizygosity or homozygosity at the Rb locus 13q14. The first hit which could occur either as a germinal or somatic event in the retinoblast is depicted at the left. In each case there would be an absence or inactivation of both RB alleles in the tumor.

Tumorigenesis. Recent evidence suggests that the two genetic events sufficient for retinoblastoma development are the loss or inactivation of both wild-type Rb alleles in 13q14 (Rb+/Rb+ → rb−/rb−). The first evidence came from a girl with a submicroscopic constitutional deletion of 13q14 in whom it could be shown that the normal 13 chromosome had been lost within her tumor cells. Since the tumor cells did not contain any genetic material from either Rb allele, it was proposed that the loss of both alleles at the Rb locus could be the two gentic events necessary for tumorigenesis. In this situation, hemizygosity at the Rb locus would be present as shown in the illustration, mechanism a, where the first hit was the submicroscopic deletion in 13q14 and the second hit was the loss of the normal 13 chromosome. This individual also provides evidence that the Rb gene is recessive, since the loss of one Rb allele was not sufficient to produce retinoblastoma, whereas the loss of both Rb alleles was apparently necessary for tumorigenesis.

The loss of chromosome 13 or a deletion of 13q14 has also been reported as a nonrandom event in tumor cells from both hereditary and nonhereditary cases. This loss of genetic material in 13q14 as a second hit again could lead to hemizygosity at the Rb locus as depicted in the illustration, mechanism a. Furthermore, using restriction-fragment-length polymorphisms specific for chromosome 13, it has been possible to show that there is frequently a nondisjunctional loss of one 13 chromosome, with duplication of the remaining 13 chromosome in retinoblastoma. If the normal 13 chromosome was lost and the 13 chromosome which contained the first hit was duplicated, this would result in homozygosity at the Rb locus, with once more a total absence of the Rb gene occurring as shown in the illustration, mechanism b. Similarly, with this technique a case of mitotic recombination has also been identified which again resulted in homozygosity for the Rb locus (see illustration, mechanism c).

Several other tumors have been identified in individuals with the nondeletion hereditary form of Rb where a deletion of 13q14 occurred in the tumor, once more resulting in homozygosity at the Rb locus (see illustration, mechanism d). Rarely it also has been found that inactivation of an Rb+ allele could be produced by a translocation of a 13 chromosome, including region 13q14, onto an inactive X chromosome, with subsequent spread of inactivation to the Rb gene as shown in the illustration, mechanism e. Lastly, it is possible that the second event in the formation of Rb could be a point or frameshift mutation of the Rb+ allele (see illustration, mechanism f). However, at present it is not possible to document such a mutation experimentally.

Other human cancer genes. Several other human tumors, including Wilms' tumor, renal cell carcinoma, neuroblastoma, and small-cell carcinoma of the lung, share certain characteristics of retinoblastoma.

Wilms' tumor is also a childhood tumor which has hereditary, nonhereditary, and chromosomal deletion forms. In the dominantly inherited form it is most often bilateral and multifocal, with an average age of diagnosis lower than that for the nonhereditary form which occurs only unilaterally. Aniridia (absence or defect of the iris) is usually found in individuals with the deletion form of Wilms' tumor, since the genes for aniridia and Wilms' tumor are linked to band 13 on the short arm p of chromosome 11 (region 11p13). Individuals have been identified who have neither aniridia nor a constitutional deletion of 11p13 but in whom a deletion including 11p13 has been found in their Wilms' tumor cells. These results suggest that a deletion or inactivation in 11p13 may be one of two genetic hits required for the development of Wilms' tumor.

Although familial renal cell carcinoma is an adult tumor, it appears at an earlier age than nonhereditary renal cell carcinoma. In addition, it often occurs bilaterally or with multiple primaries. One family of particular interest had 10 members with renal cell carcinoma who had a constitutional balanced translocation with breakpoints of 3p12 and 8q24. Twelve members with normal karyotypes did not develop cancer. Consequently, the hereditary form of renal cell carcinoma and retinoblastoma may be similar. Individuals with a 3/8 translocation (the transfer of a chromosome segment from chromosome 3 to chromosome 8) in turn may be analogous to individuals with retinoblastoma who have a deletion of 13q14, assuming that a mutation or submicroscopic deletion occurred in the 3/8 translocation at the breakpoint. In addition, an acquired translocation of 3p and 11p was also found in tumor cells from a patient with familial renal cell carcinoma who did not have a constitutional chromosomal abnormality. This individual could be analogous to cases with hereditary retinoblastoma in whom a deletion of 13q14 was found.

Although oncogenes have been implicated as possibly having a role in the etiology of neuroblastoma, this childhood tumor also has similarities to retinoblastoma. Familial cases of neuroblastoma exist, and there are a significant number of tumors with a deletion of bands p32 to pter (terminus) in chromosome 1. Statistical analysis of incidence figures also fits the hypothesis of two mutations for its origin.

Finally, there are other tumors which have specific deletions, and those which show both a dominantly inherited and noninherited pattern of susceptibility. Thus, substantial data are available to suggest that there are a group of human cancer genes which function similarly to the Rb gene.

Future studies. Whatever the function of the Rb gene is, it certainly plays a major role in the development of retinoblastoma as well as second primary cancers in those individuals who inherit the Rb gene. Future research will focus on the cloning of this gene and the elucidation of its function. DNA libraries of chromosome 13 are already available, and probes for esterase D are being made to aid in the cloning of the Rb gene. Hopefully within the next few years the task of cloning this important cancer gene will have been completed.

For background information *see* CHROMOSOME; CHROMOSOME ABERRATION; GENE; GENETICS; MUTATION; ONCOLOGY; RECOMBINATION (GENETICS) in the McGraw-Hill Encyclopedia of Science and Technology.

[WILLIAM F. BENEDICT]

Bibliography: W. F. Benedict et al., Non-random chromosomal changes in untreated retinoblastomas, *Cancer Genet. Cytogenet.*, 10:311–333, 1983; W. F. Benedict et al., Patients with 13 chromosome deletion: Evidence that the retinoblastoma gene is a recessive cancer gene, *Science*, 219:973–975, 1983; A. L. Murphree and W. F. Benedict, Retinoblastoma: Clues for human oncogenesis, *Science*, 1984; R. S. Sparkes et al., Gene for hereditary retinoblastoma assigned to human chromosome 13 by linkage to esterase D, *Science*, 219:971–973, 1983.

Optical communications

Gradient (or graded) index lenses have a number of features that are advantageous for use in optical components for optical fiber communication systems. In recent years there have been significant advances in the quality of available gradient index lenses, and components using these lenses are now in wide use.

In a conventional lens, the ray paths of the light are bent by refraction at the interfaces between the various elements, but within each element the refractive index is a constant and the ray paths are straight lines. In a gradient index lens element, the refractive index is a smooth, but not constant, function of position, and within such an element the ray paths are curved. The basic principles of gradient index optics have been understood for a long time—mirages are a consequence of a graded refractive index of the atmosphere resulting from temperature gradients. However, it is only within the past decade or so that techniques have been developed for fabricating useful index gradients in glasses, and analytical techniques have been developed for the design of optical systems using gradient index elements.

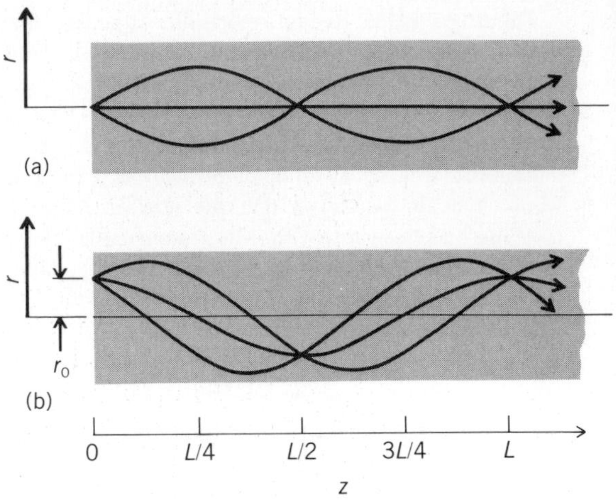

Fig. 1. Cross sections of GRIN-rods, showing ray paths for an object point (*a*) on axis and (*b*) off axis.

GRIN-rod lens. The particular type of gradient index lens that has been used in fiber components is the graded refractive index rod lens, or GRIN-rod lens. These lenses are rods with a refractive index that has its maximum value on the axis, and decreases approximately as the square of the radial distance. In such a rod the ray paths are approximately sinusoidal (Fig. 1). For an object at one end of a GRIN-rod, the rod forms an inverted image, with unit magnification, a distance $L/2$ down the rod, and an erect image after a distance L, where L is the period of the sinusoidal ray paths. At the intermediate points, $L/4$ and $3L/4$, all rays from a given object point are parallel, and thus one has a collimated beam at these points.

Lenses in fiber components. The basic building blocks used for fiber components are GRIN-rod lenses of length $L/4$, usually referred to as quarter-pitch lenses. As illustrated in Fig. 2, light from a fiber attached to one end of the lens will emerge from the other end as a parallel or collimated beam. These collimated beams are processed by conventional optical elements to attenuate, split, switch, filter them, and such; then the resulting output collimated beams pass through another (or the same) lens to focus them on the output fibers.

Lenses are necessary in most fiber components because the light emerging from the end of a fiber diverges rapidly enough that in a very short distance (typically of the order of 25 micrometers), the beam is too wide to couple efficiently to an output fiber,

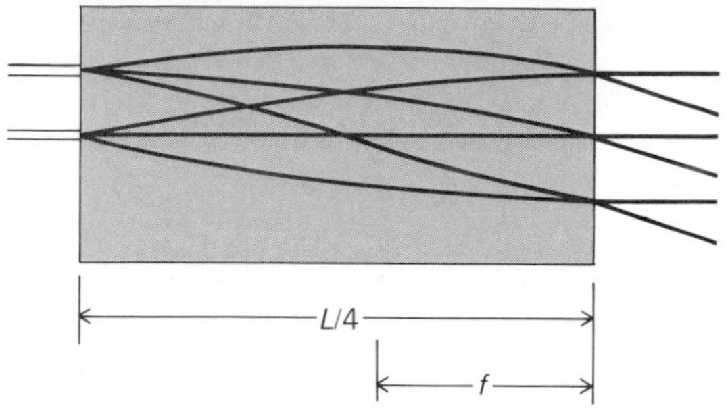

Fig. 2. Cross section of a quarter-pitch GRIN-rod lens, illustrating how light from an optical fiber is converted into a collimated beam. Distance f is the focal length of the lens.

and this distance is too small to accommodate most of the optical elements that one wants to use to process the light. In addition, some of these elements (such as interference filters) are sensitive to the angular spread of the input light beam, and would not function properly with the angular spread of the light emerging from a fiber.

Some of the characteristics of GRIN-rod lenses that make them particularly useful for fiber components are:

1. A GRIN-rod lens has planar input and output faces, and the operation of the lens does not depend on refraction at these faces. This means that the optical elements on either end of it can be glued directly to the lens face with a transparent cement. This results in a rugged, compact, stable structure with no glass-air interfaces to collect dirt or cause reflections.

2. GRIN-rod lenses are currently available with focal lengths between about 0.04 and 0.13 in. (1.1 and 3.4 mm), and numerical apertures of 0.3–0.5, which is just the required range for most fiber components.

3. The aberrations of GRIN-rod lenses are determined by the higher-order terms in the refractive index distribution, and in recent years manufacturing techniques have been improved to the point that it is possible to make lenses with negligible spherical aberration. These lenses have made it possible to produce fiber components in which the lens aberrations make a negligible contribution to the insertion loss of the component.

4. Because of their cylindrical shape, GRIN-rod lenses are considerably easier to mount and align than conventional lenses.

Fiber components using GRIN-rod lenses. The component shown in Fig. 3*a* is frequently referred to as a directional coupler, but it can perform a variety of functions. If the beamsplitter film is a simple semitransparent metal film, then part of the light input from fiber 1 will be transmitted by the film and focused on fiber 4, and part will be reflected by the film and focused on fiber 2. Thus, an input signal can be split into two output signals. The most important application of this component is for wavelength-division multiplexing, using a beam-splitter film with a wavelength-dependent reflectivity and transmission. If light at a wavelength λ_1, which is transmitted by the film, is input on fiber 1, and light at a wavelength λ_2, which is reflected by the film, is input on fiber 3, both signals will be coupled to fiber 4 (in this case fiber 2 is not used), and thus the two signals can be transmitted simultaneously over the same fiber. A similar component at the other end of fiber 4 can then be used to separate the two signals and direct them to separate detectors.

An alternate design for a wavelength multiplexer, using a diffraction grating rather than a filter, is shown in Fig. 3*b*. Considering the device as a demultiplexer, a multiplexed signal that is input on fiber 1 will be collimated by the lens and will then intercept the grating. The different-wavelength components of the input will be reflected from the grating at different angles and then pass back through the lens to be focused on separate output fibers. Note that the grating device can separate more than two channels with only a single lens and grating.

A simple fiber switch is illustrated in Fig. 3*c*. With the prism in the indicated position, light is coupled between fibers 1 and 2. When the prism is

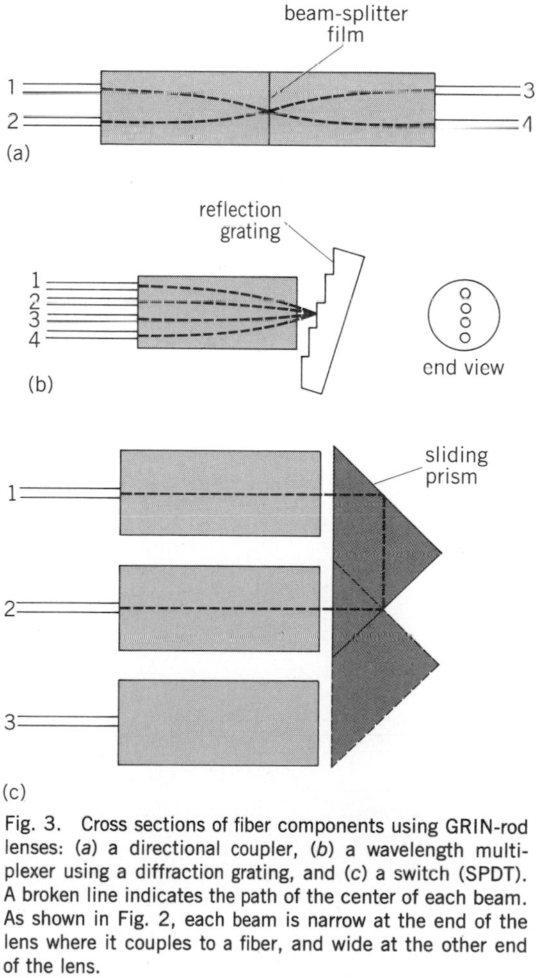

Fig. 3. Cross sections of fiber components using GRIN-rod lenses: (*a*) a directional coupler, (*b*) a wavelength multiplexer using a diffraction grating, and (*c*) a switch (SPDT). A broken line indicates the path of the center of each beam. As shown in Fig. 2, each beam is narrow at the end of the lens where it couples to a fiber, and wide at the other end of the lens.

slid to its alternate position, indicated by the broken lines, light is coupled between fibers 2 and 3.

For background information *see* LENS (OPTICS); OPTICAL COMMUNICATIONS in the McGraw-Hill Encyclopedia of Science and Technology

[W. J. TOMLINSON]

Bibliography: E. W. Marchand, *Gradient Index Optics*, 1978; W. J. Tomlinson, Applications of GRIN-rod lenses in optical fiber communication systems, *Appl. Opt.*, 19:1127–1138, April 1, 1980.

Optical pulses

Dramatic advances have taken place in the generation and application of ultrashort optical pulses with lasers. Short pulses are now being generated, and events are being measured in femtoseconds (10^{-15} s). While light can nearly travel to the Moon in 1 s, it takes 30 fs to travel a distance corresponding to one-tenth the thickness of a human hair. Recently optical pulses as short as 30 fs have been generated with newly developed laser and nonlinear pulse compression techniques.

Shortly after the invention of the laser, techniques were devised to generate short laser pulses. Since the late 1960s, progress has continued in the development of ultrashort optical pulses and measurement techniques. Only 5 years following the generation of the first optical pulses in the picosecond (10^{-12} s) range the first optical pulses less than a picosecond were generated with dye lasers. Recently, with the advent of the colliding pulse mode-locked dye laser, optical pulse widths have dipped below 0.1 ps into the femtosecond regime. Most recently, pulse compression techniques have further reduced the shortest reported optical pulse width to 30 fs.

Colliding pulse ring dye laser. The new laser configuration which has provided the means to generate the first optical pulses with a duration less than 100 fs is shown in Fig. 1. The laser cavity consists of a series of mirrors forming a ring cavity that contains only two essential elements: an optically pumped saturable gain dye (rhodamine 6G) and a saturable absorber dye (diethyloxacarbocyanine iodide) at the two focal points in the cavity. The dyes are contained in thin sheets of flowing solvent. An argon laser operating at 514.5 nanometers is used as an optical pumping source. The extreme simplicity of this configuration accounts in part for its ability to generate femtosecond optical pulses. By minimizing the amount of material in the cavity, the effects of group velocity dispersion are reduced, allowing the cavity to sustain a broad bandwidth of oscillations necessary to form a short optical pulse.

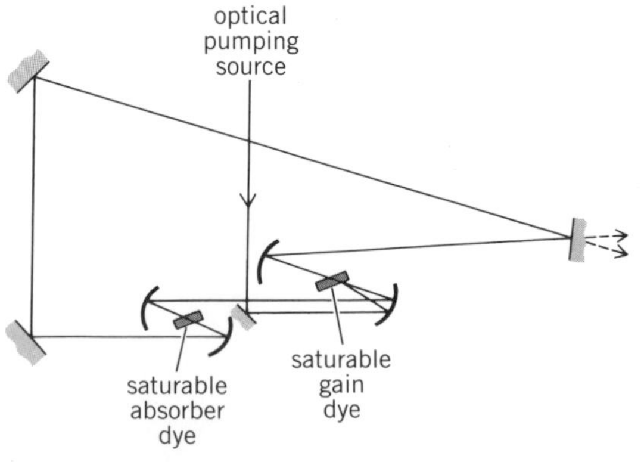

Fig. 1. Colliding pulse ring dye laser which generates optical pulses in the femtosecond time regime. (*After R. L. Fork, B. I. Greene, and C. V. Shank, Generation of optical pulses shorter than 0.1 psec by colliding pulse mode locking, Appl. Phys. Lett., 38:671–672, 1981*)

The mechanism for pulse shortening relies on the nonlinear optical properties of the organic dye molecules. When an organic dye molecule is excited by an intense light source, its optical absorption is reduced or becomes "saturated." That is, a large fraction of the molecules become excited, reducing the ground-state population and hence reducing the optical absorption. In a similar way the gain of an optical dye can be reduced when a significant number of excited molecules are stimulated to emit photons by an intense optical field.

As an optical pulse travels around the ring cavity, it is shaped each time it is amplified by the gain dye and absorbed by the saturable absorber. The leading edge of the pulse is preferentially clipped by the saturable absorber. Preferential amplification of the leading edge of the pulse, combined with cavity loss, effectively clips the rear edge of the pulse with each pass around the cavity until the limiting pulse width is achieved. The additional mechanism operative in this colliding pulse ring configuration results from the fact that there are two equally stable but oppositely directed pulses which "collide" as they meet each other traveling around the ring. The energetically most favorable place for the two pulses to meet is in the saturable absorber. Since the pulses are coherent, they can interfere and set up a standing-wave pattern in the absorber. The standing-wave pattern minimizes the energy lost to the absorber, because the field is most intense where the absorption is saturated and is weakest in the field minima where the absorption is not saturated. The shortest optical pulses are produced with a thin absorbing region that confines the standing-wave field.

Optical pulse compression. To generate even shorter pulses, optical pulse compression techniques have been devised which are analogous to electromagnetic pulse compression schemes developed in the microwave frequency range for radar. Optical pulse compression is accomplished in two steps. In the first step, a "chirp" or frequency sweep is impressed on the pulse. The pulse is then compressed by using a dispersive delay line.

A chirp can be impressed on an intense optical pulse simply by passing the pulse through an optical Kerr medium, a material whose refractive index is modified by the passage of the pulse. Very intense electrical fields comparable to those experienced by electrons in an atom are required to make a significant change in the refractive index. Almost any transparent material such as organic liquids, glass, diamond, or sapphire exhibit this effect. As the leading edge of the optical pulse rises rapidly, the refractive index is rapidly increased, leading to a time-varying phase change or frequency shift of the pulse carrier frequency. Similarly, a frequency sweep in the opposite direction occurs as the intensity of the pulse falls on the trailing edge. The result is that the pulse is sweeping in frequency, with long wavelengths in the front of the pulse and short wavelengths in the tail.

A diagram of the apparatus used for pulse compression in the femtosecond time domain is shown in Fig. 2. The incident pulse is focused with a lens into a 6-in. (15-cm) length of optical fiber a few micrometers in diameter. The electric field strengths are greatly increased when the optical pulse is focused into the fiber, and the refractive index of the fiber is modified in the manner described above, impressing a time-dependent phase change on the pulse and thereby causing a frequency sweep or chirp to be impressed on it. The pulse emerging from the optical fiber is recollimated with a lens. This pulse is broadened in both frequency and time. The optical frequency of the pulse is sweeping in time, with the pulse being redder in the front of the pulse and yellower in the rear.

To reassemble the pulse, a dispersive delay line is required. A nearly ideal dispersive delay line is a parallel pair of gratings (Fig. 2). Each wavelength of light passing through the pair is diffracted at a different angle by the first grating of the pair and recollimated by the second grating. The net effect is a wavelength-dependent optical path delay which, when properly adjusted, puts all the colors of the pulse together in time, compressing the pulse. Fortunately, the grating pair compresses the chirped pulse and provides a means of compensating other dispersive elements in the systems, such as lenses.

Pulse compression has a significant advantage over direct pulse generation in a laser cavity for the generation of optical pulses in the femtosecond time regime. As shorter pulses are generated, increasing spectral bandwidth is required. In a laser, the gain bandwidth of the lasing medium and the optical cavity determine the lasing bandwidth. In contrast, the Kerr effect is operative from the ultraviolet to the infrared regions of the spectrum. It appears feasible to compress optical pulses to a few femtoseconds, which is nearly a single optical cycle.

Limits. Fundamentally there is no limit to how short an optical pulse can be generated other than that imposed by the uncertainty principle. As stated for short optical pulses, the uncertainty principle requires that the product of the pulse width times the optical pulse bandwidth must be a constant. To generate a very short pulse requires coherence to be established over a broad range of frequencies. Typically, for a dye laser the gain bandwidth is on the order of 100 nm, which corresponds to a minimum pulse width of approximately 10 fs. Maintaining a fixed phase relationship over such a broad range of frequencies is difficult. Material and optical cavity dispersion tend to broaden the optical pulse. Linear dispersion becomes a significant consideration when making measurements with a pulse of a few femtoseconds in duration. Such a pulse will broaden significantly after it traverses 100 micrometers in almost any material. In some cases the use of pulse compression techniques described above can be used to compensate this pulse broadening, but these techniques become more difficult to apply as the pulse gets shorter.

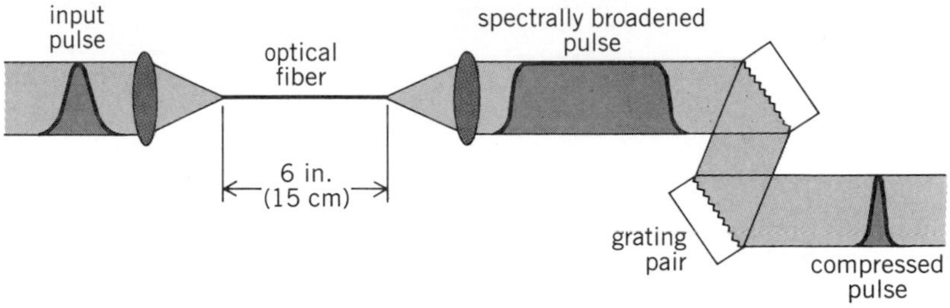

Fig. 2. Experimental configuration for compressing a femtosecond optical pulse. The incident pulse enters from the left and emerges at the right with a shortened pulse width. (*After C. V. Shank et al., Compression of femtosecond optical pulses, Appl. Phys. Lett., 40:761–763, 1982*)

The implications of generating such short bursts of light are significant for a broad range of fields, including chemistry, physics, and electronics. The transition to ever shorter times opens up new types of phenomena for investigation. The real impact of measurements in the femtosecond time regime will be to open new frontiers and investigate processes not amenable to presently available measurement techniques.

For background information *see* COHERENCE; KERR EFFECT; LASER; OPTICAL PULSES in the McGraw-Hill Encyclopedia of Science and Technology.

[C. V. SHANK]

Bibliography: R. L. Fork, B. I. Greene, and C. V. Shank, Generation of optical pulses shorter than 0.1 psec by colliding pulse mode locking, *Appl. Phys. Lett.*, 38:671–672, May 1981; C. V. Shank, Measurement of ultrafast phenomena in the femtosecond time domain, *Science*, 219:1027–1031, March 1983; C. V. Shank et al., Compression of femtosecond optical pulses, *Appl. Phys. Lett.*, 40:761–763, May 1982.

Orientation (biology)

Small mammals use a variety of sensory cues to orient at night. Vision, olfaction, vibrissal touch, audition and, in some species, echolocation enable nocturnal species to function at low light levels. Nocturnal retinas achieve their sensitivity by replacing cone receptors with elongated, densely packed rods that are highly photosensitive and are often backed by a reflecting mirrorlike structure called the tapetum. Under conditions of complete darkness, however, or in the absence of these retinal specializations, mammals must rely to a greater ex-

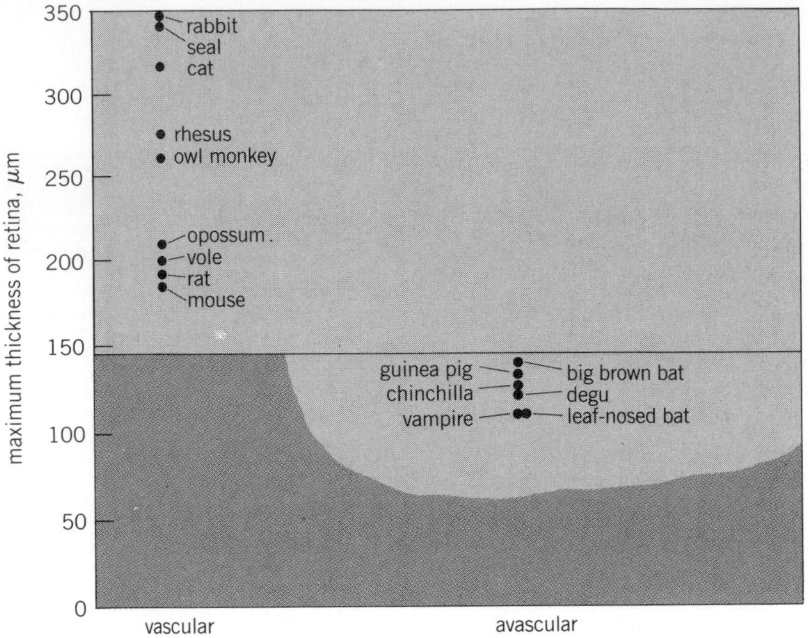

Fig. 1. Maximum retinal thickness in mammalian vascular and avascular retinas. The theoretical maximum distance for oxygen diffusion from choroid to retina in humans is 143 micrometers (gray zone). Note that mammals with avascular retinas, which depend almost entirely on the choroid for nutrition, do not exceed this thickness. (*After J. Chase, Hearing and Other Senses, Amphora Press, Groton, Connecticut, pp. 209–227, 1983*)

tent on other cues to orient at night. Recent evidence suggests that retinal avascularity in the early mammals may have precluded both the development of the tapetum and the elongation of the rods, and thus posed an anatomical limitation to nocturnal vision. On the other hand, echolocation, which is popularly thought to be restricted to bats and porpoises, has now been demonstrated in such diverse species as rats, humans, and shrews, and may well be of greater importance in the orientation of nocturnal terrestrial mammals than was previously supposed.

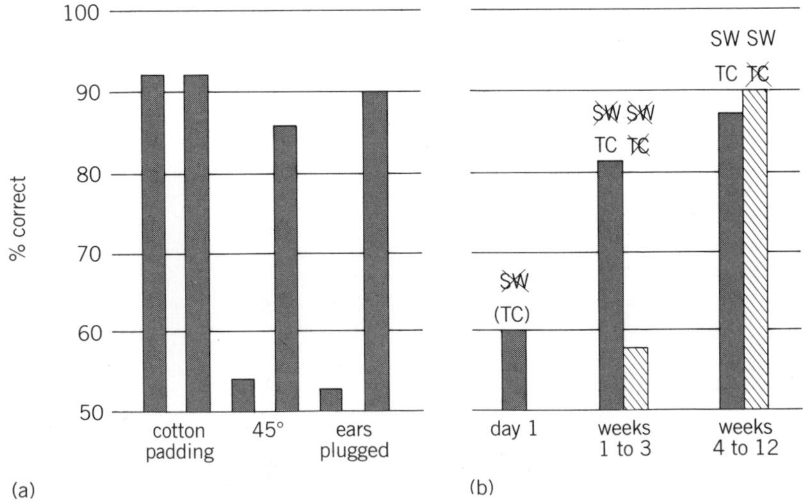

Fig. 2. Evidence of echolocation. (*a*) Blind rats successfully detect the barrier when the Y maze is padded with cotton, but not when the barrier is rotated 45° or when there is a plug (for each pair of bar graphs, the performance of control subjects is indicated by the right-hand bar). (*b*) Use of sniff-whistle (SW) and tooth chatter (TC) for echolocation. The rat was tested following unilateral section of the recurrent laryngeal nerve. Cross-hatching indicates that TC was temporarily abolished on the day of testing. On day 1 after surgery, the rat showed little TC and was lacking the whistle of the SW. TC began on day 2; by the fourth week, a modified whistle had returned to the SW. The discrimination was maintained when the rat produced either TC alone (weeks 1–3) or SW or both (weeks 4–12). (*After J. Chase, Hearing and Other Senses, Amphora Press, Groton, Connecticut, pp. 209–227, 1983*)

Retinal avascularity. Retinal avascularity imposes a nutritional constraint on the development of both the rods and the tapetum because nutrients, in particular the essential energy-producing oxygen and glucose molecules, must diffuse into the retina from the choriocapillaris which lies behind the retina. At normal metabolic rates, the choroid can supply oxygen to the retina up to a maximum distance of approximately 143 micrometers. Elongation of the rods, however, would increase retinal thickness well beyond this oxygen diffusion maximum. As Fig. 1 shows, mammalian retinas that exceed 143 μm are all vascular and, conversely, retinas that lack retinal blood vessels are restricted to thicknesses below the oxygen diffusion maximum. The only exception to this, the flying fox, has peculiar retinal mammilations that bring choroidal blood vessels well up into the retina for the direct diffusion of nutrients. The presence of a tapetum in the choroid likewise reduces the ability of the choriocapillaris to nourish the retina and, with the exception again of the flying fox, all mammalian retinas that lack blood vessels lack the tapetum. Avascular retinas share another character in common, the storage of glucose in the form of glycogen in the inner limbs of the Muller cells.

Retinal avascularity appears to be a primitive characteristic among the mammals, for it is found in both of the living monotremes, many marsupials, and certain placental groups such as the chiroptera, edentates, and some rodents. How then do these animals navigate effectively at night? While spatial memory, vibrissal touch, and olfaction are unquestionably important, several groups appear to rely extensively on echolocation or "animal sonar," notably the bats and porpoises. Echolocation is not, however, restricted to these two groups. Echolocation has been demonstrated in shrews. It has also been shown that blind humans have a modest ability to detect objects in their path by using echoes of such self-generated noises as finger snapping or cane tapping, and there is evidence suggestive of echolocation in laboratory rats as far back as the mid-1950s. It is instructive to look in detail at recent findings in the rat echolocation system which suggest that echolocation systems in terrestrial mammals would have to be designed differently from the sonar of bats and porpoises; and because of this, echolocation systems could easily go undetected even in commonly studied species. Indeed, echolocation may turn out to be far more common than was previously suspected in terrestrial mammals.

Rat echolocation. Blind rats can be trained to seek water on an elevated Y maze and will learn to select the open runway and avoid the runway that is blocked by an 8 × 9 in. (20 × 23 cm) barrier suspended 14 in. (35 cm) from the choice point. Ear-plugging such rats prevents them from making the discrimination (Fig. 2*a*), and a background of white noise will also destroy their ability to detect the barrier. The discrimination is thus clearly based on acoustic cues. The evidence suggests that the rats are indeed echolocating, and not simply noting an attenuation of ambient noise on the side with the barrier, for a 45° rotation of the barrier (which would reflect echoes *away* from the rat) destroys the discrimination. The final proof that this is an active echolocation comes from muting experiments. Blind rats that are unable to emit sounds are unable to select the open runway. Indeed, muted animals appear disoriented and are noted to vigorously clean their ears as if disturbed by the lack of acoustic imput.

Early investigators were unable to detect sound emission when rats were performing acoustic discriminations similar to those described above. It has been found that rats can use either of two different echolocation sounds, both of which are audible to the human ear yet are seldom noted under laboratory conditions (Fig. 2*b*). The more common, and also the fainter, is termed the sniff-whistle. It consists of a broadband respiratory huff and an 8-kHz whistle that seems to be produced laryngeally. The sniff-whistle is produced in trains of approximately 8 per second and is synchronous with the respiratory cycle (Fig. 3). The second sound is broadband chatter that seems to be produced by the teeth, though the tongue may also be involved in its production. It is less common and noticably louder. Both sounds are seldom noted in the laboratory because rats seldom echolocate in the light or even in the dark when in familiar territory.

However, when rats are placed in new surroundings or on an elevated platform in the dark, they will begin to sniff-whistle. The sounds become louder and more persistent when an animal is about to jump from the platform or is interested in an object lying beyond the reach of its vibrissae. If the vibrissae are numbed with Novacaine, rats sniff-whistle loudly and continuously until the anesthetic wears off. Tooth or tongue chatter is noted in animals that have been surgically muted and that can no longer produce normal sniff-whistles. It is also noted in animals with numbed vibrissae and in animals that appear to be exploring objects 3 ft (1 m) or more distant. Unlike the sniff-whistle, which is barely noticeable, tooth chatter is a loud and prominent sound that is easily localized by an observer, which may explain why rats employ it so infrequently. Functionally, the sniff-whistle and tooth or tongue chatter appear to work like the low beams and high beams of auto headlights, illuminating near and distant objects, respectively.

While rat echolocation is functional, it is clearly secondary to both visual and vibrissal orientation. Moreover, the rat's system seems to be designed to minimize the possibility that other species will detect it. Rats use echolocation only when necessary, and the sounds themselves are usually quite faint except when the animal needs information about more distant objects. Unlike aerial echolocators

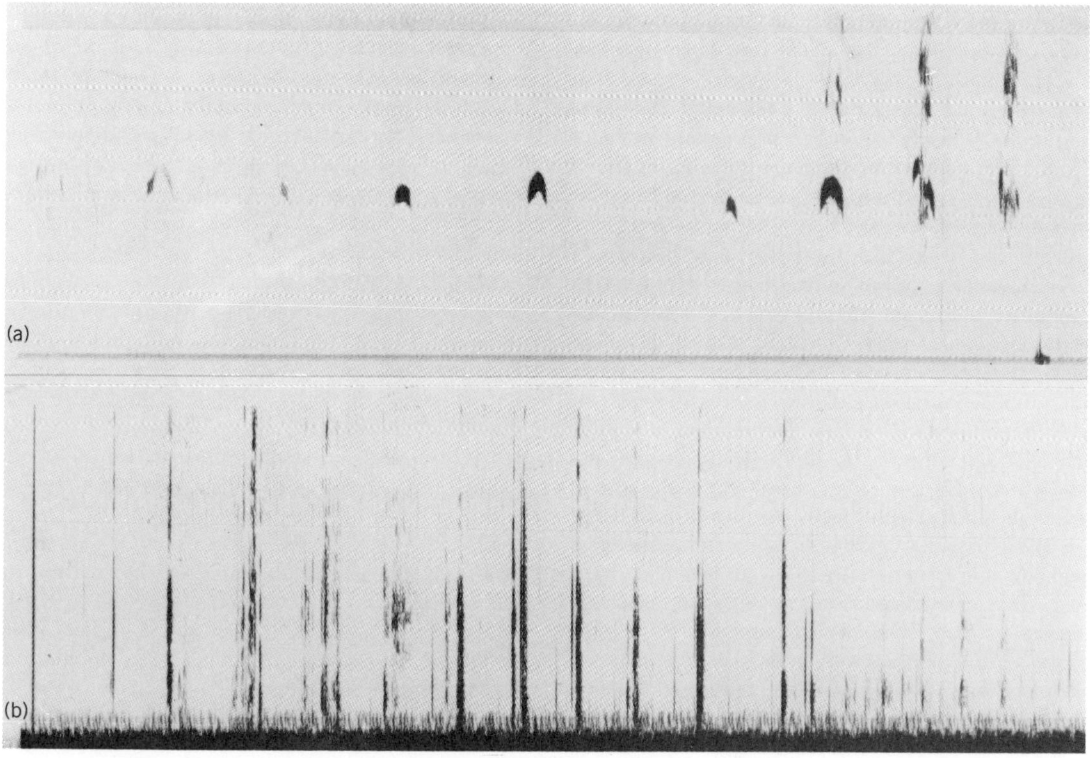

Fig. 3. Sonograms of (*a*) the sniff-whistle and (*b*) tooth chatter of an adult rat. Note that the sniff portion is too faint to appear on this recording.

such as bats, oilbirds, and swifts which face relatively less predation pressure, the terrestrial echolocator must balance the advantages of acoustic orientation against the disadvantages of betraying one's location to potential predators. If echolocation is indeed present in other terrestrial mammals, it can be predicted that they, like the rats, will have evolved a "hidden" echolocation system that is inconspicuous and difficult to detect.

Vision and echolocation: complementary senses. Vision and echolocation are in many ways complementary senses though they are partially redundant as well. Vision is silent, but of little use in complete darkness. Echolocation has a limited range and is noisy, but it is capable of detecting objects in complete darkness. In the microchiropteran bats, there is an inverse relationship between the development of vision and echolocation. Bats that feed on large objects such as fruit, flower nectar, or animal blood have large eyes and well-developed vision, but produce faint sonar cries. Bats with loud sonar cries turn out to feed largely on insects and other prey too small to be detected visually in dim light. In most species, their well-developed echolocation system is coupled with tiny eyes and poorly developed vision.

Behaviorally, there is also increasing evidence for a complementary relationship between vision and echolocation in microchiropteran bats. It has been demonstrated that bats tend to home visually when released at long distances from their roosts, but can home acoustically when released in familiar territory near the roost. It has also been reported that pallid bats tend to hunt insects acoustically at low light levels and to use vision at higher levels of illumination. While bats can orient acoustically during escape, they prefer to use visual cues in this context whenever possible. In bats, vision and echolocation both play a significant role in orientation. The question is not whether vision or echolocation are functional; both are. The interesting question is which sense is best suited for a particular purpose.

The early mammals seem to have been largely nocturnal in habit, like the modern bats and rats. If retinal avascularity was indeed the ancestral condition in mammals and limited both the development of the tapetum and the elongation of the rods, these ancestral species were faced with the dilemma of living under conditions of low ambient illumination with limited scotopic acuity. It is unclear how early echolocation appeared in the mammalian line and how widespread it is today. Moreover, the evolution of echolocation cannot be reconstructed from either the fossil remains or from an examination of living members of "primitive" groups. It is clear, however, that echolocation can serve as an important sensory backup for nocturnal mammals. An examination of the literature suggests that a number of terrestrial species may indeed be active echolocators but, like the rat, have escaped detection precisely because natural selection has favored the evolution of inconspicuous signals.

For background information *see* CHIROPTERA; EYE (VERTEBRATE); PHONORECEPTION in the McGraw-Hill Encyclopedia of Science and Technology.

[JULIA CHASE]

Bibliography: E. Buchler, The use of echolocation by the wandering *Sorex vagrans*, *Anim. Beh.*, 24:858–873, 1976; R.-G. Busnel and J. F. Fish, *Animal Sonar Systems*, 1980; J. Chase, The evolution of retinal vascularization in mammals: A comparison of vascular and avascular retinae, *J. Ophthalmol.*, 89:1518–1525, 1982; J. Chase, *Hearing and Other Senses*, pp. 209–227, 1983.

Osteichthyes

About 15 years ago, studies by James Lighthill and Theodore Wu focused on the consequences of various fish designs for thrust, drag, and efficiency. Their work emphasized the fact that fish are self-propelled flexing bodies, contrasting with the dominant opinion at the time of viewing fish as analogs of rigid vehicles, such as submarines and aircraft. Building on ideas such as those developed by Lighthill and Wu, biologists, mathematicians, and fluid dynamicists have established the locomotor performance characteristics of several fish and other aquatic vertebrates with different body and fin configurations. These researches not only have defined the functional capabilities of particular suites of structures but have also shown that design compromises are required. Optimal designs maximizing performance for particular locomotor activities are often mutually exclusive. Therefore, any tendency to evolve shapes conferring improved performance for one activity impacts negatively on performance in other areas of activity.

The functional incompatibility of optimal designs for different locomotor activities is most clearly illustrated for swimming by using the body and caudal fin. Quite different morphologies are optimal in maximizing performance for swimming at relatively constant speed, whether speeds are high or low, on the one hand, and for accelerations as in fast starts and powered turns, on the other hand. (See illustration and table).

Steady swimming. During steady swimming the body bends into a wave which travels backward over the body. To understand how this propels a fish, the body and median fins must be thought of as made up of small linked sections, or "propulsive segments." These interact along the length of the propulsive wave to accelerate water backward in the vicinity of the body by pushing "pieces of water" or water particles sideways and backward.

Thrust is equal to the rate at which momentum is ultimately discharged into the wake beyond the tip of the caudal fin (the trailing edge), and is determined by the increase in velocity imparted to the water and the mass of water accelerated by the propulsive segments. The mass of water accelerated is proportional to the square of the span, or depth, of the trailing edge. Water, however, is accelerated not only posteriorly but also laterally, so that a side force is also produced. This side force causes the

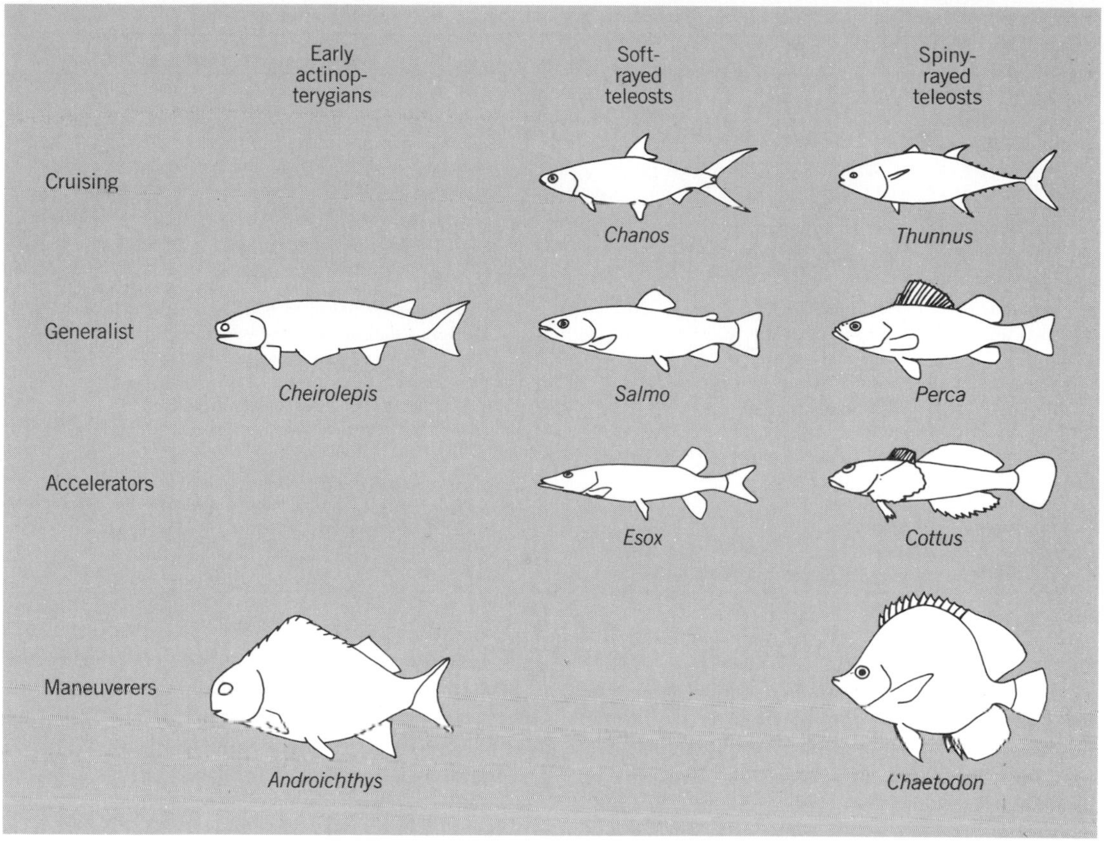

Body and fin forms of some actinopterygian fishes in three of their radiations. The radiation among now extinct early actinopterygians resulted in relatively little diversity of body forms, probably due to limitations of the skeletal structures. More specialized forms for steady swimming, acceleration, and maneuver are seen among the soft-rayed fishes. The most recent radiation of spiny-rayed fish (acanthopterygians) shows the greatest diversity and most specialized forms.

anterior of the fish, as measured at the center of mass, to move laterally, or recoil, resulting in waste of energy. The energy wastage can be minimized by reducing the area of the body and fins immediately anterior to the trailing edge where amplitudes of the propulsive wave increase rapidly. The recoil is further reduced by making the mass and depth of the body large anterior to the center of mass. Hence, the optimal shape to maximize thrust during steady swimming is a teardrop body shape attached to a deep, narrow tail by a slender caudal peduncle (see *Thunnus* in the illustration).

Performance depends on the balance between thrust and drag. In constant-speed swimming, frictional drag predominates, and is proportional to the surface area and the square of the water velocity. Drag of a self-propelled flexing body is larger than that of a rigid body, and is the highest for the largest amplitudes of the propulsive wave. Hence, drag is reduced by the same reduction in caudal area that reduces energy wastage. Drag is also reduced by the teardrop body shape. These optimal design features are typical of tunalike animals, and hence animals that possess them are said to swim in the thunniform mode (see *Thunnus* in the illustration).

Acceleration. In contrast to swimming at relatively constant speed, during accelerations the body tends to move so rapidly and through such large amplitudes that propulsive segments along the body length cannot interact. Instead, each propulsive segment independently accelerates water in its vicinity. Thrust is maximized when all portions of the body are deep, especially caudally where amplitudes are largest, in order to maximize the velocity and the total mass of accelerated water. Also, in contrast with steady swimming, resistance to motion primarily arises from inertia, and hence the extra area has little effect on drag. Instead, performance is maximized when the nonmuscle mass is minimized. Fish that simultaneously maximize thrust and minimize resistance to increase acceleration performance have not been identified so far; instead, fish appear to be resistance minimizers (for example the pike, *Esox*, in the illustration) or thrust maximizers (for example the sculpin, *Cottus*, in the illustration). The absence of simultaneous thrust maximizers and drag minimizers appears due to the need for compromises with other nonlocomotor aspects of feeding. For example, in order to obtain benthic food, sculpins have large heavy heads. As a result, inertial resistance cannot be reduced too much. Pikes may have reduced their body and fin depth anteriorly, at cost to thrust, in order to make them appear less obvious to prey.

Morphological compromise. Optimal design features for steady swimming versus acceleration are

Summary of mechanical and morphological requirements for different types of body and caudal fin swimming

Fast starts and powered turns	Constant-speed swimming
Mechanics	
Thrust is the sum of local effects along whole body length.	Mean thrust is related to the rate of momentum shedding at the caudal fin trailing edge.
Inertial resistance predominates.	Resistance is primarily due to viscous or frictional effects.
Morphology	
Large tail (or double tail) maximizes thrust, and turning moment.	Deep narrow tail maximizes thrust and minimizes drag.
Deep caudal peduncle enhanced by posterior anal and dorsal fins increases thrust and turning moments.	Small vertically flattened caudal peduncle (narrow necking) minimizes drag and recoil of the center of mass due to destabilizing side forces.
Large anterior body or depth minimize center-of-mass recoil and associated energy wastage.	Large anterior body or depth minimize center-of-mass recoil and associated energy wastage.
Flexible body allows large-amplitude lateral movements to maximize thrust and turning moments, and to minimize turning radius.	Streamlined anterior body minimizes drag but reduces body flexibility.
Large proportion of myotomal muscle mass relative to body mass (that is, minimal "dead" weight to accelerate) minimizes acceleration resistance. Large myotomal muscle mass maximizes turning moment.	Large myotomal muscle mass maximizes sprint speeds because power requirements increase with the third power of the velocity.

clearly exclusive (see table). For example, maximizing acceleration requires a large caudal area, while maximizing steady swimming performance requires this area be small. Only actinopterygian fish have some ability to mitigate the problem because their fins are collapsible. As a result, they can somewhat vary the local depth of the body to adapt it for cruising and sprinting or for accelerating.

Swimming with the body and caudal fin is not the only means of propulsion for bony fishes. Instead, the paired fins (pectorals and pelvics) and dorsal and ventrals are used for routine activities, slow swimming and precise maneuver. Reef fishes, such as the angelfishes (*Chaetodon* in the illustration) and butterfly fishes are most advanced, but again, their specialization appears to reduce performance abilities for other activities. For example, the surface area is high, adding to drag and impairing steady swimming at higher speeds using the body and caudal fin. Fish specialized for maneuver and slow swimming also tend to be short-bodied and have relatively little muscle. This would reduce acceleration abilities. In contrast, fish specialized for cruising must do so at a cost to slow swimming and maneuvering capabilities, for their fins are stiff control surfaces, lacking the flexibility needed to generate a wide range of control forces.

Because of the conflicting design requirements for different locomotor patterns, no fish can perform at high levels in all activities or excel in more than one. This has major consequences for the biology of fish with different locomotor morphologies. For example, fish specialized for steady swimming are ineffective in acceleration and powered turns, and good acceleration performance is important for avoiding predators. Therefore, thunniform fish must possess mechanisms to minimize predation risks. Adult thunniform animals are large, typically greater than 20 in. (50 cm) in length, which provides substantial protection from most aquatic predators. Unfortunately this cannot always solve the problem for the larvae and young. Therefore, thunniform bony fish spawn large numbers of pelagic eggs in warm productive waters where the young feed voraciously and grow rapidly through the vulnerable stages. Other thunniform animals (whales, dolphins, sharks, and extinct ichthyosaurs) are usually live bearers and solve the problem this way.

Foraging. The steady swimming specialists are also the largest fish, and must forage widely for food. Fish specialized for acceleration cannot exploit such widely dispersed resources and instead are the characteristic sit-and-wait ambushing piscivores. Both types of specialists share a common feature in that they eat large food items and are unable to exploit more abundant small items. These small items are exploited by small fish. Some are specialists in maneuver, but these are largely restricted to structurally complex habitats where cover protects them from predators. Most fish exploiting the abundant resources of small food items are not competitive with the maneuver specialists in cover, and as a result they feed in open waters, exposed to predators. They must be locomotor generalists, retaining adequate performance to escape the predators (for which they are a major food source), sufficient cruising abilities to find the food, and sufficient maneuverability to catch it. Examples are trout *(Salmo)* and perch *(Perca)* in the illustration.

Evolution. The various types of body and fin forms seen among living fishes have appeared gradually since the Devonian, coupled with adaptations in skeletal systems to support a wide range of large forces. Thus, successive radiation in the evolution of bony fish correlates with increasing locomotor diversity (see illustration). Among the actinopterygians, the earliest morphs are generalists from which specialists in maneuverability evolved to a small extent. Specialists for maneuverability and for steady swimming occur primarily in the most recent radiation of spiny-rayed, acanthopterygian fishes.

As a result of the need for design compromises,

recent studies on fish locomotion have profound impact on the understanding of life history strategies, behavioral ecology, and the evolution of fish.

For background information *see* OSTEICHTHYES in the McGraw-Hill Encyclopedia of Science and Technology. [PAUL W. WEBB]

Bibliography: M. J. Lighthill, *Mathematical Biofluidynamics,* Society for Industrial and Applied Mathematics, 1975; P. W. Webb, Body form, locomotion and foraging in aquatic vertebrates, *Amer. Zool.*, (in press); P. W. Webb, Locomotor patterns in the evolution of actinopterygian fishes, *Amer. Zool.*, 22:329–342, 1982; P. W. Webb and D. Weihs, *Fish Biomechanics*, 1983; T. Y. Wu, Hydrodynamics of swimming fish and cetaceans, *Adv. App. Math.*, 11:1–63, 1971.

Paleoceanography

Paleoceanography, one of the youngest branches of earth sciences, is concerned with the development of oceanic circulation during geologic time. This is a multifaceted and integrated endeavor since none of the subdisciplines is easily separated from the others. The field is holistic in character, and includes the geological history and effects of bottom, intermediate, and surface circulation patterns; planktonic and benthonic biogeographic development and rates of organic evolution; the history of biogenic productivity and dissolution and their effect on sediment distribution; and the consequences of sea-level change.

Paleoceanography was largely born of the Deep Sea Drilling Project (DSDP) and continues to be nourished by it. Because problems are of global scope, much interest exists for new sediment core material which is being supplied by expeditions of the Deep Sea Drilling Project.

A major boost for paleoceanography has resulted from development of two new coring devices now used in deep-sea drilling. The recovery of undisturbed sections of unconsolidated deep-sea sediment has long been a goal of the Deep Sea Drilling Project. Unfortunately, conventional rotary drilling usually disturbs unconsolidated sediment and renders it useless for high-resolution investigations. Pristine or largely undisturbed core sections for the entire section can now be obtained fairly routinely by using the hydraulic piston corer followed by the extended core barrel (Fig. 1). The hydraulic piston corer is used in the most unconsolidated upper part of the section, where it strokes out ahead of the nonrotating drill string, using hydraulic pressure as the driving force. When sediments become too compacted to allow the use of the hydraulic piston corer, the extended core barrel can now be employed and provides relatively undisturbed cores of high quality for high-resolution paleoceanographic studies. Core quality is maintained, despite drill rotation, because the extended core barrel drills in a less disturbed environment ahead of the main drill string.

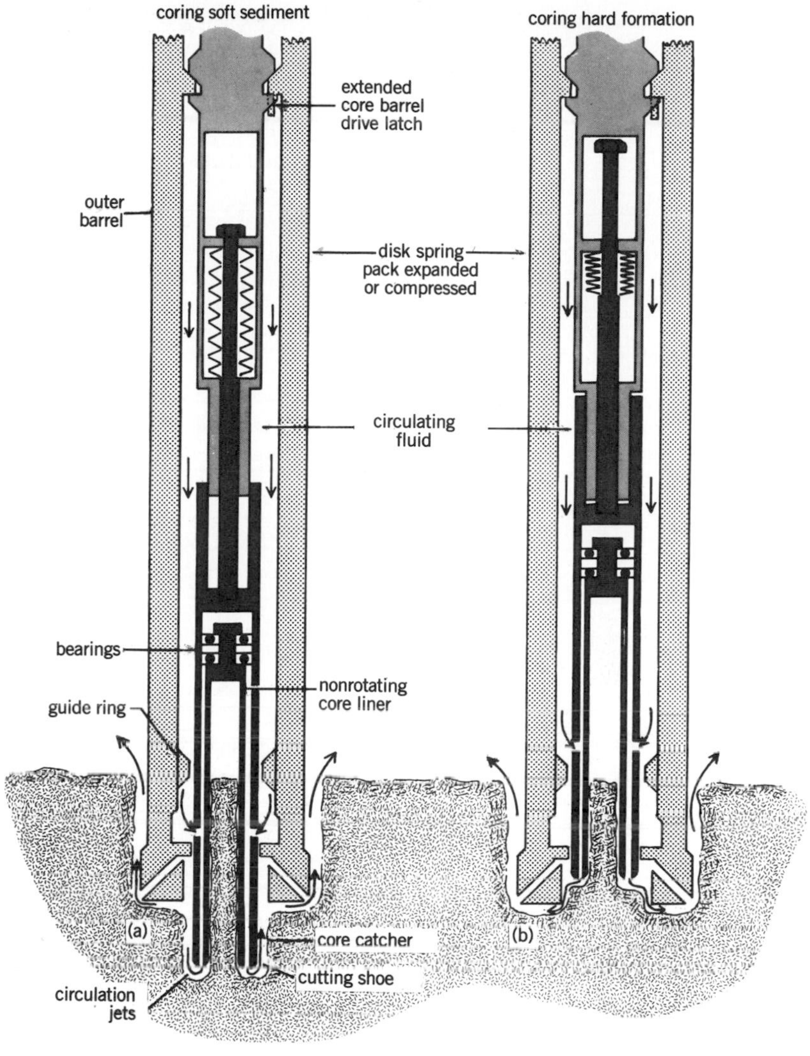

Fig. 1. Extended core barrel mechanism used at end of drill string in deep-sea drilling. (*a*) Coring soft sediment. The internal cutting core extends ahead of main drill bit. (*b*) Coring hard rocks. (*Deep Sea Drilling Project*)

Deep-ocean paleocirculation. For many years there has been much effort to better understand the history of oceanic surface waters during the late Phanerozoic. In contrast, there has been much less interest about the history and dynamics of deep-ocean circulation, a much more difficult problem because of the three-dimensional complexity of the system. This attitude is changing, with the use of benthonic foraminifera and stable isotopes remaining central to such investigations.

Benthonic foraminifera are sensitive monitors of bottom waters of both the present and the past. Different benthonic foraminiferal assemblages on the ocean floor are associated with the principal modern bottom- and deep-water masses. These associations provide criteria for reconstructing past distributions of these water masses. The traditional view that benthonic foraminiferal assemblages are distributed according to water depth has been replaced by the concept that the abundance of species is related to physical and chemical parameters of the water masses, properties that may vary spatially and temporally independently of water depth.

Fossil benthonic foraminiferal assemblages are providing useful information about past changes in deep and bottom waters in various parts of the ocean basins, especially during the Quaternary. During this interval of time, large variations are now known to have occurred in the rates of production of various bottom and intermediate waters, especially at high latitudes of the North Atlantic and the Antarctic.

Innovations have also occurred in the use of sediment parameters for study of deep-water history. Different hydrodynamic regimes seem to effect the sediment-size characteristics which can be used to distinguish between bottom currents with different velocities.

Another approach for studying bottom-water history is the utilization of variations in ^{13}C content, designated $\delta^{13}C$, in benthonic foraminifera in sedimentary sequences. Water masses of different age, oxygen content, and nutrient content will have different $\delta^{13}C$ values at different locations along its flow path. The geographic distribution of $\delta^{13}C$ in deep water is therefore the result of modern deep-water circulation patterns. Different circulation patterns through time produce different geographic distributions of $\delta^{13}C$ in the deep water. Temporal fluctuations in $\delta^{13}C$, when used in combination with other parameter changes such as benthonic foraminifera and oxygen isotopes, represent a powerful combination for the study of paleocirculation changes in the deep ocean.

Cretaceous oceans. Much interest exists in Cretaceous [65–140 million years (m.y.) ago] paleoclimates because the warm, equable conditions during that period contrast with present-day conditions. Past changes in the geography of the Earth are often cited as being important in controlling climatic change by inhibiting or accentuating oceanic heat transport to the poles. A large increase in the ability of the oceans to transport heat toward the poles has been proposed to explain Mesozoic global warmth. Climatic modeling studies show that the meridional distribution of Cretaceous temperatures cannot be successfully simulated unless additional physical feedback mechanisms, such as cloud cover, are included in the models. It has been calculated that a 2% increase of global marine solar radiation for 100 m.y. ago is sufficient to maintain warm global climates in the Cretaceous.

Cretaceous-Tertiary boundary. Catastrophic causes of the well-known terminal Cretaceous biotic extinctions have received support by the discovery of increased iridium concentrations found in clays at the Cretaceous-Tertiary boundary. Concentrations of iridium in sediments at the boundary of up to 160 times more than the background level in several areas are considered by them to have been the result of a great influx of extraterrestrial material. The scenario proposes a postimpact distribution of pulverized meteoric rock in the Earth's atmosphere for several years, resulting in both the heating and darkening of the atmosphere. The resulting suppression of photosynthesis would have had an immediate detrimental effect on the biota, as observed in the paleontological record.

It has been suggested that the extinction of large terrestrial animals was caused by atmospheric heating during cometary impact and that the extinction of marine plankton resulted from cyanide released from the fallen comet and from a catastrophic rise in the calcite compensation depth (CCD) in the ocean after detoxification of the cyanide. Some recent data suggest that at the time of extinction there was an associated reduction in productivity which would have led to a partial transfer of carbon dioxide from the oceans to the atmosphere. A resulting increase in atmospheric carbon dioxide may have created a temporary temperature rise. Some oxygen isotopic data from the Cretaceous–Tertiary boundary exhibit a temporary climatic warming episode.

The environmental and climatic effects of a meteorite impact remain uncertain. For instance, the character of land plant extinction is opposite to that expected for dust blocking and sudden heat rise, because a much higher percentage of plants at higher Northern Hemisphere latitudes became extinct compared with the tropics, where extinctions were less distinct. If the entire globe were shrouded by dust, the less hardy tropical plants should have suffered the most, rather than the least.

Early Tertiary oceans. The Paleocene (65 to 38 m.y. ago) was marked by cooling high-latitude temperatures and the development of greater latitudinal thermal contrast that eventually led to the heavily glaciated earth of the Eocene. It was mostly during the Paleocene that the loci of deep-water formation changed from low- and middle-latitude seas to the polar areas. Thus the warm, saline bottom waters of the early Tertiary were replaced by cold dense bottom waters of the later Tertiary. The latitudinal temperature gradient in the Eocene was less than half its present value: surface temperature at high latitudes was about 50°F (10°C) and a little over 68°F (20°C) at low latitudes. Thus ocean circulation must have contributed a very high proportion of the poleward transport of heat, compared with the present.

Eocene-Oligocene boundary. Significant interest has continued concerning the well-known major paleoceanographic changes associated with the terminal Eocene event (38 m.y. ago) when there was large-scale cooling of the oceans. Although there are substantial changes in fossil assemblages at this boundary, these do not approach the magnitude of the extinctions associated with the Cretaceous–Tertiary boundary. Deep-sea benthonic foraminifera exhibit no crisis in most areas, but change relatively gradually before and after the boundary. This lack of major change either reflects wide environmental tolerance of the benthonic foraminiferal species at that time or a bottom-water temperature change of lesser magnitude than generally believed.

The latest data generally support the concept of replacement of old, warm, corrosive, sluggish Eocene bottom water by young, colder, less corrosive, and more vigorously circulating bottom waters of the earliest Oligocene.

Middle Miocene event. The middle Miocene seems to represent the next crucial stage in the development of global paleoceanography, for at this time (about 14 m.y. ago) much of the Antarctic ice sheet may have formed. This event is marked by a sharp increase in the $\delta^{18}O$ values of calcareous plankton and ofbenthonic foraminifera (Fig. 2). This increase certainly reflects in part a major period of ice sheet growth, as well as a drop in surface temperatures on the Antarctic coast. Comparison of the oxygen isotopic records between high and low latitudes suggests that during the middle Miocene event, the planetary temperature gradient markedly steepened and temperatures at high and low latitudes became much less closely coupled. The existence of a widespread ice sheet from the middle miocene is supported by the presence of common and persistent ice-rafted sediments around the Antarctic continent from that time.

The increase in middle Miocene glaciation seemed to have led to a further decrease in oceanic bottom temperatures, supporting the hypothesis that part of the oxygen isotopic change resulted from temperature change, as well as from the buildup of ice. Evidence for this comes from the large changes at that time of deep-sea benthonic foraminiferal assemblages. Some paleontologists disputed that there was much Antarctic ice growth in the middle Miocene. They believe, based upon reinterpretation of oxygen isotopic data, that Antarctica has had major ice sheets at least since the Eocene and perhaps through much of the Cretaceous. Their interpretation would require that almost all Antarctic continental ice was in place prior to the middle Miocene oxygen isotopic shift, a theory that is not strongly supported by independent sedimentary evidence from the Antarctic.

Terminal Miocene Event. A surprising number of different changes are now known to have occurred during the Terminal Miocene Event (6 m.y. ago): a marked isochronous decrease of $\delta^{13}C$ isotopic ratio values (0.5–0.8‰) in marine carbonates of the Indo-Pacific region, about 6.2 m.y. ago; a widespread cooling of surface waters; a major intensification of bottom-water circulation and increased fertility of the oceans; the lowering of sea level; the isolation and desiccation of the Mediterranean; the shoaling of the Isthmus of Panama; an increase in biogenic silica productivity in the Southern Ocean; a decrease in biogenic silica sedimentation in the eastern equatorial Pacific; a switchover in relative depths of the CCD between the Atlantic and Pacific oceans; and a general increase in deep-sea accumulations rates.

The increased intensity of oceanic circulation at this time is believed to have resulted from the development of the west Antarctic ice sheet and the formation of the first major production of true Antarctic bottom water which permanently altered global abyssal circulation.

Late Pliocene events. The initiation of Northern Hemisphere ice sheets has generally been dated at 3.2 m.y., but recent data from the North Atlantic suggests that they did not accumulate until 2.5 m.y. ago. Also, oxygen isotopic evidence has been pre-

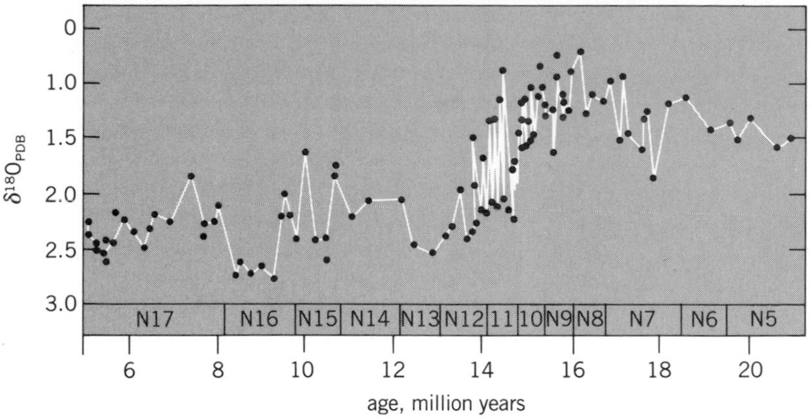

Fig. 2. Oscillations during the last 21 million years of $\delta^{18}O$ in the tests of benthonic foraminifera from a deep-drilled site near the Equator in the western Pacific (DSDP Site 289.) Oscillations to the left reflect cooler waters or increased ice accumulation on continents. The major enrichment in $\delta^{18}O$ shown at about 14 million years ago in part reflects a major buildup of ice on Antarctica. The N zones are biostratigraphic zones based upon evolution and extinction of planktonic foraminifera.

sented from several equatorial regions which suggests that the oxygen isotopic event of 3.2 m.y. ago represents only a brief excursion in ice volume. It has been bound that high-amplitude oxygen isotopic frequencies in a deep-drilled site extend back only to 2.4 m.y. ago. The age of inception of ice rafting observed in cores from high northern latitudes varies from region to region, although such material older than 2.5 m.y. ago probably represents glaciation unrelated to the development of major ice sheets.

Quaternary oceans. Quaternary (1.6 m.y. ago to the present) paleoceanographic studies have continued to focus on regional and global paleoenvironmental mapping of specific time intervals, especially during the last glaciation, on the spectral analysis of paleoenvirnomental changes in the Quaternary, and on the character of rapid deglaciations. Relationships continue to be demonstrated between orbital insolation variation and Quaternary climatic history. A high-resolution stratigraphy and chronology for Quaternary deep-sea cores has been established by maximizing coherences between orbital and oxygen isotopic signals at the three frequencies around which obliquity and precession variance are concentrated. They have identified orbital variations as the primary cause of about 60% of ice sheet variance during the last 900,000 years.

In the Pacific Ocean, mapped changes in radiolarian assemblage distributions in the Quaternary indicate that variation in intensity, direction, and mean position of the tradewinds caused marked changes in oceanic circulation patterns during the last glacial episode. The major changes in circulation seem to occur a few thousand years in advance of the intervals of increased ice sheet growth. This relationship indicates that the changes in atmospheric circulation in the tropics led and influenced the development of conditions for polar and continental ice sheet growth in the Northern Hemisphere.

In the North Atlantic it was found that during the first half of each rapid ice-growth phase of continental ice sheets in the Northern Hemisphere there were warm sea-surface temperatures in the subpolar North Atlantic (50–60°N). It is believed that the juxtaposition at high latitudes of a relatively warm ocean beside a glaciated land mass is regarded as am optimal configuration for delivering moisture needed for ice sheet growth. Spectral analysis data indicate that the oceanic moisture and sea-level act as climatic feedbcks to amplify Milankovitch (insolation) forcing of the volumetrically dominant mid-latitude ice sheets at the 23,000-year precessional cycle.

For background information *see* EXTINCTION (BIOLOGY); FORAMINIFERIDA; OCEANOGRAPHY; PALEOCEANOGRAPHY; PALEOCLIMATOLOGY; PALEOGEOLOGY in the McGraw-Hill Encyclopedia of Science and Technology. [J. P. KENNETT]

Bibliography: P. F. Ciesielski, M. T. Ledbetter, and B. B. Ellwood, The development of Antarctic glaciation and the Neogene paleoenvironment of the Maurice Ewing Bank, *Mar. Geol.*, 46:1–51, 1982; K. J. Hsu et al., Mass mortality and its environmental and evolutionary consequences, *Science*, 216(4543):249–256, 1982; T. Loutit and L. D. Keigwin, Jr., Stable isotopic evidence for latest Miocene sea-level fall in the Mediterranean region, *Nature*, 300(5888):163–166, 1982; W. F. Ruddiman et al., Oceanic evidence for the mechanism of rapid Northern hemisphere glaciation, *Quaternary Res.*, 13:33–64, 1980; N. Shackleton and A. Boersma, The climate of the Eocene Ocean, *J. Geol. Soc. London*, 138:153–157, 1981; F. Woodruff, S. M. Savin, and R. G. Douglas, Miocene stable isotope record: A detailed deep Pacific Ocean study and its paleoclimatic implications, *Science*, 212:665–668, 1981.

Paleontology

Two aspects of recent paleontological studies concern how marine fauna have yielded a better understanding of the changes in taxonomic diversity with time and of the response of this diversity to major extinctions; and how the study of preserved trackways has allowed paleontologists to reconstruct and interpret early animal life.

EVOLUTION OF DIVERSITY

Taxonomic diversity refers to the number of kinds of organisms in an ecosystem. Since the mid-nineteenth century, paleontologists have been interested in the diversity of the global ecosystem and how it changes through time. Recent studies of marine animals, which have the most extensive and best-studied fossil record, have yielded three broad generalizations about the evolution of diversity: (1) Diversity tends to increase through geologic time but does so in a stepwise rather than continuous fashion. (2) Increases in diversity are associated with expansions of different groups of organisms. (3) Most mass extinctions appear as short-term perturbations superimposed upon broader patterns of increase or stability in diversity.

These generalizations will be discussed below with respect to marine animal families. Because comprehensive data with good stratigraphic resolution have never been compiled for the immense number of species described from the fossil record, only taxa above the species level are frequently analyzed in such studies.

Phases of diversification. Figure 1 shows a graph of the diversity of recognized marine animal families through the whole of the Phanerozoic Eon. The upper curve indicates there were three main phases of diversification in the oceans:

1. A Vendian-Cambrian phase encompassing the early diversification of animals in the latest Precambrian, the rapid rise in diversity across the Precambrian-Cambrian boundary, and the slowing of diversification in the last 40 million years (m.y.) of the Cambrian.

2. A post-Cambrian Paleozoic phase comprising the great Ordovician radiations which tripled familial diversity in some 50 m.y., the subsequent 200 m.y. of near stability in diversity, and the final late Permian mass extinction which eliminated more than half of the families in the oceans.

3. A Mesozoic-Cenozoic phase encompassing the broad increase in marine diversity from the low in the Early Triassic to the all-time high in the late Cenozoic.

In its most general aspects, the pattern of diversification is the same in all three phases. As seen best in the first two phases, diversification occurs rapidly at first but then slows down so that diversity may remain nearly constant for tens to hundreds of millions of years. This general pattern conforms to the simplest models for diversification in a resource-limited ecosystem. When such a system is empty, phylogenies should grow exponentially. Exponential, or approximately exponential, growth should proceed until the system begins to become packed with species and resources become limited. Then, the potential for new species to find their own "niches" decreases, and the rate of diversification should decline. If conditions remain constant and the kinds of organisms and their ecological requirements do not change, these processes should lead eventually to long intervals during which there is little change in diversity. This is seen in the first two phases of diversification in the oceans but not in the third, which shows only a hint of slowing of diversification. Thus, it would be predicted that diversity in the oceans should increase somewhat more in future millions of years (barring any mass extinction) but should eventually stabilize as the carrying capacity of the modern oceans is reached.

Evolutionary faunas. As suggested above, taxonomic diversity should stabilize only if the kinds of animals contributing the diversity remain approximately constant. If the fossil record is studied, however, it is clear that the principal kinds of animals have not been constant. Instead, different groups of animals have expanded and declined at different times, leading to major changes in the taxonomic composition of the marine fauna. The timings of these faunal changes correspond to the beginning of new phases of diversification, so that each phase is characterized by expansion and then stabilization of a different set of higher taxa (such as classes). These sets are termed evolutionary faunas and are shown schematically in Fig. 2. Their taxonomic compositions and evolutionary histories are briefly summarized below.

The Cambrian evolutionary fauna consisted of "archaic" taxa which are now largely extinct. The most diverse class comprised the trilobites, but also important were the inarticulate brachiopods, monoplacophorans, hyolithids (probable mollusks), eocrinoids (extinct echinoderms), and various enigmatic small shelly animals. Their collective diversity is shown in Fig. 2 and by field I in Fig. 1. The Cambrian fauna expanded rapidly during the Early Cambrian and dominated the marine biota through the remainder of the Cambrian Period. In the last part of that period, however, the Cambrian fauna began a slow decline which continued through the Paleozoic Era so that after the Devonian its members were relatively unimportant in oceanic environments.

The Paleozoic evolutionary fauna consisted largely of articulate brachiopods, crinoids, corals, ostracods, cephalopods, stenolaemate bryozoans, and several other smaller classes. Their diversity is indicated by field II in Fig. 1. Most of these classes first appeared during the Cambrian Period but did not become diverse until the Ordovician. By the end of the period, members of the Paleozoic fauna dominated marine diversity. They continued to dominate the marine biota until the end of the Paleozoic Era, when they suffered severe drops in diversity during the Permian mass extinction. Thereafter, these groups never again became dominant components of the marine biota.

The Mesozoic-Cenozoic, or "Modern," evolutionary fauna principally includes marine gastropods, bivalves, and bony fishes (Osteichthyes), gymnolaemate bryozoans, rhizopods (foraminifers), marine malacostracan crustaceans, echinoids, and sharks (Chondrichthyes). Many of these classes also appeared during the Cambrian Period but during the Ordovician did not expand as rapidly as members of the Paleozoic fauna. Instead, the Modern fauna expanded slowly but steadily through the remainder of the Paleozoic Era (Fig. 2), surviving the Permian mass extinction with only a comparatively minor decline in diversity. Following the Permian, the members of the Modern fauna radiated rapidly and became dominant components of the marine biota for the whole of the Mesozoic and Cenozoic eras (as indicated by field III in Fig. 1).

Mass extinctions. The late Permian mass extinction (P in Fig. 1) is clearly of pivotal importance in the evolution of diversity in the oceans. This event, which spanned the last 15 to 20 m.y. of the Permian Period, forms the boundary between the Paleozoic and the Mesozoic-Cenozoic phases of diversification

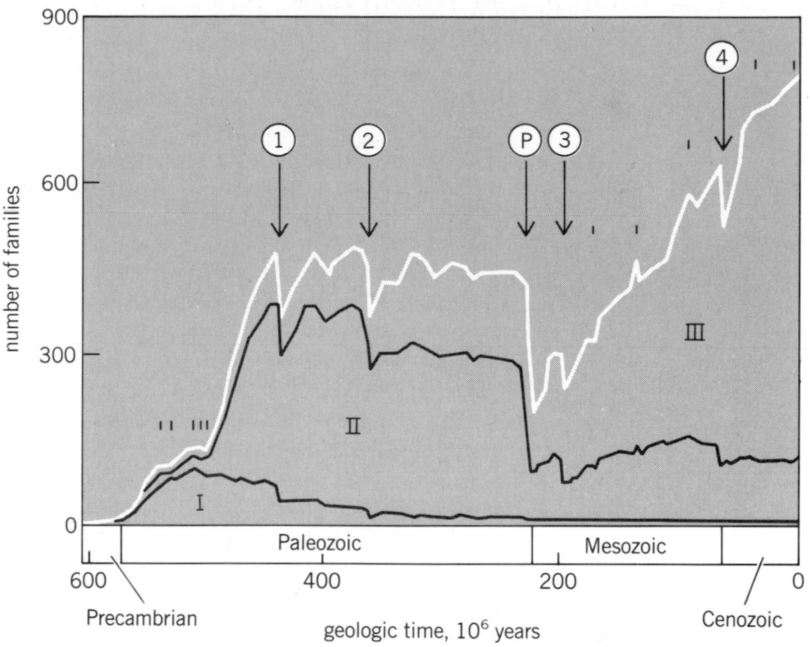

Fig. 1. The diversity of marine animal families through the Phanerozoic Eon. The white curve shows the total number of "shelly" (skeletonized) marine invertebrate and vertebrate families described from 80 stratigraphic intervals, each averaging approximately 8 million years long. The black curves break the field for total diversity into the contributions of each of the three major evolutionary faunas: I = Cambrian fauna, II = Paleozoic fauna, III = Modern fauna. The arrows and tick marks indicate the positions of mass extinctions described in the text.

as well as the changeover in dominance from the Paleozoic fauna to the Modern fauna. Approximately 52% of the families and perhaps more than 90% of the species in the oceans became extinct during this event. However, what caused this great extinction remains enigmatic. Suggested mechanisms include severe reduction of habitat area resulting from the regression of shallow seas at the end of the Permian; decline in biogeographic provinciality during the formation of the supercontinent Pangea; deterioration of the global climate as a result of huge continental air masses developing over Pangea; and changes in ocean chemistry due to orogeny, high rates of erosion, and changes in sea-floor spreading during the suturing of Pangea.

The late Permian event is the largest mass extinction in the history of life. But it is only one of approximately 15 events of accelerated extinction that have been recognized in the marine fossil record. Four other major events (1–4 in Fig. 1) caused reductions of 15 to 22% in the number of families in the oceans.

The Ashgillian event at the end of the Ordovician Period affected major reductions in the diversity of trilobites, cephalopods, articulate brachiopods, and crinoids. This event has been associated with a reduction in sea level and cooling of the global marine climate resulting from the development of continental glaciers over northern Africa (then at the South Pole).

The Frasnian event in the Late Devonian resulted

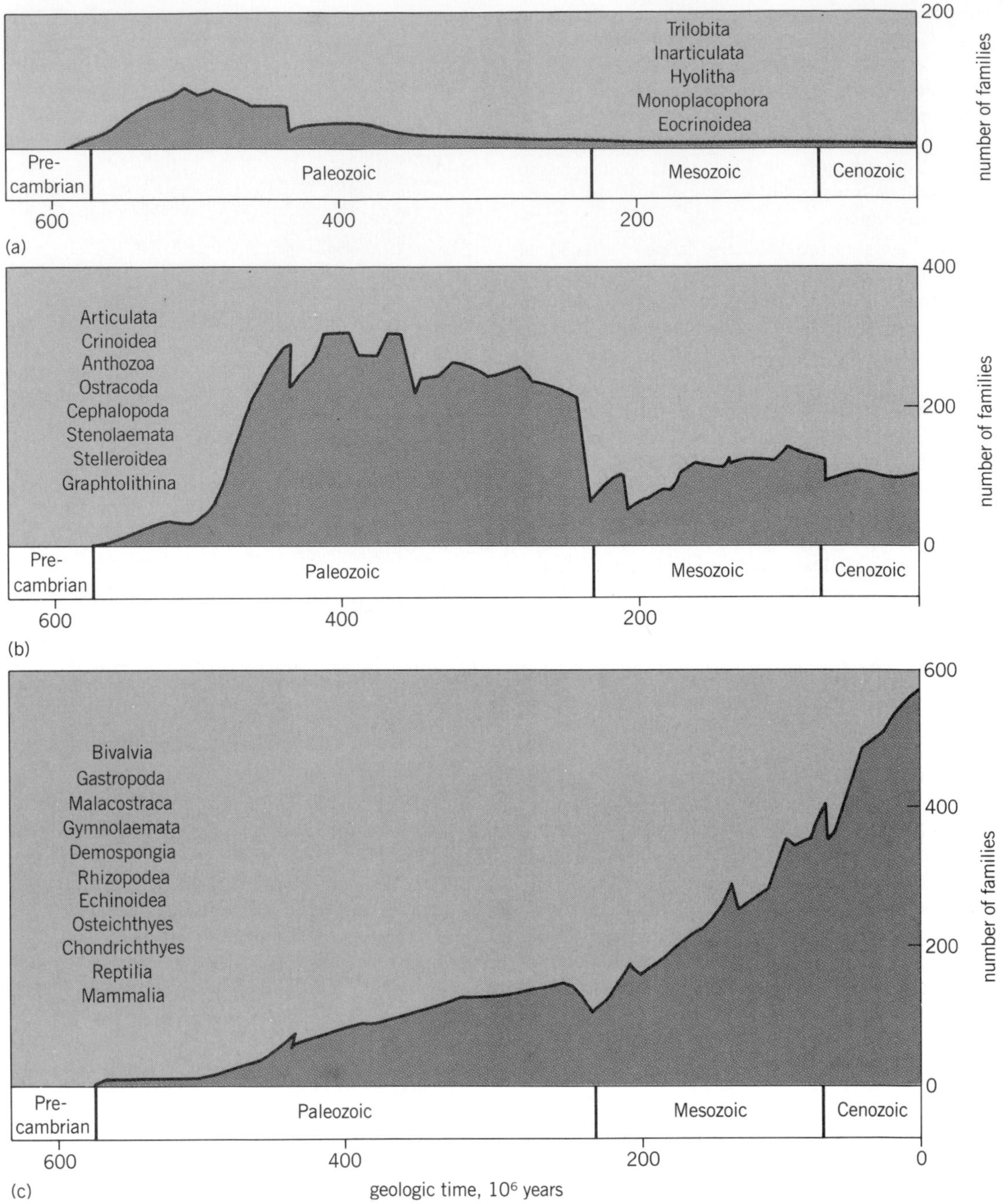

Fig. 2. The three evolutionary faunas of the Phanerozoic marine fossil record: (*a*) Cambrian; (*b*) Paleozoic; (*c*) Modern. The principal classes in each fauna are listed along with diversity curves for the total number of families through time.

in severe declines among corals, articulate brachiopods, cephalopods, placoderm fishes, and ostracods. This event has been associated with a long-term decline in provinciality as well as a possible catastrophic extraterrestrial event near the Frasnian-Famennian boundary.

The Norian event near the end of the Triassic Period approximately 30 m.y. after the great Permian mass extinction greatly affected cephalopods and marine reptiles as well as gastropods and bivalves. This event also has been associated with a major marine regression and consequent reduction in habitat area.

The Maestrichtian event at the end of the Cretaceous Period caused the extinction of ammonites as well as major declines among the marine bivalves, gastropods, echinoids, bony fishes, and sponges. On land, this event is best known for causing the final extinction of dinosaurs. Recent geochemical investigations have revealed anomalies in the concentrationsof iridium and other siderophile elements in sediments at the top of the Maestrichtian, indicating that the extinctions may have been caused by impact of a large asteroid with a diameter of approximately 6 mi (10 km).

In addition to these large mass extinctions, paleontologists have recognized at least 10 other events of unusual extinction which affected small numbers of taxonomic groups over limited geographic areas. (The positions of these events in time are indicated by the ticks above the diversity curve in Fig. 1) Five small mass extinctions, termed biomere events, eliminated endemic trilobites in North America and possibly China and Australia during the Cambrian Period. Similar small events affected European bivalves during the Early Jurassic (Toarcian Stage) and, perhaps, near the Jurassic-Cretaceous boundary. Events during the Cenomanian (Late Cretaceous Period) and late Eocene (Tertiary Period) caused severe declines among open-ocean plankton. Finally, a protracted "event" in the Pliocene and early Pleistocene epochs resulted in extinctions among benthic bivalves and gastropods as well as planktic foraminifers and algae.

All of these mass extinctions, with the exception of the late Permian event, had only temporary effects on global marine diversity. As evident in Fig. 1, the five large mass extinctions are manifested in rapid drops in familial diversity followed by immediate, if somewhat slower, rebounds. These rebounds proceeded until diversity regained previous levels of stability, as in Paleozoic events 1 and 2, or previous patterns of increase, as in Mesozoic-Cenozoic events 3 and 4. Thus, it would appear that the marine biota represents a quasi-stable system over geologic time scales. Traces of small to moderate perturbations which extinguish no more than 25% of standing families are quickly eradicated, and only very large events (the Permian mass extinction being the only example) permanently affect the course of diversification and of large-scale faunal evolution in the oceans.

[J. JOHN SEPKOSKI, JR.]

TRACKWAYS OF EXTINCT ANIMALS

Throughout the immensity of geologic time, two kinds of evidence were preserved to help paleontologists reconstruct and interpret the animal life of the past. Fossilized remains such as bones are evidence of one kind that permits reconstruction of the animal's general appearance and deduction of how it moved about and fed. The other line of evidence is the animal's footprints and trackways; these provide an insight into the animal's behavior, and permit observation of the animal "in action." The earliest documented discovery of vertebrate fossil footprints was made in 1802 by a farm boy, Pliny Moody, in the Connecticut River valley near Hadley, Massachusetts. The tracks, preserved in red sandstone of early Mesozoic age, were those of small dinosaurs. However, dinosaurs were unknown at the time, and so the footprints were believed to have been made by large birds. A quarter-century later, footprints, subsequently shown to be of Permian age, were described from a Scottish sedimentary formation by Henry Duncan, a clergyman. These were footprints of a caseasaur, a large plant-eating reptile.

Numerous other occurrences of fossil footprints began to be reported, from strata of a variety of ages. In 1841 William Logan, a Scottish geologist, found well-preserved vertebrate footprints in Carboniferous age strata at Horton Bluff, Nova Scotia. His find was disbelieved by his British peers. Logan was not vindicated until 1872 when Charles Lyell, working with John W. Dawson, discovered the fossil remains of small amphibians and reptiles in the Carboniferous coal fields at Joggins, Nova Scotia. For the balance of the nineteenth century up to about 1930, reports were published on vertebrate fossil footprints from all over the world. Thereafter, up until the recent renaissance of interest in the subject, the few papers in the field, written by amateurs, were of poor quality and generally appeared in obscure journals. Among recent important discoveries of vertebrate fossil footprints are finds made in Nova Scotia, Montana, British Columbia, and Brazil (Fig. 3). Although questions remain concerning interpretation and classification of trackways, modern research has supplied important new data on: land animals known only from their tracks; geographic and geologic ranges of species of vertebrates; behavioral habits of ancient vertebrates; prevailing environmental conditions; and the antiquity of humans' ancestors.

Fig. 3. Trackway of *Sousaichnium pricei*, a giant ignanodont dinosaur from the Sousa Formation (Jurassic), Brazil. Width of footprint is 14 in. (35 cm). (*Courtesy of G. Leonardi*)

Preservation of trackways. Among the difficulties facing the student of fossil footprints is the initial quality and character of the tracks themselves. Few high-quality impressions are found. To retain footprints, the sediment must ideally be moist, fine-grained, and cohesive. Further, the animal should traverse slowly, in order to create a detailed impression of the hand (manus) or foot (pes). In this way, accurate outlines of nails, claws, shape of pads, or the pattern of scales may be preserved. The most favorable conditions occur after high waters—the highest level of spring tides in sea or estuary, the greatest depth of river or pool during a wet season—have receded, leaving fine-grained sediment freshly exposed. Thus the footprints will solidify as the surface dries out prior to burial by the next discharge of sediments.

One important fact in favor of preservation of any particular footprint, at least under ideal conditions, is that most animals make thousands, even millions of tracks during their lifetimes. The likelihood of a few such tracks being preserved in stone greatly exceeds the likelihood of skeletal preservation of their marks. Because fine-grained sediments are readily eroded, the molds (footprint impressions) themselves may not be found. Instead the casts, composed perhaps of coarser sediment, will be discovered. Casts occur on the underside of sedimentary blocks and may not be seen unless for some reason the block has been overturned, as in a quarrying operation. Further, an important relationship exists between the size of an animal and the likelihood of its tracks being preserved and recognized. Of necessity, large animals range widely for food and visit waterholes often. Small animals whose footprints are less deeply impressed and more readily destroyed by the shifting of sediment require less food and water, and range less widely. Fossil footprints of a dinosaur are thus far more likely to be preserved than those of a lizard.

Measurements of footprints and trackways. At least three sets of footprint casts or molds in sequence are required for the measurement of a trackway. Care must be exercised to distinguish a true footprint from subsurface impressions (so-called ghost prints) which result from deformation of underlying sediment. Ideally both mold and cast should be examined; this is done by pouring artifical casts of fiberglass, rubber, or plaster of paris. In this fashion, entire trackways may be preserved for display or study purposes.

With quadrupeds, running begins with one pes followed by the manus on the same side, and by the pes and manus on the opposite side. In bipedal vertebrates, left and right foot normally alternate and are rarely set side by side.

Four kinds of measurements are essential (Fig. 4). The stride, a measure of forward movement, is taken from a fixed point on one footprint to the same point on the next footprint. The pace measures the distance between the right manus or pes and the left manus or pes, respectively. Normally the length of stride is identical for manus and pes, whereas pace may be greatly different. The third measurement is the step angle (pace angulation), which is the angle formed by joining the midpoint of three successive manus or pes prints. Lastly, trackway breadth is measured. The mark of an inefficient walker (such as an amphibian) is a broad trackway and a short stride. Conversely, a long stride and a narrow trackway indicates an efficient walker moving quickly.

In quadrupeds another important measurement is taken from the midpoint between two consecutive pes impressions and the midpoint between two similar manus impressions. This distance corresponds to that between the hipbone socket and the shoulder socket of the living animal. It provides a good estimate of body length (omitting head, neck, and tail). Measurement of the degree of overlap of the impression of the pes upon that of the manus provides an additional measurement of the distance between a quadruped's hip and shoulder. Also, the position of footprints relative to the midline of the trackway is useful in determining the gait of an animal. For example, in the case of a sprawling gait, the midline will not correspond exactly to the midpoints between paired prints. Footprints may also provide information on the soft morphology of long extinct animals—the organic tisues that decay completely under normal conditions of fossilization. For example, some of the splendidly preserved *Chirotherium* foot-

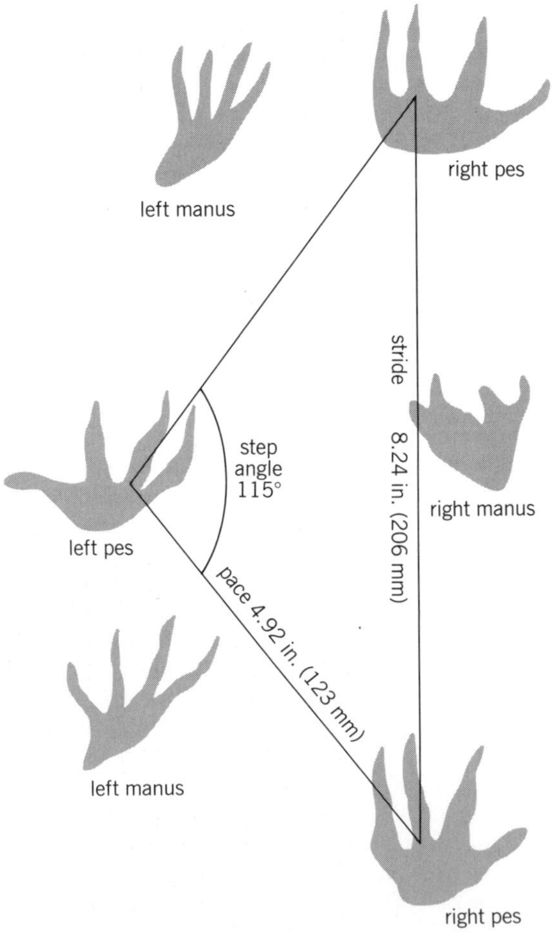

Fig. 4. Tracks of a small Carboniferous reptile showing method of systematic measurement. (*After W. A. S. Sarjeant and D. J. Mossman, Vertebrate footprints from the Carboniferous sediments of Nova Scotia: A historical review and description of newly discovered forms, Palaeogeog. Palaeoclimatol. Palaeoecol., 23:279–306, 1978*)

prints from the Triassic of England supply complete information concerning the pattern of scales on the undersurfaces of the feet. In the Upper Cretaceous of Alberta, Canada, footprints of large bipedal herbivorous dinosaurs known as hadrosaurs show that these creatures, like elephants, had pads on the soles and on the three digits of their hindfeet to cushion their great body weight.

Importance of fossil footprints. *Chirotherium*, the footprints of which were first recorded in 1834 from red sandstones of Triassic age in Thuringia, Germany, provides a classic example of an animal known only from its tracks. Acceptance of its reptilian character came with recognition of the narrowness of the trackway, and more especially with the existence of scale patterns on the best-preserved footprints. It is now recognized that the *Chirotherium* track-makers were pseudosuchian reptiles, an important group ancestral to both dinosaurs and crocodiles. A recently discovered example of an animal known only from its tracks comes from the classic locality where Logan found the first vertebrate tracks in 1841. There, at Horton Bluff, Nova Scotia, in 1964, a remarkably large, although imperfectly preserved, 300-m.y.-old trackway was discovered. It is a relatively broad trackway exhibiting short stride and clawless digits. It was probably formed by a labyrinthodont amphibian related to the giant semiaquatic predator *Eryops*.

In addition to providing information about the morphology of creatures of past times, vertebrate footprints and trackways can provide much other information. They greatly supplement the slender information provided by skeletal remains concerning the temporal and geographic distribution of vertebrate faunas. In the absence of skeletal remains of vertebrates, for example English Triassic strata, the existence of vertebrate trackways confirms that in that region no fewer than eight different types of reptiles flourished in those times. Footprints can therefore be used as a tool in stratigraphy. Thus in arid times, such as prevailed widely in the world during the late Carboniferous, Permian, and Triassic, footprints may exist where other fossils are lacking.

In 1960 footprints of herbivorous dinosaurs were discovered in the Lower Cretaceous rocks of Spitsbergen. Skeletal remains of dinosaurs are unknown from the Arctic. The presence of dinosaurs in such high latitudes indicates they may have had a wide range of climatic tolerance. Alternately, the herbivorous dinosaurs were migratory and attained these latitudes only during summer months. More important is the evidence that vertebrate footprints provide concerning individual and social behavior of the animals of past times. For example, Roland T. Bird's finds in Texas at Paluxy Creek (1939) and West Verde Creek (1954) confirm beyond doubt that giant sauropod dinosaurs could walk on land, completely unsupported by water. More recently Walter P. Coombs's study of superbly preserved trackways in the State Dinosaur Park, Rocky Hill, Connecticut, indicates that even the predatory theropod dinosaur *Eubrontes* was able to swim (Fig. 5).

Fig. 5. Swimming *Eubrontes*, a carnivorous dinosaur from Connecticut Jurassic strata. (*After D. J. Mossman and W. A. S. Sarjeant, The footprints of extinct animals, Sci. Amer., 248:75–85, January 1983*)

One important source of evidence for social behavior of dinosaurs is the occurrence of nests, complete with unhatched eggs, discovered in the 1920s in Mongolia. Another important nest discovery was made by John R. Horner in 1981 in Cretaceous sediments in Montana. Here the skeletons of 18 several-week-old hadrosaurs were discovered. Their continued association with the nest suggests social behavior among juvenile dinosaurs, and probably implies the attendance of adults. A detailed study of footprints from the Peace River Canyon in British Columbia is presently under way. Already over 1000 footprints have been mapped and measured, including the earliest known bird footprints. Splendid dinosaur trackways abound, including the entire growth series of an early hadrosaur. These show how the shape of the foot changed with growth and with increasing body weight. Evidence indicates that hadrosaurs ranged in herds, spread out on a broad front with adults preceding the juveniles. The small to medium-sized carnivorous dinosaurs also ranged in packs, possibly for hunting purposes, and were able to move quite rapidly. Clearly the stereotype picture of dinosaurs as clumsy, slow-moving, dim-witted, and essentially solitary is far from the truth.

Terrestrial record. The impressions of lobe-finned fishes that flopped across mud flats in Early Devonian times are the most ancient vertebrate trackways known. From these creatures, amphibians evolved, with characteristically broad trackways typical of rather inefficient quadruped motion. They peaked during the Carboniferous Period. Thereafter, reptiles rose to dominance in late Carboniferous and early Permian times. Diversification and increasingly efficient locomotion are shown by changes in the trackway record. In some lineages reptilian trackway narrowed and stride lengthened. A short, thumblike fifth digit developed and claws appeared. Other trackways show that by late Permian times some small reptiles adopted a bipedal gait when they moved quickly. This lineage was the precursor to the dinosaurs at the outset of the Mesozoic

Era, 225 m.y. ago.

Dinosaur footprints are the most spectacular of all vertebrate footprints. They occur in terrestrial sediments the world over. Two groups of small dinosaurs evolved into flying reptiles. Trackways of pterosaurs from the Jurassic of Arizona indicate that these creatures could also walk on all fours trailing the wing membrane behind the forefeet. Evolution of birds from dinosaurs yielded no immediate modification of the hindfeet. The synapsids, a reptilian lineage distinct from the dinosaurs, bridged the gap between primitive reptiles and the mammals. In the Late Triassic and Early Jurassic rocks of Brazil, Giuseppe Leonardi found many tracks exactly intermediate in form between reptile and mammal footprints. Dinosaur tracks vanish from geological history at the end of the Cretaceous. From the earliest Cenozoic (Paleocene) onward, a variety of mammalian footprints correspond to a great diversification among the mammals. In the Quaternary, with Mary Leakey's discovery of 3.6-m.y.-old hominid footprints on the Laetoli Plains, Kenya, humans' ancestors walked into the fossil record.

For background information *see* DINOSAUR; EXTINCTION (BIOLOGY); PALEONTOLOGY in the McGraw-Hill Encyclopedia of Science and Technology.

[DAVID J. MOSSMAN]

Bibliography: O. Kuhn, *Ichnia tetrapodorum*, in *Fossilium Catalogus*, I: *Animalia*, edited by F. Westphal, 1963; G. Leonardi, New archosaurian trackways from the Rio do Peixe Basin, Paraiba, Brazil, *Annali dell ' Universita di Ferrara*, IX: *Science Geologiche e Paleontologiche*, 5(14):239–250, 1979; J. S. Levinton, A theory of diversity equilibrium and morphological evolution, *Science*, 204: 335–336, 1979; D. J. Mossman and W. A. S. Sarjeant, The footprints of extinct animals, *Sci. Amer.*, 248(1):74–85, 1983; W. A. S. Sarjeant, Fossil tracks and impressions of vertebrates, in Robert W. Frey (ed.), *The Study of Trace Fossils: A Synthesis of Principles, Problems, and Procedures in Ichnology*, pp. 283–324, 1975; W. A. S. Sarjeant and D. J. Mossman, Vertebrate footprints from the Carboniferous sediments of Nova Scotia: A historical review and description of newly discovered forms, *Palaeogeog. Palaeoclimatol. Palaeoecol.*, 23:279–296, 1978; J. J. Sepkoski, Jr,. A factor analytic description of the Phanerozoic marine fossil record, *Paleobiology*, 7:36–53, 1981; J. J. Sepkoski, Jr., Mass extinctions in the Phanerozoic oceans: A review, in L. T. Silver and P. H. Schultz (eds.), *Geological Implications of Impacts of Large Asteroids and Comets on the Earth*, Geol. Soc. Amer. Spec. Pap. 190, pp. 283–289, 1982; J. J. Sepkoski, Jr. et al., Phanerozoic marine diversity and the fossil record, *Nature*, 293:435–437, 1981.

Particle accelerator

From the time of E. O. Lawrence's desktop cyclotron of the 1930s to the present, when the Fermilab synchrotron spreads across miles of Illinois prairie, accelerator energies have undergone a millionfold increase. Besides advancing the march of particle physics, the technology of focusing fast particles to a small point has been of use in a number of other areas, such as medicine (in cancer therapy, for example, and for producing radio-pharmaceuticals), synchrotron radiation (materials testing), and fusion (inertial confinement). While such applications are important, the accelerators described in this article are cataloged from the standpoint of particle physics, which examines the structure of matter at the smallest scale by using beams with the highest energy. *See* HEAVY-ION ACCELERATOR.

SCATTERING EXPERIMENTS

It was found in the 1920s that particles, such as electrons and protons, possess wavelike properties. The scattering of electrons in a crystal, giving a diffraction pattern characteristic of scattered waves, confirmed this. The relationship between wavelength λ and energy E for light is given roughly by the equation below, where h is Planck's constant and c is the speed of light.

$$\lambda = \frac{hc}{E} \simeq \frac{10^{-15}\ (\text{GeV-m})}{E\ (\text{GeV})}$$

In this expression, 1 GeV stands for 10^9 electron-volts, the unit of energy (and mass) used most often in high-energy physics. Thus, a particle with an energy of 100 GeV would have an equivalent wavelength of 10^{-17} m. This is much shorter than the wavelength of even the most powerful light used in a conventional microscope. Indeed, since a microscope's ability to resolve fine structure improves as the wavelength of a light source gets shorter, beams of particles can be used to "illuminate" matter at a scale billions of times smaller than the smallest organism viewed in a conventional microscope.

The static properties of elementary particles, such as mass, charge, and spin, and their dynamic properties during interactions with other particles are best studied by performing scattering experiments. By colliding one particle with another at high velocities, it may be possible to disintegrate one or both particles and see what lies within. New and exotic particles, many of which do not usually occur in nature, can be created out of the surplus kinetic energy borne by the powerful beams. The interaction forces themselves can also be studied and classified.

There are several kinds of accelerators, each suited to a particular kind of physics. Storage-ring accelerators consist of two separate beams traveling in opposite directions around a circular track. These beams are smashed head on at one or more places around the track. The center-of-mass energy for such a collision is high, but the luminosity, the number of particles which are brought to bear in the collision region, is low, owing to the relatively tenuous nature of each beam.

The other major accelerator scheme involves the collision of a beam of energetic particles against a fixed target, typically a chunk of metal. Since even a teaspoon of solid matter contains many more particles than even the most intense beam, the luminosity for a fixed-target accelerator is very high. Since the target particles are sitting at rest in the laboratory, though, the center-of-mass energy is low. Both of these factors are important: center-of-mass energy determines what kind of reactions can take place, while the luminosity determines how often reactions will occur. In general, it is desirable to have high energy and high luminosity, although in practice it is seldom possible to have both.

Most particle beams consist of protons or electrons. Other beams, such as positrons (antielectrons), antiprotons, mesons, neutrinos, and even photons, are created by crashing electrons or protons into a stationary production target and selecting out the desired secondary particles from among the collision debris.

Electrons belong to the family of leptons, pointlike particles which interact primarily via the electromagnetic force. The proton, in contrast, belongs to the hadron family and interacts primarily via the strong nuclear force. Because hadrons are composite objects, being made of quarks, the nuclear interaction between two hadrons can be very complicated. Indeed, the collision between a proton and an antiproton has been compared to smashing two Swiss watches together.

Since lepton collisions are so much "cleaner" then hadron collisions, it would seem advantageous to use only lepton beams. However, as with the energy-versus-luminosity dilemma discussed above, the issue of lepton-versus-hadron beams involves trade-offs. Electrons traveling around the circular arc of a storage ring continuously lose energy through synchrotron radiation at a rate proportional to the fourth power of the energy. The cost of counteracting this loss is a major factor in limiting the energy of circulating electron beams. Until now the principal lepton colliders, PETRA near Hamburg, West Germany, and PEP at Stanford, California, have had a maximum beam energy of about 20 GeV. To produce particles as massive as the *W* usually requires the services of protons and antiprotons, which are more immune to the attrition of synchrotron radiation, and which can deliver the highest possible center-of-mass energy.

PHYSICS AT ACCELERATORS

An accelerator can be thought of as a microscope revealing the structure of matter at the smallest scale or as a factory for producing strange, new particles. Accelerators are also time machines. Consider the high-energy collision of two particles: the ensuing miniature fireball partially recreates a tiny piece of the universe as it may have existed billions of years ago. Around 10^{-12} s after the big bang, for example, the temperature of the universe was roughly 10^{16} K or, in the equivalent energy units, about 1000 GeV. At such a time the energy density in the universe was so great that heavy particles, such as the *W* particle (with a mass of about 82 GeV), could easily materialize and then decay, and then appear once again. The *W* particle is the carrier of the weak nuclear force, just as the photon carries the electromagnetic force.

When the temperature of the universe had fallen far enough, sufficient energy could no longer be mustered to produce such a heavy dinosaur of a particle. In today's frosty universe (the temperature of the cosmic microwave background 3 K above absolute zero), the *W* appears only as a virtual particle, mediating certain kinds of radioactivity and some of the fusion reactions taking place in the Sun. That the weak force is so weak and so short-ranged is a consequence of the fact that it is borne by such a cumbersome particle.

Nevertheless, if an accelerator is powerful enough, the old days can be relived. At the CERN collider near Geneva, Switzerland, beams of protons and antiprotons collide head-on to provide a center-of-mass energy of 540 GeV. This is more than enough to produce the *W* as a physical particle in its own right. *See* INTERMEDIATE VECTOR BOSON.

Indeed, many of the experiments conducted at accelerators are designed to coax into existence particles that are not ordinarily found in nature, but which hold important places in the "standard model," the consensus overall view of particle physics that has emerged since the early 1970s. The *W* particles, the carriers of the weak force, are perhaps the best example. Their discovery at CERN provided crucial support for the Weinberg–Salam–Glashow model, a theory which comprises the electromagnetic and weak forces into a single electroweak force.

The observation of jets at the PETRA accelerator provided evidence for the existence of gluons, the carrier of the color force which acts between quarks, and lent suppport to the prevailing theory of color interactions known as quantum chromodynamics. According to quantum chromodynamics, quarks are inexorably bound together (by gluons) into groupings of two (mesons) and three (baryons, such as protons). It is these groups of quarks, known collectively as hadrons, which are observed as particles in the laboratory. Under the impact of a high-energy collision, the interquark embrace can be sundered, but only by spawning new quarks which in turn regroup into hadrons (all of which is analogous to sawing a bar magnet in half, creating not isolated poles but only two new dipoles). If the collisions are violent enough, the ensuing stream of secondary particles may emerge as collimated jets, much like volcanic eruptions from deep molten regions of the earth. *See* QUARKONIUM.

The top quark is another important particle sought by physicists. In the standard model the top is the sixth quark type. The other types, or flavors, are the up, down, strange, charm, and bottom quarks. Only the up and down quarks are the constituents of normal matter. The other quarks exist as virtual particles but can be resurrected as physical particles, if only momentarily, in high-energy collisions. Although the search for the top quark has been frustrating so far (it is perhaps too heavy to produce in existing machines), the bottom and charm quarks, or rather particles containing them, have turned up in profusion and have been the subjects of intense study in recent years. Much of this work has been done at electron-positron colliders.

Studying the *W* particle and hunting for the top quark may perhaps enjoy the highest priority, but there are many other subjects on the agenda. The following list summarizes some of the most interesting subjects that are likely to be pursued at the major accelerators in coming years:

1. Weak bosons: W^+, W^-, Z^0.
2. Top quark search.
3. Jets and states made from gluons (glueballs).
4. Particles containing bottom and charm quarks.
5. The tau neutrino (not yet seen) and neutrino oscillations.
6. Heavy leptons.
7. Tests of quantum electrodynamics (QED).
8. The search for the Higgs boson, an ingredient in the electroweak theory (supposedly endowing the *W* particle with mass) which may well exist as a particle in its own right. *See* HIGGS BOSONS.
9. Centauro events, high-energy (10^{14} eV or more) cosmic-ray showers containing a puzzling shortage of neutral pions and thought by some to represent a new physical phenomenon.
10. Particles associated with technicolor and supersymmetry, two theories that prescribe the existence of whole families of particle states.
11. Deep inelastic scattering of leptons (electrons or muons) from protons; owing to their pointlike structure and their immunity from the strong nuclear force, the lepton can penetrate far into the proton, providing information on the distribution of quarks inside the proton.

This list of subjects could change overnight. No one can anticipate exactly the physics of tomorrow. Frequently the greatest discoveries will occur in areas not included among the list of reasons for building the accelerator in the first place. On the other hand, some prognostications are superbly upheld. The electroweak theory, for example, prescribed a *W* particle with a mass of about 83 GeV. The CERN SPS synchrotron was adapted for proton–antiproton collisions, and the *W* was duly found with a mass of 81 GeV.

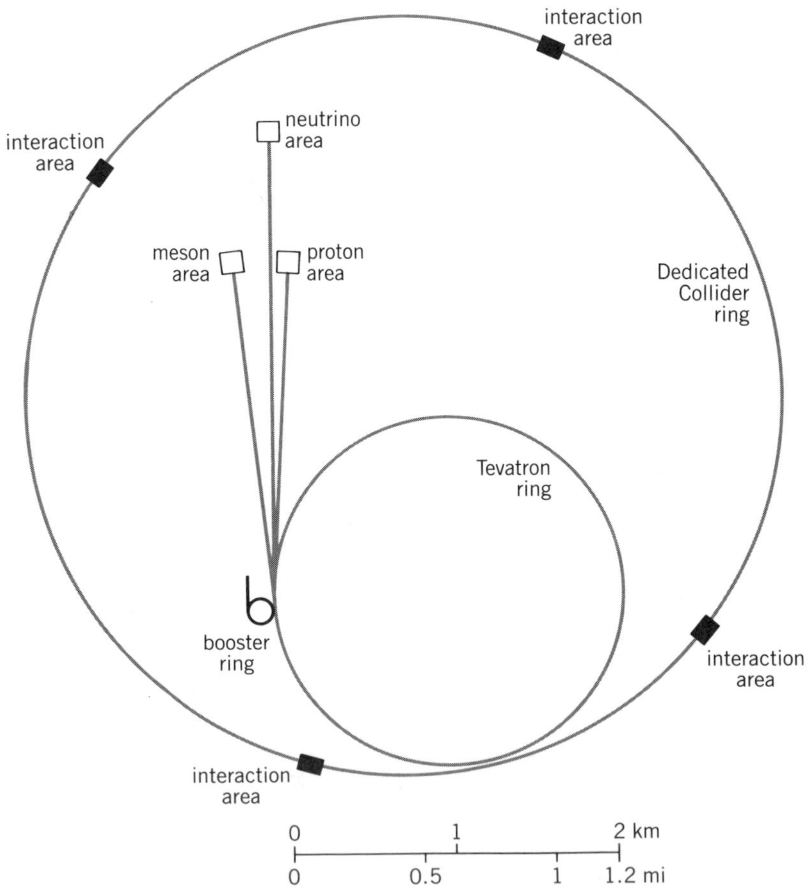

Fig. 1. Layout for the proposed Dedicated Collider at Fermilab, near Chicago. At present, protons are accelerated in the booster ring and then the Tevatron ring. The Tevatron would serve as an injector of protons and antiprotons for the larger collider.

NEW ACCELERATOR FACILITIES

The success of the discovery of the *W* particle and the expectation that, as in the past, new discoveries beckon in the energy range just over the horizon, have ensured that the bigger machines will be built, notwithstanding the constraints prevailing in government budgets. The following list of accelerator projects testifies to the strength and diversity of particle physics.

CERN. The scene of *W* and *Z* particle discoveries, CERN has long specialized in proton beams. Although there is a possibility that the proton–antiproton (p-$\bar{p}$) collider may be upgraded from 540 GeV to 900 GeV (by the year 1990), the primary construction effort will be the LEP (Large Electron Positron storage ring), a 50 GeV by 50 GeV electron–positron machine, 17 mi (27 km) in circumference, to be completed by 1988. A second phase will push the energy of each beam from 50 up to 130 GeV. At a luminosity of 3×10^{31} cm$^{-2}\cdot$s^{-1}, LEP could produce about 10^8 *Z* particles a year. The Low Energy Antiproton Ring (LEAR) is a recently commissioned machine that stores antiprotons for specialized studies. For that purpose it siphons antiprotons away from the massive proton–antiproton collider.

Fermilab. At Fermilab, near Chicago, Illinois, in the same tunnel and beneath the string of magnets constituting the old synchrotron, a second string of magnets boosts the energy of circulating protons to 1000 GeV, or 1 TeV (10^{12} electronvolts), whence the name Tevatron. Besides delivering the highest beam energy in the world, this machine (at least the second stage of magnets) is also the world's first major superconducting accelerator.

There are two principal Tevatron construction phases: (1) the extraction from the machine of 1-TeV protons (by 1984) for use in fixed-target experiments (at the meson, neutrino, and proton areas in Fig. 1), and (2) the creation (by 1986) of beams of protons and antiprotons, each with an energy of 1 TeV to be collided head-on, producing the world's largest center-of-mass energy, 2 TeV. The Fermilab proton–antiproton collider should have a luminosity of about 10^{30} $cm^{-2} \cdot s^{-1}$, about 10 times larger than that of CERN's collider.

DESY. The Deutsches Electronen Synchrotron (DESY) near Hamburg, West Germany, is the home of PETRA, a 22 GeV by 22 GeV electron–positron collider. HERA is a machine (to be completed by 1990) in which 30-GeV electrons will collide head on with 800-GeV protons, producing deep inelastic interactions of tremendous force. The proton ring in this scheme might be superconducting.

SLAC. Presently the home of PEP, an 18 GeV by 18 GeV electron–positron collider, the Stanford Linear Accelerator Center (SLAC) at Stanford, California, will next bring forth the Stanford Linear Collider (SLC), an accelerator scheme that will employ the existing linear accelerator to accelerate both electrons and positrons to energies of 50 GeV. The two beams emerge from the linear accelerator, are steered in opposite directions around a racetrack course, and collide head on with an energy of 100 GeV, enough to produce the *Z* particle. SLC should be completed by 1986, in time to challenge the LEP machine at CERN. Since the beams are discarded after a single pass (unlike ordinary storage rings, where circulating beams interact over and over again), the luminosity for SLC will be comparatively low.

CESR. At the Cornell Electron Storage Ring (CESR) in Ithaca, New York, a proposal has been made to expand the present electron–positron collider into a 50 GeV by 50 GeV machine, rivaling the LEP.

BNL. The Colliding Beam Accelerator (formerly known as ISABELLE), at Brookhaven National Laboratory (BNL), in Upton, New York, is a proposed 400 GeV by 400 GeV proton–proton (*p-p*) collider with a high luminosity (10^{33} $cm^{-2} \cdot s^{-1}$). The CBA project encountered early troubles in its ambitious superconducting magnet program and has been canceled. The 33-GeV proton synchrotron, the AGS, may be adapted to accelerate heavy ions.

Japan. At the KEK laboratory, TRISTAN, a 2-mi-circumference (3-km), 25 GeV by 25 GeV electron–positron collider, should be finished by 1987. A more ambitious version of TRISTAN, employing electron–proton collisions at a center-of-mass energy of 200 GeV (by 1989), is being contemplated.

China. The 50-GeV Peking Proton Synchrotron, a project calculated to modernize in one step the state of Chinese particle physics, has been canceled. An electron-positron collider, called BEPC, will be built in its place (by 1987).

Soviet Union. Several gigantic projects are planned. At the Serpukhov Institute near Moscow, a 3-TeV proton accelerator is due by 1990. This machine, called UNK, might later (by 1993) accommodate proton–antiproton collisions at a total energy of 6 TeV. At the Institute for Nuclear Physics at Novosibirok a novel scheme has been devised: electrons and positrons will be fired at each other by using not storage rings but opposing linear accelerators. This complex, known as VLEPP, would at first use beams with an energy of 150 GeV, to be upgraded later to 500 GeV.

FUTURE ACCELERATORS

By 1990 accelerators will routinely be sending forth 50-GeV electrons and TeV protons. To move beyond that point will require either radically new accelerator techniques or—if conventional methods are to be used—the treasuries of many nations combined. The International Committee on Future Accelerators (ICFA) was formed to consider joint contributions toward the construction of a Very Big Accelerator (VBA), a 10–20-TeV proton synchrotron, for example. The European Committee for Future Accelerators (ECFA, not to be confused with the ICFA) has contemplated constructing just such a VBA in the LEP tunnel at CERN, perhaps sometime in the early 1990s.

In the United States, the High Energy Physics Advisory Panel (HEPAP) makes recommendations to the Department of Energy as to where expenditures for particle physics research can do the most good. At an important meeting in the summer of 1983, HEPAP recommended the construction of a Superconducting Supercollider (SSC), a machine in which beams of 10–20-TeV protons would collide head on. The need to keep magnets simple and cheap would expand the overall size of the machine and necessitate the use of vast tracts of land, such as in the desert southwest (whence the nickname "desertron") or at a laboratory like Fermilab (Fig. 1).

Even if one of these next-generation schemes is built, it may be the last of its kind; the cost in money, land, electrical power, and materials is becoming too high. Consider the problem from the standpoint of the electric fields in the radio-frequency cavities used to accelerate particles: At SLAC the acceleration gradient—the energy imparted to electrons per unit frequency distance—is about 15 to 20 MeV per meter. The fields necessary for increasing this gradient a hundredfold would be large indeed. Such fields are manifested in laser light. But because the electric fields are always transverse to the laser light direction of propagation, using them to accelerate particles will require some ingenuity.

Several schemes are being studied: Near-field acceleration could, for example, use 10-micrometer radiation from a CO_2 laser, shaped by a diffraction grating so as to resemble the accelerating effect of a miniature linear accelerator. Far-field accelerators would impose an oscillating transverse component on a particle's trajectory in order to couple it to the laser radiation. In the plasma beat-wave accelerator (Fig. 2) laser light at two different frequencies is beamed through a plasma-filled chamber. If the resultant interference beat wave is in tune with the plasma's own natural oscillation, it generates traveling density waves in the plasma which could be used to accelerate a particle beam. Electrons could be sent through many stages, such as the one shown in Fig. 2, acquiring an energy of 1–2 GeV at each step. Other schemes include the inverse free-elec-

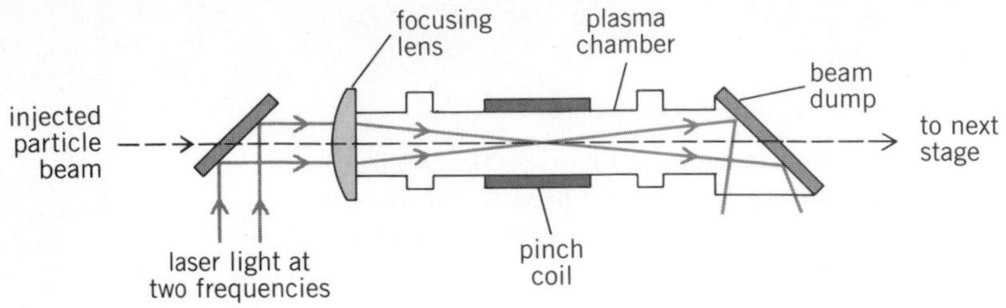

Fig. 2. Plasma beat-wave accelerator.

tron laser, inverse Cerenkov acceleration, and ionization front acceleration.

Laser accelerators may not be realized for another generation. In the meantime some physicists are returning to the study of cosmic rays, particles (some of them with multi-TeV energies) believed to be accelerated by supernovae. Others perform particle-physics experiments without an accelerator at all, moving far underground in search of proton decays, monopoles, and solar neutrinos, or setting up experiments next to nuclear reactors in order to look for neutrino or neutron oscillations. Some theorists, impatient with the relatively low energies to be had from even the best terrestrial machine, study cosmology in order to examine the output of the ultimate particle accelerator, the big bang. *See* MAGNETIC MONOPOLES; NEUTRON-ANTINEUTRON OSCILLATIONS; PROTON DECAY.

For background information *see* BIG BANG THEORY; ELEMENTARY PARTICLE; FUNDAMENTAL INTERACTIONS; PARTICLE ACCELERATOR; QUANTUM CHROMODYNAMICS; QUARKS in the McGraw-Hill Encyclopedia of Science and Technology.

[PHILLIP F. SCHEWE]

Bibliography: P. J. Channel (ed.), *Laser Acceleration of Particles*, AIP Conf. Proc. 91, 1982; R. Gustafson, *The DPF Aspen/Snowmass Summer Study*, Fermilab Report, 1982.

Perception

Human vision is particularly well suited to detect changes as well as to discern motionless objects. Two types of change, extremely important to visual perception, can be distinguished: that due to the change in position of the observer's eye, head, or body, called movements; and that due to the change in position of objects in the world, called motions.

Most recent developments have concentrated on the perception of objects in motion. Although the importance of motion has been known since Lucretius (1st century before the Christian era), the discovery that motion alone is sufficient for the identification of objects is of only recent date. Objects often have telltale motions that make recognition possible, even when the observer cannot see their substance.

Eye movements in perception. All movements create global changes on the retina. These changes occur when an observer moves through the environment, swivels the head or body, or moves the eyes. Eye movements have received much attention in vision research. Three kinds of eye movements are distinguished: saccades, pursuit movements, and micronystagmus. Saccades are the sharp and rapid movements, about 80 milliseconds in duration, that are interleaved with eye fixations, that is, periods of about 250 ms during which the eye does not move. Fixations and saccades alternate when reading text, scanning photographs or maps, or simply looking around. The purpose of saccades is to bring new information into register on the fovea, that part of the retinal surface responsible for the most acute vision. Interestingly, the world does not appear to spin around the observer during the saccades, even though textured information rapidly traverses the retina. Perception of the global movement is suppressed by a process that is incompletely understood. Part of the suppression may be due to muscular feedback of voluntary eye movements. This can be demonstrated by pressing with a finger against the corner of one eyeball through one eyelid; the world appears to move. It moves, in part, because the eye muscles were not given commands to move. Another part of suppression may be due to a phenomenon called masking: information during a fixation may simply blot out information stored during the previous saccade. Pursuit movements are smoother and slower than saccades, and can be generated only by following a moving object, such as a car or tennis ball. These movements are used to keep moving objects registered on the fovea in order to discern, identify, and track them. The third type of eye movement, micronystagmus is very small, a quivering due to tension in the eye muscles themselves. These movements occur approximately every 8 ms during fixation. Their purpose is to continually refresh the image of the visual world on the retina by jiggling the retinal image about the width of a receptor in order to keep the cells stimulated. Without these movements all visual experience ceases, and the world fades to neutral gray. Quite clearly, eye movements are necessary for perception.

Motion-detecting system. The visual system is also inherently attuned to the motions of external objects. In fact, the eye is equipped with a special motion-detecting system. The eye uses two kinds of receptor cells: the rods of the retina are responsible for the black-and-white (scotopic) vision that guides nocturnal behavior, and the cones are responsible for color (photopic) vision that guides diurnal behavior. A second important division is that between the sustained detection and the transient detection systems of ganglion cells. In the sustained detection system, the ganglion cells that receive input from receptor, bipolar, horizontal, and amacrine cell inputs respond best to near-constant stimulation; the transient detection system best to change—object motion or flicker. The populations of these cell systems differ radically in distribution across the retina: the inputs to sustained cells are located primarily in the fovea and parafovea (the region immediately surrounding the fovea); those to transient cells are moderately dense and distributed throughout the retinal surface. The fovea and parafovea together are only about 1/200 the size of the peripheral visual field of a single eye. Thus, the whole of the retina is very sensitive to motion, but only the central region of highest acuity, covering an area about the size of one's fist at arm's length, is very sensitive to relatively stationary objects. An evolutionary reason for this distribution is that in searching for prey or predator, one can detect the motion of that creature in the periphery (literally out of the corner of one's eye) and then can move one's eye to bring it into a region of much higher acuity on the fovea. The transient system seems fundamentally to be an alerting system.

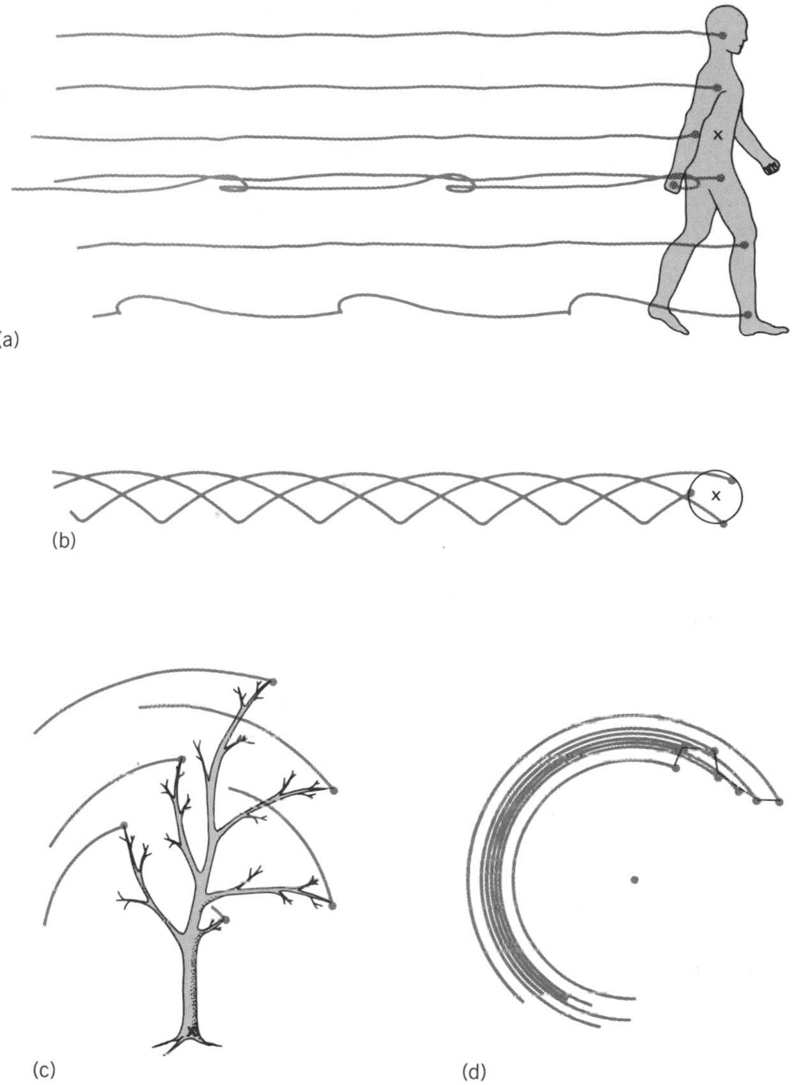

Perception of motion. (*a*) Vector traces of a walker, with lights mounted on the head, right shoulder, elbow, wrist, hip, knee, and ankle. All motion in the walker occurs around a point at mid-torso, marked X. (*b*) Vector traces of three lights mounted on the rim of a wheel, 120° apart. All motion in the wheel occurs around the axle, marked X. (*c*) Vector traces of lights mounted on a tree blowing in the wind; all motion is referred to point X, where the trunk meets the ground. (*d*) Vector traces of the stars of Ursa Major (the Big Dipper) as they trace out concentric circles over time. The center of these traces is Polaris, the North Star.

Perception of motion of objects. Unlike movements, motions create only local changes on the retina. Moreover, motions can be used for puposes beyond mere alerting; they can be used for identification as well. Different objects often have different characteristic motions. For example, people can be recognized at a long distance by their way of walking. At such distance, distinctive cues as hair length and clothing may be inperceptible, but the flow of motion is not. That viewers can identify friends by motions alone has been shown by several experiments. When small electric bulbs are placed on the joints of individuals—shoulders, elbows, wrists, hips, knees, and ankles—in an otherwise darkened room, and those individuals walk across the field of view, they can be correctly recognized more often than if a viewer were guessing randomly. Yet there is nothing visible other than a pattern of moving lights. The motion paths of such a display are shown in illustration *a*. Even when the walker is not known to the observer, the display is still perceptually rich. For example, viewers are generally able to indicate which walkers are male or female, simply from the movements of lights on joints. These results are obtained even with people who have had no experience with these experiments.

Other objects also have prototypic motions. Wheels are built so that their axles translate parallel to the surface on which they roll, and trees sway in the breeze in a manner that reveals the arborization pattern of intersections and angles of limbs with the trunk. Motions of wheels were observed by Galileo and later by psychologist Karl Duncker on rolling motion. When three lights are mounted on the rim of a wheel 120° apart, a symmetric pattern of vectors (line traces of the lights through space and time) is created in the dark, as shown in illustration *b*. In the illustration the wheel is rolling across a flat surface at right angles to the line of sight. When lights are mounted on the wheel asymmetrically, say at 0°, 90°, and 180° (but not at 270°) the configuration will look less wheellike, and more like a bouncing ball. So long as the centroid (center of gravity) of the system of lights is at the axle the display will look like a rolling wheel; if the centroid is elsewhere it generally will not, with the degree of displacement of the centroid from the axle dictating the degree to which the system looks like a rolling wheel. Another example is the motion of trees blowing in the wind. Rather than being rigid in structure as wheels are, trees are pliable. Both branches and trunks generally bend in breezes, and do so with periods similar to those of pendulums. If lights are mounted on the limbs of a tree, characteristic patterns emerge, as shown in illustration *c*.

Object motion can be useful even if the motion is extremely slow. An example is the rotation of the night sky, according to the ptolemaic model. In this situation, points of light are actually real objects—stars. In the Northern Hemisphere all rotation occurs about the celestial north pole. Polaris, or the North Star, is located at this point. As shown in illustration *d*, the stars of Ursa Major (the Big Dipper) trace out circular paths around Polaris. Migratory song birds can use such information for orientation in migratory flight. It is the motion information that is important rather than the configuration of the stars themselves. When the night sky is rotated around some other point (as one can do in a planetarium), birds orient to the new artificial north star.

In each of these cases there is a coherent event—walking people, rolling wheels, trees blowing in the wind, a rotating night sky. Coherence, in these cases, is defined by mathematical relations. All points within the moving object have a systematic relation to a defined center, called the center of moment. For the night sky it is, in the Northern Hemisphere, Polaris; for trees it is the place where the trunk meets the ground; for rolling wheels it is the axle; and for walking people it is a point within the torso, roughly where diagonal lines would intersect from left shoulder to right hip and from right shoulder to left hip. The characteristic movements around these centers of moment, and the location of these centers, guides perception of each event.

For background information *see* PERCEPTION; VISION in the McGraw-Hill Encyclopedia of Science and Technology.

[JAMES E. CUTTING]

Bibliography: D. Lee, The optic flow field: The foundation of vision, *Phil. Trans. R. Soc., London, B*, 290:169–179, 1980; R. Monty and J. Senders (eds.), *Eye Movements and Psychological Processes*, 1976; R. Walk and H. Pick (eds.), *Intersensory Perception and Sensory Integration*, 1981; A. Wertheim, W. Wagenaar, and H. Leibowitz (eds.), *Tutorials in Motion Perception*, 1980.

Periodicity in organisms

Among estuarine crabs it has been recently demonstrated that reproductive activities do not occur randomly throughout the breeding season. Instead, these activities are precisely timed to occur rhythmically in relation to lunar phase, time of day, or phase of the tide. The timing is important for increasing survival of both the adult female and her offspring.

Crab life cycle. The general life cycle of a crab begins with a larva which is free-swimming in the water column. The larva metamorphoses into a postlarval stage which settles to the bottom and grows into adulthood. Adult reproductive activity involves the following sequence of events: gonadal maturation, courtship, mating, egg laying, egg attachment to the female, egg incubation, and egg hatching. All of these events are sequentially arranged so that the eggs hatch at a precise time optimal for larval survival.

The timing of egg hatching has been extensively studied in fiddler crabs of the genus *Uca* and in the mud crab, *Rhithropanopeus harrisii*. As in most crabs, during egg incubation the female of these species has thousands of eggs attached to her posterior section or abdomen. At the time of egg hatching, the female elevates herself on her walking legs and vigorously pumps her abdomen back and forth, thereby breaking open the eggs. All of the eggs are usually hatched within a few minutes, so egg hatching can be precisely timed with respect to external events.

Estuarine cycles. These animals live in estuaries which are semienclosed bodies of coastal water connected with the open sea and diluted with fresh water from land drainage. Within an estuary there are three cycles in environmental events that occur simultaneously and affect egg hatching. First is the tidal cycle. On rising tide, oceanic water flowing into the estuary serves to increase the water depth and the salinity (amount of salt in sea water). The inflow continues until the time of high tide, after which the tide ebbs and water flows seaward. The outflow continues until the time of low tide, after which the cycle repeats. Usually, both the water depth and the salinity within the estuary decrease during the ebb tide. Most estuaries have two high and two low tides per lunar day with about 12.4 h between consecutive high tides.

The second cycle is the 24-h daily or diel cycle. Besides the change in light levels over the day, there is usually a temperature shift in which water temperatures are higher during the day and lower at night. The third cycle is the monthly lunar cycle in tidal amplitude. At the time of the new and full moon the greatest-amplitude tides (spring tides) are observed, while at the quarter phases of the moon the lowest-amplitude tides (neap tides) occur. The time between consecutive spring tides is about 14.7 days. One consequence of this cycle is that water flow in the estuary is much greater at spring tides.

Rhythms in larval release. Fiddler crabs show cycles in egg hatching related to all three of the described external cycles. A semilunar rhythm was initially suggested from nightly measurements of the abundance of newly released larvae in the water column. More larvae were present around the time of the new moon and full moon, while less were present at the quarter phases. To verify that a semilunar rhythm was in fact occurring, females with eggs were collected and kept in the laboratory under conditions of constant temperature and darkness. In this situation there were no cues to indicate tides or time of day. Each day the number of females that hatched eggs the night before was counted. Again there was a clear semilunar rhythm, as most of the crabs hatched their eggs shortly before the time of the new and full moon (Fig. 1).

While these results indicate when to expect egg hatching in the lunar month, it is equally interesting to know the exact time of hatching within the day. For these measurements, freshly collected females with eggs were kept in the laboratory under conditions of constant temperature and under the natural light-dark cycle. Egg hatching was rhythmic and occurred at the time of nighttime high tides in the field, which are followed by nocturnal ebb tides. Since hatching in the laboratory was in phase with local tides, these crabs or eggs probably have an internal clock regulating hatching.

To summarize, egg hatching by fiddler crabs generally occurs at nighttime high tides near the new and full moons. Although this is the pattern most

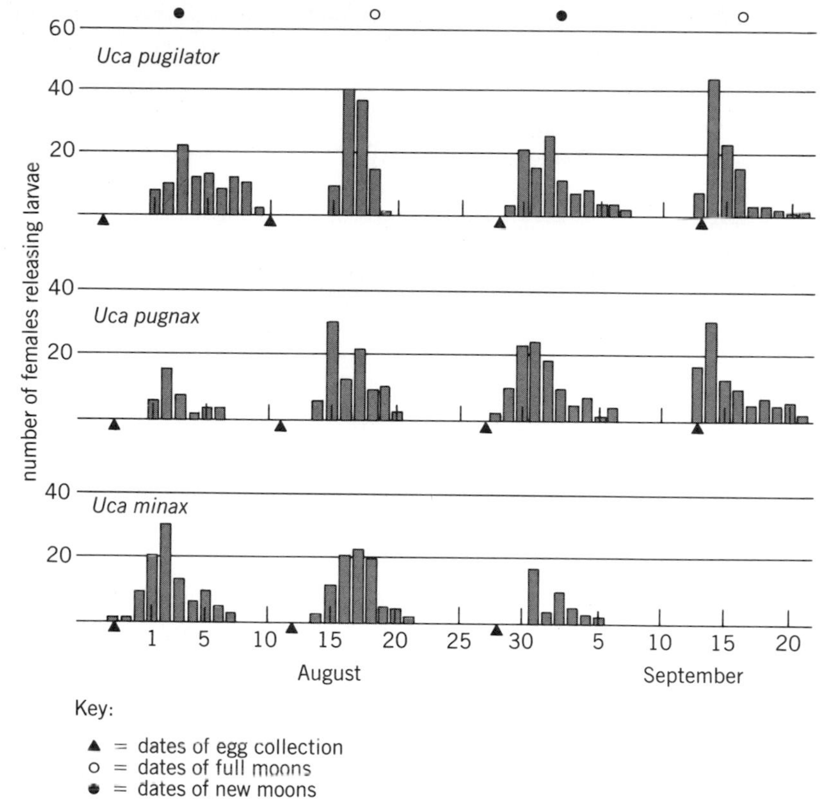

Fig. 1. Cycles in egg hatching by three *Uca* species in the laboratory. (*After J. H. Christy, Adaptive significance of semilunar cycles of larval release in fiddler crabs (genus Uca): Test of an hypothesis, Biol. Bull., 163:251–263; copyright 1982 by the Marine Biological Laboratory*)

often observed, the animals are able to make adjustments for special local conditions. For example, in Charlotte Harbor on the west coast of Florida, the tides of greatest amplitude occur during the quarter moons. The crabs living there hatch their eggs at the quarter moons.

Another case of such flexibility concerns the mud crab, *R. harrisii*. Egg hatching was compared for females from two different estuaries, one having normal tides and the other irregular tides. Females with eggs were collected from each estuary and kept in the laboratory under conditions of constant temperature and low light. Crabs from the estuary with irregular tides hatched their eggs just after the natural sunset without any relation to coastal tides (Fig. 2). In contrast, egg hatching by crabs from the estuary having regular tides generally coincided with high tide. If crabs were transplanted from conditions in one estuary to conditions in the other, the time of egg hatching also changed. So the expressed cycle in egg hatching depends upon environmental cycles in the area where the crab is living.

Since egg hatching is a short, precisely timed event, an important question is whether the timing is controlled by the female or by the developing embryos. In *R. harrisii*, if eggs are removed from the female within 2 days of hatching, they hatch at almost the same time as eggs attached to the female. In addition, upon exposure to water in which eggs hatched, females with eggs display increased abdomen pumping, the behavior observed at the time of egg hatching. Such results indicate that egg hatching is controlled by the embryo. As the eggs hatch, an active substance is released which induces abdomen pumping, which in turn serves to synchronize egg hatching.

Adaptive significance. Both the female crab and her larvae are vulnerable to predation. During egg hatching, the female must expose herself and is therefore susceptible to fish and birds which visually pursue their prey. Once in the water column the larvae themselves are vulnerable to predators which feed upon small animals (zooplankton). Unfortunately for the larvae, estuaries are nurseries for larval, postlarval, and juvenile stages of many marine and estuarine fish. These young fishes frequently feed on zooplankton.

An additional difficulty facing crabs is the highly variable nature of an estuary with respect to environmental conditions such as temperature and salinity. Developmental studies have shown that high temperature and low salinity are the most stressful conditions during larval development.

The functional advantages of the time of egg hatching can be considered relative to survival problems in estuaries. Eggs hatch at night, which is beneficial for both the female and larvae in avoiding visual predators. In addition, water temperatures are cooler at night, so the larvae are initially exposed to tolerable temperatures.

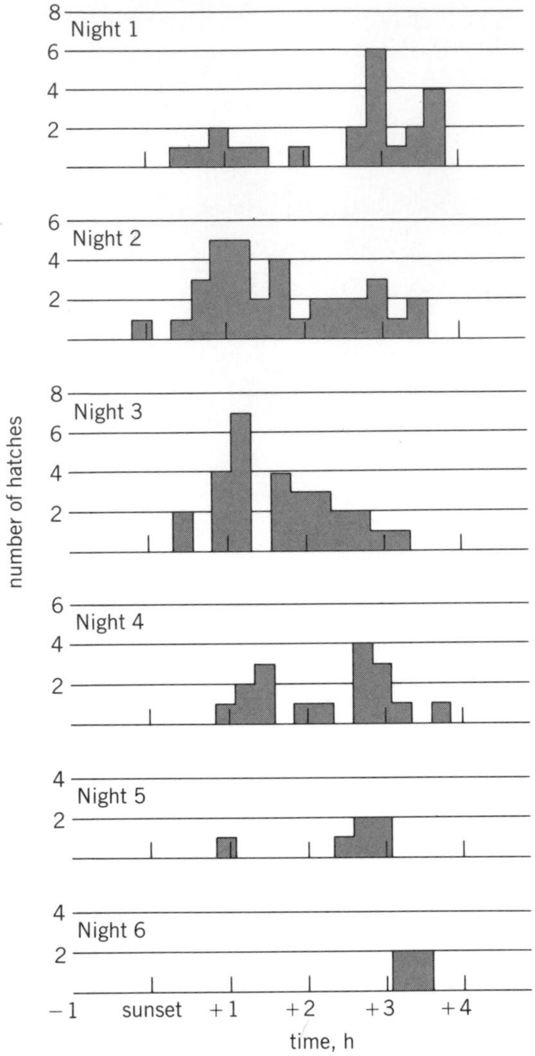

Fig. 2. Number of *Rhithropanopeus harrisii* hatching eggs on succesive nights in the laboratory relative to the time of natural sunset. (*After R. B. Forward, Jr., K. Lohmann, and T. W. Cronin, Rhythms in larval release by an estuarine crab (Rhithropanopeus harrisii), Biol. Bull., 163:287–300; copyright 1982 by the Marine Biological Laboratory*)

Eggs hatch at large-amplitude high tides, which exposes the larvae to high-salinity water that is less stressful. In addition, the larvae are initially transported seaward as the tide ebbs, exporting them from the estuary. Exportation is enhanced by the large-amplitude tides which have greater horizontal water movement and by the behavior of the larvae. Newly hatched larvae usually have a tendency to swim upward, which puts them near the surface where they are exported faster. Exportation to coastal and offshore areas where early-stage fish predators are much less abundant increases the chance of larval survival.

For background information *see* CRAB; ESTUARINE OCEANOGRAPHY; PERIODICITY IN ORGANISMS in the McGraw-Hill Encyclopedia of Science and Technology.

[RICHARD B. FORWARD, JR.]

Bibliography: M. E. Bergin, Hatching rhythms in *Uca pugilator* (Decapoda: Brachyura), *Mar. Biol.*, 63:151–158, 1981; J. H. Christy, Adaptive significance of semilunar cycles of larval release in fiddler crabs (genus: *Uca*): Test of an hypothesis, *Biol. Bull.*, 163:251–263, 1982; R. B. Forward, Jr., and K. J. Lohmann, Control of egg hatchings in the crab *Rhithropanopeus harrisii* (Gould), *Biol. Bull.*, 165:154–166, 1983; R. B. Forward, Jr., K. Lohmann, and T. W. Cronin, Rhythms in larval release by an estuarine crab (*Rhithropanopeus harrisii*), *Biol. Bull.*, 163:187–300, 1982.

Pheromones

Pheromones are substances, secreted to the outside by an animal, that affect the behavior and physiology of another individual of the same species. Although such chemical messengers are known throughout the animal kingdom, they are particularly noteworthy among insects. In many species, pheromones play a key role in pairing of the sexes, and this is perhaps best exemplified in moths. Sexual pairing in most moths species that have been studied is facilitated by female-emitted chemical signals which attract the males. Experimental evidence for the existence of moth sex pheromones was first obtained in 1900, but complete elucidation of the chemical structure of a moth sex pheromone (that of the silkworm moth, *Bombyx mori*) was not achieved until 1961. The following 15 years witnessed a concerted effort by biologists and chemists to isolate, identify, and synthesize moth pheromones. While progress continues in the area of pheromone chemistry, largely due to advances in capillary gas-liquid chromatography, mass spectrometry, and techniques for collecting chemicals, much current work is directed at the behavior associated with the release and perception of sex pheromone. Recently, attempts have been made to employ this knowledge of pheromone chemistry and associated behavior in a variety of pest control programs.

Chemistry. The original determination of the *B. mori* sex pheromone required extracts from 500,000 females. Today, however, technological developments allow the identification of chemical extracts obtained from as few as 20 individuals. Consequently, sex pheromones from several hundred moth species have been identified with great accuracy. The compounds are typically mono- or diunsaturated straight-chain hydrocarbons which are acetates, alcohols, or aldehydes (see illustration). Chain lengths vary between 12 and 21 carbon atoms. The first pheromones to be isolated from female moths were identified as single chemical compounds. However, the highly sensitive analytical techniques used today allow detection of extremely minute quantities of chemicals from female extracts or from air collected in the vicinity of a pheromone-emitting female, and most pheromones are now known to consist of blends of several compounds (multicomponent pheromones). Some researchers are also working on the emission rates of pheromones in various moth species, which appear to range from 3 to 100 nanograms per hour per individual female.

Reproductive isolation. When pheromone compounds were first isolated and described from moths, it was believed that each species utilized a unique compound and that such differences alone accounted for the reproductive isolation of sympatric moth species. However, as the number of moths for which pheromones were isolated and identified increased, findings revealed that the same compound may be found in the pheromones of several different species, and this led to a more comprehensive view of reproductive isolation in moths. It is now known that in several cases two species use the same compound as a sex pheromone, but each uses a different geometrical isomer of that compound. In other instances, in which multicomponent pheromones are involved, the pheromones of each species differ because of the presence or absence of components other than the shared compound. Finally, some species share the identical multicomponent pheromone, but are reproductively isolated because the ratios of components of the blends differ. In general, species specificity in pheromone emission by female moths corresponds with specificity in male response. For example, if females release a multicomponent pheromone consisting of a 90-to-10 ratio of compounds A and B, males will exhibit a maximum response toward that blend as opposed to other ratios of the compounds A and B.

Chemical differences in pheromones do not provide reproductive isolation in some groups of moths. Interspecific sexual attraction can be experimentally induced in some species of sesiid moths, suggesting that chemical barriers, as described above, do not exist. However, these species exhibit sexual activity at different times of the day or year, indicating that a temporal barrier serves as the mechanism of reproductive isolation.

Pheromone-induced behavior. The mechanism by which a male moth orients toward a female that is emitting pheromone involves several steps. Upon perceiving the female's pheromone, a resting male is stimulated to fly in an upwind direction with respect to the air current carrying the pheromone (positive anemotaxis). This upwind flight requires additional visual cues from the ground and stationary objects, so that the insect can judge whether it is making progress toward the pheromone source and can adjust its speed in the air current accordingly. In the absence of visual cues a male moth could fly upwind with respect to the air current, but make negative progress with respect to the ground.

When pheromone is carried downwind from its source, it creates a "plume," an elongate volume of air within which the concentration of pheromone molecules is above a threshold value for male response. A male moth that is flying upwind is able to remain within the plume by following a zigzag course. If the insect flies beyond the plume, it reverses its direction and casts back and forth in a direction perpendicular to the air current—movement that tends to return the insect to the plume. As the moth approaches the pheromone source, it enters a region of higher pheromone concentration and decreases its flight speed while casting back and forth in the plume more frequently.

In moth species utilizing multicomponent pheromones, it has been found that the various com-

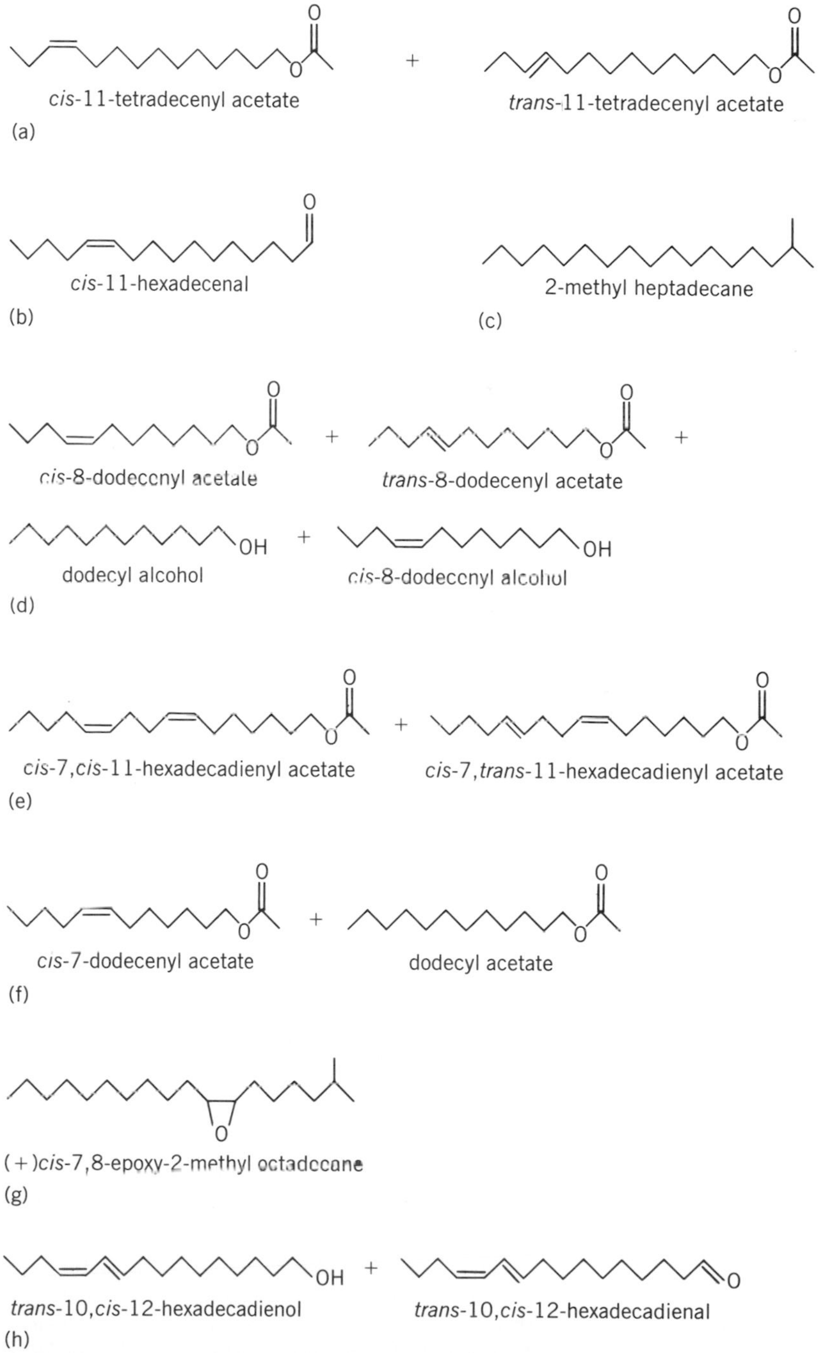

Female sex pheromones identified from moths. (*a*) European corn borer, *Ostrinia nubilalis*. (*b*) Artichoke plume moth, *Platyptilia carduidactyla*. (*c*) Tiger moth, *Holomelina nigricans*. (*d*) Oriental fruit moth, *Grapholitha molesta*. (*e*) Pink bollworm, *Pectinophora gossypiella*. (*f*) Cabbage looper, *Trichoplusia ni*. (*g*) Gypsy moth, *Lymantria dispar*. (*h*) Silkworm moth, *Bombyx mori*. (*After M. C. Birch and K. F. Haynes, Insect Pheromones, Edward Arnold Ltd., 1982*)

pounds may act together (synergism) to stimulate a certain stage of orientation in the male—for example, taking flight, positive nemotaxis, and modification of flight close to the pheromone source. Alternatively, in some species, individual pheromonal components appear to function in eliciting specific stages during orientation; for example compound A stimulates taking flight, whereas compound B stimulates positive anemotaxis.

Male pheromones. Pheromones emitted by male moths fall into two categories, courtship pheromones and long-distance pheromones. Among the moth species in which females emit pheromones in the typical manner, many species are found in which males also emit a scent, but only after coming within an inch (1 in. = 2.5 cm) or so of the female and initiating courtship. Unlike female pheromones, which are secreted from glands on the tip of the abdomen (the thoracic pheromonal glands of the female of the psychid moth *Thyridopteryx ephemeraeformis* are the only known exception), these courtship pheromones are secreted by glands situated in a variety of locations (for example, glands on the wing, scent scales on the thorax, and eversible sacs on the abdomen, called coremata), the specific location depending on the species. Courtship pheromones are believed to function in sexual selection, and their immediate effect may be to stimulate the female to remain stationary, to assume a copulatory position, or to move the final inch or so toward the male. Unlike female-emitted pheromones, courtship pheromones are generally volatile compounds. Work on the arctiid moth *Utetheisa ornatrix* shows that the biosynthesis of courtship pheromone may directly depend on specific substances acquired during larval feeding.

In a very few species of moths (for example, the galleriine waxmoths *Achroia grisella* and *Galleria mellonella*) the protocol of sexual-pair formation is reversed in that the male remains stationary and signals, and the female orients toward the male over a distance of up to several yards (1 yd = 0.9 m). These male signals are pheromones emitted from wing glands and released at extremely high rates (more than 5 micrograms per hour), and in the case of *A. grisella* an ultrasonic acoustic signal is also emitted by the male for female attraction.

Practical uses of pheromones. Various moths are serious agricultural and stored-products pests during the larval stage, and this economic importance is one factor promoting chemical identification of many moth pheromones. During the 1960s it was believed that pest populations could be controlled by synthesizing pheromones of female moths and baiting appropriately positioned traps with the synthetic compounds to ensnare multitudes of males. However, such "trap-out methods" generally have little effect on the populations of subsequent generations. Nonetheless, synthetic pheromones do show promise for controlling pest moth species by serving to detect infestations and by disrupting mating behavior. Traps baited with synthetic pheromone of a pest species can precisely indicate when and where outbreaks of a pest are beginning, thereby reducing the use of insecticide to a specific time and place. Alternatively, microcapsules containing synthetic pheromone can be evenly distributed in great numbers over an agricultural area. Pheromone released by the microcapsules disrupts the ability of males to orient toward females, and significant reductions in future populations can result. Interference with male orientation is believed to be caused by masking of the pheromone plumes of females present in the area, by providing false pheromone plumes which males follow, or by habituating males to pheromone. The disruption technique is currently being developed for the control of the pink bollworm, *Pectinophora gossypiella*, a cosmopolitan pest of cotton.

For background information *see* CHEMICAL ECOLOGY in the McGraw-Hill Encyclopedia of Science and Technology.

[MICHAEL D. GREENFIELD]

Bibliography: M. C. Birch and K. F. Haynes, *Insect Pheromones*, 1982; M. D. Greenfield, Reproductive isolation in clearwing moths (Lepidoptera: Sesiidae): A tropical-temperate comparison, *Ecology*, 64:362–375, 1983; D. A. Nordlund, R. L. Jones, and W. J. Lewis (eds.), *Semiochemicals: Their Role in Pest Control*, 1981; W. L. Roelofs, Pheromones and their chemistry, M. Locke and D. S. Smith (eds.), *Insect Biology in the Future "VBW 80"*, pp. 583–602, 1980.

Physical anthropology

In the last 100,000 years humans have occupied all continents, starting from Asia but coming perhaps originally from Africa. It is likely that most of the human racial differences arose during this period. In the last 10,000 years, with the beginning of agriculture, there were major migrations which contributed in an important way to altering the geographic distribution of genes in humans.

Spread of modern man. Human origins are obscure; the first *Homo sapiens sapiens* remains, found in South Africa, are tentatively dated at about 100,000 years ago. The earliest and most secure finds come from the Middle East and are 40,000 to 60,000 years old. Naturally, the absence of fossil finds in other areas cannot rule out the existence of early humans in other continents, or parts of them, prior to this time. But after 40,000 to 50,000 years modern man is found in all continents. In Europe, modern man replaced Neandertal man, previously the only human inhabitant of this region, between 35,000 and 40,000 years ago. Modern man possibly first arrived in America in this period from Northeastern Asia, and is first found in Australia 35,000 years ago. A clear occupation of North America occurs between 15,000 and 10,000 years ago through Behringia, a land passage between northeastern Asia and Alaska, which began to be replaced by sea around 10,000 years ago.

Until this time, modern humans were hunter-gatherers, but clearly capable of crossing seas, as is demonstrated by passage to Australia. It is only around 10,000 or 9000 years ago that an important economic development began in full strength: the introduction of plant and animal breeding. Until that time human populations could hardly reach densities above 0.1 or 0.2 inhabitant per square mile, and only in special areas. The total world population at the end of the Upper Paleolithic, around 10,000 yeas ago, was between 5 and 15 million human individuals. Even if humans originated in Africa and from there spread to Asia, it was starting from Asia that all other continents were occupied.

Formation of human races. It is probable that the low population density in the Paleolithic and the tectonic events that gave rise to the colonization of other continents generated most of the racial differences seen today. The comparisons of average genetic differences between major human ethnic stocks with those observed between the chimpanzee or gorilla and the human, or, better, those observed between more distant mammals for which separation times are known with better approximation, are consistent with the dating of 50,000–100,000 years for the divergence of human races. This is based on observations of immunological and biochemical markers and also, more recently, of mitochondrial DNA.

Compared with the history of the genus *Homo*, which goes back to 3 or 4 million years, the formation of races is a relatively recent phenomenon. This can explain why the difference between races for the average gene is small compared with that within human populations. Only about 10% of human diversity is attributable to racial differences, the rest of the variation being due to differences between individuals of the same race.

Recent population movements. The last 10,000 years have seen major changes in the human gene pool accompanying vast populaton expansions and movements, of which there exist, for the last 2000 or 3000 years, historical records. The beginnings of agriculture were highly localized: major centers were in the Middle East (development of cereals like wheat and barley as well as domestication of goats, sheep, pigs, and cattle); in Central America and perhaps the northern part of South America (corn, beans, squash); the Far East (rice, millet); and somewhat later in North Africa (millet, sorghum). The major effect of agriculture was the increase of population densities, and with it, major population explosions beginning at the centers of development of agricultural economies.

The expansion of farming technologies can be followed and dated in the archeological record. The expansions of early farming from the Middle East to Europe took a little over 3000 years, and the well-known Bantu expansion from Nigeria–Cameroon to Central and South Africa took about the same time. The expansion from the Middle East, sometimes called Neolithic, took place in the other directions as well, and is probably responsible for the spread of the Caucasian type to Europe, Arabia, India, and North Africa.

There is a remarkable agreement between gene maps, especially when the information from many genes is synthesized with appropriate statistical techniques, and maps of the diffusion of farming technologies and the artifacts that accompany them. The simplest explanation is that technical developments permitting faster population growth and spread may determine expansion of the people in directions made possible by geography, and made desirable by low levels of occupancy of the colonized areas.

The occupation of the Americas and Australia by Caucasians after the great geographic discoveries of the fifteenth century are more recent examples of demic expansion (that is, of people) made possible by technological developments, under the pressure of population increase. Another example is the spread of Indo-European-language speakers from an area in the Ukraine, beginning some 5000 years ago, which seems to have extended to almost all the area previously covered by early farming, started in the Middle East. East–West gene clines observed in Europe and northern Asia may represent the relics of many expansions from east to west from central and eastern Asia. The Mongol invasion is the last historically recorded example of these invasions, although many earlier ones in the same general direction are known. All these later expansions and movements have contributed to the highly complex geography of genes observed today, which is only somewhat simplified by restricting consideration to "aboriginal" populations, namely to those existing (and surviving) before the expansions of Caucasians in the West, of Chinese and Indians from Asia, and others.

Diversity among human races. A question of importance is the relative contribution of different evolutionary forces to the present diversity among human races. For some genes there is a positive evidence of selective effects. Perhaps the most striking example is the nearly complete fixation of a Duffy (blood group) allele in Africa; the same allele is rare in other continents. This allele is known to confer resistance to a malarial parasite, *Plasmodium vivax*. Resistance to *P. falciparum* is responsible for the prevalence of sickle cell anemia, the various thalassemias, and other genetic markers in Africa, southern Asia, or elsewhere. Almost all these genes cannot reach fixation and must remain at intermediate, usually low frequencies because individuals homozygous for them are at a serious selective disadvantage.

A correlation with the use of milk by adults has been found for lactose tolerance in Europe and Africa. Correlations with infectious diseases and with climatic conditions have been observed, but they are mostly small and there is no explanation of the mechanisms involved. Undoubtedly chance (random genetic drift) has contributed also in an important way to the variation of gene frequencies around the world, but it is difficult to evaluate its role with respect to that of selective differences.

Anthropometric differences and anthroposcopic differences (for example, height, weight, limb length, facial traits, hair shape and color, skin color) between human races are well known and form the basis of common racial stereotypes. There exists a genetic component to most of these differences, although clearly environmental effects are also present, and the mechanism of genetic inheritance is complex and poorly understood. The importance of sexual selection for these types of traits was stressed by C. Darwin; examples substantiated by data are, however, few. Anthropometric and anthroposcopic traits usually show a correlation with climate which is proportionately greater than that of genetic markers. For most of them the connection with heat and cold resistance and intensity of solar radiation is reasonably established.

For background information *see* ANTHROPOLOGY; ANTHROPOMETRY; ANTHROPOSCOPY; FOSSIL MAN; PHYSICAL ANTHROPOLOGY; POPULATION GENETICS; in the McGraw-Hill Encyclopedia of Science and Technology.

[L. L. CAVALLI-SFORZA]

Bibliography: W. F. Bodmer and L. L. Cavalli-Sforza, *Genetics, Evolution and Man*, 1976; R. C. Lewontin, The apportionment of human diversity, *Evol. Biol.*, 6:381–98, 1972; M. Nei and A. Roychoudhury, Genetic relationship and evolution of human races, *Evol. Biol.*, 14:1–60, 1982; A. Piazza, P. Menozzi, and L. L. Cavalli-Sforza, Synthetic gene frequency maps of man and selective effects of climate, *Proc. Nat. Acad. Sci. USA*, 78:2638–2642, 1981.

Pigmentation

Industrial melanism refers to the prevalence of black or dark brown forms which occur in many animal species in industrial regions. The pigments involved are often, but not necessarily, melanins. Industrial melanism was first observed in moths, and has been recorded in over 100 species, including spiders, ladybeetles, bark lice, bugs (Hemiptera), and even town pigeons. It was first observed in northwest England, one of the earliest regions to be industrialized, but many cases are to be found in parts of Europe and in the eastern United States. Interest among scientists centered on two interrelated issues: whether the frequency of melanics can be used as an indicator of environmental quality, and how the melanic frequencies are changed and the polymorphisms maintained.

Melanism in the peppered moth. The most famous industrial melanic is the peppered moth, *Biston betularia*, which has become something of a paradigm of evolutionary studies. The presence of a sooty black form of the species, called *carbonaria*, was noted by the amateur entomologists in and around Manchester, England, in the 1880s; they also recognized that this form had not been present before the middle of the nineteenth century. From then on, it was observed to spread outward from this apparent center of origin, north to Scotland and south and east to East Anglia and the London area. By 1900 it has been selected as an example for the study of the mechanism and effect of evolution, and a survey of the frequency about the country was carried out, which was published in 1906.

Various possibilities were considered to account for the existence of melanism, such as environmentally induced mutation or direct induction of black coloration by ingestion of polluted food. By 1914, however, it had been shown that the *carbonaria* form and the intermediate forms collectively called *insularia* were controlled by mendelian segregating alleles dominant to the typical pale speckled phenotype.

It was always assumed that environmental changes caused by industrialization brought about the increase in melanic frequency, although the way they do so has not always been agreed upon. The geographical pattern established by the beginning of the twentieth century remained substantially the same for another 50 years. With the introduction of smoke control legislation, however, the frequencies of melanics dropped in the vicinities of Liverpool and Birmingham, confirming the existence of a causal relation.

The population geneticist J. B. S. Haldane first used the example of *B. betularia* in 1924, when he showed that in order to produce a change in gene frequency of the magnitude observed over a period of 50 generations (the insect has an annual life cycle) it would be necessary for the selective advantage of the *carboniaria* form to be about 50%. This was an unexpectedly, and to some, unbelievably high selection pressure. The demonstration was instrumental in influencing the direction taken by the British school of ecological genetics, and the general position adopted was that selection pressures are usually high and that polymorphism, when it occurs, is the result of balancing selection.

In a series of experiments carried out in polluted industrial and unpolluted rural localities, it was shown in the 1950s that insectivorous birds attacked the peppered moths when at rest on tree trunks, and selective predation was such that the most cryptic forms in each locality had the highest survival rates. These studies did much to bring home the importance of predation as a force shaping the course of evolution, and from that time it has been generally accepted that the change in frequency which had been observed was the consequence of bird predation. Camouflage of the moths as they rest during the day on tree trunks depends on the presence and extent of epiphytic lichens. When these are abundant and diverse, the typical form is more cryptic than *carbonaria*; when lichens are absent the reverse is true, and concealment of the melanic is further enhanced if the trunk is also coated with soot. The epiphytes are intolerant of sulfur dioxide in the atmosphere, and the frequency of melanics is more closely correlated with the level of sulfur dioxide than with that of smoke pollution.

Despite the strong selection, there is no region in

the British Isles where the melanic forms have completely replaced the typical forms. In 1956 Haldane pointed out that if selection in favor of melanics as a result of bird predation was of the order of 50%, as he had earlier calculated, then the observed level of polymorphism in industrial regions would be maintained if the melanic homozygotes had a disadvantage of about 10%. It was already suggested that there may in fact be a heterozygote advantage as a result of nonvisual fitness effects. Melanics could then be suppressed in unpolluted areas and favored in polluted industrial areas by visual predation. Recent analysis of all the available data has failed to show heterozygote advantage, however, but indicates that the melanics have a nonvisual advantage over typical forms of a few percent. Like the visual advantage, this effect is dominant. A detailed field study demonstrated that the insects are highly mobile, moving on average about 1 mi (2 km) per generation. A theoretical study using these two pieces of information indicates that the pattern of melanic frequencies found in the United Kingdom could in fact be the result of a balance of visual selection and of migration between regions in which the selection differs in direction. There would therefore be no need to invoke nonvisual heterozygote advantage, and a more elementary model would explain the pattern of frequency which occurs.

It is difficult to obtain the evidence which will distinguish clearly between these population genetic theories. The visual predation theory has been criticized on the grounds that the natural resting positions of the moths when not interfered with by human investigators are not known. Whether or not the selection-migration theory stands the test of time, there are certainly many cases, both among the Lepidoptera and the other animals which display industrial melanism, where balancing nonvisual selection most probably operates. A number of moth species coexist with the peppered moth and have melanic forms. Nevertheless, the detailed frequencies often differ from that of the peppered moth and from each other, so that the patterns of selection must be assumed also to differ.

Melanism in other groups. Melanic forms occur in the ladybird beetle *Adalia bipunctata*, in which the typical pattern of black spots on a red background is reversed. This insect is widespread in Britain and Europe; the black forms are present in high numbers in industrial areas, so that technically the insect ranks as an industrial melanic. The example differs in several respects, however, from the moths. The beetles, being distasteful to birds, are less likely to be subject to predation, although some specialist predators may attack them. Black individuals have been shown to heat up more rapidly than red ones when exposed to light, and to become more active, so that the connection between melanism and industrialization may occur because industrial areas tend to be more cloudy than nonindustrial ones. In addition, current research centers on the possibility of sexual selection in which females have a preference for males of a particular type.

Yet another type of industrial melanism is seen in the town pigeon, *Columba livia*. In some urban areas there are high frequencies of dark-plumaged birds. Dark color is controlled by either of two gene loci. Males of the dark type have a reproductive advantage over pale individuals as a result of a longer breeding period and better-developed testes. Matings between unlike types appear to be favored, and the high frequencies in urban areas may be caused by the less seasonal nature of the food supply which allows the breeding season to be extended.

Conclusion. Industrial melanism is not a single phenomenon with a simple cause. In many moth species, bird predation is probably the most important factor determining frequencies in a particular place, but other, nonvisual selection is also certainly implicated. The examples of the beetle and the bird show that the different thermal properties of the pale and dark individuals may also be involved and that there may be effects on breeding behavior. In nearly all examples the result has been to establish genetic polymorphism, not the complete replacement of the preexisting types in a newly industrialized locality by melanic forms.

For background information *see* PIGMENTATION; POLYMORPHISM (GENETICS); PROTECTIVE COLORATION in the McGraw-Hill Encyclopedia of Science and Technology.

[LAURENCE M. COOK]

Bibliography: J. A. Bishop and L. M. Cook, Industrial melanism and the urban environment, *Advances in Ecological Research*, 11:373–404, 1980; B. Kettlewell, *The Evolution of Melanism*, 1973; D. R. Lees, Industrial melanism: Genetic adaptation of animals to air pollution, in J. A. Bishop and L. M. Cook (eds.) *Genetic Consequences of Man Made Change*, pp. 129–176, 1981; M. Majerus, P. O'Donald, and J. Weir, Evidence for preferential mating in *Adalia bipunctate*, *Heredity*, 49:37–49, 1982.

Piloting

Piloting is the frequent determination of a ship's position relative to geographic points. It is an art practiced by the navigator of a ship within sight of land and fixed navigational aids. Piloting enables the navigator to accurately determine the location of a ship and to avoid known hazards such as shallow water and shipwrecks. The ship's position is defined in terms of latitude and longitude. To visualize the ship's position with respect to geographic features, the navigator plots this position on a chart (nautical equivalent of a map) that has geographic features printed at their exact location. Printed charts are readily available to the public.

Navigators are seeking to use the rapid and accurate response of the computer. The principles of piloting involve problems in plane geometry and trigonometry, which a computer can solve quickly and precisely. Thus a new tool in navigation is the digital chart, which might be described as a paper chart encoded in a computer-usable format (it is also referred to as an electronic chart, and is possibly the only true electronic chart available at this time). In combination with modern electronic devices, the digital chart is used to produce a computer-generated video display which provides the navigator with an accurate pictorial presentation of the information normally gathered from a paper chart.

The paper chart is a useful tool, but contains more information than is usually necessary because the data are intended to satisfy the needs of every user, from the small-boat sailor to the supertanker captain. A digital chart is usually fabricated to show only the information important to the navigator who will use the chart. The illustration shows a representative digital chart.

Digital charts are not readily available for general use. The reasons for this are: they are expensive to produce because of the precision required; the amount of data required and the format of the data vary from application to application; and an established data base of navigational information is not yet publicly available. The lack of digital charts will impede the wide acceptance of computer-assisted piloting.

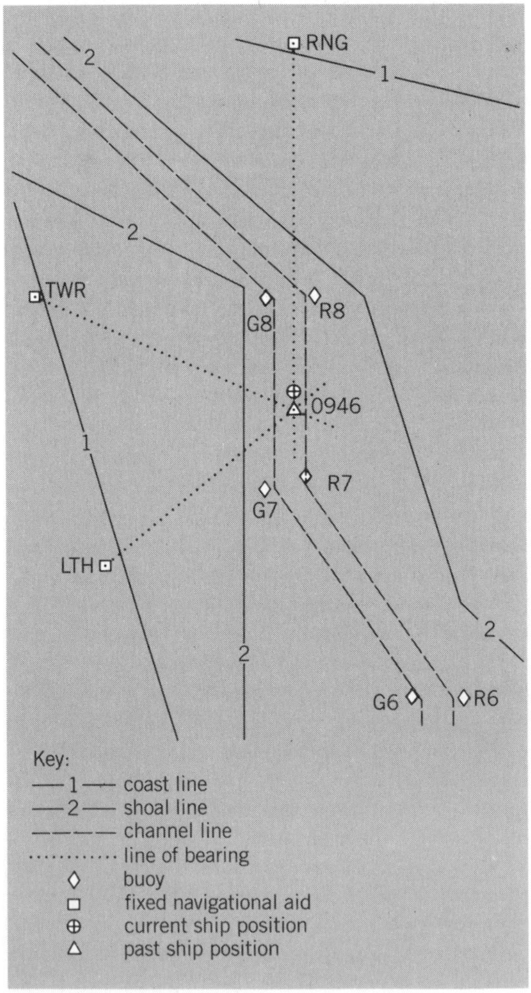

Representative digital chart for use in piloting. Colored lines can replace the numbered lines for video display. Alphanumeric labels provide identification for the navigational aids.

Presentation methods. The location of a ship is constantly changing, and traditional piloting techniques enable the navigator to determine a new position only every 3 to 5 minutes. On digital charts, however, integration of course and speed sensors allows the change in a ship's position to be determined. Therefore the ship always appears in its exact position, without time delay, on a digital chart; the error of the estimated position calculated from the sensors in the short time between positions should not be noticeable to the operator.

In a fixed-chart presentation, the details of the digital chart are fixed on the screen and the ship symbol alone moves around on the chart display. There are substantial drawbacks to this presentation. First, the scale, or the area represented by the chart, must remain fixed. Second, the chart becomes less helpful when the ship approaches the edge of the display, and there may be dangers just beyond the displayed features. Third, there are problems associated with the transition from one screen display to another, although overlapping charts may alleviate the problem.

In the fixed-location presentation, the ship's location is kept fixed on the screen while the chart moves around. This display is much more difficult to generate because the entire chart must be moved every second. However, since the ship's location is fixed on the screen, the scale can be changed at any time. This is important in relation to the fact that—depending on the location chosen for the ship symbol (usually the center)—all dangers for a specific distance (half a screen) are known. Also, the chart is always available, so there is no transition problem. If digital charts are used with radar displays, only the fixed location display can be used.

Piloting techniques. The easiest piloting technique to use with digital charts requires an external device to determine the ship's position (latitude and longitude). This can be done by using devices such as Loran C receivers, Decca receivers, navigation satellite receivers, or Omega receivers to supply data to the computer, which then automatically computes the latitude and longitude and displays the ship's position on the chart. Although many consider this to be electronic navigation rather than piloting, it does provide a determination of the ship's position relative to geographic points on the digital chart. Because of the varying accuracies of the electronic devices, this technique is used to supplement the other techniques, such as radar and visual aids.

Combining radar information with a digital chart is a second piloting technique. The scale of the radar return determines the scale to be used for the digital chart. The radar has to be presented in a television format. When the television signals for the radar and the digital chart are properly combined, the radar returns can be used to pilot by a technique called map matching. In this method, the navigator moves the land outlines from the digital chart to coincide with the radar reflections from the corresponding land mass. Since the ship is always the center of the radar return presentation, the ship's position will always be the center of the video display.

If the computer system includes a device such as a video processor which can convert the radar information into computer-usable format, radar navigational aids can be used automatically to determine the ship's position. Only one range and bearing is needed by the computer to determine the ship's position, but the accuracy or precision of the equipment or the system may cause the operator to be less confident in the position so determined. Hence, the computer must evaluate a number of radar ranges and bearings in determining the most probable ship location. Some radar manufacturers provide digital charts of limited geographic areas for use with their equipment in this manner.

The third piloting technique involves the use of visual navigational aids. Extensive operator interaction is required in this procedure, but this is usually the most accurate of the piloting techniques. The operator selects a set of three or more visual navigational aids to be used, and these must also be indicated on the digital chart. By means of an optical device the operator determines the bearing of each of the selected aids. The information is input to the computer, and the computer then displays the ship's location on the chart based upon the triangulation of the three aids. While an operator can enter bearings to visual navigation features, an automatic method may be preferred. The operator will still use an optical device to point at the navigational aid exactly and will then indicate to the computer the current bearing is associated with the indicated aid. If three or more such lines are obtained, a very accurate fix can be obtained. The type of chart shown in the illustration has been used for visual navigation. The triangle at the intersection of the three lines of bearing represents the position of the ship. The time of the position is noted beside the triangle. The ship is not at the location of the position determined by the visual navigation lines of bearing because the ship has continued moving.

The techniques discussed above are representative of the applications of digital charts in navigation. The cost of equipment, computer programs, and digital charts currently limits users to large ships. However, the time may come when digital charts will be practical for all navigation applications.

For background information *see* ELECTRONIC NAVIGATION SYSTEMS; PILOTING; RADAR in the McGraw-Hill Encyclopedia of Science and Technology.

[E. J. MEIERS, JR.]

Positron emission tomography

The measurement of brain activity has long been a challenge to scientists. The exciting development of positron emission tomography (PET scan) now allows scientists to measure noninvasively the functional activity of the living human brain. Previously, inferences about functions performed by different parts of the human brain had to be made from studies of animal brains or of cases of injury or surgery, from the detection of levels of electrical activity on the scalp through electroencephalography (EEG) usually limited to the cortical surface, or from the examination of cerebral blood flow.

For the first time, using PET scan, it is possible to obtain three-dimensional mappings of the metabolic activity of brain regions during specific psychological states or tasks. The methodology is based on a triad of technical advances. First, the development of small but highly sensitive crystal detectors and coincidence counting electronics made positron localization possible. Second, the mathematical reconstruction techniques similar to those used for computer-aided tomography allow quantitative slice images to be generated. Third, there is a mathematical model that allows quantification of regional use of brain glucose from a glucose analog autoradiography technique.

Methodology. Glucose metabolism is an indicator of brain activity since under normal circumstances cerebral tissues derive most of their energy from glucose. The amount of glucose metabolized by a specific structure at a certain point in time is related to the amount of cell firing and repolarization activity in that neuronal structure. Preliminary work in animals suggests that glucose metabolism is sensitive to changes in functional activity, including sleep, electrical stimulation, olfactory stimuli, circadian rhythms, visual and auditory activity, stress, and the administration of various pharmacological agents.

PET scan methodology involves the intravenous administration of deoxyglucose labeled with a neutron-deficient isotope containing a proton. Because of the short half-life of most common elements (oxygen-5, 2 min; nitrogen-13, 10 min; carbon-11, 20 min), fluorine-18 has been chosen as the most valuable label as it has a half-life of 110 min and therefore allows longer periods of analysis. The proton decays into a neutron, and a positron is emitted. Initially, the positron is at a very high energy level, but interacts with an electron in a very short distance and annihilates spontaneously. Energy is then released in the form of two high-energy photons (or gamma rays) that travel in opposite directions (180° apart). Multiple radiation detectors, arranged in a circle, are activated only when both photons hit detectors 180° apart simultaneously. This technique accurately determines the line segment in which the isotope resided. The data are then analyzed by a computer, which constructs three-dimensional color-coded slice images of brain structures, providing a quantitative display of glucose metabolism.

In addition to deoxyglucose, scientists are considering labeling compounds that can be used to study neuroreceptors (for example, protein synthesizer and labeled psychoactive agents such as neuroleptics and benzodiazepines). Some of these agents include ^{18}F-labeled haloperidol, ^{18}F-labeled spiroperidol, ^{11}C-labeled pimozide, and ^{11}C-labeled flunitrazepam.

Applications. The PET scan technique is already proving useful in the study of neurological and mental illness. A large number of patients and normal volunteers have been studied with PET scan techniques in several operating centers. Thus far, evidence suggests that it is possible to localize epilepsy, assess the invasiveness of tumors, and examine postinfarct damage after cerebrovascular occlusion. These findings, however, are to some extent approachable by earlier technologies. It is in the behavioral neurosciences where cerebral localization is the most uncertain that PET scan techniques have their greatest promise. In a series of studies in humans, glucose metabolism was shown to be altered in response to auditory, verbal, visual, and shock stimuli, and during periods of aphasia.

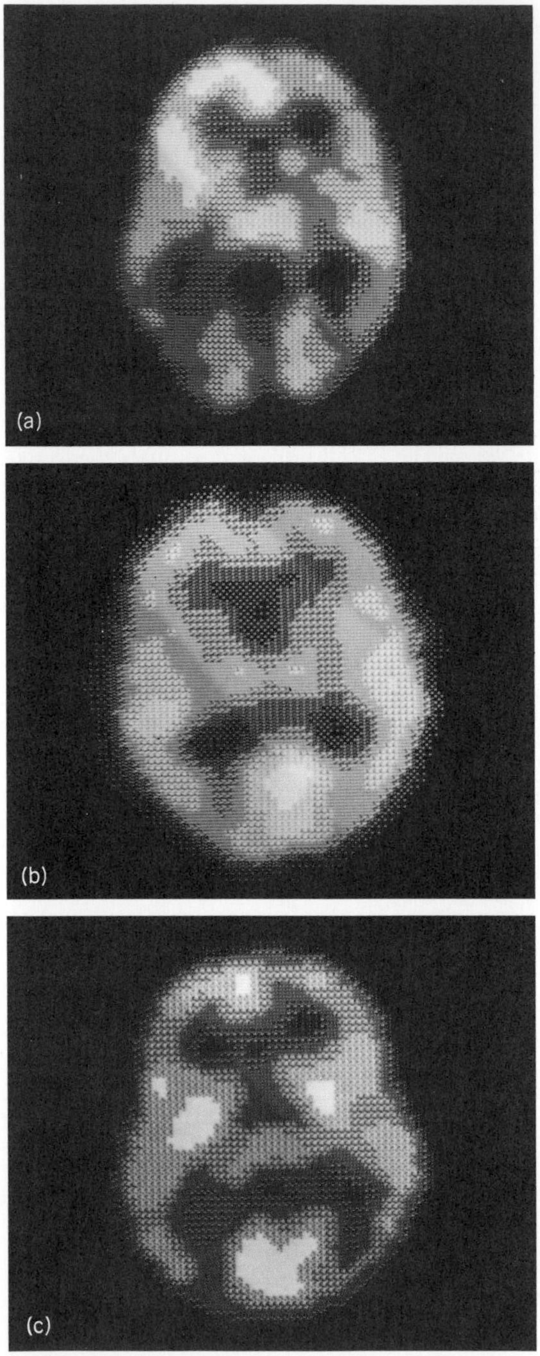

Positron emission tomographic brain scans of (*a*) normal subject, (*b*) affective disorder patient, and (*c*) schizophrenic patient while receiving shocks to the right forearm with the eyes closed. Both patients were off medication. White areas indicate glucose use (uptake of ^{18}F-deoxyglucose). The scans of both patients show the highest glucose uptake in the posterior region, where the brain's visual center resides. In contrast, the scan of the normal subject shows higher use of glucose in the frontal lobes, which are concerned with the planning and organizing of behavior.

Recent findings using PET scan techniques comparing schizophrenics to normals have been very promising. The first longitudinal study involving a 45-year-old patient with a history of schizophrenia since the age of 16 took place in 1981. Although the patient had never had neuroleptic medication before the scan, there was still a 40% decrease in frontal cortex glucose use as compared with controls. When the patient was treated with phenothiazines, there was an apparent return to normal glucose utilization levels.

In a study in 1982, local cerebral glucose use was measured with an isotope-labeled glucose analog, ^{18}F-labeled 2-fluorodeoxyglucose (FDG), in eight unmedicated schizophrenic patients and six age-matched normal volunteers. Subjects sat resting in a quiet darkened room with eyes closed after injection of the FDG. Following uptake, seven to eight horizontal brain scans parallel to a line connecting the outer canthus of the eye with the external auditory meatus were obtained. Patients with schizophrenia showed relatively lower glucose use in the frontal cortex, as compared to the occipital cortex, than did the normal controls. These results are consistent with previous blood flow studies which demonstrated decreased blood flow in schizophrenic patients during similar experimental (rest, eyes closed) conditions.

Another recent study included 16 patients with schizophrenia and 11 patients with affective disorder. Pain stimulation during the FDG uptake period was used to challenge frontal lobe activity and enhance glucose utilization differences between groups. The previous finding of a "hypofrontal" pattern in patients was confirmed (see illustration). Diminished pain responses in schizophrenics and affectively ill patients were demonstrated in both PET scan and blood flow studies, again illustrating a consistent relationship between the two techniques.

Conclusions. Although still in its early developmental stages, the PET scan technique is presently noteworthy, for it offers promise for the future investigation of specific brain structures. Furthermore, this technique holds potential for the understanding and treatment of mental illness. In fact, there is some hope that PET scan results could identify individuals who are at high risk for developing mental illness prior to the onset of actual symptoms. In combination with new clinical anatomical imaging techniques, such as electroencephalographic topography, nuclear magnetic resonance, and insights from animal data obtained with autoradiography, a level of scientific approach previously unattainable may be realized.

For background information *see* BRAIN; COMPUTERIZED TOMOGRAPHY; ELECTROENCEPHALOGRAPHY; RADIOISOTOPE (BIOLOGY); SCHIZOPHRENIA in the McGraw-Hill Encyclopedia of Science and Technology.

[MONTE S. BUCHSBAUM]

Bibliography: M. S. Buchsbaum et al., Cerebral glucography with positron tomography, *Arch. Gen. Psychiat.*, 39:251–259, 1982; M. S. Buchsbaum et al., PET image measurement in schizophrenia and affective disorder, *Ann. Neurology*, April 1984; T. Farkas et al., Regional cerebral glucose utilization in schizophrenia, *3d World Congress of Biological Psychiatry: Symposium on Cerebral Circulation and Metabolism Related to Psychopathology*, Stockholm, Sweden, 1981; L. Sokoloff et al., The [^{14}C] deoxyglucose method for the measurement of local cerebral glucose utilization: Theory, procedure and normal values in the normal and anesthetized rat, *J. Neurochem.*, 28:897–916, 1977.

Prions

Prions are small infectious pathogens that contain protein and are resistant to inactivation by most procedures which modify or hydrolyze nucleic acids. The unusual molecular properties of prions and their small size seem to distinguish them from both viruses and viroids. The most extensively studied disease caused by prions is scrapie of sheep and goats.

Prion diseases. Six diseases are probably caused by prions: scrapie of sheep and goats, transmissible mink encephalopathy (TME), chronic wasting disease (CWD) of mule deer and elk, kuru of the Fore people of New Guinea, Creutzfeldt-Jakob disease (CJD), and Gerstmann-Sträussler syndrome (GSS). The two latter diseases are found in humans throughout the world. Only in the case of scrapie has the slow infectious agent been characterized sufficiently to be classified as a prion. In the other five diseases, this classification must remain tentative until further knowledge about the molecular properties of these agents can be obtained.

All prion diseases share many features: They are confined to the central nervous system and have prolonged incubation periods ranging from 2 months to more than two decades. The clinical course of these diseases is usually rather stereotyped and progresses to death. Neurological symptomology is prominent in all these disorders. The clinical phase of prion illness is considerably shorter than the incubation period and may last for periods ranging from a few weeks to a few years. Pathologic examination of the nervous system shows a proliferation of astrocytic cells. In addition, vacuolation of the brain is found, but this is not a constant or obligatory feature of the disease.

Assays for prions. At present the only methods for measuring prion infectivity remain an incubation-time-interval assay and the end-point titration. Both methods are extremely slow because they require waiting for the onset of clinical neurological dysfunction following a prolonged incubation period in experimental animals.

The length of the incubation period varies greatly with the animal species, route of inoculation, and dose of prions. Clearly the hamster, inoculated intracerebrally with a high dose (10^7 ID_{50} units), has the shortest incubation period of all animal hosts examined. Sixty to sixty-five days after intracerebral inoculation, hamsters show tremor, ataxia, difficulty righting from a supine position, and head bobbing. Within 2–3 weeks the hamsters are dead. Because the disease develops most rapidly in the hamster, it is the preferred animal for prion research.

Molecular properties. The incubation-time-interval bioassay permitted the development of methods for purifying the prion that causes scrapie. Using partially purified fractions, procedures which hydrolyze or modify proteins were found to produce a diminution of scrapie prion infectivity; in contrast, procedures which hydrolyze or modify nucleic acids did not alter the infectivity of the prion.

Ionizing radiation, molecular sieve chromatography, and rate zonal sedimentation all indicate that the smallest or monomeric form of the prion may have a molecular weight (M_r) as low as 50,000. In agreement with this estimate, recent experimental results show that purified prions suspended in a detergent, sulfobetaine 3-14, passed through membrane filters with M_r cutoff of 100,000 but were retained by filters with a cutoff of 30,000 or less. The low molecular weight of the scrapie prion indicates that it is too small to be able to contain even a single gene.

Protein component. Radioiodination of partially purified fractions led to the discovery of a protein, designated PrP, which is a structural component of the prion. Recent studies have shown that the concentration of PrP is directly proportional to the titer of the prion. Extensive purification of prions indicates that the prion contains only one major protein, PrP. The molecular weight of PrP is 27,000–30,000. PrP is resistant to digestion by proteolytic enzymes in its native conformation but becomes sensitive to proteolysis after denaturation.

Prion rods. Electron microscopy of extensively purified fractions shows numerous clusters of rod-shaped particles. By negative staining with uranyl formate, these rods measure 10–20 nanometers in diameter and 100–200 nanometers in length (see illustration). Analogous fractions from uninoculated animals or from animals inoculated with normal-brain extracts failed to show these rods. Each rod may contain as many as 1000 PrP molecules. The monomeric, or smallest, form of the prion appears to contain three or fewer PrP molecules. To date, it has not been possible to prepare preparations which contain only single rods. The rods aggregate to form arrays or clumps of varying sizes and shapes. These observations provide ultrastructural evidence for the multiple molecular forms of prions previously reported in sucrose gradient sedimentation studies, as well as gel electrophoresis experiments. The aggregation of prions into polymeric structures of varying sizes and shapes has made purification difficult.

The ultrastructural morphology of these rods, in

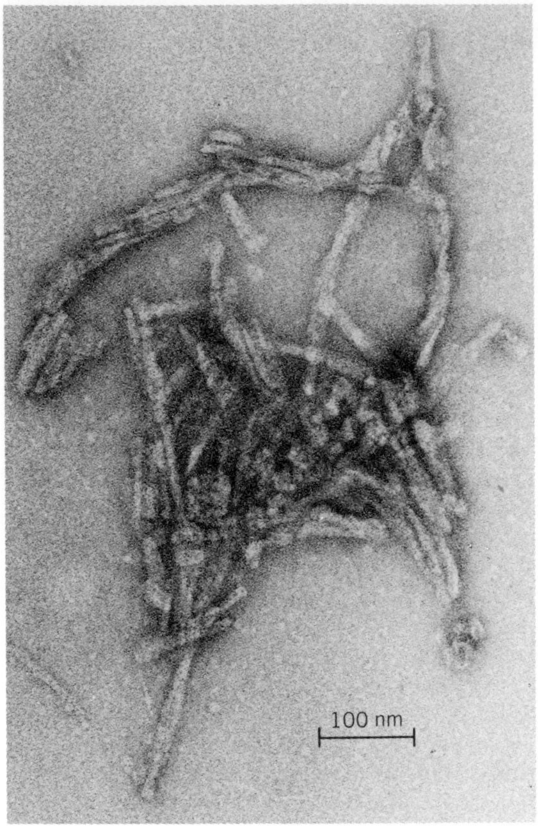

Electron micrograph of prion rods. Each rod contains as many as 1000 PrP molecules, and the rods aggregate to form clusters.

particular their assembly into arrays of different shapes and sizes, resembles that of purified amyloid. The resemblance of prion rods to amyloid is of interest because it raises the possibility that amyloid plaques found in prion diseases might, in fact, be made up of polymeric forms of prions.

Studies on the molecular structure of prions are still in an early stage of development. The mechanisms by which prions replicate are unknown, and are of great interest since the infectious particle itself is too small to contain a gene which codes for its protein component, PrP. Although there is no evidence for a nucleic acid within the prion, an oligonucleotide which is highly protected and which may act as a regulatory molecule in the replication process may be present.

Amyloid plaques. Since recent studies indicate that scrapie prions aggregate to form amyloidlike rods and that the amyloid plaques seen in prion disease may represent aggregates of prions, it is of interest to ask whether or not other diseases in which amyloid plaques are seen could be caused by prions. In Alzheimer's disease, Pick's disease, and congophilic angiopathy, the deposition of amyloid within the central nervous system is the hallmark of these disorders. In all three diseases the cause is unknown. Attempts to transmit Alzheimer's disease and Pick's disease to apes, which are a good experimental host for the human prion diseases, have not been successful to date. Alzheimer's disease is the major cause of senile dementia, which is a rapidly growing problem in many countries where the lifespan has been greatly extended in recent years. *See* ALZHEIMER'S DISEASE.

For background information *see* SCRAPIE; VIROIDS; VIRUS in the McGraw-Hill Encyclopedia of Science and Technology.

[STANLEY B. PRUSINER]

Bibliography: D. C. Bolton, M. P. McKinley, and S. B. Prusiner, Identification of a protein that purifies with the scrapie agent, *Science*, 218:1309–1311, 1982; D. C. Gadjusek, Unconventional viruses and the origin and disappearance of kuru, *Science*, 197:943–960, 1977; M. P. McKinley, D. C. Bolton, and S. B. Prusiner, A protease-resistant protein is a structural component of the scrapie prion, *Cell*, 35:57–62, 1983; S. B. Prusiner, Novel proteinaceous infectious particles cause scrapie, *Science*, 216:136–144, 1982; S. B. Prusiner et al., Scrapie prions aggregate to form amyloid-like birefringent rods, *Cell*, 35:349–358, 1983. S. B. Prusiner and W. J. Hadlow (eds.), *Slow Transmissible Diseases of the Nervous System*, vols. 1 and 2, 1979.

Propulsion

The use of microwave energy and microwave discharges for electric spacecraft propulsion systems is jointly under investigation by NASA and Michigan State University. Two thruster concepts have been identified and are under development: the microwave ion engine and the microwave plasma electrothermal engine. Although both proposed concepts make use of a microwave discharge, that is, a plasma, they belong to two more general, nonplasma classifications of electric engines. The microwave ion engine is an electrostatic ion motor having many similarities to the Kaufman engine, while the microwave plasma electrothermal engine belongs to the more general electrothermal engine category to which the better-known resistojet and dc arc jets belong. As presently envisioned, these engines would derive their electric energy from the sun, and the electric power supply would be limited to 10 kW or less. Thus engine size would be at most several horsepower (1 hp = 0.75 kW), and uses would be limited to low-thrust applications like satellite course correction and attitude control.

Both microwave thruster concepts have demonstrated scientific and technical feasibility. However, a further, more detailed test and evaluation is necessary to determine their usefulness as electric engines for spacecraft propulsion. This article describes the microwave thruster concepts and reviews their present state of development.

Microwave thruster system elements. The principal elements of electric space motors (Fig. 1) are solar cells (or a separate electric energy source), a power conditioner, a gaseous propellant, and an electric energy conversion chamber. As presently

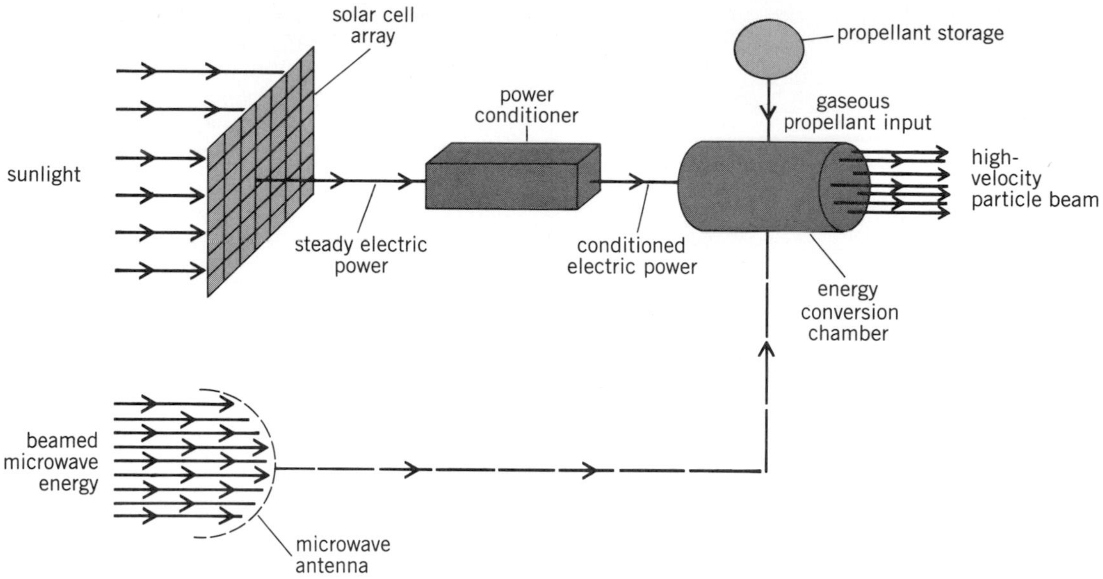

Fig. 1. The principal elements of an electric propulsion system.

envisioned, electric thrusters will use solar cell arrays to convert sunlight into direct-current (dc) electric power. The "raw" power from the solar cell panels is then converted into the proper form by the power conditioner and fed into the energy conversion chamber. The gaseous propellant is also fed into the energy conversion chamber, where the "conditioned" electric energy "energizes" and accelerates the individual molecules and ions of the propellant to high velocities. After acceleration the propellant exits the chamber as a beam of high-velocity particles, producing the required thrust.

The use of microwave energy and microwave discharges defines several of the elements of a conventional electric propulsion system. First, the power conditioner must change the solar cell energy into microwave electricity. This step can be easily performed by a magnetron which can directly convert the solar array dc power with little additional power conditioning. Also, the energy conversion chamber, referred to here as the microwave discharge chamber, must be designed to efficiently couple microwave energy to the propellant.

Finally, the use of microwave energy in the energy conversion process opens up the possibility of beaming microwave energy from a ground- or satellite-based transmitter to the satellite and the subsequent direct conversion in the microwave discharge chamber. Such a concept (shown as broken lines in Fig. 1) would not require the solar cell and power conditioning elements. Thus the mass penalty of these system elements will be replaced by the (hopefully) more modest mass of the collector (antenna) for the microwaves. Such a system may have to be modified to operate at the high microwave or millimeter frequency range to reduce collector size, but has the advantage of having the electric power supply in a central location outside the satellite, either in orbit or on Earth, beaming microwave energy when required to many different satellites.

The advantages or disadvantages of these microwave thruster concepts must be compared with similar electric engine concepts. For example, the microwave plasma ion engine should be compared with other electrostatic ion engine concepts such as the dc Kaufman ion engine and rf ion motors, and the microwave electrothermal engine must be compared with the molecular hydrogen (H_2) resistojet and the dc arcjet. In each comparison the microwave thruster has the important advantage that it produces an electrodeless discharge that can more easily provide higher electron densities and higher degrees of ionization, excitation, and dissociation than other lower-frequency and dc discharges. The lack of electrodes should simplify the design and improve lifetime and conversion efficiency of the energy conversion chamber. However, the requirement of input microwave electricity may result in a more costly and less efficient power conditioner. Since these systems are still under development, the trade-offs between the power conditioner and energy conversion chamber efficiency, system lifetime, and costs are not clear. Thus an evaluation between these two microwave propulsion concepts and other well-developed electric thrusters is not yet possible.

Microwave discharge chambers. The major innovations of these microwave thrusters have occurred in the method of plasma generation and the design of the microwave discharge chamber. The cylindrical or coaxial energy conversion chambers for the two thrusters are depicted schematically in Fig. 2. The two concepts require entirely different propellant gases and pressure-flow conditions. The ion engine requires the low pressure of 10^{-5} torr (10^{-3} pascal) and is fueled by mercury or the heavy noble gases, while the electrothermal engine favors the

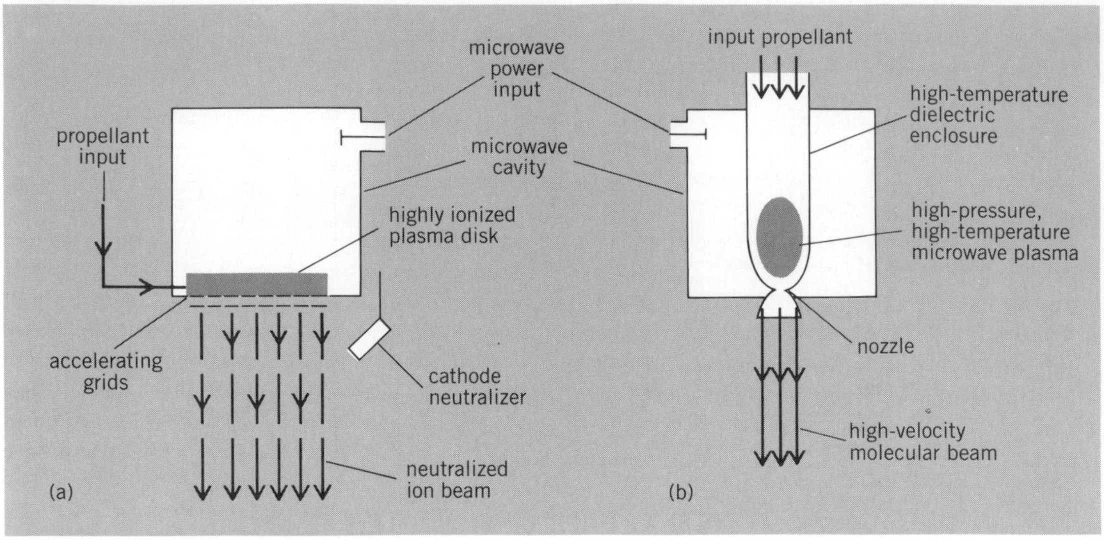

Fig. 2. Microwave discharge chamber cross sections for the (*a*) microwave ion thruster and (*b*) microwave electrothermal thruster.

high pressures of 0.5 atm (50 kilopascals) or greater with light-molecular-weight gases. Each of these chambers is specifically designed by using principles and understanding of microwave-plasma coupling technology developed earlier with 100–3000-W, continuous-wave discharges.

Microwave ion thruster. A detailed example of such a discharge chamber for the microwave ion source is shown in Fig. 3. The major components of the ion source are the cylindrical microwave cavity, the propellant input system, the quartz enclosure, the ion extraction grids, and the neutralizer cathode. The ionized gas, called a plasma, is formed in the shape of a thin disk filling the region formed by the first accelerating grid and the quartz enclosure. The disklike shape of the plasma allows the minimization of the plasma volume while maximizing the ion extraction surface.

Microwave energy at 2.45 gigahertz is coupled into the microwave cavity through a coaxial coupling port. The cylindrical microwave cavity consists of a brass cylinder and the first accelerating grid located at one end of the cavity. With the proper choice of cavity length and diameter and coupling probe depth, appropriate electromagnetic cavity modes can be excited, producing electric fields inside the cavity. The gaseous propellant flows in through the input feed tube and through an annular ring into the disk-shaped discharge zone located between the quartz enclosure and the first accelerating grid. When microwave power is coupled into the cavity, the electric fields ionize the gas located in the discharge zone, creating a disk-shaped plasma. Each gas molecule is ionized to form an ion and electron. The positively charged ions are then extracted and accelerated out the end of the cavity by an application of a high, steady electric field between the two accelerating grids. These ions travel as a beam of focused particles traveling at over 50,000 mi/h (22 km/s). The ion beam is also electrically neutralized by taking excess electrons from the plasma and emitting them from the neutralizer cathode. These electrons are then carried along with the ion beam and thus prevent ion attraction to the spacecraft surface, a condition that would prevent continued ion extraction from the grids.

A small, 2.5–3-in. (6–8-cm), 50–150-W, continuous-wave prototype engine has been designed and tested at 2.45 GHz with argon gas as the propellant and has demonstrated the feasibility of the concept. Preliminary measurements indicate that operation is possible without a dc magnetic field down to and below 10^{-4} torr (10^{-2} Pa) with gas flow rates of 10 standard cubic centimeters per minute (0.6 cubic inches per minute) of argon. Over 250 milliamperes of beam current with ion production efficiencies of 500 W/A have been achieved with a 2.5-in (6-cm) argon gas thruster. The use of confining magnetic fields should further improve overall system efficiency.

This concept can easily be scaled up to a 20-in. (50-cm) beam diameter by lowering the excitation freqency to 915 MHz. Besides its potential use in spacecraft propulsion, this concept should be useful in the earth application of ion beam materials processing since chemically active ions such as singly ionized atomic oxygen, fluorine, chlorine, and hydrogen can readily be produced without anode and cathode erosion.

Microwave electrothermal thruster. The electrothermal concept (Fig. 2*b*) converts microwave energy into heat in hydrogen and helium gases. The conversion process takes place by creating a microwave arc discharge in the flowing gaseous propellant at pressures greater than several hundred torr (1 torr = 133 Pa). The heated gas then expands thermodynamically out a nozzle, producing thrust.

Preliminary tests have shown that microwave

power coupling efficiencies to these high-pressure discharges are in excess of 96%, producing well-matched, efficient arc discharges with power densities in excess of several hundred watts per cubic centimeter. Using nitrogen (N_2) gas as the propellant and a small prototype engine, initial thruster tests with 200–500 W of continuous-wave input power have yielded energy efficiencies (measured thrust power output divided by input power) of 30–60%. These results indicate excellent promise for this concept, but further testing with hydrogen (H_2) and helium (He) gases is required.

Both microwave thruster concepts have demonstrated scientific and technical feasibility. However, a further, more detailed test and evaluation is necessary to determine the usefulness of microwave thrusters as electric engines for spacecraft propulsion.

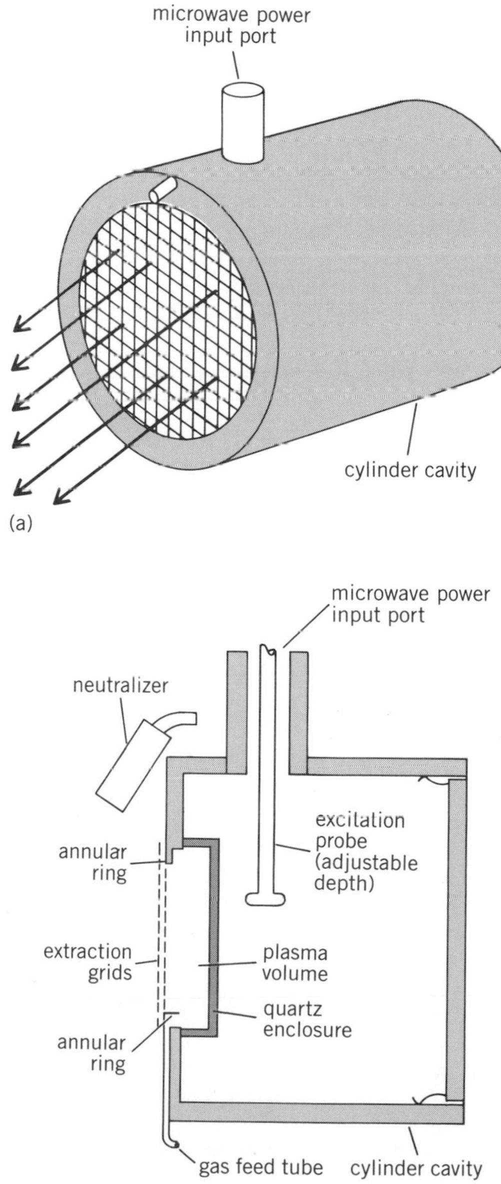

Fig. 3. A microwave ion engine in (*a*) isometric and (*b*) cross-sectional views.

For background information *see* ELECTROMAGNETIC PROPULSION; ION PROPULSION; ION SOURCES in the McGraw-Hill Encyclopedia of Science and Technology.

[JES ASMUSSEN, JR.]

Bibliography: J. Asmussen et al., The design of a microwave plasma cavity, *Proc. IEEE*, 62(1):109–117, 1974; J. R. Beattie and S. Kami, Advanced-technology 30-cm diameter mercury ion thruster, *AIAA/JSASS/DGLR 16th International Electric Propulsion Conference*, AIAA-82-1910, November 17, 1982; H. Goede et al., RF1 and ECH plasma generator development for ion thrusters, *AIAA/JSASS/DGLR 16th International Electric Propulsion Conference*, AIAA-82-1941, November 17, 1982; K. H. Groh, H. W. Lock, and H. W. Velten, Performance data comparison of the inert gas RIT 10, *AIAA/JSASS/DGLR 16th International Electric Propulsion Conference*, AIAA-82-1932, November 17, 1982.

Protein

Proteins are the most important constituents of the living organism. The first part of this article discusses recent progress in the identification of blood plasma proteins produced in association with inflammation. The second part deals with computer representations of proteins, which are powerful tools for the inference of biochemical properties of proteins.

ACUTE-PHASE PROTEINS

The acute-phase reaction (APR) follows the occurrence of an injurious stimulus, such as infection, trauma, or burn. It is typically associated with a local or general occurrence of acute inflammation. The acute-phase proteins are blood plasma proteins which exhibit a marked change in concentration in the course of the acute-phase reaction. Associated changes also occur in various other plasma and tissue components, including enzymes, hormones, and even inorganic constituents such as metals.

The acute-phase proteins are typically synthesized in the liver, in response to the stimulation of hepatocytes by interleukin-1. The latter is a monokine product of macrophages which is associated with various biological properties, including the induction of fever (leukocyte pyrogen). An experimental stimulus, for example, an injection of endotoxin in mice, is typically followed by a latent period of several hours, after which the acute-phase proteins are synthesized in increased amounts, while the synthesis of other proteins, such as albumin and transferrin, is reduced. The period of synthesis lasts for some 16 h and then subsides. Subsequently, the elevated levels of the acute-phase proteins gradually revert to normal concentrations. A similar process follows acute-phase reactions due to natural causes, such as human infection; however, the elevated serum levels of acute-phase proteins persist longer.

The most remarkable increase occurs in the concentration of the C-reactive protein and serum amyloid A component in most higher vertebrates, and that of serum amyloid P component in a few species. These proteins can increase 10- to 1000-fold during acute-phase reactions, in contrast to the much smaller changes (less than 2- to 4-fold) in the concentration of various other plasma proteins, including alpha-1 acid–glycoprotein (orosomucoid), alpha-1 antichymotrypsin, alpha-1 antitrypsin, alpha-2 macroglobulin, ceruloplasmin, complement factor 3, ferritin, fibrinogen, and haptoglobin, as well as some less well-characterized ones. While the overall course of the increase tends to be similar, the extent of the increase of each protein varies with the species, the individual, the causative agent of the acute-phase reaction, and the intensity and duration of the stimulus, so that the pattern of the reaction exhibits considerable differences.

C-reactive protein (CRP). This protein can react with and precipitate the somatic C carbohydrate of *Streptococcus pneumoniae*. Human C-reactive protein is a polypeptide with molecular weight of 21,000 that exists in serum as a characteristic, cyclic pentamer with a molecular weight of 120,000. A similar protein has been found in all the higher vertebrates that have been examined, as well as in birds and some species of fish. An analogous protein has also been found in the horseshoe crab, *Limulus polyphemus*, but its prevalence among invertebrates is undetermined. These proteins exhibit an extensive homology in primary structure as well as functional similarity.

Human serum contains approximately 800 nanograms of C-reactive protein per milliliter (the concentration ranges between 80 and 7000 ng), but the level can increase as much as 1000-fold during acute-phase reactions. Similar increases occur in many animals, including rabbits, but are much less marked in other animals, such as mice or rats. Fish that produce C-reactive protein can also exhibit an increase during acute-phase reactions, though the increase may be minimal. All animals, including the fish, that are injected with C-reactive protein–ligand complexes, or that have C-reactive protein in the serum and are injected with a ligand, exhibit a local acute inflammatory reaction. Reactivity with a ligand requires the presence of calcium, which is also required to preserve the conformational integrity of the C-reactive protein. Binding of ligands occurs apparently to at least two sites, one of which has specificity for phosphorylcholine and, to a lesser extent, other phosphate monoesters. The other site can bind carbohydrates without the phosphoryl residues, for example, depyrrylated type-4 pneumococcal polysaccharide. C-reactive protein complexes activate complement and can have an effect on clotting, phagocytosis, and the immune response.

Ligands bound by C-reactive protein are widely distributed in nature. The apparent conservation of C-reactive protein structure and function through hundreds of millions of years of evolution testifies to its biological importance.

Serum amyloid P component (SAP). This is an acute-phase reactant in some species, such as mouse and rat, but not in others, such as human or rabbit. It is a minor constituent of amyloid, a deposit that can occur in the tissues of people or animals affected with various diseases, but especially with chronic infections. Serum amyloid P component resembles C-reactive protein in its primary amino acid structure, a cyclic pentameric or decameric (double pentameric) structure visible on electron microscopic examination, and a requirement for calcium in conserving its structure and binding to ligands. It is believed to have evolved with C-reactive protein from a common precursor through gene duplication. Like C-reactive protein, it is found in the plasma of most vertebrates, including some species of fish. The SAP in plasma has a molecular weight of approximately 250,000, being composed of 10 units of molecular weight 23,000. Human serum concentration of SAP is approximately 30 micrograms per milliliter.

Serum amyloid A component (SAA). This is a polypeptide of molecular weight 12,000 that circulates in plasma in association with high-density lipoprotein with a molecular weight of approximately 160,000. Serum amyloid A component occurs in several related forms, produced by allelic genes. It is structurally analogous to a major protein component of amyloid fibrils that can develop in the tissues of subjects with chronic inflammatory diseases. In humans and many other vertebrates, the serum concentration of serum amyloid A component can increase 1000-fold during acute-phase reactions. The function of serum amyloid A component is unknown, but it is capable of inducing nonspecific immunosuppression.

Significance. It is noteworthy that the proteins which undergo changes in concentration during acute-phase reactions have various biologically important functions. It is not known whether the occurrence of a change in the level of these proteins during acute-phase reactions is selectively advantageous, and therefore conserved in evolution, or whether it is a trivial consequence of the common origin of these proteins in the hepatocyte. One could argue, however, that the capacity to bind and promote the phagocytosis of many carbohydrates (CRP, SAP), elicit immunosuppression when tissue components are released by injury (SAA), eliminate hemoglobin released during injury (haptoglobin), curtail the action of liberated proteases (alpha-1 antichymotrypsin, alpha-1 antitrypsin, alpha-2 macroglobulin, and so on), are generally beneficial in acute-phase reactions and, therefore, appropriate for evolutionary conservation.

[ALEXANDER BAUMGARTEN]

COMPUTER SURFACE GRAPHICS

Molecules are complex three-dimensional shapes formed by the interaction of atomic orbitals; however, simple approximations have been used to make models. From the time when chemists sketched their visions of chemical spatial structures on paper, the stick representation has been a favorite. As chemistry matured, the thinking turned more to representation of the space-occupying characteristics of atoms. Linus Pauling developed an atomic space–occupying model: CPK (Corey, Pauling, Koulton). Because computer graphics developed recently, the representation of molecular structures has been done only in the past 20 years. Cyrus Levinthal used a Digital PDP-1 to draw stick representations of molecules. It was not until 1976 that an effective computer representation of molecular surfaces was developed. The two methods of representation correspond in anatomical terms to the skeleton and the flesh. The choice of representation depends on what is to be illustrated.

Display techniques. The technology used to display these two modes of representation are quite different. The stick representation is generated by using a display which draws vectors (straight lines). Because the light emission of phosphors used in cathode-ray tubes decays in about 1/60 of a second, the highest-performance displays must be capable of drawing 10,000 vectors 60 times per second. Shadow mask tubes are used, so there is also a need to control the intensity of the three-color guns at each point in each vector. Since a static molecular image does not really contain all the information that scientists need in order to understand the structure of complex molecules, the displays are also equipped with matrix multipliers which perform the mathematical operations necessary to rotate the molecular image. The vectors which compose a three-dimensional object are passed in real-time through the matrix multiplier to produce the appropriate two-dimensional projection. A typical display is equipped with a bank of potentiometers which are read at each display cycle. The values are used to form the elements of the rotation and translation matrix through which the vectors must be passed.

The vector displays with color and real-time rotation are designed for this special purpose and are very expensive. With these displays, however, scientists have been able to determine the structure of proteins and nucleic acids with x-ray crystallographic methods. As the number of protein structures increased, a vector display and a related vector microfiche atlas have been used to determine the taxonomical organization of proteins.

By comparison, the technology used to display the space-filling representation of molecules is simpler and less expensive. Color cathode-ray tubes driven in the slower line-by-line mode of a conventional television set are used to form the image. A computer memory with one byte of color information for each point on the screen (called a pixel) is used to drive a simple processor which fills the screen. A relatively powerful computer must be used to determine which color is to be shown at each pixel.

Colored spheres can be generated in a simple way by decomposing the sphere into a stack of horizontal circles. The program used computes the shading of each circle and tests to determine whether a given circle is in front of or behind any circle already appearing on that line in the same position. Because the surface image of a macromolecule is inherently more complex than the related stick representation, the computation of the image is currently not in real-time (that is, less than 1/60 of a second). The surface image of a molecule of 40 atoms takes about 20 s to compute, while the equivalent image for a tetramer of hemoglobin with hydrogens (approximately 10,000 atoms) takes about 3 min (see illustration).

Applications. The computer-generated surface representations of molecules proved to be very popular because they enabled scientists to ask and answer new and very important questions about molecular structure. Part of the popularity of a medium is due to the ease with which information can be manipulated and the rapidity with which the information can be disseminated. Since data bases of macromolecules (that is, proteins and nucleic acid structures from x-ray crystallography) have been assembled for some years and computer programs for manipulating these molecules have also been developed, there remained only the technical problems of formulating and answering structural and functional questions. Since the molecular display systems are relatively expensive and in the beginning only a few were available, a pattern of collaboration was developed which encouraged scientists to come to a central facility for a few days to formulate and solve molecular graphics problems. These visits, or "fly-ins," usually result in about 100 stereo pairs of images. The image pair which differs by only a 5° rotation about the vertical axis of the molecular object produces a stereo image, which can be viewed by using an inexpensive cardboard viewer. It can be shown in a classroom by using a slightly more complicated arrangement with two projectors in which the student wears a cardboard pair of polarizing filters. The images are so realistic that many people find it hard to believe that the stereo computer images are not just photographs of the CPK plastic models. One advantage of the computer images over the plastic models is that the color assigned to each atom can be changed. New and special coloring codes for proteins and nucleic acids have been developed to bring out special structural and functional relationships.

The surfaces of macromolecules are a wonderland of texture generated by the composition of amino acids into helices and beta sheets. When one uses only the central carbon atom in each amino acid to represent a molecule, the secondary structures are easily seen. In the all-atom space-filling representation, however, only the surface can be seen. The texture of the surface is generated by the side chains of the hydrophilic amino acids, such as asparagine and glutamine (negative) and lysine and arginine (positive). In a small globular protein the surface of

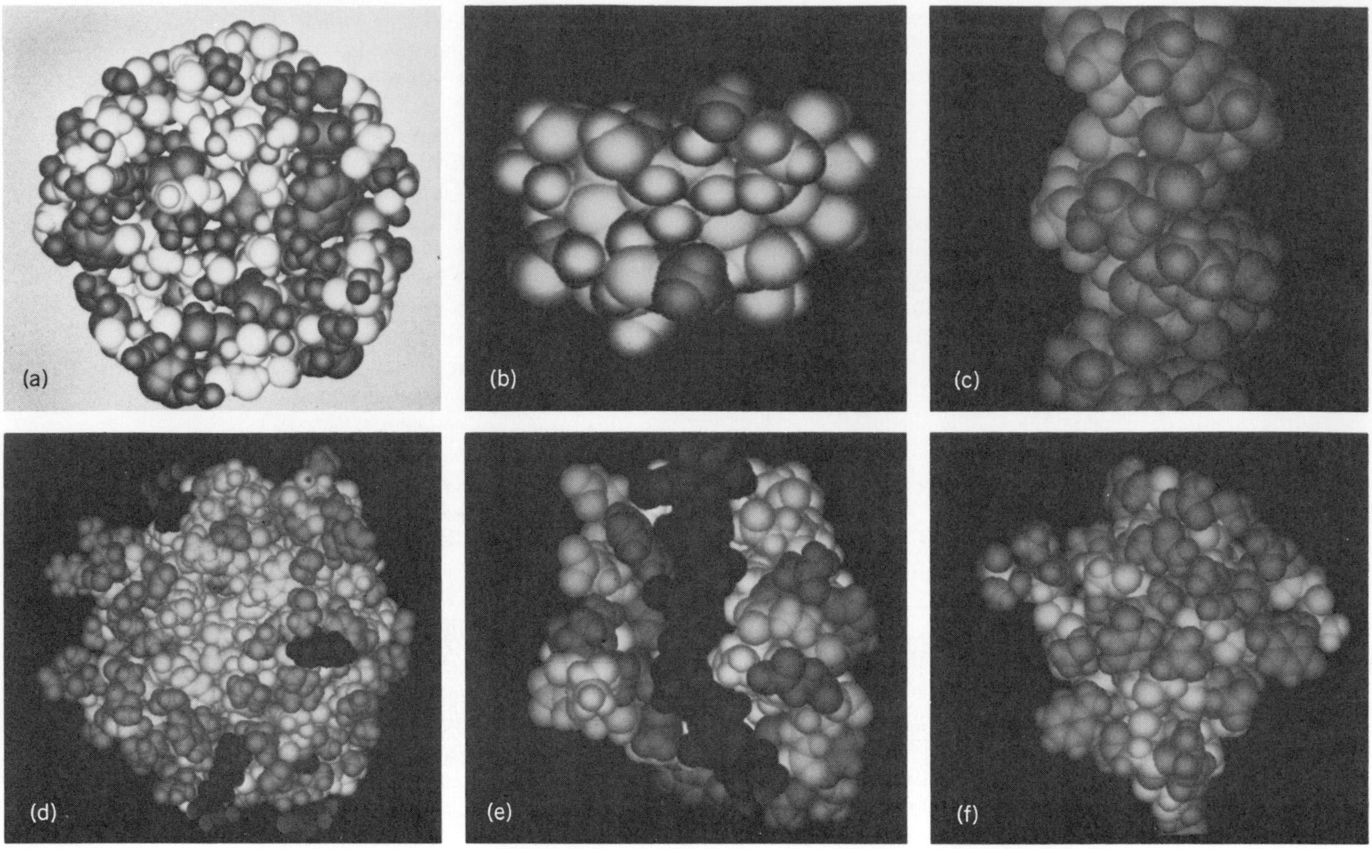

Computer presentation of proteins. (*a*) Trypsin. (*b*) Blood group LeB antigen. (*c*) Collagen. (*d*) Blood clotting factor Xa. (*e*) Chicken lysozyme complex. (*f*) Turkey insulin. (*Courtesy of Richard J. Feldman*)

the molecule is highly curved, causing the surface amino acid side chains to stick out radially into the solvent. Another component of the surface texture is the representation of the hydrogens, which are approximately equal in number to the heavy atoms. The hydrogens fill in the molecule and help the viewer to understand the microstructure. The single hydrogens attached to the rings of the tyrosine, phenylalamine, and histidine residues are visually distinct from the pairs of hydrogens along the carbon chains which form the lysine and arginine residues.

As the macromolecular surface graphics technique was developing, there was some concern that the images would be filled with too much detail. The human brain, however, is well adapted to looking at scenes of great detail. Surveying such scenes leads to interesting conclusions about the macromolecules. If the molecule is colored so that the negatively charged tips of the asparagine and glutamine residues are colored red and the positively charged tips of the lysine and arginine residues are colored blue, then from the overall view of a protein one can estimate the pH of the environment in which the protein operates. Flavodoxin, for example, has a decidedly red cast to its surface structure, while cytochrome *c* has a decidedly blue cast. Most proteins have a surface structure which is formed by a random distribution of red (−) and blue (+) residues.

Molecular graphics can be used to understand not only the static aspects of protein architecture but also the dynamic aspects of protein function. Calculations which simulate the forces between atoms in a protein can be integrated in very small time steps to produce a representation of the protein dynamics. The side chains of the surface amino acids appear to undergo waving motions, like palm trees in a hurricane. Calculations have been done on small molecules for about 10^{-10} s. Unfortunately, the computer time required to calculate the slower breathing modes of proteins as well as their functioning is still not available.

Future developments. Progress in molecular graphics is driven in equal part by advances in display and computational technology and by a constant reformulation of representation techniques and their application to new and old molecules. It has become possible in the past year or two for hardware technologists to design special-purpose computers and displays for very small markets. Routinely, new hardware is being commercially offered which is 100 times faster than an equivalent task done by software alone. Since hardware is for the moment in the ascendency, a subsequent round of software developments which will carry the representation of molecular structure and function to new heights can be expected.

For background information *see* COMPUTER GRAPHICS; PROTEIN in the McGraw-Hill Encyclopedia of Science and Technology. [RICHARD J. FELDMAN]

Bibliography: R. Allen et al. (eds.), *Marker Proteins in Inflammation*, 1982; R. J. Feldmann, *Atlas of Macromolecular Structure on Microfiche*, 1976; R. J. Feldmann and D. H. Bing, *Teaching Aids for Macromolecular Structure*, 1980; I. Kushner, J. E. Volanakis, and H. Gewurz (eds.), C-reactive protein and the plasma protein response to injury, *Ann. N.Y. Acad. Sci.*, vol. 389, 1982; J. S. Richardson, The anatomy and taxonomy of protein structure, *Advances in Protein Chemistry*, 1981.

Proton

Until a few years ago, physicists thought that all fundamental particles more massive than the proton or the electron ultimately decayed into either of these two basic constituents of matter. These two alone were thought to be immutable and not subject to radioactive decay. However, subsequent theoretical understanding, pioneered by A. Sakharov in 1967, has led to the belief that the very existence of the relatively rare proton in the vast sea of light (photons) of the universe (only one part per billion) implies that the proton is unstable. Indeed, the birth of protons in the early universe, when seen in time-reversed fashion, implies their eventual demise.

From a complementary point of view, new theories that seek to unify all fundamental forces under one rubric also lead to the decay of the proton with a predictable, but very long, mean lifetime.

Very sensitive, monumental experiments have recently been mounted to search for the decay. However, there is as yet no compelling evidence for proton decay, even though the experimental limits are considerably beyond the naive expectations of the simplest theories.

Conservation laws. Until recently physicists believed that protons were timeless, stable particles. Stability is phrased in terms of a conservation law—in this case, the conservation of baryon number (the total number of neutrons and protons).

The archetype for understanding the concept of a conserved quantum number is the conservation of electric charge. In all interactions involving electric charge, the total charge number (the difference between the numbers of positive and negative electric charges) is constant. Conservation of the quantum number associated with electric charge is a reflection of the underlying symmetry of Maxwell's equations, the basic equations of electricity and magnetism. The fundamental symmetry in these equations is known technically as gauge invariance. To calculate an electric field from an electric potential, one need not know the zero level of the electric potential energy; the ability to choose the zero point at will, without changing the physical result for the electric field, is a manifestation of the gauge invariance of the equations.

When physicists notice other numbers in nature that seem to be conserved, for example, the total number of nucleons (protons and neutrons) in a reaction, it is tempting to presume that an underlying symmetry is at work. Reactions do occur in which protons are created or destroyed. Matter plus antimatter can be made from energy: thus, an energetic proton—a particle of light—in cosmic rays or at an accelerator can materialize as a proton and an antiphoton or a neutron and an antineutron. But the sum of the baryon numbers, $+1$ for a proton or neutron and -1 for an antiproton or antineutron, remains constant. In all reactions observed to date, it appears that there is some kind of generalized electric charge number, called baryon number, which is conserved.

The same is true for the quantum number associated with electrons and their heavier cousins, muons. Here, the generalized electric charge is called lepton number. One never sees electrons created or destroyed alone. A photon may materialize into an electron and an antielectron (positron), but the sum of the lepton numbers is unchanged: lepton number in the photon is zero, and the materialized $+1$ electron and -1 positron yield zero lepton number in the final state.

Although baryon and lepton numbers are conserved, however, and behave conceptually like electric charge number, a gauge invariance (that is, a fundamental symmetry of nature like that operative in electromagnetism) responsible for their conservation has not been discovered. It is therefore possible that baryon number and lepton number only appear to be conserved at the current level of experimental sensitivity and that they are not absolutely conserved as is electric charge.

The initial belief in the stability of the proton was grounded primarily in everyday observation and enlightened speculation. A law of baryon conservation was first formulated by H. Weyl in 1929. In 1939 E. C. G. Stückelberg postulated a law of conservation of "heavy" charge, analogous to the conservation of electric charge. E. P. Wigner formulated a law of baryon stability in 1949. He remarked that electrons do not decay into neutrinos and photons, and that, by analogy, protons do not decay into positrons and photons. This means that there must be a conservation law for protons. He could not justify this law; he only posited it. Wigner did suggest the first plausible proton decay mode, into a photon and positron.

Unifying theories. The recent rebirth of interest in proton decay has been occasioned by the development of new theories that purport to unify the fundamental forces of nature. These theories envision the electromagnetic, weak, and strong forces as three manifestations of the same basic force. Each of the three forces displays a gauge invariance of the type discussed above for electromagnetism. In the unified theory, as a consequence of the overall gauge symmetry, neither baryon number B nor lepton number L alone is conserved, but a new quantum number (called fermion number, B-L). This new number assumes a role in the unified theory similar to that of electric charge number in Maxwell's subtheory. Since baryon number is no longer conserved, protons have to decay.

J. Pati and A. Salam in 1973 suggested that the three quarks (the pointlike particles that constitute a proton) might simultaneously transform into three leptons. This was expected to be an exceedingly rare event, since the probability of a single quark becoming a lepton would be small, and if that low probability is cubed, the decay rate could be almost infinitesimal. Therefore, the lifetime could be unobservably long. Pati and Salam calculated the lifetime to be 10^{29} years.

In 1974 H. Georgi and S. Glashow and, independently, H. Georgi, H. Quinn, and S. Weinberg discovered alternative reasons for proton decay and different modes of decay. The predicted lifetimes also turned out to be around 10^{29} years. In the mathematics of their "grand unification," the theories take two sets of particles that hitherto were thought to be independent—namely, quarks, which interact strongly inside the proton, and leptons, which interact only via electromagnetism and radioactive decay—and put them in the same symmetry group. At very high energies, or at very short distances, or at early times in the evolution of the universe (all of which are equivalent), quarks and leptons were members of the same family. After the big bang, the universe cooled and went through a phase transition. The various particles condensed out with different masses and different interaction properties. Nonetheless, these particles retain a primordial relationship. This means that there can be, if only rarely, transitions between them. Quarks transforming into leptons mediate proton disintegration.

Since initial experimental results in 1973, a first step in grand unification—the joining of the electromagnetic and weak interactions into the electroweak force—has been successfully completed. This unification is possible because the two interactions are both gauge-invariant. The theoretical unification posited a new weak interaction, analogous to β-decay on the one hand and electromagnetism on the other. This interaction would be mediated by a heavy neutral particle, the Z^0, the counterpart of the photon in electromagnetism. The mass of the Z^0 particle, called the intermediate boson, would be so great (about 90 times the mass of the proton) that, at the time it was postulated, it could not be produced at the highest-energy accelerators. This is part of the reason why physicists did not recognize the intimate connection between electromagnetism and radioactive decay. In 1983 the Z^0 family of particles was discovered at the CERN proton-antiproton colliding beams experiments. *See* INTERMEDIATE VECTOR BOSON.

A generalization of this unification is the more recent grand unification, in which a yet higher level force incorporates the strong force and the new electroweak force. This requires other new, heavy intermediary particles called X and Y bosons. These are so very massive that their production at accelerators cannot even be contemplated. However, like the Z^0 in 1973, their existence for even an infinitesimal time would be apparent in interactions. This fleetingly short existence is guaranteed by the Heisenberg uncertainty principle in which a very large mass, Δm, can exist for a short time, Δt, determined by the relationship $\Delta m \Delta t \sim h$, where h is Planck's constant. By spontaneously emitting one of these superheavy intermediaries, a quark inside a proton can transform into a lepton. At the same time, another quark can transform into an antiquark (Fig. 1). This antiquark and the unchanged third quark combine to form what is commonly called a pion, which subsequently decays into two photons. Both baryon number and lepton number conservation are violated, and matter is transformed into antimatter. (This decay mode is remarkably similar to that suggested by Wigner.) The great masses of the X and Y bosons explain why the transformation of quarks within a proton would be so rare.

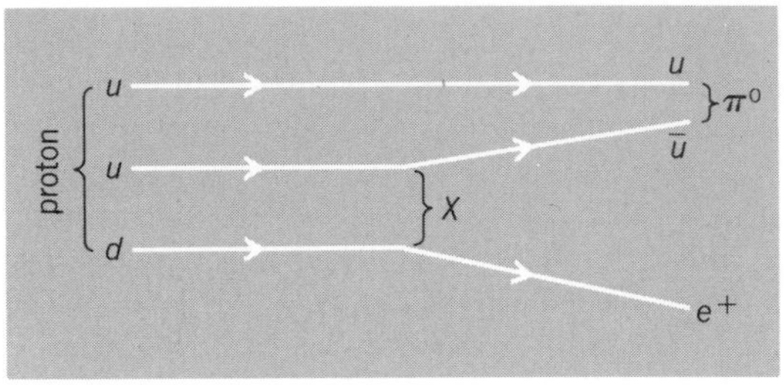

Fig. 1. Possible two-body decay mode for a proton: an up (u) and a down (d) quark transform into an anti-up quark ($\bar{u}$) and a positron, respectively. The antiquark and the unchanged third quark combine to form a pion (π^0). The exchange of a superheavy X boson causes the disintegration to occur. (*After L. R. Sulak, Waiting for the proton to decay, Amer. Sci., 70:616–625, 1982*)

Calculating the mass of the exchanged intermediary allowed theorists to predict the lifetime of the proton and set the scale for the sensitivity of the new experiments required to test the theories.

Searching for proton decay. There are two general techniques used in searching for nucleon decay. The first is independent of the particular secondary particles into which the nucleon might disintegrate. It depends only on the nucleus remaining after the disappearance of a proton or neutron from the original nucleus. The new nucleus may be one which is not ordinarily found in nature. Physicists have looked deep in the earth in environments shielded from cosmic rays, which might also induce unusual transformations of nuclei, and have searched for novel nuclei. The most sophisticated experiments using this technique have brought the lower limit on the lifetime of the proton to 10^{26} years.

To achieve higher sensitivity by ruling out cosmic-ray-induced background, the radiation emitted must be identified. This second technique (pioneered by F. Reines and M. Crouch in 1974) is exploited in several new experiments explicitly dedicated to the search for proton decay. It is advantageous to optimize a detector for certain specific decay modes. The major decay mode predicted by a broad class of unifying theories is the simplest possible reaction to observe. It is the two-body decay mode shown in Fig. 1, in which a stationary proton decays into a positron moving off in one direction and a neutral pion moving off in the opposite direction. Similarly, a decaying neutron in a nucleus is expected to emit a positron one way and a negative pion the other.

Other classes of theories lead to distinctly different decay modes. Some of these theories postulate decays which eventually produce two muons. Each of these subsequently decays into an electron. This sequence is identified by a 2-microsecond delay between the appearance of the muon and the emission of its decay electron. The detectors have generally been designed to also pick up the characteristic time decay sequence.

The new detectors must also use sufficient detector material to have on the order of 10^{33} nucleons in a completely contained volume so as to observe a few events per day at the predicted lifetime levels. For a detector material of density 62 lb/ft^3 (1 g/cm^3) this requires a cubic detector of approximately 66 ft (20 m) on a side. Because the detector has to be so massive, it must be constructed of inexpensive material. In the largest detector, that of the Irvine-Michigan-Brookhaven Collaboration (IMB), this material (water) is transparent, so that the paths of the decay products can be observed. Water also provides a unique advantage. When a charged particle travels through water faster than the speed of light in water, an electromagnetic shock wave, called Cerenkov light, is produced, which is deep blue in color and can be detected by phototubes.

Since all the particles from a decay are stopped in a massive detector, the decay of a nucleon of approximately 1 GeV mass would produce 1 GeV equivalent light. The total amount of light collected by the phototubes is summed up and compared with the proton's mass. In a two-body decay, the two daughter particles each should yield 0.5 GeV equivalent of light. The direction taken by each of the daughter particles is reconstructed to determine if the decay really produced two back-to-back tracks. All of these constraints are necessary to eliminate background that might simulate proton decay in the detector. Even though the IMB detector is very deep (2300 ft or 700 m) in the earth (Fig. 2) to shield it from the nonpenetrating component of the cosmic rays, the penetrating component (neutrinos) could induce a nuclear reaction which might look like proton decay.

The IMB detector holds 10,000 metric tons of ultrapure water. The Cerenkov light can travel through the longest distance of the detector, the diagonal of a 69-ft (21-m) cube or 98 ft (30 m), without significant loss. Phototubes, 2048 in all, sensitive to the Cerenkov light are suspended at 3.3-ft (1-m) intervals to cover all six faces of the cube. The area of each phototube is large enough to respond to light from proton decay beamed to it from anywhere in the tank.

Fig. 2. Cavity of the Irvine-Michigan-Brookhaven (IMB) proton decay experiment before it was lined with polyethylene bladders and filled with ultrapure water. Bulldozer at bottom of ramp indicates scale. Steel beams at top subsequently supported catwalk from which strings of phototubes were suspended. (*From L. R. Sulak, Waiting for the proton to decay, Amer. Sci., 70:616–625, 1982*)

The expected pattern of light produced by the 8-ft-long (2.5-m) track of a typical charged, unscattered particle from a proton decay is illustrated in Fig. 3. This track originates near the center of the 69-ft (21-m) cube. Since the Cerenkov light is produced in a cone at about 41° relative to the direction of the charged particle, a decay event at the center

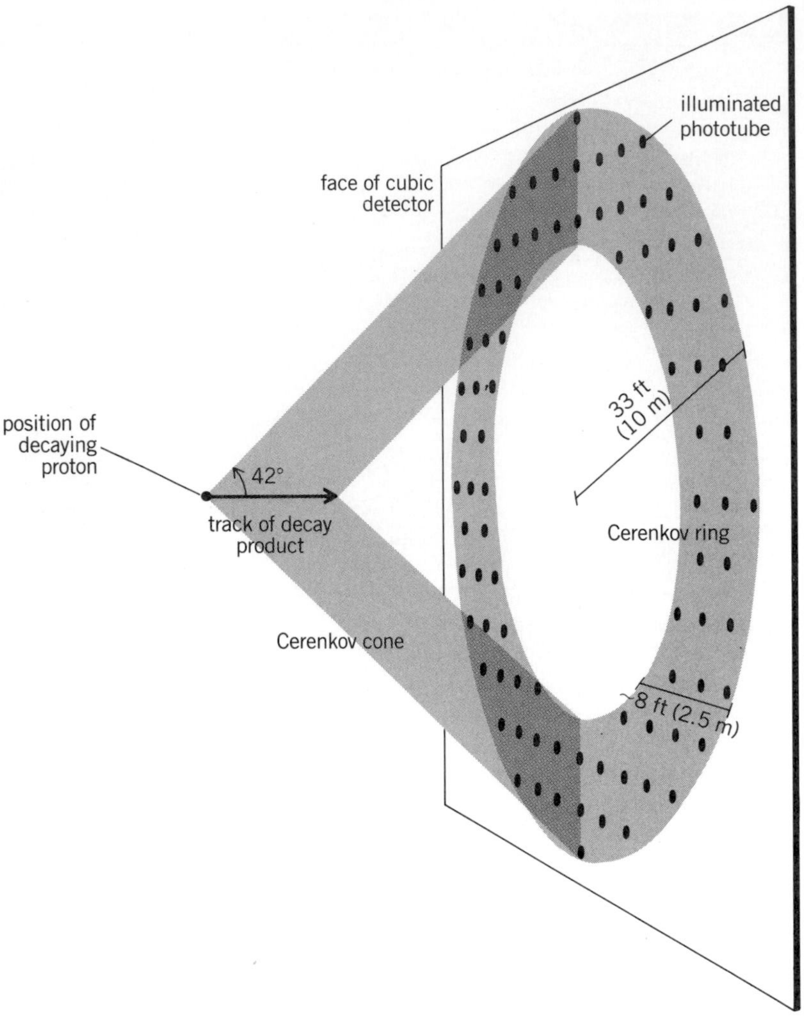

Fig. 3. Pattern of Cerenkov light produced by a particle traveling faster than the speed of light in water, in the IMB proton decay detector. (*After L. R. Sulak, Waiting for the proton to decay, Amer. Sci., 70:616–625, 1982*)

of the cube lights up a ring of tubes covering an entire face. Of the several hundred tubes on a face, typically a hundred are fired by a particle with an 8-ft-long (2.5-m) track. On the opposite wall, the other track of a two-body decay would produce a similar ring.

The width of the ring of light is a direct measure of the length of the particle track, and the radius of the ring gives the distance from the wall to the starting point of the track—that is, to the position of the decaying proton. The time of arrival of the shock wave at the wall of phototubes determines the direction in which the particle is traveling (Fig. 4). If the direction is perpendicular to the wall, as for particle a, the tubes around the edge of the circle will fire at the same time. If the particle comes in at an angle, as shown for particle b, the tubes on opposite sides of the ring will fire at different times, with the phototubes on the close end of the pattern firing much earlier than those on the ring far from the origin of decay.

The detector provides redundancy in identifying this footprint of proton decay: It is known which tubes fired and which did not. From each tube that fired, there is an electronic pulse, whose height is proportional to the intensity of light at the tube, and a time, which indicates when the shock wave hit the tube. Further, the capability of registering a second hit on each tube, up to 8 μs after the initial firing, allows identification of any muons in the detector.

As of 1983, the IMB detector had not observed any candidates for proton decay. On the other hand, the detector observes neutrino background interactions at the expected rate of one per day and with the expected characteristics. It has set the strongest limits on the proton lifetime. For the positron-pion decay mode the lifetime must be greater than 10^{32} years. For the muon-kaon decay mode, the lifetime is greater then 3×10^{31} years. These limits severely constrain the possible theories of grand unification.

Other massive detectors started taking data in 1983. The largest, the Kamioka detector in Japan, is one-third the size of IMB. Smaller detectors (1000 metric tons) in Park City, Utah, and the Frejas auto tunnel on the French-Italian border will complement

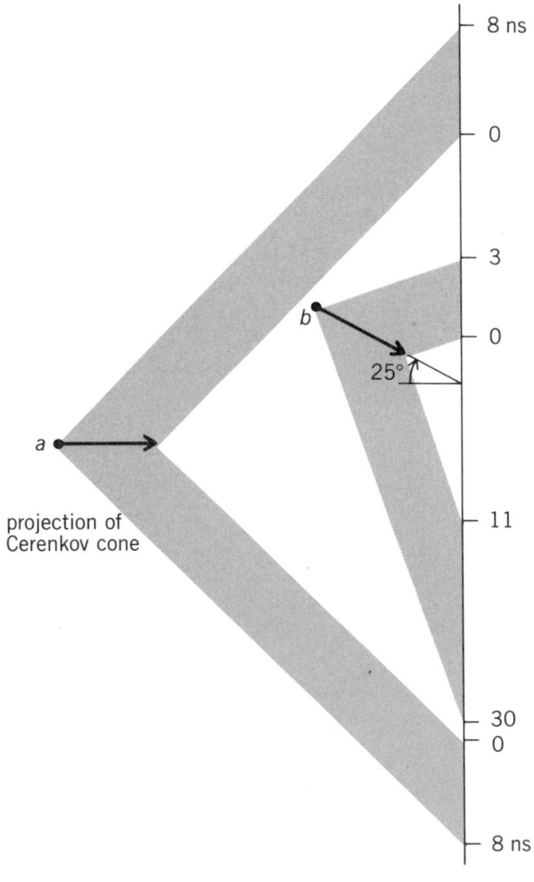

Fig. 4. Cross sections through two different Cerenkov cones in the IMB proton decay detector. Times of firing of phototubes, relative to the first phototube firing, are shown. These are used to determine the direction in which the charged particle generating the Cerenkov light is traveling. (*After L. R. Sulak, Waiting for the proton to decay, Amer. Sci., 70:616–625, 1982*)

the initial work of the very small detectors (100 metric tons) at the Mont Blanc auto tunnel on the French-Italian border and the Kolar detector in the gold fields of India.

For background information *see* CERENKOV RADIATION; FUNDAMENTAL INTERACTIONS; SYMMETRY LAWS (PHYSICS) in the McGraw-Hill Encyclopedia of Science and Technology. [LAWRENCE R. SULAK]

Bibliography: M. K. Gaillard, Toward a unified picture of elementary particle interactions, *Amer. Sci.*, 70:506–514, 1982; M. Goldhaber, P. Langacker, and R. Slansky, Is the proton stable?, *Science*, 210:851–860, 1980; L. R. Sulak, Waiting for the proton to decay, *Amer. Sci.*, 70:616–625, 1982; J. S. Trefil, How the universe began, *Smithsonian*, 14(2):32–51, 1983.

Pulsar

Pulsars are rotating neutron stars that emit brilliant flashes of electromagnetic radiation at each revolution, like beacons from a lighthouse. In the 15 years since their discovery, astronomers have found nearly 350 pulsars with pulse periods ranging from 33 ms to 4 s. In 1982 a "fast" pulsar with the short period of 1.5 ms was discovered. This new member of the cosmic menagerie has stimulated new research concerning the final state of stars, and has opened paths to experiments that use the pulsar as a precision clock.

Nature of pulsars. A systematic search for compact quasars in 1967 at Cambridge University led to the serendipitous discovery of pulsars. The subsequent identification of pulsars with neutron stars validated theories concerning the ultimate fate of massive stars that had been discussed for 30 years.

Neutron stars are formed in the collapse of stars more massive than the Sun after their nuclear fuel has been exhausted. In every star there is an inward force arising from the gravitational pull of its own mass. A normal star is held in equilibrium by an outward force resulting from the heat that is released by nuclear fusion reactions occurring in the core. The equilibrium is upset when a substantial fraction of the hydrogen fuel is converted to helium ash. The core collapses suddenly, and the overlying envelope is expelled to form a brilliant supernova. Equilibrium of the core is restored when the outward degeneracy pressure, a quantum-mechanical effect, counteracts the inward force arising from self-gravity. Ions lose their identity and merge to form a sea of neutrons, electrons, and protons that behaves like a giant atomic nucleus. The collapse leads to an increase in rotation speed and an amplification of the magnetic field, owing to the laws of conservation of angular momentum and of magnetic flux. The density of a neutron star is similar to densities in atomic nuclei, and is 10^{14} times larger than the average density of the Earth and the Sun. The surface magnetic field of a typical pulsar is 10^{12} times that on the Earth and the quiet Sun.

Pulsars are like gigantic electric generators that are continually arcing. The combination of strong magnetic fields and rapid rotation creates tremendous electrical fields in which electrons and positrons are created and accelerated to relativistic energies. These relativistic particles are responsible for the electromagnetic radiation from the pulsar. In one model, the radiation originates near the poles of a dipole field (that is, the north and the south magnetic poles). Radiation is observed from the pulsar only when one of the poles coincides with the Earth's line of sight to the star. If the dipole axis does not coincide with the rotational axis, the radiation appears to be pulsed as the beams of radiation sweep into this line of sight, like beacons from a lighthouse.

The continuous loss of accelerated particles and radiation causes the pulsar spin rate to decrease. The slowing of pulsars can be observed by measuring the arrival time of pulses over many days. The rate of slowing yields an estimate of both the pulsar age and its magnetic field intensity. Furthermore, irregularities in this rate give strong evidence for the existence of "star-quakes" resulting from the sudden adjustment of the shape of the neutron star to a more perfect equilibrium figure. These irregularities provide vital clues about the detailed physics of neutron star structure.

Surveys using large radio telescopes in the United States, Australia, and England have discovered about 350 pulsars. The galactic distribution of pulsars follows that of young stars and star-forming regions: a thin disk 10 kiloparsecs in radius. Pulsars are unique probes of the plasma and the magnetic field between stars. Only one pulsar outside the Galaxy has been discovered, in the Large Magellanic Cloud.

Discovery of 1.5-ms pulsar. Before the 1982 discovery of the 1.5-ms pulsar, astronomers thought that the maximum spin rate of pulsars would not be much faster than 30 times per second. Two arguments led astronomers to this belief: (1) The most common spin rate for pulsars was around 1 Hz, with very few exceeding the search limit of approximately 10 Hz. Searching for pulsars rotating faster than about 10 Hz was computationally difficult. (2) The neutron star formation process presented above implied that fast pulsars would be found amid their supernova debris and would be slowing rapidly. The archetype supporting this view is the Crab Nebula pulsar, which was the fastest at 30 Hz and is associated with a supernova noted by Chinese astronomers in A.D. 1054. Searches in other supernova remains, which are detectable across the entire Galaxy, had not uncovered a pulsar faster than the Crab pulsar.

The millisecond pulsar discovery story began in 1979, when D. Backer began investigating a peculiar compact radio source named 4C21.53. The source was listed in the same survey that had led to the 1967 pulsar discovery. The survey had also shown that compact radio source images were dilated by scattering along long paths of interstellar plasma. Optical images of stars are broadened by a similar scattering process in the Earth's atmosphere. The compactness of 4C21.53 was puzzling, since its location on the galactic plane implied a long path through the Galaxy. Furthermore, the intensity spectrum of 4C21.53 resembled that of pulsars rather than compact quasars. Backer reasoned that if the compact source were a nearby pulsar and not a distant quasar, then both of the above observations could be easily explained. However, the only way this object could have escaped previous pulsar searches was if its period were extremely short, less than the 10-ms limit of the surveys. Excitement about this source grew when he realized that it was probably associated with a supernovalike source.

Between 1979 and 1982 pulse searches and other investigations either were unsuccessful or added confusing facts to the observational picture. A break came in early 1982, when M. Goss and Backer obtained a radio image that pinpointed the coordinates of the compact object. The chase neared success when observations in September 1982 by S. Kulkarni at the giant radio telescope in Arecibo provided the first clue that the source, now identified as 1937+214, was a pulsar with the very short period of 1.5 ms. These observations led to a more complete study by a team of radio astronomers a month later, which showed conclusively that the source was indeed a pulsar spinning at 642 Hz.

The pulse shape of the pulsar is shown in the illustration. Six full periods of the signal are shown. The display is an average taken over several minutes, since the individual pulses are too faint to observe even with the giant Arecibo telescope. Two sharp peaks of emission are evident in each period, a main pulse and an interpulse. These two pulses are believed to arise from oppositely directed beams attached to the rapidly rotating, magnetized neutron star.

The rotation rate of the pulsar is within a factor of 2 of the rotational stability limit for neutron stars. The speed at the surface is about one-tenth the speed of light. For this reason, 1937+214 provides a natural laboratory for studying dense matter at relativistic velocities.

Observations soon indicated that the pulsar's rotation period was remarkably stable. The stability was surprising, since the energy locked up in the pulsar's rotation was comparable to that of the very young Crab pulsar and to the mechanical energy output of a supernova explosion. Arrival time measurements are continuing at the Arecibo Observatory. The period is increasing at the minute rate of 10^{-19} s/s; the pulse "clock" loses only 20 periods, or 30 ms, per year with respect to the ideal clock. For comparison, the common wristwatch may gain or lose 100 s in a year, while a laboratory atomic clock will gain or lose only a few microseconds. In fact, the pulsar clock is more accurate than its rate of slowing would indicate because the slowing is predictable. After accounting for this predictable effect, the pulsar clock only gains or loses 500 nanoseconds in a year with a period of 1.557 806 448 838 ms. As a result, the pulsar may provide the purest long-term time standard on Earth.

Origin and evolution. The tremendous rotational energy and the extremely small spindown rate of 1937+214 distinguishes it from conventional pulsars. These properties and the absence of a supernova remnant in the vicinity of the pulsar have led to the conclusion that this pulsar both has a magnetic field much weaker than the typical slow pulsar and is rather old, probably more than a million years.

An evolutionary scenario is required whereby the neutron star can obtain a rapid spin without an intense magnetic field. Two models have been discussed. The first and simplest model is that the progenitor stellar core had a weak magnetic field. Core collapse produced a rapidly spinning neutron star with a correspondingly weak field. Once spinning fast with a small field, the pulsar would remain spinning fast while all traces of the supernova event dissipated. This model is supported by the existence of weak magnetic fields in white dwarfs, which also form from the collapse of stellar cores. Opponents of this theory contend that intense magnetic fields are always present in young, hot neutron stars independent on the progenitor field.

The second model requires the initial formation of a neutron star in a binary system without disrupting the binary. Later, the secondary evolves to the red giant phase when its hydrogen fuel is exhausted. Meanwhile, the neutron star's spin decreases and its magnetic field decays. The decay of the field results from the resistive loss of the crustal currents that

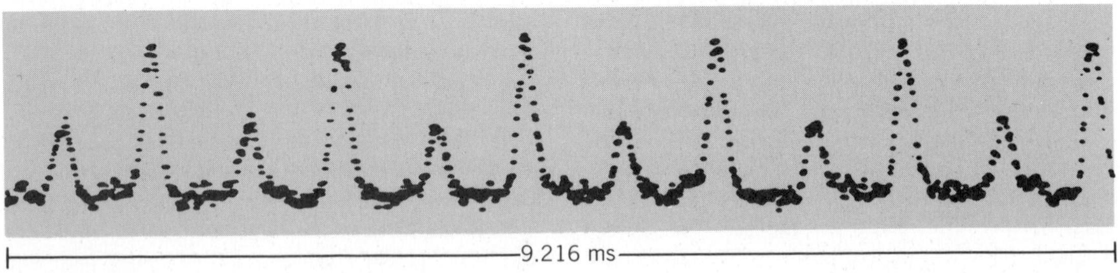

Pulse shape of the 1.5-ms pulsar discovered at the Arecibo Observatory in 1982, and observed here at a frequency of 1412 MHz. Six full periods of the signal are shown. (*From D. C. Backer et al., A millisecond pulsar, reprinted by permission from Nature, 300:615–618, copyright © 1982 Macmillan Journals Limited*)

maintain the field. If the envelope of the red giant swells to fill the binary orbit, it will transfer mass and angular momentum to the neutron star. In favorable cases the neutron star is spun up while its field remains small. After accomplishing the spin-up of its companion, this model requires a supernova explosion of the secondary to disrupt the binary. Two observations that support this scenario are the existence of x-ray-emitting binary systems, which appear to be in the low-field, mass-transfer stage discussed, and the 1983 discovery of a 6.133-ms pulsar in a binary system.

Applications of fast pulsars. The discovery of a millisecond pulsar has generated much observational and theoretical work on the death of stars. The discovery has also inspired research in areas outside traditional astronomy and astrophysics. A fundamental application of fast pulsars is the detection of gravitational waves either directly from asymmetries of the neutron star or indirectly as perturbations of the pulse arrival times. Gravitational waves are a natural by-product of modern gravitational theories. Detection of gravitational waves would provide crucial data to unification theories which attempt to explain the four basic forces as a manifestation of single basic force. A rapidly spinning object will generate gravitational waves if its shape is asymmetric with respect to the rotation axis, like a spinning, tilted cigar. An attempt to detect gravitational waves from 1937+214 has placed only uninteresting limits on the wave amplitude. Theoretical calculations suggest that 1937+214, despite its rapid spin, is not spinning fast enough to have large asymmetries. Many astronomers hope that one of the many searches that are planned or in progress at observatories in the United States, England, Australia, and India will yield a shorter-period pulsar. Discovery of a millisecond pulsar in a binary with a small separation between the binary members would allow improved tests of the general theory of relativity.

Another exciting application of short-period pulsars could be the detection of gravitational waves from a massive black hole binary in the nucleus of a distant galaxy. Gravitational waves from such a system would perturb the geometry of space along the Earth–pulsar line of sight. The pulse arrival times would be perturbed in step. Another possible source of gravitational waves could be the early universe. Timing observations of the 1.5-ms pulsar will improve the limits on the amount of gravitational energy stored in parsec-sized gravitational waves by two orders of magnitude.

The pulse arrival times are delayed by the presence of ionized gas along the line of sight. Changes in the delay can occur if small clouds of gas drift into or out of the path. Detection of these changes would lead to a new understanding of the turbulence in interstellar plasma. This effect is distinguishable from that caused by a gravity wave, since the former varies with observing frequency while the latter is frequency-independent.

Finally, the pulsar clock may lead to improvements in knowledge of the orbit of the Earth. Precise determination of the Earth's orbit is required in the realization of the pulsar clock because observations are conducted on a "platform" that moves by 20 ms daily and 500 s annually with respect to a fixed point in space. The Earth's position cannot be predicted with an accuracy better than the present pulsar clock accuracy of 500 ns for intervals exceeding 1 year. Consequently, a data base of pulsar observations extending over many years may allow a solution for improved values of the masses and positions of planets and other solar system bodies.

The discovery of fast pulsars has led to new insights concerning the final states of stars. The period stability of fast pulsars provides new opportunities for gravitational wave detection and for improvements in the model of the solar system.

For background information *see* EARTH ROTATION AND ORBITAL MOTION; GRAVITATION; NEUTRON STAR; PULSAR in the McGraw-Hill Encyclopedia of Science and Technology.

[DONALD BACKER; SHRINIVAS KULKARNI]

Bibliography: D. C. Backer et al., A millisecond pulsar, *Nature*, 300:615–618, 1982; D. C. Backer, S. R. Kulkarni, and J. H. Taylor, Timing observations of the millisecond pulsar, *Nature*, 301:314–315, 1983; R. N. Manchester and J. H. Taylor, *Pulsars*, 1977.

Quarkonium

Quarkonium is the generic name for an "atom" made up of a heavy quark and its antimatter partner, an antiquark. Quarks are presently believed to be among the elementary building blocks of matter which have mass, and an intrinsic spin angular momentum of $\frac{1}{2}\hbar$ (where $\hbar$ is Planck's constant divided by 2π). Positronium is a well-understood atom, and its general structure is similar to that of quarkonium. For positronium, made of an electron and a positron (the antiparticle of the electron), the force which holds the electron and positron together is the well-known electromagnetic force between a positively and negatively charged particle. Its constituents (electrons and positrons) can be observed in isolation. Quarkonium, though analogous in many ways to positronium, is considerably more exotic. The force at work in quarkonium is the strongest one known; now called the color force, it is thought to be the basis of all nuclear forces. In contradistinction to the electromagnetic force, the color force is presently poorly understood; also, quarks and antiquarks have never been observed in isolation. Physicists believe that quarkonium offers a unique tool with which to explore the nature of quarks and the forces between them; thus, much effort has been expended in studying quarkonium.

Spectrum of energy states. In quarkonium the bound quark and antiquark are identical in size and mass, as is the case for the electron and positron in positronium. Also, the allowed modes of motion in quarkonium are similar to those of positronium, so

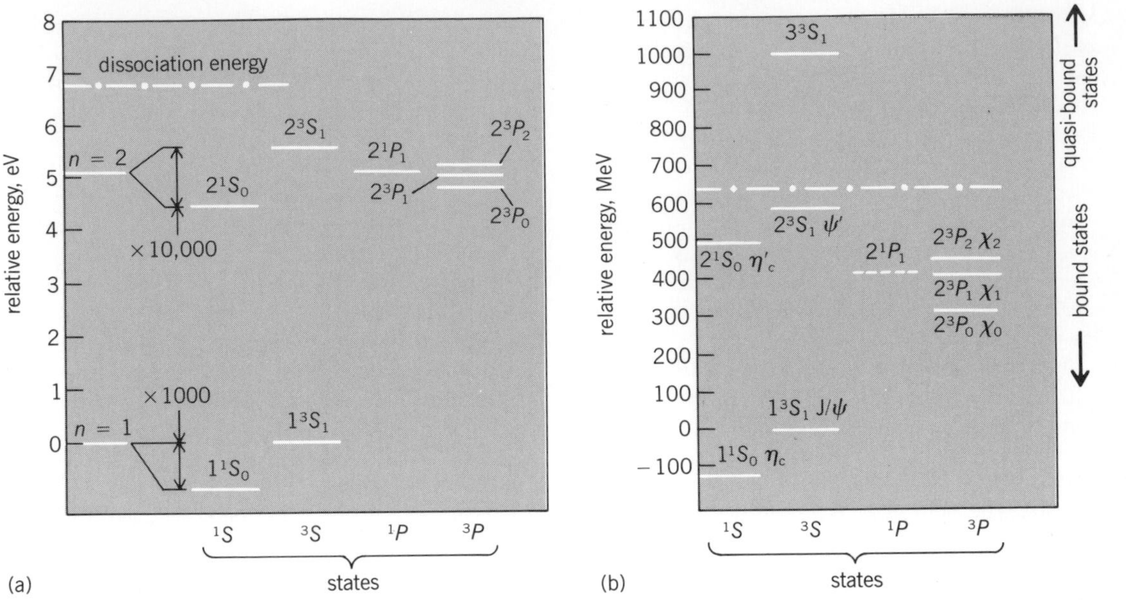

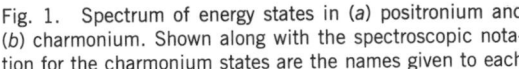

Fig. 1. Spectrum of energy states in (*a*) positronium and (*b*) charmonium. Shown along with the spectroscopic notation for the charmonium states are the names given to each particle. Observed states are indicated by solid lines; the unobserved but predicted 2^1P_1 state is indicated by broken line.

that an equivalent spectrum of energy states can be expected. Since 1974 a major effort to detect the quarkonium states has been undertaken, and well over a dozen states have been found. In a qualitative sense they correspond closely to the positronium states. On the other hand, a quarkonium system is smaller than a positronium system by a factor of 100,000, and its total mass or energy is larger by a factor of 3000 to 10,000. Moreover, from the details of the quarkonium spectrum of energy states it is apparent that the color force is not only stronger than electromagnetism but also more complex.

Figure 1 shows the spectrum of energy states for positronium, and one of the heavy quarkonium atoms made of a c (charmed) quark and its antiquark ($\bar{c}$); this $c\bar{c}$ system is called charmonium. As is seen in the figure, the spectrum of energy states is similar in positronium and charmonium, but the scale of energy differences in charmonium is greater by a factor of roughly 10^8. The energy of a state is determined by quantum-mechanical dynamics and is a function of the principal quantum number n (related to the size of the atomic orbit), and by the orientation of the particle spins and the orbital angular momentum. The dependence of these variables is concisely given in a standard "spectroscopic notation" as shown in the figure. The notation used labels a state of quarkonium as $n^{(2s+1)}\ell_j$, where n is the principal quantum number, $s\hbar$ is the sum of the spins of the two quarks in the state (0 or $\hbar$), $\ell\hbar$ is the orbital angular momentum of the two quarks in the state ($\ell = 0$ called S, $\ell = \hbar$ called P, and $\ell = 2\hbar$ called D), and $j\hbar$ is the total angular momentum of the state, $j = 0, \hbar, 2\hbar, \ldots$

In positronium the various combinations of angular momentum, for example, the 1^3S_1 state versus the 1^1S_0 state, cause only minuscule shifts in energy (shown by expanding the vertical scale), but in charmonium the shifts are much larger. All energies in Fig. 1 are given with reference to the 1^3S_1 state. At 6.8 eV, positronium dissociates into a free electron and positron. At 633 MeV above the energy of the 1^3S_1 state (called the J/ψ), charmonium becomes quasi-bound. Quarks are not observed as free particles; thus just above the dissociation energy another $q\bar{q}$ pair combines with the original $c\bar{c}$ pair to form a D and anti-D meson.

Nonrelativisitc quark systems. Originally all the hadrons known could be interpreted as combinations of just three kinds of quarks, designated up, down, and strange, or u, d, s (and antiquarks $\bar{u}$, $\bar{d}$, $\bar{s}$). It might seem that this rich variety of hadrons would offer ample opportunities to explore the force between quarks. Actually experiments with protons, neutrons, pions, and other "ordinary" hadrons can yield only indirect information about the interquark force. The reason is that the u, d, and s quarks are quite light; indeed, their mass, when expressed in energy units, is comparable to the binding energy that holds the quarks together in the hadron. As a result, the quarks in an ordinary hadron move with a speed close to the speed of light, and calculations of their properties must be done by using the complicated methods of the special theory of relativity. Up to now such calculations have been too difficult to be done reliably.

What was needed was a bound system of heavier quarks, in which the binding energy would be small compared with the quark mass. The quarks would then move much slower than the speed of light, and the complications of the theory of relativity could be ignored. Such a nonrelativistic quark system was found in 1974 with the discovery of the extraordinary J/ψ meson having a mass (in energy units) of 3095 MeV.

The J/ψ meson seemed unlikely to be any combination of u, d, or s quarks, in part because all the combinations with appropriate properties were already listed among the known hadrons. Some 10 years earlier, however, J. D. Bjorken and S. L. Glashow had speculated that there might be a fourth quark, which they had fancifully named the charmed quark (c). In 1970 Glashow and his colleagues J. Iliopoulos and L. Maiani had argued on theoretical grounds that the charmed quark must exist and should be substantially heavier than the other quarks. The J/ψ was soon recognized as a form of charmonium, that is, a meson with the quark composition $c\bar{c}$.

The discovery of charmonium stimulated a search for still heavier quarks. There was reason to think they would come in pairs, and the first two quarks after the known ones were designated bottom (b) and top (t). The first bottomonium ($b\bar{b}$) state was discovered in 1977; it is called the upsilon (Υ), and it has a mass of 9460 MeV. The toponium ($t\bar{t}$) system has not been detected. If it exists, its mass must be greater than 40,000 MeV.

In order to study quarkonium it is necessary to create it. The most effective way is through the annihilation of high-energy electrons and positrons. The high-energy annihilations take place in a device known as a storage ring, where beams of electrons and positrons circulate in opposite directions within a toroidal vacuum chamber. The electron and the positron annihilate to produce a single photon, an event forbidden in the decay of positronium because it cannot conserve both energy and momentum. These quantities must still be conserved in the high-energy annihilation, but the uncertainty principle in effect allows a momentary violation of energy conservation. Before the photon has existed long enough for its presence to be registered, it decays into two or more new particles. It is a virtual photon.

In the final state created from the decay of the virtual photon, all conservation laws must be obeyed. This condition can be met in a simple way: the virtual photon gives rise to a particle and its corresponding antiparticle. In some cases the particle-antiparticle pair is merely another electron and positron, which can then move apart and be detected. The products of the collision can also be a quark and an antiquark, however, which do not escape unencumbered. Instead, if the energies of the electron and positron are at certain values, the quark-antiquark pair becomes bound into a quarkonium atom; at most other energies additional quark-antiquark pairs materialize and become bound to the original pair. What is ultimately observed is in all cases a set of hadrons.

Probably the best quarkonium system now available to experimenters is the $b\bar{b}$ system, or bottomonium. It should be superior to charmonium because

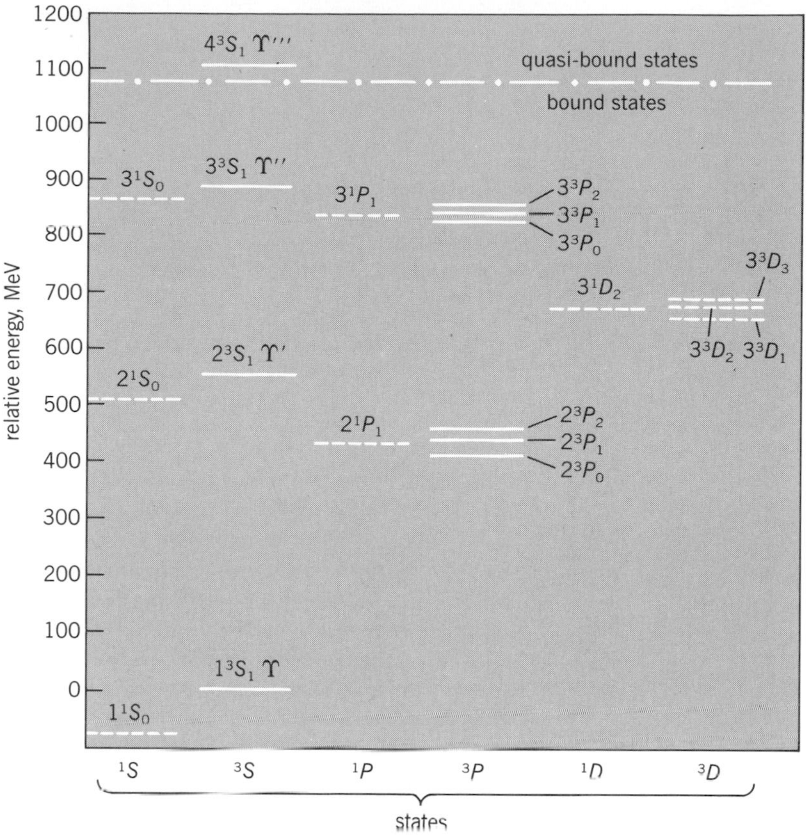

Fig. 2. Energy levels of bottomonium. Observed states are indicated by solid lines; unobserved but predicted states are indicated by broken lines.

it is appreciably heavier and is therefore a better approximation to a truly nonrelativistic system. Because of the character of the color force and the larger mass of bottomonium, the array of associated bound states should be even richer than it is in charmonium. Figure 2 shows the full richness of the bound-state spectrum expected. So far only the $^3S_1$ states and ^3P states have been observed.

Nature of the interquark force. Because the various states of quarkonium differ in the average separation between the quark and the antiquark, the energies of these systems convey information about the strength of the interquark force over a range of distances. The experiments done so far constitute measurements of the color force at distances ranging from 4×10^{-14} in. (10^{-15} m), which is roughly the size of an ordinary hadron, down to about 8×10^{-15} in. (2×10^{-16} m), a distance a fifth as large. From the measurements it is possible to construct models of how the force varies as a function of interquark distance.

Quantum chromodynamics, the current theory of quarks and their interactions, suggests that the color force may vary inversely as the square of the distance when the quarks are close together (reminiscent of Coulomb's law from electromagnetism), but may assume a constant strength for quarks that are relatively far apart. A plausible way to approximate such a force is to assume that it is merely the

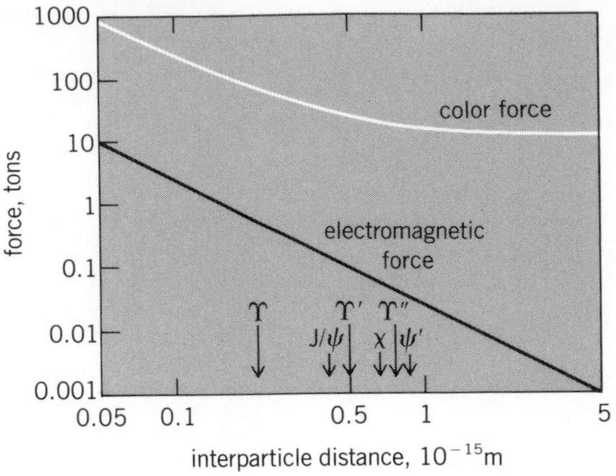

Fig. 3. Comparison of the color force between quarks with the electromagnetic force.

sum of an inverse-square force and a constant force. This model force law can be expressed mathematically by the equation $F = a/r^2 + b$, where F is the force, r is the interquark distance, and a and b are constants to be determined by experiment.

This model color force law can be tested by attempting to fit its predictions to the data for quarkonium. If the resulting curve has the right form, the values of a and b that give the best fit can then be determined. It turns out that the form of the curve is consistent with the data, and the value of b is about 16 tons (1.4×10^5 newtons). In other words, two quarks attract each other with a force of at least 16 tons regardless of how far apart they are. With this in mind, it becomes easier to understand why a quark has never been extracted from a hadron. It should be emphasized that these arguments do not prove that a constant force between quarks exists; a proof is hardly possible when the data extend only to distances of about 4×10^{-14} in. (10^{-15} m). What has been shown is that the existing data are consistent with a constant long-range force, and that if it exists, its magnitude is about 16 tons.

The value of a in the equation for the force law is more difficult to specify. The reason is that a is not actually a constant, but according to quantum chromodynamics depends to some extent on r. For the range of distances explored in the quarkonium system, a seems to have a value of approximately 1/4, compared with 1/137 for α, the constant of proportionality in Coulomb's law. Figure 3 shows some of what has been learned about the color force from the study of quarkonium, and compares the postulated force to Coulomb's law, whose inverse-square dependence on distance corresponds to a straight, sloping line on this logarithmic graph. The arrows indicate the mean-square radii of the charmonium and bottomonium states shown.

For background information *see* ELEMENTARY PARTICLE; J PARTICLE; MESON; POSITRONIUM; QUANTUM CHROMODYNAMICS; QUARKS; UPSILON PARTICLES in the McGraw-Hill Encyclopedia of Science and Technology.

[ELLIOTT D. BLOOM]

Bibliography: E. D. Bloom and G. J. Feldman, Quarkonium, *Sci. Amer.*, 246(5):66–77, 1982; E. D. Bloom and C. W. Peck, Physics with the crystal ball detector, *Annu. Rev. Nucl. Sci.*, 33:143–197, 1983; W. Chinowsky, Psionic matter, *Annu. Rev. Nucl. Sci.*, 27:393–464, 1977; N. B. Mistry, R. A. Poling, and E. H. Thorndike, Particles with naked beauty, *Sci. Amer.*, 249(1):106–115, 1983.

Quasars

On the night of March 25, 1982, a group of astronomers working in Australia obtained a spectrum of the optical counterpart of the radio source Pks 2000-330. From their spectrum analysis they concluded that the quasar identified with Pks 2000-330 had a redshift of 3.78. As the distance of a galaxy or quasar is related to its redshift, this measurement made Pks 2000-330 the most distant object known.

Until the discovery that Pks 2000-330 had a redshift of 3.78, no quasar had been found to have one greater than that of the quasar OQ172, with 3.53, measured in 1972. This apparent lack of high-redshift quasars has been interpreted as evidence for an edge to the visible universe formed by a cutoff in the space distribution of quasars. An alternative interpretation has been given in terms of selection effects that make the high-redshift quasars more difficult to identify than expected.

Identification of quasars. Quasistellar objects or quasars can be identified by their optical spectra, which show redshifted emission lines; by their ultraviolet colors which differ from the colors of field stars; and by the coincidence of a starlike object with the position of a radio source.

Photographic surveys for quasars with wide-field Schmidt telescopes find from about 100 to 200 quasars in a 6° × 6° area of the sky covered by a single photographic plate. These quasars are identified by their unusual ultraviolet color, or by low-dispersion spectra produced when the Schmidt telescope is used with an objective prism. Unfortunately, both of these survey techniques have redshift-dependent detection thresholds.

The strong emission lines in quasar spectra produce changes in the quasar colors that are redshift-dependent because the strong lines are seen at different wavelengths in the spectra of quasars at different redshifts. At redshifts greater than 2.5, the quasar's color does not distinguish it from field stars.

The redshift range of objective-prism spectra is limited to redshifts of less than 3.3 by the wavelength response of the photographic plate and by the lack of strong emission lines at wavelengths shorter than that of the Lyman-alpha line. Objective-prism surveys with red-sensitive plates have encountered difficulties with variations in the plate sensitivity with wavelength.

The optical survey work has shown that even in the limited redshift range available to the optical surveys, the radio quasars make up less than 10% of the quasars in a given region of sky. However, the radio quasars are not subject to the same redshift-dependent selection effects that beset the optical surveys, and therefore may offer the best possibility for correctly determining the true space distribution of quasars. Observing spectra of faint starlike objects coincident with radio positions is a time-consuming process that requires a large telescope and modern electronic detectors. Quasar identifications by this process require optical and radio positions with accuracies of 10^{-5} of a radian (2 arc-seconds) or better. Both OQ172 and Pks 2000-330 were identified by this method.

Distances of quasars. The relation between distance and redshift has been established for galaxies by observing that, with increasing redshift, the galaxies seen are progressively smaller and dimmer. This redshift-distance relation was predicted by the expanding-universe solution of Einstein's general theory of relativity at about the same time that the redshift-distance relation was first observed for galaxies. As the stars in a galaxy surrounding the nearby quasar 3C48 have been shown to have the same redshift as the quasar, it may be assumed that the source of the redshift seen in quasar spectra is also due to the expansion of the universe.

The distance of an object at high redshift, then, depends upon the definition of the term "distance" and upon three parameters in the pressure-free solution to Einstein's equation. The first parameter is related to the self-energy of empty space-time (the cosmological constant, usually taken to be identically zero). The second parameter is related to the radius of curvature of the universe. This can be evaluated in terms of the age of the universe or independently in terms of the expansion rate (the Hubble constant). The third parameter is related to the ratio of the kinetic energy in the expansion to the potential energy of the self-gravitation of the mass in the universe. This can be evaluated in terms of the average mass density in the universe, or independently in terms of the deceleration rate of the expansion.

For high-redshift objects, many different but equally valid definitions of "distance" are possible. For example, three valid distance measures may be defined as (1) the distance between the quasar and the Earth when the light received by the Earth was emitted by the quasar (where the quasar appears to be); (2) the distance between the quasar and the Earth when the light emitted by the quasar is received at the Earth (where the quasar is now); and (3) the distance traveled by the light from the quasar to the Earth. For nearby objects, these definitions are indistinguishable. However, because the universe expands while the light travels from the quasar to the Earth, the diferent distance measures of high-redshift objects yield different results.

The relation between distance and redshift is given by the equation below for the distance trav-

$$D = T[1 - 1/(1 + z)]$$

eled by light from the quasar to the Earth. Here, D is the distance in light-years (1 light-year = 9.46 $\times 10^{12}$ km = 5.88 $\times 10^{12}$ mi), T is the age of the universe in years, z is the redshift, and both the cosmological constant and the average mass density are taken to be zero. The distance of Pks 2000–330 as given by this equation with $T = 14.5 \times 10^9$ years is 11.5×10^9 light-years.

For background information *see* ASTRONOMICAL SPECTROSCOPY; COSMOLOGY; QUASARS; RADIO ASTRONOMY; RED SHIFT in the McGraw-Hill Encyclopedia of Science and Technology

[BRUCE A. PETERSON]

Bibliography: T. A. Boroson and J. B. Oke, Detection of the underlying galaxy in the QSO 3C48, *Nature*, 296:397–399, 1982; P. S. Osmer, Quasars as probes of the distant and early universe, *Sci. Amer.*, 246(2):126–138, February 1982; B. A. Peterson et al., PKS 2000–330, A quasi-stellar radio source with a redshift of 3.78, *Astrophys. J. Lett.*, 260:L27–L29, 1982.

Radiation damage to materials

Recent advances in research on the effects of ionizing radiation on integrated circuits include identification of the charge-trapping mechanisms at the interface of silicon and silicon dioxide in MOS integrated circuits, the development of techniques to inhibit the "latch up" phenomenon and to harden circuits to single-particle upset, and the fabrication of hardened CMOS integrated circuits.

Hole trapping and interface phenomena. Electron spin resonance measurements indicate that the well-known phenomenon of oxide hole trapping in the vicinity of a silicon–silicon dioxide (Si–SiO_2) interface occurs at oxygen vacancies. Furthermore, the density of hole traps is highest at the interface where the density of oxygen vacancies is known to be highest. Electron spin resonance measurements have also identified a strong correlation between the concentration of P_b centers (a silicon bonded to three other silicons with a "dangling bond" pointing out into a vacancy) and radiation-induced interface states. At the interface of silicon and silicon dioxide, a lattice or bond link mismatch must occur between the amorphous silicon dioxide and the crystalline silicon. This results in trivalent silicon atoms with an unpaired bond.

Electron spin resonance measurements confirm that these interface states may be unoccupied, occupied by a single electron, or occupied by two electrons. The occupancy state depends upon the oxide bias condition, because this controls the position of the Fermi level relative to the valence and conduction bands in the silicon. Experiments have determined that the interface-state density, like the density of trapped holes, increases with radiation dose. The kinetics of electron occupancy are not yet well understood.

The importance of oxide hole trapping and interface-state generation by radiation exposure has been well known in MOS integrated circuits for several years. Gate threshold voltage shifts illustrated in Fig. 1 leading to reductions in circuit speed and fanout capability in digital circuits are known consequences of exposure to ionizing radiation. At low

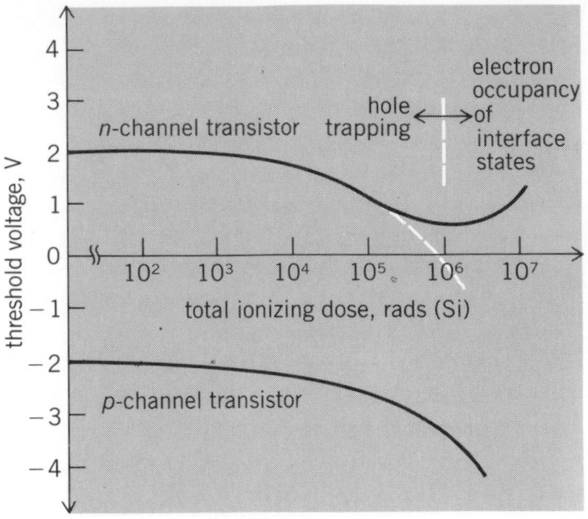

Fig. 1. Effect of radiation-induced charge buildup and interface-state production of the threshold voltage in irradiated *n*- and *p*-channel MOS transistors. 1 rad = 10^{-2} Gy.

dose levels the gate threshold shifts in *n*-channel devices are dominated by hole trapping in silicon dioxide; at high dose levels electron occupancy of interface states dominates the gate threshold shift. Therefore, at high dose exposures the gate threshold voltage of the *p*-channel and the *n*-channel may exceed the power supply voltage, whereas at medium dose exposures the *n*-channel may be driven to a depletion state where it is conducting without a voltage applied to its gate.

Although these same basic charge-trapping phenomena also occur at the silicon–silicon dioxide interface in bipolar integrated circuits, they have not generally been observed to be important at radiation exposure levels less than 10^5 rads (10^3 grays). Recently, however, bipolar integrated circuits have employed so-called recessed sidewall oxides because of their utility in reducing feature size. Increased circuit speed and higher transistor density on a chip are the benefits. These sidewall isolation regions interface with active regions of transistors while providing electrical isolation between neighboring transistors. Oxide hole trapping and silicon–silicon dioxide interface-state occupancy by electrons can produce inversions in the transistor base regions and lead to collector-to-emitter channeling along the base–oxide interface. Collector-to-collector leakage between neighboring transistors can also occur. Recent experiments have shown that bipolar integrated circuits employing recessed sidewall isolation can fail at total dose exposures less than 10^4 rads (10^2 Gy). These observations are especially significant because recessed sidewall oxides are a current design trend in bipolar integrated circuits.

Latch-up phenomena. Photocurrent generated by radiation can provide a trigger signal for parasitic *pnpn* circuits in bipolar and MOS integrated circuits. The electrical behavior of these circuits is similar to that of silicon controlled rectifiers; therefore, a low-impedance path can be triggered between the power supply and ground. Permanent damage or operational failure can occur in an integrated circuit if it stays in this "latched" state.

Experiments have shown that because of neighboring circuit interactions and conductivity modulation, radiation-induced "latch-up windows" may exist in CMOS integrated circuits. These are narrow regions of dose rate where latch-up does not occur. Models have been developed that show latch-up windows to be a special case of the classical parasitic *pnpn* latch-up phenomena.

Neutron irradiation of CMOS integrated circuits can be employed at the wafer level to inhibit latch-up. Atoms displaced from their normal lattice sites by neutrons act as recombination and trapping centers and severely reduce minority carrier lifetime. Thus, the gains of the parasitic transistors that form the *pnpn* path are reduced to the extent that the low-impedance, regenerative state of the circuit is not reached.

Experiments have demonstrated that when the feature size of CMOS integrated circuits approaches 3 micrometers or if subsequent high-temperature exposure anneals out the defects, techniques other than neutron irradiation should be employed to inhibit latch-up. An alternative is the fabrication of CMOS integrated circuits on expitaxial substrates to produce low-impedance paths that remove current from the bases of the parasitic transistors. Therefore, the parasitic silicon controlled rectifier is not permitted to reach a latched state.

Single-event phenomena. As the density of integrated circuits has gone from MSI (medium-scale integration) to VLSI (very-large-scale integration) and VHSIC (very-high-speed integrated circuit), not only has transistor size decreased, but the energy required to switch a bit on a digital gate has decreased to only 0.5 picojoule. In these circuits, a single radiation particle interacting in the appropriate region of a silicon chip can generate transient electrical signals which disrupt the operation of the circuit (single-particle upset). A heavy-ion track through the depletion region of a junction creates an ionization funnel outside the depletion region and effectively extends the junction depletion region to include the funnel (Fig. 2). Models of this phenomenon and supporting experiments demonstrate that charge generated within this funnel extension of the depletion region is collected promptly, and that charge generated within a diffusion length of the funnel diffuses to the funnel so that it is collected at later times.

In addition to advances in modeling, there have been advances in hardening integrated circuits to single-particle upset. Feedback resistors located at the output of a gate act as low-pass filters and permit CMOS integrated circuits to be hardened to upset produced by ions whose atomic weight is less than krypton. This hardening is done with negligible degradation in circuit operation.

Hardened circuits. An 8-bit microprocessor, a 2-kilobit RAM, and a 16-kilobit ROM emulating the Intel 8085, 8155, and 8355, respectively, have

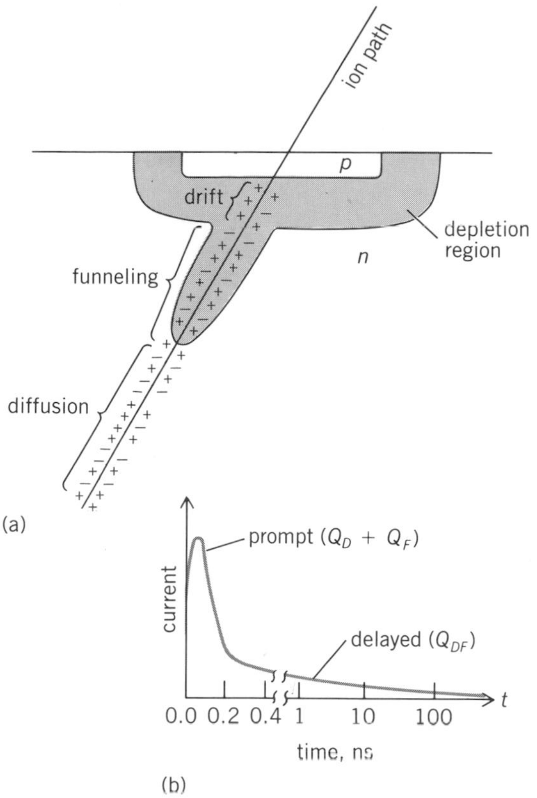

Fig. 2. Production of current pulse by a single heavy ion at a semiconductor junction. (*a*) Path of heavy ion through the junction, extending the depletion region to include a funnel. Portions of ion track giving rise to a prompt drift component Q_D, a prompt funneling component Q_F, and a delayed diffusion component Q_{DF} of current pulse are shown. (*b*) Current pulse.

been designed and built in radiation-hardened CMOS. A radiation-hardened nonvolatile 128 × 8 bit MNOS RAM with CMOS peripheral circuits and a 16-kilobyte MNOS EAROM (electrically alterable read-only memory) have been designed, fabricated, and radiation-tested. Other recent technology advances include a radiation-hardened CMOS 8-bit analog-to-digital converter and a custom CMOS large-scale integrated circuit capable of encrypting a serial data stream in accordance with the NBS Data Encryption Standard.

For background information *see* ELECTRON PARAMAGNETIC RESONANCE (EPR) SPECTROSCOPY; INTEGRATED CIRCUITS; RADIATION DAMAGE TO MATERIALS; SEMICONDUCTOR MEMORIES in the McGraw-Hill Encyclopedia of Science and Technology.

[J. A. HOOD; J. E. GOVER]

Bibliography: *IEEE Transactions on Nuclear Science*, December 1983; *Proceedings of 1983 IEEE Nuclear and Space Radiation Effects Conference*, Gatlinburg, Tennessee, July 18, 1983.

Radioactive waste management

Radioactive wastes originate almost exclusively in the nuclear fuel cycle and in the nuclear weapons program. These wastes are classified in four major categories: high-level waste (HLW), transuranic (TRU) waste, low-level waste (LLW), and uranium mill tailings. Minor waste categories, like contaminated uranium mine water or radioactive gases produced during reactor operation, will not be discussed.

The radioactivity of these wastes is commonly measured in curies (Ci). The curie, chosen to approximate the activity of 1 gram of radium-226, is equal to 3.7×10^{10} becquerels. The becquerel (Bq), the SI unit of activity (radioactive disintegration rate), is the activity of a radionuclide decaying at the rate of one spontaneous nuclear transition per second. Since the curie is a rather large unit of activity, the nanocurie (1 nCi = 10^{-9} Ci) and picocurie (1 pCi = 10^{-12} Ci) are frequently used. A common unit of nuclear generating capacity is the gigawatt (electric) [GW(e)], equal to 10^9 watts of electric power, as opposed to thermal power.

HLW includes spent fuel and reprocessing wastes. When spent fuel is reprocessed to extract plutonium and the unfissioned uranium, the resulting HLW contains most of the fission products and transuranic elements, including the residual plutonium. TRU waste, arising mainly during reprocessing and in the weapons program, is solid material contaminated to greater than 10 nCi/g (3.7×10^5 Bq/kg) with certain alpha-emitting radionuclides. Uranium mill tailings are the residues of the chemical extraction of uranium from the ore. Finally, LLW is a very broad category of wastes, covering almost every form of radioactive waste not falling into the other categories. Figures 1 and 2 show a

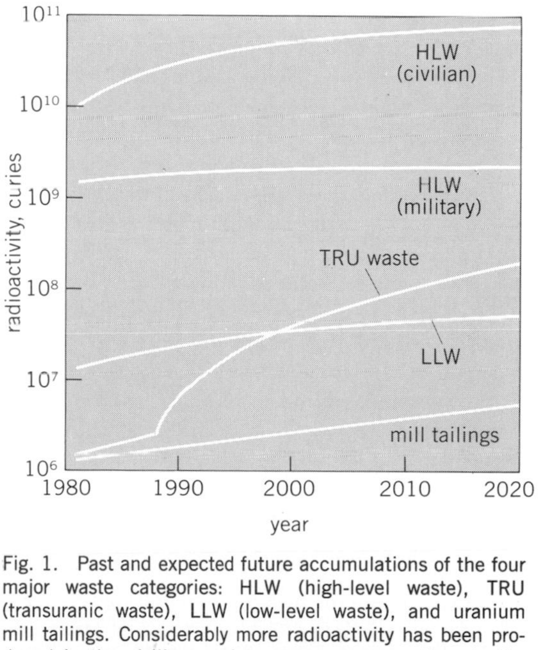

Fig. 1. Past and expected future accumulations of the four major waste categories: HLW (high-level waste), TRU (transuranic waste), LLW (low-level waste), and uranium mill tailings. Considerably more radioactivity has been produced in the civilian nuclear energy program than in the weapons program. The radioactivity in the mill tailings is the sum of the radioactivity of all uranium daughters. (*After U.S. Department of Energy, Spent Fuel and Radioactive Waste Inventories, Projections, and Characteristics, Rep. DOE/NE-0017-1, October 1982*)

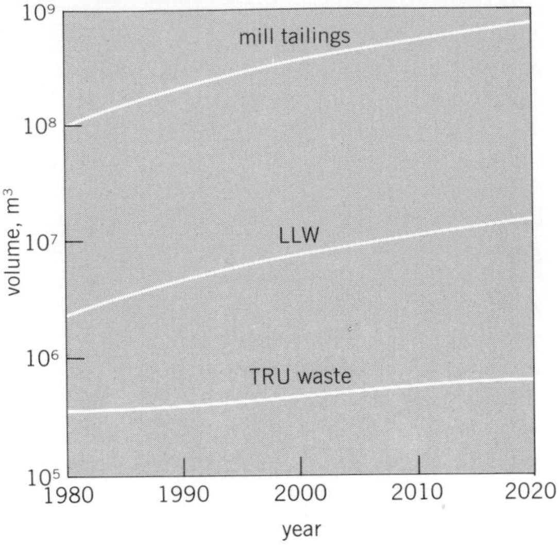

Fig. 2. Volume of the accumulated wastes shown in Fig. 1, except for that of the HLW, which is very small (< 3.5 × 10^5 ft^3 or 10^4 m^3).

comparison of the radioactivity and the volume of the different waste categories, based on a recent Department of Energy forecast, which assumes an installed civilian nuclear generating capacity of 167.4 GW(e) in the year 2000, and of 280.7 GW(e) in 2020 [1981: 54.9 GW(e)].

Since no practical methods exist to detoxify radioactive nuclides, protection against their harmful radiation must rely on their isolation from the biosphere until their radioactivity has decayed. Because each of the waste categories poses different problems, they will be discussed separately.

High-level waste. Most of the radioactivity is contained in the HLW (Fig. 1). For the first 1000 years, its toxicity is dominated by the beta- and gamma-emitting fission products (with half-lives up to 30 years); thereafter, the long-lived, alpha-emitting transuranium elements and their radioactive decay daughters are more important. Burial in geologic formations at a depth of 1600–3200 ft (500–1000 m) appears at present to be the most practical and attractive disposal method, although as of late 1983 no HLW had been disposed of in this way anywhere in the world. Most of the commercial HLW is stored as spent fuel in water-cooled basins, while the military reprocessed HLW is stored in tanks in liquid or solid form.

Geology as a predictive science is still in its infancy, and many of the parameters entering into model calculations of the long-term retention of the waste in the geologic media have to be questioned. The major single problem is the heating of the waste and its surrounding rock by the radioactive decay heat. It can accelerate the penetration of groundwater into the repository, the dissolution of the waste, and its transport to the biosphere. Much effort has been devoted to the development of improved waste forms and overpacks which promise better resistance to attack by groundwater. The suitability of rock salt, which for over 20 years had been widely considered as the preferred host rock because of its superior heat-dissipative properties and because of its plasticity (which would help to seal cracks), is now being questioned because of several shortcomings. The major drawback is that rock salt is a natural resource which has attracted a wide number of uses. It is now recognized that drilling into or through a salt formation housing a nuclear waste repository could happen. The consequences of such drilling activities add so much to the complexity of predicting the safe containment of the nuclear waste that they seem to outweigh the advantages salt might have over other rocks which are not natural resources and not water-soluble.

In the United States, suitability of one HLW disposal site in basalt, one in tuff, and several in rock salt are being studied. Selection of the repository site is not expected before 1990. The first repository for HLW will be operational after the year 2000. With a nuclear generating capacity of 300 GW(e), a new HLW repository would be needed thereafter approximately every 10 years.

Transuranic waste. Although the radioactivity of the TRU waste is considerably smaller than that of the HLW, this difference is partly offset by its higher specific radiotoxicity and its longer half-life. Consequently, disposal in a geologic repository is also envisioned for this waste category.

The rapid increase of the TRU waste inventory expected by the Department of Energy (Fig. 1) is explained by the reprocessing of spent fuel in the commercial energy program expected to begin in 1989. Of the 4400 lb (2000 kg) of TRU elements in this waste accumulated as of 1981, an estimated 2400 lb (1100 kg) was taken to Department of Energy and commercial shallow land burial sites, where the TRU elements were mixed in with LLW, and only 2000 lb (900 kg) is in retrievable storage. In addition to these TRU wastes, several million cubic meters of earth contaminated with TRU elements at concentrations exceeding 10 nCi/g (3.7 × 10^5 Bq/kg), containing several hundred kilograms of transuranic elements, is located at several Department of Energy installations. Although this earth, strictly speaking, is also TRU waste, it is not included in Figs. 1 and 2.

Present plans call for disposal of defense TRU waste in a geologic repository in bedded salt [Waste Isolation Pilot Project (WIPP), Carlsbad, New Mexico] to begin in 1989. By year 2014, 5,650,000 ft^3 (160,000 m^3) of TRU waste will have been buried there, out of a total defense and commercial accumulated volume of 22,740,000 ft^3 (644,000 m^3). These numbers show that even with the existence of one geologic repository dedicated to TRU waste disposal, waste production is expected to outpace the disposal.

Uranium mill tailings. Uranium is naturally radioactive, decaying in 14 steps to stable lead. It is a rare element, averaging 0.1–0.2% in the currently mined ore. At the mill, the rock is crushed to fine sand and the uranium is chemically extracted. The residues, several hundred thousand cubic meters for the annual fuel requirements of a 1-GW(e) reactor, are discharged to the tailings pile (Fig. 3). The tailings contain the radioactive daughters of the uranium. The long-lived isotope thorium-230 (half-life 8×10^4 years) decays into radium-226 (half-life 1.6×10^3 years), which in turn decays to radon-222 (half-life 3.8 days). Radium and radon are known to cause cancer, the former by ingestion, the latter by inhalation. Radon is an inert gas and thus can diffuse out of the mill tailings pile and into the air. It has been estimated by the Environmental Protection Agency that a person living 1600 ft (500 m) away from an unprotected tailings pile containing 280 pCi/g (1.036×10^4 Bq/kg) of ^{226}Ra would have a 30% higher chance of lung cancer from the radon gas it produces than the average person (the average radium concentration in the existing tailings piles is twice as high, and hence for these piles the estimate should be doubled). Pollution of the groundwater by radium that leached from the pile has also been observed around tailings piles, but its health effects are more difficult to estimate, since the migration in the groundwater is difficult to assess and also highly site-specific. New mill tailings piles will be built with liners to protect the groundwater, and will be covered with earth and rock to reduce atmospheric release of the radon gas. None of these measures provides protection on the time scales required for the ^{230}Th to decay. Permanent disposal methods, like chemical extraction of the thorium for disposal in a geologic repository, or burial of the tailings in deep mines, are not planned at this time.

The radioactivity contained in the mill tailings is very small relative to that of the other wastes (Fig. 1). It should be noted, however, that the total mass of ^{230}Th is, incidentally, twice as large as that of all TRU elements in the TRU waste; accumulations by 1981 were 8800 lb (4000 kg), with an activity of 80,000 Ci (3.0×10^{15} Bq) of the ^{230}Th versus 4400 lb (2000 kg) of TRU elements. It is mainly the dilution of the thorium and its daughters in the large volume of the mill tailings (Fig. 2) which reduces the health risks to individuals relative to those posed by the TRU elements in the TRU wastes. However, this advantage is partly offset by the great mobility of the chemically inert radon gas which emanates into the atmosphere from the tailings. Because of the enormous volume of the tailings, a permanent disposal—should this be considered necessary—would present serious problems.

Low-level wastes. By definition, practically everything that does not belong to one of the three categories mentioned above is considered to be low-level waste. This name is misleading because some wastes, though low in TRU content, may contain very high beta-gamma activity. For example, ion-exchange resins or activated components from reactors, which will have radioactivities measured in hundreds of Ci/m^3, may require biological shielding during handling and transport. Of the approximately 5,300,000 ft^3 (150,000 m^3) of LLW generated each year, about one-half originates in the weapons program, one-quarter in the commercial nuclear fuel cycle, and one-quarter in industry and in various institutions (hospitals, universities). While the volume originating at these various institutions is large (700,000 ft^3/year or 20,000 m^3/year), the radioactivity contained in these wastes is very small, less than 1% of the total radioactivity contained in the LLW, and is also mostly short-lived.

Fig. 3. Partial view of an operating mill tailings pile (Homestake Mining Company, Milan, New Mexico). The pile measures approximately 3000 × 3000 ft (900 × 900 m), 50 ft (15 m) high, and is entirely unprotected. In its 6×10^8 ft^3 (1.7×10^7 m^3), it contains 7000 Ci (2.6×10^{14} Bq) ^{226}Ra. The uranium extracted from this pile would have been enough to generate 90 GW(e)-year of electric energy in a light-water reactor.

The current method of LLW disposal is shallow land burial. [Prior to 1970, a total of 10^5 Ci (3.7×10^{15} Bq) of LLW generated in the United States weapons program was also disposed of by ocean dumping; up to 1980, 6×10^5 Ci (2.2×10^{16} Bq) has also been injected with grout into hydrofractured shale formations underlying the Oak Ridge National Laboratory.] At present there are five major Department of Energy shallow land burial sites and six commercial sites (only three of which are now operating). While sites in arid environments have generally performed in an acceptable manner, those located in humid environments have commonly not performed as hoped, and the fact that three of the four commercial sites in the east are no longer operating illustrates the problems they have encountered.

Besides the very large volume, the nonuniformity presents formidable problems for the LLW disposal. Current programs focus on improved waste forms, on better site-selection criteria, and on engineering improvements to the site to restrict releases. It appears inevitable that wastes containing activity of very long half-life, as they arise, for example, during the decomissioning of fuel cycle facilities, will also have to be disposed of in a geologic repository, although this would increase the disposal costs considerably.

For background information *see* NUCLEAR FUEL CYCLE; RADIOACTIVE WASTE MANAGEMENT; RADIOACTIVITY in the McGraw-Hill Encyclopedia of Science and Technology.

[ROBERT O. POHL]

Bibliography: B. W. Burton et al., *Overview Assessment of Nuclear Waste Management*, Los Alamos, Sci. Lab. Rep. LA-9395-MS, August 1982; U.S. Department of Energy, *Spent Fuel and Radioactive Waste Inventories, Projections, and Characteristics*, Rep. DOE/NE-0017-1, October 1982; U.S. Environmental Protection Agency, *Draft Environmental Impact Statement for Standards for the Control of By-product Materials from Uranium Ore Processing* (40 CFR 192), Rep. EPA 520/1-82-022, March 1983.

Reef

The term coral reef often evokes an image of luxuriant, sunlit structures formed in warm, tropical seas. Yet, scleractinian corals are known from water depths as great as 19,700 ft (6000 m) and temperatures as low as 30°F (−1°C). Other forms of coral reef occur in deep water over extensive areas of the North Atlantic, and have also been discovered in the South Atlantic and Pacific oceans, the Mediterranean Sea, and the Gulf of Mexico. The deep-water coral reefs discussed here form below the photic zone (where sunlight penetration can sustain photosynthesis), in water depths greater than 650 ft (200 m).

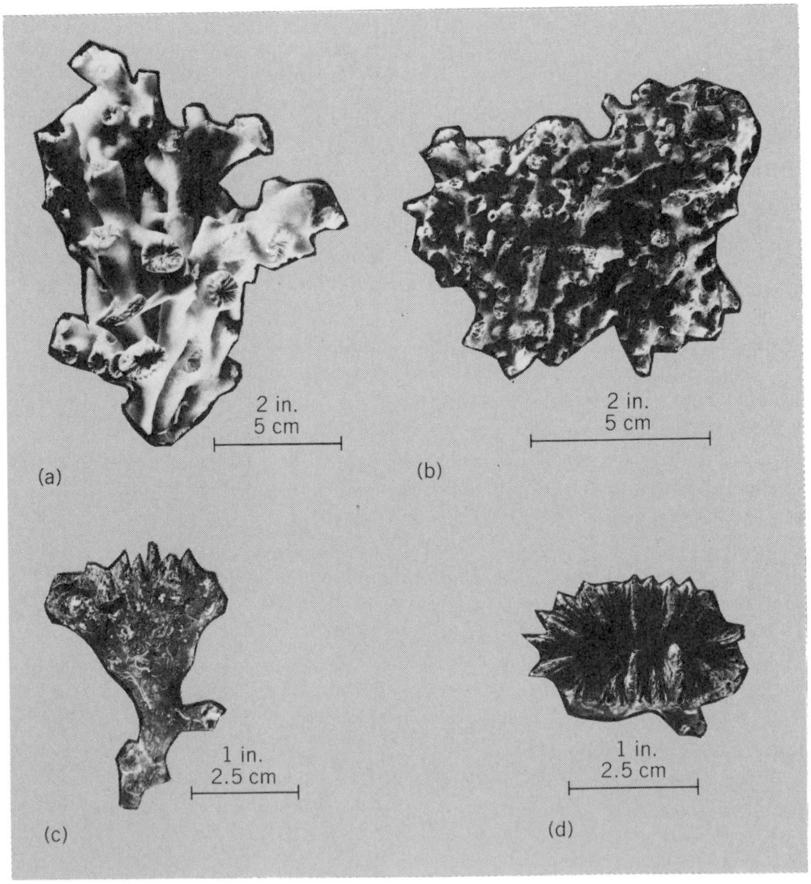

Fig 1. Deep water corals: (*a*) *Lophelia prolifera*. (*b*) *Solenosmilia variabilis*. (*c*) *Desmophyllum cristagalli*, external view and (*d*) internal view. (*Courtesy of C. R. Newton and H. T. Mullins*)

Ecology of reef-building corals. Although both deep- and shallow-water reef-building corals belong to the order Scleractinia (phylum Coelenterata, class Anthozoa), important ecological distinctions exist between the two groups. Most of the shallow-water reef-builders are hermatypes, that is, corals that contain microscopic symbiotic algae (zooxanthellae) in their gastrodermal tissues. The presence of these zooxanthellae enhances the rate at which hermatypes can secrete their skeletons, thus enabling them to form massive, laterally extensive structures, such as barrier reefs. In contrast, all deep-water corals (and some shallow-water corals) are ahermatypes, which lack symbiotic algae, and can grow only approximately one-fifth as rapidly as their hermatypic counterparts. In many shallow-water reefs, therefore, ahermatypes occupy less volume and are relegated to cryptic habitats where levels of solar illumination are insufficient to support the rapid growth of hermatypic corals. The reef-building capability of ahermatypes is thus greatest in deep waters.

Depth of occurrence. Deep-water coral reefs are presently known from depths of 650–4260 ft (200–1300 m) and are apparently most common between 1300 and 2300 ft (400 and 700 m). The deepest coral structures reported to date occur between 3300 and 4260 ft (1000 and 1300 m) on the lower continental slope north of Little Bahama Bank. As research continues, coral reefs may be found at even greater depths.

Because most reef-building ahermatypes are stenothermal, that is, they can exist only within a limited temperature range, the depth of deep-water coral reefs varies according to water temperature in a given region. The common deep-water coral *Lophelia prolifera* (Fig. 1*a*), for example, has a temperature range of 37–54°F (3–12°C). It has been reported that coral reefs constructed by *Lophelia* in Norwegian fiords are most common in 43–47°F (6 –8°C) waters at depths of only 660–990 ft (200–300 m). In contrast, in the Straits of Florida the greatest abundance of *Lophelia* mounds is between 1960 and 2300 ft (600 and 700 m), at comparable water temperatures of 42–47°F (5–8°C).

Geographic distribution. Deep-water coral reefs are patchily distributed in the North Atlantic on continental slopes and oceanic plateaus that rim the ocean basin. Their distribution shows a strong correlation with the position of major oceanic current systems (Fig. 2). The positive effect of currents on coral reef development is threefold. Strong currents apparently enhance rates of submarine lithification, and also tend to sweep the ocean floor clean of mud, so that juvenile scleractinians can find the hard substrates that are necessary for their settlement. Currents also supply necessary oxygen to bot-

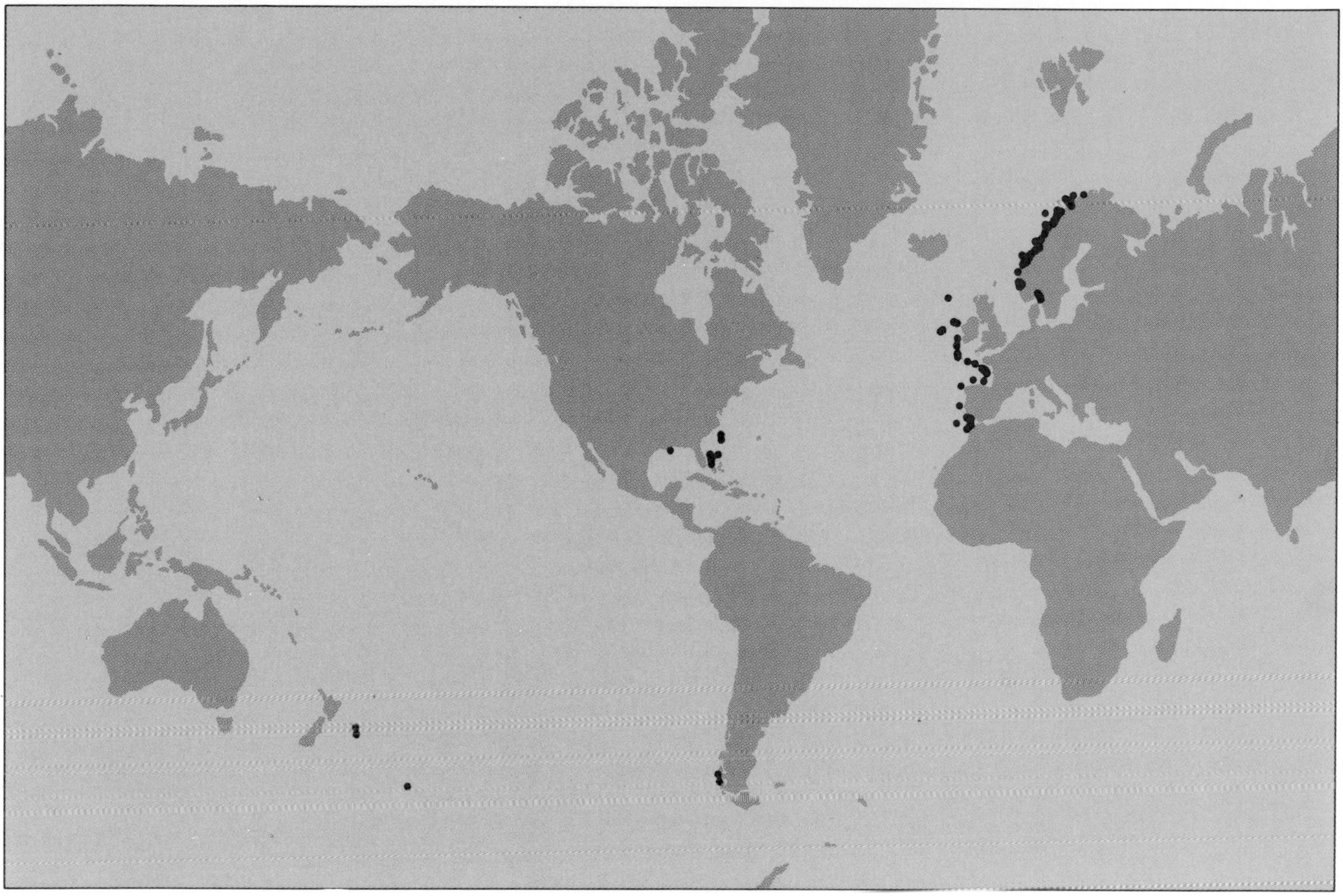

Fig. 2. Distribution of presently known deep-water coral reefs. (*Courtesy of C. R. Newton and H. T. Mullins*)

tom waters in the vicinity of deep-water reefs. In addition, oceanic currents carry in suspension organic detritus and pelagic invertebrates that probably serve as food for the deep-water ahermatypes.

Types of deep-water coral reefs. Three major types of deep-water coral reefs can be recognized, based on the occurence of *Lophelia prolifera*, *Solenosmilia variabilis*, or *Desmophyllum cristagalli* as the primary reef builders. The majority of North Atlantic deep-water coral structures are formed by the loosely branched, ahermatype *Lophelia prolifera* (Fig. 1). The complexly intertwined branches of *Lophelia* are strengthened by additional skeletal aragonite ($CaCO_3$, secreted to form the coral skeleton) and worm tubes, so that a dense coral framework is formed. Reefs of *Lophelia* are found in the Norwegian fiords, on Rockall Bank northwest of Great Britain, along the western coasts of France and Spain, off the coast of Georgia and the Carolinas, in the Florida-Bahamas region, and in the Gulf of Mexico. The *Lophelia* community typically has a low species richness of associated corals; for example, off western Europe, the only other common *Lophelia*-associated scleractinians are *Madrepora oculata* and *Dendrophyllia cornigera*.

A special case of *Lophelia*-constructed reefs is found in the Straits of Florida, where elongate coral mounds called lithoherms have formed through the combined action of colonial coral growth and submarine cementation by magnesian calcite. Coral colonies of *Lophelia* and *Enallopsammia* grow on the southern, upcurrent end of lithoherms (Fig. 3) and form a baffle for sediment, causing the mounds to accrete in an upcurrent direction. The medial and downcurrent (northern) portions of the lithoherms are mantled with concentric crusts of submarine-cemented limestone that bear a fauna of fan-shaped octocorals (class Anthozoa, order Octocorallia) and rows of crinoids (phylum Echinodermata). The lithoherms thus exhibit lateral faunal zonation, with scleractinian corals occupying the accreting, upcurrent end; octocorals attached to cemented parts of the midsection of the mounds; and crinoids on the downcurrent side, which is presently undergoing biological erosion by boring sponges.

The branched coral *Solenosmilia variabilis* (Fig. 1) is the principal frame builder of the second major type of deep-water reef community, exemplified by coral structures at 3280–4260 ft (1000–1300 m) on the lower continental slope north of Little Bahama Bank. Sixteen species of scleractinians occur in this area—the highest species richness of ahermatypes reported from any region of deep-water coral reefs. In contrast to the lithoherms, the reefs north of little Bahama Bank reveal no evidence of lithification, although initial colonies of *S. variabilis* probably col-

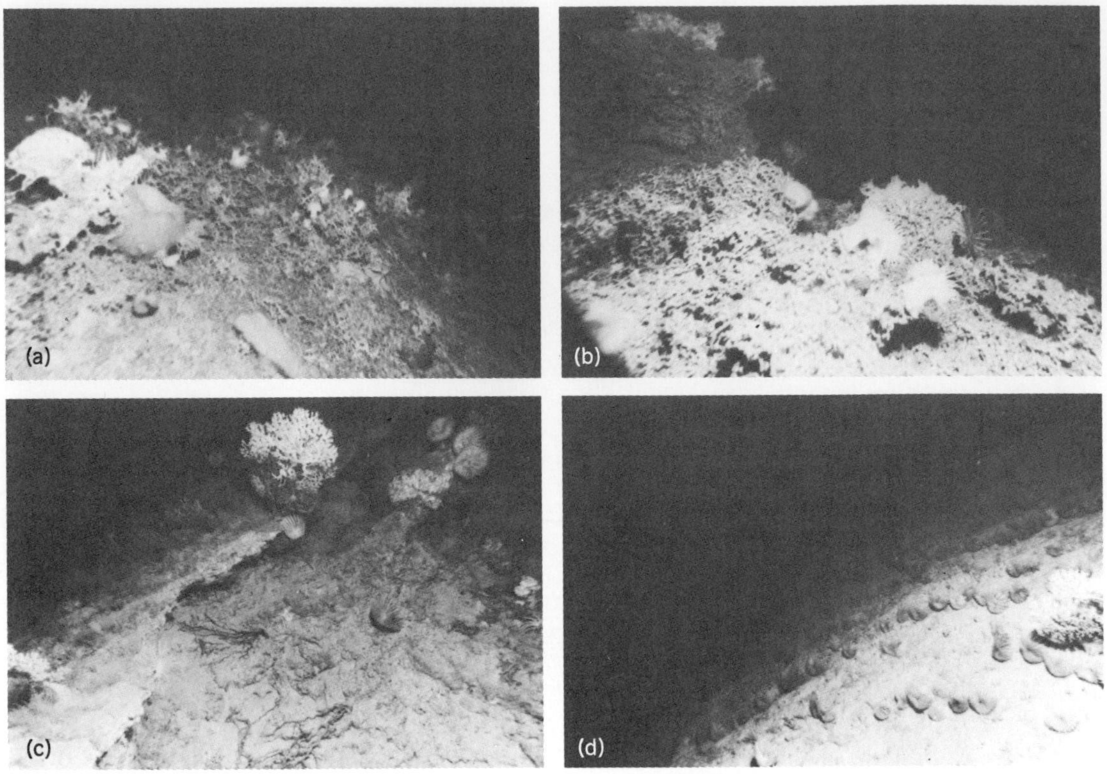

Fig. 3. Underwater photographs of lithoherms taken by the deep-sea research vessel *Alvin* at a depth of 1980–2310 ft (600–700 m) in the northern straits of Florida. (*a*) Accreting, upcurrent end of lithoherm, showing branched colonies of *Lophelia*, with sponge in left foreground. (*b*) Accreting portion of lithoherm, showing dense thicket of *Lophelia*, sea anemones, and crinoid. (*c*) Midsection of lithoherm, with octocorals attached to submarine-cemented ledges that have been undercut by submarine erosion. A few crinoids are visible on the right. (*d*) Downcurrent portion of lithoherm, showing rows of crinoids. (*Courtesy of A. C. Neumann*)

onized local hard substrates. Later colonies apparently built out over sandy areas and also served as baffles for additional sediment. Most of the ahermatypes on these reefs are solitary species that dwell in the mud-filled interstices of the colonial coral framework. The *Solenosmilia* community seems to be rarer than the *Lophelia*-dominated assemblage, as only north of Little Bahama Bank and in Chilean fiords have reefs constructed by *S. variabilis* been reported.

The pseudocolonial scleractinian *Desmophyllum cristagalli* has recently been found by S. D. Cairns and G. D. Stanley to be the primary reef builder in deep-water coral reefs in the Chilean fiords; however, very little is currently known about these *Desmophyllum*-dominated coral structures.

Associated organisms. Deep-water coral structures provide microhabitats for numerous reef-associated invertebrate species. In addition to octocorals and crinoids, cup- and fan-shaped sponges and starfish are common invertebrates on the lithoherms; however, faunal investigation of these features is not yet complete. On the unlithified coral mounds north of Little Bahama Bank, dredge hauls have yielded a host of reef-associated invertebrates, including sponges, octocorals, antipatharians, hydroids, polychaetes, bivalves, gastropods, galatheid crabs, barnacles, crinoids, and ophiuroids. The, biota of the *Lophelia* reefs and thickets of Norway has been estimated by C. Teichert to include 190 species, of which only 32 are coelenterates.

Fish are also common in deep-water coral reef ecosystems; in the Norweigian *Lophelia* communities, at least four fish species are present. Indirect evidence that the reefs north of Little Bahama Bank also support large populations of fish comes from otoliths (fish ear bones) that are common calcareous constituents of both reef and interreef sediments. Several fish species have also been observed during submersible dives on the lithoherms.

Development of deep-water coral reefs. Developmental modes and rates of accretion are not as readily observed in deep-water coral reefs as in their shallow-water counterparts; thus, ontogeny of the deep-water reefs must be reconstructed through dredged samples, submersible observations, and careful scrutiny of the few available ancient examples. After study of both modern and ancient deep-water scleractinians, D. F. Squires postulated four phases of deep-water reef development could be distinguished. The first stage involves a single colony of ahermatypes which initially colonize a local hard substrate but may later grow outward over unconsolidated sediment. Continued coral growth results in the lateral coalescence of neighboring colonies, forming a coral thicket whose crevices and interstices shelter a variety of other invertebrates. In the third, coppice, phase, fragments of dead coral colonies fall into cavities within the coral framework and also form a skirt of talus and sediment around the mound, providing an additional habitat for reef-associated invertebrates. The final phase of deep-water reef development, termed the bank stage, evolves through continued accumulation of coral talus around living ahermatype colonies. Banks are structures of relatively high vertical relief (up to 164 ft or 50 m) in which the volume of living coral is small in proportion to the surrounding apron of debris. Rates of growth of corals and of accretion of reefs in deep water remain unknown.

Future investigations. The wealth of deep-water coral reefs that have been reported from the Atlantic and other oceans underscores their biological significance as a comparison with shallow-water reef structures. Still, little is known about the factors that govern deep-water reef development, and about the accretion rates that are typical of *Lophelia-*, *Solenosmilia-*, and *Desmophyllum*-dominated coral reefs. Another question is the extent to which ecological succession is important in the history and development of deep-water coral reefs. Submersible studies will likely prove the most fruitful exploration tool for future ecological and geological investigations of deep-water reefs.

Deep-water coral structures are also of paleoecological importance: the frequency of modern deep-water coral reefs suggests that they may be more common in the fossil record than has previously been recognized. Fossil examples of deep-water scleractinian reefs have been reported from the Triassic rocks of Europe and western North America; the Cretaceous strata of New Mexico; the Tertiary rocks of Greenland, Denmark, and New Zealand; and the Pleistocene sediments of Norway. It is probable that additional fossil deep-water coral reefs will be discovered.

For background information *see* COELENTERATA; DEEP-SEA FAUNA; ECOSYSTEM; MARINE ECOSYSTEM; REEF in the McGraw-Hill Encyclopedia of Science and Technology.

[CATHRYN R. NEWTON; HENRY T. MULLINS; A. CONRAD NEUMANN]

Bibliography: S. D. Cairns and G. D. Stanley, Ahermatypic coral banks: Living and fossil counterparts, *Proceedings of the 4th International Symposium on Coral Reefs*, vol. 1, pp. 611–618, 1981; H. T. Mullins et al., Modern deep-water coral mounds north of Little Bahama Bank: Criteria for recognition of deep-water coral bioherms in the rock record, *J. Sediment. Petrol.*, 51:999–1013, 1981; A. C. Neumann, J. W. Kofoed, and G. H. Keller, Lithoherms in the Straits of Florida, *Geology*, 5:4–10, 1977; C. Teichert, Cold- and deep-water coral banks, *Amer. Ass. Petrol. Geol.*, 42:1064–1082, 1958.

Reprographics

Reprographics is the generation of graphic images (characters of various alphabets, line drawings, photographs) by using one or several of a broad spectrum of mechanical, electronic, and electromechanical devices. With the advent of computers in the field of graphics, a number of disciplines that were formerly separate, such as calligraphy, orthography, photography, lithography, planography, telegraphy, telephony, radiotelegraphy, television, and the skills of art and drafting, have been linked through the medium of electronic communications between people and machines to form an interactive communications network. This network can perform and control the cogitative and dexterous tasks and skills involved in the reproduction of text and graphic images.

Reprography has evolved over the years through the desire and need to document and distribute information. Mass distribution of information began with letterpress printing, attributed to J. Gutenberg (about 1450–1455), and progressed to offset printing (about 1796), spirit and stencil duplication (about 1900), and various forms of copying (about 1938). Sophisticated electronic techniques hastened the evolution of reprographic techniques in the 1970s; the 1980s–1990s mark the advent for possibly revolutionary changes in reprography.

ASCII. The telegraph was the first machine used for electronic communication over distance. The early telegraph used a key to interrupt an electric current flowing across a wire (line) to create the "0" state or "SPACE" condition and the "1" state or "MARK" condition, which is the idle (current-flowing) state. These states were then employed in various combinations to create the Morse code. This communications code permitted operators at either end of the telegraph line to send and receive messages. The SPACE (0)–MARK (1) states in various combinations represented the letters of the alphabet. Today, the same two states are used to represent the elements that make up the binary codes utilized by current electronic computer systems. That is, the characters 0 and 1 are binary digits, called bits, that are used to form communications codes, such as the American Standard Code for Information Interchange (ASCII), shown in the table. This is a seven-bit-plus-parity code (referred to as an eight-bit code when sent with parity) established by the American National Standards Institute to achieve compatibility between data services. This code can also be presented in octal digits (0, 1, 2, 3, 4, 5, 6, and 7), a notation using the base of eight. ASCII then has become the means for people-machine and machine-machine interactions. However, the ability of machines to understand each other has been impaired by the fact that different brands, although speaking ASCII, have been designed with different control procedures (protocols) by which they set the timing and arrange the data being communicated. That is, each manufacturer has devised protocols that govern the format and relative timing of message exchange between two communicating processes. These protocols in turn can be divided into three categories: character-oriented; byte count–oriented; and bit-oriented. Thus, manufacturers, through the proliferation of control procedures, have complicated communication processes, although,

Seven-unit codes, ACSII-1968

Binary	Octal			Binary	Octal		
0000000	000	NUL	(Blank)	1000000	100	@	
0000001	001	SOH	(Start of Header)	1000001	101	A	
0000010	002	STX	(Start of Text)	1000010	102	B	
0000011	003	ETX	(End of Text)	1000011	103	C	
0000100	004	EOT	(End of Transmission)	1000100	104	D	
0000101	005	ENQ	(Enquiry)	1000101	105	E	
0000110	006	ACK	(Acknowledge (Positive))	1000110	106	F	
0000111	007	BEL	(Bell)	1000111	107	G	
0001000	010	BS	(Backspace)	1001000	110	H	
0001001	011	HT	(Horizontal Tabulation)	1001001	111	I	
0001010	012	LF	(Line Feed)	1001010	112	J	
0001011	013	VT	(Vertical Tabulation)	1001011	113	K	
0001100	014	FF	(Form Feed)	1001100	114	L	
0001101	015	CR	(Carriage Return)	1001101	115	M	
0001110	016	SO	(Shift Out)	1001110	116	N	
0001111	017	SI	(Shift In)	1001111	117	O	
0010000	020	DLE	(Data Link Escape)	1010000	120	P	
0010001	021	DC1	(Device Control 1)	1010001	121	Q	
0010010	022	DC2	(Device Control 2)	1010010	122	R	
0010011	023	DC3	(Device Control 3)	1010011	123	S	
0010100	024	DC4	(Device Control 4-Stop)	1010100	124	T	
0010101	025	NAK	(Negative Acknowledge)	1010101	125	U	
0010110	026	SYN	(Synchronization)	1010110	126	V	
0010111	027	ETB	(End of Text Block)	1010111	127	W	
0011000	030	CAN	(Cancel)	1011000	130	X	
0011001	031	EM	(End of Medium)	1011001	131	Y	
0011010	032	SUB	(Substitute)	1011010	132	Z	
0011011	033	ESC	(Escape)	1011011	133	[	(Opening Bracket)
0011100	034	FS	(File Separator)	1011100	134	\	(Reverse Slant)
0011101	035	GS	(Group Separator)	1011101	135	]	(Closing Bracket)
0011110	036	RS	(Record Separator)	1011110	136	^	(Circumflex)
0011111	037	US	(Unit Separator)	1011111	137	___	(Underline)
0100000	040	SP	(Space)	1100000	140	‘	(Opening Single Quote)
0100001	041	!		1100001	141	a	
0100010	042	“		1100010	142	b	
0100011	043	#		1100011	143	c	
0100100	044	$		1100100	144	d	
0100101	045	%		1100101	145	e	
0100110	046	&		1100110	146	f	
0100111	047	’	(Closing Single Quote)	1100111	147	g	
0101000	050	(		1101000	150	h	
0101001	051	)		1101001	151	i	
0101010	052	*		1101010	152	j	
0101011	053	+		1101011	153	k	
0101100	054	,	(Comma)	1101100	154	l	
0101101	055	-	(Hyphen)	1101101	155	m	
0101110	056	.	(Period)	1101110	156	n	
0101111	057	/		1101111	157	o	
0110000	060	0		1110000	160	p	
0110001	061	1		1110001	161	q	
0110010	062	2		1110010	162	r	
0110011	063	3		1110011	163	s	
0110100	064	4		1110100	164	t	
0110101	065	5		1110101	165	u	
0110110	066	6		1110110	166	v	
0110111	067	7		1110111	167	w	
0111000	070	8		1111000	170	x	
0111001	071	9		1111001	171	y	
0111010	072	:		1111010	172	z	
0111011	073	;		1111011	173	{	(Opening Brace)
0111100	074	<	(Less Than)	1111100	174	|	(Vertical Line)
0111101	075	=		1111101	175	}	(Closing Brace)
0111110	076	>	(Greater Than)	1111110	176	~	(Overline (Tilde))
0111111	077	?		1111111	177	DEL	(Delete/Rubout)

fortunately, it is possible to interpret or translate the different machine protocols.

Reprographic system. In a basic reprographic system it is possible for a person to communicate with a machine by inputting a message via typewriter–optical character reader (OCR) or magnetic disk. Voice-to-digital communication, and vice versa, is now in various stages of development.

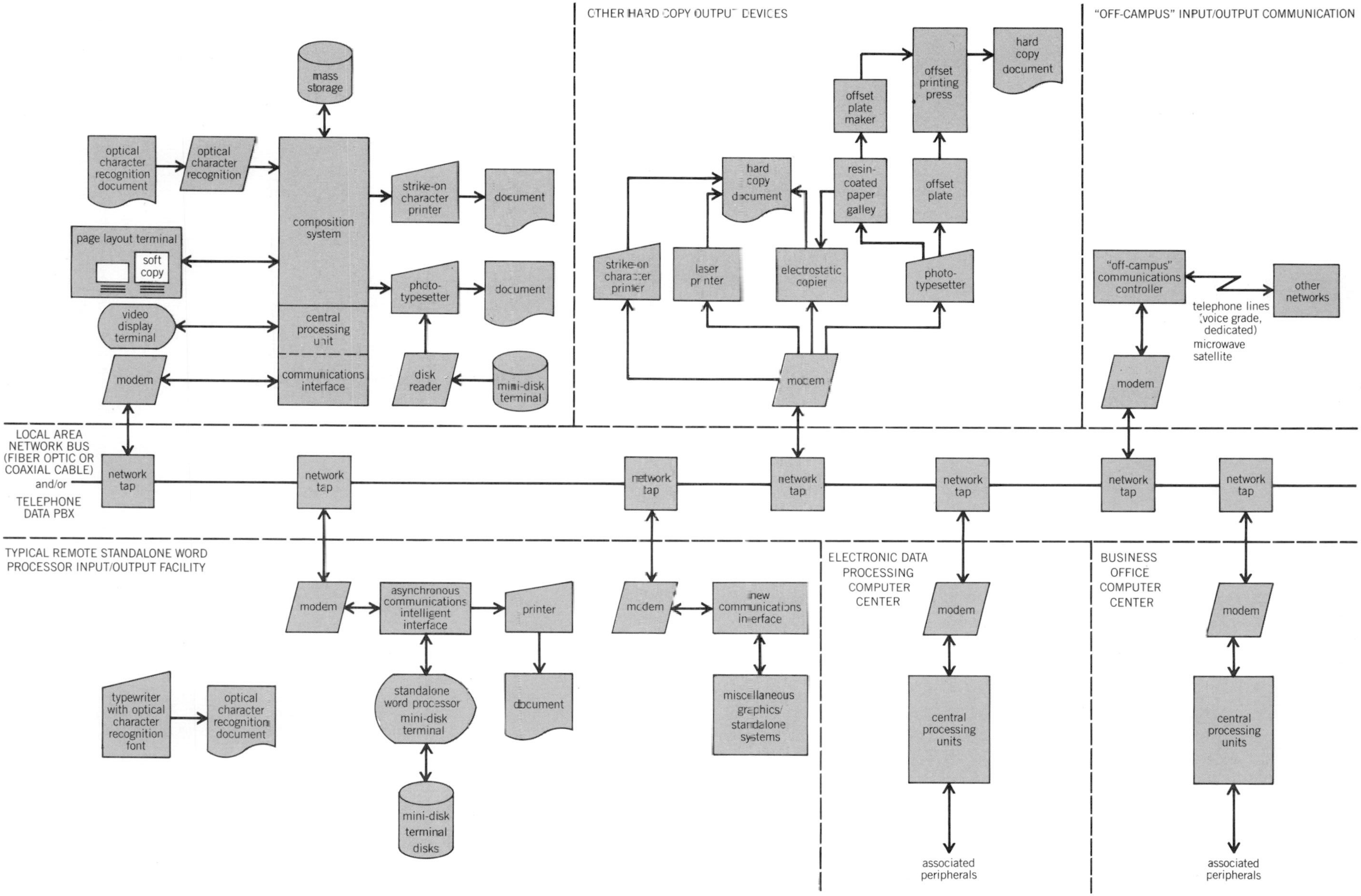

Schematic diagram of an interactive reprographic system installed at Southwest Research Institute, San Antonio, Texas. (*After W. Downing and E. Gavlick, Internal Memorandum Southwest Research Institute, June 1983*)

However, it is unlikely that applications on a wide scale will be realized before 1990.

The illustration shows a schematic diagram of a reprographic system that has been in operation since 1980. This publications installation has peripheral mainframe equipment and network interface. Three central processing units, various standalone input/output systems, several output systems, and even other network systems distantly removed can communicate over a local area network or a telephone data PBX system. The main components in such a system are: people with special training; electronic publications processing computers; electronic data processing computers; personal computers; typewriters; optical character readers (scanners); magnetic disk readers; cathode-ray tubes or video display terminals; standalone word processing units; graphics terminals; output devices, such as strike-on character printers, dot matrix impact printers, laser printers, electrostatic copiers, typesetters, and platemakers; communication interfaces (black boxes); modems; and network taps. Although not shown in the diagram, the two electronic data processing central processing units in this scheme support dot matrix printers and are capable of supporting laser printers. Also, graphics software packages are available with many current-generation central processing units. Both electronic data processing computers likewise support word processing software packages with limited applications. Software packages are available with these central processing units that permit computer-aided design, computer-aided manufacturing, and computer output microfilm.

Capability. Reprographics currently permits the composition of an entire page on a video display terminal, with subsequent reproduction in multiple copies by electronically communicating the image to a hard copy output device. Most of these hard copy devices produce documents of typewriter or better quality. The quality is better because all corrections (proofreading), changes, or copy alterations (editing) are made electronically on the video display terminals and output error-free. Typeset copy with bold, medium, light, and italic styles, various type faces (Roman, Gothic, and so on) and point sizes can be generated directly from video display terminals to either photocomposed paper galleys or offset printing plates that, in turn, produce excellent-quality reproductions. Local-area networks can interface with other such networks or national-international networks via electronic communications controllers, which in themselves are microprocessors. In large reprographic systems, digital scanners are used to "read in" photographs and line art. These then can be accessed, viewed, and arranged on page layout terminals together with the text, captions, and headlines. That is, the entire page makeup can be designed and laid out electronically and viewed as soft copy in final form before being committed to hard copy, which can be plain paper, photoreproduced paper galleys, or offset plates. The hard copies can be generated within a facility, or the entire page can be transmitted over a network for viewing remotely as soft copy on a page layout terminal or a video display terminal, or for remote reproduction of hard copies.

Reprographics has even been extended to the ends of the solar system. The pictures of the planets, especially Jupiter and Saturn, sent back by the Voyagers and Pioneers and subsequently reproduced have added considerably to documentation of the solar system. *Pioneer 10* crossed the orbit of the planet Neptune on June 13, 1983, and effectively left the solar system. NASA reports that *Pioneer 10* is still receiving commands and sending back data and that it is expected to do so for at least another 8 years. Thus, the science of reprography has extended to galactic distances.

For background information *see* CHARACTER RECOGNITION; MICROPROCESSOR; PHOTOCOPYING PROCESSES; PRINTING; TELEGRAPHY; TELETYPEWRITER; WORD PROCESSING in the McGraw-Hill Encyclopedia of Science and Technology.

[EMIL L. GAVLICK]

Bibliography: J. H. Dessauer and H. E. Clark, eds., *Xerography and Related Processes*, Focal Press, 1965; J. E. McNamara, *Technical Aspects of Data Communication*, Digital Equipment Corporation, Bedford, Massachusetts, 5th printing, July 1979; P. Waller, Public Information Officer, Press Release, NASA Ames Research Center, April 20, 1983; White Paper Report Series: *Managing Human Factors* (March 1982), *Ergonomics* (May 1982), *Word Processing* (June 1982), *Communications* (August 1982), *Reprographics* (November 1982), *Office Automation* (February 1983), *Personal Computers* (May 1983), in *Modern Office Procedures*, Penton/IPC, Inc. (Pittway Corporation).

Root

Studies involving root growth and development of plants in a soil environment have been difficult because of the lack of ready access to the roots. If the plant is disturbed in the soil, root and shoot growth processes are interrupted. Measurement of the activity of roots at all stages of plant development is necessary to the understanding of the plant–root systems, which determine yields. The plant is a complex of interactions among its parts, so that conditions affecting root growth affect the top growth and reproductive capacity. In turn, vegetative and reproductive development of the aboveground plant influences root activity. External stress on any given plant part has repercussions on the other parts and ultimately on crop yield. The dynamic interactions between the genetic characteristics of the plant, the soil, and the climatic environment have prevented the establishment of clear-cut relationships between yields and root development.

Two methods have been developed to improve the accessibility of plant roots for study. One, the minirhizotron technique, allows for study in continuous root activity in the soil environment under diverse

Fig. 1. Mini-rhizotron instrumentation for photographing plant roots includes micro-TV camera, 14-in. (36-cm) borescope, light box, video tape recorder, monitor, and 12-V battery source. (*Courtesy of D. A. Brown, Department of Agronomy, University of Arkansas*)

conditions in the field. The other, the porous-membrane root-culture technique, allows a partial environment control for the roots without altering other field conditions. These two new methods have been perfected to isolate and define the plant stresses that are caused by limiting environmental conditions.

Mini-rhizotron technique. This method was made possible in 1978 with a fiber optic duodenoscope that was modified for observing and photographing root development patterns within a soil profile with a 35-mm camera. The root was monitored by inserting the scope into a transparent Plexiglas tube embedded within the root zone. This technique permitted the observation and quantitative characterization of the root system throughout the entire growing season, which until that time had been excessively expensive, laborious, time-consuming, and nonreproducible. The scope was later modified by replacing the 35-mm camera with a micro-TV camera, video recorder, and monitor (Fig. 1).

The micro-camera is affixed to a 14-in. (36-cm) borescope with a right-angle objective lens, a lamp housing with four incandescent lamps, video tape recorder, and monitor. The unit is powered by two 12-V batteries and is completely portable for field use. The scope is inserted into the transparent observation tube and slowly pulled out of the tube. Any roots that have grown across the face of the tube are photographed and recorded on video tape, as shown in Fig. 2. Root length, area, and density are measured from a replay of the video tape into an image analyzer. Rooting patterns are measured under and to each side of the plant row six times during the growing season. The mini-rhizotron technique provides a nondestructive method for recording vertical and horizontal root development within the soil throughout the season.

Root growth of soybeans, cotton, and corn have been measured by photographing the root development patterns within a soil profile (35 in. or 89 cm depth) during the entire growing season. Root growth was greatest under the row in the vegetative stage of growth, giving way to increased root growth in the furrow during the reproductive stages. For soybeans, the percentage of the total root system developing in the furrow was 26% during the vegetative stage of growth and 67% in the latter stage of maturity. Roots under the row began to die during the early reproductive stage while the root extension into the furrow was increasing. Total root production increased through the reproductive stage; however, the rate of new root production decreased drastically. Fifty percent of the total root system was distributed within the top 12 in. (30 cm) of soil.

Four stages of root development have been related

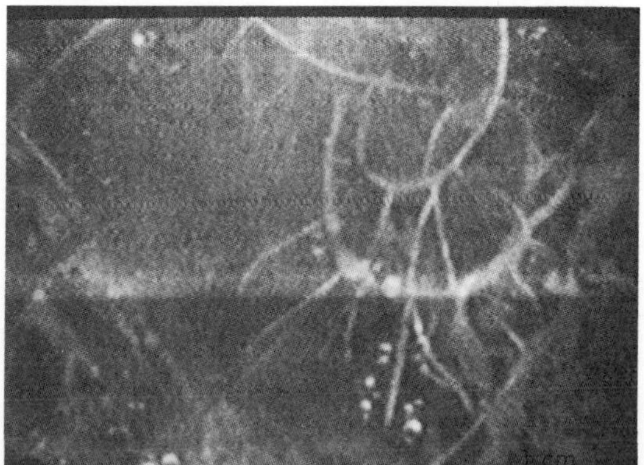

Fig. 2. Reproduction of soybean roots photographed by mini-rhizotron instrument.

to the physiological stage of shoot growth: (1) the early vegetative stage of growth is characterized by rapid root growth beneath the plant with minimum lateral development into the furrows; (2) as a plant changes from the vegetative into reproductive growth, root distribution is characterized by a slowdown in the growth rate under the plant and a rapid increase in growth into the furrow area; (3) in the mid-reproductive stage, the total amount of root growth becomes constant; however, growth under the row decreases while growth in the furrow area increases; and (4) in physiological maturity, growth under the row ceases and root losses due to decomposition are evident. Root growth in the furrow shows no significant new growth.

When new vigorously growing roots are in short supply during the reproductive stage, there is a significant physiological stress on the older roots to provide water and nutrients for the shoot portion producing the fruit. This stress becomes critical because either the older root system has extracted most of the readily available nutrients and water next to it, or the root has been unable to grow normally into a compacted or an acid subsoil with a low nutrient supply. The ability to observe root activity in place provides new possibilities for understanding how to alter the plant or the plant environment to provide greater crop yields.

Porous-membrane root-culture technique. Porous membranes that separate plant roots from actual soil contact have been used in growth chamber research where plants were grown for 4 weeks. The root growth between the membranes provides a chemical analysis of the clean (soil-free) roots. This technique enables the measurement of the suites of cations at four stages of uptake: the nonabsorbed ions in water films on the roots; the ions adsorbed onto the root surface; the cations absorbed into the roots; and the ions translocated to the aboveground portions of the plant.

In 1982, the porous-membrane technique was modified for field use (Fig. 3). This modification involved the growth of plants under field conditions,

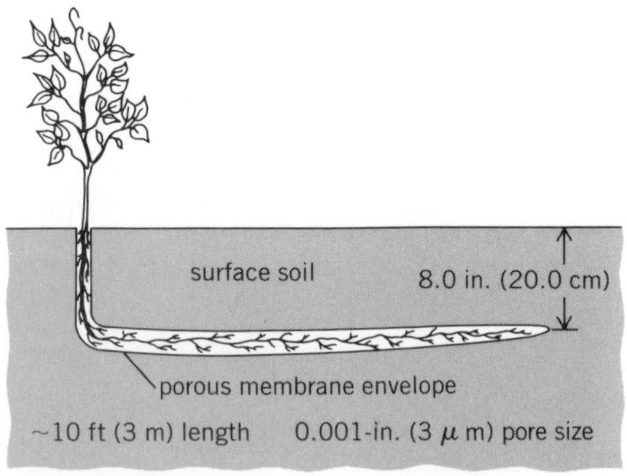

Fig. 3. Porous-membrane root-culture technique for growing plants in the field under controlled environmental stress conditions. (*Courtesy of D. A. Brown, Department of Agronomy, University of Arkansas*)

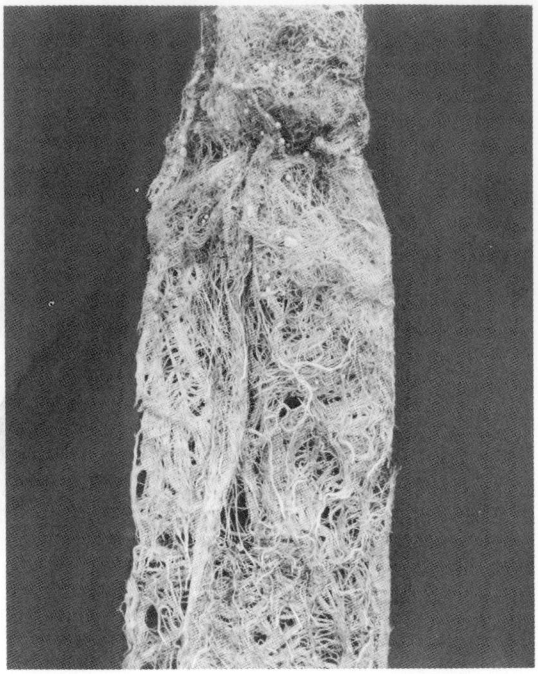

Fig. 4. Soybean root system grown in porous membrane.

with their roots restricted to a permeable-membrane envelope measuring 14 by 120 in. (35 by 300 cm) long, buried horizontally in the top soil at a depth of 10 in. (25 cm). The membrane has pores with a 3.0-micrometer diameter. One end of the membrane opens at the soil surface and permits the transplanting of a young seedling plant. The roots subsequently develop and grow within the membrane envelope in a soil environment without actual soil contact. The permeability of the membrane allows water, nutrients, and air to reach the roots in sufficient quantities for normal root growth. This technique enables the growth of plants under field conditions but without unfavorable soil effects. The entire plant and root system can be harvested at any stage of the growth in the root system by simply opening up the membrane and removing the clean (soil-free) roots (Fig. 4).

This technique also permits the measurement of changes in new root growth during plant maturity. In previous research a positive relationship was established between the rate of new root growth, nutrient absorption, and the intensity of fluorescence of the new root growth when observed under ultraviolet light (365-nanometer wavelength). By photographing the roots grown in the porous membrane under white light and then under ultraviolet light, the actively growing fraction of the total root system can be determined. This permits a calculation of the fraction of the total root system that is most actively growing during any stage of growth. Such changes can then be related to the growth and yield of soybeans under field conditions.

For plants growing under soil conditions, the new root growth is greater during the vegetative stages of growth than in the reproductive stages. The decline of new root growth in the early reproductive stage results in decreased crop yield. Membrane-grown soybeans whose roots were uninhibited by unfavorable soil conditions grew as long as 10 ft (3 m), nodulation occurred throughout the 10-ft (3-m) length, and plant leaf area was 22 ft^2 (2 m^2). The maximum yield was 1050 seedpods per plant; 51% of the pods were three-seeded. This compared with only 100 seedpods produced on a plant grown under high population density. Continuous new root growth throughout the season is required for maximum seed yield.

The capabilities provided by these two techniques will permit studies of plant–root systems that will more clearly define the interactions of plant parts. Subsequently, alterations in the environment or in the genetic characteristics or biochemical processes through plant breeding and chemical growth regulation can provide plant–root systems with the capacity for increased crop yields.

For background information *see* PLANT GROWTH; RHIZOSPHERE; ROOT in the McGraw-Hill Encyclopedia of Science and Technology.

[D. A. BROWN]

Bibliography: D. A. Brown, Dependence of crop growth and yield on root development and activity, *Proc. Plant Growth Regulator Soc.*, Monterey Park, California, July 1982; D. A. Brown and J. C. Noggle, Ion exchange in soil-plant root environments, I. Measurement of suites of cations at various stages of nutrient uptake, *Soil Sci. Soc. Amer. Proc.*, 19:131–134, 1955; J. J. Cappy and D. A. Brown, A method for obtaining soil-free, soil-solution grown plant root systems, *Soil Sci. Soc. Amer. J.*, 44:1321–1323, 1980; J. L. Sanders and D. A. Brown, A new fiber optic technique for measuring root growth of soybeans under field conditions, *Agron. J.*, 70:1073–1076, 1978.

Schizophrenia

Despite the common view of schizophrenia as a functional disorder (that is, not having an obvious physical basis, unlike syphilitic dementia for example), there is a long, controversial history of the search for the biological underpinnings of this illness. Recent technological developments in the measurement of brain structure and metabolic and biochemical activity have allowed investigators to take a new look at old ideas, with special attention to the study of negative symptoms (in contrast to positive ones, such as hallucinations and delusions). These symptoms are characterized by an absence of normal functioning, such as flat affect, social withdrawal, and poverty of speech. In some schizophrenics these negative symptoms are associated with structural abnormality, reduced frontal lobe activity, or increased platelet monoamine oxidase activity (and by implication, reduced dopaminergic activity). These findings signal a conceptual shift to the consideration of negative symptoms as the fundamental deficit in schizophrenia.

General cortical dysfunction. Compared with normals, schizophrenics manifest more soft neurological signs, such as right-left confusion, difficulty in pronouncing tongue twisters, and awkwardness in finger or foot tapping. For example, in a recent study of 20 hospitalized schizophrenic patients, 20 psychiatric controls, and 20 normal controls, schizophrenics were found to have significantly higher scores on a neuromotor examination than either of the comparison groups. The schizophrenic patients were slower and more poorly coordinated, although they did not differ in absolute strength (as measured by handgrip).

The study of cortical structure has supplied some corroborating evidence. For example, computed axial tomography (CAT scan) studies have revealed mild cortical atrophy in 6 to 60% of various samples of schizophrenic inpatients. Manifested as enlarged ventricles, this atrophy generally has been found in older, chronic patients for whom the effects of age and long-term institutionalization cannot be ruled out. Nevertheless, enlarged ventricles in schizophrenics have been found to be associated with negative symptoms, poor performance on neuropsychological testing, clinical neurological assessment, poor premorbid adjustment, and poor response to neuroleptic treatment. Cortical atrophy, as measured thus far, is not diagnostic of schizophrenia; rather, it represents one possible type of the disorder that is associated with mild but global brain damage.

Lateralized cortical dysfunction. Schizophrenia has been viewed as a disorder of the left hemisphere (which is primarily responsible for verbal processing and performance), and affective disorders as dysfunctions of the right hemisphere (which mediates nonverbal behavior and affect). Based on this premise, both neuroanatomical and functional (that is, based on performance) asymmetries have been studied. Although there is consistent evidence of left hemisphere abnormalities from both areas of research, there is as yet little understanding of the relationship between anatomical and functional findings.

Schizophrenics perform significantly worse than manics and depressives on tests mediated by the dominant hemisphere (for example, pegboard performance, sentence repetition, Wechsler Adult Intelligence Scale (WAIS) verbal IQ, and picture vocabulary), but their performance is not different from that of patients with coarse brain disease. In contrast, only those individuals with coarse brain disease differ significantly from the other groups on tests mediated by the nondominant hemisphere (for example, WAIS performance IQ, Raven's Progressive Matrices, and Benton Visual Retention).

Schizophrenics appear to be right-gazers in response to questions requiring contemplation. That is, unlike controls who look left when thinking, schizophrenics far more often look right, suggesting overinvolvement of the left hemisphere. There has long been evidence of a higher rate of left-handedness in schizophrenics than in other populations and, more recently, it has been noted that among monozygotic twins of which one is a schizophrenic, the left-handed twin is the affected one.

The assessment of sensory processing separately by hemisphere also reveals a left hemisphere abnormality, although the data are not consistent. There seems to be, for example, a left ear (hence right hemisphere) inferiority in dichotic listening speech-perception tasks, as well as a right ear (left hemisphere) superiority. *See* BRAIN.

Nonperformance measures parallel these findings. When hemispheric asymmetry is examined in electroencephalographic studies, the majority indicate a left hemisphere abnormality; this may be in the form of increased electrical activity, a difference in the dominant frequency, or as is frequently the case, greater variability in activity.

In normal right-handers, CAT scans indicate that the right frontal and left occipital lobes are wider than the left frontal and right occipital lobes, respectively. In schizophrenics there is a significantly higher rate of both frontal and occipital reversed cerebral asymmetry than that in normal individuals. That is, more schizophrenics have wider left frontal and right occipital lobes. This finding is limited to those schizophrenics with no CAT scan evidence of cortical atrophy, and is largely accounted for by reversals in the frontal lobe asymmetry. Other investigators have found lower cortical density, as measured by CAT scans, in the left frontal lobes of schizophrenics than in those of controls, lending support for a neuroanatomical asymmetry in some schizophrenics.

Metabolic studies. Studies employing the most sophisticated assessment techniques are still in the early stages of clinical application. Nevertheless, there is an apparent convergence of findings suggesting decreased metabolic activity in the frontal lobes of schizophrenics compared with that of psychiatric and normal controls. One of the first blood flow studies, conducted in 1976, compared schizophrenics with neurologically normal alcoholic men while the subjects were at rest with closed eyes. Not only did schizophrenics manifest less blood flow than the alcoholics, but also the schizophrenics exhibited decreased blood flow associated with negative symptoms, such as muteness, inactivity, and withdrawal. Positive symptoms, such as hallucinations, were associated with elevated blood flow in the postcentral region of the brain.

Similar results have been obtained in preliminary analyses using positron emission tomography (PET scan). Unmedicated schizophrenics (at rest, eyes closed) exhibit lower glucose use in the frontal cortex than do normals of the same age. As yet, no asymmetries have been found with either blood flow or PET scan techniques.

Overall, these early findings are consistent with the psychoneurological reports pointing to cortical involvement in schizophrenia. There is, however, uncertainty regarding the lateralization of cortical abnormality, although the left hemisphere in its mediation of cognitive and verbal functions is a prime candidate. Furthermore, there is growing evidence that many of the psychoneurological and cortical findings are characteristic of psychosis generally, rather than of schizophrenia specifically.

Biochemical studies. The dopamine hypothesis of schizophrenia, based on a knowledge of neuroleptic effects, animal studies, and amphetamine psychosis, states that schizophrenia is the result of the presence of excessive dopamine in the brain. After more than a decade, however, it is clear that such a simple view is not tenable. Although previous reports indicated an absolute decrease in monoamine oxidase activity, which would reflect increased dopamine, in chronic schizophrenia, the decrease has been found actually to be an effect of neuroleptic treatment. There has been somewhat better success in examining the relationship between dopaminergic activity and specific aspects of schizophrenia.

For example, a low concentration of homovanillic acid (a major dopamine metabolite whose concentration is inversely related to that of dopamine) in cerebral spinal fluid correlates with poor prognosis and first-rank symptoms (a set of delusions, such as thought insertion and thought withdrawal, assumed to be especially charactertistic of schizophrenia). High concentrations of homovanillic acid in cerebral spinal fluid may be related to a family history of schizophrenia and poor premorbid sexual adjustment. There are, however, no indications that schizophrenics have any lower absolute homovanillic acid concentrations in cerebral spinal fluid than other psychiatric or normal control groups.

There is reason to expect that a functional increase in dopamine may still account for some aspects of schizophrenia, but at the postsynaptic rather than the presynaptic stage (the focus of the previous monoamine oxidase and homovanillic acid studies). There is evidence of receptor supersensitivity in the brain of schizophrenics, which is consistent with neuroendocrinological studies of the hypothalamic-pituitary system designed to examine generalized dopaminergic hyperactivity. For example, the exaggerated response of growth hormone to injections of apomorphine in unmedicated acute and chronic schizophrenics could reflect dopamine receptor supersensitivity.

In contrast to older theories that postulate an increase in dopamine activity, there have been recent suggestions that dopaminergic underactivity underlies the negative symptoms of schizophrenia. In one study, for example, there was a significant positive correlation between platelet monoamine oxidase activity and negative symptoms in male schizophrenics; the higher the monoamine oxidase, the more negative symptoms the patients exhibited. However, there were no significant differences in absolute biochemical levels.

Conclusion. There have been major advances in knowledge of the relationship between brain, biochemistry, and behavior, but little information is available to show how these elements interact to produce schizophrenia. Simple, unidirectional causal models that posit a single etiological agent for the disease schizophrenia are untenable. Rather, examination of more delineated symptoms and signs, such as negative versus positive symptoms, is beginning to yield promising etiological leads. Perhaps certain aspects of schizophrenia, such as the loss of goal orientation, withdrawal, and apathy, are mediated by the left frontal lobe, while other aspects, such as hallucinations, find their origins in the occipital area. One lesson of the past decade is that the understanding of schizophrenia may be obtained by careful examination of specific functions rather than global syndromes.

For background information *see* BRAIN; INTERHEMISPHERIC INTEGRATION; SCHIZOPHRENIA in the McGraw-Hill Encyclopedia of Science and Technology.

[RICHARD LEWINE]

Bibliography: M. Buchsbaum et al., Cerebral glucography with positron tomography, *Arch. Gen. Psychiat.*, 39:251–259, 1982; W. Bunney, B. Garland, and M. Buchsbaum, Advances in the use of visual imaging: Techniques in mental illness, *Psychiat. Ann.*, 13:420–426, 1983; D. Newlin, B. Carpenter, and C. Golden, Hemispheric asymmetries in schizophrenia, *Biol. Psychiat.*, 16:561–582, 1981; J. Haracz, The dopamine hypothesis: An overview of studies with schizophrenic patients, *Schiz. Bull.*, 8:438–469, 1982.

Sedimentation (geology)

Accumulations of thick, gel-like muds front many shorelines of the world, from the humid tropics of the equatorial latitudes to the frozen "deserts" of the Arctic. Most coastal muds occur as intertidal mudflats and subtidal mudshoals near major rivers, and the source of muddy sediments can often be traced directly to these rivers. For example, the Mississippi River supplies muds to the coast of Louisiana, the Amazon River supplies muds to the shorelines of French Guiana, Surinam, and Guyana, and the mudflats of the Yellow Sea and Gulf of Po Hai are derived from both the Yangtze and Hwangho rivers in China. Inasmuch as nearly 50% of the world's fluvial sediment is derived from rivers in Asia, the greatest concentrations of fine-grained sediment are found in Asian countries such as India, Malaysia, Thailand, China, and Korea (see table).

The world's largest and perhaps most spectacular mudflats occur in northeastern Asia along the margins of the Yellow Sea and the Gulf of Po Hai. Here, a tide range of 10–30 ft (3–9 m), together with harsh winter-season outbreaks of cold air, create wave and current conditions of sufficient intensity to place the shelf-depth waters in a large bedshear category. Accumulation and survival of fine-grained sediments under such conditions at first appear unusual; clearly, the traditional concept that

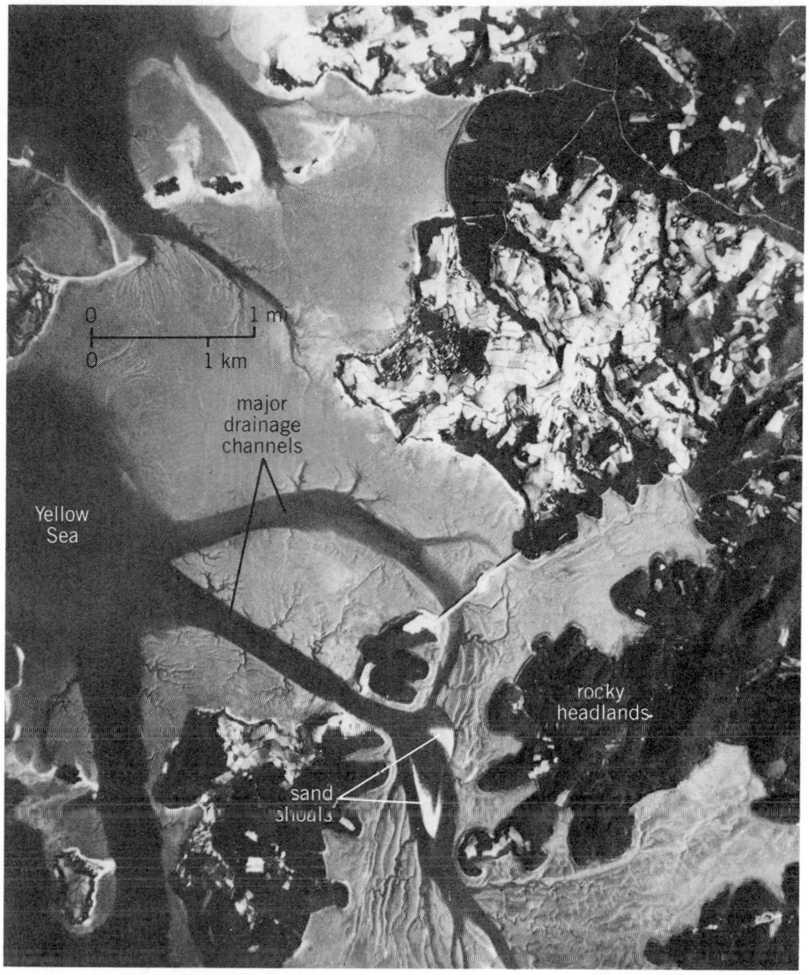

Fig. 1. Aerial photograph of intertidal mudflat on the west coast of South Korea, showing dendritic drainage channels and parabolic sand shoals. (*From J. T. Wells, Dynamics of coastal fluid muds in low-, moderate-, and high-tide-range environments, Can. J. Fish. Aquat. Sci., 40(suppl. 1):130–142, 1983*)

Major mudflat coasts of the world and their environmental settings

Location	Latitude	Climate	Tide range, ft (m)	Wind effects	Wave energy
Louisiana (U.S.A.)	29°30′N	Subtropical	1.5 (0.5)	Frontal	Low to moderate
Southwestern India (Asia)	8°00′–11°30′N	Tropical	3.3 (1)	Monsoonal	Moderate to high
Surinam–Guyana (South America)	5°00′–8°30′N	Tropical	3.3–6.6 (1–2)	Trade winds	Moderate
Netherlands–Germany (Europe)	51°30′–54°30′N	Temperate	5.0–15 (1.5–4.5)	Frontal	Moderate
China (Asia)	28°00′–40°00′N	Temperate	6.6–13 (2–4)	Monsoonal	Moderate to high
Malaysia (Asia)	2°00′–5°00′N	Tropical	10–13 (3–4)	Monsoonal	Low to moderate
South Korea (Asia)	34°30′–38°00′N	Temperate	10–30 (3–9)	Monsoonal	High

muds accumulate only in quiet, wave-free environments needs to be reexamined. Recent research aimed at addressing mudflat dynamics in large bed-shear environments indicates that, along the west coast of Korea, not only are mudflats able to provide a buffer to wave attack, but they also serve as a temporary storage facility for littoral sediments alternately accumulating and eroding. A hypothesized cycle of winter cut–summer fill may explain the large concentrations of suspended sediment frequently measured at the onset of winter monsoon winds and the largely turbid waters consistently detected by satellite sensors in the visible and near-infrared wavelength ranges.

Morphologic features. Backed by rugged headlands, conical hills, and occasional salt marshes, the west coast of South Korea is sedimentologically and morphologically complex. Exposed mudflats 10–13 ft (3–4 m) thick and 3–19 mi (5–30 km) wide overlie a basement of deeply weathered granitic rocks. Intertidal exposures display a wide range of morphologic features (Fig. 1). Most prevalent are dendritic drainage channels 3–10 ft (1–3 m) deep, which serve as conduits for carrying water and sediments seaward during falling tide. Like tributaries to a river, the smaller channels lead into large ones, and eventually to the sediment-laden waters of the Yellow Sea. Dish-shaped scour depressions several meters across and small burrows from crabs and bivalves provide additional relief to the mudflat surfaces. Patches of sand, shell, and even gravel accumulate locally and are molded by strong tidal currents into ridges and shoals, often seaward of the intertidal mudflat surfaces.

Sediment resuspension. Strong surges of polar continental air that move from northwest to southeast down the axis of the Yellow Sea inflict high-energy storm conditions on the shorelines of the Korean peninsula. These winter-season surges, which occur within a period of approximately 7 days beginning in late October, initiate sediment resuspension, vertical mixing of suspended sediment, and a wind-driven coastal current that flows to the south. As storm waves impinge on mudflat surfaces, sediments are resuspended by bottom currents, which may achieve speeds of 5 ft/s (150 cm/s). Despite the attenuating effect of muddy sediments on incoming waves, a single storm can strip away sediments deposited during many months of calm sea conditions. Vertical mixing of resuspended mud then rapidly results in a nearly uniform large concentration of suspended sediment from surface to bottom.

Sediment concentrations at the onset of winter monsoon winds may be hundreds of parts per million, typically increasing inshore, and confined to a band that parallels the shoreline. This band of turbid, destratified water ends abruptly 12–19 mi (20–30 km) offshore as a turbidity front, defined by the rapid decrease in concentration offshore from >20 to <10 ppm. Regardless of season, satellite imagery

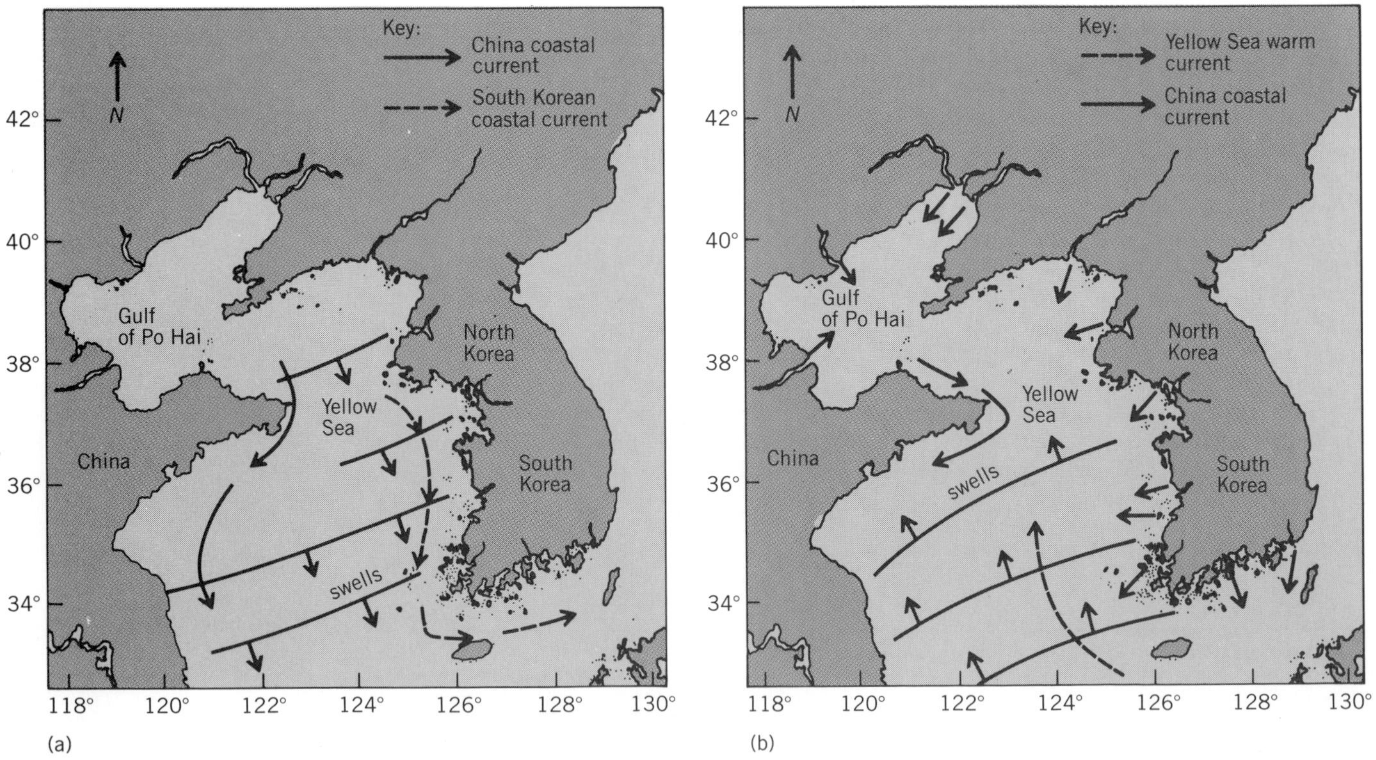

Fig. 2. Schematic representation of coastal currents: (*a*) fall/winter transport of sediment and water to the south; winds are strong northwesterlies; (*b*) spring/summer sediment replenishment from high river discharge; winds are weak and variable. (*After J. T. Wells, O. K. Huh, and Y. A. Park, Dispersal of silts and clays by winter monsoon surges in the southeastern Yellow Sea, Proceedings of the Symposium on Sedimentation on the Continental Shelf, Hangzhou, China, 1983*)

consistently shows a second turbidity front, closer inshore, that occurs at the 66 ft (20-m) bathymetric break. This second front marks the seaward boundary of the high-transport zone of a coastal mudstream, referred to as the South Korean Coastal Current.

Dispersal patterns. The wind-driven South Korean Coastal Current, setting to south during fall and winter, provides the mechanism for dispersal of resuspended sediments. The fact that suspended sediment concentrations are one to three orders of magnitude greater than "typical" continental shelf waters indicates that sediment transport rates within this coastal mudstream may be enormous, even under relatively weak currents.

On an annual time scale, a model for sedimentation suggests small sediment influx to the coast during winter, when strong northwesterly winds generate large waves in the Yellow Sea, and extremely large influx in summer, at a time when weak and variable winds from the south produce little wave activity (Fig. 2). Coastal muds are thus eroded in winter and carried south into estuaries and into the Korea Strait. In summer, muds are replenished during great river discharge and reform the band of soft material, which characteristically occurs as a series of mudflats near the coast. Thus mudflats may serve as a temporary storage facility during summer accumulation and a source during winter erosion.

Dredging records indicate that most fine-grained sediments enter estuaries along the west coast of Korea during winter months. For example, severe shoaling in the Kum River estuary (up to 8 ft or 2.5 m per year) occurs during November and December at twice the yearly rate. Thus most of the sediment accumulates in estuaries when river discharge is low and when sediments are transported south past entrances to west coast estuaries. From an engineering standpoint, analogies can be drawn to the coasts of northeastern South America and southwestern India, where case histories show harbor siltation to be tied to longshore transport of fluid muds.

The fate of the suspended coastal sediment transported south into the Korea Strait, at rates estimated to be 88–880 × 10^7 ft^3 (25–250 × 10^6 m^3) per year, is not fully known. Fragmentary evidence from satellite and sea-floor sediment distribution data indicates that some of the sediment may accumulate in shallow embayments along the south coast of the Korean peninsula, whereas the remainder may be transported into the Sea of Japan. Further, the volume transport rates are one to two orders of magnitude greater than can be explained by yield from Korean rivers alone, suggesting that muds may ultimately be derived from other parts of the Yellow Sea.

For background information *see* ESTUARINE OCEANOGRAPHY; SEDIMENTATION (GEOLOGY) in the McGraw-Hill Encyclopedia of Science and Technology. [JOHN T. WELLS]

Bibliography: J. D. Milliman and R. H. Meade, World-wide delivery of river sediment to the oceans, *J. Geol.*, 91:1–21, 1983; J. T. Wells and O. K. Huh, Tidal flat muds in the Republic of Korea; Chinhae to Inchon, *Sci. Bull. Off. Naval Res.*, Tokyo, 4:21–30, 1979; J. T. Wells, O. K. Huh, and Y. A. Park, Dispersal of silts and clays by winter monsoon surges in the southeastern Yellow Sea. *Proceedings of the Symposium on Sedimentation on the Continental Shelf*, Hangzhou, China, 1983.

Seed

Seed pathology addresses disease problems that affect seed production. A complex technology is required to produce large quantities of seeds that will given uniform stands of healthy seedlings. It involves growing, harvesting, conditioning, storing, and planting the seeds. Factors that may influence seed quality during this procedure include diseases, insects, damage due to handling, pesticide toxicity, improper storage conditions, poor weather during harvest or planting. All of these may adversely affect germination and vigor of seeds, but diseases create the additional problem of possible transmission to the succeeding crop.

Because of the extensive movement of seeds throughout the world, they are an efficient means of spreading plant pathogens between geographical regions. Since the 1920s, much effort, particularly in Europe, has gone into developing tests that will determine whether pathogens are present in seedlots. Except for plant quarantine purposes, these tests have been of limited practical value because of a lack of knowledge of relationships between test results and disease development in the field. In recent years, an increasing demand for high-quality seeds has stimulated research in all aspects of seed production. Seed pathologists have become more concerned with obtaining a better understanding of seed infection as it relates to seed quality and transmissibility of pathogens. This has required studying the seed aspect of the life cycles of pathogens as influenced by factors such as cultural practices, weather, and soil type. With this kind of information the significance of seed infection can be realistically assessed and disease control methods can be developed. Another major influence in seed pathology has been the development of new techniques to improve the detection of seed-borne bacteria and viruses. This has provided new opportunities for the control of some important seed-borne viruses and bacteria.

Infection by plant pathogens. Seeds normally are grown in a field environment where they are exposed to masses of microorganisms. Some of these are plant pathogenic fungi, bacteria, viruses, or nematodes that may become associated with seeds and cause disease. The pathogens can be present in the soil, on crop residues, and on weeds, or can have been introduced by the original seed. They can be spread to new seeds by rain, wind, insects, or handling operations. Seed infection may occur in the production field, during storage and conditioning, or after seeds are planted in the field. There are various degrees of association between pathogens and seeds. The loosest association is when pathogens are not attached to the seed. These may be infected pieces of plant debris or specialized structures of the pathogen which, essentially, are "fellow travelers" with the seedlot. Other pathogens may be attached to the outside of the seedcoat or be in the seedcoat; some are present in internal tissues of the seeds, such as the cotyledons or embryo. In general, the closer the degree of association, the greater the chance of transmission of the pathogen; however, loosely associated pathogens may be transmitted. The fungus *Sclerotinia sclerotiorum*, for example, is carried with seedlots as separate fungal structures called sclerotes and can cause serious crop losses in a broad range of crops.

Significance of seed-borne pathogens. A list of seed-borne diseases published in 1979 records almost 1500 seed-borne microorganisms on about 600 genera of agricultural, horticultural, and tree crops. These figures are misleading, however, in estimating the extent of seed-borne microorganisms as problems, particularly when seed is produced for established crop production areas where the microorganisms are known to be present.

Some perspective can be obtained by considering seed-borne microorganisms under four classes:

1. The first class consists of pathogens for which the seed is the main source of inoculum and, when seed infection is controlled, the disease is effectively controlled. For many of these pathogens, the importance of seed-born inoculum has been long recognized, and control practices have been developed. An example is lettuce mosaic virus, a serious disease of lettuce. Control is achieved by testing lettuce seeds to assure that they do not carry this virus.

2. The second class consists of major pathogens which are seed-borne, but for which the seed-borne phase of the disease is of minor importance as a source of inoculum. An example is *Leptosphaeria maculans*, the cause of blackleg of oilseed rape. In Australia, where this pathogen is a limiting factor in oilseed rape production, infected residues of rape crops are a much more important inoculum source than are infected seeds.

3. The largest group of seed-borne organisms are those that have never been shown to cause disease as a result of their presence on seeds. An example is *Chaetomium* spp. on soybeans. Rather than having detrimental effects on seeds, some of the microorganisms in this class may, in fact, be beneficial. They are known to interact on soybean seeds and could possibly be manipulated to control pathogenic fungi.

4. There is a group of microorganisms that can infect seed either in the field or in storage, causing reduction in yield and seed quality. Examples of field-infecting fungi include *Diplodia*, *Gibberella*, and *Fusarium* spp. which cause ear rots on corn. *Aspergillus* and *Penicillium* spp., both storage fungi, can invade and damage most types of seed under high-moisture storage conditions.

At present, only a small proportion of the 1500 microorganisms listed as being seed-borne can be assigned to any of these four classes. Other than the fact that the microorganisms have been shown to be associated with seeds, there usually is little information to indicate the significance of their seed-borne nature.

Control of seed diseases. Many of the strategies used to control diseases in grain crops also can be applied to seed crops, but special considerations regarding the quality of the product make disease control in the latter a more complicated matter. There also are options available in controlling seed diseases that cannot be used for grain crops.

Cultural practices. Cultural practices may be appropriate when inoculum persists in the soil or on crop residues. Burning grass-seed production fields in Oregon destroys inoculum of *Gloeotinia temulenta*, the cause of blind seed, and *Claviceps purperea*, the cause of ergot; both fungi can survive on unharvested seed. Soybean seed infection by *Phomopsis* spp. can be reduced by rotating soybean seed fields with corn. Other cultural practices such as varying planting time are sometimes effective. Winter wheat, sown early in autumn, may escape infection from bunt (caused by *Tilletia foetida*, *T. caries*) because plants are past the susceptible growth stage before spores germinate, while later-sown crops may become infected.

Resistance. Breeding for resistance to seed diseases, specifically to improve seed quality, is usually not economically feasible in temperate regions of the world, unless the disease also is an important problem in grain production fields. This approach, however, may be of more importance in developing countries where the major source of seed is that which the farmer saves from the grain crop.

Geographical isolation. Disease control has been a major consideration in locating seed production in particular geographical areas. Much seed production, therefore, is concentrated in California, Oregon, and Washington, where warm, dry conditions are unfavorable for disease development. On a smaller scale, the isolation of seed fields from other fields of the same crop within the same growing region also is of value in disease control. While this practice is primarily used to maintain varietal purity, it also serves to isolate the fields from inoculum of airborne pathogens.

Storage facilities. Storage conditions are a major consideration in maintaining seed quality. Most seed handlers appreciate the importance of correct storage conditions, but few probably realize that prevention of invasion by storage fungi (*Aspergillus* and *Penicillium* spp.) is one of the main reasons for maintaining seed moisture content below certain levels. In many tropical countries, where controlled environment storage facilities may not be available, maintenance of seed viability under conditions of high relative humidity and temperature is one of the most important limiting factors of seed production.

Seed conditioning. The importance of seed conditioning in controlling seed diseases is often overlooked. The process of cleaning and sizing seedlots automatically eliminates diseased seeds when their physical characteristics have been altered and structures of pathogens such as galls or sclerotia are present. Seed conditioning equipment has considerable potential as a means of reducing the amount of a pathogen in a seedlot to tolerable levels.

Fungicides. Perhaps the most widely used seed disease control practice is treatment of seed with fungicides. For some crops (such as corn and peanuts) the use of fungicides with broad spectra of activity against soil and seed-borne pathogens has been tried and tested over many years, and the benefits are well established. Treatment of cereal seeds with fungicides, specifically active against smuts, also have proved beneficial in some circumstances. However, with other crops, such as soybeans, the value of fungicide seed treatment has not been clearly demonstrated. This is, in part, due to a lack of knowledge of the factors that influence the efficacy of seed treatment. Application practices usually are determined by relating treatment rates to subsequent emergence and yield in a series of field tests in different locations under a variety of environmental conditions. If repeated often enough, this type of experiment may provide reasonably reliable information on application rates, but very little information is obtained about disease epidemiology. As a result, failures in control practices cannot be explained, and more fungicides tend to be used than are necessary.

Fungicides also are used as sprays on the growing seed crop to control disease on the seed. In the United States, benzimidazole fungicides have been used to control *Phomopsis* spp. infection of soybeans. In some states, however, the disease often is not serious enough to justify the treatment. A predictive method to identify fields that should be sprayed was developed in Iowa. The method uses *Phomopsis* inoculum on soybean pods to indicate the risk of serious seed disease developing. Pod-borne inoculum is measured by a simple test (Fig. 1) that uses equipment and materials obtainable at a hardware store, and with minimal training seed handlers can carry out the test themselves.

Physical and biological treatments. Physical seed treatments, using hot water or aerated steam, also are used to control seed diseases. The benefits of these treatments often have to be balanced against the damage done to seed viability, however. The use of hot water to control blackleg (caused by *Phoma lingam*) and blackrot (caused by *Xanthomonas campestris*) in high-value hybrid cabbage seed exemplifies this problem.

There has been considerable interest in recent years in treating seed with fungi and bacteria that are antagonists to seed or soil-borne pathogens. So far, however, the results have been inconsistent. One of the major problems with this approach is a lack of understanding of the ecology of the microorganisms involved. Biological seed treatment certainly has potential, but will not be widely accepted by the seed industry until these problems have been resolved.

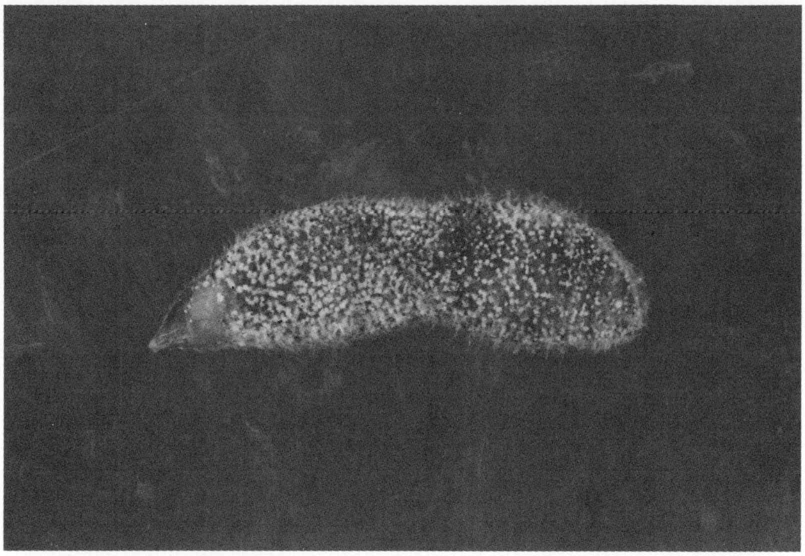

Fig. 1. Detached soybean pod tested for the presence of *Phomopsis* spp. by treating with a herbicide to induce production of fungal fruiting bodies.

Inspection. Seed diseases also can be controlled by health inspection programs. These methods are used for seedlots that have to be certified for plant quarantine purposes, or for pathogens for which tolerances of seed-born inoculum have been established for disease control.

Seed health testing. A range of methods exist for determining whether seedlots are contaminated or infected with plant pathogens. One approach is to examine seed crops while they are growing, to determine whether any diseases which have the potential to be seed-borne are present in the field. This method has the advantage of being quick and inexpensive, but it is not particularly accurate. It is, however, used extensively in the United States for seed crops grown for export. Another simple test is to examine dry seed for visual symptoms of seed-borne pathogens. This also is not accurate because only severe infections will show symptoms for many pathogens. A more sensitive test is to incubate seeds on damp blotting paper or on culture media (Fig. 2). Under these conditions, the pathogen grows out and expresses itself. Test conditions, including environmental factors such as light, temperature, and humidity, or pretreatment of the seed by freezing or surface sterilization, are usually specified for particular pathogens. Culture media, on which only certain pathogens will grow, also is a useful tool in these types of tests. Generally, incubation tests have been used only for fungal pathogens, but recent improvements in selective media have resulted in some effective tests for seed-borne bacteria. In the past, tests for seed-borne bacteria and viruses usually required growing the seeds in the laboratory or greenhouse and checking seedlings

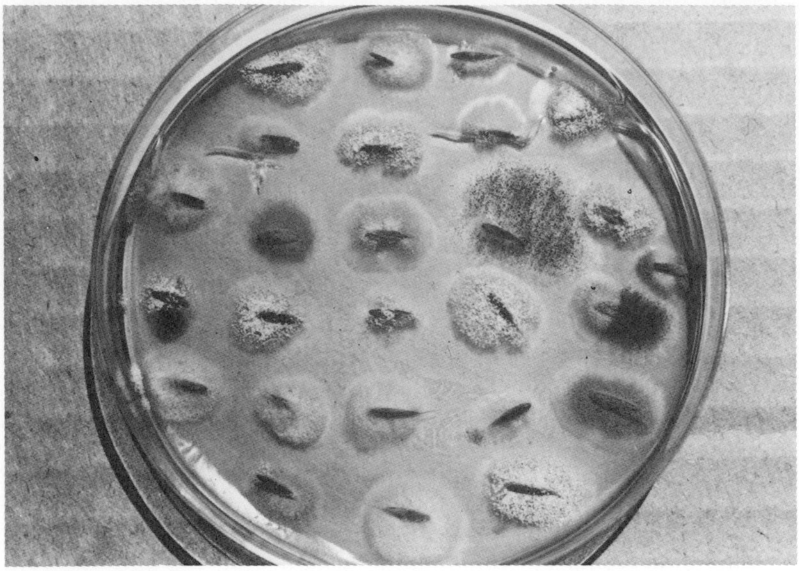

Fig. 2. Culture-plate seed health test for *Aspergillus* and *Penicillium* spp. on grass seeds.

for symptoms caused by the pathogen. Although this type of test can be quite accurate, it usually requires extensive space and also is labor-intensive.

New serological techniques have resulted in greatly improved tests for bacteria and viruses on seeds. These tests involve physicochemical reactions between blood serum and plant viruses or bacteria. The serum is obtained from animals injected with preparations of the pathogen to be tested. Seedlots suspected of carrying the pathogen are usually treated to extract the pathogen. When the serum and pathogen extract are mixed, the antibodies become associated with particles of the pathogen. In some tests the antibodies precipitate the pathogens; these are called diffusion tests. In others, the pathogen-antibody association is labeled with an enzyme, a radioisotope, or a fluorescent dye. Scanning electron microscopes also may be used to detect pathogen-antibody associations. These tests are very sensitive but are, expensive to run, requiring specialized equipment and highly trained personnel. Problems also exist in standardizing the tests because of the quality of the antiserum. However, techniques are now being developed to produce monoclonal antibodies which may facilitate the production of large quantities of high-quality antiserum.

The practical value of a seed health test lies in what it tells about the future performance of the seed. When there is little argument that a particular pathogen would cause serious crop losses if introduced into a particular geographical area, the most sensitive seed health test available should be applied to the seed to be grown there. A different approach should be taken, however, when seed is grown in an area where a disease is present. In this case, inoculum measured should be that which is most likely to be transmitted to the new crop. This could be, for example, inoculum of the pathogen internal to the seed. The most appropriate test then would be one that measures only internal inoculum and not inoculum that might be on or in the seedcoat. The most sensitive test would not be the most valuable in this context. As indicated earlier, seed health testing is of limited value in plant disease control unless the value of seed-borne inoculum in the life cycle of the pathogen is properly understood.

For background information *see* PLANT PATHOLOGY in the McGraw-Hill Encyclopedia of Science and Technology.

[DENIS C. McGEE]

Bibliography: P. Neergard, *Seed Pathology*, volumes 1 and 2, 1977; M. J. Richardson, *An Annotated List of Seed-Borne Diseases*, 3d ed., Commonwealth Agricultural Bureaux, Farnham Royal, Slough, U.K., 1979.

Semiconductor heterostructures

There has been considerable interest during the last decade in studying the properties of artificial semiconductor crystals, called superlattices, which consist of many alternating layers of two different materials. The alternating layers in these structures provide an artificially imposed periodicity in addition to the natural atomic periodicity of the individual layer materials. For layer thicknesses less than a few times 10^{-8} m, the superlattices behave as new semiconductor materials with properties that are strongly influenced by the artificially imposed period. Experimental studies of superlattices have become possible due to the development of crystal growth techniques which are capable of producing the thin layers required in these structures. Until recently, essentially all of the superlattice structures have been grown by using two crystal materials, gallium arsenide (GaAs) and gallium aluminum arsenide (GaAlAs) with lattice constants that are equal to better than one part in a thousand.

Past work on multilayered device structures with much thicker layers (greater than 10^{-6} m) has shown that lattice mismatches of greater than 0.1% result in the presence of undesirable crystal defects which unacceptably degrade the properties of the structures. The need for close lattice matching in these thicker-layered structures has considerably restricted the choices of layer materials. However, it has been known for many years that mismatches of up to several percent can be accommodated in very thin layers by elastic layer strains without the generation of mismatch defects. Recent theoretical and experimental work has emphasized that this property of solids permits high-quality superlattices to be grown from lattice-mismatched materials since the superlattice layers are typically in the required layer thickness regime for strain accommodation. These strained-layer superlattices allow considerable freedom in the choice of layer materials used to grow them so that there is a large number of potential strained-layer superlattice structures.

The flexible choice of layer materials in strained-layer superlattices, combined with the capability of influencing the strained-layer superlattice material properties through the choice of the superlattice period, implies that strained-layer superlattice properties can be varied over wide ranges. Strained-layer superlattices form a new class of artificial semiconductors with tailorable material properties. The possibility of tailoring strained-layer superlattice material properties to fit specific device applications provides considerable motivation for studying these new semiconductors.

Structural properties. The atomic structure of strained-layer superlattices is illustrated schematically in Fig. 1. When the mismatched semiconductor layers (Fig. 1*a*) are combined into a strained-layer superlattice (Fig. 1*b*), the material with the larger lattice constant uniformly contracts in the planes parallel to the interfaces, and, likewise, the material with the smaller lattice constant uniformly expands, so that the lattice constants of the strained layers in these planes (called $a^{\|}$) are equal. The final $a^{\|}$ value occurs at some intermediate value between the two unstrained lattice constants, with the fraction of the mismatch taken up by each type of layer being determined by the ratio of thicknesses of the layers and the relative stiffness of the materials. For unequal layer thicknesses and materials which are equally stiff, the thinnest of the layers takes up a larger fraction of the total mismatch, and the $a^{\|}$ value is closer to that of the unstrained lattice constant of the thicker layer. This dependence can be used to control the $a^{\|}$ value of strained-layer superlattices. The $a^{\|}$ value is the lattice constant which determines the lattice matching of the strained-layer superlattice as a whole to other semiconductors. A number of characterization techniques (transmission electron microscopy, x-ray diffraction, and ion channeling) have been used to verify the structural properties of strained-layer superlattices grown from the gallium arsenide phosphide (GaAsP) and indium gallium arsenide (InGaAs) material systems.

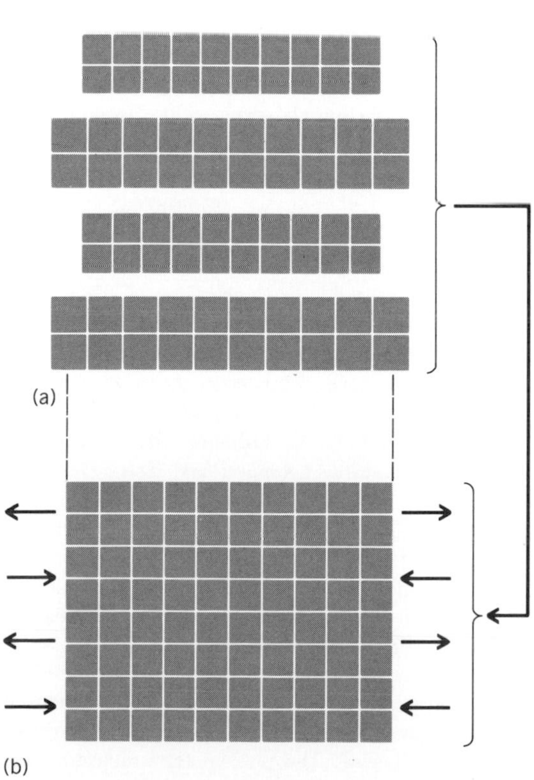

Fig. 1. Formation of strained-layer superlattice. (*a*) Two types of thin, lattice-mismatched semiconductor layers. (*b*) Formation of strained-layer superlattice by strain accommodation of the lattice mismatch.

Optical and electrical properties. An important material parameter of semiconductors is the magnitude of the band gap energy (E_g). The E_g value determines a number of other material properties, including the range of photon wavelengths which are strongly absorbed or emitted by the semiconductor. In strained-layer superlattices, the value of E_g can be tailored through the choice of the superlattice structure. For a particular pair of layer materials, the superlattice E_g is a function of the thicknesses of the layers. In general, the E_g value increases with decreasing layer thicknesses. At larger layer thicknesses, the E_g value approaches a value which depends on the pair of materials chosen. This layer thickness dependence is a quantum-mechanical effect which occurs in both closely lattice-matched superlattices and strained-layer superlattices. In the strained-layer superlattice, the layer strains also have a significant influence on the strained-layer superlattice E_g. This is a consequence of the changes in the bulk properties of the layers due to the elastic strains.

For fixed layer thicknesses, the strained-layer superlattice E_g can also be varied by changing the materials used in the layers. For example, strained-layer superlattices have been grown from alternating layers of the semiconductor gallium phosphide (GaP) and the semiconductor alloy $GaAs_xP_{1-x}$ (x and $1 - x$ label the relative amounts of GaAs and GaP, respectively). The band gap of $GaAs_xP_{1-x}$ is a monotonically decreasing function of x, and the lattice constant linearly increases with x up to a value which is 3.6% larger than that of GaP. This means that there are no pairs of GaAsP alloys with different band gaps but the same lattice constant. In contrast, strained-layer superlattices grown from alternating layers of $GaAs_xP_{1-x}$ and $GaAs_yP_{1-y}$ have E_g values and $a^{\|}$ values that can be varied independently of each other by varying the x and y values and varying the layer thicknesses of the two layer types. Theoretical band gap results for a few of these strained-layer superlattice structures are given in Fig. 2. Although these results are for structures that vary the alloy composition of only one of the layers and have only one ratio of layer thicknesses, it is clear that the E_g and $a^{\|}$ values can be varied independently. Experimental E_g and $a^{\|}$ results have been obtained for the structures in Fig. 1 with layer thicknesses greater than 6×10^{-9} m which confirm these results. The results illustrate one example of a strained-layer superlattice capability that does not

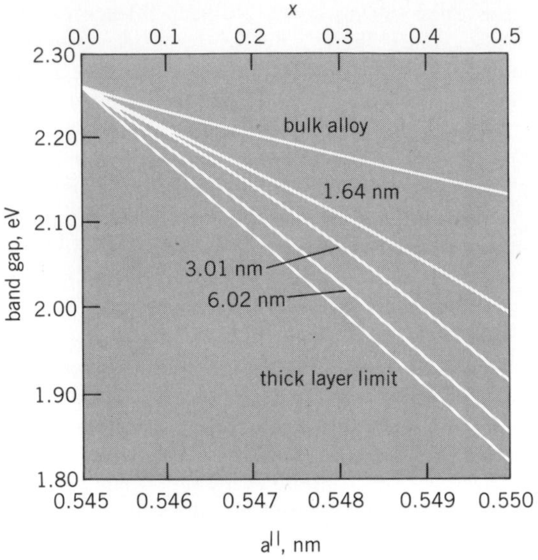

Fig. 2. Calculated band gap energies (in units of electron-volts) for $GaP/GaAs_xP_{1-x}$ strained-layer superlattices with fixed ratio of layer thicknesses (1:1) as a function of x and the GaP layer thickness. The corresponding strained-layer superlattice lattice constant in the planes parallel to the strained-layer superlattice interfaces, $a^{\|}$, is also given for each x value. The line labeled bulk alloy is the experimental band gap versus lattice constant dependence for the bulk $GaAs_{x/2}P_{1-x/2}$ alloy. (*After G. C. Osbourn, Electronic structure of $GaAs_xP_{1-x}/GaP$ strained-layer superlattices with x <0.5, J. Vac. Sci. Technol., 21(2):469–472, July-August 1982*)

exist for the materials out of which the strained-layer superlattice is grown. As a result, strained-layer superlattices have the potential of being used as the basis for a number of semiconductor device structures for which thick mismatched materials are unsuitable.

The electrical transport properties (for example, electron diffusion lengths) along the direction perpendicular to the superlattice interfaces can also be varied. This is also a quantum-mechanical effect which depends on the superlattice layer thicknesses for fixed layer materials. In general, the perpendicular diffusion lengths in the strained-layer superlattices are smaller than those of the bulk materials from which the strained-layer superlattice is grown, and decrease with increasing strained-layer superlattice layer thicknesses. Recent theoretical results indicate that the flexible choice of strained-layer superlattice layer materials and layer thicknesses allow the perpendicular diffusion lengths in strained-layer superlattices to be varied independently of both the strained-layer superlattice E_g and the strained-layer superlattice $a^{\|}$. Thus, it is possible to independently vary a transport property, an optical property, and a structural property in strained-layer superlattices. The carrier diffusion lengths in the planes parallel to the strained-layer superlattice interfaces can be equal to the values in bulk semiconductors. In fact, the first experimental studies of room-temperature and low-temperature parallel electrical transport in $In_{0.2}Ga_{0.8}As/GaAs$ strained-layer superlattices (with approximately 1.0% lattice mismatch) have been carried out, and values equal to those in high-quality bulk $In_{0.2}Ga_{0.8}As$ alloys were observed. This confirms the absence of mismatch defects in these strained-layer structures.

The artificially imposed periodicity of superlattices can also alter the strengths of optical photon emission and absorption processes from the corresponding strengths in the bulk materials which make up the strained-layer superlattice. In particular, optical transitions which cannot occur at all without the participation of lattice vibrations (indirect transitions) in bulk semiconductors can in some cases be allowed without lattice vibrations in strained-layer superlattices made from the same semiconductors. Theoretical studies have recently indicated that this can occur in certain $GaP/GaAs_xP_{1-x}$ strained-layer superlattices with x less than 0.5. Recent experimental studies of the optical properties of these GaAsP-based strained-layer superlattices indicate significant enhancements of the optical absorption strengths of transitions which would be indirect in the bulk GaAsP materials themselves.

To date, theoretical and experimental studies of strained-layer superlattice structures based on two semiconductor alloy systems (GaAsP and InGaAs) have been carried out. Future work in this new field is likely to emphasize the study of the large number of potential strained-layer superlattice structures that have not yet been examined. Studies of both the basic material properties and the variety of potential device applications of these tailorable semiconductors will be of interest.

For background information *see* CRYSTAL ABSORPTION SPECTRA; CRYSTAL STRUCTURE; SEMICONDUCTOR; SEMICONDUCTOR HETEROSTRUCTURES in the McGraw-Hill Encyclopedia of Science and Technology. [GORDON C. OSBOURN]

Bibliography: I. J. Fritz, L. R. Dawson, and T. E. Zipperian, *J. Vac. Sci. Technol.*, A1, 1983; P. L. Gourley et al., in *Proceedings of 1982 International Symposium on GaAs and Related Compounds*, Institute of Physics, London, 1983; J. W. Matthews and A. E. Blakeslee, *J. Cryst. Growth*, 32:265, 1976; G. C. Osbourn, *J. Vac. Sci. Technol.*, A1, 1983; G. C. Osbourn, R. M. Biefeld, and P. L. Gourley, *Appl. Phys. Lett.*, 41:172, 1982.

Sexually transmitted diseases

Sexually transmitted diseases comprise a large number of clinical syndromes caused by at least 20 pathogenic organisms and viruses that are transmitted by sexual contact. Some also are transmitted by nonsexual personal contact, but in each case sexual transmission is important in the overall epidemiology of the disease. The present term is technically synonymous with the term venereal disease, which historically has referred to only a few infections and, having a pejorative connotation, has been largely supplanted.

Several factors account for renewed interest in

sexually transmitted diseases during the past 15 years. Dramatic increases in the incidence of most of these diseases have resulted from changing patterns of sexual behavior and the shift from primarily male-dependent barrier contraceptives to female-dependent nonbarrier methods that may actually enhance susceptibility to the diseases and some of their complications. Their perceived clinical spectrum has broadened dramatically and now includes several conditions not previously known or widely accepted to be sexually transmitted. Similarly, sexually transmitted diseases are now recognized to assert their major long-term effects in ways not traditionally associated with venereal diseases, such as female infertility, complications of pregnancy, neonatal morbidity, and cancer. Highlights of important new knowledge in this field are summarized below.

Acquired immunodeficiency syndrome. The acquired immunodeficiency syndrome (AIDS) was first recognized in 1981, although it probably first appeared in the United States in 1978 or 1979. Characterized by various opportunistic infections (especially pneumonia due to the protozoan *Pneumocystis carinii*) and malignancies (especially Kaposi's sarcoma), AIDS is associated with profound impairment of cell-mediated immunity, involving T lymphocytes. Through 1983, over 3000 persons suffering from AIDS had been reported in the United States, over 1200 of whom had died; these numbers were expected at least to double by the end of 1984.

About 70% of AIDS cases occur in homosexual men, its occurrence being strongly correlated with numbers of sexual partners and other indices of sexual activity; this evidence of sexual transmission is corroborated by the occasional occurrence of AIDS in female sexual partners of persons with the disease. AIDS also affects intravenous drug abusers and persons with hemophilia, in whom transmission by contaminated blood, blood products, or needles is more important than sexual contact. Native Haitians, in whom sexual transmission is suspected, also are susceptible to AIDS. Fewer than 5% of persons affected fall outside these recognized risk groups. The cause of the immunologic impairment in AIDS is unknown, but an unidentified infectious agent (probably a virus) is suspected. Personal contact without overt exchange of body secretions or blood apparently confers no significant risk of transmission. *See* AIDS.

Gonorrhea. Gonorrhea remains a common sexually transmitted disease, with frequent and severe complications; about 2 million cases occur annually in the United States. Penicillin has been the mainstay of treatment since the 1940s, despite the fact that *Neisseria gonorrhoeae*, the causative organism, has become progressively resistant to this antibiotic. Such low-level resistance is mediated by chromosomal mutations that result in decreased permeability of the cell envelope to penicillin. In 1976, strains of *N. gonorrhoeae* that were absolutely resistant to penicillin were isolated simultaneously in Africa and the Philippines, and were found to produce penicillinase (β-lactamase), an enzyme that mediates the destruction of penicillin. The gene for production of penicillinase is carried on a plasmid, an extrachromosomal fragment of DNA, that may be transmitted between bacteria, facilitating the dissemination of penicillinase-producing *N. gonorrhoeae* (PPNG) in the community. By 1980, PPNG accounted for up to 70% of all isolates of *N. gonorrhoeae* in parts of Africa, Asia, and the western Pacific, so that penicillin no longer is recommended for the primary treatment of gonorrhea in much of the world. Although PPNG caused less than 1% of gonorrhea reported in the United States and Europe in 1982–1983, the incidence of PPNG infections rose rapidly through most of this period (Fig. 1). Although several other antibiotics are available for treatment of gonorrhea due to PPNG, most are more toxic or more expensive than penicillin, a critical issue in socioeconomically disadvantaged countries.

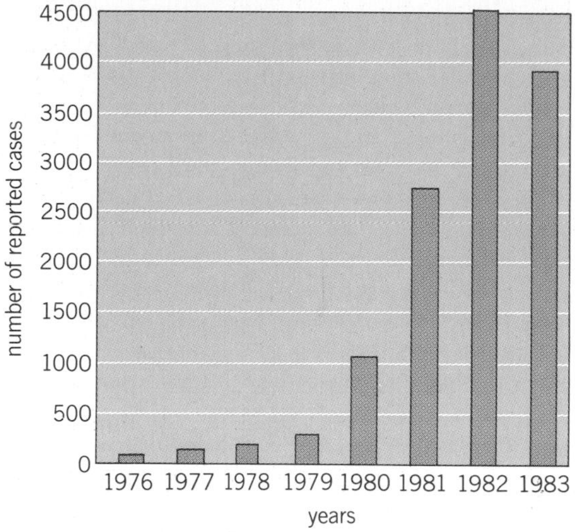

Fig. 1. Reported infections with penicillinase-producing strains of *Neisseria gonorrhoeae* in the United States, 1976–1983. (*Centers for Disease Control*)

Chlamydial infections. Infection due to *Chlamydia trachomatis*, a bacterium that requires cell culture techniques for isolation, has been shown in the past decade to be the most common sexually transmitted disease in industrialized societies. The prevalence of *C. trachomatis* infection approximates that of gonorrhea in public clinics for sexually transmitted diseases and is up to tenfold higher in family-planning and university student-health clinics; overall, chlamydial infections probably are two to three times more frequent than gonorrhea, accounting for 4 to 6 million infections annually in the United States. *Chlamydia trachomatis* causes 40–50% of cases of nongonococcal urethritis and is the most common cause of acute epididymitis in young men, which can result in sterility. It is also a common cause of acute rectal infection (proctitis) in homosexually active men. In women, chlamydial infection causes mucopurulent cervicitis, urethral infection that may mimic nonsexually transmitted bacterial urinary tract infection, and acute salpingitis (pelvic inflammatory disease), the most common cause of female infertility. The organism has been linked recently to various complications of pregnancy and to presumably precancerous changes (dysplasia) of the uterine cervix. Infants born to infected mothers are susceptible to acute conjunctivitis, pneumonia, and possibly to otitis media that may result in chronic hearing impairment. Some strains of the organism cause lymphogranuloma venereum, one of the classical venereal diseases, now uncommon in the United States and Europe. Chlamydial infections respond to treatment with the tetracycline antibiotics, erythromycin, or the sulfonamides.

Many chlamydial infections are asymptomatic or cause nonspecific signs and symptoms; for this reason, the lack of an inexpensive, readily available diagnostic test has resulted in serious underdiagnosis. The recent application of immunologic methods for rapid detection of *C. trachomatis* is therefore an important development that will greatly improve the control of this disease. Figure 2 illustrates one such test employing fluorescence microscopy to identify *C. trachomatis* in genital secretions by use of a fluorescein-conjugated monoclonal antibody to the organism.

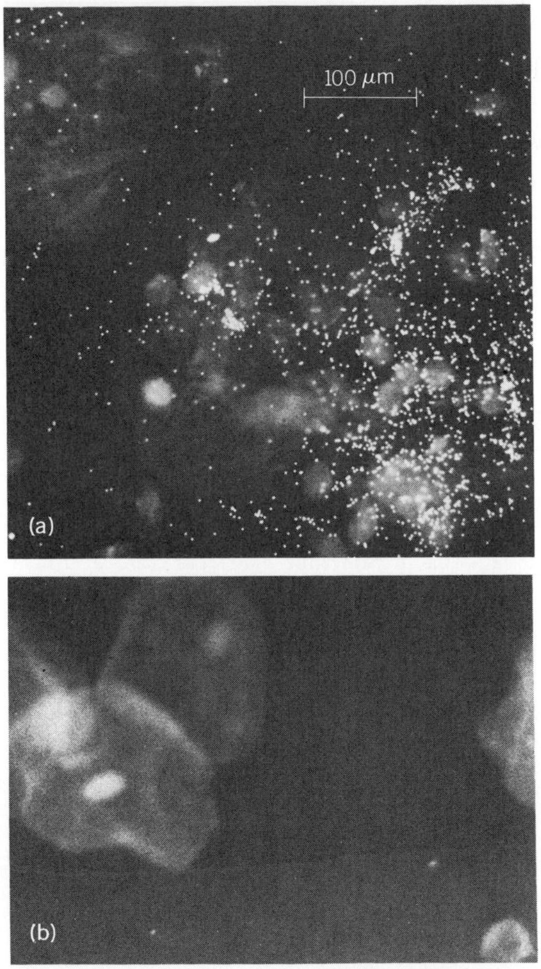

Fig. 2. Fluorescence photomicrographs of secretions from the uterine cervix. A monoclonal antibody against *Chlamydia trachomatis* was conjugated with fluorescein, a compound that fluoresces under ultraviolet light, and this reagent was applied to cervical secretions. This test gives definitive results within an hour of collection of the specimen. (*a*) Positive specimen; *C. trachomatis* elementary bodies appear as fluorescent points (in the original micrograph, bright yellow-green points on a dark red ground). (*b*) Negative specimen; only counterstained epithelial cells are seen. (*Syva Co., Palo Alto, California*)

Genital herpes. Genital infection with herpes simplex virus (usually type 2) is the third-commonest sexually transmitted disease in the United States, following chlamydial infections and gonorrhea. It is estimated that 400,000 to 600,000 new genital herpes infections are acquired annually. Although genital herpes usually is self-limiting and rarely dangerous for adults, infants born to infected mothers are susceptible to serious infections that may lead to permanent neurological impairment or death. It is estimated that at least 1000 serious neonatal herpes infections occur annually in the United States. Genital herpes has also been linked to an increased risk of squamous cell cancer of the cervix in women. These facts, the recurrent nature of the infection, and the lack of effective treatment have contributed to severe psychosexual morbidity among some affected persons and great fear of the disease in sexually active persons. The development of acyclovir, the first chemotherapeutic agent that is clinically active against genital herpes, is therefore a major advance. Applied to the skin lesions in an ointment, acyclovir ameliorates initial episodes of genital herpes but has little effect on recurrent episodes. Orally administered acyclovir is effective for both initial and recurrent episodes, and an intravenous preparation is available for severe infections in infants or adults. Other potentially effective antiviral agents also are under investigation for the treatment of genital herpes.

For background information *see* DRUG RESISTANCE; GONORRHEA; HERPES SIMPLEX; LYMPHOGRANULOMA VENEREUM; OPPORTUNISTIC INFECTION; VENEREAL DISEASE in the McGraw-Hill Encyclopedia of Science and Technology.

[H. HUNTER HANDSFIELD]

Bibliography: K. K. Holmes et al. (eds.), *Sexually Transmitted Diseases*, 1984; L. Corey et al., Genital herpes simplex virus infections: Clinical manisfestations, course, complications, *Ann. Intern. Med.*, 98:958, 1983; H. H. Handsfield, Sexually transmitted diseases, *Hosp. Prac.*, p. 99, January 1982; H. H. Handsfield et al., Epidemiology of penicillinase-producing *Neisseria gonorrhoeae* infections: Analysis by auxotyping and serogrouping, *N. Engl. J. Med.*, 306:950, 1981.

Shape memory alloys

Shape memory alloys, a new class of metallurgical materials, possess unusual mechanical deformation characteristics and other exceptional properties as well. Their intrinsic shape memory represents one of the simplest ways to generate motion, force, and work.

Metals and alloys are widely recognized for their strength and ductility. The shape of metallic objects may be changed by deformation, but once deformed, the shape of an object is usually fixed, irrespective of whether or not it is heated or cooled. In contrast, shape memory alloys can be used to make devices that change from one precise shape to another in response to temperature changes.

Characteristics. Many alloys are now known to exhibit the shape memory effect, including the alloy systems Cu-Zn, Cu-Sn, Cu-Zn-Al, Cu-Zn-Ga, Cu-Zn-Sn, Cu-Zn-Si, Cu-Al-Ni, Cu-Au-Zn, Au-Cd, Ni-Ti, Ni-Ti-Cu, Ni-Al, and Fe-Pt. These alloys have several properties in common: (1) They undergo a martensitic phase transformation from a high-temperature phase (the parent) to a low-temperature phase (known as martensite). During this transformation there is a change in crystal structure, usually from a phase of high symmetry to one of lower symmetry, and the transformation occurs in a diffusionless manner. (2) They are ordered (both the parent and the martensite). (3) The martensitic transformation is a crystallographically reversible, thermoelastic phase change.

Basically, when a specimen of one of the above alloys is deformed in the martensitic (low-temperature phase) condition, it will, after release of the external stress and upon application of heat, regain its original undistorted shape. Shape recovery occurs during reverse transformation of (deformed) martensite back to its parent phase. Provided that a certain strain (typically 6–8%) is not exceeded during deformation, shape recovery will be complete. In other words, a device made of a shape memory alloy can be conditioned to have any fixed shape in the high-temperature condition and another variable shape, obtained by deformation, in the low-temperature (martensitic) condition. This "memory" suggests interesting applications for these materials.

The substantial recovery stresses generated during reverse (martensite [deformed] → parent) transformation form the basis of a number of recently proposed designs for various solid-state heat engines. In essence, the shape memory effect represents a direct conversion of heat into mechanical energy. Roughly speaking, a stress of scale 10 is required to deform a martensitic shape memory alloy, but when a deformed specimen is heated through the reverse transformation regime the recovery stress may reach scale 100.

Crystallographic mechanisms. Martensitic transformations are shearlike or displacive in nature. Consequently the crystals of martensite which form from the parent phase take the shape of lenticular platelets, like mechanical twins. This is a low-strain-energy morphology. In general, there will be 24 different orientations of these martensite platelets. That is, the platelets will form on 24 different "habit planes," and the lattice orientation of each platelet with respect to the crystallographic axes of the parent crystal is different. These platelets form cooperatively in a self-accommodating manner.

An example is a hypothetical case where a monocrystal of the parent phase has undergone a "natural" transformation process which leads to 24 "variants" of martensite. When such a martensitic microstructure is deformed, some variants seem to consume others, hence increasing the relative volume of the former. Eventually, of the original 24 variants, only one of them survives when the memory deformation of martensite is completed. The residual variant is that particular one whose shear component is most parallel to, and thus interacts most favorably with, the directional sense of the applied stress.

A more detailed analysis would show that although 24 variants of martensite of different orientation formed initially, these various orientations are not actually independent. Some variants are twin-related to others. By means of boundary movement or additional twin formation under stress, some variants are converted (by twinning) into others. For example, the formation of twins under stress in a given variant could turn out to be the equivalent of introducing variant B within variant A, the end result being the conversion of variant A to variant B. Further, the migration of the new variant B into variant C could cause the consumption of variant C by means of twin boundary movement. By such twinning processes, most of the initially formed martensite variants are eliminated.

Ideally, these twinning processes lead to the formation of only a single variant of martensite. Because of crystallographic restrictions, considering that the martensite phase is usually of low symmetry, the crystallographic paths for the reverse transformation become highly limited, and in effect, the only way for the martensite to revert to the parent phase is by means of a specific unshearing process. This restricted reversal is the essence of the shape memory effect.

Two-way shape memory effect. The two-way shape memory effect, in which a specimen will spontaneously change shape during the formation of

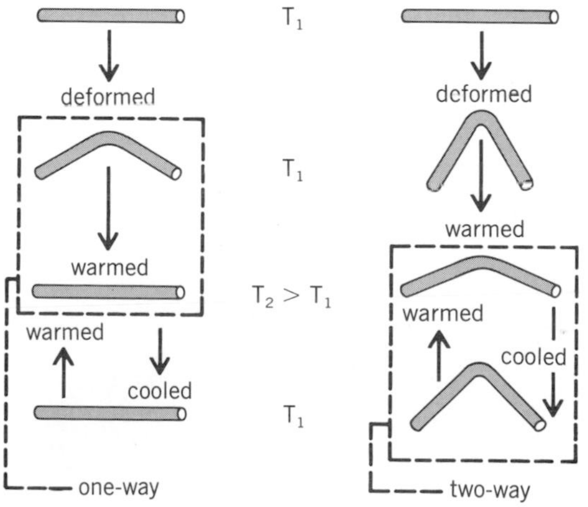

Fig. 1. Comparison of the one-way and two-way shape memory effect.

martensite upon cooling and then spontaneously undeform to the initial state during heating, may be thought of as a simple bending-unbending situation which may be repeated indefinitely. By thermomechanical treatment, stresses can be built into the parent phase which then program the martensitic transformation during cooling to occur in one direction, resulting in the formation of only a single variant. This then is essentially a shearing process, and the speciman bends during cooling. The "trained" martensite monocrystal reverts (unbends) to its parent phase by unshearing (unbending), as described above. The one-way and two-way shape memories are contrasted in Fig 1.

Applications. Both Ti-Ni and Cu-Zn-Al alloys are commercially available for shape memory designs, and various products utilizing these alloys are being marketed. Pipe couplings, electrical connectors, pins, seals, clamps, and related devices, applied in the deformed martensitic condition, are secured in place by heating. Similarly, the alloys are used to manufacture thermally activated valves, thermostats, vent shut-offs, fire door closers, and other devices. An application of the two-way shape memory effect is shown in Fig. 2. In this case, a Cu-Zn-Al shape memory spring is counterbiased by a steel spring so that its operating temperature range spans only several degrees. In the cold condition the shape memory spring is closed, but when warmed it elongates and thus opens a hinge which causes a window to open, thereby ventilating a greenhouse under warm conditions and sealing it off under cold conditions. Similarly, large-scale shape memory effect springs may be used in electrical heating systems to drive a piston-operated pump that is placed in an oil well.

Medical applications of shape memory alloys include bone plates for the fixturing of fractures, aneurysm clamps, contraceptive devices, and blood clot filters. Artificial hearts using the heat engine characteristics of shape memory alloys, and programmed pulses of electrical current, are under development. Also, orthodontic appliances of Ti-Ni alloys are currently in use and allow restoration of maloccluded teeth in a fraction of the usual time.

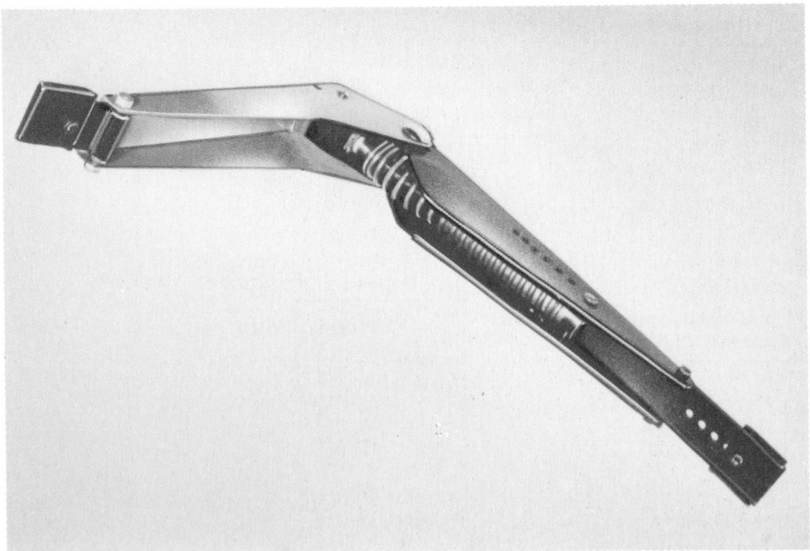

Fig. 2. Greenhouse window control incorporating a Cu-Zn-Al shape memory alloy spring.

For background information *see* ALLOY; HEAT TREATMENT (METALLURGY) in the McGraw-Hill Encyclopedia of Science and Technology.

[C. M. WAYMAN]

Bibliography: J. Perkins (ed.), *Shape Memory Effects in Alloys*, 1975; L. McD. Schetky, *Sci. Amer.*, 241:74–82, 1979; C. M. Wayman, *J. Met.*, 32:129–137, 1980; C. M. Wayman and K. Shimizu, *Met. Sci. J.*, 6:175–183, 1972.

Single sideband

The maturing telecommunications network is experiencing a shift in emphasis from conventional frequency modulation (FM) to other forms of modulation. For "long-haul" applications, both terrestrial and via satellite, spectral efficiency has important economic impact, and single-sideband amplitude modulation (SSBAM) is gaining wide acceptance. This development has taken two forms, distinguished by their applications: HiLinear (abbreviated AR, for analog radio) for multirepeater long-haul terrestrial applications; and companded single sideband (CSSB) for satellite application.

Commercial AR systems have been deployed which transport 6000 voice circuits and 3 megabits/s of data simultaneously between the same terminals and in the same 30-MHz band which previously supported 1800 voice circuits and 1.5 Mb/s with FM. The noise performance of the two systems is comparable. Similarly, CSSB has permitted increasing satellite transponder capacity from 1800 voice circuits to 7800. Importantly too, use of a single transponder by a number of separate ground terminals is now accomplished with 100% efficiency, whereas multicarrier FM loading typically involves 25–75% reduction in channel loading. The major engineering challenges to SSBAM system design are introduced below, and examples of successful system accommodations are described.

AR systems. The main functional elements of the AR system are fairly conventional, and the AR repeater (Fig. 1) has basic features which closely parallel those of a microwave repeater: down-conversion from rf to intermediate-frequency (i-f), amplification at i-f with automatic gain control (AGC), and retranslation to a frequency-shifted rf output. However, a number of basic considerations deriving from the choice of SSBAM have a major impact on the system design requirements, and lead to the inclusion in an AR repeater of elements that are less common or not found in FM repeaters: low-noise-figure microwave preamplifiers, a built-in space-diversity rf combiner option, dynamic amplitude equalization, a linearizing i-f predistorter, and locking of the microwave generator (local oscillator) to a common, high-stability synchronizing source.

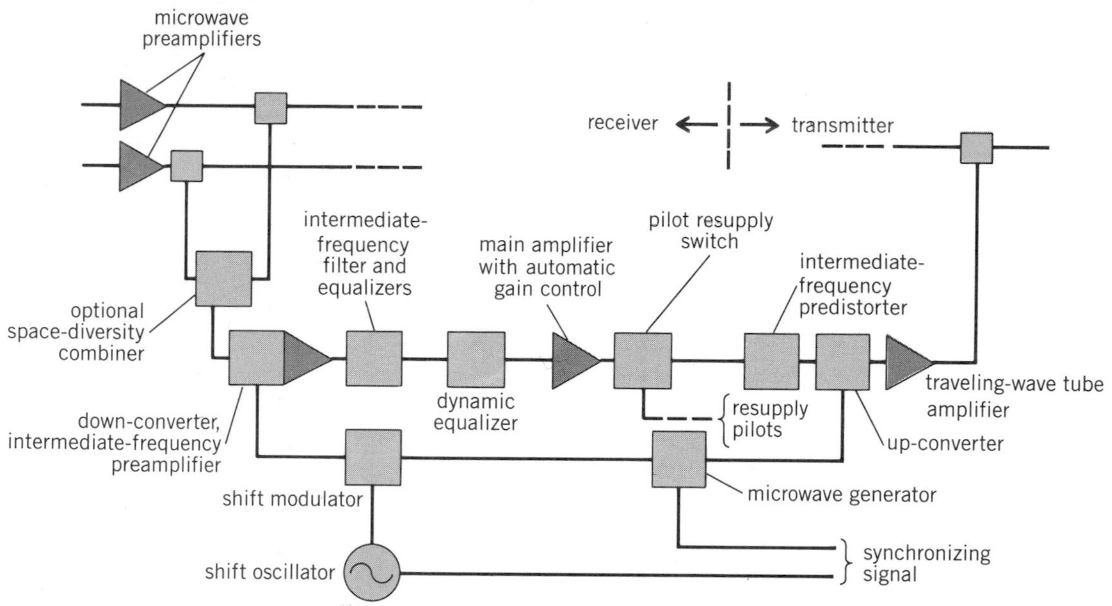

Fig. 1. Simplified block diagram of an AR repeater (transmitter–receiver) bay. The dynamic equalizer is present in main stations only.

Linearity requirements. A high degree of linearity is necessary in the active transmission elements of an AR system. Active devices may be described by a power series input–output characteristic. The only significant in-band distortion contributors, however, are the odd-order terms, the third being dominant. With typical values for an output traveling-wave-tube amplifier, the expected intermodulation noise would be 100 times greater than that tolerable from a single repeater in a standard long-haul allocation.

Reduction of this intermodulation noise is most practically obtained through the technique of predistortion. This adaptive compensation sums the signal with a properly weighted and antiphased replica of the intermodulation noise which will result from passage through the nonlinear channel. The necessary predistortion is obtained by passing a portion of the signal through a low-power nonlinear device, such as a biased diode. In practice, system improvements greater than 20 dB are obtained by this method.

Linearity is most difficult to achieve in the final power amplifier, but it can be accomplished adequately by predistortion. In the case of low-noise amplifiers, mixers, and transistor driver amplifiers, it is often possible to achieve the required linearity by operating at increased quiescent power (Fig. 2). The magnitude of intermodulation noise, *IM*, caused by nonlinear amplification is proportional to third- or higher-order powers of signal level. For a given choice of equipment, *IM* is minimized by minimizing signal level. The thermal noise contribution, *N*, on the other hand, is minimized by maximizing transmitted power. The best system solution minimizes the total (*IM* + *N*) power as indicated by the designated operating point in Fig. 2. In general, AR channel amplifiers operate with higher power drains than comparable elements in an FM system; however, the power drain per circuit served is less in an AR system.

The distortion produced by tandem nonlinear devices is coherent but, due to phase dispersion, may not be cophased. For a string of n identical nonlinear repeaters, the law of addition for intermodulation noise is defined by the equation below, where

$$W_n = K \log n + W \text{ dBrnc0}$$

W_n is the intermodulation noise of the string, W rep-

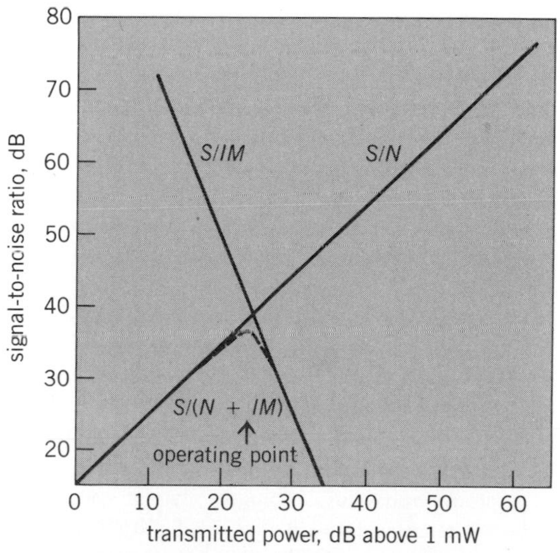

Fig. 2. Plots of the ratio of signal level to intermodulation noise (*S*/*IM*), the ratio of signal level to thermal noise (*S*/*N*), and the ratio of signal level to the sum of the two types of noise [*S*/(*N* + *IM*)], as functions of transmitted power. Both scales are logarithmic.

resents the intermodulation noise of one repeater, and K is a factor which depends on the relative phases of the noise components. The abbreviation dBrnc0 stands for decibels above reference noise, c-weighted, at the zero transmission level point; here, reference noise is 10^{-12} W, and c-weighted noise is noise whose frequency spectrum has been weighted in terms of the frequency response of human hearing. Cophased noise components add as 20 log n, random noise as 10 log n. The operative law of addition is between those extremes and will depend on the relative magnitude of the two types of contributors. Clearly, the net value of K is an extremely important factor in estimating total system noise on a radio route.

In the case of intermodulation noise contributed by repeater elements subject to predistortion correction, the residual distortion can be expected to have a random phase orientation. Nonlinear elements not (typically) subject to predistorter compensation, such as intermediate-frequency amplifiers, will generate intermodulation noise that adds systematically. Empirical results with early AR systems have indicated a K value of 17. Rearranging the signal components within the channel, typically at channel dropping stations, breaks up this coherent noise growth, and is a necessary feature of practical AR engineering.

Dispersive effects. Another basic consideration in the design of AR systems is the direct dependence of signal amplitude response to dispersive effects of the transmission medium. On typical (25-mi or 40-km) repeater sections, multipath fading can be expected to occur sufficiently often to require accommodation. Unlike FM, for which substitution of a frequency or space diversity alternate signal may be adequate, successful AM signal recovery also requires processing to compensate for in-channel amplitude dispersion. Dynamic, adaptive equalizers capable of adjusting level (over a range from +10 to −40 dB), slope dispersion (over the range ± 0.5 dB/MHz), and parabolic dispersion (over the range ± 1 to 15 dB/MHz^2) are required. In a widely used system, ±2 dB maximum deviation from perfect compensation is obtained for all but a very small fraction of time.

Absolute frequency recovery. Both voice and voice-band data applications require that customer–customer frequency offsets be controlled to within 1 Hz. Distribution of a national frequency standard has long allowed the close control of frequency offsets in multiplexers, but FM carriers are controlled only to manage intersystem interference. This is not adequate for AR; the frequency shifts which occur at each repeater must be canceled by a compensating shift at the receiving terminal. This is commonly done by phase-locking a recovered pilot tone to a local tone derived from the national standard.

Interference. The typical AR system has to operate in the interference environment caused by FM systems, digital radio systems, and other AR systems. The obvious effect of this interference is to add to the background channel noise. Other effects which can occur when two AR channels carrying different traffic interfere with each other are crosstalk and babble. If both the interfering and the desired signals are voice and are exactly aligned in frequency, the effect is called intelligible crosstalk and must be small (less than or equal to 10%) compared to thermal noise. If the two signals are increasingly offset in frequency, the interference becomes less comprehensible until, at a frequency difference of 500 Hz, the interference is judged to be unintelligible (referred to as babble) and is regarded as noise.

In order to avoid interference from cochannel FM carriers, the center of the AR message spectrum is left vacant. The noise contributed by the FM sidebands is controlled by proper selection of the spectrum gap and by transposing mastergroups along the route. The noise due to cochannel AR systems is controlled by antenna pattern isolation.

The AR system should have no significant signal-related energy outside its channel bandwidth. Thus, adjacent channel noise effects between AR channels are negligible. However, mutual problems occur when an AR and corouted FM channel are adjacent. Due to the wide bandpass of the intermediate-frequency filters needed in FM systems to reduce their own channel noise, a portion of the adjacent AR signal is coupled into the FM system and subsequently coupled back into the AR channel with a frequency offset. This energy is limited primarily by the cross-polarization discrimination of the antenna system. Estimates of these effects indicate they should be avoided if possible.

Maintenance. While AR systems do deliver the significant transport efficiencies described earlier, their successful transmission performance depends upon the proper working of adaptive equipment in every repeater. To assure their performance, and to locate troubles when they exist, it has been found desirable to implement an integrated, automated surveillance test system into the AR system. Typical measuring functions performed by such a subsystem are: measurements of the level, frequency, and slot noise of pilot tones, and measurements of intermastergroup noise, all carried out on each switch section during in-service tests; and measurement of idle noise (through tone scan), linearity, amplitude response, interference, and transmitter gain, all carried out on each switch section and each hop during out-of-service tests.

CSSB systems. Three important characteristics distinguish the long-haul microwave repeaters utilized in satellite service from those in terrestrial radio:

1. The satellite transmitter power radiated per circuit is some 7 dB greater than that utilized in AR; but the section loss (proportional to R^2/G_TA_R, where R represents hop length, G_T transmitter antenna gain, and A_R receiving antenna area) is roughly 50 dB greater.

2. Linearity requirements on the satellite transmitters are greatly relaxed relative to AR (that is, a

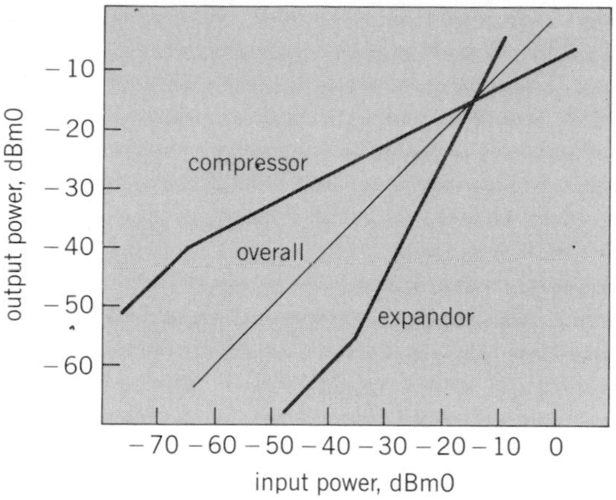

Fig. 3. Compressor, expandor, and overall characteristics of a compandor with an unaffected signal power (USP) of −15 dBm0 (decibels above 1 mW, referred to the zero transmission level point).

greater intermodulation noise–to–signal ratio is tolerable) since there is only one repeater, and therefore no 17 log *n* term in the system noise equation.

3. A single (downlink) receiver introduces the bulk of the satellite system thermal noise, whereas the typical AR receiver must share the thermal noise allocation with 150 others. (In addition, AR systems must accommodate up to 35 dB of multipath fading, which does not occur at satellite path elevation angles greater than 10°.)

Companding. The above inequalities do not completely offset each other, and for equivalent circuit performance would therefore require either a very significant reduction in channel loading or a greater tolerance for noise. The latter is indeed the case. Subjective noise improvement is achieved by introducing companding, an adaptive technique which reduces circuit noise between speech syllables and during pauses in speech, thus producing a subjective improvement in the circuit quality.

The subjective compandor advantage and potential interaction between compandor and echo canceler were studied in an extensive field trial: satellite frequency-division multiplexing–frequency modulation (FDM/FM) circuits were equipped with compandors and noise added to the transmission facility to simulate CSSB satellite transmission. Customers rated circuit quality. These ratings were compared with those resulting in a parallel experiment with customers using normal FDM/FM channels. Companded satellite circuits with noise levels up to 110,000 pWpo (picowatts, psophometrically weighted, at the zero transmission level point) were found to be subjectively equivalent to uncompanded satellite circuits with about 8000 pWpo noise. (Psophometric weighting is an international standard for weighting the noise frequency spectrum in terms of the frequency response of human hearing, which differs by about 0.5 dB from the c-weighting mentioned above.)

Introduction of compandors into the terminal equipment could alter the multichannel loading of the transmission system. The compressor component of the compandor acts as an amplifier which adjusts its gain according to the input level (Fig. 3). The higher the compandor unaffected signal power (USP; the point at which the compressor and expandor curves intersect), the higher the compressor output power for any given input below the USP. At the expandor, a higher USP value provides a larger objective noise advantage. However, this increase in objective advantage is offset by the increase in the multichannel loading, which results in increased transmission noise. These two effects cancel exactly when the noise input to the expandor falls in the nonlinear (2:1) region of the expandor. Thus, maximum compandor noise advantage can be achieved when the compandor USP is designed so that the multichannel loading does not change upon introduction of the compandors into the system and the 2:1 portion of the expandor characteristic is controlling for expected values of system noise.

Voiceband-data requirements. A widely applied noise objective for satellite-derived voice channels is 40 dBrnc0 (8900 pWpo). With 10 or more dB of subjective compandor gain, a channel noise of 50 dBrnc0 might be acceptable. However, while syllabic compandors can provide suitable voice service at these noise levels, they do not offer noise advantage for voiceband data, a continuous signal.

Since it is desirable that switched message tele-

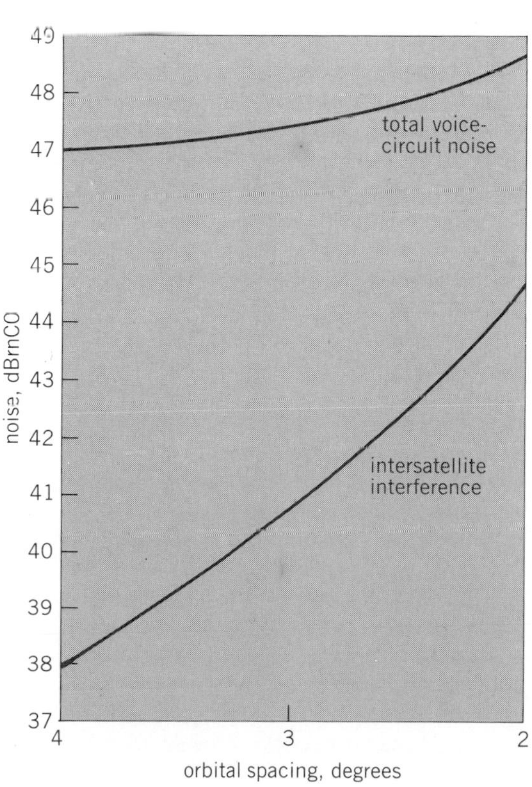

Fig. 4. Total voice-circuit and intersatellite interference noise for CSSB transmission of 6000 channels between 4-ft (12-m) Earth stations, as a function of orbital spacing.

phone service (MTS) be capable of voiceband data transport at rates up to 4800 bits per second (bps) on unconditioned lines, a study was performed which statistically simulated local loop, toll connecting trunk, and satellite connections to determine the allowable increase in satellite transmission noise for satisfactory voiceband data performance. It was found in a conservative analysis that the presence of 4800-bps voiceband data on satellite will limit this increase to about 7–10 dB.

Satellite orbital interference. The message circuit noise as a function of intersatellite spacing is shown in Fig. 4. The increased noise level limits the transponder capacity for CSSB transmission between stations with antennas 4 ft (12 m) in diameter as satellite spacing is reduced. The implementation of CSSB transmission, with the resulting increase in tolerance for noise, is a powerful means to increase the geostationary satellite orbital capacity.

For background information *see* COMMUNICATIONS SATELLITE; CROSSTALK; FREQUENCY-MODULATION RADIO; SINGLE SIDEBAND in the McGraw-Hill Encyclopedia of Science and Technology.

[ERWIN E. MULLER]

Bibliography: R. J. Brown et al., Satellite companded single sideband transmission, *1982 IEEE Global Telecommunications Conference* Miami, vol. 1, pp. A5.4.1-A5.4.5, 1982; J. Gammie and J. P. Moffatt, The AR6A single sideband microwave radio system, *1981 IEEE International Communications Conference* vol. 1, pp. 3.5.1–3.5.7, 1981.

Snow surveying

Many areas of the United States are subject to the threat of flooding from chronic and severe spring snowmelt. To help minimize loss of life and property damage, the National Weather Service issues river and flood forecasts for the nation, and hydrologists in the River Forecast Centers, using hydrologic and meteorologic data that include measurements of the water equivalent in the spring snowpack, issue river and flood forecasts for large areas of the country. Until recently, only ground-based snow measurements were available to characterize the mean areal snow water equivalent over runoff zones and river basins. The Office of Hydrology in the National Weather Service has, however, developed a technique to measure snow water equivalent from a low-flying aircraft by using natural terrestrial gamma radiation. This technique was based on theoretical work in the Soviet Union. Over large areas the operational Airborne Gamma Radiation Snow Survey Program collects real-time airborne snow water equivalent data for river basins in Minnesota and North and South Dakota which are used by river forecasters when issuing spring flood outlooks.

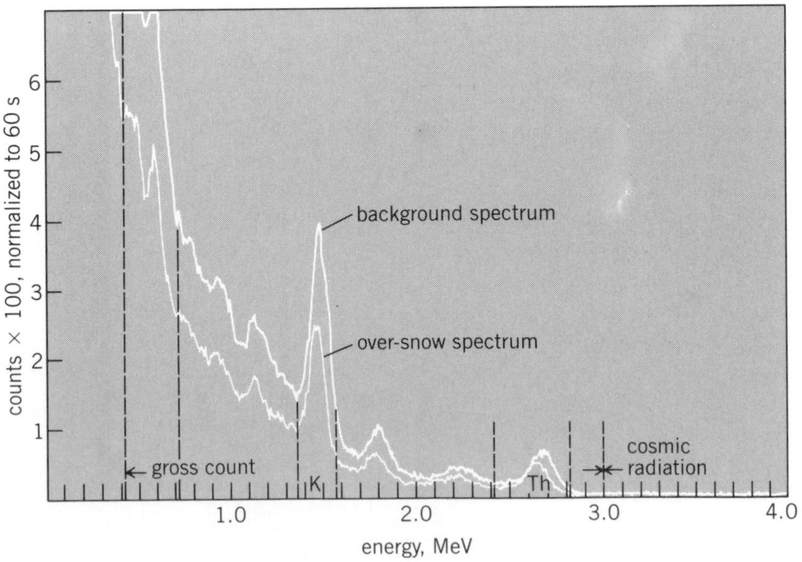

Fig. 1. Comparison of the background and oversnow radiation spectra. A snow water equivalent value is calculated by using the difference between the postassium and thorium photopeaks and the gross counts of the spectra.

Measurement technique. A low-flying aircraft measures the gamma radiation in the air as it flies over the snow cover. Some of this gamma radiation comes from radon gas and cosmic sources and some from the upper 8 in. (20 cm) of the soil. The extraneous radiation is subtracted, leaving only the gamma radiation that comes from the ground.

The natural gamma radiation from the ground is absorbed, or attenuated, by the snow cover in proportion to the mass of water in it (water equivalent). A snow water equivalent value can be calculated using background (no snow cover) radiation measurements and oversnow radiation measurements.

Typically, a network of flight lines, each 9–12 mi (15–20 km) long, is established over a river basin, and a twin-engine Aero Commander, flying at an altitude of 500 ft (150 m), measures the terrestrial gamma radiation over a path 780 ft (300 m) wide. Consequently, radiation data collected over each flight line represent mean areal measurements over approximately 2–2.5 mi^2 (4.5–6 km^2). The ability to measure mean areal snow water equivalent is attractive in comparison with the extensive ground sampling required to estimate mean values with reasonable accuracy.

The gamma radiation flux near the ground originates primarily from the natural ^{40}K, ^{238}U, and ^{232}Th radioisotopes in the soil. In a typical soil, 96% of the gamma radiation is emitted from the top 8 in. (20 cm). After a measure of the background radiation and soil moisture is made over a specific flight line, the attenuation of the radiation signal due to the snowpack overburden is used to calculate the amount of water in the snow cover. This is done by measuring the attenuation of the gamma radiation flux by using data from the potassium (K) window (1.36–1.56 MeV), the thorium (Th) window (2.41–2.81 MeV), and the gross count (GC) energy spectrum (0.41–3.0 MeV) [Fig. 1]. The potassium photopeak is consistently the strongest in the energy spectrum, and has been used successfully to measure snow water equivalent in Canada and in the United States. The gross count window accumulates an order of magnitude more counts than the K and Th photopeak windows. Consequently, gross counts are useful when measuring the variability of snow cover along a flight line or a snow cover with 6–10 in. (15–25 cm) of snow water equivalent.

The gamma radiation attenuation technique can also be used to measure soil moisture in the upper 8 in. (20 cm) when the ground is snow-free. The gamma radiation emitted from the radioisotopes in the soil is attenuated by soil moisture near the surface. Consequently, airborne radiation data collected over bare ground reflect both the radioisotope and soil moisture concentration near the surface. Recent tests show a high correlation between airborne and ground soil moisture measurements.

The principal sources of error in calculating snow water equivalent or soil moisture values using any of the three windows are incorporated in: the measurement of ground-base mean areal soil moisture used to calibrate a flight line; the measurement of air mass (that is, temperature, pressure, and radar altitude); and the limited time with which to collect radiation data. Nonetheless, the airborne technique is capable of measuring snow water equivalent with a root-mean-square error of 0.3 in. (0.8 cm) and soil moisture with a root-mean-square error of approximately 3.9% soil moisture.

Airborne detection system. The airborne detection package consists of five downward-looking thallium-activated sodium-iodide scintillation detectors; two upward-looking detectors (used to isolate the effects of the radon gas contribution); a pulse height analyzer; and a minicomputer for reducing and recording the output data onto magnetic media (Fig. 2). Temperature, pressure, and radar altitude sensors collect data that are used to calculate the air mass between the aircraft and the ground. A remote control unit is used by the system operator-navigator to control and monitor the data collection. Two microprocessors are contained in the pulse height analyzer which time, amplify, analyze, and accumulate the pulse output from the crystals. One microprocessor initializes and controls the system and formats the output, while the second is used in the analog-to-digital conversion of the temperature, pressure, and radar altitude data. All the data, including the two gamma radiation spectra (that is, from the up and down detectors), are accumulated in 1024 discrete channels, each of which is a 24-bit binary number at a given address.

The data acquisition procedure is designed to accumulate and store window data in multiple cycles of 5 s or longer. This provides the capability of analyzing the snow water equivalent or soil moisture distribution along a flight line in approximately 820-ft (250-m) segments. At the end of each flight line, the total radiation spectra (0.05–5.1 MeV) accumulated over the length of the flight line for both the up and down detectors are stored on magnetic media. These data are archived on disk for additional analysis.

Data collection and analysis. Ambient radiation data are collected by the detection system and im-

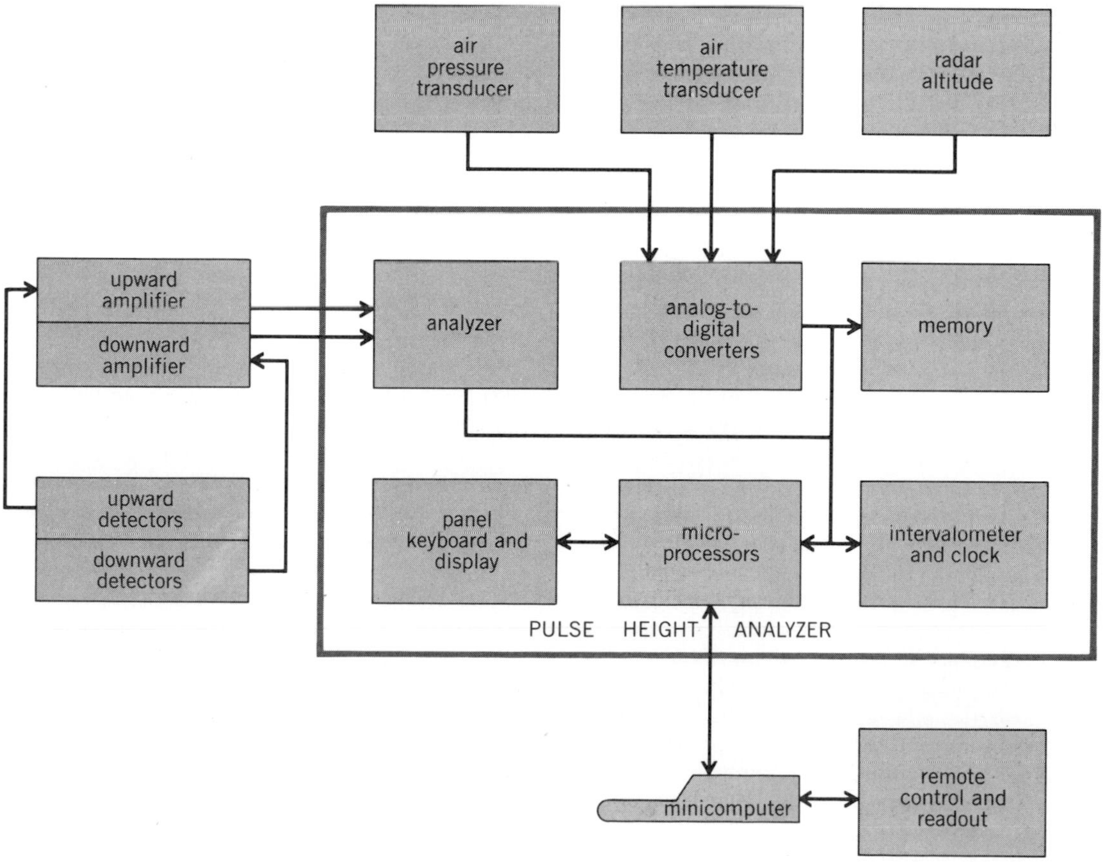

Fig. 2. Block diagram of the National Weather Service airborne gamma radiation detection system.

mediately reduced by using algorithms to describe the presence of atmospheric radon, high-energy cosmic radiation, Compton scattering effects within the radiation spectra, and extraneous background radiation contributed by the aircraft and detection system. Pressure, temperature, and radar altitude data are also recorded and used to calculate the attenuation of terrestrial radiation due to the air mass between the source and sensor (approximately 3.9 oz/in.^{2} or 17 g/cm^2 at an altitude of 500 ft or 150 m). Uncollided terrestrial radiation count rates normalized to time and air mass are used in the aircraft to calculate snow water equivalent or soil moisture values immediately following data collection for a specific flight line.

Background radiation and soil moisture data are archived in a data base and used with the current radiation data to calculate soil moisture or snow water equivalent values in the aircraft immediately after the current airborne radiation data are collected for a given flight line. In this way, real-time soil moisture or snow water equivalent values generated for 30 to 50 flight lines per day are made available to users when the aircraft lands each evening. The airborne snow survey data are transmitted in digital form over telephone lines from the field to the office in Minneapolis.

Significance. Hydrologists in the River Forecast Centers employ conceptual computer models which continuously simulate the physical states of the snowpack and soil zones by using temperature and precipitation data. Airborne snow water equivalent and soil moisture measurements are used to update, or modify, the computer simulation to more accurately reflect the true conditions at the time of the airborne observation. In addition, fall soil moisture conditions, spring snow water equivalent measurements, assumed melt patterns, estimates of future precipitation and temperature patterns, as well as other hydrologic and meteorolgic data, are used when issuing spring flood outlooks over large areas of the country. Until recently, a timely and accurate estimate of the snow water equivalent over large areas was virtually impossible. With the development of the airborne measurement technique, however, reliable real-time mean areal snow water equivalent data can be used by hydrologists to help provide accurate and timely spring flood outlooks for much of the United States.

The airborne snow measurement technique was developed over agricultural areas of the upper Midwest. Research is currently being conducted, however, to assess the capability of the technique over forested areas. Initial results indicate that the airborne technique is capable of providing reliable snow water equivalent measurements over forested areas such as the Lake Superior basin.

For background information *see* GAMMA-RAY DECTECTORS; HYDROLOGY; SNOW SURVEYING in the McGraw-Hill Encyclopedia of Science and Technology. [TOM CARROL]

Bibliography: T. R. Carroll, *Soil Science*, 132(5):358–366, 1981; T. R. Carroll, Airborne soil moisture measurement using natural terrestrial gamma radiation attenuation, *Agricultural Meteorology*, 28:19–30, 1983; E. L. Peck et al., 1971 evaluation of snow water equivalent by airborne measurement of passive terrestrial gamma radiation, *Water Resour. Res.*, 7(5); N. V. Zotimov, 1968 investigation of a method of measuring snow storage by using the gamma radiation of the earth, *Sov. Hydrology Selected Papers*, Engl. trans. no. 3.

Soil nitrogen

Soil nitrogen balance studies are characterized by the application of the principles of conservation of mass to a given soil-crop system so that nitrogen is conserved during the various biological and chemical transformations. Recent nitrogen balance studies have utilized ^{15}N-labeled compounds. These tracer studies emphasize the interaction of the labeled input within the agricultural system. Traditional nitrogen budgets consider the total nitrogen inputs and outputs, with recent studies focusing on the total nitrogen losses to the surrounding environment. These two general approaches to nitrogen balance studies are not equivalent—yet both have greatly expanded understanding of the scope and complexity of the nitrogen cycle transformations in agriculture.

The entire nitrogen cycle is involved in every nitrogen budget study. Processes such as mineralization of organic nitrogen, crop nitrogen assimilation, microbial immobilization, denitrification, ammonia volatilization, leaching, and N_2 fixation are all involved, to some degree, in nitrogen balance studies conducted under natural conditions. Furthermore, these biological processes all interact with each other and the environment over time to produce the final nitrogen budget. The specific results of any individual nitrogen balance study are therefore highly dependent upon the major nitrogen cycle processes of that particular ecosystem and the interaction of those processes with the environment. Therefore, it is not possible to make quantitative generalizations about various nitrogen cycle processes across the different studies; indeed, it is often impossible to make quantitative generalizations about the same ecosystem across different environments. Nevertheless, it is possible to make qualitative generalizations and to draw general conclusions from nitrogen budget studies that are of great general importance in understanding the soil nitrogen cycle and in managing nitrogen inputs to maximize utilization and minimize losses of nitrogen beyond the agricultural sector.

Inorganic nitrogen pool. Soils contain large amounts of organic nitrogen, but this nitrogen is a slowly reactive complex that releases only 2–5% of its nitrogen annually. The active component of the soil nitrogen is the inorganic nitrogen pool. The nitrate and ammonium pools usually constitute less than 5% of the total soil nitrogen, yet this mineral nitrogen is the focal point for the major soil nitrogen cycle processes: crop uptake, denitrification, and leaching. Crops are an effective sink for nitrogen with recoveries varying from 30 to 70%. The wide range in recovery often stems from the nitrogen management system. Highest recoveries occur when nitrogen inputs do not exceed the crop assimilation capacity, when nitrogen inputs are applied just before the crop can assimilate it, and when nitrogen inputs are placed below the soil surface. These three conditions place the nitrogen within the crop root zone in phase with the crops' nitrogen demand and assure that little is left beyond the growing season so that leaching and denitrifiction losses are minimized during the winter season.

Denitrification. This is the biological conversion of nitrate to gaseous N_2 and N_2O. Nitrogen balance studies were originally aimed at estimating these gaseous losses, and early estimates that these losses were 15% for noncropped systems and 20% for cropped systems. Recent research has shown that denitrification is a complex process that requires the presence of nitrate, an available source of carbon, a lack of oxygen, and a temperature suitable for biological activity. If one of these factors is missing, denitrification will virtually cease. Circumstances that lead to large denitrification losses are wet soils and an abundant source of carbon. Wet soils may result from soil flooding, as in paddy rice culture, which leads to large-scale, long-term denitrification; or they may result from periodic water additions, as from rains or irrigations, which can lead to small-scale, short-term denitrification losses. Denitrificaton is also known to occur in macroscopically well-aerated soils if an abundant supply of carbon is available—for example, after a manure application. Under paddy rice culture it is not uncommon to lose as much as 60% of the applied nitrogen, but more common values range from 20 to 40%. With upland crops grown on well-drained soils, denitrification losses commonly vary between 5 and 30% with 20% being a common value.

Nitrogen leaching. Losses from leaching have also been intensively studied with nitrogen balance approaches because of the environmental impact of such losses. The prerequisites for nitrogen leaching are the presence of nitrate nitrogen in the soil and the movement of water through the soil beyond the root zone. These conditions will most likely occur when water inputs exceed the soil's water-holding capacity—for example, with large rains or excess irrigations. Intermittent leaching losses may also occur after periodic water inputs if water percolates through large pores beyond the root zone; however, this type of large-pore leaching does not usually remove large masses of nitrogen since only a small fraction of the soil pore space is involved. Leaching in humid climates will most likely occur during the winter and spring months when the soil is wet and evaporation plus crop water use is small. Leaching losses in semiarid climates are much lower except for occasional large water inputs that may trigger macropore leaching events. Nitrogen leaching losses range from 5 to 50%, with common values lying between 10 and 20%. The lowest losses occur when little nitrate is left beyond the growing season, and the largest losses occur when excessive nitrogen inputs leave a large nitrate pool after the growing season. The nitrate nitrogen that is leached will affect other ecosystems beyond the agricultural sector; some may be lost by denitrification if the drainage water passes through swamps or poorly drained areas, some may appear in the groundwaters below agricultural fields, and some may appear in nearby streams fed by groundwater. The key factors in minimizing leaching losses are avoiding excess nitrogen inputs and, when possible, preventing water from percolating below the root zone—for example, through careful irrigation management.

Symbiotic nitrogen fixation. Nitrogen balance studies have also shown the importance of symbiotic N_2 fixation, a process that converts atmospheric N_2 into plant nitrogen through the bacteria in the roots of leguminous plants. Symbiotic N_2 fixation is very important because it is the dominant process that counterbalances denitrification and recycles atmospheric nitrogen directly into plants. Although this process has been known to occur for many years, there are few reliable estimates of its magnitude. Biological N_2 fixation is known to be affected by the level of soil nitrogen, pH, *Rhizobium* species, soil moisture, and other crop nutritional factors. These factors cause N_2 fixation rates to vary widely from site to site. Forage legumes, such as alfalfa and clover, generally fix the most nitrogen with values of 45–180 lb N/acre (50–200 kg/hectare), commonly credited to symbiotic fixation. Soybeans can likewise fix anywhere from 45 to 135 lb N/acre (50 to 150 kg/ha) with the lowest values occurring on soils high in available nitrogen and the highest values occurring on nitrogen-deficient soils. Some grain legumes are poor fixers; for example, dry beans usually fix less than 45 lb N/acre (50 kg/ha). Annual forage legumes, such as crimson clover, can usually fix 45–90 lb N/acre (50–100 kg/ha). Once the N_2 is fixed, it may leave the soil-crop system through grazing animals, or it may be added back into the soil organic nitrogen pool and serve as a source of nitrogen for a subsequent cereal crop. It is commonly observed that cereals following a legume need substantially less fertilizer nitrogen than those following a nonlegume.

Immobilization–mineralizaton. The conversion of mineral nitrogen to organic nitrogen and the reverse process of the production of mineral nitrogen from organic nitrogen are known as immobilization and mineralizaton, respectively. These two processes occur simultaneously in soils as organic matter is broken down into its basic nutrients and some of the basic nutrients are reutilized within the tissues of the soil microbes. The nitrate and ammonium pools intimately take part in this mineralization–immobilization turnover, although the soil microbes have a preference for ammonium. Labeled nitrogen studies have shown that anywhere between 15 and 60% of the added fertilizer nitrogen is converted into soil organic nitrogen by the microbes during a typical growing season, with a common value being about 30%. This immobilization of labeled nitrogen results from the normal mineralization–immobilization cycle and also from the decomposition of the plant root system. Immobilized ^{15}N is quite stable, and only 4–10% is remineralized annually. Annual mineralization rates normally vary between 23 and 180 lb N/acre (25 and 200 kg/ha), depending on the level of organic nitrogen, past history, soil texture, temperature, and soil moisture conditions.

Ammonia volatilization. This is another important avenue of nitrogen loss from soils. Losses are most pronounced when ammonium-producing compounds are placed on the soil surface in an alkaline environment where climatic conditions favor drying. These conditions occur most frequently in arid regions where basic soils predominate, but they may also occur near particles of urea fertilizer when it is hydrolyzed to ammonium carbonate by the ubiquitous enzyme urease. Manures spread on the soil surface provide another good environment for ammonia loss; if climatic conditions favor drying, over half the ammonium nitrogen can be lost in a week. Large ammonia losses can also occur from the flood water of rice paddies. Ammonia volatilization losses commonly vary from zero to 20%. The best way to prevent such losses is to place the nitrogen source at least 3 in. (8 cm) below the soil surface.

Secondary processes. Several other nitrogen cycle processes are involved in nitrogen balance studies, but they involve smaller nitrogen transformations and will therefore be considered as secondary processes. These include nitrogen inputs from rainfall and dry deposition of ammonia, nonsymbiotic N_2 fixation, and nitrogen losses through erosion and surface runoff. The annual nitrogen additions from precipitation and dry deposition range from 4 to 27 lb N/acre (5 to 30 kg/ha). Areas near ammonia sources, such as feedlots, commonly receive 18–36 lb N/acre (20–40 kg/ha). Nonsymbiotic N_2 fixation occurs through free-living microbes and blue-green algae. These sources contribute 4–27 lb N/acre (5–30 kg/ha) annually, depending on factors such as available carbon, sunlight, and a supply of other nutrients such as phosphorus. Nitrogen losses through erosion also vary widely, but common values would vary from 4 to 18 lb N/acre (5 to 20 kg/ha) per year. Nitrogen losses in the surface runoff are generally small and usually total less than 9 lb N/acre (10 kg/ha). In fact, it is commonly observed that nitrogen inputs through precipitation often exceed nitrogen losses through surface runoff. An exception to this would occur when a surface runoff event occurs shortly after a soluble nitrogen compound has been spread on the soil surface; under these conditions runoff losses may amount to 10% of the applied nitrogen.

Nitrogen balance priniciples have provided, and will continue to provide, a sound basis to estimate the gains, losses, and transformations of nitrogen in agricultural systems. Understanding these processes and their controlling factors will allow agricultural scientists to improve the utilization of nitrogen resources.

For background information *see* NITROGEN CYCLE; SOIL MICROORGANISMS; SOIL MINERALS, MICROBIAL UTILIZATION OF in the McGraw-Hill Encyclopedia of Science and Technology.

[JOHN J. MEISINGER]

Bibliography: F. E. Allison, The enigma of soil nitrogen balance sheets, *Adv. Agron.*, 7:213–250, 1955; F. E. Allison, The fate of nitrogen applied to soils, *Adv. Agron.*, 18:219–258, 1966; J. O. Legg and J. J. Meisinger, Soil nitrogen budgets, in Nitrogen in Agricultural Soils, *Agronomy*, 22:503–566, 1982.

Solar cell

Solar cells convert sunlight directly into electricity by use of the photovoltaic effect. Conventional solar cells consist of a single *pn* junction in a uniform-band-gap semiconducting material. In terrestrial sunlight, single-junction cells made of silicon are able to convert up to about 15% of incident solar energy to electricity, and cells made of gallium arsenide have a conversion efficiency of about 18%. Solar cells using materials with various band gaps and multiple *pn* junctions in the conversion process have the potential for significantly higher photovoltaic conversion efficiency, theoretically as high as 60%. Recent developments in cell technology using two *pn* junctions indicate an encouraging trend toward achieving practical devices with conversion efficiencies in excess of 25%.

Principles of operation. The conversion of sunlight to electricity in a solar cell involves the absorption of sunlight in a semiconductor to create free negative (electron) and positive (hole) charge pairs. The *pn* junction in a cell creates an asymmetry in the electronic structure, hence an electric field which separates the electrons and holes and causes current flow between the cell terminals. The amount of energy required to create one electron-hole pair is referred to as the band gap of the semiconductor. Photons in sunlight having energies less than the band gap of the solar cell material do not create electron-hole pairs and do not contribute to the generation of photocurrent. Photons having energies greater than or equal to the band gap create an electron-hole pair when they are absorbed, but only one pair is created regardless of the photon energy. Photon energy in excess of the band gap is dissipated as heat and is lost to the photovoltaic conversion process. Therefore, only photons with energies equal to or slightly greater than the band gap are efficiently converted to electrical energy. Since sunlight contains energy at widely varying wavelengths (300 to 2000 nanometers), the single-band-gap nature of conventional *pn*-junction cells presents a fundamental limitation to solar cell efficiency.

Theoretical analysis indicates that a substantial increase in solar cell conversion efficiency is possible by using multi-band-gap structures. Such structures are realized by using solar cells made of ma-

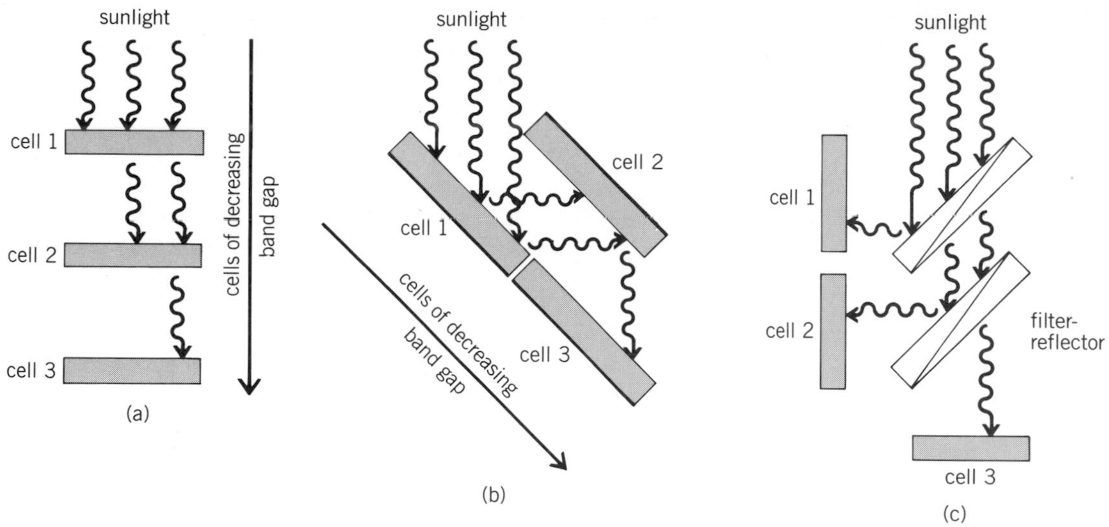

Fig. 1. Multi-band-gap solar cell configurations: *(a)* cascade; *(b)* reflective; *(c)* spectrum-splitting.

terials having various band-gap energies matched to the solar spectrum so that each cell can be designed to be optimally efficient over a limited spectral range. Three alternative configurations for multi-band-gap cells are cascade, reflective, and spectrum-splitting (Fig. 1). In the cascade cell arrangement, solar cells with decreasing band gap are stacked so that the cell facing the sun has the largest band gap. The top cell absorbs all the photons having energies at and above its band gap and transmits the remainder to the second cell. The second cell, which has a band gap less than the first cell, absorbs photons greater than its band gap and transmits less-energetic photons to the next cell. Each cell in the stack therefore absorbs a portion of the solar spectrum defined by the difference in its band gap and the preceding cell's band gap. In principle, any number of cells can be stacked in cascade. The reflective cell scheme functions much the same as the cascade stack, except that a reflector is placed on the back surface of each cell so that light not absorbed in the cell is reflected onto the next cell. The spectrum-splitting approach uses selective filter-reflectors external to the solar cells to split the solar spectrum into several photon-energy ranges and to direct each color band to cells having the optimal band gap for the range. The filter-reflectors used in this scheme are generally multilayer dichroic filters which efficiently reflect a selected narrow bandwidth of light and transmit all other wavelengths. The solar cells and filters in the spectrum-splitting approach can be arranged in any convenient band-gap order.

The individual solar cells can be interconnected in various ways to extract electric power. Series connection is the simplest circuit form of interconnection and is necessary in the cascade method if cells share a common electrode at their interface. Series connection, however, requires that each cell generate the same photocurrent for optimal operation. This requirement of photocurrent matching greatly limits the choice of band gaps of individual cells operating in multicell combinations.

Because of their complexity and resulting higher cost, multi-band-gap cells are being developed for use in concentrating solar collectors operating in the

Optimum band-gap combinations and efficiencies*

Number of cells	System efficiency, percent	Band gaps, eV										
1	32.4	1.4										
2	44.3	1.0	1.8									
3	50.3	1.0	1.6	2.2								
4	53.9	0.8	1.4	1.8	2.2							
5	56.3	0.6	1.0	1.4	1.8	2.2						
6	58.5	0.6	1.0	1.4	1.8	2.0	2.2					
7	59.6	0.6	1.0	1.4	1.8	2.0	2.2	2.6				
8	60.6	0.6	1.0	1.4	1.6	1.8	2.0	2.2	2.6			
9	61.3	0.6	0.8	1.0	1.4	1.6	1.8	2.0	2.2	2.6		
10	61.6	0.6	0.8	1.0	1.4	1.6	1.8	2.0	2.2	2.4	2.6	
11	61.8	0.6	0.8	1.0	1.2	1.4	1.6	1.8	2.0	2.2	2.4	2.6

*For ideal multi-band-gap cells operated in terrestrial sunlight at 100 suns concentration. (From A. Bennett and L. C. Olsen, Analysis of multiple-cell concentrator/photovoltaic systems, *Proceedings of the 13th IEEE Photovoltaic Specialists Conference*, pp. 868–873, 1978)

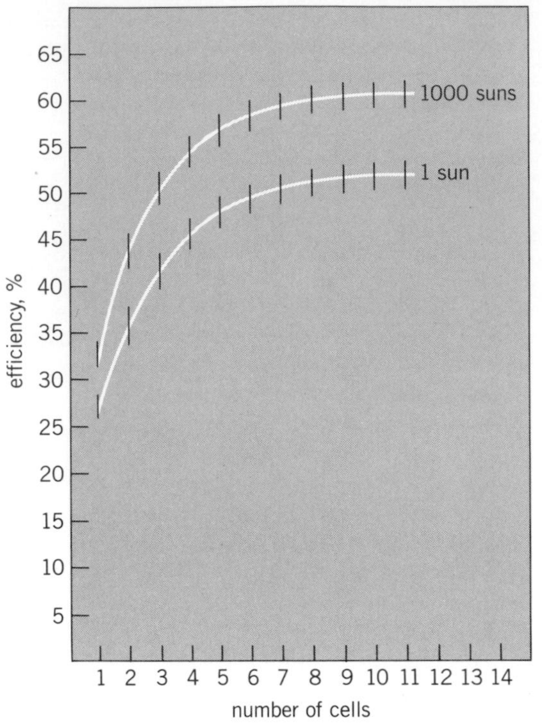

Fig. 2. Optimum efficiencies for ideal multi-band-gap cells operated in terrestrial sunlight at 1 sun and 1000 suns concentration, as a function of number of cells. *(After A. Bennett and L. C. Olsen, Analysis of multiple-cell concentrator/photovoltaic systems, Proceedings of the 13th IEEE Photovoltaic Specialists Conference, pp. 868–873, 1978).*

range of 100–1000 suns. However, there has been increasing interest in development of high-efficiency, flat-plate, nonconcentrating designs using cascade combinations of thin-film semiconductors.

The design of multi-band-gap converters is more complex than the design of single-junction cells. The optimum band gap for the individual solar cells depends on the solar spectrum, the number of cells in the system, and the method of electrical interconnection of the cells. The table and Fig. 2 show the calculated optimum band gaps and maximum conversion efficiencies for combinations of ideal solar cells. Dramatic improvement in efficiency is realized by using two- and three-cell configurations, with diminishing improvement for larger numbers of cells. A key factor limiting the performance of practical multi-band-gap cells is unwanted optical transmission and reflection losses in the cells in the cascade or reflective configurations, and in the filter-reflectors in the spectrum-splitting configuration. Such losses will limit practical multicell systems to perhaps three or four cells.

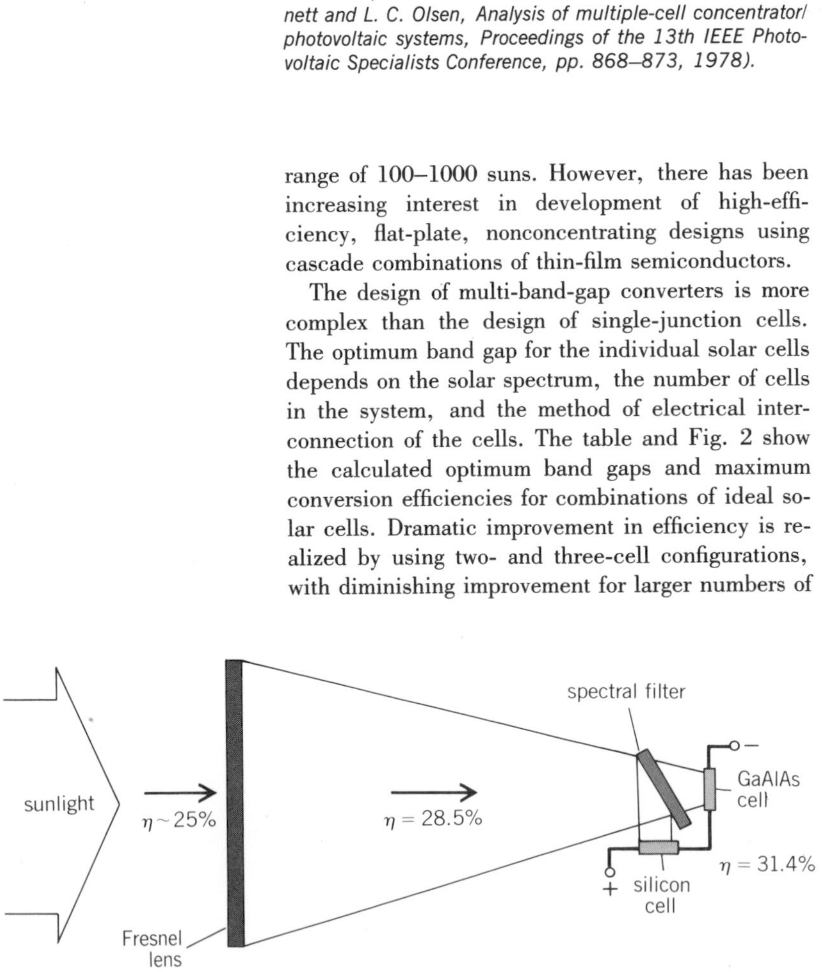

Fig. 3. An experiment concentrating spectral-splitting collector using a two-cell configuration.

Experimental results. Current research on multi-band-gap cells is focused on the simplest two-cell configurations using either the cascade or spectrum-splitting approaches. The first dramatic experimental results were obtained with a series-connected combination of silicon (band gap 1.1 eV) and gallium–aluminum arsenide (band gap approximately 1.7 eV) as shown in Fig. 3. A plastic Fresnel lens was used to concentrate sunlight at about 165 suns onto a spectrum-splitting filter which reflected photons with energy less than 1.65 eV onto the silicon cell and transmitted the remainder to the gallium–aluminum arsenide cell. The combined efficiency η of the two cells operated in series electrical connection was about 31%. Optical losses in the filter reduced this efficiency to 28.5% when referred to the solar energy incident on the filter. The overall solar-to-electric conversion efficiency, including loss in the Fresnel lens, was about 25%, a significant increase over that obtained by any previous photovoltaic collector.

Ultimately, researchers hope to develop monolithic two- or three-band-gap cascade cells having efficiencies in excess of 30%. Such cells consist of many thin layers of various *p*- and *n*-type materials with varying band gaps. For concentrator applications, structures of this complexity are within the capability of epitaxial growth technology in the gallium arsenide–based material system. For flat-plate collectors, the most promising monolithic cascade cell technology appears to be thin-film combinations of amorphous silicon–germanium alloys (band gap 1.4 eV) and amorphous silicon–carbon alloys (band gap 1.65 eV).

For background information *see* SOLAR CELL in the McGraw-Hill Encyclopedia of Science and Technology. [D. G. SCHUELER]

Bibliography: M. A. Green, *Solar Cells: Operating Principles, Technology, and System Applications*, 1982; J. C. C. Fan, B.-Y. Tsaur, and B. J. Palm, Optimal design of high-efficiency tandem cells, *Proceedings of the 16th IEEE Photovoltaic Specialists Conference*, pp. 692–701, 1982; J. J. Loferski, Theoretical and experimental studies of tandem or cascade solar cells: A review, *Proceedings of the 16th Photovoltaic Specialists Conference*, pp. 648–654, 1982; R. L. Moon et al., Multigap solar cell requirements and the performance of AlGaAs and Si cells in concentrating sunlight, *Proceedings of the 13th IEEE Photovoltaic Specialists Conference*, pp. 859–867, 1978.

Solar neutrinos

The chlorine solar neutrino experiment is the only experimental attempt that has been undertaken to observe the neutrino flux from the Sun. The results do not agree with the standard solar model, but the observed flux is only a factor of 4 lower than is expected from the standard theory, and there are nonstandard models that can account for the experimental results. New observations are needed to clarify the present results, and a number of solar neutrino detector schemes are under active consideration.

Energy production in stars. The Sun has produced energy at essentially the same rate for at least 2×10^9 years. This conclusion follows from the fact that life has existed on the Earth for this long period of time. In order to support life, the temperature and atmospheric conditions on the Earth must have remained close to those existing at the present time, and this requires that the solar luminosity has not changed very much from its present value. It was recognized in the period 1917–1920 by A. S. Eddington, H. N. Russell, and H. Shapley that all stars were at least billions of years old, but it was not understood how a star could produce energy for this long period of time. Radioactivity was discovered in 1896, and by 1920 it became clear to astronomers that energy from nuclear processes was somehow the source of stellar energy, but the mechanism was difficult to understand.

The first suggestion of the mechanism was made in 1927, when it was pointed out that at the high temperatures prevailing in the interiors of stars, positively charged nuclei could collide with sufficient energy to react by the quantum-mechanical process of tunneling through the repulsive Coulomb barrier. In 1936, H. Bethe and C. F. von Weizäcker independently proposed a detailed chain of nuclear reactions, the carbon-nitrogen cycle, in which ^{12}C and ^{14}N played the role of catalysts (Table 1). The following year, the rate of the thermal fusion reaction, $H + H \rightarrow D + e^+ + \nu$, governed by the weak process, was calculated and found to be sufficiently fast to be of importance in the interior of stars. This H-H reaction provides the first step in the proton–proton chain of reactions (Table 1). This set of reactions is believed to operate at lower temperatures than the carbon–nitrogen cycle and to play a major role in young hydrogen-rich stars like the Sun. In 1958 it was discovered that the $^3He + {}^4He \rightarrow {}^7Be + \gamma$ reaction is important and competes in a minor way with the $^3He + {}^3He \rightarrow {}^4He + 2H$ reaction for ^3He burning. The formation of ^7Be then leads to the sequential reactions of the P-PII and P-PIII branches now included in the proton-proton chain. These additional branches are not particularly important in the total energy production, but are of importance in the detection and interpretation of neutrino production in the Sun.

The nuclear processes of the proton–proton chain or the carbon–nitrogen cycle are believed to be the basic source of energy and to govern the early evolutionary development of all stars. The thermonuclear reactions serve to convert four hydrogen nuclei into a ^4He nucleus. The nuclear energy of 26 MeV per ^4He nucleus synthesized is emitted in the form of positrons (e^+), neutrinos (ν), and gamma radiation (γ) through the net reaction (1). The particular

$$4H \rightarrow {}^4He + 2e^+ + 2\nu + \text{gammas} \qquad (1)$$

nuclear reactions of these chains were chosen from laboratory studies of nuclear reactions and nuclear masses. They are the only possible normal nuclear fusion reactons for stars like the Sun, which is composed of approximately 70% hydrogen, 28% helium, and 2% heavier elements, predominantly carbon, oxygen, iron, silicon, nitrogen, magnesium and neon. Many other processes for energy production in stars have been suggested, but all have been ruled

Table 1. Nuclear reaction chains in the Sun

	Reaction	Energy, MeV	Neutrino energy, MeV
	Proton–proton chain		
P-PI	$H + H \rightarrow D + e^+ + \nu$ (99.75%)*	1.442	0–0.42 spectrum
	or $H + H + e^- \rightarrow D + \nu$ (0.25%)		1.44 line
	$D + H \rightarrow {}^3He + \gamma$	5.493	
	$^3He + {}^3He \rightarrow 2H + {}^4He$ (86%)	12.859	
P-PII	$^3He + {}^4He \rightarrow {}^7Be + \gamma$ (14%)	1.587	
	$^7Be + e^- \rightarrow {}^7Li + \nu$	0.862	0.861 line
	$^7Li + H \rightarrow \gamma + {}^8Be \rightarrow 2\,{}^4He$	17.347	
	or		
P-PIII	$^7Be + H \rightarrow {}^8B + \gamma$ (0.015%)	0.135	
	$^8B \rightarrow {}^8Be^* + e^+ + \nu$	15.079	0–14 spectrum
	$^8Be^* \rightarrow 2\,{}^4He$	2.995	
	Carbon–nitrogen cycle		
	$H + {}^{12}C \rightarrow {}^{13}N + \gamma$	1.94	
	$^{13}N \rightarrow {}^{13}C + e^+ + \nu$	1.20	0–1.20 spectrum
	$H + {}^{13}C \rightarrow {}^{14}N + \gamma$	7.54	
	$H + {}^{14}N \rightarrow {}^{15}O + \gamma$	7.29	
	$^{15}O \rightarrow {}^{15}N + e^+ + \nu$	1.73	0–1.73 spectrum
	$H + {}^{15}N \rightarrow {}^{12}C + {}^4He$	4.96	

*Relative branching ratio in percent.

out—for example, quark fusion, quarked atom reactions, magnetic monopole–catalyzed proton decay, and even a black hole at the center.

Stellar structure and evolution. The internal structure of a star is derived theoretically by starting with a mass of gas collected in space by gravitational forces. This mass is heated by the conversion of the gravitational potential into thermal energy as the star contracts ultimately into a rotating spherical body. If the star is large enough, its interior becomes sufficiently hot to generate energy by the fusion reactions of the proton–proton chain. At this stage of development the star is assumed to be highly convective and, as a result of this process, becomes uniform in composition. The mass and chemical composition determine the rate of development and the ultimate fate as a burned-out star, a supernova, or a black hole. The internal structure and evolution are derived from a set of equations: one equation describes as a function of the radius the hydrodynamic equilibrium that results from the outward pressure from gas and radiation and the inward gravitational forces; another equation describes the transport of energy as a function of the radius, chemical composition, and temperature; and a third equation describes the generation of energy that depends upon the temperature, composition, and nuclear reaction cross sections. Such evolutionary calculations have been made for stars with widely ranging masses, and the results explain satisfactorily the luminosity and surface temperatures observed for stars in clusters that presumably were formed at the same time but with different masses.

The Sun is the only star whose diameter, mass, chemical composition, and age are accurately known. Knowledge of these important parameters makes possible theoretical calculations of the Sun's internal structure, the rate of each of the fusion reactions occurring in its interior, and the expected change in luminosity with time as the Sun evolves. It is of great interest to test this basic theory of stellar evolution by observing the rate of the interior nuclear processes directly. This can, in principle, be accomplished by observing the neutrino radiation from the Sun.

Neutrino interactions. The neutrino is a member of a family of elementary particles called leptons. This family includes charged particles, the electron (e^+, e^-), the muon (μ^+, e^-), and the tauon (τ^+, τ^-), the electron neutrino ($\nu_e, \bar{\nu}_e$), the muon neutrino ($\nu_\mu, \bar{\nu}_\mu$), and the tauon neutrino ($\nu_\tau, \bar{\nu}_\tau$), where the bar indicates the antiparticle. All of these particles interact by the weak force and by the electromagnetic force if the particle has an electric charge. Since the neutrinos are neutral, they interact only by the weak processes. They can be scattered by electrons, photons, various elementary particles, and nuclei. In addition, they can be absorbed in nuclei with the emission of their charged counterpart, electron, muon, or tauon. All of these processes are very improbable. The magnitude of the interaction of a particle is usually described by the value of a cross section, that is, the target area of nucleus, electron, or particle, for the process. Neutrino interactions have cross sections in the range 10^{-37} to 10^{-50} cm^2, depending upon the particular process and the energy of the neutrino. For comparison, typical nuclear reactions have cross sections in the range 10^{-24} to 10^{-30} cm^2. The neutrino of interest in the study of the interior of the Sun is the electron neutrino (ν_e), and it will hereafter be referred to simply as the neutrino. A neutrino generated in the interior of the Sun will easily penetrate the material of the Sun without being absorbed or scattered.

Another interesting property of the electron neutrino is that it has an extremely small mass or perhaps no mass at all. During 1982–1983 there were a number of new experimental attempts to observe its mass. Only one experiment, performed in the Soviet Union, claims to have observed a mass greater than zero. This mass, expressed in energy units, is 10–30, eV, corresponding to 1/25,000 of the mass of the electron. If the neutrino has a mass, it is possible, as discussed below, that the neutrinos from the Sun could change into ν_μ or ν_τ during their passage to the Earth and influence the measurement of the solar neutrino flux. Also, if neutrinos have mass, they would play a role in the gravitational binding of galaxies and the universe.

Chlorine solar neutrino experiment. The unique property of the great penetrating power of neutrinos through matter that enables them to escape freely from a star has, of course, the contrary effect of making them extremely difficult to detect. The only operating detector that has sufficient sensitivity for observing the fluxes of neutrinos expected from the Sun is based upon a radiochemical method that uses chlorine as the target element. This detector is based on the neutrino capture process given in expression (2). The neutrino capture process is the

$$\nu + {}^{37}\mathrm{Cl} \underset{\text{decay}}{\overset{\text{capture}}{\rightleftharpoons}} {}^{37}\mathrm{Ar} + e^- \qquad (2)$$

inverse of the radioactive decay of ^{37}Ar, also shown in (2), that occurs by electron capture with a characteristic half-life of 35 days. The neutrino capture cross section of ^{37}Cl can be calculated accurately as a function of the energy of the neutrino. These calculations are based upon the theory of beta decay and the principle of detailed balancing, and require knowledge of the electron density in the neighborhood of the nucleus and the half-life of ^{37}Ar. This calculation gives the neutrino capture cross section of ^{37}Cl to produce the ^{37}Ar nucleus in its ground state. It is also possible to apply these principles to the calculation of the neutrino capture cross section to produce ^{37}Ar in various excited states, a factor of great importance for determining the expected neutrino capture rate for the more energetic neutrinos from ^{8}B decay in the Sun, in the P-PIII chain. The ability to evaluate the neutrino capture cross section for the ^{37}Cl nucleus in this detail is rather unique, and cannot be accomplished for other heavy nuclei.

The neutrino detection method depends upon us-

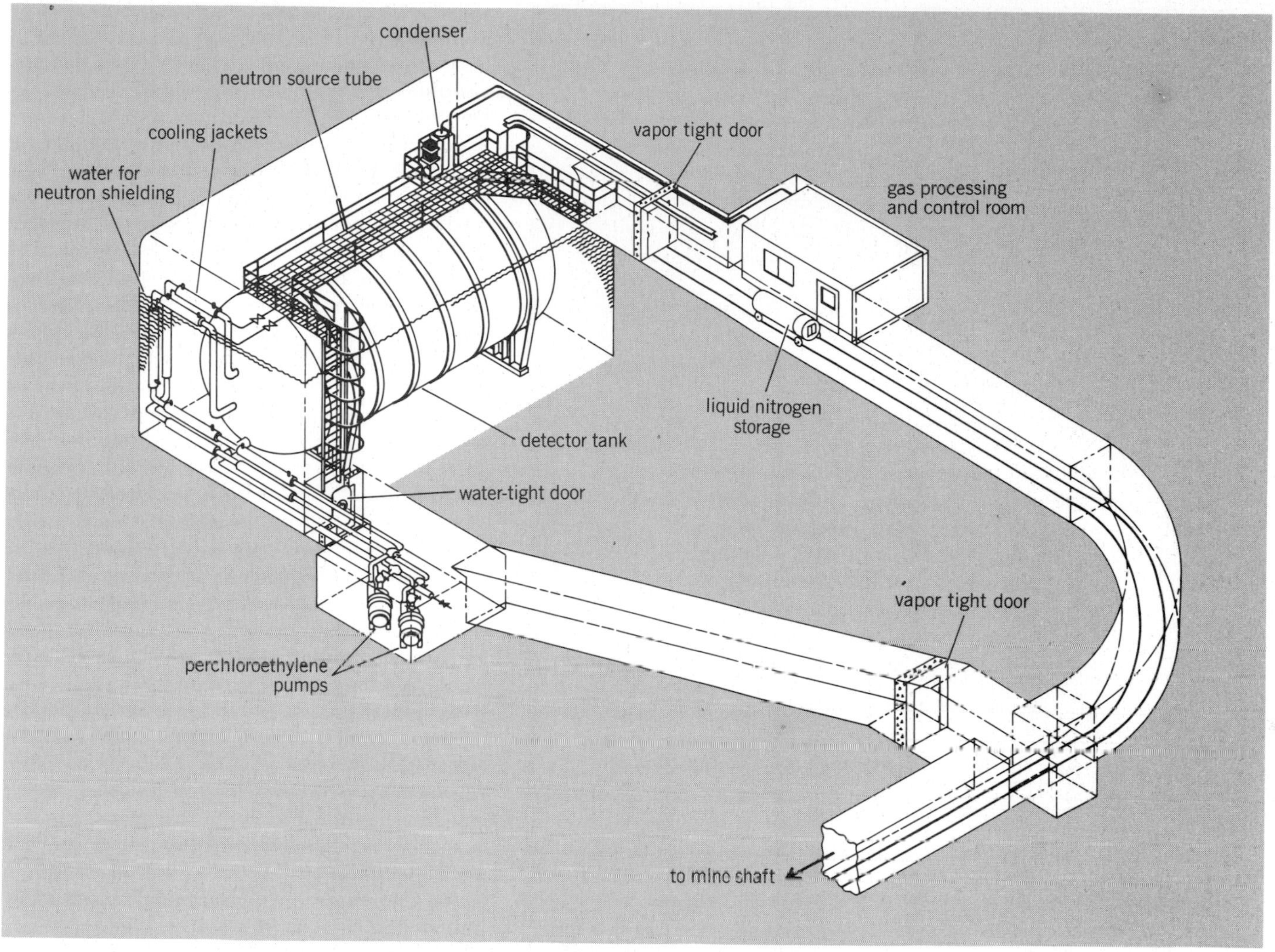

Fig. 1. Arrangement of the chlorine solar neutrino experiment in the Homestake Gold Mine at Lead, South Dakota.

ing a very large mass of chlorine-containing liquid, perchloroethylene (C_2Cl_4), and removing a small number of atoms (10–25) of radioactive argon, ^{37}Ar, that are formed by neutrino capture during a period of approximately 2 months. The ^{37}Ar is removed from the perchloroethylene in 2 days by purging the liquid with helium gas. The radioactive argon is removed from the helium gas by having it flow through a charcoal adsorber at $-319°F$ ($-195°C$). The argon recovered is purified and placed in a small proportional counter to observe the radioactive decay of ^{37}Ar.

The chlorine neutrino detector, which was built by Brookhaven National Laboratory in the Homestake Gold Mine at Lead, South Dakota (Fig. 1), was scaled to observe the flux of neutrinos expected from the Sun. The detector contains 678 tons (615 metric tons) of perchloroethylene (2.2×10^{30} ^{37}Cl nuclei) and is contained in a 100,000-gallon (400,000-liter) tank (Fig. 2). It was necessary to build the detector underground (at a depth of 4850 ft or 1480 m) to reduce the production of ^{37}Ar by cosmic rays sufficiently so that the ^{37}Ar from solar neutrinos could be observed.

When measurements were first performed in 1967–1968, it was found that the ^{37}Ar production rate was well below that predicted from theoretical models of the Sun. Refined measurements performed during 1973–1983 show that 0.44 ^{37}Ar atom is produced per day. Allowing for a small correction for cosmic-ray production, it was concluded that the ^{37}Ar production rate that could be attributed to solar neutrinos was 0.36 ± 0.05 atom per day. It is customary to express the results of solar neutrino experiments in terms of a unit called a solar neutrino unit (SNU; defined as 10^{-36} capture per target atom per second, in this case ^{37}Cl atoms). The total solar neutrino capture rate observed in the Brookhaven experiment in these units is 1.9 ± 0.3 SNU, where the error is primarily from the statistics of the recorded number of counts.

Comparison with theory. A radiochemical neutrino detector is only capable of measuring the total rate from all neutrino sources in the Sun. In Table 2 are listed the calculated fluxes of neutrinos expected from the standard solar model for all neutrino sources in the proton-proton chain and the carbon–nitrogen cycle, and their respective neutrino capture

Fig. 2. Solar neutrino detector tank in the Homestake Gold Mine at Lead, South Dakota.

cross section for the ^{37}Cl nucleus. The expected total rate for the chlorine neutrino detector is 7.8 ± 1.1 SNU. The results given are from a careful analysis of the various solar data (mass, luminosity, age, and composition), laboratory measurements of nuclear reaction rates, and the theoretically calculated opacities. The opacities are of critical importance in determining the passage of radiation through the ionized plasma in the interior of the Sun.

A major part (6.1 SNU or 80%) of the expected neutrino capture rate in the chlorine detector arises from the very low flux of energetic neutrinos from ^{8}B decay in the Sun. This result arises from the fact that these neutrinos have sufficient energy to feed a particular excited state in ^{37}Ar at 5.2 MeV, the analog state. This transition is superallowed and therefore has a high neutrino capture cross section. This feature of the chlorine experiment is of great value in understanding the interior of the Sun. The production of ^{8}B in the Sun through the P-PIII chain is a very sensitive function of the internal temperature of the Sun; the rate of the specific ^{8}B-producing reaction varies as the twenty-fifth power of the temperature. Because of this, the chlorine solar neutrino detector serves as a critical probe for the interior temperatures and thereby severely tests current understanding of the solar interior as expressed by the standard theory.

The results of the chlorine experiment are in disagreement with the standard solar model; the observed rate is about a factor of 4 below expectation. This general conclusion has been known since the initial results were obtained in 1968, and it has been difficult to understand the reason for the discrepancy. At first the chlorine experiment itself was questioned, but because of the simple nature of the chemical extraction of the rare-gas isotope ^{37}Ar and the counting of its decay process, it was doubtful that this was the explanation. A number of chemical tests were performed on the system. The results of these tests served to verify the correctness of the results.

As mentioned above, the neutrino detection efficiency of the detector depends upon the calculated neutrino capture cross sections. These calculations depend upon known priniciples and are believed to be reliable. It is possible to verify these calculations by using a neutrino source of known energy and intensity. One method would be to prepare a ^{65}Zn source in a nuclear reactor (neutrino energy = 1.34 MeV) and place it in or near the chlorine detector. A megacurie of ^{65}Zn would be required. A source of this nature would test the capture of neutrinos to form ^{37}Ar in its ground state. Another approach would be to use an accelerator to produce more en-

Table 2. Solar neutrino fluxes and cross sections for the reaction $\nu + {}^{37}Cl \rightarrow {}^{37}Ar + e^-$

Neutrino source	Flux on Earth,* $cm^{-2} \cdot s^{-1}$	Capture cross section,† cm^2	Capture rate, SNU
$H + H \rightarrow D + e^+ + \nu$	6.1×10^{10}	0	0
$H + H + e^- \rightarrow D + \nu$	1.5×10^{8}	1.56×10^{-45}	0.23
^{7}Be decay	4.3×10^{9}	2.38×10^{-46}	1.02
^{8}B decay	5.6×10^{6}	1.08×10^{-42}	6.05
^{13}N decay	4.0×10^{8}	6.61×10^{-46}	0.26
^{15}O decay	5.0×10^{8}	1.66×10^{-46}	0.08
		Total	7.64

*After J. N. Bahcall et al., *Rev. Mod. Phys*, 54:767, 1982.
†After J. N. Bahcall, *Rev. Mod. Phys.*, 50:881, 1978.

ergetic neutrinos that would test the full range of the neutrino capture cross sections. Unfortunately, because of the expense and difficulties involved in these experiments, these tests have not been performed, although considerable effort has been directed toward both of these approaches.

Possibility of neutrino oscillations. The concept of studying the solar interior by observing neutrinos is predicated upon the assumption that neutrinos of the electron type, once created, will not change form or decay during the 8 min it takes them to reach the Earth. The suggestion was made in the 1960s that it was conceivable that neutrinos could change from electron-type neutrinos into other neutrino types, a process referred to as neutrino oscillation. In 1979 it was suggested that this process could also occur when neutrinos pass through matter. If a low-energy electron neutrino oscillated into a ν_μ or ν_τ, it would not have sufficient energy to create a muon or tauon, and therefore it would not be captured. These processes are currently considered possible, and a number of experimental tests have been performed to search for these phenomena. So far, they have been negative.

Neutrino oscillations in a vacuum depend upon the neutrino having a mass, and the oscillation length is proportional to the difference in mass between the neutrino types squared and another parameter called the mixing angle θ. The square of the sine of double the mixing angle ($\sin^2 2\theta$) is a measure of the fraction of time that the neutrino is observed in a different form. If neutrinos do oscillate, the solar neutrino signal would be low, and if there are only three types of neutrinos and the mixing angle is at its maximum value, the observed flux would be reduced by a factor of 3. However, a reduction by a factor of 4 due to neutrino oscillations appears to be unlikely.

A solar neutrino experiment could be the best means of testing the survival of electron-type neutrinos because of the great distance between the Earth and the Sun. However, for such a test to be valid it would be necessary to have confidence in the solar physics and the basic fusion reaction mechanisms in the Sun. There is considerable confidence that the proton–proton reaction in the Sun does produce a flux of 6×10^{10} neutrinos per square centimeter per second. These neutrinos are below threshold for the chlorine experiment, but they could be observed by a radiochemical detector that uses gallium as a target material.

Remeasurement of reaction cross sections. Although neutrino oscillation is a conceivable explanation of the low solar neutrino capture rate in the chlorine detector, there are other more mundane explanations. One subject that received attention in 1982–1983 is the measured values of the nuclear reaction cross sections. The chlorine experiment is primarily sensitive to the flux of ^{8}B decay neutrinos, and the production of this radioactivity depends upon the successive reactions $^3He + {}^4He \rightarrow {}^7Be + \gamma$ and $^7Be + p \rightarrow {}^8\beta + \gamma$ of the P-PII and P-PIII branches. These reactions were remeasured, the first at three separate laboratories. The new values of these nuclear reaction cross sections agree within the errors with the older values, except for a single value of the $^3He(^4He, \gamma)^7Be$ reaction cross section, which is 40% lower. Incorporating these new measurements in the standard solar model calculation would give a small reduction in the theoretical neutrino capture rate in ^{37}Cl to approximately 6.5 SNU. As a result of these new investigations, there is confidence that the discrepancy between observations and theory do not arise from an error in the nuclear reaction cross sections.

Nonstandard solar models. The general consensus is that the standard solar model is probably oversimplified and does not model correctly the interior of the Sun in sufficient detail. According to the standard model, 80% of the expected rate in ^{37}Cl is accounted for by the energetic neutrinos from ^{8}B decay, and the production of this isotope is very sensitive to the internal temperatures in the Sun. One of the many attractive features of the chlorine detector is that it provides a very sensitive test of the solar model. A large effort has been devoted to theoretical models that predict low neutrino capture rates in ^{37}Cl. The general approach is to introduce a feature in the model that in some way reduces the temperatures in the central regions and thus diminishes the flux of neutrinos from ^{8}B decay.

For example, a model that assumes the interior of the Sun is continually mixed has the effect of reducing the temperature gradient. A model that permits turbulent diffusion has a similar effect if the diffusion rate is large enough. Furthermore, a model with diffusion explains the low abundance of lithium in the Sun, and is important in understanding the ^{3}He/^{4}He ratio on the solar surface. These models predict a ^{37}Cl neutrino capture rate in the range 1.5–2.4 SNU. The standard model presumes that initially the heavy-element composition throughout the Sun was the same as is observed now in its photosphere (2%). If a model is developed assuming the total heavy-element abundance was initially only 0.4%, the expected solar neutrino capture rate in ^{37}Cl would be only 2 SNU, again in agreement with the experiment. The presence of the heavy elements now observed on the solar surface could be accounted for by the later addition of cometary matter or dust. This model is not generally accepted because it implies a low helium abundance for the Sun, and a shallow convective zone.

These three models are the most appealing of the models that predict low rates for the chlorine experiment. Many other models have been suggested: for example, the Sun has a rapidly rotating core, a burned-out helium core, intense internal magnetic fields, or a core made up of heavier elements. Thus, there seem to be solar models that reasonably explain the low rate observed by the chlorine experiment; however, there is no single model that has gained general acceptance, perhaps in some degree because of some long-standing prejudices that have been established in the field of stellar evolution. It is clear that the Sun is a very dynamic object that exhibits a periodic activity cycle, differential rotation, apparent variations in its diameter, oscillations, intense flares, intense localized magnetic fields, and other phenomena that are poorly understood.

In recent years optical means of observing oscillations in the solar surface have developed rapidly. These investigations have provided a new approach to observing the interior of the Sun. The most thoroughly studied are the *p*-wave oscillations that have their maximum amplitude at the solar surface and are generally interpreted as an oscillation within the outer convective zone, although they do penetrate into the solar interior. These observations can be used to test the internal structure that is deduced by solar models. Recently *g*-mode oscillations have been resolved from the periodic occurrences of sunspot numbers. The *g*-mode oscillations have their maximum amplitude at the center of the Sun, and a study of these oscillations will give important information on the deep interior structure of the Sun.

New experiments. Clearly, new observations are needed to clarify the present results, to learn more about the neutrino spectrum from the Sun, and to give some real assurance that the Sun is indeed producing energy by hydrogen fusion.

It is highly desirable to observe the low-energy neutrinos from the primary proton–proton reaction. The flux of these neutrinos is essentially independent of the internal dynamics and processes that affect the solar model prediction of the ^{8}B production. An observation of the proton–proton reaction neutrinos would give assurance that the Sun is indeed producing energy by hydrogen fusion, and in addition would give valuable information on the fundamental properties of neutrinos. For these reasons considerable effort has been devoted to neutrino detectors with low-energy thresholds, and with sufficient sensitivity to observe the flux of these neutrinos.

It would also be important to develop a detector with directional properties to show that the Sun is indeed the source of the neutrinos. Unfortunately the only hope of establishing even a crude neutrino direction is to observe the energetic ^{8}B neutrinos by elastic scattering. A detector with this capability is beyond present technology. Within the foreseeable future the low- and medium-energy neutrino spectra from the Sun can only be studied with radiochemical detectors.

Three radiochemical solar neutrino detector schemes have been under active consideration since about 1973. For each of these detectors, Table 3 lists the percent of the neutrino capture rate that is contributed by each neutrino source in the Sun, the total neutrino capture rate in SNU, and the tons of target element needed to attain a neutrino capture rate of one per day yielding the desired radioactive isotope. The gallium detector observes predominantly the neutrinos from the proton–proton reaction, and the bromine detector observes the neutrinos from ^{7}Be decays in the Sun. The flux of neutrinos from the separate P-PI, P-PII, and P-PIII chains could be defined by observations with three radiochemical detectors, those using gallium, bromine, and chlorine. Because of the great interest in observing the proton–proton reaction, the most effort has been placed in developing a gallium detector. Two groups have independently developed this technology, and it seems likely that a gallium detector will be built in a tunnel in the Baksan valley, Soviet Union, within the next few years. At present there is little prospect of building a full-scale gallium detector in the United States.

Another way of carrying out a radiochemical solar neutrino experiment is to employ mineral grains or a massive geological deposit as a target material. In this approach, advantage would be taken of a very long exposure to accumulate a large number of atoms of a unique product that could be measured. In recent years several experiments of this nature have been suggested, including the measurement of ^{207}Pb from a thallium mineral such as lorandite, the extraction of ^{81}Kr from a salt deposit containing bromine, and the measurement of ^{97}Tc and ^{98}Tc from a molybdenum ore. A major difficulty with using a natural deposit is that background effects from trace amounts of uranium and thorium are generally a serious problem. However, these background effects apparently are small for both thallium and molybdenum ores, so that these experiments appear to be entirely feasible. The thallium experiment would be sensitive to the low-energy neutrinos from the pro-

Table 3. Comparison of radiochemical solar neutrino detectors

	Neutrino source	Gallium ^{71}Ga–^{71}Ge	Bromine ^{81}Br–^{81}Kr	Lithium ^{7}Li–^{7}Be	Chlorine ^{37}Cl–^{37}Ar
Percent of the capture rate from various solar neutrino sources for a model with rate of 1.9 SNU in ^{37}Cl	$H + H \rightarrow D + e^{+} + \nu$	81	0	0	0
	$H + H + e^{-} \rightarrow D + \nu$	2.7	17	51	12
	^{7}Be decay	14	64	8.9	23
	^{8}B decay	0.3	6.2	20	59
	^{13}N decay	1.0	4.6	3.5	1.4
	^{15}O decay	1.4	4.6	16	4.4
Total rate in SNU		88	6.8	19	1.9
Short tons (metric tons) of element needed for one neutrino cature per day	43 (39)		507 (460)	8.4 (7.6)	1610 (1460)

ton–proton reaction, and the molybdenum experiment would observe the energetic neutrinos from ^8B decay. Geological experiments have the advantage of testing for the long-term constancy of the fusion reactions in the Sun.

Neutrinos from supernovae. It is generally believed that the basic mechanism of stellar collapse that leads to a supernova event depends upon the core of a star becoming so hot that it radiates predominantly neutrinos and antineutrinos. The real proof that this theory of stellar collapse is correct would be to observe the sudden pulse of neutrinos and antineutrinos from one of these events. Unfortunately, there would be a very long wait for the event, and it would have to occur close enough to be observed. According to current views, a type II supernova arises from a star 10–20 times the mass of the Sun that prior to collapse has an iron core with a mass 1–2 times the mass of the Sun. When a collapse occurs, a sharp pulse of neutrinos lasting only a few hundredths of a second is produced by the capture of electrons by protons to form neutrons, and the star becomes a neutron star. Following this initial burst, pairs of neutrinos and antineutrinos are emitted over a period of 10–20 s. The energy of the neutrinos is in the range 10–20 MeV. The total energy emitted as neutrino radiation is of the order of 10^{53} ergs. By comparison, the Sun radiates 3×10^{35} ergs per second.

If such an event occurred, the chlorine solar neutrino experiment could observe the estimated neutrino flux if the distance were less than 30,000 light-years (2×10^{17} mi or 3×10^{17} km), the distance to the center of the Galaxy. However, this neutrino detector would not give information on the time structure of the burst. To obtain the time structure of the pulse would require a detector that could record the scattering of neutrinos from electrons, or observe their capture by hydrogen. This could be accomplished by a water Cerenkov detector or by a liquid hydrocarbon scintillator. A detector with this capability is technically feasible, but has not been built. Unfortunately, the supernova rate in the Galaxy is not known, but it is estimated to be in the range of one event every 10–50 years. If one occurred today, it could only be observed with the chlorine detector.

For background information *see* CARBON–NITROGEN–OXYGEN CYCLES; NEUTRINO; PROTON–PROTON CHAIN; SOLAR NEUTRINOS; STELLAR EVOLUTION; SUN; SUPERNOVA in the McGraw-Hill Encyclopedia of Science and Technology.

[RAYMOND DAVIS, JR.]

Bibliography: J. N. Bahcall, Solar neutrino experiments, *Rev. Mod. Phys.*, 50:881–904, 1978; J. N. Bahcall et al., Standard solar models and the uncertainties in predicted capture rates of solar neutrinos, *Rev. Mod. Phys.*, 54:767–799, 1982; C. A. Barnes et al. (eds.), *Essays in Nuclear Astrophysics*, 1982. M. M. Nieto et al. (eds.), *Science Underground*, AIP Conf. Proc. 96, 1983; R. W. Noyes, *The Sun Our Star*, 1982.

Space flight

The pace of space utilization quickened in 1983 as the United States broadened its payload capabilities with the space shuttle and the Soviet Union became operationally familiar with its new mini-space station, *Salyut 7*. Both countries further developed their skills in extravehicular activity, with the Americans performing space-walk practice and duties for the first time in 9 years.

The potential for economic competition was expanded as the French contracted to launch payloads on their Ariane rocket, including some which had been intended to fly on the United States space shuttle. In fact, there were instances in which payload owners opted to fly on one of these vehicles while also arranging for launch on the other as a back-up to ensure meeting deadlines for getting on orbit.

In a competitive move, the Soviet Union entered the comercialized booster business by offering its civilian satellite launcher, Proton, beginning in 1988, for the second-generation communications satellites of the International Maritime Satellite Organization (INMARSAT). Prices and conditions were reported to be competitive with the shuttle, and with the United States and European expendable launchers.

History was made as the *Pioneer 10* spacecraft crossed Neptune's orbit on June 13, 1983. Since the elliptical orbit of Pluto, normally the outermost planet, has taken the craft inside the orbit of Nepture until the year 2000, *Pioneer 10* had reached a distance from the Sun greater than that of any known planet. By this definition, it became the first human artifact to leave the solar system. It had taken since March 2, 1972, for *Pioneer 10* to travel the 2.8×10^9 mi (4.5×10^9 km) to Nepture's orbit. Its signals, at the speed of light, are received on Earth in 4 h 20 min. Communication time now increases at an average of about 1 min every 4 days. Pioneer's signal, sent by an 8-watt radio transmitter in a tightly focused beam of 1.5°, spreads to cover an area of more than 11×10^6 mi (1.8×10^6 km) across by the time it reaches the Earth. Here, the 210-ft-diameter (64-m) radio antennas of the Deep Space Network receive a message weakened to 2×10^{-8} watt. Even so, scientists expect to track *Pioneer* for years, hoping it will detect the heliopause, the boundary between the Sun's magnetosphere and the interstellar "wind" through which the solar system moves. Crossing of the "edge" should be signaled by an increased flux of cosmic rays expected to be detectable on the other side. There are also hopes that America's tiny survivor in unexplored territory may yield clues to the mysterious force which tugs at Uranus and Nepture, perhaps an unseen object, such as a tenth planet or a remote dark companion to the Sun.

Another milestone was the end of transmissions from Mars. Its batteries apparently dead, the *Viking Lander 1* did not call home on May 5, 1983, as it was programmed to do; nor did it respond to signals

Significant space launches in 1983

Payload	Date	Payload country or organization	Purpose and comments
IRAS	1/25/83	International	U.S., Dutch, British; infrared astronomy, U.S. Delta launched
Cosmos 1443	3/2/83	Soviet Union	Uncrewed; docked *Salyut 7*, doubling station size
NOAA-E	3/28/83	U.S.	NOAA weather, search-and-rescue beacon; U.S. Atlas-F launched
Challenger	4/4/83	U.S.	Shuttle mission 6; first flight of new shuttle; NASA's *TDRS-A* communication satellite
Soyuz T-8	4/20/83	Soviet Union	*Salyut 7* rendezvous aborted
Exosat	5/30/83	European Space Agency	X-ray observer; U.S. Delta launched
Venera 15 and 16	6/3–7/83	Soviet Union	Venus-imaging radar orbiters
Challenger	6/18/83	U.S.	Shuttle mission 7; communications satellites *Anik-C* for Canada, and *Palapa-B* for Indonesia; first U.S. woman astronaut
Soyuz T-9	6/27/83	Soviet Union	Two-person crew docked with *Salyut 7–Cosmos 1443* complex
Challenger	8/30/83	U.S.	Shuttle mission 8; *Insat 1B* (Indian communications satellite); first black American astronaut
Columbia	11/28/83	U.S.	Return of original shuttle, mission 9; *Spacelab 1* with German astronaut

from Earth. Built for a 90-day lifetime, it had functioned on the frozen, dusty, Martian surface since July 20, 1976.

Significant space launches in 1983 are listed in the table.

UNITED STATES SPACE ACTIVITY

In 1983 there were 22 United States launches carrying a total of 30 payloads which were left in orbit. Launches were grouped generally as follows: science, 3; communications, 6; weather, 3 (including one with search-and-rescue signal relay capability); crewed, 4; reconnaissance, 3; electronic intelligence, 2; and navigation, 1.

Space transportation system (STS). A new vehicle was added to the United States space shuttle fleet with the maiden flight of the *Challenger* orbiter. The concept of a family of shuttles, working in rotation, had begun. *Challenger* flew three times in 1983 and was followed by the refurbished *Columbia* making its last flight of the year. The flights were known as *STS 6* through *9*.

Challenger began life with a considerably larger cargo-carrying capacity than *Columbia*. On its maiden voyage it carried 14,000 lb (6350 kg), 40% more than *Columbia* had ever lofted. This lighter, stronger spacecraft benefitted from redesign which increased the power of its engines and reduced the weight of its external tank and two solid-fueled boosters. In the expendable fuel tank more than 10,000 lb (4500 kg) of braces, stiffeners, and other structural parts were eliminated. In the solid-fueled boosters, about 8000 lb (3600 kg) was designed out of the engine casings. Finally, in the orbiter itself, nearly 2500 lb (1100 kg) was removed by replacements such as lightweight honeycomb material for the landing gear doors. Beyond this, some 4000 lb (1800 kg) of carrying capacity were gained by increasing the rated thrust of the three main engines to 104% from the 100% of capacity used for *Columbia* missions. A significant change, in more than weight alone, was the removal of the ejection seats and heavy rails. This action is symbolic of the transition from an experimental machine, under test with a two-person crew, to an operational vehicle in regular service for carrying crews of six or more.

Shuttle mission 6. Trouble plagued the *Challenger*'s first flight. Originally scheduled for a January 20 launch from Kennedy Space Center, in Florida, the mission was delayed until April 4, 1983, because of troubles with all three of the orbiter's main engines. Inspection procedures had revealed tiny cracks in their hydrogen coolant lines. Repairs delayed the flight beyond March 20, the start of spring. This became important in resetting the hour of day for launch. The original 9:00 A.M. was no longer appropriate because it would have placed the shuttle's payload (the *TDRS-A* communications satellite) in the Earth's shadow for too long a period during the climb to geosynchronous orbit. The satellite's thermal design permitted no more than a half hour in shadow (to avoid excessive heat loss). The extended darkness period was avoided by delaying lift-off to 1:30 P.M., causing the first afternoon launch of a shuttle.

Other firsts were numerous, including reuse of the Morton Thiokol solid rockets flown on *Columbia*. The solid rocket boosters, discarded upon burnout during launch, are recovered to cut costs after parachuting into the ocean. Mission 6 even reused a parachute from an earlier launch. While *Columbia* had experienced 82 anomalies or technical problems on its maiden flight, the *Challenger* had only 22.

The prime mission was to deploy the *TDRS-A* (*Tracking and Data Relay Satellite*), the first in a trio intended to replace many ground stations, NASA's traditional communications link with earth satellites. This 5000-lb (2300-kg) spacecraft, measuring 57 ft (17.4 m) at its widest, is the largest, most complex, and most expensive ($100 million) communications satellite ever launched. Its role, at geosynchronous orbit, is to "see" satellites at lower orbits, to continuously communicate with more than a score of them at a time, and to increase the contact time previously limited by the locations of ground stations. Rotating on a spin table, the *TDRS* was spring-launched from the shuttle bay. An hour later, its Air Force booster ignited. This vehicle, the *Inertial Upper Stage* (*IUS*), malfunctioned in its second stage and began tumbling at 30 revolutions per minute. Hours of effort to command separation of *TDRS* from its booster succeeded. Engineers then designed a recovery plan to raise the satellite's elliptical orbit low point of 13,450 mi (21,650 km) to the required circular orbit of 22,300 mi (35,900 km). This was done over a period of 2 months by 39 precisely timed firings of tiny (1-lb or 4.4-newton thrust) hydrazine-fueled station-keeping jets. *See* LAUNCH VEHICLE.

Astronauts Donald Peterson and Story Musgrave performed tethered excursions into the shuttle bay for 3½ h, determining maneuverability for making repairs and inspections using the new, more flexible space suits. This activity was photographed by the other two crew members from the cabin (Fig. 1*a*, *b*, and *c*) and by Peterson (Fig. 1*d*). Peterson and Musgrave simulated a contingency operation (Fig. 1*c* and *d*) that would be required to return the tilt table for the IUS to its normal stored position in the event of failure of an automatic system designed to do so. Commander Paul Weitz and pilot Karol Bobko brought the *Challenger* to a smooth landing on hard-surface runway 22, at Edwards Air Force Base, California, after 5 days in orbit.

The shuttle orbited at 176 mi (283 km) in an orbital inclination of 28.5° to the Equator. It carried a variety of small, self-contained payloads called getaway specials. One, provided by the Japanese newspaper *Asahi Shimbun*, was designed to create snow crystals in the weightless environment. Another, from the Park Seed Company, contained 25 lb (11 kg) of common fruit and vegetable seeds for germination upon return to Earth. A third, developed by cadets of the U.S. Air Force Academy, involved six separate experiments dealing with microbiology or the properties of metal. (Canisters containing some of the getaway specials appear in the lower left foreground of Fig. 1*b*.)

Confidence in the plan to someday land a shuttle on the special runway at Kennedy Space Center was gained as a result of *Challenger*'s reentry maneuvers, which included three hypersonic S turns plus eight separate maneuver sequences involving jet firings and control surface movements. These helped further define the shuttle's flying characteristics in the critical areas of rudder effects, lateral stability, and reaction control system capabilities. A landing back to the launch base would reduce by at least 6 days the time to prepare the shuttle for its next flight.

Materials-processing research included the Lehigh University monodisperse latex processing experiment plus the biological materials separation experiment from the corporate team of McDonnel Douglas with Johnson and Johnson. Both potentially commercial experiments had flown on earlier shuttles. Physiological research supervised by physician-astronaut Musgrave included tests of eye motion, as well as eye-and-ear interaction, in search of clues for reducing space motion sickness.

Shuttle mission 7. The *Challenger* flew again from Kennedy Space Center about 9 weeks later, launching June 18, 1983, for 6-day mission. The pilot was Frederick Hauck and the commander was Robert Crippen, the first astronaut to fly a second time on the shuttle. Three mission specialists completed the first five-person crew ever to fly in space: John Fabian, physician Norman Thagard, and the first United States woman astronaut Sally Ride.

The orbital inclination was 28.5° in a circular orbit at 160 nautical miles (184 mi or 296 km). During its 97 orbits, *Challenger* deployed from the shuttle bay a Canadian communications satellite called *Anik-C2* and an Indonesian communications satellite called *Palapa-B*. A deployable payload is ejected as soon as possible following launch so that if anything forces an early termination this part of the mission will be successful. Contingency landing sites for such an event were in New Mexico, Spain, Hawaii, and Okinawa. An abort during launch would, of course, mean either a return to Kennedy Space Center or a transatlantic flight terminating in Dakar, Senegal.

Numerous experiments were carried in seven getaway-special canisters. These included crystal study and metals processing for a German Youth Fair Program; an ultraviolet spectrometer for the U.S. Air Force and the Naval Research Laboratory; Purdue University students' experiments on geotropism of plants, fluid dynamics in fluid mixtures, and recordings of nuclear particle energies; Caltech students' experiments in germinating seedlings; Camden, New Jersey, high school students' colony of hundreds of ants; ultrasensitive film from Eastman Kodak; and the Edsyn Co.'s investigation of soldering in a vacuum under microgravity.

The continuous flow electrophoresis system for separating biological materials was flown again, as was the monodisperse latex reactor from Lehigh University. In addition, there were several NASA and West German materials-processing experiments.

The *Challenger* crew made additional history by being the first to deploy a satellite from a spacecraft and later to retrieve it. The device, the largest spacecraft ever built in Europe, is a West German

Fig. 1. Extravehicular activity (EVA) of astronauts Story Musgrave and Donald Peterson in the cargo bay of the space shuttle *Challenger* during shuttle mission 6 on April 7, 1983. (*a*) Musgrave evaluating techniques required to move along the bay's edge with a bag of tools. (*b*) Musgrave (left) and Peterson floating about, restricted by tethers attached to safety slide wires. (*c*) Musgrave (left) and Peterson setting up winch operations at the aft bulkhead as a simulation for a contingency EVA in which they are to run a rope from the winch device near center through a snatch block over to the tilt table for the *Inertial Upper Stage* (*IUS*). (*d*) Musgrave running the rope through the snatch block device. (*NASA*)

platform weighing 3200 lb (1450 kg) and measuring 15 ft (4.5 m) long, called *SPAS* (*Shuttle Pallet Satellite*). By means of the multijointed, 50-ft (15-m) Canadian manipulator arm, astronauts Fabian and Ride repeatedly released and captured *SPAS*. This success brings closer the days of snaring satellites for refueling or repair or for return to Earth. The first scheduled to be so serviced is the *Solar Maximum Mission* satellite, crippled since 1980. Commander Crippen, from the aft window station, maneuvered the *Challenger* below, above, and ahead of the *SPAS*, flying in formation for hours, as far off as 1000 ft (300 m), before returning with smooth precision to snare the *SPAS*. The crew performed a different type of capture by rotating the free-flying *SPAS* while maneuvering *Challenger* and arm into position to capture. In another sequence, pilot Hauck, working at various distances from *SPAS*, fired *Challenger*'s reaction control jets to film how the exhaust plumes threw *SPAS* about. Even from 1000 ft (300 m) away, particles from *Challenger*'s jets were detected by sensors on the pallet satellite. This observation highlighted the importance of studies for avoiding contamination of the shuttle bay or nearby free-flying satellites. Television and movie cameras on *SPAS* and *Challenger* filmed the deployment and capture operations.

The *SPAS* had success with its numerous science operations. These included materials processing, a friction loss experiment, mass spectrometer shuttle environment readings, thermal studies of heat pipes, and solar cell experiments from nine countries including China. The Germans describe *SPAS* as the first reusable satellite, and they built it as a commercial vehicle on which to rent space to customers who want to fly experiments inside or outside the shuttle bay. While *SPAS* was in free drift with no perturbations, it conducted a materials-processing experiment for comparison with a similar study being done on the *Challenger*, where the materials were subject to small gravity loads from crew movements and thruster firings. The success of a *SPAS* solid-state Earth-imaging system on this flight had led to the formation of a United States–European consortium that intends to offer commercial remote sensing flights in competition with the higher-cost systems of the U.S. *Landsat* and the *SPOT* of France. The *SPAS* system, the first high-resolution, charge-coupled-device scanner to fly in space, is called MOMS (modular optoelectronic multispectral scanner).

Medical experiments were conducted on *Challenger* by physician-astronaut Thagard on the related group of symptoms known as space adaptation syndrome (S.A.S.). Involved are fatigue, headaches, problems with orienting oneself to surroundings, plus the nausea astronauts and cosmonauts have experienced for various periods. While the problem is pervasive, it is transitory, and it does not affect everyone. Thagard tested for conflict between the senses (vision versus the sense of balance). He also tested for the effects of body fluid shifting. In space, fluids migrate from the legs throughout the body (including the face, which is puffier in microgravity). There was no evidence of increased pressure inside the eyes and ears, however.

Challenger landed on June 24, 1983, at Edwards Air Force Base, forced by weather from its plan to land at the Kennedy Space Center. *Challenger* had to fly 738 mi (1188 km) cross-range to reach Edwards (from the orbit 98 reentry ground track), coming within 12 mi (19 km) of allowable cross-range limit—proving additional shuttle flexibility.

Shuttle mission 8. On August 30, 1983, *Challenger* was off again. Commander Richard Truly had previously flown on shuttle mission 2. The crew were pilot Daniel Brandenstein, physician William Thornton, and mission specialists Dale Gardner and Guion Bluford, Jr., the first black American in space. Lift-off at 2:32 A.M., making this the first night launch of the shuttle, was necessitated by the time needs of *Insat 1B*, a communications–weather satellite built by Ford Aerospace for the Indian government. *Insat 1B* had to be deployed at local dusk near the mid-Pacific for its sun sensor to obtain proper attitude information.

The night launch timing provided the astronauts with excellent photographic opportunities during extensive daylight periods over the Southern Hemisphere, compared with the darkness over these areas resulting from the previous daytime launches. New geological features were seen in Australia, and several volcanic eruptions were sighted in New Guinea.

New software instructions provided automatic guidance of the launch. For example, when the solid rockets burned at a slightly higher rate than planned, the main engines throttled back 1% to restrain acceleration through a period of maximum dynamic pressure. When the solids had been jettisoned, another improvement gimbaled the main engines so that their thrust was vectored equally around the center-of-gravity point. This provided an extra 500 lb (230 kg) of orbiter payload capability compared with the previous focusing of thrust directly through the center of gravity in the pitch axis. This flight also marked the first use of the new Morton Thiokol solid rocket motor, which provided an additional 3000 lb (13,000 newtons) of thrust. Also, its new lightweight motor cases meant 770 lb (350 kg) of payload benefit. All these features were important improvements for launching future heavy payloads.

Astronaut Bluford directed the deployment of *Insat 1B*, the sixth satellite launched from the shuttle. The 7444-lb (3377-kg) spacecraft was spin-stabilized at 40 revolutions per minute and spring-ejected at 2.5 ft/s (10.75 m/s) about 25 h after launch. The shuttle had been positioned to provide precise alignment of *Insat*'s spin axis with the needed flight path to synchronous orbit. The Morton Thiokol Star 48 solid rocket motor put *Insat* into a transfer orbit. Then, unlike previous shuttle-launched spacecraft which was used solid-propellant apogee kick motors to circularize the orbit at geosynchronous altitude, *Insat* used a liquid propulsion system. Nitrogen tetroxide and monomethyl hydrazine fueled its 100-lb-thrust (445-N) rocket engine. Final approval for deployment came from the *Insat* control center at Hassan, India. At a cost of $9 million, Lloyd's of London insured the *Insat* against loss. This was a high premium, based largely on failure of the first *Insat* launch in 1982, when the insurance payment to India was about $65 million.

Shuttle mission 8's other main goal was to aid in checking out the embattled *Tracking and Data Relay Satellite* launched on mission 6. Hopes were high from the August 7 transmission of high-quality, high-data-rate information from *Landsat 4*'s thematic mapper to the *TDRS* for relay to the ground station at White Sands, New Mexico. Transmissions from *Challenger* via *TDRS* also proved successful and built confidence for a reliable space-to-space data relay for the volumes of science and engineering data to be produced during the upcoming flight of *Spacelab 1*. The tests also showed that *TDRS* could track an orbiter automatically. Television and voice quality was noticeably better via *TDRS* as compared with transmissions through the old ground-based network.

Physician Thornton obtained medical data which convinced him that space sickness symptoms would not be a barrier to eventually flying nonastronauts on the shuttle. His physiological tests included body measurements, audiometry, eye movement, and the repeatability of physical motions. During a press conference which the astronauts held by radio from space (another first), Thornton said that he had learned more in the first 1½ h in orbit than in all his previous years of study on space adjustment problems.

Nearly 15 h of tests across 3 days were conducted with the manipulator arm maneuvering the payload flight test article (PFTA). This 7460-lb (3384-kg) barbell-shaped structure, about 20 ft (6 m) long and 15 ft (4.5 m) wide, was the largest mass ever moved by the arm. In one test, astronaut Gardner held the article over the payload bay while the reaction control jets of *Challenger* were firerd to see the effects on arm motions and how they would be absorbed by the shuttle structure. Another test successfully used the arm for the first time for television underside inspection of the orbiter.

Biological processing, using the previusly flown private-industry electrophoresis system, was conducted by astronauts Bluford and Gardner. It focused on precise separation of contaminants from live pancreas cells in hopes of producing a material capable of overcoming a human's inability to produce insulin. Intended is a product for commercial marketing in treating diabetes.

In a test of how samples of materials would react on exposure to atomic oxygen in the upper atmosphere, the *Challenger* was maneuvered to a lower orbit. The results may lead to means for preventing degradation that occurs in paints, plastic, and other substances such as the waterproofing agents on the shuttle's fabric and tile insulations.

In the shuttle's first night landing, *Challenger* touched down on September 5 at 3:40 A.M. on lighted runway 22 at Edwards Air Force Base in a hypersonic descent profile test which minimized reentry heating damage to tiles. Commander Truly reported seeing the lighted runway from an altitude of 75,000 ft (23 km) while 50 mi (80 km) from the field. He took manual control just below Mach 1 speed. With pilot Brandenstein calling altitude and airspeed, he preflared at 1750-ft (530-m) altitude, pulling *Challenger* from its 19° glide slope and 12,000 ft/min (60 m/s) descent rate to the inner glide slope of 1.5°. Movies made during reentry show a white pulsating wake trailing *Challenger.* The film shows that the wake would build and flash repeatedly at about half-second intervals from unknown causes.

Space science flights. An international effort led to the January 25 launch of *IRAS*, the *Infrared Astronomical Satellite* from Vandenberg Air Force Base, California. A joint venture of the United States, the Netherlands, and the United Kingdom, the 2370-lb (1075-kg) *IRAS* spent a 10-month lifetime successfully conducting 95% of an intended all-sky survey on infrared sources in space. Its near-polar orbit at 560 mi (900 km) was chosen to avoid contamination of the instrument by the Earth's atmosphere at lower altitudes and false readings due to proton strikes at the Van Allen radiation belts at higher altitudes. Mission duration was determined by the time it took for 125 gallons (475 liters) of liquid helium to boil off into space in the process of cooling the *IRAS*'s infrared telescopes to 2.5 K (−455°F). The deep cold environment was necessary to prevent spacecraft heat from masking the faint thermal emissions of stellar objects such as dying stars. *IRAS* surveyed practically the whole sky at four infrared wavelengths with unprecedented sensitivity. Previous infrared observations from space had used detectors cooled to about 12 K (−438°F), but the 2.5 K (−455°F) of *IRAS*'s detectors offered about a 90-fold improvement in signal-to-noise ratio. *IRAS* scanned 95% of the sky at least four times, and about 72% was surveyed six times.

The first direct evidence of a possible second solar system in the universe resulted from an *IRAS* discovery of a ring of large particles surrounding Vega, the third-brightest star in the sky. *IRAS* also discovered regions where stars seem to be coalescing from clouds of dust and gas. At the other extreme it observed ancient stars which have nearly consumed their hydrogen fuel. The spacecraft also hunted for fast-moving asteroids that may be passing close to Earth. It found at least five of these "minor planets," including one which appears to be the defunct nucleus of a comet, stripped of its volatile materials. It orbits in the trail of debris which Earth's atmosphere encounters during the well-known Geminid meteor showers. This asteroid, 1983TB, may be the parent body for the Geminid meteors. The number of actual comets discovered by *IRAS* amounted to five, and on some better-known comets *IRAS* detected tails far longer than anything seen in the visible spectrum.

Dozens of cold, point sources of infrared radiation that could not be seen in any other way were clearly visible to *IRAS*. In addition, *IRAS* found what seems to be a vast, thick ring of dust, tilted about 9° to the ecliptic, between the orbits of Mars and Jupiter, and theorized to be a cloud of debris resulting from a comet and asteroid collision. *See* INFRARED ASTRONOMY.

Other international cooperation developed rapidly when the European Space Agency (ESA) asked NASA in February 1983, with a 3-month lead time, to launch *Exosat* on the American Delta vehicle. The ESA officials had become concerned that the intended launcher, *Ariane 6*, would be delayed due to the failure of *Ariane 5*. This would have meant unknown months in storage for *Exosat*, introducing possible degradation of critical on-board components. The launch was worked into a heavy schedule, and *Exosat* was flown from Vandenberg Air Force Base on May 30 on a Delta 3914 vehicle into a highly eccentric polar orbit inclined at 72.5°. The apogee is 124,274 mi (200,000 km) and perigee is 217 mi (350 km). *Exosat* produces highly precise and detailed information on extragalactic x-ray sources correlated with observations in other wavelengths. It operates in one mode of direct pointing at an x-ray source (for up to 80 h) and in the occultation mode while the Moon passes in front of an x-ray source. A 2-year lifetime is expected, during which *Exosat* will complete 180 orbits of about 96 h each.

Satellite communications. The routine business aspects of satellite communications were highlighted by the Federal Communications Commission (FCC) in August when it issued specifications to decrease the spacing between geosynchronous satellites. The order established 2° between spacecraft operating in the Ku band (12 GHz), effective immediately, and it set a schedule for some currently orbiting satellites to move to prescribed new locations. In addition, the FCC started action on an approved plan to achieve the same 2° spacing by the 1990s for satellites operating in the C band (4 GHz), where spacing is now between 2.5° and 4°.

SOVIET SPACE ACTIVITY

In 1983 there were 98 Soviet launches carrying a total of 116 payloads. Launches were grouped generally as follows: science, 6; communications, 13; weather, 1; earth resources, 11; crew-related, 5; crewed, 2; photo reconnaissance, 27; electronic intelligence, 6; minor military, 7; early warning, 3; ocean electronic intelligence, 2; navigation, 9; and tactical communications, 6.

In an exciting year, the Soviets experienced a meteorite strike on *Salyut 7*, explosion on the launch pad, rendezvous failure, rendezvous success, orbit of Venus, scores of successful launches, development flight of a possible minishuttle test vehicle (Fig. 2), and another global scare from unplanned reentry of a nuclear-fueled satellite.

Record space flight. Retroactive news emerged about the return of the *Soyuz T-7* ferry craft from the *Salyut 7* space station. After setting a new endurance record of 211 days, cosmonauts Anatoliy Berezovoy and Valentin Lebedev were reintroduced to gravity via a hard touchdown in a windy snowstorm at night on December 11, 1982. *Soyuz T-7* rolled down a hill, tumbling the cosmonauts. Fog in the area prevented helicopters from being used, so that the crew spent 40 min waiting until a wheeled all-terrain vehicle arrived and then another 20 min until it was prepared to receive them. They spent the remaining 5 h of the night in the truck because weather was too severe for safe evacuation. Upon flight to Tyuratam on December 11, they were found to have lost several pounds and to have a reduced red blood cell count but to be in good health and adapting normally. Berezovoy's pulse rate a day after landing was 86 compared with 52–60 beats per minute prelaunch. Lebedev's comparison was 88 beats versus 70–76 beats. After 48 h on Earth, both blood pressure and pulse rates moved toward normal levels. Psychological and physiological factors helped influence the decision for the December 10 night return. The crew had become irritable with each other and had also requested a return to see the New Year come in with their families. Other reported factors showed the limited options and inflex-

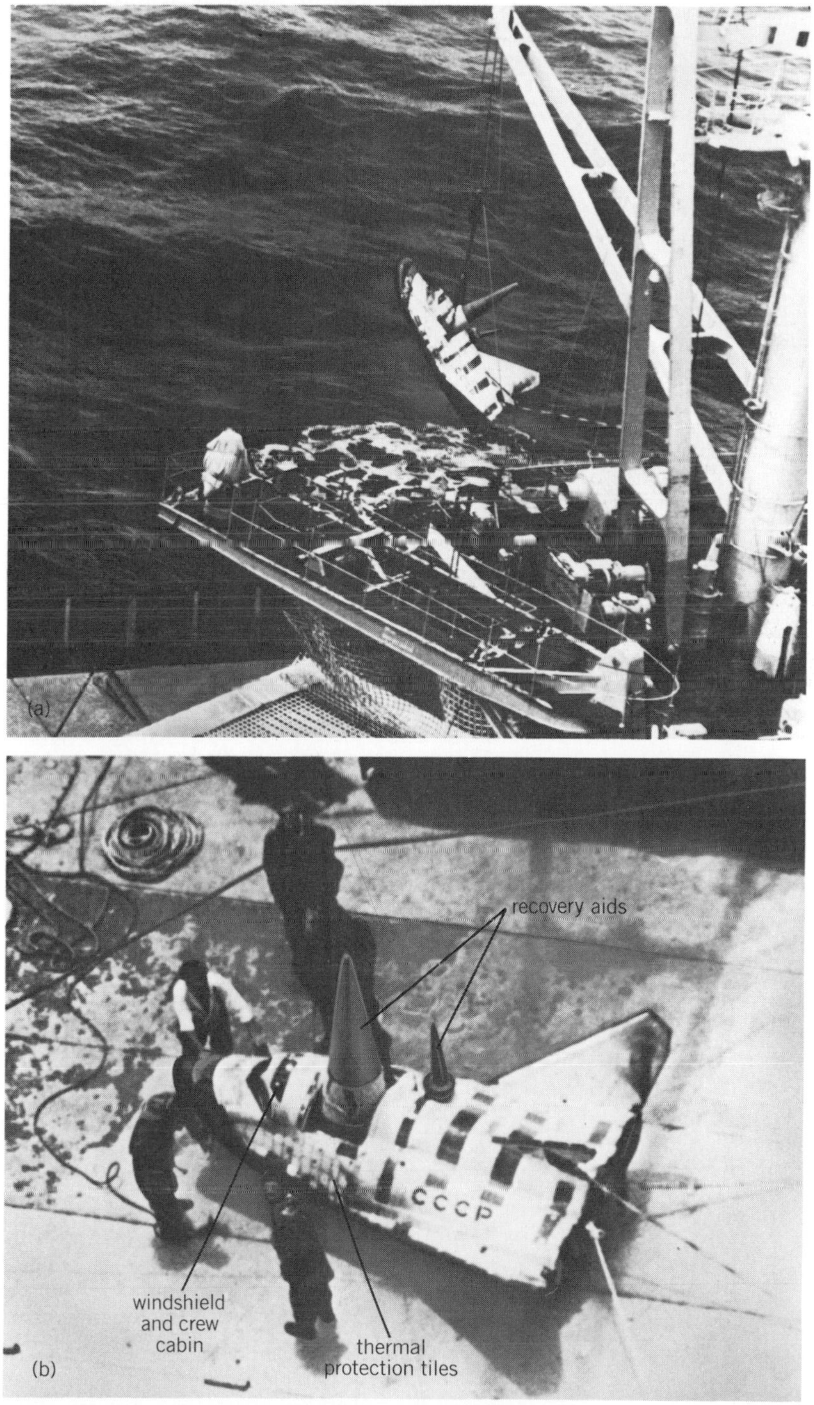

Fig. 2. Recovery of Soviet minishuttle in the Indian Ocean on March 15, 1983, viewed from an Australian Air Force aircraft. The spacecraft is a subscale test version of a future Soviet crewed spaceplane. (*a*) Spacecraft being hoisted aboard a Soviet ship, and (*b*) placed on deck. (*From Soviet spaceplane shows crew cabin, tile details, Aviat. Week Space Technol., 118(23):27, June 6, 1983*)

ibility of Soviet equipment and procedures in controlling landing times and places.

Nuclear-fueled satellite reentry. The Soviet satellite *Cosmos 1402* did not respond to an attempt to boost it and its 110-lb (50-kg) enriched uranium core into a high, 500-year storage orbit. There was widespread fear of contamination which could cause direct radiation exposure to humans or which could produce radioactive dust that might enter the food chain. The spacecraft broke up during frictional drag reentry, with some portions burning up over, or landing in, the Indian Ocean on January 23, 1983. One segment, containing the reactor core, continued to orbit until it reentered in a fiery plunge on February 7 over or into the South Atlantic. This marked the eighth nuclear-powered craft to have fallen to Earth unexpectedly. Five of these were Soviet and three were American (*Transit* in 1964, *Nimbus* in 1968, and the *Apollo 13* moon lander in 1970, all three entering the sea).

Space station activity. A heavy Proton booster on March 2, 1983, launched *Cosmos 1443* to enlarge the *Salyut 7* station for expanded crewed operations. It linked to the unoccupied station on March 10 in an important test of ability to assemble station components. Its addition was estimated to have about doubled the size of *Salyut 7*, a 42,000-lb (19,000-kg) craft about the size of a compact house trailer. The *Cosmos* addition delivered equipment and other supplies to the 200-mi-high (320-km) location. Analysts said that the two spacecraft were probably connected end to end, thus leaving two ports at opposite ends of the combination for *Soyuz* and *Progress* ferries to hook up.

On April 20, three cosmonauts were launched in *Soyuz T-8* to dock with the *Salyut 7* station, which had been unoccupied for over 4 months. The Soviets announced that they would make their quarters in the *Cosmos 1443* attachment. The crew consisted of commander Vladimir Titov and space veterans Gennadi Strekalov and Aleksandr Serebrov. Within 48 h after lift-off from the Baikonur Space Center in Kasakhstan, they were back on the ground.

Their mission was the seventh to go awry since the 30-flight *Soyuz*-to-*Salyut* series begin in 1971. Analysts determined that the failure to dock occurred because the *Soyuz* rendezvous radar antenna, a boom-held dish, would not deploy. An optically guided approach to the station was attempted but had to be aborted in fear of collision. Titov said that he could not get any closer than 500 ft (150 m) or so from the *Salyut*. Sanctioned spinning of the spacecraft failed to fling the antenna out.

On June 27, the Soviets tried again, launching *Soyuz T-9* from Tyuratam, carrying Vladimir Lyakhov and flight engineer Aleksandr Aleksandrov. The men docked the next day without incident and activated the *Salyut 7–Cosmos 1443* complex. With the addition of *Soyuz T-9*, the three docked vehicles weighed about 103,000 lb (47,000 kg) and were orbiting at an inclination of 51.6°. The capabilities of *Cosmos 1443* were finally outlined by the Soviets 4 months after it was launched. Major features included: length about 42 ft (13 m) and width about 13 feet (4 m); design for tug duties; 430 ft^2 (40 m^2) of solar array to generate 3 kilowatts of electrical power; a large fuel suppply for maneuverability or resupply of *Salyut*; Earth-return capability via a reentry module capable of carrying over 1100 lb (500 kg) of research material or bulk products manufactured in orbit; and resupply capability of several tons. The Soviets apparently plan to produce specialized versions of the habitable modules for scientific or manufacturing work in space.

A loud crack was heard aboard *Salyut 7* on July 27 when a micrometeorite or other debris damaged a window. However, the small crater (0.15 in. or 4 mm in diameter) did not threated the pressure integrity of the station. (Similar damage was suffered during shuttle mission 7 in what was apparently another rare strike. In that case, the damaged glass pane had to be replaced upon landing.) A pressurization failure would require the crew to rush into the *Soyuz* transport and shut the hatch.

A tanker spacecraft, *Progress 17*, resupplied the *Salyut* station in mid-August. before it arrived, the *Salyut* crew separated from *Cosmos 1443* and also moved *Soyuz T-9* to *Salyut's* forward docking port to accommodate *Progress 17* on the aft port. On August 23, the *Gemini*-sized reentry module from the *Cosmos 1443* tug was returned to Earth carrying 700 lb (320 kg) of film, manufactured or processed materials, and surplus hardware.

A serious propellant leak occurred aboard *Salyut 7* on September 9, when a main oxidizer line ruptured and spilled the contents of two oxidizer tanks into space. The cosmonauts donned suits and entered the *Soyuz T-9*, which was prepared for an emergency return to Earth. They reentered *Salyut 7* when extensive checking disclosed no continuing or overt hazard. Their ship was crippled, however, from the maneuvering propellant loss and with half of the 32 reaction control thrusters unusable. Meanwhile, an undisclosed but serious malfunction on the tug, *Cosmos 1443*, affected its operation, and although the Soviets had apparently wanted to redock it with the *Salyut 7*, the decision was made to discard and it was sent into a destructive reentry over the Pacific Ocean on September 19. This followed by a day the destructive reentry discard of the *Progress 17* supply tanker. Soviet space troubles then compounded. On October 27, an SL-4 booster exploded on the pad at Tyuratam while its two-person crew was being launched to safety by the rocket escape mechanism atop the vehicle. A fire had broken out at the base of the booster about a minute before lift-off. Ground controllers fired the escape system that pulled Gennadi Streakalov and Vladimir Titov to safety. These two had been part of the crew of the ill-fated *Soyuz T-8* rendezvous attempt. Their escape capsule parachuted to the ground a few thousand feet (about 1 km) from the fire. The Soviets were now confronted with the problem of two cosmonauts orbiting for 3 months in *Salyut 7* whose

Earth-return vehicle was aging. For safety purposes, the Soviets limit crewed reentry craft to about 115 days in space, based on the effects of the environment on *Soyuz* systems. The aborted launch was to have provided a fresh vehicle to the *Salyut 7* crew for a probable return in January 1984. While the reentry issue went unresolved, the Soviets launched the *Progress 18* automatic tanker on October 20 to refuel *Salyut 7*'s remaining tanks and to provide support propulsion capability. The crew pressed on with its duties, including climbing outside in early November over a 2-day period to install additional solar power arrays. The crew discarded *Progress 18* to a destructive reentry over the Pacific on November 16. They returned to Earth in the *Soyuz T-9* spacecraft on November 23, ending 149 days in space.

Venus observations. In early June, the Soviets launched two uncrewed spacecraft, *Venera 15* and *16*, toward Venus. These arrived in orbit around Venus in the second week of October and, after a month of orbit adjustment maneuvers, began obtaining high-resolution radar images of the planet's surface. The Soviets are apparently concentrating their imaging on the polar regions. Surface detail was obtained with a resolution of 0.6–1.2 mi (1–2 km) and showed impact craters, hills, major fractures, and mountain ridges. Scientists said that bizarre basalt seas were observed on the surface. In addition, Soviet media said that a type of thermal mapping was being conducted. The resolution is far better than the 12 mi (20 km) attained by the American *Pioneer Venus*, the first spacecraft to have mapped Venus by radar, but not much better than the best resolutions obtained by the Arecibo radio observatory in Puerto Rico. A Soviet broadcast said that when the first station finished working, the second orbiter would begin, perhaps an implication that the craft are battery-powered.

ASIAN SPACE ACTIVITY

There were five launches by Asian vehicles in 1983. Japan put up one scientific satellite and two communications satellites; China orbited one reconnaissance spacecraft; and India launched one Earth resources satellite.

India and Ford Aerospace recovered somewhat from the *Insat 1A* failure in 1982 by achieving full operational status for the United States–launched *Insat 1B* in October 1983. This spacecrafte provides India with services in the fields of telecommunications, direct broadcast television, and meteorology. It links 35 fixed and mobile earth stations throughout India, providing telephone trunk service, remote-area communications, and television feed uplinks. Many Indians argue that mass communications is the prime hope for progress in this vast land with only 36% literacy rate. The satellite also relays data to a service center in New Delhi from more than 100 unattended platforms gathering oceanographic, hydrological, and weather information. *See* DIRECT BROADCASTING SATELLITE SYSTEMS.

China launched its thirteenth spacecraft on August 13 into an elliptical orbit of 248 by 106 mi (400 by 170 km) inclined at about 64°. The payload, apparently a military reconnaissance satellite, ejected a reentry vehicle about 5 days later. This success moves China closer to the goal, announced in 1983, of preparing and launching its own geosynchronous communications satellite.

EUROPEAN SPACE ACTIVITY

The commercial future of Europe's *Ariane* launcher brightened considerably with the successful orbiting of two communications satellites aboard *Ariane 6* on June 16, 1983. There were no troubles in the third-stage propulsion system which had caused the loss of *Ariane 5* in September 1982 while carrying a communications satellite and a science payload. The success of the sixth flight restored confidence in an ambitious schedule for *Ariane*. A launch rate of one mission every 2 months is planned. Success was compounded on October 18 when *Ariane 7* carried the INTELSAT V communications satellite to geosynchronous orbit. The Europeans were clearly in the business of lauching spacecraft for profit.

INTERNATIONAL SEARCH AND RESCUE

On March 24, 1983, the Soviet Union launched its second satellite in the *Cospas* series of transponder-equipped vehicles which can receive signals from the emergency locator transmitters of ships or downed aircraft. It also announced construction of two new ground stations to expand coverage for the search-and-rescue system. The United States, on March 28, launched *Tiros-N*, its first satellite with such capability.

Scores of lives have been saved by the system since the Soviets put up the first *Cospas* in July 1982. A year later, Soviet and American representatives publicly agreed to continue the service through the late 1980s. They also began plans toward adding geostationary satellites to the system and to the creation of a permanent international organization to manage the program. One goal is to have a total of four polar orbiting spacecraft with search-and-rescue capability in operation at all times.

For background information *see* COMMUNICATIONS SATELLITE; INFRARED ASTRONOMY; SPACE BIOLOGY; SPACE COMMUNICATIONS; SPACE FLIGHT; SPACE PROBE; SPACE PROCESSING; SPACE SHUTTLE; VENUS; X-RAY ASTRONOMY in the McGraw-hill Encyclopedia of Science and Technology.

[CHARLES BOYLE]

Bibliography: *Aviat. Week Space Technol.*, issues from November 29, 1982, through November 28, 1983; H. S. F. Cooper, Jr., Letter from the space center, *The New Yorker*, pp. 54–103, October 3, 1983; *NASA Activities*, issues from December 1982 through November 1983; *Science News*, issues from November 28, 1982, through November 26, 1983; *Space World*, issues from December 1982 through November 1983.

Space probe

The exploration of the solar system with spacecraft started in 1962 with the launch of *Mariner 2* to Venus. In the following two decades over 40 increasingly elaborate spacecraft, launched mostly by the United States, explored Mercury, Venus, the Moon, Mars, Jupiter, and Saturn. The scientific return has been revolutionary, yielding new concepts about the evolution of planets and their atmospheres, refined ideas about the beginning of life, and insight into the first billion years of terrestrial history. This harvest is rich even though Uranus, Neptune, Pluto, and the asteroids and comets have yet to be explored. Additionally, new questions have arisen about the explored planets. Ironically, it was at the peak of the *Voyager* exploration of Jupiter and Saturn in the late 1970s and early 1980s that the United States planetary program found itself in a "going-out-of-business" mode. The last new mission to be approved (in 1977) was the *Galileo* orbiter and probe scheduled to be launched to Jupiter in 1982. A rendezvous with Halley's Comet, requiring the development of solar electric propulsion, was not approved. This opportunity thus went by default to the Soviets, Japanese, and European Space Agency, all of whom will send spacecraft to Halley's in late 1985. In the same period the proposed Venus orbiting imaging radar mission did not get into the NASA budget.

To reverse the downtrend, NASA formed the Solar System Exploration Committee (SSEC) and charged it with devising a mission implementation strategy through the year 2000, and proposing specific new missions to reinvigorate United States solar system exploration. The SSEC has concluded that there does exist a scientifically productive, technologically feasible, and more economical way to conduct planetary exploration. It structured a core program viable through the year 2000, constrained by a potential annual budget of about $300 million, but an amount deemed sufficient for a stable program.

Causes of decrease in exploration. The SSEC determined that the decrease in exploration was caused by a combination of factors. Indeed, the overall planetary budget had decreased from historical peaks, but this was recognized to be largely a symptom, not a cause. There was increasing competition from other areas of space science, especially astronomy, and the cost of individual planetary missions had increased. Cost growth came from increased sophistication of the experiments and spacecraft, the correspondingly larger launch rockets required, the longer trip times to the outer planets, inflation, and decreased overall efficiency caused by stretched development schedules. In turn, the stretched schedules were caused by budget limitations and, in the case of the *Galileo*, by delays in the space shuttle development which led to a 2-year slip. During the same period, there was a decrease in the shuttle capability to launch the *Galileo* to Jupiter, requiring a switch to a Centaur upper stage for the shuttle with a resultant further 2-year launch delay to 1986.

Reduction of costs. The core program contains several approaches to reducing mission costs.

Limited scientific objectives. Prior missions tended to be driven up in cost because of efforts to put on as many instruments as possible. This is a reasonable approach in early reconnaisance phases of exploration because there is simply not enough knowledge to pose well-structured questions. In "looking for the unknown" it is necessary to cover as broad a range of experiments as possible. Now, however, with detailed knowledge of the planets available, especially Venus and Mars, as well as the Moon, it is possible to pose specific questions which can be answered with the flight of a smaller number of instruments on simpler spacecraft.

Increased resiliency. Historically, many cost increases have been caused by spacecraft development problems. This possibility can be reduced by using existing spacecraft designs or modular designs which allow flexibility for change without concurrent increase in cost. To beat the problem of schedule delays caused by slips in launch vehicle availability or by inadequate rocket capability, only existing or firmly planned launch vehicles (for example, Centaur, IUS, and other shuttle upper stages) should be relied upon, and new technology such as solar electric propulsion should be avoided. *See* LAUNCH VEHICLE.

Increased spacecraft inheritance. A major cost-driver consists of tailoring a spacecraft design for each new mission. To reduce that, consideration must be given to using existing production-line spacecraft which are now used for Earth-orbiting missions (for example, weather and communications satellites), developing modular or reconfigurable spacecraft, using the same instrument design on several missions, and optimizing mission sequences.

Increased operations efficiency. Mission operations have historically been hand-tailored for each flight, each having its own dedicated operations crew, numbering in the hundreds for the complex *Viking* and *Voyager* type of missions. Typically a relatively large operations team is needed during the planetary encounter phase, but unfortunately that same large team must be employed during the relatively low-activity period during interplanetary cruise, which may last up to 3 years. The concept of the multimission control center allows for shared operations. By optimizing the mission sequence, say by grouping flights to the outer planets, and by automating the tedious, labor-intensive spacecraft command generation sequences, further savings are possible.

Core program. The SSEC reexamined the primary goal of planetary exploration: to determine the nature of the planets, comets, and asteroids toward an understanding of the origin and evolution of the solar system, including Earth, and toward an understanding of how the appearance of life is related to

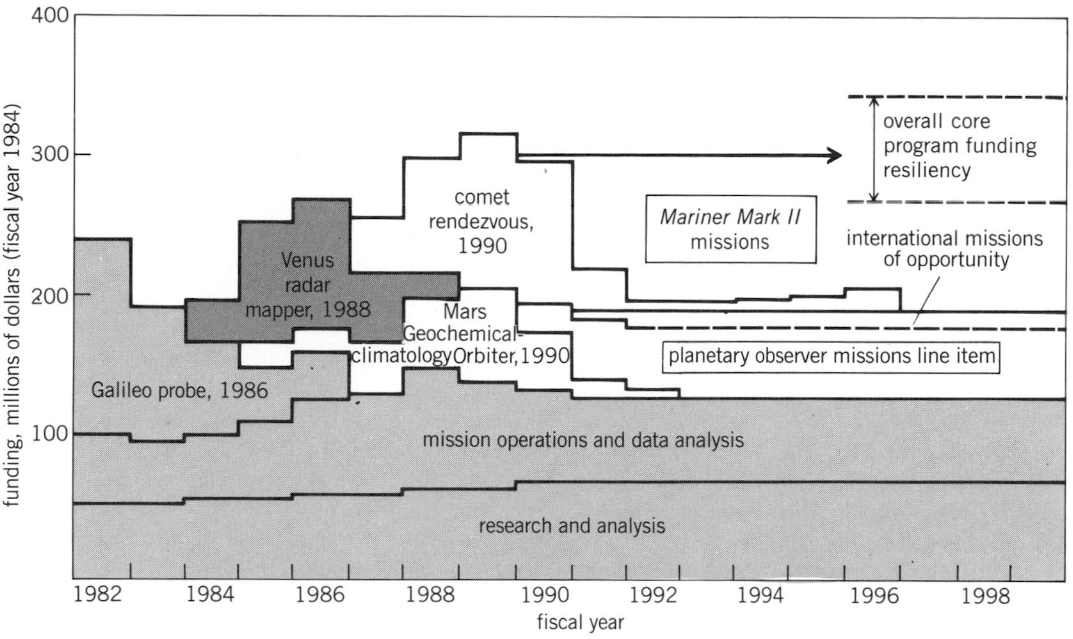

Fig. 1. Components of the core program. The *Galileo* and Venus radar mapper are a transition into the core program. The Mars geochemical/climatological mission is the first planetary observer one, and the comet mission the first *Mariner Mark II* one. Launch dates are shown. (*After NASA, Planetary Exploration Through Year 2000: A Core Program, 1983*)

the chemical history of the solar system. It was judged to be a sound goal, but the SSEC added as a secondary goal the utilization in space of resources from the Moon and Earth-crossing asteroids, believing that although accomplishment is distant, resource utilization should be considered now to the degree that it helps determine science priorities. The Space Science Board strategy of an orderly progression from reconnaissance to exploration to detailed investigation, and for a balanced program among the inner planets, outer planets, and comets and asteroids was adopted.

The core program consists of four major components (Fig. 1). Although well below historical peaks, it provides for a stable exploration program, renewing a commitment to planetary exploration. The underpinning, without which there is no future, is a strong research and analysis portion, which includes technology and instrument development for future missions, laboratory experiments, theoretical studies, ground-based telescopic observations, and future mission studies. Next is the mission operations and data analysis segment, which supports both future and current flight operations (*Voyager*, *Pioneer Venus*, and *Pioneers 6–9*) and the analysis of the great store of data obtained from all the missions to date.

Planetary observer. Planetary observers are a new mission concept which will use production-line Earth-orbital spacecraft, slightly modified, and outfitted with a limited number of experiments designed to answer well-formulated questions. The capability of Earth-orbital spacecraft and current understanding of the planets limit planetary observers to inner solar system applications. The first proposed mission is a geochemical-climatology orbiter of Mars, to be launched in 1990. The objectives are two-fold: to better understand the geologic evolution of Mars by mapping its global surface composition, topography, and gravity and magnetic fields; and to comprehend past climates by locating reservoirs of water and carbon dioxide, their daily and seasonal cycles, and the transport of these gases between Mars's surface, atmosphere, and polar caps.

The second planetary observer would be a lunar polar orbiter. Configured much like the Mars mission, it would be dedicated to a geochemical survey of the entire lunar surface, extending and improving upon the *Apollo* results and providing critical data needed for piecing together the returned lunar sample information and for assessing the potential use of resources from the Moon for future human exploration.

Additional planetary observers include a probe of the Venusian atmosphere and an exploration of Earth-crossing asteroids.

Mariner Mark II. The exploration of the outer planets, comets, and asteroids demands a spacecraft more capable than the planetary observers. For such missions, the Jet Propulsion Laboratory is designing a simplified, reconfigurable spacecraft called the *Mariner Mark II* whose modularity is expected to lower mission costs by allowing the same basic design to be used for a variety of mission types (Fig. 2), including orbiters, flyby/probes, and crude cometary sample return. By careful design of scientific observations and by use of data compression techniques, it will be possible to use lower data

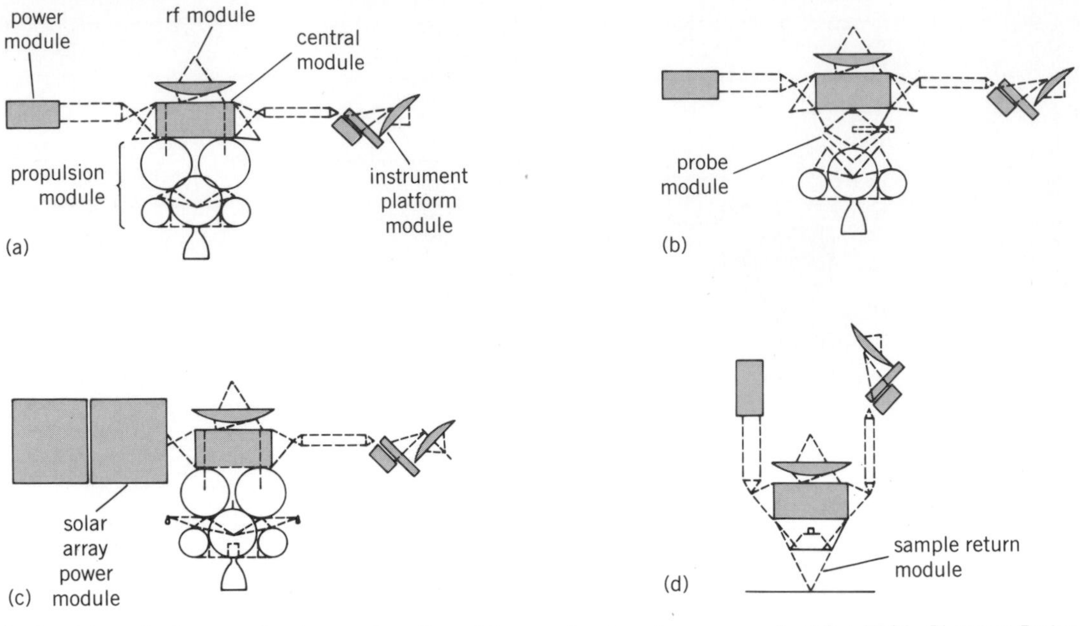

Fig. 2. Schematic diagrams of various configurations of the *Mariner Mark II* spacecraft. (*a*) Orbiter spacecraft. (*b*) Flyby/probe spacecraft. (*c*) Solar array–powered spacecraft. (*d*) Sample return spacecraft. (*After NASA, Planetary Exploration Through Year 2000: A Core Program, 1983*)

transmission rates with correspondingly smaller antennas and power systems. A multimission control center and more automation in spacecraft and ground command systems will reduce the cost of operations.

The highest-priority *Mariner Mark II* mission is a comet rendezvous with a launch in 1990. Comets, largely unexplored, are thought to consist of primitive solar system condensates (ices and solids) formed during the origin of the solar system 4.5×10^9 years ago. The ability to rendezvous is critical, to allow the close-up examination of the comet nucleus and its evolution during passage in the inner solar system. In this way the mission will be a worthy successor to the fast-flyby missions to Halley's Comet. The specific target comet would be selected after further study and by taking the Halley's results into consideration.

The second *Mariner Mark II* mission is a flyby/probe of Saturn's moon Titan. *Voyager* showed Titan's atmosphere to be rich in organic compounds with the possibility of methane oceans on its surface. An intriguing speculation is that Titan's atmosphere may look like the Earth's did before the appearance of life. Other unprioritized, candidate *Mariner Mark II* missions include atmospheric probes of Saturn and Uranus, an orbiter of Saturn, a plasmatized comet sample return, a multiasteroid orbiter and flyby, and a Mars geophysical network.

Prospects. The planetary exploration program of the United States will be reinvigorated to the degree that the proposed SSEC program is adopted. The signs are encouraging, for in 1983 Congress approved the Venus radar mission. This mission follows many precepts of the planetary observer, incorporating spare parts from other spacecraft and having simpler scientific objectives and operations than did the earlier, Venus orbiting imaging radar, which was rejected.

For background information *see* SOLAR SYSTEM; SPACE PROBE; SPACECRAFT STRUCTURE in the McGraw-Hill Encyclopedia of Science and Technology.

[NOEL W. HINNERS]

Bibliography: D. Morrison and N. W. Hinners, A program for planetary exploration, *Science*, 220:561-566, 1983; NASA, Planetary Exploration Through Year 2000: A Core Program, 1983; M. Neugebauer, Mariner Mark II and the exploration of the solar system, *Science*, 219:443–449, 1983.

Spider

The web of an orb-weaving spider is commonly composed of nonsticky radial silk threads, providing structural support like the spokes of a wheel, and a spiral of viscid silk that functions to capture the spider's prey. In addition, certain species in two families (Araneidae and Uloboridae) of orb-weaving spiders commonly add conspicuous patches or bands of ribbonlike silk to their webs. These structures are referred to as stabilimenta, so named because they were originally thought to function as structural stabilizers for the orb. Recent evidence supports a different theory of the function of stabilimenta, specifically that they act as visual markers designed to prevent birds and other large animals from accidentally crashing into and destroying the web.

Stabilimenta occur in a variety of forms and may sometimes be constructed incompletely, or not at all, by a spider. Most commonly, a stabilimentum takes the pattern of an X, a vertical stripe, or a circular patch at the center of the web.

There is no convincing evidence that stabilimenta contribute significantly to the structural integrity of an orb web. Most spiders do not add stabilimenta to

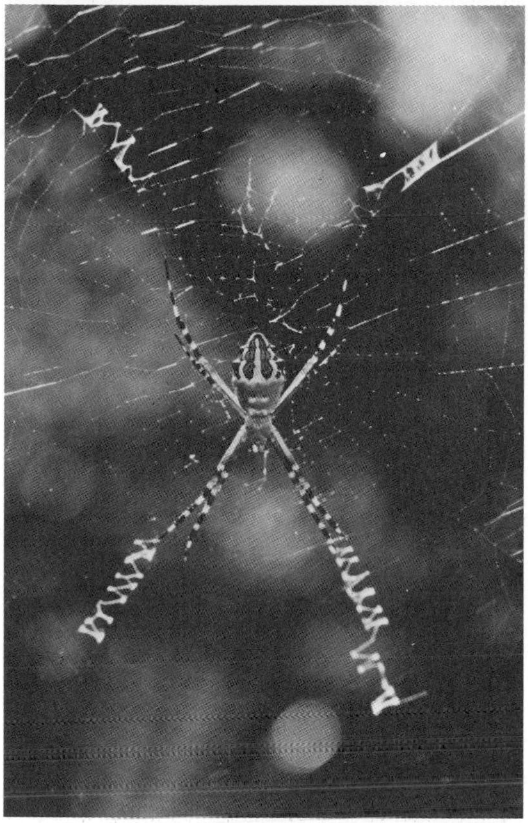

Fig. 1. Close-up of an orb-weaving spider, *Argiope florida* (Araneidae), with its X-shaped stabilimentum. The spider commonly sits in the center of its orb, head down, with its legs aligned with the stabilimentum. (*Courtesy of Thomas Eisner*)

their webs, and those that do occasionally leave them out of their spinning. Furthermore, if the stabilimentum of a web is carefully removed (by singing it with a hot needle), the resident spider continues to exhibit normal prey capture and defensive behaviors on the web, with no noticeable weakening of the web's structure.

Stabilimenta as antipredator devices. According to another, older theory, stabilimenta act as antipredator devices in two ways. First, it was proposed that the stabilimentum camouflages or conceals a spider on its web. To a human observer, however, the stabilimentum is the most conspicuous feature of an orb web and can be an aid to locating spiders. The pattern of the stabilimentum often accentuates the shape and position of the spider (Fig. 1). Second, the converse was also suggested—the stabilimentum may act in a manner analogous to the aposomatic, or warning, coloration of some noxious insects. That is, the spider might advertise its presence as prey dangerous or bad-tasting to a potential predator.

Neither of these suggestions is well supported by evidence. Predation tests with birds show that stabilimentum-building spiders are not only palatable and readily taken as prey, but that birds may even use stabilimenta to locate the spiders. On the other hand, more recent work suggests that blue jays (*Cyanocitta cristata*) prefer to feed on spiders off their webs and may shun webs with stabilimenta.

Stabilimenta as anticollision markers. Two observations suggest that stabilimenta are markers intended to prevent chance collisions between large animals, such as birds, and spider webs. First, only species that spin durable webs which are left up during the daylight hours add stabilimenta. Most orb-weaving spiders are nocturnal and take their webs down before dawn. None of these species incorporates stabilimenta into its web. Second, birds may be observed making close-range evasive moves when approaching webs with stabilimenta, flying up and over them.

This hypothesis was tested by comparing the durability of natural webs without stabilimenta and webs that had artificial equivalents of stabilimenta added. The webs used in this experiment were all of spiders that ordinarily do not add stabilimenta to their orbs; most of these species are nocturnal. After the spiders had completed spinning in the evening, they were removed from their webs, and artificial stabilimenta were added to half of the webs. Most of the artificial stabilimenta were paper strips, imitative of a common X-shaped stabilimentum, applied directly to the sticky threads of the web. To control for the possibility that the addition of paper somehow strengthened the web, some artificial stabilimenta were constructed of white plastic and suspended by black threads parallel to, but not touching, the web. The results from both conditions were the same. During the hours before dawn both marked and unmarked webs suffered comparable damage. After dawn, the unmarked webs showed significantly more damage, especially in the first hours of light when birds are most active (Fig. 2). Direct observations of unmarked control webs con-

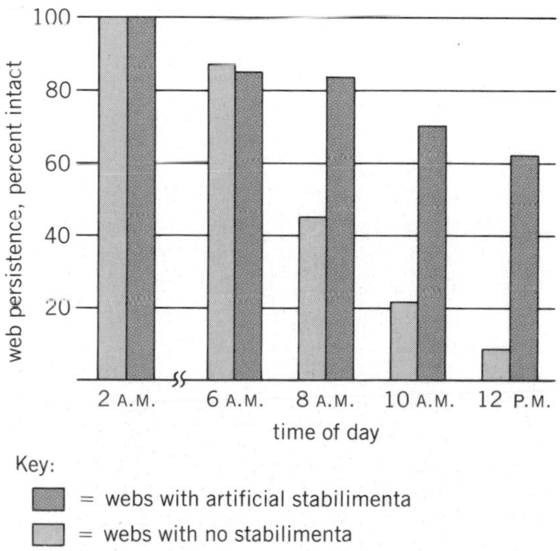

Fig. 2. Web persistence at various times of the day. (*After T. Eisner and S. Nowicki, Spider web protection through visual advertisement: Role of the stabilimentum, Science, 219:185–187, copyright 1983 by the American Associates for the Advancement of Science*)

firmed that these were in fact damaged or destroyed by flying birds.

In addition to birds, large ambulatory mammals might also be deterred from colliding with stabilimentum-marked webs, as well as large, visual insects. Some species of butterflies are sometimes observed avoiding webs with stabilimenta. Although the visual warning of a stabilimentum may result in the occasional loss of a meal for the spider, this disadvantage is probably less costly than the total destruction of a web. A spider invests not only time and energy in spinning its web, but also a considerable amount of protein in producing the silk. A spider ordinarily reingests its silk when it takes down its web, and is unable to do so when the web is ripped away by a large animal.

For background information *see* ARANEIDA in the McGraw-Hill Encyclopedia of Science and Technology. [STEPHEN NOWICKI]

Bibliography: T. Eisner and S. Nowicki, Spider web protection through visual advertisement: Role of the stabilimentum, *Science*, 219:185–187, 1983; C. C. Horton, A defensive function for the stabilimenta of two orb weaving spiders (Araneae, Araneidae), *Psyche*, 87:13–20, 1980; M. H. Robinson and B. Robinson, The stabilimentum of the orb web spider, *Argiope argentata*: An improbable defence against predators, *Can. Entomol.*, 102:641–655, 1970.

Steam turbine

Steam turbines provide most (well over 90%) of the stationary mechanical power used worldwide. Most of this power is used to drive generators and produce electricity, while the remainder drives pumps, fans, compressors, and other mechanical equipment for various industrial processes. Turbines range in size from over 1200 MW for the largest generator drives to under 1 MW for the smallest mechanical drives. This large and varied use points out the importance of turbine efficiency in world energy consumption. Even small improvements in efficiency can save tremendous amounts of fuel and have major impacts on the world economy. *See* HYDROPOWER.

The improvement of turbine efficiency has been the goal of engineers ever since initial construction of the turbine in 1883. The modern turbine represents a highly developed machine with efficiencies of over 80%. Even so, the rapid increase in the price of fuel in the last 10 years has spurred a renewed interest in further increasing turbine efficiency. Efficiency improvements are valued at \$1000–2000 per kilowatt saved when using present energy costs of \$4–5 per 10^6 Btu (10^{10} joules). Thus, an improvement of only 0.5% in efficiency for a 1000-MW generator plant will save \$1,500,000 per year in fuel costs. The design of the newer efficient steam turbines is an example of the application of more refined knowledge to a mature product, rather than minor changes to the turbine. The turbine appears unchanged physically, but the details of the major components have been refined to achieve the maximum efficiency.

Efficiency. The purpose of a steam turbine is to convert heat energy to mechanical power. Turbines are most commonly used as components in power plants based on the Rankine cycle (Fig. 1). The pump raises the pressure of cold water which is then heated and vaporized to steam in the boiler. The steam expands in the turbine, giving up its heat energy, and is finally condensed back to water and cooled in the condenser.

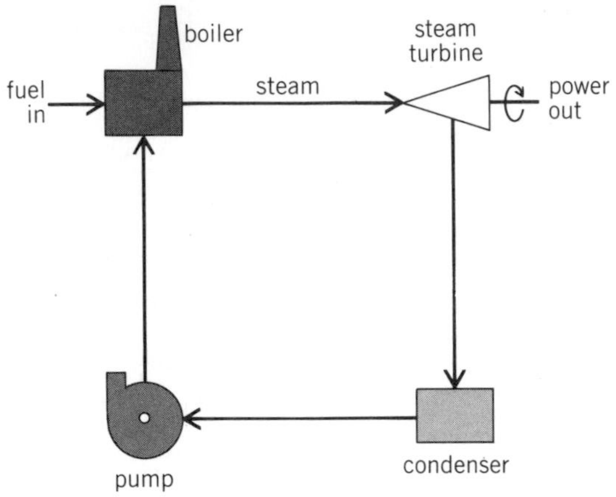

Fig. 1. Schematic of Rankine-cycle power plant.

There are many factors that affect the overall efficiency of this power-producing cycle: steam inlet temperature, discharge pressure, amount of feedwater heating, turbine efficiency, and so on. This article is limited to those factors that affect the turbine efficiency directly. Steam turbine efficiency is a measure of how well the available heat energy in the steam is converted to shaft power. Minimizing losses is equivalent to designing for high efficiency. Losses associated with various turbine areas can be grouped into the following categories: inlet and discharge losses; blade flow path losses; leakage losses; and mechanical losses.

Inlet and discharge. The turbine inlet consists of the passages necessary to carry the steam from the inlet pipe connection to the first nozzle. These passages also contain control valves to regulate the amount of steam to the turbine in order to match the load and speed requirements. Efficient turbines have passages that are designed with the lowest velocities to reduce frictional losses, as well as careful design of the flow area changes to minimize flow separation and the resulting losses. Valves, which have higher flow velocities, follow the same principle and also have flow diffusing sections following the throat to recover the velocity energy in the steam. Discharge passages must reduce the high velocity leaving the last stage and direct the flow to the exit pipe or condenser. Carefully designed flow diffusers are used to recover the energy in the blade discharge flow.

Modern efficient turbines have benefitted from the ability to calculate the detailed flow in the passages by using digital computers and to design the passages with graphic-computer methods such as computer-aided design. These calculation capabilities and an understanding of fluid mechanics result in passages designed for minimum losses.

Blading. The heart of a steam turbine is the blading because it extracts energy from the high-pressure, high-temperature steam. The blading consists of both stationary nozzles and moving blades that are attached to the rotor. The aerodynamic design of the blading plays a major role in the efficiency of a steam turbine.

The many years of experience, testing, and trial and error that have gone into the development of blade design for the steam turbine are being supplemented by advanced calculation methods, many of which are derived from research and development in aviation gas turbines and progress in computing power. Turbine blade passages are now designed by methods that calculate the complete flow field between blades. These methods allow the designer to iterate on the shape of the blades until the one with the least loss for a particular design is found. Although the effect of blade shape is subtle and not easily described, blade shape is a major factor in the efficiency of steam turbines.

An additional degree of sophistication is required in the design of the long last-stage blades of condensing turbines. These blades typically have supersonic flow fields, and shock waves must be taken into account when determining the best blade shape. Calculation methods and computing power have recently become available to calculate this type of flow. Although much has been done, further advances are expected by use of very fast computers coupled with calculation methods based on finite element methods and more accurate accounting of the fluid viscosity. Further refinement of blade and passage shape for better efficiency will result.

The aerodynamic design of blades for efficiency cannot be separated from the mechanical design, for the shapes required for efficiency must also be reliable. Thus advances in the mechanical design allow the aerodynamic designer to use larger, thinner, and narrower blades. These advances have also resulted from the development of calculation methods for the stress and vibration of blades based on finite element methods. Very long blades with very high stresses are now used to pass the high flows of condensing turbines operating at the low discharge pressure required for high efficiency.

Leakage. Efficient steam turbines make every attempt to contain the steam within the blade path where useful work can be extracted. Minimizing leakage between the rotating and stationary elements, such as across the rotor blade tips, at the base of the stationary nozzles, and where the shaft extends through the casing, is a major goal of the designer (Fig. 2).

Rotor blade tips are shrouded to prevent leakage from the pressure side to the suction side of the blade whenever mechanically possible. The leakage over the shroud is minimized by a close fit to the casing by using labyrinth teeth. The teeth are designed to run close enough to the casing to restrict the leakage, but not close enough to touch and cause mechanical damage. Materials are used that will prevent serious damage should a rub inadvertently occur.

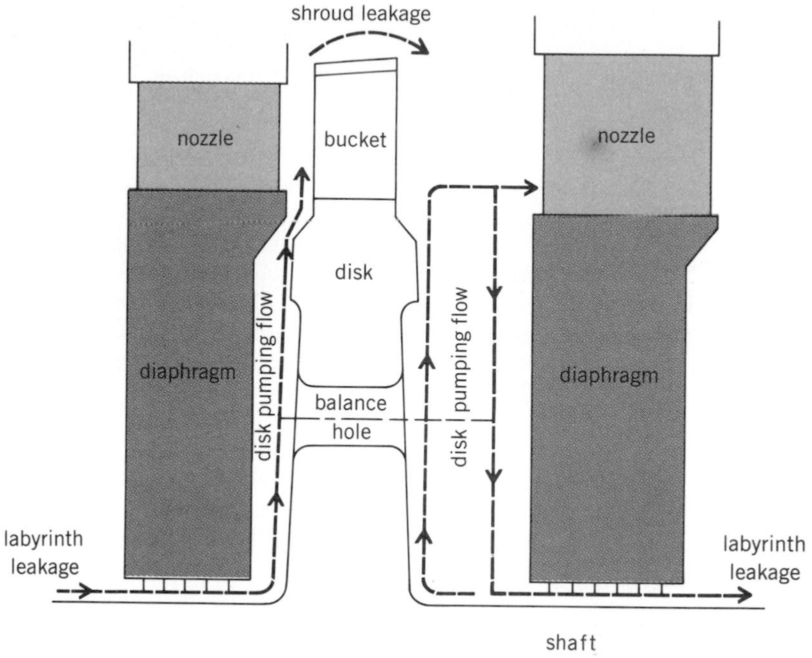

Fig. 2. Cross section of a typical steam turbine stage showing leakage paths.

Leakage between the base of the stationary nozzles and the rotor is controlled by using close-fitting labyrinth seals. Leakage is not as high as at the rotor blade tips because the diameter is smaller and the clearances are easier to control. On the other hand, this leakage is a further concern because the pumping action of the rotating disk causes the leakage flow to enter the main flow at a high velocity. This reentry of the leakage flow will disturb the main flow and cause additional losses. High-efficiency turbines have a passage shaped to ingest the leakage flow, minimizing this disturbance.

Leakage along the shaft where it goes through the casing is controlled by using close-fitting labyrinth teeth. The leakage in this area can be minimized easily because there is room for many labyrinth teeth. Special materials are used so the labyrinth teeth can rub without causing damage to the turbine.

The close clearances of all these seals require that the turbine be started and stopped carefully to prevent large differential thermal growths (expansion of the metal due to increased temperature) that would cause the seals to rub. The use of temperature sensors and computer controled startup allow the turbine to maintain the close clearances necessary for efficiency.

Mechanical factors. Mechanical losses are caused by shaft bearings and the pumps and motors required to supply oil to the turbine bearings. On large turbines these losses are a very small percentage of the output power (0.1%) and are not a major concern. However, on smaller turbines these losses can be up to 10% of the output power. Small efficient turbines use directed lubrication thrust bearings to reduce mechanical losses and increase efficiency.

For background information *see* COMPUTER-AIDED DESIGN AND MANUFACTURING; FINITE ELEMENT METHOD; POWER PLANT; RANKINE CYCLE; STEAM TURBINE.

[FRED CANOVA]

Bibliography: W. Traupel, Steam turbines, yesterday, today and tomorrow, *Institution of Mechanical Engineers Proceedings 1979*, vol. 193, no. 38.

Stratosphere

Measurements of water vapor in the stratosphere remain both difficult and challenging—not only because of the technical problems of obtaining a fast response to a trace species with instruments unavoidably contaminated by the far higher concentrations of that species existing in the lower atmosphere, but also because of the information such measurements convey regarding atmospheric circulation. Since the first frost-point hygrometer was carried into the lower stratosphere at the end of World War II by the Meterorological Research Flights (MRF) of the Royal Air Force, meteorologists have sought to explain the extremely low humidities found there—low both in terms of absolute humidity and even in the relative humidity of air capable of holding very little water vapor compared with surface air.

Early circulation model. Those first observations by the MRF recorded frost points as low as −112°F (−80°C) corresponding to relative humidities of only a few percent and to water vapor mass mixing ratios of less than 2 ppmm (parts per million by mass) compared with values of 1000 to 20,000 ppmm normally found at the surface. These led A. W. Brewer to propose a worldwide scheme of air circulation in the lower stratosphere. His model was based on the assumption that there was only one known location in the atmosphere where air could be dried to the level found by the MRF and that was at the tropical tropopause—a cold, minimum-temperature, sheet of air circling the globe above the Equator at a height near 10 mi (16 km) marking the boundary between the stratosphere above and the troposphere below and extending north-south from the mid-latitude westerly jet stream of the Northern Hemisphere to that of the Southern Hemisphere. The few observations available then indicated that the temperature dropped with altitude up to the tropopause level and increased above, and that in the tropics the minimum temperature identifying the tropopause was always in the vicinity of −112°F (−80°C).

Brewer hypothesized that on a worldwide basis air rose up through the tropical tropopause, where it was freeze-dried to a frost point (saturation point leading to deposit of frost) of −112°F (−80°C), and then spread laterally toward both poles where it descended and returned to the troposphere through the bounding polar tropopauses which were at altitudes of 6–7 mi (9–12 km) and temperatures of −40 to −58°F (−40 to −50°C). These warmer temperatures would have permitted upward diffusion of air with water vapor mixing ratios a hundred times greater than those found by the MRF. Brewer therefore reasoned that the air which had been dried by entering the stratosphere through the tropical tropopause must be descending fast enough in extratropical latitudes to prevent the upward movement of water vapor by diffusion at these latitudes.

Discordant data. Subsequent measurements of both water vapor and radioactive tracers have tended to confirm Brewer's worldwide atmospheric circulation pattern, but there have always been discordant data as well. Investigators seeking to measure stratospheric water vapor have generally found mixing ratios of 1 to 6 ppmm in the lower stratosphere, but in many cases their instruments indicated an increase with altitude above this level. Such increases were found by some even after measures were taken to eliminate the effect of outgasing of water vapor from equipment and balloon trains which could distort readings in the very dry and low-density air of the stratosphere. This problem still exists. While oxidation of methane (also ascending from the troposphere) can cause such an increase in water vapor with height, this could at most account for an increase of 2 ppmm above the mixing ratio found in the lower stratosphere.

During the Climatic Impact Assessment Program of the early 1970s, an investigation of the impact of supersonic transports on the stratosphere was undertaken. The stratosphere water vapor problem was reexamined by H. W. Ellsaesser from a budget point of view, yielding further inconsistencies. Observational data then indicated an average water vapor mixing ratio of the lower stratosphere of about 2.6 ppmm, but the saturation mixing ratio for the global mean temperature of the tropical tropopause was about 3.4 ppmm. It was inconsistent that the lower stratosphere could maintain a mixing ratio of 2.6 ppmm while all air entering through the tropical tropopause was presumed to be freeze-dried to only 3.4 ppmm. In addition, the available observations appeared to indicate a maximum stratospheric mixing ratio just above the tropical tropopause with decreases from there upward to about 12 mi (19 km) and also horizontally toward both poles. Ellsaesser concluded that these observations appeared to require a sink for removing water vapor from the stratosphere and that freeze-out in the cold pools near 12–16 mi (20–25 km) over the winter poles (particularly the one over Antarctica which gets some 36°F or 20°C colder in winter than does the one over the Arctic) appeared to be the only possible sinks compatible with the available data. While later data appear to confirm winter polar freeze-out, at least over Antarctica, this mechanism does not appear capable of removing enough water vapor from the stratosphere to take care of the budget imbalance indicated above.

Since then, further inconsistencies have been demonstrated between actual measurements and the atmospheric circulation model. NASA conducted tropical field experiments in 1977 and 1980 to determine whether the deep convective thunderstorms of the Intertropical Convergence Zone, known to penetrate the tropical tropopause in summer when the zone is near Panama, were injecting additional water vapor into the stratosphere or were instead removing water vapor from the stratosphere. Their measurements showed highest mixing ratios in the lower stratosphere at times of, and in the vicinity of, cumulonimbus penetrations of the tropopause and also showed a slow increase in lower stratospheric mixing ratios over the 30–60 days of the experiments during which penetrations were occurring. However, these experiments indicated that most (~80%) tropical water vapor soundings found that air in the upper troposphere, roughly in the height range between the levels of the polar and tropical tropopauses, was as dry or nearly as dry as that found in the lower stratosphere. These soundings gave no indication of clouds or moisture saturation, which is believed to be necessary for freeze-drying of the air, and yet indicated water vapor mixing ratios in the troposphere even lower than the saturation mixing ratio at the overlying tropical tropopause.

Drying mechanism. These data, if confirmed, indicate the existence of an unknown drying mechanism operating in the troposphere independent of, and yet equally as powerful as, the tropical tropopause freeze-drying mechanism proposed by Brewer. This possibility has as much portent for changing the understanding of atmospheric processes as did the original discovery of the aridity of the stratosphere. It suggests, first of all, that at temperatures below about −58°F (−50°C) and pressures below 200 millibars (2×10^4 pascals), water vapor does not behave as indicated by present tables for water vapor saturation pressure. This could be caused by an unidentified but ubiquitous seeding agent—perhaps sulfuric acid droplets precipitating from the Junge layer (sulfate particle layer) of the lower stratosphere. Or it could be another unique property of the water molecule, for example, a reluctance to behave as a perfect gas at temperatures below about −58°F (−50°C). Departures from the behavior of a perfect gas already appear to be indicated at much higher temperatures by the so-called e-type or continuum infrared absorption by water vapor. This phenomenon is currently believed to be due to clustering of some water vapor molecules, causing these molecules to absorb as liquid water rather than as a gas. At the much lower temperatures of the upper troposphere, such a tendency for clustering of molecules might well lead to the production of particles large enough to precipitate to lower, warmer and denser layers, thus slowly drying the upper layers of the troposphere colder than about −58°F (−50°C).

If a mechanism for drying the air in the upper tropical troposphere is confirmed, regardless of its nature, it will open up a new pathway for troposphere-to-stratosphere exchange of air—that is, tropospheric air could enter the stratosphere by moving poleward horizontally through the so-called tropopause gaps between the tropical and polar tropopauses. Such a pathway has been suggested by ozone minima near 9.3 mi (15 km) which have been traced poleward from the latitude of the westerly jet stream. Such a pathway has thus far been discounted because of the absence of the higher water vapor mixing ratios which hitherto have been believed to be representative of the upper tropical troposphere. Routine atmospheric sounding or radiosonde data generally cease to provide humidity data before reaching 200 mb (2×10^4 Pa) and thus have shed no light on this problem, which requires specialized and expensive equipment.

For background information *see* ATMOSPHERE; STRATOSPHERE; TROPOPAUSE; TROPOSPHERE in the McGraw-Hill Encyclopedia of Science and Technology. [HUGH W. ELLSAESSER]

Bibliography. A. W. Brewer, Evidence for a world circulation provided by measurements of helium and water vapour distribution in the stratosphere, *Quart. J. Roy. Meteorol. Soc.*, 75:351–363, 1949; H. W. Ellsaesser, Stratospheric water vapor, *J. Geophys. Res.*, 88:3897–3906, 1983; H. W. Ellsaesser et al., Stratospheric H_2O, *Planet. Space Sci.*, 28:827–835, 1980; G. D. Robinson, The transport of minor stratospheric constituents between troposphere and atmosphere, *Quart. J. Roy. Meteorol. Soc.*, 106:227–253, 1980.

Sudden infant death syndrome (SIDS)

A working definition of the sudden infant death syndrome (SIDS) was established in 1969 as the sudden and unexpected death of an apparently normal infant which remains unexplained after the performance of an adequate autopsy. This definition is useful in that it allows for the identification and study of a group of infants, making no assumptions about the nature of their underlying disease or the mechanism of their death. Application of this definition in practice reveals that of a group of apparently healthy infants dying suddenly and unexpectedly, 15% will manifest pathologic evidence of a disease process which is sufficient to explain the infants' death. The remaining 85% are unexplained and are classified as SIDS. In spite of the probable heterogeneity of diseases in SIDS cases, the consistent and distinctive characteristics of these infants supports the notion that many, if not the majority, represent a single disease process.

Epidemiology. The incidence of SIDS in the United States is 2.0 cases per 1000 live births. Although the rate varies with season and with geographic region, an approximate annual total of 6000–8000 cases makes SIDS the leading cause of postneonatal death in the first year of life. The epidemiologic profile of SIDS has been confirmed and filled in by several recent studies. A characteristic feature of this condition is the age distribution at the time of death, with a notable rarity of cases occurring in the first weeks of life and a peak incidence at 2–4 months of age. The other characteristics of SIDS include a higher incidence in males, prematurely born infants, multiple births, and the economically disadvantaged. More cases occur during the winter months, and this has also been found in the Southern Hemisphere.

Race affects the incidence, with the rate in Native Americans being greater than in blacks, whose rate, in turn, is greater than that in whites. The rate in Asians is lower than in whites, and this is true in both Japan and California.

Although there is a familial effect on SIDS risk, with a 10-fold increase in risk for subsequent siblings of victims, the pattern does not clearly point to a genetic mechanism. This lack of a genetic effect is also seen in twin studies, which report that monozygotic and dizygotic twin pairs have equal concordance rates. The independent effects of complex interrelated variables associated with SIDS cannot be evaluated from existing reports. However, a multicentered cooperative epidemiologic study under the direction of the National Institute of Child Health and Human Development will provide a data base sufficient to address some of these questions.

Routine postmortem findings. Reports of death scene investigations typically describe the victim as found in the crib after a time when the infant was thought to be asleep, and with no evidence of an agonal struggle. This is taken as evidence that the infants probably die during sleep, and major seizure activity did not accompany death. Autopsy findings usually include numerous hemorrhages of the thoracic viscera, pulmonary edema, and an empty bladder with wet diapers. These findings, in general, are those following death by asphyxia in young children. However, attempts to determine the sequence of events resulting in asphyxial death based on postmortem findings have so far been inconclusive.

Evidence of chronicity. A major advance in understanding SIDS emerged in the last decade as evidence accumulated supporting the idea that these babies had a chronic abnormality prior to their death. The first major contribution was a series of reports describing quantitative morphometric differences in the tissues of SIDS victims compared with control infants. SIDS victims were found to have thickened pulmonary artery walls, increased cardiac right ventricular weight, increased hepatic hematopoiesis, brown fat retention, increased adrenal medulla tissue, carotid body cell changes, and brainstem gliosis. Although not all these original studies are free of criticism, several of the findings have been confirmed and the original hypothesis of chronicity has been supported. Delayed postnatal growth, another indicator of chronic abnormality, has recently been confirmed in a large prospective study and was found to be independent of the manner of infant feeding. The weights of SIDS infants, as a group, are in the 40th percentile at birth and fall to the 20th percentile at the time of death.

Physiologic abnormalities have been documented in some SIDS victems prior to death. These have predominantly been episodes of prolonged apnea (cessation of breathing) with or without bradycardia or airway obstruction. Detailed physiologic study of some of these infants has also demonstrated unusual baseline patterns of respiratory activity during feeding and sleep.

Etiological hypotheses. The physiologic data and the postmortem evidence of asphyxial death and chronicity have been used in support of the hypothesis that SIDS is due to a chronic abnormality in the complex mechanisms which control respiration. There are several such apnea hypotheses variously proposing prolonged central apnea, airway obstruction, or failure of arousal as the predominant consequence of the underlying abnormality. A clear understanding of the relation between SIDS and these physiologic events requires more data on the variations of normal mechanisms, their nervous system control, and the circumstances under which clearly abnormal episodes occur.

Numerous other long-standing abnormalities have been suspected to precede the death of SIDS victims. These include immunologic defects or unusual responses; infectious disease; defects of metabolism, particularly involving carbohydrates; and endocrine disturbances. First, because of the heterogeneity of SIDS groups, which arises from the negative nature of the definition, some of these processes probably do account for small numbers of SIDS cases. Second, it seems likely that a chronically abnormal, predisposed infant may require a particular set of circumstances to allow that abnormality to result in sudden death. It appears that upper respiratory infections may play this role in SIDS victims, up to 60% of which had an upper respiratory infection at the time of their death.

It has been proposed that abnormalities in thyroid function are responsible for SIDS. The original observation, subsequently confirmed, noted elevated serum triiodothyronine (T-3) levels in SIDS cases compared with autopsy controls (that is, infants dying of recognized diseases). However, subsequent animal studies demonstrated that elevation of T-3 levels was a nonspecific postmortem change following sudden death. Further studies in human infants showed, when other measures of thyroid function were considered, that infants dying of recognized diseases can be separated into those with normal and those with depressed thyroid function, presumably secondary to the disease from which they died. Infants with normal thyroid function who die of recognized causes have an elevated postmortem T-3 level, like SIDS cases, while those with depressed thyroid function do not undergo such elevation. The observation that SIDS cases have an elevated T-3 postmortem level appears to indicate that their thyroid function was normal prior to death and that the difference between them and the controls is related to the disease state of the autopsy controls.

Prenatal origin. It is generally aknowledged that SIDS victims are characterized by abnormalities for some time prior to their death, and thus arises the question of when in their short life does the abnormality begin. This answer cannot be definite at present, but it appears that these infants are different from controls prior to birth. The postnatal growth disturbance is present from birth and, SIDS victims have significantly lower birth weights than controls. Large studies indicate differences in neonatal neurologic reflexes which are not related to features of the labor and delivery processes. Furthermore, a consistent pattern of materal variables is associated with an increased SIDS risk; this includes young mothers with high parity, a short interval between pregnancies, and cigarette smoking during pregnancy. The available evidence seems to suggest that SIDS is due to a chronic congenital defect in the control of basic body functions.

For background information *see* CONGENITAL ANOMALIES in the McGraw-Hill Encyclopedia of Science and Technology.

[ALFRED STEINSCHNEIDER; KEVIN WINN]

Bibliography: J. T. Tildon, L. M. Roeder, and A. Steinschneider (eds.), *The Sudden Infant Death Syndrome*, 1983; M. A. Valdes-Dapena, Sudden infant death syndrome: A review of the medical literature 1974-1979, *Pediatrics* 66:597-614, 1980.

Superconducting device

A three-terminal superconducting switch called the quiteron was proposed in 1979 and was fabricated at IBM in 1982 with advanced processing concepts. This device possesses many of the operating characteristics of a semiconductor transistor, but is based on entirely different physical principles. It makes use of the nonequilibrium superconductivity phenomenon known as quasiparticle-injection tunneling, from which the name of the device was derived acronym-wise. Transformation of the quiteron concept into an operational device and the characterization of its performance have refocused attention on the advantages of a three-terminal superconducting device.

The major impetus behind research involving superconducting electronic devices arises from their potential application in ultrahigh-performance computers. These machines require high-speed very large-scale-integration (VLSI) chips on which circuits are densely packed to avoid signal transit-time limitations. The heat generated on these densely packed, semiconductor transistor circuits is difficult to remove. An inability to dissipate additional energy from a more densely populated chip can ultimately limit computer performance.

Josephson tunnel junction. This is a high-speed low-power device that has been investigated for computer applications. Josephson devices consist of two superconducting films, for example, niobium or lead, that are separated by a thin (2-nanometer) insulating barrier, such as niobium oxide, through which electrons can tunnel from one superconductor to the other. The tunneling can be by superconducting pairs via the Josephson effect or by single quasiparticles (electrons or holes). These two tunneling phenomena result in a current–voltage characteristic with two branches. One branch is the Josephson branch and is a supercurrent; it is accompanied by no voltage drop. The supercurrent cannot exceed a certain threshold value, which is a sensitive function of magnetic field. If the threshold is exceeded, the current will be carried by the other branch, which is resistive for nonzero voltages. The resistive branch is highly nonlinear. The resistance drops considerably when the voltage exceeds that required to break the superconducting electron pairs. With a fixed dc bias, a junction may be switched from zero to a finite voltage state by a change in the applied magnetic flux. Switching takes a few picoseconds in a Josephson device, which operates at power levels of a few microwatts. This is about 1000 times lower than the power input necessary to operate a semiconductor transistor.

Although these features are attractive, the two-terminal threshold nature of the Josephson device results in a number of practical difficulties in circuit design. Integrated circuit applications require that the Josephson threshold supercurrent be well controlled. Control of the supercurrent is made difficult by an exponential dependence on the 2-nm thickness of the barrier. As a result, useful circuit elements need three or more terminals and must be constructed of junctions with other components (inductors, resistors). This limits circuit densities on a chip. In addition, Josephson junctions have low gain, which is necessary to drive other similar devices; have no convenient means of inverting an input signal, which simplifies logic design; and are "latched" into a state even after a signal has been removed.

Three-terminal devices. Three-terminal semiconductor transistors, on the other hand, have excellent circuit properties, including voltage and current gain, very large-scale-integration density potential, isolation between output and input signals, nonlatching, and inverting characteristics. If the high speed and low power of Josephson devices were to be combined with these transistor attributes, the combination of properties should provide an excellent device for applications in very densely populated integrated circuits. The concept of a three-terminal transistorlike superconducting device is as old as the field of superconducting electronics. However, fabrication and characterization of such a device generally require advanced superconducting device processing. The necessary processing sophistication has only recently been developed as part of Josephson tunnel junction technology.

Gray device. In 1978, K. Gray fabricated and characterized a novel tunneling device that he called a superconducting transistor. In contrast to the Josephson device, Gray's device made use of nonequilibrium superconductivity for its operation. This device was formed by stacking three superconducting layers separated by two tunnel barriers to give a two-junction structure. Gray used aluminum films for the superconducting electrodes because of ease of fabrication and because they offered a convenient time scale (100 nanoseconds) to investigate nonequilibrium phenomena. The middle aluminum electrode is common to both junctions. It forms an injector junction with one of the outer electrodes, and a collector junction with the other. The injector junction is used to create a nonequilibrium state in the middle electrode by injection of a low level of quasiparticles at energies just greater than the junction energy gap. This disturbs the equilibrium that exists in a superconductor between the electron pairs, the quasiparticles, and the phonons. Nonequilibrium effects are maximized, relative to heating, by making the middle film very thin (30 nm). The collector junction, whose relative resistance is very low, is used to collect the excess quasiparticles in the middle electrode. The collection process can involve several excursions of quasiparticles across the collector junction for each injected quasiparticle and result in a current gain. The current gain, defined as the excess collector current compared with the injector current, is obtained if the quasiparticle tunneling rate is large compared with the rate of recombination of electron pairs. Although current gains of 3 to 4 have been achieved, device operation with aluminum electrodes is slow. In addition, the injection current is at a higher voltage than the extraction voltage, which results in a voltage loss. Gray believes that this device could be made significantly faster with other superconducting materials.

Quiteron. The basis for operation of the quiteron can be traced to nonequilibrium superconductivity studies and to Gray's earlier three-terminal nonequilibrium superconducting device. The device structures are similar. Three superconducting layers, preferably composed of different materials, are separated by two ultrathin (about 2-nm) insulating layers to form a double junction device. The quiteron also operates by injection of quasiparticles into a common, thin middle electrode to create a nonequilibrium state. However, the quiteron differs from Gray's device by being shifted much further from equilibrium during operation and by using the nonlinearity of the quasiparticle current–voltage branch.

In the quiteron, quasiparticles are heavily injected into the middle electrode. This causes electron pairs to be broken until, at the limit, the properties of the electrode become those of a normal metal rather than of a superconductor. As the electrode properties approach those of a normal metal, the superconducting energy gap approaches zero. The middle electrode must be thinner than the other electrodes, so that its superconducting properties are the ones that are primarily affected by the heavy quasiparticle injection. The heavy quasiparticle injection is through the injector junction, which is biased at a voltage greater than its energy gap. As the gap vanishes in the middle layer, the acceptor junction is effectively converted from a superconductor–insulator–superconductor (S-I-S) junction into a normal insulator–superconductor (N-I-S) junction. This conversion lowers the onset voltage of the sharp nonlinearity of the acceptor current–voltage characteristic. The quiteron device exploits this modification of the tunneling characteristics to change the acceptor from high resistance to low resistance in the vicinity of the onset voltage.

Biased appropriately, the quiteron operates as an inverter. With low quasiparticle injection across the injector junction (input 0), the acceptor junction passes very little current because it is biased below its energy gap voltage and it is a superconductor–insulator–superconductor junction. If a second device is used as an external load and is put in parallel with the acceptor, the current and power applied across the acceptor junction will be shunted through this external load (output 1). If an intense quasiparticle current is injected through the injector junction (input 1), the acceptor becomes a normal metal–insulator–superconductor junction and permits most of the current to flow through it with almost no power delivered to the external load (output 0).

At its present level of development, the quiteron requires an elaborate fabrication process. It possesses device properties which include a power (that is, the product of current times voltage) gain of 2, nonlatching switching, natural signal inversion, and a potential for high packing density. Its speed is greater than 100 picoseconds, and its operating power is about 10 times higher than that obtained from Josephson devices, but some improvements may be possible by further reduction in device size and redesign of the device to minimize heating effects. A magnetic field of sufficient strength must be imposed on the device to inhibit supercurrent tunneling. This field is increased as the tunnel junction size is decreased.

For integrated circuit applications, additional power gain, to about 10, is required. Besides the increase in device operating speed and the decrease in operating power, a modified device must still be nonlatching, have inverting characteristics, have output isolated from input, and have very large-scale-integration density potential. Practical questions remain about the degree of isolation of the control electrode from the output, the relationship between the impedances of the input and output, and the manufacturability of the experimental structure.

For background information *see* JOSEPHSON EFFECT; INTEGRATED CIRCUITS; SUPERCONDUCTING DEVICES; TRANSISTOR; TUNNELING IN SOLIDS in the McGraw-Hill Encyclopedia of Science and Technology. [STANLEY I. RAIDER]

Bibliography: S. M. Faris et al., *IEEE Trans. Magnet.*, MAG-19:1293, 1983; K. E. Gray, *Appl. Phys. Lett.*, 32:392, 1978; J. Matisoo, *Sci. Amer.*, 242:50, May 1980; S. I. Raider and R. E. Drake, *IEEE Trans. Magnet.*, MAG-17:229, 1981.

Superconductivity

Superconductors that contain elements with large magnetic moments (or large spin) are called magnetic superconductors. The simple fact that the diamagnetic effect of the superconductivity shields the magnetic moments suggests that superconductors tend to oppose the formation of magnetic order. Furthermore, the magnetism also tends to oppose the formation of the superconducting state because the scattering of conduction electrons by the spin magnetic moments tends to break the Cooper pairs. These considerations raise important questions regarding the interplay between magnetism and superconductivity, and the study of magnetic superconductors attracted much attention as early as 1956.

Interactions. In magnetic superconductors, there are three kinds of interactions in addition to those in nonmagnetic superconductors. They are (1) the $\vec{\sigma} \cdot \vec{S}$ interaction (the interaction between the spin $\vec{\sigma}$ of conduction electrons and the localized spin $\vec{S}$); (2) the dipole interaction, $\vec{b} \cdot \vec{S}$ (where $\vec{b}$ is the magnetic induction); and (3) the exchange interaction, $\vec{S}(x)\gamma_0(x - y)\vec{S}(y)$, between localized spins at locations x and y, which does not include the effects of the above two interactions. The first two interactions modify the exchange interaction among the localized spins; thus, the exchange interaction coupling γ_0 must be replaced with the total exchange interaction coupling γ.

Early studies. Until recently, most of the experiments which searched for magnetic superconductors were performed by using alloys with magnetic imputities, and it was found that the presence of the $\vec{\sigma} \cdot \vec{S}$ interaction easily quenched the superconductivity. The first serious theoretical study of superconductors with magnetic impurities, made by A. A. Abrikosov and L. P. Gor'kov in 1960, clearly showed that when the concentration of these impurities becomes sufficiently large they act to break the Cooper pairs and hence quench the superconductivity. Since then, the effects of magnetic impurities have been widely studied, but until recently most of the studies focused on the $\vec{\sigma} \cdot \vec{S}$ effect, ignoring the effect of the dipole interaction.

The magnetic properties of magnetic superconductors are controlled mainly through the total exchange interaction coupling γ among the localized spins. Part of this exchange interaction is mediated by the conduction electrons through the $\vec{\sigma} \cdot \vec{S}$ interaction and is called the RKKY interaction. The superconducting electron energy gap suppresses the RKKY coupling at low k (momentum) and creates a maximum of RKKY coupling at a certain nonzero value of k, favoring a kind of spin-periodic phase. This phenomenon, theoretically predicted in 1959, is called the Anderson–Suhl effect. The effect is weakened when there is a strong spin-orbit coupling of the conduction electrons. In the mixed state or in a thin film under an external field, localized magnetic moments tend to polarize the conduction electron spin through the $\vec{\sigma} \cdot \vec{S}$ interaction, creating an energy gap between the two spin states of the conduction electrons. This tends to suppress the superconductivity.

Rare-earth compounds. The discovery in 1977 that certain ternary compounds containing rare-earth elements become superconducting dramatically opened a new era in the experimental and theoretical study of magnetic superconductors. A particularly systematic study has been made on two groups of compounds: the Rh_4B_4 group, with formulas of the type $RERh_4B_4$ (where RE represents a rare-earth element); and the Chevrel compounds of the type $REMo_6X_8$ (where X represents sulfur or selenium). The fact that many of these compounds become superconducting suggests that the $\vec{\sigma} \cdot \vec{S}$ interaction is weak. Recent band-theoretical calculations indeed lend support to the relative weakness of this interaction. The conduction electrons are mainly the $4d$ electrons of rhodium (Rh) or molybdenum (Mo), while the localized spin moment is carried by the $4f$ electrons of the rare-earth atoms. The $\vec{\sigma} \cdot \vec{S}$ interaction is called the *d-f* interaction. (This interaction is sometimes referred to as the *s-f* interaction, in analogy with the *s-d* interaction.) The Rh_4B_4 or Mo_6X_8 units form clusters that surround the rare-earth atoms with a relatively large separation. This large separation is the primary reason for the weakness of the *d-f* interaction. Since the *d-f* interaction is weak, the electromagnetic effect will play an important role and should therefore be considered.

Electromagnetic shielding effect. Recent theoretical studies have shown that in the Meissner state the persistent current tends to shield the spin magnetic moment, substantially suppressing the total exchange coupling $\gamma(k)$ at low momentum. This creates a maximum of $\gamma(k)$ at a certain nonzero value of k, which tends to induce a spin-periodic phase below a certain temperature (say, T_p), unless this phase is overpowered by some other phase such as normal ferromagnetism. This phenomenon should not be confused with the previously mentioned Anderson-Suhl effect; the electromagnetic shielding effect is caused by the dipole interaction while the Anderson-Suhl effect is created by the *d-f* interaction. The presence of electromagnetic shielding effect has been confirmed by ultrasonic attenuation experiments in mixed states of $E_rRh_4B_4$.

Electron energy gap. The *d-f* interaction modifies the electron energy gap Δ through the self-energy induced by the correlation function χ of the localized spins. When χ for the normal state is used and the scattering recoil momentum of conduction electrons is ignored, the singularity of χ at a temperature T equal to T_m, the normal Curie temperature, creates an infrared divergence that makes Δ vanish at T_m. However, recent calculations have shown that when the scattering recoil momentum of conduction electrons is carefully taken into account, the infrared divergence is eliminated and Δ remains finite at $T = T_m$ unless the *d-f* interaction is extremely strong. Furthermore, when χ for the superconducting state is used, χ has no singularity at $T = T_m$, though it has a singularity at $T = T_p$, the critical temperature for the spin-periodic phase caused by the shielding effect, which makes Δ vanish at $T = T_p$. (The temperature T_p is less than T_m.) However, Δ can remain finite at $T = T_p$ when a certain amount of magnetic anisotropy of the material is present. Thus, Δ does not necessarily vanish at $T = T_m$ or at $T = T_p$ unless *d-f* interaction is strong.

Reentrant phenomenon. Among the rare-earth ternary compounds mentioned above, $ErRh_4B_4$ and $HoMo_6S_8$ are of particular interest because of their reentrant behavior; when the temperature is cooled below T_c, the superconducting critical temperature, a ferromagnetic normal phase appears at a temperature T_{c2} of approximately 1 K. The temperature behavior of the ac susceptibility and resistivity of $ErRh_4B_4$ (Fig. 1) clearly exhibits the reentrant phenomenon. The hysteresis at T_{c2} in Fig. 1 shows that the transition at this temperature is first order. A periodic spin structure with a long period, approximately 10 nanometers for $ErRh_4B_4$ and approximately 20 nanometers for $HoMo_6S_8$, was observed in neutron diffraction experiments in a small range of temperature above T_{c2}, and a similar experiment with a single crystal of $ErRh_4B_4$ clearly confirmed the presence of a spin-sinusoidal phase (Fig. 2). Since the *d-f* interaction is weak and there is a strong spin-orbit coupling, the origin of this spin-sinusoidal phase is not the Anderson-Suhl effect but the electromagnetic shielding effect. The neutron scattering experiments showed also that a ferromagnetic component coexists with the spin-sinusoidal part. This suggests the coexistence of spatially different domains of ferromagnetic and spin-sinusoidal superconducting phase that appear at $T_{c2} < T \lesssim T_p$.

A measurement of energy gap by tunneling experiments for $ErRh_4B_4$ has shown that the order parameter does not vanish even at a temperature immediately above T_{c2}. This is consistent with the fact that the phase transition at T_{c2} is of the first order.

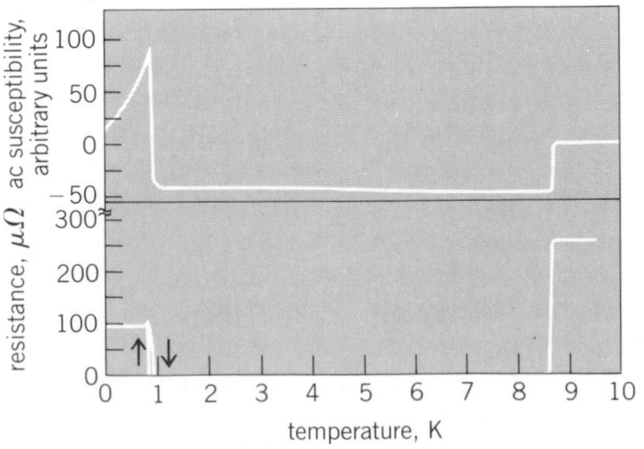

Fig. 1. Ac susceptibility and resistivity of $ErRh_4B_4$ as a function of temperature. (*After W. A. Fertig et al., Destruction of superconductivity at the onset of long range magnetic order in the compound ErRh₄B₄, Phys. Rev. Lett., 38:987–990, 1977*)

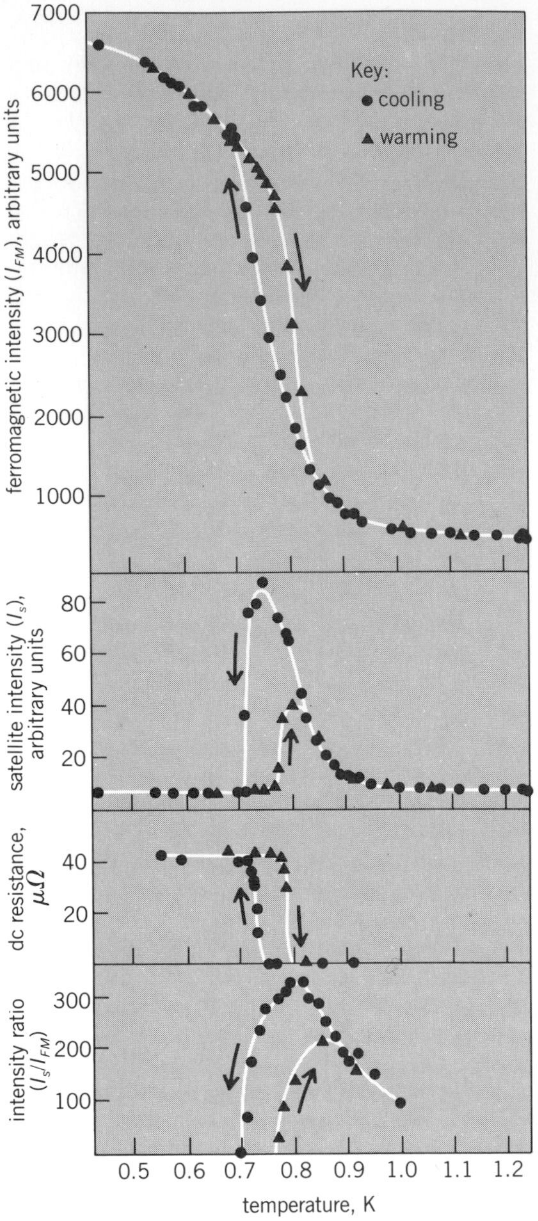

Fig. 2. Temperature behavior of neutron scattering intensity of a single crystal of $ErRh_4B_4$ near the transition temperature T_{c2}. Peak in the satellite intensity indicates presence of a periodic spin structure. 1 kilooersted(kOe) = 7.96 × 10^4 A/m. (*After S. K. Sinha et al., Study of coexistance of ferromagnetism and superconductivity in single-crystal ErRh₄B₄, Phys. Rev. Lett., 48:950–953, 1982*)

Mixed states. From detailed studies of critical fields and magnetization curves, it can be shown theoretically that, roughly speaking, when the *d-f* interaction is ignored a magnetic superconductor with Ginzburg-Landau parameter κ behaves like a nonmagnetic one with $\kappa' = \kappa\sqrt{1 + 4\pi\chi}$, where χ is the magnetic susceptibility. Since χ increases (κ' decreases) with decreasing temperature, a magnetic

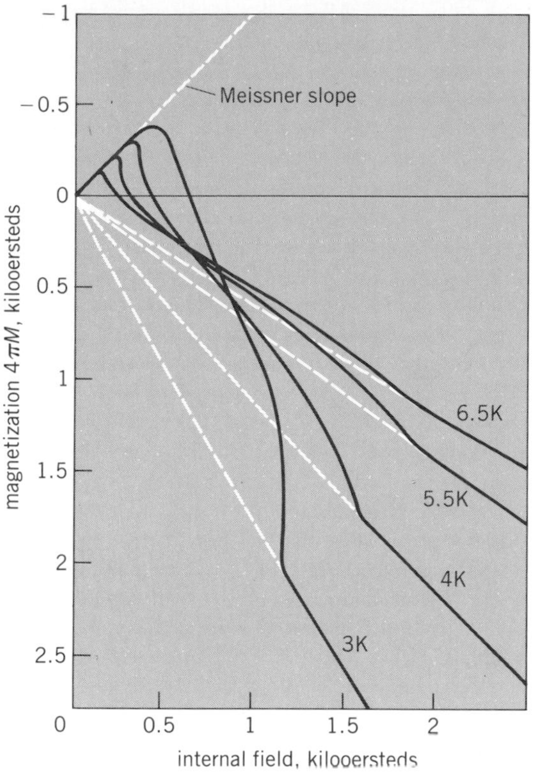

Fig. 3. Magnetizaton versus the internal field for the *a* axis (easy axis) direction of a single crystal of $ErRh_4B_4$ at various temperatures between 3 and 6.5 K. (*After F. Behroozi et al., Observation of a first order phase transition in single-crystal $ErRh_4B_4$ at H_{c2}, Phys. Rev., B27:6849–6852, 1983*)

superconductor with decreasing temperature tends to make a transition from type II to type I. The theoretical prediction of this phenomenon agreed with a subsequent experiment on a polycrystal of $ErRh_4B_4$. However, the real situation seems to be more complex still because a recent experiment with a single crystal of $ErRh_4B_4$ revealed that the phase transition at the upper critical field H_{c2} seems to be of first order. This may be an effect of the *d-f* interaction. The magnetization curves obtained from this experiment are shown in Fig.3.

The d-f interaction. Although the *d-f* interaction in $RERh_4B_4$ and $REMo_6X_8$ is known to be weak, it does exist and influences the experimental data in many ways, including depression of T_c, modification of the energy gap Δ, and modification of the magnetization curves in the mixed state. Recently, the *d-f* interaction has attracted considerable attention because the data obtained for the single crystal of $ErRh_4B_4$ in the above experiment indicates the need to include it, even though it is considered to be weak.

Antiferromagnetic superconductors. The above discussion is concerned mainly with the unusual behavior of ferromagnetic superconductors. Many of the magnetic superconductors comprising rare-earth ternary or pseudoternary compounds are antiferromagnetic. Furthermore, the antiferromagnetic phase coexists with the superconducting phase because the Cooper pairs experience only the average moment of the antiferromagnetic state, which is practically zero. A theoretical analysis of antiferromagnetic superconductors has been made and can explain most, but not all, of the H_{c2} versus T curves, which can assume a variety of forms depending on the compound. A particular feature of antiferromagnetic superconductors is that they usually have a very high H_{c2}. This may have important practical applications.

Prospects. The experimental study of magnetic superconductors has now become very active; many new materials have been discovered that exhibit a variety of unusual behavior, and many theoretical predictions await confirmation. Since a magnetic moment can support a vortex current and a vortex current can in turn induce a magnetic moment, it is theoretically possible that vortices may be self-consistently created in certain samples without an external field. There has been no definite experimental proof for this, although there is one set of experimental data for $(Ce_{1-x}Gd_x)Ru_2$ that could be interpreted as indicating the presence of self-induced vortices. The compounds which contain cerium tend to exhibit a behavior that suggests the Kondo effect or lattice Kondo effects may play an important role. In the cases of Y_4Co_3 (which may be of itinerant electron ferromagnetic type) and $Ho(Rh_{0.3}Ir_{0.7})_4B_4$, the magnetic transition takes place before the superconducting transition (T_{c2} is greater than T_c). A period of rapid development in the physics of magnetic superconductors seems in prospect.

For background information *see* KONDO EFFECT; MAGNETISM; PHASE TRANSITIONS; SUPERCONDUCTIVITY in the McGraw-Hill Encyclopedia of Science and Technology.

[HIROOMI UMEZAWA]

Bibliography: F. Behroozi et al., Observation of a first order phase transition in a single-crystal $ErRh_4B_4$ at H_{c2}, *Phys. Rev.*, B27:6849–6852, 1983; M. Ishikawa and Ø. Fisher, Destruction of superconductivity by magnetic ordering in $Ho_{1.2}Mo_6S_8$, *Solid State Commun.*, 23:37–39, 1977; W. A. Fertig et al., Destruction of superconductivity at the onset of long range magnetic order in the compound $ErRh_4B_4$, *Phys. Rev. Lett.*, 38:987–990, 1977; S. K. Sinha et al., Study of coexistence of ferromagnetism and superconductivity in single-crystal $ErRh_4B_4$, *Phys. Rev. Lett.*, 48:950–953, 1982.

Supercritical fields

Recent experiments involving collisions of heavy atoms have provided evidence for the spontaneous emission of positrons. This observation indicates that a breakdown of the neutral vacuum is taking place, a massive phase transition in the ground state of quantum electrodynamics. The spectrum of the emitted positrons also suggests that long-lived giant nuclear molecules are formed in these collisions.

Phenomena in large-Z atomic systems. Atomic systems with a nuclear charge Z much greater than 100 exhibit a number of unique features not other-

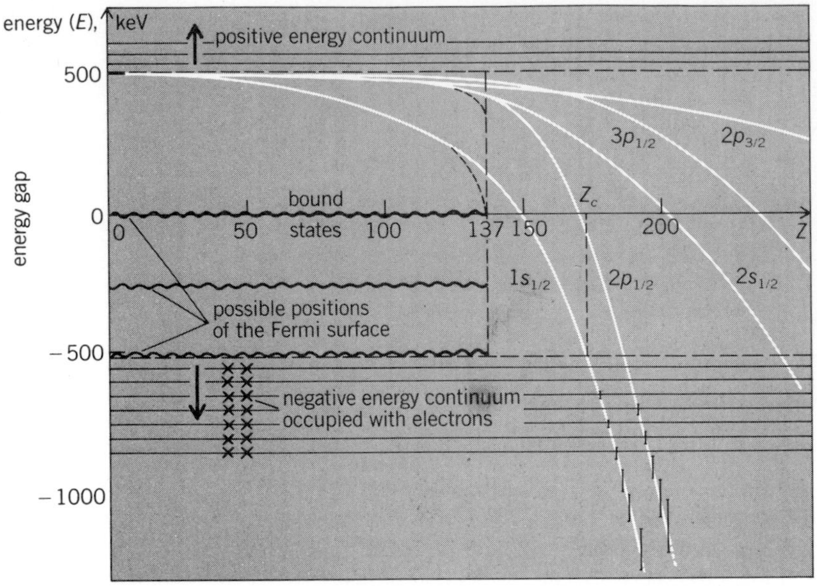

Fig. 1. Energy spectrum of a Dirac electron in the central field of an extended nucleus with charge Z. The various shells, energy levels at which electrons orbit the nucleus, are indicated by curved lines.

wise found in nature. The binding energies of electrons around hypothetical nuclei with Z up to about 200 are shown in Fig. 1. For $Z > 150$, the binding energy of K-shell ($1s_{1/2}$) electron exceeds its rest energy m_0c^2, where m_0 is the electron rest mass and c is the speed of light; that is, adding an electron to the atom actually diminishes the total mass of the system.

At the critical charge $Z_c \approx 170$–173, the binding energy reaches twice the electron rest mass, the threshold for the spontaneous creation of an electron–positron pair. This signals the transition to a new ground state of quantum electrodynamics: for $Z > Z_c$ the static electric field surrounding a bare nucleus spontaneously creates two electrons to fill the vacant K shell, while two positrons are emitted to balance the overall charge. Thus, the ground state is no longer a bare nucleus but one that is surrounded by two electrons. Since the new ground state carries charge, it is called a charged vacuum.

Thus, atomic systems with large Z offer the unique opportunity to study a phase transition, that is, a discontinuous change, in the ground state of a relativistic quantum field theory by laboratory experiments. For a long time it was not clear whether modifications in the single-particle picture introduced by interactions with virtual electron–positron pairs (vacuum polarization) and virtual photons (self-energy) might change the value of the critical charge and, possibly, prevent the binding energy from exceeding $2m_0c^2$ altogether. Recent accurate calculations of these corrections to first order in the fine-structure constant α (approximately 1/137), but to all orders in the Coulomb field, characterized by the value of $Z\alpha$, have shown the following results. At $Z = Z_c$, vacuum polarization causes an increase in the binding energy by 10.7 keV, whereas self-energy corrections reduce the binding by 11.0 keV. The net effect is a tiny change of 0.3 keV in the total binding energy of 1022 keV at the critical charge, which may be neglected for all practical purposes.

Study of quasimolecules. The study of such superheavy atomic systems is complicated by the fact that, in spite of considerable efforts, no stable nucleus with Z much greater than 100 has been found. The essential step toward an experimental realization of the phenomena discussed above was made in 1968, when it was suggested that, during the subbarrier collisions of two very heavy atoms, with nuclear charges Z_1 and Z_2, a supercritical nuclear charge, $Z = Z_1 + Z_2$, could be assembled for a sufficiently long interval of time. For example, a uranium–uranium collision yields a total $Z = 92 + 92 = 184$. The argument for this idea was based on the observation that the relative velocity of the two nuclei is of the order of $0.1c$, where c is the speed of light, while the K electrons move with a velocity essentially equal to the velocity of light due to the strong binding. Thus it was expected that the K-shell electrons could follow the nuclear motion and adjust to the instantaneous relative position of the nuclei, circling about both Coulomb centers. Such a collision system is called a quasimolecule or, in the

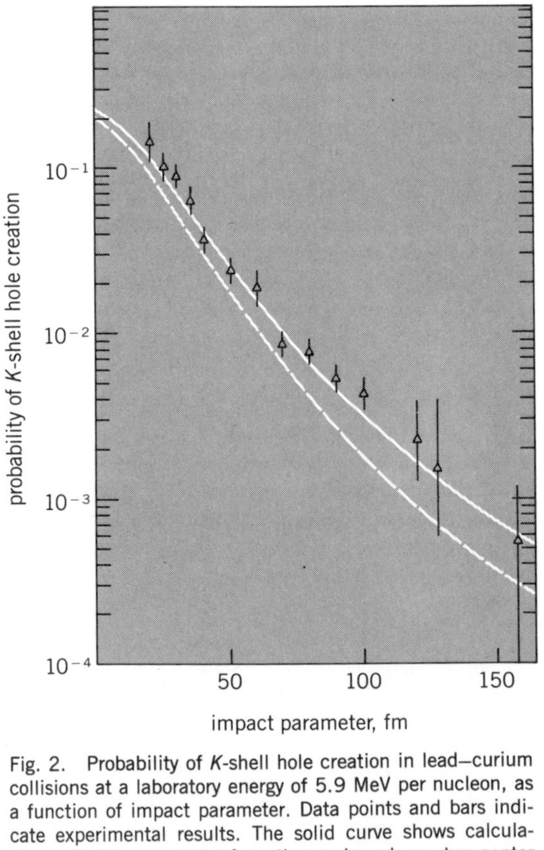

Fig. 2. Probability of K-shell hole creation in lead–curium collisions at a laboratory energy of 5.9 MeV per nucleon, as a function of impact parameter. Data points and bars indicate experimental results. The solid curve shows calculations of the parameter-free theory based on two-center Dirac–Hartree–Fock states containing the electron–electron interaction. The calculations shown by the broken curve are based on standard two-center Dirac levels without the electron–electron interaction.

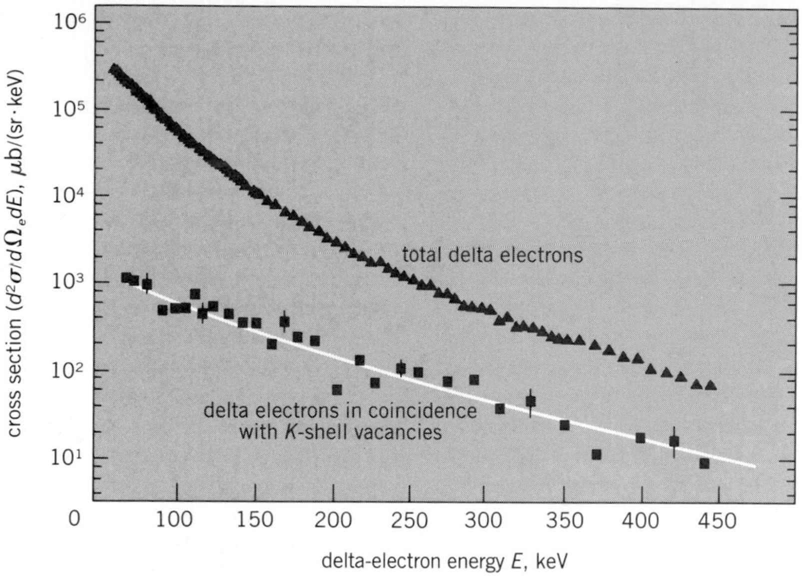

Fig. 3. Cross sections for the total production of delta electrons, and for production of delta electrons in coincidence with K-shell vacancies, for the iodine–lead system at a laboratory energy of 500 MeV, plotted as a function of the kinetic energy of the delta electrons. Data points indicate experimental results. Curve shows calculations of the parameter-free theory. 1 microbarn (μb) = 10^{-34} m^2.

limit when the internuclear distance is much smaller than the K-shell radius, a quasiatom. This expectation has been confirmed in great detail by numerous studies of the x-rays (quasimolecular x-rays), K-vacancy production, and delta-electron emission from heavy-ion collisions.

Some indication of the agreement between theory and experiments in this field is given in Fig. 2, which shows the impact parameter dependence of the K-shell hole-production probability for the lead–curium system, and in Fig. 3, which shows a delta-electron spectrum for the iodine–lead system. The excellent agreement of the experimental results with the parameter-free theory is remarkable. Since many other details of the experiments (energy dependence, Z_1+Z_2 dependence, and so forth) are reproduced by the theory, these processes can be said to be quantitatively understood.

Observation of spontaneous positrons. Of even greater significance is the observation of sharp line structures in the positron spectra of uranium–uranium and uranium–curium collisions at energies close to the Coulomb barrier. Two experimental groups have independently established these positron lines. According to an idea proposed in 1978, the spontaneous vacuum decay should indeed produce a sharp positron emission line, if the two colliding ions "stick" together long enough.

By investigating the gamma-ray and delta-electron spectra, experimentalists have been able to show that the observed positron line does indeed stem from the decay of the neutral vacuum into a charged vacuum. The argument is as follows: If the sharp line structure were of trivial origin, such as internal conversion in one of the separated atoms, either an observable x-ray line in the photon spectrum or a significant structure in the delta-electron spectrum should appear, superimposed on a smooth background. However, such effects have not been observed.

A typical positron spectrum, together with the theoretical calculations, is shown in Fig. 4. It dem-

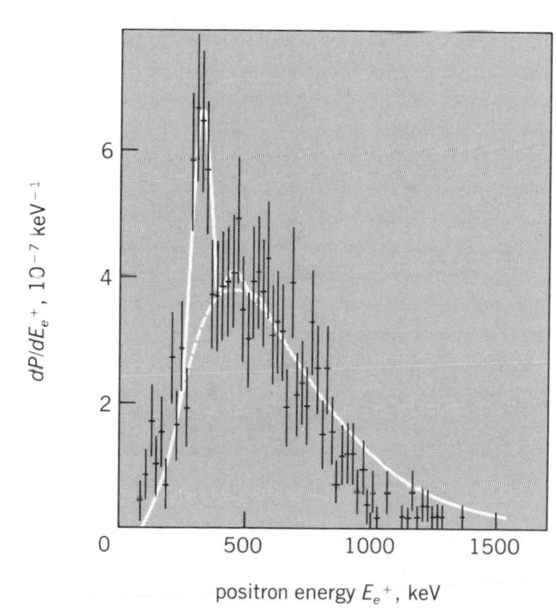

Fig. 4. Spectrum of emitted positrons in uranium–curium collisions with backward-scattered ions at a laboratory energy of 5.9 MeV per nucleon. dP/dE_{e^+} is the probability that a positron will be produced within a unit energy interval. Crosses indicate experimental results. The solid curve indicates the theoretical prediction of the total positron yield (spontaneous plus dynamical positrons); the broken curve indicates the predicted yield of dynamical positrons only.

onstrates both the remarkable positron line structure and the excellent agreement between theory and experiment. In fact, the position of the positron line structure also agrees with theoretical calculations of the position of pockets in the nucleus–nucleus potential. This suggests that a giant nuclear molecule of rather long lifetime (greater than 10^{-19} s) has indeed been formed.

This is a fundamental discovery: It shows that the ground state of a field theory, quantum electrodynamics, undergoes a massive phase transition and, furthermore, that rather long-lived giant nuclear systems (most likely giant nuclear molecules) with charges around $Z = Z_1 + Z_2 \approx 170–190$ are formed. Study of the Raman-type fine structure of the positron spectra is a promising technique for investigating the nuclear structure of these systems.

For background information *see* NUCLEAR MOLECULE; PHASE TRANSITIONS; QUANTUM ELECTRODYNAMICS; QUASIATOM in the McGraw-Hill Encyclopedia of Science and Technology.

[WALTER GREINER]

Bibliography: E. Berderman et al., Narrow structure in positron spectra from U + U collisions, *Sci. Rep.*, GSI 83–1, p. 147, 1982; H. Bokemeyer et al., Study of positron creation in overcritical HI-collision physics, *Sci. Rep.*, GSI 83–1, 146, J. S. Greenberg and W. Greiner, Search for the sparking of the vacuum, *Phys. Today*, 35(8):24–32, August 1982; W. Greiner, *Quantum Electrodynamics of Strong Fields, NASI at Lahnstein/Rhein 1981*, 1983; J. Reinhardt et al., The decay of the vacuum in the field of superheavy nuclear systems, *Zeitschr. Phys.*, A303:173–188, 1981.

Supersymmetry

The last 20 years have seen an unprecedented expansion in the use of symmetry considerations in the study of problems in physics. Supersymmetry is the latest and most elaborate application. Although it was originally introduced in the early 1970s for studying problems in elementary particle physics, its use and applications have now spread to other fields of physics. In the last few years, experimental examples of supersymmetry have been discovered in the spectra of atomic nuclei.

Definition. Physicists believe that all constituents of matter have a property called angular momentum. According to quantum theory, angular momentum can be either an integer or half-integer multiple of Planck's constant, $\hbar$. Particles for which the angular momentum is an integer multiple of $\hbar$ are called bosons; those for which the angular momentum is a half-integer multiple of $\hbar$ are called fermions. Up to the early 1970s, when applying symmetry considerations to the study of phenomena in physics, physicists always considered symmetry transformations relating either bosons to bosons or fermions to fermions. In the early 1970s it was suggested that there may exist symmetries relating bosons to fermions, and vice versa. Because of their remarkable properties, and in order to distinguish them from the usual symmetries, these symmetries were called supersymmetries.

Mathematical formalism. The study of symmetries requires the use of a mathematical formalism, known as theory of group transformations. Most applications in physics rely on the theory of Lie groups, which constitute a special type of group of transformations first studied by S. Lie and others at the end of the nineteenth century. The mathematical properties of these groups can be concisely stated by writing down the commutation relations satisfied by the generators G_α of the corresponding infinitesimal transformations, Eq. (1), where the commutator

$$[G_\alpha, G_\beta] = \sum_\alpha c^\gamma_{\alpha\beta} G_\gamma \qquad (1)$$

of two operators A and B is defined by $[A,B] = AB - BA$, and the coefficients $c^\gamma_{\alpha\beta}$ specify the group. The study of supersymmetries requires the introduction of a new mathematical formalism, the theory of graded groups of transformations or supergroups. While the mathematical theory of Lie groups has been known for more than 80 years, that of graded Lie groups is still being developed. The main difference between graded and nongraded Lie groups is that in the graded case there are two types of generators, the bosonic, G_α, and the fermionic, F_i. These generators satisfy commutation relations more elaborate than those of nongraded groups, Eqs. (2). In these expressions, in addition to com-

$$\begin{aligned} [G_\alpha, G_\beta] &= \sum_\gamma c^\gamma_{\alpha\beta} G_\gamma \\ [G_\alpha, F_i] &= \sum_j f^j_{\alpha i} F_j \\ \{F_i, F_j\} &= \sum_\alpha g^\alpha_{ij} G_\alpha \end{aligned} \qquad (2)$$

mutators, there now appear anticommutators. The anticommutator of two operators A and B is defined as $\{A,B\} = AB + BA$. The coefficients $c^\gamma_{\alpha\beta}$, $f^j_{\alpha i}$, g^α_{ij} again specify the group.

Supersymmetry in particle physics. Supersymmetry was originally formulated for applications to this field of physics. Early ideas about supersymmetry, put forward in 1971, went to a large extent unnoticed. The subject was rediscovered in 1973 and placed into a systematic perspective in 1973 and 1974, in an approach that led to the construction of theories of elementary particles with well-defined relations among masses and coupling constants of the fundamental particles in the theory. Subsequently, supersymmetries were used in attempts to construct a unified theory of all known interactions in physics: the gravitational, the weak, the electromagnetic, and the strong interactions. Particularly important here is the attempt to include the gravitational interaction, since there is at present no workable theory of gravitation consistent with the principles of quantum mechanics. Theories of gravitation that make use of supersymmetry have been called supergravity theories. The most elementary example of a supergravity theory, constructed in 1976, assumes the existence of a field particle with angular momentum 2 (graviton) and a second one with angular momentum $\frac{3}{2}$(gravitino). Many other theories have been constructed since then. Despite all these attempts, no experimental verification of supersymmetric theories in elementary particles physics has yet been found.

Supersymmetry in nuclear physics. Supersymmetry has also been applied to the study of problems in nuclear physics. The early ideas on this subject were presented in 1980 and 1981. Atomic nuclei are composed of neutrons and protons. In medium-mass and heavy nuclei, two protons or two neutrons appear to bind into pairs. The pairs have angular momenta that are integer multiples of $\hbar$ and thus are bosons. In nuclei with an even number of protons and neutrons (even-even nuclei), all particles are paired. However, in nuclei with an even number of protons and an odd number of neutrons, or vice versa (even-odd nuclei), complete pairing is impossible and in addition to pairs (bosons) there are unpaired neutrons or protons. These unpaired particles are fermions. If the theory describing these nuclei is required to have supersymmetry, a set of relations is obtained linking together properties of even-even and even-odd nuclei. The most important relation is that linking the excitation spectra of atomic nuclei, that is, the set of their allowed quantum-mechanical energy levels. The first experimentally found example of supersymmetry in nuclei is shown in the illustration. Since the discovery of this example, several others have been found, in the same and in other regions of the periodic table. A general feature of the experimental examples found so far is that supersymmetry in nuclei appears to be somewhat broken (see illustration). This is not unexpected, since supersymmetry is a rather complex type of symmetry requiring very stringent conditions on the interactions between the constituent particles. Despite the partial breaking, the fact that the experimental situation can be described, at least approximately, by invoking purely symmetry concepts is a remarkable result. This result is now being exploited systematically in the study of complex atomic nuclei.

Outlook. Experimental examples of supersymmetry have been found in nuclear physics. The implications of this discovery in the search for supersymmetry in other fields of physics, especially in elementary particle physics, are not yet clear. Supersymmetries observed in nuclei are of a different type than those mostly sought in elementary particle physics, since the bosons employed in the description of atomic nuclei are composite particles (pairs). Most supersymmetric theories in elementary particle physics employ bosons which are not composite. Recently, supersymmetry has been applied to the problem of electrons moving in a type II superconductor. In this application, the bosons are also composite particles (pairs of electrons). The fact that no experimental verification of supersymmetric theories with noncomposite bosons at present exists indicates either that the experimental search has not proceeded far enough or that the only realizable types of supersymmetry are those with composite bosons.

In conclusion, the discovery of supersymmetry in nuclear physics has led to major advances in this field and has shown that this new and elaborate type of symmetry exists in nature. However, the scope

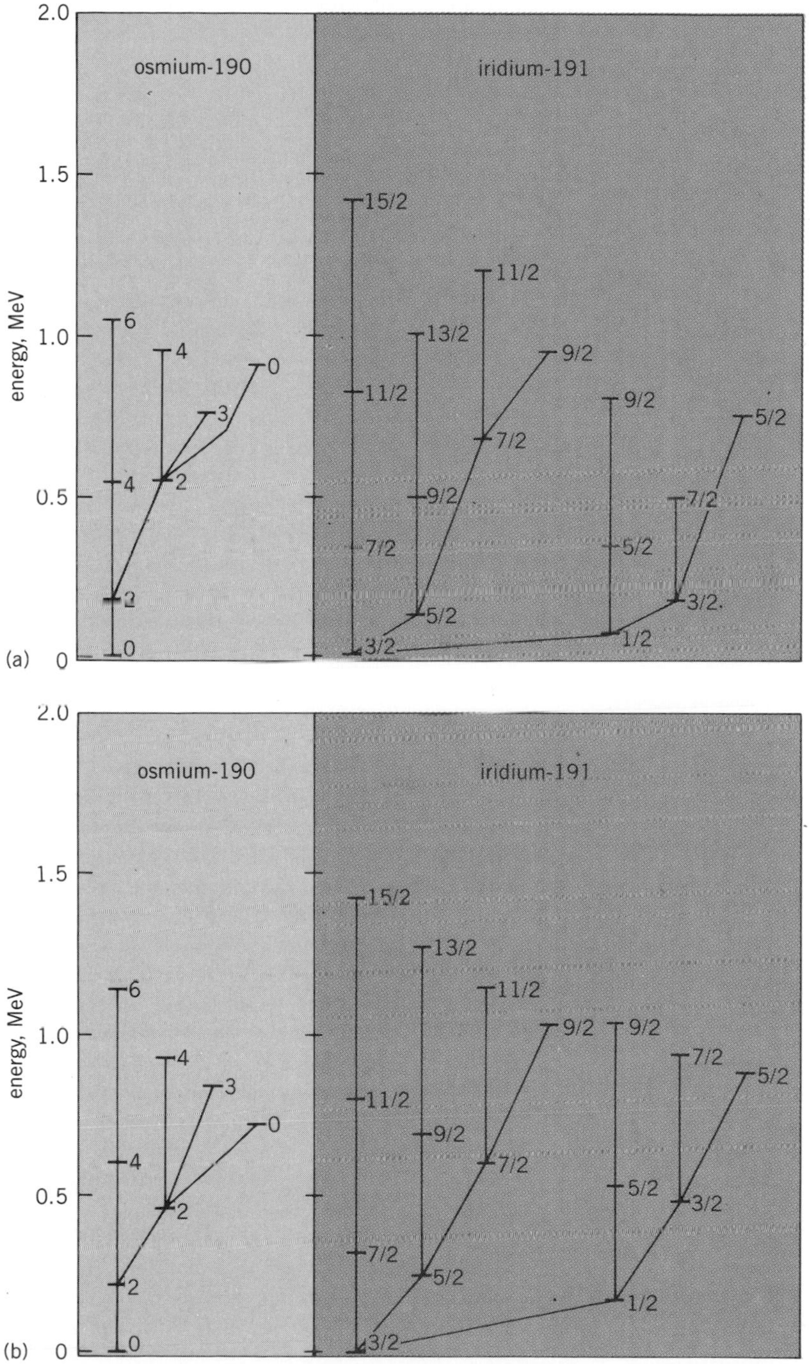

Examples of a dynamic supersymmetry in nuclear physics: energy spectra of osmium-190, composed of 76 protons and 114 neutrons, and iridium-191, composed of 77 protons and 114 neutrons. (*a*) Observed spectra; (*b*) theoretically predicted spectra. Each allowed energy level is labeled by the value of the total angular momentum *J*. The lines between levels indicate decay from the upper to the lower level by emission of intense electromagnetic radiation. The extent to which the observed and predicted spectra agree is evidence of supersymmetry in these nuclei; the extent to which they disagree is indicative of partial breaking of the supersymmetry.

and limitation of the concept of supersymmetry remain to be determined.

For background information *see* ELEMENTARY PARTICLE; FUNDAMENTAL INTERACTIONS; NUCLEAR STRUCTURE; SYMMETRY LAWS (PHYSICS) in the McGraw-Hill Encyclopedia of Science and Technology. [FRANCESCO IACHELLO]

Bibliography: F. Iachello, Supersymmetry in nuclei, *Amer. Sci.*, 70(3):294–299, May 1982; D. Z. Freedman and P. van Nieuwenhuizen, Supergravity and the unification of the laws of physics, *Sci. Amer.*, 238(2):126–143, February 1978; B. G. Levi, Nuclei may exhibit supersymmetry, *Phys. Today*, 33(9):21–22, September 1980; G. B. Lubkin, Interacting-boson model emphasizes symmetry group, *Phys. Today*, 31(7):17–20, July 1978.

Surface physics

The study of surfaces using low-energy monoenergetic positron beams is developing rapidly. Positrons are the antimatter partners of electrons. They have the same mass m (1/1836 of a proton mass) and intrinsic angular momentum ($\hbar/2$, where $\hbar$ is Planck's constant divided by 2π). However, the positron has a unit positive charge which is the opposite of the electron's charge. Electrons are presently used in a number of techniques to study surfaces, and in each of these spectroscopies positrons may be substituted for electrons with the advantage that the probing particle is now distinguishable from the myriad of electrons of the object under study.

A positron can annihilate with an electron, and this provides a unique way to remove electrons from a solid. Information about the state of an electron is carried away by the annihilation quanta, usually two photons of energy approximately mc^2 (where c is the speed of light) or 511,000 eV. The total momentum and energy of the photons can be measured, and this gives information about the electron state if the positron comes to rest before the annihilation. In fact, most of the time the positron does thermalize before annihilation because the slowing-down time is about 100 times shorter than the positron annihilation lifetime in a solid. A positron also can form a bound state with an electron. These matter-plus-antimatter objects analogous to hydrogen atoms are known as positronium atoms, and were first observed in 1951.

Techniques for generating a beam of low-energy positrons will be discussed in this article, and current knowledge about the unique interactions of positrons at surfaces and how these interactions can be used to delineate the characteristics of surfaces in some instances will then be briefly outlined. The field is partially developed because it is only about 5 years since the positron-beam technique was first combined with the ultrahigh-vacuum techniques needed for the study of clean surfaces.

Positron beams. Positrons with energies of the order of 10^6 eV are obtained from the β-decay of some radioactive nuclei (such as ^{22}Na, ^{58}Co, ^{68}Ge, ^{64}Cu) or from the electron-positron-photon showers that result from stopping high-energy electron beams (10^7 eV) in a thick target. These positrons are too energetic to give much information about systems with atomic binding energies (a few eV). Because annihilation rates are orders of magnitude less than slowing-down rates, the annihilation of positrons in a sample does tell something about the chemical or physical states of the sample. However, the relatively long annihilation lifetime (10^{-10} s) also makes it possible to stop the positrons in a solid and to extract some of them before they decay. Simply putting a solid surface next to a β^+ (positron) emitter results in the emission of a few (1 in 10^8) slow positrons (approximately 1-eV kinetic energy). This scheme works much better if the moderator surface is a well-annealed single crystal so that the implanted and thermalized positrons can diffuse back to the surface. It is also essential that the moderator have a negative positron affinity. This can be achieved by using a relatively high-density metal with a clean surface which has possibly been treated by coating it with a monolayer or so of an adsorbate that increases the electron work function. Figure 1 shows the configuration of a radioactive source and slow positron moderator. The source and moderator are located in an ultrahigh-vacuum chamber (10^{-10} torr or 10^{-8} pascal). The positrons are transported away by an electric field and are formed into a beam by means of conventional electron optics or are confined by a guiding axial magnetic field.

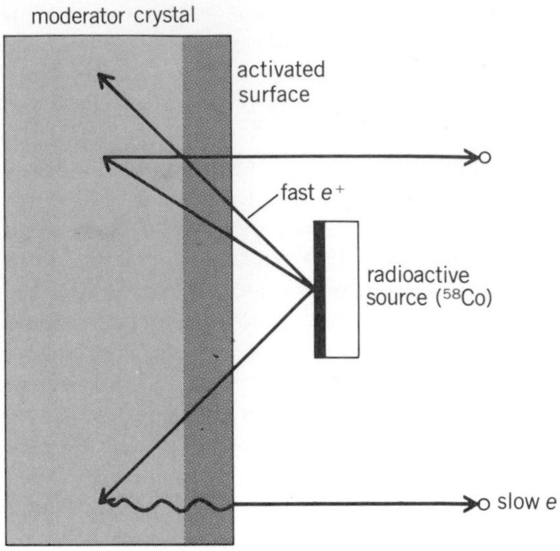

Fig. 1 Configuration for obtaining slow positrons (e^+) by moderating the fast positrons from a radioactive source using an activated single crystal surface.

Positron surface interactions. Figure 2 shows some of the things that can happen when a positron strikes a solid surface. The incident positron beam may be diffracted by the regular array of surface atoms. The positrons may penetrate deeply into the solid and thermalize. Subsequently they will diffuse through the crystal until they annihilate with an electron, get trapped by a defect in the crystal, or reach the surface. At the surface the positrons are trapped in a surface state by their image potential interaction, or they are ejected as a free low-energy positron or as an atom of positronium (chemical symbol Ps). These three processes have similar probabilities of occurrence at a number of metal surfaces. The surface positrons may be boiled away as positronium atoms if the solid is hot enough. A positron which is deflected back to the surface before thermalizing can escape as a fast positron, as an energetic positronium atom possibly in one of its excited states, or as the positronium negative ion, a positronium atom with an extra electron attached.

Some of these processes are being used in experiments to study the atomic physics of positronium and in scattering experiments. On the other hand, a number of the processes are sensitive to the type and condition of the solid surface being used as a target. Some of the possibilities for exploiting these interactions to analyze surfaces are listed below. None of these has yet been developed into a routine analysis tool.

Positron work function. Implanted positrons are ejected from some metal surfaces with a definite energy (an eV or so) that can be measured precisely. These surfaces are said to have a negative work function for positrons, ϕ_+. Theoretically ϕ_+ is the sum of an attractive bulk term primarily from the negative electron cloud that forms around the positron, and a repulsive surface term Δ from the electric dipole layer due to the metallic electrons extending out beyond the last layer of ions. The electron work function ϕ_- is the least amount of energy required to remove an electron from a metal surface. The work functions ϕ_+ and ϕ_- change by equal and opposite amounts when Δ is changed by the absorption of small quantities of contaminants on a surface.

Positron emission spectroscopy. It has been found that positrons are emitted from a negative positron work function surface primarily in a direction normal to the surface with a roughly thermal spread in angle and energy. If the outgoing positrons interact with an adsorbed layer of molecules, some of the positrons lose discrete amounts of energy ΔE_i corresponding to the excitation of the molecules. Peaks then appear in the positron energy spectrum at energies $-\phi_+ - \Delta E_i$. It is possible that at low temperatures these spectra will have better resolution than those of the corresponding electron method known as electron energy loss spectroscopy (EELS).

Positronium velocity spectroscopy. When positronium is formed at a surface, it leaves behind a hole in the Fermi sea of electrons. For a simple metal the energy and momentum spectrum of the outgoing positronium is proportional to the density of electron states near the surface at a corresponding energy and momentum dictated by conservation of the total energy and the momentum parallel to the surface. For example, an aluminum surface emits positronium with a relatively flat distribution that is sharply cut off for energies greater than 2.6 eV. This energy is the negative of the positronium work function,

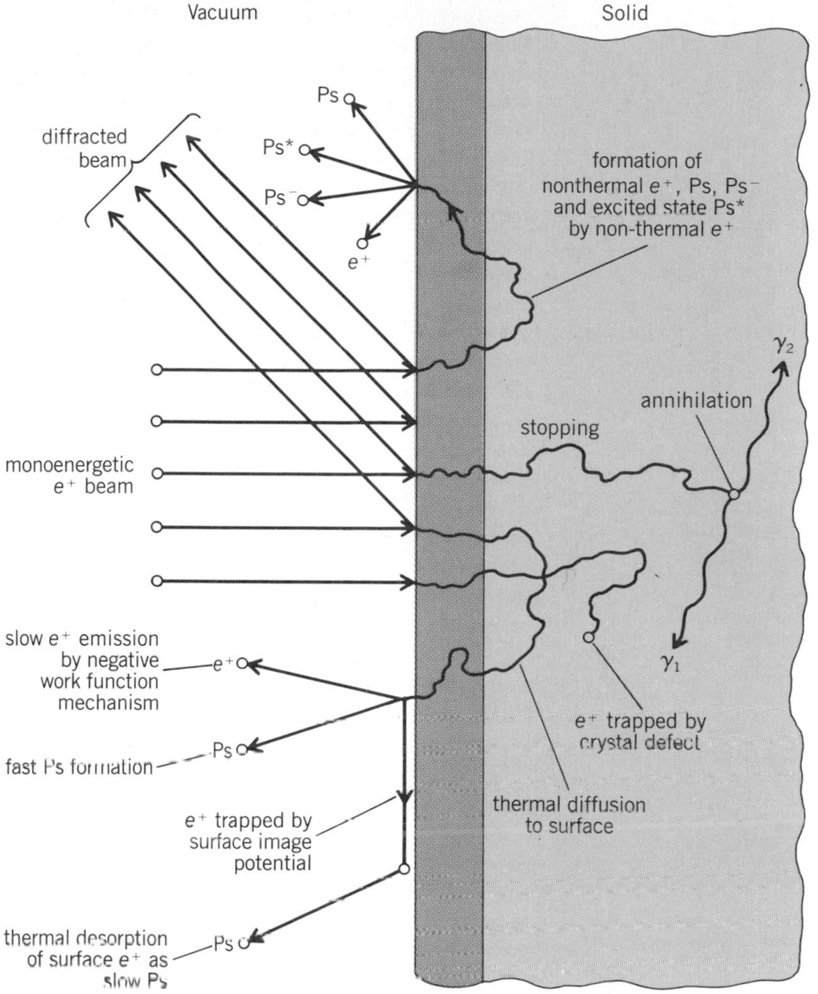

Fig. 2 Some of the processes that occur when a monoenergetic slow positron beam impinges upon a metal surface in ultrahigh vacuum.

$-\phi_{Ps} = \frac{1}{2}R_\infty - \phi_+ - \phi_-$ where ϕ_+ and ϕ_- are the energies expended in removing a positron and an electron from the bulk solid, and $\frac{1}{2}R_\infty$ is the energy gained by forming them into an atom of positronium. Since ϕ_{Ps} involves the sum of ϕ_+ and ϕ_-, it is insensitive to contaminant layers that change the dipole potential Δ. However, the positronium velocity spectrum is very sensitive to surface contamination. It is expected that this spectroscopy will be a useful way to study surface electrons. In particular, surface magnetism can be studied if the probing positron beam is polarized. For beams derived from β-decay this is the case, provided steps are taken not to destroy the large polarization inherent in the β-decay process. It has been found that positron beams with up to 60% polarization can be produced.

Positron diffraction. The kinematics of positron diffraction are identical to those of the familiar low-energy electron diffraction (LEED). However, the dynamics of positron diffraction are much different and promise to make this one of the most useful of the positron surface-analysis techniques. Positrons are repelled from the ion cores of a solid, and they thus avoid the regions where the potential is strongest. Spin-orbit effects are much reduced, and it should be a good approximation to treat the positron as a spinless particle. Furthermore, the potential is much better understood for a positron because of the absence of the exchange terms which are needed in electron scattering.

In a LEED or LEPD (P for positron) experiment, a narrow beam of electrons or positrons strikes the target surface, and the diffracted beams in certain directions permitted by the regular array of surface atoms are observed. The symmetry of the diffraction pattern is used to determine the type of surface lattice, but the intensities of the diffracted beams give information on the arrangement of atoms that make up the unit cell. The interpretation of these intensities requires an extensive calculation because the probing particles are strongly interacting and scatter many times. The utility of LEED is presently limited partially by the amount of computer time required. The simplicity of the positron problem would allow the computation to be done an order of magnitude faster.

Positron depth profiling. Positrons of energy E stop at a mean depth roughly proportional to $E^{3/2}$. The spread in stopping depths is comparable to the mean depth, but this still allows useful information to be obtained about the presence and depth of interfaces, vacancies, voids, bubbles, dislocations, and other defects below the surface. Positrons are generally trapped by such defects and exhibit different annihilation characteristics that signal their presence. Depths of the order of 10 μm can be probed by a 100,000-eV positron beam.

Positron surface state. Positrons trapped by the image potential well at a surface have an annihilation lifetime longer than bulk positrons, about 5×10^{-10} s. It should be possible to measure the momentum of the annihilation radiation from surface positrons to study the two-dimensional surface electron charge density. Surface positrons can be desorbed as positronium having a thermal velocity distribution by heating the sample. The surface positrons are expected to be mobile. When very high density of surface positrons is obtained (say 1 per 150 nanometers) they should be observed to be spontaneously desorbed as positronium molecules (a positronium molecule, Ps_2, is formed through the chemical binding of two positronium atoms). The Ps_2 binding energy of 0.2 eV can be sufficient to overcome the individual Ps surface binding energies for suitably chosen sample surfaces.

Future techniques. Plans are under way to make very high-intensity slow positron sources. It would then be possible to study positronium diffraction, for example. When a positronium atom of a few tens of eV strikes a surface, it should break up with high probability. The diffraction problem then becomes a weak scattering one, and the interpretation of Ps diffraction intensities would be simple. Various high-resolution positron microprobes are also possible: a positron microscope for imaging defects, and possibly a positronium or Ps^- microscope for extreme surface sensitivity.

For background information *see* ELECTRON DIFFRACTION; POSITRON; POSITRONIUM; SURFACE PHYSICS; WORK FUNCTION (ELECTRONICS) in the McGraw-Hill Encyclopedia of Science and Technology.

[ALLEN P. MILLS, JR.]

Bibliography: P. G. Coleman, S. C. Sharma, and L. M. Diana (eds.), *Positron Annihilation*; A. P. Mills, Jr., Surface analysis and atomic physics with slow positron beams, *Science*, 218:335–340, 1982.

Swine production

Pigs have been a source of human food for centuries; today their number is growing faster than that of the human population. Progress in nutrition, genetics, reproduction, husbandry, and disease control has greatly increased efficiency of pork production. The lean modern pig not only receives favor with consumers, but it is produced at a lower cost than its obese predecessor. Lard yield per carcass declined from about 14% in 1960 to about 5% in 1980. Modern pigs require 3–3½ lb of feed from birth to market to produce 1 lb of live weight gain compared with more than 4 lb 50 years ago.

The early sexual maturity, prolificacy, and short gestation period of pigs compared with other food animals, and associated favorable heritability of the traits involved, makes possible relatively rapid changes in growth efficiency and body composition. Advances in health care have reduced veterinary costs associated with disease treatment and have resulted in more emphasis on herd health and disease prevention programs; recombinant DNA techniques have led to the development of a new vaccine against foot-and-mouth disease.

Interactions of economics and the biology of the pig have produced dramatic changes in swine housing and feeding. Pigs are commonly raised in a carefully controlled environment arranged to include a unit housing pregnant sows, a farrowing unit where sows give birth, a nursery, and a growing-finishing unit. All units have slatted floors for easy removal of wastes, and most units in temperate zones have a ventilation system. Most farrowing and nursery units have supplemental heating units. Many swine herds are maintained relatively free of disease by measures that limit animal and human traffic; such herds are termed specific-pathogen-free (SPF).

Recent breakthroughs in research in nutrition, genetics, and reproduction have had a major impact on swine production and are expected to result in further important advances in swine management and production.

Nutrition. Feed represents 55–85% of the total cost of commercial pork production. The economics of swine feeding is dependent on local feedstuff availability, competition for the same foodstuff for use by humans or other animals, and the price of purchased protein, mineral, and vitamin supplements.

The pig has a digestive system similar to that of humans so that, like the chicken, it is in direct competition with the human population for available food supplies. Nutrient requirements of pigs of different ages and stages of production (gestation, lactation, growth) have been determined with high precision, so that diets now are formulated with computer technology by commercial feed manufacturers, and in many cases, by individual pork producers. Not only are synthetic vitamins and high-purity mineral salts routinely used, but synthetic amino acids such as lysine are used to provide a more nutritionally balanced mixture of amino acids, allowing lower total protein level in the diet. Although unprocessed maize and cereal grains make up a major part of most pig diets, many by-product foodstuffs such as milling by-products (wheat bran, corn bran), distillery by-products (distillers' grains, distillers' solubles), and slaughter plant by-products (blood meal, meat meal, bone meal) constitute increasingly larger parts of computer-formulated least-cost diets. Such by-products would have little value if not used for livestock, or costs of disposing of them would increase significantly the market price of the final product. Forages, such as alfalfa, can be used effectively in pig diets, especially for pregnant females.

Gene-splicing techniques may soon be applied in nutrition by changing the microfloral activities of the gastrointestinal tract of pigs to produce cellulase, the enzyme that unlocks the glucose contained in cellulose, the most abundant of plant constituents. Such a development would change drastically the constraints on pig production imposed by the inability of the digestive tract to assimilate large amounts of these complex carbohydrates.

Amino acids may be widely available soon from synthetic sources as another result of gene splicing. The availability of such amino acids at economically competitive prices could result in significant annual savings of protein, because the metabolic needs of the animal for protein and amino acids could be met by a lower total content of the diet associated with more efficient utilization of a well-balanced mixture of amino acids.

New feed resources will also play a major role in the future of pork production. Plant breeders continue to produce new varieties of high-lysine maize, barley, and oats which show promise of reducing protein supplementation. New knowledge of the physiology of digestion and sites of absorption of amino acids from the gastrointestinal tract has provided a better basis for assessing amino acid bioavailability. Several microbial protein sources, including dried bacteria, yeast cells, and algae, offer potential for expanded use in pork production. A species of blue-green algae, *Arthrospira platensis*, is grown in Taiwan and other tropical regions on the effluent from fermented swine wastes; its filamentous structure allows harvest by gravity filtration, thereby lowering production costs. Its high protein content and apparent relative freedom from toxic constituents favor its expanded use as a swine feed.

Genetics. Population geneticists in government, university, and private laboratories, by capitalizing on sophisticated modeling techniques made possible by computer technology and by introducing genetically diverse breeding stock into crossbreeding systems, are making rapid strides in improving heritable production traits, and are actively engaged in such research for application in commercial pork production.

Methods of accurately estimating body composition of live animals, such as ultrasonic backfat measurements and other noninvasive techniques, and of predicting subsequent body composition of individual animals by appropriate measurements early in life offer possibilities for more effective genetic selection for desired traits such as increased muscling. Genetic engineering by recombinant DNA techniques (gene splicing) shows promise of allowing mass production of hormones such as insulin and growth hormones for use in swine production.

The functional basis for genetic variation in economically important traits, such as growth and lactation, relates to hormones, enzymes, and various intracellular processes; yet the genetic bases for these relationships are not well understood. Current research on growth of muscle cells in culture tubes and other artificial-environment approaches to studies of protein and fat synthesis and degradation offer possibilities for providing more complete knowledge of the metabolic controls of muscle accretion in pigs.

Reproduction. Current research efforts in nutritional, genetic, and endocrine factors associated with high prenatal mortality show promise of making it feasible to induce superovulation without a concomitant increase in fetal deaths. The use of exogenous hormones and other agents to synchronize ovulation among groups of gilts has been possible for some time. The practice has not been adopted because, until recently, boar semen freezing techniques have not been available. Synchronized ovulation, coupled with artificial insemination with frozen semen from genetically superior boars and the timed induction of parturition with administration of prostaglandins, offers a potentially attractive reproductive management package for use in large pig production units.

An improvement in reproductive efficiency must be accompanied by improved methods of perinatal care. Pigs, after receiving colostrum, which provides antibody protection, can be reared successfully from 2 or 3 days of age on liquid diets or on dry diets. Combining the technology which would allow birth of an average of 15 to 20 viable pigs per litter with the early weaning of part of the litter at 2 to 3 days of age could result in an increase in the number of pigs marketed per breeding female from the present norm of 14–18 to an average of 24 or more pigs yearly.

Embryo transfer techniques are available. Initial applications probably will be found in reducing costs of movement of superior germ plasm over great distances.

Goals and prospects. Attainable goals in production traits to allow the sustained competitive position

Attainable goals for some economically important production traits

Trait	Present	Goal
Number of litters per sow annually	1.8	2.2
Number of pigs marketed per sow annually	13	24
Percentage loss of pigs, birth to weaning	15–30	5–10
Number of pigs raised per litter	7.4	11
Percentage of fat in edible carcass	41	32
Pounds of feed per pound of body weight gain, birth to market	3.5	2.5
Months of age to market	5–6	4–5
Average backfat thickness at market, in. (cm)	1.6 (4)	.8 (2)

of pork as a significant source of animal protein in the coming decades are suggested in the table. Each trait listed is directly related to efficiency of the total pork production enterprise. Since feed represents 55–85% of total cost of production, a reduction from the current 3.5 lb to a projected 2.5 lb of feed required per pound of live weight gain would represent significant progress in lowering pork production costs on an industry-wide basis. Such a change presupposes no change in the reliance on high-concentrate feed resources as the major source of nutrients. If technology provides a means of utilizing large portions of fibrous by-product feedstuffs noncompetitive with humans, the incentive for improved efficiency of feed utilization of the magnitude suggested will be diminished. Such factors as international trade policies, changes in purchasing power among consumers in developing countries, and sustained increases in global capacity for feed grain and protein supplement production probably will have an impact on future pork demand at least as great as the impact effected by advances in biology and husbandry.

For background information *see* ANIMAL FEEDS; BLEEDING (ANIMAL); SWINE PRODUCTION in the McGraw-Hill Encyclopedia of Science and Technology. [WILSON G. POND]

Bibliography: H. T. Fredeen and B. G. Harmon, The swine industry: Changes and challenges, *J. Animal Sci.*, 57(suppl. 2):100–118, 1983; J. L. Krider, J. H. Conrad, and W. E. Carroll, *Swine Production*, 5th ed., 1982; W. G. Pond, Modern pork production, *Sci. Amer.*, 248:96–103, 1983; W. G. Pond and J. H. Maner, *Swine Production and Nutrition*, 2d ed., 1984.

Systolic arrays

Very high-performance computer systems must rely heavily on parallelism, since there are severe physical and technological limits on the ultimate speed of any single processor. The systolic array concept allows effective use of a very large number of processors in parallel for important application areas such as signal processing. It is a kind of array or matrix operation on elements of a signal or of data.

This article describes the principle of systolic arrays, and illustrates the ideas by considering a simple systolic array design digital filtering of electrical signals or of data manipulation. The range of problems that can be solved by systolic arrays, as well as implementation alternatives for systolic array processors, is also discussed.

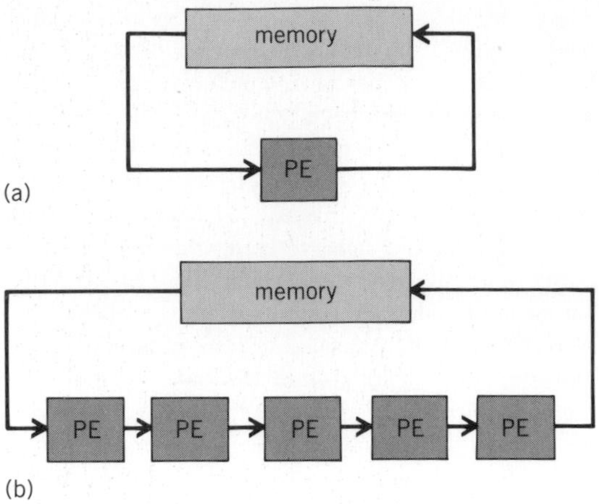

Fig. 1. Processor architecture. (*a*) Conventional processor with one processing element (PE). (*b*) Systolic array processor with array of PEs. (*After H. T. Kung, Why systolic architecture?, Computer, 15:37–46, 1982*)

Principle. Systolic arrays are suited for front-end processing that deals with large amounts of data obtained directly from sensors. Although processing of this kind usually requires much computing power, it is highly regular and many operations can be performed simultaneously, or in parallel. The systolic array architecture exploits this regularity and parallelism to meet the computation requirement efficiently.

The principle of a systolic array architecture (Fig. 1) is: by replacing a single processing element (PE) with an array of processing elements or systolic cells a higher computation throughout can be achieved without increasing the input/output bandwidth. The function of the memory is analogous to that of the heart; it "pulses" data through the array of cells. The crux of this approach is to ensure that once a data item is brought out from the memory it can be used effectively at each cell it passes while being pumped from cell to cell along the array. Being able to use each input data item several times is one of the many advantages of a systolic array. Other advantages include modular expandability, simple and regular data and control flows, the use of simple and uniform cells, the use of efficient fault-tolerant schemes, and the elimination of global data communication. These properties are highly desirable for very large-scale-integration (VLSI) implementations. Indeed, the advances in VLSI technology have been a major motivation for recent interest in systolic arrays.

Digital filtering implementation. An important application of systolic arrays is in digital filtering implementation. Given inputs x_i and weights w_j, the filtering problem is to compute outputs y_i defined by $y_i = w_1x_i + w_2x_{i+1} + \ldots + w_kx_{i+k-1}$. Figure 2 depicts a typical systolic filtering array with $k = 3$ weights. Weights are preloaded into the array, one for each cell. During computation, both partial results for y_i and inputs x_i flow from left to right, but the former move twice as fast as the latter. More precisely, each x_i stays inside every cell it passes for one cycle, and thus each x_i takes twice as long to march through the array as does a y_i. One can check that each y_i is initialized to zero before entering the leftmost cell, and is able to accumulate all its terms while marching to the right. For example, y_1 accumulates w_3x_3, w_2x_2, and w_1x_1 in three consecutive cycles at the leftmost, middle, and rightmost cells, respectively.

From the user's point of view, the filtering can be performed by simply pumping the inputs into the left end cell and then collecting the results from the right end. A unique property of the systolic approach is that as the number of cells is increased the system performance improves proportionally; in general, the performance of a systolic array is bounded only by problem size, which for the filtering example is the number of weights.

Scope of application. A large number of systolic array designs have been developed and used to solve an extremely broad range of problems. In the areas of signal and image processing, systolic arrays have been used in various types of filtering, convolution, correlation, interpolation, and resampling operations, in computing discrete Fourier transforms, and in encoding and decoding for error correction. Matrix arithmetic operations carried out by systolic arrays include matrix–vector and matrix–matrix multiplication, solution of various types of linear systems, least-squares computation, singular-value decomposition, and eigenvalue computation. Nonnumeric applications include the implementation of data structures (such as stacks, queues, dictionaries), dynamic programming, polynomial and multiprecision integer arithmetic, and Monte Carlo simulation.

Several theoretical frameworks for the systematic design of systolic arrays are being developed.

Implementation alternatives. Existing systolic array processors exhibit a large variety of implementation alternatives. They can be classified along two dimensions: flexibility and interconnection topology.

Flexibility. A systolic array processor may implement a single systolic design or a variety of them. Systolic array processors can be classified into five categories according to their degree of flexibility: single-purpose systolic arrays; multipurpose systolic arrays; processors based on nonprogrammable building blocks; processors based on programmable building blocks; and programmable systolic arrays.

Single-purpose systolic arrays are reasonable if one or more of the following conditions hold: the performance of the processor is of ultimate importance and its use is well understood; the processor will be used in large quantities despite the fact that it is single-purpose; and the design and implementation cost of the processor is low.

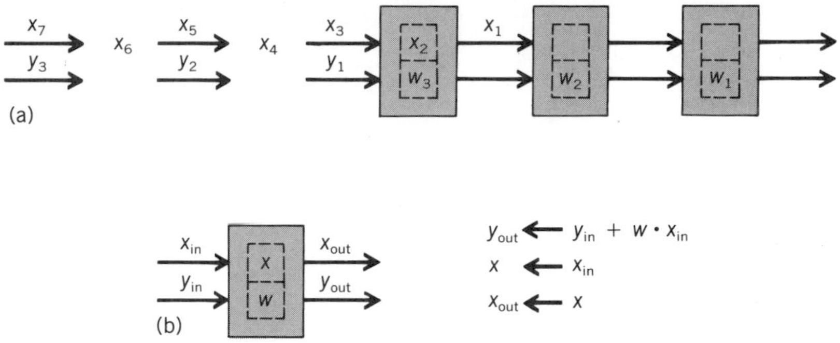

Fig. 2. Application of systolic arrays to digital filtering. (*a*) Systolic array for filtering of 3 weights. (*b*) Operations carried out by each cell of the array during each cycle. (*After H. T. Kung, Why systolic architecture?, Computer, 15:37–46, 1982*)

Multipurpose systolic array processors can implement a predefined set of computations. The approach is based on the observation that many systolic designs, like those for digital filtering, and matrix operations can be excuted on systolic arrays of very similar structures.

Some common functions such as multiply–accumulate are performed by cells of a large number of systolic arrays. Thus it is possible to construct building-block processors capable of executing a few predefined, commonly used functions, and then connect them to form a variety of systolic processors of different sizes and shapes.

In the programmable building-block approach, building blocks can be programmed to implement a large family of systolic cells. Of course, the programmable approach is not as efficient as the nonprogrammable one, because of the overhead for supporting the programmability. Nevertheless, it meets the need of implementing those systolic arrays, none of which is important enough to warrant individual implementation as custom hardware, but which in aggregate can justify the cost of an array processor. Moreover, in some systolic designs, the instruction that the cell executes in a cycle depends, in a complicated way, on the inputs to the cell and the cell's state during that cycle. The programmable approach seems to be the only effective way to implement those complicated, data-dependent systolic arrays.

In programmable systolic array processors, a fixed number of programmable processing elements are connected together in a certain manner, possibly with other control circuits. Programmable systolic arrays are more flexible than the multipurpose systolic arrays in the sense that the processing elements are programmable, and sometimes even their interconnections can be configured by software control before a computation starts.

Interconnection topology. Systolic arrays typically call for simple and regular array interconnections between their processing elements. Interconnections for arrays with n processing elements can be classified into three classes according to the number of computations performed for each input/output oper-

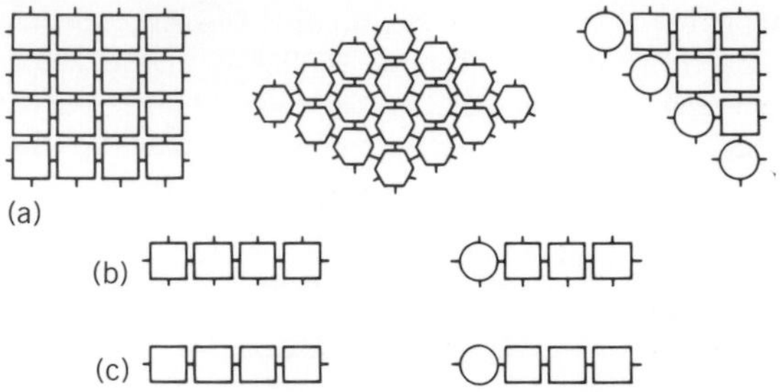

Fig. 3. Typical interconnection schemes for systolic arrays. (*a*) Two-dimensional systolic arrays. (*b*) Degenerate two-dimensional systolic arrays. (*c*) Linear or one-dimensional systolic arrays.

ation: two-dimensional systolic arrays, degenerate two-dimensional systolic arrays, and linear or one-dimensional systolic arrays.

For two-dimensional systolic arrays (Fig. 3*a*), a number on the order of n processing elements perform computations at each cycle, while a number on the order of $\sqrt{n}$ boundary processing elements perform input/output. Thus, the computation over input/output ratio is on the order of $\sqrt{n}$. Examples are systolic arrays for matrix arithmetic.

Degenerate two-dimensional systolic arrays (Fig. 3*b*) consist of only a small (constant) number of rows or columns of processing elements. At each cycle, on the order of n processing elements perform computations, and on the order of n processing elements perform input/output. Thus, the computation over input/output ratio is on the order of 1. Examples are systolic arrays for the solution of triangular linear systems and orthogonal transformations.

For linear or one-dimensional systolic arrays (Fig. 3*c*), on the order of n processing elements perform computations at each cycle, but only the two end processing elements perform input/output. Thus, the computation over input/output ratio is on the order of n. Examples are various systolic filtering arrays such as the one discussed above.

For situations where the input/output bandwidth between the host system and the systolic array is a major limiting factor for achieving high performance, linear arrays are preferable to two-dimensional arrays, and degenerate two-dimensional arrays are least desirable. Linear arrays have the additional advantage that they can usually be safely synchronized by a simple, global clock. To avoid the excessive input/output bandwidth required by degenerate two-dimensional arrays, it may sometimes be cost-effective to associate each processing element with a local memory, where data required by the processing element can be held.

For background information *see* DIGITAL COMPUTER; INTEGRATED CIRCUITS in the McGraw-Hill Encyclopedia of Science and Technology.

[H. T. KUNG]

Bibliography: A. L. Fisher et al., Architecture of the PSC: A programmable systolic chip, *Proceedings of the 10th Annual Symposium on Computer Architecture*, pp. 48–53, Stockholm, Sweden, June 1983; H. T. Kung, Why systolic architectures?, *Computer*, 15:37–46, 1982; H. T. Kung and M. S. Lam, Fault tolerance and two-level pipelining in VLSI systolic arrays, *Proceedings of MIT Conference on Advanced Research in VLSI*, pp. 74–83, January 1984; H. T. Kung, and C. L. Leiserson, Systolic arrays (for VLSI), in C. A. Mead and L. A. Conway, *Introduction to VLSI Systems*, pp. 271–292, 1980.

Tectonics

Sedimentary basins are large depressions in the Earth's crust that are filled with more than about 0.6 mi (1 km) of sediment. In recent years, geological and geophysical studies of basins have focused on the processes that initiate the depressions and maintain them for tens or even hundreds of millions of years. These studies have employed new methods that quantify the subsidence history of basins by using data from stratigraphic sections measured through the basin fill. The results are providing clues about the origin of the depressions and the mechanisms that control their subsidence. Sedimentary basins can now be classified genetically on the basis of subsidence mechanisms. It also appears possible in certain types of basins to distinguish between the subsidence mechanisms and eustatic

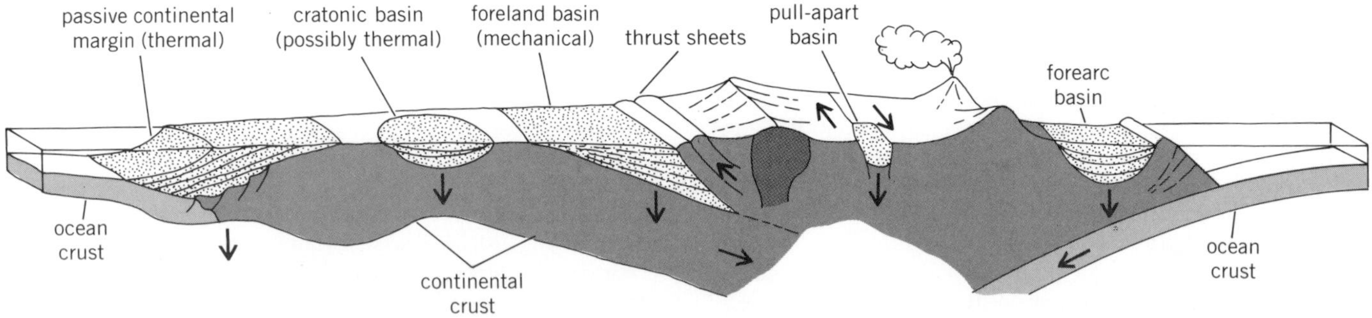

Fig. 1. Diagrammatic examples of typical sedimentary basins in relation to their geologic setting. (*After G. Bond and W. C. Pitman, III, The evolution of sedimentary basins, Lamont-Doherty Geological Observatory Yearbook 1981–1982*)

(global) changes in sea level that have occurred during the geologic past.

Types of sedimentary basins. Sedimentary basins conventionally have been classified according to the tectonic setting in which they occur, that is, the relation of the basin to surrounding geologic features. Some of the most common types of basins are shown in relation to their tectonic settings in Fig. 1. Recent geophysical and geological studies of basins have led to a new classification which places basins into one of two genetic categories: those formed by mechanical subsidence mechanisms and those initiated by processes that lead to thermal subsidence mechanisms. One of the best examples of a basin formed by mechanical subsidence is the foreland basin (Fig. 1). This type of basin develops between the stable interior of a continent and an orogenic belt (a linear region which has been subject to compressional deformation, usually resulting in the formation of mountains). It is thought that the margin of the continental crust is bent downward, somewhat like a springboard, under the weight of thrust sheets that form as a result of compression along the edge of the orogenic belt (Fig. 1). The resulting basin extends as much as 125–250 mi (200–400 km) into the stable continental interior beyond the edge of the mountain belt, and subsides as much as 1.8–2.4 mi (3–4 km) below sea level. The rates of subsidence in this type of basin vary greatly through time, probably owing to changes in rates of convergence between the orogenic belt and the continental plates. Examples of active foreland basins include the Ganges Basin in northern India and the Persian Gulf in Arabia.

A good example of a type of basin formed by thermal subsidence mechanisms is the passive continental margin (Fig. 1). This type of basin forms in rift belts where the lithosphere, or solid portion of the Earth which normally extends to a depth of about 78 mi (125 km), has been stretched. The stretching eventually causes the continent to rift, and the two pieces drift apart as new ocean floor forms between them. During the stretching process the lithosphere in the rift belt is heated. After the continent splits apart, the rifted edges or margins of the new continents begin to cool by vertical and lateral conduction of the heat produced by rifting. The cooling causes the lithosphere in the margins to contract and subside. This subsidence, which continues for about 200 million years until thermal equilibrium is restored, produces the passive margin type of basin. Geophysical models for thermal contraction of passive margins have shown that for about 80 million years after the continents begin to drift apart, the form of thermal subsidence is a linear function of the square root of time. After about 80 million years the subsidence decays exponentially with a time constant of about 50 million years. The rate of subsidence, however, is directly proportional to the amount of extension of the lithosphere that occurred during the stretching phase. Modern examples of the passive margin type of basin include the margins of eastern North America, eastern South America, and the eastern, southern, and western margins of Africa.

The pull-apart basin, a combined mechanical and thermal basin, is formed by shear stresses between large faults. The forearc basin is formed between an island arc and a trench. Its origin is uncertain, but it may be a mechanical basin.

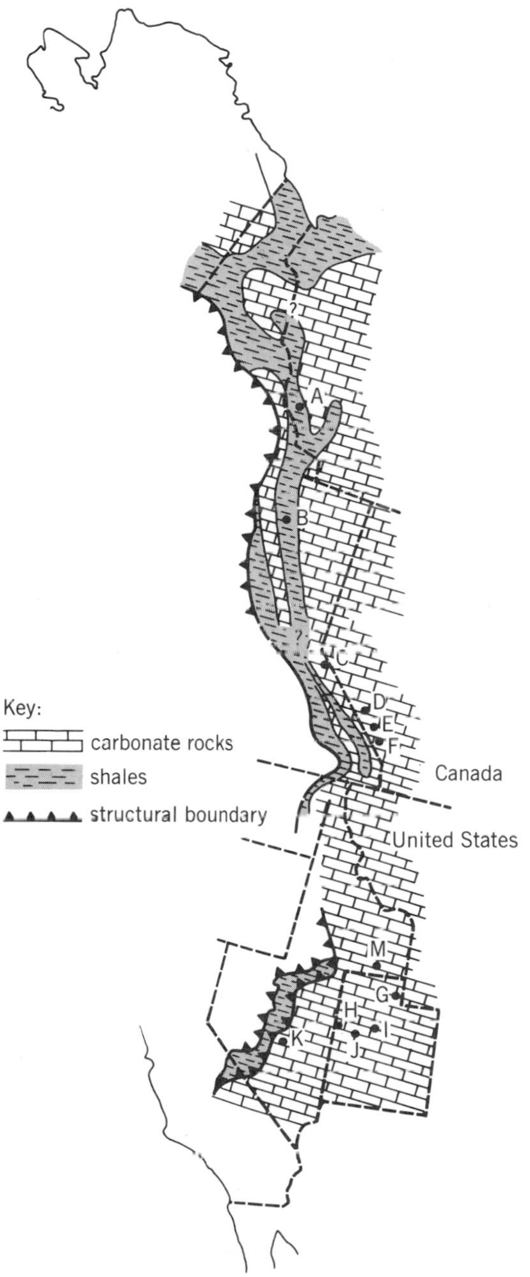

Fig. 2. Location of strata deposited in the ancient passive margin of western North America. Solid dots (A–M) indicate locations of stratigraphic sections of the passive margin strata used to construct the tectonic subsidence curves in Fig. 3. The structural boundary separates nuclear North America from terranes that have collided with its western margin since early Paleozoic time. (*After G. Bond and W. C. Pitman, III, The evolution of sedimentary basins, Lamont-Doherty Geological Observatory Yearbook 1981–1982*)

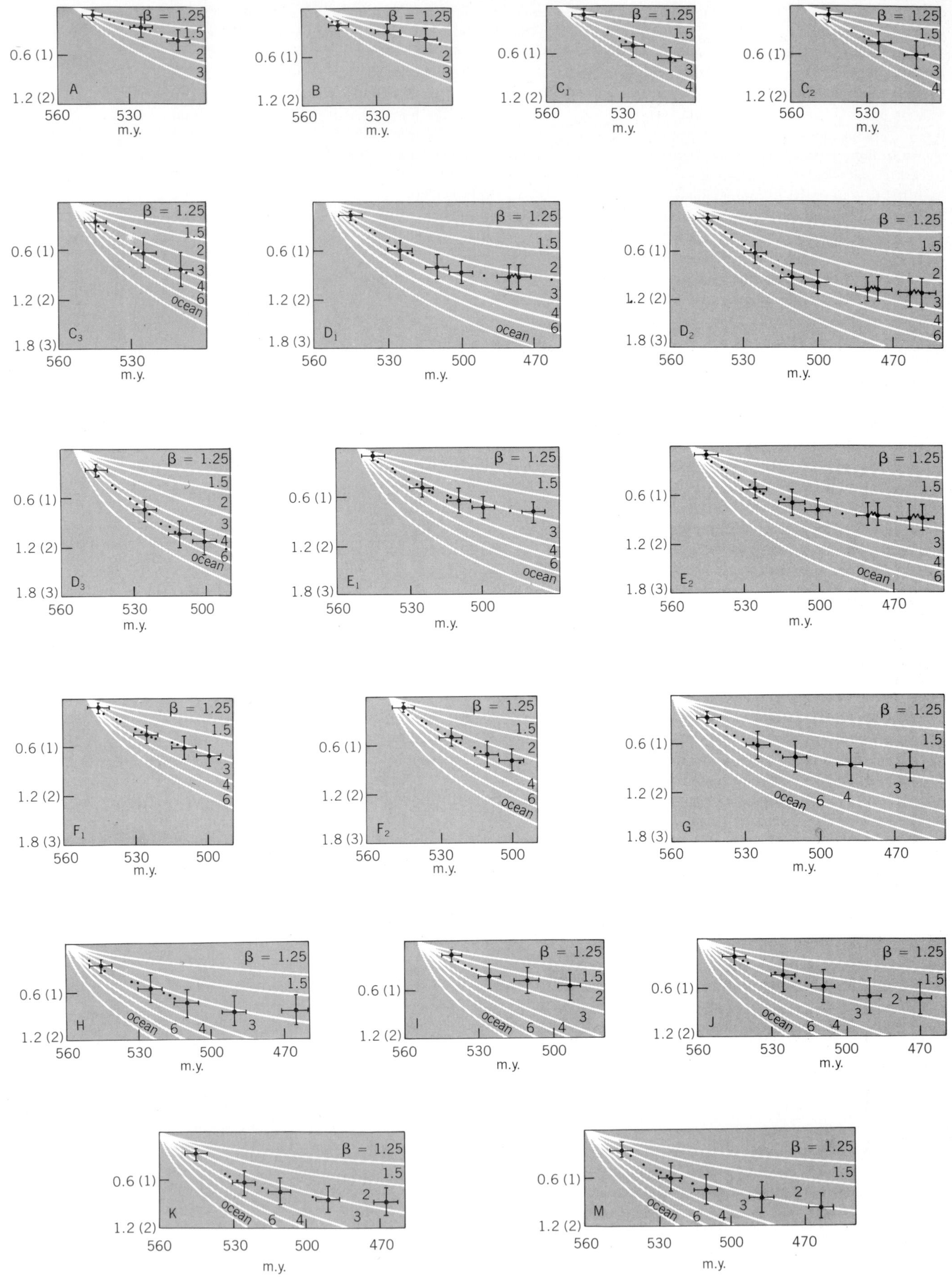

A
B
C1
C2
C3
D1
D2
D3
E1
E2
F1
F2
G
H
I
J
K
M
β = 1.25
1.5
2
3
4
6
ocean
0.6 (1)
1.2 (2)
1.8 (3)
560
530
500
470
m.y.

Fig 3. Tectonic subsidence for sections (A–M) in Fig. 2 compared with subsidence curves calculated from a model of postrift thermal contraction for passive margins (solid curves). Subscripts identify sections from the localities indicated by letters that are too close together to indicate separately. Vertical bars indicate magnitude of error due to compaction corrections; horizontal bars are magnitude of error due to uncertainty in stratigraphic correlations. The term β indicates the amount of extension of the lithosphere during the stretching (rifting phase). The abbreviation m.y. indicates age in millions of years. The good fit between the points for tectonic subsidence and the thermal contraction model curves is evidence that the subsidence mechanism in the ancient margin was thermal. All vertical scales represent depth in miles (kilometers). (*After G. Bond and W. C. Pitman, III, The evolution of sedimentary basins, Lamont-Doherty Geological Observatory Yearbook 1981–1982*)

Identification of subsidence mechanisms. The mechanical or thermal subsidence that forms a sedimentary basin, referred to in general as the tectonic subsidence, is distorted by nontectonic processes, the most important of which are compaction, which is the expulsion of water from the sedimentary layers, and the additional subsidence caused by the weight of accumulating sediment. The effects of these nontectonic processes must be quantitatively removed from the measured thickness of the sediment column before tectonic subsidence curves can be constructed and the subsidence mechanism identified. The effects of compaction are removed by calculating the original thicknesses and densities of sedimentary layers by using either porosity-depth profiles from wells drilled into the basin or empirical porosity-depth equations derived from published data. Additional subsidence due to the weight of sediment is calculated and removed by using equations that express an isostatic equilibrium between the thickness, elevation, and density of the sedimentary layers, the density of the mantle (material below the lithosphere), and the strength of the lithosphere. A graph of the tectonic subsidence versus time produces the tectonic subsidence curve, the shape of which can be interpreted directly in terms of mechanical or thermal subsidence mechanisms.

Figure 2 shows the location of strata deposited in the ancient passive margin of western North America.

Examples of the tectonic subsidence curves for thermal and mechanical basins are given in Figs. 3 and 4. Tectonic subsidence curves with a distinct thermal form (Fig. 3) were constructed from an ancient passive margin in western North America that

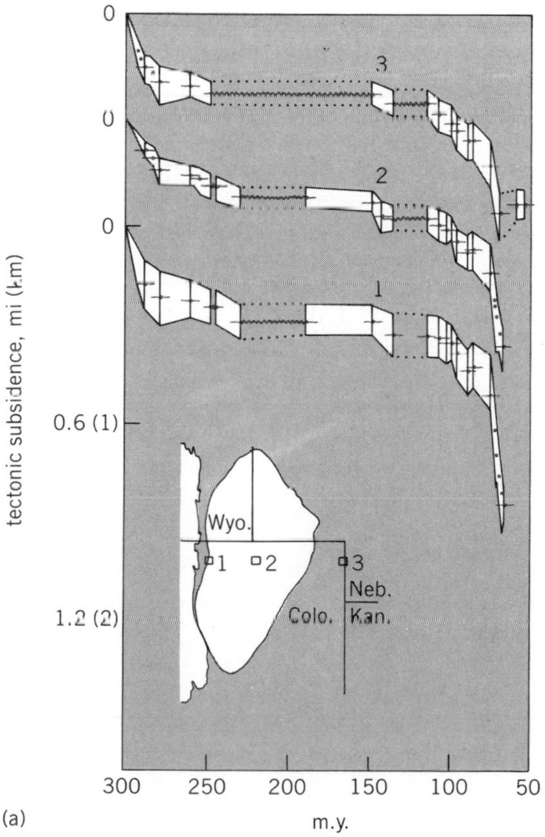

(a)

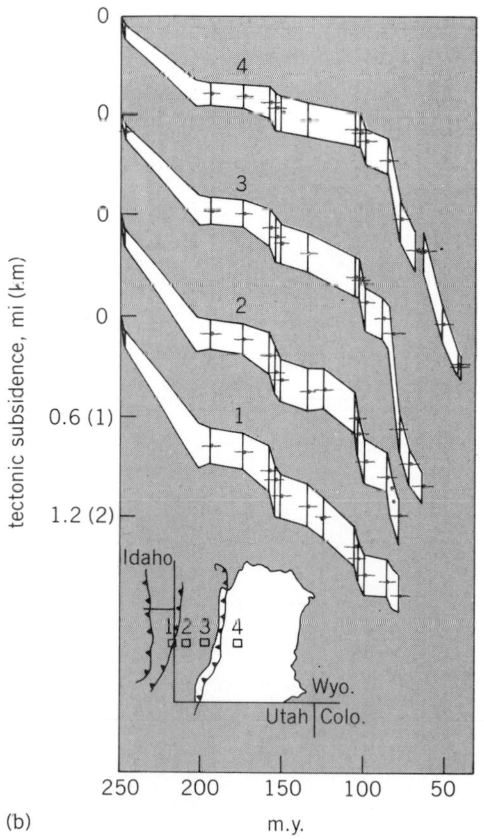

(b)

Fig. 4. Tectonic subsidence calculated from two foreland basins in western North America: (*a*) Denver Basin and (*b*) Green River Basin. The abrupt changes in the rate of subsidence are characteristic of a mechanical subsidence mechanism produced by episodes of thrusting in the orogenic belt to the west. The value on the vertical axis, tectonic subsidence, is equivalent to depth in miles (kilometers).

subsided between about 550 and 450 million years ago. The tectonic subsidence curves with a characteristic mechanical form (Fig. 4) are from two foreland basins in western North America that formed between about 160 and 50 million years ago during a period of major orogeny in the mountain belt to the west. Figure 3 shows that the curves from the passive margin have a smoothly decaying form that corresponds closely to the form of curves calculated from a model of postrift thermal contraction, whereas the curves from the foreland basin (Fig. 4) have a much more complex form characterized by alternating episodes of rapid and slow rates of subsidence. The episodes of rapid subsidence are broadly correlative with the major periods of thrust faulting in the orogenic belt to the west. The contrast between the smoothly decaying form of subsidence in the passive margin and the episodic subsidence in the foreland basin is one of the most characteristic differences between thermal and mechanical mechanisms of basin subsidence.

Sea-level changes. In certain cases, the form of tectonic subsidence curves can be used to identify eustatic (global) sea-level changes. So far, the most encouraging results have come from passive margins. Sea-level changes during the past 160 million years have been identified from tectonic subsidence curves for the modern passive margins off the east coast of the United States and off the northwest coast of Africa. Much earlier changes in sea level, between about 540 and 480 million years, have been inferred from the tectonic subsidence curves constructed from the ancient passive margin in western North America.

Figure 5 is an example of an inferred global or eustatic sea-level change (ΔSL) calculated from tectonic subsidence curves for the ancient passive margin in western North America (Fig. 3). The calculation of ΔSL is based on the observation that the tectonic subsidence for the ancient margin cannot be fit exactly to the model subsidence curves (see Fig. 3). If the misfit is entirely a result of a eustatic change in sea level, the downward deflection of the tectonic subsidence from the model curves between 545 and about 510 million years records a sea-level rise, that is, an increase in water depth relative to that predicted by the thermal subsidence model. Similarly, the upward deflection of the tectonic subsidence between about 510 and 480 million years records a sea-level fall, that is, a decrease in water depth relative to that predicted by the subsidence model. The sea-level change (ΔSL) in Fig. 5 is calculated from the magnitude of the misfit between the tectonic subsidence and the model curves. The result supports the assumption of a eustatic sea-level change because the curves for ΔSL have approxi-

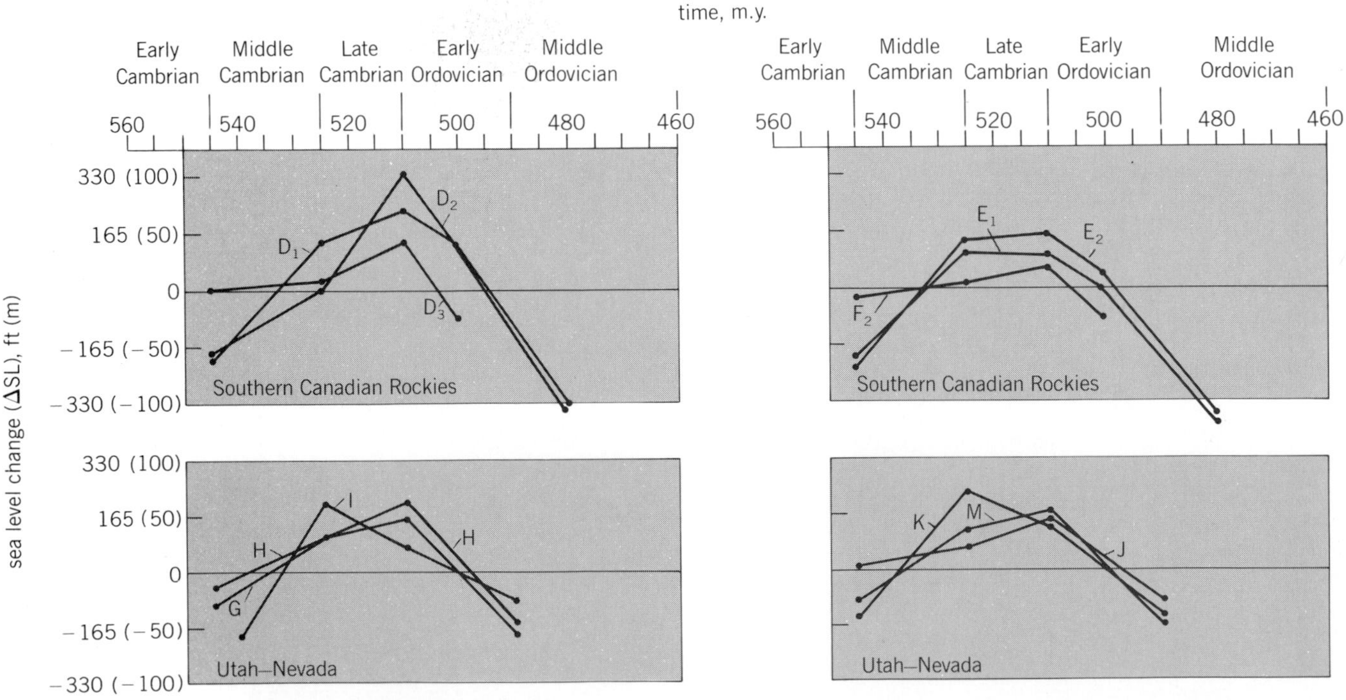

Fig. 5. Apparent sea-level curves that were calculated from the misfit observed in Fig. 3 between the tectonic subsidence for the early Paleozoic miogeosyncline in western North America and the subsidence predicted by a model of postrift thermal contraction in a passive margin. Letters D–M correspond to curves D–M in Fig. 3 and locations D–M in Fig. 2. Subscripts refer to different sections from the locality designated by the letter that are too close together to indicate separately in Fig. 2. The zero line corresponds to the position of the model curves relative to the points for tectonic subsidence in Fig. 3. The zero line does not correspond to present-day sea level. Note the similarity in form and timing of the apparent sea-level change (ΔSL) in the 12 locations extending along the ancient passive margin. This is considered good evidence that the curves represent a eustatic or global sea-level rise and fall between about 540 and 480 million years ago.

mately the same timing and a similar magnitude over a distance of more than 930 mi (1500 km) along the ancient passive margin. In addition, the inferred sea-level rise between about 540 and 510 million years ago corresponds closely in time to a major episode of marine flooding that has been recognized on several continents, and the inferred sea-level fall between 510 and 480 million years ago occurs at the same time as a major episode of emergence of the continents. Additional studies are currently under way to determine if the same sea-level curve can be derived from subsidence curves for ancient passive margins elsewhere in North America and on other continents.

For background information *see* BASIN; CONTINENTAL MARGINS; STRUCTURAL GEOLOGY; TECTONOPHYSICS in the McGraw-Hill Encyclopedia of Science and Technology.

[GERARD C. BOND]

Bibliography: C. Beaumont, Foreland basins, *Roy. Astron. Soc. Geophys.*, 54:291–329, 1981; G. C. Bond and M. A. Kominz, Construction of tectonic subsidence curves for the early Paleozoic miogeocline, Southern Canadian Rocky Mountains: Implications for subsidence mechanisms, age of breakup and crustal thinning, *Geol. Soc. Amer. Bull.*, in press; D. McKenzie, Some remarks on the development of sedimentary basins, *Earth Planet. Sci. Lett.*, 40:25–32, 1978; A. B. Watts, The U.S. Atlantic continental margin: Subsidence history, crystal structure and thermal evolution, *Amer. Ass. Petrol. Geol. Continuing Educ. Course Note Ser.*, no. 19, 1981.

Telephone

An increasing number of large businesses are acquiring private automatic branch exchange (PABX) systems, replacing in many cases manual or semiautomatic (PBX) systems. The newer systems are controlled by computers (microprocessors) and contain features that were heretofore impractical. This article reviews the development of telephone switching systems, leading to theadvent of computer-controlled systems, and discusses the application of computer control to the PABX and to related systems.

Development of switching systems. In the early years of the telephone, as more people started using the service, a problem arose because each call was routed by hand. An operator had to connect a wire from a subscriber jack to the called party jack. The longer distance the call, the more people became involved in placing it. The cure was the automatic exchange. This device automatically routed a telephone call based on a number dialed by the subscriber.

As more subscribers and more cities installed telephones, the number sequences became more complex. The first numbers consited of four digits which indicated the routing in a local switch. This local switch is called a CO (central office). Three digits were added as the system allowed direct dialing between COs. The three digits identified the CO to which the called party was connected. A digit "1" was added to indicate that the call was a toll call to a city located within the same region. As the system grew, three-digit numbers called area codes were added. The system has grown to the point that country codes are now available to dial many places throughout the world. Many cities in the United States can reach a foreign number by dialing 011, the country code, city code, and the subscriber's number.

The original switches that handled the dial information determined the call routing by comparing the received digits to a wire logic comparator circuit. As the dialing information expanded, the use of wire logic became impractical. With the advent of electronics, dialing information came to be handled by simple electronic logic circuits. As the dialing information and routing became more complex, the exchanges were designed by using computers for logic control. Computers can be updated by data tables that update their memories, where as logic circuits, whether wire or electronic, must be rewired. This feature allowed the exchanges to be updated more quickly, giving them greater flexibility.

Computers began to open the field to new and more powerful switching systems. The exchange required only a small percentage of the available computer power, leaving spare computing power available for new features.

Computers provided a new way to assign subscriber telephone numbers. In the early days of telephone systems the subscriber number indicated the actual wire position in the CO to which the subscriber telephone was connected. To change the location of the subscriber without changing the subscriber number required relocating the switch input wires to another pair of wires coming from the new location. This was time-consuming and in many cases impractical. In a computer-controlled switch, two numbers are used. The first number is the subscriber telephone number. The second number is the junction number (the location on the switch where the subscriber telephone is connected). To change the location of the subscriber, an update to the routing data table is the only operation required. No rewiring is needed at the CO. This usually means that a subscriber location can be changed from one location to another in a just a second. The exchange can now change a subscriber location to a different location and return it at a present time. This feature is available to subscribers that are connected to an electronic switching system (ESS), a type of computer-controlled switching system.

PABX features. As the telephone became more popular, it became more important for businesses to equip themselves with large automated telephone systems called private automatic branch exchanges. As businesses grew, it was discovered that they required only a small number of outside lines and a larger percentage of internal cross-links (a cross-link is used to connect telephones together in a

PABX). It became too costly to provide each employee with a private line to the CO.

The PABX was then used to provide interoffice communications as well as interface to the CO. The computer-controlled PABX allowed many features to be incorporated in the office, including call forwarding, camp on, hold, conference call, call pickup, call waiting, speed dialing, hunt group, and hot-line operation.

Call forwarding allows a telephone call to be transferred to a different number under a programmable set of conditions. The conditions could be transferred if busy, no answer, busy and no answer, or during a certain time of day.

Camp on is used to call a party back and place a call to a number that was busy.

The call hold feature allows single-line phones to place a call on hold, allowing the subscriber to place another call or answer a call that is waiting.

A conference call allows the subscriber to add more than two persons to a telephone conversation. The ESS performs this function by making multiple subscriber connections.

Call pickup is used by one subscriber to answer calls directed to another subscriber. In an office, this feature could be used by a person to answer a call that is directed to another phone.

Call waiting is a feature that provides a tone or other indication, while the subscriber is in a telephone conversation, to tell the subscriber that a call is waiting.

Speed dialing is a feature provided in microprocessor PABXs that allows the subscriber to enter a two-, three-, or four-digit code to dial a preprogrammed number. The microprocessor PABXs can also be set up to be programmed for a special set of speed dialing numbers that are available to only certain subscribers. All of these attributes are contained in the data look-up table that the microprocessor refers to each time a call routing operation is required.

Hunt group is supplied as a means of finding alternate lines to be used to complete a call. An incoming hunt group can transfer incoming calls to different extensions based on a preprogrammed set of routing rules. If a line is busy, the PABX will hunt for the next extension available. If a call is not answered by a certain number of rings, programmed into the data table, the PABX will route the call to another extension. Outgoing hunt groups could use different types of long-distance telephone services based on expense to place the call and the availability of a circuit on the various services. This alternated path routing is determined by a preprogrammed set of rules contained in the PABXs data table.

The PABX can be equipped with a hot-line service, which can be assigned to certain VIP telephones in the system. When a VIP telephone places a hot-line call, the calling party is connected to the called party even if the called party line is busy with another call.

The microprocessor PABX provides new features because it is controlled not by hardware but by software. Software can be changed by reprogramming ROMs (read only memories) rather than by changing wiring, which is required to make a hardware change. The use of microprocessor systems also allows information-processing functions to be incorporated in the switching processor. These functions could be used to check a dialed number for errors before placing a call, or to determine the most economical way of placing the call.

Data transmission. Future telephone systems will be able to transmit data as well as receive information from standard telephones. Several telephone systems are available that can forward digits to a larger computer system that, in turn, can produce synthesized speech to provide information without the use of a special data terminal.

Digital telephone systems. With the introduction of digitally controlled switches, a new fully digital PABX was designed. This PABX system provides faster telephone routing. (Tone systems require at least 10 milliseconds to determine a valid dialed digit whereas digital systems can send a dialed digit in several microseconds.) Digital PABXs also provide vitually noise-free operation. This greatly improves the clarity of a telephone conversation.

The first telephone systems were analog. Analog refers to a method of converting voice energy to an electrical signal that varies in amplitude and frequency. This signal can be sent over a wire and reproduced at the receiving end by modulation of the air to reproduce the sound pressure and frequently of the original voice energy. The digital PABX converts this analog waveform into 1's and 0's (1 = on state, 0 = off state). When the analog signal is converted, a digital signal is produced that is the value of the analog waveform, in binary (on and off data), at the instant the measurement was made. Several advantages are gained through the conversion of the analog signal to digital. The analog signal can be distorted by noise whereas the digital signal is immune to noise. The digital telephone system can provide virtually noise-free long-distance telephone calls. The digital system also allows several conversations to be sent over a pair of wires that could be used by only one subscriber in an analog system.

Digital telephone systems provide better telephone service for less cost. Rural America has been the largest recipient of this new technology. Remote areas of the United States can now be provided with telephone service as economically as the big cities because only one pair of wires needs to be run between subscribers.

To provide more economical service, call concentrators have been installed in the field. These units connect many subscribers to a small number of telephone lines. The microprocessor is used in the field to interface the subscriber telephone with telephone lines going to the CO. The average home uses the telephone for only a very short period in each month. In the analog system, each pair of wires going to a subscriber was used by only that subscriber. With the digital call concentrator, a set of 10 lines can handle the requirements of as many as 30 subscribers.

Other microprocessor applications. Microprocessor-controlled switches can provide other features besides call routing. The PABXs can provide the operator with the telephone number of the calling party. When the telephone system is used in a motel, it can identify the room number of the calling party. The system can automatically accept a charge card number. The system is limited only by the avalable microprocessor memory and the imagination of the designers. As digital memory becomes less expensive, telephone systems will become even more capable.

Future microprocessor telephone systems will provide subscribers with features that will enhance their communications needs. The mobile cellular telephone system will allow a person to be contacted by telephone anywhere in the United States. Each subscriber will be assigned a special telephone number. The telephone will contain a small transmitter that will send out a weak radio-frequency signal. This signal will be picked up by a receiver located close by. When the subscriber moves to a different location, the cellular telephone system will inform the next receiver located in another nearby location. The telephone call will be automatically routed from site to site as required to maintain the telephone conversation. This will mean that subscribers can take their telephones anywhere and still be able to receive telephone calls directed to them. With the use of satellites to cover areas not covered by local receivers, a subscriber will never be out of touch with his or her telephone calls. This concept is currently being applied to domestic commercial airplanes. *See* MOBILE RADIO.

For background information *see* DATA COMMUNICATIONS; MICROPROCESSOR; PULSE MODULATION; SWITCHING SYSTEMS (COMMUNICATIONS); TELEPHONE PRIVATE BRANCH EXCHANGE (PBX); TELEPHONE SERVICE; VOICE RESPONSE in the McGraw-Hill Encyclopedia of Science and Technology.

[RICHARD H. COLEMAN]

Tooling

Recent developments in tooling involve the use of new coating materials on cutting tools. This article discusses two of these materials, titanium nitride and silicon nitride. Thin titanium nitride coatings can substantially increase the lives of high-speed tools, while silicon nitride holds promise of providing cutting tools that can be used at elevated temperatures as well as at high cutting speeds.

Titanium nitride coatings. Thin layers of titanium nitride can be applied to high-speed steel cutting tools in order to substantially increase the service life of the tool. Research shows that the inert titanium nitride reduces the sticking friction between the cutting tool and the metal chip being produced. Consequently the temperature of the cutting tool is greatly reduced, and the cutting speed (for a given tool life) can be increased, in some conditions by as much as 200%. Titanium nitride coatings are particularly useful for improving the service life of drills. The chips being machined move along the coated drill flutes more easily. At present, the application of titanium nitride layers approximately doubles the cost of the drill, but because speeds can be significantly increased there is a net saving in final production costs.

High-speed steels. Although cemented carbide cutting tools have replaced high-speed steel in many turning and milling operations, high-speed steels continue to play an important role in production, especially for cutting operations in which sudden impacts or torques may be imposed on the cutting tool. Drilling, tapping, and sawing operations are typical examples. Now that titanium nitride layers can be applied to these steels, their application may increase. Essentially, high-speed steels are high-carbon steels containing significant percentages of tungsten or molybdenum, as well as chromium, vanadium, and sometimes cobalt. When heat-treated for use as cutting tools, the structures consist of alloyed martensite strengthened by precipitation-hardening carbides formed during tempering. High-speed steels can be manufactured by conventional casting followed by working and shaping, or by powder met-

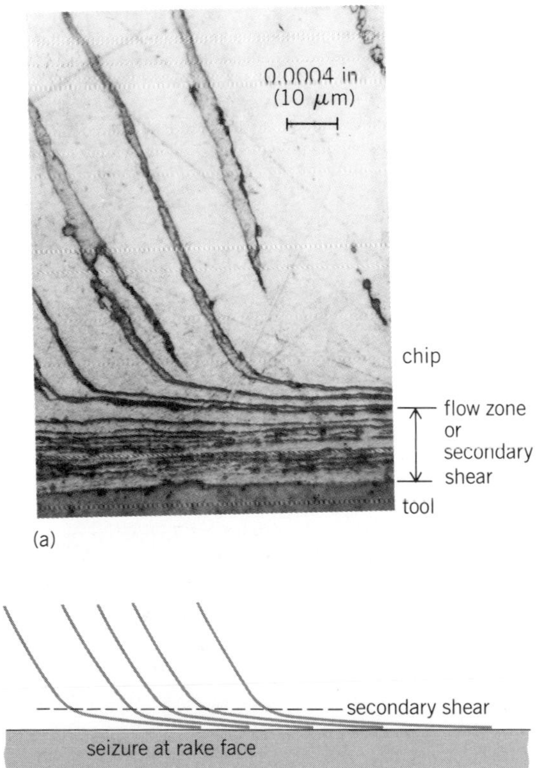

Fig. 1. Metallographic section showing the welding effect between chip and tool in the metal cutting process: (*a*) photograph and (*b*) diagram.

allurgy techniques. Powder metallurgy high-speed steels have greater homogeneity than the wrought products. In either case, the uncoated cutting tool consists of an iron-based matrix with dissolved and some undissolved primary carbides.

Seizure effect. During any metal machining process, very high normal loads are applied by the chip to the face of the tool. The chip moves over the tool with a high rubbing velocity (also cleaning the tool very effectively), and consequently the temperatures formed are very high. All these conditions encourage solid-state welding between the chip and tool, especially when commercial steels are being machined with high-speed steel tools. Despite the different alloys in the steel chip and the steel tool, there is a natural propensity for the iron-based materials to exhibit solid-state welding. Figure 1 shows that in commercial cutting the frictional interactions are so intense that a flow zone, or sticking zone, occurs in the lower face of the chip. This sticking friction or seizure is similar to the events that occur when an automobile engine seizes, or the processes that occur in the friction welding (inertia welding) operation.

Anything that can be done to reduce the seizure effect is beneficial to tool life. For example, free machining additions (lead or manganese sulfide) can be added to the work material to reduce its shear strength in the flow zone and thus reduce the stresses and temperatures applied to the tool. Alternatively, inert coatings can be applied to the tool, and when this is done the steel chip cannot weld as strongly to the tool. There are a variety of coatings that have been developed to achieve this. Aluminum oxide, titanium carbide, and titanium nitride are three of the more common coatings that have been developed.

Coating methods. Physical vapor deposition is at present the most common way of applying the titanium nitride to the high-speed steel substrate. It is a low-temperature process, and it can be applied to the tools without upsetting the previous heat treatment and tempering, carried out to achieve the optimum properties of the underlying high-speed steel substrate. There are a variety of methods for creating a reactive metal vapor that, under vacuum, will adhere to the substrate. Both evaporation and sputtering processes can be used, and then, by applying a suitable field environment, the coating adheres to the substrate by ion plating.

Tool life improvement. The improvement of tool life with inert coatings has been well documented. Recent drilling experiments involved the use of titanium nitride–coated drills on AISI 4340 steel. For example, to obtain a fixed tool life of 100 in. (250 cm) of cut material, it was shown that uncoated high-speed steel drills can be used at 50 surface feet (5 surface meters) per minute, whereas under the same cutting conditions titanium nitride–coated tools can be used at 200 surface feet (60 surface meters) per minute for the same tool life. Thus, although the coating process approximately doubles the cost of the drill, these dramatic increases in cutting speed lead to such an improved throughput of parts production that the overall economy of the manufacturing process is increased.

Some of the reasons behind these improvements in tool life are shown in Fig. 2. A metallographic technique has been developed which uses a loss of temper in the high-speed steel tools to monitor the temperature changes. This technique has been used to investigate the different temperature profiles in uncoated and coated high-speed steel turning tool inserts. Figure 3 shows the coating adhering to the surface of the high-speed steel substrate and the loss-of-temper effect in the high-speed steel tool. This is the darkened arc; it correlates with the temperature contours in Fig. 2. When machining AISI 1045 steel, the results in these tests show that the cutting speed can be approximately doubled (from 148–328 ft or 45–100 m per minute) with coated tools, while still maintaining much the same temperature profile within the tool.

The results can also be evaluated by considering the cutting forces on the tool. The cutting forces show that the titanium nitride coatings reduce the frictional forces on the tool. This is associated with lower temperatures and rates of wear. The coatings

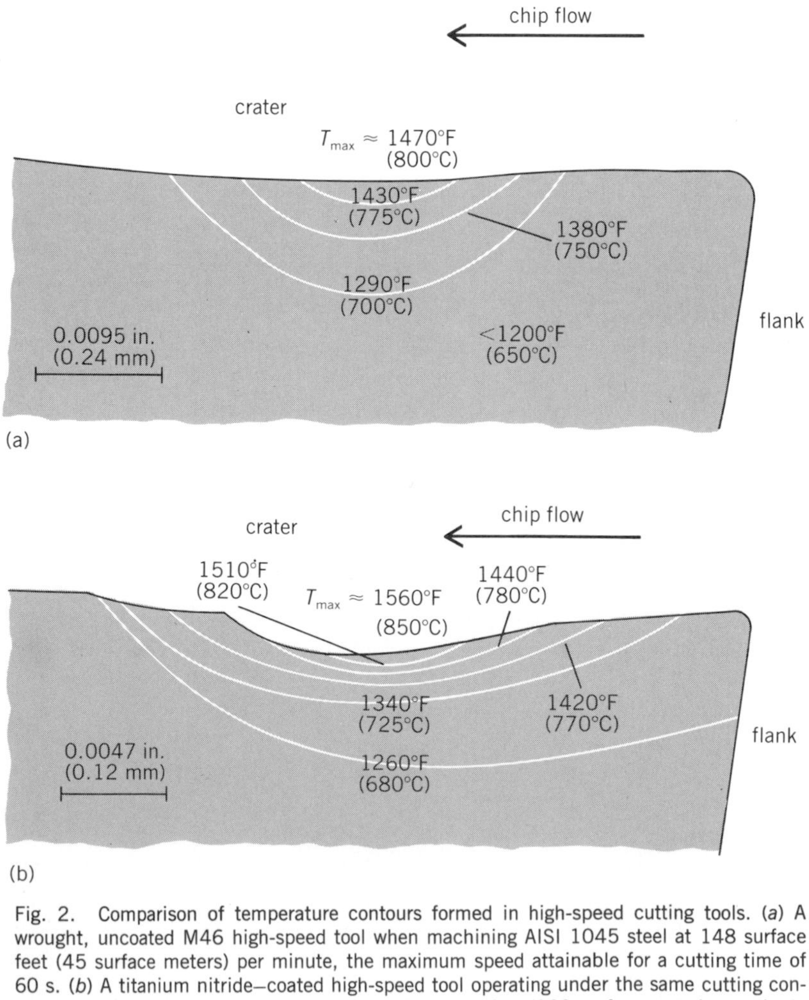

Fig. 2. Comparison of temperature contours formed in high-speed cutting tools. (*a*) A wrought, uncoated M46 high-speed tool when machining AISI 1045 steel at 148 surface feet (45 surface meters) per minute, the maximum speed attainable for a cutting time of 60 s. (*b*) A titanium nitride–coated high-speed tool operating under the same cutting conditions; the cutting speed is increased to 328 surface feet (100 surface meter) per minute for the 60-s time.

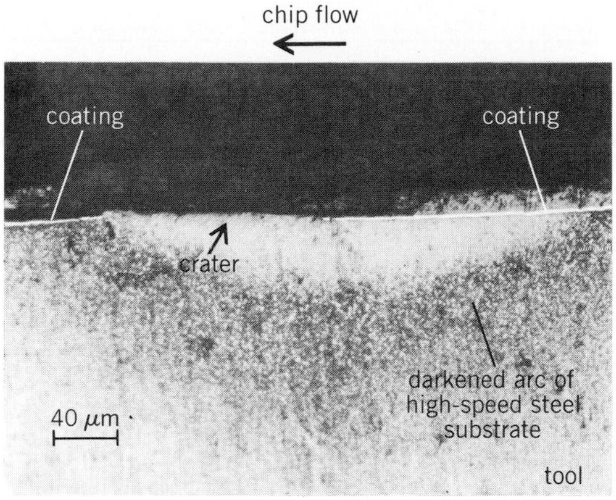

Fig. 3. Metallographic section through a coated high-speed steel cutting tool. The coating is the thin white line. Toward the end of this test the coating has been worn away in the crater area, and at this point temperatures begin to accelerate rapidly.

themselves do not act as a low-conductivity thermal barrier. Calculations show that the coating is so thin that it cannot prevent the heat from flowing through the surface of the tool into the bulk. The contribution of the titanium nitride is to reduce the friction with the chip and thereby reduce the heat flux at the top surface of the coating. The titanium nitride is significantly harder than the underlying high-speed steel substrate, and thus tool wear by abrasion is reduced. Its inertness, from a chemical point of view, prevents any possible diffusion of tool material into the chip. All wear processes are influenced by temperature, and thus the reduction of temperature is a key feature.

It can be shown that at the upper limit of cutting speed the high shear stresses applied to the coating begin to break down its coherency in the region of maximum temperature and stress. A crater may be formed in the tool, a small distance (0.02–0.04 in. or 0.5–1 mm) away from the cutting edge. This region is indicated with an arrow in Fig. 3. It has been found that as the coating begins to break down, tool wear then accelerates dramatically; once the high-speed steel substrate and the chip are in intimate contact, the temperatures in the tool rise very rapidly. However, the presence of some coating on the front part of the rake face is always beneficial. Even when coated drills are used, the performance of drills that have been reground and have lost their coating on the drill end is still improved over uncoated drills. The coating remaining in the flutes still reduces the temperature in the drill and hence improves life. Titanium nitride coatings are an important contribution to the performance of cutting tools, especially high-speed steel drills.

[PAUL K. WRIGHT]

Silicon-nitride-based cutting tools. The development cycle for cutting tool materials has run the gamut from very tough materials (steel) to extremely hard materials (diamond). This trend, to a great extent, is the direct result of the demand for higher productivity, which can be achieved either by cutting at higher speed or by extending tool life. Since the 1900s an exponential increase in productivity capability, as measured by the cutting speeds available, has been achieved (Fig. 4). It is anticipated that this trend will not continue. Therefore, the need for cutting tool materials with elevated temperature strength and chemical inertness, capable of operating at high cutting speeds, is becoming critical. Ceramic materials are the prime candidates to fulfill these requirements.

Today, the predominant materials used for cutting tools are tungsten carbide and cobalt (WC–Co) based cemented carbides. The application range of cemented carbides has been extended to a maximum of approximately 1500 surface feet (460 surface meters) per minute with alumina-coated (Al_2O_3) tools. Although convertional ceramic cutting tool materials (mainly alumina-based) have been available for over 40 years, and can attain cutting speeds well over 1500 surface feet (460 surface meters) per minute, they have found limited success due to their inherently poor thermal shock resistance and low fracture strength. On the other hand, although silicon nitride (Si_3N_4) has long been recognized as one of the toughest ceramic materials available, its potential as a cutting tool material has not been fully realized.

Silicon nitride materials. Silicon nitride is a compound of highly convalent bond character and as such requires additions of sintering aids to achieve theoretical densities. For this purpose, many compounds (mainly oxides and nitrides) have been investigated, yielding a number of materials with a wide spectrum of properties. The development of Si_3N_4 as potential material for cutting tools has been proceeding for over a decade, but it is only in the last few years that Si_3N_4-based cutting tools have found commerical applications. These tools can be categorized into three basic groups; Si_3N_4 [Si_3N_4 + sintering aid (MgO, Y_2O_3)]; SiAlON [Si_3N_4 + AlN

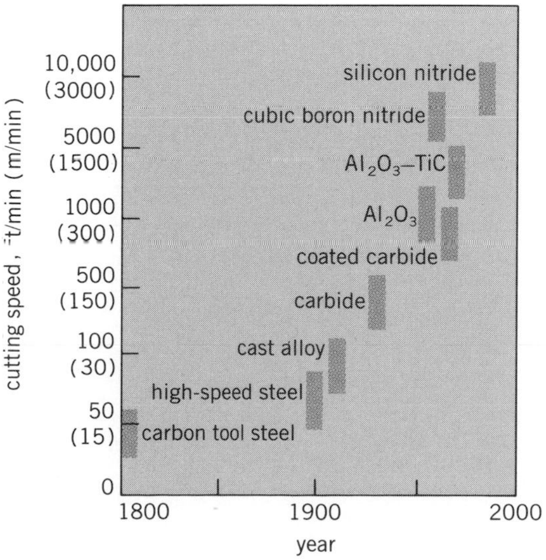

Fig. 4. Change in productivity via the introduction of new cutting tool materials.

Hardness and fracture toughness of ceramic cutting tool materials

Material	Years of introduction	Fracture toughness (K_{IC}), meganewtons/$m^{3/2}$	Hardness (H), giganewtons/m^2
Al_2O_3 sintered	1930–1931	2–2.5	15.0
Al_2O_3 hot pressed	1944–1945	—	—
Al_2O_3 + TiC hot pressed	1968–1970	3–3.5	17.0
Si_3N_4 + Y_2O_3 hot pressed	1960–1965	4–5.0	12.0
Typical SiAlON	1968–1975	4–5.0	12.5
Si_3N_4 + Y_2O_3 sintered	1972–1974	4–6.0	12.0
Si_3N_4 + Y_2O_3 + TiC composite	1979–1981	4–6.0	15.0

+ SiO_2 + Al_2O_3]; and composite [Si_3N_4 + sintering aid (MgO, Y_2O_3) + dispersed phase (TiC, and so on)].

The table compares the hardness and fracture toughness of Al_2O_3- and Si_3N_4-based cutting tool materials. Although all the Si_3N_4-based cutting tool materials are tougher, only the composite materials yield hardnesses approaching those of the Al_2O_3-based materials. In spite of their moderately lower hardness, these Si_3N_4 materials have exhibited higher abrasive wear resistance than the Al_2O_3-based materials. This is consistent with the empirical model that predicts the abrasion wear resistance of ceramic cutting tools to be directly proportional to the value of the expression $K_{IC}{}^{3/4}H^{1/2}$, where K_{IC} is fracture toughness and H the hardness of the cutting tool materials. Because these critical mechanical properties have been found to maintain their relative ranking at elevated temperatures, it is possible to predict the performance ranking of ceramic cutting tools in turning of gray cast iron, since in this case tools wear basically by abrasion (Fig. 5). In the machining of steels, it has been found that chemical interaction, diffusion, and adhesion predominate, and therefore the model is not directly applicable.

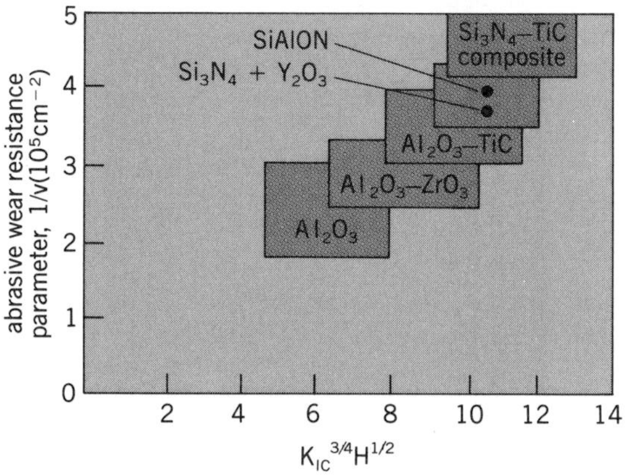

Fig. 5. Graph showing abrasive wear resistance of ceramic cutting tools as a function of hardness H and fracture toughness K_{IC}. Here ν = the volume of material removed per unit length of travel in an abrasive wear test.

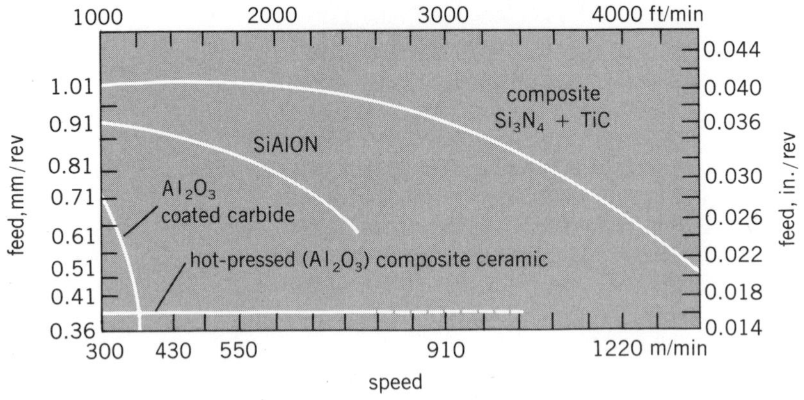

Fig. 6. Comparison between proven application range for gray cast iron machining of various cutting tool materials.

Potential. The potential for remarkable improvements in productivity (that is, increases in cutting speeds by as much as 500%) in gray cast iron through the application of Si_3N_4-based cutting tool materials has been clearly demonstrated by test results. Extensive data, especially on the composite Si_3N_4-based tools, have not only demonstrated considerable improvements in tool life over state-of-the-art ceramic tools, but also higher reliability due to their superior fracture toughness and shock resistance. These tool materials operate at high cutting speeds normally associated with ceramics and exhibit high impact resistance, leading to feed rates approaching those of ceramic-coated carbide (Fig. 6). The thermal shock resistance of Si_3N_4-based composites is sufficient to allow coolants to be employed during turning operations, a situation not sustained by Al_2O_3.

The potential of Si_3N_4-based materials to machine superalloys has yet not been fully realized, but initial results are very promising. Productivity rate increase up to 400% with consistent and predictable tool life has been obtained over conventional cemented carbides and Al_2O_3-based materials.

Although the present state-of-the-art Si_3N_4-based ceramic cutting tool materials have proven to be superior for machining cast iron and superalloys, their use in other applications (that is, steels, aluminum, copper, and so on) can be visualized, and are in the development stages.

Conclusion. While in their infancy, silicon nitride–based cutting tool materials have demonstrated that they possess the potential to meet the present and future challenges of high-productivity machining. Furthermore, due to Si_3N_4's excellent high-temperature stability, oxidation and thermal shock resistance, and the possibility of tailoring the structure and properties of the silicon nitride–based composite materials, these cutting tool materials may be second only to tungsten carbide–cobalt cemented carbides in their range of applications with a definite advantage in the attainment of higher productivity.

For background information *see* HIGH-TEMPERATURE MATERIALS; STEEL; TOOLING in the McGraw-Hill Encyclopedia of Science and Technology.

[V. K. SARIN; S. T. BULJAN; J. T. SMITH]

Bibliography: R. D. Baker, High-velocity cutting tools: Application guidelines, *Biennial International Machine Tool Technical Conference Proceedings, McLean, Virginia*, National Machine Tool Builders Association, pp. 487–528, 1982; S. T. Buljan and V. K. Sarin, Improved productivity through application of silicon nitride cutting tools, *Carbide Tool J.*, 14:40–46, 1982; W. E. Henderer, Performance of titanium nitride coated high speed steel drills, *Proceedings of the 11th North American Manufacturing Research Conference*, University of Wisconsin–Madison, pp. 337–341, 1983; V. K. Sarin and S. T. Buljan, *Advanced Silicon Nitride Based Ceramic for Cutting Tools*, SME Tech. Pap. MR83-189; E. M. Trent, *Metal Cutting*, 1977; P. K. Wright and E. M. Trent, Metallographic methods for determining temperature gradients in cutting tools, *J. Iron Steel Inst.*, 211(5):364–368, 1973.

Transducer

Recently, a new class of acoustic sensors has been demonstrated, and various sensor types are under development. These fiber optic sensors are based upon changes which take place in the characteristics of light traveling through an optical fiber when it is exposed to acoustic energy.

An optical fiber is a thin (approximately 100 micrometers in diameters), glass, wirelike waveguide formed from high-purity glass in such a way that light travels slightly slower at its center than at its edges. As a result, light coupled into the center of the fiber becomes "trapped" and can travel through great lengths of such fiber (in excess of tens of kilometers). Since its origin in the early 1970s, fiber optic technology has rapidly expanded into communication networks (such as telephonics) and is projected to eventually replace most conventional copper signal transmission lines. *See* OPTICAL COMMUNICATIONS.

Several years ago, researchers observed that optical fibers could be made susceptible to external perturbations such as acoustic vibrations. They found that by optimizing the material jacket or the mechanical mounting of the fiber, large sensitivity to acoustic waves could be obtained. Sensitive fiber optic acoustic sensor devices soon followed.

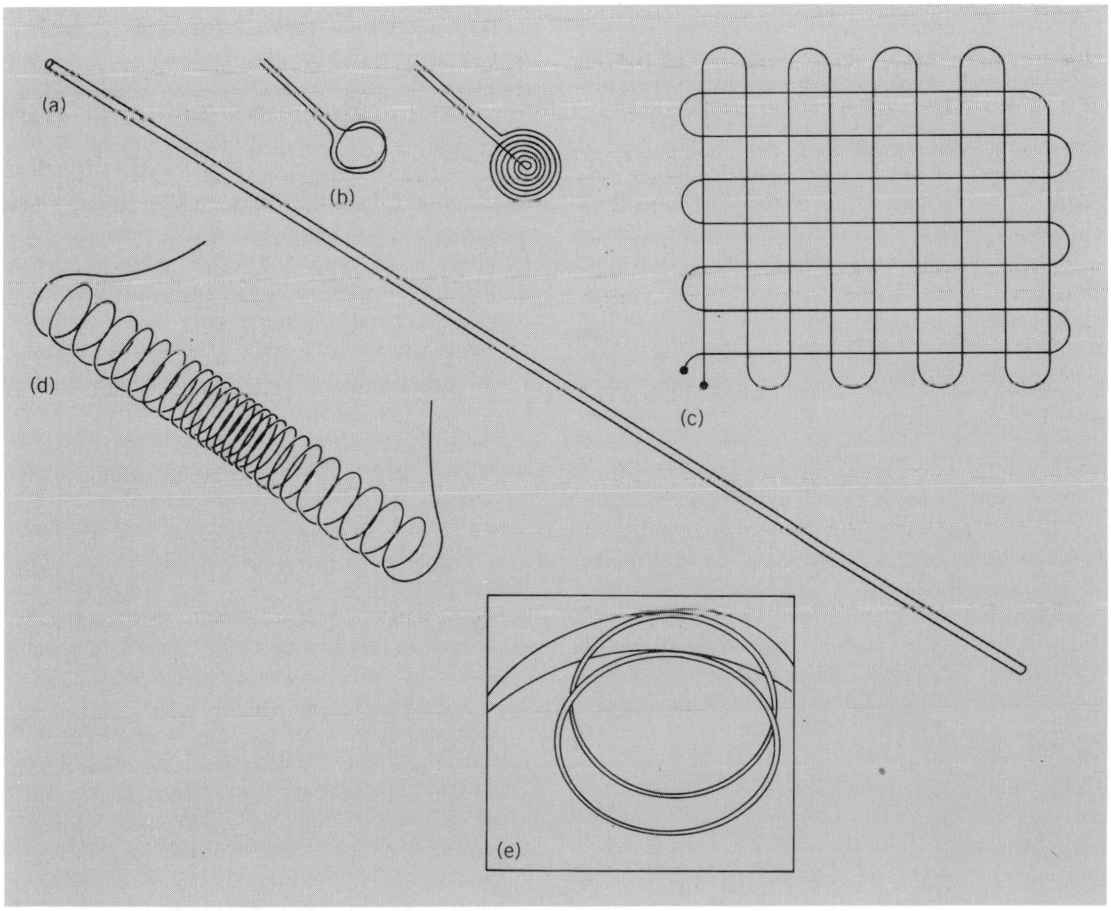

Fig. 1. Various fiber optic sensor arrangements, illustrating geometric versatility. (*a*) Line array. (*b*) Small loops and disks. (c) Area sensor. (*d*) Shaded element. (e) Gradient element.

Advantages of fiber optic sensors. Fiber optic sensors have a number of attractive properties. The sensing portion of the device as well as the signal leads back to some platform or station can be entirely optical, that is, devoid of electronic signals or power. As such, these elements are immune to electromagnetic interference, a problem which plagues conventional electric devices. There is no crosstalk between various sensor leads since, unlike conventional electric cables, the optical field does not extend outside the core of the fiber. Because there are no electric signals at the sensor, it can be used safely in hazardous environments such as explosive atmospheres. Dramatic sensitivities are possible, typically increasing linearly with the length of fiber used to form the sensor. *See* ELECTROMAGNETIC COMPATIBILITY.

Perhaps the most intriguing possibilities of fiber optic sensors lie in their geometric versatility which derives from the fact that the transduction takes place in a thin, long, bendable fiber. Figure 1 depicts some examples of useful geometries. Small, lightweight loops and disks can be readily formed whose sensitivity and environmental dependence can be tailored by appropriate choice of the fiber coating material (Fig. 1*b*). It is possible to cover large areas with a single sensor by laying the fiber in various patterns over the particular structure involved (Fig. 1*c*). In many applications, large signal-to-noise gains can be realized by spatially shading the sensor response, and this can be done quite simply with fiber optic sensors (Fig. 1*d*). As a final example, simple sensors which respond to the pressure gradient—rather than pressure—can be obtained by appropriate arrangement of signal and reference fibers (Fig. 1*e*).

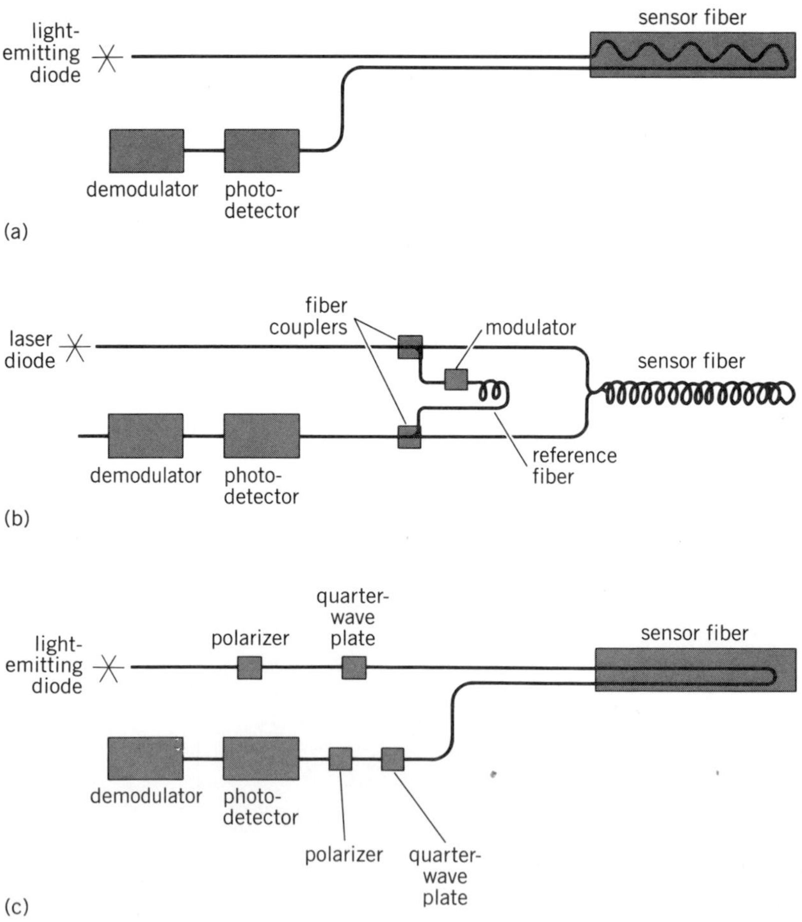

Fig. 2. Basic fiber optic acoustic sensor detection techniques. (*a*) Intensity detection. (*b*) Phase detection. (c) Polarization detection.

Types of detection. Although there is a wide variation in the manner in which various groups have implemented fiber optic acoustic sensors, all such sensor types can be broadly categorized into intensity, phase, or polarization detection devices. Each of these and the necessary components are diagrammed in Fig. 2.

Intensity detection. From a component viewpoint, intensity devices (Fig. 2*a*) are simple. Light from an incoherent source (such as a light-emitting diode) is coupled into an optical fiber. Typically, a multimode fiber is used which is mounted in such a way that acoustic waves change the mode power distribution exiting the fiber. This mode power modulation can be a redistribution of propagating core modes or a transition from core modes to radiation (loss) modes. In the former case, an appropriate mode filter is employed after the fiber. The light level transmitted through the fiber is then monitored by means of a photodiode. If T is the optical transmission coefficient for the specified optical modes through the device, and W_0 is the input light level to the fiber, then W_0T is the light power incident on the detector. If an acoustic wave of pressure ΔP changes the transmission by ΔT, the detector signal current i_s is given by Eq. (1), where q is the detector quantum efficiency, e is the electronic charge, h is Planck's constant, ν is the light frequency, and $\Delta T/\Delta P$ is the transduction coefficient of the sensor. The rms shot noise in a band Δf is given by Eq. (2).

$$i_s = \frac{qeW_0}{h\nu}\left(\frac{\Delta T}{\Delta P}\right)\Delta P \qquad (1)$$

$$i_N^2 = 2\Delta f e^2 q W_0 T/h\nu \qquad (2)$$

Thus, the minimum detectable pressure (for signal-to-noise of 1) is given by Eq. (3). Insertion of reasonable values for the parameters in Eq. (3) indicates that pressures as low as 10^{-9}lbf/in.^{2} (10^{-5} pascal) should be detectable. This is comparable to the threshold for human hearing.

$$P_{\min} = \left(\frac{2Th\nu\Delta f}{qW_0}\right)^{1/2}\left(\frac{\Delta T}{\Delta P}\right)^{-1} \qquad (3)$$

Phase detection. In a fiber phase sensor (Fig. 2*b*), light from a stable, coherent diode laser is coupled into a single-mode fiber. This light is split equally by an evanescent field fiber coupler into a sensing fiber and a reference fiber. The modulator shown in the reference arm indicates provision to phase- or frequency-modulate the light transmitted through that fiber or to frequency-modulate the light from the laser source. The two beams are subsequently recombined by a second fiber coupler and photodetected. Under the proper conditions, a signal results in the photodetector output which is proportional to the difference in optical phase, $\Delta\phi$, induced by the acoustic wave between the reference and sensing beams.

For homodyne operation (no external frequency or phase modulation of the reference beam), the photodetector output signal is given by Eq. (4). Here Ω

$$i_s = W_0 T \frac{qe}{h\nu} \left\{ \sum_{n=0}^{\infty} J_{2n+1}(\Delta\phi) \cdot \sin([2n + 1]\Omega t) \sin \phi_0 - \sum_{n-1}^{\infty} J_{2n}(\Delta\phi) \cdot \cos(2n\Omega t) \cos \phi_0 \right\} \quad (4)$$

is the acoustic frequency, t is the time, J_n is an integer order Bessel function, and ϕ_0 is the "static" phase difference between sensing and reference beams in the absence of the acoustic signal. As can be seen by examination of Eq. (4), stability of ϕ_0 or, equivalently, of path length, between reference and sensing arms is essential, and ϕ_0 should be $\pi/2$ to achieve maximum output at Ω. For most applications, environmental conditions prevent path length stability, and various demodulation schemes have been utilized to obtain stable signals. These include heterodyne–frequency-modulation discrimination with a Bragg modulator, phase-stabilized feedback loop with a fiber stretcher, synthetic heterodyning with an ac phase modulator, laser frequency modulation with path length mismatch, and passive quadrature detection.

To estimate the threshold detectability, consider Eq. (4) for the small-signal case ($\Delta\phi$ much less than 1) with $\phi_0 = \pi/2$. The photodetector current for a pressure wave of ΔP is given by Eq. (5), where

$$i_s = W_0 T \frac{qe}{h\nu} \left(\frac{\Delta\phi}{\Delta P} \right) \Delta P \quad (5)$$

$(\Delta\phi/\Delta P)$ is the transduction coefficient of the sensor, namely the change in optical phase per unit pressure induced in the sensing fiber by the acoustic wave. Again assuming a signal-to-noise ratio of 1 and a noise level as given by Eq. (2) yields Eq. (6).

$$P_{\min} = \left(\frac{2Th\nu\Delta f}{qW_0} \right)^{1/2} \left(\frac{\Delta\phi}{\Delta P} \right)^{-1} \quad (6)$$

Inserting reasonable values for these parameters indicates that with 300 ft (100 m) of sensing fiber, 10^{-11}lbf/in.^{2} (10^{-7} pascal) could be detected. These levels are significantly below those achievable with conventional devices.

Polarization detection. Fiber optic acoustic sensors based upon polarization effects for transduction (Fig. 2*c*) are, in essence, a special case of the phase interferometers in that orthogonal polarization states are compared after traveling through the same fiber, rather than parallel polarization phases being compared after transit through different fibers. On the one hand, environmental effects are minimized since both states travel in the same fiber. On the other hand, the fiber must be fabricated or mounted in such a way that the phases of orthogonal polarization states are modulated unequal amounts by the acoustic wave.

Fiber optic acoustic transduction. Unlike conventional acoustic transduction principles (for example, piezoelectricity), there is a great degree of flexibility with which optical fiber transduction effects can be optimized. This is especially true in the case of interferometric phase devices. Here, the acoustic transduction effect is simply the change of the travel time of light propagating in the optical fiber. This results from the combined effects of direct fiber length changes and strain-induced changes in the refractive index of the glass. The magnitude of each of these effects can be controlled by choosing the thickness and elastic parameters of the material deposited over the glass fiber. A wide variety of materials can be readily deposited, including plastics, rubbers, epoxies, metals, ceramics, and other glasses, to arrive at the specific response required. Sensor fibers with high, low, broadband, narrow-band, temperature-stable, and pressure-stable acoustic sensitivities have been demonstrated.

Intensity transduction principles by comparison are quite diverse, including not only those involving effects within an optical fiber itself, but also "hybrid" schemes such as coupling from one fiber to another and transmission through mechanical apertures. Almost all of those involving transmission effects within a fiber, however, utilize microbend mode coupling in a fiber. This technique utilizes multimode fibers which are bent axially in a sinusoidal fashion by a mechanical deformer. Transduction effects are maximized by designing specific spatial bending periodicities which match the inverse differential wave number of the coupled modes. Unlike the case for phase transduction, it is not yet possible to apply simple coatings to microbending fibers to tailor the transduction effect.

Polarization transduction has received the least research attention. A "hybrid" device has been demonstrated which utilizes a birefringent polyester material inserted between the input and output ends of two optical fibers. Another device utilizes an unbroken optical fiber in which the required birefringence is created by twisting the fiber about its axis. Finally, it has been demonstrated that at sufficiently high frequencies, ultrasonic waves can themselves induce birefringence and can thus be detected by utilizing a conventional nonbirefringent fiber.

For background information *see* OPTICAL FIBERS; PRESSURE TRANSDUCER; SOUND; TRANSDUCER; in the McGraw-Hill Encyclopedia of Science and Technology.

[J. A. BUCARO]

Bibliography: J. A. Bucaro et al., Fiber optic acoustic transduction, *Physical Acoustics*, W. P. Mason and R. N. Thurston (eds.), vol. 16, ch. 7,

1982; T. G. Giallorenzi, Optical communications research and technology: Fiber optics, *Proc. IEEE*, 66(7):774–780, July 1978; T. G. Giallorenzi, Optical fiber sensor technology, *IEEE J. Quant. Electr.*, QE18(4):626–665, April 1982.

Transuranium elements

Some atoms of elements 107 and 109, the heaviest detected so far, have been produced and identified by a new separation and detection technique sensitive enough to detect single decaying nuclei. Stable isotopes of chromium and iron were fused with bismuth to produce the new isotopes. The fused systems were made by cold fusion, a reaction where only one neutron is emitted. The production rates are as small as 1 out of 10^{11} nuclear reactions.

Identification of single decaying nuclei. The search for rare isotopes requires highly sensitive, universal, rapid detection methods. At the heavy-ion accelerator UNILAC, in Darmstadt, West Germany, a new method has been developed which is adapted to the production of new nuclei by fusion and which fulfills these criteria more than any previous technique (Fig. 1). In the fusion of nuclei in an incident beam with other nuclei in a stationary target, the conservation of momentum requires that all fused systems travel in a small cone surrounding the beam direction with a velocity smaller than that of the projectiles inducing the reaction. The fused nuclei, which are highly charged ions traveling with a few percent of the velocity of light, are separated from these projectiles by a velocity filter, a combination of electric and magnetic fields which can separate ions in a preselected velocity band. This "separator for heavy-ion reaction products" (SHIP) can suppress the primary beam by factors up to 10^{16}, and generates a beam consisting of those particles passing the filter in a given velocity window. This beam contains the fused systems, a background of degraded beam particles at a rate of about 100/s, and eventually other reaction products which have the preset velocity. The separation time is the time of flight through the filter and amounts to about 0.5 microsecond. The separation is independent of the chemical properties of the fused systems, which leave the SHIP with energies between 20 and 100 MeV. Forty percent of the fused systems are transported to a detector position.

The fused nuclei are detected by a system which makes full use of the fact that the energy obtained in the nuclear reaction has been conserved. The velocity of all particles in the selected-velocity bin of SHIP is measured by a time-of-flight detector. In the focal plane, the energy, time of arrival, and position of each particle are registered by an array of position-sensitive surface barrier detectors into which the particles are implanted. The radioactive decays of the implanted reaction products are registered by the same detector system. Energy, time of emission, and position of alpha particles and fission products following the implantation are registered. Time correlations of events in the position windows defined by the implanted reaction product allow reconstruction of the decay chain. The background in best cases is small enough to correlate still decays with half-lives up to 30 min. Events followed by decay chains can be detected unambiguously, allowing the identification of single decaying atoms, even if the production rates are of the order of one per week, a rate corresponding to a cross section 10^{-39} m^2.

Discovery of elements 107 and 109. All the identified elements beyond atomic number 100 can be synthesized in reactions between stable isotopes. Soviet scientists at Dubna were the first to use this method in place of the former technique of bombardment of actinide targets, which had been used by American scientists at Berkeley to synthesize elements through atomic number 106. The Soviet workers showed in 1974 that fermium can be made through the fusion of lead and argon. It was shown at Dubna and Darmstadt that all previously known elements can be made by bombarding ^{208}Pb and ^{209}Bi targets with beams of ^{40}Ar, ^{48}Ca, ^{50}Ti, ^{54}Cr, and ^{58}Fe. At Darmstadt it was shown that the excitation energy of the systems formed in the fusion reactions is low, between 15 and 25 MeV, and most of the isotopes are produced through evaporation of only one or two neutrons from the fused system. The fusion process is much colder than for actinide-based reactions. The reactions make it possible to produce the very fragile elements beyond atomic number 106, which have fission barriers between 1.5 and 2.5 MeV.

Figure 2 shows decay chains of the isotopes ^{262}107 and ^{266}109. For element 107, the "eka-rhenium" which was searched for in the prefission era, seven events have been found in the reaction $^{54}\text{Cr} + {}^{209}\text{Bi} \rightarrow {}^{262}107 + n$. For element 109, one event has been seen in the reaction $^{58}\text{Fe} + {}^{209}\text{Bi} \rightarrow {}^{266}109 + n$. The isotopes produced are odd-odd nuclei, which have a highly increased stability toward spontaneous fission. They are still alpha-active, thus allowing the application of the chain analysis technique. The alpha energies of the two isotopes

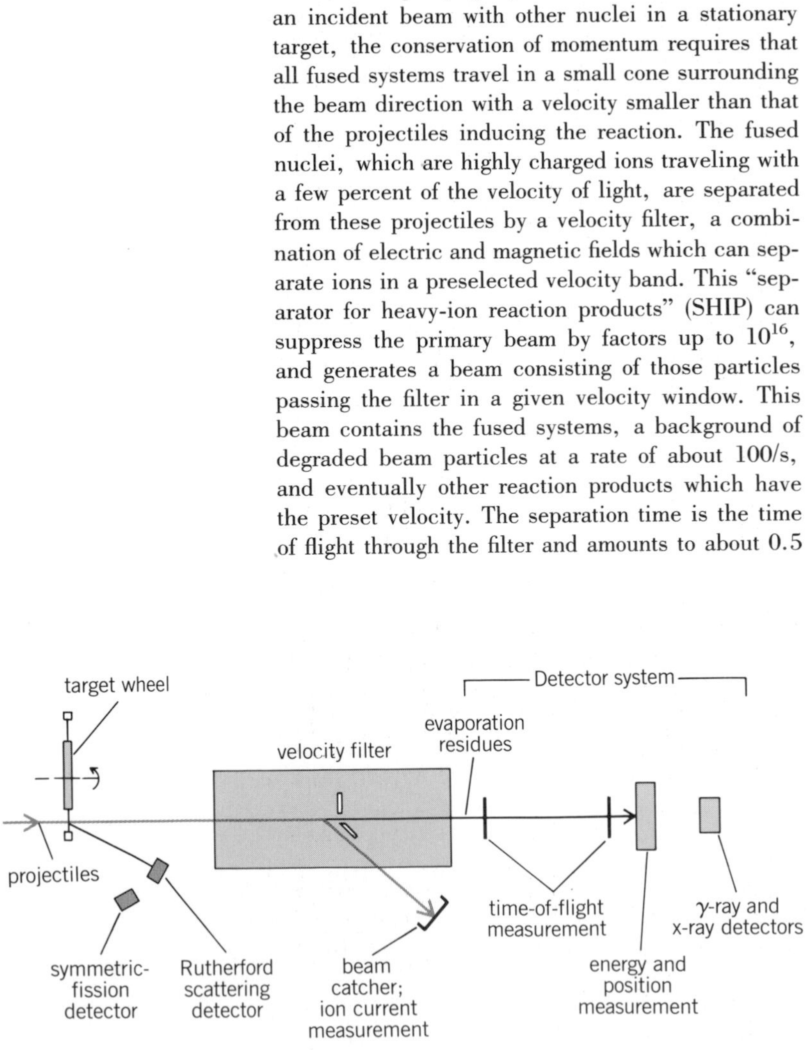

Fig. 1. Experimental setup to detect single decaying atoms produced through fusion.

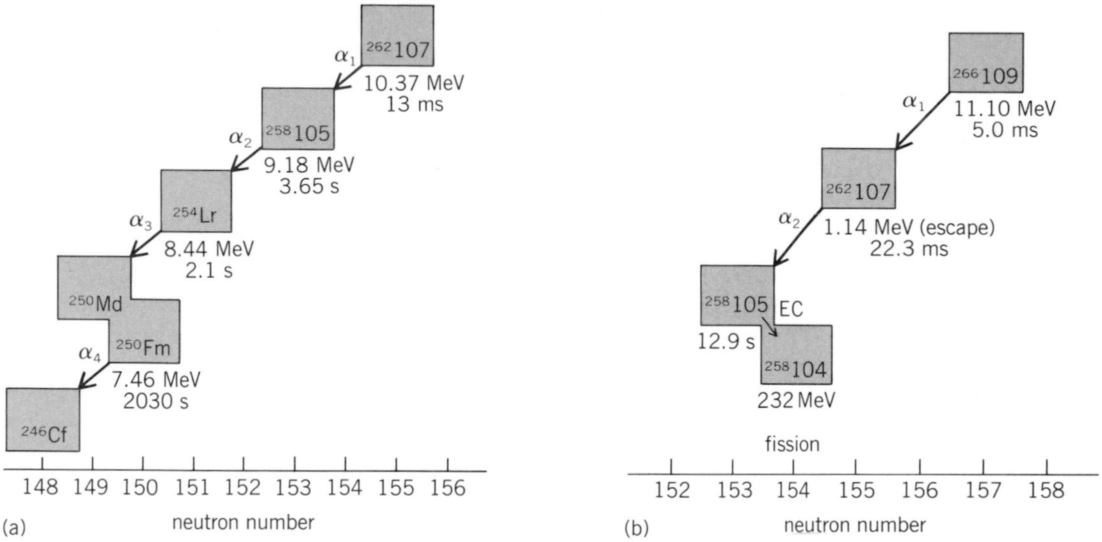

Fig. 2. Decay chains observed in reactions of ^{54}Cr and ^{58}Fe with ^{209}Bi targets leading to two new elements. Disintegration energy, correlation times between successive events, and type of decay are given for each step. (*a*) Decay chain of the isotope $^{262}107$. (*b*) Decay chain of the isotope $^{266}109$.

found are 10.4 MeV and 11.1 MeV, their half-periods 5 ms and 3 ms, and their production cross sections 200 picobarns (2×10^{-38} m^2) and less than 10 picobarns (10^{-39} m^2). To produce one surviving nucleus of element 109, 10^{11} nuclear encounters between ^{58}Fe and ^{209}Bi nuclei at a well-defined energy are necessary.

For background information *see* NUCLEAR REACTION; TRANSURANIUM ELEMENTS in the McGraw-Hill Encyclopedia of Science and Technology.

[P. ARMBRUSTER]

Bibliography: G. Münzenberg et al., The velocity filter SHIP, a separator of unsowed heavy ion fusion products, *Nucl. Instr. Meth.*, 161:65–82, 1979; G. Münzenberg et al., Identification of element 107 by alpha correlation chains, *Z. Phys.*, A300:107–108, 1981; G. Münzenberg et al., Observation of one correlated alpha decay in the reaction 58-Fe on 209-Bi→267-109, *Z. Phys.*, A309:89–90, 1982.

Typewriter

Electronic typewriters are computer-based and enhance typing function through the use of microprocessor technology (microchip computers) with small memories and simple commands. Such systems were introduced in 1978.

Product groups. Electronic typewriters are designed to substitute for and ultimately replace two separate, but related, product groups.

1. An electronic typewriter is an improved, albeit more expensive, kind of machine. Microcomputer technology offers some additional functions and greater ease of use.

2. An electronic typewriter can be thought of as a low-cost word processor, offering many word-processing functions at a lower price than that of a dedicated word-processing system.

Since the word-processing functions of electronic typewriters are necessarily limited by their small memories and lack of adequate video displays, add-on products are now available which can convert electronic typewriters into higher-function word processors. Such products typically offer more storage, a partial page display, and an additional microprocessor with software for added functionality. Such add-ons can survive in the marketplace only as accessories to already-sold electronic typewriters now that the price of many low-level word processors is below that of most electronic typewriters plus add-on product combinations.

Electronic typewriter functions. The electronic typewriter typically offers a set of augmented typing functions. Such functions often include (but are not limited to): centering text between margins or about a point; setting up tabs for tables (and calculating correct tab positions automatically); remembering multiple formats (margins and tab settings); remembering words and phrases for faster typing; addressing an envelope from a memory-stored address; and permitting the editing (revision) of short text segments (ranging from a partial line to several pages).

Many electronic typewriter models offer no visual display to facilitate the process of text entry and revision. The most common type of display offered on electronic typewriters is generally a partial-line variety; such displays give only a very primitive indication of what is happening "now" and what will happen "next."

Some of the most recent and most expensive electronic typewriters offer higher functionality, such as removable media to permit large sets of documents to be sequentially revised and stored; and communication functions, permitting the electronic typewriter to be employed as a terminal (typically a relatively unsophisticated ASCII or TTY terminal). Some experts believe that nonvideo terminals are useful only for the most limited output functions.

Prospects. The notion that an electronic typewriter is a low-cost word processor has been largely supplanted by the advent of the personal computer. The personal computer offers superior processor capacity (and therefore substantially more functionality), together with the full hardware of a word processor (for example, a partial page display and removable high-density diskette media), and enhanced word-processing software is now becoming available. Its user acquires at once a much more powerful, multifunction, programmable system that offers greater growth potential for both the user and the work station.

Organizations which do not use much word processing often consider themselves to be not ready for computer systems. They will find that the electronic typewriter clearly offers some enhanced funtionality and an opportunity for secretaries or typists to learn how to work with a typing device with memory and editing capabilities. As some electronic typewriters cost no more than ordinary electric machines the days of the nonelectronic typewriter are clearly numbered.

Firms already using word processing or other front-office computing will find that the electronic typewriter is not the best available investment; a combination of personal computer work stations and office automation systems will serve them better.

Electronic typewriters are expected to supplant electric typewriters before the end of the 1980s. The market for such devices should continue into the 1990s until the increasing automation of the office makes it clear that any further investment in work stations without much more powerful processing and a usefully sized display make the workers and their organizations less productive.

For background information *see* MICROCOMPUTER; MICROPROCESSOR; TYPEWRITER; WORD PROCESSING in the McGraw-Hill Encyclopedia of Science and Technology.

[AMY D. WOHL]

Underwater navigation

LAUNS is an acronym for local area underwater navigation system. Its design was motivated by the stringent maintenance requirements that apply to North Sea structures, which make the cost of inspection a significant factor in offshore operations. Various sophisticated equipment must be deployed, including a number of vessels, a spread of diving bells, equipment for cleaning the structure, various television assemblies, and underwater cameras. The whole operation requires continual backup and support to meet the required safety standards. There is increasing use of automatic equipment to replace the diver and make the operation less dependent on the so-called weather window (in general, the period between April and October for North Sea operations).

LAUNS. Navigation plays a key role in this activity. The complete specification for the navigation system is extensive, but the main requirements are that it should (1) be a multiuser system (new users should not impair the system's performance, nor should the system have a user saturation level); (2) provide a common frame of reference for both underwater and surface vessels to a distance of 0.6 mi (1 km) from the platform; and (3) make it possible to position a remotely controlled vehicle so that defects in the structure can be photographed with sufficient precision for measurements to be made from the photograph for design checks and analysis.

The second of these requirements must satisfy the needs of dynamic positioning systems for surface vessels, since anchoring is severely restricted or not allowed in the vicinity of a platform. In order to maintain an acceptably small watch-keeping area in a specified set of meteorological conditions, navigation data must be supplied to a vessel's dynamic positioning system at intervals that do not exceed 1 s.

These requirements imply a passive system. Given the anomalous nature of sound propagation underwater, a hyperbolic system was considered to be the best choice. This has the further advantage of allowing a variety of processing techniques, from direct hyperbolic conversions to more sophisticated methods of estimating the direct ranges to the acoustic transmitters. The technique varies according to the processing capabilities and equipment configuration carried by the users. The system operates at frequencies in the 40–100-kHz range and is deployed as a permanent installation providing a continuously available navigational frame of reference for all phases of the inspection process and all the users involved.

Principles of operation. A LAUNS chain consists of a number (up to 10) of acoustic transmitters. Each station is frequency coded. The cables from the platform to the transmitters (Fig. 1) are capacity-matched so that all 10 stations are simultaneously triggered. Each transmitting station is identified to the receiver by its frequency and, as soon as it is received, its time of arrival is noted by the receiver's clock. Since this clock is free-running and not referenced to a clock on the platform, this logged time of arrival is not significant on its own. The differences between the logged times of arrival of the various triggers, however, are significant, and each time difference defines a unique hyperboloid of two sheets with the two relevant transmitters as foci. If n stations have been received (n is equal to or less than 10), there are $\frac{1}{2}n(n-1)$ time differences, but only some of the possible sets of $n-1$ of these are independent.

This set of $n-1$ time differences forms a "redundant" set on which it is possible to use estima-

Fig. 1. Configuration of LAUNS.

tion techniques to derive a "most probable position." There are many sets of $n - 1$ independent lines of position (in fact, n to the power of $n - 2$ of them), and they do not all give the same most probable position, but if n is greater than 5, the physical properties of the system are such that these differences are small.

Greater accuracies can be achieved by estimation and filtering, but the raw data processed by simple averaging techniques give positional accuracies of ± 3 ft (1 m) in the X and Y (east and north) components. Depth (the Z component) is also a by-product of the computation but, with the transmitters on the sea bed, this computed value of depth is generally worthless and a separate depth sensor is used for submersible vehicles and divers. The computed depth must be corrected for tidal variations. In the case of surface vessels, only the tidal variations need to be entered. It is also necessary to make provision for rejection criteria since on occasion some of the time differences may be polluted by the presence of local hot spots in the water, severe tunneling, or temperature inversions. The accuracy degrades as the distance from the net increases and is a function of the geometry of the net.

The rate at which the LAUNS chain is triggered is the system data cycle and is governed by the required operating range of the system and the speed of acoustic propagation in sea water. A 1-s data rate is used, allowing unambiguous system operation to a maximum slant range of 0.9 mi (1.5 km). Under certain conditions, transmitter cycles overlap, but the ambiguities generated by the overlap are resolved by using the present estimate of the receiver's position. The system operates in the 40–70-kHz band.

System calibration. When the transducer array is laid on the seabed, initial calibration is performed by monitor hydrophones on the platform itself. Furthermore, each seabed transducer can act as both a transmitter and a receiver, and the system performs self-checking by transmitting to the other seabed transducers. This function is stepped round each one in turn and, since the cables to each transducer are capacitatively matched, it is possible to measure the exact time taken by the acoustic signal to travel from any transducer to any other that is acoustically open to it. This large amount of redundant data can be used to determine the relative position of each transducer with accuracies of better than 20 in. (50 cm). This can be combined with known fixed hydrophones on the rig to give absolute positions in terms of rig coordinates. Further fixed hydrophones on the rig can then give estimates of the average speed of acoustic propagation. The station positions derived from this self-check are compared to the stored values to confirm that each station is within its positional tolerance.

Applications. The most important application of LAUNS is as an underlying frame of reference for the inspection and maintenance of offshore structures. In this context there are equipment variants for surface ships (from which the operation is controlled), diving bells, the divers themselves, and remotely controlled vehicles. Steel structures consist of a large number of similar or identical nodes and members so that structural ambiguities become a problem. Concrete structures present a large featureless expanse. The primary purpose of LAUNS is to identify positively the portion of the structure being examined. The subsea environment is an extremely hostile one, and even experienced divers can become totally disoriented. It is only with great difficulty that they can perform the exacting work required of them.

LAUNS not only provides a means of structural identification but ensures that data can be compared between one inspection cycle and the next on the same portion of the structure. Increasingly, remotely controlled vehicles are replacing divers, and this makes it possible to use photographic evidence directly as a means of measurement of the development of hairline cracks and other structural flaws. To do this, a remotely controlled vehicle with a camera must be positioned in a known attitude relative to the structure on successive occasions. This is achieved by positioning the remotely controlled vehicle by means of LAUNS to a precision of ±3 ft (1 m) from the required location on the structure. This is close enough for the picture matching and autotrack camera systems on the remotely controlled vehicle itself to position the vehicle in a known attitude relative to the structure. During this phase, the remotely controlled vehicle will be operating at engine powers that prevent acoustic reception until it has docked. The effect is purely local to the remotely controlled vehicle.

The diver navigation unit is another feature of the system. Here, the diver steers a small camera frame to which both power and data are supplied by cables that form part of an umbilical. This in turn connects to the surface via the diving bell umbilical, or directly to the control station. Its function is to guide a diver to any required position that can be specified in terms of LAUNS coordinates. Video data can also be transmitted. The position at which the diver is required to be is inserted into the LAUNS processor either on the bell or on the surface vessel, and the position of the diver navigation unit is sensed by a transducer in the unit itself. This transducer-sensed position is relayed back to the originating processor together with depth from the unit's self-contained depth sensor. This enables the immediate computation of the distance to go and the depth change required.

The unit also displays command information (Fig. 2) regarding the line joining its present position to the destination (the relative position vector). As soon as any movement is made, a new relative position vector is computed, and the command information changes accordingly. Navigation is thus from waypoint to waypoint along routes selected by the diver so as to avoid unforeseen obstacles. The way-

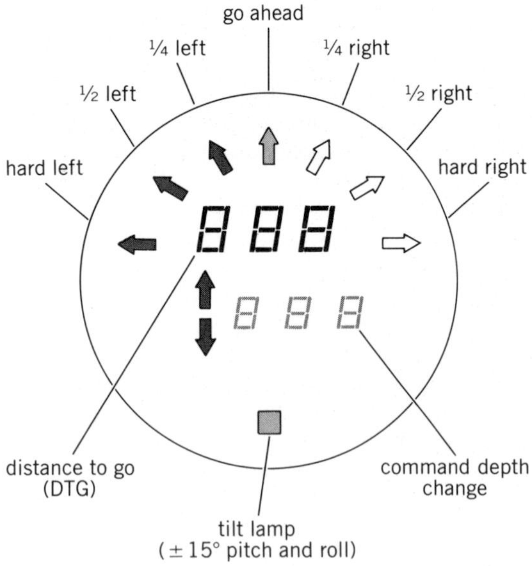

Fig. 2. Diver navigation unit (DNU) display.

points are defined by the operations controller and inserted through the master console. The whole operation can be monitored on a plotting table or a visual display unit.

Trials and demonstrations at various locations have shown that the diver navigation unit operates extremely successfully at the 1-s data cycle. One of the most remarkable features of the system is the ease with which divers and remotely controlled vehicle pilots learn to use the system.

For background information *see* UNDERWATER NAVIGATION in the McGraw-Hill Encyclopedia of Science and Technology. [M. G. PEARSON]

Bibliography: M. Baker et al., *Offshore Inspection and Maintenance, the Implications of the North Sea Experience*, Financial Times Ltd., 1978; T. W. A. Harris, *Survey*, 10:12–14, 1981; Lloyds Register of Shipping, *Guidelines for Underwater Surveys of Offshore Installations*; M. G. Pearson, *J. Navig.*, 35:374–386, 1982.

User friendly systems

User friendly systems create effective human-computer interactions by matching the system design to the users' skills and needs. The user interface must be friendly not just to the people but to their way of working, that is, it must be ergonomic. Awareness of these human-factors requirements for interactive systems has greatly expanded in the last few years, fueled by advances in recognition of individual differences, in new interaction styles, and in the process of design and implementation. *See* ERGONOMICS.

User differences. At one time, direct interaction with computer systems was limited to professional programmers, and early user interfaces reflected their needs (or level of tolerance). As interactive systems came to be used by diverse classes of people, there was growing recognition that each group of users had its own requirements. Recent systems are designed to allow for the frequency with which people will encounter the system, the amount of learning they will experience, the conceptual and cognitive skills they bring to the interface, and the particular characteristics of each task environment.

For an infrequent user of an interactive system, it is essential that the system provide ongoing information about the options currently available, to limit the knowledge the person must retain. The mechanics of use must be kept simple. More regular users will tolerate increased complexity if it gives them greater discretion over operations. Frequent users of several different services will want some measure of integration among them, so that patterns of use have consistent effects.

The best systems provide a growth path as knowledge increases. Long-term users are expected to progress from an early focus on the ease of learning a system to concern for the power and ease of use in their tasks. Later mastery will make efficiency of use a major interest. Finally, expert users will be most concerned with the ease of extending and customizing the user interface.

This learning sequence requires a corresponding sequence of different interaction styles. For example, the first interface may display available options (a menu), but the users will be able to directly generate instructions in a command language. Further learning may lead to more abbreviated commands and creation of new operations (macros) from a sequence of frequent commands.

Long-term use does not guarantee such skill acquisition. This sequence of development must support individual learning styles as well. For example, some people will require strategically timed intervention (by the system or another user) to move to a more effective interaction style.

Full mastery of a powerful user interface requires a suitable conceptual model that organizes how the system works and how it can be used to accomplish tasks. This is another potential difference between users: many people will continue to use recipes of commands without developing more understanding or a good mental map. Today's best systems explicitly build a conceptual framework by displaying operation context (the objects and processes available) and encouraging user excursions to explore new facilities.

Differences in task structure are also now recognized as important. Two people who appear to be engaged in the same task, like document preparation, may have distinct needs. Interface designers must allow for differences in distractions or interruptions, in the duration and purpose of their work, in control over how and when it is performed, and in the amount of satisfaction they derive from it.

Advances in interaction styles. Improvements in computing technology have dramatically lowered the cost of processor cycles. Expanded use of computers has meant that development cost for both hardware and software can be spread over larger unit volumes. These developments have allowed user interface designers to create cost-effective products to address the differing user needs outlined above. Key advances include increased information flow in the human-computer dialog, minimization of user time with more system processing, and inventive interaction styles which have expanded designers' imagination and users' expectations.

When processor cycles were seen as a scarce resource, interactive users typed one-line statements at a keyboard and received a line-oriented response. With more processing power, the system response expanded to a page-size screen of information; part of the display processing intelligence was moved to the user's terminal. More recently, the user interface has acquired the ability to show several screen windows into different ongoing activities. The screen display has expanded from lines of characters to special forms and pictograms. The net effect has been a large increase in the communication bandwidth between system and user.

Increased system and terminal or work-station processing has enabled designers to concentrate on minimizing user interaction time. The most obvious time-saver has been less reliance on typing skills:

keyboards now commonly have function keys to invoke frequent commands and sometimes to select menu items, and a hand-controlled screen pointer (a mouse or a puck). Speaking instructions to the machine is still not a commercial option except in very limited situations, but command languages are available which resemble natural languages in grammar and flexibility.

The use of English-like statements optimizes user time (especially for novices) at a cognitive rather than physical level. Progress in this area will ultimately be more valuable than optimizing mechanical actions. For instance, systems displaying an explicit conceptual model aid the user's thinking by clarifying the current status and available activities.

On-line assistance is another area where interface designs have advanced. Instead of stating only problem symptoms when users make an error, an ergonomic interface provides better assistance by reporting current status of suspended activities, a diagnosis of causes for the mistake, and a set of possible corrective actions. Simple errors like spelling slips or missing punctuation can often be corrected automatically; more complex situations require that the system contain knowledge about the semantic structure of operation sequences. When users ask for help, a good interface makes available specific information about present conditions, as well as a facility for exploring further system capabilities.

The most creative of the current interaction styles combine these elements into natural, powerful, and consistent user interfaces: a natural query language for a data base; a text formatter which allows experimentation to display different typefaces; a word processor which moves sections visually between documents; an electronic spreadsheet which updates total rows dynamically as entries change; or an electronic filing cabinet which zooms in or out of structural levels of detail. The users feel that they are directly manipulating the objects in their daily tasks in powerful and natural ways which amplify (rather than replace) human intelligence.

User interface design process. Awareness of user needs, availability of cost-effective processing, and imaginative examples of interaction styles are vital components for user friendly systems. An ergonomic user interface requires in addition an effective development process which orders the necessary design decisions, supports implementation with productivity tools, and encourages testing and tuning of human-factors quality.

The interface design process sequences the designer's choices to ensure that the more fundamental decisions are given priority. The developer may begin with functional requirements, product constraints (like a limited display size for a low-cost briefcase computer), and informal scenarios of use. However, when the design is to be formalized, the specification should begin with a thorough task analysis and definition of target users. Then conceptual models are designed for various stages of learning. Dialog styles and operation semantics are decided, followed by details of language and format. Physical items like key placement and color ranges are specified last. This design cycle proceeds in a top-down fashion, in which the fundamental—and least easily changed—decisions are made first and influence all later choices.

Implementation tools are just beginning to emerge to increase the productivity of user interface development. Certain tools such as screen formatting packages and parser-generators have been in common use. At a more complex level, tools are becoming available which provide features like menu presentations, handling of incomplete commands, and context display. By ensuring common components for much of the user interaction, these high-level dialog managers present an integrated user interface across different applications. They also can support generic operations which form a common nucleus for each command language, and user profiles to give parts of the interaction individualized attributes.

While functional testing of computer systems is a standard next step after implementation, human-factors testing is another recent development. Behavioral science research has begun to address user interfaces, so that a growing body of empirical results can be applied during designs. But each interactive system is a unique invention and needs to be tested for ease of learning and use. Ease of learning can be reviewed with scenarios and prototypes, using subjects from the target user population. Ease of use, as a later concern, requires at least a demonstration version which can be used over a period of time. More complete information can be obtained from field trial use in a real working environment. Software monitors are usually employed to collect and analyze field trial usage, yielding data on command frequency, error-prone operations, and so forth.

Increased understanding of ergonomic requirements, technology advances, and improved development cycles all have contributed to better interactive systems. Yet the next generation of user interfaces is already taking shape, as designers continue to seek graceful cooperation in human-computer interaction.

For background information *see* COMPUTER; COMPUTER GRAPHICS; HUMAN-FACTORS ENGINEERING; MULTIACCESS COMPUTER in the McGraw-Hill Encyclopedia of Science and Technology.

[TOM CAREY]

Bibliography: J. A. Larson, *Tutorial: End User Facilities in the 1980's*, IEEE Computer Society, 1982; *Proceedings of ACM Conference on Human Factors of Computer Systems*, 1982/1983; B. Shackel (ed.), *Man-Computer Interaction: Human Factors Aspects of Computers and People*, 1981.

Vertebrata

The morphological features shared by vertebrates and other deuterostomes (organisms that are characterized by bilateral symmetry, indeterminate embryonic cleavage, and ciliated larvae) gave the first indication that vertebrates must have arisen phylogenetically from some group of early deuterostomes.

Shared derived characters of vertebrates*

Character	Embryonic origin	Associated functions
Nervous system		
Cranial nerves with peripheral sensory ganglia	NC, P	DH, G
Trunk nerves with peripheral sensory ganglia	NC	DH
Peripheral motor ganglia	NC	G, F
Forebrain	NC?	DH, F
Paired special sense organs		
Olfactory organs	P	DH
Eyes (accessory structures)	NC? (P)	DH
Labyrinthine organs	P	DH
Lateral line mechanoreceptors	P	DH
Lateral line electroreceptors	P	DH
Gustatory organs	NC, P	DH
Pharyngeal and alimentary systems		
Cartilaginous pharyngeal bars	NC	DH, F, G
Branchiomeric muscle	MH	DH, F, G
Smooth muscle of gut	MH	F
Chromaffin cells, adrenal cortex	NC	DH, F
Circulatory system		
Muscularized aortic arches	NC	G
Muscular heart	MH	G
Vasoreceptors of aortic arches	NC	G
Skeletal system		
Anterior neurocranium and sensory capsules	NC	DH, F
Dermal armor and derivatives	NC	DH
Calcitonin cells	NC	G

*NC, neural crest; P, epidermal placodes; H, hypomere; MH, muscularized hypomere; DH, food detection and ingestion; G, gas exchange; F, food processing.

Most discussions of the origin of vertebrates have focused on these shared primitive features, and the failure to categorize how vertebrates differ from other living deuterostomes has masked the evolutionary and functional shifts that accompanied the origin of vertebrates. Although all parts of the vertebrate body differ from those of other deuterostomes, the organization of the vertebrate head differs most. Many of these differences arise embryologically from muscularization of the hypomere, neural crest, and epidermal placodes.

The table lists most of the derived features unique to vertebrates. One must understand the origin of these features and their transitional forms in order to explain the origin of vertebrates. Because the unique vertebrate features involve multiple organ systems, it is likely that these organs arose as a result of changes in the genetic "program" that determines the early embryonic pattern of tissue interaction and differentiation.

Embryonic development of shared features. The shared derived features of vertebrates listed in the table arise embryonically from hypomere, neural crest, and epidermal placodes. The hypomere (the ventral unsegmented portion of the mesoderm) of the vertebrate trunk forms the smooth muscle of the gut and striated muscle of the heart, whereas in the head the hypomere forms striated branchiomeric muscle of the pharynx. Although other deuterostomes possess a hypomeric division of the mesoderm, it forms connective tissue rather than muscle. Water is moved through the pharynx, and particles through the gut by the action of the endodermal cilia. The neural crest is a transitory embryonic neuroectodermal tissue of vertebrates that initially forms adjacent to the neural plate and subsequently migrates to specific locations throughout the body, where its cells differentiate into a number of tissue types. In the trunk, the neural crest forms the sensory ganglia of the spinal nerves, peripheral visceral motor neurons, melanocytes, and the chromaffin tissue of the adrenal organs. In the head, however, the neural crest forms skeletal tissues such as cartilage, bone, dentine, and enamellike tissues, as well as the sensory ganglia of some of the cranial nerves and all peripheral visceral motor neurons.

Unlike the neural crest, epidermal placodes, which are thickenings of the ectoderm, arise only in the head. These embryonic structures give rise to the lateral line and inner ear organs and nerves, as well as to the sensory ganglia of the cranial nerves that innervate taste buds and chemoreceptors of the carotid and aortic bodies.

Origin of neural crest and placodes. Although developmental studies demonstrate that the shared derived features of vertebrates arise from a few embryonic tissues, the origin of the neural crest and epidermal placodes is not obvious. All of the derived features are involved in gas exchange or the detection, ingestion, and processing of food. Hypomeric muscle appears to have arisen from primitive, hypomeric myoepithelial cells, which are widespread among many deuterostomes. Such muscularization allows movement of materials through the gut of larger organisms in which ciliary currents would be less effective. Both neural crest and epidermal placodes are specifically involved in sensory and integrative functions. This suggests that the neural crest and epidermal placodes are somehow related. Both tissues are migratory derivatives of ectoderm that form sensory neurons and special sense organs. Both produce or induce proteoglycans (mucopolysaccharides) as extracellular matrices.

Placodes differ from the neural crest in that they arise only in the head and form only sensory receptors or sensory neurons that become part of the cranial nerve ganglia. Placodes form within the ectoderm prior to the migration of neural crest cells. Thus, placodes do not appear to be a specialized portion of the neural crest, nor do they appear to be induced by the neural crest. Both the neural crest and placodes are probably parallel derivatives of an earlier phylogenetic tissue that formed sensory, integrative, and motor cells in other deuterostomes. Large portions of the sensory, integrative, and motor elements of other deuterostomes are located in an epidermal plexus. This plexus exhibits variable degrees of mid-dorsal condensation and invagination in other deuterostomes.

In addition, these organisms also possess a perivisceral plexus. As all neurons in metazoans are believed to arise only from ectodermal tissue, it is likely that the elements of the perivisceral neural plexus arise and migrate, like the neural crest and placodes, from the ectoderm. These observations suggest that parts of the nervous system and special sense organs of vertebrates arose phylogenetically by condensation and hypertrophy of the epidermal nerve plexus of earlier deuterostomes.

Origin of vertebrate features. Both the neural crest and epidermal placodes form special sense organs and many portions of the cranial nerves. The structures in vertebrates that arise embryonically from the neural crest and epidermal placodes appear to be homologous to portions of the epidermal nerve plexus of other deuterostomes. In the head, the neural crest also forms connective, skeletal, and muscular tissue—tissues that in the trunk are usually formed from mesoderm. The origin of vertebrates appears to have been associated with a shift from passive to active predation, apparently accomplished by utilizing mesodermal tissue to form new head muscles as well as a muscular heart. The new head muscles were used to pump water across the pharyngeal gills for oxygen extraction, and also allowed more active prey capture. The new muscular heart formed part of an extensive modification of the circulatory system that increased the volume of blood flowing through the gills and to the rest of the body. Portions of the epidermal nerve plexus were involved in monitoring and controlling these new muscular organs and formed new sense organs as well. Many of the skeletal tissues of the head—such as cartilage, bone, dentine, and enamellike tissues—appear to have arisen in association with new sense organs, and only secondarily provided mechanical support for many of the head tissues. Thus the origin of vertebrates involved fundamental changes in gas exchange and distribution, as well as the modification of many of the associated structures for the capture of larger prey items.

The origin of parts of the vertebrate nervous system and special sense organs from the epidermal nerve plexus of earlier deuterostomes accounts for the role of the neural crest and placodes in the embryonic development of these features, but it does not account for the role of the neural crest in the embryonic formation of skeletal elements in the head of vertebrates. It is most likely that these tissues arose as transduction enhancement tissues associated with special sense organs and only secondarily became involved as mechanical support tissues. The earliest fossil vertebrates (ostracoderms) possessed all of the skeletal components seen in the head of modern vertebrates, as well as an extensive ossified integumentary armor. The external surface of this armor consisted of an outer layer of enamellike tissue and an inner layer of dentine. These layers were in turn supported by a deeper layer of bone. A large number of flask-shaped depressions (pore-canal system) pitted the outer surface of the integumentary armor of ostracoderms. These depressions appear to have housed electroreceptors sensitive to weak electric fields generated by other organisms. The crystalline materials forming enamellike and dentine tissues have the highest electrical resistivity of any vertebrate material, and it appears that they initially arose in phylogeny to increase the sensitivity of electroreception and only secondarily came to provide protective and supportive functions. Enamel and dentine are extremely brittle, and it is possible that bone initially provided support for these materials, as well as providing storage for calcium ions.

Similarly, cartilage may have arisen phylogenetically to provide mechanoreceptive functions and may only secondarily have acquired support functions. The head and body of most modern fishes are characterized by ciliated receptors of the lateral line system (neuromasts) that are highly sensitive to water movement. The sensitivity of these neuromasts is enhanced by a cap of proteoglycans similar to those forming cartilage. Other deuterostomes possess similar ciliated sensory cells on the body surface as well as on the pharyngeal surface. Thus cartilage cells (chondrocytes) may have arisen phylogenetically as elements of a perivisceral nerve plexus that were highly sensitive to mechanical distortions of the pharyngeal wall. The sensitivity of these cells may have been enhanced by the secretion of a proteoglycan cap that was secondarily selected for support functions and its initial sensory function was lost. Both bone and cartilage spread from integumental and pharyngeal sites, involving the capsules of the special sense organs and braincase as well as numerous sites within the trunk. The presence of these materials in other head and trunk regions can have occurred only if these tissues evolved many times independently in early vertebrates, or if the capacity to deposit proteoglycans and crystalline matrices shifted among cell lineages. The latter is the more likely case.

Vertebrate ancestors. Analysis of the shared derived features of vertebrates suggests that vertebrates arose from ciliated filter feeders which used muscular propulsion similar to that in modern cephalochordates. Vertebrate ancestors probably possessed a notochord and segmented trunk muscles for locomotion. Their pharynx was supported by a skeleton of collagenous bars, and they fed by transporting a mucous strip that was moved by ciliated endoderm. Gases used in respiration diffused across the pharyngeal and skin surfaces. Thus the ancestor lacked both a central heart and gill capillary networks.

The shift to the vertebrate condition primarily involved modification of the pharynx, permitting use of larger prey items. Such an adaption would have expanded the range of available nutrients and would also have established advantages for increased size and metabolic output. An improved gas exchange mechanism was brought about by muscularization of the hypomere, leading to new pharyngeal muscles that allowed deformation of the pharynx. The collagenous pharyngeal bars were replaced by cartilaginous bars that had the advantage of elastic recoil, thus storing much of the energy generated during pharyngeal deformation. The wall of the gut became muscularized, thus increasing its capacity to transport larger prey.

The circulatory system acquired capillary beds beneath the gill epithelia, and development of a muscular heart and aortic arches completed the suite of changes that allowed for increased metabolic output. These changes and the advantages they provided also involved the development of new sensory, integrative, and motor controls which were centralized in the expanded neural tube of the head.

Modification of the integument to enhance electroreception by the deposition of crystalline matrices occurred late in the transition to vertebrates. Thus ossification and the fossil record of the new hard tissues are indicative of later stages in the transition to vertebrates. The new sense organs and much of the integrative machinery of the vertebrate brain apparently developed from portions of the epidermal nerve plexus anterior to the existing neural tube. Thus much of the vertebrate head is an addition rather than a modified portion of the head of other deuterostomes.

For background information *see* ANIMAL EVOLUTION; CHORDATA; NEURAL CREST; VERTEBRATA in the McGraw-Hill Encyclopedia of Science and Technology.

[R. GLENN NORTHCUTT; CARL GANS]

Bibliography: A. D'Amico-Martel and D. M. Noden, Contributions of placodal and neural crest cells to avian cranial peripheral ganglia, *Amer. J. Anat.*, 166:445–468, 1983; C. Gans and R. G. Northcutt, Neural crest and the origin of vertebrates: A new head, *Science*, 220:268–274, 1983; N. LeDouarin, *The Neural Crest*, 1982; R. G. Northcutt and C. Gans, The genesis of neural crest and epidermal placodes: A reinterpretation of vertebrate origins, *Quart. Rev. Biol.*, 58:1–28, 1983.

Vibration

Elastic structures subjected to external excitation undergo vibration. In many cases, vibration is to be expected and can be tolerated, provided it does not exceed a certain level. This is the case with structures supporting rotating machinery or with vehicles in motion such as automobiles and aircraft. If the level increases, however, vibration can cause discomfort, or even damage to the structure or the equipment supported by the structure.

To prevent undesirable vibration, one can adjust the system parameters, such as the stiffness and damping, in particular the latter. Indeed, damping has been used to reduce the vibration of machinery, appliances, vehicles, and even massive structures such as multistory buildings. Damping represents a passive means of suppressing vibration, however, and has many limitations. For example, except for the case of heavy damping, reduction of the vibration amplitude to a tolerable level can extend over an unacceptably long period of time. Moreover, damping of any kind may not be feasible. A case in point is an antenna dish in a spacecraft. Such an antenna tends to be very flimsy, so that the fundamental frequency of the structure is low. Moreover, any internal structural damping is very light, and the addition of damping material is not feasible. Hence, any disturbance is likely to produce a persisting vibration of the antenna, during which time the signal distortion prevents the antenna from performing its task.

It follows that passive vibration suppression is not always satisfactory, which points to active vibration suppression as the only alternative. Active vibration suppresssion involves feedback control of the vibratory motion, whereby the forces designed to reduce the vibration depend on the system response, generally on the system displacements and velocities.

The idea of feedback control can be traced to ancient times. Early examples of feedback controls are characterized by the fact that they were designed to control a single quantity. The development of feedback control systems, and in particular of techniques for the analysis and design of such systems, did not begin in earnest until the twentieth century, primarily during and after World War II. Added impetus was provided by the development of the digital computer. Feedback control of structures is a more recent development. It has been applied extensively to aerospace structures and to a lesser extent to civil structures.

Elastic structures are basically distributed-parameter systems, as the mass and stiffness are distributed throughout the entire structure. Distributed-parameter systems have an infinite number of degrees of freedom. At times, structures are modeled as discrete systems through a process known as discretization. The most common discretization procedures consist of lumping of the mass and stiffness, and series discretization in conjunction with truncation. Discrete systems have a finite number of degrees of freedom.

Feedback control. The behavior of systems is governed by their eigenvalues, or poles. If all the eigenvalues have negative real parts, the motion of the system decays with time, and the system is said to be asymptotically stable. If the eigenvalues have nonpositive real parts, that is, if some of the eigenvalues have negative real parts and the balance have zero real parts, some oscillation persists as time passes, and the system is said to be merely stable. If at least one of the eigenvalues possesses a positive real part, the motion diverges and the system is unstable. Active vibration suppression implies the application of feedback control forces so as to render a merely stable or an unstable system asymtotically stable. Hence, the role of feedback control is to modify the system eigenvalues so that all the poles of the controlled system have negative real parts.

In the case of linear controls, the feedback control forces are proportional to the system velocities and displacements, where the constants of proportionality are known as control gains. The presence of forces proportional to the velocities brings to mind viscous damping forces, which exhibit the same characteristic. Hence, feedback control in which the forces are proportional to velocities can be interpreted as externally imposed damping forces.

The eigenvalues of a system for which the control forces do not depend on the state (displacements and velocities) are known as open-loop poles, and those of a system subjected to feedback control are called closed-loop poles. Similarly, the equations of motion in the presence of feedback control forces are called closed-loop equations. The term closed-loop can be traced to the block diagram describing the system. In the case of feedback control, the diagram has the form of a loop in which the feedback forces close the loop. The term open-loop is used for forces not closing the loop.

Modal control of distributed systems. Most techniques for the control of elastic systems fall in the general class of modal control. The idea of modal control simply implies that the vibration of an elastic system can be controlled by controlling the modes of vibration. There is a marked difference between modal control of distributed systems and discrete systems. This can be traced to the ability (or lack of ability) to compute the control gains, that is, the coefficients of the state variables.

The motion of a distributed elastic system is governed by a partial differential equation, and the solution of this equation is subject to given boundary conditions. The solution can be expressed as a linear combination of space-dependent modal functions multiplied by time-dependent modal coordinates. For flexible structures, the modes possess the orthogonality property, which implies that each and every mode can be excited independently of any other mode. This permits the reduction of the partial differential equation to an infinite set of second-order ordinary differential equations for the modal coordinates, where the equations are known as modal equations. Each modal equation contains a forcing term called modal control. For linear state-feedback control, the modal control forces can be expressed as linear combinations of the modal displacements and modal velocities, where the coefficients are called modal control gains. Feedback control can be best visualized by recalling that elastic restoring forces are proportional to displacements and viscous damping forces are proportional to velocities. Hence, the feedback control forces can be interpreted as externally imposed restoring forces and damping forces, and the modal control gains can be interpreted as associated stiffness and damping coefficients, respectively. In general, the feedback modal control forces depend on all the modal coordinates and velocities, so that modal controls tend to recouple the modal equations. This type of control is called coupled control. Coupled control cannot be designed for distributed systems because of the inability to compute the control gains.

A viable alternative to coupled control is to design the modal controls so that the modal control for the rth mode depends only on the rth coordinate and the rth velocity. This type of control retains the independence of the modal equations and is known as natural control. For natural control, the modal gains do not include cross products and can be computed with great ease. In fact, they can be computed so as to satisfy certain optimal performance criteria. Then, the distributed control can be synthesized in the form of a linear combination of modal controls. Implementation of such control requires a distributed actuator force. In this regard, electrostatic forces may be considered. It is more likely, however, that implementation of the control will be carried out only approximately by means of a finite number of discrete actuators, where the discrete actuators are designed so as to mimic the distributed actuator.

Modal control of discretized systems. Another alternative to designing distributed controls is to discretize the structure, perhaps by the finite element method. The equations of motion of the discretized system have the appearance of the equations of a discrete system. Indeed, they are defined by $n \times n$ mass and stiffness matrices and are in terms of n-dimensional displacement and control vectors. The equations can once again be reduced to modal form, except that the necessary expansion is in terms of orthogonal modal vectors instead of space-dependent modal functions. Here too, the modal controls can be such that the rth modal control depends only on the rth modal displacement and velocity, leading to natural control. In the latter case, n actuators are necessary to implement the control, where the gains can be selected so as to render the control optimal. If the modal controls depend on all the modal displacements and velocities, the closed-loop equations of motion are coupled. As in the case of distributed systems, the control itself is referred to as coupled. Coupled control can be carried out with fewer actuators than n, but the control is not globally optimal. Moreover, difficulties can be expected in computing the gains, particularly for high-order systems.

For background information *see* CONTROL SYSTEMS; FINITE ELEMENT METHOD; MECHANICAL VIBRATION; VIBRATION; VIBRATION DAMPING in the McGraw-Hill Encyclopedia of Science and Technology.

[LEONARD MEIROVITCH]

Bibliography: H. H. E. Leipholz (ed.), *Proceedings of the IUTAM Symposium on Structural Control*, 1979; L. Meirovitch (ed.), *Proceedings of the 1st VPI&SU/AIAA Symposium on Dynamics and Control of Large Flexible Spacecraft*, 1977, *2d Symposium*, 1979, *3d Symposium*, 1981.

Volcano

Extensive eruptions of El Chichón in Mexico in 1982 were followed by spectacular eruptions of Kilauea in Hawaii, beginning in early 1983. These events provided scientists with singular opportunities to study volcanic processes and their ramifications.

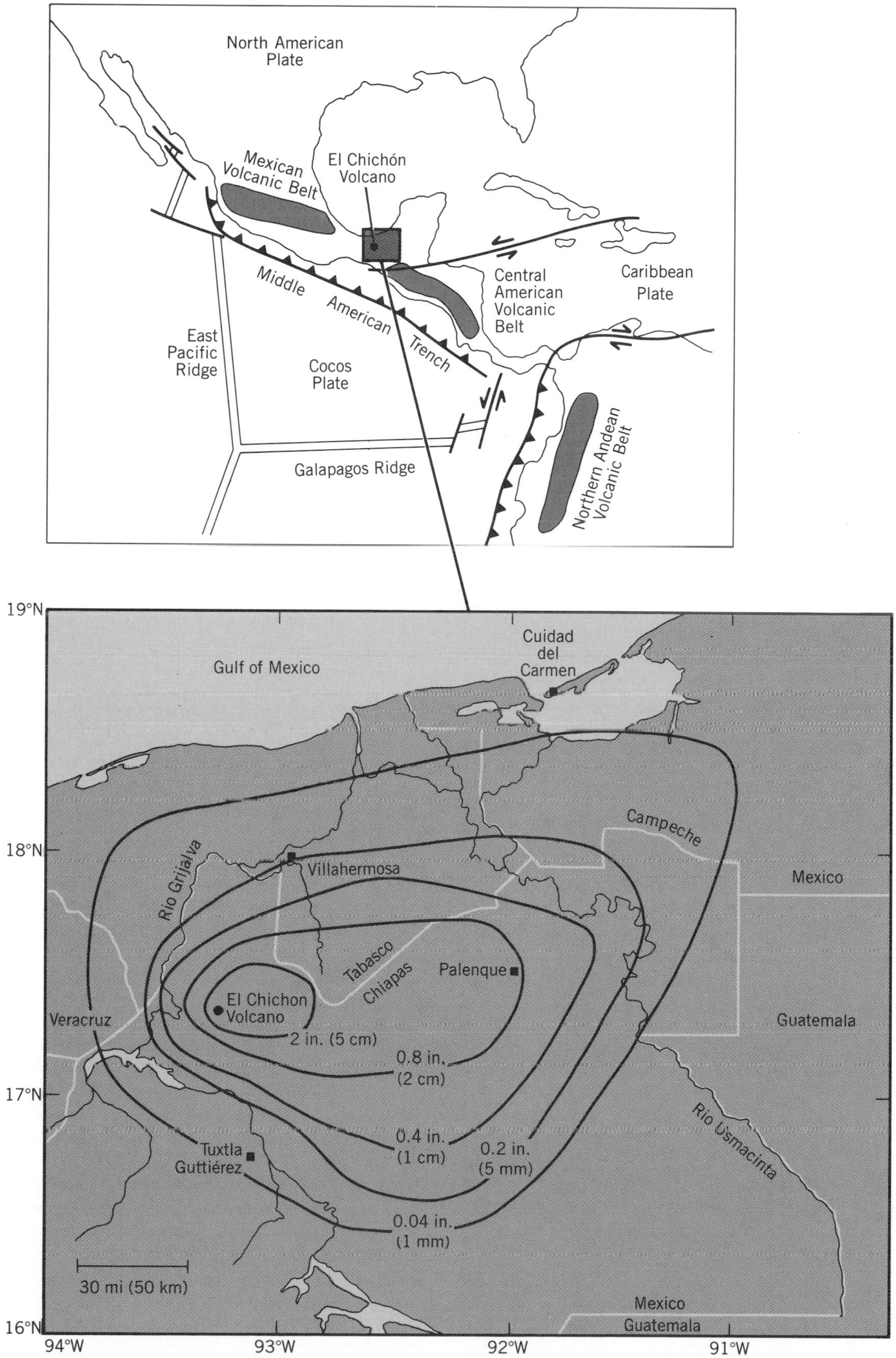

Fig. 1. Area of the El Chichón eruption. A generalized geologic location map is shown at top, and at bottom an enlargement of the shaded area indicating contoured thickness for fresh, compacted ash; major rivers are indicated.

EL CHICHÓN

In late March and early April of 1982, a sequence of three major pyroclastic eruptions took place in southeastern Mexico at El Chichón volcano, which showed no previous record of activity. The eruptions were relatively small in volume, involving only about 0.07 mi^3 (0.3 km^3) of andesitic magma. They were extraordinary, however, in that the magma was unusually rich in sulfur. The large magmatic sulfur content was expressed mineralogically by 1 vol % crystals of anhydrite ($CaSO_4$), an uncommon igneous mineral, in the 1982 pumices. Accidental fragments of the underlying volcanic pile, brought to the surface by the eruptions, contain potassic alteration assemblages and sulfide mineralization, suggesting that a magmatic–hydrothermal ore deposit may underlie the volcano. The large sulfur content of the magma also produced an unusually large stratospheric cloud consisting principally of micrometer-sized droplets of H_2SO_4, which may significantly influence the Earth's climate in coming years.

Fig. 2. Changes in El Chichón volcano. (*a*) Summit prior to the 1982 eruption (*photograph by Reneé Canul, Commission Federal Electricidad, Mexico*). (*b*) New crater left by the eruption as it appeared in June 1982 (*photograph by Wendell A. Duffield, USGS*)

Geologic setting. El Chichón volcano is an unimposing topographic feature in the northern Chiapas highlands, an area of dense tropical vegetation. It forms part of a diffuse volcanic zone filling the gap between the Mexican and Central American volcanic belts (Fig. 1). El Chichón lies 217 mi (350 km) inland from the Middle America Trench, where the Cocos Plate is subducting northeastward beneath the North American and Caribbean plates. An inclined seismic zone, presumably corresponding to the top of the subducting Cocos Plate, dips about 30° northeastward beneath Chiapas to underlie El Chichón at a depth of about 93 mi (150 km). The andesitic magmas of El Chichón and other continental margin and island arc volcanoes are generally thought to be crystal fractionation products of basalts generated in subduction zones through complex interactions between the underthrust plate and the overlying mantle wedge.

Surprisingly, El Chichón was not known to geologists until 1928, when it was first described by F. K. G. Mullerried. It is a small composite volcano, 2.5 × 3 mi (4 × 5 km) in diameter, and rests on a Jurassic-to-Miocene sequence of carbonates, lutites, arenites, and evaporites. These sediments have been deformed into northwest-trending anticlines. Before the recent eruptions, the summit of El Chichón (4430 ft or 1350 m) was marked by a central dome filling an elliptical crater 6200 × 3000 ft (1900 × 900 m; Fig 2*a*). Shortly before the 1982 eruptions, there was vigorous fumarolic activity in the crater from both the central dome and the surrounding moat, including strong alteration of the moat andesite, and local patches of native sulfur. The moat was strewn with charred logs and the ground was hot to touch.

Eruption chronology. Following a month of intense shallow-level seismic activity, the first eruption began on March 28, 1982, at 11:30 P.M., and was over by 6:00 the next morning. The cloud top, determined by infrared satellite imaging, rose to 11 mi (17 km), penetrating into the stratosphere by approximately 0.6 mi (1 km). The pumice- and ashfall blanket extended several hundred miles to the east (Fig. 1), closing airports in the nearby cities of Villahermosa and Tuxtla Guttierez and forcing the evacuation of tens of thousands of people. Occasional small eruptions occurred during the following days. The second major blast came at 8:00 P.M. on April 3, and the largest and final eruption followed the next morning at 5:30. These eruptions also penetrated the stratosphere and produced extensive fallout blankets to the east of the volcano. Sometime during the night of April 3–4, pyroclastic flows and surges descended the flanks of the volcano on all sides, overrunning several villages. The eruptions destroyed the former crater dome, leaving a new crater 0.6 mi (1 km) in diameter, consisting of coalesced explosion pits (Fig 2*b*). Combined ash thicknesses for the 1982 eruptions are shown in Fig. 1, with the 0.04 in. (1-mm) line extending over 155 mi (250 km) to the east of El Chichón. The combined volumes of the eruptions represent only 0.07 mi^3 (0.3 km^3) of magma, similar to the volume involved in the May 18, 1980, eruption of Mount St. Helens, Washington.

Nature of ejecta. Fresh pumices collected shortly after the eruptions are andesitic in composition, with 56 wt % SiO_2, 1.2 wt % SO_3, and relatively large alkali contents ($Na_2O + K_2O$). They contain about 50 wt % crystals (plagioclase > hornblende > augite > magnetite > anhydrite > apatite > sphene > pyrrhotite) enclosed in a vesiculated glassy matrix. Pumices resampled in February 1983, following the heavy rains of the previous summer, contain only 0.2 wt % SO_3 and just traces of anhydrite. Anhydrite is not usually considered to be an igneous mineral, but its occurrence in the fresh Chichón pumices, where it forms euhedral crystals up to 0.01 in. (0.3 mm) long, leaves no doubt that it crystallized from the magma prior to eruption. Anhydrite may be considerably more common in igneous rocks than previously thought, with the misconception resulting from the rapid hydration and dissolution of anhydrite in vesicular pumices at the Earth's surface. In the Chichón area, where rainfall exceeds 13 ft (4 m) per year, near-complete removal of anhydrite took less than 10 months. Thick, sulfate-bearing evaporite beds underlie the volcano, and stable-isotope studies are in progress to constrain the importance of assimilation of these beds in the generation of the high magmatic sulfur content of the 1982 andesite.

In addition to the new magma, represented by fresh pumices, the eruptions of El Chichón brought many accidental fragments of the former crater dome and the underlying volcanic pile to the surface. These accidental fragments include many that show strong potassic alteration of the original volcanic mineral assemblages and contain veinlets of pyrite (FeS_2), anhydrite, and chalcopyrite ($CuFeS_2$). The textures and mineral assemblages in these fragments are similar to those found in alteration halos surrounding porphyry–copper and other magmatic–hydrothermal ore deposits. The large calculated preeruptive sulfur (>2 wt %) and water (>7 wt %) contents of the 1982 Chichón andesite, the coexistence in the magma of primary anhydrite and pyrrhotite, and the potassic alteration and sulfide mineralization in dense, accidental lithic clasts suggest that El Chichón may represent the surface expression of an actively forming, mineralized magmatic–hydrothermal system.

Stratospheric cloud. Following a major pyroclastic eruption, the great majority of particulate silicate material falls rapidly back to the earth, forming the ash-fall blanket. Erupted gases and micrometer-sized silicate particles can, however, enter the stratosphere and remain suspended for years. In the case of El Chichón, the ash-fall blanket fell to the east, but the stratospheric winds took the upper levels of the cloud to the west. Satellite imagery permitted tracking of the April 4 cloud as it circled the globe, reaching Hawaii on April 9, Japan on April 18, and completing its first circuit on April 26. Since then, the cloud has been circling every 10 days or so, smearing into a nearly continuous band at lower northern latitudes, as it slowly disperses northward. The Chichón cloud is considerably more massive and extends to considerably higher levels than other eruption-related stratospheric clouds that have been investigated. Recent studies of such clouds have shown that although silicate particles do exist, the main mass is represented by aerosol droplets of sulfuric acid, less than 1 micrometer in diameter, formed through combination of volcanic SO_2 gas with water vapor. These stratospheric acid droplets can influence global climate through backscattering of incident sunlight. Past studies of the interrelationships between volcanism and climatology have assumed that climatic influence could be related directly to measureable eruptive volume. The 1982 eruptions of El Chichón have seriously weakened that assumption, showing that a volumetrically small eruption may cause a significant climatic effect if the magma is sufficiently sulfur-rich.

[JAMES LUHR]

KILAUEA

A major eruption of Hawaii's Kilauea volcano began in early January 1983. By the end of July, six eruptive episodes had produced about 3.5×10^9 ft^3 (10^8 m^3) of lava from East Rift Zone vents 9–12 mi (15–20 km) from the summit caldera (Fig. 3 and 4).

In addition to their detailed observations of the eruption, geophysicists at the United States Geological Survey's Hawaiian Volcano Observatory use a variety of techniques to monitor the underground movement of magma. Kilauea's magma rises from a depth of tens of miles into a shallow holding chamber beneath the summit caldera. As the magma chamber fills, the ground surface above it swells slightly. Conversely, net drainage of the chamber is

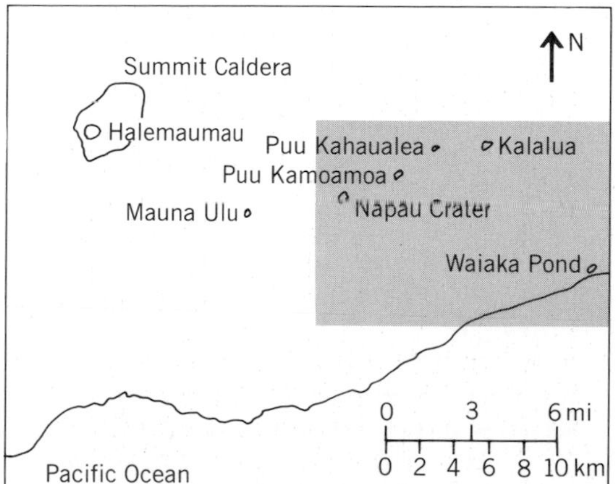

Fig. 3. Index map of Kilauea's summit region and East Rift Zone. The inset defines the location of the area detailed in Fig. 4. (*Hawaiian Volcano Observatory, USGS*)

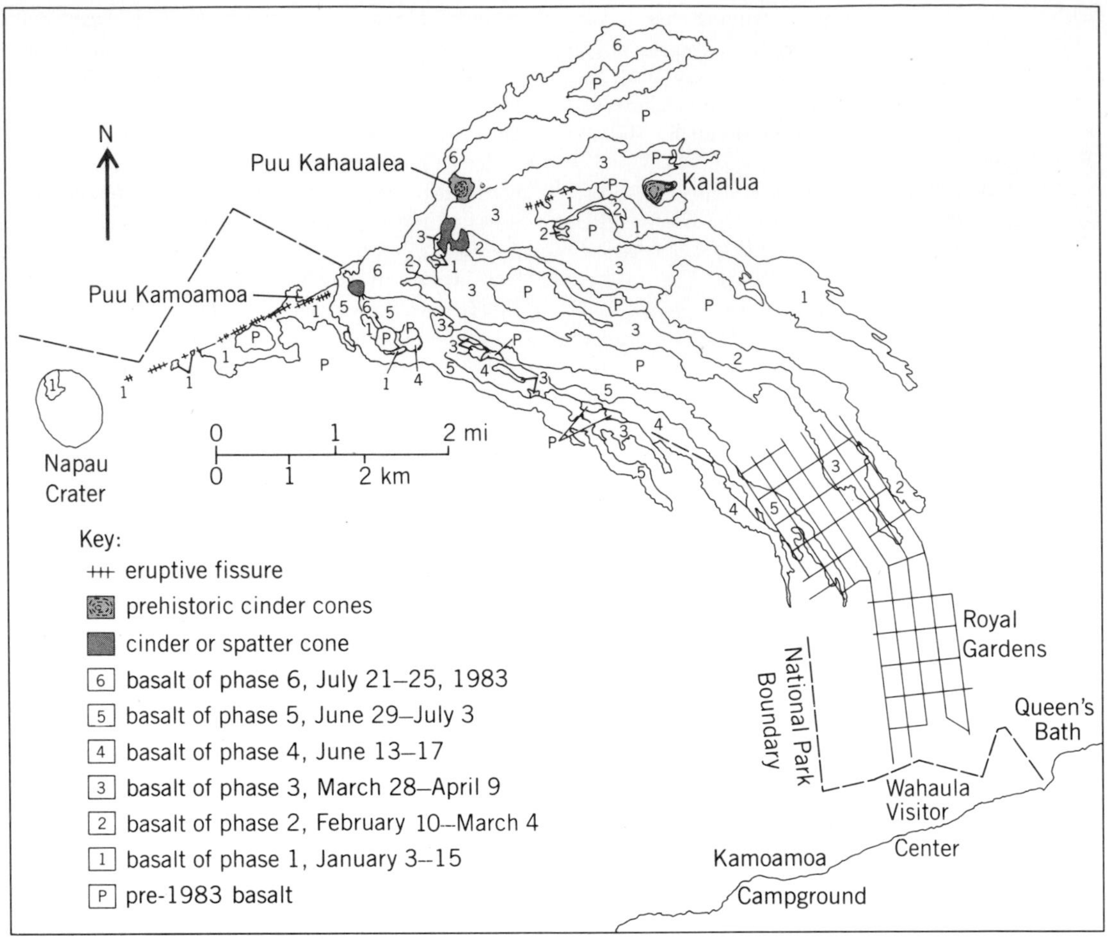

Fig. 4. Map showing eruption fissures and lava flows produced January–July 1983. (*Hawaiian Volcano Observatory, USGS*)

reflected by a slight deflation at the surface. By measuring the surface changes (Fig. 5) it is possible to estimate changes in the amount of magma within the chamber at a given time and the rate of change in magma volume. Numerous seismographs record harmonic tremor, a characteristic seismic signal produced by moving magma (Fig. 5), as well as small earthquakes associated with subsurface fracturing often induced by magmatic pressure. Magma leaving the summit chamber sometimes moves upward to be erupted within the summit caldera (most recently in September 1982), but numerous eruptions have also occurred from the East and Southwest rifts, fissure zones that extend miles from the summit area. As the eruptions occurred, they were under constant observation by the Hawaiian Volcano Observatory, where a chronology of the eruption was prepared.

Phase 1 (January 2–23). In the weeks before the 1983 eruption, seismographs recorded increasing numbers of microearthquakes in the East Rift Zone. Early on January 2, a swarm of small earthquakes and harmonic tremor started about 6 mi (10 km) southeast of the caldera, then intensified and migrated about 5 mi (9 km) downrift in the next few hours. About 30 min after the beginning of the earthquake swarm, tiltmeters started to show slow summit deflation, indicating drainage of magma from the summit chamber. The deflation rate accelerated in the early afternoon as seismicity concentrated in a zone extending 2 mi (3 km) east from Napau Crater (Fig. 4), where the eruption started just after midnight on January 3. Lava fountaining that fed small lava flows continued for about 9½ h as a system of fissure vents extended about 4 mi (6 km) downrift. After a 4½-h pause, activity resumed for about an hour at the downrift end of the vent system.

Renewed eruptive activity started shortly before noon on January 5, and numerous episodes of vigorous lava production lasted from 2 min to 18¾ h during the next 10 days. Much of the lava that was erupted drained into an open crack 300 ft (about 100 m) from the east end of the vents. During parts of January 7–8, the main eruptive center shifted temporarily to vents ½ mi (about 1 km) farther downrift. These vents produced lava fountains 250–300 ft (80–100 m) high and a lava flow nearly 4 mi (6 km) long. Through January 8, rapid summit deflation was accompanied by intense shallow seismicity as magma drained from the summit chamber into the East Rift. Major lava production ended on January

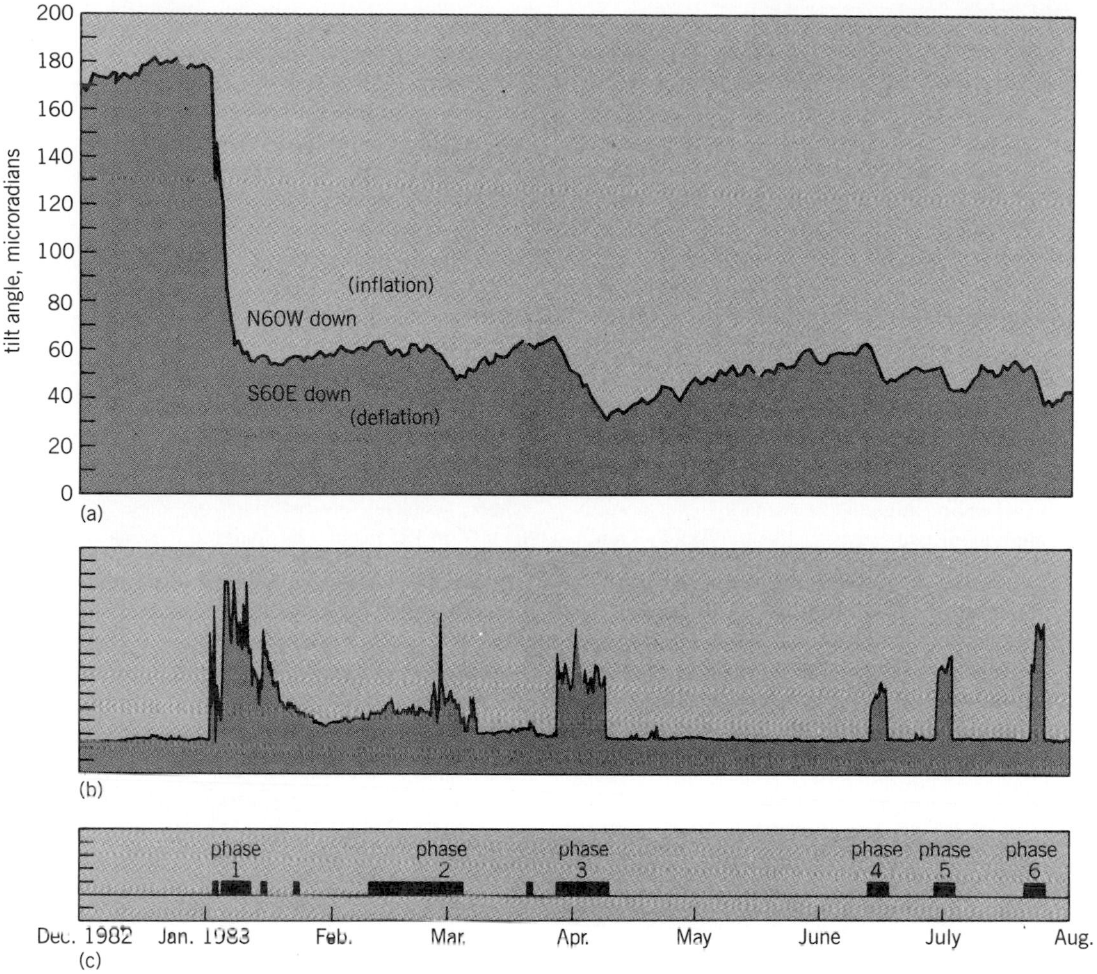

Fig. 5. Graphs showing (*a*) changes in tilt at the Kilauea summit at Uwekahuna, (*b*) relative amplitude of harmonic tremor on the middle East Rift, and (*c*) periods of lava production, December 1982–July 1983. The vertical axes in (*b*) and (*c*) are dimensionless. (*Hawaiian Volcano Observatory, USGS*)

15, but after an 8-day hiatus, a small quantity of lava was extruded during the evening of January 23.

Phase 2 (February 10–March 4). Activity was largely limited to incandescence and emission of burning gases through early February. The eruption resumed on February 10 when increased production of lava spatter was observed at the vents that had been active on January 7–8. By February 12 a second small spatter cone had formed, and a glowing crack extended tens of feet northeast of the two cones. Through February 24, intermittent low lava fountains and small lava flows formed a flat-topped shield about 600–700 ft (200 m) long and 300 ft (100 m) wide, capped by a line of spatter cones. Activity intensified early on February 25, but the strongest lava emission shifted 300 ft (about 100 m) uprift and activity ceased at the shield. The main lava fountain was commonly 150–250 ft (40–80 m) high and 100 ft (30 m) in diameter at its base, rising from a lava pond 200 ft (60 m) across. Lava spilled from the south side of the pond and followed the path of the large January 7 flow at first, but was diverted southeast by early February 27. On March 2, the lava entered the sparsely populated Royal Gardens subdivision (Fig. 5), where it destroyed two dwellings before lava production stopped March 4.

Phase 3 (March 28–April 9). Lava fountaining was reported early March 28 just inside the Hawaii Volcanoes National Park boundary, about 1 mi (2 km) uprift from the site of the main late-February to early-March lava production. Vigorous activity at this vent continued until late March 30, and lava flowed nearly 3 mi (5 km) to the southeast along the park boundary.

About 1 mi (2 km) downrift, activity began March 28 from the late-February to early-March vent, and two lava fountains eventually formed. The more vigorous north fountain, commonly 300 ft (100 m) high and at times estimated to be 1000 ft (300 m) high, was continuously active by late March 29. Lava moved slowly about 2 mi (3 km) to the northeast. Lava production from the south fountain became continuous on April 4, feeding three successive flows that advanced 300 to more than 600 ft (100–200 m) per hour and at times faster than 660 ft (200 m) per hour. The longest of these flows entered the

Royal Gardens subdivision on April 8, destroyed six structures, and continued to advance through at least April 11, two days after the supply of new lava from the vent stopped. Pele's hair (a form of volcanic spun glass traditionally thought to be hair from the Hawaiian volcano goodess Pele) fell 10 mi (17 km) away. When lava fountaining ended on April 9, harmonic tremor decreased significantly and summit deflation stopped.

Phase 4 (June 13–17). Harmonic tremor, at a very low level between eruptive phases, began to build early on June 13, and lava fountains were first reported during the late morning, from the vent near the Hawaii Volcanoes National Park boundary that had been active during March 28–30. Lava production quickly concentrated at the northeast end of the vent area, building a steep-sided spatter cone 100–150 ft (30–40 m) high. Lava cascaded over a spillway one-half to two-thirds of the way up the south side of the cone. A low fountain, up to about 60 ft (20 m) high, emerged from the surface of the lava pond that filled the cone to the level of the spillway. Most of the lava moved southeast at about 100–600 ft (30–200 m) per hour, flowing 4.7 mi (7.5 km) along the National Park boundary on top of and beside the late-March lava (Fig. 5). The flow entered the Royal Gardens subdivision only locally, and no homes were destroyed. Harmonic tremor declined rapidly in the early afternoon of June 17 as lava production ended abruptly.

Phase 5 (June 29–July 3). About midmorning on June 29, harmonic tremor began to increase again, and an hour later a pool of lava was seen slowly rising in the main June 13–17 vent. By early afternoon lava was cascading over the spillway and moved southeastward over earlier flows. Advancing at average rates of 250–550 ft (80–165 m) per hour, the flow front entered the Royal Gardens subdivision on July 1, where it burned and crushed seven dwellings and cut off four others from road access. Periods of stagnation up to a few hours long alternated with rapidly moving surges that advanced the flow front by 300–1000 ft (100–300 m) in 30 min. A second vent on the west flank of the cone sent small lava flows to the north and northeast until mid to late afternoon on June 30, when it began to feed a flow that extended 3 mi (5 km) along the southwest edge of the March–April and mid-June lavas (Fig. 5).

Fountains rose to as much as 150 ft (50 m) from lava ponds within the vents. As harmonic tremor amplitude dropped dramatically early on July 3, the vents stopped emitting lava, but the main flow continued to advance for more than 3 h, finally halting 5 mi (8 km) from the vent (Fig. 5).

Phase 6 (July 21–25). Renewed activity from the same cone was first observed about dawn of July 21. Lava cyclically filled and drained from its funnel-shaped interior until midafternoon the next day, when the pond filled to a depth of about 60 ft (20 m) and lava spilled over the rim to begin feeding flows to the north, northeast, and southeast. Fountains played continuously from the surface of the lava pond, reaching heights of 150–500 ft (50–150 m) on July 23. Blocked from advancing southeast by flows and vent deposits from earlier phases, the main flow moved northeast at as much as 650 ft (200 m) per hour, reaching 4 mi (6 km) from the vent before activity stopped on July 25 (Fig. 5).

As of early August, eruptive activity had not resumed. However, low-level harmonic tremor was continuing as it had between previous active phases, suggesting that the East Rift remained "open" to the magma draining from the summit chamber. Chemical and microscopic analysis indicates that all of the lava erupted in 1983 is material that had previously been intruded into the East Rift, stored there for several years, then pushed to the surface by magma leaving the summit chamber. From tilt measurements, geophysicists calculate that about 3.5×10^9 ft^3 (10^8 m^3) of magma has entered the East Rift from the summit chamber during the 1983 eruption.

For background information *see* ANHYDRITE; CLIMATOLOGY; MAGMA; PUMICE; VOLCANO; VOLCANOLOGY in the McGraw-Hill Encyclopedia of Science and Technology.

[LINDSAY McCLELLAND; EDWARD W. WOLFE]

Bibliography: P. E. Damon and E. Montesinos, Late Cenozoic volcanism and metallogenesis over an active Benioff zone in Chiapas, Mexico, *Ariz. Geol. Soc. Digest*, 11:155–168, 1978; W. A. Duffield et al., Storage, migration, and eruption of magma at Kilauea Volcano, Hawaii, 1971–1972, *J. Volcanol. Geotherm. Res.*, 13:273–307, 1982; J. B. Gill, Orogenic andesites and plate tectonics, 1981; G. A. MacDonald, A. T. Abbot, and F. L. Peterson, *Volcanoes in the Sea: The Geology of Hawaii*, 2d ed., 1983; M. P. Ryan, R. W. Koyanagi, and R. S. Fiske, Modelling the 3-dimensional structure of macroscopic magma transport systems: Application to Kilauea Volcano, Hawaii, *J. Geophys. Res.*, 86(B9):7111–7129, 1981; H. Sigurdsson et al., Eruption column collapse, surge and pyroclastic flows of El Chichón, *Am. Geophys. Union. Trans. EOS*, 63(45):1126, 1982; D. A. Swanson, W. A. Duffield, and R. S. Fiske, *Displacement of the South Flank of Kilauea Volcano: The Result of Forceful Intrusion of Magma into the Rift Zones*, USGS Prof. Pap. 963, 1976.

Waste energy recovery

Millions of gallons of combustible liquid wastes and organic chemical compounds are generated each year in the United states. Portions of these are recycled, or destroyed by conversion to harmless carbon dioxide and water by the processes of combustion or thermal oxidation, thus eliminating their potentially hazardous form. The remainder are placed in landfills or improperly disposed of at high cost, without eliminating the hazards. Most of the heat released by combustion can be recovered by heat exchange to steam, liquids, or air, replacing costly fuel. The value of the recovered heat is an economic benefit which helps pay the cost of disposal.

The Resource Conservation and Recovery Act of 1976 regulates the "cradle-to-grave" disposition of wastes. A list, Principal Organic Hazardous Compounds, has been prepared which identifies substances that could be hazardous when improperly disposed of. These compounds must be listed on manifests which are part of the permitting system intended to follow them to their ultimate disposal. Many of these materials are flammable liquid wastes and toxic chemicals, which are suitable for thermal destruction under certain conditions. Combustible liquids may be classified as wastes if they cannot be used in their present form, and if they contain contaminants or materials which are unsuitable or hazardous for disposal.

Thermal destruction. Systems used to thermally destroy wastes consist of storage and pumping systems, burners, combustion air fans, primary and secondary combustion or mixing chambers, boilers and exchangers, and induced-draft fans and stacks to discharge the gaseous products above ground level (Fig. 1). Emission controls may be required.

Petroleum wastes containing primarily carbon and hydrogen have heating values ranging from 16,000 to 21,000 Btu/lb (38–49 megajoules/kg). Alcoholic compounds and solvents, containing a large fraction of oxygen, have a dry heating value of about 10,000 Btu/lb (24 MJ/kg), or less when contaminated with water. Aqueous wastes may contain about 5% organic contaminants, which may contribute enough heat to evaporate the water. Halogenated wastes containing chlorine, bromine, and fluorine may have heating values as low as 4,000 Btu/lb (9.4 MJ/kg). If the heating value of the waste is insufficient to sustain combustion, auxiliary fuel or waste may have to be cofired to maintain flame stability and assure temperatures necessary for destruction of the waste.

In order to burn, liquid wastes must be atomized into small droplets which will vaporize rapidly, and injected into a high-temperature environment in the presence of oxygen. Light oils and solvents are readily atomized; heavy sludges can be heated or thinned, or may need special equipment such as rotary kilns to provide suitable conditions for combustion. Combustion of liquid and gaseous wastes requires evaporation of liquid fuel and water, raising the fuel and air to the ignition temperature, mixing the reactants properly, and providing some time (up to several seconds) to react. Autoignition temperatures vary with the chemical nature of the substance, and range from 600 to 1100°F (320 to 590°C) for most common combustible liquids and gases. Final combustion temperatures depend on the heating value of the fuel, moisture content, and the amount of air supplied.

The chlorides in the products of combustion, primarily hydrochloric acid (HCl), can be removed by wet scrubbers which dissolve acids, by dry scrubbers which adsorb the HCl onto dry particles collected by a fabric filter or other device, or by adding alkaline materials to the waste stream of the flame, thus forming harmless salts, for example NaCl.

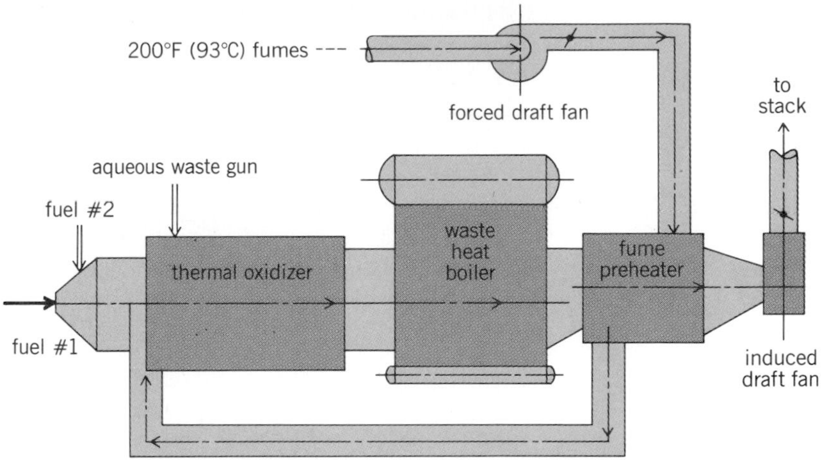

Fig. 1. Thermal oxidizer for destruction of several waste streams, with waste heat boiler, fume preheater, and forced- and induced-draft fans. (*After F. Hasselriis, Design and operation of a versatile pollution control liquid waste thermal destruction system with maximum energy recovery, J. Energy Resource Technol., June 1983*).

Flame destruction and thermal oxidation. Flame oxidation, flameless thermal oxidation, and pyrolysis take place during the destruction process. The temperatures achieved after flame combustion range from 1500 to 3200°F (820 to 1800°C). The radicals released under these conditions are extremely effective in the destruction of most organic compounds. In either flameless or flame oxidation conditions, compound destructibility is a function of the chain propagation and branching rates, which in turn are a function of the bond dissociation energy required for the particular chain propagation reactions. In the presence of flames at temperatures in excess of 1100°F (590°C), combustion of gases can occur in milliseconds, followed by a sudden increase in temperature resulting from the release of heat.

Burners are designed to atomize the fuels, and to introduce air in one or more stages to create highly turbulent combustion. The combustion chamber, usually refractory-lined to resist high temperatures, provides additional mixing, temperature, and time to destroy the trace of hydrocarbons left over from the flame zone. Those organic molecules which do not meet oxygen molecules may be pyrolized rather than oxidized, forming products of incomplete combusion, which may be destroyed beyond the flame by thermal oxidation. Sufficient and properly distributed oxygen must be provided to carry out the oxidation of the compounds present, plus enough additional excess air to provide oxygen in the right proportions to match the fuel in all locations of the flame and beyond. Insufficient oxygen reduces the flame temperature due to the partial combustion, and excessive combustion air reduces the flame temperature because of the large amount of nitrogen which accompanies the oxygen, both of which absorb heat from the reaction.

Wastes can be injected separately into the flameless region of a furnace and raised to sufficient temperature, and the necessary oxygen supplied at the right time and location, to assure destruction of the wastes. Successful incineration of waste is a function of various physical and chemical properties of the waste feed, including surface tension, viscosity, volatile content or evaporation rate, and water content, as well as the incinerator design parameters, such as design of the fuel nozzle or the burner and furnace configuration.

Thermal destruction of difficult compounds. Halogens, including chlorides, bromides, and fluorides, form chemical compounds which are highly stable in their normal form, such as polychlorinated biphenyls, dioxins, and furans, and are irritating or toxic to humans and other forms of life. They are also relatively difficult to destroy. While totally destroyed in flames within a range of satisfactory oxygen concentrations, any that may escape the flame due to nonideal conditions must be destroyed under suitable temperatures and oxygen levels.

The temperatures to achieve required destruction levels can be determined in the laboratory. Figure 2*a* shows test data obtained in purely thermal (nonflame) destruction of a number of typical compounds plotted as the logarithm of the destruction efficiency versus the temperature to which the gases are subjected. The data fall on straight lines which can be described mathematically in terms of the chemical properties and rate constants of the substance. The lines virtually intersect 100% compound (zero destruction) at approximately the autoignition temperature, and fall at an angle reflecting the kinetic constants of the substance. The temperature at which 99.99% destruction will occur (T99.99) can be determined experimentally or from empirical equations. Biphenyl, dibenzofurans and dioxins, and vinyl chloride follow similar destruction curves, showing T99.99 below 1380°F (750°C) for residence times over 0.5 s. Figure 2*b* shows that residence time has a comparatively small influence on destruction efficiency: increasing residence time from 1 to 2 s reduces the temperature required to destroy vinyl chloride by about 36°F (20°C).

Heavy polychlorinated biphenyls, which have been used because of their fire-resistant properties, are more difficult to destroy. Decachlorobiphenyl and hexachlorobenzene show similar destruction

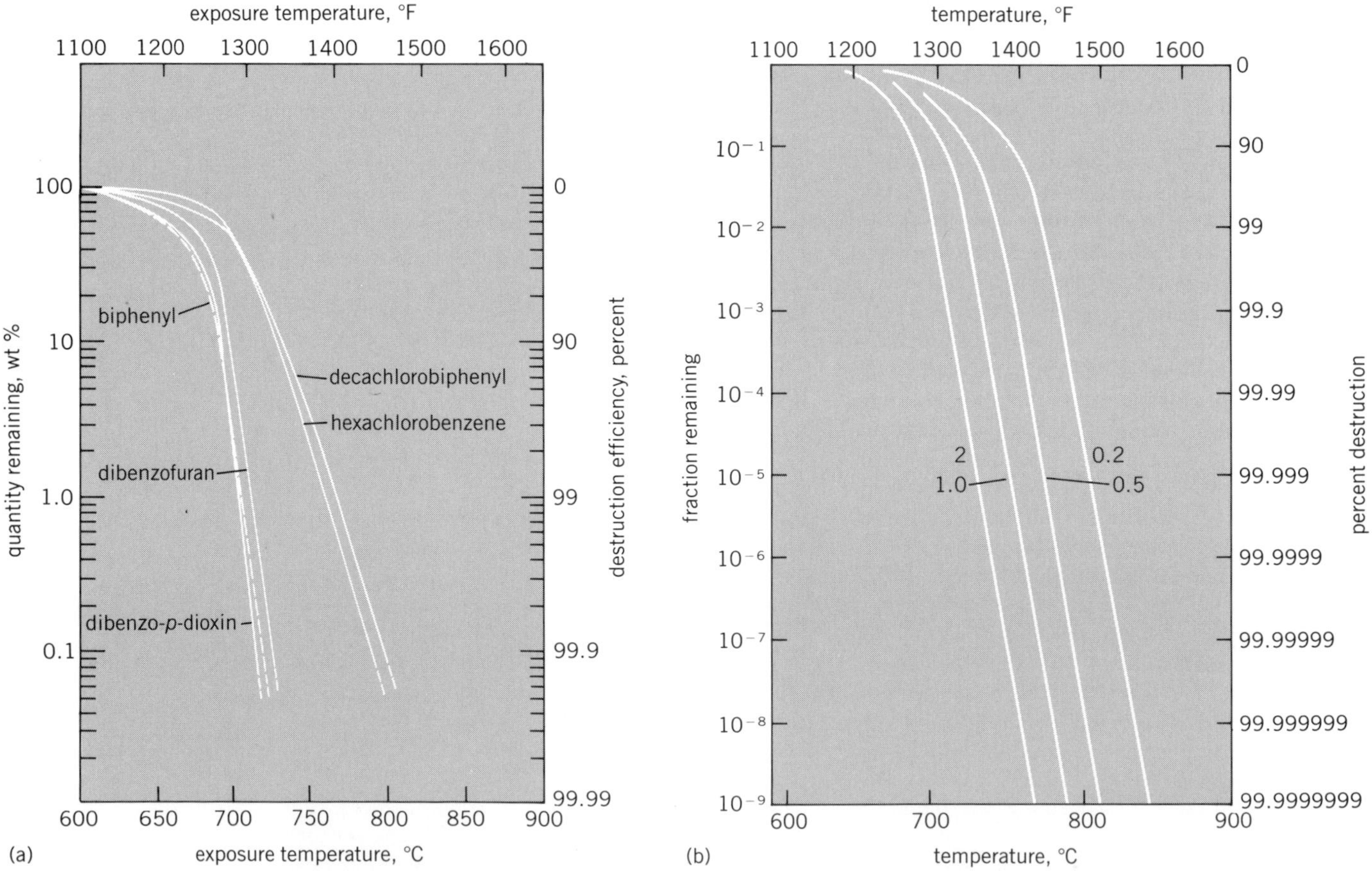

Fig. 2. Graphs showing the destruction efficiency of thermal oxidation. (*a*) Thermal destruction profiles for individual compounds as a function of exposure temperature [in air, residence time: 1.0 ± 0.1 s] (*after B. Dellinger et al., Laboratory determinations of high temperature decomposition behavior of industrial organic materials, 75th Annual Meeting of the Air Pollution Control Association, New Orleans, 1982*). (*b*) Destruction of vinyl chloride as a function of retention time and temperature, residence time indicated by numbers at the curves (*after K. C. Lee et al., Revised model for the prediction of the time-temperature requirements for thermal destruction of dilute organic vapors and its usage for predicting compound destructibility, 75th Annual Meeting of the Air Pollution Control Association, New Orleans, 1982*).

curves, and a T99.99 of 1530°F (830°C).

The effect of variations in oxygen concentration of the combustion atmosphere, and retention time on the destruction of pentachlorobiphenyl are shown in Fig. 3. As the oxygen concentration is reduced, additional temperature is required to achieve the same destruction efficiency. T99.99 is achieved at 1400°F (760°C) in air, 1530°F (830°C) with 2.5% oxygen. In pure nitrogen a temperature of 1830°F (1000°C) may be required. If furnace temperatures are not high enough, some chlorinated compounds can form intermediate compounds which may require temperatures still higher than those required by the parent compounds.

Recovery of heat energy. The recovery of heat from the high-temperature gaseous products of waste combustion reduces the amount of expensive prime fuels, such as oil and gas, which would otherwise be required to produce needed energy. The savings thus help to pay for the systems needed to store, mix, filter, sediment, pump, and feed the wastes to the boilers, as well as emission control and ash disposal costs.

The burners, furnace, and recovery equipment must be designed so that the products of combustion are maintained at high enough temperatures and are sufficiently well mixed to assure the destruction of difficult compounds before the gases are cooled down by heat recovery surfaces. Otherwise, products of incomplete combustion would be formed and escape into the atmosphere.

When wastes are destroyed in refractory-lined furnaces (thermal oxidizers), the minimum temperature maintained in the furnace must be high enough to assure destruction of the critical difficult compounds in the waste. Higher temperatures, up to the refractory limit, may be used to permit delivery of additional heat to the heat recovery system, as required, provided that sufficient oxygen is also provided. Fumes generated by the associated processes may be used as combustion air for the thermal oxidizer, subject to the same limits.

Many chemical manufacturers have installed incinerators (or thermal oxidizers) within their facilities (on site), incorporating heat recovery, and thus reducing their overall purchased fuel costs as well as the costs of transportation and disposal of liquid wastes off site. The continuous demand for heat in many chemical processing plants and petroleum refineries may provide a steady market for the recovered heat.

A number of facilities have been built which accept wastes from off-site sources. The heat recovered from destruction of these wastes can be used for distillation and other on-site processing purposes. However, the irregular nature of the waste stream and the need for heat limits the feasibility of heat recovery.

Thermal oxidizers coupled to boilers. Some industrial plants have installed thermal oxidizers coupled to boilers. These serve the multiple purposes of incinerating odorous or contaminated fumes from treating ovens and other processing equipment while generating all of the steam and some power for their facilities. These plants have burned not only their own wastes but also large amounts of wastes shipped by distant waste processors and producers. One facility near Philadelphia has burned about 3,000,000 gallons (10,000,000 liters) of imported wastes per year while supplying 30,000–70,000 lb/h (10,000–30,000 kg/h) of steam to a plant producing phenolic laminates.

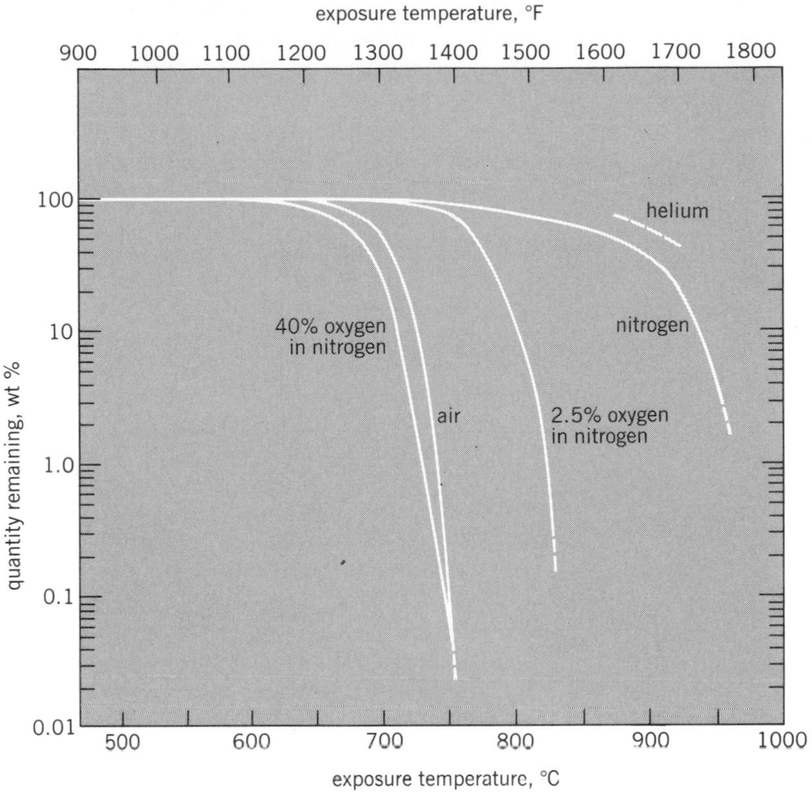

Fig. 3. Decomposition of pentachlorobiphenyl in different gaseous atmospheres showing effect of oxygen concentration on temperature required for compound destruction at 2 s residence time, except for helium with 1.2 s residence time. (*After B. Dellinger et al.*)

Waste burning in conventional boilers. Clean waste liquids can be burned in conventional boilers. Since they are likely to vary considerably in heating value and viscosity, interfering with control of fuel/air ratios, more sophisticated control methods for fuel and air controls are necessary. If only a small percentage of waste is burned, the ability to control fuel/air ratios in the boiler is not seriously affected, but as waste disposal rates increase, control becomes more critical and difficult. When mixed wastes are burned, with varying heating values, a primary means of controlling the feed of waste to the furnace is the furnace temperature itself, thus assuring total destruction of the principal or most difficult-to-burn substances. A carbon monoxide meter sampling the stack gases can be used to trim the combustion air to achieve optimum destruction efficiencies.

Many wastes can be blended with fuel oil and

burned in industrial furnaces, thus serving as an effective and economical means of disposal and minimizing disturbance to the combustion systems. When burning chlorinated compounds, it is necessary to determine the ash and halogen content, as well as the ability of the burner, furnace, and boiler configuration to cope with these contaminants, so that under all operating conditions destruction of difficult compounds is assured and products of incomplete combustion are kept to acceptable limits. Emissions controls may be required to add removal efficiency to destruction efficiency.

The carbon monoxide and total hydrocarbons measured in the stack gas emissions of waste burners (as well as conventional fuel burners) are good indications of the efficiency of combustion. Low carbon monoxide serves as evidence that difficult wastes are indeed being destroyed. This is especially important when wastes are burned in boilers with little if any refractory to maintain high combustion temperatures prior to cooling the gases by the boiler tubes.

Heat recovery equipment to recover heat from the products of combustion is similar if not identical to that used to recover heat from conventional fuels. When the wastes liberate ash, especially abrasive ash, and ash which has a low enough melting point to have slagging and fouling tendencies, the tubes must be arranged in straight rows to minimize erosion and permit cleaning of the tubes to maintain heat transfer rates and prevent plugging of gas passages. Finned tubes can be used where not subjected to excessive fouling from dust particles.

For background information *see* ENERGY CONVERSION in the McGraw-Hill Encyclopedia of Science and Technology.

[FLOYD HASSELRIIS]

Bibliography: B. Dellinger et al., Laboratory determinations of high temperature decomposition behavior of industrial organic materials, *75th Annual Meeting of the Air Pollution Control Association, New Orleans*, 1982; F. Hasselriis, Design and operation of a versatile pollution control/liquid waste thermal destruction system with maximum energy recovery, *J. Energy Resourc. Technol.*, June 1983; K. C. Lee et al., Revised model for the prediction of the time–temperature requirements for thermal destruction of dilute organic vapors and its usage for predicting compound destructibility, *75th Annual Meeting of the Air Pollution Control Association, New Orleans*, 1982; W. Tsang and W. Shaub, *Chemical Processes in the Incineration of Hazardous Waste*, National Bureau of Standards, 1981.

Whistling

Although the acoustic mechanism underlying whistling is still not completely understood, sufficient empirical data exist to reveal aerodynamic and acoustic similarities between systems of widely differing geometry, and to make it possible to predict the presence, frequency, and amplitude of whistles with an increasing degree of accuracy. The basic mechanism involves an instability of a fluid flow that converts the energy from the steady flow to oscillatory energy. One or more feedback paths sustain and control the oscillation, resulting in the typical whistle sound with a relatively high amplitude at a narrow band of frequencies.

The particular form that the flow instability will take, and the type of control exerted by the feedback, are both heavily dependent on the geometry of the system. Experimentation with different geometries has led to some general rules predicting where instabilities will form and the type of symmetry they will exhibit. The flow rate at which a given geometry will generate instabilities must generally also be determined empirically. The frequencies at which a given geometry will whistle can be predicted more easily than the range of flow rates within which these whistles will occur, but the extension of such formulations to systems with different geometries is not always obvious.

Whistles have been investigated for over a century, resulting in several practical applications, such as better control of unwanted whistles, and the invention of new whistles. Mathematical formulations of the instability and feedback mechanisms have been developed that make it possible to relate seemingly disparate phenomena. More recently this has led to a thorough analysis of the tone production of organ pipes and to the extension of results to the relatively inaccessible system of human whistling.

Acoustic mechanism. An orifice tone arises when air flows through a single constriction (Fig. 1). It is known experimentally that air going through a con-

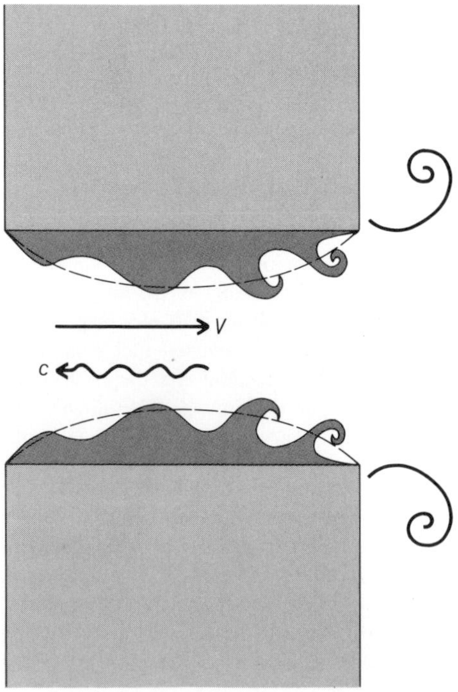

Fig. 1. Diagram of the acoustic mechanism of the orifice tone. Air flows through the constriction at a velocity V; the acoustic waveform generated by the vortices at the outlet travels back to the inlet at the speed of sound, c. (*After C. Shadle, Experiments on the acoustics of whistling, Phys. Teacher, 21(3):148–154, 1983*)

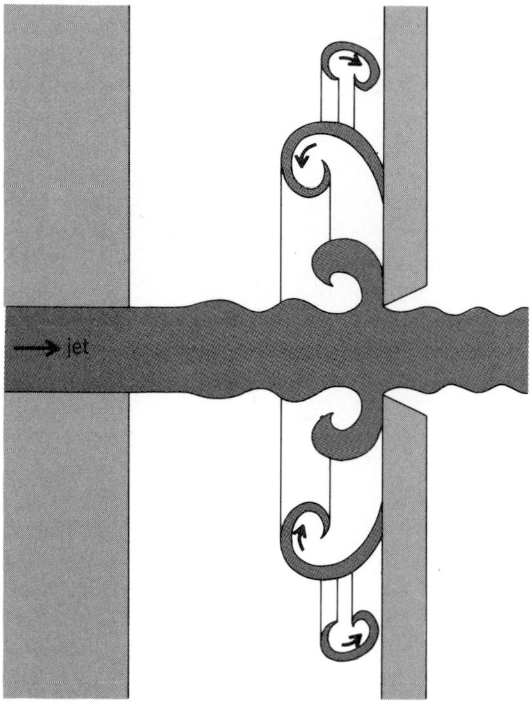

Fig. 2. Cross-sectional view of the hole-tone system, showing the ring vortices as revealed by smoke. (*After R. C. Chanaud and A. Powell, Some experiments concerning the hole and ring tone, J. Acoust. Soc. Amer., 37(5):902–911, 1965*)

striction will form a vena contracta, that is, a jet of air with a cross-sectional area less than that of the constriction. Thus the jet is surrounded by a quiescent annulus of air. It has been shown both theoretically and experimentally that a disturbance of the boundary between two regions of differing airflow velocity will, under certain conditions, propagate downstream with an exponentially increasing displacement. The greater the difference between the flow velocities, the faster the exponential rises; this means the perturbations will be amplified more when the flow rate through the orifice is higher. When they are large enough, these perturbations will "roll up" into ring vortices, which are convected downstream.

The widening of the constriction presents a discontinuity to the vortices, which generate acoustic pressure fluctuations as they roll out the end. This acoustic waveform travels upstream and perturbs the jet at the inlet, where it is most unstable, and this disturbance in turn propagates downstream with increasing amplitude. Although the initial perturbations of the boundary layer occur at random intervals, within a few cycles the positive feedback of the acoustic waveform emphasizes one characteristic frequency and its harmonics above all others.

An orifice tone will not occur unless the inlet has abrupt edges. This apparently provides a definite location for the flow to detach from the walls and thus potentially to become unstable. An unstable shear layer is common to all geometric configurations that give rise to whistles. The aeolian tone is produced by air flowing past a cylinder. Two layers are formed on the downstream side of the cylinder; for a flow rate in the unstable range, these layers alternately curl up into vortices, applying an alternating force to the cylinder. The effect of this force was shown most dramatically in the Tacoma (Washington) Bridge disaster (1940), where the fluctuating forces coupled into the resonances of the bridge. The sounds heard when the wind blows through porch railings or tree branches are examples of aeolian tones.

The edge tone occurs when a jet of air exiting from a nozzle strikes a wedge and sets up a pattern of alternate vortices. The shape of the wedge is not critical; it can be a thin plate on edge, a cylinder, a wire, or a wedge of any angle through 180°. The vortex street forms even in the absence of the wedge, but dies out in a much shorter distance. The wedge creates paths for both aerodynamic and acoustic feedback that sustains the oscillation. The sounds produced by organ pipes and other wind instruments, and by blowing across the top of a bottle, all depend on the edge tone mechanism.

A hole tone occurs when air flows through concentric holes in two closely spaced parallel plates. Through the use of flow visualization techniques, ring vortices have been observed in the space between the two plates (Fig. 2). The hole tone is much less sensitive than the orifice tone to the shape of the hole through which the air passes. There may be walls on either side; if present, they establish additional feedback paths. The whistling teakettle and, apparently, the human whistle both depend on the hole tone mechanism. The configuration for a typical human whistle includes reduced crosssectional areas above the tip of the tongue and at the lips (Fig. 3). Although flow visualization techniques

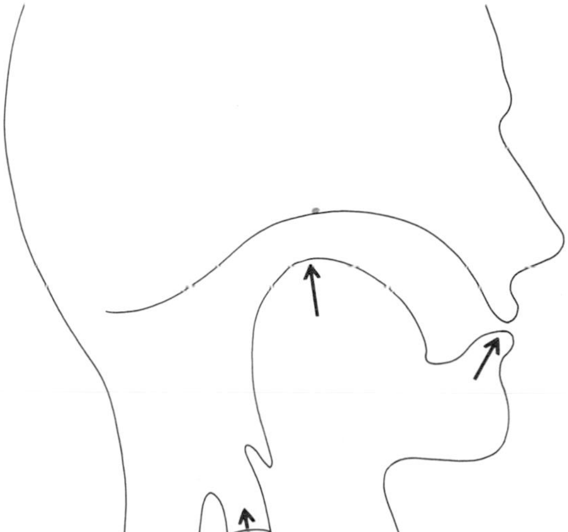

Fig. 3. Cross-sectional view of the human vocal tract in a typical whistle configuration, showing certain geometric similarities to the hole-tone configuration. The arrows point to regions of reduced cross-sectional area.

cannot be used inside a vocal tract, it seems plausible that the first constriction generates an unstable jet and the second constriction reinforces the instabilities, producing ring vortices between the tongue and lips.

Range of instability. Conditions under which boundary layers are unstable can be predicted theoretically for simple geometries, but are difficult to compute for more complex shapes. Further, flows are generally unstable at a much lower flow rate than predicted, due to upstream disturbances. The degree of instability can be specified by using the Reynolds number (Re), a dimensionless parameter defined by the equation below, where ν = kine-

$$Re = \frac{Vd}{\nu}$$

matic viscosity of the fluid (0.15 cm^2/s for air), V = a representative velocity, and d = a representative dimension. As V or d is increased (retaining the same geometrical shape, though not necessarily the same size), Re increases to and beyond Re_{crit}, the critical Reynolds number at which the flow becomes unstable. Re_{crit} varies from one geometrical configuration to another, and for some configurations a different Re_{crit} exists for each mode of instability. For a circular jet, Re_{crit} is on the order of 10^1 to 10^2 depending on the exact mode of instability. The regions of occurrence of hole and edge tones all fall within the unstable regions for a circular jet, although each kind of tone operates over a different range of Re. For human whistling, Re is found to be well above Re_{crit} for a circular jet; in view of the irregularities in the vocal tract, this indicates that the flow is probably fully turbulent.

A more precise method of calculating unstable regions involves measuring the admittance function of the jet in an organ pipe. This function, which is the ratio of the acoustic flow out of the mouth of the pipe to the acoustic pressure just inside the mouth, forms a spiral which alternates between stable (energy-subtracting) and unstable (energy-adding) regions for the jet. Since the position on this spiral is determined by the frequency of the jet oscillations and the blowing pressure, regions of instability can be predicted for this particular geometry much more precisely than with the Reynolds number.

Frequency prediction of whistles. The presence of a whistle and, to some extent, how readily it occurs and settles into steady state are determined by the stability of the shear layer of the jet. The frequency and amplitude of the whistle, once established, depend on the feedback paths. When there are no resonators present, the feedback path generally consists of the convection of the vortices downstream from the point of generation, and the propagation of the acoustic wave upstream from the discontinuity at which it was generated. Since the vortex convection velocity depends linearly on the mean flow velocity, the frequency of a whistle produced without resonators rises nearly linearly with velocity. This has been shown to hold for aeolian, orifice, hole, ring, and edge tones. For the hole, ring, and edge tones, the frequency is also inversely dependent on the distance between the plates or between the jet and the edge or ring; since these form the two ends of the feedback path, a greater distance increases the time required to traverse the loop, thus lowering the whistle frequency. Different states are possible which correspond to different integral numbers of wavelengths packed into the feedback path. The whistle tone jumps abruptly from one state to the next as the flow velocity increases, causing a frequency jump, with a hysteresis loop.

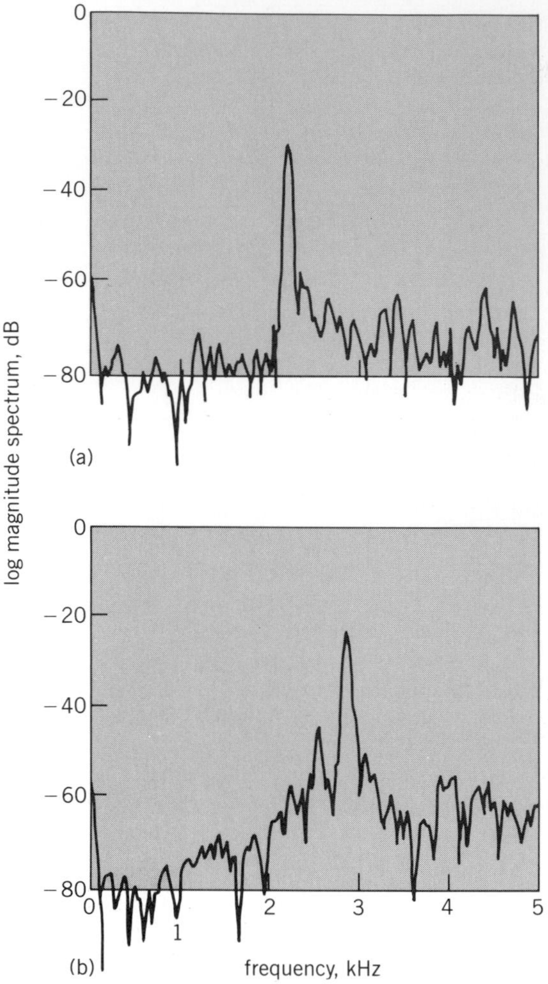

Fig. 4. Log-magnitude spectra taken of 25-millisecond portions of (*a*) a regular whistle near high end of subject's range and (*b*) an overblown whistle, produced by using a faster flow velocity for the same vocal tract configuration as in *a*. (*After C. Shadle, Experiments on the acoustics of whistling, Phys. Teacher, 21(3):148–154, 1983*)

It has been found that whistle frequencies alter substantially to match the natural resonances when cavities or reflectors are added to a system, and tend to cluster at the intersection of the cavity resonance frequencies and the frequencies proportional to velocity predicted for the system without resonators. This type of frequency prediction has been refined for an apparatus in which flow passed over a cavity, striking an edge. Using hot-wire anemometers, the displacement of the shear layer across the cavity opening was measured, and was used to derive forward-and-backward admittance functions similar to those measured for the jet in an organ pipe. This method can be used to predict not only frequency but also bandwidth of the resulting tones, giving good agreement with the observed spectra.

Through observation of the growth and decay progression of the edge tones as flow rate increased, evidence has been found for the capture of frequency-mobile shear tones by the cavity resonances, with enhancement of the cavity resonance possible under favorable phase conditions. Similar progressions have been observed for laminar and turbulent flow, though resonances are excited at different flow speeds in the two cases, due to different velocity profiles. The growth-and-decay progression for turbulent flow is similar to that observed for human whistles. Figure 4 shows spectra for two different flow rates through the same vocal tract configuration. The two graphs depict a frequency jump between neighboring vocal tract resonances; when the whistle is coupled into the upper resonance, the lower resonance is still clearly excited, at a lower amplitude, by the background turbulent rumble. The evidently correct application to human whistling of results from a more accessible system bodes well for the eventual comprehension of whistles produced with more exotic vocal tract configurations.

For background information *see* FLUID-FLOW PROPERTIES; KARMAN VORTEX STREET; MUSICAL INSTRUMENTS in the McGraw-Hill Encyclopedia of Science and Technology.

[CHRISTINE H. SHADLE]

Bibliography: R. C. Chanaud and A. Powell, Some experiments concerning the hole and ring tone, *J. Acout. Soc. Amer.*, 37(5):902–911, 1965; S. A. Elder, R. M. Farabee, and F. C. Demetz, Mechanisms of flow-excited cavity tones at low Mach number, *J. Acoust. Soc. Amer.*, 72(2):532–549, 1982; N. H. Fletcher and S. Thwaites, The physics of organ pipes, *Sci. Amer.*, 248(1):94–103, 1983; D. K. Holger, T. A. Wilson, and G. S. Beavers, The amplitude of edgetone sound, *J. Acoust. Soc. Amer.*, 67(5):1507–1511, 1980.

Wing

To avoid the sonic boom, overland flights of aircraft must be limited to a subsonic ground speed, that is, approximately 750 mi/h (1200 km/h). Since the speed of sound at flight altitudes is about 100 mi/h (160 km/h) slower than at ground level, the airplane must fly supersonically in its own medium in order to reach this overland speed limit. Under standard conditions the flight Mach number at the sonic boom speed limit is about 1.15. However, with strong headwinds aloft and a high temperature at ground level the Mach number may be as high as 1.3. Under such conditions the waves made by the airplane will be refracted by wind and temperature gradients, become vertical, and disappear before reaching the ground. The sonic boom speed limit still permits a gain of some 200 mi/h (320 km/h) over current jet transports.

Fuel consumption. An airplane designed for such service should be capable of efficient flight at a variety of speeds, from subsonic to supersonic. Unfortunately flight at supersonic speed entails some loss in fuel efficiency, though the loss at transonic speeds need not be as great as the loss at higher supersonic speeds. It seems possible that the increased utilization of the aircarft and the time saved for the passengers could make up for a moderate increase in fuel consumption. This article discusses an unusual design for an airplane which is the most fuel-efficient for flight in the transonic speed range—an airplane with a pivoted olique wing.

The efficiency of current aircraft designed for subsonic speeds is remarkably high. Thus a Boeing 747 cruising at Mach .8 in the thin air at 40,000 ft (12,000 m) can get 60 passenger miles per gallon (26 km/liter), about the same as an automobile but 10 times as fast. The Antonov AN 22, with its long narrow straight wing and 20-ft-diameter (6-m) propellers could do even better, but it is somewhat slower (Mach .7) and its productivity is less.

Aerodynamic efficiency. The energy required to propel an airplane on a trip of reasonable length is basically the drag times the distance. Hence the important parameter characterizing the aerodynamic efficiency of an airplane is its lift-to-drag ratio

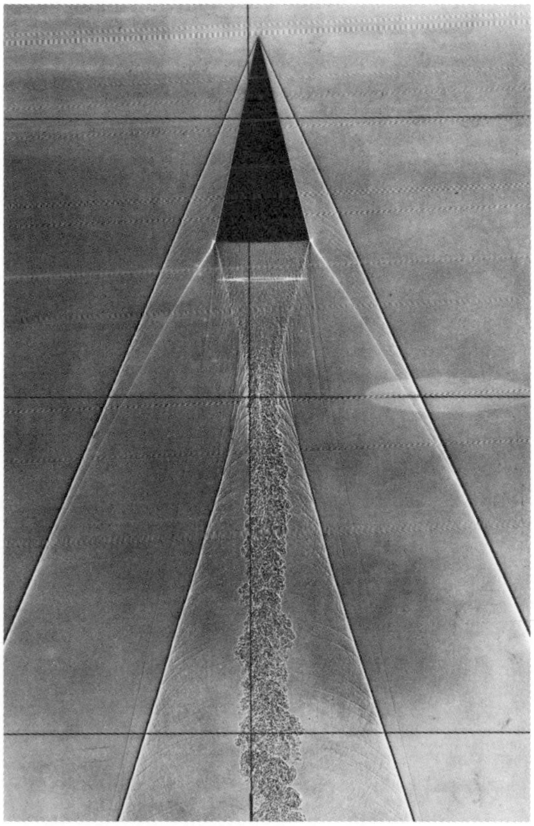

Fig. 1. Shock waves produced by a cone in a supersonic airstream.

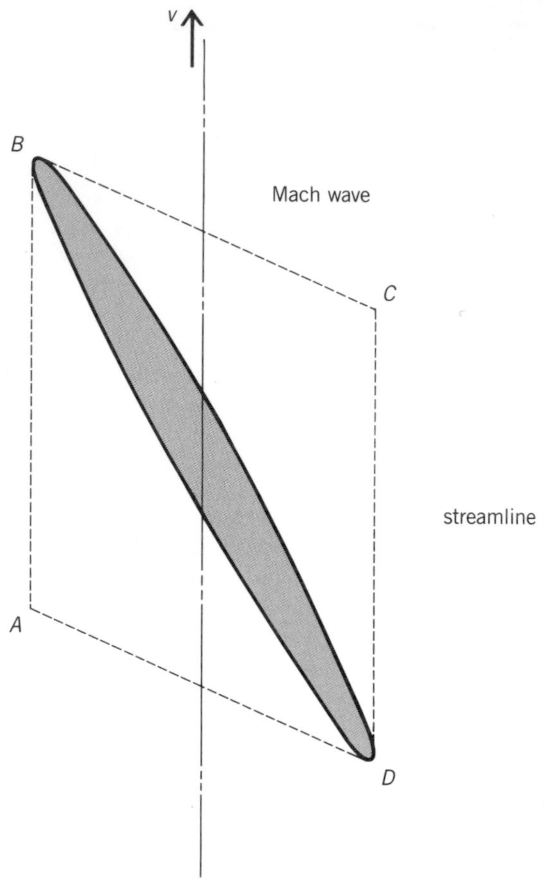

Fig. 2. Wing shape having minimum drag for area *ABCD*.

(*L*/*D*). A sailplane having long narrow wings and a small body can achieve an *L*/*D* greater than 40 to 1. Such efficiency could be maintained right up to Mach .7 if the flight altitude were increased to keep the drag from increasing with speed. However, above a Mach number of .7 the straight wing begins to develop shock waves, the flow becomes unsteady, and the *L*/*D* falls precipitously. Beyond this speed the wing must be swept or turned at an oblique angle to the flow in an effort to recapture the favorable properties of shock-free subsonic flow. If the swept or oblique wing is sufficiently long and narrow, it will behave as if it were flying at a reduced speed, equal to the component of the flight velocity in the direction perpendicular to its long axis, that is, $V_{\text{effective}} = V \cos \beta$, where β is the angle of sweep. Thus flying at Mach 1 with 45° of sweep, the component Mach number will be .7, below the critical Mach number of the wing sections.

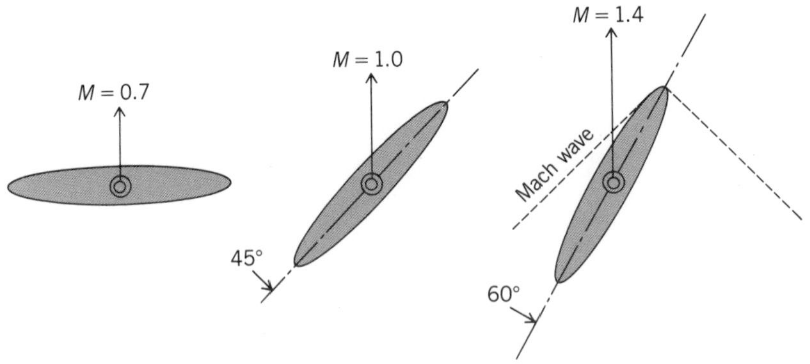

Fig. 3. Ideal wing shape at three different speeds.

At subsonic speeds it is convenient to divide the drag into two components. The friction drag is approximately proportional to the wetted area of the surface. The vortex drag is inversely proportional to the square of the wing span. A well-known mathematical problem in the theory of wings at subsonic speeds is the determination of the wing shape giving the smallest drag for a given total lift and a given span. The solution given by the theory is a long narrow elliptic planform perpendicular to the flight direction.

At speeds below the speed of sound the air in the region ahead moves to accommodate the oncoming wing. At supersonic speeds, however, no signal can travel ahead and the air meets the oncoming wing with a shock. Figure 1 shows the bow shock wave made by a body of revolution in a supersonic airstream. It came as a surprise when it was learned that the shock wave made by a supersonic airplane can reach all the way to the ground from 60,000 ft (18,000 m). The energy required to maintain this wave system is responsible for the wave drag.

Wing shape. By the early 1950s the theory of supersonic flow had progressed to the point where it was possible to consider wings and bodies having a minimum total drag, including the wave drag. Figure 2 shows the solution given by the theory for the wing shape having the highest ratio of lift to drag at supersonic speed. Here, two Mach waves (at the correct Mach angle to the flow) and any two parallel streamlines of the flow are laid out. The question then is: Of all possible distribution of a given total lift and a given total volume within this area, what particular distribution will have the smallest drag? Surprisingly, the theory gives again a long narrow ellipse. However, the ellipse must be turned at an angle, keeping its long axis well behind the Mach wave from its leading tip. Now the sweep angle of the Mach wave is such that the component of velocity normal to the wavefront is just the speed of sound. Hence placing the long axis of the wing behind the Mach wave means that the wing must have subsonic sweep. At Mach 1.4 the wave angle is 45°. With a wing sweep of 60° the component Mach number perpendicular to its long axis will be .7, below the critical Mach number of the wing sections; that is, the wing will behave for the most part as if it were flying subsonically (Fig. 3). Following the preference for bilateral symmetry, one might think of bending the forward half of the oblique wing back to form a conventional swept-back wing (or, alternatively, to bend the rear half forward to create a swept-forward wing). Unfortunately such a change would increase the transonic wave drag by an order of magnitude.

Thus aerodynamic theory indicates that the optimum wing shape is the same for both subsonic and supersonic speeds. The wing has simply to be turned about a vertical axis to adapt to the speed. This simple solution has its limits, however. At Mach 1.4 the sweep angle required is already 60°. At Mach 2 the required angle is 70°, and the gain in *L/D* over shapes such as the delta becomes marginal. Hence the oblique wing appears best adapted to the transonic speed range, from Mach .8 to 1.4.

It was not until early 1970 that definitive wind tunnel experiments were made to test the theory of the oblique wing. The tests covered the range from Mach .6 to 1.4 and were performed on a wing with semielliptical planform having a 10-to-1 axis ratio and made of solid steel. For support, the wing was mounted on a small body. Figure 4 summarizes the results of this test, which showed higher *L/D* ratios than had previously been obtained.

Following these wind tunnel tests, the Boeing Company undertook a study of possible application of the oblique wing to a passenger transport. The new design was supposed to carry 200 passengers across the United States at a speed just under the sonic boom speed limit. Boeing engineers considered four possible designs for the same task: a fixed delta wing, a fixed swept-back wing, a wing with variable sweep, and a pivoted oblique wing. The study showed the oblique wing design to be significantly better than the others in fuel economy, gross weight, operating cost, and so on. However, for reasons explained above, the fuel economy at these speeds cannot be quite as good as that of a definitely subsonic airplane. At the time the study was completed, the price and scarcity of fuel had increased to the point where it seemed better to stay with conventional designs flying at Mach .8.

Flight characteristics. A major stumbling block in application of the oblique wing has been lack of knowledge of its stability and controllability in flight. NASA has carried out flight tests with an airplane known as the AD-1, a small subsonic jet made of foam and fiberglass (Fig. 5). The AD-1 carries a single pilot, who by pushing a switch can turn the wing to any desired angle up to 60°. The controls are perfectly conventional, and no provision was made to correct the cross-coupling of motions that occurs in flight. The flight test pilots have reported little difficulty in control at wing angles up to 45°. At 60° wing angle it becomes difficult to make coordinated turns at steep bank angles. The chief difficulty here is the flexibility of the wing, which acts to resist turning in the direction of the forward tip and to promote turning in the other direction.

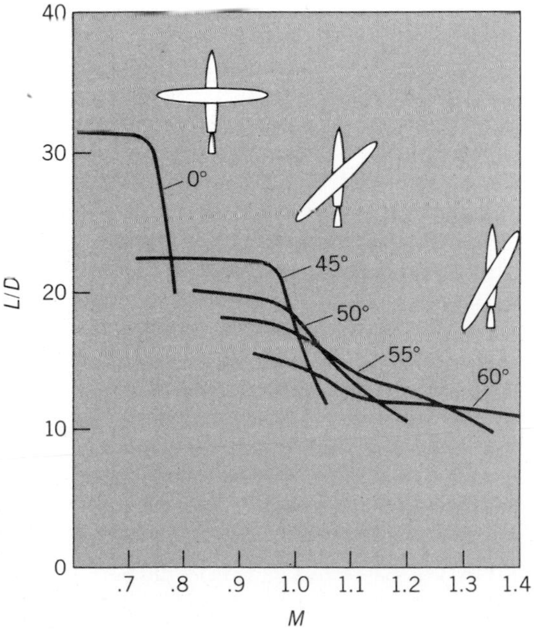

Fig. 4. Wind tunnel test of oblique winged model at supersonic speed. (*NASA Ames Research Center*)

Fig. 5. The AD-1 in flight with the wing at an extreme sweep angle (60°). (*NASA*)

Commercial oblique winged transport. Historically, increases of air speed have led to lower travel costs. Thus a single medium-sized jet can carry many more passengers across the Atlantic in a year's time than the largest ocean liner, and at lower fares. Whether or not commercial transport aircraft will be permitted to exceed the present limit of Mach .8 will depend on the price and availability of fuel. Evidently, taking the big step, going to Mach 2 and increasing the fuel consumption by a factor of 3 or 4, turned out to be uneconomical. The very fast supersonic transport has other problems in addition to its fuel consumption. It cannot utilize its full speed on overland flights but must slow down to transonic speed, below the sonic boom speed limit. The oblique winged transonic transport, designed for this speed, is much more fuel-efficient. Moreover, the oblique winged transport can easily change its wing sweep in flight, adapting to a range of Mach numbers, including .8. Because of this and other advantages, the oblique winged transport may in the future find a place among the commercial aircraft of the world.

For background information *see* AERODYNAMICS;

SHOCK WAVE; SUPERSONIC FLIGHT; TRANSONIC FLIGHT in the McGraw-Hill Encyclopedia of Science and Technology.

[ROBERT T. JONES]

Bibliography: Boeing Commercial Airplane Co., *Oblique Wing Transonic Transport Configuration Development*, NASA CR-151928, D6-75793; R. M. Hicks and J. P. Mendoza, Oblique wing: Sonic boom, NASA TM X 62, 247, February 1973; R. T. Jones, New design goals and a new shape for the SST, *Astronaut. Aeronaut.*, December 1972; R. T. Jones, The oblique wing: Aircraft design for transonic and low supersonic speeds, *Acta Astronaut.*, 4:99–109, 1977; R. T. Jones, Theoretical determination of the minimum drag of airfoils at supersonic speeds, *J. Aeronaut. Sci.*, vol. 19, no. 12, December 1952; R. T. Jones, *Wing Planforms for High Speed Flight*, NASA Tech. Rep. 863, 1947; W. P. Nelms, Jr., Applications of oblique wing technology, *American Institute of Astonautics and Aeronautics: Aircraft Systems and Technology Meeting*, Dallas, September 27–29, 1976.

Zeolite

Zeolite catalysts are characterized by small (1.0 nanometer) and uniform pores. If most catalytic sites are located inside this pore structure and if the pores are small, the reaction rates and selectivities are determined by the dimensions and configurations of the reactants, products, or transition states.

In general, one type of molecule will react preferentially and selectively in a shape-selective catalyst if its diffusivity is at least one or two orders of magnitude higher than that of competing molecular types.

The advantages of shape-selective catalysis are: continuous conversion of undesirable impurities to harmless substances or smaller molecules that can be removed easily, increase in selectivity of a desired product, and reduction of coke formation.

The role of diffusivity was recently verified by determining the molecular diffusivities in a zeolite at steady-state and actual reaction conditions. Some of the newest applications of shape-selective catalysis are catalytic dewaxing and the production of gasoline from methyl alcohol. Recently, modified zeolites with greatly improved selectivity have been developed. Deuterium nuclear magnetic resonance spectroscopy has given evidence of restriction of molecular motion of paraxylene inside a type of zeolite known as ZSM-5. Potentially the most significant new development is the discovery and synthesis of a series of aluminophosphate molecular sieves.

Properties. Zeolites are crystalline aluminosilicates. The small pores and cavities account for their "molecular sieving" properties. Ion exchange allows the introduction of cations with various catalytic properties. Zeolites in which most or all cationic sites are occupied by protons can have a very high number of very strong acid catalytic sites. If the overwhelming majority of the catalytic sites are confined within the pore structure and if the pores are small, the fate of the reactant molecules and the probability of forming product molecules are determined mostly by molecular dimensions and configurations. In shape-selective catalysis, only molecules whose dimensions are less than a critical size can enter the pores, have access to internal catalytic sites, and react there. Furthermore, only molecules that can leave appear in the final product.

Pore diameters in zeolites*

Number of tetrahedra† in ring	Maximum free diameter, nm	Example‡
6	0.28	
8	0.43	Erionite, zeolite A
10	0.63	ZSM-5, ferrierite
12	0.80	Zeolites L and Y, mordenite
18	1.5	Not yet observed

*From S. M. Csicsery, Shape-selective catalysis in zeolites, ACS Meeting, Seattle, March 20–25, 1983, *Preprints of the Division of Fuel Chemistry of the American Chemical Society*, 28(2):116–126, 1983.

†SiO_4 or AlO_4.

‡A, L, and Y refer to various structural types of synthetic zeolites.

Pore diameters depend on the number of SiO_4 or AlO_4 tetrahedra in a ring (see table). The pore size may limit the entry of the reacting molecule, the departure of the product molecule, or the formation of certain transition states. This makes it possible to distinguish the reactant, product, or restricted transition-state-type selectivities.

Reactant- and product-type selectivities. Reactant selectivity occurs when only part of the reactant molecules are small enough to diffuse through the catalyst pores (Fig. 1*a*). Product selectivity occurs when some of the products formed within the pores are too bulky to diffuse out as observed products. They are either converted to less bulky molecules (for example, by equilibration) or eventually deactivate the catalyst by blocking the pores (Fig. 1*b*).

Shape-selective catalysis most frequently involves acid-catalyzed reactions such as isomerization,

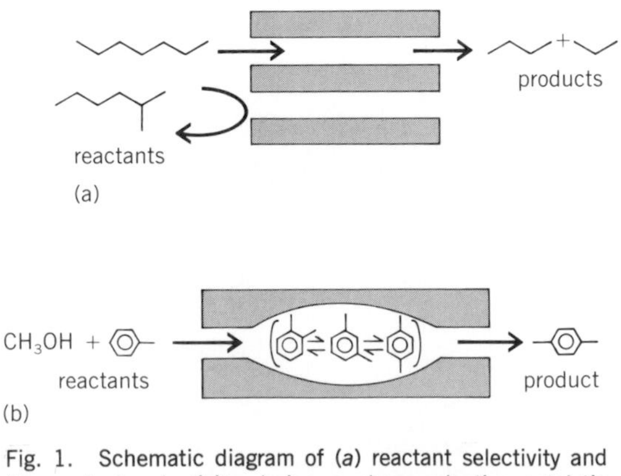

Fig. 1. Schematic diagram of (*a*) reactant selectivity and (*b*) product selectivity during a shape-selective catalytic chemical reaction. (*After S. M. Csicsery, Shape-selective catalysis in zeolites, ACS Meeting, Seattle, March 20–25, 1983, Preprints of the Division of Fuel Chemistry of the American Chemical Society, 28(2):116–126, 1983*)

cracking, and dehydration. Acid-catalyzed reactivities of primary, secondard, and tertiary carbon atoms differ. Tertiary carbon atoms react much more easily than secondary carbon atoms. Primary carbon atoms do not form carbonium ions under ordinary conditions and therefore do not react. In most cases, isoparaffins crack and isomerize much more rapidly than normal paraffins. Because normal paraffins diffuse faster through the pores of small- and medium-pore-size zeolites than isoparaffins, this order is reversed in most shape-selective acid catalysis. Normal paraffins react faster under these conditions than branched ones, which sometimes do not react at all.

Restricted transition-state-type selectivity. Restricted transition-state selectivity occurs when certain reactions are prevented because the corresponding transition state would require more space than available in the cavities. Neither reactant nor potential-product molecules are prevented from diffusing through the pores. Only the formations of the transition state is hindered. Most examples involve bimolecular transition states. For example, in the transalkylation of dialkylbenzenes, one alkyl group is transferred from one aromatic ring to another. Thermodynamic equilibrium favors the 1,3,5-trialkylbenzene isomers, and therefore they are the predominant transalkylation products over most acid catalysts. However, it has been found that over H-mordenite, 1,3,5-trialkylbenzenes are absent from the product. In the narrow, confined pores of H-mordenite the space available cannot accommodate the bimolecular transition state leading to the 1,3,5-trialkylbenzene isomers (Fig. 2). The other trialkylbenzene isomers can form in H-mordenite as their transition states are smaller.

Xylene isomerization is usually accompanied by transalkylation, forming trimethylbenzenes and toluene. Transalkylation involves a bimolecular transition state. The space available in the H-ZSM-5 molecular sieve is less than in H-mordenite, providing no room for any bimolecular transition state; xylene isomerization can proceed without trimethylbenzene formation. This has important practical consequences: xylene yields are significantly higher with H-ZSM-5 than they were with the older, non-shape-selective catalysts. Furthermore, since trialkylbenzenes are catalyst poisons, the life of the new shape-selective H-ZSM-5 isomerization catalyst is much longer than that of the old types. As a result, today most commercial xylene isomerization plants use H-ZSM-5 catalysts. In addition, restricted transition-state-type selectivity in ZSM-5 (a pentasil zeolite) molecular sieves prevents most coke precursors from polymerizing and forming coke. This has great commercial significance because certain reactions can be performed in ZSM-5 without metal hydrogenation components and high hydrogen pressure. On ZSM-5, most coke is deposited on the outer surface of the crystallites, whereas in most other molecular sieves the greater part of the coke forms within the pores. A simplified scheme of these phenomena is shown in Fig. 3. Activity is barely affected in the first case (Fig. 3*a*), while it decreases rapidly in the second (Fig. 3*b*).

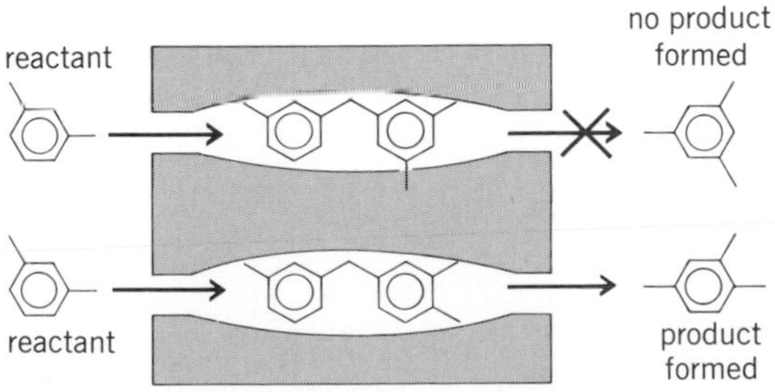

Fig. 2. Schematic diagram of restricted transition-state selectivity. (*After S. M. Csicsery, Shape-selective catalysis in zeolites, ACS Meeting, Seattle, March 20–25, 1983, Preprints of the Division of Fuel Chemistry of the American Chemical Society, 28(2):116–126, 1983*)

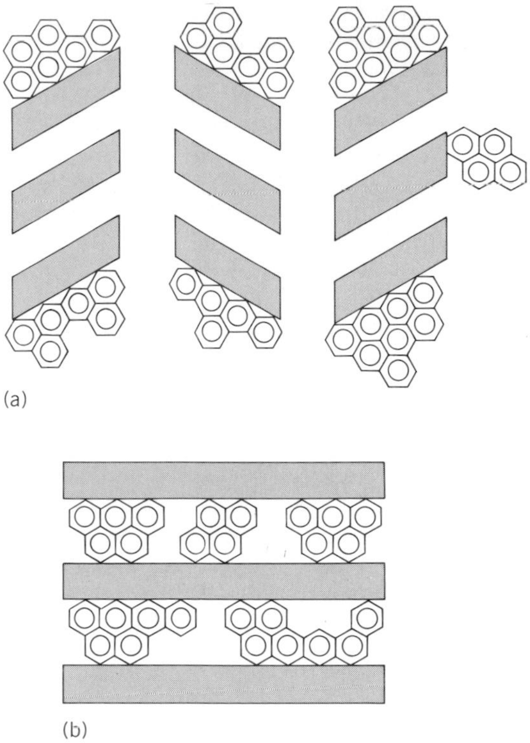

Fig. 3. Simplified schematic diagram of coke formation in (*a*) pentasil and (*b*) mordenite (and other large-pore) zeolites. Note that most of the coke is formed on the crystalline surfaces in the first case and within pores in the second. (*After S. M. Csicsery, Shape-selective catalysis in zeolites, ACS Meeting, Seattle, March 20–25, 1983, Preprints of the Division of Fuel Chemistry of the American Chemical Society, 28(2):116–126, 1983*)

Improvement of shape selectivity. It is possible to further improve the shape selectivity of molecular sieves. One method reduces the number of active sites on the external surface of the zeolite crystallites, for example, by neutralizing external sites with large molecules that cannot diffuse into the pores. Other methods involve modifying the internal structure. For example, in the reaction: toluene + methanol → paraxylene, the paraxylene selectivity of ZSM-5 was increased from 46 to 97% by treatment with phosphorus, boron compounds, or polymers. This has significance for commercial applications as paraxylene is the intermediate of an important industrial chemical, terephthalic acid. Product selectivity, restricted transition-state-type selectivity, and molecular traffic control may all contribute to this reaction and several others in which paraxylene is formed above its equilibrium concentration. Reactant and product selectivities are mass-transfer-limited and therefore affected by crystallite size, whereas restricted transition-state selectivity is not. This method has been used experimentally to distinguish reactant and product-type selectivities and restricted transition-state-type selectivity. Observed rates depend on the intrinsic, uninhibited rate constant and, if mass transfer is limiting, on the diffusivities of the reactant (or product) molecules and on catalyst particle size.

Molecular traffic control. It has been suggested that in zeolites with intersecting pores of more than one size, for example, ZSM-5, reactant molecules may preferentially enter the catalyst through one of the pore systems while the products diffuse out through the other. Such molecular traffic control would minimize counterdiffusion and thus increase reaction rates.

For background information *see* CATALYSIS; MOLECULAR SIEVE; ZEOLITE in the McGraw-Hill Encyclopedia of Science and Technology.

[SIGMUND M. CSICSERY]

Bibliography: D. W. Breck, *Zeolite Molecular Sieves*, 1974; S. M. Csicsery, The cause of shape-selectivity of transalkylation in mordenite, *J. Catal.*, 23:124–130, 1971; S. M. Csicsery, Shape-selective catalysis, in J. A. Rabo (ed.), *Zeolite Chemistry and Catalysis*, ACS Monogr. 171, pp. 680–713, 1976; W. O. Haag, R. M. Lago, and P. B. Weisz, Transport and reactivity of hydrocarbon molecules in a shape-selective zeolite, *Faraday Gen. Discuss.*, 72:317–330, 1982, and *Selectivity in Heterogeneous Catalysis*, University of Nottingham, England, September 14–16, 1981; P. B. Weisz et al., Catalysis by crystalline aluminosilicates II. Molecular-shape selective reactions, *J. Catal.*, 1:307–312, 1962; P. B. Weisz et al., Intracrystalline and molecular shape-selective catalysis by zeolite salts, *J. Phys. Chem.*, 64:382, 1960; P. B. Weisz et al., Molecular shape-selective catalysis, *Pure Appl. Chem.*, 52:2091–2103, 1980.

McGRAW-HILL YEARBOOK OF SCIENCE AND TECHNOLOGY

List of Contributors

List of Contributors

A

Ackroyd, Dr. Frederick. Massachusetts General Hospital, Department of Surgery, Harvard Medical School, Shriners Burns Institute. ARTIFICIAL SKIN—coauthored.

Aldridge, Dr. Richard J. Department of Geology, University of Nottingham. CONODONT—coauthored.

Altamuro, Vincent M. President, Robotics Research, Division of VMA, Inc., Toms River, New Jersey. ROBOTICS—feature.

Armbruster, Dr. P. Gesellschaft für Schwerionenforschung mbH, Darmstadt, West Germany. TRANSURANIUM ELEMENTS.

Asmussen, Dr. Jes, Jr. Professor of Electrical Engineering, College of Engineering, Department of Electrical Engineering and Systems Science, Michigan State University. PROPULSION.

Awramik, Dr. S. M. Department of Geological Sciences, University of California, Santa Barbara. CELL (BIOLOGY).

B

Backer, Dr. Donald C. Radio Astronomy Laboratory, University of California, Berkeley. PULSAR—coauthored.

Bada, Dr. Jeffrey L. Scripps Institution of Oceanography, Amino Acid Dating Laboratory, University of California, San Diego. ARCHEOLOGICAL CHEMISTRY.

Bain, Dr. Roger J. Associate Professor of Geology, Department of Geology, University of Akron. GYPSUM.

Barger, Prof. Vernon D. Department of Physics, University of Wisconsin. HIGGS BOSON.

Bar-Itzhack, Dr. I. Y. Associate Professor, Aeronautical Engineering Department, Technion–Israel Institute of Technology, Haifa. INERTIAL GUIDANCE SYSTEM.

Bartram, James F. Middletown, Rhode Island. ACOUSTIC SIGNAL PROCESSING.

Bastian, Robert K. Environmental Scientist, Office of Water, U.S. Environmental Protection Agency, Washington, D.C. NATURAL WASTE TREATMENT—feature, coauthored.

Baumgarten, Dr. Alexander. Director, Clinical Immunology Laboratory, Yale–New Haven Hospital, New Haven, Connecticut. IMMUNOLOGY; PROTEIN—in part.

Benedict, Dr. William F. Clayton Ocular Oncology Center, Children's Hospital of Los Angeles. ONCOLOGY.

Benforado, Jay. Senior Scientist, Office of Research and Development, U.S. Environmental Protection Agency, Washington, D.C. NATURAL WASTE TREATMENT—feature, coauthored.

Bloom, Prof. Elliott D. Stanford Linear Accelerator Center, Stanford University. QUARKONIUM.

Bond, Dr. Gerard C. Lamont-Doherty Geological Observatory of Columbia University, Palisades, New York. TECTONICS.

Boyle, Charles. NASA Goddard Space Flight Center, Greenbelt, Maryland. SPACE FLIGHT.

Briggs, Dr. Derek E. G. Geology Department, University of London, Goldsmiths' College. CONODONT—coauthored.

Brody, Dr. Aaron L. Manager, Market Development, Valley Forge Marketing and Research Center, Container Corporation of America, Oaks, Pennsylvania. FOOD ENGINEERING—in part.

Brown, Dr. D. A. Professor of Soil Science, Agronomy Department, University of Arkansas. ROOT.

Browning, Dr. Keith A. Chief Meteorological Officer, Radar Research Laboratory, Royal Signals and Radar Establishment, Worcester, England. NOWCASTING.

Brownlee, Dr. Donald E. Department of Astronomy, University of Washington, Seattle. INTERPLANETARY MATTER.

Bucaro, Dr. J. A. Naval Research Laboratory, Washington, D.C. TRANSDUCER.

Buchsbaum, Dr. Monte S. Professor of Psychiatry, Department of Psychiatry and Human Behavior, California College of Medicine, University of California, Irvine. POSITRON EMISSION TOMOGRAPHY.

Buljan, Dr. S. T. GTE Laboratories, Waltham, Massachusetts. TOOLING—in part.

Büning, Dr. Jürgen. Zoologisches Institut, Universität Münster, West Germany. INSECT.

Buntschuh, Robert F. RCA, Government Systems Division, Astro-Electronics, Princeton, New Jersey. DIRECT BROADCASTING SATELLITE SYSTEMS.

Burke, Dr. John F. Massachusetts General Hospital, Department of Surgery, Harvard Medical School, Shriners Burns Institute. ARTIFICIAL SKIN—coauthored.

C

Cabrera, Prof. Blas. Department of Physics, Stanford University. MAGNETIC MONOPOLES.

Canova, Fred. Worthington Group, McGraw-Edison Corporation, Wellsville, New York. STEAM TURBINE.

Carbonara, Dr. Robert S. Associate Manager, Process Metallurgy Section, Battelle Columbus Laboratories, Columbus, Ohio. METALLURGY.

Carey, Prof. Tom. University of Guelph (Ontario, Canada), College of Physical Science, Department of Computing and Information Science. USER FRIENDLY SYSTEMS.

Carroll, Dr. Tom. National Weather Service, National Oceanic and Atmospheric Administration, U.S. Department of Commerce. SNOW SURVEYING.

Carson, Prof. Hampton L. Department of Genetics, University of Hawaii. GENETICS.

Cavalli-Sforza, Dr. L. L. Professor of Genetics, Department of Genetics, Stanford University School of Medicine. PHYSICAL ANTHROPOLOGY.

Chase, Dr. Julia. Department of Biological Sciences, Barnard College. ORIENTATION (BIOLOGY).

Cheryan, Dr. Munir. Associate Professor, College of Agriculture, Department of Food Science, University of Illinois. FOOD ENGINEERING—in part.

Clarkson, Dr. Euan N. K. Grant Institute of Geology, University of Edinburgh. CONODONT—coauthored.

Coleman, Richard H. Manager, Baseband Engineering, COMSAT General Corporation, Washington, D.C. TELEPHONE.

Cook, Dr. Laurence M. Department of Zoology, University of Manchester, England. PIGMENTATION.

Corigliano, Dr. Horace J. Applied Magnetics, Magnetic Head Division, Goleta, California. ELECTRONIC WARFARE—in part.

Cox, Prof. F. E. G. Zoology Department, King's College, London, England. MEDICAL PARASITOLOGY.

Csicsery, Dr. Sigmund M. Chevron Research Company, Richmond, California. ZEOLITE.

Cutting, Dr. James E. Department of Psychology, Cornell University. PERCEPTION.

D

Damuth, Dr. John E. Senior Research Associate, Lamont-Doherty Geological Observatory of Columbia University, Palisades, New York. MARINE SEDIMENTS.

Davis, Dr. Raymond, Jr. Brookhaven National Laboratory, Associated Universities, Inc., Department of Chemistry, Upton, New York. SOLAR NEUTRINOS.

Delmer, Dr. Deborah P. Principal Scientist, Arco Plant Cell Research Institute, Dublin, California. CELL WALLS (PLANT).

DeMaster, Dr. David J. Department of Marine, Earth and Atmospheric Sciences, School of Physical and Mathematical Sciences, North Carolina State University. MARINE SEDIMENTS—in part.

Dickey, Dr. John S., Jr. Department of Geology, Syracuse University. EARTH.

Draper, Dr. Clifton W. Laser Studies Group, Western Electric Company, Princeton, New Jersey. LASER ALLOYING.

Dykhuizen, Dr. Daniel. Department of Genetics, Washington University Medical School, St. Louis, Missouri. CHEMOSTAT.

E

Ellsaesser, Dr. Hugh W. Lawrence Livermore Laboratory, Livermore, California. STRATOSPHERE.

Epel, Dr. David. Department of Biological Sciences, Hopkins Marine Station, Pacific Grove, California. FERTILIZATION.

Erb, Dr. Karl A. Nuclear Division, Oak Ridge National Laboratory, Oak Ridge, Tennessee. COULOMB EXCITATION.

F

Feldmann, Dr. Richard J. Computer Specialist, Department of Computer Research, Department of Health and Human Services, Public Health Service, National Institutes of Health. PROTEIN—in part.

Forward, Dr. Richard B., Jr. Associate Professor, Zoology, Duke University Marine Laboratory, Pivers Island, Beaufort, North Carolina. PERIODICITY IN ORGANISMS.

Foushee, Harvey C. Corporate Engineering, Western Electric Company, New York, New York. ERGONOMICS.

Fox, William K. President, Standard Havens Research Corporation, Kansas City, Missouri. FUEL.

G

Gall, Dr. Joseph G. Department of Embryology, Carnegie Institution, Baltimore, Maryland. CHROMOSOME.

Gans, Dr. Carl. Division of Biological Sciences, University of Michigan. VERTEBRATA—coauthored.

Gavlick, Emil L. Council of Reprographic Executives, Southwest Research Institute, San Antonio, Texas. REPROGRAPHICS.

Gibson, Dr. Arthur C. Department of Biology, University of California, Los Angeles. CACTUS.

Gonzalez, Prof. R. C. Department of Electrical Engineering and Computer Science, University of Tennessee. COMPUTER VISION.

Gover, Dr. J. E. Sandia National Laboratories, Albuquerque, New Mexico. RADIATION DAMAGE TO MATERIALS—coauthored.

Green, Dr. Daniel W. E. Harvard-Smithsonian Center for Astrophysics, Cambridge, Massachusetts. COMET.

Green, Prof. Robert E., Jr. Chairman, Materials Science and Engineering, Johns Hopkins University. NONDESTRUCTIVE TESTING.

Greenfield, Dr. Michael D. Department of Biology, University of California, Los Angeles. PHEROMONES.

Greenwood, Dr. James R. Director, Public Health Laboratory, County of Orange Health Care Agency, Santa Ana, California. GARDNERELLA.

Greiner, Prof. Walter. Institute for Theoretical Physics, University of Frankfurt, West Germany. SUPERCRITICAL FIELDS.

Grover, Kenneth M. GSA-International Hydropower, Croton Falls, New York. HYDROPOWER.

Gupta, Dr. Umesh C. Research Branch, Agriculture Canada, Prince Edward Island. MOLYBDENUM.

H

Haenlein, Prof. George F. W. Animal Science Department, University of Delaware. GOAT.

Handsfield, Dr. H. Hunter. STD Program Director, Sexually Transmitted Disease Control Program, Harborview Medical Center, Seattle, Washington. SEXUALLY TRANSMITTED DISEASES.

Hardyck, Dr. Curtis. School of Education, University of California, Berkeley. BRAIN.

Hartnoll, Dr. Richard. Department of Marine Biology, University of Liverpool, Port Erin, Isle of Man. ECOLOGICAL INTERACTIONS—in part.

Hasselbrack, Sally A. Boeing Company, Seattle, Washington. AIRPLANE.

Hasselriis, Dr. Floyd. Consulting Engineer, Forest Hills, New York. WASTE ENERGY RECOVERY.

Hawthorne, Dr. F. C. Department of Earth Sciences, University of Manitoba, Winnipeg. MINERAL.

Haxby, Dr. W. F. Lamont-Doherty Geological Observatory of Columbia University, Palisades, New York. GEOTECTONIC IMAGERY.

Hayes, William C. "Electrical World," McGraw-Hill Publications Company, New York, New York. ELECTRICAL UTILITY INDUSTRY.

Heiken, Dr. Grant. Earth and Space Sciences Division, Los Alamos National Laboratory, Los Alamos, New Mexico. GEOTHERMAL POWER—coauthored.

Hinners, Dr. Noel W. Goddard Space Flight Center, Greenbelt, Maryland. SPACE PROBE.

Holder, Dr. Nigel. Department of Anatomy, Kings College, University of London, England. DEVELOPMENTAL BIOLOGY.

Hood, Dr. J. A. Sandia National Laboratories, Albuquerque, New Mexico. RADIATION DAMAGE TO MATERIALS—coauthored.

House, Dr. William A. Research Physiologist, U.S. Plant, Soil and Nutrition Laboratory, USDA-ARS-NER-NAA, Ithaca, New York. NUTRITION.

Huckenholz, Prof. Dr. Hans G. Professor of Petrology and Director of the Mineralogisch-Petrographisches Institut der Ludwig Maximilians Universität, München, West Germany. GARNET.

I

Iachello, Dr. Francesco. A. W. Wright Nuclear Structure Laboratory, Yale University. SUPERSYMMETRY.

J

Johnson, Dr. A. A. Professor of Materials Science, Speed Scientific School, University of Louisville. DENTAL MATERIALS—coauthored.

Jones, Dr. R. N. Department of Agricultural Botany, School of Agricultural Science, University College of Wales. CHROMOSOME—in part.

Jones, Dr. Robert T. Senior Research Associate, NASA–Ames Research Center, Moffett Field, California; and Consulting Professor, Stanford University. WING.

K

Kan-Mitchell, Dr. J. School of Medicine, University of Southern California, Los Angeles. IMMUNOTHERAPY—coauthored.

Keiser, Dr. Bernhard E. Keiser Engineering, Inc., Vienna, Virginia. ELECTROMAGNETIC COMPATIBILITY.

Kennett, Dr. J. P. Department of Oceanography, University of Rhode Island. PALEOCEANOGRAPHY.

Knight, Dr. W. E. Research Leader, Forage Research Unit, U.S. Department of Agriculture, Delta States Area, Crop Science Research Laboratory. CLOVER—coauthored.

Kulkarni, Dr. Shrinivas. Radio Astronomy Laboratory, University of California, Berkeley. PULSAR—coauthored.

Kung, Prof. H. T. Professor of Computer Science, Department of Computer Science, Carnegie-Mellon University. SYSTOLIC ARRAYS.

Kutler, Dr. Paul. Computer Aerodynamics, NASA, Ames Research Center, Moffett Field, California. AERODYNAMICS.

L

Lacy, John A. Eastman Kodak Company, Rochester, New York. MICROGRAPHICS.

Lewine, Dr. Richard. Illinois State Psychiatric Institute, Chicago. SCHIZOPHRENIA.

Lippincott, Dr. James A. College of Arts and Sciences, Section of Biological Sciences, Northwestern University. CROWN GALL.

Lorian, Dr. Victor. Department of Microbiology, Bronx-Lebanon Hospital Center, New York, New York. ANTIMICROBIAL AGENTS—in part.

Luhr, Dr. James F. Geology Department, Franklin and Marshall College, Lancaster, Pennsylvania. VOLCANO.

M

McClelland, Lindsay. Smithsonian Institution, Washington, D.C. VOLCANO—in part.

MacDonald, Verne H. Technical Staff, Cellular Systems Engineering and Advanced Development Department, Bell Telephone Laboratories, Holmdel, New Jersey. MOBILE RADIO.

McGee, Dr. Denis C. Department of Plant Pathology, Iowa State University. SEED.

McGuire, Dr. W. S. Crop Science Department, Oregon State University. CLOVER—coauthored.

Macher, Dr. Abe M. Staff Fellow, Microbiology Service, Department of Clinical Pathology, Clinical Center, National Institutes of Health, Bethesda, Maryland. ACQUIRED IMMUNE DEFICIENCY SYNDROME (AIDS).

McSween, Dr. Harry Y., Jr. Department of Geological Sciences, University of Tennessee. METEORITE.

Mahon, Joseph B. Director, Space Transportation Support Division, NASA Headquarters, Washington, D.C. LAUNCH VEHICLE—coauthored.

Mahowald, Dr. Anthony P. Department of Biology, Indiana University. CELL DIFFERENTIATION.

Marchalonis, Dr. John J. Chairman, Department of Biochemistry, Medical University of South Carolina, Charleston. IMMUNOLOGICAL PHYLOGENY.

Meehan, Dr. Richard T. Department of Internal Medicine, University of Iowa Hospital and Clinics. HIGH-ALTITUDE ILLNESS.

Meiers, E. J., Jr. Commandant G-TES-1, U.S. Coast Guard, Washington, D.C. PILOTING.

Meirovitch, Prof. Leonard. University Distinguished Professor, Department of Engineering Science and Mechanics, Virginia Polytechnic Institute and State University. VIBRATION.

Meisinger, Dr. John J. Soil Scientist, Soil Nitrogen and Environmental Chemistry Laboratory, Agricultural Environmental Quality Institute, U.S. Department of Agriculture, Agricultural Research Service. SOIL NITROGEN.

Menge, Dr. Bruce A. Department of Zoology, Oregon State University. ECOLOGICAL INTERACTIONS—in part.

Millar, I. C. Maritime Technology Committee, MEERB, Department of Industry, London, England. MARINE NAVIGATION—in part.

Mills, Dr. Allen P., Jr. Bell Laboratories, Murray Hill, New Jersey. SURFACE PHYSICS.

Mitchell, Dr. M. S. Comprehensive Cancer Center, Kenneth Norris, Jr. Cancer Research Institute, University of Southern California, Los Angeles. IMMUNOTHERAPY—coauthored.

Mitchell, Prof. Rodger. Zoology Department, Ohio State University. AGROECOSYSTEMS.

Mohapatra, Prof. Rabindra N. Physics Department, University of Maryland. NEUTRON-ANTINEUTRON OSCILLATIONS.

Morgan, Dr. Walter L. Consultant, Communications Center of Clarksburg (Maryland). COMMUNICATIONS SATELLITE.

Mortelmans, Dr. Kristien E. Director, Microbial Genetics Department, Toxicology Laboratory, Life Sciences Division, Stanford Research Institute. MUTAGENS AND CARCINOGENS.

Mossman, Dr. David J. Department of Geology, Mount Allison University, Sackville, New Brunswick, Canada. PALEONTOLOGY—in part.

Muller, Dr. Erwin E. Bell Laboratories, Holmdel, New Jersey. SINGLE SIDEBAND.

Mullins, Dr. Henry T. Department of Geology, Syracuse University. REEF—coauthored.

Mundinger, Dr. Paul C. Department of Biology, Queens College, CUNY. BIRDS.

Murnick, Dr. D. E. Bell Laboratories, Murray Hill, New Jersey. ATOMIC STRUCTURE AND SPECTRA—coauthored.

Murphy, Dr. Eugene F. (Retired) Director, Office of Technology Transfer, Veterans Administration. BIOENGINEERING—feature.

N

Neumann, Dr. A. Conrad. Department of Geology, Syracuse University. REEF—coauthored.

Newton, Dr. Cathryn R. Department of Geology, Syracuse University. REEF—coauthored.

Northcutt, R. Glenn. Division of Biological Sciences, University of Michigan. VERTEBRATA—coauthored.

Nowicki, Dr. Stephen. Section of Neurobiology and Behavior, Division of Biological Sciences, Cornell University. SPIDER.

O

Ormrod, Dr. John H. Applied Physics Division, Chalk River Nuclear Laboratories, Atomic Energy of Canada, Ltd., Chalk River, Ontario. CYCLOTRON.

Osbourn, Dr. Gordon C. Compound Semiconductor Research Division, Sandia National Laboratories, Albuquerque, New Mexico. SEMICONDUCTOR HETEROSTRUCTURES.

P

Patel, Dr. C. K. N. Executive Director, Research, Physics Division, Bell Laboratories, Murray Hill, New Jersey. ATOMIC STRUCTURE AND SPECTRA—coauthored.

Pearce, Dr. Cedric. Radioisotope Laboratory, University of Illinois. ANTIBIOTIC.

Pearson, Dr. M. G. Royal Institute of Navigation, Royal Geographic Society, London, England. UNDERWATER NAVIGATION.

Pedigo, Prof. Larry P. Professor of Entomology, Department of Entomology, Iowa State University. INTEGRATED PEST MANAGEMENT—feature.

Peterson, Dr. Bruce A. Mount Stromlo and Siding Spring Observatories, Australian National University, Canberra. QUASARS.

Pettitt, Dr. Roland. Earth and Space Sciences Division, Los Alamos National Laboratory, Los Alamos, New Mexico. GEOTHERMAL POWER–coauthored.

Pfeffer, Dr. Robert A. Harry Diamond Laboratories, Woodridge Research Laboratories, Woodbridge, Virginia. ELECTROMAGNETIC PULSE (EMP).

Pionke, Dr. Harry B. Northeast Watershed Research Center, U.S. Department of Agriculture, Agriculture Research Service, University Park, Pennsylvania. ACID RAIN—coauthored.

Pitt, C. W. Department of Electronic and Electrical Engineering, University College, London, England. MOLECULAR ENGINEERING.

Pohl, Prof. Robert O. Laboratory of Atomic and Solid State Physics, Cornell University. RADIOACTIVE WASTE MANAGEMENT.

Pond, Dr. Wilson G. Department of Animal Science, University of Nebraska. SWINE PRODUCTION.

Price, Dr. Donald L. Director, Neuropathology Department, and Professor, Departments of Pathology, Neurology, and Neuroscience, Johns Hopkins University School of Medicine, Baltimore, Maryland. ALZHEIMER'S DISEASE.

Prusiner, Dr. Stanley B. Department of Neurology, School of Medicine, University of California, San Francisco. PRIONS.

Puckett, Lanny. Program Manager, Electronic Systems, Sperry Corporation, Reston, Virginia. MARINE NAVIGATION—in part.

Pugh, Dr. Howel G. Lawrence Berkeley Laboratory, University of California, Berkeley. HEAVY-ION ACCELERATOR.

R

Rafaels, Umberto. Director, Marketing Communications, Local Digital Distribution Company, Rockville, Maryland. DATA COMMUNICATIONS.

Raider, Dr. Stanley I. Thomas J. Watson Research Center, International Business Machines, Yorktown Heights, New York. SUPERCONDUCTING DEVICE.

Raymond, Dr. Kenneth N. Department of Chemistry, University of California, Berkeley. BIOMIMETIC CHEMISTRY—coauthored.

Richards, Prof. J. Ian. School of Mathematics, University of Minnesota. NUMBER THEORY.

Ringlee, Dr. Robert J. Power Technologies, Inc., Schenectady, New York. ELECTRIC POWER SYSTEMS.

Robinson, Dr. F. Neville H. Clarendon Laboratory, Oxford University, England. CHAOTIC BEHAVIOR.

Rodgers, Steven J. Department of Chemistry, University of California, Berkeley. BIOMIMETIC CHEMISTRY—coauthored.

Rohlf, Dr. James W. Assistant Professor of Physics, High Energy Physics Laboratory, Harvard University. INTERMEDIATE VECTOR BOSON.

Rowley, Dr. John. Earth and Space Sciences Division, Los Alamos National Laboratory, Los Alamos, New Mexico. GEOTHERMAL POWER—coauthored.

Royer, Dr. G. P. Director, Biotechnology Division, Corporate Research Department, Standard Oil (Indiana), Naperville, Illinois. ENZYME.

S

Sanders, Prof. T. H., Jr. Department of Materials Engineering, Purdue University. ALUMINUM ALLOYS.

Sarin, Dr. V. K. Senior Staff Scientist, GTE Laboratories, Waltham, Massachusetts. TOOLING—in part.

Schewe, Dr. Phillip F. Public Information Division, American Institute of Physics, New York, New York. PARTICLE ACCELERATOR.

Schnabel, Dr. Ronald R. Northeast Watershed Research Center, U.S. Department of Agriculture, Agriculture Research Service, University Park, Pennsylvania. ACID RAIN—coauthored.

Schueler, Dr. D. G. Sandia National Laboratories, Albuquerque, New Mexico. SOLAR CELL.

Schuman, Dr. Gerald E. High Plains Grasslands Research Station, U.S. Department of Agriculture, Agricultural Research Service, Cheyenne, Wyoming. LAND RECLAMATION—coauthored.

Sepkoski, Dr. J. John, Jr. Department of Geophysical Sciences, University of Chicago. PALEONTOLOGY—in part.

Shadle, Dr. Christine H. Research Laboratory of Electronics, Massachusetts Institute of Technology. WHISTLING.

Shank, Dr. C. V. Bell Laboratories, Holmdel, New Jersey. OPTICAL PULSES.

Smith, Dr. J. T. GTE Laboratories, Waltham, Massachusetts. TOOLING—in part.

Sock, Alan S. Litton Industries, College Park, Maryland. ELECTRONIC WARFARE—in part.

Solem, Dr. Johndale C. Los Alamos National Laboratories, Los Alamos, New Mexico. MICROHOLOGRAPHY.

Steinschneider, Dr. Alfred. President, American Sudden Infant Death Syndrome Institute, Atlanta, Georgia. SUDDEN INFANT DEATH SYNDROME (SIDS)—coauthored.

Strom, Dr. Stephen E. Five College Astronomy Department, University of Massachusetts. INFRARED ASTRONOMY.

Sulak, Prof. Lawrence R. Physics Department, University of Michigan. PROTON.

T

Taylor, Dr. E. M., Jr. High Plains Grasslands Research Station, U.S. Department of Agriculture, Agricultural Research Service, Cheyenne, Wyoming. LAND RECLAMATION—coauthored.

Tomasz, Dr. Alexander. Department of Microbiology, Rockefeller University. ANTIMICROBIAL AGENTS—in part.

Tomlinson, W. J. Central Services Organization, Holmdel, New Jersey. OPTICAL COMMUNICATIONS.

U

Umezawa, Prof. Hiroomi. Killam Memorial Professor of Science, and Professor of Physics, Theoretical Physics Institute, University of Alberta. SUPERCONDUCTIVITY.

V

Vaux, Dr. Henry J., Jr. Professor of Resource Economics, College of Natural and Agricultural Sciences, Citrus Research Center and Agricultural Experiment Station, Department of Soil and Environmental Sciences, University of California, Riverside. IRRIGATION (AGRICULTURE).

von Fraunhofer, Dr. J. A. Professor of Biomaterials Science, University of Louisville, Health Sciences Center. DENTAL MATERIALS—coauthored.

W

Wayman, Prof. C. M. Department of Metallurgy and Mining Engineering, University of Illinois, Urbana. SHAPE MEMORY ALLOYS.

Webb, Dr. Paul W. School of Natural Resources, University of Michigan. OSTEICHTHYES.

Weiner, Dr. S. A. Manufacturing Processes Laboratory, Manufacturing Systems and Machining Department, Ford Motor Company, Redford, Michigan. EVAPORATIVE CASTING.

Wells, Dr. John T. Associate Professor, Coastal Studies Institute, Center for Wetland Resources, Louisiana State University. SEDIMENTATION (GEOLOGY).

Wild, Dr. Jack W. Deputy Director, Space Transportation Support Division, NASA Headquarters, Washington, D.C. LAUNCH VEHICLE—coauthored.

Winn, Dr. Kevin. American Sudden Infant Death Syndrome Institute, Atlanta, Georgia. SUDDEN INFANT DEATH SYNDROME (SIDS)—coauthored.

Wisnosky, Dennis E. Group Vice President, GCA Corporation, Industrial Systems Group, Naperville, Illinois. COMPUTER INTEGRATED MANUFACTURING (CIM).

Wohl, Amy D. President, Advanced Office Concepts Corporation, Bala Cynwyd, Pennsylvania. TYPEWRITER.

Wolfe, Dr. Edward. Hawaiian Volcano Observatory, U.S. Geological Survey. VOLCANO—in part.

Wright, Prof. Paul K. Department of Mechanical Engineering, Carnegie-Mellon University. TOOLING—in part.

Z

Zeilik, Prof. Michael. Department of Physics and Astronomy, University of New Mexico. ARCHEOASTRONOMY—feature.

Zen, Dr. E-an. Geological Survey, U.S. Department of the Interior, Reston, Virginia. MUSCOVITE.

Zuelow, Dale L. Supervisor of Manufacturing Research, Lockheed Corporation, Burbank, California. MACHINING OPERATIONS.

McGRAW-HILL YEARBOOK OF SCIENCE AND TECHNOLOGY

Index

Index

Asterisks indicate page references to article titles.

A

B

C

D

E

F

G

H

I

J

K

L

M

N

Q

R

S

T

U

V

W

X–Z